D1604530

WITHDRAWN

COLD SPRING HARBOR SYMPOSIA ON QUANTITATIVE BIOLOGY

VOLUME LXII

COLD SPRING HARBOR SYMPOSIA ON QUANTITATIVE BIOLOGY

VOLUME LXII

Pattern Formation during Development

COLD SPRING HARBOR LABORATORY PRESS
1997

COLD SPRING HARBOR SYMPOSIA ON QUANTITATIVE BIOLOGY VOLUME LXII

©1997 by Cold Spring Harbor Laboratory Press
International Standard Book Number 0-87969-535-8 (cloth)
International Standard Book Number 0-87969-536-6 (paper)
International Standard Serial Number 0091-7451
Library of Congress Catalog Card Number 34-8174

Printed in the United States of America
All rights reserved

COLD SPRING HARBOR SYMPOSIA ON QUANTITATIVE BIOLOGY
Founded in 1933 by
REGINALD G. HARRIS
Director of the Biological Laboratory 1924 to 1936

Previous Symposia Volumes

I (1933) Surface Phenomena
II (1934) Aspects of Growth
III (1935) Photochemical Reactions
IV (1936) Excitation Phenomena
V (1937) Internal Secretions
VI (1938) Protein Chemistry
VII (1939) Biological Oxidations
VIII (1940) Permeability and the Nature of Cell Membranes
IX (1941) Genes and Chromosomes: Structure and Organization
X (1942) The Relation of Hormones to Development
XI (1946) Heredity and Variation in Microorganisms
XII (1947) Nucleic Acids and Nucleoproteins
XIII (1948) Biological Applications of Tracer Elements
XIV (1949) Amino Acids and Proteins
XV (1950) Origin and Evolution of Man
XVI (1951) Genes and Mutations
XVII (1952) The Neuron
XVIII (1953) Viruses
XIX (1954) The Mammalian Fetus: Physiological Aspects of Development
XX (1955) Population Genetics: The Nature and Causes of Genetic Variability in Population
XXI (1956) Genetic Mechanisms: Structure and Function
XXII (1957) Population Studies: Animal Ecology and Demography
XXIII (1958) Exchange of Genetic Material: Mechanism and Consequences
XXIV (1959) Genetics and Twentieth Century Darwinism
XXV (1960) Biological Clocks
XXVI (1961) Cellular Regulatory Mechanisms
XXVII (1962) Basic Mechanisms in Animal Virus Biology
XXVIII (1963) Synthesis and Structure of Macromolecules
XXIX (1964) Human Genetics
XXX (1965) Sensory Receptors
XXXI (1966) The Genetic Code
XXXII (1967) Antibodies
XXXIII (1968) Replication of DNA in Microorganisms
XXXIV (1969) The Mechanism of Protein Synthesis
XXXV (1970) Transcription of Genetic Material
XXXVI (1971) Structure and Function of Proteins at the Three-dimensional Level
XXXVII (1972) The Mechanism of Muscle Contraction
XXXVIII (1973) Chromosome Structure and Function
XXXIX (1974) Tumor Viruses
XL (1975) The Synapse
XLI (1976) Origins of Lymphocyte Diversity
XLII (1977) Chromatin
XLIII (1978) DNA: Replication and Recombination
XLIV (1979) Viral Oncogenes
XLV (1980) Movable Genetic Elements
XLVI (1981) Organization of the Cytoplasm
XLVII (1982) Structures of DNA
XLVIII (1983) Molecular Neurobiology
XLIX (1984) Recombination at the DNA Level
L (1985) Molecular Biology of Development
LI (1986) Molecular Biology of *Homo sapiens*
LII (1987) Evolution of Catalytic Function
LIII (1988) Molecular Biology of Signal Transduction
LIV (1989) Immunological Recognition
LV (1990) The Brain
LVI (1991) The Cell Cycle
LVII (1992) The Cell Surface
LVIII (1993) DNA and Chromosomes
LIX (1994) The Molecular Genetics of Cancer
LX (1995) Protein Kinesis: The Dynamics of Protein Trafficking and Stability
LXI (1996) Function & Dysfunction in the Nervous System

Front Cover (Paperback): GHY03: *Drosophila* third instar wing imaginal disc triple labeled for the expression of wing patterning genes. (*Green*) Vestigial expression; (*blue*) apterous expression; (*red*) CiD. (Image courtesy of J. Williams, S. Paddock, and S. Carroll, University of Wisconsin.)

Back Cover (Paperback): GHY02: Confocal imaging of pattern formation in *Drosophila*. (*Top*) Triple-labeled embryo at the cellular blastoderm stage showing the expression of three segmentation genes: *hairy* (*red*), *Krüppel* (*green*), and *giant* (*blue*). (Image courtesy of J. Langeland, S. Paddock, and S. Carroll, University of Wisconsin.) (*Bottom*) Triple-labeled third instar wing imaginal disc with nested domains of Spalt (*purple* and *light blue*) and Omb (*green* and *yellow*) expression in the vestigial domain (*red*) (for details, see Kim et al. 1997 [*Nature 388:* 304].) (Image courtesy of J. Kim and S. Carroll, University of Wisconsin.)

Authorization to photocopy items for internal or personal use, or the internal or personal use of specific clients, is granted by Cold Spring Harbor Laboratory Press, provided that the appropriate fee is paid directly to the Copyright Clearance Center (CCC). Write or call CCC at 222 Rosewood Drive, Danvers, MA 01923 (508-750-8400) for information about fees and regulations. Prior to photocopying items for educational classroom use, contact CCC at the above address. Additional information on CCC can be obtained at CCC Online at http://www.copyright.com/

All Cold Spring Harbor Laboratory Press publications may be ordered directly from Cold Spring Harbor Laboratory Press, 10 Skyline Drive, Plainview, NY 11803-2500 (Phone: 1-800-843-4388 in Continental U.S. and Canada). All other locations: (516) 349-1930. FAX: (516) 349-1946. E-mail: cshpress@cshl.org. For a complete catalog of all Cold Spring Harbor Laboratory Press publications, visit our World Wide Web Site http://www/cshl.org/

Symposium Participants

ABATE-SHEN, CORY, Center for Advanced Biotechnology and Medicine, UMDNJ-Robert Wood Johnson Medical School, Piscataway, New Jersey

ADDIVINOLA, FRANK, Lab. of Biochemistry, National Institutes of Health, Bethesda, Maryland

AIZAWA, SINICHI, Lab. of Morphology, Institute of Molecular Embryology and Genetics, Kumamoto University School of Medicine, Kumamoto, Japan

AKSAN, ISIL, Eukaryotic Transcription Laboratory, Marie Curie Research Institute, Oxted, United Kingdom

AMEIXA, CLARA, Dept. of Molecular Haematology, Institute of Child Health, London, United Kingdom

AMIEUX, PAUL, Dept. of Pharmacology, University of Washington, Seattle

ANDERSON, DAVID, Dept. of Biology, Howard Hughes Medical Institute, California Institute of Technology, Pasadena, California

ANDERSON, KATHRYN, Dept. of Molecular Biology, Memorial Sloan-Kettering Cancer Institute, New York, New York

ARTINGER, KRISTIN, Cardiovascular Research Center, Massachusetts General Hospital, Harvard Medical School, Charlestown, Massachusetts

ARUGA, JUN, Lab. of Molecular Neurobiology, Institute of Physical and Chemical Research, Tsukuba, Japan

AVILION, ARIEL, Div. of Developmental Genetics, National Institute for Medical Research, London, United Kingdom

BACH, INGOLF, Dept. of Medicine, University of California at San Diego, La Jolla

BACHVAROVA, ROSEMARY, Dept. of Cell Biology, Cornell University Medical College, New York, New York

BAKER, BRUCE, Dept. of Biological Sciences, Herring Laboratories, Stanford University, Stanford, California

BAKER, NICK, Dept. of Molecular Genetics, Albert Einstein College of Medicine, Bronx, New York

BALINT-KURTI, PETER, Lab. of Cellular and Developmental Biology, National Institutes of Health, Bethesda, Maryland

BALLY-CUIF, LAURE, CNRS, Ecole Normale Superieure, Paris, France

BANG, ANNE, Dept. of Molecular Neurobiology, Salk Institute for Biological Studies, La Jolla, California

BAO, ZHENG-ZHENG, Dept. of Genetics, Harvard Medical School, Boston, Massachusetts

BARTELS, JANET, Dept. of Biology, Yale University, New Haven, Connecticut

BAYLIES, MARY, Memorial Sloan-Kettering Cancer Institute, New York, New York

BAZAN, FERNANDO, Dept. of Molecular Biology, DNAX Research Institute, Palo Alto, California

BEACHY, PHILIP, Dept. of Molecular Biology and Genetics, Johns Hopkins University School of Medicine, Baltimore, Maryland

BECKENDORF, STEVEN, Dept. of Molecular and Cell Biology, University of California, Berkeley

BEDDINGTON, ROSA, Dept. of Developmental Genetics and Mammalian Development, National Institute for Medical Research, London, United Kingdom

BEGLEY, DALE, Dept. of Bioinformatics,The Jackson Laboratory, Bar Harbor, Maine

BEHRINGER, RICHARD, Dept. of Molecular Genetics, M.D. Anderson Cancer Center, University of Texas, Houston

BELLIVEAU, MICHAEL, Dept. of Genetics, Harvard Medical School, Boston, Massachusetts

BENNETT, RANDY, Dept. of Zoology, Brigham Young University, Provo, Utah

BERGEMANN, ANDREW, Dept. of Pathology, Mount Sinai School of Medicine, New York, New York

BERKOWITZ, LAURA, Dept. of Biology, Indiana University, Bloomington

BERTOLIN, RA, Consorzio Milano Ricerche, Milan, Italy

BHANDT, PURIMA, Dept. of Molecular Biology and Genetics, Johns Hopkins University School of Medicine, Baltimore, Maryland

BIGGIN, MARK, Dept. of Molecular Biophysics and Biochemistry,Yale University, New Haven, Connecticut

BLADER, PATRICK, CNRS, IGBMC, Strasbourg, France

BLITZ, IRA, Dept. of Developmental and Cell Biology, University of California, Irvine

BLOOM, THEODORA, *Current Biology*, London, United Kingdom

BORGHESANI, PAUL, Dept. of Neurology, Beth Israel-Deaconess Medical Center, Harvard Medical School, Boston, Massachusetts

BORYCKI, ANNE-GAELLE, Dept. of Cell and Developmental Biology, University of Pennsylvania, Philadelphia

BOWLES, JOSEPHINE, Centre for Cellular and Molecular Biology, University of Queensland, Brisbane, Australia

BRANDLI, ANDRE, Institute of Cell Biology, Swiss Federal Institute of Technology, Zurich, Switzerland
BRICKMAN, JOSHUA, Lab. of Mammalian Development, National Institute for Medical Research, London, United Kingdom
BRODERS, FLORENCE, CNRS, Institut Curie, Paris, France
BRODY, THOMAS, *The Interactive Fly*, Washington, D.C.
BROIHIER, HEATHER, Dept. of Developmental Genetics, New York University Medical Center, New York, New York
BUMCROT, DAVID, Ontogeny, Inc., Cambridge, Massachusetts
BUSCHER, DIRK, Institute for Molecular Biology, Hannover Medical School, Hannover, Germany
CANN, MARTIN, Dept. of Pharmacology, Cornell University Medical College, New York, New York
CAPECCHI, MARIO, Eccles Institute of Human Genetics, University of Utah, Salt Lake City
CARROLL, SEAN, Lab. of Molecular Biology, University of Wisconsin, Madison
CASCIO, STEPHANIE, Dept. of Neurosurgery, Rhode Island Hospital, Brown University, Providence, Rhode Island
CASEY, ELENA, Dept. of Developmental Biology, National Institute for Medical Research, London, United Kingdom
CHAN, YEE-MING, Dept. of Biochemistry, University of California, San Francisco
CHAZAUD, CLAIRE, CNRS, IGBMC, Strasbourg, France
CHEAH, KATHRYN, Dept. of Biochemistry, University of Hong Kong
CHEN, ELIZABETH, Dept. of Developmental Biology and Biological Sciences, Stanford University, Stanford, California
CHEN, JAU-NIAN, Cardiovascular Research Center, Massachusetts General Hospital, Harvard University, Charlestown, Massachusetts
CHEN, SIMING, Dept. of Pharmacology, Cornell University Medical College, New York, New York
CHEN, XIN, Dept. of Cell Biology, Harvard Medical School, Boston, Massachusetts
CHEN, ZHOUFENG, Dept. of Biology, California Institute of Technology, Pasadena
CHENG, YAN, G.W. Hooper Foundation, University of California, San Francisco
CHIA, WILLIAM, Institute of Molecular and Cell Biology, National University of Singapore, Singapore
CHITNIS, AJAY, Cardiovascular Research Center, Massachusetts General Hospital, Harvard Medical School, Charlestown, Massachusetts
CHO, KEN, Dept. of Developmental Biology, University of California, Irvine
CONNORS, STEPHANIE, Dept. of Cell and Molecular Biology, University of Pennsylvania School of Medicine, Philadelphia
CORBIN, JOSHUA, Brookdale Center for Molecular Biology, Mount Sinai School of Medicine, New York, New York
COROMINAS, MONTSERRAT, Dept. of Genetics, University of Barcelona, Spain
COSTANTINI, FRANK, Dept. of Genetics and Development, Columbia University College of Physicians & Surgeons, New York, New York
CREASE, DEVANAND, Wellcome-CRC Institute of Cancer and Developmental Biology, Cambridge, United Kingdom
CROSIER, PHILIP, Dept. of Molecular Medicine, University of Auckland School of Medicine, Auckland, New Zealand
CZERNY, THOMAS, Research Institute of Molecular Pathology, Vienna, Austria
DAHMANN, CHRISTIAN, Dept. of Zoology, University of Zurich, Switzerland
DANILOV, VLATKO, Institut fur Genetik, Forschungszentrum Karlsruhe, Karlsruhe, Germany
DAUBAS, PHILIPPE, Dept. of Molecular Biology, CNRS, Institut Pasteur, Paris, France
DAVIS, NICOLE, Dept. of Genetics, Harvard Medical School, Cambridge, Massachusetts
DE ROBERTIS, EDDY, Dept of Biological Chemistry, University of California School of Medicine, Los Angeles
DEARDORFF, MATTHEW, Dept. of Medicine, University of Pennsylvania School of Medicine, Philadelphia
DEAROLF, CHARLES, Dept. of Pediatrics, Massachusetts General Hospital, Boston, Massachusetts
DENG, CHUXIA, Lab. of Biochemistry and Metabolism, NIDDK, National Institutes of Health, Bethesda, Maryland
DESPLAN, CLAUDE, Howard Hughes Medical Institute, Rockefeller University, New York, New York
DOBENS, LEONARD, Dept. of Cutaneous Biology Research, Massachusetts General Hospital, Charlestown, Massachusetts
DOEBLEY, JOHN, Dept. of Plant Biology, University of Minnesota, St. Paul
DRIEVER, WOLFGANG, Dept. of Developmental Biology, University of Freiburg, Freiburg, Germany
DROUIN, JACQUES, Dept. of Molecular Genetics, Institut de Recherches Cliniques de Montreal, Quebec, Canada
DYSON, STEVEN, Wellcome-CRC Institute of Cancer and Developmental Biology, Cambridge, United Kingdom
ENGELKAMP, DIETER, Cell Genetics Section, Medical Research Council, Edinburgh, Scotland, United Kingdom

GRIGORI ENIKOLOPOV, Beckman Neuroscience Center, Cold Spring Harbor Laboratory, Cold Spring Harbor, New York
EPHRUSSI, ANNE, Developmental Biology Program, European Molecular Biology Laboratory, Heidelberg, Germany
EPSTEIN, DOUGLAS, Dept. of Developmental Genetics, Skirball Institute, New York University Medical Center, New York, New York
EZER, SINI, Dept. of Medical Genetics, University of Helsinki, Finland
FAINSOD, ABRAHAM, Dept. of Cellular Biochemistry, Hebrew University-Hadassah Medical School, Jerusalem, Israel
FARMER, SUSAN, Dept. of Biochemistry and Molecular Genetics, University of Alabama, Birmingham
FERGUSON, CHIP, Dept. of Molecular Genetics and Cell Biology, University of Chicago, Chicago, Illinois
FERGUSON, CHRISTINE, Dept. of Craniofacial Development, United Medical and Dental Schools, Guy's Hospital, London, United Kingdom
FINLEY, MICHAEL, Dept. of Cell Biology and Physiology, Washington University School of Medicine, St. Louis, Missouri
FLEISSNER, ERWIN, Div. of Science and Mathematics, Hunter College, City University of New York, New York
FODE, CAROL, CNRS, IGBMC, Strasbourg, France
FOUQUET, BERNADETTE, Cardiovascular Research Center, Massachusetts General Hospital, Charlestown, Massachusetts
FRANK, DALE, Dept. of Biochemistry, Technion, Haifa, Israel
FRENZ, DOROTHY, Dept. of Otolaryngology, Anatomy, and Structural Biology, Albert Einstein College of Medicine, Bronx, New York
FUSHIMI, DAISUKE, Dept. of Biophysics, Kyoto University, Kyoto, Japan
GAMMILL, LAURA, Dept. of Biology, Whitehead Institute for Biomedical Research, Massachusetts Institute of Technology, Cambridge, Massachusetts
GILBOA, LILACH, Dept. of Neurobiochemistry, Tel Aviv University, Ramat Aviv, Israel
GOLDMAN, DEVORAH, Dept. of Anatomy, University of California, San Francisco
GOODMAN, COREY, Dept. of Molecular and Cell Biology, Howard Hughes Medical Institute, University of California, Berkeley
GORFINKIEL, NICOLE, Centro de Biologia Molecular-CSIC, Facultad de Ciencias, Universidad Autonoma de Madrid, Spain
GOTOH, YUKIKO, Dept. of Virus Research, Kyoto University, Kyoto, Japan
GRINBLAT, YEVGENYA, Whitehead Institute for Biomedical Research, Massachusetts Institute of Technology, Cambridge, Massachusetts
GRINDLEY, JUSTIN, Dept. of Cell Biology, Vanderbilt University School of Medicine, Nashville, Tennessee
GROSSNIKLAUS, UELI, Delbrück Laboratory, Cold Spring Harbor Laboratory, Cold Spring Harbor, New York
GRUNWALD, DAVID, Dept. of Human Genetics, University of Utah, Salt Lake City
GULISANO, MASSIMO, Dept. of Developmental Neurobiology, United Medical and Dental Schools, Guy's Hospital, London, United Kingdom
GURDON, JOHN, Dept. of Zoology, Molecular Embryology Research Unit, Cancer Research Campaign, University of Cambridge, Cambridge, United Kingdom
HAFFTER, PASCAL, Max-Planck-Institute, Tübingen, Germany
HARIHARAN, ISWAR, Cancer Center, Massachusetts General Hospital, Harvard Medical School, Charlestown, Massachusetts
HARLAND, RICHARD, Dept. of Molecular and Cell Biology, Div. of Biochemistry and Molecular Biology, University of California, Berkeley
HARRIS, SARAH, Dept. of Molecular Genetics, University of Glasgow, Anderson College, Lanarkshire, Scotland, United Kingdom
HARRIS, STEPHEN, Dept. of Medicine, University of Texas Health Science Center, San Antonio
HARVEY, RICHARD, Dept. of Molecular Biology, Walter and Eliza Hall Institute of Medical Research, Royal Melbourne Hospital, Victoria, Australia
HAUBENWALLNER, SABINE, Dept. of Molecular Genetics, University and Biocenter, Vienna, Austria
HAUSDORFF, SHARON, Dept. of Cancer Biology, Harvard Medical School, Boston, Massachusetts
HAYASHI, SHIGEO, Genetic Stock Research Center, National Institute of Genetics, Mishima, Japan
HENDRICKSON, MARY, Dept. of Genetics and Development, Columbia University, New York, New York
HENKEMEYER, MARK, Center for Developmental Biology, University of Texas Southwestern Medical Center, Dallas
HERAULT, YANN, Dept. of Zoology and Animal Biology, University of Geneva, Geneva, Switzerland
HERSKOWITZ, Ira, Dept. of Biochemistry, University of California, San Francisco
HOGAN, BRIGID, Dept. of Cell Biology, Howard Hughes Medical Institute, Vanderbilt University School of Medicine, Nashville, Tennessee
HOLDENER, BERNADETTE, Dept. of Biochemistry and Cell Biology, State University of New York, Stony Brook
HOLLEMANN, THOMAS, Dept. of Developmental Biochemistry, University of Goettingen, Goettingen, Germany

HOODLESS, PAMELA, Prog. in Developmental Biology, Hospital for Sick Children, Toronto, Canada

HOPKINS, NANCY, Dept. of Biology, Center for Cancer Research, Massachusetts Institute of Technology, Cambridge, Massachusetts

HORB, MARKO, Dept. of Biochemistry and Cell Biology, State University of New York, Stony Brook

HORVITZ, H. ROBERT, Dept. of Biology, Howard Hughes Medical Institute, Massachusetts Institute of Technology, Cambridge, Massachusetts

HOSKINS, SALLY, Dept. of Biology, City College of New York, New York

HSIAO, EDWARD, Dept. of Molecular Biology and Genetics, Johns Hopkins University School of Medicine, Baltimore, Maryland

HU, GEZHI, Center for Advanced Biotechnology and Medicine, UMDNJ-Robert Wood Johnson Medical School, Piscataway, New Jersey

HUTSON, RICHARD, Lab. of Pathology, National Cancer Institute, National Institutes of Health, Bethesda, Maryland

IMONDI, RALPH, Dept. of Neuroscience, Albert Einstein College of Medicine, Bronx, New York

INGHAM, PHILIP, Developmental Genetics Program, The Krebs Institute, University of Sheffield, Sheffield, United Kingdom

ISAACS, HARRY, School of Biology and Biochemistry, University of Bath, Bath, United Kingdom

ITASAKI, NOBUE, Dept. of Neurobiology, Medical Research Council, National Institute for Medical Research, London, United Kingdom

ITO, NORIKO, Dept. of Molecular and Cell Biology, University of California, Berkeley

ITOH, KEIJI, Molecular Medicine Unit, Beth Israel Hospital, Harvard Medical School, Boston, Massachusetts

IZPISÚA-BELMONTE, JUAN CARLOS, Lab. of Gene Expression, The Salk Institute for Biological Studies, La Jolla, California

JACOBS, DAVID, Dept. of Biology, University of California, Los Angeles

JAN, YUH NUNG, Dept. of Physiology, University of California, San Francisco

JEANNOTTE, LUCIE, Centre de Recherche de L'Hotel-Dieu de Quebec, Quebec, Canada

JENSEN, JAN, Dept. of Developmental Biology, Hagedorn Research Institute, Gentofte, Denmark

JESSELL, THOMAS, Dept. of Biochemistry and Molecular Genetics, Howard Hughes Medical Institute, Columbia University College of Physicians & Surgeons, New York, New York

JOHNSON, MICHELLE, Lab. of Pediatric Surgical Research, Massachusetts General Hospital, Harvard Medical School, Boston, Massachusetts

JOYNER, ALEXANDRA, Developmental Genetics Program, Skirball Institute, New York University Medical Center, New York, New York

KAMMANDEL, BIRGITTA, Dept. of Molecular Cell Biology, Max-Planck-Institute, Goettingen, Germany

KAPRIELIAN, ZAVEN, Dept. of Pathology and Neuroscience, Albert Einstein College of Medicine, Bronx, New York

KAWAMURA, KOKI, Dept. of Anatomy, Keio University School of Medicine, Tokyo, Japan

KELLY, ANNEMARIE, Medical Research Council, Institute of Hearing Research, Nottingham, United Kingdom

KENYON, CYNTHIA, Dept. of Biochemistry and Biophysics, University of California, San Francisco

KESSLER, DANIEL, Dept. of Cell and Developmental Biology, University of Pennsylvania School of Medicine, Philadelphia

KHAN, TASLIMA, Dept. of Developmental Biology, National Institute for Medical Research, London, United Kingdom

KIM, YONGSOK, Lab. of Molecular Cardiology, National Institutes of Health, Bethesda, Maryland

KIMURA, SHIOKO, Lab. of Metabolism, National Cancer Institute, National Institutes of Health, Bethesda, Maryland

KING, NICOLE, Dept. of Molecular and Cellular Biology, Environmental Health, University of Cincinnati, Cincinnati, Ohio

KONDO, SHIGERU, Dept. of Medical Chemistry, Kyoto University School of Medicine, Kyoto, Japan

KONDOH, HISATO, Dept. of Molecular and Cellular Biology, Osaka University, Osaka, Japan

KOOPMAN, PETER, Dept. of Molecular and Cellular Biology, University of Queensland, Brisbane, Australia

KOSEKI, HARUHIKO, Div. of Molecular Immunology, Center for Biomedical Science, Chiba University School of Medicine, Chiba, Japan

KOUTSOURAKIS, MANOUSSOS, Dept. of Cell Biology and Genetics, Erasmus University, Rotterdam, The Netherlands

KRASNOW, MARK, Dept. of Biology, Stanford University School of Medicine, Stanford, California

KROLL, KRISTEN, Dept. of Cell Biology, Harvard Medical School, Boston, Massachusetts

KRUMLAUF, ROBB, Dept. of Developmental Neurobiology, Medical Research Council, National Institute for Medical Research, London, United Kingdom

KUEHN, MICHAEL, Experimental Immunology Branch, National Cancer Institute, National Institutes of Health, Bethesda, Maryland

KUME, SHOEN, Mikoshiba Calciosignal Net Project, ERATO, Japan Science and Technology Corp., Tokyo, Japan

KUO, CHAY, Committee on Genetics, University of Chicago, Chicago, Illinois

KUROIWA, ATSUSHI, Div. of Biological Science, Nagoya University, Nagoya, Japan

LACY, ELIZABETH, Dept. of Molecular Biology, Memorial Sloan-Kettering Cancer Center, New York, New York

LAKE, ROBERT, Dept. of Cell Biology, Howard Hughes Medical Institute, Yale University, New Haven, Connecticut

LANCTOT, CHRISTIAN, Dept. of Molecular Genetics, Institut de Recherches Cliniques de Montreal, Quebec, Canada

LANE, CONNIE, Dept. of Biology, University of California, Santa Barbara

LARRAIN, JUAN, Dept. of Molecular and Cellular Biology, Catholic University of Chile, Santiago, Chile

LECUIT, THOMAS, Developmental Biology Program, European Molecular Biology Laboratory, Heidelberg, Germany

LEE, KYU-HO, Dept. of Cardiology and Pediatrics, Children's Hospital, Harvard Medical School, Boston, Massachusetts

LEHMANN, RUTH, Developmental Genetics Program, Skirball Institute, New York University Medical Center, New York, New York

LETTICE, LAURA, Human Genetics Unit, Medical Research Council, Western General Hospital, Edinburgh, Scotland, United Kingdom

LEVIN, MARGARET, Dept. of Biological Sciences, Stanford University, Stanford, California

LEVINE, MICHAEL, Div. of Genetics, University of California, Berkeley

LEVY-STRUMPF, NAOMI, Dept. of Molecular Genetics, Weizmann Institute of Science, Rehovot, Israel

LEWANDOSKI, MARK, Dept. of Anatomy, University of California, San Francisco

LEYNS, LUC, Dept. of Biological Chemistry, Howard Hughes Medical Institute, University of California, Los Angeles

LI, HUASHUN, Dept. of Anatomy and Neurobiology, Washington University School of Medicine, St. Louis, Missouri

LOCKWOOD, WENDY, Dept. of Biology, University of Michigan, Ann Arbor

LOGAN, MALCOLM, Dept. of Genetics, Harvard Medical School, Boston, Massachusetts

LOOMIS, CYNTHIA, Dept. of Dermatology and Cell Biology, New York University Medical Center, New York, New York

LOPEZ-DEE, ZENAIDA, Institute of Biology, University of the Philippines, Guezon, Philippines

LOSICK, RICHARD, Dept. of Cellular and Developmental Biology, Harvard University, Cambridge, Massachusetts

LOWE, CHRISTOPHER, Dept. of Ecology and Evolution, State University of New York, Stony Brook

MA, HONG, Delbrück Laboratory, Cold Spring Harbor Laboratory, Cold Spring Harbor, New York

MA, QIUFU, Div. of Biology, California Institute of Technology, Pasadena

MAHANTHAPPA, NAGESH, Ontogeny, Inc., Cambridge, Massachusetts

MAJUMDAR, ARINDAM, Renal Unit, Massachusetts General Hospital, Charlestown, Massachusetts

MALIK, SUNDEEP, Dept. of Pharmacology and Physiology, University of Rochester, Rochester, New York

MANNERVIK, MATTIAS, Dept. of Medical Immunology and Microbiology, Karolinska Institute, Uppsala, Sweden

MAROTEAUX, LUC, CNRS, IGBMC, Strasbourg, France

MARTIENSSEN, ROBERT, Delbrück Laboratory, Cold Spring Harbor Laboratory, Cold Spring Harbor, New York

MARTIN, GAIL, Dept. of Anatomy, University of California, San Francisco

MARTIN-BLANCO, ENRIQUE, Dept. of Zoology, University of Cambridge, Cambridge, United Kingdom

MATISE, MICHAEL, Dept. of Developmental Genetics, Skirball Institute, New York University Medical Center, New York, New York

MATSUI, HIDEO, Dept. of Medicine and Pharmacology, VA Medical Center, University of Tennessee, Memphis

MATSUNO, KENJI, Dept. of Neuroanatomy, Osaka University Medical School, Osaka, Japan

MATTESON, PAUL, Dept. of Molecular Biology, Lewis Thomas Laboratory, Princeton University, Princeton, New Jersey

MAYES, CARYL, Developmental Biology Program, European Molecular Biology Laboratory, Heidelberg, Germany

MCCORMICK, BETSY, Fred Hutchison Cancer Research Center, Seattle, Washington

MCNEILL, HELEN, Dept. of Biological Sciences, Stanford University, Stanford, California

MECHTA, FATIMA, Dept. of Molecular Biology of Development, CNRS, Ecole Normale Superieure, Paris, France

MEHRA, ARUN, Dept. of Anatomy and Cell Biology, University of Toronto, Toronto, Canada

MEIJLINK, FRITS, Hubrecht Laboratory, Netherlands Institute for Developmental Biology, Utrecht, The Netherlands

MELTON, DOUGLAS, Dept. of Molecular and Cellular Biology, Howard Hughes Medical Institute, Harvard University, Cambridge, Massachusetts

MERLINO, GLENN, Dept. of Molecular Biology, National Cancer Institute, National Institutes of Health, Bethesda, Maryland

MEYEROWITZ, ELLIOT, Dept. of Biology, California Institute of Technology, Pasadena

MEYERS, ERIK, Dept. of Anatomy, University of California, San Francisco
MINOWADA, GEORGE, Dept. of Anatomy, University of California School of Medicine, San Francisco
MITCHELL, PAMELA, Dept. of Pharmacology, University of Zurich, Zurich, Switzerland
MIURA, NAOYUKI, Dept. of Biochemistry, Akita University School of Medicine, Akita, Japan
MOORE, ADRIAN, Human Genetics Unit, Medical Research Council, Western General Hospital, Edinburgh, Scotland, United Kingdom
MOORE, JAMES, Delbrück Laboratory, Cold Spring Harbor Laboratory, Cold Spring Harbor, New York
MOOS, MALCOLM, Lab. of Developmental Biology, Center for Biologics Evaluation and Research, Food and Drug Administration, Bethesda, Maryland
MORAIS DA SILVA, SARA, Dept. of Developmental Genetics, National Institute for Medical Research, London, United Kingdom
MORGAN, BRUCE, Dept. of Cutaneous Biology Research, Massachusetts General Hospital, Harvard Medical School, Charlestown, Massachusetts
MURCIA, NOEL, Dept. of Biology, Oak Ridge National Laboratory, Oak Ridge, Tennessee
MYAT, MONN MONN, Dept. of Immunology, Rockefeller University, New York, New York
MYSOREKAR, INDIRA, Dept. of Molecular Biology and Pharmacology, Washington University School of Medicine, St. Louis, Missouri
NASCONE, NANETTE, Dept. of Cell Biology, Harvard Medical School, Boston, Massachusetts
NASTOS, ARISTOTELIS, Institute of Cell Biology, University Hospital of Essen, Essen, Germany
NATHANS, JEREMY, Dept. of Molecular Biology and Genetics, Howard Hughes Medical Institute, Johns Hopkins University School of Medicine, Baltimore, Maryland
NEUHAUS, HERBERT, Dept. of Bone and Connective Tissue Biology Research, Genetics Institute, Inc., Cambridge, Massachusetts
NGO-MULLER, VALERIE, Dept. of Cell and Molecular Biology, Tulane University, New Orleans, Louisiana
NGUYEN, HANH, Dept. of Medicine and Cardiology, Albert Einstein College of Medicine, Bronx, New York
NICOLAS, JEAN-FRANCOIS, Dept. of Molecular Biology of Development, Institut Pasteur, Paris, France
NIEHRS, CHRISTOF, Div. of Molecular Embryology, Deutsches Krebsforschunzentrum, Heidelberg, Germany
NILSON, LAURA, Dept. of Molecular Biology, Princeton University, Princeton, New Jersey
NISHIDA, YASUYOSHI, Div. of Biological Science, Nagoya University Graduate School of Science, Nagoya, Japan
NISWANDER, LEE, Memorial Sloan-Kettering Cancer Institute, New York, New York
NOJI, SUMIHARE, Dept. of Biological Science and Technology, University of Tokushima, Tokushima, Japan
NUSSE, ROEL, Dept. of Developmental Biology, Howard Hughes Medical Institute, Stanford University, Stanford, California
NYBAKKEN, KENT, G.W. Hooper Foundation, University of California, San Francisco
OAKEY, REBECCA, Div. of Genetics, Children's Hospital of Philadelphia, Philadelphia, Pennsylvania
O'CONNOR, MICHAEL, Dept. of Molecular Biology and Biochemistry, University of California, Irvine
OHLMEYER, JOHANNA, Dept. of Biological Sciences, Columbia University, New York, New York
OHYAMA, KYOJI, Dept. of Anatomy, Keio University School of Medicine, Tokyo, Japan
OKADA, AMI, Dept. of Biological Sciences, Stanford University, Stanford, California
OLIVER, GUILLERMO, Dept. of Genetics, St. Jude Children's Research Hospital, Memphis, Tennessee
OLSON, ERIC, Dept. of Molecular Biology and Oncology, University of Texas Southwestern Medical Center, Dallas
ORNITZ, DAVID, Dept. of Molecular Biology and Pharmacology, Washington University School of Medicine, St. Louis, Missouri
OVERBEEK, PAUL, Dept. of Cell Biology, Howard Hughes Medical Institute, Baylor College of Medicine, Houston, Texas
PAI, JIH-TUNG, Dept. of Molecular Genetics, University of Texas Southwestern Medical Center, Dallas
PENNATI, ROBERTA, Dept. of Biology, University of Milan, Milan, Italy
PERA, EDGAR, Dept. of Molecular Cell Biology, Max-Planck-Institute, Goettingen, Germany
PEREIRA, FRED, Dept. of Cell Biology, Baylor College of Medicine, Houston, Texas
PERRIMON, NORBERT, Dept. of Genetics, Harvard Medical School, Boston, Massachusetts
PHILPOTT, ANNA, Dept. of Cell Biology, Harvard Medical School, Boston, Massachusetts
PHIMISTER, BETTE, *Nature Genetics*, New York, New York
PICCOLO, STEFANO, Dept. of Biological Chemistry, Howard Hughes Medical Institute, University of California, Los Angeles
PITUELLO, FABIENNE, Centre de Biologie du Developpement, CNRS, Université Paul Sabatier, Toulouse, France
POIRIER, FRANCOISE, ICGM, CNRS, INSERM, Paris, France

PUJADES, CRISTINA, Dept. of Molecular Biology of Development, CNRS, Ecole Normale Superieure, Paris, France

PUOTI, ALESSANDRO, Dept. of Biochemistry, University of Wisconsin, Madison

QUATRANO, RALPH, Dept. of Biology, University of North Carolina, Chapel Hill

RADICE, GLENN, Dept. of Obstetrics and Gynecology, University of Pennsylvania, Philadelphia

RAMAMURTHY, BASKAR, Delbrück Laboratory, Cold Spring Harbor Laboratory, Cold Spring Harbor, New York

RANKIN, CHRISTOPHER, Dept. of Molecular Biology and Genetics, Johns Hopkins University School of Medicine, Baltimore, Maryland

RAO, NAGARAJA, Dept. of Cancer Research, Eli Lilly & Co., Indianapolis, Indiana

RATNER, DAVID, Dept. of Biology, Amherst College, Amherst, Massachusetts

RAZ, EREZ, Cardiovascular Research Center, Massachusetts General Hospital, Charlestown, Massachusetts

RENNEBECK, GABRIELA, Dept. of Zoology, University of Texas, Austin

RIUS, RICARDO, NHLBI, LBG, National Institutes of Health, Bethesda, Maryland

RIVERA-POMAR, ROLANDO, Dept. of Molecular Developmental Biology, Max-Planck-Institute, Goettingen, Germany

ROBBINS, DAVID, G.W. Hooper Foundation, University of California, San Francisco

ROBERTSON, ELIZABETH, Dept. of Cellular and Developmental Biology, Harvard University, Cambridge, Massachusetts

ROSA, FREDERIC, CNRS, INSERM, Ecole Normale Superieure, Paris, France

ROSEN, MITCHELL, Developmental Biology Branch, RTD-NHEERL, U.S. Environmental Protection Agency, Research Triangle Park, North Carolina

ROSEN, VICKI, Genetics Institute, Inc., Cambridge, Massachusetts

ROSENTHAL, ARNON, Dept. of Developmental Biology, Genentech, Inc., South San Francisco, California

ROSENTHAL, NADIA, Cardiovascular Research Center, Massachusetts General Hospital, Charlestown, Massachusetts

ROSSANT, JANET, Samuel Lunenfeld Research Institute, Mount Sinai Hospital, Toronto, Ontario, Canada

ROSSNAGEL, KARIN, Max-Planck-Institute, Tübingen, Germany

ROTH, SIEGFRIED, Dept. of Biochemistry, Max-Planck-Institute, Tübingen, Germany

ROVESCALLI, ALLESSANDRA, Lab. of Biochemistry and Genetics, National Heart, Lung, and Blood Institute, National Institutes of Health, Bethesda, Maryland

ROWE, PETER, Children's Medical Research Institute, Wentworthville, New South Wales, Australia

ROWITCH, DAVID, Dept. of Molecular and Cellular Biology, Harvard University, Cambridge, Massachusetts

RUBIN, GERALD, Dept. of Molecular and Cell Biology, Howard Hughes Medical Institute, University of California, Berkeley

RUNKO, ERIK, Dept. of Neuroscience, Albert Einstein College of Medicine, Bronx, New York

RUPP, RALPH, Dept. of Biology, Friedrich-Miescher Laboratorium, Tübingen, Germany

RUTHER, ULRICH, Institute for Molecular Biology, Hannover Medical School, Hannover, Germany

RYAN, AIMEE, Dept. of Medicine, University of California at San Diego, La Jolla

SAKA, YASUSHI, Div. of Developmental Biology, National Institute for Medical Research, London, United Kingdom

SAMAKOVLIS, CHRISTOS, Center for Molecular Pathogenesis, Umea University, Umea, Sweden

SAMPATH, KARUNA, Institute of Molecular Agrobiology, Singapore

SARGENT, MICHAEL, Dept. of Developmental Biology, National Institute for Medical Research, London, United Kingdom

SASAKI, HIROSHI, Dept. of Molecular and Cell Biology, Osaka University, Osaka, Japan

SATO, SHERYL, Dept. of Genetics and Biochemistry, NIDDK, National Institutes of Health, Bethesda, Maryland

SATOH, NORIYUKI, Dept of Zoology, Kyoto University, Kyoto, Japan

SAXTON, TRACY, Samuel Lunenfeld Research Institute, Mount Sinai Hospital, Toronto, Ontario, Canada

SCHAUERTE, HEIKE, Max-Planck-Institute, Tübingen, Germany

SCHIER, ALEXANDER, Dept. of Developmental Genetics and Cell Biology, Skirball Institute, New York University Medical Center, New York, New York

SCHMID, BETTINA, Dept. of Cell and Developmental Biology, University of Pennsylvania, Philadelphia

SCHULTHEISS, THOMAS, Dept. of Biological Chemistry and Molecular Pharmacology, Harvard Medical School, Boston, Massachusetts

SCHWEITZER, RONEN, Dept. of Genetics, Harvard Medical School, Boston, Massachusetts

SCOTT, MATT, Dept. of Developmental Biology, Stanford University School of Medicine, Stanford, California

SEMBA, ICHIRO, Craniofacial Development Section, National Institutes of Health, NIAMS, Bethesda, Maryland

SETTLES, A. MARK, Delbrück Laboratory, Cold Spring Harbor Laboratory, Cold Spring Harbor, New York

SHEIKH, HUMA, Dept. of Developmental Neurobiology, United Medical and Dental Schools, Guy's Hospital, London, United Kingdom

SHEN, MICHAEL, UMDNJ-Robert Wood Johnson Medical School, Piscataway, New Jersey

SHILO, BENNY, Dept. of Molecular Genetics, Weizmann Institute of Science, Rehovot, Israel

SHISHIDO, EMIKO, National Institute for Basic Biology, Okazaki, Aichi, Japan

SHOJI, WATARU, Dept. of Biology, University of Michigan, Ann Arbor

SHUM, LILLIAN, Craniofacial Development Section, NIAMS, National Institutes of Health, Bethesda, Maryland

SIEGEL, VIVIAN, *Cell*, Cambridge, Massachusetts

SIROTKIN, HOWARD, Dept. of Molecular Genetics, Albert Einstein College of Medicine, Bronx, New York

SIVASANKARAN, RAJEEV, Zoological Institute, University of Zurich, Zurich, Switzerland

SIVE, HAZEL, Whitehead Institute for Biomedical Research, Massachusetts Institute of Technology, Cambridge, Massachusetts

SMIDT, MARTEN, Dept. of Medical Pharmacology, University of Utrecht, Utrecht, The Netherlands

SMITH, JIM, Dept. of Developmental Biology, National Institute for Medical Research, London, United Kingdom

SONG, JIHWAN, Cardiovascular Research Center, Massachusetts General Hospital, Charlestown, Massachusetts

SPICER, DOUGLAS, Dept. of Biological Chemistry and Molecular Pharmacology, Harvard Medical School, Boston, Massachusetts

SPRADLING, ALLAN, Dept. of Embryology, Howard Hughes Medical Institute, Carnegie Institution of Washington, Baltimore, Maryland

SRIVASTAVA, DEEPAK, Dept. of Molecular Biology and Oncology, University of Texas Southwestern Medical Center, Dallas

STEMPLE, DEREK, Div. of Developmental Biology, National Institute for Medical Research, London, United Kingdom

STEWART, DAVID, Cold Spring Harbor Laboratory, Cold Spring Harbor, New York

STILLMAN, BRUCE, James Laboratory, Cold Spring Harbor Laboratory, Cold Spring Harbor, New York

STRUHL, GARY, Dept. of Neurobiology, Howard Hughes Medical Institute, Columbia University, New York, New York

STRUMPF, DAN, Dept. of Molecular Genetics, Weizmann Institute of Science, Rehovot, Israel

SU, YI-CHI, Dept. of Molecular Pathogenesis, New York University Medical Center, New York, New York

SUGIMOTO, HANA, Div. of Medical Sciences, Harvard University, Cambridge, Massachusetts

SUPP, DOROTHY, Dept. of Developmental Biology, Children's Hospital of Cincinnati Research Foundation, Cincinnati, Ohio

SYMES, KAREN, Dept. of Biochemistry, Boston University School of Medicine, Boston, Massachusetts

TABATA, TETSUYA, Institute of Molecular and Cellular Biosciences, University of Tokyo, Tokyo, Japan

TABIN, CLIFFORD, Dept. of Genetics, Harvard Medical School, Boston, Massachusetts

TADA, MASAZUMI, Div. of Developmental Biology, National Institute for Medical Research, London, United Kingdom

TAKAHASHI, YOSHIKO, Dept. of Bioscience, Kitasato University, Sagamihara, Japan

TAKEICHI, MASATOSHI, Dept. of Biophysics, Faculty of Science, Kyoto University, Kyoto, Japan

TAM, PATRICK, Embryology Unit, Children's Medical Research Institute, Wentworthville, New South Wales, Australia

TESSIER-LAVIGNE, MARC, Div. of Anatomy, University of California, San Francisco

THAYER, JEANETTE, Dept. of Anatomy, Midwestern University, Glendale, Arizona

THESLEFF, IRMA, Institute of Biotechnology, University of Helsinki, Helsinki, Finland

THOMAS, ADRI, Dept. of Molecular Cell Biology, University of Utrecht, Utrecht, The Netherlands

THOMAS, LINDA, National Institute of Dental Research, National Institutes of Health, Bethesda, Maryland

THOMAS, PAUL, Lab. of Mammalian Development, National Institute for Medical Research, London, United Kingdom

THOMAS, SHEILA, Dept. of Cancer Biology, Beth Israel-Deaconess Medical Center, Harvard Medical School, Boston, Massachusetts

THOMSEN, GERALD, Dept. of Biochemistry and Cell Biology, State University of New York, Stony Brook

TIAN, HUI, Dept. of Molecular Genetics, University of Texas Southwestern Medical Center, Dallas

TIAN, JINGDONG, Dept. of Biochemistry and Cell Biology, State University of New York, Stony Brook

TIRET, LAURENT, Dept. of Molecular Biology, CNRS, Institut Pasteur, Paris, France

TORRES, MONICA, Dept. of Pharmacology, Howard Hughes Medical Institute, University of Washington, Seattle

TOYAMA, REIKO, Lab. of Molecular Genetics, National Institutes of Child Health and Human Development, National Institutes of Health, Bethesda, Maryland

TSUDA, LEO, Dept. of Biological Chemistry, University of California, Los Angeles

UENO, NAOTO, Dept. of Development, Div. of Morphogenesis, National Institute for Basic Biology, Okazaki, Japan

UTSET, MANUEL, Dept. of Physiology, Howard Hughes Medical Institute, University of California, San Francisco

VIELLE, JEAN-PHILIPPE, Delbrück Laboratory, Cold Spring Harbor Laboratory, Cold Spring Harbor, New York

VINCENT, ALAIN, Centre de Biologie du Developpement, CNRS, Université Paul Sabatier, Toulouse, France

VOGT, THOMAS, Dept. of Molecular Biology, Lewis Thomas Laboratory, Princeton University, Princeton, New Jersey

WEINBERG, ERIC, Dept. of Biology, Goddard Laboratories, University of Pennsylvania, Philadelphia

WEINSTEIN, MICHAEL, Lab. of Biochemistry and Metabolism, NIDDK, National Institutes of Health, Bethesda, Maryland

WHARTON, KRISTI, Dept. of Molecular Biology, Cell Biology, and Biochemistry, Brown University, Providence, Rhode Island

WILKINSON, DAVID, Div. of Developmental Neurobiology, Medical Research Council, National Institute for Medical Research, London, United Kingdom

WILLIAMS, TREVOR, Dept. of Biology, Yale University, New Haven, Connecticut

WILLIAMSON, TONI, Dept. of Cell Biology, Ludwig Institute for Cancer Research, La Jolla, California

WIMMER, ERNST, Howard Hughes Medical Institute, Rockefeller University, New York, New York

WRIGHT, CHRISTOPHER, Dept. of Cell Biology, Vanderbilt University School of Medicine, Nashville, Tennessee

WU, CHIUNG-YUAN, Dept. of Cell Biology, Vanderbilt University School of Medicine, Nashville, Tennessee

WUNDERLE, VERONIQUE, Dept. of Genetics, Human Molecular Genetics, University of Cambridge, Cambridge, United Kingdom

WUNDERLICH, WINFRIED, Institute of Molecular Genetics, University of Vienna, Vienna, Austria

WURST, WOLFGANG, Dept. of Mammalian Genetics, GSF-Research Center, Munich, Germany

YANAGAWA, SHIN-ICHI, Dept. of Viral Oncology, Institute for Virus Research, Kyoto University, Kyoto, Japan

YANG, MING, Delbrück Laboratory, Cold Spring Harbor Laboratory, Cold Spring Harbor, New York

YAWORSKY, PAUL, Mayo Clinic, S.C. Johnson Medical Research Center, Scottsdale, Arizona

YELICK, PAMELA, Dept. of Cytokine Biology, Forsythe Dental Center, Boston, Massachusetts

YELON, DEBORAH, Dept. of Biochemistry and Biophysics, University of California, San Francisco

YISRAELI, JOEL, Dept. of Anatomy and Cell Biology, Hebrew University Medical School, Jerusalem, Israel

YU, RUTH, Lab. of Biochemistry, Dept. of Virus Research, Kyoto University, Kyoto, Japan

ZAPPAVIGNA, VINCENZO, Lab. of Gene Expression, DIBIT H San Raffaele, Milan, Italy

ZECCA, MYRIAM, Zoological Institute, University of Zurich, Zurich, Switzerland

ZEITLINGER, JULIE, European Molecular Biology Laboratory, Heidelberg, Germany

ZELTSER, LORI, Dept. of Developmental Neurobiology, United Medical and Dental Schools, Guy's Hospital, London, United Kingdom

ZHANG, CHAOHUI, Dept. of Molecular Oncology, Genentech, Inc., South San Francisco, California

ZHANG, JIAN, Dept. of Cell Biology, University of Miami School of Medicine, Miami, Florida

ZHANG, YAN, Dept. of Biological Sciences, Columbia University College of Physicians & Surgeons, New York, New York

ZHANG, YUE, Dept. of Pharmacology, State University of New York, Stony Brook

ZHAO, RENBIN, Dept. of Molecular Biology and Genetics, Johns Hopkins University Medical School, Baltimore, Maryland

ZHU, WENCHENG, Dept. of Molecular Genetics, Wadsworth Center, State University of New York, Albany

ZIPURSKY, LARRY, Dept. of Biological Chemistry, Howard Hughes Medical Institute, University of California, Los Angeles

ZON, LEONARD, Dept. of Hematology and Oncology, Howard Hughes Medical Institute, Children's Hospital, Boston, Massachusetts

First row: G. Martin; N. Hopkins, E. Fleissner; J. Gurdon
Second row: R. Lehmann; R. Beddington, N. Rosenthal; L. Jeannotte
Third row: I. Herskowitz, T. Grodzicker, R. Losick; R. Krumlauf, M. Tessier-Levigne
Fourth row: S. Carroll, M. Levine

First row: D. Stemple; A. Spralding; P. Ingham; S. Carroll
Second row: R. Nusse, J. Witkowski; I. Herskowitz, B. Stillman
Third row: A. Nastos, A. Moore; R. Losick, E. Robertson
Fourth row: G. Rubin, D. Melton, M. Levine

First row: J. Watson, M. Scott; H. Ma, S. Kunes; L. Gammill
Second row: H. Ma, J. Nathans, M. Levine; E. Meyerowitz; I. Thesleff
Third row: Enjoying the view at the beach; E. Robertson; B. Hogan; J. Smith

First row: D. Melton; J. Watson, M. Capecchi; P. Beachy
Second row: Picnic at Airsley
Third row: V. Siegel, M. Scott; R. Beddington, P. Tam

Foreword

One of the great scientific accomplishments of the past decade was the recognition that the mechanisms used for patterning of tissues and organs during development are remarkably similar among species. What works for flies and frogs also serves human beings very well as embryos acquire their form and identity. To celebrate these marvelous discoveries, the 62nd Symposium in this series focused on pattern formation during development, with a particular emphasis on evolutionarily conserved mechanisms and molecules.

This outstanding series of meetings began in 1933, in the midst of economic depression. Each meeting lasted for 5 weeks while scientists were in residence at Cold Spring Harbor, and there were no time limits on the length of presentations or discussions. This "experiment in scientific procedure," reflecting the apparent tranquility of academic life in those days, lasted for 8 years. The second Symposium, entitled Aspects of Growth, included several papers dealing with differentiation and development.

For the 1941 Symposium, the new Director of the Laboratory, Milislav Demerec, condensed the program into 2 weeks, and his second Symposium focused on The Relation of Hormones to Development, with 17 presentations. In 1948, Demerec changed the meeting to 8 days, held in early June, and this has been the arrangement until this year. Given the dramatic changes in the pace of scientific discovery in the biological sciences, the modern methods for rapid dissemination of information, and the fact that in many families, both spouses have demanding careers, scientists find it difficult, if not impossible, to spend more than a week at a meeting. Thus, I have shortened the program to 5 days, still considered by some to be a long time for a meeting. This was done after much thought, keeping in mind the goals of this particular series and indeed of all the meetings at Cold Spring Harbor Laboratory. The idea is not to present a series of observations from a collection of individuals, but to provide sufficient time during the meeting for the experts in a field of biology to discuss the issues of the day and to determine how progress might be accomplished.

During 8 days in 1954, the Symposium focused on The Mammalian Fetus: Physiological Aspects of Development. The science was descriptive, but some ideas presented were the forerunners of the modern approaches to understanding development. Salome Glueckshon-Waelsch discussed the T (Brachyury) locus and a genetic approach to understanding development that foreshadowed the powerful approaches to come. Then after a long hiatus between Symposia on this topic, Joe Sambrook organized in 1985 the 50th Symposium on the Molecular Biology of Development. The Symposium celebrated the conversion of developmental biology from a descriptive, anatomically based science to a mechanism-driven science of great interest.

In the dozen years since, we have learned much about the exquisite mechanisms that specify development, but most importantly, the evolutionary conservation of the pathways that govern development has emerged. We are no longer bound by technology, as in the early days of these Symposia, and modern technology has taken us to new heights unimagined even 12 years ago.

The field of developmental biology is now very large, and coupled with the decision to shorten the program, the inevitable difficult decisions about the selection of speakers and topics had to be made. I would have liked to include the important topic of imprinting, for example, but perhaps this could be the topic of its own Symposium in the near future. I thank Brigid Hogan and Gerry Rubin for much valuable advice and help with the scope of the meeting and potential speakers. The formal scientific program consisted of 64 oral presentations and a remarkable 233 poster presentations, and the meeting attracted 391 participants. Introductory talks on the first evening were given by Mario Capecchi, Brigid Hogan, Roel Nusse, and Gail Martin, and the Reginald G. Harris Lecture was presented by John Gurdon. Sean Carroll presented a fascinating Dorcas Cummings Lecture to visiting scientists and the local community on the formation and evolution of animal body patterns.

Essential funds to run this meeting were obtained from the National Institute of Child Health and Human Development and the National Institute of Neurological Disorders and Stroke, both branches of the National Institutes of Health. In addition, financial help from the Corporate Sponsors of our meetings program is essential for these Symposia to remain a success and we are most grateful for their continued support. These sponsors are: Alafi Capital Company; Amgen Inc.; BASF Bioresearch Corporation; Becton Dickinson and Company; Boehringer Mannheim Corporation; Bristol Myers Squibb Company; Chiron Corporation; Chugai Research Institute for Molecular Medicine, Inc.; Diagnostic Products Corporation; The Du Pont Merck Pharmaceutical Company; Forest Laboratories, Inc.; Genentech, Inc.; Glaxo Wellcome Inc.; Hoechst Marion Roussel; Hoffmann-La Roche Inc.; Johnson & Johnson; Kyowa Hakko Kogyo Co., Ltd.; Life Technologies, Inc.; Eli Lilly and Company; Merck Genome Research Institute; Novartis Pharma Research; OSI Pharmaceuticals, Inc.; Pall Corporation; The Perkin-Elmer Corporation; Applied Biosystems Division; Pfizer Inc.; Pharmacia & Upjohn, Inc.; Research Genetics, Inc.; Schering-Plough Corporation; SmithKline Beecham Pharmaceuticals; Wyeth-Ayerst Research; and Zeneca Group plc.. Plant Corporate Associates are American Cyanamid Company; Kirin Brewery Co., Ltd.; Monsanto Company; Pioneer Hi-Bred International, Inc.; and Westvaco Corporation.

I thank the staff, including Diane Tighe and Mary Smith, in our meetings and courses office under the talented direction of David Stewart, for their efficient and outstanding organization of this meeting. Mary Horton and Wendy Crowley handled the various grant applications, and Herb Parsons and his staff provided flawless audiovisual assistance. The organization of this meeting relied on great help from my assistant Delia King. It was again a pleasure to work with the Laboratory Press, under the direction of John Inglis, particularly Dorothy Brown and Joan Ebert.

Bruce Stillman
December 1997

Contents

Tissue Specification

Patterning and Transcription

Signaling in Organogenesis

Neural Induction and Pathfinding

Germ Plasm Assembly and Germ Cell Migration in *Drosophila*

C. Rongo,[1,2] H. Tarczy Broihier,[2] L. Moore,[2] M. Van Doren, A. Forbes, and R. Lehmann[3]
Skirball Institute, New York University Medical Center, New York, New York 10016

In most organisms, primordial germ cells are set aside from somatic cells early in development. Germ cells follow a stereotyped differentiation program that leads to the production of egg and sperm. In many organisms, germ cells are formed in a specialized cytoplasm, the germ plasm that is synthesized during oogenesis and deposited in the egg. During embryogenesis, germ cells migrate through the developing embryo to reach mesodermal cell populations that contribute the somatic component of the gonad. In this paper, we summarize and discuss recent advances in our understanding of the genetics of germ plasm assembly and germ cell differentiation in *Drosophila.*

GERM PLASM ASSEMBLY AND GERM CELL DETERMINATION

In *Drosophila*, germ plasm is assembled during oogenesis at the posterior pole of the growing oocyte. Germ plasm contains the posterior determinant *nanos* (*nos*), which is required for abdomen formation, and the electron-dense polar granules that are thought to harbor yet largely unknown factors required for the formation of the pole cells, the primordial germ cell (PGCs) (Illmensee and Mahowald 1974; Illmensee et al. 1976; Wang and Lehmann 1991). Although many genes have been identified that have a role in germ plasm assembly, only the *oskar* (*osk*) gene has been shown to be both necessary and sufficient to direct all steps required for germ cell formation (Ephrussi et al. 1991; Kim-Ha et al. 1991; Ephrussi and Lehmann 1992; for review, see Rongo and Lehmann 1996).

The *osk* gene product is a limiting factor in determining the number of pole cells that form, the amount of *nos* RNA localized, and the amount of Nos protein synthesized (Ephrussi and Lehmann 1992; Smith et al. 1992; Gavis and Lehmann 1994). Mislocalization of *osk* RNA to the anterior pole leads to ectopic germ plasm assembly at the anterior (Ephrussi and Lehmann 1992). At this ectopic site, germ cells form, *nos* RNA becomes localized and translated, and a second abdomen develops in mirror image to the posterior abdomen. This result demonstrates that the location of *osk* RNA within the oocyte defines where germ cells form.

Drosophila ovaries are composed of strings of maturing egg chambers. Each egg chamber consists of the oocyte and its 15 sister cells, the nurse cells, surrounded by somatic follicle cells (for review, see Mahowald and Kambysellis 1980; Spradling 1993). *osk* RNA is transcribed in the nurse cells and then transported into the oocyte through large intercellular bridges, the ring canals, by a microtubule-dependent mechanism (Ephrussi et al. 1991; Kim-Ha et al. 1991; Clark et al. 1994; Pokrywka and Stephenson 1994). During stage 8 of oogenesis, *osk* RNA becomes concentrated at the posterior pole of the oocyte, a process that also requires correct polarization of the microtubule network and the function of the Staufen (Stau) protein as well as tropomyosin II (Fig. 1) (Ephrussi et al. 1991; Kim-Ha et al. 1991; St. Johnston et al. 1991; Clark et al. 1994; Theurkauf 1994; Erdélyi et al. 1995; González-Reyes et al. 1995; Roth et al. 1995). *osk* and three other genes (*vasa* [*vas*], *pipsqueak* [*pip*], *tudor* [*tud*]) are required for the assembly of the germ plasm (for review, see Rongo and Lehmann 1996).

Osk protein is required to anchor *osk* RNA to the posterior pole throughout oogenesis and early embryogenesis (Markussen et al. 1995; Rongo et al. 1995). Oocytes mutant for any of three "delocalizing alleles" of *osk* localize *osk* RNA and Stau protein up to stage 10 of oogenesis, but they do not maintain localization into embryogenesis (Fig. 2) (Ephrussi et al. 1991; Kim-Ha et al. 1991; St Johnston et al. 1991; Rongo 1996). These three delocalizing alleles are caused by nonsense mutations that terminate the protein in its amino-terminal half. In contrast, all known missense mutations of *osk* do not affect *osk* RNA and Stau protein localization (Fig. 2) (Ephrussi et al. 1991; Kim-Ha et al. 1991; St. Johnston et al. 1991; Rongo 1996). Sequences responsible for both the initiation and maintenance of *osk* RNA localization at the posterior pole reside within the *osk* 3′-untranslated region (UTR), and Osk and Stau proteins are required to maintain the localization of *osk* RNA or a reporter RNA containing the *osk* 3′UTR (Kim-Ha et al. 1993; Rongo et al. 1995). These results imply a positive feedback model in which Osk separately conducts multiple functions, including germ plasm assembly and *osk* RNA localization maintenance (Fig. 1).

Two proteins of 69 kD ($p69^{Osk}$) and 54 kD ($p54^{Osk}$) molecular mass are generated from a single species of *osk*

[1]*Present addresses*: University of California, Department of Molecular and Cell Biology, 555 Life Sciences Addition, Berkeley, California 94720; [2]Department of Biology, Massachusetts Institute of Technology, Cambridge, Massachusetts, 02139; [3]Howard Hughes Medical Institute.

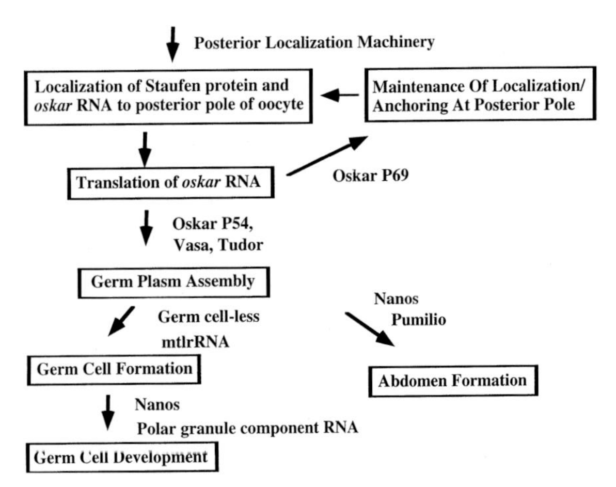

Figure 1. Pathway of germ plasm assembly. The abbreviated pathway focuses on a description of the role of Osk and downstream factors in germ plasm assembly and germ cell development (for more detailed review, see Rongo and Lehmann 1996). Processes are boxed and gene products required in respective process are next to arrow. See text for abbreviations.

RNA (Ephrussi et al. 1991; Kim-Ha et al. 1991; Markussen et al. 1995; Rongo et al. 1995). The relative contribution of each of these proteins was tested by inserting sequences encoding 100 bases of the *Xenopus* β-globin 5′UTR into *osk* genomic sequences just upstream of the first AUG codon to produce $p69^{Osk}$ and upstream of the second AUG codon to produce $p54^{Osk}$. Females that carry the the P[β*gAUG1*] transgene produce predominantly the $p69^{Osk}$ isoform and very little $p54^{Osk}$, whereas females that carry P[β*gAUG2*] transgene produce only $p54^{Osk}$ (Rongo 1996).

To determine whether the $p69^{Osk}$ and $p54^{Osk}$ isoforms could form abdomen, determine germ cells, and maintain *osk* RNA localization, we introduced the β*gAUG1* and β*gAUG2* transgenes into the background of the nonsense mutations osk^{54}/osk^{84} and assayed their ability to function in each of three processes: *osk* RNA localization, germ plasm formation (as measured by the number and frequency of germ cell formation), and *nanos* RNA localization and translation (as assayed by the number of abdominal segments formed) (Fig. 2). We find that $p69^{Osk}$ is required for the maintenance of wild-type levels of *osk* RNA at the posterior of embryos but does not complement the abdominal defects nor the pole cell defects. In contrast, one copy of the β*gAUG2* transgene complements the abdomen and pole cell defects of an *osk* mutant and can give rise to fertile flies. However, the transgenic RNA is poorly maintained during embryogenesis. These results suggest that $p54^{Osk}$ primarily functions in germ plasm formation, including abdomen and pole cell formation, whereas $p69^{Osk}$ is primarily involved in the maintenance of *osk* RNA localization. These findings are consistent with those of Markussen et al. (1995) and Breitweiser et al. (1996). These authors showed that the shorter isoform ($p54^{Osk}$) of *osk* is sufficient to assemble functional germ plasm and that this protein form is a component of the polar granules and associates specifically with germ plasm components such as Vasa. Furthermore, the longer Osk isoform ($p69^{Osk}$) is able to provide function for the maintenance of RNA localization and can interact with Stau but is not a component of the germ plasm and does not interact with Vas protein (Markussen et al. 1995; Breitweiser et al. 1996).

Although the two Osk protein isoforms fulfill different

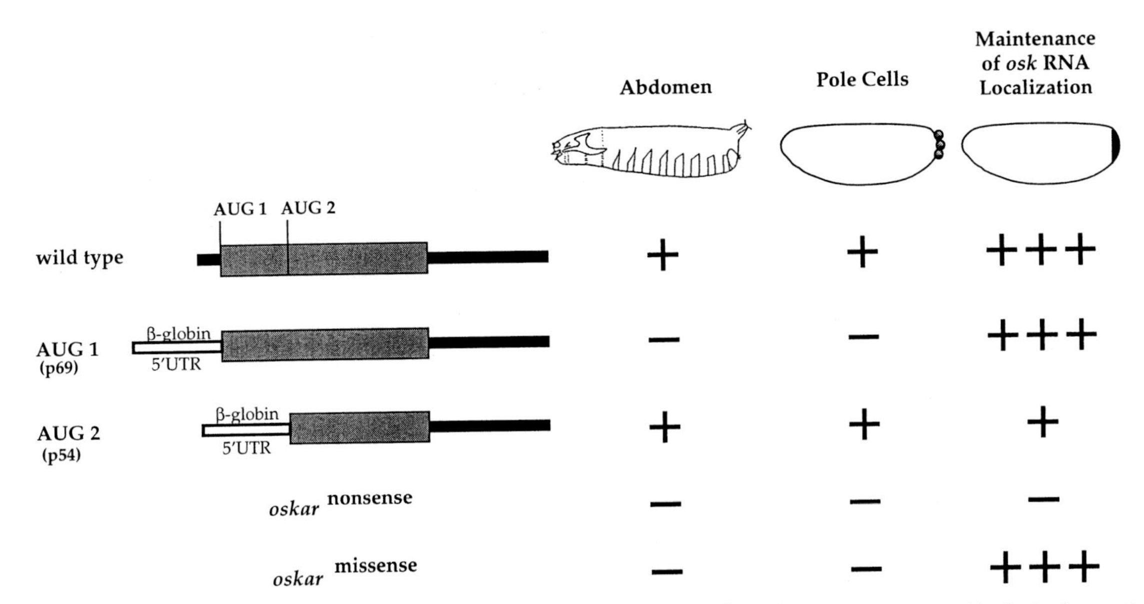

Figure 2. Oskar isoforms have different functions. For each transgenic construct, three phenotypes were tested in the background of a nonsense mutation for *osk*: osk^{54}/osk^{84}: (1) Abdomen formation assayed by number of abdominal segments formed and amount of *nos* RNA localized, (2) pole cells: assayed by percentage of embryos with pole cells and average number of pole cell/embryos, and (3) maintenance of *osk* RNA localization was assayed by in situ hybridization to stage-10 ovaries and embryos. The phenotypes shown are those of a single copy of the transgene. Missense mutation for *osk* is: osk^{166}/osk^{54} or osk^{CE3}/osk^{84}. For details, see Rongo (1996).

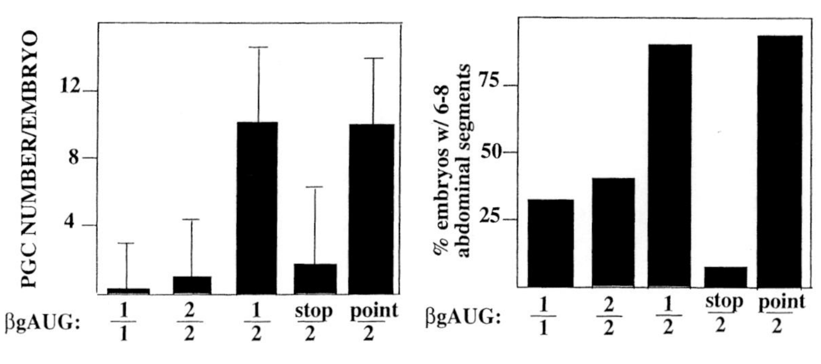

Figure 3. Oskar isoforms cooperate. The number of pole cells formed was determined by staining embryos from transgenic females with anti-Vas antibody. The background for the β*gAUG1*/β*gAUG1*, β*gAUG2*/β*gAUG2*, β*gAUG1*/β*gAUG2* was osk^{54}/osk^{84}. On average, 50 embryo were counted for the number of pole cells (PGCs). For abdomen formation, hatch rates were determined for each genotype, and cuticle preparations were prepared from unhatched embryos. Cuticles were classified as 0–1, 2–5, and 6–8 abdominal segments, respectively. The hatch rates for the different genotypes are β*gAUG1*/β*gAUG1* = 54%, β*gAUG2*/β*gAUG2* = 49%; β*gAUG1*/β*gAUG2* = 93%; β*gAUG2*/osk^{54}/osk^{84}= 32%; β*gAUG2*/osk^{166}/osk^{84} = 90%. The controls osk^{54}/osk^{84} and osk^{166}/osk^{54} produced only embryos with no abdominal segments and no PGCs. stop indicates osk^{54}/osk^{84} t; point osk^{166}/osk^{54} or osk^{CE3}/osk^{84}. Between 100 and 600 cuticles were analyzed for each experiment. For details, see Rongo (1996).

functions, they do cooperate in the assembly of the germ plasm: We find that $p69^{Osk}$ potentiates $p54^{Osk}$ function by fully maintaining *osk* RNA localization as a positive feedback. *osk* mutants carrying one copy of β*gAUG1* and one copy of β*gAUG2* localize more *nos* RNA and Vas protein, form more abdominal segments, and specify more germ cells than two copies of either transgene (Fig. 3). The enhancement of β*gAUG2* activity by β*gAUG1* is also achieved by introducing β*gAUG2* into *osk* missense mutants that completely lack activity required for abdomen and pole cell formation, but retain the ability to maintain RNA localization, but not when introduced in the background of *osk* nonsense mutations (Figs. 2 and 3). Thus, the mechanism of β*gAUG2* enhancement by β*gAUG1* occurs through maintenance of *osk* RNA localization.

A feedback model may explain the interaction between the two Osk isoforms (Fig. 1). *osk* RNA becomes initially localized dependent on the polarity of the microtubule network and the function of Stau protein. Upon localization, *osk* RNA is translated to produce two Osk protein isoforms. The $p69^{Osk}$ isoform stabilizes the initial localization by anchoring *osk* RNA and Stau protein to the posterior pole, allowing the concentrated synthesis of the $p54^{Osk}$ isoform, which functions to assemble the germ plasm. High levels of localized Osk protein ensure that downstream components of the germ plasm such as *nos* and Vas are properly localized.

MATERNAL CONTROL OF GERM CELL FORMATION AND DEVELOPMENT

Most aspects of early germ cell development, such as the precocious formation of germ cells at the posterior pole prior to cellularization of the somatic cells of the embryo are likely directed by factors produced during oogenesis and localized to the posterior pole. Furthermore, germ cells are transcriptionally silent during the early stages of development (Zalokar 1976; Seydoux and Dunn 1997; M. Van Doren, A. Williamson, and R. Lehmann, unpubl.). In contrast to genes like *osk* and *vas*, which are required for the assembly of germ plasm and affect both germ cell formation and abdomen formation, one would expect that genes which act downstream from germ plasm assembly should specifically affect germ cell formation at the posterior pole or germ cell migration during embryogenesis.

At least three genes, *germ cell less* (*gcl*), *polar granule component* (*pgc*), and *large mitochondrial ribosomal* RNA (*mtlr*RNA), have been identified as candidate genes specifically involved in the formation or function of the primordial germ cells but not in other aspects of germ plasm such as polar granule formation or *nos* RNA localization (Nakamura et al. 1996; Jongens et al. 1992; Kobayashi and Okada 1989). A direct functional analysis of these genes has been hampered by the lack of mutations in these genes. The *pgc* gene codes for an untranslated RNA and *mtlr*RNA is transcribed by mitochondria. Thus, isolation of mutations using conventional screens for point mutations may not generate mutations in these genes. However, experiments in which the levels of the respective gene products have been reduced have led to the conclusion that *gcl* and *mtlr*RNA are involved in germ cell formation, whereas *pgc* is required in the further development of germ cells (Jongens et al. 1994; Nakamura et al. 1996).

In addition to these candidate genes, mutations in *nos* have revealed a dual role of *nos* during early embryogenesis. *nos* RNA is localized to the posterior pole and is translated from this localized source to form a posterior to anterior protein gradient (Wang et al. 1994). This gradient of Nos protein is required to permit the expression of abdomen-specific gap genes such as *knirps*. Nanos protein, together with the RNA-binding protein Pumilio, represses translation of the maternally provided, uniformly distributed RNA of the transcription factor *hunchback* (*hb*) in the prospective abdominal region (Struhl 1989; Hulskamp et al. 1989; Irish et al. 1989; Wharton and Struhl 1991; Murata and Wharton 1995; Zamore et al. 1997). In the absence of *nos* or *pumilio* function, *hb* RNA is translated throughout the embryo and represses transcription of abdomen-specific genes such that no abdomen forms (Fig. 4A) (Struhl 1989; Hulskamp et al. 1989; Irish et al. 1989).

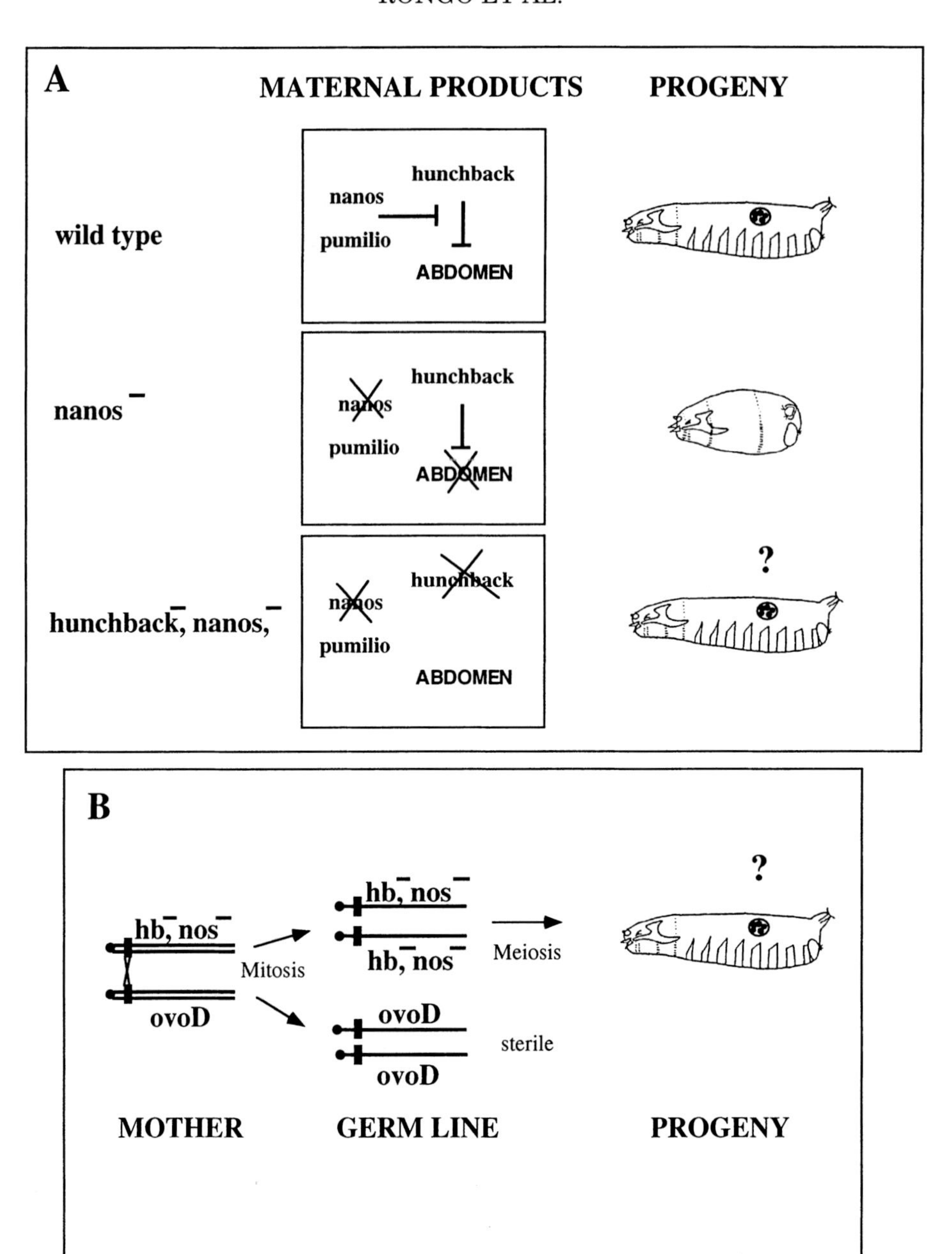

Figure 4. Simultaneous removal of *hunchback* and maternal *nanos* allows observation of the behavior of *nos* mutant germ cells in a normally developing embryo. (*A*) Outline of rationale and phenotypes; (*B*) germ line clones were introduced in the female germ line in *trans* to the dominant female sterile mutation *ovoD* that blocks oogenesis at an early stage. Only germ cells where recombination occurred will give rise to progeny. For the *ovoD*, Flp technique, see Chou and Perrimon (1992).

Once pole cells form at the posterior pole, maternal Nos protein is sequestered into the germ cells. Incorporation of Nos protein into the pole cells is not required for germ cell formation, since pole cells form in embryos from *nos* mutant females. In the absence of *nos* function, embryos fail to form any tissues normally derived from the abdomen. Since mesoderm cells located in abdominal segments 4 to 7 (parasegments 10–12) act as guidance for germ cell migration and become part of the embryonic gonad, it is difficult to assess the developmental potential of germ cells in embryos derived from *nos* mutant females. To produce embryos that lack maternal Nos protein but form normal somatic structures, we examined the progeny of females that lacked both Nanos and maternal Hunchback in their germ line (Fig. 4B). Germ cells lacking maternal Nos fail to migrate correctly (Forbes and Lehmann 1998). Instead of transversing from the midgut into the mesoderm, mutant germ cells remain on the midgut (Fig. 7C, see below). The morphology of the germ cells is altered and embryos with pole cells lacking *nos* develop into sterile adults that lack all germ line in the ovary and testis (Forbes and Lehmann 1998). The failure of germ cells lacking maternal *nos* activity to migrate could be a consequence of the premature differentiation of the mutant germ cells. Kobayashi et al. (1996) have shown that in the absence of maternal *nos*, certain P-element enhancer traps normally expressed in the germ cells following gonad coalescence are transcribed prematurely. This suggests that in some respects, *nos* mutant germ cells behave as if they are already associated with the somatic gonad while they are still on the surface of the gut. In analogy to its role in abdomen formation, the presence of Nos in germ cells may allow the expression or translation of genes required for normal germ cell migration by preventing the premature synthesis of gene products required for a later program of germ cell development.

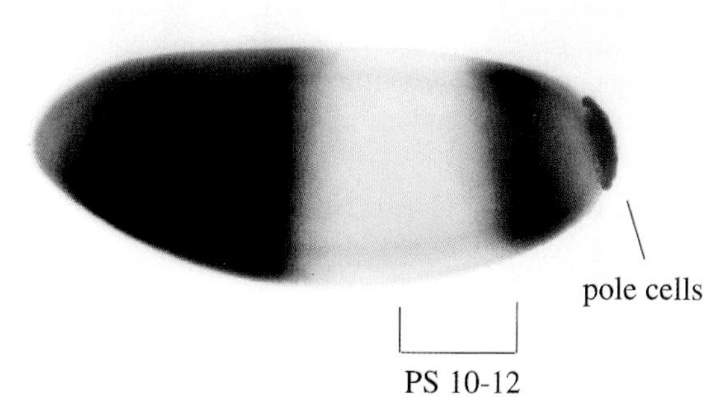

Figure 5. Germ line and somatic components originate from different locations in early embryos. Embryo double labeled for Nos *(brown)* and Hb *(blue)*. The region from where the gonadal mesoderm originates according to fate maps is highlighted. PS10–12 correspond to the region between the posterior part of abdominal segment 4 and the anterior part of abdominal segment 7. Orientation of embryos is dorsal up and anterior to left.

GERM CELL MIGRATION

The embryonic and adult gonad consists of germ cells that give rise to sperm and egg and somatic support cells which allow gonadogenesis to occur (for review, see Williamson and Lehmann 1996). Primordial germ cells (PGCs) form at the posterior pole of the early embryo, whereas the somatic components of the embryonic gonad are derived from mesodermal cells located in abdominal segments A4–A7 (parasegments [PS] 10–12) (Fig. 5). Proper gonadogenesis requires the correct migration of germ cells toward the somatic gonadal precursors and subsequent association of germ cells and gonadal mesoderm during embryogenesis. In addition to the components of the gonad that form during embryogenesis, fertility of the adult also requires the coordinate association of the developing gonad with imaginal discs that contribute the external genitalia.

We have initiated genetic screens to identify zygotically active genes required for germ cell migration and gonad formation in the *Drosophila* embryo. Phenotypic analysis of the mutants identified so far by us and others show that the process of gonad assembly can be dissected into discrete functional steps where each step is defined by specific gene functions (summarized in Fig. 6): (1) Migration of germ cells through the posterior midgut; (2) migration of germ cells along the dorsal surface of the midgut; (3) migration of germ cells away from the midgut and into lateral mesoderm; (4) association of germ cells with somatic gonadal precursors; (5) alignment of germ cells with somatic gonadal precursors; and (6) gonad coalescence.

Migration of Germ Cells through the Posterior Midgut

Ultrastructural studies have shown that during the movement of the germ cells through the posterior midgut, apical junctions dissolve in the blind end of the midgut, and intercellular gaps form through which the germ cells migrate (Jaglarz and Howard 1995; Callaini et al. 1995). Mutations affecting the development of the posterior midgut suggest that these gaps are required for germ cell movement through this tissue. *Serpent* (*srp*) and *huckebein* (*hkb*) are both genes required for the proper differentiation of the midgut, and mutations in them cause a transformation of part of the midgut into a more hindgut-like tissue that forms a tight epithelium compared to the "permeable" mesenchyme of the posterior midgut. In *srp* and *hkb* mutant embryos, germ cells remain inside a pocket formed by the enlarged hindgut primordium (Fig. 7) (Reuter 1994; Jaglarz and Howard 1995; Rehorn et al. 1996; Warrior 1994).

Migration Along the Dorsal Surface of the Midgut

Once outside of the gut wall, germ cells move along its basal surface to its dorsal side. Several lines of evidence suggest that movement of germ cells toward the dorsal side of the midgut requires specific interactions between germ cells and the endoderm but does not depend on mesodermal cues. First, PGCs move toward the dorsal side in embryos mutant for *twist* (*twi*) or *snail* (*sna*) that lack all mesoderm (Fig. 8). Second, germ cells that lack maternal Nos protein largely fail to move toward the dorsal side and do not enter the overlying mesoderm (Forbes and Lehmann 1998). Thus, germ cell autonomous functions are needed for this movement. Finally, this directed migration of the germ cells along the basal surface of the midgut is affected in embryos mutant for *wunen* (*wun*). In *wun* mutant embryos, germ cells do not remain associated with the gut but scatter throughout the embryo (Fig. 7B). *Wun* encodes a transmembrane protein that has sequence homology with the enzyme type-2 phosphatidic acid phosphatase (PAP2). The expression pattern of this gene along the ventral part of the midgut, as well as ectopic expression experiments, suggests that Wun acts by repelling germ cells away from other areas of the gut (Zhang et al. 1996, 1997). These results indicate that movement to the dorsal side may require active migration by the PGCs on the midgut. Alternatively, changes in the adhesive property of the germ cells and PAP2 expressing and nonexpressing tissues may be sufficient to cause the relocation of germ cells on the endoderm.

Movement of Germ Cells from Endoderm Toward Mesoderm

Once germ cells have moved toward the dorsal side, they transverse from the midgut primordium into the mesoderm. This process depends on the normal differentiation of two cell populations: the caudal visceral mesoderm which provides a "bridge" that connects the midgut and the lateral mesoderm and a specialized cell population within the dorsolateral mesoderm that will give rise to the somatic gonadal precursors and acts as an attractant for PGCs (Fig. 9). Two genes, *zinc-finger homeodomain factor-1* (*zfh-1*) and the *Drosophila* homolog of the mouse T gene, *brachyenteron* (*byn*), both affect the migration of the caudal visceral mesoderm cells, such that these are not juxtaposed to the germ cells as they leave the endoderm (Singer et al. 1996; Broihier et al. 1998). As a consequence, many germ cells fail to leave the gut and remain associated with the midgut rudiment throughout further development.

Two genes, *zfh-1* and *tinman* (*tin*) are required for the specification and development of the gonadal mesoderm (Boyle et al. 1997; Broihier et al. 1998; Moore et al. 1998). The gonadal mesoderm derives from bilateral clusters of about 32 cells within the lateral mesoderm of each segment of PS10–12 (Boyle et al. 1997). Embryos mutant for either *zfh-1* or *tin* have a reduced number of gonadal mesoderm precursors. However, in embryos that lack both *zfh-1* and *tin* gene function, gonadal mesoderm precursors fail to form altogether (Fig. 8) (Broihier et al. 1998). This result suggests that the two genes have partially overlapping functions in specification of gonadal mesoderm precursors. The effect on germ cell migration is

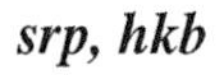

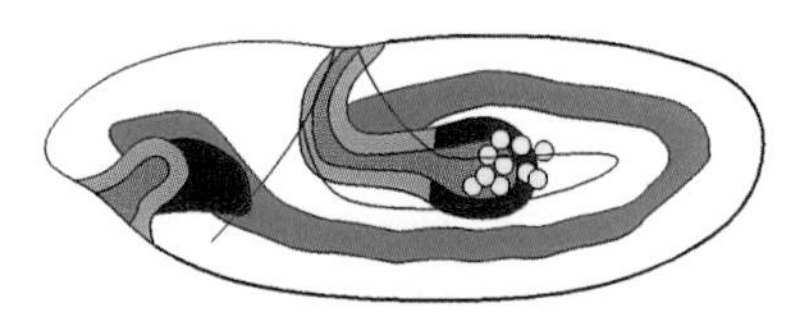

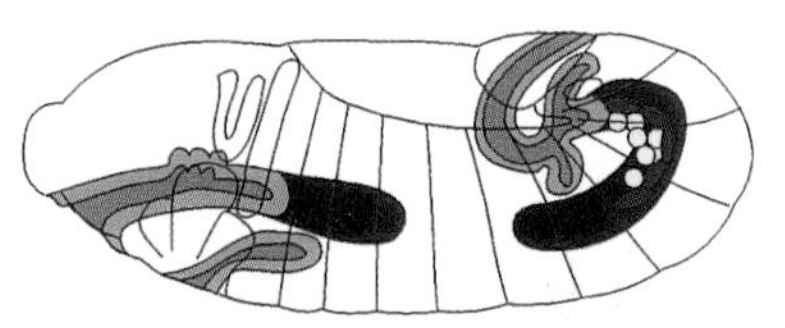

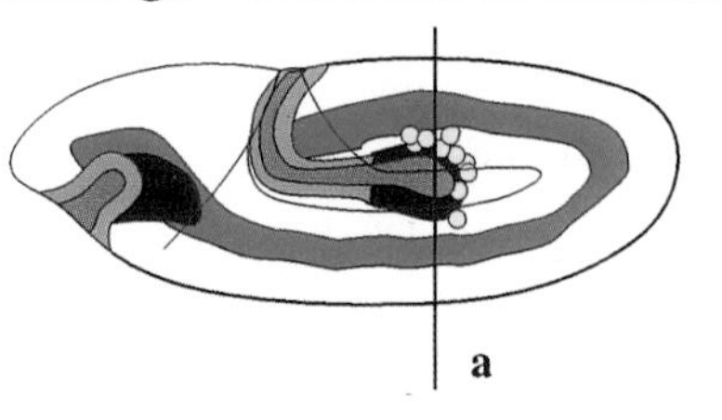

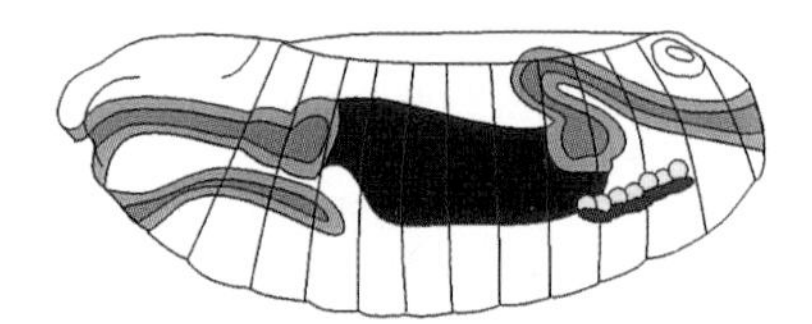

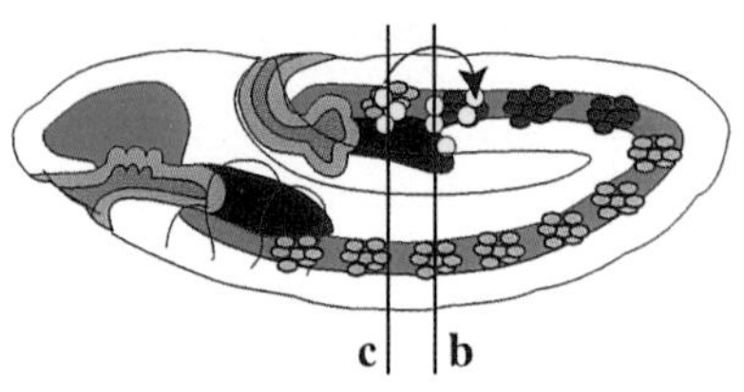

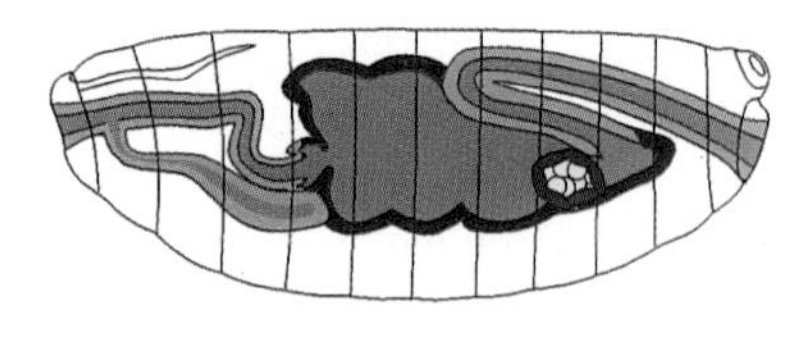

Figure 6. Genetics of germ cell migration. Genes that affect a particular step are shown next to the respective stage. Stages are from top left to bottom right: early stage 10, late stage 10, stage 11, stage 12, stage 13, stage 14. Two genes (*zfh-1* and *tin*) are found at two positions in the pathway. For *zfh-1*, this positioning indicates its role in caudal visceral mesoderm migration and gonadal mesoderm development. For *tin*, this positioning reflects the early role of *tin*, which is only revealed in a *zfh-1* mutant background, and the late role of *tin* in somatic gonadal mesoderm differentiation. The role of *heartless* (*htl*) in gonad formation is not clear, since *htl* affects the normal dorsolateral migration of the invaginating mesoderm along the ectoderm (Beiman et al. 1996; Gisselbrecht et al. 1996). Mutations in *htl* reduce the number of gonadal mesoderm cells found in late embryos. In addition, gonadal mesoderm cells are irregularly shaped in *htl* mutants, suggesting an additional defect in gonadal mesoderm differentiation similar to that seen in *foi*. (*Green*) mesoderm; (*yellow*) PGCs; (*red*) posterior and anterior midgut rudiment; (*blue*) hindgut and foregut; (*pink/brown*) clusters of lateral mesoderm: *pink* for clusters that give rise to only fatbody and *brown* for clusters that give rise to gonadal mesoderm and fatbody. (*Bottom panel, left*) Arrow denotes the "lagging" germ cells described in text; (*a*, *b*, and *c*) level of cross sections shown in Fig. 9. Orientation of embryos is dorsal up and anterior to left. (Adapted from Moore et al. 1998; embryo drawings courtesy of V. Hartenstein.)

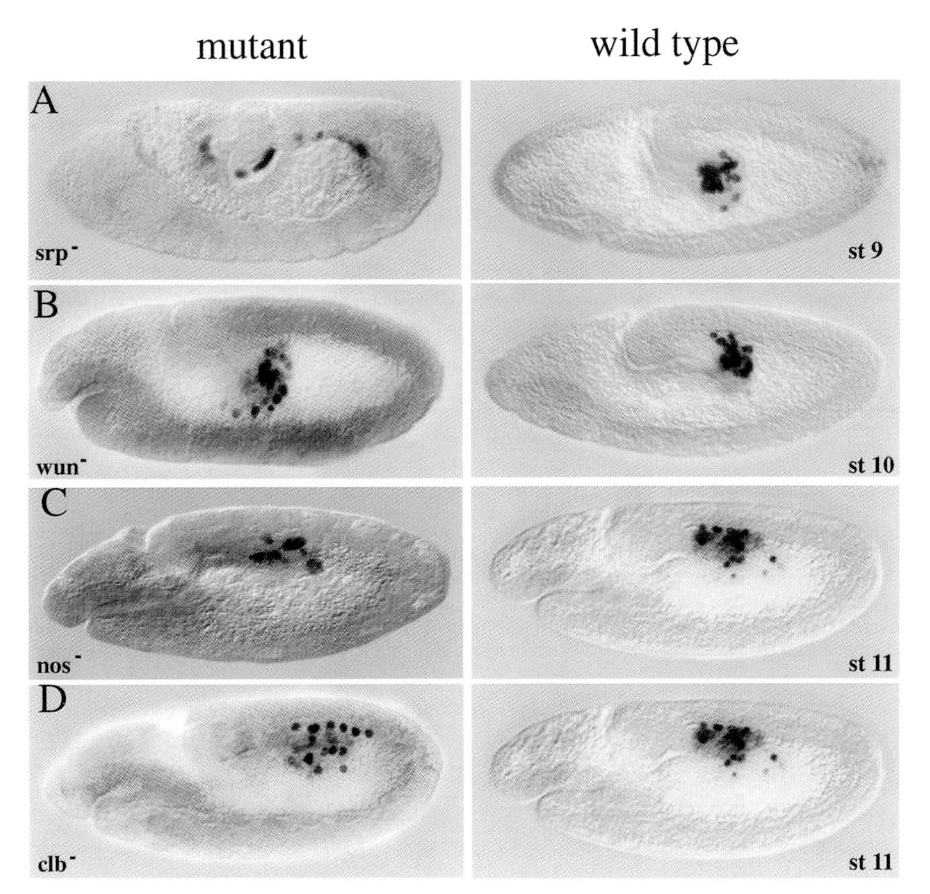

Figure 7. Mutant phenotypes ordered according to the earliest stage of PGC migration affected. (*Left*) Mutant embryos; (*right*) wild-type embryo of stage where defect is first detectable. Staining of PGCs with Vasa antibody. Orientation of embryos is dorsal up and anterior to left. (Adapted from Moore et al. 1998.)

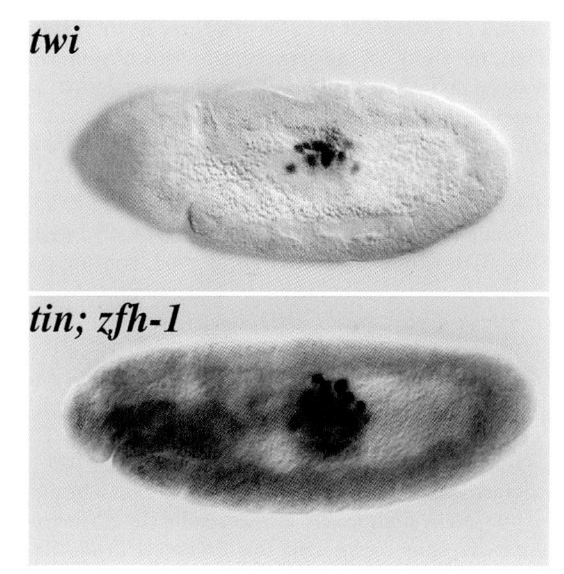

Figure 8. Embryos that lack all mesoderm have a phenotype similar to that of embryos double mutant for *zfh-1* and *tin*. Embryos are at stage 11 of embryogenesis; however, germ cells remain on dorsal side of gut. Staining of PGCs was done with Vasa antibody.

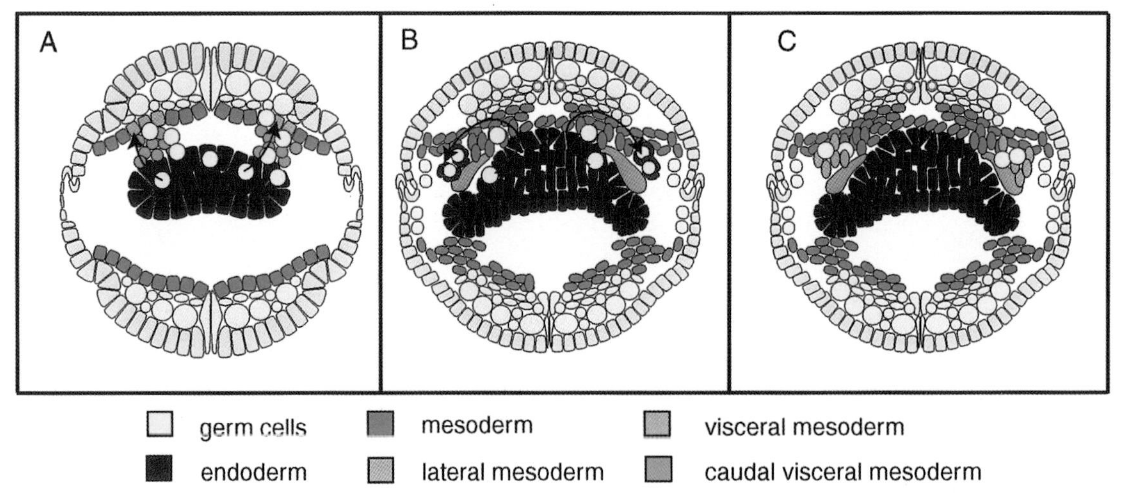

Figure 9. Cross section through embryos at different stages and at different anteroposterior levels reveals different migratory routes of germ cells. (*A*) During late stage 10, germ cells move along the caudal visceral mesoderm. (*B*) During stage 11, germ cells move directly from the gut into the mesoderm. To reach the somatic gonadal precursors, germ cells must migrate around the visceral mesoderm at this stage. (*C*) Germ cells that migrated from the midgut into mesoderm of PS13 migrate anteriorly within the mesoderm to reach the somatic gonadal precursors. Cross sections are dorsal up; because of germ band extension, similar structures are seen in the dorsal and ventral "halves" of the embryo. (Adapted from Broihier et al. 1998; embryo drawings courtesy of V. Hartenstein.)

observed earlier in *zfh-1* mutants than in *tin* mutants; it is thus likely that early during development *zfh-1*$^+$ function can partially compensate for the loss of *tin* function (Fig. 10).

In single *zfh-1* mutants, germ cells fail to leave the surface of the gut and do not associate with the lateral mesodermal cells. Lost germ cells in these mutants often remain attached to the gut, or scatter within the mesoderm and ectoderm (see Fig. 7). In the *tin, zfh-1* double mutants, germ cells do not leave the posterior midgut (see Fig. 8). Thus, *zfh-1* and *tin* are necessary in directing the germ cells away from the endoderm into a specific mesodermal region. A similar phenotype has been observed in mutants lacking *twist (twi)* and *snail (sna)* activity (Fig. 8) (Warrior 1994). However, although in *twi* and *sna* mutant embryos mesoderm fails to form altogether, mesodermal cell types form in the *zfh-1, tin* double mutants. Thus, the ability of the germ cells to detach from the endoderm epithelium and move into the mesodermal layer depends on signals emanating from the mesoderm. The nature of the attractant is yet unknown. The *columbus* (*clb*) gene is a good candidate to be involved in the production of this attractant. *clb* mutant embryos have a PGC migration phenotype very similar to that of *zfh-1* mutants (Fig. 7D). However, in contrast to *zfh-1*, somatic gonad formation is unaffected in *clb* mutant embryos (Moore et al. 1998).

Association with Somatic Gonadal Precursors

During stage 11 of embryogenesis, when germ cells contact gonadal mesoderm precursors, Zfh-1 protein is expressed at high levels in these cells. In *zfh-1* mutant embryos, some germ cells that do detach from the gut remain in a region of the mesoderm posterior to where gonadal mesoderm is normally specified. These germ cells may correspond to "lagging" germ cells that are observed in wild-type embryos to migrate from the midgut into mesoderm posterior to PS12 (Fig. 10) (Broihier et al. 1998). However, in contrast to wild-type development where these germ cells can navigate anteriorly within the mesoderm until they reach gonadal mesoderm, in *zfh-1* mutant embryos, germ cells remain where they entered the mesoderm. This observation suggests that somatic gonadal precursors attract germ within the mesoderm over a distance. Other germ cells continue to migrate past lateral mesoderm into the ectoderm in *zfh-1* mutant embryos. This phenotype is not observed in *abdA* mutants (see below) where somatic gonadal precursors fail to form but *zfh-1* is expressed at low levels in all mediolateral clusters, suggesting that *zfh-1* function is required for the adherence between germ cells and mesodermal cells.

Alignment of Germ Cells with Somatic Gonadal Precursors

The homeotic genes *abdA* and *AbdB* are required for gonad assembly (Cumberledge et al. 1992; Warrior 1994; Boyle and DiNardo 1995; Greig and Akam 1995). Germ cells are able to move through the posterior midgut and initially find lateral mesoderm in *abdA* mutants. However, germ cells fail to maintain their specific association with the mesoderm since somatic gonadal precursors are not specified; consequently, germ cells disperse in the posterior of the embryo (Fig. 10A) (Brookman et al. 1992; Cumberledge et al. 1992; Boyle and DiNardo 1995; Moore et al. 1998). Zfh-1 protein is expressed at high levels in those mediolateral clusters of cells that give rise to the gonadal mesoderm, whereas lower levels of Zfh-1 mark clusters at similar positions in the other para-

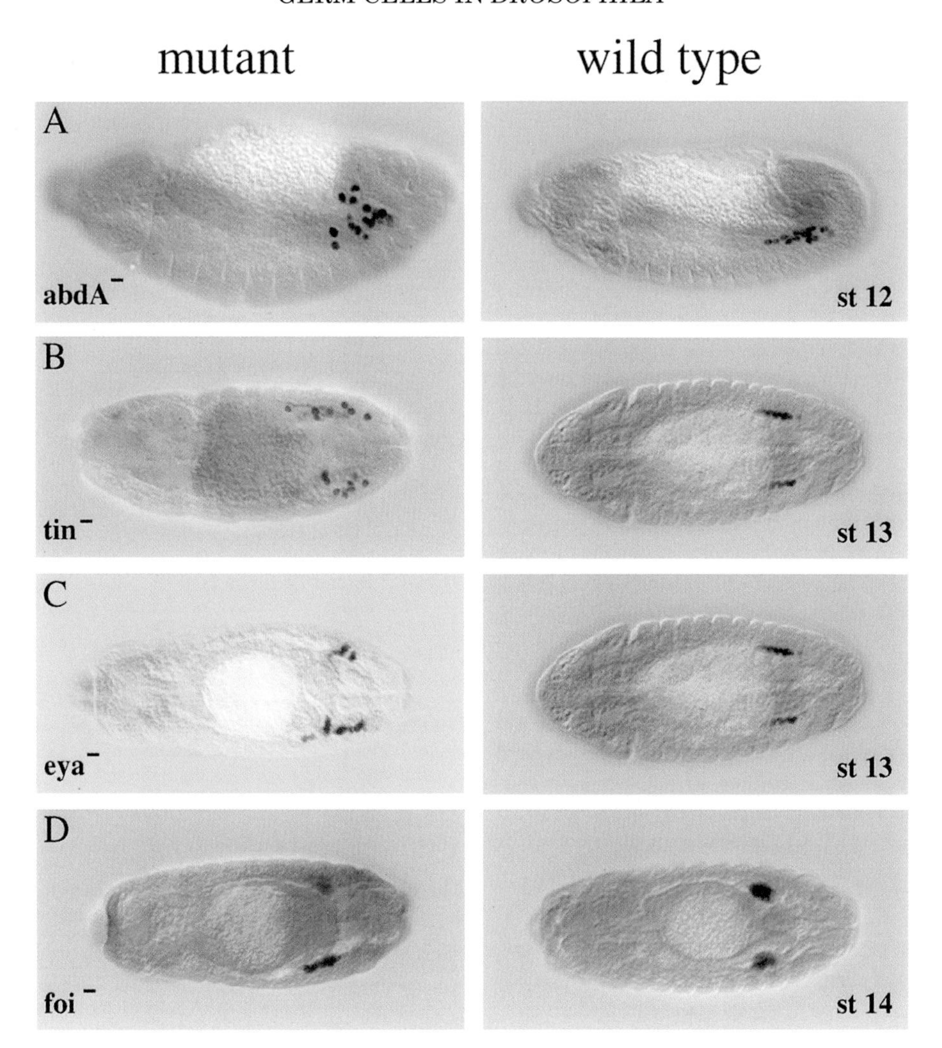

Figure 10. Mutant phenotypes ordered according to the earliest stage of PGC migration affected. (*Left*) Mutant embryos; (*right*) wild-type embryo of stage where defect is first detectable. Staining of PGCs with Vasa antibody. Orientation of embryos is dorsal up and anterior to left in *A*, frontal views in *B–D*. (Adapted from Moore et al. 1998.)

segments. These cell clusters give rise to fatbody (L. Moore and R. Lehmann, unpubl.). In *abdA* mutants, Zfh-1 protein is expressed initially in all clusters of mesodermal cells; however, its expression is not up-regulated in PS10–12. Coincidentally, other markers of the somatic gonadal precursors, such as Eyes absent (Eya also referred to as Clift) protein, are not expressed, and the cells that would normally give rise to gonadal mesoderm instead express markers characteristic of fatbody (L. Moore and R. Lehmann, unpubl). Thus, *abdA* (in PS10–12) and *AbdB* (restricted to PS12) may act as genetic switches that control the specification of precursors for the somatic gonad versus fatbody.

Although embryos double mutant for *tin* and *zfh-1* have a dramatic and early effect on germ cell migration, embryos only mutant for *tin* do not display an effect on germ cell migration until germ band retraction. In *tin* mutant embryos, germ cells migrate through the gut epithelium and remain associated with somatic gonadal precursors throughout germ band retraction (Fig. 10B). The germ cells attempt to line up along with the somatic gonadal precursors, and the alignment deteriorates as development ensues, resulting in a complete failure of gonad coalescence (Moore et al. 1998). *tin* is required for proper development of the somatic gonadal precursors (Boyle et al. 1997). By the end of embryogenesis, most gonadal mesoderm markers are no longer expressed in embryos lacking *tin* function. It is unclear why *tin* mutants show such a relatively late germ cell migration defect, given their striking effect on expression of markers for gonadal mesoderm development. One possible explanation rests on the finding described above that *tin* and *zfh-1* fulfill partially overlapping functions in gonad formation. This hypothesis would suggest that during the early stages of germ cell migration when germ cells leave the endoderm and move toward the mesoderm and when they initially align with the lateral mesoderm, *zfh-1* function alone is sufficient. However, differentiation of the gonadal mesoderm requires both *zfh-1* and *tin*.

Gonad Coalescence

Analysis of mutations in *eya/Clift* as well as those in a novel gene *fear of intimacy* (*foi*) suggests that these two genes act at a stage when germ cells and somatic gonadal precursors coalesce to form the embryonic gonad (Boyle et al. 1997; Moore et al. 1998; M. Van Doren and R. Lehmann, unpubl.). In *eya* mutant embryos, somatic gonadal precursors become specified as judged from the initially normal Zfh-1 expression pattern but fail to differentiate into gonadal mesoderm (Broihier et al. 1998). As a consequence, germ cells line up correctly and remain attached to mesodermal cells during germ band retraction but disperse at stage 14 when, during normal development, germ cells and somatic gonadal precursors located in PS11 and PS12 move anteriorly to coalesce into the embryonic gonad (Fig. 10C) (Boyle et al. 1997; H.T. Broihier, L. Moore, and R. Lehmann, unpubl.).

A defect in *foi* mutant embryos becomes apparent at a very late stage of embryogenesis: Germ cells and somatic gonadal precursors remain in a line, instead of the characteristic round shape normally found in gonads by stage 14. Somatic gonadal precursors form in normal number, but their morphology and shape are aberrant, suggesting that the defect lies within the somatic component of the gonad rather than the germ cells (Moore et al. 1998; M. Van Doren and R. Lehmann, unpubl.). In wild-type embryonic gonads, somatic gonadal precursors are tightly associated with one another and with the encapsulated germ cells. This is in sharp contrast to that seen in *foi* mutants, where the somatic gonadal precursors seem to be incapable of making close contacts with one another or with the germ cells (Fig. 10D).

SUMMARY

Many aspects of germ cell behavior, migration, and gonad formation are shared between vertebrate and invertebrate species. For example, a specialized germ plasm has been observed in many species including *Caenorhabditis elegans* and *Xenopus*. Furthermore, the fact that Vasa marks germ cells in many species suggests that even certain molecular aspects of germ cells may be common between different organisms. In most organisms, germ cells initially form at a location away from their target mesodermal tissues and have to migrate to reach the mesoderm. Further genetic studies will reveal the extent to which molecular aspects of germ cell migration and gonad formation are conserved.

ACKNOWLEDGMENTS

We thank past and present members of our lab for contributions and lively discussion that furthered and often questioned our understanding of germ plasm. Work on germ plasm assembly is supported by a grant from the National Institutes of Health to R.L. R.L. is a Howard Hughes Medical Institute investigator. A.F. is supported by a fellowship from the Human Frontiers Program, and M.V.D. is supported by a fellowship from the American Cancer Society.

REFERENCES

Beiman M., Shilo B.-Z., and Volk T. 1996. *heartless*, a *Drosophila* FGF receptor homolog, is essential for cell migration and establishment of several mesodermal lineages. *Genes Dev.* **10:** 2993.

Boyle M. and DiNardo S. 1995. Specification, migration and assembly of the somatic cells of the *Drosophila* gonad. *Development* **121:** 1815.

Boyle M., Bonini N., and DiNardo S. 1997. Expression and function of *clift* in the development of somatic gonadal precursors within the *Drosophila* mesoderm. *Development* **124:** 971.

Broihier H.T., Moore L., Van Doren M., Newman S., and Lehmann R. 1998. *zfh-1* is required for germ cell migration and gonadal mesoderm development in *Drosophila. Development* **124:** 4.

Brookman J.J., Toosy A.T., Shashidhara L.S., and White R.A.H. 1992. The 412 retrotransposon and the development of gonadal mesoderm in *Drosophila. Development* **116:** 1185.

Breitwieser W., Markussen F.H., Horstmann H., and Ephrussi A. 1996. Oskar protein interaction with Vasa represents an essential step in polar granule assembly. *Genes Dev.* **10:** 2179.

Callaini G., Riparbelli M.G., and Dallai R. 1995. Pole cell migration through the gut wall of the *Drosophila* embryo: Analysis of cell interactions. *Dev. Biol.* **170:** 365.

Chou T.B. and Perrimon N. 1992. Use of a yeast site-specific recombinase to produce female germline chimeras in *Drosophila. Genetics* **131:** 643.

Clark I.E., Giniger H., Ruohola-Baker, L.Y. Jan, and Y.N. Jan. 1994. Transient posterior localisation of a kinesin fusion protein reflects anteroposterior polarity of the *Drosophila* oocyte. *Curr. Bio.* **4:** 289.

Cumberledge S., Szabad J., and Sakonju S. 1992. Gonad formation and development requires the abd-A domain of the bithorax complex in *Drosophila melanogaster. Development* **115:** 395.

Ephrussi A. and Lehmann R. 1992. Induction of germ cell formation by *oskar. Nature* **358:** 387.

Ephrussi A., Dickinson L.K., and Lehmann R. 1991. *Oskar* organizes the germ plasm and directs localization of the posterior determinant *nanos. Cell* **66:** 37.

Erdélyi M., Michon A.-M., Gulchet A., Glotzer J.B., and Ephrussi A. 1995. Requirement for *Drosophila* cytoplasmic tropomyosin in *oskar* mRNA localization. *Nature* **377:** 524.

Forbes A. and Lehmann R. 1998. Nanos and Pumilio have critical roles in the development and function of *Drosophila* germline stem cells. *Development* **124:** 4.

Gavis E.R. and Lehmann R. 1994. Translational regulation of *nanos* by RNA localization. *Nature* **369:** 315.

Gisselbrecht S., Skeath J.B., Doe C.Q., and Michelson A. 1996. *heartless* encodes a fibroblast growth factor receptor (DFR1/DFGF-R2) involved in the directional migration of early mesodermal cells in the *Drosophila* embryo. *Genes Dev.* **10:** 3003.

González-Reyes A., Elliott H., and St. Johnston D. 1995. Polarization of both major body axes in *Drosophila* by *gurken-torpedo* signalling. *Nature* **375:** 654.

Greig S. and Akam M. 1995. The role of homeotic genes in the specification of the *Drosophila* gonad. *Curr. Biol.* **5:** 1057.

Hulskamp M., Schroder C., Pfeifle C., Jäckle H., and Tautz D. 1989. Posterior segmentation of the *Drosophila* embryo in the absence of a maternal posterior organizer gene. *Nature* **338:** 629.

Illmensee K. and Mahowald A.P. 1974. Transplantation of posterior polar plasm in *Drosophila*. Induction of germ cells at the anterior pole of the egg. *Proc. Natl. Acad. Sci.* **71:** 1016.

Illmensee K., Mahowald A.P., and M. R. Loomis. 1976. The ontogeny of germ plasm during oogenesis in *Drosophila. Dev. Biol* **49:** 40.

Irish V., Lehmann R., and Akam M. 1989. The *Drosophila* posterior-group gene *nanos* functions by repressing *hunchback* activity. *Nature* **338:** 646.

Jaglarz M.K. and Howard K.R. 1995. The active migration of *Drosophila* primordial germ cells. *Development* **121:** 3495.

Jongens T.A., Hay B., Jan L.Y., and Jan Y.N. 1992. The *germ cell-less* gene product: A posteriorly localized component necessary for germ cell development in *Drosophila. Cell* **70:** 569.

Jongens T.A., Ackerman L.D., Swedlow J.R., Jan L.Y., and Jan Y.N. 1994. *germ cell-less* encodes a cell type-specific nuclear pore-associated protein and functions early in the germ-cell specification pathway of *Drosophila. Genes Dev.* **8:** 2123.
Kim-Ha J., Smith J.L., and Macdonald P.M. 1991. *oskar* mRNA is localized to the posterior pole of the *Drosophila* oocyte. *Cell* **66:** 23.
Kim-Ha J., Webster P.J., Smith J.L., and Macdonald P.M. 1993. Multiple RNA regulatory elements mediate distinct steps in localization of *oskar* mRNA. *Development* **119:** 169.
Kobayashi S. and Okada M. 1989. Restoration of pole-cell-forming ability to u.v.-irradiated *Drosophila* embryos by injection of mitochondrial lrRNA. *Development* **107:** 733.
Kobayashi S., Yamada M., Asaoka M., and Kitamura T. 1996. Essential role of the posterior morphogen *nanos* for germline development in *Drosophila. Nature* **380:** 708.
Mahowald A.P. and Kambysellis M.P. 1980. Oogenesis. In *The genetics and biology of* Drosophila (ed. M. Ashburner and T.R.F. Wright), p. 141. Academic Press, London.
Markussen F.H., Michon A.M., Breitwieser W., and Ephrussi A. 1995. Translational control of *oskar* generates Short OSK, the isoform that induces pole plasm assembly. *Development* **121:** 3723.
Moore L., Broihier H.T., Van Doren M., Lunsford L., and Lehmann R. 1998. Identification of genes controlling germ cell migration and embryonic gonad formation in *Drosophila. Development* **124:** 4.
Murata Y. and Wharton R.P. 1995. Binding of pumilio to maternal hunchback mRNA is required for posterior patterning in *Drosophila* embryos. *Cell* **80:** 747.
Nakamura A., Amikura R., Mukai M., Kobayashi S., and Lasko P.F. 1996. Requirement for a noncoding RNA in *Drosophila* polar granules for germ cell establishment. *Science* **274:** 2075.
Pokrywka N.J. and Stephenson E.C. 1994. Localized RNAs are enriched in cytoskeletal extracts of *Drosophila* oocytes. *Dev. Biol.* **166:** 210.
Rehorn K.P., Thelen H., Michelson A.M., and Reuter R. 1996. A molecular aspect of hematopoiesis and endoderm development common to vertebrates and *Drosophila. Development* **122:** 4023.
Reuter R. 1994. The gene *serpent* has homeotic properties and specifies endoderm versus ectoderm within the *Drosophila* gut. *Development* **120:** 1123.
Rongo C. 1996. "The role of RNA localization and translational regulation in *Drosophila* germ cell determination." Ph.D. thesis, Massachusetts Institute of Technology, Cambridge, Massachusetts.
Rongo C. and Lehmann R. 1996. Regulated synthesis, transport and assembly of the *Drosophila* germ plasm. *Trends Genet.* **12:** 102.
Rongo C., Gavis E.R., and Lehmann R. 1995. Localization of *oskar* RNA regulates *oskar* translation and requires oskar protein. *Development* **121:** 2737.
Roth S., Neuman-Silberberg F.S., Barcelo G., and Schüpbach T. 1995. cornichon and the EGF receptor signaling process are necessary for both anterior-posterior and dorsal-ventral pattern formation in *Drosophila. Cell* **81:** 967.
Seydoux G. and Dunn M.A. 1997.Transcriptionally repressed germ cells lack a subpopulation of phosphorylated RNA polymerase II in early embryos of *Caenorhabditis elegans* and *Drosophila melanogaster. Development* **124:** 2191.
Singer J.B., Harbecke R., Kusch T., Reuter R., and Lengyel J.A. 1996. *Drosophila brachyenteron* regulates gene activity and morphogenesis in the gut. *Development* **12:** 3707.
Smith J.L., Wilson J.E., and Macdonald P.M. 1992. Overexpression of *oskar* directs ectopic activation of *nanos* and presumptive pole cell formation in *Drosophila* embryos. *Cell* **70:** 849.
Spradling A.C. 1993. Developmental genetics of oogenesis. In *The development of* Drosophila melanogaster (ed. M. Bate and A. Martinez Arias), p. 1. Cold Spring Harbor Laboratory Press, Cold Spring Harbor, New York.
St. Johnston D., Beuchle D., and Nüsslein-Volhard C. 1991. *Staufen*, a gene required to localize maternal RNAs in the *Drosophila* egg. *Cell* **66:** 51.
Struhl G. 1989. Differing strategies for organizing anterior and posterior body pattern in *Drosophila* embryos. *Nature* **338:** 741.
Theurkauf W.E. 1994. Premature microtubule-dependent cytoplasmic streaming in *cappuccino* and *spire* mutant oocytes. *Science* **265:** 2093.
Wang C. and Lehmann R. 1991. *nanos* is the localized posterior determinant in *Drosophila. Cell* **66:** 637.
Wang C., Dickinson L.K., and Lehmann R. 1994. The genetics of *nanos* localization in *Drosophila. Dev. Dyn.* **199:** 103.
Warrior R. 1994. Primordial germ cell migration and the assembly of the *Drosophila* embryonic gonad. *Dev. Biol.* **166:** 180.
Wharton R.P. and Struhl G. 1991. RNA regulatory elements mediate control of *Drosophila* body pattern by the posterior morphogen *nanos. Cell* **67:** 955.
Williamson A. and Lehmann R. 1996. Germ cell development in *Drosophila. Ann. Rev. Cell Dev. Biol.* **12:** 365.
Zalokar M. 1976. Autoradiographic study of protein and RNA formation during early development of *Drosophila* eggs. *Dev. Biol.* **49:** 425.
Zamore P.D., Williamson J.R., and Lehmann R. 1997. The Pumilio protein binds RNA through a conserved domain that defines a new class of RNA-binding proteins. *RNA* **3:** (in press).
Zhang N., Zhang J., Cheng Y., and Howard K. 1996. Identification and genetic analysis of *wunen*, a gene guiding *Drosophila melanogaster* germ cell migration. *Genetics* **143:** 1231.
Zhang N., Zhang J., Purcell K.J., Cheng Y., and Howard K. 1997. The *Drosophila* protein Wunen repels migrating germ cells. *Nature* **385:** 64.

moderate levels of Tub-Osk protein (Fig. 1A, lane 8) partially rescues this defect (Table 1) (Markussen et al. 1995). In addition, embryos from vas^{PD}/vas^{D1} females lack the entire abdomen and never hatch, but expression of Tub-Osk protein does not rescue this defect, and 99% of the embryos show the *vas* maternal-effect phenotype (Table 1). The level of Vas is normal in *osk* mutants (Hay et al. 1990), and the level of Tub-Osk protein is similar in both *osk* and *vas* mutant backgrounds (Fig. 1A); this suggests that in this context, Vas has an essential function in pole plasm independent of its requirement in Osk accumulation. Therefore, it seems that the primary cause of the pole plasm defect in *vas* is not the reduced level of total Short Osk. Rather, it appears that there is a requirement for Vas in addition to translation of Short Osk, such as in translation of other mRNAs (Gavis et al. 1996) and/or phosphorylation of Short Osk. Although our data show a correlation between lack of Osk phosphorylation and the failure to form pole plasm in *osk* and *vas,* we do not know in which way Short Osk phosphorylation affects the activity of Osk. One possibility is that phosphorylation of Short Osk enhances its activity in *osk* translation and/or pole plasm formation. Posterior identity of the oocyte is determined by a posterior extracellular signal transmitted by a pathway involving protein kinase A (Lane and Kalderon 1994), and Short Osk could be a target for this pathway at the posterior pole. In addition to localization of the *osk* mRNA, this could contribute to restrict Osk activity to the posterior pole.

ACKNOWLEDGMENTS

We thank members of the Ephrussi Lab for comments on the manuscript and especially Anne-Marie Michon for her additional help with the figure. F.-H.M. was supported by a predoctoral fellowship from EMBL and a postdoctoral fellowship from the Norwegian Research Council. W.B. was supported by a grant of the Boehringer Ingelheim Fonds and a predoctoral fellowship from EMBL.

REFERENCES

Breitwieser W., Markussen F.-H., Horstmann H., and Ephrussi A. 1996. Oskar protein interaction with Vasa represents an essential step in polar granule assembly. *Genes Dev.* **10:** 2179.

Curtis D., Lehmann R., and Zamore P.D. 1995. Translational regulation in development. *Cell* **81:** 171.

Ephrussi A., and Lehmann R. 1992. Induction of germ cell formation by *oskar*. *Nature* **358:** 387.

Ephrussi A., Dickinson L.K., and Lehmann R. 1991. *oskar* organizes the germ plasm and directs localization of the posterior determinant *nanos*. *Cell* **66:** 37.

Gavis E.R., Lunsford L., Bergsten S.E., and Lehmann R. 1996. A conserved 90 nucleotide element mediates translational repression of *nanos* RNA. *Development* **122:** 2791.

Hay B., Jan L.H., and Jan Y.N. 1990. Localization of vasa, a component of *Drosophila* polar granules, in maternal-effect mutants that alter embryonic anteroposterior polarity. *Development* **109:** 425.

Kennelly P.J., and Krebs E.G. 1991. Consensus sequences as substrate specificity determinants for protein kinases and protein phosphatases. *J. Biol. Chem.* **266:** 15555.

Kim-Ha J., Kerr K., and Macdonald P.M. 1995. Translational regulation of *oskar* mRNA by bruno, an ovarian RNA-binding protein, is essential. *Cell* **81:** 403.

Kim-Ha J., Smith J. L., and Macdonald P.M. 1991. *oskar* mRNA is localized to the posterior pole of the *Drosophila* oocyte. *Cell* **66:** 23.

Lane M.E. and Kalderon D. 1994. RNA localization along the anteroposterior axis of the *Drosophila* oocyte requires PKA-mediated signal transduction to direct normal microtubule organization. *Genes Dev.* **8:** 2986.

Lehmann R. and Nüsslein-Volhard C. 1986. Abdominal segmentation, pole cell formation, and embryonic polarity require the localized activity of *oskar*, a maternal gene in *Drosophila*. *Cell* **47:** 141.

Lindsley D.L. and Zimm G.G. 1992. *The genome of* Drosophila melanogaster. Academic Press, San Diego, California.

Markussen F.-H., Michon A.-M., Breitwieser W., and Ephrussi A. 1995. Translational control of *oskar* generates Short Osk, the isoform that induces pole plasm assembly. *Development* **121:** 3723.

Rongo C. and Lehmann R. 1996. Regulated synthesis, transport and assembly of the *Drosophila* germ plasm. *Trends Genet.* **12:** 102.

Rongo C., Gavis E.R., and Lehmann R. 1995. Localization of *oskar* RNA regulates *oskar* translation and requires Oskar protein. *Development* **121:** 2737.

Smith J., Wilson J., and Macdonald P.M. 1992. Overexpression of *oskar* directs ectropic action of *nanos* and presumptive pole cell formation in *Drosophila* embryos. *Cell* **70:** 849.

Spradling A.C. 1993. Developmental genetics of oogenesis. In *The development of* Drosophila melanogaster (ed. M. Bate and A. Martinez-Arias), p. 1. Cold Spring Harbor Laboratory Press, Cold Spring Harbor, New York.

St. Johnston D. 1995. The intracellular localisation of messenger RNAs. *Cell* **81:** 161.

Controls of Cell Fate and Pattern by 3′ Untranslated Regions: The *Caenorhabditis elegans* Sperm/Oocyte Decision

A. Puoti,[1] M. Gallegos,[1] B. Zhang,[1] M.P. Wickens,[1] and J. Kimble[1,2]
[1]*Department of Biochemistry, University of Wisconsin-Madison, Madison, Wisconsin 53706;* [2]*Howard Hughes Medical Institute, Department of Medical Genetics and Laboratory of Cell and Molecular Biology, University of Wisconsin-Madison, Madison, Wisconsin 53706*

Most multicellular creatures comprise numerous cell types organized into complex tissues and organs. How are distinct cell types governed to adopt tissue- and organ-specific patterns? We have approached this fundamental problem in the germ line of the nematode *Caenorhabditis elegans*. Specifically, we have investigated the molecular controls specifying the pattern of gamete differentiation in hermaphrodites.

C. elegans can develop as either of two sexes: XX embryos develop as self-fertilizing hermaphrodites, whereas XO embryos become male. A hermaphrodite is essentially a female that can regulate her sex-determining genes to produce both sperm and oocytes. The hermaphrodite and male sexes differ in nearly all body tissues and organs. Specification of an embryo as hermaphrodite or male relies on a genetic pathway that regulates sexual fates in both somatic and germ line tissues (for review, see Meyer 1997). The first step in the pathway is assessment of the ratio of X chromosomes to sets of autosomes. The second step involves signal transduction to coordinate virtually all cells to follow the same fate. This second step may be likened to the community effect in vertebrates (Gurdon et al. 1993). Intriguingly, the signal transduction pathway that regulates sexual fate in *C. elegans* is a variant of the *hedgehog* pathway of flies and vertebrates: The receptor, TRA-2, is a homolog of the membrane protein called patched (the *hh* receptor) (Kuwabara et al. 1992), and the downstream transcription factor, TRA-1, is a homolog of the zinc finger protein called Ci in flies and GLI in vertebrates (Zarkower and Hodgkin 1992).

In the hermaphrodite germ line, germ cells make two major cell fate decisions. First, a germ cell either divides mitotically to amplify the pool of germ cells or enters the meiotic cell cycle. This germ line decision between mitosis and meiosis is governed by a Lin-12/Notch-related signal transduction pathway (for review, see Kimble and Simpson 1997). Second, a germ cell differentiates as either sperm or oocyte. This paper focuses on the generation of the pattern of sexual fates in the hermaphrodite germ line (Fig. 1). Before any overt gamete differentiation, the germ line expands from 2 to about 200 cells during the first three larval stages (L1–L3) (Fig. 1, top). During L4, spermatogenesis begins in the proximal-most regions of the two U-shaped germ line tubes to generate approximately 160 sperm in each gonadal arm (Fig. 1, middle). (In the gonad, "proximal" means closer to the mid-ventrally located vulva.) During late L4 or young adulthood, spermatogenesis stops and oogenesis ensues (Fig. 1, bottom). The adult germ line continues to produce oocytes, which can be either self-fertilized using the animal's own sperm or cross-fertilized by male sperm. Because the specification of both sperm and oocytes occurs within an XX animal, the regulation that achieves both sexual fates is independent of the X/A ratio.

The production of sperm in an otherwise female animal bears on two issues of broad biological interest. First, how are distinct fields generated within a uniform group of cells? Normally, the sex determination pathway directs all cells in the organism to follow one of two sexual fates uniformly. How is this pathway regulated to generate both male and female gametes in a single organism? The second issue concerns how regulatory pathways evolve to generate species-specific differences. In the case addressed here, how does a hermaphrodite/male species evolve from a female/male species? An understanding of the underlying molecular machinery that specifies sperm and then oocytes in the *C. elegans* hermaphrodite germ line is essential to approach either question. Progress in this arena has relied on the analysis of dominant gain-of-function mutations in two sex-determining genes; *tra-2* and *fem-3* (Figs. 2 and 3). As described in the following sections, posttranscriptional regulation of *tra-2* has a central role in the onset of spermatogenesis in the hermaphrodite germ line, whereas posttranscriptional regulation of *fem-3* is key for the switch from spermatogenesis to oogenesis. Furthermore, the regulatory circuitry that achieves appropriate *tra-2* and *fem-3* expression levels is beginning to emerge as both genetic and molecular candidates for *trans*-acting factors are identified (Fig. 4).

REGULATION OF THE ONSET OF SPERMATOGENESIS BY THE *tra-2* 3′ UNTRANSLATED REGION

The onset of spermatogenesis in the hermaphrodite germ line relies on repression of the *tra-2* gene by regulatory elements located in the *tra-2* 3′ untranslated region

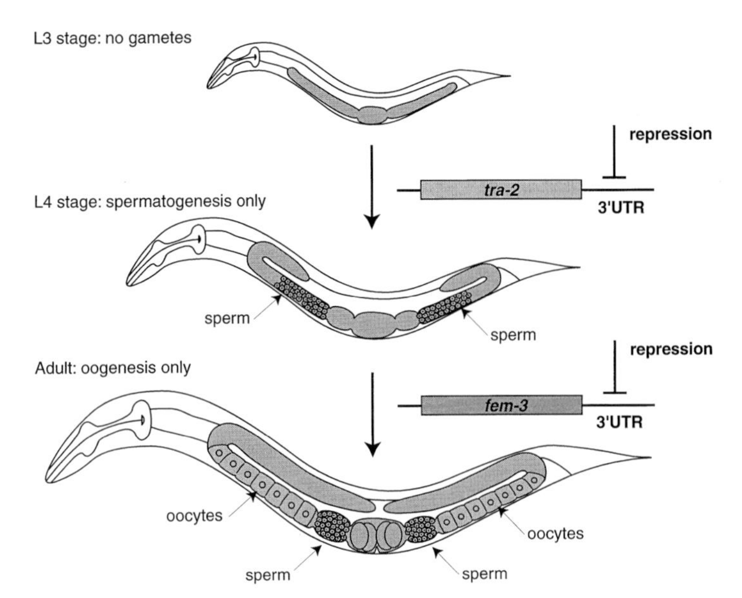

Figure 1. Development of the hermaphrodite germ line. Anterior is to the left, and dorsal is to the top in each worm diagram; the complete gonad is shaded. At the L3 stage (*top*), no gametes have been made; at L4 (*middle*), spermatogenesis has begun proximally; and as the animal molts into adulthood (*bottom*), oogenesis begins. The onset of spermatogenesis is regulated by 3′UTR repression of the female-promoting gene *tra-2* and the switch from spermatogenesis to oogenesis by 3′UTR repression of the male-promoting gene *fem-3*. See text for explanation and references.

(UTR). Normally, *tra-2* promotes female development (Hodgkin and Brenner 1977). However, rare gain-of-function (gf) *tra-2* mutations transform XX animals from hermaphrodites into females (Doniach 1986; Schedl and Kimble 1988). Thus, XX *tra-2(gf)* mutants make no sperm but instead produce oocytes continuously (Fig. 2A). They are therefore defective in the regulation required for onset of hermaphrodite spermatogenesis.

tra-2(gf) mutations disrupt the translational repression of *tra-2* mRNA (Goodwin et al. 1993). All *tra-2(gf)* mutations affect one or both of two identical 28-nucleotide sequences, which are arranged as a direct repeat in the center of the 3′UTR (Fig. 2B). These direct repeat elements (DREs) mediate negative regulation of *tra-2* translation. Indeed, one DRE mediates partial repression, whereas two are required for full repression, suggesting the existence of a translational rheostat.

The *trans*-acting regulatory machinery that represses *tra-2* translation has not yet been identified, although progress has been made. An activity, called DRF (for *DR*E binding *f*actor), in crude worm extracts binds specifically to the *tra-2* DRE (Goodwin et al. 1993). Furthermore, mutations in the *laf-1* gene appear to act upstream of *tra-2* and can derepress expression of a reporter transgene that is regulated by the *tra-2* 3′UTR (Goodwin et al. 1997). An idea that is consistent with all data available is that *laf-1* encodes a component of DRF and that *laf-1*/DRF represses *tra-2* via its 3′UTR (Fig. 4). A second

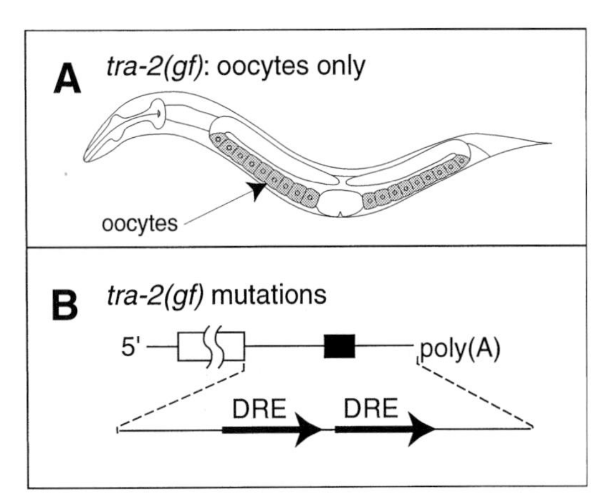

Figure 2. The *tra-2* gene is regulated by elements in its 3′UTR to achieve the onset of spermatogenesis. (*A*) Diagram as in Fig. 1. XX animals bearing dominant gain-of-function mutations in *tra-2* do not make sperm but instead produce only oocytes. These animals are therefore equivalent to females. (*B*) Molecular basis of *tra-2(gf)* mutations. Within the 3′UTR are two direct repeat elements (DREs). Each *tra-2(gf)* mutation disrupts at least one DRE. The strongest *tra-2(gf)* mutation removes both DREs, whereas weaker *tra-2(gf)* mutations remove only one DRE. See text for explanation and references.

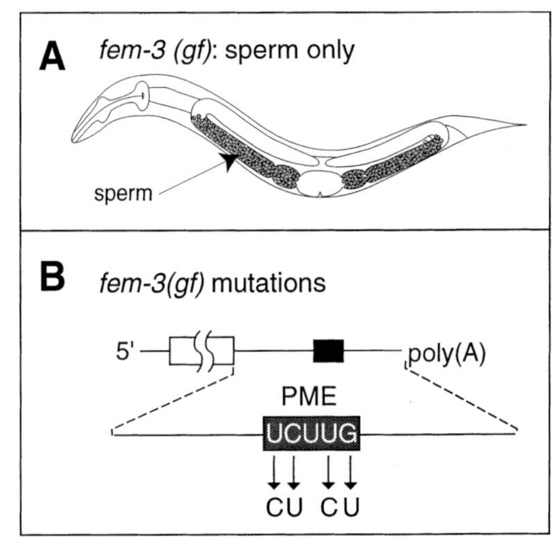

Figure 3. The *fem-3* gene is regulated by elements in its 3′UTR to achieve the switch from spermatogenesis to oogenesis. (*A*) Diagram as in Fig. 1. XX animals bearing dominant gain-of-function mutations in *fem-3* do not make oocytes but instead produce only sperm. These animals therefore have a Mog phenotype (for masculinization of the germ line). (*B*) Molecular basis of *fem-3(gf)* mutations. Within the 3′UTR, single base changes or point mutations define a *cis*-acting regulatory element (PME). See text for explanation and references.

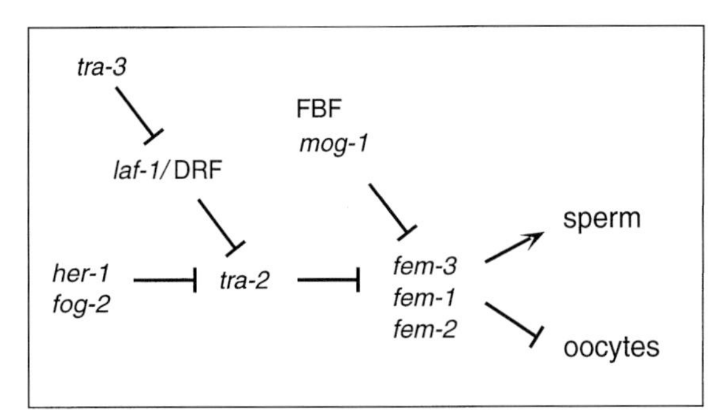

Figure 4. Regulatory pathway to generate a hermaphrodite germ line. *her-1* and *fog-2* negatively regulate *tra-2*; *tra-2*, in turn, negatively regulates the *fem* genes; the *fem-3* genes promote spermatogenesis and inhibit oogenesis. To achieve the onset of spermatogenesis, *laf-1*/DRF inhibits *tra-2* through its 3′UTR; furthermore, the *tra-3* gene can repress *laf-1*/DRF activity. To achieve the switch from spermatogenesis to oogenesis, FBF and the *mog* genes repress *fem-3* via its 3′UTR. See text for further explanation and references.

genetic candidate for the regulator of *tra-2* is *gld-1*. This gene encodes an RNA-binding protein and phenotypically affects the sperm/oocyte switch, among other effects on germ line development (Francis et al. 1995; Jones and Schedl 1995).

A genetic pathway for translational repression of *tra-2* mRNA has been proposed (Goodwin et al. 1997). The *tra-3* gene, which encodes a calpain-like protease (Barnes and Hodgkin 1996), appears to regulate DRF: In wild-type embryos, DRF is absent, whereas in *tra-3* null mutant embryos, DRF is present (Goodwin et al. 1997). Therefore, wild-type *tra-3* is required for the absence of DRF in embryos. Furthermore, *tra-3* appears to act upstream of *laf-1* in a genetic hierarchy for *tra-2* regulation (Fig. 4). A simple molecular hypothesis is that the DRF repressor is a substrate for the TRA-3 protease.

REGULATION OF THE SWITCH FROM SPERM TO OOCYTES BY THE *fem-3* 3′UTR

The switch from spermatogenesis to oogenesis in the hermaphrodite germ line relies on repression of the *fem-3* gene by regulatory elements located in its 3′UTR. Normally, *fem-3* promotes male development (Hodgkin 1986; Barton et al. 1987). Loss-of-function *fem-3* mutants cannot make sperm but instead make oocytes only. In contrast, rare gain-of-function *fem-3* mutants make only sperm in a typical hermaphrodite soma (Fig. 3A) (Barton et al. 1987). Thus, XX *fem-3(gf)* mutants are defective in the regulation required for the hermaphrodite switch from spermatogenesis to oogenesis.

The *fem-3(gf)* mutations disrupt a posttranscriptional repression of *fem-3* mRNA (Ahringer and Kimble 1991). All *fem-3(gf)* mutations affect a small region in the center of its 3′UTR. Indeed, 17 of 19 mutations are point mutations in a small five-nucleotide region (Fig. 3B). This point mutation element (PME) identifies a site of negative regulation; the size of the complete element may well exceed the five nucleotides identified by mutations.

The ability of the *fem-3* 3′UTR to regulate gene expression has been confirmed by transgenic reporter experiments (M. Gallegos et al., in prep.). The reporter transgenes place the *lacZ*-coding region plus a *fem-3* 3′UTR under control of a heat shock promoter. This promoter drives expression in somatic tissues, but not the germ line. A transgene carrying the wild-type *fem-3* 3′UTR expresses β-galactosidase at a lower level than that carrying a gain-of-function mutant *fem-3* 3′UTR. Therefore, the *fem-3* 3′UTR mediates repression in somatic tissues, as assayed by this transgene, as well as in the germ line for the sperm/oocyte switch. On the basis of these data, the regulatory machinery should be present throughout the animal. Consistent with this conclusion, a PME-specific binding activity is present in both somatic and germ line tissues (M. Gallegos et al., in prep.).

We have taken both genetic and molecular approaches to identify the *trans*-acting regulatory machinery for *fem-3* 3′UTR repression, as described in the next two sections.

mog GENES AND *fem-3* 3′UTR REGULATION

Several genes have been identified genetically as candidate *trans*-acting regulators of the *fem-3* 3′UTR (Graham and Kimble 1993; Graham et al. 1993). In general, recessive mutations in *trans*-acting factors are predicted to have the same phenotype as dominant mutations in the *cis*-acting regulatory element. We refer to the phenotype of the *fem-3(gf)* mutations as Mog (for *m*asculinization *o*f the *g*erm line). Six *mog* genes have been identified by recessive mutations. One of these genes, *mog-1*, has been characterized genetically in detail (Graham and Kimble 1993). By genetic criteria, the null phenotype for *mog-1* is Mog, and its position in the genetic hierarchy is consistent with its being a *fem-3* repressor (see Fig. 4).

In addition to their Mog phenotype, most *mog* mutants also have a smaller than normal germ line, suggesting that these genes also may be required for normal germ line proliferation. Furthermore, all *mog* mutants are required maternally for embryonic viability. This effect on embryogenesis may result from a defect in oogenesis such that imperfect embryos are made or may reveal a requirement for *mog* activity during embryogenesis. In either case, the *mog* genes appear to be required for several aspects of development, including the sperm/oocyte switch.

The *mog* genes have been implicated in regulation of the *fem-3* 3′UTR by reporter experiments (M. Gallegos et al., in prep.). Whereas a transgene carrying the wild-type *fem-3* 3′UTR is normally repressed, it becomes derepressed when placed in a *mog* mutant background. This derepression is observed for mutations in each of the *mog* genes, but not in various other genetic backgrounds, including other mutants that masculinize the germ line (*gld-1*[*Mog*]) or affect other aspects of somatic or germ line development. Therefore, the *mog* genes have a dramatic effect on regulation of the transgene, which supports the hypothesis that they may be directly involved in the *trans*-acting machinery that represses via the *fem-3* 3′UTR. Furthermore, the *mog* genes appear to be critical for *fem-3* 3′UTR regulation both in the germ line, as evidenced by the Mog phenotype, and in somatic tissues, as seen in the transgenic assays.

Two *mog* genes have been cloned (A. Puoti and J. Kimble, in prep.). Surprisingly, *mog-1* and *mog-5* are related in that they both encode proteins of the DEAH-box family of putative RNA helicases. MOG-1 and MOG-5 share 35% identity at the amino acid level and are possible homologs of yeast Prp16p and Prp22p, respectively. Both Prp16p and Prp22p are involved in pre-mRNA splicing (Company et al. 1991; Ruby and Abelson 1991; Schwer and Guthrie 1991; Umen and Guthrie 1995). MOG-1 and MOG-5 both contain a series of serine-arginine (SR) dipeptides typical of general splicing factors such as U2AF and SF2/ASF (Fu 1995). Intriguingly, the SR domains are not present in either Prp16p or Prp22p from yeast but are conserved in HRH1, the human homolog of Prp22p (Ono et al. 1994). Therefore, the SR region may be specific to higher eukaryotes. Three of four *mog-1* alleles contain nonsense mutations, consistent with genetic evidence that these alleles are null. Furthermore, one missense mutation in *mog-1* and another in *mog-5* alter amino acids that are well conserved throughout the DEAH family (A. Puoti and J. Kimble, in prep.).

Northern blot analysis shows that both *mog-1* and *mog-5* mRNAs are expressed in the *C. elegans* germ line, consistent with their role in germ line sex determination (A. Puoti and J. Kimble, in prep.). Steady-state levels of *mog-1* and *mog-5* mRNA peak in L4 at the time of the sperm/oocyte switch. They are also detected at lower levels in somatic tissues. MOG-1 and MOG-5 have, respectively, three and four putative nuclear localization signals in their amino-terminal portion, suggesting that they may function in the nucleus. Indeed, a *mog-1*::GFP translational fusion protein is detected in the nuclei of somatic tissues. (The absence of MOG-1::GFP from the germ line is likely a technical artifact because transgenes do not express well in the germ line.) The finding that *mog-1* is expressed in somatic tissues supports the reporter experiments described above, in which *mog-1* was found to regulate *lacZ* expression in somatic tissues. Furthermore, it suggests that *mog-1* has a function in somatic tissues in addition to its function in the sperm/oocyte decision in the germ line.

FBF AND *fem-3* 3′UTR REGULATION

As an alternate approach to identification of a *trans*-acting regulator of the *fem-3* 3′UTR, we used the yeast RNA three-hybrid system. Specifically, we screened a *C. elegans* cDNA library for PME-specific binding proteins (Zhang et al. 1997). The RNA three-hybrid system employs two hybrid proteins and one hybrid RNA (SenGupta et al. 1996). In our screen, one hybrid protein consisted of the LexA DNA-binding domain fused to the MS2 coat RNA-binding domain. The RNA hybrid, which was used as "bait," harbored the cognate MS2 RNA and a 37-nucleotide stretch of the *fem-3* 3′UTR carrying the PME. The second hybrid protein was encoded by a library of *C. elegans* cDNAs fused to a GAL4 activation domain. Transcriptional activation of reporter genes is selected in yeast, and dependence on the RNA intermediate is monitored with an *ADE2* marker carried on the RNA plasmid (Zhang et al. 1997).

One protein was identified that binds specifically to the PME; this protein is dubbed FBF-1 (for *fem-3-b*inding *f*actor). Using the sequence of *fbf-1* cDNA, we scanned the *C. elegans* genomic sequence and screened a *C. elegans* cDNA library. In this way, we found two nearly identical genes, *fbf-1* and *fbf-2*. Both FBF-1 and FBF-2 proteins bind specifically to the PME of the *fem-3* 3′UTR. The coding regions of *fbf-1* and *fbf-2* are 93% identical at the nucleotide level, and the predicted proteins are 91% identical at the amino acid level. The simplest hypothesis is that *fbf-1* and *fbf-2* are redundant. Consistent with this hypothesis, neither *fbf-1* nor *fbf-2* corresponds to a previously identified *mog* gene.

We examined *fbf-1* and *fbf-2* RNAs and proteins using Northern blots and immunofluorescence (Zhang et al. 1997). Only one major *fbf-1* and *fbf-2* RNA is seen during development: It is restricted to the germ line and peaks during the fourth larval stage. Using anti-FBF antibodies, we found that FBF protein is cytoplasmic and limited to the germ line. Therefore, *fbf* appears to be expressed at the right time and in the appropriate place to play a key part in the posttranscriptional regulation required for the sperm/oocyte decision.

The functions of *fbf-1* and *fbf-2* were investigated by the technique of RNA-mediated interference (RNAi) (Zhang et al. 1997). RNAi is a reverse genetic approach that can be used to investigate the loss-of-function phenotype of a molecularly identified gene. In RNAi, either sense or antisense RNA corresponding to a gene of interest is injected into the hermaphrodite germ line, and the corresponding endogenous gene is silenced in the injected animal's progeny (Guo and Kemphues 1995). This technique has been used primarily for silencing of maternal effect genes (Rocheleau et al. 1997 and references therein) but is often effective for zygotic genes as well (Zhang et al. 1997; J. Kimble et al., unpubl.).

Injection of RNA that targets either *fbf-1* or *fbf-2* had a dramatic effect in the progeny of the injected animal: The germ line contained only sperm—no oocytes were observed (Zhang et al. 1997). In addition to this sexual transformation, some animals had a smaller than normal germ line, but no lethality or other morphological defects were observed. Given the high sequence identity between *fbf-1* and *fbf-2*, it seems likely that the RNA directed against either one influences both genes. We conclude that removal of *fbf-1* and *fbf-2* eliminates the sperm/oocyte switch.

Epistasis experiments were done next to investigate the position of *fbf* in the hierarchy of sex-determining genes (Zhang et al. 1997). To this end, *fbf-1* RNA was injected into single mutants homozygous for either *fog-2*, *her-1*, or *fem-3*. All three genes normally promote male development; *fog-2* and *her-1* genes act upstream of *fem-3* (see Fig. 4). After injection, progeny were then examined for germ line masculinization. The germ lines of both *fog-2* and *her-1* mutants were masculinized by *fbf* RNA, whereas the germ line of *fem-3* mutants remained female. Therefore, if *fbf* acts in the linear pathway, its position is downstream from *fog-2* and *her-1* and upstream of *fem-3* (see Fig. 4). This position is consistent with its proposed role as a negative regulator of *fem-3*.

The FBF protein is homologous to Pumilio from *Drosophila* (Zhang et al. 1997). Central to both FBF and Pumilio are eight repeats with the same core consensus amino acids (Barker et al. 1992; Zhang et al. 1997). Deletion experiments with FBF show that these eight repeats are essential for binding the *fem-3* PME, as assayed by the yeast RNA three-hybrid system. A search for other proteins with similar repeats reveals examples in *Saccharomyces cerevisae, Schizosaccharomyces pombe, C. elegans, Drosophila*, and humans. We conclude that FBF and Pumilio are founding members of a new family of RNA-binding proteins.

COORDINATION OF *tra-2* AND *fem-3* TO REGULATE GERM LINE PATTERN

The hermaphrodite germ line generates sperm and then oocytes, with sperm proximally and oocytes distally (see Fig. 1, bottom). In the previous discussion, we focused on the role of posttranscriptional regulation of *tra-2* and *fem-3* in regulating germ line pattern. However, this discussion is not complete without mention of the roles of the TRA-2 and FEM-3 proteins in specification of germ line fates. The TRA-2 protein promotes female development, including oogenesis (Hodgkin and Brenner 1977), whereas the FEM-3 protein promotes male development, including spermatogenesis (Hodgkin 1986; Barton et al. 1987). The functional relationship between TRA-2 and FEM-3 proteins appears to be regulatory: TRA-2 is thought to be a negative regulator of FEM-3 (Hodgkin 1986). TRA-2 encodes a membrane protein and appears to relay intercellular signaling (Kuwabara et al. 1992), whereas FEM-3 is novel (Ahringer et al. 1992). Recent studies demonstrate that the intracellular domain of TRA-2 binds FEM-3, suggesting that FEM-3 repression by TRA-2 may be direct (A. Spence and P. Kuwabara, pers. comm.). The FEM-3-binding activity of TRA-2 and the lack of sequence motifs typical of enzymatic activity in TRA-2 (e.g., a kinase domain) together suggest that TRA-2 may repress FEM-3 by sequestration.

The levels of TRA-2 and FEM-3 appear to be poised in a delicate balance in the hermaphrodite germ line: *tra-2(gf)* mutants are thought to make excess TRA-2 protein, which could swamp out available FEM-3, and *fem-3(gf)* mutants may make excess FEM-3, which could free FEM-3 from the available TRA-2. Intriguingly, *tra-2(gf)*; *fem-3(gf)* double mutants can possess a virtually normal hermaphrodite germ line with sperm and then oocytes (Barton et al. 1987; Schedl and Kimble 1988). The strength of the individual *tra-2(gf)* or *fem-3(gf)* allele is critical to the ultimate phenotype of the double mutant. For example, a strong *tra-2(gf)* allele is epistatic to a weak *fem-3(gf)* allele but is neutralized by a strong *fem-3(gf)* allele. Perhaps when the allelic strengths of the *gf* alleles are matched, the levels of TRA-2 and FEM-3 are comparable, albeit higher than normal, and the balance between these two proteins is restored.

The ability of the *tra-2(gf)*; *fem-3(gf)* double mutant to develop a normal hermaphrodite germ line demonstrates that the *tra-2/fem-3* 3′UTR controls can be bypassed to generate the sperm/oocyte pattern. We suggest that this pattern relies not only on 3′UTR controls, but also on additional factors. Because the balance of these two products appears to be critical, one can envision numerous other controls of gene expression that might impact pattern formation. A simple idea is that controls other than those acting through the DREs and PME ensure that *fem-3* is expressed before *tra-2* during germ line development but that as the germ line grows, *tra-2* levels catch up with those of *fem-3* and override its activity. We speculate that an alternative mechanism may normally reinforce the 3′UTR controls to ensure the proper pattern of sperm and then oocytes. Because mutations in the other control have not yet been identified, despite the application of strong genetic selections to search for them, these putative controls may involve a composite of regulatory influences that cannot be eliminated by mutation of a single gene.

EVOLUTION OF HERMAPHRODITISM FROM UNDERLYING MATERNAL CONTROLS

C. elegans reproduces as a hermaphrodite/male strain: Hermaphrodites can produce embryos by either self-fertilization or mating with males. However, closely related nematode species (e.g., *Caenorhabditis remanei*) reproduce by a female/male strategy (Fitch and Thomas 1997). Because many branches of the nematode phylogenetic tree possess both hermaphrodite/male and female/male species, transitions between these two modes appears to be relatively easy. Similarly, *C. elegans* can be transformed from a hermaphrodite/male species to a female/male strain by mutations in one of several genes: *tra-1* (Hodgkin 1983), *tra-2* (Doniach 1986), or *fog-2* (Schedl and Kimble 1988).

The simplest hypothesis is that hermaphrodite/male species evolved from ancestral female/male species. The fundamental difference between hermaphrodite/male and female/male species resides in the regulation permitting sperm in the hermaphrodite germ line. We suggest that this regulation may be a modification of underlying controls of maternal mRNAs. Regulation by 3′UTR elements is fundamental to the control of maternal mRNAs in many organisms, including *C. elegans*, *Drosophila*, and *Xenopus* (for review, see Wickens et al. 1996). The primary function of the *fem-3* gene is to promote male development throughout the animal: It is required for specification of XO embryos as males in addition to its requirement for spermatogenesis. The *fem-3* gene acts maternally, and its maternal mRNA is subject to posttranscriptional regulation (Hodgkin 1986; Barton et al. 1987; Ahringer et al. 1992). Perhaps FBF and the PME are ancestral maternal controls that silence *fem-3* in the germ line of both hermaphrodite and female animals. In females, this silencing would be absolute: No sperm would be produced. However, in hermaphrodites, this silencing has been modified so that *fem-3* is transiently expressed.

Now that specific regulatory circuits have been defined, at least in broad outline, one can explore their generality. For example, do other hermaphroditic species use the same regulatory circuit as *C. elegans* to produce sperm transiently or have they instead exploited other mechanisms? And how great a role do posttranscriptional controls have in modifying regulatory pathways to achieve species-specific differences—not only in the nematode germ line, but more generally in the development of other tissues and phyla? The answers to these questions should emerge over the next few years.

ACKNOWLEDGMENTS

This work was supported by research grants from the National Institutes of Health and the National Science Foundation to J.K. and M.W. A.P. was an EMBO postdoctoral trainee, and M.G. was a National Institutes of Health predoctoral trainee. J.K. is an investigator of the Howard Hughes Medical Institute.

REFERENCES

Ahringer J. and Kimble J. 1991. Control of the sperm-oocyte switch in *Caenorhabditis elegans* hermaphrodites by the *fem-3* 3′ untranslated region. *Nature* **349:** 346.

Ahringer J., Rosenquist T.A., Lawson D.N., and Kimble J. 1992. The *Caenorhabditis elegans* sex determining gene *fem-3* is regulated post-transcriptionally. *EMBO J.* **11:** 2303.

Barker D., Wang C., Moore J., Dickinson L., and Lehmann R. 1992. *Pumilio* is essential for function but not for distribution of the *Drosophila* abdominal determinant Nanos. *Genes Dev.* **6:** 2312.

Barnes T.M. and Hodgkin J. 1996. The *tra-3* sex determination gene of *Caenorhabditis elegans* encodes a member of the calpain regulatory protease family. *EMBO J.* **15:** 4477.

Barton M.K., Schedl T.B., and Kimble J. 1987. Gain-of-function mutations of *fem-3*, a sex-determination gene in *Caenorhabditis elegans. Genetics* **115:** 107.

Company M., Arenas J., and Abelson J. 1991. Requirement of the RNA helicase-like protein PRP22 for release of messenger RNA from spliceosomes. *Nature* **349:** 487.

Doniach T. 1986. Activity of the sex-determining gene *tra-2* is modulated to allow spermatogenesis in the *C. elegans* hermaphrodite. *Genetics* **114:** 53.

Fitch D.H.A. and Thomas W.K. 1997. Evolution. In C. elegans *II* (ed. D.L. Riddle et al.), p. 815. Cold Spring Harbor Laboratory Press, Cold Spring Harbor, New York.

Francis R., Barton M.K., Kimble J., and Schedl T. 1995. *gld-1*, a tumor suppressor gene required for oocyte development in *Caenorhabditis elegans. Genetics* **139:** 579.

Fu X.-D. 1995. The superfamily of arginine/serine-rich splicing factors. *RNA* **1:** 663.

Goodwin E.B., Okkema P.G., Evans T.C., and Kimble J. 1993. Translational regulation of *tra-2* by its 3′ untranslated region controls sexual identity in *C. elegans. Cell* **75:** 329.

Goodwin E.B., Hofstra K., Hurney C.A., Mango S., and Kimble J. 1997. A genetic pathway for regulation of *tra-2* translation. *Development* **124:** 749.

Graham P.L. and J. Kimble. 1993. The *mog-1* gene is required for the switch from spermatogenesis to oogenesis in *Caenorhabditis elegans. Genetics* **133:** 919.

Graham P.L., Schedl, T., and Kimble J. 1993. More *mog* genes that influence the switch from spermatogenesis to oogenesis in the hermaphrodite germ line of *Caenorhabditis elegans. Dev. Genet.* **14:** 471.

Guo S. and Kemphues K.J. 1995. *par-1,* a gene required for establishing polarity in *C. elegans* embryos, encodes a putative Ser/Thr kinase that is asymmetrically distributed. *Cell* **81:** 611.

Gurdon J.B., Lemaire P., and Kato K. 1993. Community effects and related phenomena in development. *Cell* **75:** 831.

Hodgkin J. 1983. Two types of sex determination in a nematode. *Nature* **304:** 267.

———. 1986. Sex determination in the nematode *C. elegans:* Analysis of *tra-3* suppressors and characterization of *fem* genes. *Genetics* **114:** 15.

Hodgkin J. and Brenner S. 1977. Mutations causing transformation of sexual phenotype in the nematode *Caenorhabditis elegans. Genetics* **86:** 275.

Jones A.R. and Schedl T. 1995. Mutations in *gld-1*, a female germ cell-specific tumor suppressor gene in *Caenorhabditis elegans*, affect a conserved domain also found in Src-associated protein Sam68. *Genes Dev.* **9:** 1491.

Kimble J. and Simpson P. 1997. The LIN-12/Notch signaling pathway and its regulation. *Annu. Rev. Cell Dev. Biol.* **13:** 333.

Kuwabara P.E., Okkema P.G., and Kimble J. 1992. *tra-2* encodes a membrane protein and may mediate cell communication in the *Caenorhabditis elegans* sex determination pathway. *Mol. Biol. Cell* **3:** 461.

Meyer B.J. 1997. Sex determination and X chromosome dosage compensation. In C. elegans *II* (ed. D.L. Riddle et al.,), p. 209. Cold Spring Harbor Laboratory Press, Cold Spring Harbor, New York.

Ono Y., Ohno M., and Y. Shimura. 1994. Identification of a putative RNA helicase (HRH1), a human homolog of yeast Prp22. *Mol. Cell. Biol.* **14:** 7611.

Rocheleau C. E., Downs W.D., Lin R., Wittmann C., Bei Y., Cha Y.-H., Ali M., Priess J.R., and Mello C.C. 1997. Wnt signaling and an APC related gene specify endoderm in early *C. elegans* embryos. *Cell* **90:** 707.

Ruby S.W. and Abelson J. 1991. Pre-mRNA splicing in yeast. *Trends Genet.* **7:** 79.

Schedl T. and J. Kimble. 1988. *fog-2*, a germ-line-specific sex determination gene required for hermaphrodite spermatogenesis in *Caenorhabditis elegans. Genetics* **119:** 43.

Schwer B. and C. Guthrie. 1991. PRP16 is an RNA-dependent ATPase that interacts transiently with the spliceosome. *Nature* **349:** 494.

SenGupta D., Zhang B., Kraemer B., Pochart P., Fields S., and Wickens M. 1996. A three-hybrid system to detect RNA-protein interactions in vivo. *Proc. Natl. Acad. Sci.* **93:** 8496.

Umen J.G. and C. Guthrie. 1995. The second catalytic step of pre-mRNA splicing. *RNA* **1:** 869.

Wickens M., Kimble J., and Strickland S. 1996. Translational control of developmental decisions. In *Translational control* (ed. J. Hershey et al.), p. 411. Cold Spring Harbor Laboratory Press, Cold Spring Harbor, New York.

Zarkower D. and Hodgkin J. 1992. Molecular analysis of the *C. elegans* sex-determining gene *tra-1*: A gene encoding two zinc finger proteins. *Cell* **70:** 237.

Zhang B., Gallegos M., Puoti A., Durkin E., Fields S., Kimble J., and Wickens M.P. 1997. A conserved RNA binding protein that regulates sexual fates in the *C. elegans* hermaphrodite germ line. *Nature* (in press).

The *Drosophila* Germarium: Stem Cells, Germ Line Cysts, and Oocytes

A.C. SPRADLING, M. DE CUEVAS, D. DRUMMOND-BARBOSA, L. KEYES, M. LILLY, M. PEPLING, AND T. XIE
Howard Hughes Medical Institute Laboratories, Department of Embryology, Carnegie Institution of Washington, Baltimore, Maryland 21210

Drosophila oogenesis (for review, see King 1970; Mahowald and Kambysellis 1980; Spradling 1993) provides an outstanding opportunity to study several fundamental aspects of developmental biology (see, e.g., Ray and Schupbach 1996). A major advantage is cellular simplicity. Each ovary is composed of 15 to 16 ovarioles: strings of progressively more mature egg chambers, tipped with a special region known as the germarium where the egg chambers originate (Fig. 1). The linear organization of ovarioles and their multiplicity ensure that all stages of oogenesis are readily available and easily recognized. Each egg chamber develops independently of its neighbors; cellular interactions involve only the constituent cells: a germ line cyst consisting of a single oocyte connected to 15 nurse cells via a system of intercellular bridges, and a surrounding monolayer of up to 650 somatic follicle cells (Fig. 1B). Egg chamber construction begins with germ line cyst formation in the anterior third (region 1) of the germarium. Completed 16-cell cysts move posteriorly through region 2a, acquire a monolayer of profollicle cells in region 2b, and then bud (in region 3) from the germarium as a new egg chamber. Well before budding is completed, the 16 cyst cells have differentiated into an oocyte and 15 nurse cells.

Drosophila females produce eggs rapidly throughout most of adult life. Indeed, to reproduce at an equivalent rate by weight, a human would have to deliver a full-term baby approximately every 3 hours, day and night. Such rapid egg production ultimately depends on distinct populations of germ line and somatic stem cells that reside within each germarium. The small number, discrete location, and high activity of ovarian stem cells make the germarium an extremely attractive system for studying how cellular proliferation within mature tissues is regulated. In contrast to vertebrate stem cells within basal epidermal layers, hair follicles, bone marrow, or intestinal crypts, the interactions of ovarian stem cells with their neighbors can be readily analyzed anatomically, genetically, and at the molecular level (see Lin and Schagat 1997).

Germ line cysts are formed at an early stage of gametogenesis in a wide range of organisms (for review, see de Cuevas et al. 1997). Male gametes from virtually all animal groups develop synchronously within such interconnected cysts, and female gametes in diverse species also originate in such clusters. Most such cysts appear to form in a similar manner. One daughter of a germ line stem cell differentiates as a cyst-forming cell or cystoblast. The cystoblast subsequently undergoes a series of synchronous, rapid divisions in which cytokinesis remains incomplete, so that all progeny cystocytes become interconnected via ring canals formed at the site of former cleavage furrows. Much remains to be learned about the mechanism of cyst formation and about cyst function. In most organisms, stem cell division and cyst formation are confined to embryonic gonads, complicating their analysis. Consequently, the *Drosophila* germarium, where stem cell division and cyst formation continue throughout adult life, provides an excellent place for such studies.

METHODS

Drosophila *strains.* The mutant strains and constructs used have been described previously: *oskar*^{301} (Margolis and Spradling 1995); *wg-lacZ, ts-hh, hs-hh* (Forbes et al. 1996b); *hts* alleles (Yue and Spradling 1992); *cycE*^{1672} (Lilly and Spradling 1996); *pumilio* alleles (Lin and Spradling 1997); and *orb*^{dec} (Christerson and McKearin 1996). All flies were maintained at 22–25°C on standard medium. The strains used to generate *a-spec*$^-$ clones (de Cuevas et al. 1996; Lee et al. 1997) and *ptc*$^-$ clones (Forbes et al. 1996b) were also described. Enhancer trap lines were generated using the PZ element as described previously (Spradling 1993).

Antibodies. Rabbit antibody 354 or mouse monoclonal antibody 323 was used to visualize α-spectrin (Byers et al. 1987). Antibodies to Hh and Arm protein were prepared as described previously (Forbes et al. 1996a,b). Antibodies recognizing Profilin (Lynn Cooley: Verheyan and Cooley 1994), Sxl (Paul Schedl: Bopp et al. 1993), Vasa (Paul Lasko: Lasko and Ashburner 1988), and cyclin E (H. Richardson: Richardson et al. 1993) were generous gifts. TRITC-, FITC-, or Cy3-conjugated secondary antibodies were purchased from Jackson ImmunoResearch Laboratories and used at a 1 : 200 dilution.

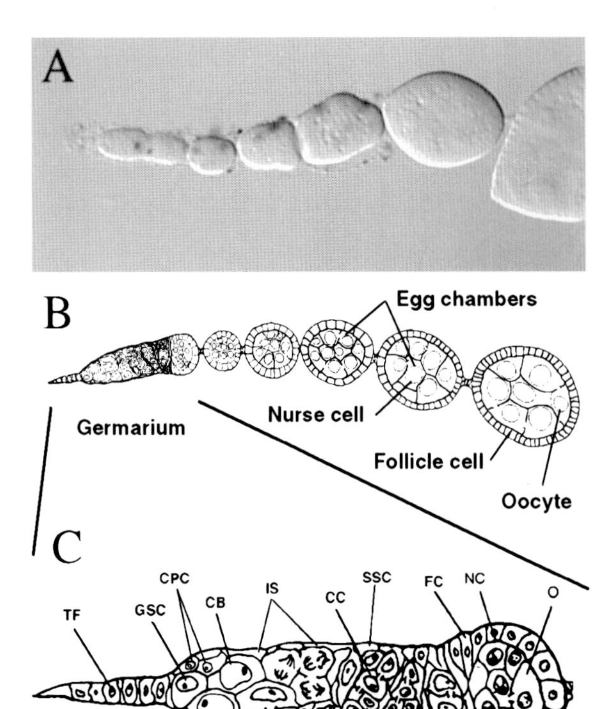

Figure 1. *Drosophila* oogenesis and the germarium. A single ovariole from a *Drosophila* ovary is shown in *A,* accompanied by a diagram (*B*) of a section through a similar ovariole. (Adapted from King 1970.) At the anterior (*left*) lies the germarium followed by a succession of sequentially more mature egg chambers it has produced, each consisting of a 16-cell germ line cyst surrounded by follicle cells. In *C,* the germarium is illustrated at higher magnification to illustrate the anatomical relationships of its constituent cell types. The 300–400 germarial cells include at least five germ line cell types: stem cells (GSC), cystoblasts (CB), cystocytes (CC), oocytes (O), and nurse cells (NC). Somatic cell types include terminal filament cells (TF), cap cells (CPC), inner sheath cells (IS), somatic stem cells (SSC), and follicle cells (FC). The four recognized subregions are indicated below: Region 1 contains the GSC and dividing germ line cysts. In region 2a, completed 16-cell cysts have not yet acquired follicle cells, whereas in region 2b, inwardly migrating follicle cells surround the cysts, which acquire a lens shape and span the width of the germarium. At the end of region 2b, a population of budding follicle cells separates the newly completed egg chamber from the germarium. (Adapted from de Cuevas and Spradling 1996.)

Immunostaining and fluorescence microscopy. Ovaries were dissected in saline solution, fixed at room temperature, washed, and then incubated overnight in primary antibodies at the appropriate dilution as described previously (de Cuevas et al. 1996). After washing, ovaries were incubated with secondary antibodies, washed further, and mounted for microscopy using a Zeiss Laser Scan Microscope. X-Gal staining was carried out as described previously (Margolis and Spradling 1995).

RESULTS

Early Germ Line Cells Associate Intimately with a Succession of Specific Somatic Cell Populations

Two to three germ line stem cells (GSC) reside near the anterior tip of the germarium (Fig. 2A) as deduced on morphological grounds (King 1970; Carpenter 1975) and verified by lineage marking (Fig. 2B) (Wieschaus and Szabad 1979; Margolis and Spradling 1995) and laser ablation (Lin and Spradling 1993). Like other early germ line cells, GSCs are large cells with round nuclei that contain relatively sparse cytoplasm and few membranous organelles. Two subpopulations of somatic cells associate intimately with GSCs, suggesting the potential for regulatory interactions. The anterior surface of each GSC lies adjacent to one of the basal cells of the terminal filament (Lin and Spradling 1995). This distinctive stack of about ten coin-shaped somatic cells has been proposed to organize ovariole differentiation (King 1970) and regulate the rate of GSC division (Lin and Spradling 1993). More laterally, GSCs contact a second distinct cell group: the cap cells (Forbes et al. 1996a). The existence of basal and cap cells was only revealed by studies of gene expression in the anterior germarium. For example, *engrailed* is expressed throughout the terminal filament and in cap cells (Fig. 2F), whereas *armadillo* and *wingless* expression is confined to cap cells (Fig. 2G,H). Despite examination of thousands of enhancer trap lines, no gene specific for GSCs has yet been identified.

GSCs divide approximately parallel to the anteroposterior axis of the germarium. These divisions are asymmetric because fusome material (see below) is distributed differentially between the daughter cells, which acquire different fates (Lin and Spradling 1997). The anterior daughter retains its attachment to basal and cap cells and remains a stem cell. The posterior daughter rapidly acquires the cystoblast fate and greatly up-regulates *bag-of-marbles* (*bam*) expression (Fig. 2C). As the cystoblast moves posteriorly, contacts are established with another somatic cell type, the inner sheath cells (Fig. 2I) (Margolis and Spradling 1995). Inner sheath cells extend long processes that contact germ line cysts throughout regions 1 and 2a (see King 1970).

Cystoblasts undergo four synchronous rounds of division within approximately 24 hours while in region 1, but increase only about fourfold in volume. Completed 16-cell cysts spend nearly 2 days passing through region 2a, after which they begin to contact the somatic stem cells (SSC) and their progeny—the prefollicle cells (Fig. 2A). Cell lineage tracing determined that exactly two SSCs are present in most ovarioles just anterior to the region 2a/2b boundary (Fig. 2J). Their location has been pinpointed by comparison with other cell groups in double-labeling experiments (Fig. 2E). As they pass the SSCs, germ line cysts become lens-shaped, lose contact with the inner sheath, and acquire a layer of prefollicle cells. Thus, germ line cysts are passed from one subset of follicle cells to the next during their trip through the germarium, each of which may provide new signals. Additional subpopula-

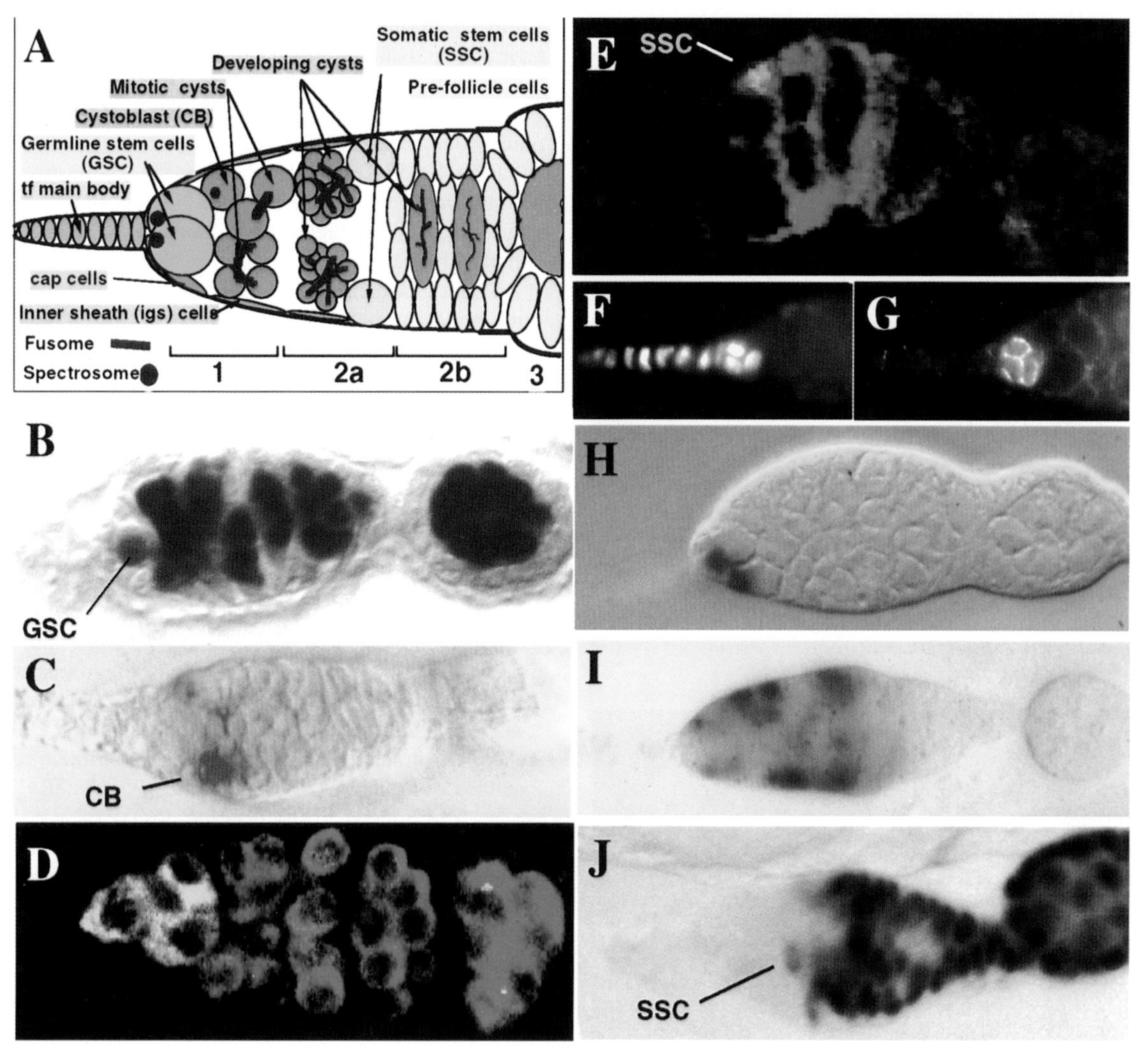

Figure 2. Germarial cell types. Each panel shows a germarium at about 400× magnification with its anterior end on the left. (*A*) By omitting some of the closely packed cells and color coding, the location of the different cell types is more clearly illustrated. The position and structure of spectrosomes and fusomes (*red*) within germ line cells are also illustrated. (*B*) A germarium containing a *lacZ*-marked GSC has been stained to reveal approximately eight progeny cysts (a typical number). Because all the cysts are labeled, only the marked GSC was active in this germarium. (*C*) A germarium hybridized in situ to reveal *bam* RNA, a marker for the cystoblast and early germ line cysts. (Adapted from McKearin and Spradling 1990.) (*D*) A germarium stained for Vasa (*green*) and Sxl (*red*). Vasa is abundant in the cytoplasm of all germ line cells. Sxl is also cytoplasmic until the four-cell cyst stage, but becomes nuclear in older cysts. (*E*) An SSC clonally marked with *lacZ* (*red*) can be seen to lie just anterior to the inwardly migrating prefollicle cells, which express high levels of Profilin (*green*). Some progeny follicle cells of the SSC can be seen in the stage-2 egg chamber to the right. (*F*) Engrailed expression in terminal filament and cap cells. (*G*) Armadillo expression in cap cells. (*H*) *wingless* expression in cap cells revealed by X-gal staining of a *wg-lacZ* strain. (*I*) *patched* expression in inner sheath cells revealed by X-gal staining of a *ptc-lacZ* strain. (*F, G, H,* and *I,* Adapted from Forbes et al. 1996b.) (*J*) A germarium containing a *lacZ*-marked SSC was stained to reveal all prefollicle cells. Because all the prefollicle cells are labeled, only one SSC was active in this germarium. (*B* and *J,* Adapted from Margolis and Spradling 1995.)

tions of follicle cells are involved in polarizing the new egg chamber and budding it from the germarium in region 3; however, these events are beyond the scope of this paper.

Translational Regulation in Early Germ Line Development

Screens for female sterile mutations identified two major classes of genes with opposite effects on early germ cell development (Fig. 3). Stem cell maintenance genes are required to retain a steady supply of germ line cysts. Females bearing a novel class of mutations in the *pumilio* gene ($pum^{ovarette}$ mutations) are unable to maintain a normal germarial structure (Lin and Spradling 1997). Instead of retaining active stem cells adjacent to terminal filament and cap cells, each ovariole houses only one or two germ line derivatives (Fig. 3B). These can take the form of relatively normal egg chambers or aberrant aggregates of undifferentiated, disconnected germ cells. When these

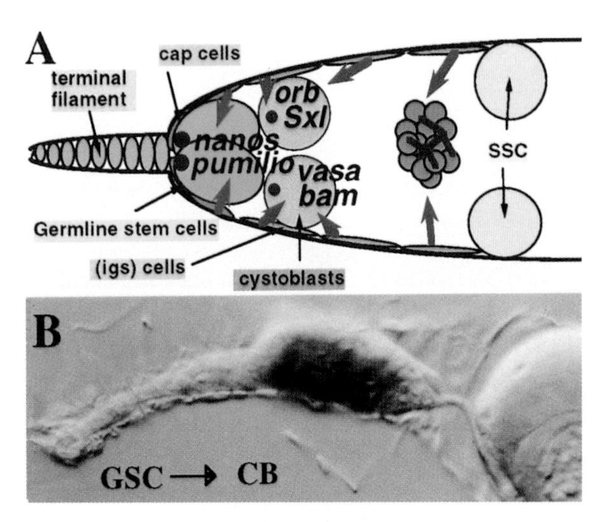

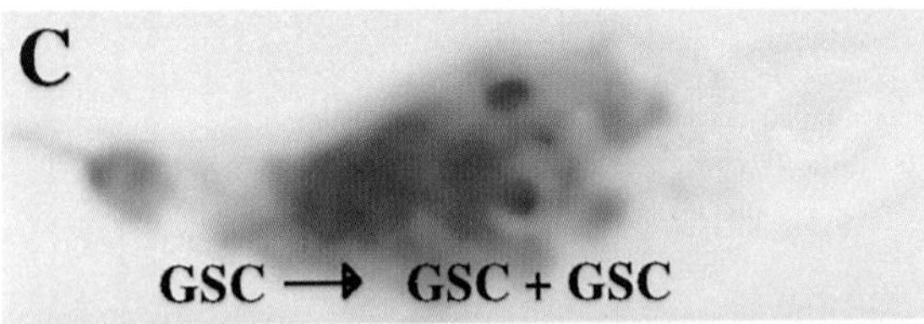

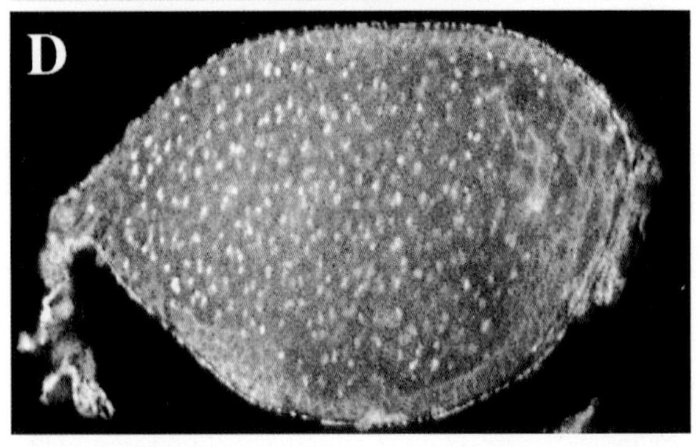

Figure 3. Translational regulation and signaling control early germ line development. (*A*) The anterior half of a germarium is diagrammed to illustrate key cell types, genes, and signals that regulate germ line cyst production. Green arrows and red arrows indicate hypothetical signals from cap cells and inner sheath cells, respectively, that likely also influence germ line development in regions 1 and 2a. (*B*) A $pumilio^{2003}$ germarium stained for Vasa reveals that GSCs are not maintained. Two cysts that developed from faulty GSCs can be seen posterior to the empty, tube-like germarium. Magnification, 200×. (Adapted from Lin and Spradling 1997.) (*C*) A tumorous germarium from an orb^{dec} mutant female stained with X-gal to reveal enhancer trap expression in the proliferating germ line cells. (*D*) A tumorous germarium from a bam^{1} mutant female stained for α-spectrin to reveal the spectrosomes (*round circles*) in the proliferating germ line cells (H. Lin and A. Spradling, unpubl.). In both *C* and *D*, early germ line (or possibly CB) daughter cells fail to differentiate as cystocytes and continue to divide like their mother.

nonfunctional egg chambers leave the germarium, they are not replaced. The germarium assumes the appearance of a thin tube, similar to that seen when the germ line is prevented from ever developing (e.g., in the progeny of osk^{301} mothers). Thus, $pum^{ovarette}$ stem cells cannot be maintained and instead differentiate into cystoblast-like cells.

A similar phenotype has been reported for a class of strong *nanos* mutations, although the effects on stem cells have not been analyzed in detail (Lehmann and Nüsslein-Volhard 1991). Nanos is known to cooperate with Pumilio in regulating the early patterning of the embryonic abdomen and in pole cell migration (Murata and Wharton 1995; Kobayashi et al. 1996). These proteins bind to conserved sequences called NREs within specific target mRNAs and repress their translation (Murata and Wharton 1995).

A larger group of mutations, the tumorous ovary mutations, appear to have the opposite effect. Instead of disappearing, germ line cells accumulate excessively but fail to form normal 16-cell cysts in females mutant for *orb* (Fig. 3C), *bam* (Fig. 3D), some *vasa* alleles, or whose germ lines lack *Sxl*. Unlike Bam (Ohlstein and McKearin 1995), Orb, Vasa, and Sxl are expressed in GSCs and in developing cysts (Fig. 2D) (Lautz et al. 1994). All of these genes have now been implicated in translational regulation. Orb is strongly related to the cytoplasmic polyadenylation element binding protein of vertebrates (CPEB) (Hake and Richter 1994). CPEB regulates stored mRNAs after fertilization, and controls meiotic maturation by activating c-*mos* mRNA (see Stebbins-Boaz et al. 1996). Vasa structurally resembles EIF-4A, a major translation initiation factor (Lasko and Ashburner 1988). Vasa is required for the production of Osk protein from localized *osk* mRNA (Kim-Ha et al. 1995). Finally, *Sxl* has been postulated to equalize the dosage of X-linked genes in the early germ line by binding to U-rich stretches in the 3′ untranslated regions (UTRs) of mRNAs from X-linked genes (Kelley et al. 1993). In the absence of these genes, sequestered mRNAs might not be regulated appropriately in cystoblasts and their progeny, thereby locking cells in a proliferative state.

Signals between Germ Line and Somatic Cells also Mediate Cyst Development

Egg chamber production requires that cysts and prefollicle cells be produced in a highly coordinated fashion and at rates determined by external factors such as the availability of nutrients and suitable egg-laying sites. Consequently, some of the most important controls, including those upstream of translational cues, are likely to depend on signals transmitted between subpopulations of somatic cells and developing germ line cells. Cystoblasts differ from stem cells by contacting inner sheath rather than cap and terminal filament base cells (see Fig. 3A, pink and green arrows). If these somatic cells send signals that are important for cyst differentiation, then genes expressed in cap or inner sheath cells may exist that influence cyst development. Recently, several candidates for such genes have been identified (T. Xie and A. Spradling, unpubl.).

Signaling in the reverse direction, from germ line to soma, is already known to be important (Fig. 4) (Goode et al. 1996). Inner sheath cells depend on signals from underlying germ line cells to survive. These cells cover about 50% of the length of germaria from well-fed wild-type females, extending from the cap cells to the SSCs

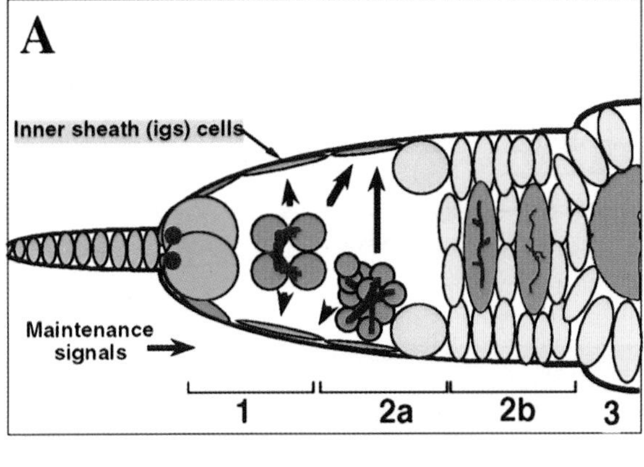

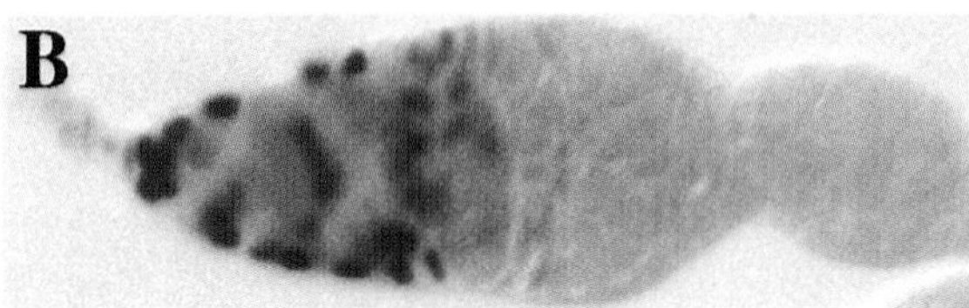

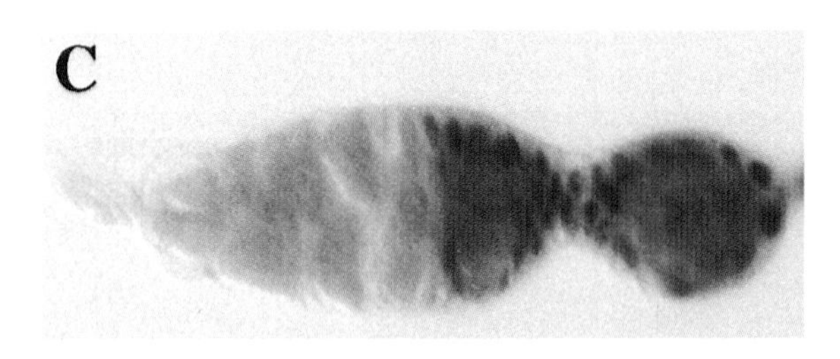

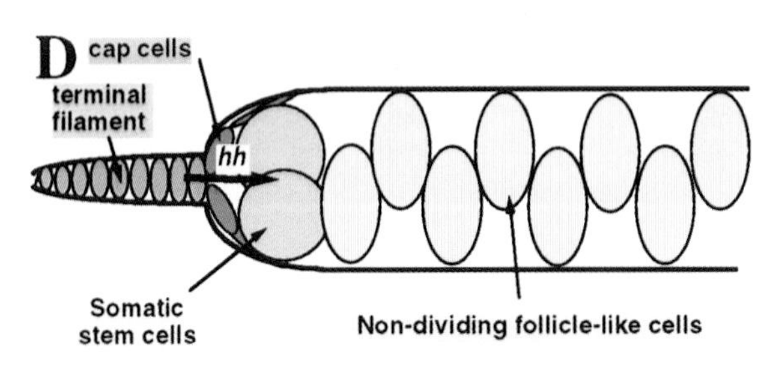

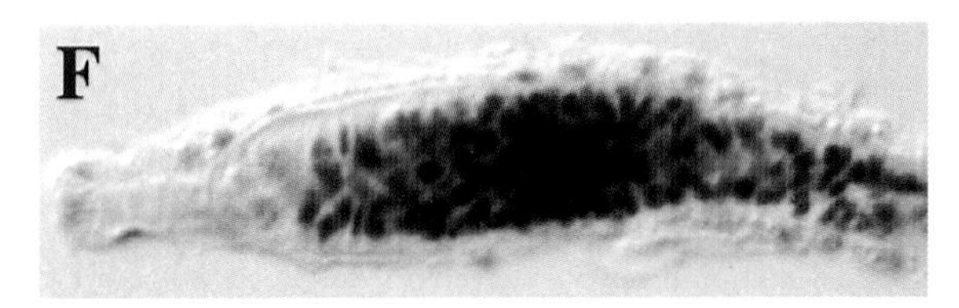

Figure 4. Germ line cells maintain the inner sheath. (*A*) Maintenance signals (*black arrows*) sent by developing germ line cysts (*purple*) and received by the somatic inner sheath cells (*pink*). In wild type (*A–C*), a normal population of inner sheath and cap cells (*B*) and prefollicle cells (*C*) is maintained. In osk^{301} germaria (*D–F*), germ line cells are absent. Inner sheath cells are not observed, but cap cells remain (*E*). SSC progeny are found throughout the interior of the germarium (*F*). PZ1444 (*B,E*) and PZ2954 (*C,F*) enhancer trap expression is shown. (Adapted from Margolis and Spradling 1995.)

(Fig. 4B). However, this organization is hardly a fixed anatomical feature of the germarium, as is sometimes depicted. Indeed, inner sheath cells are not found in germaria that never contained a germ line (Fig. 4D,E). Cells expressing follicle cell markers (Fig. 4C) continue to be produced but now are found more anteriorly than in a normal germarium (Fig. 4F). Inner sheath cells also disappear when germ line cells are lost after eclosion (cf. Fig. 3B and Fig. 4E,F) (T. Xie and A. Spradling, unpubl.). This suggests that region-1 and -2a germ line cells send signals that are required to maintain the inner sheath (Fig. 4A, black arrows). These cells are known to send signals mediated by *braniac*, *egghead,* and the epidermal growth factor (EGF) receptor only slightly later in development, which are required for the proper budding of egg chambers (Goode et al. 1996). The requirement for germ line signals to maintain the inner sheath cells suggests that this aspect of the germarium is dynamic. The size of regions 1 and 2a may simply reflect the number of cysts currently developing within these regions. If so, the distance between the germ line and somatic stem cells may vary depending on the developmental state of the germarium.

Regulating Somatic Cell Production

Two somatic stem cells satisfy the ovariole's requirement for a continuous supply of follicle cells (Margolis and Spradling 1995). Although the products of SSC division differ in function, physical asymmetries in the segregation of specific molecular components during SSC divisions have not been observed. Fusomes (see below) are not detected, and SSC progeny cells do not divide synchronously, although many eventually become joined by small ring canals that resemble early germ line bridges (see Robinson and Cooley 1996). Thus, available evidence suggests that somatic and germ line stem cells are controlled by very different mechanisms.

Many of the somatic stem cell progeny begin a program of rapid division (doubling time 9.6 hr), undergo morphological changes characteristic of an epithelial-mesenchymal transition, and express high levels of profilin (Fig. 2E). These prefollicle cells leave the surface epithelial layer and migrate inward between successive germ line cysts and surround each cyst as a cellular monolayer. For reasons that are not clear, the cyst changes its shape just before and as it acquires this follicular coating. Perhaps because of continuing growth, one of the two cysts that have been advancing side by side in region 2a begins to pull ahead. By region 2b, each cyst spans the entire width of the germarium and has acquired a characteristic lens shape. The pattern of cystocyte interconnections is likely to influence cyst shape and oocyte position at this time, but whether the enveloping follicle cells also contribute is not clear.

Germaria can produce egg chambers at very different rates depending on available nutrients without accumulating an excess of either germ line or somatic cells. Thus, the relative production and stability of germ line and somatic cells must be well-regulated. One of the most important signals controlling follicle cell production is the *hedgehog* pathway (Fig. 5A) (Forbes et al. 1996a). *hedgehog* is expressed primarily in the terminal filament and cap cells (Fig. 5B, C). When *hedgehog* signaling is reduced, the size of the germarium shrinks dramatically over the course of several days (Fig. 5D). Egg chamber production decreases and eventually ceases if reduced levels of Hh persist. The few chambers that do bud under these conditions often contain multiple cysts (Fig. 5D). All of these effects are consistent with a strong relative reduction in the rate of somatic cell compared with germ cell production.

The opposite effect is observed when *hh* signaling is increased (Fig. 5E). The germarium swells to two or three times its normal size, and egg chambers bud off with a large excess of follicle cells that are eventually shed, greatly bloating the interfollicular stalks. The timing of increased cell production argues that the responsive cells are located in region 2b. Later follicle cells are not affected, even though they normally continue to divide for several more days. Thus, only the somatic stem cells and/or some of their immediate progeny are likely to respond to *hh*-mediated signals. As with other Hh-responsive cells, the effects on follicle cells appear to be mediated by *patched*, because *ptc*$^-$ clones behave like cells exposed to excess Hh (Fig. 5F). However, unlike most other Hh effects in *Drosophila*, neither *wg* nor *dpp* mediates this pathway in follicle cells (Forbes et al. 1996b).

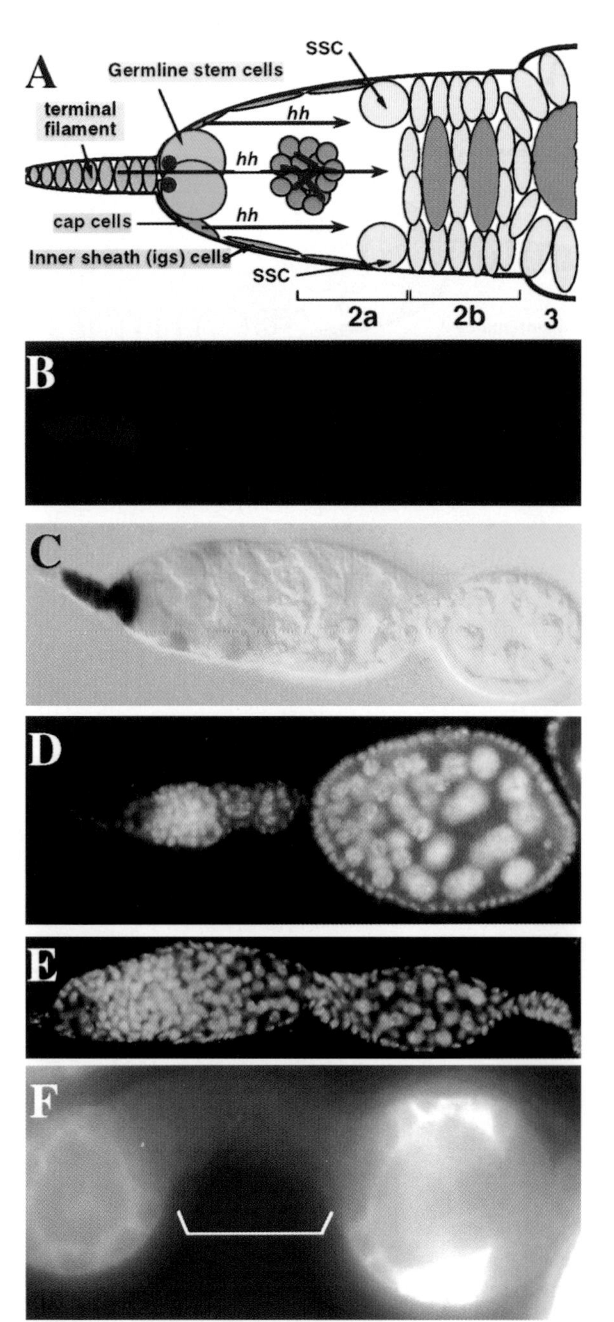

Figure 5. *Hedgehog* signaling regulates prefollicle cell proliferation. Illustrated are *hh*-mediated signals (*black arrows*) sent by terminal filament (*pink*) and cap cells (*green*) that influence the division of SSCs and some of their progeny (*yellow*). Anti-Hh antibody staining (*B*) and *hh-lacZ* expression (*C*) show that Hh is expressed predominantly in the terminal filament and cap cells. When Hh activity is reduced by shifting *hh*^{ts} flies to the nonpermissive temperature (D), the germarium shrinks and the few remaining follicle cells envelop multiple germ line cysts. When Hh activity is increased using an *hsp70-hh* construct (*E*), the germarium greatly expands and excess follicle cells bud off newly formed egg chambers. Similar overproliferation of follicle cells occurs within clones (bracket) of *ptc*$^-$ cells (*F*). (Adapted from Forbes et al. 1996a,b.)

Fusome Mediates Germ Line Cyst Formation

The actin and microtubule cytoskeletons are likely to have central roles in many of the regulatory pathways described above (for review, see Cooley and Theurkauf 1994). Indeed, the large size and unusual biology of germ line cells make *Drosophila* oogenesis a favorable venue for studying morphogenetic mechanisms that lie downstream from intercellular signals. Early germ cells use one mechanism that has been known since the early part of this century. Developing germ line cysts contain a large cytoplasmic organelle called the fusome that spans all the constituent cystocytes (Fig. 6A–C) (for review, see Telfer 1975; de Cuevas et al. 1997; McKearin 1997). The existence of such an organelle during these stages is highly conserved among diverse insects (see Büning 1995). One pole of the mitotic spindle in each cystocyte is always associated with the fusome (Fig. 6A), and the organelle segregates asymmetrically at each division into only one of the daughters, along with all preexisting ring canals. The structure of the fusome during the cystocyte cell cycles is dynamic; new branches arise after each mitosis is completed (Fig. 6C). In females, the fusome disappears by the time a new egg chambers buds from the germarium.

The membrane skeleton proteins α-spectrin and β-spectrin and *hu-li tai shao* (*hts*) protein (adducin) are major fusome constituents (Lin et al. 1994; de Cuevas et al. 1996). In most cells, the spectrin-based membrane skeleton is located underneath the plasma membrane, where it strengthens the lipid bilayer and anchors integral membrane proteins (for review, see Kusumi and Sako 1996). This organization is seen in most vertebrate and *Drosophila* cell types examined by antibody staining, in-

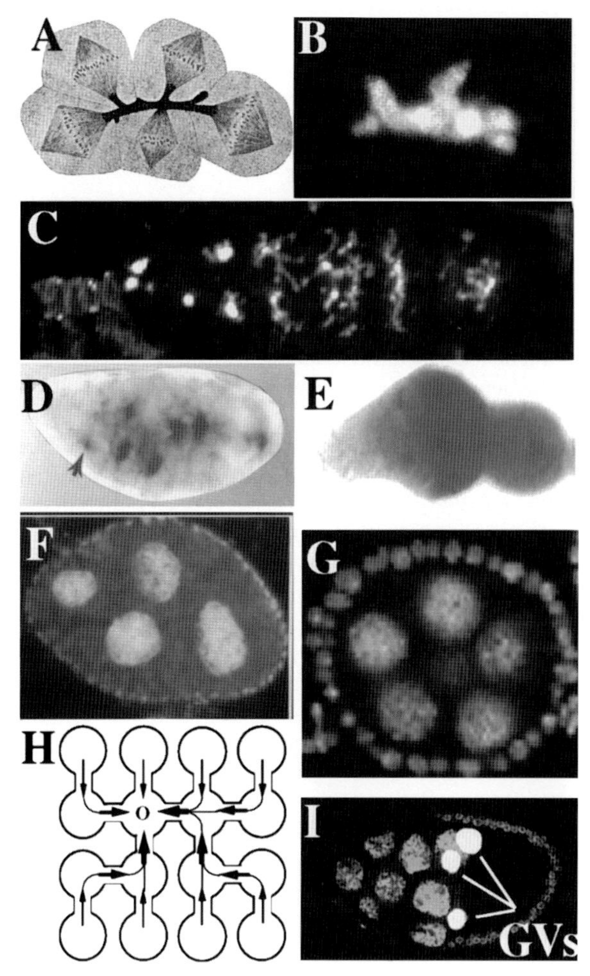

Figure 6. Structure and function of the fusome. (*A*) Drawing of synchronous mitoses in an eight-cell wasp cyst (Reprinted from Maziarski 1913) illustrating the relationship between spindle poles and the fusome. (*B*) A fusome revealed by antibody staining. (*C*) Spectrosomes and fusomes within the germ line cells of a germarium are seen clearly because of the presence of an *α-spec⁻* clone in the somatic cells. (Adapted from de Cuevas et al. 1996.) (*D*) Germarium showing the accumulation of *hts* mRNA within the future oocyte of each 16-cell germ line cyst. (Reprinted, with permission, from Yue and Spradling 1992.) (*E*) *hts* RNA remains uniformly distributed in *hts* mutant cysts. (From L. Yue and A.C. Spradling, unpubl.) (*F*) An *hts* mutant egg chamber showing characteristic reduction in cyst cell number and absence of an oocyte. (*G*) Egg chambers whose germ cells lack α-spectrin resemble *hts* mutant chambers. (Adapted from de Cuevas et al. 1996.) (*H*) Model of polarized transport through ring canals. In a wild-type cyst, the downstream cell always contains four ring canals and differentiates into the oocyte. (*I*) An egg chamber produced by a $cycE^{1672}$ female that contains 3 germinal vesicles and 13 nurse cells, illustrating the importance of cell cycle regulators in oocyte determination. The germinal vesicles (GVs) are labeled with anti-cyclin E antibody. (Adapted from Lilly and Spradling 1996.) *F, G,* and *I* were stained with DAPI or propidium iodide to label nuclear DNA.

cluding all the ovarian somatic cells. In contrast, germ cells lack submembranous staining, a fact that may help explain their round shape. Instead, these proteins are concentrated in the fusome (Fig. 6B). All three membrane skeletal proteins label the entire fusome at all stages where it is present. In addition, several other more transient fusome constituents have been identified. Ankyrin is detected preferentially on spectrosomes and early fusomes (de Cuevas et al. 1996). One form of Bam protein labels spectrosomes and fusomes in dividing cystocytes but then declines in amount (Ohlstein and McKearin 1995). CycA protein associates transiently with spectrosome and cystocyte fusomes during G_2 of the cell cycle (M. Lilly et al., unpubl.). Finally, Ptc is observed coincident with just the core region of degenerating fusomes (Forbes et al. 1996b).

Fusome Is Required for Cyst Formation and Oocyte Determination

The role of the fusome has been clarified by the analysis of mutations in its membrane skeleton components. Strong *hts* mutations (Fig. 6F) (Yue and Spradling 1992; Lin et al. 1994), or null α-spectrin germ line clones (Fig. 6G) (de Cuevas et al. 1996), both have a similar effect on developing cysts. The number of cystocytes is reduced and no longer corresponds to a power of two, indicating that cystocyte divisions are no longer synchronized. Because the fusome spans all the cells of the cyst, and is observed to accumulate large amounts of CycA protein in G_2, the interaction of CycA and the fusome has been proposed to be responsible for synchronizing cystocyte cell cycles (M. Lilly et al., unpubl.). The fusome may ensure that CycA becomes activated in all cystocytes simultaneously, driving them synchronously into M phase. Moreover, the fusome likely influences other cystocyte cell cycle parameters. Normally, these divisions are accelerated relative to cellular mass increase, i.e., cystocytes become smaller after each round of mitoses. In mutants that compromise the fusome, the cleavage-like character of these divisions appears to be reduced or lost. Thus, in hts^1 females, cysts average about four cells but appear no smaller overall when they bud from the germarium than a corresponding 16-cell wild-type cyst.

The fusome also appears to be critically important for oocyte formation. A variety of mRNAs and proteins are transported via polarized microtubules into the presumptive oocyte beginning shortly after cyst formation is complete (Fig. 6D) (see Cooley and Theurkauf 1994; St. Johnston 1995). Several genes, including *egalitarian* and *Bic-D*, are required for differential transport and for oocyte formation (see Mach and Lehmann 1997). The fusome appears to have a crucial early role in establishing and polarizing the microtubule network (Lin and Spradling 1995). In the absence of a fusome due to loss of *hts* or α-spectrin, oocyte formation is blocked (Fig. 6F,G), and differential transport is severely compromised (Fig. 6E). Recently, the minus-end-directed microtubule motor cytoplasmic dynein was shown to be required for cyst polarization and oocyte determination (McGrail and Hays 1997). Interestingly, defects were seen in the orientation of cystocyte mitotic spindles and in fusome structure in the mutant clones, rather than simply an absence of transport along an intact microtubule system. This provides further evidence that establishing a properly polarized microtubule network within germ line cysts requires early fusome-mediated events. Whether cytoplasmic dynein also functions as a motor to transport specific components from the nurse cells to the oocyte in early cysts remains to be addressed.

What is the critical component that must accumulate in the oocyte to specify an oocyte fate? Previous candidate genes for such an oocyte determinant such as *egl* and *Bic-D* now appear to encode proteins involved in transport or localization rather than the determinants themselves (Mach and Lehmann 1997). Perhaps the strongest clue as to the nature of oocyte specification comes from studies of the hypomorphic $cycE^{1672}$ mutation (Lilly and Spradling 1996). In $cycE^{1672}$ homozygotes, two or even three of the 16 cystocytes differentiate as oocytes (Fig. 6I), as judged from nuclear morphology and function. Lowering CycE levels is likely to alter the length and timing of cystocyte cell cycles, suggesting that fine details of these cell cycles are critical for the decision between oocyte and nurse cell. Consistent with this model, entry into meiosis is the first step in oocyte differentiation. All 16 cystocytes normally begin meiosis, since they undergo a premeiotic S phase, and several continue into meiotic prophase. Normally, however, only one cell remains in the meiotic cycle. The other 15 exit meiotic prophase, commence endocycles, and differentiate as nurse cells.

The cell cycle changes induced by $cycE^{1672}$ somehow assist cells in maintaining the meiotic cycle even though their upstream position on the microtubule network prevents them from permanently accumulating the cyst's transported products. Indeed, some of the extra germinal vesicles show small increases in ploidy in later egg chambers, suggesting that meiosis is not fully maintained and that some endocycles have taken place. *encore*, another gene required for cystocyte cell cycles, is also needed to prevent the normal germinal vesicle from undergoing similar changes (Hawkins et al. 1996).

DISCUSSION

Germarium as a System for Analyzing Cellular Proliferation

The *Drosophila* germarium provides outstanding opportunities for understanding how multiple signaling pathways and mechanisms collectively act to regulate cellular proliferation and differentiation. The somatic stem cells resemble those found in vertebrate epithelia in at least two respects. First, both *Drosophila* SSC and the stem cells of the human basal epidermis appear to use the *hedgehog/patched* pathway as a major regulator of proliferation (Johnson et al. 1996). Second, like all known vertebrate stem cells, the SSCs are not morphologically distinctive but can be recognized only by lineage marking experiments. It may be that genetic markers of such stem cells eventually will be found. However, it remains a real possibility that stem cell properties are not intrinsic but are a response induced in a susceptible undifferentiated cell to a highly specialized microenvironment organized by differentiated surrounding cells. The *Drosophila* germarium provides an extremely powerful system for identifying such stem cell "niches" because functional stem cells can be localized accurately relative to adjacent cells. The importance of specific signaling pathways and anatomical organization can then be tested by manipulating the strength, location, and nature of the signals in these regions.

Translational Regulation in the Germ Line

A relative absence of specific transcriptional regulation and reliance on translational mechanisms is emerging as a characteristic of some germ line cells (Curtis et al. 1995). Translational regulation is implicated extensively in the formation and functioning of the "germ plasm" at the posterior pole of the egg (for review, see Rongo and Lehmann 1996). Many of the same genes required in the early germarium, including *pumilio*, *nanos*, *orb*, and *vasa* are needed later for the assembly and/or function of pole plasm. Early embryonic germ cells remain transcriptionally quiescent in both *Drosophila* (Zalokar 1976) and *Caenorhabditis elegans* (Seydoux et al. 1996) embryos. It may be that the specific transcription factors required for gene activation are incompatible with the pluripotent state of germ cells or that a quiescent state minimizes the exposure of germ cells to vertically transmitted parasites. Instead, germ cells appear to rely on regulated translation of long-lived maternally inherited mRNAs for an unusually long period of time during embryogenesis. It has been shown that nanos is needed for germ cells to associate properly with gonadal mesoderm (Kobayashi et al. 1996).

Our analyses of the $pum^{ovarette}$ class mutations suggests that germ cells rely on translational regulation for a longer period than previously thought. The nature of the genes that disrupt early germ cell differentiation suggests that translational regulation remains important not only in germ line stem cells, but also during cystocyte divisions. Such a situation would help explain several observations. It provides a reason for Sxl protein to remain cytoplasmic in germ line stem cells and early cysts, where it has been proposed to carry out dosage compensation by translationally repressing X-linked genes (Kelley et al. 1993). The chromatin-based mechanism used in somatic cells might not function appropriately in conjunction with the transcriptional machinery characteristic of these early germ cells.

Several early embryos regulate the activity of stored maternal mRNAs by controlling the length of their poly(A) tails (for review, see Wickens 1992). This mechanism is also used in *Drosophila* after fertilization (Sallés et al. 1994). Because mutations in the CPEB homolog *orb* can block germ line differentiation and produce tumorous ovaries, it is plausible that changes in polyadenylation are important in stem cells or early cysts. It is particularly interesting that several aspects of the meiotic cell cycle appear to be controlled by CPEB, including release from meiotic arrest upon fertilization by up-regulating c-*mos* mRNA translation (Gebauer and Richter 1996; Stebbins-Boaz et al. 1996).

Germ Line Cyst Formation: A Conserved Process Mediated by the Fusome

The active association of the fusome with germ line cyst formation is a widespread phenomenon that has been known for nearly 100 years (Telfer 1975; Büning 1995; de Cuevas et al. 1997). The favorable biology and genetics available in *Drosophila* have now begun to reveal how this organelle controls the cystocyte cell cycles and selects a cell to become the oocyte. There are two basic classes of mechanisms the fusome could use to synchronize cystocyte cycles and accelerate their G_1-S transitions. The fusome may possess the capacity to transport cell cycle regulatory factors relatively quickly throughout the cyst. If materials moved at the rate of typical microtubule motors, then it would take only a short time to move a gene product throughout the cyst, as is seen in completed 16-cell cysts. However, microtubule-based transport has several drawbacks as a synchronization mechanism. The polarity of the microtubules would have to be mixed to equalize the concentration of critical products throughout the cyst, complicating the specification of a single polarity directed toward the oocyte later. Alternatively, the fusome might control a mechanism that equalized the phosphorylation state of cell cycle regulatory proteins such as a cycA/cdc2 complex. Perhaps, in addition to CycA, proteins that regulate the activation and degradation of cyclin/cdks are associated with this structure. We hope to address some of these questions in the near future by analyzing mosaic cysts in which some cystocytes lack key fusome components or candidate regulators.

One of the most interesting open questions concerns how the fusome templates microtubule polarity so that 16-cell cysts can rapidly begin to transport materials differentially to the presumptive oocyte. It seems likely that the answer to this question will require a better understanding of the intricate choreography of fusome growth and contraction during each cycle. How does the fusome associate with the centrosome and orient cystocyte spindles? Is polarity templated and stored after each division or is an inactive organizing center deposited before or during the first division that can be activated and rapidly generate a polarized cytoskeleton using materials cached within the fusome arms? Ultimately, we can look forward to an understanding of why female gametes so frequently derive from interconnected cysts during early developmental stages.

ACKNOWLEDGMENTS

This work was supported by the Howard Hughes Medical Institute, and by National Institutes of Health grant GM-27875 to A.C.S. The authors particularly thank previous co-workers who have helped to analyze the structure and function of the germarium: Dennis McKearin, Lin Yue, Haifan Lin, Jon Margolis, and Zandy Forbes.

REFERENCES

Bopp D., Horabin J.I., Lersch R.A., Cline T.W., and Schedl P. 1993. Expression of the Sex-lethal gene is controlled at multiple levels during *Drosophila* oogenesis. *Development* **118:** 797.

Büning J. 1995. *The insect ovary*. Springer-Verlag, Berlin.

Byers T.J., Dubreuil R., Branton D., Kiehart D.P., and Goldstein L.S.B. 1987. *Drosophila* spectrin. II. Conserved features of the α-subunit are revealed by analysis of cDNA clones and fusion proteins. *J. Cell Biol.* **105:** 2103.

Carpenter A.T.C. 1975. Electron microscopy of meiosis in *Drosophila melanogaster* females. I. Structural arrangement and temporal change of the synaptonemal complex in wild-type. *Chromosoma* **51:** 157.

Christerson L.B. and McKearin D.M. 1996. Orb is required for anteroposterior and dorsoventral patterning during *Drosophila* oogenesis. *Genes Dev.* **8:** 614.

Cooley L. and Theurkauf W.E. 1994. Cytoskeletal functions during *Drosophila* oogenesis. *Science* **266:** 590.

Curtis D., Lehmann R., and Zamore P.D. 1995. Translational regulation in development. *Cell* **81:** 171.

de Cuevas M., Lee J.K., and Spradling A.C. 1996. α-Spectrin is required for germline cell division and differentiation in the *Drosophila* ovary. *Development* **122:** 3959.

de Cuevas M., Lilly M., and Spradling A.C. 1997. Germ line cyst formation in *Drosophila*. *Annu. Rev. Genet.* (in press).

Forbes Z., Lin H., Ingham P., and Spradling A.C. 1996a. *hedgehog* is required for the proliferation and specification of somatic cells prior to egg chamber formation in *Drosophila*. *Development* **122:** 1125.

Forbes Z., Spradling A.C., Ingham P., and Lin H. 1996b. The role of segment polarity genes during early oogenesis in *Drosophila*. *Development* **122:** 3283.

Gebauer F. and Richter J.D. 1996. Mouse cytoplasmic polyadenylylation element binding protein: An evolutionarily conserved protein that interacts with the cytoplasmic polyadenylylation elements of c-*mos* mRNA. *Proc. Natl. Acad. Sci.* **93:** 14602.

Goode S., Melnick M., Chou T.B., and Perrimon N. 1996. The neurogenic genes *egghead* and *brainiac* define a novel signaling pathway essential for epithelial morphogenesis during *Drosophila* oogenesis. *Development* **122:** 3863.

Hake L.E. and Richter J.D. 1994. CPEB is a specificity factor that mediates cytoplasmic polyadenylation during *Xenopus* oocyte maturation. *Cell* **79:** 617.

Hawkins N.C., Thorpe J., and Schüpbach T. 1996. *encore*: A gene required for the regulation of germ line mitosis and oocyte differentiation during *Drosophila* oogenesis. *Development* **122:** 281.

Johnson R.L., Rothman A.L., Xie J., Goodrich L.V., Bare J.W., Bonifas J.M., Quinn A.G., Myers R.M., Cox D.R., Epstein Jr. E.H., and Scott M.P. 1996. Human homolog of *patched*, a candidate gene for the basal cell nevus syndrome. *Science* **272:** 1668.

Kelley R.L., Solovyeva I., Lyman L.M., Richman R., Solovyev V., and Kuroda M.I. 1993. Expression of *msl-2* causes assembly of dosage compensation regulators on the X chromosomes and female lethality in *Drosophila*. *Cell* **81:** 867.

Kim-Ha J., Kerr K., and Macdonald P.M. 1995. Translation regulation of oskar by bruno, an ovarian RNA-binding protein, is essential: mRNA is localized to the posterior pole of the *Drosophila* embryo. *Cell* **81:** 403.

King R.C. 1970. *Ovarian development in* Drosophila melanogaster. Academic Press, New York.

Kobayashi S., Yamada M., Asaoka M., and Kitamura T. 1996. Essential role of the posterior morphogen Nanos for germline development in *Drosophila*. *Nature* **380:** 708.

Kusumi A. and Sako Y. 1996. Cell surface organization by the membrane skeleton. *Curr. Opin. Cell Biol.* **8:** 566.

Lasko P. and Ashburner M. 1988. The product of the *Drosophila* gene *vasa* is very similar to eukaryotic initiation factor 4A. *Nature* **335:** 611.

Lautz V., Chang J.S., Horabin J.I., Bopp D., and Schedl P. 1994. The *Drosophila* Orb RNA-binding protein is required for the formation of the egg chamber and establishment of polarity. *Genes Dev.* **8:** 598.

Lee J.K., Brandin E., Branton D., and Goldstein L.S. 1997. α-

Spectrin is required for ovarian follicle monolayer integrity in *Drosophila melanogaster. Development* **124:** 353.
Lehmann R. and Nüsslein-Volhard C. 1991. The maternal gene *nanos* has a central role in posterior pattern formation of the *Drosophila* embryo. *Development* **112:** 679.
Lilly M. and Spradling A.C. 1996. The *Drosophila* endocycle is controlled by *cyclin E* and lacks a checkpoint ensuring S phase completion. *Genes Dev.* **10:** 2514.
Lin H. and Schagat T. 1997. Neuroblasts: A model for the symmetric division of stem cells. *Trends Genet.* **13:** 33.
Lin H. and Spradling A.C. 1993. Germline stem cell division and egg chamber development in transplanted germaria. *Dev. Biol.* **159:** 140.
———. 1995. Fusome asymmetry and oocyte determination in *Drosophila. Dev. Genet.* **16:** 6.
———. 1997. A novel group of *pumilio* mutations affects the asymmetric division of germline stem cells in the *Drosophila* ovary. *Development* **124:** 2463.
Lin H., Yue L,. and Spradling A. 1994. The *Drosophila* fusome, a germline-specific organelle, contains membrane skeletal proteins and functions in cyst formation. *Development* **120:** 947.
Mach J.M. and Lehmann R. 1997. An Egalitarian-Bicaudal D complex is essential for oocyte specification and axis determination in *Drosophila. Genes Dev.* **11:** 423.
Mahowald A.P. and Kambysellis M.P. 1980. Oogenesis. In *The genetics and biology of* Drosophila (ed. M. Ashburner and T.R.F. Wright), vol. 2, p. 141. Academic Press, New York.
Margolis J. and Spradling A.C. 1995. Identification and behavior of epithelial stem cells in the *Drosophila* ovary. *Development* **121:** 3797.
Maziarski S. 1913. Sur la persistance des résidues fusoriaux pendent les mombreuses générations cellularies au cours de l'ovogénese de *Vespa vulgaris*. L. *Arch. Zelforsch.* **10:** 507.
McGrail M. and Hays T.S. 1997. The microtubule motor cytoplasmic dynein is required for spindle orientation during germline cell divisions and ooctye differentiation in *Drosophila. Development* **124:** 2409.
McKearin D. 1997. The *Drosophila* fusome, organelle biogenesis and germ cell differentiation: If you build it . . . *BioEssays* **19:** 147.
McKearin D.M. and Spradling A.C. 1990. *bag-of-marbles:* A *Drosophila* gene required to initiate both male and female gametogenesis. *Genes Dev.* **4:** 2242.
Murata Y. and Wharton R.P. 1995. Binding of Pumilio to maternal *hunchback* mRNA is required for posterior patterning in *Drosophila* embryos. *Cell* **80:** 747.
Ohlstein B. and McKearin D. 1995. A role for the *Drosophila bag-of-marbles* protein in the differentiation of cytoblasts from germline stem cells. *Development* **121:** 2937.
Ray R.P. and Schupbach T. 1996. Intercellular signaling and the polarization of body axes during *Drosophila* oogenesis. *Genes Dev.* **10:** 1711.
Richardson H.E., O'Keefe L.V., Reed S.I., and Saint R. 1993. A *Drosophila* G_1-specific cyclin E homolog exhibits different modes of expression during embryogenesis. *Development* **119:** 673.
Robinson D. and Cooley L. 1996. Stable intercellular bridges in development: The cytoskeleton lining the tunnel. *Trends Cell Biol.* **6:** 474.
Rongo C. and Lehmann R. 1996. Regulated synthesis, transport and assembly of the *Drosophila* germ plasm. *Trends Genet.* **12:** 102.
Sallés F.J., Lieberfarb M.E., Wredan C., Gergen J.P., and Strickland S. 1994. Coordinate initiation of *Drosophila* development by regulated polyadenylation of maternal messenger RNAs. *Science* **266:** 1996.
Seydoux G., Mello C.C., Pettitt J., Wood W.B., Priess J.R., and Fire A. 1996. Repression of gene expression in the embryonic germ lineage of *C. elegans. Nature* **382:** 713.
Spradling A. 1993. Developmental genetics of oogenesis. In *Development of* Drosophila melanogaster (ed. M. Bate and A. Martinez Arias), p. 1. Cold Spring Harbor Laboratory Press, Cold Spring Harbor, New York.
Stebbins-Boaz B., Hake L.E., and Richter J.D. 1996. CPEB controls the cytoplasmic polyadenylation of cyclin, Cdk2 and c-*mos* mRNAs and is necessary for oocyte maturation in *Xenopus. EMBO J.* **15:** 2582.
St. Johnston D. 1995. The intracellular localization of messenger RNAs. *Cell* **81:** 161.
Telfer W.H. 1975. Development and physiology of the oocyte-nurse cell syncytium. *Adv. Insect Physiol.* **11:** 223.
Verheyen E.M. and Cooley L. 1994. Profilin mutations disrupt multipleactin-dependent processes during *Drosophila* development. *Development* **120:** 717.
Wickens M. 1992. Forward, backward, how much, when: Mechanisms of poly(A) addition and removal and their role in early development. *Semin. Dev. Biol.* **3:** 399.
Wieschaus E. and Szabad J. 1979. The development and function of the female germline in *Drosophila melanogaster,* a cell lineage study. *Dev. Biol.* **68:** 29.
Yue L. and Spradling A.C. 1992. *hu-li tai shao*, a gene required for ring canal formation during *Drosophila* oogenesis, encodes a homolog of adducin. *Genes Dev.* **6:** 2443.
Zalokar M. 1976. Autoradiographic study of protein and RNA formation during early development of *Drosophila* eggs. *Dev. Biol.* **49:** 425.

Genetic Characterization of *hadad*, a Mutant Disrupting Female Gametogenesis in *Arabidopsis thaliana*

J.M. Moore,[1] J.-P. Vielle Calzada, W. Gagliano,[2] and U. Grossniklaus
Cold Spring Harbor Laboratory, Cold Spring Harbor, New York 11724; [1]Graduate Program in Genetics, State University of New York, Stony Brook, New Yrok 11794

The plant life cycle alternates between a diploid and a haploid generation, the spore-producing sporophyte and the gamete-producing gametophyte. Unlike in animals, where meiotic products differentiate directly into gametes, the haploid spores of plants divide mitotically to form a multicellular haploid organism. The differentiation of gametes occurs later in the development of the gametophyte. In lower plants, gametophytes constitute the dominant phase of the life cycle, are free living, and are highly differentiated for interactions with the environment (Kenrick 1994). In contrast, the gametophytes of flowering plants consist of a small number of cells and develop within the reproductive organs of the flower. Male gametophytes (pollen or microgametophyte) are produced in the anthers and usually consist of only three cells (a vegetative cell and two sperm cells). The female gametophyte (embryo sac or megagametophyte) most often consists of seven cells and develops within the ovule, a specialized organ derived from the placental tissue of the ovary wall.

Megasporogenesis and megagametogenesis are highly conserved processes in angiosperms. In *Arabidopsis* as in most higher plants, monosporic development produces an embryo sac of the *Polygonum* type (Maheswari 1950). In the developing ovule, a single sporophytic cell differentiates into a megaspore mother cell and undergoes meiosis to form a tetrad of megaspores. Only one meiotic product survives, divides three times in a syncytium, and cellularizes to form the mature female gametophyte consisting of an egg cell and two synergids at the micropylar pole, three antipodals at the chalazal pole, and a centrally located, binucleate central cell (Fig. 1). After fertilization of both the egg and central cell, the ovule develops into a seed. This involves the coordinate development of both ferilization products, zygote and endosperm, as well as the surrounding sporophytic cell layers that develop into the seed coat. Despite their clonal origin, the eight nuclei of the embryo sac differentiate along four alternative developmental pathways. Cell specification may involve local signaling processes either between the gametophytic cells or from surrounding sporophytic tissues. One of the most striking features is the polar nature of megasporogenesis and megagametogenesis. Development of the embryo sac and the differentiation of its constituent cells occur along the chalazal-micropylar (proximodistal) axis of the ovule. The highly polar nature of this developmental process suggests the involvement of positional information within the embryo sac and/or local signaling events in the ovule (Willemse 1981; Reiser and Fischer 1993). Alternatively, cell lineage constraints could be involved in cell fate decisions as suggested by the stereotypic nuclear division patterns in a wide range of species (Schnarf 1936; Folsom and Cass 1990; Russell 1993; Huang and Sheridan 1994). Although many examples illustrate the importance of position and signaling for plant development (Steeves and Sussex 1989; van den Berg et al. 1995; Weigel and Doerner 1996), little is known about the underlying molecular mechanisms. The megagametophyte with its polar organization and small number of distinct cell types is ideally suited to address fundamental aspects of plant development such as the role of positional information, cell lineage and cell-cell communication for plant morphogenesis, and cellular differentiation.

Much attention has been given to the structure and development of the female gametophyte for more than a century. Whereas many studies have focused on morphological descriptions (Hofmeister 1849; Strasburger 1879; Cass and Jensen 1970; Willemse and van Went 1984; Russell 1985; Mogensen 1988; Huang and Russell 1992; Schneitz et al. 1995; Christensen et al. 1997), little emphasis has been given to molecular and genetic approaches (Vollbrecht and Hake 1995; Nadeau et al. 1996). Genetic analysis has mainly focused on the characterization of female sterile mutants that disrupt the development of the sporophytic tissues of the ovule surrounding the megagametophyte (Robinson-Beers et al. 1992; Lang et al. 1994; Léon -Klosterziel et al. 1994; Modrusan et al. 1994; Ray et al. 1994; Gaiser et al. 1995; Reiser et al. 1995; Eliott et al. 1996; Klucher et al. 1996). Whereas these studies have led to the formulation of genetic models for ovule morphogenesis (Angenent and Colombo 1996; Baker et al. 1997; Schneitz et al. 1997), the genetic basis and molecular mechanisms controlling the development of the female gametophyte are almost completely unknown.

[2]*Present address:* Graduate Program in Natural Resources, Ohio State University Columbus, Ohio 43210-1085.

 © 1997 Cold Spring Harbor Laboratory Press 0-87969-535-8/97.

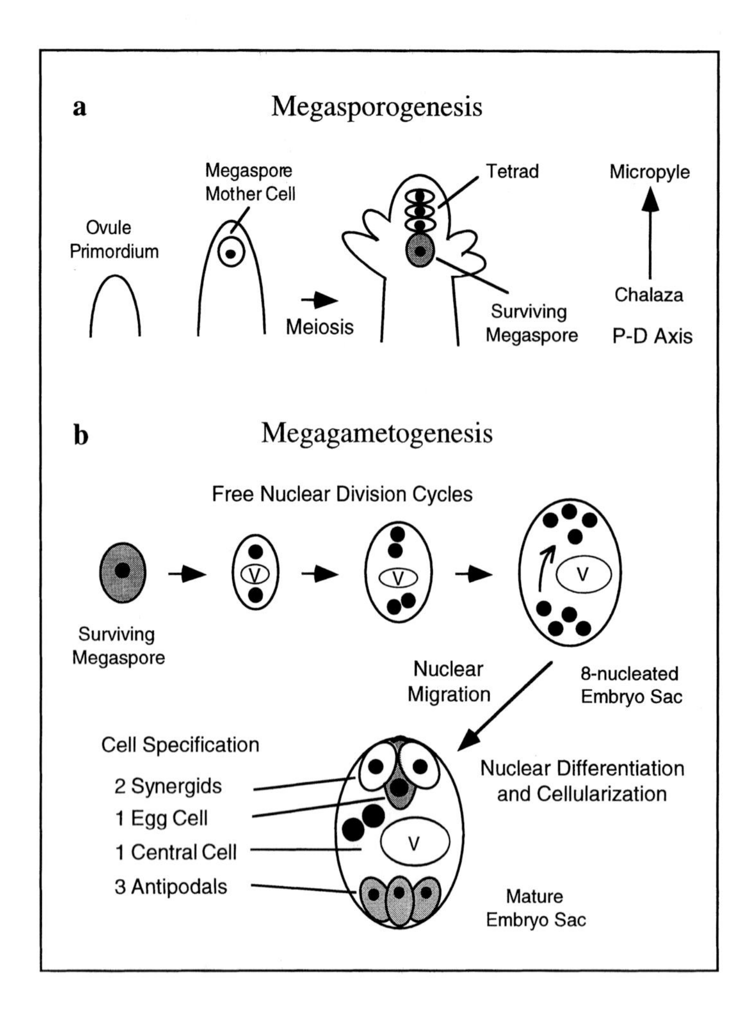

Figure 1. Diagram of megasporogenesis and megagametogenesis. (*a*) Megasporogenesis. A single megaspore mother cell differentiates in the nucellus and undergoes meiosis to produce four megaspores. Only the most chalazal one survives and gives rise to a megagametophyte of the *Polygonum* type. (*b*) Development of the megagametophyte. The surrounding sporophytic tissue of the ovule is not shown. Megagametogenesis can be subdivided into distinct developmental events: nuclear division cycles, nuclear migration, nuclear differentiation and fusion, cellularization, and cell differentiation. Three free nuclear divisions give rise to an eight-nucleated embryo sac. One of the chalazal nuclei migrates to the micropylar pole. Cellularization forms the mature embryo sac with its four distinct cell types. The two polar nuclei of the central cell fuse prior to fertilization (not shown). (V) Vacuole.

Transmission studies of chromosomal deletions (Patterson 1978; Coe et al. 1988; Buckner and Reeves 1994; Vizir et al. 1994) and deficiency analysis in maize (Vollbrecht and Hake 1995) suggest that a large number of loci essential for female gametogenesis are dispersed throughout the genome. Nevertheless, gametophytic mutants have rarely been isolated. Only two *Arabidopsis* mutants, *Gametophyte factor 1* (*Gf1*) and *prolifera* (*prl*), are known to affect megagametogenesis (Rédei 1965; Springer et al. 1995). *Gf1* is not transmitted through the female gametophyte and shows reduced transmission through the male. Mutant *Gf1* megagametophytes arrest at the one-nuclear stage and do not initiate the mitotic division cycles (Christensen et al. 1997). *prl*, as a member of the *MCM2-3-5* family and a putative component of DNA replication licensing factor, is an essential gene required in all dividing cells. In maize, megagametophytes carrying *indeterminate gametophyte* (*ig*) or the *r-X1* deficiency undergo abnormal mitotic divisions and are transmitted through the female gametophyte at a reduced frequency (Kermicle 1971; Lin 1978 1981; Weber 1983; Huang and Sheridan 1996). Embryo sacs mutant for *lethal ovule* (*lo1* and *lo2*) do not produce viable seeds (Singleton and Mangelsdorf 1940; Nelson and Clary 1952). *lo2* gametophytes arrest during the mitotic phase of megagametogenesis (Sheridan and Huang 1997).

Here, we describe the rationale of a genetic screen aimed at the identification of mutants required for megagametogenesis which is based on an insertional mutagenesis system in *Arabidopsis thaliana* (Springer et al. 1995; Sundaresan et al. 1995). We focus on the genetic and phenotypic characterization of *hadad* (*hdd*), a mutant that predominantly affects the female gametophyte but also shows reduced transmission through the male. Wild-type *HDD* activity is required for normal progression through the mitotic division cycles of the haploid gametophytes, and mutant embryo sacs arrest after one or two divisions or progress very slowly through megagametogenesis and eventually degenerate.

EXPERIMENTAL PROCEDURES

Plant material. Plants were grown on CustomBlen plus low 1 soil (Scotts) containing 1.36 kg/m^3 of 12-12-12 slow-release fertilizer (Osmocote). The systemic insecticide Marathon (Olympic Horticultural) was added to 1.36 kg/m^3, and the soil was rinsed before planting. Plants were cultivated either in an indoors growth facility at an average temperature of 19°C under F20T12 Daylight and F20T12 Plant Light tubes (Sylvania) in a 4:1 ratio in continuous light of approximately 220 lumens/m^2 at pot height or in a growth incubator (Percival) at 20°C under F48T12 Cool White high-output light tubes (Sylvania) in a 16-hour photoperiod of approximately 480 lumens/m^2 at pot height. Plants were watered using a subirrigation system. The wild-type strains were *A. thaliana* (L.) Heynh. var. Landsberg (*erecta* mutant; L*er*) and *A. thaliana* (L.) Heynh. var. Columbia (Col). Insertional mutagenesis using enhancer- and gene-trap *Ds* elements was performed as described previously (Sundaresan et al. 1995).

Ovule preparations and microscopy. For whole-mounted specimens, the carpel valves of siliques at various stages of development were removed using hypodermic needles (1-cc insulin syringes U-100 28G1/2, Becton-Dickinson). Dissected carpels were fixed in FAA (10:7:2:1 of ethanol [95%]:dH$_2$O:formaldehyde [37%]: acetic acid) for 2 hours and were transferred either directly or progressively into a drop of Herr's solution (Herr 1971) on a slide. After further dissection, they were allowed to clear at least overnight before microscopical inspection. Specimens were viewed on a Leica DMR microscope with differential interference contrast (DIC) optics. Photographs were taken on Techpan film (Kodak). For sectioned material, siliques were fixed in 3% glutaraldehyde for 2 hours, rinsed in 50 mM cacodylate buffer (Electron Microscopy Sciences), and postfixed in 2% OsO_4 (in cacodylate buffer). After dehydration in an ethanol series, specimens were embedded in Spurr's media as described previously (Spurr 1969). Specimens were sectioned on an Ultracut E ultramicrotome (Reichert-Jung) and observed on a Leica DMR microscope under brightfield. Photographs were taken on Royal Gold 400 film (Kodak) using a CB12 daylight filter. Slides or negatives were scanned on a 35 Sprintscan slide scanner (Polaroid) and processed for publication using Adobe Photoshop 3.0 (Adobe Systems).

Transmission and segregation analysis. For transmission analysis, *hdd*/+ plants were reciprocally crossed to Col wild-type plants. Since *hdd* is in L*er* background, the *erecta* marker allowed us to distinguish progeny of a cross from possible contaminants resulting for self-fertilization. Seeds resulting from these crosses or self-fertilization (segregation analysis) were sterilized by ethanol immersion for 10 minutes, followed by 30% bleach immersion for 5 minutes with three rinses in sterile water. They were then plated on 0.7% agar media (Difco) containing 4.4 g/liter Murashige-Skoog salts (Carolina Biological) and 10 g/liter Sucrose (Gibco BRL). The pH was adjusted to 5.7 with KOH. After autoclaving, 50 mg/liter kanamycin (Sigma) was added, which allowed us to follow the kanamycin resistance marker present on the *Ds* element linked to *hdd.*

RESULTS

An analysis of the genetic and molecular basis of gametophyte development has been difficult because of the inaccessability of the embryo sac and the small number of cells involved. Since gametophyte development depends on expression of the haploid genome, mutants affecting the gametophytic phase can in principle be identified by the occurrence of defective gametophytes produced by a plant heterozygous for such a mutation. Mutations in genes essential for gametogenesis will produce 50% defective and 50% normal gametophytes in a heterozygote (or heterozygos sector of an M1 plant). Thus, female gametophyte lethal mutations are expected to display a semisterile phenotype since only half of the ovules will produce viable seeds. In contrast, male gametophyte lethal mutants are fully fertile due to an excess of wild-type pollen relative to the number of ovules available for fertilization. A subset of these mutations can be recognized based on 50% pollen abortion, a feature that led to the isolation of a number of male gametophytic mutants in maize and *Arabidopsis* (see, e.g., Singleton and Mangelsdorf 1940; Rédei 1964; Laughnan and Gabay-Laughnan 1983; Xu et al. 1995; Chen and McCormick 1996). Because mutant alleles of gametophytically required loci are not transmitted to the next generation by defective gametophytes, such mutations are characterized by non-Mendelian segregation. Disruption of a gene essential to the development or function of the gametophyte in one sex but not affecting it in the other is expected to segregate the mutant phenotype in one half (rather than one quarter) of the progeny from a heterozygote. Distorted segregation has been used to demonstrate the gametophytic requirement of several mutants affecting pollen viability in maize and *Arabidopsis* (see, e.g., Correns 1902; Manglesdorf and Jones 1926; Rédei 1964; Demerec 1929; Bianci and Lorenzoni 1975; Ottaviano et al. 1988; Nelson 1994; Chen and McCormick 1996).

Insertional Mutagenesis Screen for Mutants Disrupting Megagametogenesis

We have used a recently established insertional mutagenesis system in *Arabidopsis* to identify genes required in the gametophytic phase of the life cycle. The enhancer detection and gene-trap system (O'Kane and Gehring 1987; Skarnes 1990; Wilson et al. 1990) developed at Cold Spring Harbor Laboratory is based on the *Ac*/*Ds* transposable element from maize (Springer et al. 1995; Sundaresan et al. 1995). We performed a two-step screen for mutants that show both characteristics of a mutation disrupting megagametogenesis: semisterility and non-Mendelian segregation. Among close to 5000 transposants usually carrying single randomly distributed *Ds* elements, we identified about 150 plants with reduced fertility (J. Moore et al., unpubl.). A reduction in seed set can be caused by a variety of reasons such as poor growth conditions, sporophytically required female sterile mutations with low penetrance, megagametophyte lethal mutations, and reciprocal translocations. In plants heterozygous for a reciprocal translocation, half the spores will carry a chromosomal imbalance that leads to gametophyte lethality (Sree-Ramulu and Sybenga 1979,1985; Koornneef et al. 1982; Patterson 1994; Ray et al. 1997).

These different possibilities can be distiguished genetically by characterizing the inheritance of the trait.

Whereas plants producing fewer seeds due to environmental effects or mutations in sporophytically required genes show normal Mendelian inheritance, gametophytic mutants display a distorted segregation among their progeny. Since the transposons used are marked with a kanamycin resistance marker, their segregation pattern can easily be followed in seedlings. Analyzing the segregation pattern of the transposon rather than semisterility allows us to immediately focus on lines with a tight linkage of phenotype and *Ds* insertion which are likely to be tagged. Gametophytic mutants not transmitted through the megagametophyte but normally through the male segregate for semisterile and wild-type plants in a ratio of 1:1. The same inheritance of semisterility is observed in translocation heterozygotes. In contrast, kanamycin resistance will segregate in a Mendelian fashion whether unlinked to or directly associated with the translocation breakpoint. Thus, segregation distortion is a powerful means to distinguish between various causes of reduced fertility and directly indicates whether a mutation is likely to be tagged by the *Ds* insertion. We characterized the segregation of kanamycin resistance in 126 lines with reduced fertility and found that 59 of them showed a non-Mendelain segregation pattern consistent with a gametophytic defect (J. Moore and U. Grossniklaus, unpubl.). Reduced fertility and distorted segregation strongly suggest that these mutations affect the development or function of the female gametophyte, although they may affect male gametogenesis to some degree as well. Because the disrupted genes are required for normal fertility, we name megagametophyte lethal mutants after gods and godesses associated with fertility cults of different ethnic groups.

hdd Heterozygotes Are Semisterile and Show Distorted Segregation

In our screen for transposants with reduced fertility, we found that the gene-trap transposant GT 1925 produces only about half the seeds as wild type. Whereas in wild type, about 95% of the ovules become fertilized and develop into seeds, in GT 1925, approximately 50% of the ovules arrest (Table 1). Arrested ovules do not initiate seed development, degenerate, and leave a characteristic empty space in the seed pod, the phenotypic trait we screened for (Fig. 2). To determine the cause of reduced fertility in GT 1925, we analyzed the inheritance of the *Ds* element by following the segregation of kanamycin-resistance in the progeny. GT 1925 segregated kanamycin-sensitive and -resistant seedlings in a ratio of 1.8:1.0 as compared to 1:3 expected for normally segregating insertions (Table 2). The non-Mendelian inheritance strongly suggests that semisterility in GT 1925 is caused by a disruption of megagametogenesis. We named this megagametophyte lethal mutation *hadad* (*hdd*) after the the god of fertility of the Aramaeans.

The degree of segregation distortion does not only depend on a lack or reduction of transmission through the female and/or male gametophyte, but also on the number of *Ds* elements present. Although the system used here in general produces single *Ds* insertions, two or more insertions are sometimes recovered (Sundaresan et al. 1995; J. Moore and U. Grossnikaus, unpubl.). To determine the number of *Ds* elements in GT 1925, we performed a Southern blot analysis on progeny of GT 1925 (data not shown). In addition to an internal band derived from the *Ds* insertion, we detected only one border fragment using an appropriate probe derived from *Ds,* suggesting that a single *Ds* element was present in the original transposant. Among 73 plants analyzed, the *Ds* insertion always cosegregated with semisterility and segregation distortion. The tight linkage between the gametophytic defect and the transposon strongly suggests that the *Ds* insertion disrupted the wild-type *HDD* gene. However, an unequivocal confirmation that *HDD* is molecularly tagged will require a reversion analysis.

Table 1. Seed Phenotype in *hdd*/+ and Wild-type Plants

	Mean per silique	S.D.	N (ovules)	%
hdd				
undeveloped	33.3	5.5	1567	50
normal	32.4	5.2	1522	49
aborted	0.1	0.3	6	1
total	65.8	7.1	3095	100
Ler				
undeveloped	4.6	1.7	55	6
normal	65.3	6.8	783	93
aborted	0.6	1.5	6	1
total	70.5	5.0	844	100

Green or dry siliques of *hdd* heterozygotes and L*er* wild-type plants were opened and the seeds were classified as undeveloped ovules, normal, and aborted seeds. The mean number of each class per silique and the standard deviation (S.D.) are given. N (ovules) is the total number of ovules or seeds scored in this class.

hdd Affects Predominantly the Female Gametophyte

A female-specific gametophyte lethal which is normally transmitted through the male will segregate kanamycin-sensitive to -resistant seedlings in a ratio of 1:1. An additional reduction of transmission through the male would increase the number of kanamycin-sensitive seedings at the expense of the resistant ones. The distorted segregation ratio of kanamycin-sensitive to -resistant seedlings being 1.8:1.0 in *hdd* suggests that transmission of the mutant allele is also reduced through the male gametophyte. The female defect is highly penetrant since consistently 50% of the ovules fail to initiate seed development. To determine to which extent *hdd* affects the development of the male gametophyte, we performed reciprocal outcrosses to wild type (Table 3). Indeed, male transmission was reduced by 51%, indicating a male requirement for *HDD*. Female transmission was drastically reduced to 3% but not completely abolished. These transmission frequencies are consistent with the segregation distortion observed in the original transposant. On the ba-

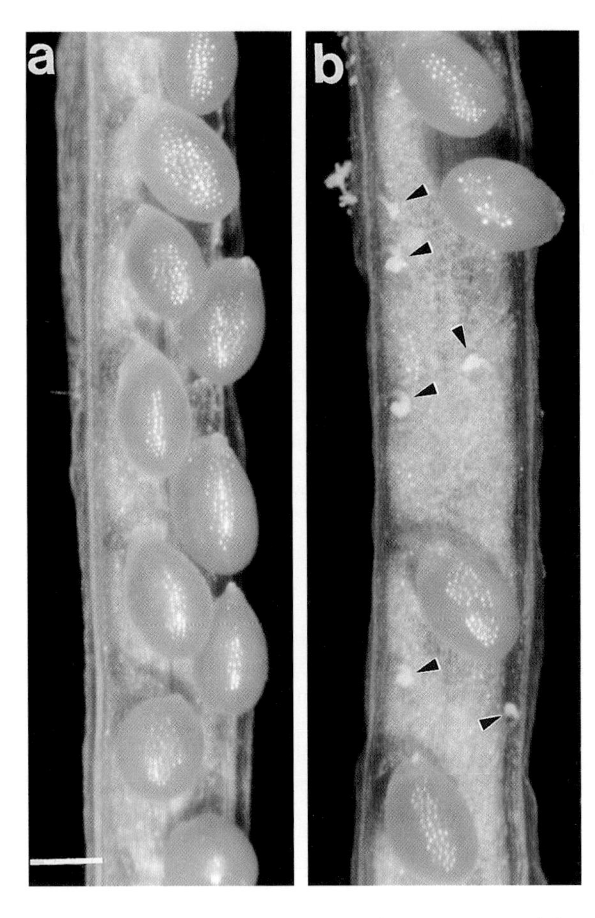

Figure 2. Semisterile phenotype in *hdd* heterozygotes. (*a*) Immature wild-type silique with normally developing seeds. (*b*) Immature silique of an *hdd* heterozygote containing normal seeds and undeveloped ovules. Ovules that remain unfertilized (*arrowheads*), and do not initiate seed development, degenerate and leave an empty space in the seed pod. Bar, 100 μm.

sis of the percentage of functional *hdd* gametophytes produced, we expect a ratio of kanamycin-sensitive to -resistant seedlings of 1.9:1.0.

Although we observed always more kanamycin-sensitive than -resistant seedlings indicating a consistent male effect, the segregation ratio varied in different experiments ranging from 1.1:1.0 to 4.0:1.0 (Table 2). Within each experiment, the standard deviation was somewhat higher but comparable to the one observed with normally segregating *Ds* insertions. This suggests that the large variation between experiments was due to fortuitous variability of environmental conditions under which the plants were grown for different experiments. Female transmission rates were highly reproducible: Among 32 outcrosses, 27 showed no transmission at all, 3 produced a single seed carrying the mutant *hdd* allele, and 1 each produced two and three kanamycin-resistant seedlings, respectively. The seed phenotype of plants grown under different conditions was also very consistent. In contrast, male transmission varied greatly between crosses, with transmission frequencies ranging from about 20% to 100% with an average of 49%. Taken together, these results suggest that the megagametophytic defect in *hdd* mutants is highly penetrant, whereas the male defect is variable and is strongly influenced by environmental conditions.

hdd Embryo Sacs Arrest during Megagametogenesis

To elucidate the function of *HDD* in megagametogenesis, we characterized the phenotype of gametophytes carrying the mutant *hdd* allele in cleared and sectioned specimens (Figs. 3 and 4). To facilitate the interpretation of the mutant phenotype, we first describe the wild-type development of the embryo sac in some detail (see also Schneitz et al. 1995; Christensen et al. 1997). Within a developing ovule, a single subepidermal cell enlarges, differentiates into a megaspore mother cell, and undergoes meiosis to produce four megaspores (Hill and Lord 1994). In *Arabidopsis*, the two meiotic nuclear divisions occur before cytokinesis, leading to the formation of tetrads with a linear or multiplanar arrangement (Webb and Gunning 1990; Schneitz et al. 1995). Only the chalazal-most megaspore survives (functional megaspore), whereas the other three undergo programmed cell death and degenerate (Fig. 3a). The functional megaspore gives rise to the mature embryo sac by three consecutive mitotic divisions that occur in a syncytium. After the first division, the nuclei migrate to opposing poles of the developing megagametophyte (Fig. 3b), and a large vacuole forms in its center. As the embryo sac enlarges, the integuments grow to envelop the nucellus. Asymmetric growth of the integuments gives the ovule its characteristically curved shape. The nuclei at each pole undergo two synchronous divisions to form the four- and eight-nucleated embryo sac (Fig. 3c,d). A single nucleus at the chalazal pole starts migrating toward the micropylar pole and gives rise to one of the two polar nuclei. Cellularization leads to the formation of seven cells: an egg cell and two synergids at the micropylar pole, three antipodals at the chalazal pole, and a central cell harboring the two polar nuclei.

The two synergids and the egg cell are arranged in triangular configuration at the micropylar pole to form the egg apparatus (Webb and Gunning 1988; Mansfield et al. 1991). The egg cell is located at a slightly chalazal position with respect to the synergids (Fig. 3e). One of the synergids typically degenerates prior to fertilization, but the moment for initiation of the degenerative process varies (Russell 1992). The large central cell is highly vacuolated and contains the polar nuclei originating from opposite poles. They will fuse prior to fertilization to form the diploid nucleus of the central cell (Fig. 3e). Three antipodal cells differentiate at the chalazal pole; they usually degenerate after the fusion of the two polar nuclei (Christensen et al. 1997). Double fertilization of both the egg and central cell gives rise to the zygote and the triploid primary endosperm nucleus. The cytoplasm in the zygote reorganizes (Jensen 1968; Olson and Cass 1981; Russell 1993), and the zygote elongates but does not divide for some time. In contrast, the primary en-

Table 2. Segregation Analysis of the *Ds* Element in *hdd* Mutants

	Kan^S (%)	Kan^R (%)	Segregation ratio	S.D. (%)	N (seedlings)	N (plants)
GT 1925	64.1	35.9	1.8:1.0	n.a.	590	1
A	59.2	40.8	1.5:1.0	7.4	608	10
B	58.3	41.7	1.4:1.0	2.9	3,496	7
C	52.3	47.7	1.1:1.0	4.3	2,898	4
D	69.7	30.3	2.3:1.0	7.9	940	9
E	79.9	20.1	4.0:1.0	4.0	882	5
F	60.3	39.7	1.5:1.0	5.3	618	6
Total	59.0	41.0	1.4:1.0	9.6	10,032	42
G	25.5	74.5	1.0:2.9	1.1	2,152	8
H	24.7	75.7	1.0:3.1	2.2	1,422	5

Segregation of kanamycin resistance in seedlings was analyzed in *hdd* heterozygotes. GT 1925 was the originally isolated transposant that showed semisterility and non-Mendelian segregation. Experiments A through F represent plants derived from GT 1925 that were grown at different times in different locations. The percentage of kanamycin-resistant (Kan^R) and -sensitive (Kan^S) seedlings is the mean of several plants (N[plants]) grown under the same conditions. The standard deviation (S.D.) is indicated. In experiments G and H, plants segregating the kanamycin resistance marker in a normal Mendelian fashion are shown for comparison. N (seedlings) is number of seedlings scored. n.a. indicates not applicable.

dosperm nucleus divides syncytially a few times before the first division of the embryo occurs (Fig. 3f). Subsequent seed development depends on the coordinate morphogenesis of embryo, endosperm, and the integumental cell layers forming the seed coat.

To determine the function of *HDD* in female gametogenesis, we analyzed the phenotype of ovules in *hdd* heterozygotes. These plants segregate ovules containing mutant and wild-type megagametophytes in a ratio of 1:1, allowing for a direct comparison of wild-type and mutant gametophytes within the same silique. Using cleared and sectioned material, we analyzed the developmental progression of wild-type and *hdd*-defective ovules before and after pollination. In all ovules, meiosis resumes normally, as three megaspores degenerate and a single haploid megaspore differentiates in the center of the young nucellus (Fig. 4a). No developmental defects are found before the first division of the viable megaspore (Fig. 4b). Defective embryo sacs arrest at variable stages of megagametogenesis, either after the first mitotic division of the functional megaspore or after additional mitotic divisions. Ovules containing a two-nucleated megagametophyte before or after vacuolization and expansion can be found in pollinated pistils at stages when wild-type ovules contain early globular embryos (Fig. 4c). The establishment of the micropylar-chalazal polarity that characterizes the developing megagametophyte after the first mitotic division is not affected. In arrested megagametophytes containing more than two nuclei, aberrant patterns of cellularization can occur at the micropylar or chalazal pole (Fig. 4e); however, cellularization does not result in the differentiation of specific cell types (synergids, egg cell, antipodals). Some defective embryo sacs undergo all three mitoses and form an eight-nucleated gametophyte, but the distribution of the nuclei is abnormal (Fig. 3g).

Megagametophytes containing odd numbers of nuclei were also observed, suggesting that mitotic divisions are not necessarily synchronized. The occurrence of asynchronous divisions is further illustrated by the presence of embryo sacs that had four nuclei at the micropylar pole but only two at the chalazal pole (Fig. 3h). As a consequence of limited megagametophyte growth and expansion, portions of nucellar cells found within the region defined by the endothelium are not re-absorbed (Fig. 4c,d). The rest of the sporophytic tissue is structurally normal and indistinguishable from wild type. Several days after pollination, the remnants of degenerating nuclei can still be identified in arrested megagametophytes (Fig. 3i).

Mitotic Progression of *hdd* Embryo Sacs Is Severely Delayed

Phenotypic characterization strongly suggested that *HDD* is required for normal progression through the mitotic division cycles. *hdd* mutant gametophytes were arrested at various mitotic stages in siliques containing wild-type gametophytes that had already initiated embryogenesis. To gain insight into the temporal progression of mutant gametophytes, we phenotypically characterized megagametogenesis in wild-type and *hdd* heterozygous plants from the tetrad stage through early embryogenesis (Fig. 5). To draw quantitative conclusions, usually more than 200 ovules of each developmental phase were scored. The stage of the ovules analyzed was determined using sporophytic characteristics to allow a comparison of the embryo sac at various stages between wild-type and *hdd* heterozygous plants (for details of the developmental phases, see legend to Fig. 5). The development of the sporophytic tissues of the ovule appears to be largely independent of megagametogenesis (Hülskamp et al. 1995; Schneitz et al. 1997; J. Moore and U. Grossniklaus, unpubl.) such that the developmental stage of an ovule can be determined even if it does not contain an embryo sac. Since ovules within a gyneocium develop synchronously (with the exception of ovules at the ends of the carpels which are often delayed), individual gyneocia were assigned to one of seven developmental phases based on sporophytic criteria (Fig. 5a–g). The stage to which megagametogenesis had progressed was determined for each embryo sac (stages ranging from

Table 3. Transmission of *hdd* through Male and Female Gametophytes

	Kan^S	Kan^R obs.	N(seedlings)	Kan^R exp.	Transmission (%)
hdd/+ X Col	513	8	521	260.5	3.1
Col X *hdd*/+	800	258	1058	529	48.8

Transmission of *hdd* through male and female gametophytes was determined by following the kanamycin resistance marker associated with *hdd*. *hdd* heterozygous females were crossed to Col wild-type plants (*hdd*/+ x Col) to determine female transmission, and the reciprocal cross was performed to determine male transmission (Col x *hdd*/+). The number of observed kanamycin-resistant seedlings (Kan^R obs.) was divided by the number of expected kanamycin-resistant seedlings (Kan^R exp.) based on Mendelian segregation. N (seedings) is the number of seedlings scored. Kan^S is kanamycin-sensitive seedlings.

tetrad stage to early seed development, see legend to Fig. 5). For each developmental phase, the percentage of embryo sacs at a given stage of megagametogenesis in wild-type and *hdd* heterozygotes was plotted as a histogram (Fig. 5) allowing a quantitative comparison of the developmental profile.

The histograms in Figure 5 reflect the general morphogenetic progression of the embryo sac over the course of gametogenesis as ovule development proceeds. In wild type (white bars), there is a continuous development from the tetrad stage to the mature embryo sac with some variation from ovule to ovule (Fig. 5a–c). Then the female gametophytes appear to halt their developmental progression at the mature stage such that by phase (d), the vast majority of the embryo sacs (about 85%) have reached maturity awaiting fertilization (Fig. 5d). Since fertilization events are not synchronized, there is a considerable variation of developmental progression in the early phase of seed development (Fig. 5e,f). Whereas in more than 40%, the endosperm has undergone four nuclear division cycles by phase (f), more that 30% of the ovules are still at the mature stage, leading to a biphasic distribution (Fig. 5f). In the next phase, this biphasic distribution disappears such that in more than 92% of the ovules, the first division of the zygote has occurred (Fig. 5g).

The developmental profile of *hdd* heterozygotes (black bars) looks very similar to that of wild type in early phases of development (Fig. 5a,b). In fact, the embryo sacs appear to be slightly advanced in their development as compared to wild type (Fig. 5a). This may be due to a sampling effect or caused by the mutation. At later phases of development, there is a clear delay in the development of about 50% of the gametophytes, which leads to a biphasic distibution of embryo sac stages that becomes apparent when the majority of wild-type gametophytes reach maturity and is maintained throughout development thereafter (Fig. 5d–g). Developmental arrest of embryo sacs is first detectable in phase (c) when only 39% of the embryo sacs in *hdd* heterozygotes have progressed past the two-nuclear stage as compared to 79% in wild type (Fig. 5c). About half of the gametophytes lag behind and are at the one- or two-nuclear stage of megagametogenesis. This delay becomes more prominent in the next developmental phase when only 42% of the gametophytes in the mutant have completed the mitotic cycles as compared to 85% in wild type. The vast majority of the remaining ovules in *hdd* heterozygotes (42%) are apparently arrested at the two- and four-nuclear stage (Fig. 5d). As development proceeds, about half of the ovules contain embryo sacs that progress together with wild type. In phases (f) and (g), for instance, 50% and 51% of all ovules scored have reached the mature stage or initiated seed development as compared to 96% and 92% in wild type, respectively. The remaining megagametophytes arrest or degenerate, and some undergo delayed mitotic divisions.

Ovules harboring degenerated gametophytes occur at a frequency of about 5% in wild type. In *hdd* heterozygotes, the frequency of embryo sacs that degenerate or contain uneven numbers of nuclei increases to 13% in phase (d), suggesting that some of the gametophytes arrested at the two- or four-nuclear stage degenerate at this point (Fig. 5d). A second increase in the number of degenerating and aberrant embryo sacs is observed during early phases of seed development when this class comprises 30% of the gametophytes scored (Fig. 5g). In the same period, the number of gametophytes that have not reached the mature stage decreases from 38% in phase (f) to 20% in phase (g), suggesting that most of the embryo sacs which have not yet completed the mitotic division cycles degenerate. Taken together, these results strongly suggest that the wild-type HDD product is required for normal progression through the nuclear division cycles. Whereas half of the ovules that contain megagametophytes carrying a wild-type *HDD* allele progress normally through gametogenesis and early seed development, the embryo sacs carrying a mutant copy of *hdd* arrest predominantly at the two- and four-nuclear stage. They subsequently degenerate or undergo delayed and irregular mitotic divisions but usually do not form mature functional embryo sacs. The results of this phenotypic characterization is completely consistent with the highly penetrant semisterility and the almost complete lack of transmission of the mutant *hdd* allele through the female gametophyte.

DISCUSSION

A Combination of Semisterility and Non-Mendelian Segregation Is a Stringent Test for Mutants Disrupting Female Gametogenesis

Plants have evolved a characteristic life strategy with alternating generations and the absence of a distinct germ line, features that have important implications for the development of the gametes and embryogenesis (Walbot

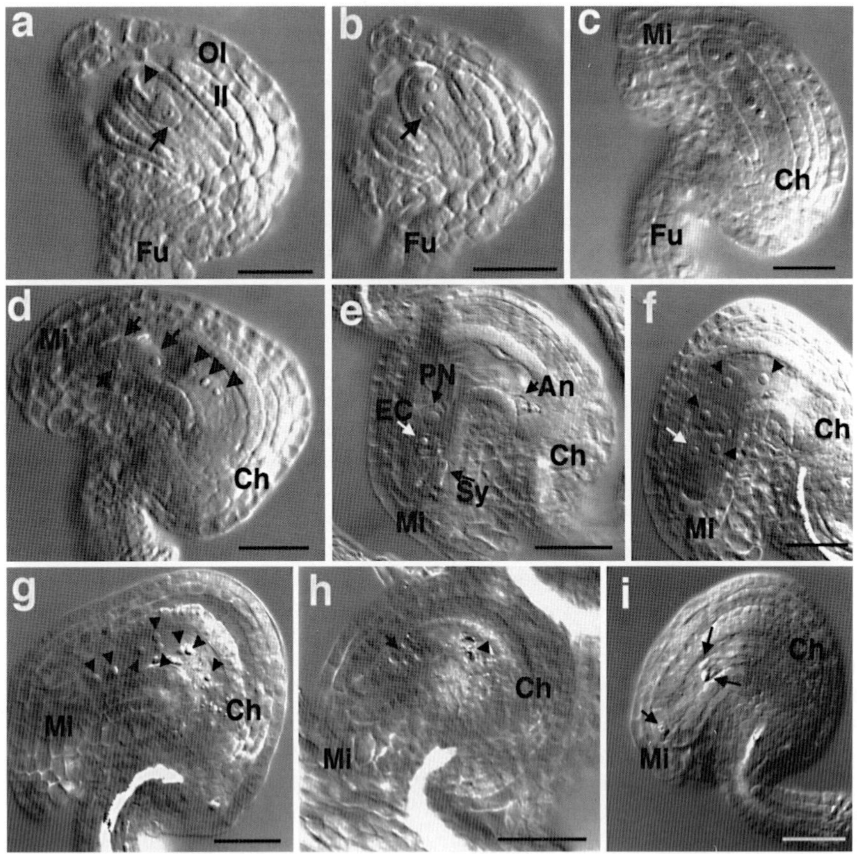

Figure 3. Megagametophyte development in cleared ovules of wild type (*a–f*) and *hdd*/+ (*g–i*) plants. (*a*) Viable megaspore (*arrow*) and remnants of three degenerated megaspores (*arrowhead*) at the initiation of megagametogenesis. (*b*) Two-nucleated megagametophyte (*arrow*). (*c*) Four-nucleated megagametophyte. (*d*) Eight-nucleated noncellularized megagametophyte (only six out eight nuclei are visible). The nuclei at the micropylar pole (*arrowheads*) will give rise to the egg apparatus and the binucleated central cell; the three nuclei at the chalazal pole (arrows) will give rise to the antipodals. (*e*) Cellularized megagametophyte showing a synergid (Sy), the egg cell (EC), the fused diploid polar nucleus (PN), and two out of three antipodals (An). (*f*) Fertilized megagametophyte. The zygote (*white arrow*) will give rise to the embryo, and mitotic divisions of free nuclei in the central cell (*arrowheads*) will result in the formation of the endosperm. (*g*) Defective *hdd* ovule containing an arrested noncellularized megagametophyte with eight nuclei (*arrowheads*). (*h*) Arrested megagametophyte in which mitotic divisions at the micropylar (*arrow*) and chalazal (*arrowhead*) poles were asynchronous. (*i*) Degeneration of nuclei (*arrows*) in an arrested 4-nucleated megagametophyte. (Ch) Chalazal pole; (Fu) funiculus; (II) inner integument; (Mi) micropylar pole; (OI) outer integument. Bars: (*a–c*) 10 μm; (*d*) 9 μm; (*e–f*) 11 μm; (*g–i*) 13 μm.

1996). For instance, mutations in essential genes cannot be transmitted through the haploid gametophytic phase. Therefore, many genes having crucial roles in developmental processes cannot be identified in screens that concentrate on later stages of development. On the other hand, the haploid nature of the gametophytes allows for identification of gametophytic mutants directly in mutagenized M1 plants. A mutation disrupting female gametogenesis results in a semisterile phenotype. Although the haploid nature of the megagametophyte offers attractive opportunities for chemical mutagenesis aimed at the identification of semisterile mutants, such screens have proven difficult due to the large number of aborted gametophytes and embryos that are produced after efficient mutagenesis. This results in a very high background of partial sterility making it difficult to identify true gametophytic mutants (A. Ray; G. Drews; both pers. comm.). Alternatively, distorted segregation can be used to identify chemically induced gametophytic mutants if multiply marked chromosomes are available. Such screens are promising and allow the identification of mutants affecting both male and female gametogenesis (M. Hülskamp, pers. comm.). However, they are limited to particular chromosomal regions for which easy scorable markers are available. Moreover, such screens do not allow a distinction between mutations disrupting gametogenesis and chromosomal rearrangements (e.g., inversions and translocations) that lead to an aberrant segregation pattern posing problems for a screen aimed at the identification of gametophytically required loci (A. Ray, pers. comm.).

Here, we describe a two-step strategy based on an insertional mutagen that combines both traits associated with megagametophyte lethality, semisterility, and non-Mendelian segregation. Semisterility is directly scored in the original transformant or transposant, and the inheritance pattern is then characterized in the progeny of putative semisterile plants. The use of an insertional mutagen such as a transposon or T-DNA construct solves some of the problems associated with chemical mutagenesis.

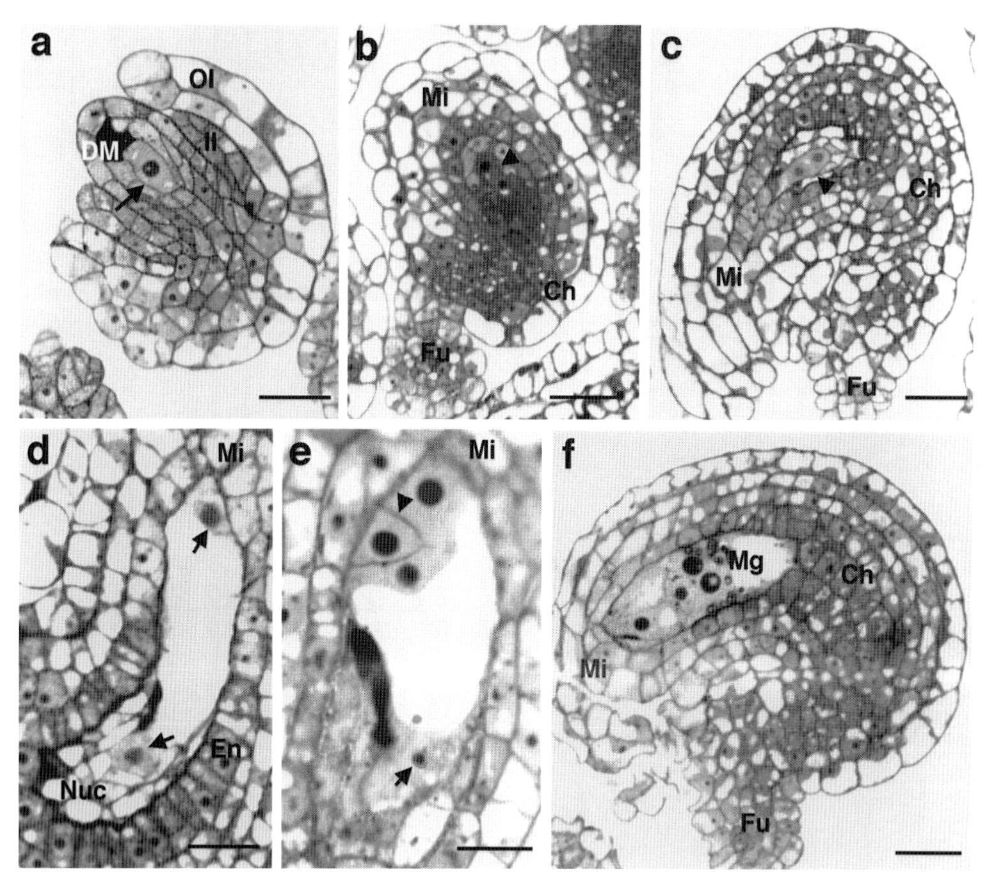

Figure 4. Megagametophyte development in sectioned ovules of *hdd*/+ plants. (*a*) The viable megaspore (VM) in the young ovule at the initiation of megagametogenesis; one of the degenerated megaspores (DM) is visible. (*b*) Wild-type developing ovule containing a two-nucleated megagametophyte (*arrowhead*) before vacuolization. (*c*) Defective mature ovule where the megagametophyte (*arrowhead*) is arrested at the two-nucleated stage prior to vacuolization. (*d*) Arrested two-nucleated megagametophyte after vacuolization; the two degenerating nuclei (*arrows*) are located at the micropylar (Mi) and chalazal poles; a portion of nonreabsorbed nucellus (Nuc) is visible within the region defined by the endothelium (En) (*e*) Aberrant cellularization in a five-nucleated megagametophyte. Three nuclei showing signs of cellularization (*arrowhead*) are located at the micropylar pole, and one is located at the chalazal pole (*arrow*); a second chalazally located nucleus is not visible in this section. (*f*) Wild-type ovule showing a normally cellularized megagametophyte (Mg). (Ch) Chalazal pole; (Fu) funiculus; (II) inner integument; (Mi) micropylar pole; (OI) outer integument. Bars: (*a*) 5 μm; (*b*) 7 μm; (*c*) 8 μm; (*d–e*) 6 μm; (*f*) 8 μm.

First, the high lethality and inviability associated with ethylmethanesulfonate (EMS) treatment is circumvented, allowing the screening of healthy plants for reduced fertility. Second, easily scorable markers present on the insertion are directly linked to insertional mutants and allow for a rapid and accurate assay of segregation in the progeny. Futhermore, this allows one to focus immediately on cadidates that are likely to be tagged, facilitating a subsequent molecular isolation of the affected gene. A search for non-Mendelian segregation in T-DNA collections has proven to be a powerful means to identify gametophytic mutants affecting either sex (R. Howden, S.-K. Park, and D. Twell; G. Drews; both pers. comm.). However, T-DNA mutagenesis quite often generates chromosomal rearrangements that may result in distorted segregation that is not caused by a gametophyte lethal, a problem that so far has not been observed with transposon-based systems. In summary, a combined screen for semisterility and distorted segregation allows the efficient identification of transposon-induced mutants that predominantly affect the female gametophyte. However, given the nature of transposon systems available in *Arabidopsis* to date, saturation mutagenesis is not feasible and a genetic analysis of female gametogenesis as outlined here is necessarily limited to a representative subset of gametophytic mutants.

HDD Is Required for Progression through the Mitotic Phase of Gametogenesis

Our phenotypic characterization of *hdd* heterozygous plants strongly suggested that the HDD wild-type product is required for normal progression through the nulcear division cycles in the female gametophtye. Embryo sacs carrying the mutant *hdd* allele arrest after one or two divisions and eventually degenerate. Some of mutant megagametophytes undergo additional mitotic cycles and often show an aberrant number of nuclei and asynchronous divisions at the two poles, suggesting that *HDD* is involved in the coordination of the division cycles. Mu-

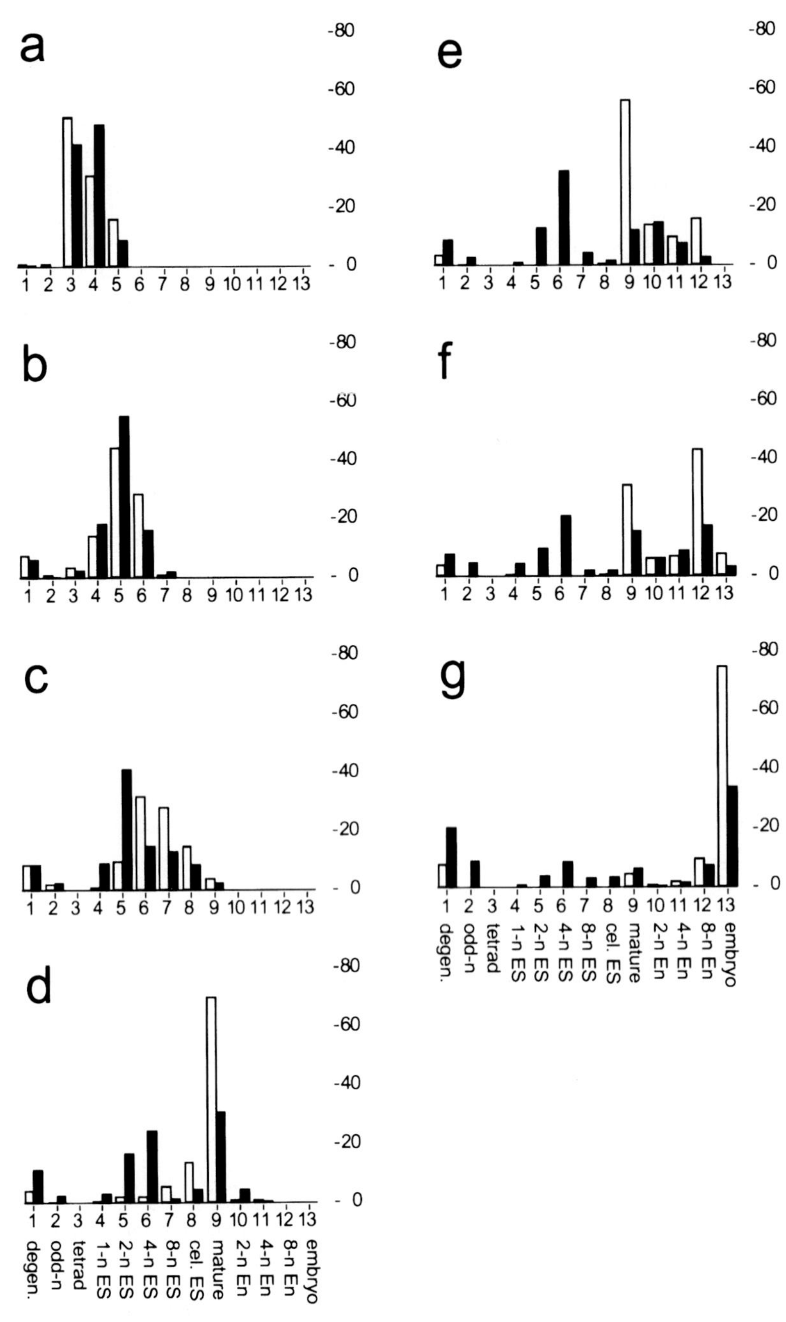

Figure 5. Histograms representing the distribution of developmental stages of embryo sacs in wild type (*white bars*) and *hdd*/+ (*black bars*) plants at various phases of ovule development. Ovule development was subdivided into seven phases based on morphological criteria of the sporophyte, and the developmental stage to which each megagametophyte had progressed was classified. The bars represent the percentage of embryo sacs found at each stage. Each of the developmental phases (*a–g*) encompasses two subsequent stages as defined by Schneitz et al. (1995): (*a*) stages 2-V and 3-I [N(wt) = 283, N(*hdd*) = 244]; (*b*) stages 3-II and 3-III [N(wt) = 306, N(*hdd*) = 372]; (*c*) stages 3-IV and 3-V [N(wt) = 259, N(*hdd*) = 213]; (*d*) stages 3-VI and 4-I [N(wt) = 556, N(*hdd*) = 316]; (*e*) stages 4-II and 4-III [N(wt) = 321, N(*hdd*) = 117]; (*f*) stages 4-IV and 4-V [N(wt) = 246, N(*hdd*) = 378]; (*g*) stages 4-VI and subsequent initiation of embryogenesis [N(wt) = 113, N(*hdd*) = 417]. Megagametophytes within each phase were classified by the number, position, and size of nuclei in the embryo sac (ES) into the following developmental stages (stages as defined by Christensen et al. [1997] are indicated by brackets): (1) degenerated embryo sacs lacking nuclei; (2) embryo sacs containing nuclei of aberrant number and/or position; (3) ovules having completed meiosis (tetrad stage); (4–7) embryo sacs containing 1 (FG1), 2 (FG2,3), 4 (FG4), or 8 (FG5) syncytial nuclei; (8) cellularized megagametophyte (FG6); (9) mature embryo sac with fused polar nuclei (FG7,8); (10–12) embryo sacs with 2, 4, and 8 (or more) endosperm (En) nuclei; (13) embryo sacs where the zygote has undergone the first division. N(wt) is the number of ovules scored in wild type; N(*hdd*) is the number of ovules scored in *hdd*/+ plants

tant embryo sacs arrested at the two-nuclear stage usually show a normal distribution of the nuclei along the chalazal-micropylar axis, indicating that polarity of the first division and subsequent migration is not affected. In contrast, mutant embryo sacs arrested later in gametogenesis show an aberrant distribution and migration of the nuclei. The failure of mutant embryo sacs to undergo normal nuclear division and migration appears to be the main defect in *hdd* mutants that leads to semisterility and an almost complete lack of transmission through the female gametophyte.

The observed phenotype shows certain similarities to *lo2*, a megagametophyte lethal mutation in maize. *lo2* embryo sacs show a defect in nuclear division and migration and arrest predominantly at the one- and two-nucleated stage, although some *lo2* megagametophytes undergo all three division cycles (Sheridan and Huang 1997). However, the nuclei in *lo2* mutants often enlarge dramatically, whereas the nuclear morphology in *hdd* mutants prior to degeneration is not obviously different from that of wild type. As observed for *hdd*, transmission of *lo2* is extremely low through the female. *lo2* is readily transmitted through the male, but transmission through the pollen can also be reduced (Nelson and Clary 1952; Sheridan and Huang 1997). In contrast to *lo2*, *hdd*, and *Gf1*, which all show a strong reduction of transmission through the female and also have an effect on pollen viability, transmission of *prl* is only reduced to 50% in the female and there is no effect on male transmission (Nelson and Clary 1952; Rédei 1965; Springer et al. 1995). In *hdd*, the degree of male transmission appears to be strongly influenced by environmental conditions as it is also often observed in sporophytic mutants affecting male fertility (R. Pruitt, pers. comm.). In contrast, the female defect in *hdd* shows little variability.

Interestingly, some of the mutant embryo sacs cellularize prematurely before the mitotic division cycles are completed (Fig. 4e), suggesting that nuclear division and cellularization are regulated independently. However, these cells do not differentiate into a particular gametophytic cell type, suggesting that cell specification may depend on the correct spatial context and/or the presence of neighboring cells. Premature cellularization and asynchronous divisions at the poles have also been observed in segmental deletions in maize (Vollbrecht and Hake 1995), suggesting that several loci including *hdd* are involved in the spatial and temporal coordination of cellularization, nuclear division, and migration. Whereas these processes are normally tightly coordinated, they are uncoupled in *hdd* mutant embryo sacs.

Does *HDD* Have a Gametophyte-specific or General Role for Mitotic Progression?

Because the gametophytic phase of the life cycle is haploid, screens for mutants disrupting male or female gametogenesis will uncover both genes essential to cellular processes in all cells and genes with specific roles for the development of male and female gametophytes. In a genetic screen as outlined here, truly essential genes with an absolute requirement for cell viability can only be recovered as hypomorphic mutations with residual transmission through one of the two gametophytes. It is likely that a screen which focuses on mutants that show low transmission through the female (semisteriles) but considerable transmission through the male (segregation ratios close to 1:1) will eliminate many basic factors that have housekeeping functions. On the other hand, such a screen will identify many genes involved in basic cell biological processes such as the control of cell division, the orientation of division planes, and the formation of cell walls, all events that can be studied in the relatively simple context of the embryo sac.

Nuclear migration and the relative position of the division planes appear to have an important role in cell determination of the gametophytic cells (Cass et al. 1985; Russell 1993). At present, we do not know whether *hdd* has a role in the regulation of mitosis only during female (and male) gametogenesis and or throughout plant development. To date, we have not recovered plants homozygous for *hdd*, but due to the low frequency of transmission through the female gametophyte, a larger number of plants will have to be analyzed to determine whether rare homozygous embryos are viable or abort as was observed for *prl* homozygotes (Springer et al. 1995). Molecular characterization of *hdd* and an analysis of its expression pattern will allow conclusions about its role in coordinating mitosis with plant development and differentiation.

ACKNOWLEDGMENTS

We are grateful to Bob Pruitt for stimulating discussions and helpful suggestions. Special thanks go to our summer students Bill Wagner and Nikhil Laud for technical support, and to Tim Mulligan of CSHL's Agricultural Field Station for taking care of the plants. We also thank Gary Drews, Martin Hülskamp, Ross Howden, S.-K. Park, Bob Pruitt, Animesh Ray, and David Twell for permission to cite unpublished data. We are thankful to V. Sundaresan, Rob Martienssen, and co-workers for allowing us to use their starter lines for insertional mutagenesis prior to publication. This work was supported by the Cold Spring Harbor Laboratory President's Council, National Science Foundation grant MCB-9723948 to U.G., and the Robertson Research Foundation. U.G. was supported by fellowships of EMBO, the Human Frontiers Science Program, and the Janggen-Poehn Foundation.

REFERENCES

Angenent G.C. and Colombo L. 1996. Molecular control of ovule development. *Trends Plant Sci.* **1:** 228.

Baker S.C., Robinson-Beers K., Villanueva J.M., Gaiser J.C., and Gasser C.S. 1997. Interactions among genes regulating ovule development in *Arabidopsis thaliana*. *Genetics* **145:** 1109.

Bianchi A. and Lorenzoni C. 1975. Gametophytic factors in *Zea mays*. In *Gamete competition in plants and animals* (ed. D.L. Mulcahy), p. 257. North-Holland, Amsterdam.

Buckner B. and Reeves S.L. 1994. Viability of female gametophytes that possess deficiencies for the region of chromosome 6 containing the *Y1* gene. *Maydica* **39:** 247.

Cass D.D. and Jensen W.A. 1970. Fertilization in barley. *Am. J. Bot.* **57:** 62.
Cass D.D., Peteya D.J., and Robertson B.L. 1985. Megagametophyte development in *Hordeum vulgare.* 1. Early megagametogenesis and the nature of wall formation. *Can. J. Bot.* **63:** 2164.
Chen Y.-C.S. and McCormick S. 1996. *sidecar pollen*, an *Arabidopsis thaliana* male gametophytic mutant with aberrant cell division during pollen development. *Development* **122:** 3243.
Christensen C.A., King E.J., Jordan J.R., and Drews G.N. 1997. Megagametogenesis in *Arabidopsis* wild type and the *Gf* mutant. *Sex Plant Reprod.* **10:** 49.
Coe E.H., Neuffer M.G., and Hoisington D.A. 1988. The genetics of corn. In *Corn and corn improvement* (ed. G.F Sprague and J.W. Dudley), p. 81. American Society for Agronomy, Madison, Wisconsin.
Correns C. 1902. Scheinbare Ausnahmen von der Mendel'schen Spaltungsregel für Bastarde. *Ber. Dtsch. Bot. Ges.* **20:** 157.
Demerec M. 1929. Cross sterility in maize. *Z. Indukt. Abstammungs-Verebungsl.* **50:** 281.
Eliott R.C., Betzner A.S., Huttner E., Oakes M.P., Tucker W.Q.J., Gerentes D., Perez P., and Smyth D.R. 1996. *AINTEGUMENTA*, an *APETALA2*-like gene of *Arabidopsis* with pleiotropic roles in ovule development and floral organ growth. *Plant Cell* **8:** 155.
Folsom M.W. and Cass D.D. 1990. Embryo sac development in soybean: Cellularization and egg apparatus expansion. *Can. J. Bot.* **68:** 2135.
Gaiser J.C., Robinson-Beers K., and Gasser C.S. 1995. The *Arabidopsis SUPERMAN* gene mediates asymmetric growth of the outer integument of ovules. *Plant Cell* **7:** 333.
Herr J.M., Jr. 1971. A new clearing-squash technique for the study of ovule development in angiosperms. *Am. J. Bot.* **58:** 785.
Hill J.L. and Lord D.R. 1994. Wild type flower development: Gyneocial initiation. In Arabidopsis: *An atlas of morphology and development* (ed. J. Bowman), p. 158. Springer Verlag, New York.
Hofmeister W. 1849. *Die Entstehung des Embryo der Phanerogamen.* F. Hofmeister, Leipzig.
Huang B.-Q. and Russell S.D. 1992. The female germ unit. *Int. Rev. Cytol.* **140:** 233.
Huang B.-Q. and Sheridan W.F. 1994. Female gametophyte development in maize: Microtubular organization and embryo sac polarity. *Plant Cell* **6:** 845.
———. 1996. Embryo sac development in the maize *indeterminate gametophyte1* mutant: Abnormal nuclear behavior and defective microtubule organization. *Plant Cell* **8:** 1391.
Hülskamp M., Schneitz K., and Pruitt R.E. 1995. Genetic evidence for long-range activity that directs pollen tube guidance in *Arabidopsis*. *Plant Cell* **7:** 57.
Jensen W.A. 1968. Cotton embryogenesis: The zygote. *Planta* **9:** 346.
Kenrick P. 1994. Alternation of generations in land plants: New phylogenetic and morphological evidence. *Biol. Rev.* **69:** 293.
Kermicle J.L. 1971. Pleiotropic effects on seed development of the *indeterminate gametophyte* gene in maize. *Am. J. Bot.* **58:** 1.
Klucher K.M., Chow H., Reiser L., and Fischer R.L. 1996. The *AINTEGUMENTA* gene of *Arabidopsis* required for ovule and female gametophyte development is related to the floral homeotic gene *APETALA2*. *Plant Cell* **8:** 137.
Koornneef M., Dresselhuys H.C., and Sree-Ramulu K. 1982. The genetic identification of translocations in *Arabidopsis*. *Arabidopsis Inf. Serv.* **19:** 93.
Lang J.D., Ray S., and Ray A. 1994. *sin1*, a mutation affecting female fertility in *Arabidopsis*, interacts with *mod1*, its recessive modifier. *Genetics* **137:** 1101.
Léon-Klosterziel K.M., Keijzer C.J., and Koornneef M. 1994. A seed shape mutant in *Arabidopsis* that is affected in integument development. *Plant Cell* **6:** 385.
Laughnan J.R. and Gabay-Laughnan S.J. 1983. Cytoplasmic male sterility in maize. *Annu. Rev. Genet.* **17:** 27.
Lin B.-Y. 1978. Structural modifications of the female gametophyte associated with the *indeterminate gametophyte* (*ig*) mutant in maize. *Can. J. Genet. Cytol.* **20:** 249.
———. 1981. Megagametogenetic alterations associated with the *indeterminate gametophyte* (*ig*) mutation in maize. *Rev. Bras. Biol.* **41:** 557.
Maheswari P. 1950. *An introduction to the embryology of angiosperms*. McGraw-Hill, New York.
Mangelsdorf P.C. and Jones D.F. 1926. The expression of mendelian factors in the gametophyte of maize. *Genetics* **11:** 423.
Mansfield S.G., Briarty L.G., and Erni S. 1991. Early embryogenesis of *Arabidopsis thaliana*. I. The mature embryo sac. *Can. J. Bot.* **69:** 447.
Modrusan Z., Reiser L., Feldman K.A., Fischer R.L., and Haughn G.W. 1994. Homeotic transformation of ovules into carpel-like structures in *Arabidopsis*. *Plant Cell* **6:** 333.
Mogensen H.L. 1988. Exclusion of male mitochondria and plastids during syngamy in barley as a basis for maternal inheritance. *Proc. Natl. Acad. Sci.* **85:** 2594.
Nadeau J.A., Zhang X.S., Li J., and O'Neill S.D. 1996. Ovule development: Identification of stage specific and tissue specific cDNAs. *Plant Cell* **8:** 213.
Nelson O.E. 1994. The gametophyte factors of maize. In *The maize handbook* (ed. M. Freeling and V. Wallbot), p. 496. Springer Verlag, New York.
Nelson O.E. and Clary G.B. 1952. Genic control of semi-sterility in maize. *J. Hered.* **43:** 205.
O'Kane C. and Gehring W.J. 1987. Detection *in situ* of genomic regulatory elements in *Drosophila*. *Proc. Natl. Acad. Sci.* **84:** 9123.
Olson A.R. and Cass D.D. 1981. Changes in megagametophyte structure in *Papaver nudicaule* following in vitro placental pollination. *Am. J. Bot.* **68:** 1338.
Ottaviano E., Petroni D., and Pé M.E. 1988. Gametophytic expression of genes controlling endosperm development in maize. *Theor. Appl. Genet.* **75:** 252.
Patterson E.B. 1978. Properties and uses of duplicate deficient chromosome complements in maize. In *Maize breeding and genetics* (ed. D. Walden), p. 693. Wiley, New York.
———. 1994. Translocations as genetic markers. In *The maize handbook* (ed. M. Freeling and V. Wallbot), p. 361. Springer Verlag, New York.
Ray A., Robinson-Beers K., Ray S., Baker S.C., Lang J.D., Preuss D., Milligan S.B., and Gasser C.S. 1994. *Arabidopsis* floral homeotic gene *BELL* (*BEL1*) controls ovule development through negative regulation of *AGAMOUS* gene (*AG*). *Proc. Natl. Acad. Sci.* **91:** 5761.
Ray S., Park S.-S., and Ray A. 1997. Pollen tube guidance by the female gametophyte. *Development* **124:** 2489.
Rédei G.P. 1964. A pollen abortion factor. *Arabidopsis Inf. Serv.* **1:** 10.
———. 1965. Non-Mendelian megagametogenesis in *Arabidopsis*. *Genetics* **51:** 857.
Reiser L. and Fischer R.L. 1993. The ovule and the embryo sac. *Plant Cell* **5:** 1291.
Reiser L, Modrusan Z., Margossian L., Samach A, Ohad N., Haughn G.W., and Fischer R.L. 1995. The *BELL1* gene encodes a homeodomain protein involved in pattern formation in the *Arabidopsis* ovule primordium. *Cell* **83:** 735.
Robinson-Beers K., Pruitt R.E., and Gasser C.S. 1992. Ovule development in wild type *Arabidopsis* and two female-sterile mutants. *Plant Cell* **4:** 1237.
Russell S.D. 1985. Preferential fertilization in *Plumbago:* Ultrastructural evidence for gamete-level recognition in an angiosperm. *Proc. Natl. Acad. Sci.* **82:** 6129.
———. 1992. Double fertilization. *Int. Rev. Cytol.* **140:** 357.
———. 1993. The egg cell: Development and role in fertilization and early embryogenesis. *Plant Cell* **5:** 1349.
Schnarf K. 1936. Contemporary understanding of embryo-sac development among angiosperms. *Bot. Rev.* **2:** 565.
Schneitz K., Hülskamp M., and Pruitt R.E. 1995. Wild-type ovule development in *Arabidopsis thaliana:* A light microscope study of cleared whole-mount tissue. *Plant J.* **7:** 731.

——. 1997. Dissection of sexual organ ontogenesis: A genetic analysis of ovule development in *Arabidopsis thaliana*. *Development* **124:** 1367.

Sheridan W.F. and Huang B.-Q. 1997. Nuclear behavior is defective in the maize (*Zea mays* L.) *lethal ovule2* female gametophyte. *Plant J.* **11:** 1029.

Singleton W.R. and Mangelsdorf P.C. 1940. Gametic lethals on the fourth chromosome of maize. *Genetics* **25:** 366.

Skarnes W.C. 1990. Entrapment vectors: A new tool for mammalian genetics. *Biotechnology* **8:** 27.

Springer P.S., McCombie W.R., Sundaresan V., and Martienssen R.A. 1995. Gene trap tagging of *PROLIFERA*, an essential *MCM2-3-5*-like gene in *Arabidopsis*. *Science* **268:** 877.

Spurr A.R. 1969. A low-viscosity epoxy resin embedding medium for electron microscopy. *J. Ultrastruct. Res.* **26:** 31-43

Sree-Ramulu K. and Sybenga J. 1979. Comparison of fast neutrons and X-rays in respect to genetic effects accompanying induced chromosome aberrations and analysis of translocations in *Arabidopsis thaliana*. *Arabidopsis Inf. Serv.* **16:** 27.

———. 1985. Genetic background damage accompanying reciprocal translocations induced by X-rays and fission neutrons in *Arabidopsis* and *Secale*. *Mutat. Res.* **149:** 421.

Strasburger E. 1879. *Die Angiopsermen und die Gymnospermen*. Jena.

Steeves T.A. and Sussex I.M., Eds. 1989. *Patterns in plant development*. Cambridge University Press, United Kingdom.

Sundaresan V., Springer P.S., Volpe T., Haward S., Jones J.D.G., Dean C., Ma H., and Martienssen R.A. 1995. Patterns of gene action in plant development revealed by enhancer trap and gene trap transposable elements. *Genes Dev.* **9:** 1797.

van den Berg C., Willemsen V., Hage W., Weisbeek P., and Scheres B. 1995. Determination of cell fate in the *Arabidopsis* meristem by directional signaling. *Nature* **378:** 62.

Vizir I.Y., Anderson M.L., Wilson Z.A., and Mulligan B.J. 1994. Isolation of deficiencies in the *Arabidopsis* genome by γ-irradiation of pollen. *Genetics* **137:** 1111.

Vollbrecht E. and Hake S. 1995. Deficiency analysis of female gametogenesis in maize. *Dev. Genet.* **16:** 44.

Walbot V. 1996. Sources and consequences of phenotypic and genotypic plasticity in flowering plants. *Trends Plant Sci.* **1:** 27.

Webb M.C. and Gunning B.E.S. 1988. The microtubular cytoskeleton during embryo-sac development in *Arabidopsis thaliana*. In *Pollination'88* (ed. R.B. Knox et al.), p. 62. University of Melbourne, Australia.

———. 1990. Embryo sac development in *Arabidopsis thaliana*. I. Megasporogenesis, including the microtubular cytoskeleton. *Sex. Plant Reprod.* **3:** 244.

Weber D.F. 1983. Monosomic analysis in diploid crop plants. In *Cytognetics of crop plants* (ed. P.K. Gupta and U. Sinha), p. 352. Macmillan, New Delhi, India.

Weigel D. and Doerner P. 1996. Cell-cell interactions: Taking cues from the neighbors. *Curr. Biol.* **6:** 10.

Willemse M.T.M. 1981. Polarity during megasporogenesis and megagametogenesis. *Phytomorphology* **31:** 124.

Willemse M.T.M. and van Went J.L. 1984. The female gametophyte. In *Embryology of angiosperms* (ed. B.M. Johri), p. 159. Springer Verlag, New York.

Wilson C., Bellen H.J., and Gehring W.J. 1990. Position effects on eukaryotic gene expression. *Annu. Rev. Cell Biol.* **6:** 679.

Xu H., Knox R.B., Taylor P.E., and Singh M.B. 1995. *Bcp1*, a gene required for male fertility in *Arabidopsis*. *Proc. Natl. Acad. Sci.* **92:** 2106.

Establishment of Cell Type in a Primitive Organism by Cell-specific Elimination and Proteolytic Activation of a Transcription Factor

A. HOFMEISTER AND R. LOSICK
Department of Molecular and Cellular Biology, Harvard University, Cambridge, Massachusetts 02138

Understanding how a dividing cell gives rise to progeny cells that assume distinct fates and acquire specialized functions is a fundamental challenge in development. Because cell fate and cell specialization are largely determined at the level of gene expression, an understanding of the establishment of cell fate requires an elucidation of the mechanisms that cause progeny cells to follow dissimilar programs of gene expression. In some cases, extrinsic cues, such as proximity to an external signaling molecule, set in motion a chain of events that triggers dissimilar gene expression between the products of cytokinesis (Horvitz and Herskowitz 1992). In other instances, however, differential gene expression arises from the asymmetric inheritance from the parental cell of factors that govern gene expression. For example, the asymmetric distribution of the transcription factors Prospero in *Drosophila melanogaster* (Doe and Spana 1995; Hirata et al. 1995; Knoblich et al. 1995) and ASH-1 in *Saccharomyces cerevisiae* (Bobola et al. 1996; Sil and Herskowitz 1996) causes progeny cells to express different sets of genes. Here, we describe the mechanisms by which differential gene expression is established following cell division during the developmental process of spore formation in *Bacillus subtilis*. Earlier work had shown that cell fate and differential gene expression in this bacterium are chiefly governed by two transcription factors called σ^F and σ^E that are synthesized in the parental cell but do not become active in directing gene expression until after division, when the two factors each become active in different progeny cells (for review, see Losick and Stragier 1992; Stragier and Losick 1996). In this paper, we principally consider the σ^E factor, which is subject to two independent regulatory mechanisms: One that causes the factor to localize differentially to one progeny cell and another that couples its activation to the completion of cytokinesis.

Spore formation in *B. subtilis* takes place within a structure known as the sporangium. Initially, the sporangium consists of a single cell bounded by a cell wall. The earliest distinctive morphological manifestation of sporulation, and a hallmark of differentiation in this organism, is the formation of a septum near one pole of the sporangium (Piggot and Coote 1976). This asymmetrically positioned septum partitions the sporangium into two, dissimilar-sized progeny cells called the forespore (the small progeny cell) and the mother cell. The progeny cells do not separate. Rather, they remain side by side, bounded within the cell wall of the sporangium. Later in development, the forespore is wholly engulfed by the mother cell, a configuration in which the sporangium becomes a cell-within-a-cell. The two progeny have different fates: The forespore is destined to become the spore, whereas the mother cell contributes to the development of the spore but is ultimately discarded by lysis.

The fates of the progeny cells are determined by the differential activation of σ^F and σ^E, which establish dissimilar programs of gene expression in the two sporangial compartments involving additional transcription factors at later stages of development (Stragier and Losick 1996). The σ^F and σ^E factors are synthesized in the predivisional sporangium, but they are inactive in directing gene expression prior to polar septation. Instead, their activation is delayed until after asymmetric division, when σ^F directs gene expression in the forespore and σ^E in the mother cell. Thus, temporal as well as spatial controls govern the establishment of cell fate by restricting the activities of σ^F and σ^E to the right time and the right place. This raises two questions: How are the transcription factors activated in a cell-type-specific fashion? What are the mechanisms that delay their activation until after cell division?

The σ^F factor is regulated at the level of its activity by a pathway involving proteins called SpoIIAB, SpoIIAA, and SpoIIE. SpoIIAB is an antagonist of σ^F that reversibly binds to the transcription factor, thereby trapping it in an inactive complex (Duncan and Losick 1993; Min et al. 1993; Alper et al. 1994; Diederich et al. 1994; Duncan et al. 1995, 1996; Lewis et al. 1996). Release of σ^F from the SpoIIAB/σ^F complex occurs when SpoIIAB becomes sequestered in an alternative protein complex with SpoIIAA. Formation of the SpoIIAB/SpoIIAA complex is, in turn, governed by the phosphorylation state of SpoIIAA. When SpoIIAA is phosphorylated, it cannot bind to SpoIIAB and hence σ^F is inactive. Conversely, when SpoIIAA is not phosphorylated, SpoIIAB is sequestered in a complex with SpoIIAA, thereby releasing active σ^F. The phosphorylation of SpoIIAA is determined by the opposing activities of a kinase and a phosphatase. Interestingly, the protein kinase is SpoIIAB, a dual function protein that is capable both of phosphorylating SpoIIAA and of inhibiting σ^F (Min et al. 1993). Dephosphorylation of SpoIIAA-P is catalyzed by SpoIIE, a membrane-bound phosphatase (Duncan et al. 1995).

The key to understanding the cell-type-specific activation of σ^F is therefore the question of how dephosphorylation of SpoIIAA-P is achieved selectively in the forespore. A clue comes from the discovery that at the time of asymmetric division the SpoIIE phosphatase is located exclusively in the sporulation septum (Arigoni et al. 1995). This important finding demonstrates that the pathway governing the activation of σ^F is directly tied to septum formation and provides a plausible explanation for the dependence of σ^F activation on cell division. The further question of how SpoIIE-mediated dephosphorylation of SpoIIAA-P occurs preferentially in the forespore is not understood, although several possibilities have been considered (Arigoni et al. 1995; Glaser et al. 1997).

Next, we turn to σ^E, the principal subject of this paper. The σ^E factor is produced in the predivisonal sporangium but does not become active in directing gene expression until after polar division when its activity is confined to the mother cell (Driks and Losick 1991; Harry et al. 1995). Spatial and temporal mechanisms conspire to restrict σ^E-directed gene expression to the right place and the right time. Results described here and elsewhere (Pogliano et al. 1997) show that the transcription factor is eliminated from the forespore chamber of the postseptation sporangium, a finding that helps explain the compartmentalization of σ^E-directed gene expression. In addition, however, σ^E is subject to a timing mechanism that delays its activiation until after polar division (Margolis et al. 1991; Zhang et al. 1996). This timing mechanism operates at the level of the proteolytic activation of the inactive proprotein precursor, pro-σ^E, to σ^E. The primary product of the gene for σ^E has an NH_2-extension of 27 amino acids that blocks the transcriptional activity of the σ factor (LaBell et al. 1987). The proteolytic conversion of pro-σ^E to mature σ^E in the mother cell is tied to the activation of σ^F in the forespore via a signal transduction pathway that operates across the closely opposed membranes that separate the two sporangial compartments (Karow et al. 1995; Londoño-Vallejo and Stragier 1995). The pathway consists of the signaling protein, SpoIIR, which is produced in the forespore under the control of σ^F, and SpoIIGA, a membrane-bound protein that is likely to be the proprotein processing enzyme (Hofmeister et al. 1995; Karow et al. 1995; Londoño-Vallejo and Stragier 1995). Here, we discuss data that indicate that SpoIIR is exported from the forespore into the intermembrane space between the forespore and the mother cell where it activates the putative processing enzyme SpoIIGA, thereby triggering proteolytic activation of pro-σ^E within the mother cell. Because the activation of σ^E is tied to the activation of σ^F, the signal transduction pathway couples the activation of σ^E to the formation of the sporulation septum and thereby delays the onset of σ^E-directed gene expression until after asymmetric divsion (Margolis et al. 1991; Zhang et al. 1996).

RESULTS

Absence of σ^E from the Forespore Prior to Its Activation

To investigate the mechanism by which σ^E activity is confined to the mother cell, we examined the distribution of the σ^E protein in postdivisional sporulating cells using immunofluorescence microscopy (Pogliano et al. 1997). A prerequisite for this study was the ability to readily distinguish the forespore from the mother cell compartment of individual postdivisional sporangia. This was possible by the use of 4′,6-diamidino-2-phenylindole (DAPI), which stains the forespore chromososome as a round and highly condensed nucleoid as compared to the elongate and more diffuse nucleoid of the mother cell (Fig. 1D, G, and J, blue) (Setlow et al. 1991). The mouse monoclonal antibody we used to immunostain σ^E binds to both the inactive and the active form of the transcription factor (Trempy et al. 1985), which we could therefore not distinguish in this study and which we thereby refer to as pro-σ^E/σ^E.

In postdivisional sporangia, pro-σ^E/σ^E immunostaining (Fig.1B, red) was observed to coincide with the mother cell nucleoid (Fig.1D, blue) and could not be detected in the forespore. When the same cells were stained with rabbit anti-σ^F antibodies, fluorescence from immunostaining of σ^F (Fig. 1C, green) was detected similarly in the forespore and the mother cell. Therefore, pro-σ^E/σ^E seems to be absent from the forespore of postdivisional sporangia, whereas σ^F is present in both the forespore and the mother cell. The lack of pro-σ^E/σ^E from the forespore was unexpected since the gene encoding pro-σ^E is known to be transcribed in predivisional cells (Kenney and Moran 1987) and since pro-σ^E/σ^E immunostaining can be detected prior to asymmetric division (not shown). The fact that both transcription factors differ in their postdivisional pattern of localization reflects the fact that their cell-type-specific activation is controlled by different mechanisms.

The localization of pro-σ^E/σ^E exclusively in the mother cell was similar to the localization that had been described for a reporter of σ^E activity (Driks and Losick 1991; Harry et al. 1995). In confirmation of this observation, sporulating cells were coimmunostained with mouse anti-pro-σ^E/σ^E antibodies and rabbit antibodies that bind to β-galactosidase synthesized from a gene fusion (*spoIID-lacZ*) whose transcription depends on σ^E. Indeed, pro-σ^E/σ^E (Fig. 1E, red) showed a staining pattern that paralleled that observed for β-galactosidase (Fig. 1F, green) in that both proteins were present in the mother cell compartment of the sporangia but absent from the majority (91%) of the forespores (Table 1). Therefore, the compartmentalization of σ^E-directed gene expression may involve a mechanism that excludes pro-σ^E/σ^E from the forespore.

Persistence and Activation of σ^E in the Forespore of a *spoIIIE* Mutant

A clue as to whether the absence of pro-σ^E/σ^E from the forespore is in fact sufficient for restricting σ^E activity to

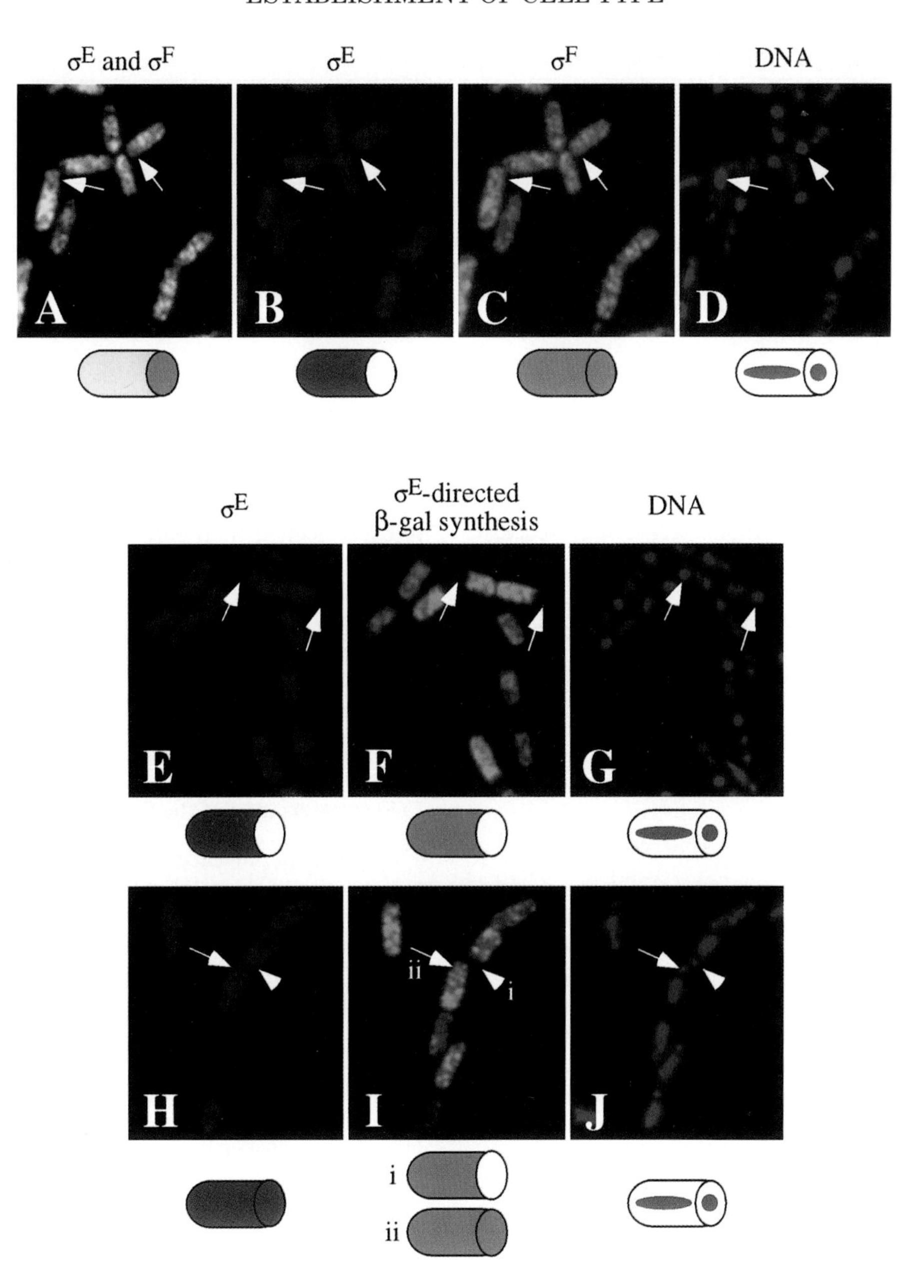

Figure 1. Immunolocalization of σ^F, pro-σ^E/σ^E, and σ^E activity in sporulating wild-type and *spoIIIE* mutant cells of *B. subtilis*. Cells were harvested 2 hr after the induction of sporulation and prepared for immunofluorescence microscopy as described recently (Pogliano et al. 1997). The activity of σ^E was visualized by immunostaining β-galacosidase synthesized from a *spoIID-lacZ* fusion that had been inserted into the chromosome at the *amyE* locus. Where the red and green fluorophores overlap, as in the doubly exposed image shown in panel *A*, a yellow to orange color is visible. Arrows point to the forespores and are oriented perpendicularly to the long axis of the postdivisional developing cells. (*A–D*) Arrows point to wild-type cells that display pro-σ^E/σ^E immunostaining in the mother cell and σ^F immunostaining in both the forespore and the mother cell. (*E–G*) Wild-type cells showing pro-σ^E/σ^E immunostaining (*red*) and σ^E activity (*green*) in the mother cell are indicated by arrows. (*H–J*) Sporulating cells of the *spoIIIE::spc* mutant strain fall into two classes. One class (*i;* marked by the arrowhead) displays pro-σ^E/σ^E immunostaining (*red*) in both the forespore and mother cell, whereas σ^E activity (*green*) can only be detected in the mother cell. The other class (*ii; arrow*) shows pro-σ^E/σ^E immunostaining and σ^E activity in both the forespore and the mother cell.

the mother cell came from the analysis of a strain carrying a null mutation in the *spoIIIE* gene (*spoIIIE::spc*). Note that the forespore nucleoids of these *spoIIIE* mutant cells are substantially smaller than those of wild-type cells (compare J with G in Fig.1, blue), reflecting the previously described DNA translocation defect of this mutant (Wu and Errington 1994). Strikingly, the *spoIIIE* mutation allowed pro-σ^E/σ^E to persist in the forespore of sporulating cells (Fig. 1H, red). More than 89% of the postdivisional cells were found to immunostain for pro-σ^E/σ^E in both the forespore and the mother cell as compared to only 9% in wild-type cells (Table 1). Significantly, σ^E activity (Fig. 1I, green) was observed in the forespore of some *spoIIIE* mutant cells that allowed pro-

Table 1. Sporulating *spoIIIE* Mutant Cells Contain Pro-σ^E/σ^E and σ^E Activity in Both the Forespore and the Mother Cell

Strain	Relevant genotype	Pro-σ^E/σ^E and σ^E activity in mother cell only (%)	Pro-σ^E/σ^E in mother cell and forespore (%)	
			σ^E activity in mother cell only	σ^E activity in mother cell and forespore
BZ184	+	91	9	0
KJP131	*spoIIIE::spc*	11	55	34

Cells were collected 2 hr after the initiation of sporulation. A total of 509 wild-type and 327 *spoIIIE::spc* sporulating cells were scored. The activity of σ^E was visualized by immunofluorescence microscopy localizing β-galactosidase produced from the *spoIID-lacZ* fusion. Pro-σ^E/σ^E protein was monitored by immunolocalization.

σ^E/σ^E to persist in the small sporangial chamber. Thus, in a substantial proportion (38%) of the *spoIIIE* mutant cells, the continued presence of pro-σ^E/σ^E in the forespore resulted in the misactivation of σ^E in the forespore. Taken together, our immunofluorescence microscopical data suggest that the depletion of pro-σ^E/σ^E from the forespore of sporulating cells contributes to the establishment of mother-cell-specific transcription. The absence of pro-σ^E/σ^E from the forespore, however, may not be sufficient for the establishment of mother-cell-specific gene expression because its presistence in the forespore of *spoIIIE::spc* mutant sporulating cells did not always lead to σ^E activity.

Signal-dependent Proteolytic Activation of pro-σ^E

The σ^E factor is separately regulated at the level of the proteolytic activation of pro-σ^E. To gain insight into the biochemical mechanism of intercellular signal transduction for the postdivisional proteolytic activation of σ^E, we (Hofmeister et al. 1995) devised an assay to monitor the signal-dependent conversion of pro-σ^E to mature σ^E. Recipient cells of *B. subtilis* were generated that can be induced to synthesize both pro-σ^E and the putative protease, SpoIIGA, during vegetative growth. In these cells, pro-σ^E was not proteolytically processed to σ^E (not shown), thus confirming genetic inferences that SpoIIGA is by default inactive and becomes active only in response to a signal produced in the forespore (Stragier et al. 1988; Margolis et al. 1991; Losick and Stragier 1992). We could therefore attempt to identify a factor that, when added to recipient cells, is capable of activating SpoIIGA for intracellular processing of pro-σ^E to σ^E.

A likely candiate for the processing-promoting factor was the product of the *spoIIR* gene. Expression of *spoIIR* had been shown to be sufficient for triggering SpoIIGA-dependent pro-σ^E processing (Londoño-Vallejo and Stragier 1995). In addition, the *spoIIR* gene product had been proposed to carry an amino-terminal signal sequence which makes it competent for secretion from the forespore (Karow et al. 1995). We therefore generated donor cells of *B. subtilis* that can be induced for the synthesis of σ^F during vegetative growth knowing that in these cells σ^F would direct the transcription of *spoIIR*. Recipient cells were then incubated with conditioned medium of donor cell cultures, and the conversion of pro-σ^E to σ^E was monitored by Western blot analysis with the mouse monoclonal antibodies that bind to pro-σ^E and σ^E.

Figure 2 shows that the signal-dependent pro-σ^E processing assay faithfully reproduces the in vivo process of intercellular signal transduction in that it requires the presence of SpoIIGA and is dependent on SpoIIR. Conditioned medium from donor cell cultures of a *spoIIR* deletion mutant failed to stimulate the SpoIIGA-mediated conversion of pro-σ^E to σ^E (Fig. 2c), suggesting that the processing-promoting factor is indeed SpoIIR or is at least dependent on SpoIIR. We conclude that extracellularly added SpoIIR is capable of signaling the intracellular processing of pro-σ^E to σ^E in the presence of the integral membrane protease SpoIIGA.

Characterization of the Signaling Protein SpoIIR

The signal-dependent pro-σ^E processing assay enabled us to purify and characterize SpoIIR as a functional protein. Previously, we had noted that synthesis of SpoIIR under the control of inducible promotors in either vegetative *B. subtilis* or *Escherichia coli* cells did not yield the quantities required for a successful purification (Hofmeister et al. 1995). In an attempt to increase the level of expression of *spoIIR*, we synthesized SpoIIR without its putative signal sequence as a fusion protein to the signal sequence of the secreted protein subtilisin, which then mediated the secretion of SpoIIR into the culture medium of an overproducing *B. subtilis* strain. The stability of SpoIIR in conditioned medium was augmented by using an overproducing strain that carried mutations in the genes that code for the prevalent extracellular proteases. Conditioned medium (1 liter) from cultures of the overproducing strain yielded an average of 1 mg of purified SpoIIR after ammonium sulfate fractionation, anion exchange chromatography, and gel filtration.

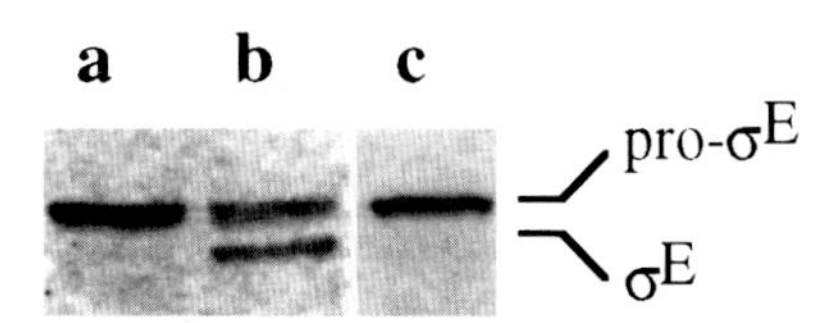

Figure 2. Signal-dependent pro-σ^E processing in recipient cells of *B. subtilis*. Recipient cells were engineered to produce either (*a*) pro-σ^E alone or (*b,c*) SpoIIGA and pro-σ^E. These cells were then incubated at 37°C for 30 min with conditioned medium of (*a,b*) donor cells and (*c*) the congenic *spoIIR* deletion strain that had been induced to synthesize σ^F. Proteins were subjected to SDS-PAGE, and Western blot analysis was performed with the monoclonal antibody that binds to pro-σ^E and σ^E.

The protein profile after gel filtration of enriched SpoIIR in Figure 3A shows that SpoIIR elutes in two peaks. One peak consists of homogeneous SpoIIR and corresponds to a molecular mass of 44 kD indicative of a homodimer. The other peak contains SpoIIR together with additional proteins and corresponds to a molecular mass of 160 kD. The high-molecular-mass complex is stable upon rechromatography, suggesting that native SpoIIR might exist as a multimeric complex as well as a homodimer. Since both peaks contained pro-σ^E processing activity (Fig. 3B), we are left to speculate whether SpoIIR exerts its function as a homodimer or as a higher-molecular-mass complex (see Discussion).

Interaction of pro-σ^E with SpoIIGA

To obtain further evidence in support of the notion that SpoIIGA is the specific protease for pro-σ^E processing, we performed in vivo chemical cross-linking with *B. subtilis* cells that had synthesized both pro-σ^E and SpoIIGA or, as a negative control, pro-σ^E alone. Cells were incubated in the presence or in the absence of the homobifunctional *N*-hydroxy-succinimidyl-ester cross-linker ethylene glycobis(succinimidylsuccinate) (EGS). Cellular proteins were then separated by SDS-PAGE and subsequently analyzed by Western blot analysis with the mouse monoclonal antibody that binds to pro-σ^E and σ^E. As shown in Figure 4d, addition of EGS caused the appearance of antibody-reactive protein complexes of higher molecular masses than pro-σ^E only in cells that had been engineered to produce pro-σ^E and SpoIIGA. EGS did not lead to the appearance of these or other protein complexes in cells that had been engineered to produce pro-σ^E alone (Fig. 4b). In the absence of EGS, no protein complexes could be detected (Fig. 4a,c). Therefore, pro-σ^E forms SpoIIGA-dependent protein complexes in the presence of EGS, suggesting that pro-σ^E interacts with SpoIIGA. This finding is consistent with the idea that SpoIIGA is a protease that catalyzes the conversion of pro-σ^E to σ^E.

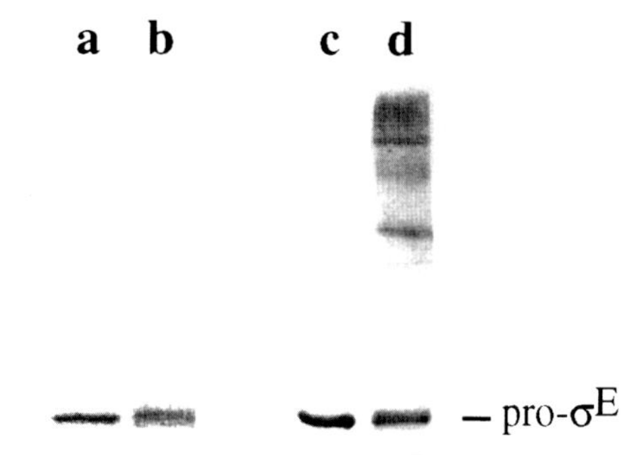

Figure 4. Interaction between pro-σ^E and SpoIIGA. The figure shows the Western blot analysis of proteins after in vivo cross-linking of vegetative *B. subtilis* cells that had been engineered to synthesize either (*a,b*) pro-σ^E alone or (*c,d*) pro-σ^E and SpoIIGA. Vegetative *B. subtilis* cells were incubated at 4°C for 2 hr (*a,c*) in the absence of chemical cross-linker or (*b,d*) in the presence of 2 mM EGS [ethylene glycobis(succinimidylsuccinate)]. Proteins were then subjected to SDS-PAGE and Western blot analysis was carried out with the monoclonal antibody that binds to pro-σ^E and σ^E.

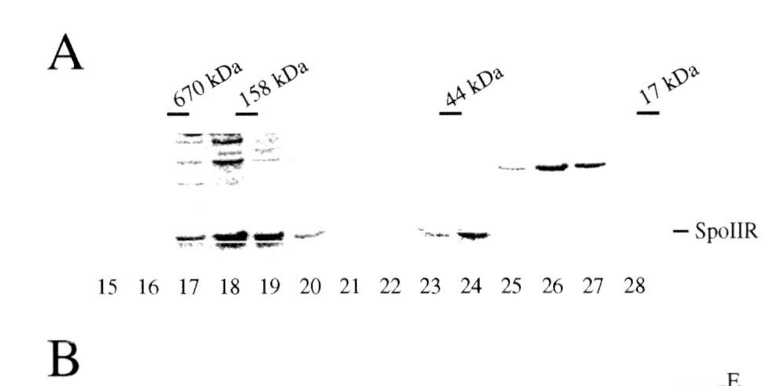

Figure 3. Purification and characterization of SpoIIR by gel filtration. (*A*) Protein profile obtained by gel filtration of enriched SpoIIR on a Superdex 75 column after SDS-PAGE and subsequent silver staining. (*B*) Western blot analysis of pro-σ^E processing in recipient cells after incubation with each one of the fractions 15–28. The pro-σ^E processing assays were preformed at 37°C for 30 min. Proteins were subjected to SDS-PAGE, and Western blot analysis was performed with the monoclonal antibody that binds to pro-σ^E and σ^E.

DISCUSSION

Spore development in *B. subtilis* is regulated at the transcriptional level by the sequential activation of several transcription factors each in only one of the two cell types (Losick and Stragier 1992). The transcription factors that establish differential gene transcription selectively in the forespore and in the mother cell are both synthesized in the predivisional cell but remain inactive until after asymmetric division. The challenge in understanding development in this primitive organism is therefore to elucidate the mechanisms that govern the cell-type-specific activation of these two trancription factors.

One of the two transcription factors that establish cell-type-specific gene transcription is synthesized as an inactive proprotein, pro-σ^E, in the predivisional cell (Kenney and Moran 1987; LaBell et al. 1987). The experiments presented here argue that mother-cell-specific activation of σ^E is controlled by at least two mechanisms, one of which is responsible for eliminating σ^E from the forespore. We have shown that pro-σ^E/σ^E is absent from the forespore following asymmetric division. This absence contributes to the establishment of mother-cell-specific gene transcription because a *spoIIIE::spc* mutant that failed to deplete pro-σ^E/σ^E from the forespore was capable of activating σ^E in this cell type. Our observation that the persistence of pro-σ^E/σ^E in the forespore of *spoIIIE::spc* mutant cells did not always result in σ^E-directed gene expression suggests that the elimination of pro-σ^E/σ^E from the forespore may not be the only mechanism responsible for confining σ^E-directed gene transcription to the mother cell. Nevertheless, our findings with the *spoIIIE* mutant indicate that cell-specific elimination of σ^E contributes significantly to the compartmentalization of gene expression.

Interestingly, the *spoIIIE* mutant which we describe as

being deficient in the depletion of pro-σ^E/σ^E from the forespore has previously been characterized as being defective in chromosome segregation during sporulation (Wu and Errington 1994). During sporulation, polar division precedes full segregation of a chromosome into the forespore. Initially, after division, only the replication origin-proximal region of the chromosome is present in the forespore. Next, in a process that depends on SpoIIIE, which is located in the septum, the remainder of the chromsome is translocated into the small chamber of the sporangium. This finding raises the possibility that the establishment of cell fate is somehow tied to chromosome segregation. Conceivably, the chromosomal location of a gene could govern its spatial transcription pattern during chromosome translocation. If, for instance, a gene whose transcription depends on its translocation into the forespore encodes a protein that is essential for eliminating pro-σ^E/σ^E, a failure to segregate the chromosomes could indirectly lead to the persistence of pro-σ^E/σ^E in the forespore. Alternatively, the SpoIIIE protein itself could, directly or indirectly, be involved in eliminating pro-σ^E/σ^E from the forespore by controlling either its specific degradation or its asymmetric segregation to the mother cell.

The pattern of pro-σ^E/σ^E localization was distinct from the one we observed for the transcription factor σ^F in that σ^F was found to be similarly present in the mother cell and in the forespore. Thus, the forespore-specific activation of σ^F is achieved by mechanisms that do not involve the differential localization of the transcription factor itself. Rather, it has been suggested that the localization to the asymmetrically positioned division septum of the SpoIIE phosphatase that indirectly controls σ^F activity is crucial in establishing σ^F-directed gene transcription in the forespore (Arigoni et al. 1995). Therefore, σ^F seems to be regulated exclusively at the level of its activity, whereas σ^E is regulated by its subcellular localization as well as at the level of its activity.

The proteolytic activation of σ^E is controlled via a signal transduction pathway that operates across the membranes of the forespore and the mother cell (Hofmeister et al. 1995; Karow et al. 1995; Londoño-Vallejo and Stragier 1995). This intercellular signal transduction pathway is thought to involve three proteins: SpoIIR, SpoIIGA, and pro-σ^E. We have demonstrated that SpoIIR is an extracellular signaling protein capable of triggering the intercellular SpoIIGA-dependent conversion of pro-σ^E to σ^E. A factor in conditioned medium from *B. subtilis* cells that had been engineered to produce σ^F during growth signaled pro-σ^E processing in *B. subtilis* cells that had been induced to synthesize SpoIIGA and pro-σ^E during growth. The processing-promoting factor is SpoIIR itself because processing was dependent on an intact copy of the *spoIIR* gene in donor cells and processing could be achieved with SpoIIR that had been purified from conditioned medium of *B. subtilis* cells engineered to overproduce the signaling protein. Together with the fact that SpoIIR is produced under the control of σ^F exclusively in the forespore (Karow et al. 1995; Zhang et al. 1996) and that SpoIIR carries an amino-terminal sequence characteristic of secreted proteins (Karow et al. 1995), these results indicate that SpoIIR is secreted from the forespore into the intermembrane space between the forespore and the mother cell membranes where it signals the SpoIIGA-dependent processing of pro-σ^E in the mother cell.

Consistent with the idea that SpoIIGA is a protease that catalyzes the conversion of pro-σ^E to σ^E, we have obtained evidence suggesting that pro-σ^E interacts with SpoIIGA. In vivo chemical cross-linking yielded pro-σ^E-specific complexes only in *B. subtilis* cells that had been induced to synthesize both pro-σ^E and SpoIIGA during growth. We propose that SpoIIGA is a receptor/protease that is activated on the outside surface of the membrane to catalyze the intracellular proteolytic processing of pro-σ^E to active σ^E.

How might SpoIIGA be activated for pro-σ^E processing? One possibility is that SpoIIR exerts its effect indirectly either by catalyzing the formation of a specific ligand or by interacting with an as yet unidentified protein thereby unleashing SpoIIGA for pro-σ^E processing or inducing the formation of a heteromultimeric protease complex. Alternatively, the signaling protein SpoIIR could directly interact with the putative transmembrane receptor domain of SpoIIGA to activate its intracellular protease domain. In this context, it is interesting that our gel filtration experiments characterized SpoIIR as a functional homodimer as well as a functional oligomeric complex. Ligand-induced dimerization of SpoIIGA has been suggested to provide a mechanism for generating a functional protease domain (Stragier et al. 1988) because a SpoIIGA monomer contains only one of the two DSG sequence motifs necessary to form the catalytic site of aspartic proteases (Pearl and Taylor 1987). A SpoIIR dimer could cause SpoIIGA to homodimerize to an active protease by directly interacting with its putative transmembrane receptor domain.

Whatever the detailed mechanisms for signaling pro-σ^E processing, the intercellular signal transduction pathway is an important checkpoint in spore development because it ensures that pro-σ^E processing in the mother cell occurs subsequent to the activation of σ^F in the forespore. In this way, the intercellular signal transduction pathway might indirectly contribute to the establishment of mother cell fate by delaying pro-σ^E processing until after the elimination of pro-σ^E/σ^E from the forespore.

ACKNOWLEDGMENTS

We thank W. Haldenwang and L. Duncan for their gift of antibodies and K. Pogliano for strains and discussions. This work was supported by National Institutes of Health grant GM-18568 to R.L. A.H. was a postdoctoral fellow of the Alexander von Humboldt Foundation.

REFERENCES

Alper S., Duncan L., and Losick R. 1994. An adenosine nucleotide switch controlling the activity of a cell type-specific transcription factor in *B. subtilis*. *Cell* **77:** 195.

Arigoni F., Pogliano K., Webb C., Stragier P., and Losick R.. 1995. Localization of protein implicated in establishment of cell type to sites of asymmetric division. *Science* **270:** 637.

Bobola N., Jansen R.-P., Shin T.H., and Nasmyth K. 1996. Asymmetric accumulation of Ash1p in postanaphase nuclei depends on a myosin and restricts yeast mating-type switching to mother cells. *Cell* **84:** 699.

Diederich B., Wilkinson J.F., Magnin T.,. Najafi S.M.A, Errington J., and Yudkin M. 1994. Role of interactions between SpoIIAA and SpoIIAB in regulating cell-specific transcription factor σ^F of *B. subtilis*. *Genes Dev.* **8:** 2653.

Doe C.Q. and Spana E.P. 1995. A collection of cortical crescents: Asymmetric protein localization in CNS precursor cells. *Neuron* **15:** 991.

Driks A. and Losick R. 1991. Compartmentalized expression of a gene under the control of sporulation transcription factor σ^E in *Bacillus subtilis*. *Proc. Natl. Acad. Sci.* **88:** 9934.

Duncan L. and Losick R. 1993. SpoIIAB is an anti-sigma factor that binds to and inhibits transcription by regulatory protein σ^F from *Bacillus subtitis*. *Proc. Natl. Acad. Sci.* **90:** 2325.

Duncan L., Alper S., and Losick R. 1996. SpoIIAA governs the release of the cell-type specific transcription factor σ^F from its anti-sigma factor SpoIIAB. *J. Mol. Biol.* **260:** 147.

Duncan L., Alper S., Arigoni F., Losick R., and Stragier P. 1995. Activation of cell-specific transcription by a serine phosphatase at the site of asymmetric division. *Science* **270:** 641.

Glaser P., Sharpe M.E., Raether B., Perego M., Ohlsen K., and Errington J. 1997. Dynamic, mitotic-like behavior of a bacterial protein required for accurate chromosome partitioning. *Genes Dev.* **11:** 1160.

Harry E., Pogliano K., and Losick R. 1995. Use of immunofluroescence to visualize cell-specific gene expression during sporulation in *Bacillus subtilis*. *J. Bacteriol.* **177:** 3386.

Hirata J., Nakagoshi H., Nabeshima Y.-I., and Matsuzaki F. 1995. Asymmetric segregation of the homeodomain protein Prospero during *Drosophila* development. *Nature* **377:** 627.

Hofmeister A.E.M., Londoño-Vallejo J.-A., Harry E., Stragier P., and Losick R. 1995. Extracellular signal protein triggering the proteolytic activation of a developmental transcription factor in *B. subtilis*. *Cell* **83:** 219.

Horvitz H.R. and Herskowitz I. 1992. Mechanisms of asymmetric cell division: Two Bs or not two Bs, that is the question. *Cell* **68:** 237.

Karow L.M., Glaser P., and Piggot P.J. 1995. Identification of a gene, *spoIIR*, which links the activation of σ^E to the transcriptional activity of σ^F during sporulation in *Bacillus subtilis*. *Proc. Natl. Acad. Sci.* **92:** 2012.

Kenney T.J. and Moran Jr. C.P. 1987. Organization and regulation of an operon that encodes a sporulation-essential sigma factor in *Bacillus subtilis*. *J. Bacteriol.* **169:** 3329.

Knoblich J.A., Jan L.Y., and Jan Y.N. 1995. Asymmetric segregation of Numb and Prospero during cell division. *Nature* **377:** 624.

LaBell T.L., Trempy J.E., and Haldenwang W.G. 1987. Sporulation-specific σ factor σ^{29} of *Bacillus subtilis* is synthesized from a precursor protein, P^{31}. *Proc. Natl. Acad. Sci.* **84:** 1784.

Lewis P.J., Magnin T., and Errington J. 1996. Compartmentalized distribution of the proteins controlling the prespore-specific transcription factor σ^F of *Bacillus subtilis*. *Genes Cells* **1:** 881.

Londoño-Vallejo J.-A. and Stragier P. 1995. Cell-cell signaling pathway activating a developmental transcription factor in *Bacillus subtilis*. *Genes Dev.* **9:** 503.

Losick R. and Stragier P. 1992. Crisscross regulation of cell-type-specific gene expression during development in *Bacillus subtilis*. *Nature* **355:** 601.

Margolis P., Driks A., and Losick R. 1991. Establishment of cell type by compartmentalized activation of a transcription factor. *Science* **254:** 562.

Min, K.-T., Hilditch K.T., Diederich B., Errington J., and Yudkin M.D. 1993. σ^F, the first compartment-specific transcription factor of *B. subtilis*, is regulated by an anti-sigma factor that is also a protein kinase. *Cell* **74:** 735.

Pearl L.H. and Taylor W.R. 1987. A structural model for the retroviral proteases. *Nature* **329:** 351.

Piggot P.J. and Coote J.G. 1976. Genetic aspects of bacterial endospore formation. *Bacteriol. Rev.* **40:** 908.

Pogliano K., Hofmeister A.E.M., and Losick R. 1997. Disappearance of the σ^E transcription factor from the forespore and the SpoIIE phosphatase from the mother cell contributes to establishment of cell-specific gene expression during sporulation in *Bacillus subtilis*. *J. Bacteriol.* **179:** 3331.

Setlow B., Magill N., Febbroriello P., Nakhimousky L., Koppel D.E., and Setlow P. 1991. Condensation of the forespore nucleoid early in sporulation of *Bacillus* species. *J. Bacteriol.* **173:** 6270.

Sil A. and Herskowitz I. 1996. Identification of an asymmetrically localized determinant, Ash1p, required for lineage-specific transcription of the yeast *HO* gene. *Cell* **84:** 711.

Stragier P. and Losick R. 1996. Molecular genetics of sporulation in *Bacillus subtilis*. *Annu. Rev. Genet.* **30:** 297.

Stragier P., Bonamy C., and Karmazyn-Campelli C. 1988. Processing of a sporulation sigma factor in *Bacillus subtilis*: How morphological structure could control gene expression. *Cell* **52:** 697.

Trempy J.E., LaBell T.L., Ray G.L., and Haldenwang W.G. 1985. P^{31}, a σ^{29}-like RNA polymerase binding protein of *Bacillus subtilis*. In *Molecular biology of microbial differentiation* (ed. J.A. Hoch and P. Setlow), p. 193. American Society for Microbiology, Washington, D.C.

Wu L.J. and Errington J. 1994. *Bacillus subtilis* SpoIIIE protein required for DNA segregation during asymmetric cell division. *Science* **264:** 572.

Zhang L., Higgins M.L., Piggot P.J., and Karow M.L. 1996. Analysis of the role of prespore gene expression in the compartmentalization of mother cell-specific gene expression during sporulation of *Bacillus subtilis*. *J. Bacteriol.* **178:** 2813.

Building Organs and Organisms: Elements of Morphogenesis Exhibited by Budding Yeast

I. HERSKOWITZ
Department of Biochemistry and Biophysics, School of Medicine, University of California, San Francisco, San Francisco, California 94143-0448

During development of a multicellular organism, several processes contribute to the formation of organs and organisms of characteristic shape. Morphogenesis is influenced by the total number of cells involved in a structure, by their arrangement relative to each other, by cell shape, and by other specialized characteristics of the cells such as their ability to produce and respond to signaling molecules. A given structure can contain many different types of cells, produced as a result of asymmetric cell divisions that yield cells of different cellular fates.

Although yeast is a single-celled organism, features of its life cycle have made it possible to study many of the processes that give shape to multicellular organisms. One process that yeast does not exhibit is cell migration, although it is able to organize and orient its cell shape in a manner like that of migrating cells (Chenevert 1994). This paper summarizes these processes, describes the responsible molecular machinery, and places these processes in the context of the life cycle of yeast.

CELL SPECIALIZATION IS DETERMINED BY HOMEODOMAIN PROTEINS AND OTHER TRANSCRIPTIONAL REGULATORS THAT ACT ON KNOWN TARGETS

The context for understanding the morphogenetic processes exhibited by yeast is provided by its specialized cell types. Yeast has three cell types, **a** and α cells, typically haploid, and the **a**/α cell, which is typically diploid (Herskowitz 1989; Johnson 1992). **a** and α cells are specialized to mate with each other and form the **a**/α diploid. To facilitate the mating process, each of the haploid cell types produces a mating pheromone, α-factor by α cells and **a**-factor by **a** cells, which induce differentation in the mating partner. More specifically, α-factor induces **a** cells to arrest in the G_1 phase of the cell division cycle, undergo morphological alteration (growth toward the mating partner to form a pear-shaped "shmoo"), and express genes necessary for cell fusion and other aspects of mating.

α cells produce α-factor and respond to **a**-factor, and **a** cells produce **a**-factor and respond to α-factor. All of these differences between **a** and α cells as well as others (see Herskowitz 1988; Herskowitz et al. 1992) are governed by a single locus, the mating-type locus, which codes for regulatory proteins (Fig. 1). The mating-type locus of α cells, *MAT*α, codes for two regulatory proteins, α1, which activates transcription of α-specific genes, and α2, a homeodomain protein that represses transcription of **a**-specific genes. In **a** cells, the **a**-specific genes are transcribed because of the absence of α2, and the α-specific genes are not transcribed because of the absence of α1. A special feature of **a**/α cells—in fact, the molecular determinant for this cell type—is the presence of a novel repressor, **a**1-α2, which is a heterodimer comprising the homeodomain protein coded by *MAT***a** (**a**1) and the homeodomain protein coded by *MAT*α (α2) (Li et al. 1995). Combinatorial regulatory proteins such as **a**1-α2, comprising two homeodomain proteins, are also known to have important roles in fungal development (Herskowitz 1992; Banuett 1995) and in development of *Drosophila* and other metazoans (Mann and Chan 1996). The yeast cell monitors its ploidy (haploid vs. diploid) by assessing whether this novel regulatory protein, encoded by two different alleles of a given locus, is present. **a**1-α2, like α2, is a repressor, but it has a different set of binding sites and thus a different set of genes that are repressed. The genes repressed by α2 (the **a**-specific genes) and the genes repressed by **a**1-α2 (the haploid-specific genes) are thus direct targets of homeodomain proteins. Among the large number of genes turned off by **a**1-α2 (Fig. 2) is *MAT*α*1*, which thereby ensures that **a**/α cells do not express α-specific genes.

As described below, the three cell types differ from each other in many respects that are relevant to understanding morphogenesis; in particular, their responsiveness to the mating pheromones, as well as in their cell division plane and cellular polarity.

PROLIFERATION OF YEAST CAN BE CONTROLLED BY A CYCLIN-DEPENDENT KINASE INHIBITOR, FAR1, WHOSE ACTIVITY IS CONTROLLED BY THE PHEROMONE RESPONSE PATHWAY

As part of the mating process, the mating pheromones of the haploid mating-type cells cause their mating partners to arrest in the G_1 phase of the cell cycle (see Herskowitz 1995). The mating pheromones are peptides: α-factor is 13 amino acids in length, and **a**-factor is a co-

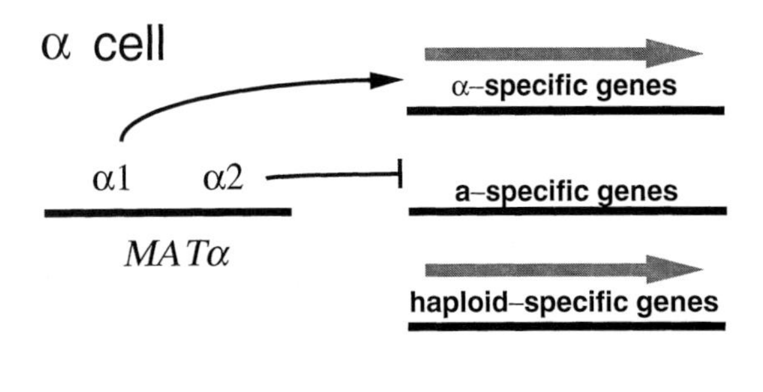

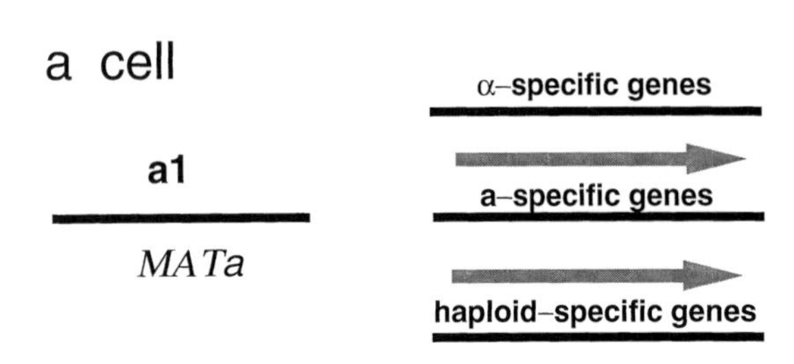

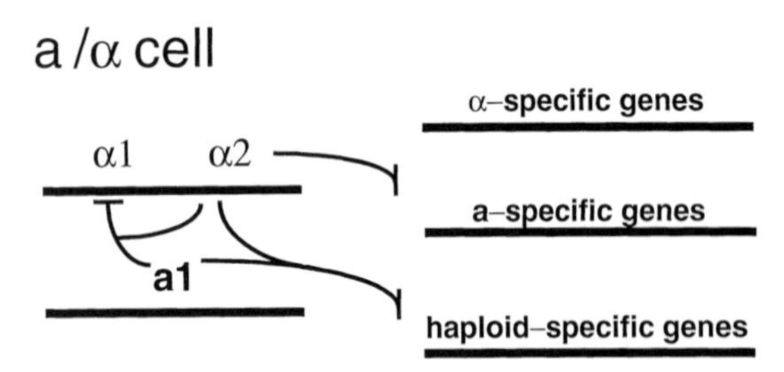

Figure 1. Control of yeast cell specialization by regulatory proteins encoded by the mating-type locus. α-specific genes are expressed only in α cells; **a**-specific genes are expresssed only in **a** cells; haploid-specific genes are expressed in both **a** and α cells but not in **a**/α cells. Lines from the mating-type locus indicate activation or repression of transcription from the target genes by direct action on the upstream regulatory regions of these genes. Thick arrow over gene sets indicates transcription. α2 and **a**1 are homeodomain proteins (see text).

valently modified peptide of 12 amino acids. These pheromones act on serpentine receptors that trigger a MAP kinase cascade and cause activation of a transcription factor, STE12. This protein turns on a large set of genes, such as *FUS1*, that are necessary for cell fusion and other steps in the mating process. Most importantly, with respect to control of cell number, STE12 increases transcription of a gene called *FAR1*, whose product is responsible for causing cell cycle arrest (Fig. 3). FAR1 causes cell cycle arrest in G_1 by inhibiting activity of key cyclin-dependent species necessary for progression from G_1 to S, in particular, the catalytic subunit CDC28 associated with G_1 cyclins CLN1 and CLN2 (Peter and Herskowitz 1994).

The size of a population of cells during development of a multicellular organism could be established by governing synthesis of a cell division regulator such as **a**-factor or α-factor or by controlling response to it. For example, a

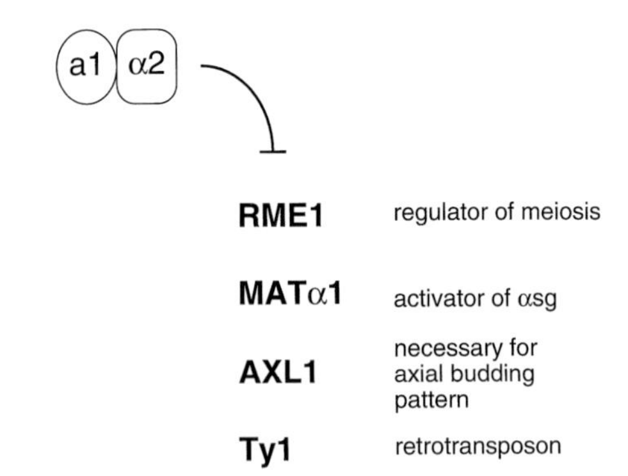

Figure 2. The combinatorial homeodomain protein **a**1-α2 represses transcription of a diverse set of genes. Four representative genes are shown. Others include genes coding for the heterotrimeric G protein involved in the pheromone response pathway (*GPA1*, *STE4*, and *STE18*) (see Herskowitz et al. 1992).

clone of **a** cells would be able to increase in cell number until encountering a sufficiently high level of α-factor produced by nearby α cells. This type of local interaction has been invoked by Garcia-Bellido and colleagues to account for coordinated growth that occurs within compartment borders during development of the *Drosophila* wing disk (Milán et al. 1996). Recent studies have shown that a FAR1 analog of *Drosophila*, the Dacapo protein, has a role in arresting cell proliferation during fruit fly development (de Nooij et al. 1996; Lane et al. 1996). Because Dacapo is a member of the p21/p27 family of cyclin-dependent kinase inhibitors, it would not be surprising if as-

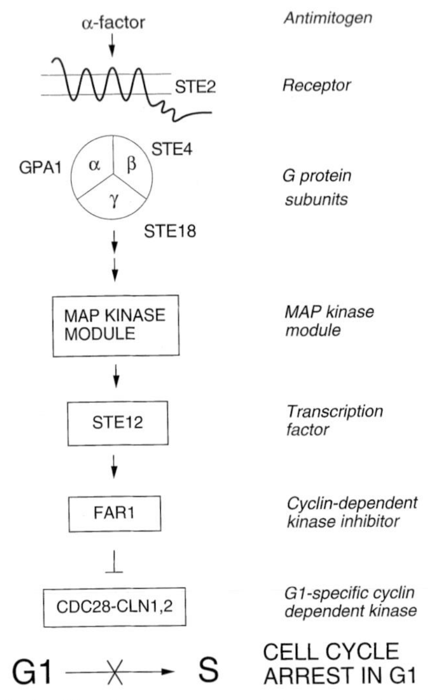

Figure 3. Proliferation of yeast cells is governed by FAR1, a cyclin-dependent kinase inhibitor, whose activity is regulated by the pheromone response pathway. α-factor and **a**-factor are antimitogenic growth and differentiation factors that coordinate the process of mating. For additional details, see text and Herskowitz (1995).

pects of mammalian morphogenesis were also governed by FAR1-like proteins.

YEAST CELL POLARITY DETERMINES THE ARRANGEMENT OF CELLS RELATIVE TO EACH OTHER

Budding yeast chooses a single site on its cell surface from which to produce a new (daughter) cell (see Chant and Pringle 1991; Chant 1994). This is the position at which the actin cytoskeleton is organized and is thought to be determined by the localization of actin-organizing proteins such as CDC42 and BEM1. Strikingly, the position chosen differs in different cell types: **a** and α cells exhibit one pattern, whereas **a**/α cells exhibit another. More specifically, **a** and α cells form buds adjacent to the site where they previously budded (the "axial" budding pattern). In contrast, daughter **a**/α cells always bud away from their mother cell, whereas the mother cell can bud either toward or away from the daughter cell (the "bipolar" budding pattern) (Fig. 4).

These different choices for cell polarization result in differences in cell division planes exhibited by **a** or α cells and **a**/α cells. In budding yeast, the position of the new bud determines the orientation of the mitotic spindle. Thus, successive cell divisions of an **a**/α cell occur in approximately the same direction, resulting in an elongated clone of cells (Fig. 5A). In contrast, cell divisions of an **a** or α cell occur at an angle to each other (Fig. 5B), resulting in a compact arrangement of progeny cells.

We now understand at least part of the molecular basis for these differences. As noted above, the characteristic regulatory feature of **a**/α cells is that they contain the novel repressor, **a**1-α2. It was therefore hypothesized that the bipolar budding pattern of **a**/α cells results from repression of some product necessary for the axial budding pattern exhibited by **a** and α cells (Chant and Herskowitz 1991). The machinery responsible for the budding pattern has been identified in a series of mutant hunts, which reveal three groups of genes and encoded proteins.

One group (*BUD1*, *BUD2*, and *BUD5*) is required for the correct budding pattern in all cell types (see Herskowitz et al. 1995). Mutants lacking any of these genes

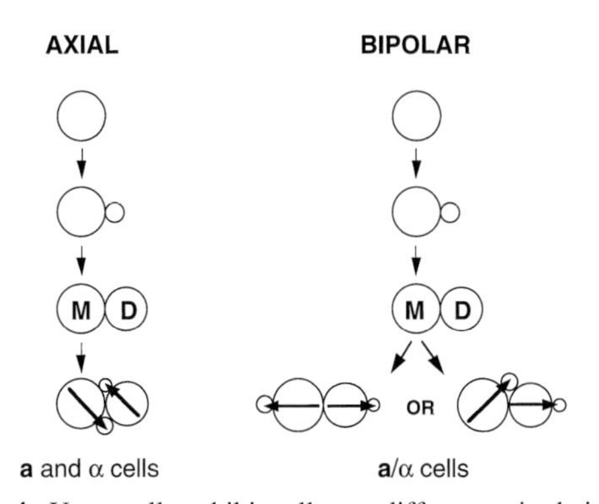

Figure 4. Yeast cells exhibit cell-type differences in their budding pattern. (M) Mother cell; (D) daughter cell. Arrows indicate the axis of cell division (see Fig. 5). (Reprinted, with permission, from Chant and Herskowitz 1991 [copyright Cell Press].)

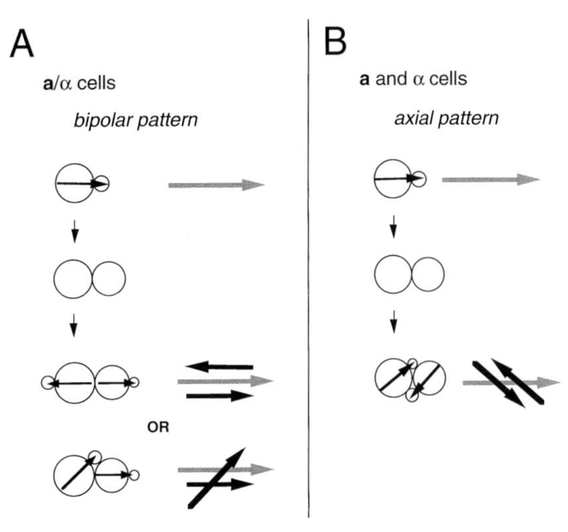

Figure 5. Yeast cells exhibit cell-type differences in their axis of cell division. Long arrows to the right of the cells indicate the axis of cell division. The axis of cell division of the original cell is shown by a lighter arrow and reproduced after the next round of cell division to contrast the different patterns exhibited by **a**/α versus **a** or α cells.

exhibit a random budding pattern. These genes code for a function module consisting of a RAS-related GTPase (BUD1), its guanine nucleotide exchange protein (BUD5), and its GTPase-activating protein (BUD2). These proteins are believed to be involved in a macromolecular assembly process that brings proteins such as BEM1 and CDC24 (the exchange protein for CDC42) to the nascent budding site (Herskowitz et al. 1995; Park et al. 1997).

A second group (*BUD3*, *BUD4*, *AXL1*, and *AXL2* [aka *BUD10*]) is required for the correct budding pattern only in **a** and α cells (Chant and Herskowitz 1991; Fujita et al. 1994; Halme et al. 1996; Roemer et al. 1996). **a** or α cells lacking any of these products exhibit a bipolar budding pattern. It has been proposed that these proteins are involved in choosing an axial landmark. BUD3, BUD4, and AXL2/BUD10 are located at the mother-bud neck and thus appear to be associated with the landmark (Chant et al. 1995; Halme et al. 1996; Roemer et al. 1996; Sanders and Herskowitz 1996). In contrast, AXL1 might be involved in preventing use of a bipolar landmark.

A third group of genes (*BNI1*, *BUD6*, *BUD7*, *BUD8*, and *BUD9*) is required for the correct budding pattern only in **a**/α cells (Zahner et al. 1996). **a**/α cells lacking any of these products exhibit either a random or axial pattern. **a** or α cells lacking any of these products exhibit a normal, axial pattern. It is proposed that these proteins are involved in creating a bipolar landmark (Pringle et al. 1995; Zahner et al. 1996). This landmark is thought to be located at the distal tip of daughter cells and at both ends of mother cells. Most notably, the BUD8 protein has been localized to the distal tip of daughter cells (H. Harkins and J. Pringle, pers. comm.).

The distinctive budding pattern of **a**/α cells was hypothesized to result from repression of one or more genes required for the axial budding pattern characteristic of **a**

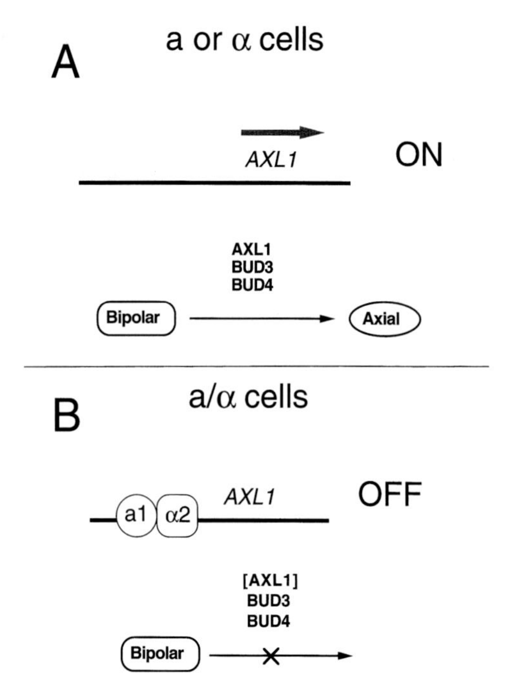

Figure 6. The morphogenetic difference exhibited by haploid (**a** or α) cells versus diploid (**a**/α) cells results from repression of *AXL1* by a homeodomain protein, **a**1-α2. (*Panel A*) *AXL1* is transcribed in **a** or α cells and functions with *BUD3*, *BUD4*, and *AXL2* (*BUD10*) to confer the axial budding pattern. (*Panel B*) In **a**/α cells, *AXL1* is repressed by **a**1-α2 (Fujita et al. 1994). Consequently, **a**/α cells exhibit a bipolar budding pattern.

and α cells (Chant and Herskowitz 1991), and thus we tested whether *BUD3* or *BUD4* was repressed in **a**/α cells. These genes are expressed in **a**/α cells, and their proteins localize normally (Chant et al. 1995; Sanders and Herskowitz 1996). In contrast, *AXL1* is repressed in **a**/α cells (Fujita et al. 1994). The distictive budding pattern of **a**/α cells thus can be explained by the action of a homeodomain protein (**a**1-α2) that represses synthesis of a morphogenetic regulatory protein, AXL1 (Fig. 6).

YEAST CELL SHAPE CAN BE REGULATED AND THEREBY CONTRIBUTE TO MORPHOGENETIC DIFFERENTIATION

In addition to controlling the plane of cell division, yeast exhibits other kinds of morphological differentiations (Kron and Gow 1995). There are at least two examples of morphological differentiation that yeast exhibits in response to external cues. First, when yeast cells of the **a**/α cell type are grown under nitrogen limitation, they exhibit pseudohyphal differentiation (Gimeno et al. 1992; Kron et al. 1994). Under these conditions, yeast cells bud in a unipolar manner, are strikingly elongated, and remain attached to each other; in addition, the pseudohyphae grow into the agar and remain attached to it.

A second example of morphological differentiation is the changes induced by mating pheromones. In this case, production of mating pheromone by one partner causes the other cell to form a projection that is oriented toward the mating partner (Chenevert 1994; Leberer et al. 1996; Valtz and Herskowitz 1996).

YEAST CELLS EXHIBIT ASYMMETRIC CELL DIVISION TO GENERATE PROGENY CELLS WITH DIFFERENT CELL FATES

During development of multicellular organisms, many cell divisions yield daughter cells with different fates. How mitotic division produces cells of different cell fates has been one of the classic challenges for cell and developmental biologists working on a wide variety of organisms.

In the bacterium *Bacillus subtilis*, nutritional starvation causes the bacterial cell to divide asymmetrically, yielding cells of two different sizes. This **structural** difference becomes transformed into a **functional** difference in that the smaller cell becomes the spore and the larger cell becomes the mother cell. It has been proposed that the different size of the two compartments leads to a difference in activity of a phosphatase that ultimately activates a transcription factor in one compartment and not in the other (Duncan et al. 1995).

Yeast cells likewise exhibit cell divisions that yield structurally distinct cells that exhibit functional differences. In the case of yeast, mother cells and daughter cells are structurally different in their history and composition. These cells differ functionally in that the mother cell transcribes the *HO* gene, whereas the daughter cell does not. This mode of regulation is termed "asymmetric" or "mother-daughter" control. This mode of regulation was discovered by analyzing the process by which yeast cells carrying the *HO* gene change their mating type, which revealed a striking stem-cell lineage (Strathern and Herskowitz 1979).

Yeast cells carrying the *HO* gene are able to change their cell type from **a** to α or from α to **a** by a programmed

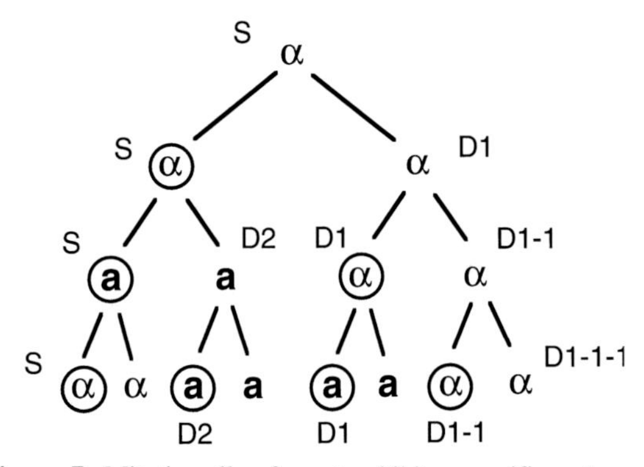

Figure 7. Mitotic cells of yeast exhibit a specific pattern of switching of mating types. The diagram represents mitotic divisions beginning with a single spore cell (S), which carries the *HO* gene, whose expression allows mating-type switching. The initial S cell divides to produce itself (S) and its first daughter (D1). The lineage of first daughters (D1, D1-1, D1-1-1) is a stem cell lineage. Cells that have undergone at least one cell division (*circled*) are competent to exhibit mating-type switching because they transcribe the *HO* gene. In contrast, daughter cells are unable to transcribe *HO* (see text). Cells that are competent to switch mating types do so in at least 75% of cell divisions. Incompetent cells switch in less than 0.1% of cell divisions.

genetic rearrangement (Herskowitz et al. 1992). Beginning with an α cell, it is possible to follow the changes from α to **a** and the reverse by analyzing individual cells for response to α-factor (Hicks and Herskowitz 1976). Following cells in this manner revealed certain rules of switching, in particular, that each cell division yields one cell with competence to switch mating types and one cell that is not competent to switch mating type (Fig. 7) (Strathern and Herskowitz 1979). The succession of daughter cells is invariably incompetent for switching and thus is a stem-cell lineage.

What is it about mother cells that allows them to transcribe the *HO* gene? What is it about daughter cells that prevents them from transcribing the *HO* gene?

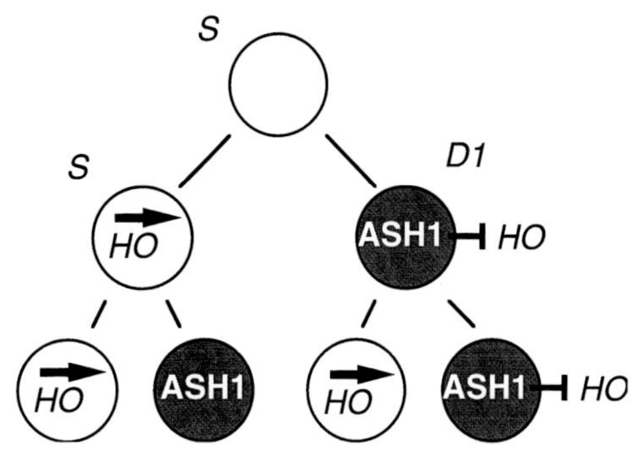

Figure 8. The ASH1 protein is an asymmetrically localized determinant of cell fate. Daughter cells (*shaded*) contain the ASH1 protein, which represses transcription of the *HO* gene. In *ash1* deletion mutants, both mother and daughter cells switch in 94–95% of cell divisions (Sil and Herskowitz 1996).

ASYMMETRIC CELL FATE RESULTS FROM LOCALIZATION OF A REPRESSOR PROTEIN TO THE DAUGHTER CELL NUCLEUS

To identify the regulator responsible for mother-daughter regulation, Anita Sil screened for mutants in which *HO* was expressed in both mother and daughter cells (Sil and Herskowitz 1996). This hunt (and similar studies by Bobola et al. 1996) identified a mutation in what was termed the *ASH1* gene (asymmetric synthesis of *HO*): Inactivation of *ASH1* allows daughter cells to transcribe the *HO* gene as efficiently as mother cells.

The ASH1 protein is localized to the nucleus and, most remarkably, preferentially in daughter cells (Bobola et al. 1996; Sil and Herskowitz 1996): Approximately 73% of cells with a large bud that have two nuclei exhibit asymmetric distribution of ASH1, with ASH1 present in daughter cells and not in mother cells. Another 15% of large-budded cells that had undergone nuclear division exhibited ASH1 in both mother and daughter nuclei, but with a predominance in the daughter cell nucleus. Daughter cells thus contain the negative regulatory protein, ASH1, and do not transcribe *HO*, whereas mother cells lack ASH1 and can transcribe *HO* (Fig. 8).

How does ASH1 protein become localized to daughter cell nuclei? *ASH1* is transcribed in a cell-cycle-dependent manner, in the predivisional cell containing two nuclei during M phase (Bobola et al. 1996). Several possibilities can be imagined by which ASH1 protein becomes localized. *ASH1* might be transcribed only in daughter cell nuclei, and its protein remain localized in the daughter cell compartment. Another possibility is that *ASH1* is transcribed in both nuclei, but its mRNA is transported into the daughter compartment or stabilized there, or selectively translated there. Finally, it is possible that ASH1 protein is synthesized in both mother and daughter compartments but is stable only in the daughter cell compartment.

The observation that ASH1 protein is delocalized in an *ASH1* strain deleted for its 3′-untranslated region (UTR) (Sil 1996) raised the possibility that localization of ASH1 protein is mediated by the *ASH1* mRNA. It is known from work with localized RNA species from *Drosophila* and *Xenopus* that the 3′ UTR mediates RNA localization and translation (St. Johnston 1995; Bassell and Singer 1997). A direct test of the localization hypothesis has awaited development of a facile technique for in situ localization of mRNA in yeast, which has now been accomplished. Using this technique, it has recently been observed that the ASH1 mRNA is localized in a cap at the distal pole of daughter cells in postanaphase cells with large buds (Long et al. 1997; Takizawa et al. 1997).

The next step in understanding the relationship between cell structure (the difference between mother and daughter cells) and cell function (localization of the ASH1 transcript and subsequent repression of *HO* transcription) is to understand how the ASH1 transcript becomes localized. Nasmyth and colleagues have isolated several mutants that are defective in localizing ASH1 protein (Jansen et al. 1996). These mutants fail to exhibit proper localization of ASH1 RNA (Long et al. 1997; Takizawa et al. 1997). We reasoned that the cellular structure involved in marking the daughter cell for bipolar budding might also be involved in directing or anchoring ASH1 mRNA and therefore tested localization of ASH1 protein in these mutants. Most notably, we found that ASH1 protein is inefficiently localized in a mutant defective in the *BUD8* gene (A. Sil, pers. comm.). We have recently found that the ASH1 transcript is also localized inappropriately in mutants lacking *BUD8* (A. Sil and P. Takizawa, pers. comm.).

As noted above, the BUD8 protein has been shown to associate with the distal tip of the bud, i.e., with the daughter cell compartment. We thus propose that the ASH1 transcript might be anchored or directed to the distal tip by association with the BUD8 protein or with another protein bound to BUD8 (A. Sil and P. Takizawa, pers. comm.).

Earlier we noted that the budding pattern of yeast is different in the different cell types, but we did not comment on the possible biological significance for this difference. Interestingly, the rationale for controlling the budding pattern and the plane of cell division may be related to the biological basis for the switching pattern.

To link the phenomena of budding pattern and the pattern of mating-type switching, it is first necessary to note that mating-type switching yields stable **a**/α diploid cells. Consider the behavior of a haploid α spore cell that con-

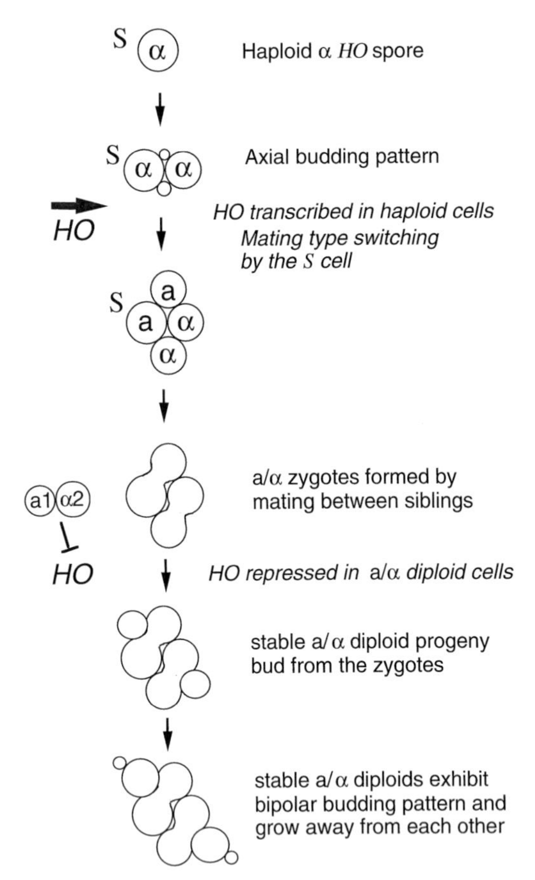

Figure 9. The pattern of mating-type switching and the budding pattern act together to facilitate mating of haploids and to avoid competition among diploids. The diagram depicts germination of a single spore cell carrying the *HO* gene. Although sporulation of laboratory **a**/α strains typically produces four viable haploid spores, sporulation in nature can yield fewer than four spores and inefficient germination, thereby leaving a single viable meiotic product (the "lonely spore" hypothesis; Herskowitz 1988). The process of mating-type switching leads to production of stable **a**/α diploids from **a** or α haploids because transcription of *HO* is turned off by **a**1-α2 after the **a**/α diploid has been formed (Jensen et al. 1983).

tains the *HO* gene (Fig. 9). After its first cell division, both progeny cells remain with α mating type. In the next cell division, however, one cell is competent to switch mating type and does so with high efficiency to yield two **a** cells. Thus, at the four-cell stage, two **a** and two α cells are present. The axial budding pattern exhibited by haploid cells results in the formation of a compact microcolony in which these cells are adjacent to each other. The process of mating to form an **a**/α cell is thus facilitated (Nasmyth 1982; Hartwell 1991): Two **a**/α diploids are produced. There are two important consequences for the formation of an **a**/α cell, both of which result from the presence of the novel regulatory protein, **a**1-α2, the molecular signature of this cell type. First, **a**1-α2 turns off transcription from the *HO* gene, so that further mating-type switching does not occur (Jensen et al. 1983). Second, **a**1-α2 turns off transcription of the *AXL1* gene, which causes a change in the budding pattern to bipolar (Fujita et al. 1994). Thus, the process of mating-type switching results in the production of stable (**a**/α) diploid cells beginning with haploid cells, the process of diploidization discovered many years ago (Winge and Roberts 1949). Once the diploid cell has formed, it need not grow near its siblings for further mating. Quite the contrary, it may be advantageous for the diploid cells to grow away from each other to avoid nutritional competition. Thus, the bipolar pattern exhibited by **a**/α cells may help diploid cells avoid competing with each other for nutrients (Hartwell 1991). The rationale for diploidization may simply be that this cell type has two copies of every yeast gene, but it is also the case that the **a**/α cell exhibits numerous other phenotypic differences that might be advantageous to the organism (see Herskowitz 1988).

WHAT A SINGLE CELL CAN DO

We have considered a variety of different phenomena that have roles in the life cycle of yeast and that provide mechanistic insights into morphogenesis. We have seen how cell specialization determined by homeodomain proteins acting on defined cell targets controls many aspects of a cell's behavior that are relevant to morphogenesis: The yeast cell types differ in their response to antimitogenic growth and differentiation factors that influence their state of proliferation and are programmed to divide in different division planes depending on their cell type. We have also seen that mitotically growing yeast cells exhibit asymmetric cell division to produce a specific cell lineage. These asymmetric cell divisions yield cells with differing competencies for changing from one haploid cell type to another and thus also ultimately contribute to morphogenesis. As I have noted elsewhere (Herskowitz 1989), in trying to unravel the logical and molecular basis of development in multicellular organisms, it may be instructive to consider what a single cell can do.

ACKNOWLEDGMENTS

I thank Heidi Harkins, John Pringle, Anita Sil, and Pete Takizawa for allowing me to cite unpublished work and Flora Banuett for comments on the figures. The work described here has been supported by research grants from the National Institutes of Health, the American Cancer Society, the National Science Foundation, the Markey Program in Biological Sciences, and the Herbert W. Boyer Fund, all of whom are gratefully acknowledged.

REFERENCES

Banuett F. 1995. Genetics of *Ustilago maydis*, a fungal pathogen that induces tumors in maize. *Annu. Rev. Genet.* **29:** 179.

Bassell G. and Singer R.H. 1997. mRNA and cytoskeletal filaments. *Curr. Opin. Cell Biol.* **9:** 109.

Bobola N., Jansen R.-P., Shin T.H., and Nasmyth K. 1996. Asymmetric accumulation of Ash1p in postanaphase nucleic depends on a myosin and restricts yeast mating-type switching to mother cells. *Cell* **84:** 699.

Chant J. 1994. Cell polarity in yeast. *Trends Genet.* **10:** 328.

Chant J. and Herskowitz I. 1991. Genetic control of bud site selection in yeast by a set of gene products that constitute a morphogenetic pathway. *Cell* **65:** 1203.

Chant J. and Pringle J.R. 1991. Budding and cell polarity in *Saccharomyces cerevisiae*. *Curr. Opin. Genet. Dev.* **1:** 342.
Chant J., Mischke M., Mitchell E., Herskowitz I., and Pringle J.R. 1995. Role of Bud3p in producing the axial budding pattern of yeast. *J. Cell Biol.* **129:** 767.
Chenevert J. 1994. Cell polarization directed by extracellular cues in yeast. *Mol. Biol. Cell* **5:** 1169.
de Nooij J.C., Letendre M.A., and Hariharan I.K. 1996. A cyclin-dependent kinase inhibitor, Dacapo, is necessary for timely exit from the cell cycle during *Drosophila* embryogenesis. *Cell* **87:** 1237.
Duncan L., Alper S., Arigoni F., Losick R., and Stragier P. 1995. Activation of cell-specific transcription by a serine phosphatase at the site of asymmetric division. *Science* **270:** 641.
Fujita A., Oka C., Arikawa Y., Katagai T., Tonouchi A., Kuhara S. and Misumi, Y. 1994. A yeast gene necessary for bud-site selection encodes a protein similar to insulin-degrading enzymes. *Nature* **372:** 567.
Gimeno C.J., Ljungdahl P.O., Styles C.A., and Fink G.R. 1992. Unipolar cell divisions in the yeast *S. cerevisiae* lead to filamentous growth: Regulation by starvation and *RAS*. *Cell* **68:** 1077.
Halme A., Michelitch M., Mitchell E.L., and Chant J. 1996. Bud10p directs axial cell polarization in budding yeast and resembles a transmembrane receptor. *Curr. Biol.* **6:** 570.
Hartwell L. 1991. Yeast embryology. Pathways of morphogenesis. *Nature* **352:** 663.
Herskowitz I. 1988. Life cycle of the budding yeast *Saccharomyces cerevisiae*. *Microbiol. Rev.* **52:** 536.
——— 1989. A regulatory hierarchy for cell specialization in yeast. *Nature* **342:** 749.
——— 1992. Yeast branches out. *Nature* **357:** 190.
——— 1995. MAP kinase pathways in yeast: For mating and more. *Cell* **80:** 187.
Herskowitz, I., Rine J., and Strathern J. 1992. Mating-type determination and mating-type interconversion. In *The molecular and cellular biology of the yeast* Saccharomyces*: Gene expression* (ed. E.W. Jones et al.), vol. 2, p. 583. Cold Spring Harbor Laboratory Press, Cold Spring Harbor, New York.
Herskowitz I., Park H.-O., Sanders S., Valtz N., and Peter M. 1995. Programming of cell polarity in budding yeast by endogenous and exogenous signals. *Cold Spring Harbor Symp. Quant. Biol.* **60:** 717.
Hicks J.B. and Herskowitz I. 1976. Interconversion of yeast mating types. I. Direct observations of the action of the homothallism (*HO*) gene. *Genetics* **83:** 245.
Jansen R.-P., Dowzer C., Michaelis C., Galova M., and Nasmyth K. 1996. Mother cell-specific *HO* expression in budding yeast depends on the unconventional myosin Myo4p and other cytoplasmic proteins. *Cell* **84:** 687.
Jensen R., Sprague, Jr. G.F., and Herskowitz I. 1983. Regulation of yeast mating-type interconversion: Feedback control of *HO* gene expression by the mating type locus. *Proc. Natl. Acad. Sci.* **80:** 3035.
Johnson A. 1992. A combinatorial regulatory circuit in budding yeast. In *Transcriptional regulation* (ed. S.L. McKnight and K.R. Yamamoto), p. 975. Cold Spring Harbor Laboratory Press, Cold Spring Harbor, New York.
Kron S.J. and Gow N.A. 1995. Budding yeast morphogenesis: Signaling, cytoskeleton and cell cycle. *Curr. Opin. Cell Biol.*. **7:** 845.
Kron S.J., Styles C.A., and Fink G.R. 1994. Symmetric cell division in pseudohyphae of the yeast *Saccharomyces cerevisiae*. *Mol. Biol. Cell* **5:** 1003.
Lane M.E., Sauer K., Wallace K., Jan Y.N., Lehner C.F., and Vaessin H. 1996. Dacapo, a cyclin-dependent kinase inhibitor, stops cell proliferation during *Drosophila* development. *Cell* **87:** 1225.
Leberer E., Chenevert J., Leeuw T., Harcus D., Herskowitz I., and Thomas D.Y. 1996. Genetic interactions indicate a role for Mdg1p and the SH3 domain protein Bem1p in linking the G-protein mediated yeast pheromone signaling pathway to regulators of cell polarity. *Mol. Gen. Genet.* **252:** 608.
Li T., Stark M.B., Johnson A.D., and Wolberger C. 1995. Crystal structure of the MAT**a**1/MATα2 homeodomain heterodimer bound to DNA. *Science* **270:** 262.
Long R.M., Singer R.H., Meng X., Gonzalez I., Nasmyth K., and Jansen R.-P. 1997. Mating type switching in yeast controlled by asymmetric localization of *ASH1* mRNA. *Science* **277:** 383.
Mann R.S. and Chan S.-K. 1996. Extra specificity from *extradenticle*: The partnership between HOX and PBX/EXD homeodomain proteins. *Trends Genet.* **12:** 258.
Milán M., Campuzano S., and García-Bellido A. 1996. Cell cycling and patterned cell proliferation in the wing primordium of *Drosophila*. *Proc. Natl. Acad. Sci.* **93:** 640.
Nasmyth K. 1982. Molecular genetics of yeast mating type. *Annu. Rev. Genet.* **16:** 439.
Park H.-O., Bi E., Pringle J.R., and Herskowitz I. 1997. Two active states of the Ras-related Bud1/Rsr1 protein bind to different effectors to determine yeast cell polarity. *Proc. Natl. Acad. Sci.* **94:** 4463.
Peter M. and Herskowitz I. 1994. Direct inhibition of the yeast cyclin-dependent kinase Cdc28-Cln by Far1. *Science* **265:** 1228.
Pringle J.R., Bi E., Harkins H.A., Zahner J.E., De Virgilio C., Chant, J., Corrado K., and Fares H. 1995. Establishment of cell polarity in yeast. *Cold Spring Harbor Symp. Quant. Biol.* **60:** 729.
Roemer T., Madden K., Chang J., and Snyder M. 1996. Selection of axial growth sites in yeast requires Axl2p, a novel plasma membrane glycoprotein. *Genes Dev.* **10:** 777.
Sanders S.L. and Herskowitz I. 1996. The Bud4 protein of yeast, required for axial budding, is localized to the mother/bud neck in a cell cycle-dependent manner. *J. Cell Biol.* **134:** 413.
Sil A. 1996. "Identification and analysis of an asymmetrically localized determinant of cell fate in *Saccharomyces cerevisiae*." Ph.D. thesis. University of California, San Francisco.
Sil A. and Herskowitz I. 1996. Identification of an asymmetrically localized determinant, Ash1p, required for lineage-specific transcription of the yeast *HO* gene. *Cell* **84:** 711.
St. Johnston D. 1995. The intracellular localization of messenger RNAs. *Cell* **81:** 161.
Strathern J.N. and Herskowitz I. 1979. Asymmetry and directionality in production of new cell types during clonal growth: The switching pattern of homothallic yeast. *Cell* **17:** 371.
Takizawa P.A., Sil A., Swedlow J.R., Herskowitz I., and Vale R.D. 1997. Actin-dependent localization of an RNA encoding a cell-fate determinant in yeast. *Nature* **389:** 90.
Valtz N. and Herskowitz I. 1996. Pea2 protein of yeast is localized to sites of polarized growth and is required for efficient mating and bipolar budding. *J. Cell Biol.* **135:** 725.
Winge Ø. and Roberts C. 1949. A gene for diploidization in yeasts. *C.R. Trav. Lab. Carlsberg Ser. Physiol.* **24:** 341.
Zahner J.E., Harkins H.A., and Pringle J.R. 1996. Genetic analysis of the bipolar pattern of bud site selection of the yeast *Saccharomyces cerevisiae*. *Mol. Cell. Biol.* **16:** 1857.

Cortical Asymmetries Direct the Establishment of Cell Polarity and the Plane of Cell Division in the *Fucus* Embryo

R.S. QUATRANO
Department of Biology, University of North Carolina, Chapel Hill, North Carolina 27599-3280

Cell morphogenesis in plants requires that the site of cell expansion and the plane of cell division be closely regulated to generate the observed diversity of cell types. The same coordination between cell division and expansion is believed to be responsible for tissue and organ morphogenesis, since plant cells are immobile and fixed relative to each other. The mechanisms operative in plant cells that are responsible for morphogenesis require the localization of the macromolecular complexes controlling cell expansion or cell division to one or more cortical sites or domains (Fowler and Quatrano 1997). The sites where these domains are assembled are determined by directional cues originating from intracellular or extracellular sources, often in the form of gradients. Hence, to understand the cellular and molecular basis of plant cell morphogenesis, the fundamental questions that need to be addressed are: How do plant cells perceive external gradients and process the signal into cortical asymmetries? What is the nature of these asymmetries and how are they assembled? How is this localized complex stabilized at the site and how do these cortical domains direct local expansion and/or orient the plane of cell division?

These questions are not unique to plant cells. A general pattern seems to be emerging from a number of prokaryotic and eukaryotic systems in which a cortical site, identified by some internal or external cue, serves as an assembly site for a cytoskeletal complex, which then becomes anchored at that site and serves to direct local morphogenesis (Fowler and Quatrano 1997). In this volume, several of these diverse systems (e.g., bacteria, yeast, and insects) are also discussed with respect to these fundamental questions and present an important evolutionary perspective. However, I will focus on plant cell morphogenesis and concentrate on the interactions between the cytoskeleton, plasma membrane, and cell wall in generating stable cortical domains in zygotes of the brown alga *Fucus*.

ZYGOTES OF THE FUCALES: MODELS TO STUDY THE ESTABLISHMENT OF CORTICAL ASYMMETRIES

Numerous examples exist in which polar cells direct local cell expansion or the orientation of cell division during vascular plant development, e.g., root hairs, root cortical cells, microspores, cells of the stomatal complex, and zygotes. However, in most cases, to identify the cue(s) and the signaling pathway that establishes the polar axis in these cells, to determine the nature of the cortical asymmetry that is established in response to the signal, to elucidate how the asymmetry is stabilized or fixed, or to understand how this structural localization may affect subsequent determination of cell fate is not amenable to experimental approaches at the cellular level. Using zygotes of the Fucales, such as *Fucus* and *Pelvetia*, the question can be experimentally approached of how molecular asymmetries are established with respect to external spatial cues, and hopefully, this information can be applied to higher plant as well as other eukaryotic systems.

Eggs of the Fucales are large (75–100 μm in diameter) and apolar, lack a large central vacuole, and along with sperm, are released from mature plants under laboratory conditions. Fertilization occurs in a defined seawater medium, resulting in the symmetrical secretion of a cell wall which allows the diploid zygote to adhere to the substratum within a few hours (Quatrano 1982). This attachment allows for easy manipulation of solutions and for orienting external gradients to study the basis of how gradients, such as light and other vectors, establish cortical asymmetries (Fig. 1a). At 14 hours after fertilization, the secretory apparatus is redirected from a symmetrical pattern to a defined cortical site where a localized bud or rhizoid is formed (Fig. 1b). This polar growth of the rhizoid (or localized cell expansion) is followed by an asymmetric cell division at 24 hours, with the cell plate and its associated vesicles forming in a division plane perpendicular to the rhizoid growth axis. At this stage, vesicles containing cell wall material are being directed not only to the site of tip growth (i.e., the rhizoid), but also to the site of cell plate formation. As a result of this oriented division, the two-celled embryo is composed of a smaller, more elongated rhizoid cell and a larger, more rounded thallus cell (Fig. 1c). These cells differ not only in shape, but also in organellar content and molecular composition, both in the cytoplasm and in the cell wall (Goodner and Quatrano 1993; Fowler and Quatrano 1995).

The development of the early embryo occurs in synchrony and in the absence of surrounding cells. The surface domain identified for polar growth of the zygote is not believed to be preset in the egg (Jaffe 1958) but rather is initially set by the site of sperm entry (Knapp 1931) and

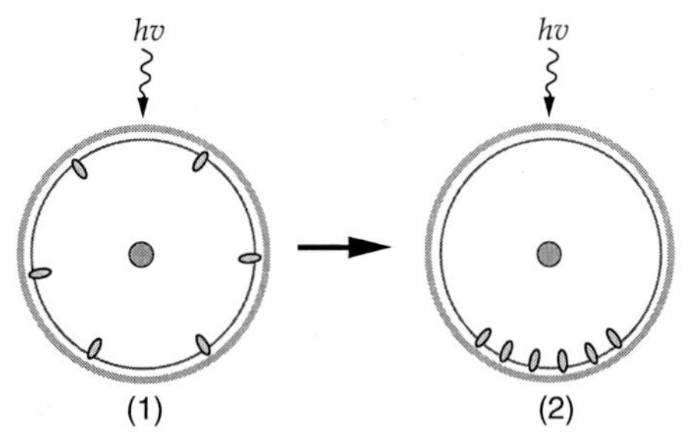

a. An external light gradient is perceived and processed into a cortical asymmetry by the actin-dependent translocation of plasma membrane components.

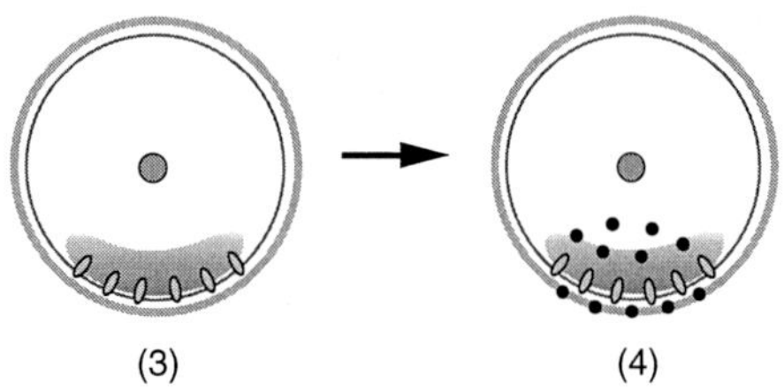

b. The local accumulation of plasma membrane and cortical components forms a target site for Golgi vesicle secretion required to stabilize the molecular asymmetry in the membrane.

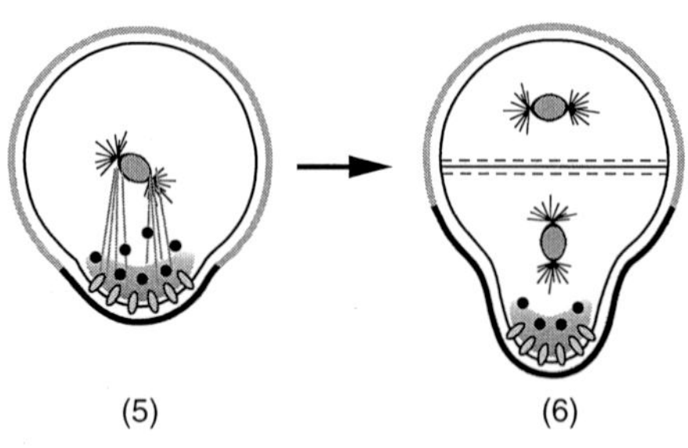

c. Golgi-derived components secreted into the cell wall are required for polar growth and for the proper orientation of the division apparatus.

Figure 1. Stages in the establishment of a polar axis and an oriented cell division in brown algal zygotes that are oriented in a unilateral light gradient (*hv*). (*a*) A cortical asymmetry is first observed in the plasma membrane by 6 hr after fertilization. An example is the translocation of dihydropyridine (DHP) receptors (*ovals*) from a symmetrical distribution (*1*) to the shaded quadrant of the zygote (*2*). This cortical asymmetry is labile and can be repositioned to another site by changing the direction of the light. (*b*) The colocalization F-actin and Ca^{++} in the cortex beneath the DHP receptors generates a locally specialized cortical domain (*shaded area*) (*3*). This unique local environment leads to secretion of a specific population of Golgi vesicles, the F granules (*dark circles*), at the cortical site, resulting in the deposition of their contents into the plasma membrane and cell wall (*4*). This secretion stabilizes the plasma membrane asymmetries in the form of an axis stabilizing complex which fixes the cortical domain as the site for polar growth. (*c*) The fixed asymmetry in the plasma membrane and cell wall (dark cell wall at the emerging tip) is maintained during polar growth and has a role in orienting the asymmetric division plane (*5*), perhaps by interacting with microtubules to alter spindle orientation through centrosome rotation. Cell wall/membrane molecules, locally deposited by targeted secretion in only the rhizoid cell wall (*6*), may also affect the determination of rhizoid and thallus traits of the daughter cells.

does not require any other gradient after fertilization (R. Quatrano, unpubl.). This initial site for subsequent polar growth is not fixed at the time of sperm entry, however. The polar axis can be realigned with various gradients. For example, when zygotes of *Fucus* or *Pelvetia* are subjected to a gradient of light during the first several hours after fertilization, the rhizoid always forms on the shaded side of the light gradient several hours later. Most importantly for the analysis of the mechanisms involved, the polarity of the axis is not irreversibly established until about 10 hours after fertilization. During the time between 4 and 10 hours, the direction of the light can be changed several times, with the resulting polar growth (and site of polar secretion) and orientation of the plane of division always determined by the last light gradient. Hence, within the first 10 hours after fertilization, external gradients imposed on the zygote can realign an axis of symmetry, initially set by sperm entry, and establish a polarity of that axis by redirecting the secretory apparatus to produce localized cell expansion. In this system, then, the signal, as well as the cortical site toward which the secretory apparatus and growth will be directed, can be identified and analyzed. Furthermore, the polar axis which results is responsible for the differentiation of the first two cells of the embryo and the developmental axis of the whole organism (Fowler and Quatrano 1995; Quatrano and Shaw 1997).

Much of our recently published work on this system has focused on three questions: What is the nature of the molecules that are localized in the cortex and serve as the target site for polar vesicle secretion? What are the processes required to irreversibly stabilize the axis of polarity established by a light gradient and what proteins may be involved? What is the nature of the positional information at the cortex or in the cell wall that plays a part in orienting the plane of the first cell division?

CORTICAL MARKERS FOR POLAR VESICLE SECRETION

Cytological markers have been identified to monitor the orientation of the polar axis in response to a light gradient. These include free calcium and dihydropyridine (DHP) receptors (Shaw and Quatrano 1996a), actin (Kropf et al. 1989), and a specifically stained population of secretory vesicles (Shaw and Quatrano 1996b). Using these probes, we can monitor the temporal and spatial localization of a cortical site(s) relative to the direction(s) of unilateral light imposed during the first 10 hours after fertilization. Fluorescently labeled dihydropyridine (FL-DHP) can be used as a vital stain in *Fucus* zygotes to follow the distribution of DHP receptors during the establishment of cell polarity (Shaw and Quatrano 1996a). Whether these DHP receptors actually represent calcium channels (Knaus et al. 1992) has not been determined, but the localization of DHP receptors is clearly a marker for plasma membrane components (Shaw and Quatrano 1996a). We found that distribution of FL-DHP in the plasma membrane is symmetrical within the first several hours after fertilization (Fig. 1a,1). However, in response to a unilateral light gradient during the first 6 hours after fertilization, DHP receptors accumulated on the shaded side of the zygote and predicted the site of future polar outgrowth of the rhizoid (Fig. 1a,2). If the light gradient was changed by 180° after the initial asymmetry of FL-DHP was observed, we could monitor the redistribution of DHP receptors to the opposite pole, i.e., the new site for rhizoid outgrowth. In the presence of cytochalasin, neither the asymmetry of FL-DHP (Shaw and Quatrano 1996a) nor the orientation of the rhizoids to the shaded side of the light gradient (Quatrano 1973) was observed. On the basis of the site, timing, photo-reversibility, and actin dependence of the asymmetric localization of DHP receptors, we concluded that FL-DHP is a vital probe for visualizing the orientation of the polar axis in response to a light gradient (Shaw and Quatrano 1996a). FL-DHP can now be used to monitor in real time the orientation of the polar axis in a large population of developing zygotes. We can now ask: Are the receptors for vectors other than light (e.g., exogenous electric fields) linked to the localization of FL-DHP, and if so, are they linked through the same or a parallel response pathway? What are the intermediates in the light response pathway that result in the localization of DHP receptors? Using FL-DHP as a probe, we can add to the medium or microinject into zygotes various inhibitors, antibodies, or analogs of potential signaling intermediates to determine their possible role in the orientation of the polar axis. Hence, the initial asymmetry that we detect in a light gradient is the actin-dependent translocation of existing plasma membrane components (Fig. 1a).

The asymmetric distribution of DHP receptors coincides temporally and spatially with an increase in intracellular calcium, as measured by microinjected calcium green dextran (Shaw and Quatrano 1996a), as well as an F-actin patch or cap (Kropf et al. 1989). All three components comprise the target site for future polar secretion and growth (Fig. 1b,3). A structural complex that may have similar properties is the yeast Exocyst, a bud-tip-localized complex of several proteins that are specifically required for secretion in yeast (TerBush et al. 1996).

POLARIZED SECRETION STABILIZES THE POLAR AXIS ESTABLISHED BY A LIGHT GRADIENT

We have also utilized toluidine blue O (TBO), a dye specific to sulfated polysaccharides when used at a pH below 2 (Brawley and Quatrano 1979), as a marker for the Golgi vesicles that are directed to the site for polar growth (F granules). Two components that are found in F granules include the sulfated fucan F2 and a protein that cross-reacts with an antibody to human vitronectin (Wagner et al. 1992). We have cloned a gene whose protein product has epitopes similar to those of vitronectin as it is recognized by antibodies to human vitronectin and a heparin-binding peptide, and can bind F2 (C. Taylor et al., unpubl.). This F2-binding protein, with an mRNA of approximately 6 kb, has no sequence homology with any known protein, except for several basic amino acid domains which are similar to the heparin-binding domains in human vitronectin. We determined by TBO staining that the local secretion of F2 into the cell wall at the target site (Fig. 1b,4) occurs with the same kinetics in a population of zygotes as does the fixation of the polar axis, i.e., the irreversible stabilization of the polarity such that an orienting light vector cannot change the orientation of the polar axis (Shaw and Quatrano 1996b).

To further confirm this correlation between secretion and the fixation of the polar axis, we disrupted Golgi-mediated secretion by treating zygotes continually with brefeldin-A (BFA). In the presence of BFA, the fixation of the light-induced polar axis in *Fucus* zygotes, as well as polar growth of the rhizoid, was prevented. However, BFA did not interfere with the initial orientation of the axis (i.e., the establishment of the target site as assayed by localization of DHP receptors and F-actin) or with the timing of the first cell division. Hence, based on the timing of F2 secretion into the wall, and the inhibition of polar axis fixation in the presence of BFA, we concluded that local secretion of the contents of F granules transforms the labile target site into the fixed growth site for the rhizoid (Fig. 1b) (Shaw and Quatrano 1996b).

Previous work showed that microfilaments (Quatrano 1973) and an intact cell wall (Kropf et al. 1988) are also required for the fixation of the polar axis. F granules are neither targeted to a cortical site in the presence of cytochalasin nor are their contents assembled into the cell wall when the zygote is continually in the presence of a mixture of wall-degrading enzymes (Kropf et al. 1988). Coupled with the BFA results, we now believe that the critical event for stabilizing the polar axis is actually the localized secretion of F granules at the site of polar growth. In the absence of F granule secretion, components of the target site that are assembled in response to a light gradient are not sufficient for axis fixation.

Evidence for the role of polarized secretion in axis fixation must now be incorporated into our model for the axis stabilizing complex (ASC) (Quatrano 1990), a hypothetical structural unit that serves to stabilize molecular asymmetries in the plasma membrane at the target site for vesicle deposition (Fig. 2). This model was originally based on the requirements for an intact cell wall and actin cytoskeleton to stabilize the polar axis. We proposed that to stabilize the polar axis, a direct coupling of the actin cytoskeleton with the cell wall, through some transmembrane protein, was required (Kropf et al. 1988). This model is similar to that of focal adhesions in mammalian cells: It includes an extracellular matrix component (e.g., vitronectin and its associated sulfated heparin) that binds to a receptor peptide in a transmembrane protein (e.g., integrin), which links to the actin cytoskeleton through actin-binding proteins (e.g., vinculin) (Burridge and-Chrzanowska-Wodnicka 1996). Our revised model now includes a step in the assembly of the ASC that allows for the secretion of F granules, providing the necessary plasma membrane/cell wall components to stabilize the molecular asymmetries in the plasma membrane, e.g., the extracellular ligand(s) for the postulated ASC transmembrane receptor (Fowler and Quatrano 1995). However, there is no direct evidence yet for a physical link between the cytoplasmic actin network and the cell wall at the time when the axis will no longer respond to reorienting incident light.

What genes might be involved in the transport and secretion of the F granules in *Fucus*? The requirements for the temporal and spatial establishment of the polar axis in

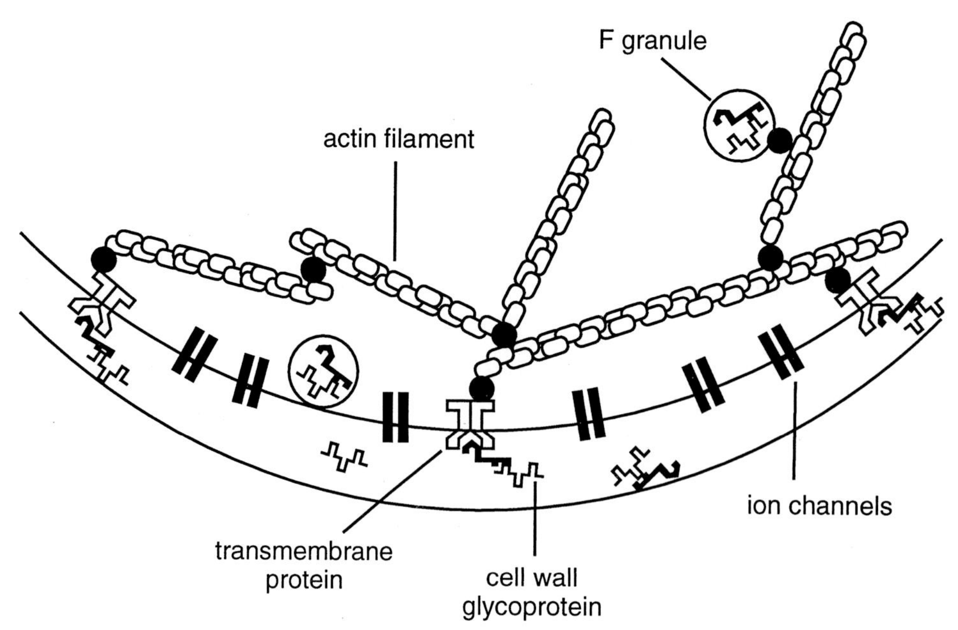

Figure 2. A schematic diagram of the cortical domain that represents the fixed site of polar growth in *Fucus* zygotes. A postulated transmembrane protein links the actin cytoskeleton with cell wall glycoproteins stabilizing plasma membrane components (e.g., DHP receptors) that accumlated at that site in response to a unilateral light gradient (see Fig. 1b,4). The actin filaments provide the tracks for transport of the localized F granules whose contents are secreted into the cell wall and are required to complete the axis stabilizing complex and to orient the plane of cell division. (Reprinted, with permission, from Goodner and Quatrano 1993.)

Fucus are similar to the stages of bud-site selection, polarity establishment, and bud growth in yeast (Chant and-Stowers 1995; Fowler and Quatrano 1997). Polarized growth of the yeast bud also involves the selective targeting of proteins to a patch on the cell surface at the predetermined bud site. Protein delivery to this cortical domain is probably mediated by an actin/myosin-dependent translocation of vesicles to the bud site and appears to be regulated by GTPase molecular switches that specify movement to and docking of vesicles to the targeting patch. The *ras* superfamily of GTPases includes members of the rac/rho/CDC42 family that regulate the assembly of the actin cytoskeleton at localized sites at the plasma membrane in yeast (Chant and Stowers 1995), as well as the rab GTPase family, which includes the *rab/ypt1/sec4* genes, which regulates vesicle transport and fusion in yeast and mammals (Drubin and Nelson 1996; Roemer et al. 1996). Proteins in these families are localized to the cytoplasmic regions of polar bud growth in yeast and homologs have recently been isolated from *Fucus*. Preliminary results indicate that a rac-like GTPase from *Fucus* can interact with certain signaling intermediates (e.g., STE20p) in the bud formation pathway in yeast (J. Fowler et al., unpubl.). These results, and ongoing investigations involving microinjection of recombinant proteins and antibodies, as well as immunolocalization of these gene products in *Fucus* during rhizoid outgrowth, may indicate that similar pathways operate to localize cortical complexes in these two organisms. Furthermore, the localization of glucan synthase at the tip of the emerging yeast bud (Drgonová et al. 1996; Qadota et al. 1996) and the requirement for new cell wall synthesis during rhizoid elongation in *Fucus* suggest that an alginate or cellulose synthase may also be localized in the emerging rhizoid.

THE RHIZOID CELL WALL MAY CONTAIN POSITIONAL INFORMATION THAT HELPS DIRECT THE ORIENTATION OF THE FIRST CELL DIVISION

Our first indication that a cell wall/plasma membrane domain might possess positional information required for the orientation of the first cell division in *Fucus* came during our experiments with BFA. Although continual incubation of zygotes in BFA did not appreciably delay cell division, we noticed that the orientation of first division plane was not always oriented perpendicular to the light gradient (Shaw and Quatrano 1996b). Furthermore, since BFA prevented secretion and rhizoid outgrowth, the resulting division in BFA-treated zygotes was symmetric rather than asymmetric, resulting in an embryo with two cells of identical shape. This raised the interesting possibility that targeted secretion into the cell wall was providing not only material for polar outgrowth of the rhizoid, but also some type of positional cues for orienting the plane of the first cell division. This result highlighted the critical role of targeted secretion not only in localized cell expansion, but also in orienting the plane of cell division, both key components for cellular morphogenesis in plants.

These results now allowed us to approach the question of whether the abnormal positioning of the cell plate influenced the site of polar growth relative to the light gradient. Since the site of polar growth is established but not fixed in the presence of BFA, we asked: Does the position of the cell plate in BFA-treated zygotes alter the site of future polar growth, or is this cortical site maintained on the shaded side of the light gradient independent of the plane of cell division? While in a unilateral light gradient, zy-

Asymmetric Segregation of the *Drosophila* Numb Protein during Mitosis: Facts and Speculations

J.A. KNOBLICH, L.Y. JAN, AND Y.N. JAN
Departments of Physiology and Biochemistry, Howard Hughes Medical Institute, University of California, San Francisco, San Francisco, California 94143-0724

During the development of a multicellular organism, many different cell types are generated from a single precursor cell. Generation of this diversity requires that cells be able to divide into two different daughter cells. Two mechanisms are possible (Horvitz and Herskowitz 1992): The two daughters are initially identical but become different by interacting with each other, by interacting with neighboring cells, or by responding to diffusible factors released from those neighboring cells. Alternatively, the mother cell can divide asymmetrically into two daughter cells that are committed to different developmental pathways already at the time of their birth. Such asymmetric cell divisions can be generated if the mother cell is capable of segregating determinants preferentially into one of its two daughter cells so as to initiate a particular developmental pathway in this daughter cell, but not its sister cell. Although many examples of cell fate determination by cell-cell interaction or by diffusible factors have been studied extensively and are now fairly well understood at the molecular level, we have begun only recently to understand the mechanisms of asymmetric cell division.

Information on the mechanisms of asymmetric cell division has emerged from studies in *Saccharomyces cerevisiae* (Chang and Drubin 1996), *Caenorhabditis elegans* (Guo and Kemphues 1996a), and *Drosophila melanogaster*, as well as studies of prokaryotic organisms such as *Caulobacter crescentus* and *Bacillus subtilis* (Shapiro and Losick 1997). Here, we focus on asymmetric cell divisions during the development of the fruit fly *Drosophila melanogaster*. In *Drosophila*, formation of both the peripheral nervous system (PNS) and the central nervous system (CNS) requires asymmetric cell divisions. An external sensory (ES) organ, one type of sensory structure found in the PNS, consists of four cells that arise from a single sensory organ precursor (SOP) cell in two rounds of asymmetric cell division (Fig. 1a,b) (Bodmer et al. 1989). The SOP cell divides asymmetrically into a IIA cell and a IIB cell. The IIA cell gives rise to the two outer cells, the hair cell and the socket cell, whereas the IIB cell gives rise to the two inner cells, the neuron and the sheath cell. Asymmetric cell divisions also take place in the CNS: Most neuroblasts, the CNS founder cells, divide asymmetrically into a larger apical daughter cell and a smaller basal ganglion mother cell (GMC). The apical cell retains neuroblast characteristics and continues to divide in a stem-cell-like fashion, whereas the GMC divides only once into two cells that later both differentiate into neurons.

The protein Numb has been shown to have an important role in the asymmetric divisions of both SOP cells and neuroblasts. *numb* was originally identified as a mutant that leads to cell fate changes in ES organs and other sensory structures of the PNS (Uemura et al. 1989). In a *numb* mutant, the SOP cell divides into two IIA cells and gives rise to twice the number of outer cells and no inner cells. Conversely, the overexpression of *numb* leads to the opposite cell fate transformation: The SOP cell divides into two IIB cells and gives rise to duplicated inner cells and no outer cells (Rhyu et al. 1994). A similar function for *numb* has been demonstrated for certain neuroblasts in the CNS (Spana et al. 1995). Immunocytochemistry using an antibody against the Numb protein revealed that Numb is a membrane-associated protein (Rhyu et al. 1994). Although the protein is homogeneously distributed around the cell membrane during interphase, it redistributes dramatically in mitotic neuroblasts and SOP cells. Starting in late prophase, the Numb protein forms a cortical crescent and concentrates in the membrane area overlying one of the two spindle poles; in telophase, the protein specifically enters only one of the two daughter cells (Rhyu et al. 1994; Knoblich et al. 1995). Thus, Numb fulfills the criteria for a segregating determinant in the asymmetric cell divisions of the *Drosophila* nervous system. A similar asymmetric localization has been demonstrated for the nuclear homeodomain transcription factor Prospero (Hirata et al. 1995; Knoblich et al. 1995; Spana and Doe 1995). During mitosis, Prospero is found at the cell membrane, where it colocalizes with Numb. In telophase, Numb and Prospero enter the same daughter cell and, after mitosis, Prospero translocates into the nucleus, whereas Numb stays at the cell membrane. Localization of Numb and Prospero occurs independently, but the tight correlation suggests a similar mechanism.

To understand asymmetric cell divisions in *Drosophila*, we will have to identify and characterize the cellular mechanism that enables neuroblasts and SOP cells to specifically segregate proteins such as Numb and Prospero into one of their two daughter cells. Here, we discuss some of the more recent findings on the asymmetric localization of Numb and Prospero. We also pre-

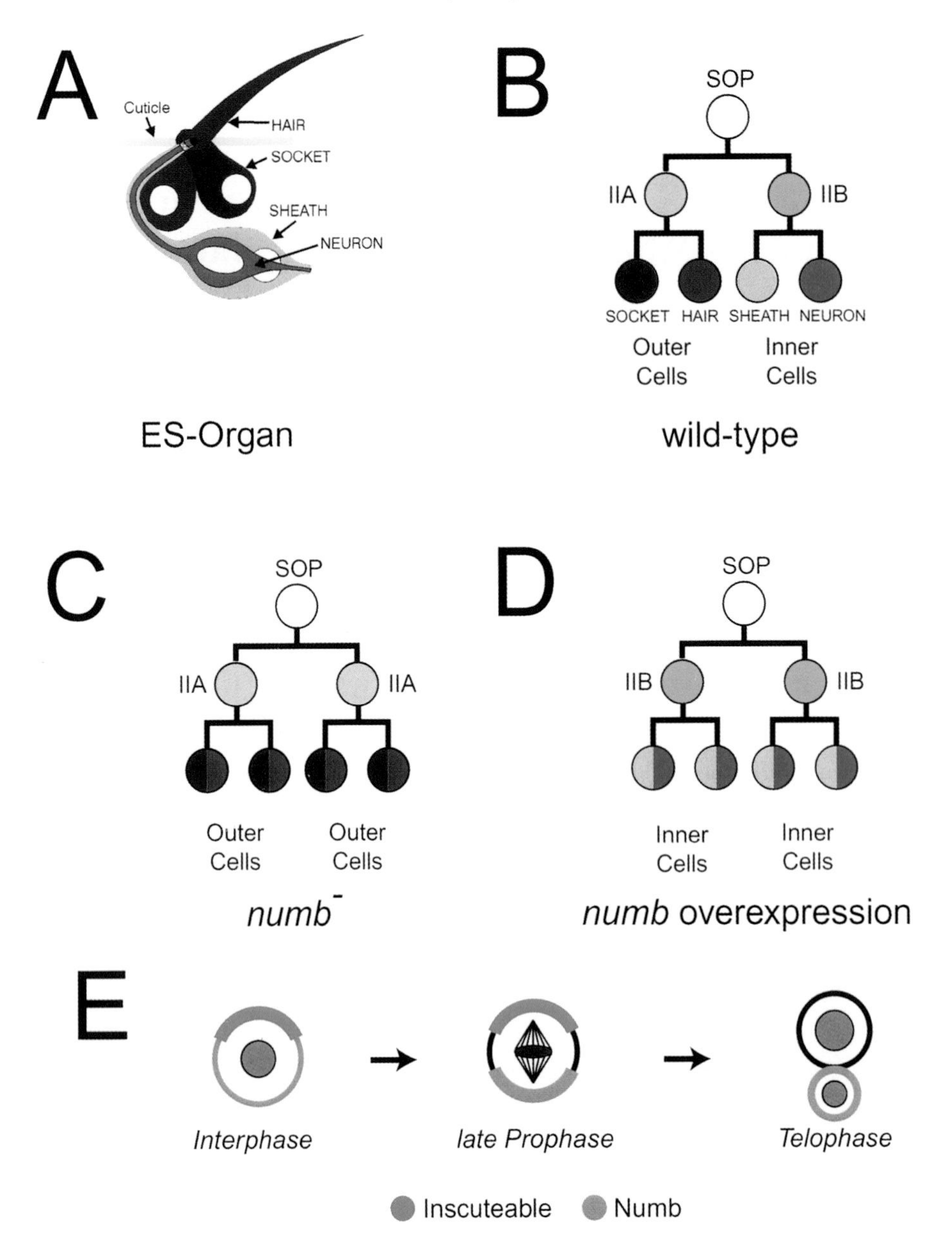

Figure 1. (*A*) Schematic view of a *Drosophila* external sensory (ES) organ. ES organs consist of two outer cells (hair and socket) and two inner cells (neuron and sheath). (*B*) In a wild-type embryo, the sensory organ precursor (SOP) cell divides into the IIA cell and the IIB cell. The IIA cell gives rise to the two outer cells, whereas the IIB cell gives rise to the two inner cells. (*C*) In a *numb* mutant, the SOP cell divides into two IIA cells, which give rise to four outer cells and no inner cells. (*D*) When *numb* is overexpressed, the SOP cell divides into two IIB cells, which give rise to four inner cells and no outer cells. It should be noted that absence and overexpression of numb also lead to defects in the IIA and IIB cell divisions (Uemura et al. 1989; Rhyu et al. 1994) (*E*) Asymmetric localization of Numb and Inscuteable in neuroblasts. In late interphase cells, Inscuteable is localized asymmetrically to the apical cell cortex, but Numb is distributed evenly around the cell membrane. In late prophase, Numb localizes to the basal membrane area opposite the Inscuteable crescent. In telophase, Numb preferentially enters one of the two daughter cells, but asymmetric localization of Inscuteable is no longer detectable.

sent new data on the identification of the Numb localization domain and speculate about possible mechanisms for Numb localization.

Numb LOCALIZATION IS CORRELATED WITH SPINDLE ORIENTATION AND REQUIRES THE *inscuteable* GENE

The asymmetric localization of Numb and Prospero is tightly correlated with the position of one of the centrosomes and the orientation of the mitotic spindle (Knoblich et al. 1995; Spana and Doe 1995). Although the Numb and Prospero crescents in different cell types can have different orientations with respect to the anteroposterior, dorsoventral, and apical-basal axes, they always form in the membrane area overlying one of the two spindle poles. However, the orientation of the mitotic spindle and the localization of Numb and Prospero are independent events: No defects in spindle orientation are observed in *numb* or *prospero* mutants, and neither Numb nor Prospero localization is affected by disruption of the mitotic spindle using the microtubule drug colcemid (Knoblich et al. 1995). These observations have led us to propose that an organizer of asymmetric cell division governs several independent events during asymmetric cell divisions in the *Drosophila* nervous system and provides positional information for both spindle orientation and protein segregation (Knoblich et al. 1995).

The organizer hypothesis has recently been supported by the identification and characterization of the *inscuteable* gene (Kraut and Campos-Ortega 1996; Kraut et al. 1996). Inscuteable provides a molecular link between spindle orientation and Numb and Prospero localization: In wild-type embryos, neuroblasts usually divide with their mitotic spindle oriented along the apical-basal axis and the Numb and Prospero crescents form over the basal centrosome. In *inscuteable* mutants, however, the orientation of the mitotic spindle and the division plane become random and Numb and Prospero are either not asymmetrically localized or their crescents form at random positions (Kraut et al. 1996). Furthermore, the tight correlation between the orientation of the spindle and the orientation of the crescent is lost. Thus, the *inscuteable* gene is required for both spindle orientation and asymmetric protein localization in *Drosophila* neuroblasts. Interestingly, *inscuteable* can also induce a reorientation of the mitotic spindle in cells that normally do not divide along the apical-basal axis: When *inscuteable* is ectopically expressed in epithelial cells of the ectoderm, which normally divide with their mitotic spindles oriented parallel to the epithelial surface, these cells rotate their spindles by 90° and divide perpendicularly to the surface (Kraut et al. 1996). However, *inscuteable* is not sufficient to induce Numb localization in these cells: In both wild-type and transgenic embryos, the Numb protein is present in ectodermal epithelial cells, but during mitosis, it does not specifically segregate into one of the two daughter cells.

Like Numb and Prospero, the Inscuteable protein is localized asymmetrically in neuroblasts (Kraut et al. 1996). However, there are some characteristic differences: In contrast with Numb and Prospero, which become localized in prophase of mitosis and are not asymmetrically localized in interphase cells, the Inscuteable protein is already asymmetrically localized before mitosis. Furthermore, whereas the Numb and Prospero crescents form on the basal side of neuroblasts, Inscuteable localizes to the opposite, apical side. Taken together, these results suggest that in neuroblasts, the Inscuteable protein recognizes some positional information in late interphase and localizes to the apical side of the cell cortex, where it functions in setting up an axis of asymmetry. When the cell enters mitosis, this axis is then recognized both by the machinery that orients the mitotic spindle and by the machinery for Numb and Prospero localization.

What is the nature of the positional information that directs Inscuteable localization? Several results suggest that Inscuteable recognizes epithelial apical-basal polarity for its localization. When ectopically expressed in epithelial cells of the *Drosophila* prospective epidermis, the hindgut epithelium, or the salivary gland epithelium, the Inscuteable protein always localizes to the apical membrane area irrespective of the cells' orientation along the major body axes (Kraut et al. 1996; J.A. Knoblich and Y.N. Jan, unpubl.). Exclusively apical localization is no longer observed when *inscuteable* is ectopically expressed in the prospective epidermis of *crumbs* mutant embryos (Tepass et al. 1990; Wodarz et al. 1993), which have defects in epithelial apical-basal polarity (J.A. Knoblich and Y.N. Jan, unpubl.). Furthermore, SOP cells, which usually divide with their spindle oriented parallel to the epithelial plane (rather than along the apical-basal axis), do not require *inscuteable* for spindle orientation or Numb and Prospero localization (Kraut et al. 1996; J.A. Knoblich and Y.N. Jan, unpubl.), suggesting that other molecules direct and coordinate these processes along the dorsoventral and anteroposterior axes.

At least two mechanisms are possible by which Inscuteable localized to one pole of the cell could direct the localization of Numb and Prospero to the other pole of the cell: Inscuteable could repel Numb and Prospero from the apical membrane area and thus move the proteins to the basal side of the cell. This seems unlikely, both because the Numb and Prospero crescents and the Inscuteable crescent do not appear to touch each other (Kraut et al. 1996) and because Inscuteable localization precedes Numb/Prospero localization. Alternatively, Inscuteable localization could induce the polarization of some cytoskeletal element, which then is used for the transport of Numb and Prospero. Microtubules are an obvious candidate, but drug experiments have shown that they are not required for asymmetric localization (Knoblich et al. 1995). Actin, in contrast, seems to play a role: After disruption of the actin cytoskeleton with the drug cytochalasin D, the Numb and Prospero crescents frequently form at incorrect positions (Knoblich et al. 1995) and are no longer correlated with the orientation of the mitotic spindle (Knoblich et al. 1997). Treatment with the more potent actin drugs latrunculin A and B completely abolishes Numb and Prospero localization (Knoblich et al. 1997). A function of Inscuteable in polarizing the actin cytoskeleton is therefore plausible, although it remains possible that there are other mechanisms for directing Numb and Prospero localization (see below).

THE AMINO TERMINUS OF THE NUMB PROTEIN DIRECTS ITS ASYMMETRIC LOCALIZATION

Whereas Inscuteable directs the asymmetric localization of Numb to the basal side of a dividing cell, it is unlikely that Inscuteable itself is a component of the Numb localization machinery. To identify this machinery, it would be helpful to know the functional domains of the Numb protein that mediate asymmetric localization. We have therefore carried out a deletion analysis of the Numb protein (Knoblich et al. 1997). Numb has no overall homology to other known proteins. However, the protein contains an amino-terminal phosphotyrosine binding

(PTB) domain (Bork and Margolis 1995) and the very amino terminus of Numb matches the consensus for *N*-myristoylation (Uemura et al. 1989). We have constructed a series of Numb deletions and have determined their subcellular distribution both in RNA injection experiments and in transgenic flies expressing these deletion constructs. The results are summarized in Figure 2. Most of the Numb deletion constructs behave identically to the full-length protein: In all cells, the proteins are found exclusively at the cell membrane, and in dividing neural precursor cells, they localize asymmetrically into a crescent overlying the basal one of the two spindle poles. However, deletion of the very amino terminus of the Numb protein or the amino-terminal half of the PTB domain affects the subcellular localization in a characteristic and reproducible way: In both interphase and mitotic cells, the proteins are found mostly in the cytoplasm, and during mitosis, no signs of asymmetric localization can be detected. Thus, the amino terminus of the Numb protein is required for both cell membrane association and asymmetric localization. To test a possible function of *N*-myristoylation in the membrane localization of the Numb protein, we have mutated the second amino acid of the Numb protein from glycine to alanine, a mutation that destroys the *N*-myristoylation signal and has been shown to completely abolish *N*-myristoylation in other proteins. In RNA injection experiments, this mutation has no effect on membrane association or asymmetric localization, indicating that *N*-myristoylation is not essential for Numb localization. We have also determined whether the amino terminus of the Numb protein can induce asymmetric localization of a heterologous protein. Although fusion of the first 41 amino acids of Numb to the LacZ protein results in a mostly cytoplasmic fusion protein, fusions containing the first 76 or 119 amino acids of Numb are localized to the cell membrane with no signs of asymmetric localization during mitosis. Only a fusion protein containing the first 226 amino acids of Numb, including the complete PTB domain, is localized asymmetrically in dividing neural precursor cells.

The membrane localization domain of the Numb protein is thus located at the very amino terminus: The amino-terminal-most 76 amino acids of the protein are both required and sufficient for localization to the cell membrane. The situation for asymmetric localization is more complex. Although deletion of the Numb PTB domain does not affect its asymmetric localization, this domain is necessary for asymmetric localization when

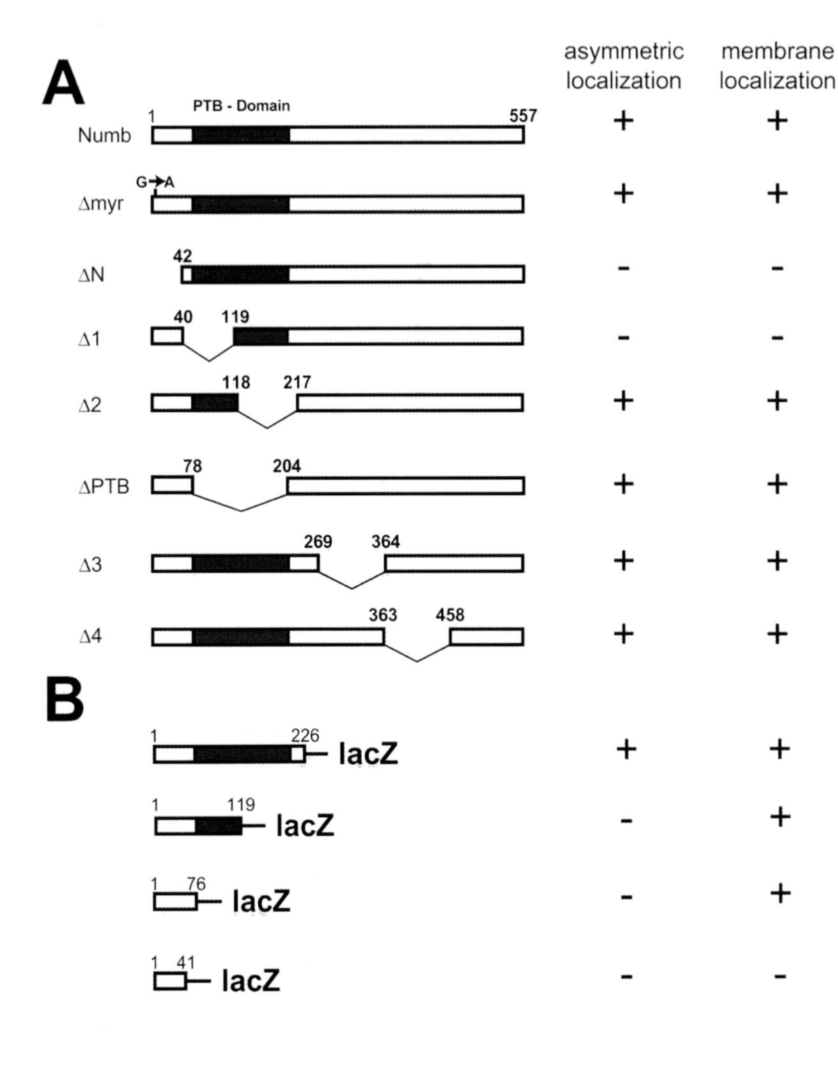

Figure 2. The amino terminus of the Numb protein directs its asymmetric localization. (*A*) Seven deletion constructs and one point mutant of the Numb protein were assayed in RNA injection experiments for their ability to localize asymmetrically in dividing neural precursor cells and to segregate preferentially into one daughter cell. The results for ΔN and ΔPTB were verified in transgenic flies expressing these deletion constructs. (*B*) Four fusion constructs of amino-terminal fragments of the Numb protein and LacZ were tested for their ability to localize asymmetrically in RNA injection experiments and transgenic flies. The details of these experiments are published elsewhere (Knoblich et al. 1997).

amino-terminal fragments of Numb are fused to LacZ. Several explanations are possible: The Numb localization domain could be located at the amino terminus of the protein but might not be able to function correctly, when fused to LacZ, either because correct folding requires other parts of Numb or because steric problems inhibit binding of essential factors. Alternatively, Numb might contain two redundant localization domains, including one within the PTB domain and another in a more carboxy-terminal region of the protein. This could account for the observation that the PTB domain is only required for asymmetric localization in LacZ fusions of Numb without the carboxyl terminus.

POSSIBLE MECHANISMS FOR THE ASYMMETRIC LOCALIZATION OF NUMB

No Numb-associated protein has yet been shown to be involved in asymmetric localization. We can therefore only speculate about the mechanism that leads to the asymmetric localization of Numb in mitotic cells. Below, we present and discuss three models for Numb localization that involve directed transport along cytoskeletal elements, binding to a prelocalized anchor protein, or capping as the underlying mechanism (Fig. 3).

The directed transport model. In this model, the Numb protein is bound to a molecular motor either constitutively or in a cell-cycle-dependent fashion. During prophase, movement of this motor along some cytoskeletal element transports Numb to the basal side of neuroblasts. The fact that microtubules are not required for the asymmetric localization of Numb (Knoblich et al. 1995) leaves actin or some other cytoskeletal structure as possible candidates for this cytoskeletal element. In fact, actin/myosin-based transport has been suggested as the mechanism for transport of determinants during asymmetric cell divisions in other organisms. During *C. elegans* development, the asymmetric localization of the proteins PAR1, PAR2, and PAR3 (Etemad-Moghadam et al. 1995; Guo and Kemphues 1995; Boyd et al. 1996) require NMY-2, a nonmuscle myosin II (Guo and Kemphues 1996b). In yeast, Ash1p is produced in only one of the two daughter cells (Bobola et al. 1996; Sil and Herskowitz 1996); this asymmetry requires the type V myosin She1p/Myo4p (Jansen et al. 1996). However, in *Drosophila*, the process of Numb localization is relatively insensitive to actin drugs. Cytochalasin D at a concentration that completely inhibits cytokinesis does not inhibit Numb crescent formation, even though the Numb crescent frequently forms at aberrant positions (Knoblich et al. 1995). Complete disruption of actin filaments with the more powerful drugs latrunculin A and B, in contrast, inhibits Numb localization (Knoblich et al. 1997) and leaves open the possibility of directed transport along actin fibers. The function of Inscuteable in this model would be to direct the movement of the motor. Inscuteable could either control the polarity of the cytoskeletal element used for movement or directly act on the motor and influence directionality or speed.

The anchor model. In this model, the Numb protein is freely diffusible within the cell membrane in interphase cells, but it aquires affinity for a prelocalized anchor dur-

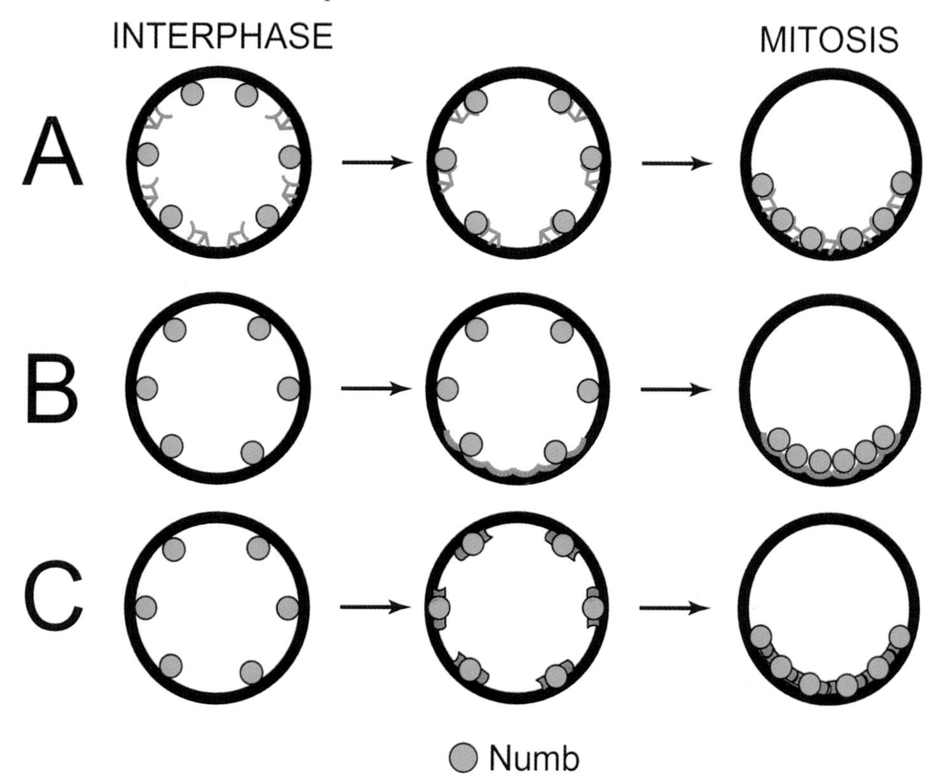

Figure 3. Three models for the asymmetric localization of the Numb protein during mitosis. (*A*) The directed transport model. (*B*) The anchor model. (*C*) The capping model. See text for details.

ing mitosis. Lateral diffusion of membrane-associated proteins is a rapid process, and within a short time, all Numb protein will be captured by the localized anchor. The time difference between Inscuteable and Numb localization could be explained by assuming that Inscuteable repels an anchor molecule from the apical membrane area in late interphase, and subsequently Numb acquires affinity for this molecule during prophase. This model is reminiscent of some aspects of *oskar* RNA localization to the posterior pole of the *Drosophila* oocyte. Even though the initial localization of *oskar* RNA is thought to involve microtubule-dependent transport (Theurkauf et al. 1993; Pokrywka and Stephenson 1995; Clark et al. 1997), both localization during later stages of oogenesis and the maintenance of *oskar* RNA localization seem to be mediated by binding of the RNA to a localized anchor (Glotzer et al. 1997). The requirement for tropomyosin in this process (Erdelyi et al. 1995) suggests involvement of the actin cytoskeleton, even though the localization is insensitive to cytochalasins (Pokrywka and Stephenson 1995).

The capping model. When membrane proteins that are normally distributed homogeneously in the cell membrane are cross-linked by bivalent antibodies, they aggregate into large patches and ultimately form a cap on one side of the cell (de Petris and Raff 1973). Either directed membrane flow or directed flow of the membrane actin cytoskeleton has been suggested as the underlying mechanism (Bretscher 1996). However, capping so far has been observed only in cells capable of locomotion, and the speed of the process is slow compared with Numb localization. Thus, this model entails a novel, capping-like movement of Numb that is directed by Inscuteable.

The data available do not allow us to distinguish between these three models. It is also possible that Numb localization occurs by a combination of these mechanisms. In fact, the results from our deletion analysis would be compatible with the existence of two localization domains that mediate Numb localization by two different localization mechanisms. Given the feasibility of genetic analysis in *Drosophila*, more information to distinguish these different possibilities should become available soon.

ACKNOWLEDGMENTS

We thank the members of the Jan Lab for helpful discussions and critical comments on the manuscript. J.A.K. is supported by the Howard Hughes Medical Institute and L.Y.J. and Y.N.J. are HHMI investigators.

REFERENCES

Bobola N., Jansen R.P., Shin T.H., and Nasmyth K. 1996. Asymmetric accumulation of Ash1p in postanaphase nuclei depends on a myosin and restricts yeast mating-type switching to mother cells. *Cell* **84:** 699.

Bodmer R., Carretto R., and Jan Y.N. 1989. Neurogenesis of the peripheral nervous system in *Drosophila* embryos: DNA replication patterns and cell lineages. *Neuron* **3:** 21.

Bork P. and Margolis B. 1995. A phosphotyrosine interaction domain. *Cell* **80:** 693.

Boyd L., Guo S., Levitan D., Stinchcomb D.T., and Kemphues K.J. 1996. PAR-2 is asymmetrically distributed and promotes association of P granules and PAR-1 with the cortex in *C. elegans* embryos. *Development* **122:** 3075.

Bretscher M.S. 1996. Getting membrane flow and the cytoskeleton to cooperate in moving cells. *Cell* **87:** 601.

Chang F. and Drubin D.G. 1996. Cell division: Why daughters cannot be like their mothers. *Curr Biol* **6:** 651.

Clark I.E., Jan L.Y., and Jan Y.N. 1997. Reciprocal localization of Nod and kinesin fusion proteins indicates microtubule polarity in the *Drosophila* oocyte, epithelium, neuron and muscle. *Development* **124:** 461.

de Petris S. and Raff M.C. 1973. Normal distribution, patching and capping of lymphocyte surface immunoglobulin studied by electron microscopy. *Nat. New Biol.* **241:** 257.

Erdelyi M., Michon A.M., Guichet A., Glotzer J.B., and Ephrussi A. 1995. Requirement for *Drosophila* cytoplasmic tropomyosin in *oskar* mRNA localization. *Nature* **377:** 524.

Etemad-Moghadam B., Guo S., and Kemphues K.J. 1995. Asymmetrically distributed PAR-3 protein contributes to cell polarity and spindle alignment in early *C. elegans* embryos. *Cell* **83:** 743.

Glotzer J.B., Saffrich R., Glotzer M., and Ephrussi A. 1997. Cytoplasmic flows localize injected *oskar* RNA. *Curr. Biol.* **7:** 326.

Guo S. and Kemphues K.J. 1995. *par-1*, a gene required for establishing polarity in *C. elegans* embryos, encodes a putative Ser/Thr kinase that is asymmetrically distributed. *Cell* **81:** 611.

———. 1996a. Molecular genetics of asymmetric cleavage in the early *Caenorhabditis elegans* embryo. *Curr. Opin. Genet. Dev.* **6:** 408.

———. 1996b. A non-muscle myosin required for embryonic polarity in *Caenorhabditis elegans*. *Nature* **382:** 455.

Hirata J., Nakagoshi H., Nabeshima Y., and Matsuzaki F. 1995. Asymmetric segregation of the homeodomain protein Prospero during *Drosophila* development. *Nature* **377:** 627.

Horvitz H.R. and Herskowitz I. 1992. Mechanisms of asymmetric cell division: Two B's or not two B's, that is the question. *Cell* **68:** 237.

Jansen R.P., Dowzer C., Michaelis C., Galova M., and Nasmyth K. 1996. Mother cell-specific *HO* expression in budding yeast depends on the unconventional myosin myo4p and other cytoplasmic proteins. *Cell* **84:** 687.

Knoblich J.A., Jan L.Y., and Jan Y.N. 1995. Asymmetric segregation of Numb and Prospero during cell division. *Nature* **377:** 624.

———. 1997. The N-terminus of the *Drosophila* Numb protein directs membrane association and actin-dependent localization. *Proc. Natl. Acad. Sci.* (in press).

Kraut R. and Campos-Ortega J.A. 1996. *inscuteable*, a neural precursor gene of *Drosophila*, encodes a candidate for a cytoskeleton adaptor protein. *Dev. Biol.* **174:** 65.

Kraut R., Chia W., Jan L.Y., Jan Y.N., and Knoblich J.A. 1996. Role of *inscuteable* in orienting asymmetric cell divisions in *Drosophila*. *Nature* **383:** 50.

Pokrywka N.J. and Stephenson E.C. 1995. Microtubules are a general component of mRNA localization systems in *Drosophila* oocytes. *Dev. Biol.* **167:** 363.

Rhyu M.S., Jan L.Y., and Jan Y.N. 1994. Asymmetric distribution of numb protein during division of the sensory organ precursor cell confers distinct fates to daughter cells. *Cell* **76:** 477.

Shapiro L. and Losick R. 1997. Protein localization and cell fate in bacteria. *Science* **276:** 712.

Sil A. and Herskowitz I. 1996. Identification of asymmetrically localized determinant, Ash1p, required for lineage-specific transcription of the yeast *HO* gene. *Cell* **84:** 711.

Spana E.P. and Doe C.Q. 1995. The Prospero transcription factor is asymmetrically localized to the cell cortex during neuroblast mitosis in *Drosophila*. *Development* **121:** 3187.

Spana E.P., Kopczynski C., Goodman C.S., and Doe C.Q. 1995. Asymmetric localization of Numb autonomously determines sibling neuron identity in the *Drosophila* CNS. *Development* **121:** 3489.

Tepass U., Theres C., and Knust E. 1990. *crumbs* encodes an EGF-like protein expressed on apical membranes of *Drosophila* epithelial cells and required for organization of epithelia. *Cell* **61:** 787.

Theurkauf W.E., Alberts B.M., Jan Y.N., and Jongens T.A. 1993. A central role for microtubules in the differentiation of *Drosophila* oocytes. *Development* **118:** 1169.

Uemura T., Shepherd S., Ackerman L., Jan L.Y., and Jan Y.N. 1989. *numb*, a gene required in determination of cell fate during sensory organ formation in *Drosophila* embryos. *Cell* **58:** 349.

Wodarz A., Grawe F., and Knust E. 1993. *crumbs* is involved in the control of apical protein targeting during *Drosophila* epithelial development. *Mech. Dev.* **44:** 175.

On the Roles of *inscuteable* in Asymmetric Cell Divisions in *Drosophila*

W. CHIA, R. KRAUT, P. LI, X. YANG, AND M. ZAVORTINK
Institute of Molecular and Cell Biology, National University of Singapore, Singapore 119076

Asymmetric cell divisions (for review, see Horvitz and Herskowitz 1992) in which one stem cell divides to produce two progeny that adopt distinct cellular identities are utilized in a variety of developmental contexts (for review, see Posakony 1994; Jan and Jan 1995) and organisms including vertebrates (Chenn and McConnell 1995), invertebrates (Sternberg and Horvitz 1984), and plants (for review, see Stauger and Doonan 1993) to generate cellular diversity. The generation of asymmetric sibling cell fates can be mediated extrinsically, through Notch signaling (for review, see Artavanis-Tsakonas et al. 1995); intrinsically, through asymmetrically localized and segregated protein determinants (for review, see Doe and Spana 1995; Campos-Ortega 1996); or both mechanisms can work together (Guo et al. 1996; Spana and Doe 1996).

Here, we are concerned with how intrinsic determinants effect asymmetric cell divisions in the developing *Drosophila* nervous system. Two proteins, Numb (Uemura et al. 1989; Rhyu et al. 1994) and Prospero (Hirata et al. 1995; Knoblich et al. 1995; Spana and Doe 1995), fulfill some of the criteria expected for intrinsic determinants of asymmetric cell division. Both proteins are produced by *Drosophila* neural stem cells, asymmetrically localized during stem cell mitosis, and both partition preferentially to only one of the two postmitotic daughter cells. Numb is localized asymmetrically as a cortical crescent during division of sensory organ precursor cells in the developing peripheral nervous system (PNS). Numb subsequently segregates into one of the two daughter cells (Rhyu et al. 1994) where it remains associated with the cellular cortex. *numb* is required to effect distinct sibling cell fates. In *numb* loss-of-function mutants, both sibling cells adopt the fate of the progeny which normally fails to inherit Numb (Uemura et al. 1989); when Numb is overexpressed, both siblings adopt the identity of the cell which normally inherits Numb (Rhyu et al. 1994). *numb* also appears to act in an analogous manner to mediate the distinct cellular identities of the sibling neurons derived from the central nervous system (CNS) midline precursor, MP2 (Spana et al. 1995). Numb functions by directly interacting with and suppressing Notch activity in the daughter cells that inherit Numb in both the SOP (Frise et al. 1996; Guo et al. 1996) and MP2 cell lineages (Spana et al. 1995; Spana and Doe 1996).

Although Numb acts as an intrinsic determinant in the asymmetric cell divisions of SOPs in the PNS and MP2 in the CNS, *numb* loss of function does not result in severe defects in CNS development, and there is no convincing evidence that it has a general role in mediating the asymmetric cell division of neuroblasts. Prospero (Pros), on the other hand, is required in ganglion mother cells (GMCs) to activate genes required to specify GMC identity (Doe et al. 1991) and to repress genes that are normally expressed in neuroblasts (Vaessin et al. 1991). *pros* is transcribed in all neuroblasts and encodes a homeodomain-containing transcription factor (Doe et al. 1991; Vaessin et al. 1991; Matsuzaki et al. 1992). Pros protein localization is both cell-type- and cell-cycle-dependent. In neuroblasts, Pros is associated with the cellular cortex, as an apical crescent in late interphase and as a basal cortical crescent throughout mitosis (Hirata et al. 1995; Knoblich et al. 1995; Spana and Doe 1995). At the end of telophase, Pros is partitioned into the more basal daughter cells that will become the GMCs. A 119-amino-acid region of Pros (residues 825–943) appears to be sufficient to impart asymmetric localization and segregation when fused in-frame to β-galactosidase. Shortly following cell division, Pros is released from the cortex and translocates into the nuclei of the GMCs (Hirata et al. 1995; Knoblich et al. 1995; Spana and Doe 1995).

If the asymmetrically localized protein determinants are to be partitioned into just one progeny following cell division, the processes of mitotic spindle orientation and protein localization must be coordinated in the progenitor cell. In wild-type neuroblasts, not only does Pros colocalize with Numb throughout mitosis, but the position of the Numb and Pros crescents is tightly correlated with the orientation of the mitotic spindle, such that the basal protein crescents always overlie one of the two centrosomes (Knoblich et al. 1995; Spana and Doe 1995). Consequently, Numb and Pros are always segregated to the more basal daughter, the GMC. Nevertheless, this strong correlation between protein localization and mitotic spindle orientation can be decoupled following drug treatment, suggesting that these processes are independent (Knoblich et al. 1995). One possible mechanism by which the two independent processes of protein localization and mitotic spindle orientation might normally be tightly coupled might be because they both respond to the same positional information. So which molecule might be providing or interpreting this positional information?

A second issue, raised by the localized Numb and Pros (and Inscuteable) protein crescents, is whether their respective RNAs might also be asymmetrically localized. A mechanism that is generally employed in a variety of cell types to localize proteins to specific subcellular regions or compartments is to localize their mRNAs (for review, see St. Johnston 1995). The importance of mRNA localization as a mechanism to localize proteins has been particularly elegantly demonstrated by the genetic analysis of the *Drosophila* body pattern (for review, see St. Johnston and Nüsslein-Volhard 1992). For example, the formation of an anterior Bicoid morphogen gradient necessary for the formation of the head and thorax is effected through the localization of *bicoid* mRNA to the anterior pole of the egg (Driever and Nüsslein-Volhard 1988a,b); similarly, the localization of *oskar* mRNA to the posterior pole of the oocyte defines the site of accumulation of both Osk protein and the posterior determinant Nanos necessary for abdominal development (Ephrussi et al. 1991; Kim-Ha et al. 1991; Lehmann and Nüsslein-Volhard 1991; Wang and Lehman 1991).

Here, we summarize the evidence that the gene *inscuteable* (*insc*) acts to organize the asymmetric cell division of neuroblasts by providing the positional information required to mediate several independent processes, including protein localization, RNA localization, and mitotic spindle orientation.

NOTEWORTHY FEATURES OF THE DEDUCED INSCUTEABLE PROTEIN SEQUENCE

P-element insertions near the 5′-noncoding region of *insc* were independently isolated in two separate screens. AB44 was identified (A. Beerman and J.A. Campos-Ortega, pers. comm.) in a screen for enhancer trap lines that express β-galactosidase in embryonic neuroblasts. Similarly, P180 was identified in a screen of P-element lethal lines which affect the number of RP2 neurons (X. Yang and W. Chia, unpubl.) in a antibody screen using the Anti-Even-Skipped antibody (Frasch et al. 1986; Patel et al. 1994). Cloning and analysis of both genomic and cDNA from the region of the AB44 insertion have shown that the transcription unit associated with a 4.2-kb embryonic transcript is disrupted by the AB44 insertion (Kraut and Campos-Ortega 1996), whereas P180 is inserted upstream of this transcription unit (S. Bahri and W. Chia, unpubl.). Imprecise excision events induced by mobilizing AB44 (Kraut and Campos-Ortega 1996) and P180 (S. Bahri, unpubl.) have produced small deletions that remove all ($insc^{P72}$) or part ($insc^{P49}$ and $insc^{E70}$) of the genomic region encoding the 4.2-kb transcript. Subsequent to the analyses of these mutants, we realized that *insc* was in fact allelic to two previously described mutations, *not enough muscles* (Burchard et al. 1995) and *fata-morgana* (Kania et al. 1995).

Sequence analysis of the 4.2-kb *insc* transcript has shown that it encodes an 859-amino-acid deduced protein (Kraut and Campos-Ortega 1996). The following are the main features of the deduced sequence: (1) A putative binding site for a WW domain, PPPPPPY at residues 236–242 (a WW domain is a module that, like the SH3 domains, binds proline-rich motifs but prefers a run of prolines followed by a tyrosine [see, e.g., Staub et al. 1996]). (2) A putative nuclear localization sequence, RRGVFFNDAKIERRRYL at residues 405–421; however, the significance of this sequence is unclear since we have never observed nuclear localization of Insc protein in any context. (3) At the carboxyl terminus of Insc is a putative PDZ-binding site, QESFV-COOH at residues 855–859. Many PDZ domain-containing proteins localize to submembranous sites (see, e.g., Ponting et al. 1997), which is consistent with the observed subcellular localization of Insc (see below). However, the putative PDZ-binding site does not appear to be essential for the apical localization of Insc or for its ability to cause realignment of the mitotic spindle along the apical/basal axis when expressed in epithelial cells (see below); an ectopically expressed Insc protein containing a Myc epitope tag at its carboxyl terminus (which should disrupt the PDZ-binding site) can nevertheless localize to the apical surface and cause reorientation of mitotic spindle (M. Zavortink, unpubl.). (4) Kraut and Campos-Ortega (1996) also point out that Insc shares 23% identity and 45% similarity with human ankyrin; in addition, a sequence motif that is related to the so-called ankyrin repeats found in many proteins is repeated five times in Insc, at residues 258–290, 329–362, 449–481, 496–528, and 541–573. The region of Insc containing the ankyrin repeat motifs appears to be required for mitotic spindle orientation and for Insc apical localization (M. Zavortink, unpubl.).

THE INSCUTEABLE PROTEIN EXPRESSION PATTERN

Insc is expressed in many cell types; some types are known to express and localize Numb and/or Pros (Kraut and Campos-Ortega 1996; Kraut et al. 1996). Here, we summarize the embryonic expression pattern. Insc expression is first observed in the procephalic neurogenic region (PNR) of the ectoderm (in the head) shortly following gastrulation. The Insc-expressing ectodermal cells in the head comprise a subset of the ectodermal cells that form a monolayered epithelium. Whereas most of the cells in this epithelium divide with their mitotic spindles oriented parallel to the surface of the epithelium, the Insc-expressing cells of the PNR, which will give rise to the larval brain, divide with their mitotic spindles oriented perpendicular to this surface. Both Numb and Pros segregate into the basal daughter cell (Kraut et al. 1996). Later during embryogenesis, Insc is expressed in neuroblasts in the developing segmented CNS. In contrast to Insc-expressing cells in the PNR, neuroblasts delaminate from the ectoderm prior to mitosis. However, like the Insc-expressing cells of the PNR, neuroblasts in the segmented portion of the CNS also express Numb and Pros and they divide with their mitotic spindles oriented perpendicular to the apical surface of the embryo, with Numb and Pros segregating into the more basal daughter cell. Insc is also expressed in the developing PNS in the sensory organ precursors that also asymmetrically localize Numb and

Pros (Knoblich et al. 1995; Spana et al. 1995). In addition, Insc expression is seen in precursor cells in the midgut primordium (Kraut and Campos-Ortega 1996), which probably correspond to the cells described by Spana and Doe (1995) and Hirata et al. (1995) that asymmetrically localize and segregate Pros. Insc is expressed in muscle progenitor cells (A. Carmena et al., unpubl.) which do not express Pros. The role of *insc* in asymmetric cell divisions has been best characterized in neuroblasts and in the cells of the PNR.

INSC PROTEIN IS ASYMMETRICALLY LOCALIZED IN A CELL-CYCLE-DEPENDENT MANNER IN THE THE ECTODERMAL CELLS OF THE PNR AND IN NEUROBLASTS

The basal cortical localization of Numb and Pros is cell-cycle-regulated and occurs only during mitosis. Basal protein crescents first become apparent during prophase and persist until telophase, at which time both proteins segregate into the basal daughter cell (Hirata et al. 1995; Knoblich et al. 1995; Spana and Doe 1995). The first indication that *insc* might be involved in the localization of Numb and Pros was the observation that the Insc protein was itself asymmetrically localized in neuroblasts, as protein crescents on the apical side of the cortex, opposite to the side where Numb and Pros are localized (Kraut and Campos-Ortega 1996). The cell cycle dependence of Insc localization in neuroblasts has been examined by double staining wild-type embryos with an antibody that specifically recognizes Insc and a DNA stain (Lundell and Hirsh 1994) that enables the progression of these cells through mitosis to be followed (Fig. 1). Insc protein crescents become apparent only when neuroblasts delaminate from the ventral ectoderm. Staining is first seen in the form of a stalk which represents the apical surface of delaminating neuroblasts that retain contact with the epithelial surface (Fig. 1a). Since neuroblasts replicate their DNA prior to delamination (Hartenstein et al. 1994), delaminating neuroblasts are late interphase cells. In fully delaminated interphase neuroblasts, Insc remains associated with the apical cell cortex in the form of a protein crescent (Fig. 1b). Shortly following delamination, neuroblasts enter mitosis and divide, with their mitotic spindle oriented perpendicular to the apical surface of the embryo (Hartenstein et al. 1994; Spana and Doe 1995). The apical Insc crescent remains during prophase and metaphase (Fig. 1c,d), but becomes hard to detect by late anaphase (Fig. 1e). We do not know whether this is due to delocalization or degradation of the protein. Thus, unlike Numb and Pros, Insc does not appear to be preferentially segregated into one of the daughter cells. Insc localization in cells of the PNR shows a similar cell cycle dependence (Kraut et al. 1996).

The dependence of Insc localization on elements of the cytoskeleton has been examined by treating embryos with cytochalasin D and colcemid (Kraut et al. 1996). Disruption of actin filaments by treatment with cytochalasin D results in the failure to form Insc crescents in mitotic (metaphase) neuroblasts (Fig. 1f–g). This effect produced by cytochalasin D is probably not an artifact caused by cell death or damage since Pros was still asymmetrically localized as cortical crescents in the same cells (Fig. 1g). However, the position of the Pros crescent is incorrect, i.e., not basal, in about half the neuroblasts scored. Thus, with respect to Pros localization, cytochalasin-D treatment produces effects similar to *insc* loss of function (see below), consistent with the notion that Insc must be apically localized in order to

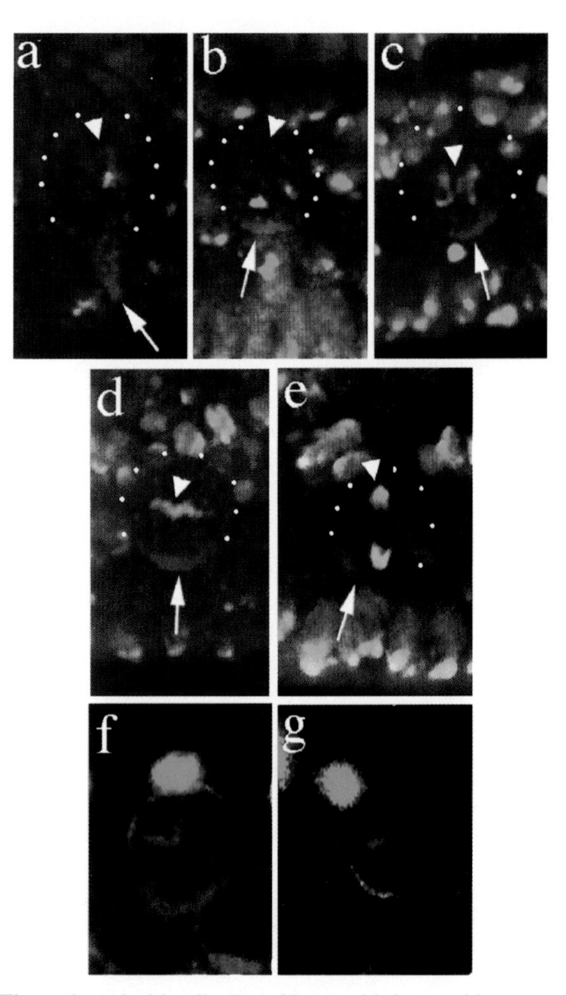

Figure 1. Apical localization of Inscuteable in neuroblasts precedes entry into mitosis and is dependent on the actin cytoskeleton. (*a–e*) Parasagittal optical sections through the ventral ectoderm of stage-9–10 embryos stained for Insc protein (*red, arrows*) and DNA (*green, arrowheads*). Apical is down. Neuroblasts are outlined by dots. (*a*) Delaminating neuroblast; (*b*) delaminated late interphase neuroblast; (*c*) prophase neuroblast; (*d*) metaphase neuroblast; (*e*) anaphase neuroblast (see text); (*f–g*) cytochalasin-D treatment disrupts Insc localization. After treatment with 1 mg/ml cytochalasin D for 30 min, Insc (*red*) becomes delocalized in metaphase cells (DNA in *blue*), whereas Pros (*green*) is still localized into a crescent (*g*). Identical treatment without the drug does not disrupt Insc localization (*f*). After cytochalasin-D treatment, Insc is delocalized in 20 of 21 pro- or metaphase neuroblasts, but localized in 21 of 22 control neuroblasts. Pros is asymmetrically localized in all of the drug-treated and control neuroblasts. However, the Pros crescents are misoriented in 47% of the neuroblasts after cytochalasin-D treatment (14% in the controls). (Reprinted, with permission, from Kraut et al. 1996 [copyright Macmillan].)

function. Destruction of microtubules with colcemid has no apparent affect on the formation of Insc apical crescents. Thus, Insc localization is dependent on the intact microfilament and precedes both mitosis and the asymmetric localization of Numb and Pros. It is interesting to note that although the localization of determinants in oocytes appears to be dependent on the intact microtubule, the asymmetric localization of a variety of localized proteins found in the somatic stem cells of a range of organisms appears to depend on intact microfilament (see, e.g., Goodner and Quatrano 1993; Guo and Kemphues 1996).

A REQUIREMENT FOR *insc* IN THE ASYMMETRIC LOCALIZATION OF NUMB AND PROSPERO

Mitotic neuroblasts in *insc* mutant embryos exhibit defects in Numb and Pros localization (Figs. 2 and 3). In wild-type neuroblasts, Numb and Pros colocalize as basal crescents during mitosis and partition into the more basal GMC daughter (Fig. 2a–c, g–i). Quantitation of the defects seen with Pros localization in *insc* null mutants has shown that 24% of the dividing mutant neuroblasts fail to form a cortical crescent with Pros distributed throughout their cortex; Pros cortical crescents do form in the majority (76%) of mutant mitotic neuroblasts; however, these crescents are localized to random positions on the cortex (Figs. 2d–f and 3b). Similar defects in Numb localization are also seen in the mutant neuroblasts (Fig. 2j–1).

The effect of *insc* loss of function on protein localization in the cells of the PNR which normally express Insc is more dramatic (Kraut et al. 1996). Prior to mitosis in wild-type embryos, these cells are morphologically indistinguishable from other ectodermal cells which do not express Insc. However, in contrast to most other ectodermal cells, these normally Insc-expressing cells divide with their mitotic spindle oriented perpendicular to the apical surface of the embryo. Both Numb and Pros localize asymmetrically as basal crescents and segregate into the basal daughter cell. The cells in the PNR of *insc* null embryos fail to localize Numb and Pros during mitosis. This mutant phenotype occurs with complete expressivity in these cells. Both proteins are localized throughout the cell cortex, but no Numb or Pros crescents can be detected and the proteins segregate into both daughter cells. These observations indicate that the correct localization of Numb and Pros in neuroblasts and cells of the PNR requires *insc*.

A ROLE FOR *insc* IN MITOTIC SPINDLE ORIENTATION

In wild-type neuroblasts, there is a tight correlation between the position of the colocalized Numb and Pros crescents and the orientation of the mitotic spindle

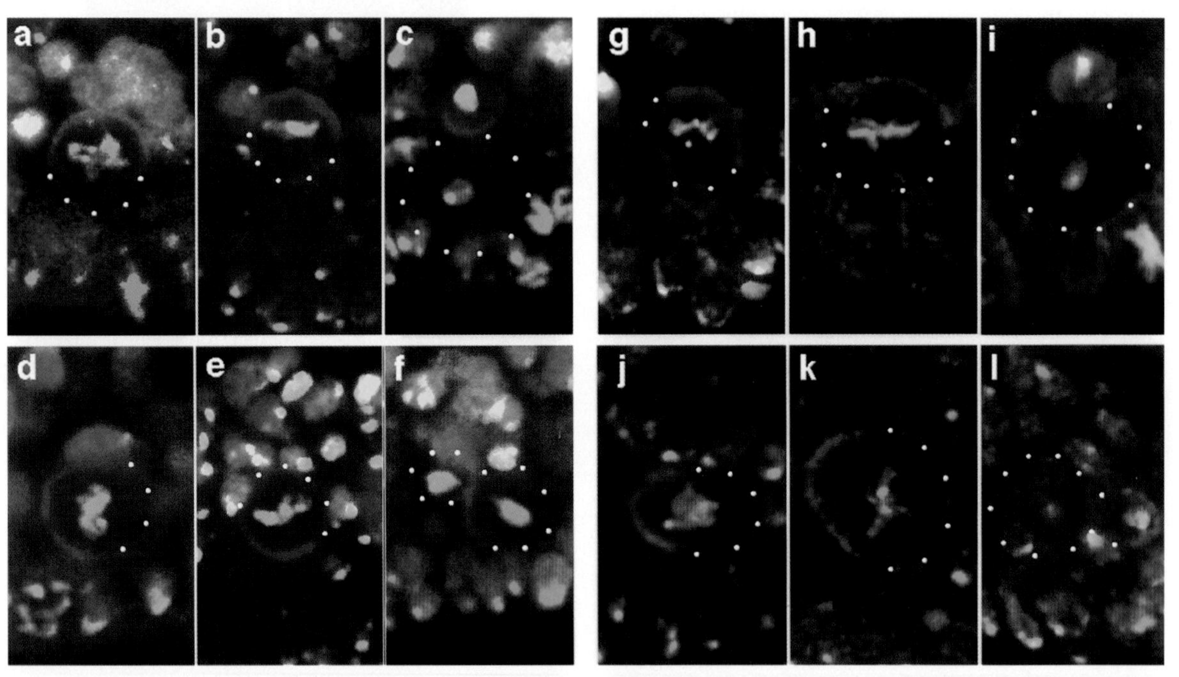

Figure 2. Basal cortical localization of Prospero and Numb in neuroblasts requires *inscuteable*. (*a–l*) Parasagittal optical sections through the ventral ectoderm of stage-10–11 embryos showing different neuroblasts (*dotted line*) at various stages of division. Apical is at the bottom. (*a–c*) Pros (*red*) and DNA (*green*) in wild-type neuroblasts. During mitosis (*a–c*), Pros is localized basally (*up*) at prophase (*a*) and metaphase (*b*) and segregates into the basally located ganglion mother cell at telophase (*c*). Note that divisions are oriented apicobasally. (*d–f*) Pros (*red*) and DNA (*green*) in *insc*^{P49} neuroblasts. A Pros crescent is found at random, nonbasal locations (*d,e,f*) or Pros is delocalized and found throughout the cortex (not shown). Note that division is misoriented and Pros sometimes fails to become localized to the ganglion mother cell as in *f*. (*g–i*) Numb (*red*) and DNA (*green*) in wild-type neuroblasts. Like Pros, Numb is also localized basally during mitosis and segregates into the ganglion mother cell (*i*). (*j–l*) Numb (*red*) and DNA (*green*) in *insc*^{P49} neuroblasts. Numb, like Pros, is not always basally localized (*j,k,l*) and division is likewise misoriented (*l*). Similar results are obtained for *insc*^{P72}. (Reprinted, with permission, from Kraut et al. 1996 [copyright Macmillan].)

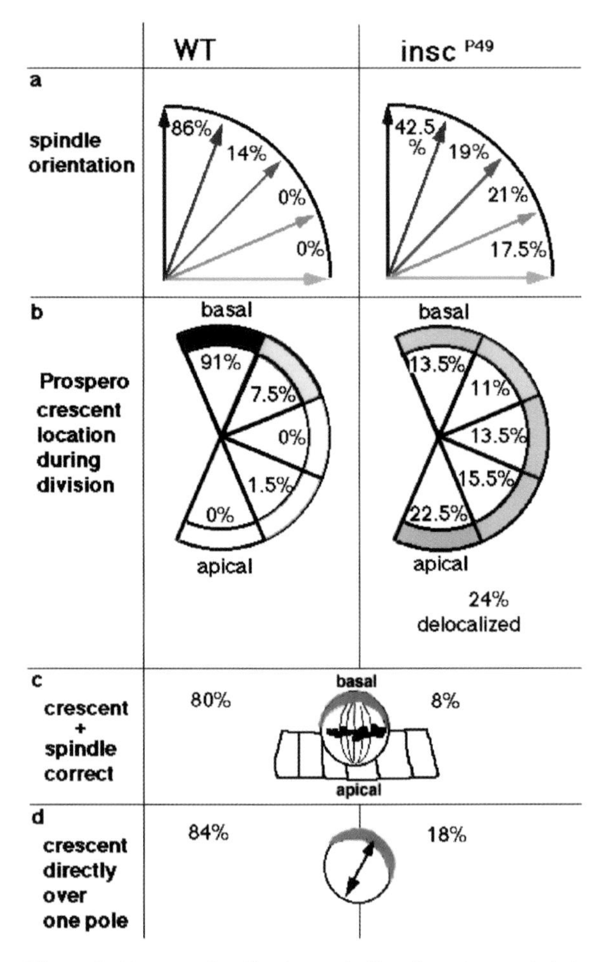

Figure 3. Prospero localization, spindle orientation, and their correlation are perturbed in dividing mutant neuroblasts. (*a*) In wild-type neuroblasts, most mitotic spindles are oriented perpendicular to the plane of the overlying ectoderm ($n = 50$); in $insc^{P49}$ embryos, spindle orientation is less ordered ($n = 52$). (*b*) Pros crescents in wild-type neuroblasts are primarily basally oriented ($n = 68$); in mutant neuroblasts, Pros often fails to form a crescent (24%) and crescents that do form are oriented randomly ($n = 111$). (*c*) In 80% of wild-type neuroblasts ($n = 50$), the division axis is apical-basal and the Pros crescent is localized directly over the basal pole; in $insc^{P49}$, only 8% of the neuroblasts show this configuration ($n = 52$). (*d*) In mutant neuroblasts, the orientation of the division axis and the localization of the Pros crescent are no longer tightly coupled. In wild type, the Pros crescent is localized directly over one spindle pole (regardless of the division axis) in 84% of dividing neuroblasts ($n = 50$) vs. only 18% ($n = 52$) in mutants. (Reprinted, with permission, from Kraut et al. 1996 [copyright Macmillan].)

(Knoblich et al. 1995; Spana and Doe 1995). An analysis of mitotic spindle orientation and Pros localization in dividing neuroblasts of mutant embryos has indicated that *insc* not only is required for both processes, but is also required to coordinate these processes. The orientation of the mitotic spindle in dividing neuroblasts can be deduced by double staining embryos for centrosomes and DNA. In wild-type neuroblasts, one of the two centrosomes migrates to the basal side after delamination (Spana and Doe 1995), and the mitotic spindle is aligned perpendicular to the epithelium (Fig. 4a,b).

In *insc* mutant embryos, while the centrosomes still migrated to opposite poles, their mitotic spindle failed to become oriented along the apical basal axis in most mutant neuroblasts (Fig. 4c,d). Statistical analyses reveal that spindle orientation is less ordered in mutant neuroblasts (Fig. 3a). Moreover, whereas in wild-type neuroblasts the Pros crescent overlies one of the spindle poles, this rarely occurs in mutant neuroblasts (Fig. 3c,d). Hence, the tight correlation between the position of the Pros crescent and the orientation of the mitotic spindle, seen in wild-type neuroblasts, also requires *insc*.

A requirement for *insc* for apical/basal orientation of the mitotic spindle has also been seen in the ectodermal cells of the PNR (Kraut et al. 1996). In wild-type embryos, in contrast to neuroblasts, the mitotic spindle is established parallel to the epithelial surface in the Insc-expressing cells of the PNR. Early in mitosis, however, the spindle reorients by 90°, leading to its alignment along the apical-basal axis (Foe 1989). This reorientation of the mitotic spindle in these cells does not occur in *insc* mutants (Kraut et al. 1996). These observations demonstrate that *insc* is required for the apical/basal orientation of mitotic spindles in both neuroblasts and cells of the PNR. Additional experiments have shown that *insc* also has an instructive role in orienting mitotic spindles of ectodermal cells. The GAL4 system (Brand and Perrimon 1993)

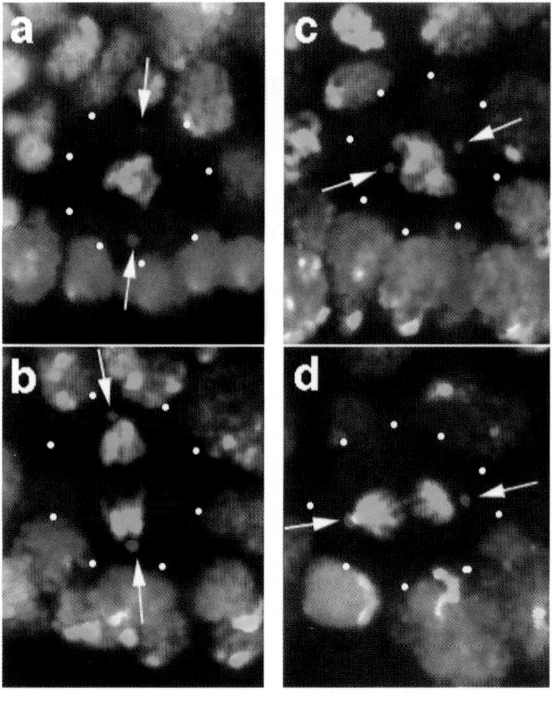

Figure 4. Correct orientation of the mitotic spindle in neuroblasts requires *inscuteable*. (*a–d*) Parasagittal optical sections through the ventral ectoderm of stage-10–11 wild-type (*a,b*) and $insc^{P49}$ (*c,d*) embryos, double stained for centrosomes (*red, arrows*) and DNA (*green*), showing the orientation of mitosis in dividing neuroblasts. Apical is down. (*a,b*) In wild-type embryos, mitotic spindles are aligned in an apical-basal or nearly apical-basal orientation in essentially all neuroblasts. (*c,d*) In *insc* mutants, spindles are oriented more randomly (see *a*). Note that in *insc* mutants, as in wild type, the centrosomes migrate to opposite poles of the cell. (Reprinted, with permission, from Kraut et al. 1996 [copyright Macmillan].)

was used to express *insc* ectopically in the expression pattern of the pair rule gene *hairy* (Kraut et al. 1996). This enables Insc to be expressed in ectodermal cells which do not normally express Insc; these cells do not reorient their mitotic spindle and divide with their spindle aligned parallel to the epithelial surface. In embryos in which Insc is ectopically expressed in these normally Insc-negative ectodermal cells, the ectopically produced protein is found on the apical cell cortex; moreover, these cells will reorient their mitotic spindle and divide with their spindle aligned perpendicular to the epithelial surface (Kraut et al. 1996). Therefore, Insc expression in ectodermal cells early during embryogenesis appears to be both necessary and sufficient for reorientation of their mitotic spindle and division plane.

BOTH *insc* AND *pros* RNA*s* ARE ASYMMETRICALLY LOCALIZED IN NEUROBLASTS

Since Insc and Pros proteins are localized as cortical crescents on the apical/basal regions of the NB cell cortex in a cell-cycle-dependent manner, it seemed logical to ask whether this asymmetric localization of the proteins is also reflected at the level of their respective RNAs. The location of *pros* and *insc* RNAs at various stages of the neuroblast cell cycle has been determined by whole-mount in situ hybridization experiments (Tautz and Pfeifle 1989) with single-stranded anti-sense RNA probes using stage-10–11 wild-type embryos (P. Li et al., unpubl.). The embryos were doubly labeled with a DNA stain (Lundell and Hirsh 1994) to follow the progression of neuroblasts through mitosis. The results indicate that both *pros* and *insc* RNAs are also asymmetrically localized with a cell cycle dependence that parallels the localization of the respective proteins they encode. Hence, both the Pros protein (Fig. 5a) and *pros* RNA (Fig. 5d) are localized as crescents to the apical cortex during late interphase, although in the case of the RNA, the localization can be apical but is apparently cytoplasmic rather than cortical. From prophase to telophase, both Pros protein (Fig. 5b,c) and RNA (Fig. 5e,f) are localized as basal cortical crescents. Ultimately, both the *pros* RNA and protein segregate preferentially into the more basally located daughter cell, the GMC.

The *insc* RNA (Fig. 6c), like the protein it encodes (Fig. 6a), is also asymmetrically localized to the apical cortex of neuroblasts at interphase. However, in contrast to the Insc protein that remains apically localized from prophase (Fig. 6b) to metaphase, *insc* RNA is found in the cytoplasm when neuroblasts enter mitosis (Fig. 6d). Therefore, *insc* RNA is cortical only during late interphase, whereas *pros* RNA is associated primarily with the cell cortex even during mitosis.

INSC IS REQUIRED FOR THE LOCALIZATION OF *pros* RNA TO THE BASAL CORTEX OF NEUROBLASTS DURING MITOSIS

Since we have previously suggested (Kraut et al. 1996) that *insc* might be providing or interpreting positional information required for processes associated with the asymmetric neuroblast cell division, such as localization of Pros (and Numb) protein to the basal cortex and orientation of the mitotic spindle along the apical/basal axis, it seems possible that *insc* might be playing an analogous part for *pros* RNA localization. This possibility was examined by comparing *pros* RNA localization between wild-type and *insc* mutant neuroblasts at various stages of the cell cycle. In wild-type neuroblasts, *pros* RNA localization changes from the apical cortex to the basal cortex between late interphase and prophase (Fig. 5d,e) which mirrors the change in the Pros protein localization (see Fig. 5a–b); this basal cortical localization of *pros* RNA remains throughout mitosis with the RNA preferentially segregating to the GMC at telophase (Fig. 5f). In *insc* mutant neuroblasts, this dramatic change in the localization of the *pros* RNA during the interphase to prophase transition does not occur, and *pros* RNA remains primarily on the apical cortex during mitosis (Fig. 5g–i). These observations suggest that whereas the apical localization of *pros* RNA during interphase does not require *insc* function, the basal localization during mitosis does. Therefore, in addition to its requirement for correct spindle orientation and protein localization, *insc* also has a role in the localization of *pros* RNA.

These observations suggest that the apical localization and the basal localization of *pros* RNA are mediated by two independent mechanisms. The apical localization during interphase is mediated by an *insc* (and *staufen*, see below)-independent mechanism. In contrast, the basal localization of *pros* RNA during mitosis requires *insc* (and *staufen*, see below). These observations also imply that *insc* is required, probably indirectly, for moving *pros* RNA to the basal cortex during the start of mitosis. In the absence of *insc* function, the *pros* RNA fails to move and stays associated with the apical cortex. The finding that *pros* RNA continues to be localized primarily to the apical cortex when *insc* is absent has several implications. First, the *insc*-independent mechanism that operates to effect localization to the apical cortex during interphase is normally overridden by the *insc*-dependent process during mitosis. Second, cues distinct from those provided by *insc* must be used to read the apical basal axis of the neuroblast during interphase; this earlier cue may act, along with apical vectorial transport from the nucleus, to localize initially a variety of RNAs to the apical region during interphase (see, e.g., Davis and Ish-Horowicz 1991; Davis et al. 1993; Francis-Lang et al. 1996). Conceivably, the apical cortex of NBs may serve as an initial distribution point for proteins and RNAs that may ultimately be targeted to a different final destination. Finally, since the interphase apical RNA localization is largely unchanged in mutant NB, it is possible to eliminate models in which *pros* RNA is transported along astral microtubules stabilized at the apical surface by interaction with Insc.

A ROLE FOR STAUFEN IN THE LOCALIZATION OF *pros* RNA

It seems likely that Insc is not directly involved in the basal localization of *pros* RNA, because when the RNA

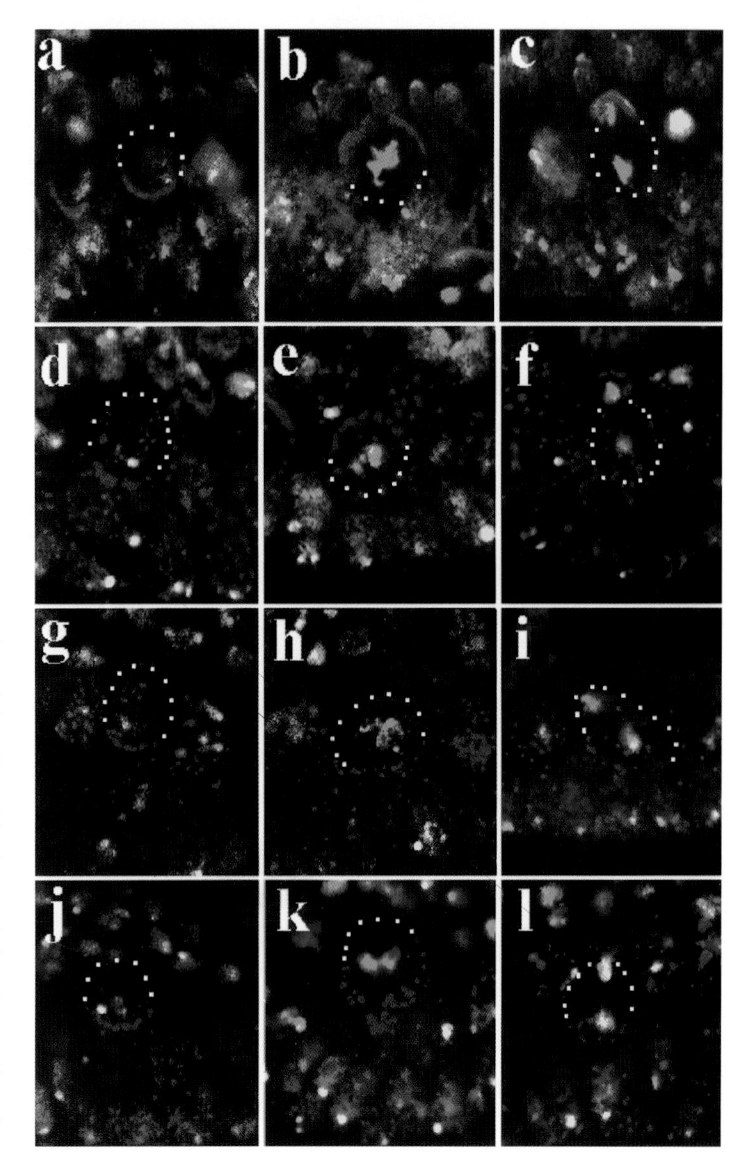

Figure 5. *insc* and *stau* are required for the change in *prospero* RNA localization from the apical to the basal cortex as neuroblasts enter mitosis. In all panels, *white dots* delineate the boundary of the cells, and apical is toward the bottom. DNA staining is in *green*. In wild-type neuroblasts, Pros protein (*a–c, red*) can be seen as an apical cortical crescent during late interphase (*a*); however, in mitotic neuroblasts (*b*, prophase and *c*, telophase), Pros protein crescents localize to the basal cortex. The localization of *pros* RNA (*red, d–l*) is also cell-cycle-dependent and reflects the Pros protein localization in wild-type embryos (*d–f*). In interphase neuroblasts (*d*), *pros* RNA is apically localized; however, as neuroblasts enter mitosis, *pros* RNA becomes basally localized starting at prophase (*e*). The basal localization remains throughout mitosis until at telophase, the cortically localized *pros* RNA segregates primarily into the more basally located daughter, the GMC (*f*). *pros* RNA does not localize to the basal cortex of neuroblasts in *insc* (*g–i*) and *stau* (*j–l*) mutant embryos. In mutant interphase neuroblasts, the apical localization of *pros* RNA is unchanged (*g,j*); however, in contrast to wild-type neuroblasts, *pros* RNA remains on the apical cortex of mutant neuroblasts from prophase (*h,k*) to telophase (*i,l*).

is basally localized, Insc protein is on the opposite side of the neuroblast. Experiments based on the yeast two-hybrid screen (Finley and Brent 1994) using various portions of Insc as bait have identified Staufen (Stau) as a molecule that interacts with Insc; deletion analyses using both yeast two-hybrid and in vitro protein-binding assays have shown that the two proteins interact via their respective carboxy-terminal regions (P. Li et al., unpubl.). Stau and Insc proteins are both cortically localized in neuroblasts, and during interphase, the two proteins are largely colocalized on the apical cortex of neuroblasts (P. Li et al., unpubl.). Stau is a double-stranded RNA-binding protein that has been implicated in mRNA transport in the oocyte (see below), and we have shown that it can bind to the 3′-untranslated region (UTR) of *pros* RNA in vitro. Hence, a role for *stau* in *pros* RNA localization can be rationalized.

A possible role for *stau* in *pros* RNA localization has been examined and the results are summarized in Figure 5j–l. RNA in situ experiments have shown that the apical cortical localization of *pros* RNA in neuroblasts is unaffected in embryos lacking both maternal and zygotic *stau* (Fig. 5j). However, the change in *pros* RNA localization from the apical cortex at interphase to the basal cortex at prophase fails to occur in mutant neuroblasts (Fig. 5k). Further analysis has shown that only the zygotic component of *stau* is required for the apical to basal change in *pros* RNA localization, consistent with previous observations (St. Johnston et al. 1991) which indicated that maternal Stau protein becomes undetectable in somatic tissues by the time of neurogenesis. The *stau* loss-of-function phenotype is indistinguishable from that seen in *insc* mutants with respect to *pros* RNA localization. Hence, we can conclude that the apical cortical localization of *pros* RNA during interphase requires neither *insc* nor *stau* function; the basal cortical localization during mitosis requires both *insc* and *stau* functions. However, *stau* is not required for the other functions that are mediated by *insc*. In the absence of *stau*, Pros (and Numb) localization and mitotic spindle

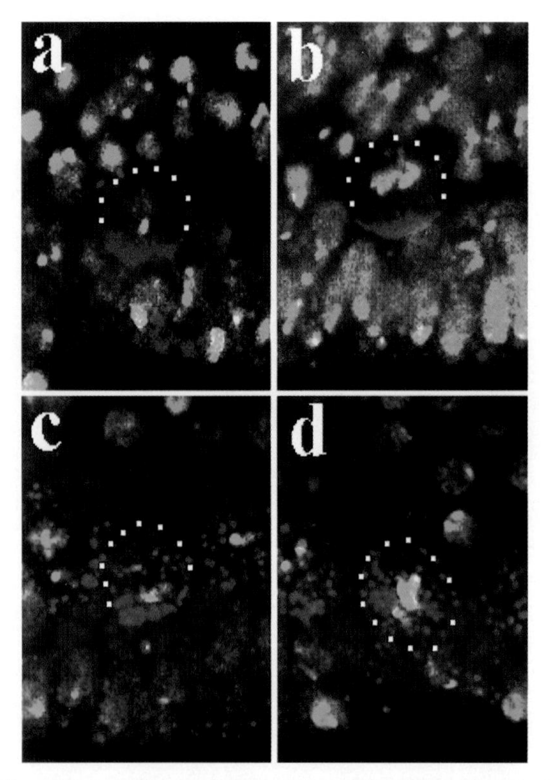

Figure 6. *insc* RNA and Insc protein both localize to the apical cell cortex in wild-type interphase neuroblasts. In all panels, *white dots* delineate the boundary of the cells, and apical is toward the bottom. DNA staining is in *green*. In wild-type embryos (*a–d*), Insc protein (*red, a* and *b*) is localized as an apical cortical crescent during late interphase (*a*) and prophase (*b*). *insc* RNA (*red, c* and *d*) is localized as an apical cortical crescent only during interphase (*c*); during prophase, it is found in the cytoplasm (*d*).

orientation in neuroblasts remain wild type in *stau* null embryos (P. Li et al., unpubl.). Therefore, we envision *stau* as a downstream partner of *insc* involved in effecting only one aspect, that of RNA localization, of the several processes that require *insc*.

Since *insc* is unlikely to be directly involved in changing the localization of *pros* RNA (see above), can such a role be envisioned for *stau*? Stau has been implicated in having a direct role in the localization and transport of *bicoid* and *oskar* RNAs in the oocyte. Localization of *bcd* mRNA to the anterior pole is a complex multistep process that requires *stau* for the final step (St. Johnston et al. 1989, 1991). Signals located in the 3′ UTR of *bcd* can recruit Stau to form ribonucleoprotein (RNP) particles which are subsequently transported in a process that requires intact microtubules (Ferrandon et al. 1994). *stau* is also required for the anterior to posterior transport of *osk* mRNA in the oocyte (for review, see St. Johnston 1995). Consistent with a direct role in mRNA transport, Stau contains five copies of a double-stranded RNA-binding motif, binds to double-stranded RNA in vitro (St. Johnston et al. 1992), and colocalizes with *osk* mRNA to the posterior and *bcd* mRNA to the anterior of the egg (St. Johnston et al. 1991; Ferrandon et al. 1994). It is therefore appealing to think that Stau might have a similar role to directly transport *pros* RNA from the apical to basal cortex as neuroblasts enter mitosis. However, the available evidence does not rule out the possibility that Stau might have an indirect role; for example, a complex including Insc and Stau might be required to facilitate the interaction of *pros* RNA with other factors directly involved in its transport.

SUMMARY

Three processes requiring *insc*, protein localization, RNA localization, and mitotic spindle orientation, appear to be tightly correlated but nevertheless independent. The apical cortical localization of Insc protein occurs prior to mitosis and precedes the basal localization of Pros, Numb, and *pros* RNA. In *insc* null mutants, neuroblasts divide in more random orientations and fail to correctly localize Pros, Numb, and *pros* RNA to their correct position during mitosis. Our results suggest that asymmetric localization of Insc during interphase functions to establish positional information for spindle orientation and the basal localization of Numb, Pros, and *pros* RNA during mitosis. Insc probably interacts with a number of interacting partners in order to effect its various functions. One of its partners, Stau, is involved exclusively in the process of RNA localization.

ACKNOWLEDGMENT

We thank Jose Campos-Ortega for his generosity with unpublished reagents at the start of our efforts to understand the role of *insc* in asymmetric cell divisions; Pat O'Farrell for his insight and advice; Juergen Knoblich, Uttam Surana, and our colleagues in the Chia Lab for some interesting discussions. We also thank Renate Renkawitz-Pohl and Hugo Bellen for bringing *not enough muscles* and *fata-morgana* to our attention.

REFERENCES

Artavanis-Tsakonas A., Matsuno K., and Fortini M.E. 1995. Notch signalling. *Science* **268**: 225.

Brand A. and Perrimon N. 1993. Targeted gene expression as a means of altering cell fates and generating dominant phenotypes. *Development* **118:** 401.

Burchard S., Paululat A., Hinz U., and Renkawitz-Pohl R. 1995. The mutant *not enough muscles* reveals reduction of the *Drosophila* embryonic muscle pattern. *J. Cell Sci.* **108:** 1443.

Campos-Ortega J.A. 1996. Numb diverts Notch pathway off the Tramtrack. *Neuron* **17:** 1.

Chenn A. and McConnell S.K. 1995. Cleavage orientation and the asymmetric inheritance of Notch1 immunoreactivity in mammalian neurogenesis. *Cell* **82:** 631.

Davis I. and Ish-Horowicz D. 1991. Apical localisation of pair-rule transcripts require 3′ sequences and limits protein diffusion in the *Drosophila* blastoderm embryo. *Cell* **67:** 927.

Davis I., Francis-Lang H., and Ish-Horowicz D. 1993. Molecular mechanisms of intracellular transcript localisation and export in early *Drosophila* embryos. *Cold Spring Harbor Symp. Quant. Biol.* **58:** 793.

Doe C.Q. and Spana E. 1995. A collection of cortical crescents: Asymmetric protein localisation and CNS precursor cells. *Neuron* **15:** 991.

Doe C.Q., Chu-LaGraff Q., Wright D.M., and Scott M.P. 1991. The *prospero* gene specifies cell fate in the *Drosophila* central nervous system. *Cell* **65:** 451.

Driever W. and Nüsslein-Volhard C. 1988a. A gradient of Bicoid protein in *Drosophila* embryos. *Cell* **54:** 83.

———. 1988b. The Bicoid protein determines position in the *Drosophila* embryo in a concentration dependent manner. *Cell* **54:** 95.

Ephrussi A., Dickinson L.K., and Lehmann R. 1991. *oskar* organizes the germ plasm and directs localisation of the posterior determinant *nanos*. *Cell* **66:** 37.

Ferrandon D., Elphick L., Nüsslein-Volhard C., and St. Johnston D. 1994. Staufen protein associates with the 3′ UTR of *bicoid* mRNA to form particles that move in a microtubule dependent manner. *Cell* **79:** 1121.

Finley R.L., Jr. and Brent R. 1994. Interaction trap cloning with yeast. In *Gene probes: A practical approach* (ed. D. Hames and D. Glover). IRL Press, Oxford.

Foe V. 1989. Mitotic domains reveal early commitment of cells in *Drosophila* embryos. *Development* **107:** 1.

Francis-Lang H., Davis I., and Ish-Horowicz D. 1996. Asymmetric localisation of *Drosophila* pair-rule transcripts from displaced nuclei: Evidence for directional nuclear transport. *EMBO J.* **15:** 640.

Frasch M., Hoey T., Rushlow C., Doyle H., and Levine M. 1986. Characterisation and localisation of the even-skipped protein of *Drosophila. EMBO J.* **6:** 749.

Frise E., Knoblich J.A., Younger-Shepherd S., Jan L.Y., and Jan Y.N. 1996. The *Drosophila* Numb protein inhibits signaling of the Notch receptor during cell-cell interaction in sensory organ lineage. *Proc. Natl. Acad. Sci.* **93:** 11925.

Goodner B. and Quatrano R.S. 1993. Fucus embryogenesis, a model to study the establishment of polarity. *Plant Cell* **5:** 1471.

Guo M., Jan L.Y., and Jan Y.N. 1996. Control of daughter cell fates during asymmetric division: Interaction of Numb and Notch. *Neuron* **17:** 27.

Guo S. and Kemphues K. 1996. Molecular genetics of asymmetric cleavage in the early *Caenorhabditis elegans* embryo. *Curr. Opin. Genet. Dev.* **6:** 408.

Hartenstein V., Younossi-Hartenstein A., and Lekven A. 1994. Delamination and division in the *Drosophila* neuroectoderm: Spatialtemporal pattern, cytoskeletal dynamics, and common control by neurogenic and segmeny polarity genes. *Dev. Biol.* **165:** 480.

Hirata J., Nakagoshi H., Nabeshima Y., and Matsuzaki F. 1995. Asymmetric segregation of a homeoprotein, prospero, during cell division in neural and endodermal development. *Nature* **377:** 627.

Horvitz H.R. and Herskowitz I. 1992. Mechanisms of asymmetric cell divisions: Two Bs or not two Bs, that is the question. *Cell* **68:** 237.

Jan Y.N. and Jan L.Y. 1995. Maggot's hair and bug's eye: Role of cell interactions and intrinsic factors in cell fate specification. *Neuron* **14:** 1.

Kania A., Salzberg A., Bhat M., D'Evelyn D., He Y., Kiss I., and Bellen H. 1995. P-element mutations affecting embryonic peripheral nervous system development in *Drosophila melanogaster. Genetics* **139:** 1663.

Kim-Ha J., Smith J.L., and Macdonald P.M. 1991. *oskar* mRNA is localised to the posterior pole of the *Drosophila* oocyte. *Cell* **66:** 23.

Knoblich J.A., Jan L.Y., and Jan Y.N. 1995. Localisation of Numb and Prospero reveals a novel mechanism for asymmetric protein segregation during mitosis. *Nature* **377:** 624.

Kraut R. and Campos-Ortega J.A. 1996. *inscuteable*, a neural precursor gene of *Drosophila*, encodes a candidate for a cytoskeleton adaptor protein. *Dev. Biol.* **174:** 65.

Kraut R., Chia W., Jan L.Y., Jan Y.N., and Knoblich J.A. 1996. Role of *inscuteable* in orienting asymmetric cell divisions in *Drosophila. Nature* **383:** 50.

Lehmann R. and Nüsslein-Volhard C. 1991. The maternal gene *nanos* has a central role in the posterior pattern formation of the *Drosophila* embryo. *Development* **112:** 679.

Lundell M.J. and Hirsh J. 1994. A new visible light DNA fluorochrome for confocal microscopy. *BioTechniques* **16:** 434.

Matsuzaki F., Koizumi K., Hama C., Yoshioka T., and Nabeshima Y. 1992. Cloning of the *Drosophila prospero* gene and its expression in ganglion mother cells. *Biochem. Biophys. Res. Commun.* **182:** 1326.

Patel N., Condron B., and Zinn K. 1994. Pair-rule expression patterns of *even-skipped* are found in both short and long germ beetles. *Nature* **367:** 429.

Ponting C.P., Phillips C., Davies K., and Blake D.J. 1997. PDZ domains: Targeting signalling molecules to sub-membranous sites. *BioEssays* **19:** 469.

Posakony J.W. 1994. Nature versus nurture: Asymmetric cell divisions in *Drosophila* bristle development. *Cell* **76:** 415.

Rhyu M.S., Yan L.Y., and Jan Y.N. 1994. Asymmetric distribution of Numb protein during division of the sensory organ precursor cell confers distinct fates to daughter cells. *Cell* **76:** 477.

Spana E. and Doe C.Q. 1995. The Prospero transcription factor is asymmetrically localised to the cell cortex during neuroblast mitosis in *Drosophila. Development* **121:** 3187.

———. 1996. Numb antagonises Notch signaling to specify sibling neuron cell fate. *Neuron* **17:** 21.

Spana E., Kopczynski C., Goodman C.S., and Doe C.Q. 1995. Asymmetric localisation of Numb autonomously determines sibling neuron identity in the *Drosophila* CNS. *Development* **121:** 3489.

Staub O., Dho S., Henry P.C., Correa J., Ishikawa T., McGlade J., and Rotin D. 1996. WW domains of Nedd4 bind to the proline-rich PY motifs in the epithelial Na^+ channel deleted in Liddle's syndrome. *EMBO J.* **15:** 2371.

Stauger C. and Doonan J. 1993. Cell division in plants. *Curr. Opin. Cell Biol.* **5:** 226.

Sternberg P.W. and Horvitz H.R. 1984. The genetic control of cell lineage during nematode development. *Annu. Rev. Genet.* **18:** 489.

St. Johnston D. 1995. The intracellular localisation of mRNAs. *Cell* **81:** 161.

St. Johnston D. and Nüsslein-Volhard C. 1992. The origin of pattern and polarity in the *Drosophila* embryo. *Cell* **68:** 201.

St. Johnston D., Beuchle D., and Nüsslein-Volhard C. 1991. *staufen*, a gene required to localised maternal RNAs in the *Drosophila* egg. *Cell* **66:** 51.

St. Johnston D., Brown N.H., Gall J.G., and Jantsch M. 1992. A conserved double-stranded RNA-binding domain. *Proc. Natl. Acad. Sci.* **89:** 10979.

St. Johnston D., Driever W., Berleth T., Richstein S., and Nüsslein-Volhard C. 1989. Multiple steps in the localisation of *bicoid* mRNA to the anterior pole of the *Drosophila* oocyte. *Development Suppl.* **107:** 13.

Tautz D. and Pfeifle C. 1989. In situ hybridisation to embryos with nonradioactive probes. In Drosophila: *A laboratory manual* (ed. M. Ashburner), p. 194. Cold Spring Harbor Laboratory Press, Cold Spring Harbor, New York.

Uemera T., Shepherd S., Ackerman L., Jan L.Y., and Jan Y.N. 1989. *numb*, a gene required in determination of cell fate during sensory organ formation in *Drosophila* embryos. *Cell* **5:** 349.

Vaessin H., Grell E., Wolff E., Bier E., Jan L.Y., and Jan Y.N. 1991. *prospero* is expressed in neuronal precursors and encodes a nuclear protein that is involved in the control of axonal outgrowth in *Drosophila. Cell* **67:** 941.

Wang C. and Lehmann R. 1991. Nanos is the localised posterior determinant in *Drosophila. Cell* **66:** 637.

Maternal Genes with Localized mRNA and Pattern Formation of the Ascidian Embryo

S. YOSHIDA, Y. SATOU, AND N. SATOH
Department of Zoology, Graduate School of Science, Kyoto University, Sakyo-ku, Kyoto 606-01, Japan

It has been 110 years since, in 1887, the French biologist Laurent Chabry carried out blastomere destruction experiments with ascidian eggs, the first such effort in the history of embryology. Chabry destroyed one blastomere of a 2-cell *Ascidiella* embryo and found that the remaining blastomere continued to cleave as if it were half of the whole embryo and eventually formed a half-larva instead of a complete dwarf larva. After obtaining a similar result using a 4-cell embryo, he reached the conclusion that ascidian embryos could not compensate for missing parts and that the developmental pattern was thus a "mosaic." Such a mosaic nature of ascidian embryogenesis is dependent on maternal information that is prelocalized in the egg cytoplasm and segregated into certain lineages of blastomere. Since then, ascidian eggs and embryos have provided an appropriate experimental system to investigate the molecular nature of localized maternal factors and their roles in cell specification and pattern formation (Jeffery and Swalla 1990; Satoh 1994; Satoh et al. 1996).

The fertilized egg of ascidians develops quickly into a tadpole larva. The ascidian tadpole consists of about 2600 cells, which constitute a small number of distinct tissues including epidermis, central nervous system with two sensory organs, spinal (nerve) cord, endoderm, mesenchyme, notochord, and muscle (cf. Satoh 1994). The lineage of these embryonic cells is completely described up to the gastrula stage (Conklin 1905; Nishida 1987). In addition to classic descriptive and experimental studies (Reverberi 1971), recent studies have obtained convincing evidence of maternal factors or determinants responsible for the differentiation of muscle (Nishida 1992; Marikawa et al. 1994), epidermis (Nishida 1994a), and endoderm (Nishida 1993), factors for the establishment of the anteroposterior (A/P) axis of the embryo (Nishida 1994b), and those for the initiation of gastrulation (Jeffery 1990; Nishida 1996). In particular, the posterior-vegetal cytoplasm of the fertilized egg or the so-called myoplasm contains muscle determinants, factors for the A/P axis establishment, and those for the initiation of gastrulation. Several studies have explored the molecular nature of such localized factors (Nishikata et al. 1987; Swalla et al. 1991, 1993; Swalla and Jeffery 1995). In this study, we also attempted an isolation of cDNA clones for maternal genes with localized messages.

ISOLATION OF cDNA CLONES FOR MATERNAL GENES WITH LOCALIZED mRNA

As listed in Table 1, we have isolated cDNA clones for novel maternal genes with localized mRNA from the ascidian *Ciona savignyi* (Yoshida et al. 1996; Y. Satou and N. Satoh, in prep.). The genes were named *posterior end mark (pem)*, because, as shown in Figure 1, the transcript is initially concentrated in the posterior-vegetal cytoplasm of the fertilized egg, and later the distribution of the transcript marks the posterior end of the developing embryo.

pem

The *pem* cDNA clone was isolated with *C. savignyi* egg fragments. Centrifugation of the unfertilized eggs yielded four types of fragments: a large nucleated red fragment and small enucleated black, clear, and brown fragments. Fusion experiments using these fragments revealed that maternal factors for muscle and endoderm differentiation and for the A/P axis establishment are preferentially separated into the black fragments (Marikawa et al. 1994). In addition, the results of UV-irradiation experiments using the black fragments suggested that maternal mRNAs are associated with the activities of these factors (Marikawa et al. 1995). A differential screening of cDNA libraries of black and red fragments yielded a cDNA clone for *pem* (Yoshida et al. 1996). A Northern blot analysis of the poly(A)$^+$ RNA of the red and black fragments revealed a predominance of *pem* mRNA in the black fragments.

The *pem* cDNA predicts a protein of 374 amino acids with a calculated molecular mass of 41 kD (Fig. 2). The amino acid sequence, however, showed no significant similarity to known proteins. It contains no motifs specific for transcriptional factors, integral membrane proteins, or signaling molecules.

As shown in Figure 1A–F, the localization and distribution of *pem* mRNA mark the posterior end of developing embryos. In unfertilized eggs (Fig. 1A), the mRNA was detected in the peripheral cytoplasm except for the animal pole region in which the nucleus is present. In ascidian eggs, fertilization evokes a dynamic rearrangement of the egg cytoplasm called ooplasmic segregation, yielding the establishment of the dorsoventral (D/V) and

Table 1. Maternal Genes with Localized mRNA in the Eggs of the Ascidian *Ciona savignyi*

Gene	Size of cDNA (bp)	Predicted protein (no. of amino acids)	Consensus sequence (motifs and domains)	Accession number
pem	1886	374	no	D83482
pem-2	3266	820	SH3, GDS domains of CDC24 family	AB001770
pem-3	1980	465	KH domains and ring finger	AB001769
pem-4	2281	516	C2H2-type zinc fingers	AB001771
pem-5	2688	768	no	AB001772
pem-6	1646	202	no	AB001773

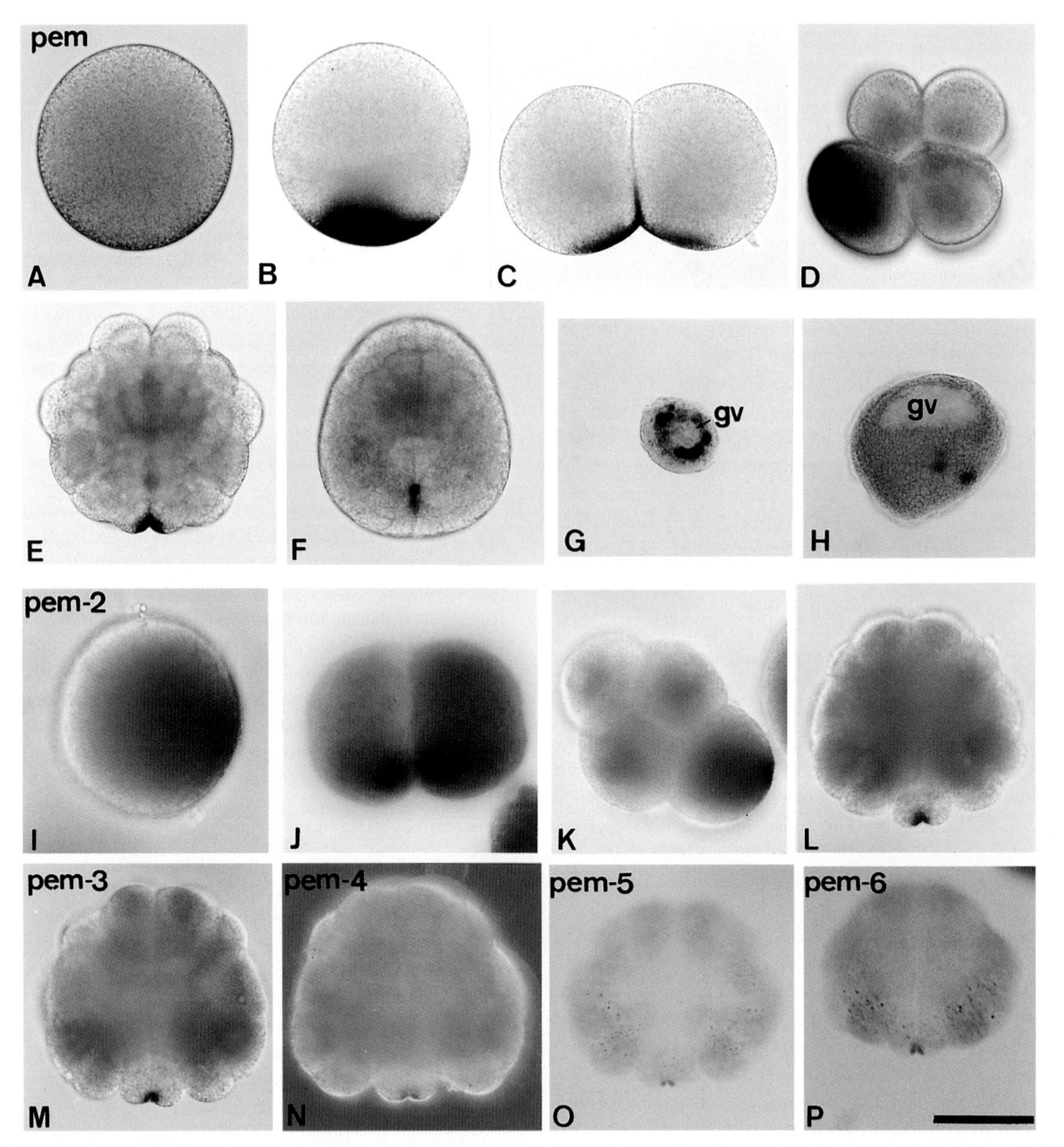

Figure 1. The distribution of maternal mRNA of *pem. pem-2, pem-3, pem-4, pem-5,* and *pem-6* mark the posterior end of developing embryos, as revealed by whole-mount in situ hybridization. Scale bar in P represents 100 μm for all photographs. (*A–F*) Distribution of *pem* maternal mRNA. (*A*) An unfertilized egg, lateral view. The animal pole is up and the vegetal pole is down. (*B*) A fertilized egg after completion of the first phase of ooplasmic segregation, lateral view. (*C*) A 2-cell embryo, animal pole view. (*D*) An 8-cell embryo, lateral view. (*E*) A 32-cell embryo, vegetal pole view. (*F*) An embryo at the neural plate stage, vegetal pole view. (*G,H*) The occurrence and distribution of *pem* mRNA during the process of oogenesis. Oocytes at the early (*G*) and late (*H*) stages of oogenesis. (gv) Germinal vesicle. (*I–L*) Distribution of *pem-2* maternal mRNA. (*I*) A fertilized egg after completion of the second phase of ooplasmic segregation, lateral view. (*J*) A 2-cell embryo, animal pole view. (*K*) An 8-cell embryo, lateral view. (*L*) A 32-cell embryo, vegetal pole view. (*M–P*) Distribution of maternal mRNA of *pem-3* (*M*), *pem-4* (*N*), *pem-5* (*O*), and *pem-6* (*P*) at the 32-cell stage, vegetal pole view.

```
PEM       1:MKMVVPLIESQNVPLVKASPHLPLKRMAACPASQHTIAVPKRSRLTAASEAHLRHLSYNM 60
Ci-PEM   18:MKVVIPMIESQNVPLVRASPHLPVKRMAPCPSSQHFVAVPKRIRLTASSEAHLRQSSYSL 77

PEM      61:AMASPNVVASSRHYGLTASPNPQTPRGLSVVPSNSIRPGNCCGKENLCFGSATSS-TGFS 119
Ci-PEM   78:AMSSPNMVV-AHNYVATPSNISTPNRGLSFVPINTIRSVNCCAKENLCFGQ-TSSATGLS 135

PEM     120:TPPYGSSKPVTYNYAPSASRRSSLKSLVRRSKKRSPYKSRISNKSKRPEKIPEDRVQDSF 179
Ci-PEM  136:TPPYGSQTQNT-HFVQSASRRGSLKQLVRKSKKKSPYKNRVSRKVKRQINIVEESSQDSL 194

PEM     180:QSQPKALVVYNEFLRQLESENQIKNSGKSQNTTKSSGRYSRGFTPSVISTPQHNISPPSS 239
Ci-PEM  195:RGQPKALVVYNEFLRQLESENQGKCTTKTLNTTRASAKYSRGFTPSVISTPQHNASPPLS 254

PEM     240:RESIHPE-STSDYFSAPSSG-A-----SAEDSFVDFLKYIEIGVSSDRRKRLLLNET-SD 291
Ci-PEM  255:REISQIEAATSDYFSAPNSGSASSKISSADDSFVDFLRYIEIGVARN-DKLLLLNEIRNQ 313

PEM     292:-L---KT--SAESS-SSRQ-TKPAKKRFVISTPESEQLSDVRNNNLKSPILPGMYPIGAP 343
Ci-PEM  314:RLNEDKSKFSSGVSLSTQNSSKSVRKPLLTSTPQSDQQSDVINNNRKATALPGMYPIGAP 373

PEM     344:RHASSPVNFKQPA-VSAGGASEFTSAVWRPWC                             374
Ci-PEM  374:RYASSPITFKQQQMVAAGGRIEYNSAVWRPWC                             405
```

Figure 2. Comparison of amino acid sequences of PEM from *Ciona savignyi* and its homolog (*Ci*-PEM) of *C. intestinalis*. (*Shaded boxes*) Identical amino acid sequences.

A/P axes of the embryo. The ooplasmic segregation consists of two distinct phases, the first of which involves rapid movement of the peripheral cytoplasm including the myoplasm (the cytoplasm to be segregated into muscle-lineage blastomeres) to form a transient cap near the vegetal pole of the egg. After the first phase of ooplasmic segregation, the *pem* transcript was found to be concentrated near the vegetal pole (Fig. 1B). During the second phase, the myoplasm shifts from the vegetal pole region to a new position near the subequatorial zone of the egg and forms a crescent, which is a landmark of the posterior side of the future embryo. After the second phase of ooplasmic segregation, the *pem* transcript moved to the subequatorial region to form a crescent-like structure. The cleavage of ascidian eggs is bilaterally symmetrical. The animal pole view of the 2-cell stage embryo shown in Figure 1C indicates that *pem* mRNA was localized in the rather narrow peripheral cytoplasm of the posterior region of the blastomeres. The second cleavage furrow is at a right angle to the first. At the 4-cell stage, the *pem* transcript was restricted to the posterior blastomere, near the posterior-most region and also along the second cleavage furrow. The third cleavage is latitudinal and results in a slightly unequal division. In the 8-cell stage, the embryo distribution of the *pem* transcript was restricted to the posterior region of B4.1 cells (a pair of posterior vegetal blastomeres in the bilaterally symmetrical embryo; Fig. 1D).

The region with positive hybridization signals became narrower as the development proceeded. At the 16-cell stage, *pem* mRNA was found in the posterior-most region of the embryo. Positive hybridization signals for the *pem* transcript were detected in the posterior cytoplasm of B6.3 in the 32-cell stage embryo (Fig. 1E), then in B7.6 of the 64-cell stage embryo. At the gastrula stage, the *pem* transcript was seen in a few invaginating posterior cells (Fig. 1F), and finally, positive hybridization signals were found in a few cells of the ventral side of the tail of the tailbud embryo, presumably endodermal strand cells.

We also examined the occurrence and distribution of *pem* mRNA during the period of oogenesis. As shown in Figure 1G, in oocytes at the early stage of oogenesis, hybridization signals were seen around the germinal vesicle as granular structures. As oogenesis proceeded, the granular structures became separate from the germinal vesicle to the peripheral cytoplasm of oocytes (Fig. 1H). In the full-grown oocytes, such granular structures were no longer detectable. The ascidian egg is enclosed by the chorion and follicle cells. In addition, there are many test cells in the perivitelline space. The distribution of *pem* transcripts in the process of oogenesis reveals that *pem* mRNA is produced by the oocyte itself.

It is worth mentioning here that the early process of the localization of *pem* mRNA described above is distinguishable from that of the myoplasm. As noted above, the localization domain of *pem* transcripts in fertilized eggs and early embryos up to the 8-cell stage overlaps with the myoplasmic domain, although the *pem* domain is narrower than the myoplasmic domain. Moreover, experiments with inhibitors of cytoskeletal components suggested that the cytoskeletal components required for the localization of myoplasm are also required for the correct localization of *pem* transcripts (data not shown). The myoplasm is a unique cytoskeletal domain which consists of several cytoskeletal proteins. It has been shown that, during oogenesis, the component of myoplasm is broadly distributed around the germinal vesicle and then translocated to the peripheral cytoplasm as the oocyte maturates (Swalla et al. 1993). Therefore, the localization of *pem* mRNA during oogenesis is distinct from that of the myoplasm. It is likely that some cytoskeletal machinery other than the myoplasm is responsible for the correct localization of *pem* mRNA and in turn plays a part in the establishment of the polarity of the ascidian egg.

pem-2

cDNA clones for *pem-2, pem-3, pem-4, pem-5,* and *pem-6* have been isolated from a cDNA library of fertilized egg mRNAs subtracted with gastrula mRNAs of *C. savignyi*. The library was estimated to contain about 1000 independent clones. From the library, clones were randomly selected and their nucleotide sequences were determined from the 3′ end to prevent the further analysis of the same clones. Each clone was then examined for the localization of corresponding mRNA by whole-mount in situ hybridization. We have examined 102 clones to date and found that among them 7 cDNA clones are of mitochondrial genes, 2 clones are of cytoplasmic actin gene(s), 3 are of α-tubulin gene(s), 1 clone is of a β-tubulin gene, and 1 is of a TFIIB gene. In addition, we were able to find 5 independent clones that contained mRNAs that were localized in the posterior end of the 8-cell and 32-cell embryos. Because the localization patterns of these clones resembled that of the *pem* cDNA clone, we designated these clones as *pem-2*, *pem-3*, *pem-4*, *pem-5*, and *pem-6*, in the order of their isolation (Table 1). The results of Northern blot analysis as well as in situ hybridization suggested that among these localized messages, *pem* mRNA is the most abundant, and the amount of mRNA was lessened in the order of *pem-2*, *pem-3*, *pem-4*, *pem-5*, and *pem-6*.

Two kinds of cDNA clones were isolated for *pem-2*. One was 3260 bp long and the other 2950 bp long. The two clones differed from each other in their 3′ untranslated region (UTR) sequence length. The *pem-2* cDNA had a single open reading frame encoding a polypeptide of 820 amino acids (Table 1). Database searches indicated that the predicted PEM-2 protein contained a signal for nuclear localization, an src homology 3 (SH3) domain, and a consensus sequence of the guanine nucleotide dissociation stimulators (GDSs; also known as guanine nucleotide exchange factors or guanine nucleotide releasing proteins) that have been classified as members of the CDC24 family. The CDC24 family GDSs are thought to be specific to Rho/Rac proteins which are involved in the organization of the cytoskeleton; the family includes Dbl, Vav, Bcr, rasGRF, and ect2 (Boguski and McCormick 1993). This suggests that PEM-2 is a new member of the CDC24 family. Many proteins of this family contain additional functional domains, including pleckstrin homology (PH), src homology 2 (SH2), and SH3 domains. PEM-2 also has an SH3 domain. This finding provides further support that PEM-2 is a member of the CDC24 family. These characteristic motifs suggest that the PEM-2 has a complex function in embryogenesis.

Similar to *pem*, the localization and segregation of *pem-2* mRNA marked the posterior end of developing embryos (Fig. 1I–L). In the unfertilized eggs, the hybridization signal appeared in the peripheral cytoplasm but its localization was not so conspicuous. After the first phase of ooplasmic segregation, the *pem-2* hybridization signal became stronger near the vegetal pole, and after the second phase of ooplasmic segregation, the *pem-2* transcript moved to the subequatorial region to form a rather broad crescent-like structure (Fig. 1I). The *pem-2* transcript was localized at the posterior-vegetal cytoplasmic region of the 2-cell (Fig. 1J) and 4-cell embryos. At the 8-cell stage, *pem-2* mRNA was restricted to the very narrow posterior region of B4.1 cells (Fig. 1K). At the 32-cell stage, the *pem-2* mRNA was found in the posterior-most region of the embryo, only in the posterior cytoplasm of the B6.3 cells (Fig. 1L). At the tailbud stage, hybridization signals were detectable in two cells of the endodermal strand.

pem-3

As shown in Figure 1M, the localization and segregation of *pem-3* mRNA also marked the posterior end of developing embryos, although the *pem-3* transcript became undetectable by the neurula stage, presumably reflecting a lower amount of the mRNAs. The cDNA for the *pem-3* gene was 1980 bp long. The cDNA had a single open reading frame that predicted a polypeptide of 465 amino acids (Table 1). The predicted PEM-3 protein contained two KH domains. The KH domains of PEM-3 showed extensive similarity to those of *Caenorhabditis elegans* MEX-3, suggesting that *pem-3* is a candidate homolog of *C. elegans mex-3*. *C. elegans mex-3* regulates blastomere identity in the early embryos (Draper et al. 1996). In addition, PEM-3 also contained the consensus sequence of the ring finger or C3HC4 zinc finger motif. These results indicated that PEM-3 is a probable RNA-binding protein.

pem-4

As in the cases of the other *pems*, the localization and segregation of *pem-4* mRNA also marked the posterior end of developing embryos (Fig. 1N). The cDNA for the *pem-4* gene was 2281 bp long and had a single open reading frame encoding a polypeptide of 516 amino acids (Table 1). The predicted PEM-4 protein contained three C2H2-type zinc finger motifs, suggesting that PEM-4 functions as a transcriptional factor.

pem-5

The cDNA for the *pem-5* gene was 2688 bp long and had a single open reading frame encoding a polypeptide of 758 amino acids (Table 1). The predicted PEM-5 protein did not contain any consensus motif, nor did it show any similarity to known proteins. As shown in Figure 1O, the localization and segregation of *pem-5* mRNA also marked the posterior end of developing embryos; although, presumably reflecting a lesser amount of *pem-5* transcript, its distribution pattern was not as evident until the 8-cell stage and was much narrower at later stages than those of *pem* and *pem-2*.

pem-6

The cDNA for *pem-6* was 1646 bp long and had a single open reading frame encoding a polypeptide of 202 amino acids (Table 1). Database searches indicated that none of the reported proteins had similarity to the predicted PEM-6 protein. The localization and segregation of *pem-6* mRNA also marked the posterior end of developing embryos (Fig. 1P). However, as in the case of the *pem-5* transcript, the *pem-6* distribution pattern was not as evident until the 8-cell stage and was much narrower at later stages than those of *pem* and *pem-2*.

PUTATIVE FUNCTION OF PEM

The Overexpression of PEM Affects the Development of the Anterior and Dorsal Structures of the Larva

The characteristic distribution of *pem* mRNA implies a significant role of the gene in embryogenesis. To explore its putative function, we injected synthetic capped *pem* mRNA into *C. savignyi* fertilized eggs to examine the effects of the overexpression of the PEM protein. As controls, we injected synthesized *lacZ*-RN3 RNA or synthesized *pem* mRNA with a frameshift mutation. Eggs from both controls developed into morphologically normal tadpole larvae (Fig. 3A) except for a very few larvae with abnormalities. However, the microinjection of *pem* mRNA into fertilized eggs resulted in a disturbance of the development of the anterior and dorsal structures of the larva (Fig. 3B).

The injection of *pem* mRNA into single fertilized eggs did not alter the cleavage pattern. Gastrulation took place normally, followed by neurulation and the formation of tailbud embryos. Tadpole-type larvae hatched as usual. As shown in Figure 3B, the tail region of the experimental larvae looked normal. However, a deficiency in the development of the anterior and dorsal trunk region was evident. The most marked effect of the *pem* mRNA microinjection was a lack of the adhesive organ (Fig. 3B). At the anterior end of ascidian larvae, there is an adhesive organ that consists of three adhesive papillae arranged in a triangular field (Fig. 3A). Since the papillae secrete ad-

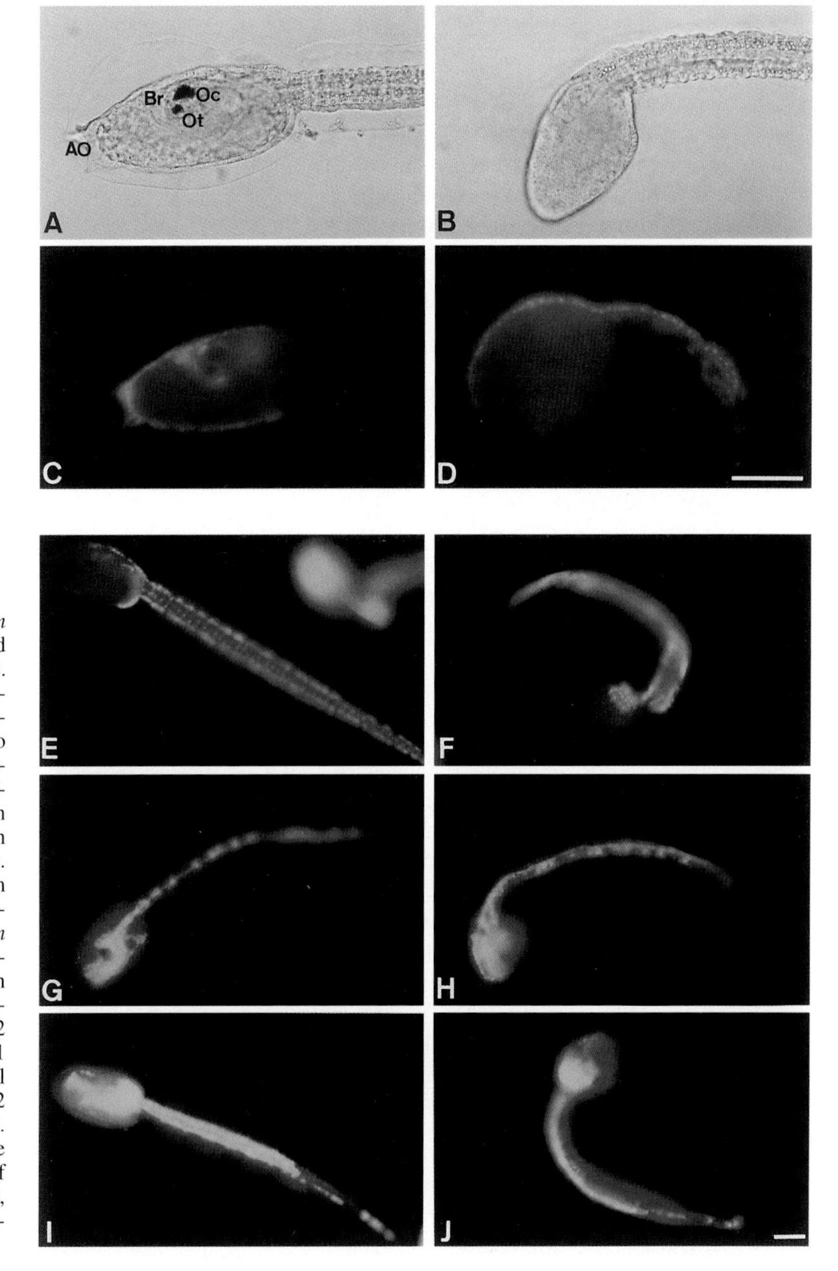

Figure 3. The overexpression of *pem* affects the formation of the anterior and dorsal structures of the tadpole larvae. Synthetic *pem* mRNA was microinjected into fertilized eggs, and the injected eggs were allowed to develop into tadpole larvae. (*A*) A control larva developed from an egg injected with *lacZ*-RN3 mRNA, showing the trunk region of the larva with the adhesive organ (AO), otolith (Ot), and ocellus (Oc). (Br) Brain. Bar in *D* represents 100 μm for *A–D*. (*B*) An experimental larva developed from an egg injected with *pem* mRNA showing lack of the adhesive organ, sensory pigment cells and brain vesicle. (*C–J*) Effects of *pem* overexpression on the distribution of a4.2 (*C,D*), b4.2 (*E,F*), A4.1 (*G,H*), and B4.1 (*I,J*) progenitor cells. (*C,E,G,I*) Control larvae showing progenitor cells of a4.2 (*C*), b4.2 (*E*), A4.1 (*G*), and B4.1 (*I*). (*D,F,H,J*) PEM-overexpressed larvae showing changes in the distribution of the progenitor cells of a4.2 (*D*), b4.2 (*F*), A4.1 (*H*), and B4.1 (*J*). Bar in *J* represents 100 μm for *E–J*.

hesives that are used to effect settlement at the onset of metamorphosis, this organ may be homologous to the sucker or cement gland of amphibian tadpoles. Almost all of the larvae developed from eggs injected with 125 pg *pem* mRNA failed to form the adhesive organ. Therefore, the anterior surface of the larva was smooth (Fig. 3B).

Another marked deficiency induced by the *pem* mRNA injection was evident in the sensory pigment cells of the brain vesicle (the anterior portion of the central nervous system of ascidian larvae). The brain of *Ciona* larvae contains two sensory organs: the otolith and ocellus (Satoh 1994). The microinjection of 125 pg of synthesized *pem* mRNA resulted in the formation of no pigment cells (Fig. 3B). When these affected structures were compared, the damage was more evident in the more anterior structures. In addition, sections of deficient larvae revealed a lack of the brain vesicle (data not shown). The trunk region of the deficient larva was occupied by endoderm cells.

Because the injection of *pem* mRNA with a frame-shifted mutation did not affect the formation of the anterior and dorsal structures of the larva, it is highly likely that the maternal message functions after translation.

Distribution of a4.2 Descendants in PEM-overexpressed Embryos

As described above, the effect of the overexpression of the *pem* gene was restricted to the anterior and dorsal structures of the larva; the ventral trunk as well as the tail looked normal in the *pem*-overexpressing larvae. The disappearance of the anterior and dorsal structures suggests that the presumptive cells of these structures change their developmental fates to other cell types or that the presumptive cells remain undifferentiated somewhere in the embryo. Most of the affected structures are derived from a pair of a4.2 (anterior and animal blastomeres) progenitor cells of the 8-cell embryo. We examined these possibilities by tracing the lineage of the progenitor cells.

The injection of a lineage tracer into either right a4.2 or left a4.2 revealed striking effects of *pem* overexpression on the distribution of the a4.2 descendant cells. In the control larvae, the anterior and middle portions of the trunk (including the adhesive organ, epidermis, and brain vesicle with two sensory organs) were labeled (Fig. 3C), confirming that these structures are formed by a4.2 descendant cells. The labeled region never crossed over the boundary that divides the middle and posterior portions of the larval trunk (Fig. 3C). The a4.2 descendant cells do not contribute to the posterior portion of the central nervous system (visceral ganglion and spinal cord) that extends into the tail region. In the experimental larvae with *pem* overexpression, labeled epidermal cells and neuronal cells were evident in the anterior portion of the tail (Fig. 3D). This suggests that *pem* overexpression does not affect the differentiation of a4.2-derived neuronal cells, but may change the patterning of the nervous system, causing the translocation of a4.2 descendants posteriorly to the region that is usually occupied by the visceral ganglion and spinal cord derived from b4.2 and A4.1. Together with the posterior localization of the *pem* transcript, the *pem* gene may have an important role in the patterning of the embryo.

In contrast, the results of the injection of the lineage tracer into b4.2 (posterior and animal blastomere), A4.1 (anterior and vegetal), or B4.1 (posterior and vegetal) progenitor cells showed that the overexpression of *pem* had almost no effect on the distributions of the b4.2 (Fig. 3E,F), A4.1 (Fig. 3G,H), or B4.1 descendant cells (Fig. 3I,J) in the larvae. The b4.2 progenitor cells contribute mainly to the epidermis of the anterior-dorsal trunk and tail region (Fig. 3E), as well as the dorsal part of the central nervous system and muscle cells at the tip of the tail, which is not obvious in the figure. The distribution of b4.2 progenitor cells in the *pem* mRNA-injected larvae was not affected, although the epidermis of the anterior-dorsal trunk region was not seen in the experimental larvae (Fig. 3F). The A4.1 progenitor cells mainly give rise to the notochord, the posterior portion of the central nervous system of the tail, and muscle cells at the tip of the tail (Fig. 3G). Endoderm cells of the trunk region are also derived from this blastomere (Fig. 3G). A similar labeling pattern was found in the A4.1 progenitor cells in the experimental larvae (Fig. 3H), suggesting no obvious changes in the development of the A4.1 progenitor cells in the experimental larvae. The B4.1 progenitor cells contribute mainly to muscle, the endodermal strand of the tail, and a few notochord cells at the tip of the tail (Fig. 3I). Endoderm and mesenchyme cells of the trunk region are also derived from this blastomere. There was no detectable change in the B4.1 progenitor cells in the experimental larvae (Fig. 3J). This is consistent with the above-mentioned finding that the formations of endoderm, mesenchyme, muscle, and notochord appeared to be unaffected by the overexpression of the *pem* gene.

Lithium Counteracts the Effects of PEM Overexpression

pem mRNA is localized in the posterior-most cytoplasm of the B4.1 cells of the 8-cell embryo, and its overexpression causes the translocation of the anterior-dorsal neural structure to the posterior position, suggesting that the anterior neuronal cells acquired characteristics of the posterior neuronal cells by receiving PEM signals. It is likely that *pem* is required for the determination of the posterior character in the nervous system of the ascidian embryo. However, it remains uncertain whether *pem* overexpression truly causes posteriorization or causes a simple translocation of the anterior neuronal cells.

In amphibian embryos, the administration of lithium is known to cause an expansion of the dorso-anterior structure, leading to the duplication of axial structures or, in extreme cases, entirely dorsalized embryos lacking ventral structures (Kao et al. 1986). To explore the mechanism of patterning of ascidian embryo and the possible cascade of *pem* function, we examined the effect of lithium treatment in *pem*-overexpressed embryos. We found that lithium treatment could rescue the anterior and dorsal structures in *pem*-overexpressed embryos; i.e., the anterior-most adhesive organ as well as the dorsal brain with sensory pigment cells was formed in the *pem*-overexpressed larvae. These results suggest that *pem* has a role in the establishment of the anterior and dorsal patterning of the ascidian embryo via a signaling cascade that lithium mimics. There is growing evidence that the endogenous targets of lithium are factors involved in the *wnt* signaling cascade (Klein and Melton 1996) and that factors involved in the pathway are required for the anterior-dorsal axis formation (Heasman et al. 1994; He et al. 1995; Sokol et al. 1995). Therefore, it is an intriguing research question as to whether this cascade is also involved in the pattern formation, in relation to the function of the *pem* gene, in ascidian embryos.

PEM Functions in Vertebrate Embryos

Another question arises as to whether PEM also functions in vertebrate embryos. To examine this possibility, we injected synthetic *pem* mRNA into *Xenopus* eggs. When the *pem*-RNA-injected embryos were examined at the tailbud stage, the resultant embryos showed the loss of the anterior-dorsal structures, similarly to the case of *pem* mRNA injection into ascidian eggs. They lacked the adhesive organ and dorsal nervous system (data not shown). Therefore, vertebrate *pem* homologs may function in their embryos as well.

Isolation of *pem* Homolog from a Closely Related Ascidian Species

Despite such a possible conservation of PEM function, attempts to isolate its homolog from *Xenopus* cDNA with *pem* cDNA as a probe or from another ascidian, *Halocynthia roretzi*, which belongs to another order of ascidians, have been unsuccessful. Therefore, as the first step to isolate a vertebrate *pem* homolog, we isolated a homolog from a very closely related ascidian species, *Ciona intestinalis*. As shown in Figure 2, the comparison of the PEM amino acid sequence with that of *Ci*-PEM demonstrated several regions with conserved sequences. In particular, the amino acid sequences of a few central regions as well as the carboxyl terminus were highly conserved; this information may be useful for further attempts to isolate vertebrate *pem* homologs.

DELETION OF THE B6.3 BLASTOMERE DID NOT AFFECT THE PATTERN FORMATION OF THE EMBRYO

The clones we have isolated thus far show the localization of corresponding mRNA in the posterior end of the developing ascidian embryo. In addition, despite such localization of *pem* mRNAs, their cDNA sequences did not contain the localization signal sequence-motif in the 3′ UTR as reported in some other localized messages (Ding and Lipshitz 1993). At the 32-cell stage, a pair of very small blastomeres are formed at the posterior-most region of the vegetal hemisphere. This blastomere is named B6.3, and it contains the developmental fates giving rise to muscle, the endodermal strand, and mesenchyme. Interestingly, all of the *pem* mRNAs appeared to be localized in this cell, as shown in Figure 1. To examine a possible role of B6.3 in the function of *pems* and the pattern formation of the embryo, we killed the cell shown with a fine tungsten needle. Such B6.3-deprived embryos were allowed to develop into tadpole larvae. However, the resultant embryos developed into morphologically normal tadpole larvae. The deletion of B6.3 did not disturb the pattern formation, including that of the anterior and dorsal structures. This suggests that all of the *pem* mRNAs are translated into proteins, which are not restricted to a certain region but distributed rather widely in the posterior region of the egg and embryo. This result also suggests that PEMs have completed their function before the 32-cell stage. In addition, this concentration of *pem* mRNAs into the B6.3 blastomere suggests machinery that attracts and anchors mRNAs localized to the posterior end of the embryo as a whole.

ACKNOWLEDGMENTS

We thank Dr. P. Lemaire for pBluescript-RN3 vector. This research was supported by a grant-in-aid for Specially Promoted Research (07102012) from the Ministry of Education, Science, Sports and Culture, Japan to N.S. S.Y. and Y.S. were supported by predoctoral fellowships from the Japan Society for the Promotion of Science for Japanese Junior Scientists with research grants 7-5137 and 8-6806, respectively.

REFERENCES

Boguski M.S. and McCormick F. 1993. Proteins regulating Ras and its relatives. *Nature* **366:** 643.

Chabry L. 1887. Contribution a l'embryologie normale et teratologique des Ascidies simples. *J. Anat. Physiol.* **23:** 167.

Conklin, E.G. 1905. The organization and cell lineage of the ascidian egg. *J. Acad. Nat. Sci.* **13:** 1.

Ding D. and Lipshitz H.D. 1993. Localized RNAs and their functions. *BioEssays* **15:** 651.

Draper B.W., Mello C.C., Bowerman B., Hardin J., and Priess J.R. 1996. MEX-3 is a KH domain protein that regulates blastomere identity in early *C. elegans* embryos. *Cell* **87:** 205.

He X., Saint-Jeannet J.-P., Woodgett J., Varmus H., and Dawid I. 1995. Glycogen synthase kinase-3 and dorsoventral patterning in *Xenopus* embryos. *Nature* **374:** 617.

Heasman J., Crawford A., Goldstone K., Garner-Hamrick P., Gumbiner B., McCrea P., Kintner C., Yoshida-Noro C., and Wylie C. 1994. Overexpression of cadherins, and underexpression of β-catenin inhibit dorsal mesoderm induction in early *Xenopus* embryos. *Cell* **79:** 791.

Jeffery W.R. 1990. Ultraviolet irradiation during ooplasmic segregation prevents gastrulation, sensory cell induction, and axis formation in the ascidian embryo. *Dev. Biol.* **140:** 388.

Jeffery W.R. and Swalla B.J. 1990. The myoplasm of ascidian eggs: A localized cytoskeletal domain with multiple roles in embryonic development. *Semin. Dev. Biol.* **1:** 373.

Kao K.R., Masui Y., and Elinson R.P. 1986. Lithium-induced respecification of pattern in *Xenopus laevis* embryos. *Nature* **322:** 371.

Klein P.S. and Melton D.A. 1996. A molecular mechanism for the effect of lithium on development. *Proc. Natl. Acad. Sci.* **93:** 8455.

Marikawa Y., Yoshida S., and Satoh N. 1994. Development of egg fragments of the ascidian *Ciona savignyi*: The cytoplasmic factors responsible for muscle differentiation are separated into a specific fragment. *Dev. Biol.* **162:** 134.

———. 1995. Muscle determinants in the ascidian egg are inactivated by UV irradiation and the inactivation is partially rescued by injection of maternal mRNAs. *Roux's Arch. Dev. Biol.* **204:** 180.

Nishida H. 1987. Cell lineage analysis in ascidian embryos by intracellular injection of a tracer enzyme. III. Up to the tissue restricted stage. *Dev. Biol.* **121:** 526.

———. 1992. Regionality of egg cytoplasm that promotes muscle differentiation in embryo of the ascidian, *Halocynthia roretzi. Development* **116:** 521.

———. 1993. Localized regions of egg cytoplasm that promote expression of endoderm-specific alkaline phosphatase in embryos of the ascidian *Halocynthia roretzi. Development* **118:** 1.

———. 1994a. Localization of egg cytoplasm that promotes differentiation to epidermis in embryos of the ascidian *Halocynthia roretzi. Development* **120:** 235.

———. 1994b. Localization of determinants for formation of the anterior-posterior axis in eggs of the ascidian *Halocynthia roretzi. Development* **120:** 3093.

———. 1996. Vegetal egg cytoplasm promotes gastrulation and is responsible for specification of vegetal blastomeres in embryos of the ascidian *Halocynthia roretzi. Development* **122:** 1271.

Nishikata, T., Mita-Miyazawa I., Deno T., and Satoh N. 1987. Monoclonal antibodies against components of the myoplasm of eggs of the ascidian *Ciona intestinalis* partially block the development of muscle-specific acetylcholinesterase. *Development* **100:** 577.

Reverberi G. 1971. Ascidians. In *Experimental embryology of marine and fresh-water invertebrates* (ed. G. Reverberi), p. 507. Elsevier, North-Holland, Amsterdam.

Satoh N. 1994. *Developmental biology of ascidians.* Cambridge University Press, New York.

Satoh N., Makabe K.W., Katsuyama Y., Wada S., and Saiga H. 1996. The ascidian embryo: An experimental system for studying genetic circuitry for embryonic cell specification and morphogenesis. *Dev. Growth Differ.* **38:** 325.

Sokol S., Klingensmith J., Perrimon N., and Itoh K. 1995. Dorsalizing and neuralizing properties of Xdsh, a maternally expressed *Xenopus* homologue of *dishevelled. Development* **121:** 1637.

Swalla B.J. and Jeffery W.R. 1995. A maternal RNA localized in the yellow crescent is segregated to the larval muscle cells during ascidian development. *Dev. Biol.* **170:** 353.

Swalla B.J., Badgett M.R., and Jeffery W.R. 1991. Identification of a cytoskeletal protein localized in the myoplasm of ascidian eggs: Localization is modified during anural development. *Development* **111:** 425.

Swalla B.J., Makabe K.W., Satoh N., and Jeffery W.R. 1993. Novel genes expressed differentially in ascidians with alternate modes of development. *Development* **119:** 307.

Yoshida S., Marikawa Y., and Satoh N. 1996. *posterior end mark,* a novel maternal gene encoding a localized factor in the ascidian embryo. *Development* **122:** 2005.

Related Signaling Networks in *Drosophila* That Control Dorsoventral Patterning in the Embryo and the Immune Response

L.P. WU AND K.V. ANDERSON
Molecular Biology Program, Sloan-Kettering Institute, New York, New York 10021

As described in other chapters in this volume, many signaling pathways are used for related developmental processes in animals as different as *Drosophila* and mice, but, in addition, a single pathway may be used repeatedly during the life of a single organism to mediate communication between cells in very different developmental events. When a particular signaling pathway is used in similar contexts in two very different organisms, such as the use of the Hedgehog signaling pathway in the development of limbs in *Drosophila* and in vertebrates, it is tempting, and perhaps correct, to think that the ancestor of both flies and vertebrates used Hedgehog signals to pattern its limbs. However, at the moment, there is no clear logic that explains why and how a particular signaling pathway is used repeatedly in different contexts during the life of a single organism.

Genetic studies in *Drosophila* should provide an ideal starting point for understanding the repeated use of signaling pathways in different cellular contexts. The Toll-Dorsal pathway is an example of a genetically defined multi-use pathway, which defines the earliest dorsoventral (D/V) asymmetry in the *Drosophila* embryo (Chasan and Anderson 1993), and is expressed repeatedly during *Drosophila* development in a variety of tissues (Hashimoto et al. 1991). Perhaps the use of the Toll pathway that is most different from its function in early embryonic axis determination is its activation in the fat body of larvae and adults as a part of the immune response of the animal to infection (Ip et al. 1993; Lemaitre et al. 1995 1996). To understand this signaling pathway in this very different context, we have begun a genetic analysis of the *Drosophila* immune response and how the components that control early embryonic patterning are used in this physiological response.

CONSERVED AND DIVERGENT EARLY AXIS-DETERMINING MECHANISMS

There appears to be a conserved process that establishes the D/V pattern of both *Drosophila* and vertebrate embryos at the gastrula stage, just after the earliest axis-determining events that are controlled by the Toll-Dorsal pathway in *Drosophila* (Fig. 1) (Ferguson 1996). In *Xenopus*, a gradient of the bone morphogenetic protein 4 (BMP4) defines cell fate along the D/V axis, and in *Drosophila*, an activity gradient of Decapentaplegic (Dpp) , the homolog of BMP4, also defines D/Vcell fates. In the *Drosophila* embryo, the *dpp* gene is expressed on the dorsal side of the embryo and the *short gastrulation* (*sog*) gene is expressed ventrally. The Sog product inhibits Dpp activity, thereby establishing the gradient of Dpp activity (Holley et al. 1995; Biehs et al. 1996). In the *Xenopus* embryo, BMP4 is expressed ventrally and Chordin, a homolog of Sog, is expressed dorsally (Sasai et al. 1994; Schmidt et al. 1995). Chordin binding to BMP4 prevents the binding of BMP4 to its cell surface receptor, and thereby helps establish the gradient of BMP4 activity in the embryo (Piccolo et al. 1996).

Despite this similarity of patterning mechanisms at the gastrula stage, the available evidence suggests that the mechanisms that set up the region-specific expression of *dpp* and *sog* in *Drosophila* are not the same as those that control regional expression of *BMP4* and *chordin* in *Xenopus*. In *Xenopus*, cortical rotation during the first cell cycle defines the D/V axis of the embryo (Vincent et al. 1986). The molecular events triggered by cortical rotation are not completely clear, but they appear to include dorsal activation of β-catenin (Miller and Moon 1996; Rowning et al. 1997), which eventually leads to the dorsal activation of *chordin* and dorsal repression of *BMP4* expression.

In *Drosophila*, there is no cortical rotation after fertilization and no early role for β-catenin (Armadillo) in axis determination. Instead, a very different set of events leads to asymmetric expression of *dpp* and *sog* in the *Drosophila* embryo. A set of maternal effect genes control the production of a gradient of the transcription factor Dorsal, a member of the Rel/NF-κB family, in the nuclei of the blastoderm embryo (Chasan and Anderson 1993). Among other transcriptional targets, the Dorsal protein in ventral and lateral nuclei activates the transcription of *sog* and represses the transcription of *dpp*, leading to their region-specific expression in the gastrula-stage embryo (Huang et al. 1993; François et al. 1994).

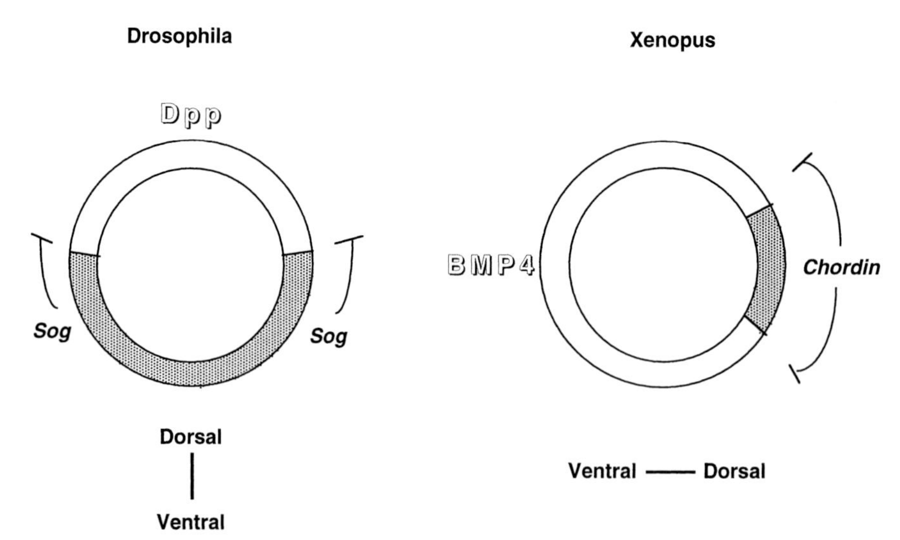

Figure 1 Common signals control D/V axis determination at the gastrula stage in *Drosophila* and *Xenopus*. Gradients of the homologous growth factors Dpp and BMP4 determine cell fates along the D/V axes of *Drosophila* and *Xenopus* embryos. In *Drosophila*, the Dpp gradient is established by regionalized transcription of *dpp* on the dorsal side and its antagonist *sog* on the ventral side of the embryo. Nuclear Dorsal protein in nuclei on the ventral side of the embryo controls the regionalized transcription of these two molecules: Dorsal activates the transcription of *sog* and represses the transcription of *dpp*. In *Xenopus*, the BMP4 gradient is established by regionalized transcription of *BMP4* on the ventral side and its antagonist *chordin* (which is homologous to *sog*) on the dorsal side of the embryo. The direct regulators of region-specific transcription of *BMP4* and *chordin* are not known.

THE TOLL PATHWAY DEFINES THE FIRST D/V POLARITY OF THE *DROSOPHILA* EMBRYO

In the *Drosophila* embryo, the gradient of nuclear Dorsal protein is controlled by a genetically defined signaling pathway that is activated ventrally by an unidentified extracellular cue and relays the signal to control Dorsal nuclear localization (Fig. 2) (Chasan and Anderson 1993; Belvin and Anderson 1996). The extracellular portion of the pathway relies on successive proteolytic activation steps in which four different members of the trypsin family of serine proteases act in sequence and ultimately lead to the ventral proteolytic activation of the uniformly distributed precursor of the Spätzle protein. Cleaved Spätzle is thought to be the ligand that activates the receptor, Toll, on the ventral side of the embryo. Activation of the membrane receptor Toll leads to a series of events in the cytoplasm, including activation of the Tube and Pelle proteins. Activation of Pelle, a cytoplasmic serine/threonine kinase, leads, directly or indirectly, to the degradation of the IκB-like protein Cactus. Degradation of Cactus liberates Dorsal from its cytoplasmic complex with Cactus and allows Dorsal to translocate into nuclei.

This cytoplasmic signaling pathway from Toll to Dorsal is very similar to the events that lead to the activation of mammalian NF-κB (a homolog of Dorsal) in response to the cytokine interleukin-1. Activation of the interleukin-1 receptor (IL-1R), whose cytoplasmic domain is similar to that of Toll, leads to binding and phosphorylation of IRAK, a serine/threonine kinase related to Pelle (Cao et al. 1996), and eventually causes the degradation of IκB, a homolog of Cactus, in a cytoplasmic IκB-NF-κB complex. Degradation of IκB allows NF-κB to move into nuclei, where it is active as a transcription factor (Fig. 2).

At this point in time, there is no evidence that the Toll signaling pathway has a conserved function in early D/V patterning in vertebrate embryos. Targeted mutations in mammalian Rel proteins (the homologs of Dorsal) affect various aspects of immune cell function but do not cause early embryonic patterning defects (Beg et al. 1995; Sha et al. 1995; Weih et al. 1995). However, it is possible that other uncharacterized Rel family members could play a part in patterning or that overlapping functions of the Rel proteins obscure an early embryonic function.

CONSERVATION OF THE TOLL SIGNALING PATHWAY IN THE INNATE IMMUNE RESPONSE

Despite the evidence suggesting that the Toll pathway does not have a conserved function in embryonic patterning, the similarity of the Toll and IL-1R signaling pathways implies that the biochemical steps in this pathway have been conserved in evolution. The available evidence suggests that the function of this pathway may also have been conserved in at least one context, as a part of the innate immune responses of both *Drosophila* and mammals.

In all animals, the first rapid response to infection by pathogens is mediated by the innate immune response. In both insects and vertebrates, the innate immune system recognizes general surface features of microbes, such as the complex polysaccharides in the cell walls, and responds by activating macrophages and by secreting specific proteins and peptides into the circulation that kill the pathogens (Fig. 3). In mammals, macrophages engulf in-

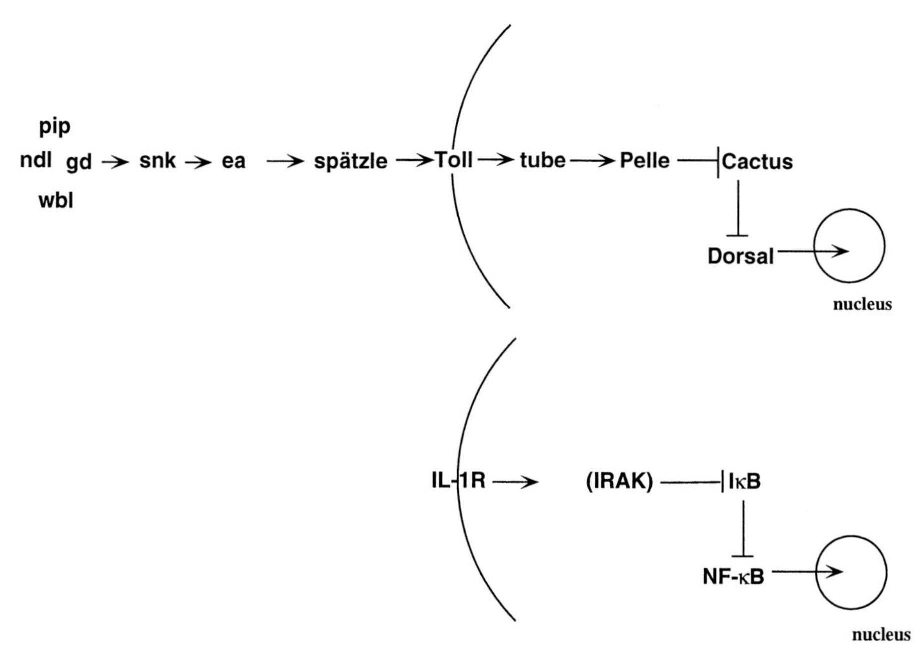

Figure 2. The Toll signaling pathway as defined in the *Drosophila* embryo and the homologous mammalian IL-1 receptor signaling pathway. Genetic analysis has defined the steps in the Toll signaling pathway: A series of extracellular events leads to the activation of the receptor, Toll, on the ventral side of the embryo; activation of Toll leads to the degradation of Cactus in the cytoplasmic Cactus-Dorsal complex and the release of Dorsal to the nucleus. The events in the cytoplasmic portion of this signaling pathway are similar to the events activated by the binding of the cytokine IL-1 to its receptor: IL-1R activation activates IRAK (which is related to Pelle); activation of the cytoplasmic pathway leads to degradation of IκB (which is homologous to Cactus) in the cytoplasmic IκB-NF-κB complex and the release of NF-κB (which is homologous to Dorsal) to the nucleus.

vading pathogens and secrete cytokines that activate other aspects of the immune response (Sipe 1985). Similarly, macrophage-like cells of insects are activated by infection and participate in phagocytosis and encapsulation of invading pathogens (Götz and Boman 1985). A second major branch of the mammalian innate immune response is the production of the acute-phase proteins by the liver and their secretion into the circulation; these acute-phase proteins are involved directly in killing bacteria. Analogously, the insect fat body (the counterpart to the liver) produces and secretes a set of antimicrobial peptides in response to infection (Hultmark 1993; Hoffmann 1995).

The transcription factors of the Rel/NF-κB family have key roles in the innate immune response in both mammals and insects (Bauerle and Henkel 1994). In mammalian macrophages, bacterial lipopolysaccharide (LPS) triggers an uncharacterized signaling pathway that activates NF-κB (Kopp and Ghosh 1995). As a consequence, cytokines including IL-1 and tumor necrosis factor-α (TNF-α) are secreted by macrophages. IL-1 and TNF-α then both activate signaling pathways that lead to nuclear localization of NF-κB in other macrophages and in the liver, where NF-κB activates transcription of acute-phase proteins, including components of complement.

The best-characterized aspect of the *Drosophila* immune response is the rapid induction of transcription of genes encoding antimicrobial peptides in the fat body (Hoffmann 1995). The first clue that Rel proteins might be involved in the *Drosophila* immune response was the identification of κB elements, sites where Rel proteins could bind, in the promoters of a number of the antimicrobial peptide genes (Reichhart et al. 1992; Engström et al. 1993). This in turn led to the discovery that all three known *Drosophila* members of the Rel/NF-κB family, *dorsal*, *Dif* (*D*orsal-related *i*mmunity *f*actor), and *relish*, are expressed in the larval fat body, and all three are activated in response to infection (Reichhart et al. 1993; Lemaitre et al. 1995; Dushay et al. 1996). When larvae are infected by bacteria, Dorsal and Dif move rapidly from the cytoplasm to the nuclei of fat body cells (Ip et al. 1993; Reichhart et al. 1993; Lemaitre et al. 1995). Nuclear translocation of Dorsal in the fat body requires the activities of the *Toll*, *tube*, and *pelle* genes (Lemaitre et al. 1995). Thus, the same signaling pathway used in the early embryo is activated by infection.

Even more remarkable than the use of similar signaling pathways in the innate immune responses of mammals and *Drosophila* is the use of related molecules in plant disease resistance responses. Among the large group of genes that has been identified recently from a variety of plants that mediate recognition and responses to pathogens, several protein families have been identified that have homology with components of the Toll signaling pathway (Staskawicz et al. 1995; Baker et al. 1997; Cao et al. 1997; Ryals et al. 1997). These proteins include domains with leucine-rich repeats like those in the Toll extracellular domain, proteins with domains homologous to the Toll and IL-1R cytoplasmic domains, serine/threonine kinases related to Pelle and IRAK, and ankyrin-repeat proteins similar to Cactus and IκB. The presence of

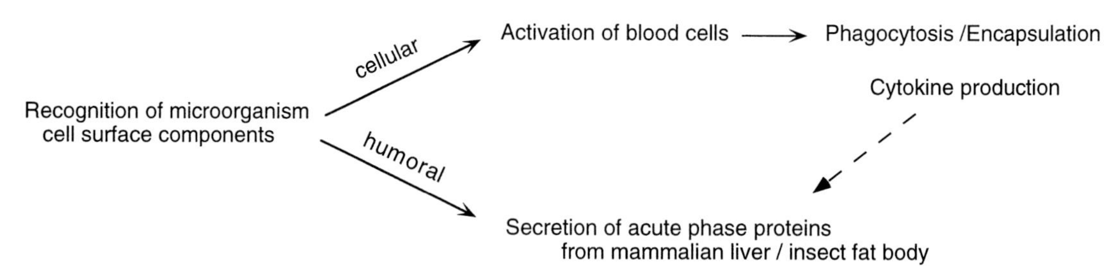

Figure 3. In both mammals and *Drosophila*, there are both cellular and humoral branches of the innate immune response. The cellular response involves the activation of macrophages and the humoral response is the secretion of peptides and proteins that directly attack microbes from the liver or the fat body.

these domains in plant disease resistance genes, insect immune response genes, and genes mediating mammalian innate immune responses raises the possibility that the first function of these proteins in evolutionary history was as part of an ancestral immune response.

THE FUNCTION OF THE TOLL AND OTHER SIGNALING PATHWAYS IN THE *DROSOPHILA* IMMUNE RESPONSE

The biochemical data suggest that the *Drosophila* Rel proteins could mediate transcription of genes induced in response to infection. Dif and Dorsal are both present in nuclei after infection. All three *Drosophila* Rel proteins are capable of binding the κB sites in the promoters of genes encoding antimicrobial peptide genes. All three can activate the transcription of the *cecropin A1* gene in *Drosophila* blood cell lines, and both Dorsal and Dif have been shown to activate transcription of the *diptericin* gene (Ip et al. 1993; Petersen et al. 1995; Dushay et al. 1996; Gross et al. 1996).

The in vivo function of the Rel proteins in the immune response is not so simple, however. Even though Dorsal moves from the fat body cytoplasm to nuclei in response to infection, mutant larvae that lack Dorsal show normal induction of both *cecropin A1* and *diptericin*. No mutations in *Dif* or *relish* have been reported, so the function of the Rel proteins in the immune response was not immediately clear.

A major step forward in determining the function of the Toll pathway in the immune response was the demonstration that adult flies that lack *spätzle*, *Toll*, *tube*, or *pelle* do not induce the transcription of the antifungal peptide gene, *drosomycin*, to normal levels (Lemaitre et al. 1996). In wild-type *Drosophila*, *drosomycin* RNA, like the transcripts of the antibacterial peptide genes, is rapidly induced in response to infection either by bacteria or by fungi. In contrast, in *spätzle*, *Toll*, *tube*, or *pelle* mutant adult flies, the normal induction of *drosomycin* does not take place. The *cactus* gene also participates in this response: In *cactus* mutant animals, *drosomycin* is expressed even in the absence of infection. Thus, Cactus is a negative regulator of the Toll signaling pathway in the immune response, just as it is in the early embryo. Other genes required in the early embryo are not required for *drosomycin* induction: Mutations in both genes that act upstream (*gastrulation defective, snake, easter*) and the gene at the end of the pathway (*dorsal*) have no effect on expression of the antifungal peptide gene.

These studies demonstrate that the Toll signaling pathway has a clear function in the immune response, but they also point out that many key components have not been identified. For instance, there are presumably unidentified regulators that activate *spätzle* and a Rel protein other than Dorsal that mediates the activation of *drosomycin*. The function of Dorsal in the immune response is mysterious; although it is clearly activated as a part of the immune response, it is not necessary for the transcription of either antibacterial or antifungal peptides.

The most apparent gap in our understanding of the *Drosophila* immune response is how the antibacterial peptide genes are induced in response to infection. Although the presence of functional κB elements in the promoters of the antibacterial peptide genes is good evidence that these genes can be regulated by the Rel transcription factors, larvae and adults that lack Toll show the normal induction of transcription of the antibacterial peptide genes in response to infection (Lemaitre et al. 1995, 1996; Gross et al. 1996). Therefore, other unknown signaling pathways must also be activated in response to infection. Because of these gaps in our understanding, we are using a genetic approach to define the events activated by infection.

A GENETIC SCREEN FOR *DROSOPHILA* IMMUNE RESPONSE GENES

A prerequisite for the genetic identification of components that mediate the *Drosophila* immune response is the development of simple, rapid assays for aspects of the immune response that can be used to test the immune responses of animals that carry random newly induced mutations. We have developed simple assays that make it possible to test several aspects of the immune response (L.P. Wu and K.V. Anderson, in prep.), including the activation of transcription of a gene encoding one of the antibacterial peptides that is induced in response to infection, *diptericin*. A reporter gene in which the *diptericin*

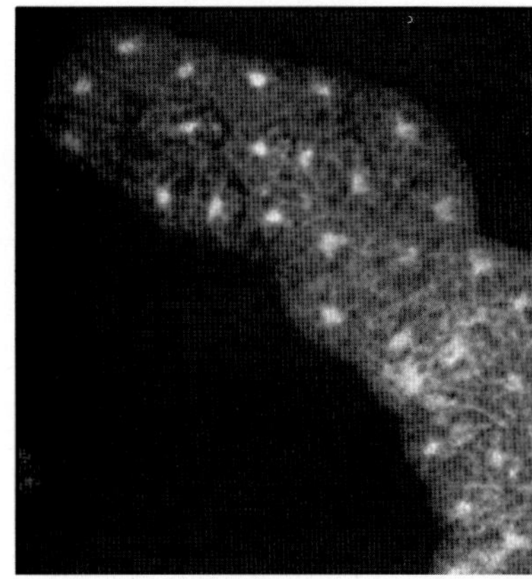

Figure 4. Dif protein resides in the cytoplasm of fat body cells in uninfected wild-type larvae. By 30 min after infection, most of the Dif protein has moved into the nuclei of the fat body cells, as described previously (Ip et al. 1993). Dorsal protein moves into fat body nuclei with a similar time course (data not shown; Lemaitre et al. 1995).

promoter is fused to the coding regions of β-galactosidase was constructed by Reichhart et al. (1992). This reporter gene, like the endogenous *diptericin* gene, is not normally expressed in the larval fat body but is rapidly induced in response to infection.

We have used the *diptericin-lacZ* reporter gene to screen for new ethylmethanesulfonate (EMS)-induced recessive mutations on the *Drosophila* third chromosome that prevent the normal induction of this gene in response to infection (L.P. Wu and K.V. Anderson, in prep.). To identify mutants, F3 larvae homozygous for newly induced mutations and for the *diptericin-lacZ* reporter gene were injected with a solution of *Escherichia coli*. After 3 hours, the fat bodies from the larvae from each line were dissected and stained for β-galactosidase activity. From this screen, we identified 57 mutations that prevent the expression of *diptericin-lacZ* after infection.

ANALYSIS OF REL PROTEIN ACTIVATION IN THE IMMUNE-RESPONSE-DEFICIENT MUTANTS

On the basis of preliminary complementation analysis, we have identified approximately 40 genes on the third chromosome that are required for induction of *diptericin* expression. To help classify the mutants and identify those genes that are most likely to provide new insights into the signaling pathways activated by infection, we have examined whether the mutations interfere with the nuclear translocation of the Rel proteins Dorsal and Dif in response to infection.

In wild-type larvae, both Dorsal and Dif proteins are translocated into fat body nuclei within 30 minutes after infection (Fig. 4). All the mutants that we have examined that fail to express *diptericin-lacZ* change Dif nuclear localization in response to infection, but some of these mutants do not affect Dorsal nuclear localization. For instance, in larvae homozygous for a mutation in *ird4*, Dorsal shows the normal response to infection, moving rapidly into nuclei, but Dif remains completely cytoplasmic. From these results, we infer that nuclear Dif is required for the activation of *diptericin* expression.

PARALLEL PATHWAYS CONTROL NUCLEAR LOCALIZATION OF REL PROTEINS IN RESPONSE TO INFECTION

Thus far, we have identified mutations in four different genes that prevent Dif nuclear localization in response to infection (class I mutations); at least two of these genes do not alter Dorsal nuclear localization (L.P. Wu and K.V. Anderson, in prep.). These mutations probably define components of a new signaling pathway that controls the activation of Dif in response to infection. Previous studies had shown that mutations which inactivate components of the Toll pathway (*Toll*, *tube*, and *pelle*) prevent Dorsal nuclear localization after infection (Lemaitre et al. 1995). We have found that Dif is localized to the nucleus normally in *Toll*$^-$or *pelle*$^-$ mutants. Thus, infection must activate two independent signaling pathways that control Rel protein localization in the fat body cells: The class I genes control Dif nuclear localization and the Toll pathway controls Dorsal nuclear localization.

Mutations in *cactus* cause constitutive nuclear localization of Dorsal in both embryos and the fat body (Roth et al. 1991; Lemaitre et al. 1995). We have found that Dif is also constitutively nuclear in *cactus*$^-$ fat body cells. Therefore, the two signaling pathways both use Cactus to retain both Dorsal and Dif in the cytoplasm. On the basis of what is known about the regulation of Cactus in the embryo, it is likely that the two signaling pathways both act by targeting Cactus for degradation. Thus, Cactus is targeted for degradation in response to a specific pathway depending on the Rel protein in the complex.

CONSTITUTIVE DIF NUCLEAR LOCALIZATION IN CLASS II MUTANTS

We have identified mutations in a second set of genes that prevent *diptericin* induction (class II mutations) (L.P. Wu and K.V. Anderson, in prep.). In these mutants, Dif is present in the fat body nuclei both before and after infection, but even though Dif is nuclear, *diptericin-lacZ* is not transcribed. This could indicate that these mutants lack some other factor, in addition to Dif, required for *diptericin* transcription. Alternatively, the presence of nuclear Dif prior to infection in these mutants suggests that there is some other defect in these animals, and perhaps chronic nuclear Dif initiates a feedback loop that prevents Dif from acting as a transcriptional activator on this promoter.

We think it likely that the class II mutants lead to chronic activation of the immune response through one of a number of possible mechanisms. One possibility is that these mutations affect the integrity of the larval tissues, leading to exposure of abnormal surfaces and chronic activation of the immune response. Mutations in at least two of the class II genes do not cause lethality, which argues against a lack of tissue integrity in those animals. Another possibility is that these genes encode negative regulators of steps in the Dif signaling pathway. In this scenario, double-mutant analysis with mutations that block Dif nuclear localization and those with chronic nuclear Dif should help define the order of events in the signaling pathway that controls Dif nuclear localization.

SIGNALING PATHWAYS IN THE INNATE IMMUNE RESPONSE

On the basis of the analysis of mutants that affect the *Drosophila* immune response, we have identified a new signaling pathway that controls activation of Dif, a Rel protein, in the immune response. In contrast to the early embryo where a single Rel pathway is activated, at least two parallel pathways activate two different Rel proteins in the fat body immune response. Furthermore, although only 12 genes are known that act in the early embryonic signaling pathway, our data suggest that there are more genes that affect Rel protein activation in the larval immune response. Some of the genes that we have identified appear to affect Dif activation specifically and probably define a new signaling pathway that controls the activation of this Rel family protein in the immune response. We expect that other genes identified in the immune response screen will identify components common to both Rel signaling components, perhaps including the uncharacterized subunits of the IκB kinase (Chen et al. 1996). Such genes would be required for the maternal Toll signaling pathway but would have been selectively overlooked in the maternal effect screens in which homozygous mutants were required to survive to adult stages. Further analysis should allow us to identify those common components and make it possible to determine the function of those genes in D/V patterning. Analysis of other mutants identified in the screen could well define additional signaling pathways activated as a part of the immune response.

THE EVOLUTIONARY ORIGIN OF THE SIGNALING PATHWAY

As we learn more about the use of the Rel signaling pathways in particular contexts in development and in immune responses, it becomes tempting to contemplate the original function of the pathway during evolution and how it became used for such apparently different ends as specifying an array of embryonic cell types and mediating responses to pathogens. With the current data, the simplest hypothesis is that a Toll/Rel signaling pathway played a part in self-defense in an ancestor that gave rise to both plants and animals and that function, although it has undergone substantial modification in the ensuing billion years, has been preserved in modern organisms. In plants, a serine/threonine kinase very similar to those involved in plant pathogen responses is used in the control of cell proliferation in meristems (Clark et al. 1997). At some point during plant evolution, it appears that a disease resistance gene became employed to control this aspect of development. Similarly, we imagine that at some point during animal evolution, the Toll/Dorsal signaling pathway became harnessed to events that activate it on the ventral side of the embryo and to downstream targets that control D/V cell fates, including *dpp* and *sog*. It seems quite likely that the recruitment of the Toll/Dorsal pathway to early development took place at some point during the early evolution of insects, after the divergence of invertebrates and vertebrates. But the definitive test of when this pathway acquired its axis determination function awaits a thorough understanding of the events that initiate axis determination in vertebrates, including mammals, as well as in other arthropods.

Very recently, human genes more closely related to Toll than is the IL-1R have been identified, and a potential role for one of these genes in the immune response has been described (Medzhitov et al. 1997). With these human homologs in hand, it will be possible to test whether these genes serve a role in development in mammals. These new perspectives should help us think about the original function of this pathway and when it became exploited for regulation of early development.

ACKNOWLEDGMENTS

This work was supported by National Institutes of Health grant GM-35437 and National Science Foundation grant MCB-9604323 to K.V.A. L.P.W. is a Leukemia Society of America Special Fellow.

REFERENCES

Baker B., Zambryski P., Staskawicz B., and Dinesh-Kumar S.P. 1997. Signaling in plant-microbe interactions. *Science* **276:** 726.

Bauerle P.A. and Henkel T. 1994. Function and activation of NF-κB in the immune system. *Annu. Rev. Immunol.* **12:** 141.

Beg A.A., Sha W.C., Bronson R.T., Ghosh S., and Baltimore D.

1995. Embryonic lethality and liver degeneration in mice lacking the RelA component of NF-κB. *Nature* **376:** 167.
Belvin M.P. and Anderson K.V. 1996. A conserved signaling pathway: The *Drosophila* Toll-Dorsal pathway. *Annu. Rev. Cell Dev. Biol.* **12:** 393.
Biehs B., Francois V., and Bier E. 1996. The *Drosophila short gastrulation* gene prevents Dpp from autoactivating and suppressing neurogenesis in the neuroectoderm. *Genes Dev.* **10:** 2922.
Cao H., Glazebrook J., Clarke J.D., Volko S., and Dong X. 1997. The *Arabidopsis NPR1* gene that controls systemic acquired resistance encodes a novel protein containing ankyrin repeats. *Cell* **88:** 57.
Cao Z., Henzel W.J., and Gao X. 1996. IRAK: A kinase associated with the interleukin-1 receptor. *Science* **271:** 1128.
Chasan R. and Anderson K.V., eds. 1993. Maternal control of dorsal-ventral polarity and pattern in the embryo. In *The development of* Drosophila melanogaster (ed. M. Bate and A. Martinez-Arias), pp. 387. Cold Spring Harbor Laboratory Press, Cold Spring Harbor, New York.
Chen Z.J., Parent L., and Maniatis T. 1996. Site-specific phosphorylation of IκB-α by a novel ubiquitination-dependent protein kinase activity. *Cell* **84:** 853.
Clark S.E., Williams R.W., and Meyerowitz E.M. 1997. The *CLAVATA1* gene encodes a putative receptor kinase that controls shoot and floral meristem size in *Arabidopsis*. *Cell* **89:** 575.
Dushay M., Asling B., and Hultmark D. 1996. Origins of immunity: *Relish*, a compound Rel-like gene in the antibacterial defense of *Drosophila*. *Proc. Natl. Acad. Sci.* **93:** 10343.
Engström Y., Kadalayil L., Sun S.C., Samakovlis C., Hultmark D., and Faye I. 1993. κB-like motifs regulate the induction of immune genes in *Drosophila*. *J. Mol. Biol.* **232:** 327.
Ferguson E.L. 1996. Conservation of dorsal-ventral patterning in arthropods and chordates. *Curr. Opin. Genet. Dev.* **6:** 424.
François V., Solloway M., O'Neill J., Emery J., and Bier E. 1994. Dorsal-ventral patterning of the *Drosophila* embryo depends on a putative negative growth factor encoded by the short gastrulation gene. *Genes Dev.* **8:** 2602.
Götz P. and Boman H.G. 1985. Insect immunity. In *Comprehensive insect Physiology* (ed. G. Kerkut and L. Gilbert), p. 453. Pergammon Press, Oxford, United Kingdom.
Gross I., Georgel P., Kappler C., Reichhart J.M., and Hoffmann J.A. 1996. *Drosophila* immunity: A comparative analysis of the Rel proteins dorsal and Dif in the induction of the genes encoding diptericin and cecropin. *Nucleic Acids Res.* **24:** 1238.
Hashimoto C., Gerttula S., and Anderson K.V. 1991. Plasma membrane localization of the *Toll* protein in the syncytial *Drosophila* embryo: Importance of transmembrane signaling for dorsal-ventral pattern formation. *Development* **111:** 1021.
Hoffmann J.A. 1995. Innate immunity of insects. *Curr. Opin. Immunol.* **7:** 4.
Holley S.A., Jackson P.D., Sasai Y., Lu B., De Robertis E.M., Hoffmann F.M., and Ferguson E.L. 1995. A conserved system for dorsal-ventral patterning in insects and vertebrates involving *sog* and *chordin*. *Nature* **376:** 249.
Hultmark D. 1993. Immune reactions in *Drosophila* and other insects: A model for innate immunity. *Trends Genet.* **9:** 178.
Huang J.D., Schwyter D.H., Shirokawa J.M., and Courey A.J. 1993. The interplay between multiple enhancer and silencer element defines the pattern of decapentaplegic expression. *Genes Dev.* **7:** 694.
Ip Y.T., Reach M., Engström Y., Kadalayil L., Cai H., Gonzalez-Crespo S., Tatei K., and Levine M. 1993. Dif, a dorsal-related gene that mediates an immune response in *Drosophila*. *Cell* **75:** 753.
Kopp E.B. and Ghosh S. 1995. NF-κB and Rel proteins in innate immunity. *Adv. Immunol.* **58:** 1.
Lemaitre B., Nicolas E., Michaut L., Reichhardt J.M., and Hoffman J.A. 1996. The dorsoventral regulatory gene cassette *spätzle/Toll/cactus* controls the potent antifungal response in *Drosophila* adults. *Cell* **86:** 973.
Lemaitre B., Meister M., Govind S., Georgel P., Steward R., Reichhart J.M., and Hoffmann J.A. 1995. Functional analysis and regulation of nuclear import of Dorsal during the immune response in *Drosophila*. *EMBO J.* **14:** 536
Medzhitov R., Preston-Hurlburt P., and Janeway C.A. 1997. A human homologue of the *Drosophila* Toll protein signals for activation of adaptive immunity. *Nature* **388:** 394.
Miller J.R. and Moon R.T. 1996. Signal-transduction through β-catenin and specification of cell fate during embryogenesis. *Genes Dev.* **10:** 2527.
Petersen U.M., Bjorklund G., Ip Y.T, and Engström Y. 1995. The dorsal-related immunity factor, Dif, is a sequence-specific trans-activator of *Drosophila cecropin* gene expression. *EMBO J.* **14:** 3146.
Piccolo S., Sasai Y., Lu B., and De Robertis E.M. 1996. Dorsoventral patterning in *Xenopus*: Inhibition of ventral signals by direct binding of chordin to BMP-4. *Cell* **86:** 589.
Reichhart J.M., Georgel P., Meister M., Lemaitre B., Kappler C., and Hoffmann J.A. 1993. Expression and nuclear translocation of the rel/NF-κB-related morphogen dorsal during the immune response of *Drosophila*. *C.R. Acad. Sci.* **316:** 1218.
Reichhart J.M., Meister M., Dimarcq J.L., Zachary D., Hoffman D., Ruiz C., Richards G., and Hoffmann J.A. 1992. Insect immunity: Developmental and inducible activity of the *Drosophila* diptericin promoter. *EMBO J.* **11:** 1469.
Roth S., Hiromi Y., Godt D., and Nüsslein-Volhard C. 1991. *cactus*, a maternal gene required for proper formation of the dorsoventral morphogen gradient in *Drosophila* embryos. *Development* **112:** 371.
Rowning B.A., Wells J., Wu M., Gerhart J.C., Moon R.T., and Larabell C.A. 1997. Microtubule-mediated transport of organelles and localization of β-catenin to the future dorsal side of *Xenopus* eggs. *Proc. Natl. Acad. Sci.* **94:** 1224.
Ryals J., Weymann K., Lawton K., Friedrich L., Ellis D., Steiner H.Y., Johnson J., Delaney T.P., Jesse T., Vos P., and Uknes S. 1997. The *Arabidopsis* NIM1 protein shows homology to the mammalian transcription factor inhibitor I-κB. *Plant Cell* **9:** 425.
Sasai Y., Lu B., Steinbeisser H., Geissert D., Gont L.K., and De Robertis E.M. 1994. *Xenopus chordin:* Anovel dorsalizing factor activated by organizer-specific homeobox genes.*Cell* **79:** 779.
Schmidt J.E., Suzuki A., Ueno N., and Kimelman D. 1995. Localized BMP-4 mediates dorsal/ventral patterning in the early *Xenopus* embryo. *Dev. Biol.* **169:** 37.
Sha W.C., Liou H.C., Tuomanen E.I., and Baltimore D. 1995. Targeted disruption of the p50 subunit of NF-κB leads to multifocal defects in immune responses. *Cell* **80:** 321.
Sipe J.D. 1985. Cellular and humoral components of the early inflammatory reaction. In *The acute-phase response to injury and infection: Research monographs in cell and tissue physiology* (ed. A.H. Gordon and A. Koj), p. 3. Elsevier, Amsterdam.
Staskawicz B.J., Ausubel F.M., Baker B.J., Ellis J.G., and Jones J.D. 1995. Molecular genetics of plant disease resistance. *Science* **268:** 661.
Vincent J.P., Oster G.F., and Gerhart J.C. 1986. Kinematics of gray crescent formation in *Xenopus* eggs: The displacement of subcortical cytoplasm relative to the egg surface. *Dev. Biol.* **113:** 484.
Weih F., Carrasco D., Durham S.K., Barton D.S., Rizzo C.A., Ryseck R.P., Lira S.A., and Bravo R. 1995. Multiorgan inflammation and hematopoietic abnormalities in mice with a targeted disruption of RelB, a member of the NF-κB/Rel family. *Cell* **80:** 331.

Nodal Signaling and Axis Formation in the Mouse

I. Varlet, J. Collignon,[1] D.P. Norris, and E.J. Robertson
Department of Molecular and Cellular Biology, Harvard University, Cambridge, Massachusetts 02138

Members of the transforming growth factor-β (TGF-β) superfamily of secreted growth factors function as key signaling molecules guiding morphogenetic processes throughout development (for review, see Hogan 1997). Work from our lab has focused on the *nodal* gene, originally identified due to its close linkage with the 413.d proviral integration site (Zhou et al. 1993; Conlon et al. 1994). Homozygous *413.d* mutant embryos lacking *nodal* expression fail to initiate primitive streak formation and are arrested at the gastrulation stage of development (Conlon et al. 1994).

The mouse *nodal* gene represents the prototype of a discrete subfamily of TGF-β proteins. To date, *nodal* homologs have been cloned from chick, *Xenopus*, and zebrafish (Jones et al. 1995; Smith et al. 1995; Joseph and Melton 1997; C. Wright, pers. comm.). Interestingly, multiple *Xenopus* and zebrafish *nodal* homologs have been identified, and experiments in *Xenopus* indicate they have distinct functional activities. For example, both *Xnr-1* and *Xnr-2* act as potent inducers of dorsal mesoderm cell types in manipulated frog embryos and can also completely rescue UV-ventralized embryos (Jones et al. 1995), whereas *Xnr-3* appears to be required for cell migration and movement (Smith et al. 1995). Strikingly, several aspects of the nodal signaling pathway are highly conserved between higher and lower vertebrates. Thus, recent studies have shown that the mouse *nodal* gene is transiently expressed in a discrete population of lateral plate mesoderm on the left side of the early somite stage embryo, where its expression appears to be functionally correlated with the direction of axial rotation and the establishment of the definitive left-right (L/R) axis (Collignon et al. 1996; Lowe et al. 1996). Similarly, both the chick *Cnr-1* (Levin et al. 1995) and *Xenopus Xnr-1* (Lowe et al. 1996) genes are asymmetrically expressed in left lateral mesoderm, suggesting that a conserved molecular pathway determines the vertebrate L/R body axis.

We have used gene targeting techniques to introduce a *lacZ* reporter gene under control of the endogenous *cis*-regulatory elements into the mouse *nodal* locus (Collignon et al. 1996). The increased resolution afforded by LacZ staining has allowed us to document a highly dynamic pattern of *nodal* expression during early stages of mouse development (Collignon et al. 1996; Varlet et al. 1997). Here, we review a series of genetic and mosaic experiments specifically designed to address the role of the nodal signaling pathway in tissue populations in which the gene is transiently expressed.

To distinguish nodal activities in the embryonic ectoderm and the primitive endoderm lineages of the embryo, we have examined reciprocal chimeras in which the endoderm cells are either genetically wild type or *nodal*-deficient. Collectively, these experiments show unique roles for nodal signaling in both the epiblast lineage and primitive endoderm during gastrulation and establishment of the anteroposterior (A/P) axis (Varlet et al. 1997). A particularly intruiging finding is that the rostral-most regions of the developing central nervous system (CNS) are consistently missing in embryos in which the primitive endoderm lacks nodal, thereby demonstrating an essential role for the primitive endoderm lineage in the establishment of anterior pattern during mouse gastrulation.

To determine where in the cascade of events determining the L/R body axis the *nodal* gene is likely to function, the *nodal.lacZ* allele has been introduced into mutant backgrounds known to affect the establishment of body situs. Asymmetric expression of *nodal* was found to be perturbed in both the *inv* and *iv* mutations (Collignon et al. 1996; Lowe et al. 1996), suggesting that *nodal* acts downstream from these mutations. Moreover, we have shown that in embryos carrying the *nodal.lacZ* allele and which are also heterozygous for a mutation in the transcription factor HNF3β, the asymmetric pattern of *nodal* is also highly disturbed, leading to the development of situs defects in later-stage embryos (Collignon et al. 1996). Here, we report that a proportion of these double heterozygous embryos also develop severe anterior defects including holoprosencephaly, accompanied by craniofacial defects of varying severity. It thus seems likely that this genetic interaction results in defects in both the formation of anterior midline structures including the prechordal plate and patterning of neural crest derivatives.

MATERIALS AND METHODS

Mouse strains and embryonic stem cells. The *nodal*^{lacZ} allele was maintained on a 129/Sv background, and animals were genotyped as described previously (Collignon et al. 1996). Mice carrying the 413.d retroviral integration were maintained on a 129/Sv background and genotyped by polymerase chain reaction (PCR) as described by Conlon et al. (1994). HNF3β mutant ani-

[1]*Present address:* Department of Developmental Neurobiology, Guy's Hospital, London SE1 9RT, England.

mals were maintained and genotyped as described by Ang and Rossant (1994). For analysis of nodal.LacZ expression during development, embryos were collected from matings between heterozygous *nodal*^{lacZ} males and 129/Sv or CD-1 (Charles River, Wilmington, MA) females and stained for β-galactosidase activity as described previously (Hogan et al. 1994), using either X-gal or Salmon-gal (Biosynth) as the substrate. Embryos were postfixed in 4% paraformaldehyde, embedded in paraffin wax and sectioned at 9 μm. Sections were dewaxed using standard procedures, mounted in either 80% glycerol or Permount, and photographed under Normarski optics. Embryonic stem (ES) cell lines carrying either the BT-5 or ROSA26 *lacZ* reporter gene were prepared as described previously (Varlet et al. 1997). The experiments presented here were carried out using two independent *lacZ*$^{+}$ wild-type ES cell lines (R26.1 and BT5.7) and two *lacZ*$^{+}$*413.d* homozygous *nodal*-deficient ES cell lines (ER.1 and ER.4).

Generation and analysis of chimeras. Chimeras were generated by blastocyst injection as described previously (Bradley 1987). Blastocysts were collected from matings between *413.d* heterozygous animals or from outbred CD-1 strain animals (Charles River, Wilmington, MA). For production of chimeric embryos with differing ES cell contributions, the number of ES cells injected into the blastocoel cavity was varied from between 2 and 4 cells up to a maximum of 12–14 cells. Following transfer into pseudopregnant foster females, the manipulated embryos were recovered at day 10.5 of development. Embryos were fixed and processed for either β-galactosidase staining or in situ hybridization. In experiments using embryos from *413.d* intercrosses, the genotype of the host blastocyst was determined retrospectively by PCR analysis of DNA samples prepared from the endodermal fraction of individual visceral yolk sacs.

Whole-mount in situ hybridization and immunohistochemistry. Whole-mount in situ hybridization using digoxygenin-labeled RNA probes was performed as described by Wilkinson (1992). Additionally, a fluorescein-labeled RNA probe detected with Magenta-Phos substrate (Biosynth) was used for double labeling as described by Levin et al. (1995). The *Shh*, *En-1*, *Ox-2*, *Krox-20,* and *lefty* probes have been described previously (Echelard et al. 1993; Davis and Joyner 1988; Ang et al. 1994; Wilkinson et al. 1989; Meno et al. 1996). The *Wnt-8b* probe was kindly provided by Scott Lee and Andrew McMahon (S. Lee and A. McMahon, unpubl.). To evaluate Shh expression, embryos stained for β-galactosidase activity were subsequently stained with antibodies raised

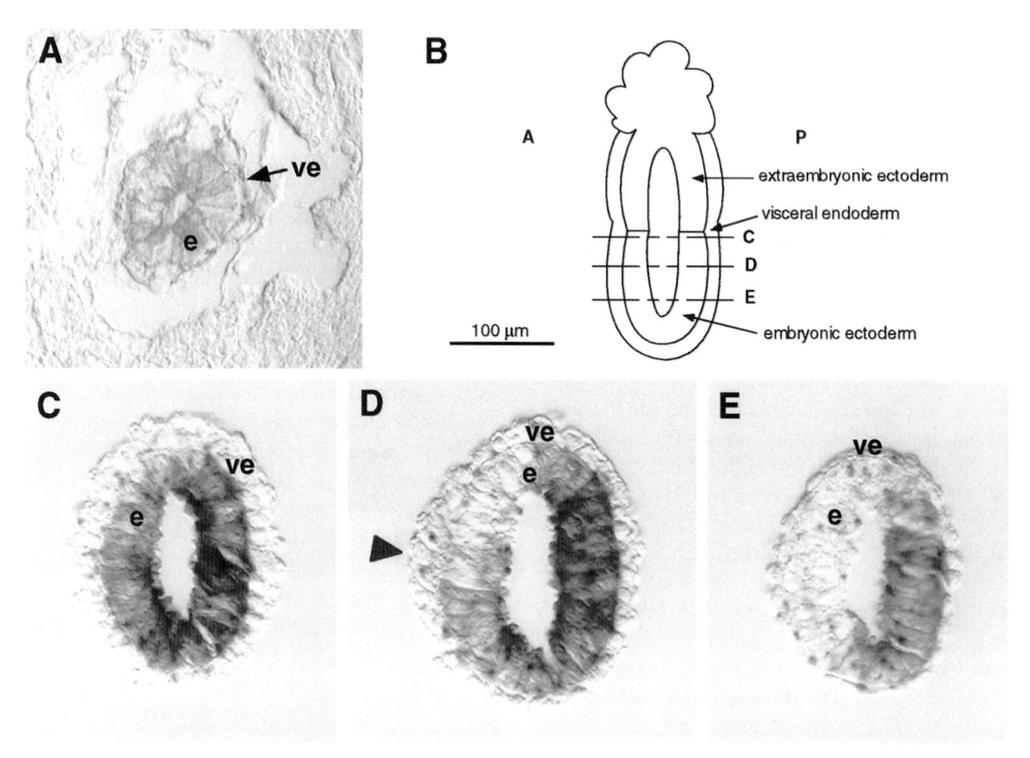

Figure 1. Dynamic pattern of *nodal.lacZ* expression during gastrulation. Embryos heterozygous for the *nodal*^{lacZ} allele recovered at 5.5, 6.0, and 6.5 dpc were stained with X-gal. (*A*) Transverse section through a 5.5 dpc embryo within the deciduum. Low levels of expression are seen in both the visceral endoderm and the epiblast. (*B*) Schematic diagram of a 6.0 dpc embryo (prior to streak formation) showing the appropriate locations of the transverse sections presented in *C–E*. Higher levels of nodal expression are broadly distributed in the epiblast adjacent to the extra embryonic junction (*C*). In more distal regions, LacZ becomes localized to the presumptive posterior side of the embryo (*D,E*). Low levels of expression are seen throughout the visceral endoderm *(C–E)* and in the anterior epiblast *(D,E*). A distinct patch of thickened visceral endoderm (*red arrow*) can be seen at the prospective anterior side of the embryo in the section shown in *D*. The anterior side of the embryo is toward the left in *C–E*. Abbreviations: (e) Embryonic ectoderm; (m) embryonic mesoderm; (ps) primitive streak; (ve) visceral endoderm; (xe) extraembryonic ectoderm; (xm) extraembryonic mesoderm; (A) anterior side of the embryo; (P) posterior side of the embryo.

against the 19-kD form of the protein as described previously (Bumcrot et al. 1995; Martí et al. 1995). Following photography, embryos were processed for sectioning.

RESULTS

Nodal Expression Domains in Early Mouse Embryos

To determine the onset of *nodal* expression, embryos carrying the *nodal*^{lacZ} allele (Collignon et al. 1996) were examined for LacZ activity from 3.5-day postcoitum (dpc) onward. No detectable activity was observed at the blastocyst stage (data not shown). However, shortly after implantation (5.5 dpc), low levels of staining were detected throughout the embryonic ectoderm and associated overlying primitive endoderm (Fig. 1A). By approximately 6.0 dpc, just prior to overt streak formation, LacZ activity was detected in the embryonic epiblast and primitive endoderm (Fig. 1B). Low but uniform LacZ expression was seen throughout the visceral endoderm cell population (Fig. 1B–E), and was most noticeable in a morphologically distinct region of anterior endoderm lying approximately 10–15 μm distal to the junction of the embryonic and extraembryonic tissues, which retains a more thickened character. However, the more obvious staining seen in this population likely reflects increased cell density since analysis of numerous independent embryos revealed LacZ-positive cells in all regions of the visceral endoderm, including the proximal region overlying the extraembryonic ectoderm. In contrast, within the embryonic ectoderm, nodal activity was highest in the posterior proximal quadrant, suggesting gene activity is down-regulated on the prospective anterior side of the embryo. Interestingly, expression remained radially symmetrical at levels coincident with or just distal to the region where the embryonic and extraembryonic cell populations abut (Fig. 1C). However, 10–15 μm distal to this region corresponding to the level at which a distinct thickening of the anterior visceral endoderm is visible, expression is clearly strongest on the prospective posterior side of the egg cylinder (Fig. 1D). Similarly, toward the distal tip of the egg cylinder, LacZ expression was markedly higher on the posterior side (Fig. 1E). These results demonstrate that the primitive streak forms at the site of highest *nodal* activity within the ectoderm.

The pattern of *nodal* expression changes rapidly during the next few hours of development. Within the ectoderm, *nodal* expression was rapidly lost as the streak elongates distally, and by late streak stages, only a few cells, strictly confined to the posterior side of the embryo, show weak LacZ staining (Fig. 2A) (Collignon et al. 1996). Previous studies indicated that *nodal* transcripts are most prevalent

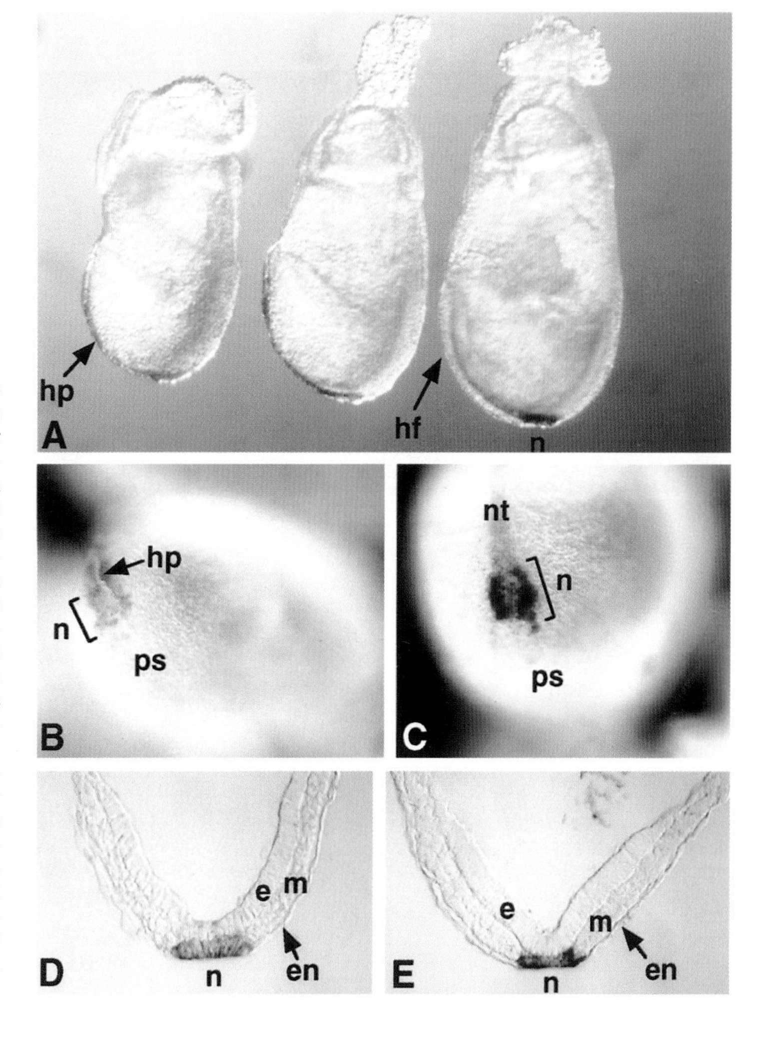

Figure 2. Nodal and Shh expression during node formation. (*A*) Lateral views of 7.5 dpc embryos heterozygous for the *nodal*^{lacZ} allele stained for β-galactosidase activity. LacZ staining is progressively down-regulated in the primitive streak (left-hand side embryo) and becomes localized to the node by the headfold stage (right-hand side embryo). Cells of the midline are stained by anti-Shh antibody (*brown*). (*B*) Whole-mount analysis of a late streak stage embryo (ventral view) showing Shh (*brown*) and nodal (*blue*) expressing cells in the distal tip of the primitive streak during the initial stages of node formation. Shh expression extends anterior of the streak in the cells of the head process. (*C*) Following the formation of a distinctive node (headfold stage), cells at the periphery of the notochordal plate express nodal in a domain adjacent to and partially overlapping with Shh expressing cells. (*D,E*) Frontal sections through the node region of embryos of a late streak and a presomite embryo, respectively, showing persisting nodal expression in the notochordal plate. Abbreviations: (e) Ectoderm; (en) endoderm; (hf) headfold; (hp) head process; (m) mesoderm; (n) node; (nt) notochord; (ps) primitive streak.

in the region of the mature node at 7.5 dpc (Zhou et al. 1993; Conlon et al. 1994). To extend these observations, we have further investigated the origin of this node-specific domain of expression. Earlier at the allantoic bud stage, LacZ expression appears in a discrete cell population at the distal tip of the streak (Fig. 2B). To further localize these cells, LacZ-stained embryos representing late streak to early headfold stages were incubated with antibodies specific for the 19-kD form of Sonic hedgehog (Shh) protein (Bumcrot et al. 1995). At these stages, Shh expression has been detected in midline cells of the prechordal mesoderm and in the forming notochordal plate and notochord (Fig. 2) (Martí et al. 1995). We found that *nodal*-expressing cells first appear at the most caudal limit of the Shh expression domain. A few hours later, following the appearance of a morphologically distinctive node, *nodal*-expressing cells were exclusively confined to the outermost ventral layer of the notochordal plate and were present at the lateral-most aspect of the Shh expression domain (Fig. 2C–F) (Collignon et al. 1996). In sum, *nodal* is expressed prior to primitive streak formation. Expression continues during the initial stages of streak induction and is then rapidly down-regulated as the streak elongates. Subsequently, *nodal* expression is detected in a small subset of node progenitors, and following the formation of the morphologically distinct node becomes restricted to the edges of the notochordal plate.

By early somite stages, *nodal* expression becomes markedly asymmetrical with respect to the prospective L/R axis. Consistently more LacZ-positive cells are present on the left side of the node as opposed to the right (Collignon et al. 1996). Additionally, at the 3–4-somite stage, an asymmetric domain of nodal becomes apparent in the somatic and splanchnic mesoderm of the lateral plate on the left side of the embryo (Fig. 3A,B). This asymmetric expression is evident before the embryo displays overt morphological asymmetry, but following embryonic turning, nodal expression is lost from the lateral plate (Fig. 3C). At 9.5–10.5 days of development, *nodal.lacZ* expression is restricted to a region of the dorsal midline of the forebrain and in the base of the second branchial arch (Fig. 3D). We failed to detect *nodal.lacZ* activity in embryonic tissues at later embryonic stages.

Nodal Expression Is Required in Both the Primitive Endoderm and Epiblast for Normal Gastrulation

The gastrulation defect displayed by *nodal*-deficient embryos could potentially reflect an essential role for nodal signaling in the primitive endoderm, the embryonic ectoderm, or both cell lineages. To distinguish these possibilities, we undertook a mosaic analysis using wild-type and *nodal*-deficient ES cells. It is known that ES cells display a marked developmental bias when introduced into recipient blastocysts, almost exclusively colonizing the

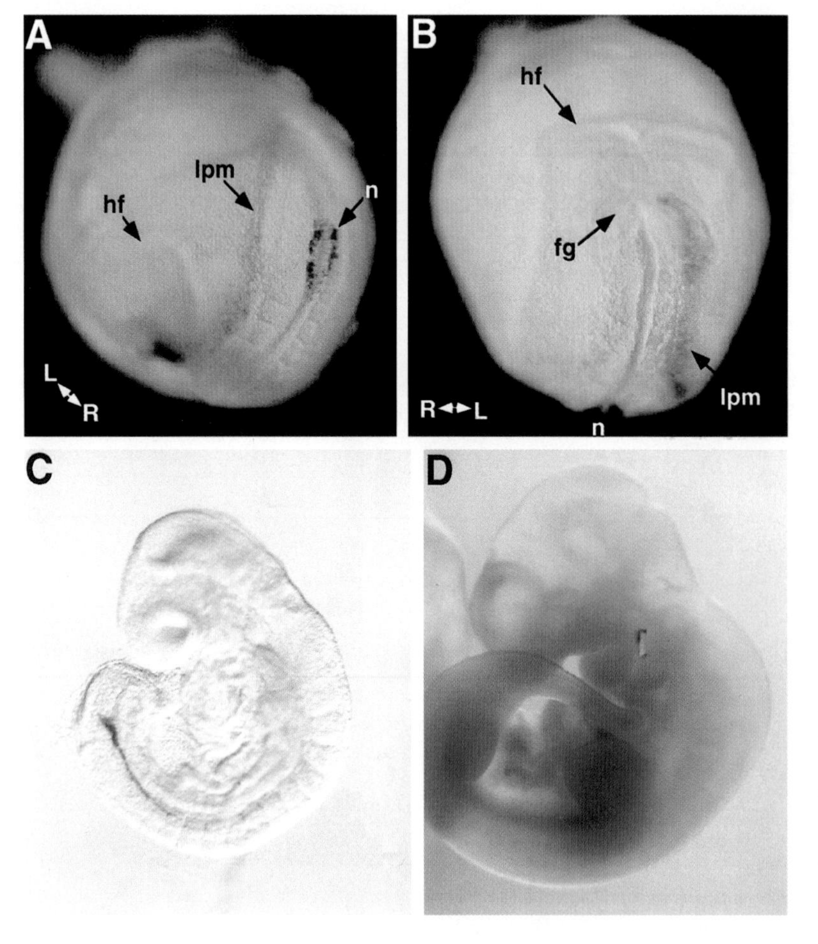

Figure 3. Expression of *nodal.lacZ* between 8.5 and 9.5 days of development. (*A,B*) Asymmetric *nodal.lacZ* expression in lateral plate mesoderm, viewed as ventral and frontal views, respectively, at the 4-somite stage. In addition to the mature node, nodal expression is observed in a band of mesodermal cells present on the left side of the embryo, with a rostral limit at the position of the forming heart. Nodal expression is present in both the somatic and splanchnic mesoderm. (*C*) Following the process of embryonic turning, nodal expression is no longer detectable in the lateral plate mesoderm. *nodal.lacZ* activity is retained in cells of the tail bud. (*D*) At 9.5 days of development, *nodal.lacZ* activity is confined to the dorsal midline of the telencephalon, and to a small goup of cells located between the first and second branchial arches.

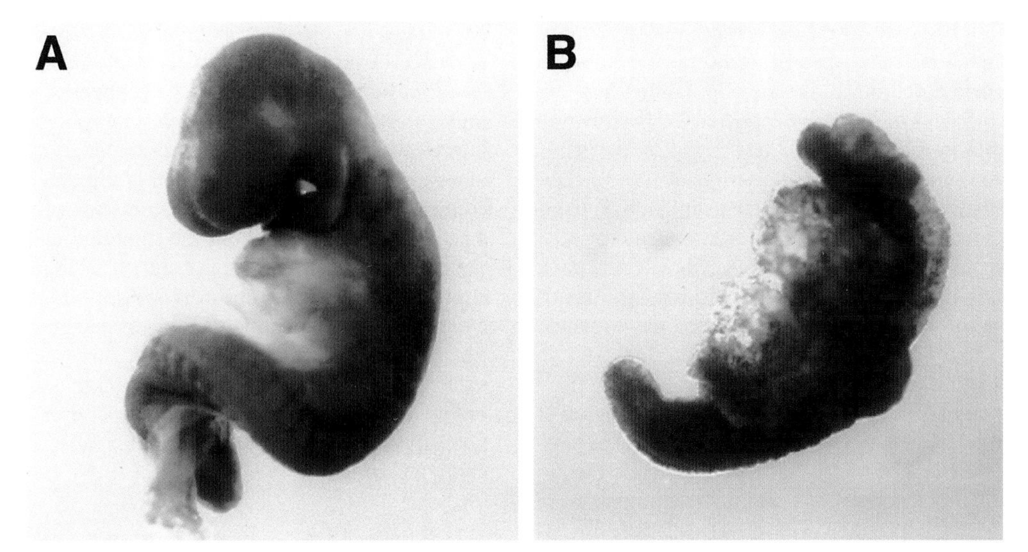

Figure 4. Chimeras in which the primitive endoderm is wild-type and the ES cells are nodal-deficient show no defects in formation of the anterior aspects of the axis (*A*). In comparison, extensively wild-type chimeras in which the primitive endoderm is *nodal*-deficient exhibit anterior defects at 10.5 days (*B*). In both cases, the respective ES cell derivatives are distributed uniformly along the A/P and D/V axes.

embryonic epiblast (Beddington and Robertson 1989; Varlet et al. 1997). Thus, mosaic embryos contain ES cell derivatives largely confined to the embryonic portion, whereas the primitive endoderm and its derivatives are of host origin. In the present experiments, we used ES cell lines carrying ubiquitously expressed *lacZ* transgenes to simultaneously mark and follow the fates of the injected ES cells.

To test the role of nodal activity in the primitive endoderm and epiblast, a series of reciprocal chimeras were constructed. Thus, wild-type ES cells were injected into *nodal*-deficient blastocysts to generate chimeric conceptuses in which the primitive endoderm was composed exclusively of *nodal*-deficient cells. Alternatively, *nodal*-deficient ES cells injected into wild-type blastocysts gave chimeras in which the primitive endoderm was genetically wild type.

We initially tested whether the introduction of wild-type ES cells could rescue the gastrulation defect in nodal-deficient *413.d* mutant embryos. Increasing numbers of wild-type $lacZ^+$ ES cells were injected into host blastocysts recovered from intercross matings between *413.d* heterozygous animals. Chimeras were identified by LacZ staining of 10.5 dpc conceptuses. All of these were judged to have undergone gastrulation, since they were developing within a distinctive visceral yolk sac (VYS). Approximately 15% of the chimeras proved to be derived from *413.d* mutant blastocysts. These embryos had clearly gastrulated to produce extraembryonic mesodermal tissue but were all morphologically abnormal, and smaller in comparison to their littermates. Interestingly, the extent of development of the A/P axis in these rescued animals correlated strongly with the degree of colonization by wild-type $lacZ^+$ ES cells. When the wild-type contribution was 10% or less, the chimeric embryos resembled simple cylinders of cells lacking overt morphological hallmarks. More strongly colonized embryos (10–30% wild-type contribution) displayed a more robust A/P axis with morphologically distinct anterior and posterior structures including a neural tube and somites. In contrast, extensively colonized chimeras (wild-type contribution 30% or higher) were relatively well developed (Fig. 4B). In the majority of extensive chimeras, the trunk and caudal-most structures appeared grossly normal (Figs. 4B and 5A). However, we observed pronounced abnormalities affecting the development of anterior neural structures. For example, as shown in Figure 4B, the neural structures rostral to the otic vesicle are clearly defective.

Interestingly, none of a large number of chimeras generated using nodal-mutant ES cells, in which the majority of the epiblast derivatives were *nodal*-deficient, showed overt abnormalities when examined at the mid-streak to early headfold stage (not shown). To exclude the possibility that chimeras containing a large component of *nodal*-mutant cells develop anterior patterning defects at later stages, the embryonic development of both classes of chimera was compared at 10.5 dpc. In marked contrast to the clear anterior defects present in extensively chimeric embryos obtained from *nodal*-mutant blastocysts (Fig. 4B), all of the chimeras examined in which $lacZ^+$ *nodal*-deficient ES cell derivatives comprised greater than 80% of the embryo showed correct patterning of the anterior CNS structures (Fig. 4A). Collectively, these experiments allow us to conclude that nodal activity in the cells of the epiblast is required to initiate gastrulation. Moreover, the anterior defects documented in chimeras generated from *nodal*-mutant blastocysts can be attributed to loss of nodal signaling in the primitive endoderm during early gastrulation.

Molecular Characterization of the Anterior Defects in Chimeras

To more clearly delineate the nature of the anterior truncations described above, day-10.5 chimeras were further analyzed by assessing the expression of *Krox-20*, *En-1*, and *Wnt-8b* mRNAs, specific markers of hindbrain, midbrain, and forebrain tissue subpopulations, respectively. At 9.5 days of development, *Krox-20* is known to be expressed in rhombomeres 3 and 5 and by day 10 is confined to rhombomere 5 (Wilkinson et al. 1989). *Wnt-8b* marks the dorsal region of the telencephalon, and expression extends rostrally into the diencephalon at 10.5 days of development (S. Lee and A. McMahon, pers. comm.). In double-labeling in situ hybridization experiments, control chimeric embryos showed the expected patterns of *Krox-20* and *Wnt-8b* expression (Fig. 5A). In contrast, extensively rescued chimeras generated from *nodal* mutant blastocysts express *Krox-20* in the region of the otic vesicle, but there was no evidence for the expression of *Wnt-8b* (Fig. 5A). To further characterize the anterior neural tissue populations, additional chimeras were analyzed for the expression of *En-1*. At 10.5 dpc, *En-1* is strongly expressed throughout a ring of neural tissue at the hindbrain-midbrain junction (Davis and Joyner 1988). In keeping with the *Krox-20* results, none of the poorly rescued chimeras analyzed showed evidence for expression of *En-1* mRNA in the developing CNS, although a specific hybridization signal was present in the developing somites and limb buds. However, in extensively rescued chimeras, the hindbrain-midbrain region, or isthmus, had clearly been induced as evidenced by the presence of a characteristic stripe of *En-1* expression (data not shown). In poorly rescued chimeras, the anterior regions of the embryo were highly morphologically abnormal, with very little tissue present anterior to the otic vesicles. To establish the character of this anterior region, chimeras were assessed for the expression of *Otx-2*, normally expressed throughout the midbrain and forebrain regions (Ang et al. 1994). As shown in Figure 5B, the chimeras displayed a robust hybridization signal in the anterior region, indicating that this tissue retains an anterior neural character. Collectively, these gene marker studies show that whereas posterior regions of the midbrain encompassing the hindbrain-midbrain boundary do form, the rostral-most neural structures fail to form in this class of chimera.

Genetic Interactions between Nodal and HNF3β

We and others previously provided genetic evidence supporting an important role for the nodal-signaling pathway in the establishment of normal body situs. Thus, the sidedness of nodal expression in the lateral plate mesoderm is perturbed in both the *inv* (Collignon et al. 1996; Lowe et al. 1996) and *iv* mutant backgrounds (Lowe et al. 1996). Similar effects are seen on the expression of *lefty*,

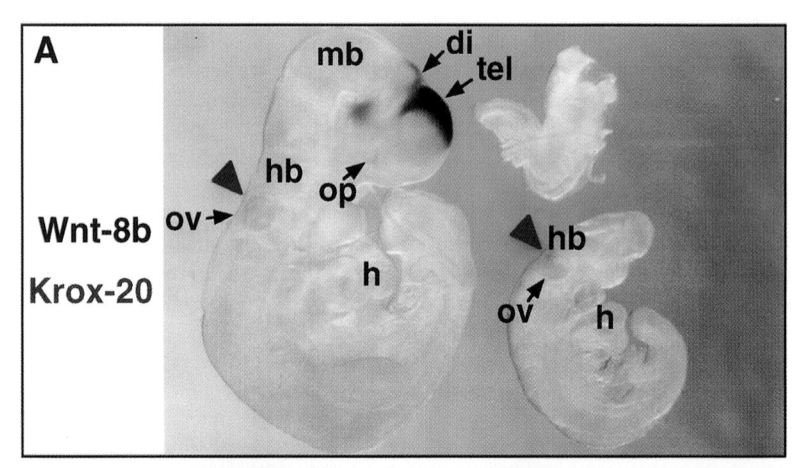

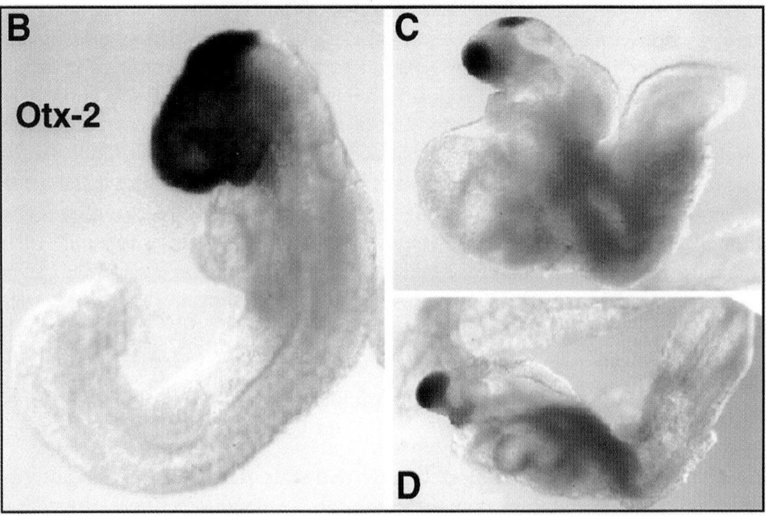

Figure 5. Anterior neural defects found in chimeras at 10.5 dpc. (*A*) Whole-mount in situ hybridization analysis of *Wnt-8b* (*purple stain*) and *Krox-20* (*light pink stain, arrowhead*) expression in wild type (*left*) and two rescued chimeras (*right*). Wnt-8b expression is absent in both rescued chimeras. *Krox-20* expression in r5 is detected in the chimera exhibiting the most extensive rescue. (*B,C,D*) Otx-2 expression in wild-type (*B*) embryos and in two rescued chimeras (*C,D*).

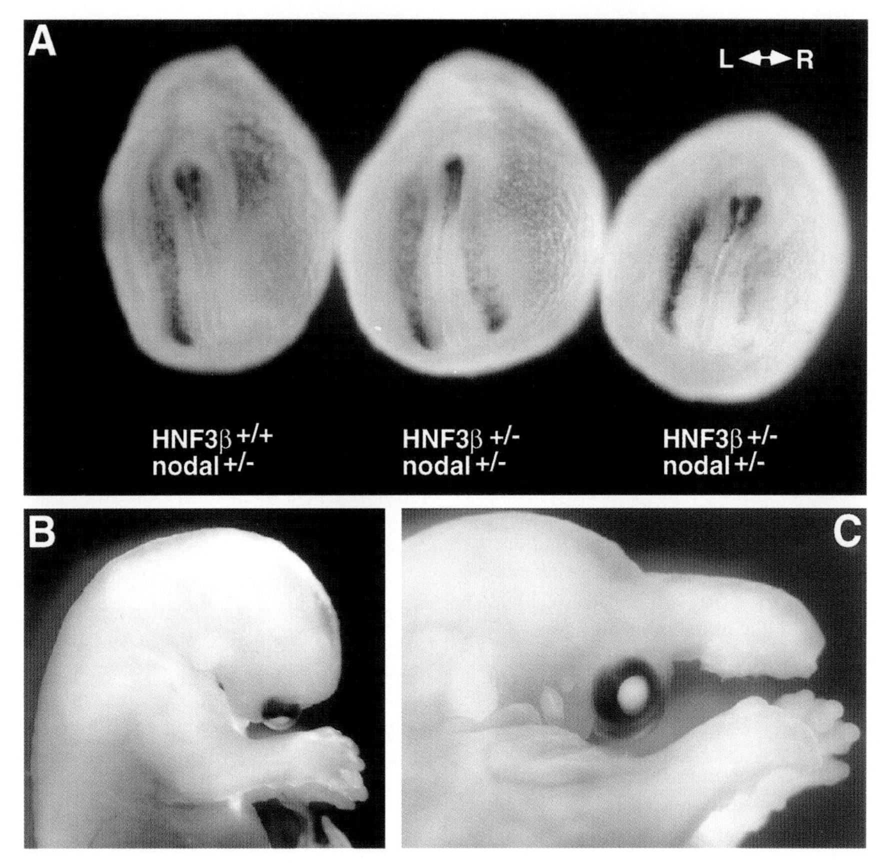

Figure 6. Genetic interactions between *nodal* and *HNF3β*. (*A*) Misexpression of *nodal.lacZ* in double heterozygous embryos stained at the 8–12-somite stage. Views of two double heterozygous embryos (*right)* and a HNF3*β* wild-type embryo (*left*) showing bilateral and left-sided LacZ expression, respectively. (*B,C*) At 15.5 days, approximately half of the double heterozygous mice are characterized by severe craniofacial defects including cyclopia (*B*) and holoprosencephaly (*C*).

a novel member of the TGF-β superfamily also normally expressed on the prospective left side of the body axis (Meno et al. 1996). These observations suggest that both growth factor genes contribute to the generation of normal heart and body situs.

We previously uncovered a second genetic interaction between nodal and HNF3β, a member of the forkhead family of transcriptional factors essential for mouse gastrulation and node formation (Ang and Rossant 1994; Weinstein et al. 1994). Thus, embryos heterozygous for mutations in both HNF3β and nodal showed bilateral LacZ staining in lateral plate mesoderm (Fig. 6A) (Collignon et al. 1996). We have extended these experiments to test whether expression of *lefty* is similarly disturbed in this genetic background. Whole-mount in situ hybridization experiments using a *lefty* cDNA probe demonstrate that many of these double heterozygous embryos also show aberrant expression patterns. These include bilateral expression patterns or complete loss of expression (I. Varlet and E. Robertson, unpubl.). These experiments indicate that *nodal* and *lefty* genes are likely to be regulated by common mechanisms. Transgenic experiments using genomic fragments from the *nodal* locus linked to a minimal promoter driving a *lacZ* reporter sequence have allowed us to determine that the elements responsible for directing nodal expression to the left lateral plate mesoderm reside within the body of the gene (D. Norris and E. Robertson, unpubl.).

The functional consequence of the striking deregulation of asymmetric expression of *nodal* in the nodal$^{+/-}$; HNF3β$^{+/-}$ double heterozygotes is that a proportion of the embryos develop situs defects, affecting positioning of the abdominal viscera and the heart. Moreover, of a panel of embryos dissected at day 15.5 of development, a number were found to have developed severe craniofacial defects, including cyclopia and holoprosencephaly (Fig. 6B). Genotyping of affected individuals confirmed that, without exception, they were double heterozygotes. However, only approximately half of the double heterozygotes display abnormalities at this stage of development. In keeping with this observation, genotyping of liveborn offspring from these crosses shows that the remaining double heterozygotes are viable and fertile. Preliminary histological analysis of the severely affected class of embryos indicates they present with a highly complex and variable phenotype. Thus, it appears that defects in both the formation of ventral and dorsal midline structures, together with neural crest patterning defects (J. Collignon and G. Koentges, unpubl.), contribute to the development of this distinctive phenotype.

DISCUSSION

Experiments presented here underscore the important part played by the nodal signaling pathway in multiple tissue sites during development. In keeping with the phenotypic abnormalities originally described in the 413.d strain, the *nodal* gene is first expressed in the pregastrulation stage of development. The chimera experiments indicate that very low levels of nodal activity provided by a minor population of wild-type cells are sufficient to alleviate the block to gastrulation in *nodal*-deficient epiblast tissue. Thus, the introduction of a small number of *lacZ*$^+$ wild-type ES cells is sufficient to allow a *nodal*-mutant embryo to develop a well-elaborated A/P axis. These results suggest that localized nodal signaling within the embryonic ectoderm is necessary to promote the formation of the primitive streak.

Shortly after implantation, *nodal* is transiently expressed in the layer of primitive endoderm cells that invests the epiblast tissue. Cell-marking and mosaic experiments have shown that the primitive endoderm lineage contributes exclusively to the extraembryonic tissues (for review, see Tam and Beddington 1992). Beyond possible roles in supporting the growth of the underlying ectoderm, the primary function of the visceral endoderm remains poorly understood. Because *nodal* is expressed in both the endoderm and ectoderm lineages, we devised an experimental strategy that allowed us to test whether *nodal* expression was essential for any aspects of endoderm function. These studies show that the transient expression of *nodal* in the visceral endoderm is not required for the initiation of gastrulation. Strikingly, nodal signaling in this cell layer appears to be essential to confer correct anterior patterning of the neural plate at later stages of development. Thus, mosaic embryos composed largely of wild-type cells developing in combination with mutant primitive endoderm lack the most rostral aspects of the axis. Conversely, mosaic embryos composed largely of *nodal*-mutant ectoderm derivatives developing in conjunction with wild-type primitive endoderm form a normal A/P axis. We found that the forebrain, and possibly anterior midbrain, structures fail to form in the absence of *nodal*-expressing primitive endoderm. As induction of the forebrain, and possibly regions of the midbrain in the vertebrate CNS, is contingent on signals provided by the prechordal plate tissue (for review, see Shimamura et al. 1995), it seems likely that formation of this tissue is adversely affected in these chimeras. A number of morphological studies have shown that the prechordal plate forming at the rostral midline of the embryo is composed of a complex population of closely associated endoderm and anterior mesoderm cells (Poelmann 1981; Sulik et al. 1994). Although the embryonic origins of the endoderm cells that contribute to the prechordal plate have yet to be established, our experiments demonstrate that *nodal* expression in the primitive visceral endoderm prior to gastrulation is required for the correct morphogenesis of this population of cells, and hence the establishment of correct anterior positional identity in the developing neural tissue.

Recently, evidence has been provided for important inductive interactions between the primitive endoderm and underlying ectoderm during early embryogenesis. Thus, expression of the homeobox gene *Hesx1/Rpx* (Thomas and Beddington 1996), first detected at the start of gastrulation, is confined to a small anterior domain of primitive endoderm, but a few hours later becomes apparent in the underlying anterior ectoderm. This second domain of ectodermal *Hesx1* expression is in part dependent on signals from the endoderm, since expression in the ectoderm is lost or severely depleted by physical removal of the anterior endoderm at earlier stages (Thomas and Beddington 1996). A similar finding has been reported for *Otx-2*. Thus, a *lacZ* reporter allele of *Otx-2* is initially activated in the visceral endoderm layer of *Otx-2* mutant embryos but then fails to be expressed in the underlying anterior ectoderm (Acampora et al. 1995), suggesting that expression of *Otx-2* itself in the anterior endoderm is normally required for its subsequent induction in the ectoderm. In keeping with these findings, a series of elegant tissue recombination experiments have endorsed an important role for anterior mesendoderm populations in patterning the developing neural plate (Ang and Rossant 1993; Ang et al. 1994).

Mosaic embryos developing within a *nodal*-deficient visceral endoderm exhibit a distinctive physical constriction between the embryonic and extraembryonic regions (Varlet et al. 1997) similar to that seen in *HNF3β*-, *Otx-2*-, and *Lim-1*-deficient mutants. Since these mutants also exhibit defects in the formation of anterior regions of the neural axis, it is tempting to speculate that these molecules participate in a common pathway which is initiated prior to gastrulation by signals provided from the primitive endoderm. Evidence is now accumulating that the primitive endoderm is patterned with respect to the prospective A/P axis at early stages of postimplantation mouse development (Thomas and Beddington 1996). It will be interesting to examine whether *nodal*, expressed throughout the visceral endoderm of the pregastrulation stage embryo, participates in setting up this marked regional identity within the endoderm.

A variety of experiments in mouse, chick, and *Xenopus* have endorsed an important role for the nodal-signaling pathway in the development of L/R body asymmetry (for review, see Varlet and Robertson 1997). However, the mechanism whereby the initial L/R axis is specified remains to be determined, although manipulations in *Xenopus* (Lohr et al. 1997) and chick embryos (Levin et al. 1995) have implicated signals from the midline and node, respectively, as potential sources regulating the expression of *nodal*. Here, we show a genetic interaction between *nodal* and *HNF3β*, both of which are coexpressed in the primitive endoderm and node of the embryo. Thus, in a proportion of doubly heterozygous embryos, the asymmetric distribution of both *nodal* and *lefty* mRNAs is altered. In keeping with previous data, these disturbances are associated with a high incidence of situs defects in later-stage embryos. Unexpectedly, however, numerous double heterozygous embryos develop very severe craniofacial defects, in the most extreme cases presenting with holoprosencephaly and cyclopia. Similar anterior defects have been described in mouse embryos lacking the *Shh* gene (Chiang et al. 1996). However, in contrast to the *Shh* mutants, here we show that the remainder of the body axis develops normally. It seems likely that the genetic interaction between *HNF3β, nodal,* and *lefty* is indirect. Whatever the mechanism, collectively, this interaction appears to impact on aspects of the Shh signaling pathway, leading to subtle defects in production of midline structures such as the rostral regions of the notochord and the prechordal plate, known to confer patterning to the anterior aspects of the CNS. Further studies are required to unravel the defects in tissue interactions and concomitant alterations in gene expression patterns that underlie the situs and anterior craniofacial defects arising in these embryos.

ACKNOWLEDGMENTS

We thank Andy McMahon, Scott Lee, Eliza Marti, and David Bumcrot for kindly providing the Wnt8b probe and Shh antibodies. These experiments were supported by a grant from the National Institutes of Health (HD-25208) to E.J.R and by postdoctoral fellowships from the HFSP (I.V.), EMBO (J.C.), and the Jane Coffin Childs Foundation (D.P.N.).

REFERENCES

Acampora D., Mazan S., Lallemand Y., Avantaggiato V., Maury M., Simeone A., and Brûlet P. 1995. Forebrain and midbrain regions are deleted in *Otx2*$^{-/-}$ mutants due to a defective anterior neuroectoderm specification during gastrulation. *Development* **121:** 3279.

Ang S.-L. and Rossant J. 1993. Anterior mesendoderm induces mouse *Engrailed* genes in explant cultures. *Development* **118:** 139.

———. 1994. *HNF-3β* is essential for node and notochord formation in mouse development. *Cell* **78:** 561.

Ang S.-L., Conlon R.A., Jin O., and Rossant J. 1994. Positive and negative signals from mesoderm regulate the expression of mouse *Otx2* in ectoderm explants. *Development* **120:** 2979.

Beddington R.S.P. and Robertson E.J. 1989. An assessment of the developmental potential of embryonic stem cells in the midgestation embyro. *Development* **105:** 733.

Bradley A. 1987. Production and analysis of chimeric mice. In *Teratocarcinomas and embryonic stem cells: A practical approach* (ed. E. J. Robertson), p. 131. IRL Press, Oxford, United Kingdom.

Bumcrot D.A., Takada R., and McMahon A.P. 1995. Proteolytic processing yields two secreted forms of Sonic hedgehog. *Mol. Cell. Biol.* **15:** 2294.

Chiang C., Litingtung Y., Lee E., Young K.E., Corden J.L. Westphal H., and Beachy P.A. 1996. Cyclopia and defective axial patterning in mice lacking Sonic hedgehog gene function. *Nature* **374:** 356.

Collignon J., Varlet I., and Robertson E. J. 1996. Relationship between asymmetric *nodal* expression and the direction of embryonic turning. *Nature* **381:** 155.

Conlon F.L., Lyons K.M., Takaesu N., Barth K.S., Kispert A., Herrmann B., and Robertson E.J. 1994. A primary requirement for *nodal* in the formation and maintenance of the primitive streak in the mouse. *Development* **120:** 1919.

Davis C.A. and Joyner A.L. 1988. Expression patterns of the homeo box-containing genes *En-1* and *En-2* and the proto-oncogene *int-1* diverge during mouse development. *Genes Dev.* **2:** 1736.

Echelard Y., Epstein D.J., St. Jacques B., Shen L., Mohler J., McMahon J.A., and McMahon A.P. 1993. Sonic hedgehog, a member of a family of putative signaling molecules, is implicated in the regulation of CNS polarity. *Cell* **75:** 1417.

Hogan B.L.M. 1997. Bone morphogenetic proteins: Multifunctional regulators of vertebrate development. *Genes Dev.* **10:** 1580.

Hogan B., Beddington R., Costantini F., and Lacy E. 1994. *Manipulating the mouse embryo: A laboratory manual*, 2nd edition. Cold Spring Harbor Laboratory Press, Cold Spring Harbor, New York.

Jones C.M., Kuehn M.R., Hogan B.L.M., Smith J.C., and Wright C.V.E. 1995. Nodal-related signals induce axial mesoderm and dorsalize mesoderm during gastrulation. *Development* **121:** 3651.

Joseph E.M. and Melton D.A. 1997. *Xnr-4*, a *Xenopus* nodal-related gene expressed in the Spemann organizer. *Dev. Biol.* **183:** 367.

Levin M., Johnson R.L., Stern C.D., Kuehn M., and Tabin C. 1995. A molecular pathway determining left-right asymmetry in chick embryogenesis. *Cell* **82:** 803.

Lohr J.L., Danos M.C., and Yost H.J. 1997. Left-right aysmmetry of a nodal-related gene is regulated by dorsoanterior midline structures during *Xenopus* development. *Development* **124:** 1465.

Lowe L.A., Supp D.M., Sampath K., Yokoyama T., Wright C.V.E., Potter S.S., Overbeek P., and Kuehn M.R. 1996. Conserved left-right asymmetry of nodal expression and alterations in murine *situs inversus*. *Nature* **381:** 158.

Martí E., Takada R., Bumcrot D.A., Sasaki H., and McMahon A.P. 1995. Distribution of Sonic hedgehog peptides in the developing chick and mouse embryo. *Development* **121:** 2537.

Meno C., Saijoh Y., Jujii H., Ikeda M., Yokoyama M., Toyoda Y., and Hamada H. 1996. Left-right asymmetric expression of the TGFβ family member lefty in mouse embryos. *Nature* **381:** 151.

Poelmann R.E. 1981. The head process and the formation of definitive endoderm in the mouse embryo. *Anat. Embryol.* **162:** 41.

Shimamura K., Hartigan D.J., Martinez S., Puelles L., and Rubenstein J.R. 1995. Longitudinal organization of the anterior neural plate and neural tube. *Development* **121:** 3923.

Smith W.C., McKendry R., Ribisi Jr. S., and Harland R.M. 1995. A *nodal*-related gene defines a physical and functional domain within the Spemann organizer. *Cell* **82:** 37.

Sulik K., Dehart D.B., Inagaki T., Vrablic J.L., Gesteland K., and Schoenwolf G.C. 1994. Morphogenesis of the murine node and notochordal plate. *Dev. Dyn.* **201:** 260.

Tam P.P.L. and Beddington R.S.P. 1992. Establishment and organization of germ layers in the gastrulating mouse embryo. *Ciba Found. Symp.* **165:** 27.

Thomas P. and Beddington R.S.P. 1996. Anterior primitive endoderm may be responsible for patterning the anterior neural plate in the mouse embryo. *Curr. Biol.* **6:** 1487.

Varlet I. and Robertson E.J. 1997. Left-right asymmetry in vertebrates. *Curr. Opin. Genet. Dev.* **7:** 519.

Varlet I., Collignon J., and Robertson E.J. 1997. Nodal expression in the primitive endoderm is required for the specification of the anterior axis during mouse gastrulation. *Development* **124:** 1033.

Weinstein D.C., Ruiz i Altaba A., Chen W.S., Hoodless P., Prezioso V.R., Jessell T.M., and Darnell Jr. J.E. 1994. The winged-helix transcription factor *HNF-3β* is required for notochord development in the mouse embryo. *Cell* **78:** 575.

Wilkinson D.G. 1992. Whole-mount in situ hybridization of vertebrate embryos. In *In situ hybridization: A practical approach* (ed. D. G. Wilkinson), p. 75. IRL Press, Oxford, United Kingdom.

Wilkinson D.G., Bhatt S., Chavrier P., Bravo R., and Charnay P. 1989. Segment-specific expression of a zinc-finger gene in the developing nervous system of the mouse. *Nature* **337:** 461.

Zhou X., Sasaki H., Lowe L., Hogan B.L.M., and Kuehn M.R. 1993. *Nodal* is a novel TGF-β-like gene expressed in the mouse node during gastrulation. *Nature* **361:** 543.

Axis Duplication and Anterior Identity in the Mouse Embryo

P. THOMAS, J.M. BRICKMAN, H. PÖPPERL,[1] R. KRUMLAUF, AND R.S.P. BEDDINGTON
[1]*Deutsches Krebsforschungszentrum, Abteilung Angewandte Tumorvirologie, Im Neuenheimer Feld 242, D-69120 Heidelberg, Germany;* [2]*MRC National Institute for Medical Research, The Ridgeway, Mill Hill, London NW7 1AA, United Kingdom*

The origin of the anteroposterior (A/P) axis in vertebrate embryos remains unclear, not least because most emphasis in the past has been placed on explaining the provenance of the vertebrate organizer, and it has been tacitly assumed that definition of the rostrocaudal axis is a direct consequence of establishing dorsoventral (D/V) polarity and dependent on ensuing organizer activity. In *Xenopus,* many of the cell biological and molecular mechanisms involved in determining the location of the organizer, as well as some of the signaling interactions underlying its unique patterning properties have been identified (for review, see Sive 1993; Kessler and Melton 1994; Miller and Moon 1996). Although this work provides an increasingly coherent and compelling explanation for the determination of the earliest embryonic axis, the D/V axis, it does not explain how D/V polarity can be utilized to establish an orthogonal A/P axis. Clearly, once gastrulation is under way, the organizer and its derivatives influence A/P pattern (Hemmati-Brivanlou et al. 1990; Sharpe and Gurdon 1990; Saha and Grainger 1992), but this may explain refinement of A/P pattern, rather than explain how, for example, nascent neurectoderm and gut endoderm acquire anterior character de novo.

In the *Drosophila* embryo, the D/V and A/P axes are established using separate molecular pathways (for review, see St. Johnston and Nüsslein-Volhard 1992; Chasan and Anderson 1993). Initially, there is little if any interaction between the signaling pathway responsible for setting up D/V polarity and those genes comprising the anterior, posterior, and terminal classes responsible for A/P pattern of the zygote. Mutants, such as *Toll*, demonstrate that D/V polarity can be ablated while the A/P axis is preserved, and homozygous *bicoid* or *nanos* embryos show the converse. Therefore, there is no a priori reason vertebrates may not also establish their A/P and D/V axes independently, the two cooperating in pattern formation only once gastrulation has commenced.

In the mouse, extremely little is known about the origin of any of the definitive axes of the embryo, and again A/P pattern has been considered a direct consequence of gastrulation: the products of the primitive streak and organizer bestowing pattern on the overlying ectoderm. Studying the origin of the A/P axis in mammals is difficult because an unequivocal axis is only evident at gastrulation when the embryo has implanted in the uterus and is no longer amenable to direct observation or manipulation. Morphological asymmetry of the blastocyst has been described and histological analysis has been used to argue that this asymmetry is preserved during implantation such that what was the longest side of the blastocyst with respect to its proximodistal axis always attaches either to the right or left uterine wall and that this presages the formation of the primitive streak, which also always forms next to either the right or left uterine wall (Smith 1985). However, although postimplantation embryos do exhibit consistent morphological asymmetry in the extraembryonic region, lineage labeling indicates that there is only a modest correlation between this asymmetry and the future plane of primitive streak formation, and it certainly does not predict the polarity of the A/P axis (Gardner et al. 1992). Recently, more elaborate tracing techniques have been used to show that the axis of bilateral symmetry in the blastocyst coincides with the animal vegetal axis of the fertilized egg (Gardner 1997). This raises the possibility that information present in the oocyte could influence the subsequent definition of axes in the embryo (Gardner 1997), but such a hypothesis crucially requires that there be a causal link between bilateral symmetry in the blastocyst and the site of primitive streak formation at gastrulation, and evidence for this is still missing.

We report here the results of two rather different studies in the mouse, one documenting early asymmetric gene expression and cell fate in the primitive endoderm lineage of the mouse prior to gastrulation and the other assessing the effects of ectopic *Wnt8* expression (Pöpperl et al. 1997) on axis formation. Together, these two studies reinforce the notion that anterior identity is established prior to overt gastrulation and therefore independently of organizer function (Thomas and Beddington 1996).

MATERIALS AND METHODS

Generation of transgenic mice. The transgene used to misexpress *Cwnt8* (Pöpperl at al. 1997) was constructed in pBluescript KS$^+$ and consisted of a 4.3-kb human β-actin promoter, including intron 1 in the 5′-untranslated region (Ng et al. 1985), a 1.29-kb *Eco*RI–*Xmn*I fragment containing the *Cwnt8*-coding region (Hume and Dodd 1993), and a 0.3-kb SV40 polyadenylation sequence. Linear, vector-free DNA fragments (1 ng/μl) were injected into a pronucleus of 0.5-day post coitum (dpc) zygotes obtained from crosses between F_1 hybrids (CBA × C57BL6). Some transgenics were made in zygotes derived from crosses between F_1 (CBA × C57BL6) females and F_1 (CBA × C57BL6) males homozygous for a transgene comprising the *lacZ* gene under control of the *Hoxb1 rhombomere 4* enhancer (Pöpperl et al. 1995). Visceral yolk sac or other extraembryonic tissues were used to provide DNA for polymerse chain reaction (PCR) analysis to identify those embryos that contained *β-actin-Cwnt8* insertions. Primer pairs specific for the transgene (*Cwnt8/SV40pA*: 5′-GGCGTTCCTCGTGCATAGTCGG-3′; 5′-GATGAGTTTGGACAAACCAC-3′) and internal control primers to *myogenin* (5′-CCAAGTTGGTGTCAAAAGCC-3′; 5′-CTCTCTGCTTTAAGGAGTCAG-3′) were used in the PCR analysis.

DiI labeling and orthotopic grafting of postimplantation embryos. Embryos that were to be labeled or serve as recipients for grafts were derived from F_1(CBA × C57BL6) intercrosses, dissected from the uterus on the sixth to ninth day of gestation, and Reichert's membrane-reflected. Orthotopic anterior ectoderm grafts were prepared and grafted as described previously (Beddington 1981), donor tissue having been recovered from 8-day embryos carrying a transgene that expresses *lacZ* ubiquitously and constitutively in midgestation embryonic tissues (Beddington et al. 1989).

For all labeling experiments, micropipettes and DiI-labeling solution were prepared as described previously (Beddington 1987; Serbedzija et al. 1989). DiI labeling of 8.5-dpc rostral neurectoderm was carried out by hand essentially as described by Wilson and Beddington (1996); the labeling pipette was inserted through the visceral yolk sac and amnion, and the mouth of the pipette was pushed gently against the dorsal surface of the rostralmost prosencephalon in the region of the midline. A small quantity of dye was aspirated before flushing the surface of the cranial neural folds with M2 medium. For DiI labeling of pregastrulation embryos, single embryos were transferred individually to hanging drops of M2 medium in a micromanipulation chamber (Gardner 1978), and labeling was carried out using a Leitz micromanipulator, the embryos being immobilized with a holding pipette. Approximately six cells of the distalmost visceral endoderm of 5.5-dpc embryos were labeled by apposing the injection pipette to the distal tip of the egg cylinder and slowly expelling DiI solution for 30 seconds using a de Fonbrune suction and force pump. After labeling, embryos were washed once in M2 medium and either immediately fixed in 4% paraformaldehyde at 4°C (to assess the efficacy of labeling) or transferred to culture.

Embryos (8.5 dpc) were cultured according to the method of Beddington (1987), but younger embryos were placed in siliconized (Repelcote) embryological dishes (Raymond Lamb) containing Dulbecco's modified Eagle's medium (DMEM) + 10% rat serum. All embryos were cultured for 24 hours at 37°C in a humidified atmosphere of 5% CO_2 in air and then fixed either in 4% paraformaldehyde at 4°C (DiI-labeled) or in 0.2% glutaraldehyde (orthotopic grafts). The distribution of DiI-labeled cells in intact embryos was examined in an Axiophot (Zeiss) compound microscope as described previously (Wilson and Beddington 1996), and embryos were photographed using P1600 film (Kodak).

Whole-mount in situ hybridization. Embryos were fixed in 4% paraformaldehyde in phosphate-buffered saline (PBS) for about 24 hours at 4°C, dehydrated through a methanol series, and processed as described by Wilkinson (1992). RNase was omitted in hybridizations using the heterologous rat *vhh1* probe. Digoxigenin-labeled antisense riboprobes used to detect the expression of *T, Cwnt8, Hesx1, Shh, Wnt1, cerberus-like*, and *Hex* were generated using the appropriate RNA polymerase according to manufacturer's instructions (Boehringer Mannheim) from the linearized plasmids listed below.

Brachyury (T) probe: A plasmid containing a 2.0-kb full-length cDNA encoding mouse *T* (gift from B. Herrmann; Wilkinson et al. 1990).

CWnt8 probe: Plasmid *pGEMwnt8* (Hume and Dodd 1993) containing a 1.7-kb *Eco*RI fragment of a cDNA clone spanning the entire coding region.

Hesx1 probe: The riboprobe was transcribed from a 394-bp *Alu*I fragment (Thomas and Beddington 1996).

Shh probe: A plasmid containing an *Xho*I fragment of a rat *vhh1* cDNA (a gift from T. Jessell; Roelink et al. 1994).

Wnt1 probe: Plasmid *dnWnt-1 CS2*$^+$ (a gift from R.T. Moon; Hoppler et al. 1996), containing 915 bp of the mouse *Wnt1* cDNA.

Hex probe: Antisense *Hex* probe was transcribed from a *Hex* cDNA clone (a gift from L. Wiedemann) and spanned positions 291–818 (EMBL Acc. No. Z21524)

cerberus-like probe: IMAGE consortium EST sequence from a mouse embryonic region cDNA library (Harrison et al. 1995) comprising 487 bases (available through Gene Bank [AA120122]). The sequence contains 72 bp of 5′ UTR and 414 bp of predicted coding sequence.

Histochemical and immunohistochemical staining. Histochemical staining to detect β-galactosidase activity was performed as described previously (Whiting et al. 1991). An anti-T antibody (a gift from Dr. B. Herrmann) was used on whole embryos either directly or after in situ hybridization using the conditions described by Kispert and Herrmann (1994). Embryos were processed for wax histology according to the method of Beddington (1994), 7-μm serial sections were cut (Bright 6030 Microtome), and dewaxed, and the sections were mounted in DPX (BDH, Ltd.).

RESULTS

Expression Pattern of *Hesx1* and the Fate of Cells Expressing *Hesx1*

The possibility that visceral endoderm might constitute a highly patterned tissue (perhaps capable of bestowing pattern on the overlying epiblast/ectoderm) prior to or coincident with the onset of gastrulation was raised by the

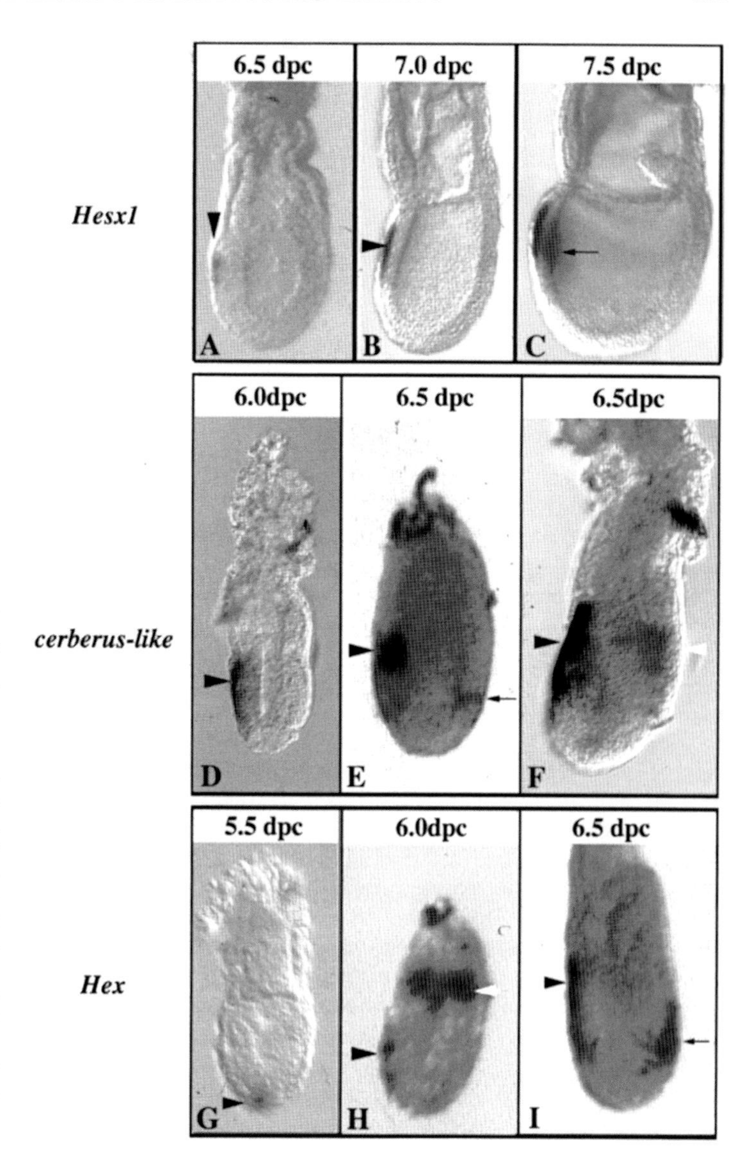

Figure 1. Asymmetric gene expression in the egg cylinder. (*A–C*) *Hesx1* expression in the anterior visceral endoderm (AVE; *A* and *B, closed arrowhead*) and later in the adjacent ectoderm (*C, closed arrow*). (*D-F*) *cerberus-like* expression in the AVE (*closed arrowheads*) of a pregastrulation embryo (*D*), during early gastrulation (*E,F*) when transcripts are also evident in the node (*E,* arrow) and are clearly on the side of the embryo opposite to the primitive streak marked by *T* expression (*F, white arrowhead*). (*G-I*) *Hex* expression. (*G*) At 5.5 dpc, *Hex* transcripts (*closed arrowhead*) are localized to a few visceral endoderm cells at the distal tip of the egg cylinder. (*H*) At 6.0 dpc, when *T* expression is symmetrical at the embryonic extraembryonic junction (*open arrowhead*), *Hex* transcripts are evident in the visceral endoderm only on one side of the egg cylinder. (*I*) During early gastrulation, *Hex* transcripts are present in the AVE (*closed arrowhead*) and in cells emerging from the node (*arrow*). (*A–C,* Reprinted with permission, from Thomas and Beddington 1997.)

early expression pattern of *Hesx1* (Fig. 1A–C) (Hermesz et al. 1996; Thomas and Beddington 1996). *Hesx1* transcripts are first detected in a small patch of anterior endoderm at the onset of gastrulation (Fig. 1A) when derivatives of the primitive streak have yet to reach the anterior aspect of the egg cylinder. Lineage analysis demonstrated that this endoderm was visceral in character (Thomas and Beddington 1996), belonging to the primitive endoderm lineage and thus not destined to contribute to the definitive gut endoderm of the embryo (Gardner and Rossant 1979). This endodermal expression domain remains more or less constant as the primitive streak elongates, and by the full-length streak stage, transcripts begin to appear in the ectoderm immediately adjacent to this patch of endoderm, expression in the two tissue layers then remaining in almost complete register until early somite stages (Fig. 1B,C). However, although the domain of *Hesx1* expression in the endoderm layer appears not to change during gastrulation, by the time the headfolds begin to form (Fig. 1C), visceral endoderm has been replaced by definitive endoderm emerging from the streak. These cell movements in the context of the expression profile of *Hesx1* during gastrulation suggested that reciprocal patterning interactions may occur between the anterior endoderm and ectoderm throughout gastrulation and that such interactions were initiated by the extraembryonic endodermal tissue lineage.

At 8.5 dpc, *Hesx1* transcripts are found in the rostral extreme of the developing central nervous system (CNS) (Fig. 4H), and a day later, they were found in the ventral diencephalon (Hermesz et al. 1996; Thomas and Beddington 1996). To ascertain whether the domain of *Hesx1* expression in the ectoderm corresponds to the progenitor population of the prosencephalon, and subsequently ventral diencephalon, lineage analysis was undertaken to determine by orthotopic grafting the fate of those ectodermal cells that express *Hesx1* at the full-length streak stage (Fig. 2A). Of seven embryos that received a graft of transgenic anterior ectoderm, six had developed normally to the early somite stage after 24 hours in culture. All six embryos contained transgenic cells in the rostralmost neural folds (Fig. 2A), and only in two embryos was there

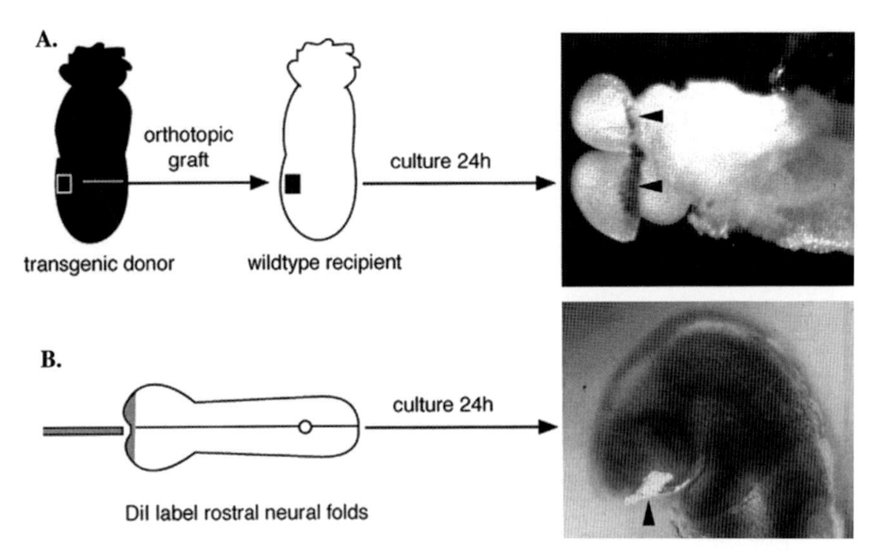

Figure 2. The fate of *Hesx1*-expressing ectoderm cells. (*A*) Anterior ectoderm, corresponding to the region expressing *Hesx1*, from a 7.5-dpc *lacZ* expressing transgenic embryo was grafted into an equivalent position in a wild-type embryo of the same age. After 24 hr of development in culture, transgenic cells (*closed arrowheads*) were restricted to the rostral extreme of the neural folds (the prospective prosencephalon). (*B*) The rostral neural folds of early somite stage embryos (8.5 dpc) were labeled on their dorsal surface with DiI, and after 24 hr in culture, fluorescent progeny were restricted to the ventral diencephalon (*closed arrowhead*).

evidence of additional surface ectoderm labeling. In all cases, the graft-derived cells remained relatively coherent and showed considerable expansion in a mediolateral as opposed to A/P direction. The fate of this most rostral aspect of the prosencephalon at the early somite stage was then determined using DiI labeling (Fig. 2B). Ten embryos were labeled, and three were inspected immediately after labeling. All three embryos contained the DiI label restricted to a rostromedial region of the prosencephalic neurectoderm (data not shown). All seven embryos placed in culture for 24 hours developed normally, but two embryos proved to be devoid of Di-labeled cells. The remaining five embryos contained a relatively coherent patch of fluorescent cells located in the ventral aspect of the diencephalon (Fig. 2B). Therefore, it would appear that unlike the situation in visceral endoderm, *Hesx1* expression in the ectoderm may be heritable and constitutes a lineage-specific marker for the progenitors of the ventral diencephalon.

Early Expression Pattern of *cerberus-like* and *Hex* Genes

The *cerberus* gene in *Xenopus* encodes a secreted protein that can induce additional head structures if expressed ectopically. It is normally expressed in the non-involuting deep endomesoderm of the dorsal half of the *Xenopus* embryo, and once gastrulation is under way, its transcripts are restricted entirely to anterior endoderm tissue (Bouwmeester et al. 1996). A mouse clone (*cerberus-like*) encoding a peptide fragment of at least 138 amino acids with 54% similarity and 34% identity to *Xenopus cerberus* was used here as a probe. Transcripts were first detected in the inner cell mass of hatched blastocysts (4.5 dpc; data not shown). By 6.0 dpc, *cerberus-like* transcripts were restricted to the visceral endoderm on one side of the prestreak egg cylinder (Fig. 1D). At the onset of gastrulation, double in situ analysis with *T* (Fig. 1F) showed that *cerberus-like* transcripts were on the opposite side of the embryo from the primitive streak and thus present in anterior visceral endoderm (AVE). *Cerberus-like* transcripts were also detectable transiently in the vicinity of the node during the early stages of gastrulation (Fig. 1E), presumably marking emerging gut endoderm cells, since *cerberus-like* transcripts persist until the early somite stage in the rostralmost definitive gut endoderm (data not shown). Thus, the *cerberus-like* gene appears to have an expression pattern comparable to its counterpart in *Xenopus* in that during gastrulation, transcripts are found exclusively in anterior endoderm tissue. Later, unlike what has been reported for *Xenopus cerberus* (Bouwmeester et al. 1996), *cerberus-like* transcripts are seen in three stripes corresponding to a domain in the anterior presomitic mesoderm, the somite in the process of forming and the most recently formed somite, for the duration of somitogenesis (data not shown).

Hex encodes a divergent homeodomain protein that is not affiliated to any of the established classes of homeobox genes (Bedford et al. 1993). *Hex* transcripts are first detected in hatched blastocysts where they are restricted to the primitive endoderm population at the interface between the inner cell mass and the blastocoel cavity (data not shown). During implantation, *Hex* expression remains restricted to the visceral endoderm, and at 5.5 dpc, it is detected in a small group of cells at the distal tip of the egg cylinder (Fig. 1G) and in slightly older embryos in a unilateral patch of visceral endoderm immediately proximal to the distal tip (data not shown). At 6.0 dpc, double in situ analysis of *Hex* and *T*, which we have previously shown to be expressed in a proximal symmetrical ring of epiblast cells prior to gastrulation (Thomas and Beddington 1996), confirmed that asymmetrical *Hex* expression in the primitive endoderm preceded primitive streak formation (Fig. 1H). During the early stages of gastrulation (6.5 dpc), *Hex* transcripts remain localized to a strip of anterior medial visceral endoderm extending about two thirds of the way down the egg cylinder (Fig. 1I). A second domain of *Hex* transcripts in the region of the node is also evident at this stage (Fig. 1I), and at later stages, *Hex* is expressed in anterior gut endoderm and its derivatives (data not shown). Therefore, both *Hex* and *cerberus-like* are first expressed in the AVE prior to primitive streak formation.

Fate of Distal 5.5-dpc Visceral Endoderm Cells

To examine how the distribution of *Hex* transcripts becomes asymmetrical, the fate of the distal endoderm population that expresses *Hex* apparently symmetrically at 5.5 dpc was determined using DiI as a lineage tracer (Fig. 3). Inspection of nine embryos immediately after labeling revealed that between three and six distal endoderm cells were labeled. However, the frequency of entirely unlabeled embryos (15/34) among those examined, either immediately after labeling ($n = 9$) or those that had developed normally after a further 24 hours in culture ($n = 25$), indicated that the overall efficiency of labeling was only 56%. Remarkably, in all 14 embryos that did contain DiI-labeled endoderm cells after further development in culture, the labeled cells invariably formed a coherent patch that extended up only one side of the egg cylinder (Fig. 3); a radial or bilateral distribution of DiI cells was not observed. Furthermore, in all four embryos that had developed a recognizable primitive streak, the labeled patch of endoderm was on the opposite side of the egg cylinder from the nascent streak (Fig. 3). Therefore, it would appear that cell movement or growth in the visceral endoderm layer is such that cells originally at the distal tip of the egg cylinder are invariably displaced anterioward.

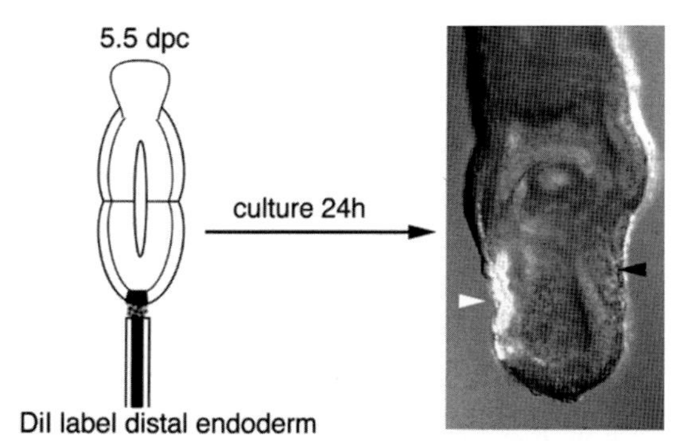

Figure 3. Tracing cell movement in the 5.5-dpc visceral endoderm. Up to six visceral endoderm cells were labeled at the distal tip of cylinder, and after 24 hr of further development in culture, the fluorescent progeny formed a coherent patch of cells extending up the anterior side of the egg cylinder (*open arrowhead*) on the side of the embryo opposite from the nascent primitve streak (*closed arrowhead*).

Misexpression of *Cwnt8* Causes Incomplete Axis Duplication and Anterior Truncations

Ectopic expression of the Wnt1 class of proteins in *Xenopus* causes complete axis duplication when expressed early and anterior truncation of the axis if expressed after the mid blastula transition (Smith and Harland 1991; Sokol et al. 1991; Christian and Moon 1993; Moon et al. 1993). The effects of a member of this class, *Cwnt8*, on axis formation in the mouse was examined using transgenesis to ectopically express the gene. Putative transgenic embryos were analyzed between 6.5 and 9.5 dpc, and transgenesis was confirmed either by whole-mount in situ hybridization using a *Cwnt8* probe or by PCR analysis of extraembryonic tissues. Analysis of 6.5–7.5-dpc embryos demonstrated that *Cwnt8* was ectopically expressed and that transcripts in a few embryos could be detected in all tissues of the conceptus. However, when transgenic embryos were compared, it was evident that *Cwnt8* expression was neither consistent nor ubiquitous, although the highest level of transcripts always occurred in visceral endoderm cells (Fig. 4A). This variability can be ascribed to position effects influencing the transgene at different integration sites in the genome and is probably compounded by mosaicism due to the time of integration.

Approximately 80% of all *Cwnt8* transgenic embryos (80/105) were morphologically abnormal, and these abnormalities could be classified into distinct categories. Overt axis duplication within a single amnion occurred in about 23% of transgenic embryos (24/105; Fig. 4D,E). Since duplications always occurred within a single amnion, they must have arisen from the production of more than one primitive streak within a single egg cylinder, rather than from an earlier event (Kaufman 1992). Most transgenic embryos at the egg cylinder stage (6.5–7.5 dpc) were abnormally constricted at the embryonic-extraembryonic junction (24/37; 64.9%), and frequently, the rostral half of the cylinder was deformed (Fig. 4A,C). Apart from axis duplication, the predominant phenotype observed between the headfold stage and 25 somite stage was truncation of anterior structures (38/68; 55.9%). These phenotypes were never observed in transgenic embryos containing other *β-actin* expression constructs or unrelated transgenes (Pöpperl et al. 1995; data not shown).

Analysis of 8.5–9.5-dpc transgenic embryos showed that axis duplication was manifest either as two axes in opposing orientation giving a head-to-head duplication (Fig. 4E) or as parallel axes (Fig. 4D). No axial duplication which included the formation of two complete heads was observed, and there was always a pronounced reduction in the amount of neurectoderm rostral to the anterior limit of the two notochords (Fig. 4D). Duplicated parallel axes tended to merge caudally (Fig. 4D), but surprisingly their rostral extemities also appeared to be fused, being conjoined by an eptihelial fold (Fig. 4D). Where head-to-head duplications occurred, the two axes were always fused rostrally (Fig. 4E), and although each axis contained a notochord, no somites were discernible in the

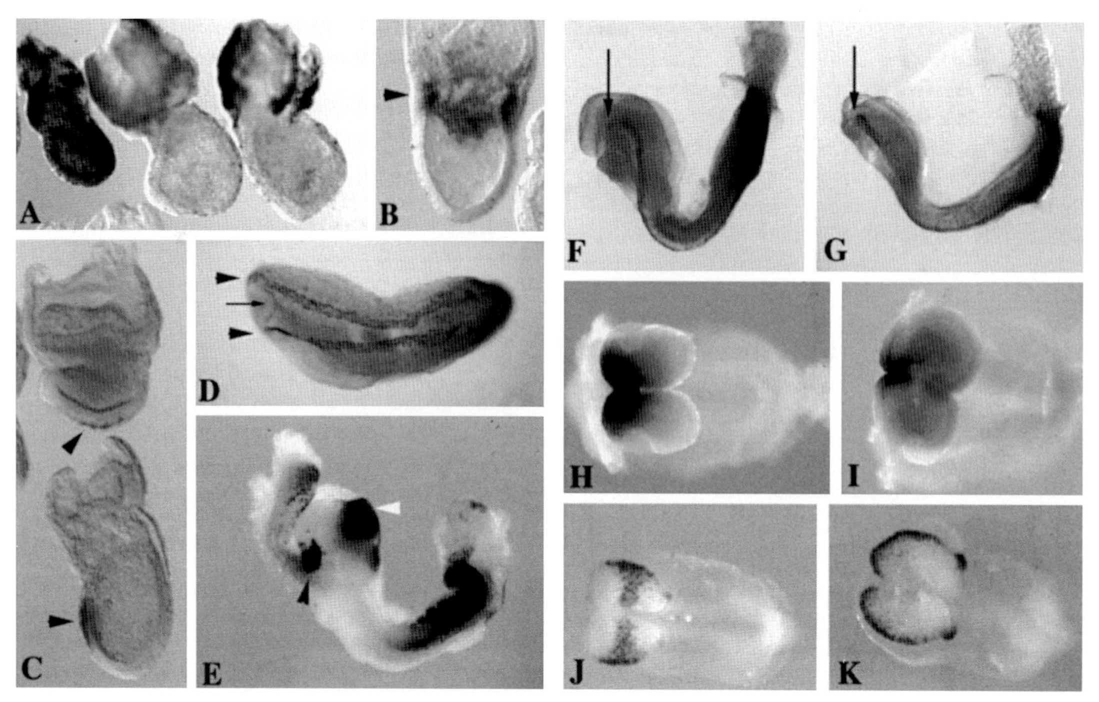

Figure 4. Ectopic expression of *Cwnt8* causes axis duplication and anterior truncation. (*A*) Expression of the *Cwnt8* transgene in 7.5-dpc embryos. Transcripts predominate in the visceral endoderm, and the two embryos on the right show the characteristic junctional constriction and abnormal morphology. (*B*) 7.5-dpc transgenic embryo showing circumferential expression of *T* in the proximal epiblast. (*C*) A transgenic embryo (*top*) at 7.5 dpc showing a single domain of distally displaced *Hesx1* expression (*arrowhead*). A control embryo showing the normal domain of *Hesx1* expression (*arrowhead*) at this stage is shown at the bottom. (*D*) Ventral view of a transgenic embryo at 8.5 dpc with parallel axis duplication showing two notochords, marked by anti-T antibody (*closed arrowheads*). The notochords extend to the rostral limit of the axes which appear to be conjoined by a single epithelial fold (*arrow*). (*E*) Lateral view of a transgenic embryo at 9.5 dpc generated in the *Hoxb1 r4 enhancer* line showing head-to-head duplication of the axis, the caudal end of the secondary axis being top left. Both primary (*open arrowhead*) and secondary (*closed arrowhead*) axes contain a discrete domain of β-galactosidase activity indicative that each axis contains a rhombomere 4. (*F–K*) Anterior development in 8.5-dpc dorsalized transgenic embryos (*G,I,K*) compared with control embryos (*F,H,J*). (*F*) Normally, the rostral limit of the notochord marked by anti-T antibody (*arrow*) coincides with the midbrain/forebrain boundary, but in a transgenic embryo (*G*), it extends almost to the rostral limit of the axis and the heart is absent. (*H*) *Hesx1* transcripts mark prospective forebrain in a control embryo. (*I*) The domain of *Hesx1* expression is severely reduced in a transgenic embryo. (*J*) *Wnt1* expression in the prospective midbrain of a control embryo. (*K*) The *Wnt1* domain in a transgenic embryo is expanded and extends to the rostral limit of the CNS, although transcripts are absent from the most rostral medial portion of the anterior neurectoderm. (*A–K*, Reprinted with permission, from Pöpperl et al. 1997.)

second axis. Axis duplication induced by ectopic *Cwnt8* in the transgenic line of mice containing the *Hoxb1 rhomobomere (r) 4 enhancer* driving a *lacZ* reporter showed unequivocally that the secondary axis at 9.5 dpc included the hindbrain, since a second rhombomere 4 was present (Fig. 4E). However, both the primary and secondary axes showed a deficit in cranial neurectoderm rostral to the hindbrain (Fig. 4E).

At egg cylinder stages (7.5 dpc), two or more primitive streaks, and occasionally two nodes, could be discerned using *T* transcripts or T protein as a marker (Pöpperl et al. 1997; data not shown). In one case, *T* expression was observed as a discontinuous circumferential ring abutting the extraembryonic region (Fig. 4B), indicating that a streak might be able to form anywhere on the circumference of the proximal epiblast. Where two streaks occurred on opposite sides of the egg cylinder, the incipient headfolds were displaced to the distal aspect of the conceptus. Likewise, when detectable, *Hesx1*, the earliest marker for prospective forebrain and foregut (see above), was often displaced from its normal anterior position to a distal domain in 7.5-dpc transgenic embryos (Fig. 4C). Importantly, no duplication of the domain of *Hesx1* expression was observed in any embryo. These data suggest that head-to-head duplications may arise if two streaks form on opposite sides of the egg cylinder but that parallel duplicated axes ensue if the two streaks are closer together. However, in neither case does streak duplication lead to complete rostral duplication.

The development of anterior structures was examined more rigorously in 8.0–9.0-dpc *Cwnt8* transgenic embryos that lacked overt duplications but showed features of dorsalization (i.e., an enlarged notochord). As well as being the most common phenotype exhibited by transgenic embryos, it was more straightforward to examine anterior development in the absence of axial duplications and fusions. The notochord extended much closer to the anterior limit of the embryo than normal in these dorsalized embryos (Fig. 4F,G), which indicates that at least part of the axial mesoderm underlying the prosencephalon was missing since this tissue does not normally express *T* (Hermann 1991). In fact, most of the prospective forebrain domain was missing since the expression domain of *Hesx1* was greatly diminished (Fig. 4H,I) or absent. On the other hand, the midbrain seemed to be expanded because *Wnt1* transcripts that are normally restricted to the midbrain and dorsal aspect of the hindbrain (McMahon et al. 1992; Parr et al. 1993) extended almost to the anterior limit of the neuroectoderm (Fig. 4J,K). However, the rostral medial domain where *Hesx1* expression tended to persist (Fig. 4I,K) was not invaded by *Wnt1* transcripts. Thus, the midbrain (and possibly hindbrain) domains were expanded at the expense of the prosencephalon.

Defective foregut and heart development was seen in some transgenic embryos; three embryos at 9.5 dpc lacked a morphologically recognizable foregut and heart, and there was no evidence of heart tube formation in a further four 8.5-dpc embryos (Fig. 4G). Since both the heart and foregut are descended from rostral tissues during gastrulation, their reduction or absence may also reflect a form of anterior truncation.

DISCUSSION

The notion that all A/P patterns in the mouse embryo are dictated exclusively by products of the primitive streak during gastrulation is no longer tenable in the light of the experiments presented here and those recently reported by other investigators. Although purely descriptive, the expression patterns of a number of genes now demonstrate that A/P pattern exists prior to primitive streak formation and that it appears to emerge first in the visceral endoderm layer, an epithelium that envelopes the founder tissue of the fetus (the epiblast) but does not itself contribute to the fetus (Gardner and Rossant 1979; Cockroft and Gardner 1987; Lawson et al. 1987). Even though the onset of *Hesx1* expression in AVE appears to coincide with streak formation (Fig. 1A) (Hermesz et al. 1996; Thomas and Beddington 1996), it is highly unlikely that any product of the streak could induce *Hesx1* expression in such a small group of outer cells on the opposite side of the embryo from the nascent streak and separated from it by the proamniotic cavity. Even more compelling is the expression of the *cerberus-like* gene that is clearly expressed in AVE at least 12 hours before streak formation (Fig. 1D–F). The expression pattern of this gene at 6.0 dpc is reminiscent of that described for an antigen (VE-1) recognized by an antibody of unknown epitope specificity (Rosenquist and Martin 1995). More importantly, several genes, which are known from mutational studies to affect anterior development of the embryo, such as *Otx2* (Acampora et al. 1995; Ang et al. 1996), *nodal* (Varlet et al. 1997; E. Robertson, pers. comm.), *HNF3β* (Ang and Rossant 1994; Weinstein et al. 1994; E. Robertson, pers. comm.), and *Lim1* (Shawlot and Behringer 1995), are all expressed first in visceral endoderm prior to gastrulation.

The expression pattern of *Hex* reveals the earliest asymmetry in gene expression related to A/P patterning yet identified (Fig. 1G–I). *Hex* is expressed in implanting blastocysts in the primitive endoderm cells as they form at the surface of the inner cell mass. Less than a day later, its transcripts are detected in a small group of endoderm cells at the distal tip of the egg cylinder, and thus its expression seems to be maintained in only a small subset of primitive endoderm descendants, a population that faithfully marks one pole of the proximodistal axis (Fig. 5). How such restriction in expression is brought about is not obvious, although since the primitive endoderm becomes multilayered within the blastocyst (Gardner 1983), a sustained and intimate association with epiblast could play a part. Cell lineage analysis shows that these distal cells are destined to move proximally as a coherent population and that their route from distal to proximal precisely coincides with the future anterior side of the embryo (Figs. 3 and 5). Therefore, it would appear that a proximodistal axis established in the blastocyst is translated into an A/P axis by virtue of vectorial movement within the visceral endoderm layer. As *Hex* expression seems to be maintained in the cells as they move, this gene marks the future anterior side of the embryo about 24 hours before gastrulation starts. What provides the impetus for such vectorial movement? One might speculate that differential growth could be responsible for nonrandom movement of the endoderm epithelium and that the asymmetry of implantation (Smith 1985) is normally sufficient to ensure such nonuniform growth. Alternatively, the asymmetry documented in the blastocyst (Smith 1985), which has now been shown to correlate with the animal/vegetal axis of the egg (Gardner 1997), may provide the requisite bias. Since the visceral endoderm layer overlying the epi-

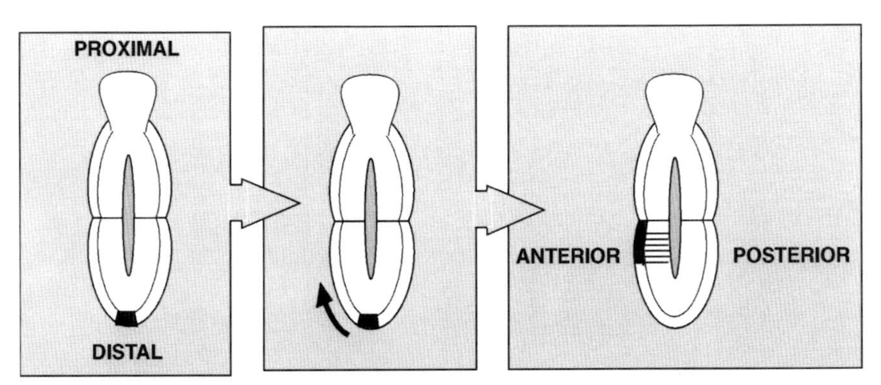

Figure 5. Diagram illustrating the proposed "rotation" of the visceral endoderm which translates a proximal distal axis (marked by *Hex* transcripts at the distal pole; *black*) into an A/P axis. (*Black stripes*) Anterior ectoderm.

blast contains less than 100 cells at 5.5 dpc (Snow 1977), a very small bias in growth may be all that is required to convert a proximodistal axis into an A/P one. Therefore, any influence that generates some differential growth in the embryonic portion of this tissue may be sufficient to bring about this conversion. This may explain why at a low frequency, blastocysts explanted into culture can sometimes gastrulate and form an apparently normal A/P axis (Hsu 1979), even though the process of egg cylinder formation in vitro is abnormal (Wu et al. 1981).

Although it is now incontrovertible that the pregastrulation AVE has a unique molecular character that distinguishes it from other regions, the evidence that it may be instrumental in bestowing pattern on the embryo itself, or the nature of the pattern it can impart, is less clear. Two possibilities exist. Products of the streak and node are responsible for all rostrocaudal pattern in the embryo, and the pattern observed in the AVE is purely coincidental and has no role in patterning the embryo. Alternatively, AVE does serve to define anterior identity in the embryonic region, either by establishing an anterior terminus that is required for subsequent patterning by products of the streak or by directly inducing definitive anterior embryonic tissues such as the heart, foregut, and forebrain. That many of the genes expressed in the AVE are the same as those later expressed in the anterior mesendoderm emerging from the node (Fig. 1E, e.g., *cerberus-like;* Fig. 1I, *Hex*) suggests that if it can bestow anterior pattern, then it may do so by inducing specific tissues.

The sequential expression of *Hesx1* first in a small domain of AVE and then in the immediately adjacent ectoderm (Fig. 1A–C) certainly resembles an inductive interaction. Moreover, if this AVE is removed during early gastrulation, the ectoderm remains viable and proliferates, but little or no *Hesx1* expression occurs in the ectoderm. If endoderm in the same position is removed at late streak stages, some *Hesx1* expression although much reduced always persists in rostromedial neurectoderm (Thomas and Beddington 1996). This indicates that AVE does have a function beyond simply maintaining ectoderm integrity and growth. The ectoderm cells expressing *Hesx1* are clearly fated to be the most rostral neurectoderm (Fig. 2) (Beddington 1981; Tam 1989; Lawson and Pedersen 1992), and heterotopic transplantation has identified anterior ectoderm as the least developmentally labile ectoderm in the embryo by the late primitive streak stage because it shows a strong bias toward ectoderm differentiation wherever it is placed (Beddington 1982). This indicates that a decisive interaction may have occurred by the late primitive streak stage to restrict its fate. However, the interaction that restricts the potency of anterior ectoderm could still be instigated by the mesendoderm emanating from the organizer, since this will have largely replaced AVE by the late streak stage.

The most compelling evidence to implicate AVE in anterior patterning of the embryo comes from chimeras containing a mixture of wild-type and *nodal*$^-$/*nodal*$^-$ cells (Varlet et al. 1997). Embryonic stem (ES) cells injected into the blastocyst rarely colonize the primitive endoderm lineage (Beddington and Robertson 1989), and it is therefore possible to make chimeras that contain predominantly wild-type cells in the primitive endoderm but mostly *nodal*$^-$/*nodal*$^-$ cells in the embryo. Even though intact *nodal*$^-$/*nodal*$^-$ mutants fail to gastrulate, such chimeric embryos now proceed to advanced fetal stages and show reasonably normal anterior development. In contrast, in the converse experiment where the visceral endoderm is largely mutant and the embryo wild type, the anterior part of the axis is truncated. Therefore, wild-type visceral endoderm is essential for normal anterior development.

Further evidence that AVE may be required for anterior development comes from examining the axis-inducing properties of the mouse organizer (or node). Heterotopic grafts of the organizer can generate axis duplications, but organizers from neither late primitive streak stage embryos (Beddington 1994) nor early gastrulation stage embryos (P. Tam, pers. comm.) produce duplication of the forebrain. Ectopic *Cwnt8* expression can duplicate the primitive streak and the node in mouse (Fig. 4) (Pöpperl et al. 1997), much as ectopic expression of *Xwnt8* induces a secondary organizer in *Xenopus* (Christian and Moon 1993; Smith and Harland 1991; Sokol et al. 1991). However, unlike the case in *Xenopus*, these duplications in mouse never include the forebrain region or heart (Fig. 4D,F). Dorsalized embryos lacking overt duplications also showed signs of anterior truncation (Fig. 4F–K). Since it is known that expression of *Xwnt8* after the mid blastula transition in *Xenopus* causes anterior truncation (Moon et al. 1993), it is possible that the failure of *Cwnt8* to cause forebrain duplication is due to a later, secondary posteriorization of the induced axis. However, even at egg cylinder stages, there was no evidence for more than one prospective forebrain region (Fig. 4C). Furthermore, both head-to-head and parallel duplicated axes were invariably fused rostrally (Fig. 4D, E) indicating that they always shared a common anterior terminus. One could argue that rostral fusion is an inevitable consequence of the constraints imposed by egg cylinder architecture, but this cannot be the case because vincristine sulfate administered during gastrulation does generate twins with complete axes (Kaufman and O'Shea 1978). Therefore, at present, it would seem that with the exception of vincristine, all molecular and mechanical perturbations causing duplication of the mouse organizer result only in duplication of tissue underlain by notochord. This again is consistent with the AVE being responsible for establishing anterior identity initially and the node-derived axial mesendoderm serving subsequently to maintain this anterior identity and embellish anterior pattern. Such a scenario would explain why embryos homozygous for a null mutation in *HNF3β*, which fail to develop a recognizable node and do not produce axial mesendoderm, nonetheless elaborate remarkably normal anterior pattern (Ang and Rossant 1994; Weinstein et al. 1994).

A problem remains in reconciling the ability of ectopic *Wnt8* and early organizer grafts to induce complete axis duplications in *Xenopus* with their inability to do so in mouse. One possibility is that mammals have adopted an

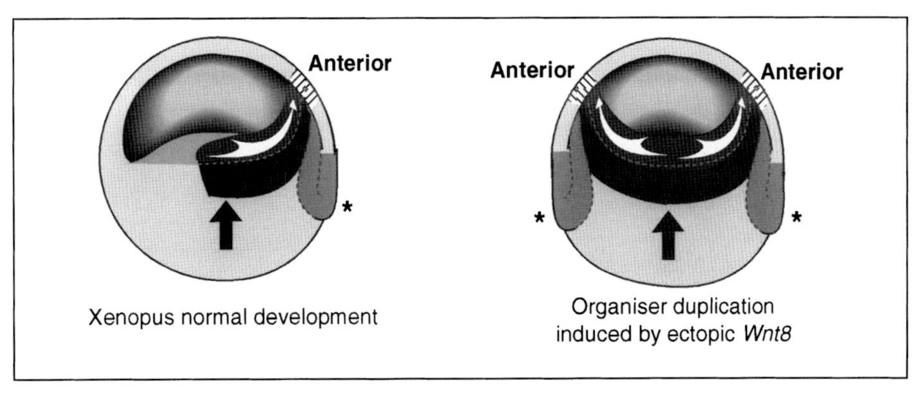

Figure 6. Diagram illustrating how duplication of the axis by ectopic expression of *Xwnt8* may not require duplication of an anterior signal. In the normal embryo (*left panel*), the deep mesendoderm cells (*black*) are already internal at the time gastrulation commences and they migrate rostrally in front of the prechordal cells (*dark gray*) which do ingress at the dorsal blastopore lip (*asterisk*; see Bouwmeester et al. 1996). Ectopic expression of *Xwnt8* (*right panel*) causes duplication of the organizer (*asterisks*) and its prechordal derivatives (*gray*). If the initial signal that provides deep endomesoderm with its anterior character is located relatively centrally in the embryo (*large closed arrow*), then the presence of two organizers instigating gastrulation movements may serve simply to divert the migration of deep endomesoderm toward both the native and induced organizer. This would cause complete rostral duplication (*black stripes*) without duplication of the initial anterior signal.

independent anterior signaling mechanism. Alternatively, the geometry of the mouse egg cylinder may have revealed the presence of an independent anterior patterning process simply because the proamniotic cavity physically separates it from the organizer, whereas this is not the case in *Xenopus*. In *Xenopus*, the cells that express *cerberus*, and thus may be responsible for head induction, are the deep endomesoderm cells in the dorsal half of the embryo that abut the organizer but are distinct from those that will form prechordal plate (Bouwmeester et al. 1996). The *cerberus* expressing cells, which do not ingress, will eventually give rise to the most anterior gut derivatives and heart (Bouwmeester et al. 1996). If the *cerberus-like* gene in mouse is also functionally homologous to *Xenopus cerberus*, then its expression initially in the AVE (Fig. 1D) highlights the topographical differences between the two embryos. Clearly, early organizer grafts in *Xenopus* will usually include *cerberus*-expressing cells, whereas mouse grafts will not. If the original signal responsible for *cerberus* induction in *Xenopus* emanates from deep vegetal cells in a more central location in the *Xenopus* blastula, then duplication of the organizer by Wnt8 may simply serve to divert the movement of *cerberus*-expressing cells such that they now move toward both the native and secondary organizer. This would not require explicit duplication of the original anterior signal (Fig. 6). Therefore, it is conceivable that an independent mechanism for the initial specification of anterior identity exists in vertebrates other than mammals. If this is so, and it resides in the equivalent of the AVE, then in chick embryos this would be the anterior hypoblast cells and in zebrafish, the syncytial yolk cell.

ACKNOWLEDGMENTS

We thank Mike Jones for his helpful discussions regarding *Xenopus* development and Richard Harland for drawing our attention to the *cerberus-like* clone in the EST database. R.S.P.B. is an international scholar of the Howard Hughes Medical Institute. H.P. was supported by EMBO and HFSP postdoctoral fellowships; J.B. is an HFSP long-term fellow.

REFERENCES

Acampora D., Mazan S., Lallemand Y., Avataggiato V., Maury M., Simeone A., and Brûlet P. 1995. Forebrain and midbrain regions are deleted in $Otx2^{-/-}$ mutants due to a defective anterior neurectoderm specification during gastrulation. *Development* **121:** 3279.

Ang S.-L. and Rossant J. 1994. *HNF-3β* is essential for node and notochord formation in mouse development. *Cell* **78:** 561.

Ang S.-L., Jin O., Rhinn M., Daigle N., Stevenson L., and Rossant J. 1996. A targeted mouse Otx2 mutation leads to severe defects in gastrulation and formation of axial mesoderm and to deletion of rostral brain. *Development* **122:** 243.

Beddington R.S.P. 1981. An autoradiographic analysis of the potency of embryonic ectoderm in the 8th day postimplantation mouse embryo. *J. Embryol. Exp. Morphol.* **64:** 87.

———. 1982. An autoradiographic analysis of tissue potency in different region of the embryonic ectoderm during gastrulation in the mouse. *J. Embryol. Exp. Morphol.* **69:** 265.

———. 1987. Isolation, culture and manipulation of post-implantation mouse embryos. In *Mammalian development: A practical approach.* (ed. M. Monk), p. 43. IRL Oxford.

———. 1994. Induction of a second neural axis by the mouse node. *Development* **120:** 613.

Beddington R.S.P., and Robertson E.J. 1989. An assessment of the developmental potential of embryonic stem cells in the midgestation mouse embryo. *Development* **105:** 733.

Beddington R.S.P., Morgenstern J., Land H., and Hogan A. 1989. An in situ transgenic enzyme marker for the midgestation mouse fetus and the visualization of inner cell mass clones during ealy organogenesis. *Development* **106:** 37.

Bedford F.K., Ashworth A., Enver T., and Wiedemann L. 1993). *HEX*: A novel homeobox gene expressed during haematopoeisis and conserved between mouse and human. *Nucleic Acids Res.* **21:** 1245.

Bouwmeester T., Kim S.-H., Sasai Y., Lu B., and De Robertis E.M. 1996. Cerberus is a head-inducing secreted factor ex-

pressed in the anterior endoderm of Spemann's organizer. *Nature* **382:** 595.
Chasan R. and Anderson K.V. 1993. Maternal control of dorsal-ventral polarity and pattern in the embryo. In *The development of* Drosophila melanogaster (ed. M. Bate and A. Martinez Arias), vol. 1, p. 387. Cold Spring Harbor Laboratory Press, Cold Spring Harbor, New York.
Christian J.L. and Moon R.T. 1993. Interactions between Xwnt-8 and Spemann organizer signaling pathways generate dorsoventral pattern in the embryonic mesoderm of *Xenopus*. *Genes Dev.* **7:** 13.
Cockroft D.L. and Gardner R.L. 1987. Clonal analysis of the developmental potential of 6th and 7th day visceral endoderm cells in the mouse. *Development* **101:** 143.
Gardner R.L. 1978. Production of chimeras by injecting cells or tissue into the blastocyst. In *Methods in mammalian reproduction* (ed. J.C. Daniel), p. 137. Academic Press, New York.
———. 1983. Origin and differentiation of extraembryonic tissues in the mouse. *Int. Rev. Exp. Pathol.* **24:** 63.
———. 1997. The early blastocyst is bilaterally symmetrical and its axis of symmetry is aligned with the animal-vegetal axis of the zygote in the mouse. *Development* **124:** 289.
Gardner R.L. and Rossant J. 1979. Investigation of the fate of 4.5 day post coitum mouse inner cell mass cells by blastocyst injection. *J. Embryol. Exp. Morphol.* **52:** 141.
Gardner R.L., Meredith M.R., and Altman D.G. 1992. Is the anterior-posterior axis of the fetus specified before implantation in the mouse? *J. Exp. Zool.* **264:** 437.
Harrison S.M., Dunwoodie S.L., Arkell R.M., Lehrach H., and Beddington R.S.P. 1995. Isolation of novel tissue-specific genes from cDNA libraries representing the individual tissue components of the gastrulating mouse embryo. *Development* **121:** 2479.
Hemmati-Brivanlou A, Stewart R.M., and Harland R.M. 1990. Region-specific neural induction of an engrailed protein by anterior notochord in *Xenopus*. *Science* **250:** 800.
Hermesz E., Mackem S., and Mahon K.A. 1996. *Rpx*: A novel anterior-restricted homeobox gene progressively activated in the prechordal plate, anterior neural plate and Rathke's pouch of the mouse embryo. *Development* **122:** 41.
Herrmann B.G. 1991. Expression pattern of the *Brachyury* gene in whole-mount T^{ws}/ T^{ws} mutant embryos. *Development* **113:** 913.
Hoppler S., Brown J.D., and Moon R.T. 1996. Expression of a dominant-negative Wnt blocks induction of *MyoD* in *Xenopus* embryos. *Genes Dev.* **10:** 2805.
Hume C.R. and Dodd J. 1993. *Cwnt-8C*: A novel *Wnt* gene with a potential role in primitive streak formation and hindbrain organization. *Development* **119:** 1147.
Hsu Y.C. 1979. In vitro development of individually cultured whole mouse embryos from blastocyst to early somite stage. *Dev. Biol.* **68:** 453.
Kaufman M.H. 1992. Disposition of the extra-embryonic membranes associated with various types of twinning. In *The atlas of mouse development*, p. 479. Academic Press, London.
Kaufman M.H., and O'Shea K.S. 1978. Induction of monozygotic twinning in the mouse. *Nature* **276:** 707.
Kessler D.S. and Melton D.A. 1994. Vertebrate embryonic induction: Mesodermal and neural patterning. *Science* **266:** 596.
Kispert A. and Herrmann B.G. 1994. Immunohistochemical analysis of the Brachyury protein in wild-type and mutant mouse embryos. *Dev. Biol.* **161:** 179.
Lawson K.A. and Pedersen R.A. 1987. Cell fate, morphogenetic movement and population kinetics of embryonic endoderm at the time of germ layer formation in the mouse. *Development* **101:** 627.
———. 1992. Clonal analysis of cell fate during gastrulation and early neurulation in the mouse. *Ciba Found. Symp.* **165:** 3.
McMahon A.P., Joyner A.L., Bradley A., and McMahon J.A. 1992. The midbrain-hindbrain phenotype of *Wnt-1*$^{-}$/*Wnt-1*$^{-}$ mice results from stepwise deletion of *engrailed*-expressing cells by 9.5 days *postcoitum*. *Cell* **69:** 1.
Miller J.R. and Moon R.T. 1996. Signal transduction through β-catenin and specification of cell fate during embryogenesis. *Genes Dev.* **10:** 2527.
Moon R.T., Christian J.L., Campbell R.M., McGrew L.L., DeMarais A.A., Torres M., Lai C.-J., Olson D.J., and Kelly G.M. 1993. Dissecting Wnt signaling pathways and Wnt-sensitive developmental processes through transient misexpression analyses in embryos of *Xenopus laevis*. *Development* (suppl.) p., 85.
Ng S.-Y., Gunning P., Eddy R., Ponte P., Leavitt J., Shows T., and Kedes L. 1985. Evolution of the functional human *β-actin* gene and its multi-pseudogene family: Conservation of noncoding regions and chromosomal dispersion of pseudogenes. *Mol. Cell. Biol.* **5:** 2720.
Parr B.A., Shea M.J., Vassileva G., and McMahon A.P. 1993. Mouse *Wnt* genes exhibit discrete domains of expression in the early embryonic CNS and limb buds. *Development* **119:** 249.
Pöpperl H., Schmidt C.W., Hume C., Dodd J., Krumlauf R., and Beddington R.S.P. 1997. Misexpression of *Cwnt8C* in the mouse induces an ectopic embryonic axis and causes a truncation of the anterior neurectoderm. *Development* (in press).
Pöpperl H., Bienz M., Studer M., Chan S.K., Aparacio S., Brenner S., Mann R.S., and Krumlauf R. 1995. Segmental expression of *Hoxb-1* is controlled by a highly conserved autoregulatory loop dependent upon *exd/pbx*. *Cell* **81:** 1031.
Roelink H., Augsburger A., Heemskerk J., Korzh V., Norlin S., Ruiz i Altaba A., Tanabe Y., Placzek M., Edlund T., Jessell T.M., and Dodd J. 1994. Floor plate and motor neuron induction by *vhh-1*, a vertebrate homolog of *hedgehog* expressed by the notochord. *Cell* **76:** 761.
Rosenquist T.A. and Martin G.R. 1995. Visceral endoderm-1 (VE-1): An antigen marker that distinguishes anterior from posterior embryonic visceral endoderm in the early post-implantation mouse embryo. *Mech. Dev.* **49:** 117.
Saha M.S. and Grainger R.M. 1992) A labile period in the determinaiton of the anterior-posterior axis during early neural development in *Xenopus*. *Neuron* **8:** 1003.
Serbedzija G.N., Bronner-Fraser M., and Fraser S.E. 1989. A vital dye analysis of the timing and pathways of avian trunk neural crest cell migration. *Development* **106:** 809.
Sharpe C.R. and Gurdon J.B. 1990. The induction of anterior and posterior neural genes in *Xenopus laevis*. *Development* **109:** 765.
Shawlot W. and Behringer R.R. 1995. Requirement for *Lim1* in head-organizer function. *Nature* **374:** 425.
Sive H.L. 1993. The frog prince-ss: A molecular formula for dorsoventral patterning in *Xenopus*. *Genes Dev.* **7:** 1.
Smith L.J. 1985. Embryonic axis orientation in the mouse and its correlation with blastocyst relationships to the uterus. II. Relationships from 4.5 to 9.5 days. *J. Embryol. Exp. Morphol.* **89:** 15.
Smith W.C. and Harland R.M. 1991. Injected *Xwnt-8* RNA acts early in *Xenopus* embryos to promote formation of a vegetal dorsalizing center. *Cell* **67:** 753.
Snow M.H. L. 1977. Gastrulation in the mouse: Growth and regionalization of the epiblast. *J. Embryol. Exp. Morphol.* **42:** 293.
Sokol S., Christian J.L., Moon R.T., and Melton D.A. 1991. Injected *Wnt* RNA induces a complete body axis in *Xenopus* embryos. *Cell* **67:** 741.
St. Johnston D. and Nüsslein-Volhard C. 1992. The origin of pattern and polarity in the *Drosophila* embryo. *Cell* **68:** 201.
Tam P.P.L. 1989. Regionalisation of the mouse embryonic ectoderm: Allocation of prospective ectodermal tissues during gastrulation. *Development* **107:** 55.
Thomas P. and Beddington R. 1996. Anterior primitive endoderm may be responsible for patterning the anterior neural plate in the mouse embryo. *Curr. Biol.* **6:** 1487.
Varlet I., Collignon J., and Robertson E.J. 1997. *nodal* expression in the primitive endoderm is required for specification of the anterior axis during mouse gastrulation. *Development* **124:** 1033.
Weinstein D.C., Ruiz i Altaba A., Chen W.S., Hoodless P., Prezioso V.R., Jessell T.M. and Darnell Jr. J.E. 1994. The

winged-helix transcription factor HNF-3β is required for notochord develoment in the mouse embryo. *Cell* **78:** 575.
Whiting J., Marshall H., Cook M., Krumlauf R., Rigby P.W.J., Scott D., and Allemann R.K. 1991. Multiple spatially specific enhancers are required to reconstruct the pattern of *Hox-2.6* gene expression. *Genes Dev.* **5:** 2048.
Wilkinson D.G., Ed. 1992. Whole mount in situ hybridisation of vertebrate embryos. In *In situ hybridisation*, p. 75. IRL Press, Oxford, United Kingdom.
Wilkinson D.G., Bhatt S., and Herrmann B.G. 1990. Expression pattern of the mouse *T* gene and its role in mesoderm formation. *Nature* **343:** 657.
Wilson V. and Beddington R.S.P. 1996. Cell fate and morphogenetic movement in the late mouse primitive streak. *Mech. Dev.* **55:** 79.
Wu T.C, Wan Y.J., and Damjanov I. 1981. Positioning of inner cell mass determines the development of mouse blastocysts in vitro. J. *Embryol. Exp. Morphol.* **65:** 105.

FGF Signaling in Mouse Gastrulation and Anteroposterior Patterning

J. ROSSANT, B. CIRUNA, AND J. PARTANEN
Samuel Lunenfeld Research Institute, Mount Sinai Hospital, Toronto, Ontario, Canada, M5G 1X5, and Department of Molecular and Medical Genetics, University of Toronto

In several different species, fibroblast growth factors (FGFs) have been shown to be involved in many different aspects of cell behavior, including promotion of mitogenesis, cell migration, and induction of specific cell fates. We have been interested in the roles that FGF signaling might have in the induction and patterning of the developing mesoderm lineage in the vertebrate embryo. Studies in *Xenopus laevis* first implicated FGFs in mesoderm development. Exogenous basic FGF, when added to animal cap explants, is able to induce mesoderm from tissue normally fated to become ectoderm (Kimelman and Kirschner 1987; Paterno et al. 1989; Slack et al. 1989). The introduction of a dominant-negative FGF receptor into *Xenopus* embryos at various stages of development has demonstrated that FGF signaling is required for the expression of several early mesodermal markers, for the induction of posterior and ventral mesoderm, and for proper mesodermal maintenance during gastrulation (Amaya et al. 1991, 1993; Kroll and Amaya 1996).

These kinds of studies indicate a general role for FGF signaling in mesoderm development, but they do not specify the relative roles of the different receptors and ligands that might be expressed in the intact embryo. In vertebrates, there are more than ten different FGFs and four different receptors, each with differently spliced isoforms. In the mouse, it has been shown that FGFs 3, 4, 5, and 8 are expressed within the primitive streak in spatial and temporal patterns, which would be consistent with a role for FGF signaling in mesoderm induction or in regulating processes of fate determination (Wilkinson et al. 1988; Haub and Goldfarb 1991; Hebert et al. 1991; Niswander and Martin 1992; Crossley and Martin 1995). So far, the published mutations in relevant FGFs have not shed much light on their possible roles in gastrulation. Mutations in *Fgf3* and *Fgf5* led to late defects in ear and tail development and in the hair growth cycle, respectively (Mansour et al. 1993; Hebert et al. 1994). Targeted disruption of *Fgf4* resulted in peri-implantation lethality prior to mesoderm formation (Feldman et al. 1995). None of these studies definitively exclude any of the FGFs from acting at gastrulation, since there is clear overlap in expression and known cross-specificity in ligand-receptor interactions (Ornitz et al. 1996).

Given the large number of ligands and the extensive cross-talk in the FGF system, we have focused on trying to assess the role of FGF signaling in mesoderm development by studying the roles of the receptors, which are fewer in number and also restricted in their expression. *Fgfr1* is first expressed throughout the primitive ectoderm. In mid-streak-staged embryos, *Fgfr1* expression is concentrated in the posterior mesoderm lateral to the primitive streak and is maintained in the migrating mesodermal wings; headfold stage embryos show strong expression in both neurectoderm and the developing paraxial mesoderm (Orr-Urtreger et al. 1991; Yamaguchi et al. 1992). Targeted mutation of *Fgfr1* resulted in embryonic lethality between day 7.5 and day 9.5 of development, with defects first manifesting themselves at the onset of gastrulation (Deng et al. 1994; Yamaguchi et al. 1994). Homozygous *Fgfr1* mutant embryos showed complex morphological abnormalities, including thickening of the posterior streak suggestive of defects in cell behavior within the primitive streak. Paraxial mesoderm of *Fgfr1* mutant embryos was much reduced, yet axial mesoderm was still present and possibly expanded, suggesting that FGFR1 might also have a role in patterning the mesoderm populations arising at different positions along the streak (Deng et al. 1994; Yamaguchi et al. 1994). However, it was difficult to distinguish primary defects associated with the *Fgfr1* mutation from secondary defects resulting from grossly abnormal morphogenetic movements at gastrulation. We have performed mosaic analysis by aggregating homozygous $Fgfr1^{-/-}$ embryonic stem (ES) cells with wild-type embryos and show that defects in primary mesodermal patterning may be secondary to an initial deficiency in the ability of mutant cells to traverse the primitive streak.

FGFS AND ANTEROPOSTERIOR PATTERNING

In addition to a role in the initial formation of mesoderm at the primitive streak, experiments with *Xenopus* embryos have implicated a function for FGFs in assigning anteroposterior (A/P) positional values in the developing mesoderm and neurectoderm. Ectopic application or expression of FGFs has been observed to result in an anterior expansion of the expression patterns of some posteriorly expressed *Hox* genes and of the caudal-related gene *Xcad3* (Ruiz i Altaba and Melton 1989; Cho and De Robertis 1990; Kolm and Sive 1995; Pownall et al. 1996). On the other hand, when a dominant-negative FGF receptor is expressed in *Xenopus* embryos, repression or posteriorization of *Hox* and *Xcad3* gene expression has been observed. This repression of *Hox* gene expression can be overcome by ectopic expression of *Xcad3* (Pownall et al. 1996), supporting a model where FGFs regulate *Hox* genes through induction of caudal-related transcription factors (Subramanian et al. 1995). However, different members of the *Hox* gene family may respond differently to FGF signals (Kroll and Amaya 1996). Furthermore, the morphogenetic abnormalities seen in embryos expressing the dominant-negative receptor complicate the interpretation of altered gene expression in terms of a simple posteriorizing effect.

The mouse has served as a valuable model system for studying the A/P patterning mechanisms of vertebrates, because of the relative ease in generating mutations in putative patterning genes. Formal proof for the involvement of *Hox* genes in vertebrate A/P patterning has come from the studies of mouse *Hox* mutants generated by gene targeting (Capecchi 1996). Targeted mutagenesis of mouse caudal-related *Cdx* genes also first suggested their involvement as positive regulators of *Hox* gene expression (Subramanian et al. 1995; Chawengsaksophak et al. 1997). However, to date, there has been no clear genetic evidence to implicate FGF signaling in A/P patterning in the mouse. Expression of FGFR1 not only in the primitive streak, but also after gastrulation in the presomitic mesoderm and the early developing somites suggests that FGFR1 might be involved in various stages of the establishment and maintenance of A/P patterning. However, the early lethality and lack of somite formation in the FGFR1 mutant embryos precluded an analysis of any possible later role for FGFR1 in A/P patterning of the paraxial mesoderm. The ability to engineer subtle alterations in the FGFR1 signaling pathway and to detect their consequences on embryonic patterning (using the morphological landmarks of the vertebral column) has allowed us to study the function of FGF signaling in A/P patterning in the mouse. Our results show that alterations in signaling through FGFR1 can result in changes in vertebral identities independent of defects in the production and segmentation of paraxial mesoderm. Thus, our results are consistent with the suggested involvement of FGF signaling in positive regulation of *Hox* gene expression.

RESULTS AND DISCUSSION

Chimeric Analysis of *Fgfr1* Mutant ES Cells

Diploid chimeric embryos were generated by aggregating an 8–10-cell clump of either *Fgfr1*$^{+/-}$ or *Fgfr1*$^{-/-}$ embryonic stem (ES) cells with CD-1 8-cell embryos, using the standard morula aggregation technique (Nagy et al. 1993). ES cells were derived from *Fgfr1* mutant strains that also contained the ROSA26 insertion that results in ubiquitous expression of β-galactosidase (Ciruna et al. 1997). Aggregates were transferred into the uteri of CD-1 foster mothers, chimeric embryos were dissected at early to mid gestational stages, and the contribution of ES cells and CD-1 cells to the embryo was determined by whole-mount β-galactosidase staining.

Analysis of chimeras at 9.5 to 10.5 days of gestation revealed that *Fgfr1*$^{-/-}$ cells were deficient in their ability to contribute to gut, cephalic mesenchyme, heart, and somites but were evenly interspersed with wild-type cells in all other cell layers (Fig. 1). This deficiency was not absolute—some mutant cells were observed in all structures, suggesting that *Fgfr1* is not absolutely required for specification of any particular mesoderm cell fate, a conclusion previously reached from the analysis of the development of teratomas from *Fgfr1* mutant ES cells (Deng et al. 1994). Examination of chimeras at earlier stages showed that the failure of mutant cells to contribute to these specific mesoderm populations was largely due to their failure to populate the precursors of these cells at the primitive streak stage. Mutant cells were deficient in contributing to mesoderm cells passing through the early primitive streak and tended to accumulate at the posterior end of the embryo. Heart and cephalic mesenchyme are among the first populations of mesoderm cells to migrate through the streak (Parameswaran and Tam 1995), and, in chimeras, wild-type cells consistently contributed to these early populations at the expense of mutant cells (Fig. 2).

In contrast to the anterior mesoderm and endodermal populations, the limb bud, lateral mesoderm, and allantois of chimeric embryos were well colonized by *Fgfr1*$^{-/-}$ cells. Thus, mutant cells were ultimately able to traverse the posterior streak. This skewed distribution of *Fgfr1*$^{-/-}$ mesoderm might indicate a differential requirement for FGFR1 at various A/P levels of the primitive streak. *Fgf4* and *Fgf5* are expressed predominantly within the anterior two thirds of the egg cylinder and streak (Haub and Goldfarb 1991; Hebert et al. 1991; Niswander and Martin 1992) and could therefore be responsible for anterior streak-specific FGFR1 signaling. A null mutation in *Fgf8*, which is expressed throughout the streak, appears to have a more severe problem in generating mesoderm cells at the streak than do *Fgfr1* mutants (Lewandoski et al., this volume), suggesting that there may also be overlapping requirements for FGF receptors in integrating the different FGF signals emanating from the streak.

The aberrant morphogenetic movement of *Fgfr1* mutant cells at the streak could be the result of defects in either cell-cell adhesion or migration. Analysis of mutant cell distribution in chimeras with high mutant contribution showed that some mutant cells were capable of undergoing the epithelium-mesenchyme transition at the streak but failed to move away from the streak as effectively as wild-type cells. This is suggestive of a defect in cell migration. Cell–extracellular matrix (ECM) interactions are necessary for cell migration, and mouse embryos homozygous for mutations in fibronectin (a glycoprotein component of the ECM) and in focal adhesion kinase (a nonreceptor protein tyrosine kinase thought to mediate integrin signaling) both show phenotypes similar to those of homozygous *Fgfr1*$^{-/-}$ embryos (George et al. 1993; Furuta et al. 1995). A role for FGFR signaling in cell migration has been established in other organisms

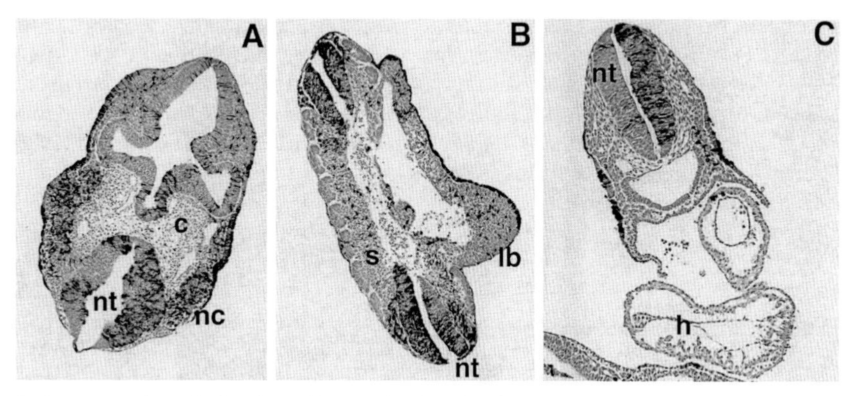

Figure 1. Histological sections through β-galactosidase stained *Fgfr1*$^{-/-}$↔CD-1 chimeric embryos at E9.5. *Fgfr1* mutant cells (*stained dark*) contribute strongly to the neural tube and neural crest lineages of chimeric embryos, but they appear to be deficient at contributing to cephalic mesenchyme (*A*), somitic mesoderm (*B*) and the heart (*C*). (nt) Neural tube; (nc) neural crest; (c) cephalic mesenchyme; (S) somites; (lb) limb bud; (h) heart.

(Reichman-Fried et al. 1994; DeVore et al. 1995). In particular, mutations of the *Drosophila* FGF-R2 gene, *heartless* (*htl*), show phenotypes strikingly similar to those observed in this study (Fig. 2): Invaginated mesodermal cells remain aggregated along the ventral midline and fail to migrate in a dorsolateral direction (Beiman et al. 1996; Gisselbrecht et al. 1996). Because these progenitors fail to acquire position-specific inductive cues, *htl* embryos show a reduction in cardiac, visceral, and dorsal somatic muscle fates. However, *htl* mesodermal precursors remain competent to receive inductive signals, and mesodermal fates can be rescued by ectopically expressed decapentaplegic protein (Beiman et al. 1996; Gisselbrecht et al. 1996). This is analogous to the results obtained here which demonstrate that mutant *Fgfr1* cells can still form all types of mesoderm but fail to migrate properly through the primitive streak to a position where they can receive signals to specify mesodermal fates (Tam et al. 1997). Therefore, there may be a conserved role for FGFR signaling in the morphogenesis of mesoderm formation at gastrulation.

Defects in traversing the primitive streak could also involve altered adhesive properties of *Fgfr1*$^{-/-}$cells. Mutant cells that accumulated within the streak of *Fgfr1*$^{-/-}$↔CD-1 chimeras maintained a columnar epithelial morphology, arguing that *Fgfr1*$^{-/-}$ cells were failing to undergo fully an epithelial to mesenchymal transition (EMT). EMT events are thought to be regulated by a family of calcium-dependent cell adhesion molecules (the cadherins). E-cadherin, for example, is expressed in all cells of the early egg cylinder but is down-regulated at gastrulation within the primitive streak and nascent mesodermal populations (Damjanov et al. 1986). Burdsal et al. (1993) have shown that function perturbing antibodies against E-cadherin can force an EMT in cultured epiblast tissue: Epiblast cells lose cell-cell contacts, flatten, and assume a mesenchymal morphology. Therefore, there appears to be a causal relationship between loss of cadherin function and EMT. FGFR1 may regulate the expression or function of these adhesion molecules. Indeed, exogenous FGF has been shown to cause mesenchymal transformation of cultured epithelial cell lines

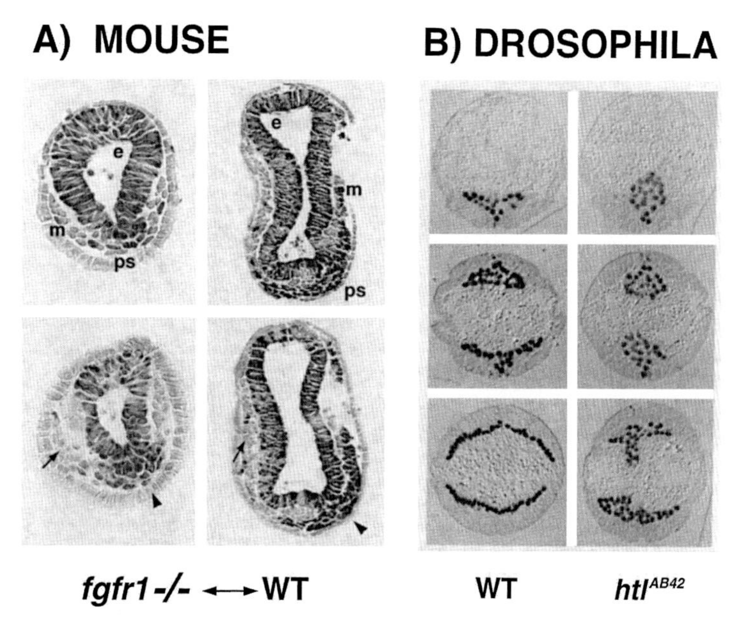

Figure 2. The behavior of both murine *Fgfr1* mutant and *Drosophila heartless* mutant mesodermal precursor cells suggests a conserved role for FGFR signaling in the morphogenesis of mesoderm formation at gastrulation. (*A*) Transverse sections of β-galactosidase stained *Fgfr1*$^{-/-}$↔CD-1 chimeras at E7.5; primitive streak is toward the bottom. *Fgfr1* mutant cells accumulate along the primitive streak of chimeric embryos (*arrowheads*) and show defects in lateral migration. As a result, the mesodermal wings are composed largely of wild-type cells (*arrows*). (e) Epiblast; (m) mesodermal wings; (ps) primitive streak. (*B*) *Twist*-stained wild-type and *heartless Drosophila* embryos at different developmental stages; ventral is toward the bottom. In wild-type embryos, mesodermal cells invaginate, disperse, and spread laterally along the inner surface of the ectoderm. In *htl* mutants, however, mesodermal cells remain clustered close to the ventral midline and fail to migrate in a dorsolateral direction. (Reprinted, with permission, from Gisselbrecht et al. 1996.)

(Boyer et al. 1992), and this FGF-induced EMT has been associated with the cellular redistribution of E-cadherin. There is also some evidence that cell adhesion molecules may act upstream of FGF signaling by binding to and activating FGF receptors (for review, see Green et al. 1996). Further study will be needed to determine the relationship between FGFR1 signaling and regulation of cell adhesion at gastrulation.

Generation and Analysis of an Allelic Series of *Fgfr1* Mutations

To create hypomorphic or site-specific alleles of the *Fgfr1* gene, a series of point mutations as well as neo-cassette insertions were introduced into the *Fgfr1* locus by gene targeting (Fig. 3). These alleles fall into three categories: (1) The protein coding capacities of the two exons (exons 6 and 7) encoding alternatively spliced isoforms of FGFR1 (isoforms IIIc and IIIb; Johnson et al. 1991) were separately disrupted by introducing stop codons into these exons (Fig. 3A); (2) alleles with a neo cassette in a sense orientation in introns 7 and 15 were generated (Fig. 3C); and (3) one of the autophosphorylation sites and the phospholipase-Cγ1 docking site of FGFR1 (Y766; Mohammadi et al. 1991) was inactivated by mutation into phenylalanine (Fig. 3B).

Embryos homozygous for an allele with a point mutation in the IIIc exon displayed a phenotype very similar to those of the embryos homozygous for the putative null alleles (Deng et al. 1994; Yamaguchi et al. 1994). They showed a greatly reduced amount of *Mox-1*-positive paraxial mesoderm, which was not organized into somites. On the other hand, mice homozygous for an allele where the exon IIIb was inactivated were viable and fertile. We conclude that IIIc is the dominant isoform and responsible for the majority of FGFR1 functions. However, in a situation where the overall activity of the receptor is reduced, the mutation in the IIIb exon further enhances the phenotype. Thus, in the case of *Fgfr1,* there appears to be a certain amount of redundancy between the two isoforms.

The neo-cassette insertions into introns 7 and 15 produced alleles (n7 and n15) that showed qualitatively similar but quantitatively different phenotypes. When homozygous, both n7 and n15 caused neonatal lethality, defects in craniofacial and limb patterning, as well as abnormal development of the A/P axis. In both cases, posterior truncations were observed: In the n7 homozygotes, there were defects in tail development, whereas n15 homozygotes displayed more severe posterior deletions that often extended into the lumbro-sacral level. The tail and limb defects were more severe in neonates homozygous for a IIIbn allele, which carries both the neo insertion in intron 7 and a point mutation in the IIIb exon. In addition to the posterior deletions, skeletal analysis of n7 and n15 homozygotes revealed transformations in vertebral identities predominantly in an anterior direction. These alterations were seen at all levels of the vertebral column and were not confined to the regions of axis truncation. A commonly seen change was an anterior transformation of the first lumbar vertebra into a rib-bearing vertebra (Fig. 4, left), and transformations were also apparent at the cervical level. The n7 and n15 alleles are likely to be hypo-

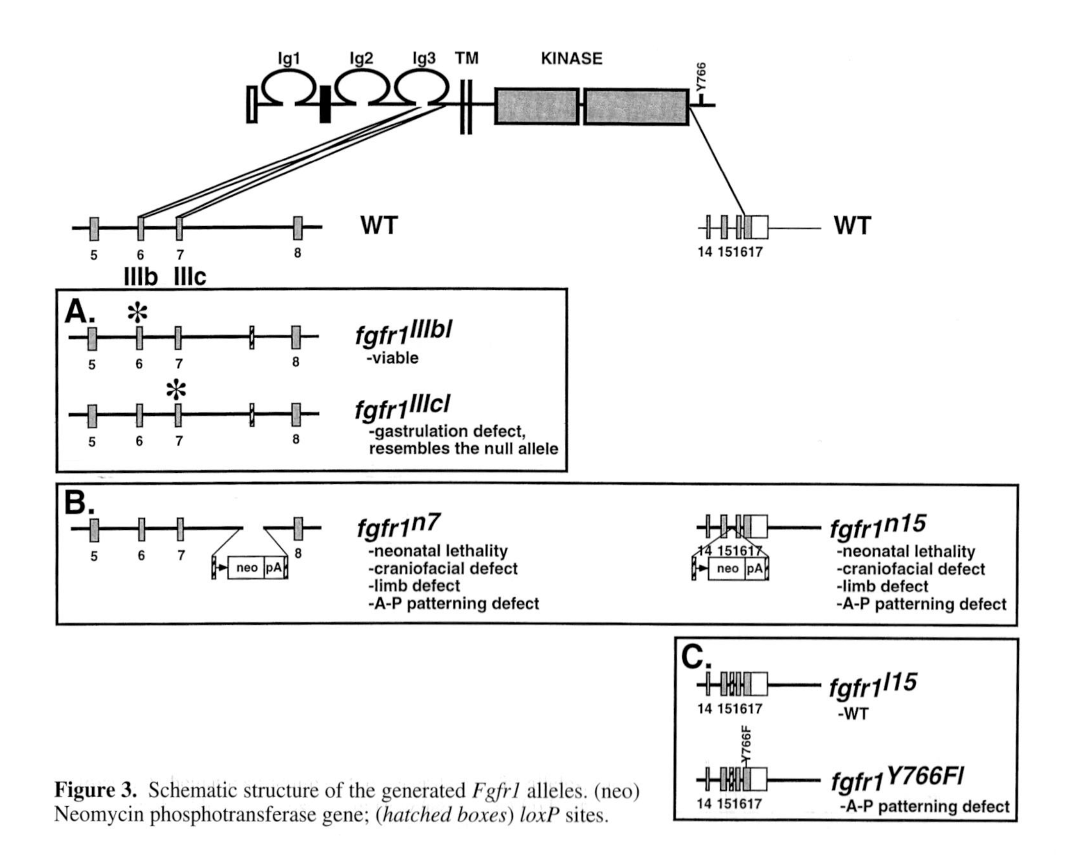

Figure 3. Schematic structure of the generated *Fgfr1* alleles. (neo) Neomycin phosphotransferase gene; (*hatched boxes*) *loxP* sites.

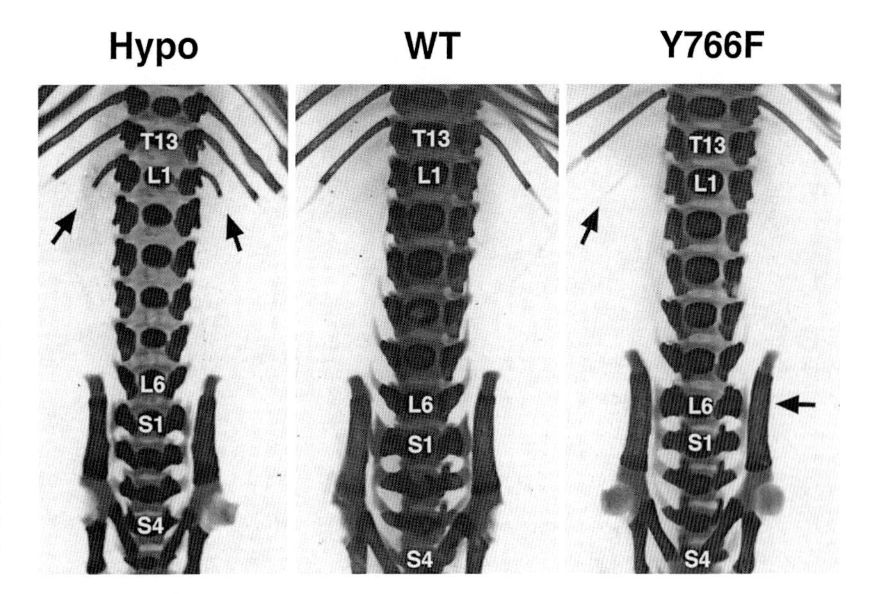

Figure 4. Homeotic transformations at the lumbar level of *Fgfr1* mutants. (*A*) Skeleton of a newborn *Fgfr1*n7/n7 mouse. Arrows indicate an extra pair of ribs on the first lumbar vertebra (L1). (*B*) Wild type. (*C*) Skeleton of a newborn *Fgfr1*Y766F/Y766F mouse. Arrows indicate the loss of ribs on the last thoracic vertebra (T13). In addition, the last lumbar vertebra (L6) is fused to the iliac bones.

morphic alleles, producing reduced levels of normal FGFR1 protein. The neo cassette (containing its own polyadenylation signal) may cause aberrant splicing and/or premature termination of transcription, leading to reduced mRNA levels, as has been shown with a similar neo-casette insertion into the N-*myc* locus (Moens et al. 1993). Based on genetic criteria, n7 is a hypomorphic allele (the phenotype of n7/null is more severe than n7/n7).

A series of posterior truncations somewhat similar to that of the *Fgfr1* mutants is seen with *Wnt-3a/vestigial tail* alleles (Takada et al. 1994; Greco et al. 1996). However, no homeotic transformations were observed in the *Wnt-3a* mutants, suggesting that WNT-3a and FGF signals have distinct functions during axis elongation. In addition, this difference argues further against the possibility that the transformations observed in *Fgfr1* hypomorphs are secondary to an axis elongation defect.

Mice homozygous for the *Fgfr1* allele with the Y766F mutation were viable and did not show obvious craniofacial, limb, or tail defects. Interestingly, however, skeletal analysis of newborn Y766F homozygotes revealed frequent transformations throughout the vertebral column. In contrast to the n7/n15 homozygotes, these transformations were exclusively in the posterior direction. For example, Y766F homozygotes often showed a posterior transformation of the last thoracic vertebra into a lumbar phenotype (Fig. 4, right). Furthermore, the Y766F allele appeared to be semi-dominant and does not behave genetically as a hypomorph. These observations together suggest that the Y766F mutation creates a gain-of-function allele encoding an overactive receptor. Interestingly, the phosphorylation of Y766 in FGFR1 has been implicated in the internalization and degradation of the receptor (Sorokin et al. 1994). In addition, a similar autophosphorylation site capable of binding PLC-γ1 negatively regulates the biological activity of the *Drosophila* torso receptor tyrosine kinase (Cleghon et al. 1996). Thus, it is possible that in the Y766F mutants, a receptor down-regulation signal (maybe involving PLC-γ1) is disrupted, leading into extended and amplified signaling through FGFR1. The fact that hypomorphic and gain-of-function alleles of *Fgfr1* have opposing effects on A/P patterning argues for a direct role for *Fgfr1* in this process.

Correct establishment of vertebrate *Hox* gene expression leading into development of distinct structures along A/P axis is an interesting but still poorly understood phenomenon. This complex process appears to be regulated at multiple levels (van der Hoeven et al. 1996). Opening of the chromatin of *Hox* complexes may be regulated by the polycomb and trithorax family members (Yu et al. 1995; Shumacher et al. 1996) and is presumably permissive for later gene activation. Translation of this competence into actual *Hox* gene activation appears to be regulated by intercellular signals, which act to activate progressively *Hox* genes along the A/P axis. Two extensively studied candidates for such signals are retinoic acid and FGFs. The direct involvement of retinoic acid in the regulation of some aspects of *Hox* gene expression in mice has been supported by gene targeting studies of the retinoic acid receptors and deletion of retinoic acid response elements of *Hoxa1* (Lohnes et al. 1993; Dupe et al. 1997). The data presented here provide the first genetic evidence of involvement of FGF signaling in A/P patterning of the mesoderm, and the opposite phenotypes observed in putative hypomorphic and gain-of-function alleles of *Fgfr1* provide strong support for a crucial role for FGFR1 in patterning the body axis (Fig. 5). In *Xenopus*, it has been proposed that FGF may act to activate posterior *Hox* gene expression via up-regulation of the Caudal genes (Pownall et al. 1996), which have been shown to regulate directly at least one *Hox* gene in mammals (Subramanian et al. 1995). However, the exact mechanisms leading to homeotic transformation in the *Fgfr1* mutants is still not clear, and we are currently examining how and when the patterns of expression of the *Hox* genes are altered in these mutants.

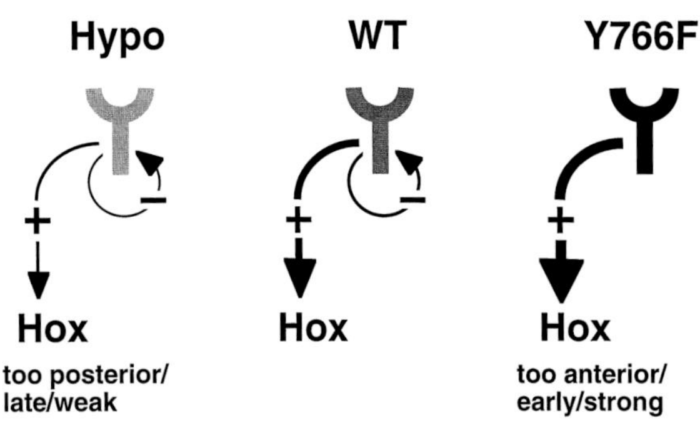

Figure 5. A model for FGFR1 function during A/P patterning and the effects of hypomorphic and Y766F mutations on it.

CONCLUSION

By using a combination of mosaic analysis and generation of an allelic series of mutations, we have been able to extend our understanding of the role of FGFR1 in early mouse development. We show that signaling through FGFR1 is required for correct epithelium-mesenchymal transition at the primitive streak, elongation of the axis and somite formation, and A/P patterning of somite derivatives. There are still other aspects of the hypomorphic mutations and of the chimeras that have not yet been fully explored. These include apparent defects in limb patterning and possible roles for FGFR1 in internal tissue development. Clearly, more analysis is needed to dissect the complex pathways of action of FGFs in development.

ACKNOWLEDGMENTS

J.R. is a Terry Fox Research Scientist of the National Cancer Institute of Canada and an International Research Scholar of the Howard Hughes Medical Institute, and B.G.C. is a scholar of the Natural Sciences and Engineering Research Council of Canada. This work was supported by a program project from the NCIC with funds from the Terry Fox Foundation.

REFERENCES

Amaya E., Musci T.J., and Kirschner M.W. 1991. Expression of a dominant negative mutant of the FGF receptor disrupts mesoderm formation in *Xenopus* embryos. *Cell* **66:** 257.

Amaya E., Stein P.A., Musci T.J., and Kirschner M.W. 1993. FGF signalling in the early specification of mesoderm in *Xenopus*. *Development* **118:** 477.

Beiman M., Shilo B.Z., and Volk T. 1996. *heartless*, a *Drosophila* FGF receptor homolog, is essential for cell migration and establishment of several mesodermal lineages. *Genes Dev.* **10:** 2993.

Boyer B., Dufour S., and Thiery J.P. 1992. E-cadherin expression during the acidic FGF-induced dispersion of a rat bladder carcinoma cell line. *Exp. Cell Res.* **201:** 347.

Burdsal C.A., Damsky C.H., and Pedersen R.A. 1993. The role of E-cadherin and integrins in mesoderm differentiation and migration at the mammalian primitive streak. *Development* **118:** 829.

Capecchi M.R. 1996. Function of homeobox genes in skeletal development. *Ann. N.Y. Acad. Sci.* **785:** 34.

Chawengsaksophak K., James R., Hammond V.E., Kontgen F., and Beck F. 1997. Homeosis and intestinal tumours in *Cdx2* mutant mice. *Nature* **385:** 84.

Cho K.W. and De Robertis E.M. 1990. Differential activation of *Xenopus* homeobox genes by mesoderm-inducing growth factors and retinoic acid. *Genes Dev.* **4:** 1910.

Ciruna B.G., Schwartz L., Harpal K., Yamaguchi T.P., and Rossant J. 1997. Chimeric analysis of fibroblast growth factor receptor-1 (*Fgfr1*) function: A role for FGFR1 in morphogenetic movement through the primitive streak. *Development* **124:** 2829.

Cleghon V., Gayko U., Copeland T.D., Perkins L.A., Perrimon N., and Morrison D.K. 1996. *Drosophila* terminal structure development is regulated by the compensatory activities of positive and negative phosphotyrosine signaling sites on the Torso RTK. *Genes Dev.* **10:** 566.

Crossley P.H. and Martin G.R. 1995. The mouse *Fgf8* gene encodes a family of polypeptides and is expressed in regions that direct outgrowth and patterning in the developing embryo. *Development* **121:** 439.

Damjanov I., Damjanov A., and Damsky C.H. 1986. Developmentally regulated expression of the cell-cell adhesion glycoprotein cell-CAM 120/80 in peri-implantation mouse embryos and extraembryonic membranes. *Dev. Biol.* **116:** 194.

Deng C.X., Wynshaw-Boris A., Shen M.M., Daugherty C., Ornitz D.M., and Leder P. 1994. Murine FGFR-1 is required for early postimplantation growth and axial organization. *Genes Dev.* **8:** 3045.

DeVore D.L., Horvitz H.R., and Stern M.J. 1995. An FGF receptor signaling pathway is required for the normal cell migrations of the sex myoblasts in *C. elegans* hermaphrodites. *Cell* **83:** 611.

Dupe V., Davenne M., Brocard J., Dolle P., Mark M., Dierich A., Chambon P., and Rijli F.M. 1997. In vivo functional analysis of the Hoxa-1 3′ retinoic acid response element (3′RARE). *Development* **124:** 399.

Feldman B., Poueymirou W., Papaioannou V.E., DeChiara T.M., and Goldfarb M. 1995. Requirement of FGF-4 for postimplantation mouse development. *Science* **267:** 246.

Furuta Y., Ilic D., Kanazawa S., Takeda N., Yamamoto T., and Aizawa S. 1995. Mesodermal defect in late phase of gastrulation by a targeted mutation of focal adhesion kinase, FAK. *Oncogene* **11:** 1989.

George E.L., Georges-Labouesse E.N., Patel-King R.S., Rayburn H., and Hynes R.O. 1993. Defects in mesoderm, neural tube and vascular development in mouse embryos lacking fibronectin. *Development* **119:** 1079.

Gisselbrecht S., Skeath J.B., Doe C.Q., and Michelson A.M. 1996. *heartless* encodes a fibroblast growth factor receptor (DFR1/DFGF-R2) involved in the directional migration of early mesodermal cells in the *Drosophila* embryo. *Genes Dev.* **10:** 3003.

Greco T.L., Takada S., Newhouse M.M., McMahon J.A., McMahon A.P., and Camper S.A. 1996. Analysis of the *vestigial tail* mutation demonstrates that *Wnt-3a* gene dosage regulates mouse axial development. *Genes Dev.* **10:** 313.

Green P.J., Walsh F.S., and Doherty P. 1996. Promiscuity of fibroblast growth factor receptors. *BioEssays* **18:** 639.

Haub O. and Goldfarb M. 1991. Expression of the fibroblast growth factor-5 gene in the mouse embryo. *Development* **112:** 397.

Hebert J.M., Boyle M., and Martin G.R. 1991. mRNA localization studies suggest that murine FGF-5 plays a role in gastrulation. *Development* **112:** 407.

Hebert J.M., Rosenquist T., Gotz J., and Martin G.R. 1994. FGF5 as a regulator of the hair growth cycle: Evidence from targeted and spontaneous mutations. *Cell* **78:** 1017.

Johnson D.E., Lu J., Chen H., Werner S., and Williams L.T. 1991. The human fibroblast growth factor receptor genes: A common structural arrangement underlies the mechanisms for generating receptor forms that differ in their third immunoglobulin domain. *Mol. Cell. Biol.* **11:** 4627.

Kimelman D. and Kirschner M. 1987. Synergistic induction of mesoderm by FGF and TGF-β and the identification of an mRNA coding for FGF in the early *Xenopus* embryo. *Cell* **51:** 869.

Kolm P.J. and Sive H.L. 1995. Regulation of the *Xenopus* labial

homeodomain genes, *HoxA1* and *HoxD1:* Activation by retinoids and peptide growth factors. *Dev. Biol.* **167:** 34.
Kroll K.L. and Amaya E. 1996. Transgenic *Xenopus* embryos from sperm nuclear transplantations reveal FGF signaling requirements during gastrulation. *Development* **122:** 3173.
Lohnes D., Kastner P., Dierich A., Mark M., LeMeur M., and Chambon P. 1993. Function of retinoic acid receptor γ in the mouse. *Cell* **73:** 643.
Mansour S.L., Goddard J.M., and Capecchi M.R. 1993. Mice homozygous for a targeted disruption of the proto-oncogene *int-2* have developmental defects in the tail and inner ear. *Development* **117:** 13.
Moens C.B., Stanton B.R., Parada L.F., and Rossant J. 1993. Defects in heart and lung development in compound heterozygotes for two different targeted mutations at the N-*myc* locus. *Development* **119:** 485.
Mohammadi M., Honegger A.M., Rotin D., Fischer R., Bellot F., Li W., Dionne C.A., Jaye M., Rubinstein M., and Schlessinger J. 1991. A tyrosine-phosphorylated carboxy-terminal peptide of the fibroblast growth factor receptor (Flg) is a binding site for the SH2 domain of phospholipase Cγ1. *Mol. Cell. Biol.* **11:** 5068.
Nagy A., Rossant J., Nagy R., Abramow-Newerly W., and Roder J.C. 1993. Derivation of completely cell culture-derived mice from early-passage embryonic stem cells. *Proc. Natl. Acad. Sci.* **90:** 8424.
Niswander L. and Martin G.R. 1992. Fgf-4 expression during gastrulation, myogenesis, limb and tooth development in the mouse. *Development* **114:** 755.
Ornitz D.M., Xu J., Colvin J.S., McEwen D.G., MacArthur C.A., Coulier F., Gao G., and Goldfarb M. 1996. Receptor specificity of the fibroblast growth factor family. *J. Biol. Chem.* **271:** 15292.
Orr-Urtreger A., Givol D., Yayon A., Yarden Y., and Lonai P. 1991. Developmental expression of two murine fibroblast growth factor receptors, *flg* and *bek. Development* **113:** 1419.
Parameswaran M. and Tam P.P. 1995. Regionalisation of cell fate and morphogenetic movement of the mesoderm during mouse gastrulation. *Dev. Genet.* **17:** 16.
Paterno G.D., Gillespie L.L., Dixon M.S., Slack J.M., and Heath J.K. 1989. Mesoderm-inducing properties of INT-2 and kFGF: Two oncogene-encoded growth factors related to FGF. *Development* **106:** 79.
Pownall M.E., Tucker A.S., Slack J.M., and Isaacs H.V. 1996. *eFGF*, *Xcad3* and *Hox* genes form a molecular pathway that establishes the anteroposterior axis in *Xenopus. Development* **122:** 3881.
Reichman-Fried M., Dickson B., Hafen E., and Shilo B.Z. 1994. Elucidation of the role of breathless, a *Drosophila* FGF receptor homolog, in tracheal cell migration. *Genes Dev.* **8:** 428.
Ruiz i Altaba A. and Melton D.A. 1989. Interaction between peptide growth factors and homoeobox genes in the establishment of antero-posterior polarity in frog embryos. *Nature* **341:** 33.
Shumacher A., Faust C., and Magnuson T. 1996. Positional cloning of a global regulator of anterior-posterior patterning in mice. *Nature* **383:** 250.
Slack J.M., Darlington B.G., Gillespie L.L., Godsave S.F., Isaacs H.V., and Paterno G.D. 1989. The role of fibroblast growth factor in early *Xenopus* development. *Development* (suppl.) **107:** 141.
Sorokin A., Mohammadi M., Huang J., and Schlessinger J. 1994. Internalization of fibroblast growth factor receptor is inhibited by a point mutation at tyrosine 766. *J. Biol. Chem.* **269:** 17056.
Subramanian V., Meyer B.I., and Gruss P. 1995. Disruption of the murine homeobox gene *Cdx1* affects axial skeletal identities by altering the mesodermal expression domains of *Hox* genes. *Cell* **83:** 641.
Takada S., Stark K.L., Shea M.J., Vassileva G., McMahon J.A., and McMahon A.P. 1994. *Wnt-3a* regulates somite and tailbud formation in the mouse embryo. *Genes Dev.* **8:** 174.
Tam P.P., Parameswaran M., Kinder S.J., and Weinberger R.P. 1997. The allocation of epiblast cells to the embryonic heart and other mesodermal lineages: The role of ingression and tissue movement during gastrulation. *Development* **124:** 1631.
van der Hoeven F., Zakany J., and Duboule D. 1996. Gene transpositions in the HoxD complex reveal a hierarchy of regulatory controls. *Cell* **85:** 1025.
Wilkinson D.G., Peters G., Dickson C., and McMahon A.P. 1988. Expression of the FGF-related proto-oncogene *int-2* during gastrulation and neuregulation in the mouse. *EMBO J.* **7:** 691.
Yamaguchi T.P., Conlon R.A., and Rossant J. 1992. Expression of the fibroblast growth factor receptor *fgfr-1/flg* during gastrulation and segmentation in the mouse embryo. *Dev. Biol.* **152:** 75.
Yamaguchi T.P., Harpal K., Henkemeyer M., and Rossant J. 1994. *fgfr-1* is required for embryonic growth and mesodermal patterning during mouse gastrulation. *Genes Dev.* **8:** 3032.
Yu B.D., Hess J.L., Horning S.E., Brown G.A., and Korsmeyer S.J. 1995. Altered *Hox* expression and segmental identity in Mll-mutant mice. *Nature* **378:** 505.

Lineage and Functional Analyses of the Mouse Organizer

P.P.L. Tam, K.A. Steiner, S.X. Zhou, and G.A. Quinlan
Embryology Unit, Children's Medical Research Institute, Wentworthville, NSW 2145, Australia

A comparison of the fate maps of the mouse gastrula with those of other vertebrate gastrulae reveals a remarkable convergence of the basic body plan during early embryogenesis (Lawson et al. 1991; Tam and Quinlan 1996). The homology of the body blueprint and the mounting evidence that an essentially similar repertoire of gene functions is involved in the delineation of tissue pattern and lineage specification strongly advocate for a conservation of the genetic and morphogenetic process of embryogenesis during evolution (Arendt and Nübler-Jüng 1997; Tam and Behringer 1997).

In the zebrafish, frogs, and birds, a group of cells in the gastrula expresses a common set of genes that encode transcription factors and secreted growth factors (Lemaire and Kodjabachian 1996). This population is able to induce the formation of a whole new body axis made up primarily of host tissues if transplanted to an ectopic site in another embryo (Hornbruch et al. 1979; Gilbert and Saxen 1993; Shih and Fraser 1995; Storey et al. 1995). Because the host tissues recruited to the induced axis are often properly assembled into neural and somitic structures, this population is said to act as an organizer. This functional attribute of the organizer is consistent with its being a source of signal(s) for patterning of the embryonic tissues during development. In the mouse embryo, an inductive activity comparable to that of amphibian and avian organizers has been demonstrated for the node of the late gastrula (Beddington 1994). Heterotopic transplantation of the node results in the induction of a partial axis containing neural and somitic tissues derived from the host. The induced axis apparently does not include any anterior (cranial) structures, suggesting that the activity that organizes the head may be found in an organizer present in earlier stage embryos.

In the early mouse gastrula, several organizer-specific genes are expressed in a group of cells in the posterior epiblast that is positioned between the progenitors of the extraembryonic and embryonic mesoderm (for review, see Tam and Behringer 1997). Cells at this site have been shown to contribute descendants that colonize the notochord and head process of the early-somite-stage embryo (Lawson et al. 1991; Quinlan et al. 1995). Therefore, this epiblast cell population in the early gastrula displays a cell fate similar to that of the node (Beddington 1994; Sulik et al. 1994). Whether a genealogical relationship exists between this early epiblast population and the node of the late gastrula is not known. This early epiblast population has not yet been tested directly for its activity in tissue patterning.

This study aimed to analyze the lineage and functional properties of the potential organizer in the early mouse gastrula. Fate-mapping and cell-tracking experiments have identified the cell population in the posterior epiblast of the early mouse gastrula that gives rise to tissues typically derived from the node of the late gastrula. This cell population was tested for its patterning activity by transplantation to heterotopic sites in the early- and the late-gastrula stage embryos.

EXPERIMENTAL PROCEDURES

Fate-mapping and Cell-tracking Experiment

Early-primitive-streak (PS) stage embryos were explanted from uterus of pregnant mice at 6.5 days postcoitum. A line of transgenic mice expressing an X-linked *HMG-nls-lacZ* transgene was used as the source of donor embryos. Cells were isolated from these embryos and transplanted to nontransgenic host embryos of the ARC strain. In the experimental chimeras, cells descended from the transplanted population can be tracked by the expression of the *lacZ* transgene (Tam and Tan 1992).

In this experiment, tissue fragments within a zone 50–100 μm from the proximal border of the epiblast on the posterior side of the early-PS stage embryo (Fig. 1a) were dissected from the *lacZ*-transgenic embryo. This epiblast fragment (the P-Ep fragment) was dissociated into clumps of 5–20 cells which were transplanted orthotopically to the equivalent site in the host embryo also at the early-PS stage (Fig. 1a[III]). The embryos were cultured for 44–46 hours until they reached the early somite stage. The distribution of the graft-derived cells in the host tissues was assessed on histological sections of the embryos.

The movement of cells descended from the P-Ep population was tracked during mesoderm formation. The cells were labeled by the microinjection of a carbocyanine dye (DiI or DiO, Molecular Probes) into the P-Ep region of the early-PS stage embryo (Fig. 1e[I]). The position of the labeled cells was scored at 12–24 hours after labeling by confocal microscopy (Quinlan et al. 1995). In the mid-PS stage embryos, cells in the distal region of the mesoderm (Fig. 1e[II]) and the epiblast (Fig. 1e[III]) were labeled differently in the same embryo by two car-

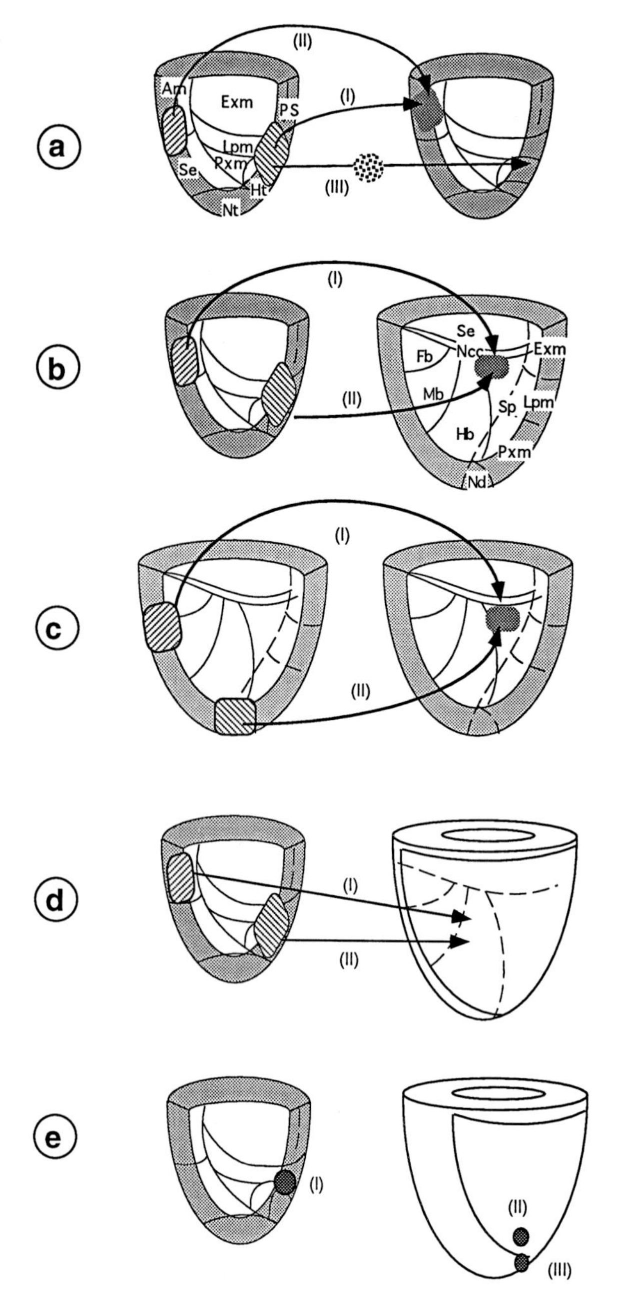

Figure 1. The experimental strategy for testing the cell fate and patterning activity of epiblast/ectoderm cell population (see text for details of the experiment).(*a*) Testing the patterning activity of (I) the anterior epiblast and (II) the posterior epiblast (P-Ep) of the early-PS stage synchronic host embryo. (III) Assessing the cell fate of the P-Ep population by orthotopic transplantation. (*b*) Testing the patterning activity of (I) the anterior epiblast and (II) the P-Ep cells of early-PS stage embryo in the late-PS stage host embryo. (*c*) Testing the patterning activity of (I) the anterior ectoderm and (II) the node of the late-PS stage embryo in synchronic host embryo. (*d*) Testing the inductive activity of (I) the anterior epiblast and (II) the P-Ep cells of early-PS stage embryo by transplanting the tissue into the cranial mesoderm of the late-PS stage embryo. (*e*) In situ labeling by carbocyanine dye of cell populations in the early- and mid-PS stage embryos to trace the movement of the descendants of the P-Ep cells. (I) In the early-PS stage embryo, a group of epiblast cells localized midway down the posterior side of the egg cylinder was labeled. In the mid-PS stage embryo, (II) cells in the most distal region of the mesoderm and (III) in the midline epiblast anterior to the margin of the mesoderm were labeled simultaneously with different dyes to determine the relative movement of the two populations during in vitro development.Only the right half of the epiblast or the ectoderm is shown. The area occupied by the progenitors of different tissue lineages is demarcated by lines mapped onto the epiblast and the ectoderm. For the host embryo of strategy (*d*), the domains of the precursor of the brain parts and the spinal cord in the ectoderm underneath the cranial mesenchyme are shown by dashed lines. Abbreviations for tissue types: (Am) Amnion ectoderm; (Exm) extraembryonic mesoderm; (Fb) forebrain; (Hb) hindbrain; (Ht) heart; (Mb) midbrain; (Nd) node; (Ncc) neural crest cells; (Nt) neuroectoderm; (Lpm) lateral plate mesoderm; (PS) primitive streak; (Pxm) paraxial mesoderm; (Se) surface ectoderm; (Sp) spinal cord.

bocyanine dyes. The localization of the two labeled populations was analyzed after 12 hours of development by confocal microscopy.

Testing the Patterning Activity of Epiblast Fragments

Experimental strategy. Fragments of the epiblast of the early-PS stage embryo were transplanted to heterotopic sites in the early- and late-PS stage embryos. The embryos were examined at 5–6 hours after transplantation for an assessment of the molecular and morphogenetic response of the host and the graft tissues. The embryo was also examined after 24–46 hours of culture for the formation of secondary axis and ectopic embryonic structures and for gene expression by whole-mount in situ hybridization.

Isolation of tissue fragments. Fragments of epiblast were dissected from the same P-Ep region as for the cell fate study. Tissues were also taken from the anterior epiblast (A-Ep) which is fated for the ectoderm of the amnion and the surface ectoderm (Lawson et al. 1991; Quinlan et al. 1995). The node region was isolated from the anterior region of the primitive streak of the late-PS stage embryo. The anterior ectoderm of the late-PS embryo that is fated to form the forebrain and midbrain was dissected free of the mesoderm and the endoderm. The various types of tissue fragments were trimmed down to about 25–30 μm for transplantation.

Transplantation. In the first series of transplantation experiments, the P-Ep fragment of the early-PS stage embryo was grafted to the anterior region of the embryo at the same developmental stage (Fig. 1a[I]). For a comparison of the effect of transplantation, the A-Ep fragment was also transplanted to the same anterior site (Fig. 1a[II]). In the second series, the P-Ep of the early-PS stage embryo was transplanted to the lateral region of the late-PS stage embryo (Fig. 1b[II]) at the site where tissues are fated to form the lateral mesoderm, surface ectoderm, and the neural tube caudal to the hindbrain level. In parallel experiments, A-Ep fragments of the early-PS stage embryo (Fig. 1b[I]), and the anterior ectoderm and the node of the late-PS stage embryo (Fig. 1c[I,II]) were also transplanted to the lateral region of the late-PS stage embryo. In the third series of experiments, the effect of the A-Ep and P-Ep fragments of the early-PS stage embryo on the morphogenesis of the neural plate was tested. The tissue fragments were grafted to the prospective cranial mesenchyme of the hindbrain of the late-PS stage embryo (Fig. 1d).

Analysis of results. At the time of collection, embryos were fixed in 4% paraformaldehyde and stained by X-gal (Progen) or M-gal (Biosynth AG) reagents to reveal the *lacZ* activity of the transplanted tissues. The specimens were processed either for histology (all three series) or for whole-mount in situ hybridization and immunostaining (the second and third series). The expression of tissue- or region-specific genes was studied: *Brachyury* (primitive streak and axial mesoderm), *Lim1* (anterior mesendoderm and lateral mesoderm), *Hnf3β* (node and axial mesoderm), *Otx2* (prospective anterior neuroectoderm), *Hoxb1* (posterior germ layers in the late gastrula), *Mox1* (paraxial mesoderm), and *NCAM* (neural tissues). Embryos in the third series of experiments were tested at the early-somite stage for the expression of the *Otx2* mRNA (forebrain and midbrain) and the *En2* protein (midbrain and hindbrain) activity.

RESULTS

The P-Ep Fragments Contain the Progenitors of the Node Derivatives

After 2 days of culture, 46 (of 73) host embryos contained *lacZ*-expressing cells derived from the orthotopically transplanted P-Ep population. Two different patterns of distribution of graft-derived cells were found (Table 1). In most embryos, graft-derived cells were found in the head process, the notochord, the notochordal plate, and floor plate of the neural tube (Fig. 2a–c). Colonization of the floor plate and the axial mesoderm were often found in the same segmental region of the embryo (Fig. 2d). Graft-derived tissues were also found in the paraxial and lateral mesoderm, heart, and endoderm. In some embryos, graft-derived cells colonized mainly the paraxial and lateral mesoderm (Table 1). There was no consistent segmental colocalization of the graft-derived cells in the neural tube and the paraxial or lateral mesoderm. The discrepancy in tissue colonization of the two groups of embryos may be due to the variation of the site of transplantation. Cells grafted in a slightly more lateral site were among the progenitors of the paraxial and lateral mesoderm and thus may not contribute to midline structures.

Cells from the P-Ep Population Reach Anterior Midline via Migration in the Mesoderm and Anterior Extension

Twelve hours after dye labeling at the early-PS stage, descendants of P-Ep cells were found in the distal region near the anterior end of the primitive streak of the now mid-PS stage embryo ($n = 15$). Cells were localized in the midline epiblast or in the adjacent mesodermal layer, or at both sites (Fig. 2e). After another 12–24 hours of de-

Table 1. Developmental Fate of Cells of the Posterior Epiblast of the Early-primitive-streak Stage Mouse Embryo

		Number of graft-derived cells in														
		neurectoderm		paraxial mesoderm					midline mesoderm				endoderm			
No. of embryos	Total cell no.	FP	NE	CRM	SOM	PSM	INM	LPM	HPM	NCD	ND	HT	FGN	TRN	YS	PS
With contribution to the floor plate of the neural tube and the midline mesoderm																
20	1209	**54**	0	390	235	70	20	4	**125**	**88**	**34**	42	87	36	8	16
	mean = 60.5 ± 12.0	**4.4%**			57.5%		1.9%			**20.4%**		3.4%		10.8%		1.2%
Without contribution to the floor plate of the neural tube and the midline mesoderm																
9	733	0	28	142	140	140	20	190	0	0	0	0	7	32	34	0
	mean = 81.6 ± 14.6		3.8%		57.6%		28.8%							9.9%		

Abbreviations: FP = floor plate, NE = neural plate, CRM = cranial mesoderm, SOM = somites, PSM = presomitic mesoderm; INH = intermediate mesoderm, LPM = lateral plate mesoderm, HPM = head process mesoderm, NCD = notochord, ND = node, HT = heart, FGN = foregut endoderm, TRN = trunk endoderm, YS = yolk sac endoderm, PS = primitive streak. Bold type indicates the node-derivatives.

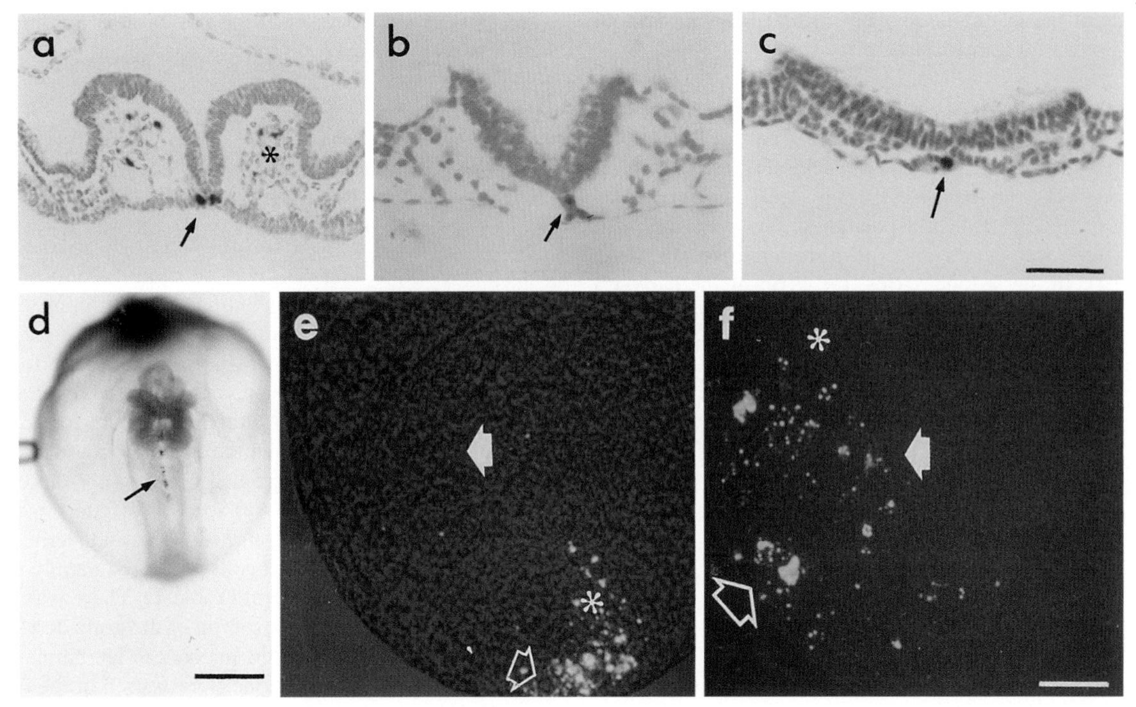

Figure 2. (*a–f*) The distribution of the P-Ep cells of early-PS stage embryo following cell transplantation and in situ labeling. Host embryos cultured for 46 hr to the early-somite stage showing the localization of the graft-derived cells (stained blue by X-gal and indicated by arrows) in (*a*) the anterior axial mesoderm (head process), cranial mesenchyme (asterisk), (*b*) notochord, and (*c*) notochordal plate and the prospective floor plate. Bar, 100 μm (*a–c*). (*d*) The localization of graft-derived cells (arrow) along the embryonic midline. Bar, 1 mm. (*e*) Labeled cells descending from the P-Ep were localized in the distal mesoderm (asterisk) and distal epiblast of the embryo (open arrow) at 12 hr after labeling. (*f*) Labeled cells originated from the distal region of the mesoderm (DiI: red label, asterisk) and the distal midline epiblast (DiO; green label, open arrow) were localized to the proximal and anterior region of the early-headfold stage embryo at 24 hr after labeling. In *e* and *f*, the solid arrow on the lateral region of the embryo points anteriorly. Bar, 20 μm (*e, f*)

velopment, labeled cells were distributed to the anteroproximal region of the mesoderm (the prospective heart and cranial mesoderm) and they also spread along the anterior midline of the embryo. To further analyze the migration of the P-Ep descendants, cells in seven mid-PS stage embryos were labeled at the distal mesoderm (Fig. 1e[II]) and epiblast (Fig. 1e[III]) with different dyes. During development, the labeled mesodermal cells migrated anteroproximally and contributed to the anterior paraxial mesoderm and the midline mesoderm. Descendants of the labeled epiblast cells were found mostly in the midline mesoderm (Fig. 2f). In summary, cells derived from the P-Ep population were distributed to both the axial and cranial mesenchyme by displacement along the midline of the embryo and migration with the mesoderm.

P-Ep Tissues of the Early-PS Stage Embryo Disrupts Anterior Development of the Host Embryo

Transplantation of the P-Ep cells to the anterior region of the early-PS stage embryos resulted in a diverse range of development. Of the 32 embryos examined, 8 (25%) showed normal development to the early-somite stage. In contrast, 7 (60%) of the 12 early-PS stage host embryos that received anterior epiblast graft developed normally. In both groups of embryos, the graft-derived cells were found primarily in the amnion and the surface ectoderm of the host embryo, as would be predicted from the normal fate of the cells in the anterior epiblast. About 37% of the embryos with P-Ep grafts and 20% of those with A-Ep grafts showed retarded development and reached the presomite headfold stage.

Of particular interest is the finding that in about 50% of the embryos receiving the P-Ep graft, profound disruption of neural tube morphogenesis was displayed (Fig. 4a). Neural folds and somites were absent, but allantois was present. The main bulk of graft-derived cells remained in the anterior end of the embryo (Fig. 4a). The graft-derived cells colonized the rudimentary neural plate and the adjacent surface ectoderm. Graft-derived cells spread anteriorly to the yolk sac, reminiscent of allantoic formation. There was a marked constriction of the anterior region of the embryo.

Gene Expression in the Graft and the Response of the Late-PS Stage Host Tissues (Table 2)

Early-PS stage P-Ep graft (Fig. 1b[II]). Within 5–6 hours after transplantation, the P-Ep grafts had integrated well into the tissues of the host embryo. When analyzed by whole-mount in situ hybridization, the grafted tissues were found to express the *Brachyury* (Fig. 3a), *Lim1* (Fig. 3b), *Hnf3β* (Fig. 3c) and *Otx2* (Fig. 3d) genes. An up-regulation of the *Brachyury* expression was found in the host tissue between the graft and the primitive streak, which is reminiscent of a newly organized primitive streak (Fig. 3a). In contrast, an up-regulation of *Lim1* expression was found in the anterior germ layer tissues (Fig. 3b). No novel expression of the *Hnf3β* (Fig. 3c), *Otx2* (Fig. 3d), or *Hoxb1* (data not shown) genes was found.

Late-PS stage node graft (Fig. 1c[II]). The node that was grafted to the lateral region of the late-PS host embryo expressed the *Brachyury* and *Lim1* genes (Fig. 3e). No graft showed any *Otx2* expression. In the host embryos, an up-regulation of *Brachyury* expression was found in the primitive streak and the allantois. Germ layer tissues located between the graft and the primitive streak exhibited novel *Brachyury* expression, reminiscent of an ectopic primitive streak. In the host embryos, *Lim1* activity that is found normally in the prospective lateral mesoderm remained unchanged. However, in some cases, the presence of the node graft caused an up-regulation of *Lim1* expression in the anterior germ layer tissues and the primitive streak (Fig. 3e). No host tissues showed any novel expression of the *Otx2*.

Early-PS stage A-Ep (Fig. 1b[I]) and late-PS stage anterior ectoderm (Fig. 1c[I]). *Brachyury* expression was found in two of ten grafts, *Lim1* in two of five grafts, and *Otx2* was not expressed in any anterior epiblast grafts. The late-PS stage anterior ectoderm grafts did not express *Brachyury* or *Lim1* genes. There was no change in the host pattern of expression in any of these genes.

Early-PS Stage P-Ep Displayed Patterning Activity Like the Late-PS Stage Node (Table 3)

In the majority of the host embryos, the P-Ep tissues extensively colonized the host neural tube, paraxial, and lateral mesoderm. Formation of an extra axis or an expanded neural tube was found in about 25% of the samples. In one case, a partial duplication of the headfold and the foregut of the host embryo was found. The graft-de-

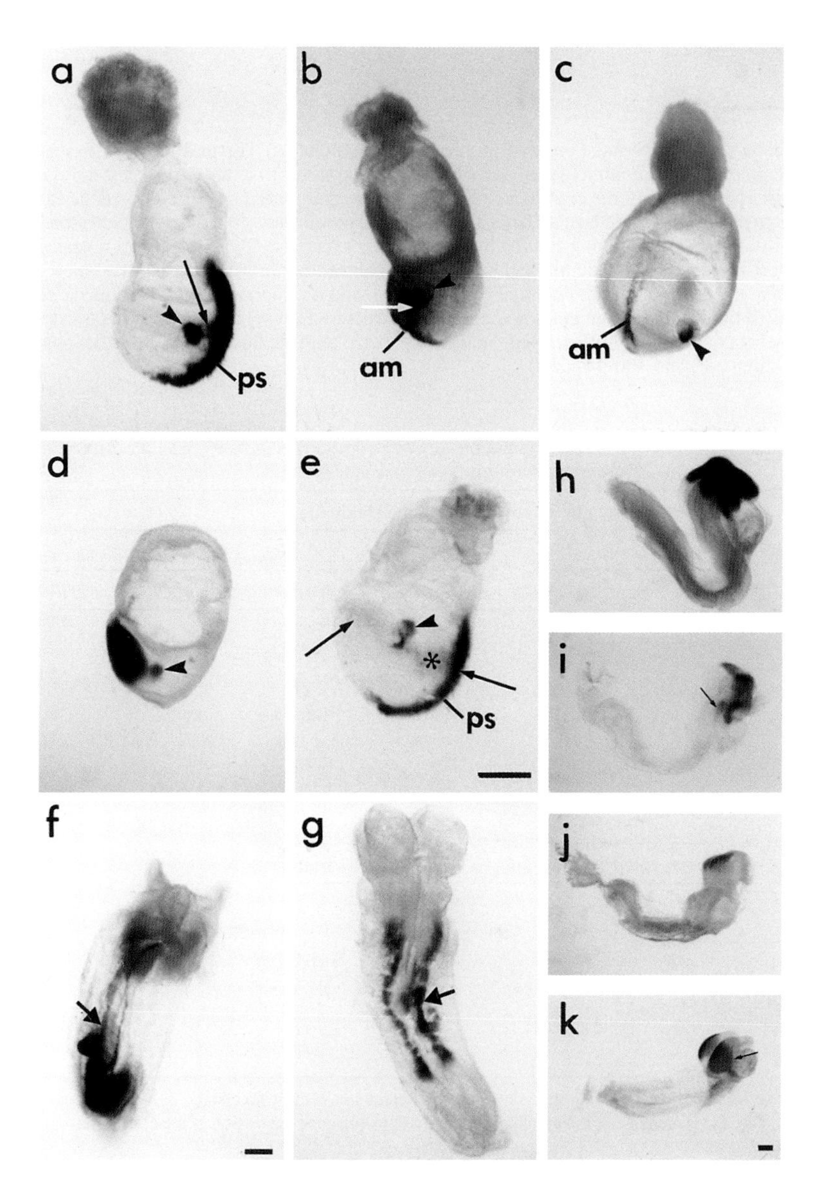

Figure 3. (*a–e*) Immediate response of the host tissues and the graft (arrowheads) in embryos at 5–6 hr after transplantation of P-Ep cells of the early-PS stage or the node of late-PS stage embryos. (*a*) *Brachyury* expression in the P-Ep transplant and the adjacent host tissues (arrow). (*b*) *Lim1* expression in the P-Ep graft and the anterior host tissues (white arrow), in addition to the normal expression in anterior midline tissue. (*c*) Expression of *Hnf3β* in the P-Ep graft, with no change in the normal expression in the anterior axial tissues of the host. (*d*) *Otx2* expression in the P-Ep graft with no change in the normal expression in the anterior tissues of the host. (*e*) Novel expression of *Lim1* in the anterior germ layers (arrow) and the primitive streak (arrow). Asterisk indicates the expected *Lim1* expression at this stage of development. In *a–e*, anterior side of the embryo is to the left. Bar, 100 μm (*a–e*). (*f,g*) Whole-mount in situ hybridization showing (*f*) the presence of an extra *Brachyury* expressing axial structure (arrow) and (*g*) the supernumery *Mox1* expressing somites (arrow). (*f,g*) Anterior (headfolds) is to the top. Bar, 100 μm (*f,g,h*). (*h–k*) Changes in the expression of the *Otx2* and *En2* in the forebrain and midbrain induced by the P-Ep graft. Normal patterns of expression are shown in *h* and *j*. (*i*) The transplanted P-Ep tissues in the cranial mesenchyme diminished the expression of *Otx2* in the midbrain but induced ectopic expression in the hindbrain and lateral surface ectoderm (thin arrow). (*k*) The P-Ep graft induced an expanded domain of *En2* expression in the midbrain and the adjacent surface ectoderm (thin arrow). In *h–k*, anterior (headfolds) is to the right side. Abbreviations: (am) anterior midline; (ps) primitive streak. Bar, 100 μm (*i,j,k*).

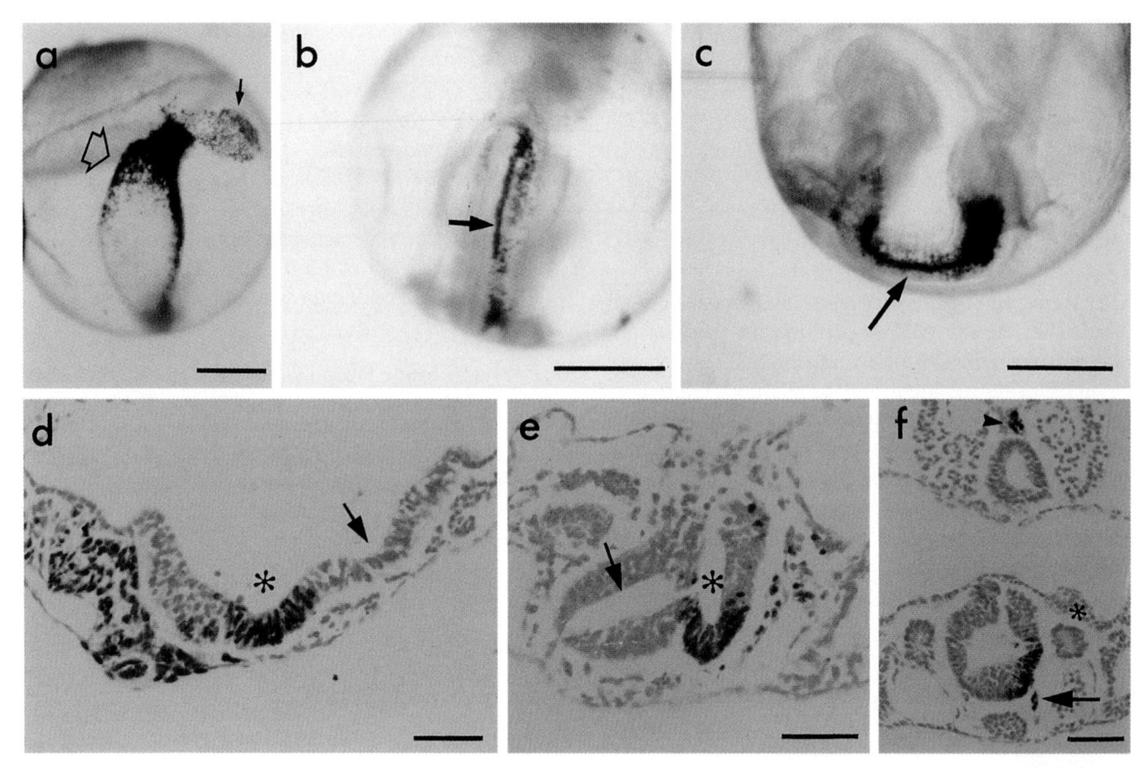

Figure 4. The patterning activity of the P-Ep cells of early-PS stage embryo on the host body plan. (*a*) The graft-derived tissues (stained blue with X-gal) occupy the anterior region of the host embryo (open arrow) that received the graft at the early-PS stage and was then cultured for 46 hr in vitro. There is no morphogenesis of the cranial neural plate and the graft-derived cells colonize the extraembryonic tissues (solid arrow). When grafted to the late-PS stage host embryo, the grafted tissues formed (*b*) elongated axial structures (arrow) and (*c*) additional paraxial (somitic) tissues (arrow) alongside the host embryonic axis. Histological analyses revealed the presence of (*d*) a duplicated neural plate (asterisk; host neural plate marked by the arrow) and extra graft-derived paraxial mesodermal tissues, (*e*) secondary neural tube (asterisk; and host neural tube marked by the arrow), and (*f*) enlarged neural tube made up of both host and graft tissues. In *f*, the graft-derived tissues formed an extra notochord like structure (arrow) and induced the formation of an extra row of somites (asterisk) from the host tissue. The graft-derived "notochord" fused with the host notochord in the posterior region of the embryo (arrowhead). Bars: 1 mm (*a–c*) and 100 μm (*d–f*).

Table 2. Response to the Host Tissues and the Graft following Transplantation of Embryonic Fragments to the Late Mouse Gastrula at 7.5 Days Postcoitum

	Gene expression by whole-mount in situ hybridization							
	T		*Lim1*		*Hnf3*β		*Otx2*	
Transplants	host	graft	host	graft	host	graft	host	graft
Analyzed 5–6 hours after grafting								
Early-PS								
Posterior epiblast	+(72)	+(85)	+(60)	+(69)	n.c.	+(60)	n.c.	+(90)
Anterior epiblast	n.c.	±(17)	n.c.	±(40)	n.d.	n.d.	n.c.	–
Late-PS								
Node	+(42)	+(93)	+(63)	+(63)	n.d.	n.d.	n.c.	–
Anterior ectoderm	n.c.	–	n.c.	–	n.d.	n.d.	n.d.	n.d.
	T		*Moxl*		*NCAM*			
	host	graft	host	graft	host	graft		
Analyzed 24–26 hours after grafting								
Early-PS								
Posterior epiblast	n.c.	+(90)	+(33)	+(33)	+(65)	+(65)		
Anterior epiblast	n.c.	±	n.c.	–	n.c.	–		
Late-PS								
Node	n.d.	n.d.	+(60)	+(60)	+(38)	+(75)		
Anterior ectoderm	n.d.	n.d.	n.c.	–	n.c.	–		

Expression pattern: + = expressed in graft and host tissues associated with the graft, – = no expression, n.c. = no change in host expression pattern, n.d. = not determined. Numbers in parentheses are the percentage of the number of embryos showing gene expression in the graft, or ectopic expression in the host tissues.

Table 3. Organizing Activity and Differentiation of Tissue Grafts in Host Mouse Embryos Cultured for 24 Hours following Grafting at the Late-primitive-streak Stage

No. of embryos analyzed	Induction of		Incorporation into host tissues										Self differentiation to	
			axial					localized						
	extra axis	neural tube expansion	NE	SE	M+E	MES	NCD	ECT	M+E	MES	M+N	NCD	cyst	epithelioid
6.5-day posterior epiblast (P-Ep)														
43	5	6	3	2	5	8	1	2	3	2	1	0	0	5
	11(25.5%)				19(44.1%)					8(18.6%)				5(11.6%)
6.5-day anterior epiblast (A-Ep)														
16	0	0	1	0	1	0	0	3	2	4	0	0	1	5
					2(12.5%)					9(56.3%)				6(37.5%)
7.5-day node														
58	13	2	2	6	8	7	3	2	8	0	1	0	0	8
	15(25.8%)				**23(39.7%)**					14(24.1%)				8(13.7%)
7.5-day anterior ectoderm														
7	0	0	0	0	0	0	0	3	0	1	0	0	1	2
										4(57.2%)				3(42.8%)

Abbreviations: NE = neuroepithelium, SE = surface ectoderm, M+E = mesoderm and ectoderm, MES = mesoderm, NCD = notochord, M+N = mesoderm and endoderm. Bold type indicates significant differences between grafts of the same developmental age.

rived tissues typically formed an elongated structure parallel to part of the host axis (Fig. 4b,c) and often fuzed with the host axis posteriorly (Figs. 3f and 4f). It resembled the condensed axial mesoderm (Fig. 4d,f) and expressed the *Brachyury* gene (Fig. 3f). A duplicated neural tube or neural plate, composed mostly of host tissues, was often found associated with the ectopic graft-derived axial structure (Fig 4d,e). In some cases, the induced neuroepithelium also expressed *NCAM* activity (Table 2). In the new axis, an extra mass of paraxial mesoderm (both host and graft-derived) was present (Fig. 4e), and in some cases, extra somites were formed (Fig. 4f). The extra mesodermal tissues were found to express *Mox1* activity (Fig. 3g; Table 2). A similar pattern of tissue colonization and induction of axial and neural structures was found in the embryos receiving grafts of the late-PS stage node (Table 3). The induced tissues also expressed *Mox1* and *NCAM* (Table 2). The node graft was capable of inducing secondary axial structures more frequently than the P-Ep graft (Table 3).

A-Ep or anterior ectoderm did not induce any secondary axis or alter the pattern of *Brachyury, Mox1,* or *NCAM* expression in the host tissues. Most grafts remained as clumps or colonized nearby host tissues. About 40% of the grafts differentiated into cystic structures.

The P-Ep Fragment Changes the Expression of Two Anterior Neural Genes

The P-Ep fragment remained as a tightly packed cell clump or an epithelial cyst when transplanted to the paraxial mesoderm of the upper hindbrain. The graft has induced ectopic expression of both the *Otx2* and *En2* genes. *Otx2* transcripts that are normally localized in the forebrain and midbrain (Fig. 3h) were found to be more patchy and occasionally in the upper hindbrain and the surface ectoderm adjacent to the graft (Fig. 3i). Ectopic expression was confined to the tissue adjacent to the graft and not in the tissues on the contralateral side of the neural axis. There was an expansion of *En2* activity in the midbrain and the surface ectoderm adjacent to the graft (compare Fig. 3j,k), concomitant with the reduction in *Otx2* expression (Fig. 3i). Transplants of A-Ep tissues of the early-PS stage embryo to the cranial mesenchyme of the late-PS embryo did not induce any change in the expression of the *Otx2* and *En2* genes.

DISCUSSION

Cell Fate Suggests the Presence of the Node Progenitors in Early Mouse Gastrula

In the early-PS stage mouse gastrula, several organizer- or axial mesoderm-specific genes such as *EphA2, Gsc, Hnf3β, Lim1, nodal* and *Brachyury* are expressed in localized areas in the posterior region of the epiblast (Blum et al. 1992; Sasaki and Hogan 1993; Ang and Rossant 1994; Ruiz and Robertson 1994; Shawlot and Behringer 1995; Herrmann 1996; Thomas and Beddington 1996; Varlet et al. 1997). Previous analysis of cell lineages (Lawson et al. 1991) and our fate-mapping study have shown that descendants of cells in this region of the epiblast contribute to the midline neuroectoderm, notochord, and the head process. These are typical derivatives of the node of late-PS stage embryos (Beddington 1994; Sulik et al. 1994). Descendants of this population also colonize the node of the early-somite stage embryo (this study). Thus, both the cell fate and the genetic activity suggest that some of the epiblast cells in the posterior region of the early-PS stage embryo are progenitors of the node. We have further shown that the P-Ep cells are distributed to the distal mesoderm and the midline epiblast during the first 12 hours of gastrulation. Descendants of cells from both sites of the mid-PS stage embryo eventually colonize the anterior axial mesoderm. These findings suggest that both anterior extension and migration in the mesoderm of P-Ep derived cells may be instrumental to the formation of the axial mesoderm.

Results of our study show that the P-Ep population may contain other mesodermal tissues such as the paraxial (cranial and somitic) mesoderm, the heart, and the lateral mesoderm. This finding is concordant with the prospective fate of cells in the adjacent posterior regions of the epiblast previously discovered by Lawson et al. (1991) and Parameswaran and Tam (1995). However, by late-gastrula stage, the node displays a more restricted cell fate and contributes predominantly to floor plate and the notochord. A comparison of the developmental fate of cells in the organizer of other vertebrate gastrulae reveals that the intermingling of precursors of axial mesoderm with those of neural, somitic, and endodermal progenitors is a common characteristic of the organizer of the early gastrula of the zebrafish and the frog (Table 4 and references therein). With development of the axis, the cellular composition of the *Xenopus* organizer changes with regard to its contribution to different cranio-caudal segments of the chordamesoderm and the association with more caudal mesoderm recruited from the lateral marginal zone (Lane and Keller 1997). Parallel to the changes in cell fate, the organizer cells also change their genetic activity (Vodicka and Gerhart 1995). The developmental changes in cell fate and molecular characteristics suggest that the organizer is a dynamic cell population that may display different patterning activity during embryogenesis.

The P-Ep Cells Display Organizer Activity

The patterning activity of the P-Ep cells that give rise to the node and node derivatives has been tested by heterotopic grafting to primitive-streak stage embryos. When transplanted to the anterior site in the early-PS stage embryo, the graft-derived cells in about 50% of the embryos have totally replaced the host tissues in the anterior region of the embryo. This is accompanied by a reduction in area of the prospective cranial neural plate and an atypical displacement of graft-derived cells to the yolk sac which is reminiscent of the formation of yolk sac mesoderm and allantois. The precise identity of these graft-derived tissues and whether this may represent a posterior transformation of the anterior embryonic structures is not known.

Similar to the node, the P-Ep population may induce the formation of extra neural and somitic tissues in the late-PS stage host embryos and differentiate into tissues that structurally resemble axial mesoderm and also express the notochord-specific gene. P-Ep cells also contribute significantly to the induced neural tissues and paraxial mesoderm. This may reflect the heterogeneous cellular composition of the P-Ep population, in contrast to the node which contributes mainly to the notochord and the neural tube in the duplicated embryonic axis (Beddington 1994).

The P-Ep cells and the node displayed similar molecular characteristics after they were transplanted in an ectopic environment. P-Ep grafts when transplanted to the late-PS stage embryo express the *T, Lim1, Otx2,* and *Hnf3β* genes. The node grafts express *T and Lim1,* although not *Otx2*. Results of our study have shown that P-Ep grafts can elicit ectopic *Lim1* and *T* expression in the host tissues. Specifically, *Lim1* expression is enhanced in the anterior germ layer tissues in addition to the normal activity in the midline mesoderm. *T* expression is up-regulated in the host tissues on the posterior side of the graft and extends into the host primitive streak. Node that is transplanted to the late-PS stage host elicits a similar ec-

Table 4. Comparison of the Cell Fate of the Organizer of Vertebrate Gastrulae Revealed by In Situ Labeling or Cell Transplantation

Cell population	Ectoderm	Cell fate/Tissue contribution: axial mesoderm	paraxial mesoderm	Endoderm
Zebrafish (50% epiboly, 6 hours) dorsal margin epiblast including embryonic shield	neural tube tissue including the floor plate	trunk notochord and tail hypochord	head mesenchyme and muscles of trunk and tail	
Xenopus (stage 10) dorsal blastopore lip		notochord	head mesenchyme, somites	pharyngeal endoderm, archenteron wall
Chick (HH stage 3–9) Hensen's node	floor plate	prechordal plate, head process, notochord	cells in the medial portion of the somites	gut endoderm
Mouse				
Posterior epiblast (early-primitive-streak stage) (% total cellular contribution)	floor plate (4.4%)	head process, notochord plate and notochord (20.4%)	head mesenchyme and somites (57.5%)	foregut endoderm, trunk endoderm (10.8%)
Node (late-primitive-streak stage)	floor plate from dorsal part	notochord and node from ventral part		gut endoderm from ventral part

References: Zebrafish (Shih and Fraser 1995; Melby et al. 1996); *Xenopus* (Smith and Slack 1983; Bouwmeester et al. 1996; Lane and Keller 1997); chick (Selleck and Stern 1991, 1992); mouse: posterior epiblast (this paper); node (Beddington 1994; Sulik et al. 1994).

topic expression of the *T* and *Lim1* genes. In some embryos, there is a significant up-regulation of the *Lim1* activity in the primitive streak. It is not known how these induced gene activities may be translated during the patterning of the host tissues into a second axis, but the findings strongly suggest that the P-Ep and node tissues have established a new anteroposterior polarity in the body plan of the host embryo. *Lim1* activity is reputed to be associated with the organization of the head structures and *T* is required for posterior axis formation. Our finding suggests that the P-Ep and the node of the mouse gastrula should possess both head and trunk organizing activity.

It is not clear at this stage whether or not the P-Ep cells can organize the entire body axis, as would be expected from an early embryonic organizer. Like the node transplant, none of the tissues induced by the P-Ep population seem to display any significant anterior characteristics (P.P.L. Tam, unpubl.). There may be a genuine lack of activity to specify anterior structures. Alternatively, the lack of anterior differentiation may be due to the incompetence of the late-PS stage germ layers to respond to P-Ep activity. By interacting the P-Ep fragments with the cranial neuroectoderm, we have shown that the expression of two brain-specific genes may be altered by the presence of the P-Ep fragments. The P-Ep fragments in the cranial mesenchyme do not show the extent of self-differentiation as in lateral embryonic sites (Inagaki and Schoenwolf 1993). The expression of inductive activity may therefore be independent of the morphogenesis of the organizer (Hatta and Takahashi 1996).

The Ontogeny of the Gastrula Organizers

Experimental evidence obtained from this study so far points to the existence of an organizer in the early mouse gastrula. We propose to name this organizer as the early gastrula organizer to indicate its lineage relationship and functional similarity to the node of the late mouse gastrula. In the pregastrulation avian embryo, the organizer-specific *goosecoid* gene is expressed in cells localized closed to the Koller's sickle. These cells are then displaced anteriorly during the extension of the primitive streak and finally reside in the Hensen's node. When tested by heterotopic transplantation, the *goosecoid*-expressing cells can induce the formation of ectopic axial structures similar to that induced by the Hensen's node (Izpisúa-Belmonte et al. 1993). Therefore, in both the mouse and avian gastrula, an organizer that displays similar gene activity (De Robertis et al. 1994) and patterning activity (Izpisúa-Belmonte et al. 1993) is formed before or soon after the primitive streak is established in the posterior region of the embryo.

It is not known how the organizer is specified during embryogenesis. In the avian embryo, the posterior marginal zone has been shown to be critical to the formation of the primitive streak (Stern 1990; Eyal-Giladi et al. 1994). Recently, a protein that is homologous to the *Xenopus* Vg1 has been shown to be expressed in the tissues of the posterior marginal zone, and ectopic expression of this protein induced the formation of an extra primitive streak in the avian gastrula (Selerio et al. 1996). Since appropriate *goosecoid* activity is detected in tissues at the anterior region of the ectopic primitive streak, it is likely that the organizer is specified concomitantly with the formation of the primitive streak. It is not known if the primitive streak and the organizer in the mouse gastrula are also induced in a similar manner. By analogy of the topography of the body plan, the epiblast and extraembryonic ectoderm located proximal to the early gastrula organizer are the most probable equivalent to the posterior marginal zone. Future studies may be focused on the analysis of gene activity and direct experimental manipulation to test the inductive activity of these tissues.

ACKNOWLEDGMENTS

We thank Peter Rowe, Bruce Davidson, Anne Camus, and Richard Behringer for reading the manuscript; Siew-lan Ang, Richard Behringer, Bernhard Herrmann, Janet Rossant, and Alex Joyner for riboprobes and En2 antibody. Our work is supported by the Human Frontier of Science Program, the National Health and Medical Research Council (NHMRC) of Australia, and Mr. James Fairfax. P.P.L.T. is a Principal Research Fellow of the N.H.M.R.C.

REFERENCES

Ang S.-L. and Rossant J. 1994. *HNF-3β* is essential for node and notochord formation in mouse development. *Cell* **78:** 561.

Arendt D. and Nübler-Jüng K. 1997. Dorsal or ventral: Similarities in fate maps and gastrulation patterns in annelids, arthopods and chordates. *Mech. Dev.* **61:** 7.

Beddington R.S.P. 1994. Induction of a second neural axis by the mouse node. *Development* **120:** 603.

Blum M., Gaunt S.J., Cho K.W.Y., Steinbeisser H., Blumberg B., Bittner D., and De Robertis E.M. 1992. Gastrulation in the mouse: The role of the homeobox gene *goosecoid*. *Cell* **69:** 1097.

Bouwmeester T., Kim S.-H., Sasai Y., Lu B., and De Robertis E.M. 1996. Cerberus is a head-inducing secreted factor expressed in the anterior endoderm of Spemann's organizer. *Nature* **382:** 595.

De Robertis E.M., Fainsod A., Gont L.K., and Steinbeisser H. 1994. The evolution of vertebrate gastrulation. *Development* (suppl.), p. 117.

Eyal-Giladi H., Lotan T., Levin T., Avner O., and Hochman J. 1994. Avian marginal zone cells function as primitive streak inducers only after their migration into the hypoblast. *Development* **120:** 2501.

Gilbert S. and Saxen L. 1993. Spemann's organizer: Models and molecules. *Mech. Dev.* **41:** 73–89.

Hatta K. and Takahashi Y. 1996. Secondary axis induction by heterospecific organizers in Zebrafish. *Dev. Dyn.* **205:** 183.

Herrmann B.G. 1996. The role of *Brachyury* in the axial development and evolution of vertebrates. In *Mammalian development* (ed. P . Lonai), p. 29. Harwood Academic, Amsterdam.

Hornbruch A., Summerbell D., and Wolpert L. 1979. Somite formation in the early chick embryo following grafts of Hensen's node. *J. Embryol. Exp. Morphol.* **51:** 51.

Inagaki T. and Schoenwolf G.C. 1993. Axis development in avian embryos: The ability of Henson's node to self differentiate as analysed with heterochronic grafting experiments. *Anat. Embryol.* **188:** 1.

Izpisúa-Belmonte J.-C., De Robertis E.M., Storey K.G., and Stern C.D. 1993. The homeobox gene *goosecoid* and the origin of organizer cells in the early chick blastoderm. *Cell* **74:** 645.

Lane M.C. and Keller R. 1997. Microtubule disruption reveals that Spemann's organizer is subdivided into two domains by the vegetal alignment zone. *Development* **124:** 895.

Lawson K.A., Meneses J.J., and Pedersen R.A. 1991. Clonal analysis of epiblast fate during germ layer formation in the mouse embryo. *Development* **113:** 891.

Lemaire P. and Kodjabachian L. 1996. The vertebrate organizer: Structure and molecules. *Trends Genet.* **12:** 525.

Melby A.E., Warga R.M., and Kimmel C.B. 1996. Specification of cell fates at the dorsal margin of the zebrafish gastrula. *Development* **122:** 2225.

Parameswaran M. and Tam P.P.L. 1995. Regionalisation of cell fate and morphogenetic movement of the mesoderm during mouse gastrulation. *Dev. Genet.* **17:** 16.

Quinlan G.A., Williams E.A., Tan S-S., and Tam P.P.L. 1995. Neuroectodermal fate of epiblast cells in the distal region of the mouse egg cylinder: Implication for body plan organization during early embryogenesis. *Development* **121:** 87.

Ruiz J. and Robertson E. J. 1994. The expression of the receptor protein tyrosine kinase gene, *eck,* is highly restricted during early mouse development. *Mech Dev.* **46:** 87.

Sasaki H. and Hogan B.L.M. 1993. Differential expression of multiple fork head related genes during gastrulation and axial pattern formation in the mouse embryo. *Development* **118:** 47.

Selerio E.A.P., Connolly D.J., and Cooke J. 1996. Early developmental expression and experimental axis determination by the chicken *Vg1* gene. *Curr. Biol.* **6:** 1476.

Selleck, M.A.J. and Stern C.D. 1991. Fate mapping and cell lineage analysis of Hensen's node in the chick embryo. *Development* **112:** 615.

———. 1992. Commitment of mesoderm cells in Hensen's node to the chick embryo to notochord and somite. *Development* **114:** 403.

Shawlot W. and Behringer R.R. 1995. Requirement for *Lim1* in head-organizing function. *Nature* **374:** 425.

Shih J. and Fraser S.E. 1995. Distribution of tissue progenitors within the shield region of the zebrafish gastrula. *Development* **121:** 2755.

Smith J.C. and Slack J.M.W. 1983. Dorsalization and neural induction: Properties of the organizer in *Xenopus laevis. J. Embryol. Exp. Morphol.* **78:** 299.

Stern C.D. 1990. The marginal zone and its contribution to the hypoblast and primitive streak of the chick embryo. *Development* **109:** 667.

Storey K.G., Selleck M.A.J., and Stern C.D. 1995. Neural induction and regionalisation by different subpopulations of cells in Hensen's node. *Development* **121:** 417.

Sulik K., DeHart D.B., Inagaki T., Carson J.L.M., Vrablic T., Gesteland K., and Schoenwolf G.C. 1994. Morphogenesis of the murine node and notochordal plate. *Dev. Dyn.* **201:** 260.

Tam P.P.L. and Behringer R.R. 1997. Mouse gastrulation: The formation of a mammalian body plan. *Mech. Dev.* (in press).

Tam P.P.L. and Quinlan G.A. 1996. Mapping vertebrate embryos. *Curr. Biol.* **6:** 106.

Tam P.P.L. and Tan S.S. 1992. The somitogenetic potential of cells in the primitive streak and the tail bud of the organogenesis-stage mouse embryo. *Development* **115:** 703.

Thomas P. and Beddington R.S.P. 1996. Anterior primitive endoderm may be responsible for patterning the anterior neural plate in the mouse embryo. *Curr. Biol.* **6:** 1487.

Varlet I., Collignon J., and Robertson E.J. 1997. *nodal* expression in the primitive endoderm is required for specification of the anterior axis during mouse gastrulation. *Development* **124:** 1033.

Vodicka M.A. and Gerhart J.C. 1995. Blastomere derivation and domains of gene expression in Spemann's organizer of *Xenopus laevis. Development* **121:** 3505.

Goosecoid and *Goosecoid*-related Genes in Mouse Embryogenesis

M. Wakamiya,[1] J.A. Rivera-Pérez,[1] A. Baldini,[2] and R.R. Behringer[1]
[1] *Department of Molecular Genetics, The University of Texas M.D. Anderson Cancer Center, Houston, Texas 77030;* [2]*Department of Molecular and Human Genetics, Baylor College of Medicine, Houston, Texas 77030*

The generation and analysis of loss-of-function mutations in the mouse have been instrumental in the identification of genes, including *HNF3β, Lim1, nodal*, and *Otx2*, that are essential during gastrulation to pattern the mouse embryo (Zhou et al. 1993; Ang and Rossant 1994; Weinstein et al. 1994; Acampora et al. 1995; Matsuo et al. 1995; Shawlot and Behringer 1995; Ang et al. 1996). One surprising finding has been the lack of any gastrulation or axial patterning defects in mice with mutations in the homeobox-containing gene *goosecoid* (*gsc*) (Rivera-Pérez et al. 1995; Yamada et al. 1995).

gsc was originally identified in a screen for homeobox-containing genes from a *Xenopus* dorsal lip cDNA library (Blumberg et al. 1991). Injection of *gsc* mRNA into *Xenopus* embryos can cause the formation of a secondary body axis, mimicking the activity of Spemann's organizer (Cho et al. 1991). *gsc* transcripts are detected in the organizer regions of various vertebrate embryos (Cho et al. 1991; Blum et al. 1992; Izpisúa-Belmonte et al. 1993; Stachel et al. 1993; Schulte-Merker et al. 1994). In the mouse, *gsc* has been reported to be expressed in a biphasic pattern during embryogenesis (Blum et al. 1992; Gaunt et al. 1993). Initially, *gsc* is detected between embryonic day 6.4 (E6.4) and E6.8 (Blum et al. 1992) in the anterior region of the primitive streak (Fig. 1a). *gsc* transcripts are next detected at E10.5 (Gaunt et al. 1993) in craniofacial structures, the limbs, and the ventrolateral body wall (Fig. 1b). The expression pattern of *gsc* suggests an important role in gastrulation and craniofacial, limb, and ventrolateral body wall development. Whereas craniofacial and rib cage defects are observed in *gsc*-null mice, no abnormalities have been detected during gastrulation and axis formation or in limb development. In addition, transplant experiments with *gsc*-null ovaries demonstrate that no maternally derived ovarian *gsc* transcript or Gsc protein provides *gsc* function for axis formation (Rivera-Pérez et al. 1995). Other possible explanations for the lack of gastrulation defects in *gsc*-null embryos could be that subtle alterations occur during gastrulation that were not previously detected. Furthermore, it is also possible that other genes may compensate for *gsc* during gastrulation in its absence.

Recently, novel *gsc*-related genes have been reported in mouse, human, and chick (Galili et al. 1997; Gottlieb et al. 1997; Lemaire et al. 1997). The mouse *gscl* gene was discovered by DNA sequence analysis of a region of mouse chromosome 16 that is syntenic to the DiGeorge and velocardiofacial syndrome (DGS/VCFS) minimal critical region (Galili et al. 1997). The human *GSCL* gene has also been isolated and maps to the same region (Gottlieb et al. 1997). DGS/VCFS patients present with craniofacial and heart abnormalities caused by a haploinsufficiency of a gene or genes deleted in human 22q11.2. The observation that *gsc*-null mice develop craniofacial defects makes *gscl* an exciting candidate gene for causing some of the abnormalities observed in the DGS/VCFS. In addition, the identification of *gscl* suggests that it may compensate for *gsc* during gastrulation in *gsc*-null embryos.

We have examined *gsc*-null embryos to uncover potentially subtle changes during gastrulation. However, no transient delay in the development of *gsc*-null embryos during gastrulation was detected, suggesting that *gsc*-null embryos do not have obvious defects in gastrulation. In addition, we have examined the temporal and tissue expression patterns of *gsc* and *gscl* during embryogenesis and in adult tissues. The data indicate that *gscl* is a candidate gene that may compensate for *gsc* during gastrulation and may contribute to the development of the DGS/VCFS.

METHODS

Mice. Two different *gsc* null alleles (gsc^f and gsc^r) that produce the same mutant phenotype were generated previously by gene targeting in mouse embryonic stem (ES) cells (Rivera-Pérez et al. 1995). Each allele can be distinguished by Southern blot (Rivera-Pérez et al. 1995) or by polymerase chain reaction (PCR) (J.A. Rivera-Pérez et al., in prep.). C57BL/6J × 129SvEv (B6 × 129) F_1 $gsc^f/+$ females and B6 × 129 F_1 $gsc^r/+$ males were interbred to generate B6 × 129 F_2 gsc^f/gsc^r mutant embryos. The embryos from this cross were dissected at 2:00 pm on the seventh day of gestation and staged according to the method of Downs and Davies (1993). The embryos were subsequently genotyped by PCR, taking care to exclude maternal tissue.

Whole-mount in situ hybridization. Whole-mount in situ hybridization was performed according to the proto-

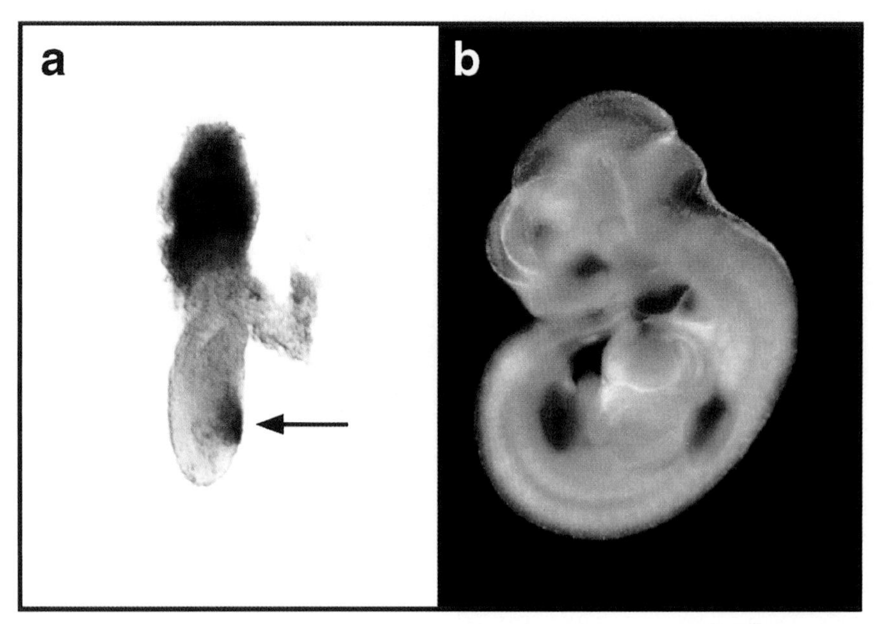

Figure 1. Spatial distribution of *gsc* transcripts by whole-mount in situ hybridization. (*a*) E6.5 embryo. *gsc* transcripts are detected in the anterior region of the primitive streak (arrow). (*b*) E10.5 embryo. *gsc* transcripts are detected in the nasal process, the first and second branchial arches, the proximal limb buds, and ventrolateral body wall.

col of Conlon and Rossant (1992), using a mouse *gsc* probe (Shawlot and Behringer 1995).

Reverse transcription-polymerase chain reaction (RT-PCR). Total RNA (3.0 μg) (or 2.3 μg for E6.5 embryos) was reverse-transcribed using an oligo(dT) primer and amplified for 30 cycles using mouse *gsc* primers: sense 5′-GGCACCTCCTCGGACTACGG-3′ (exon 1) and antisense 5′-ATGTAGGGCAGCATCTGGTG-3′ (exon 2) and for 35 cycles mouse *gscl* primers: sense 5′-ACCATCTTCAGCGAGGAGCA-3′ (exon 2) and antisense 5′-AGTTTTCTTAGTCCCAGGCA-3′ (exon 3). For adult tissues, RT-PCR products were subjected to Southern blotting using ^{32}P-labeled probes containing *gsc-* or *gscl*-coding sequences. The integrity of the RNA and the efficiency of the reverse transcription for adult tissues were monitored by amplification of hypoxanthine-guanine phosphoribosyl transferase (HPRT). The sequences of the HPRT primers were: 5′-GTTGAGAGATCATCTCCACC-3′ and 5′-AGCTATGATGAA CCAGGTTA-3′.

RESULTS

Gastrulation Appears Normal in *gsc*-null Embryos

Although *gsc* is precisely expressed during gastrulation in the anterior region of the primitive streak (Fig. 1a), *gsc*-null mice gastrulate and are born without axial defects (Rivera-Pérez et al. 1995; Yamada et al. 1995). One possible explanation for these results could be that *gsc*-null mice might have subtle defects in gastrulation. Perhaps gastrulation initiates later in *gsc*-null embryos in comparison to wild-type embryos. If this were true, one would predict that the *gsc*-null embryos would then regulate and catch up developmentally with their wild-type littermates because at later stages, *gsc*-null embryos are morphologically indistinguishable from their wild-type littermates. Therefore, we examined in more detail the development of *gsc*-null embryos during gastrulation.

gsc heterozygotes were interbred and the resulting embryos on the seventh day of development were dissected, morphologically staged, and genotyped. Five litters (7–10 embryos per litter) from heterozygous crosses representing a total of 40 embryos (11 wild-type, 20 heterozygotes, and 9 mutants) were examined. The stages of the embryos within any specific litter varied from the late primitive streak stage to the late allantoic bud stage (Downs and Davies 1993). Genotype analysis of the morphologically staged embryos did not reveal any correlation with the developmental stage of the embryos and their *gsc* genotype. Thus, by these criteria, *gsc*-null embryos develop in vivo like wild-type embryos during gastrulation.

gsc and *gsc-like* Genes in the Mouse, Human, and Chick

Recently, novel *gsc*-related genes have been reported in mouse, human, chick, and *Drosophila* (Goriely et al. 1996; Hahn and Jäckle 1996; Galili et al. 1997; Gottlieb et al. 1997; Lemaire et al. 1997). Figure 2 shows an alignment of the homeodomain regions encoded by *gsc* and the newly described *gsc*-related genes from these vertebrate species (Blum et al. 1992; Izpisúa-Belmonte et al. 1993; Blum et al. 1994; Galili et al. 1997; Gottlieb et al. 1997; Lemaire et al. 1997) and *Drosophila* (Goriely et al. 1996; Hahn and Jäckle 1996). The homeodomains encoded by the mouse, human, and chick *gsc* genes are identical. The homeodomains of mouse gscl and human GSCL are 97% identical, differing by only two amino acids. Mouse *gscl*

```
Mouse Gsc    KRRHRTIFTDEQLEALENLFQETKYPDVGTREQLARKVHLREEKVEVWFKNRRAKWRRQKR
Human GSC    ------------------------------------------------------------- 100%
Chick GSC    ------------------------------------------------------------- 100%

Mouse Gscl   T-------SE---Q---A--VQNQ--------R--VRIR----R-------------H--- 74%
Human GSCL   T-------SE---Q---A--VQNQ----S---R--GRIR----R-------------H--- 73%
Chick GSX    T--------E---Q---T--HQNQ----I---H--NRI--K--R-------------H--- 75%

Dros  Gsc    ---------E----Q--AT-DK-H----VL-----LR-D-K--R-------------K--- 76%
```

Figure 2. Alignment of the homeodomains encoded by *gsc* and *gsc*-related genes in mouse (Blum et al. 1992; Galili et al. 1997), human (Blum et al. 1994; Gottlieb et al. 1997), chick (Izpisúa-Belmonte et al. 1993; Lemaire et al. 1997), and *Drosophila* (Goriely et al. 1996). The percent identity relative to mouse *gsc* is indicated to the right of each sequence.

and human *GSCL* map to regions of the mouse and human genomes, respectively, that are syntenic (Gottlieb et al. 1997). Thus, mouse *gscl* and human *GSCL* are likely homologs of each other. The mouse gscl and human GSCL homeodomains are 73% and 72% identical with the homeodomains of mouse gsc and human GSC, respectively.

A *gsc*-related gene (*GSX*) has recently been reported in the chick (Lemaire et al. 1997). The homeodomain of GSX is 74% identical to the homeodomain of chick GSC (Izpisúa-Belmonte et al. 1993). Interestingly, the homeodomains of mouse gscl and human GSCL differ by approximately ten amino acids in comparison to the homeodomain of chick GSX, suggesting the possibility that *gscl/GSCL* and *GSX* may not be homologs of each other and that there may be multiple *gsc*-like genes in vertebrates. The identification of *gscl* in the mouse suggests that this gene might compensate for *gsc* in *gsc*-null mice during gastrulation.

Expression of Mouse *gsc* and *gscl* during Embryogenesis and in Adult Tissues

Previous reports have suggested that *gsc* is expressed in a biphasic pattern, with expression initially detected in a narrow window of development during gastrulation, then later beginning at approximately E10.5 to at least E14.5 (Blum et al. 1992; Gaunt et al. 1993). However, using RT-PCR, we found that *gsc* was expressed at all embryonic stages examined from E6.5 to birth (Fig. 3). In addition, *gsc* transcripts were detected in ES cells and extraembryonic tissues. Recent whole-mount in situ hybridization studies of *gsc* in the mouse confirm that *gsc* is expressed between E6.5 and E10.5 in ventral forebrain neuroepithelium, prechordal mesoderm, and foregut endoderm (Filosa et al. 1997). Interestingly, *gscl* was also detected by RT-PCR at all stages of embryogenesis examined in this study, notably, during gastrulation at E6.5 (Fig. 3).

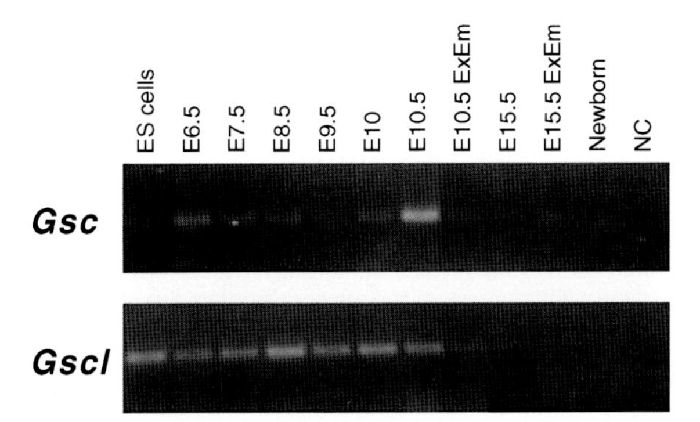

Figure 3. RT-PCR analysis of *gsc* and *gscl* expression during mouse embryogenesis. Ethidium-bromide-stained gel. (Exem) Extraembryonic tissues; (NC) negative control.

Although most interest in *gsc* has focused upon its embryonic expression pattern and roles during embryogenesis, its expression pattern in adult tissues has been ignored. Therefore, we examined *gsc* expression in a wide variety of tissues from adult mice by RT-PCR (Fig. 4). Transcripts were detected in eye, skin, thymus, thyroid region, tongue, stomach, ovary, and testis. Weak expression was also detected in brain, heart, skeletal muscle, esophagus, large and small intestine, and adrenal. We also examined *gscl* expression in the same series of adult tissues (Fig. 4). *gscl* transcripts were abundant in brain, eye, thymus, thyroid region, stomach, bladder, and testis. Weak expression was also detected in tongue and ovary but notably not in heart. Thus, *gsc* and *gscl* are coexpressed in a subset of adult organs but are also uniquely expressed in other organs.

DISCUSSION

Gain-of-function experiments in *Xenopus* clearly demonstrate that ectopic expression of *gsc* can induce the formation of a secondary body axis (Cho et al. 1991), yet loss-of-function mutations in the mouse indicate no essential role for this transcription factor during gastrulation or axial patterning (Rivera-Pérez et al. 1995; Yamada et al. 1995). The detailed examination of gastrulation stage *gsc*-null embryos showed no differences with wild-type or heterozygous embryos, indicating that gastrulation proceeded normally without *gsc*. We previously suggested that the levels of *gsc* expression might respond to environmental cues to maintain correct levels of organizer activity and this remains a possibility (Rivera-Pérez et al. 1995). However, it is also possible that other genes may compensate for *gsc* in its absence. Indeed, although *gsc*-null mice have normal limbs, severe limb abnormalities are observed when *gsc/Mhox* double-mutant mice are generated (Martin et al. 1995; J. Martin, pers.

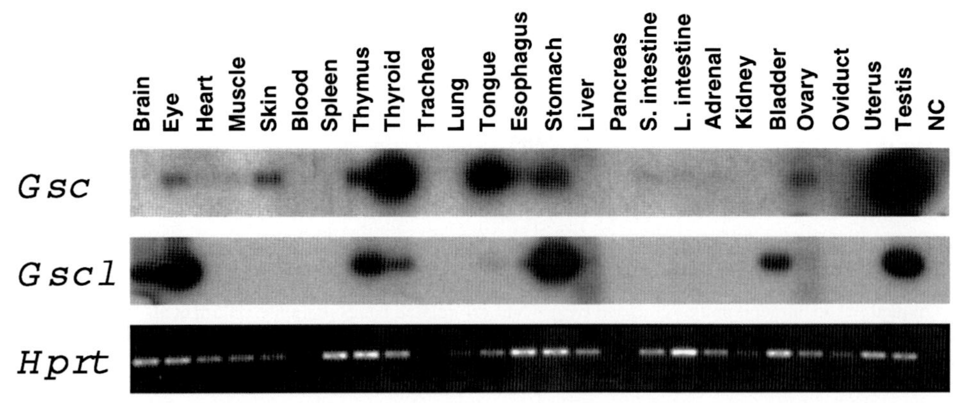

Figure 4. RT-PCR analysis of *gsc* and *gscl* expression in adult mouse organs. Southern blot of *gsc* and *gscl* RT-PCR products. Ethidium-bromide-stained gel of *Hprt* RT-PCR products to assess the integrity of the mRNA. (NC) Negative control.

comm.). *HNF3β* and *gsc* have recently been shown to be coexpressed in all three germ layers in the anterior regions of headfold to early somite stage embryos (Filosa et al. 1997). In addition, *HNF3β* $^{+/-}$, *gsc* $^{-/-}$ mutant mice have defects in the development of the forebrain, optic vesicles, neural tube, branchial arches, and heart that are not observed in either *HNF3β* $^{+/-}$ mice or *gsc* $^{-/-}$ mice (Filosa et al. 1997). Therefore, other genes may provide *gsc* function during gastrulation in the absence of *gsc*.

The identification of *gsc*-related genes such as *gscl* suggests that it may compensate for *gsc* during gastrulation in *gsc*-null mice. Recently, a *gsc* gene (*D-gsc*) was reported in *Drosophila* (Goriely et al. 1996; Hahn and Jäckle 1996). The homeodomain of *D-gsc* is 76% identical to the vertebrate gsc homeodomains (Fig. 2). *D-gsc* mRNA was shown to rescue axial development of UV-irradiated *Xenopus* embryos, indicating that *D-gsc* was a functional *gsc* homolog (Goriely et al. 1996). Interestingly, the homeodomain of mouse gscl shares approximately the same amount of amino acid identity (74%) relative to the homeodomain of mouse gsc (Galili et al. 1997). It will be important to test the activity of *gscl* in the *Xenopus* gain-of-function assay.

For *gscl* to remain a viable candidate gene that can compensate for *gsc,* it should be coexpressed at the same time and the same tissues as *gsc*. We found that *gsc* and *gscl* were coexpressed during gastrulation and at later stages of embryogenesis. Previously, *gscl* was detected in E12 mouse embryos by RT-PCR (Galili et al. 1997). In the chick, *GSC* and *GSX* are initially coexpressed within the primitive streak, but this expression segregates later during gastrulation with *GSX* expression persisting in the neural plate (Lemaire et al. 1997). It will be important to determine the spatial pattern of *gscl* in wild-type gastrula stage embryos.

The expression patterns of developmental control genes in adult organs are typically ignored. However, such information can be useful for a global understanding of the function of a gene in fundamental cellular processes. Our RT-PCR studies revealed a number of tissues not previously known to express *gsc*. The expression of *gsc* detected by RT-PCR in these adult tissues has also been corroborated by the detection of β-galactosidase staining in the same tissues of mice carrying a *lacZ* reporter introduced into the *gsc* locus by gene targeting in mouse ES cells (M. Wakamiya and R. Behringer, in prep.). Our studies have also revealed that *gsc* and *gscl* are also coexpressed in a subset of adult tissues, suggesting that these genes may interact after birth in these organs. One of the organs that coexpressed *gsc* and *gscl* was the testis that had previously been shown to express human *GSCL* (Gottlieb et al. 1997). *gsc* and *gscl* were also uniquely expressed in a subset of adult tissues, suggesting that they may also act exclusive of each other in these tissues.

gscl has attracted great interest from human geneticists interested in the molecular basis of the DGS/VCFS because it maps to the region that defines this genetic defect (Galili et al. 1997; Gottlieb et al. 1997). Individuals that are haploid for deletions of chromosome 22q11.2 have hypoplasia of the parathyroid gland and thymus, cardiac outflow defects, and craniofacial abnormalities. Interestingly, we detected *gscl* expression in the thyroid region and the thymus. However, no *gscl* transcripts were detected in adult heart, although *gscl* may have been expressed in the heart earlier during embryogenesis. Ultimately, the deletion of the mouse *gscl* gene by targeted mutagenesis in ES cells should provide important information regarding the role of *gscl* in embryogenesis, its contribution to the development of the DGS/VCFS, and its relationship to *gsc* during gastrulation.

ACKNOWLEDGMENTS

We thank Bill Shawlot for the picture of *gsc* expression in the mouse gastrula.

REFERENCES

Acampora D., Mazan S., Lallemand Y., Avantaggiato V., Maury M., Simeone A., and Brûlet P. 1995. Forebrain and midbrain regions are deleted in *Otx2* $^{-/-}$ mutants due to a defective anterior neuroectoderm specification during gastrulation. *Development* **121:** 3279.

Ang S.-L. and Rossant J. 1994. *HNF3-β* is essential for node and notochord formation in mouse development. *Cell* **78:** 561.

Ang S.-L., Jin O., Rhinn M., Daigle N., Stevenson L., and Rossant J. 1996. A targeted mouse *Otx2* mutation leads to severe defects in gastrulation and formation of axial mesoderm and to deletion of rostral brain. *Development* **122:** 243.

Blum M., Gaunt S.J., Cho K.W.Y., Steinbeisser H., Blumberg B., Bittner D.A., and De Robertis E.M. 1992. Gastrulation in the mouse: The role of the homeobox gene *goosecoid*. *Cell* **69:** 1097.

Blumberg B., Wright C.V.E., De Robertis E.M., and Cho K.W.Y. 1991. Organizer-specific homeobox genes in *Xenopus laevis* embryos. *Science* **253:** 194.

Cho K.W.Y., Blumberg B., Steinbesser H., and De Robertis E.M. 1991. Molecular nature of Spemann's organizer: The role of the *Xenopus* homeobox gene *goosecoid*. *Cell* **67:** 1111.

Downs K.M. and Davies T. 1993. Staging of gastrulation mouse embryos by morphological landmarks in the dissecting microscope. *Development* **118:** 1255.

Filosa S., Rivera-Pérez J.A., Gómez A.P., Gansmuller A., Sasaki H., Behringer R.R., and Ang S.-L. 1997. *goosecoid* and *HNF-3β* genetically interact to regulate neural tube patterning during mouse embryogenesis. *Development* **124:** 2843.

Galili N., Baldwin H.S., Lund J., Reeves R., Gong W., Wang Z., Roe B.A., Emanuel B.S., Nayak S., Mickanin C., Budarf M.L., and Buck C.A. 1997. A region of mouse chromosome 16 is syntenic to the DiGeorge, velocardiofacial syndrome minimal critical region. *Genome Res.* **7:** 17.

Gaunt S.J., Blum M., and De Robertis E.M. 1993. Expression of the mouse *goosecoid* gene during mid-embryogenesis may mark mesenchymal cell lineages in the developing head, limbs and body wall. *Development* **117:** 769.

Goriely A., Stella M., Coffinier C., Kessler D., Mailhos C., Dessain S., and Desplan C. 1996. A functional homologue of *goosecoid* in *Drosophila*. *Development* **122:** 1641.

Gottlieb S., Emanuel B.S., Driscoll D.A., Sellinger B., Wang Z., Roe B., and Budarf M.L. 1997. The DiGeorge syndrome minimal critical region contains a *goosecoid*-like (*GSCL*) homeobox gene that is expressed early in human development. *Am. J. Hum. Genet.* **60:** 1194.

Hahn M. and Jäckle H. 1996. *Drosophila goosecoid* participates in neural development but not in body axis formation. *EMBO J.* **15:** 3077.

Izpisúa-Belmonte J.C., De Robertis E.M., Storey K.G., and Stern C.D. 1993. The homeobox gene *goosecoid* and the origin of organizer cells in the early chick blastoderm. *Cell* **74:** 645.

Lemaire L., Roeser T., Izpisúa-Belmonte J.C., and Kessell M. 1997. Segregating expression domains of two *goosecoid* genes during the transition from gastrulation to neurulation in chick embryos. *Development* **124:** 1443.

Martin J.F., Bradley A., and Olson E.N. 1995. The *paired*-like homeobox gene *MHox* is required for early events of skeletogenesis in multiple lineages. *Genes Dev.* **9:** 1237.

Matsuo I., Kuratani S., Kimura C., Takeda N., and Aizawa S. 1995. Mouse *Otx2* functions in the formation and patterning of rostral head. *Genes Dev.* **9:** 2646.

Rivera-Pérez J.A., Mallo M., Gendron-Maguire M., Gridley T., and Behringer R.R. 1995. *goosecoid* is not an essential component of the mouse gastrula organizer but is required for craniofacial and rib development. *Development* **121:** 3005.

Schulte-Merker S., Hammerschmidt M., Beuchle D., Cho K.W.Y., De Robertis E.M., and Nüsslein-Volhard C. 1994. Expression of zebrafish *goosecoid* and *no tail* gene products in wild-type and mutant *no tail* embryos. *Development* **120:** 843.

Shawlot W. and Behringer R.R. 1995. Requirement for *Lim1* in head-organizer function. *Nature* **374:** 425.

Stachel S.E., Grunwald D.J., and Myers P. 1993. Lithium perturbation and *goosecoid* expression identify a dorsal specification pathway in the pregastrula zebrafish. *Development* **117:** 1261.

Weinstein D.C., Ruiz i Altaba A., Chen W.S., Hoodless P., Prezioso V.R., Jessell T.M., and Darnell J.E., Jr. 1994. The winged-helix transcription factor *HNF-3β* is required for notochord development in the mouse embryo. *Cell* **78:** 575.

Yamada G., Mansouri A., Torres M., Stuart E.T., Blum M., Schultz M., De Robertis E.M., and Gruss P. 1995. Targeted mutation of the murine *goosecoid* gene results in craniofacial defects and neonatal death. *Development* **121:** 2917.

Zhou X., Sasaki H., Lowe L., Hogan B.L.M., and Kuehn M.R. 1993. Nodal is a novel TGF-β-like gene expression in the mouse node during gastrulation. *Nature* **361:** 543.

Cell Response to Different Concentrations of a Morphogen: Activin Effects on *Xenopus* Animal Caps

J.B. Gurdon, K. Ryan, F. Stennard, N. McDowell, A.M. Zorn, D.J. Crease, and S. Dyson
Wellcome CRC Institute, and Department of Zoology, University of Cambridge, Cambridge CB2 1QR, England

INTRODUCTION

By long-range signaling, we refer to any situation in which cells signal to other cells not in contact with them. It is convenient to exclude from this definition signaling by hormones because these are transmitted in the blood circulation. We are interested in those cases in which signaling takes place over distances of up to 500 microns and usually over a few to about 20 cell diameters. Signaling of this kind is of particular interest because it is intimately connected with the concept of a morphogen gradient. According to this concept, molecules, usually proteins, are secreted by a cell or cells in one position in an embryo and spread from this region to create a concentration gradient throughout a field of adjacent cells. These cells are believed to sense the concentration of the morphogen around them and to make qualitatively different responses, such as the activation of different genes, according to morphogen concentration. The morphogen concept is likely to prove particularly important in development because one developmental event, namely, the activation of a morphogen-encoding gene in one part of an embryo, can generate several different types of gene activities, and hence cell types, in nearby cells; furthermore, these new cell types will have a defined spatial relationship to each other. There is a long history of interest in morphogen gradients, especially from the theoretical point of view (Turing 1952; Wolpert 1969, 1989). Recent experimental contributions include work on invertebrates (Lawrence and Struhl 1996) and on vertebrate embryos (Dosch et al. 1997; Tonegawa et al. 1997).

The work described here makes use of an experimental system in which activin acts as a morphogen on *Xenopus* blastula animal cap cells and which enables us to analyze several aspects of the mechanism of long-range morphogen action. There are several reasons for believing that activin signaling has an important function in early *Xenopus* development. First, it is the single most active known molecule in its ability to induce embryonic ectoderm tissue to form mesodermal cell types. It is also the effective component of cultured cell media and of embryo tissues that have strong mesoderm-inducing activities (Smith et al. 1990; Thomsen et al. 1990; Tiedemann et al. 1992). Second, activin is present as protein in *Xenopus* embryos (Asashima et al. 1991). Third, dominant-negative receptors inhibit mesoderm formation. This result was originally established by Hemmati-Brivanlou and Melton (1992), although it later emerged that the construct used, a complete type II activin receptor lacking its intracellular domain, inhibited signaling by activin as well as by related transforming growth factor-β (TGF-β) molecules including Vg1 and BMP4. Recently, another type of dominant-negative type II activin receptor, lacking both the intracellular and transmembrane domains, has been found, at the appropriate concentration, to block signaling by activin but not by Vg1 or any other signaling molecules known to be active in early embryos. This construct greatly delays mesodermal gene activation and causes early developmental abnormalities (Dyson and Gurdon 1997). We conclude that activin has a necessary function in early *Xenopus* development independent of any function attributable to Vg1 or BMP4. Activin gene knockout experiments have indicated that activin is not essential for early mouse development (Matzuk et al. 1995), but is essential in the *Medaka* fish (Wittbrodt and Rosa 1994). Current evidence therefore suggests that activin has an important role in early *Xenopus* development. It seems likely to be a component of the dorsovegetal (Nieuwkoop) signaling center, with mesoderm-inducing activity from the 250-cell to late blastula stages.

Reasons for attributing a morphogen-like activity to activin derived initially from experiments of Green and Smith (1990) and of Green et al. (1992). They showed that the treatment of dissociated *Xenopus* animal cap cells elicited several different types of mesodermal gene activity according to the concentration of activin to which they were exposed. The actual concentration of activin that caused these cells of normal ectodermal fate to switch to mesodermal or mesoderm-dependent gene expression was extraordinarily low, in the range of 20–40 pM (Green et al. 1992). Although these experiments gave convincing evidence that activin has concentration-dependent effects, they did not exclude the possibility that this was achieved by a heterogeneous cell population, in which individual cells could respond to different activin concentrations, rather than by a situation in which each cell can respond in different ways to various concentrations of activin. To address this question, we devised an in situ analysis of activin effects on blastula animal cap cells (Gurdon et al. 1994). We implanted beads loaded with activin protein into intact animal caps and analyzed the effect of activin signaling by in situ hybridization with various mesodermal gene probes. We recognized a low concentration response by activation of *Xbrachyury* (*Xbra*; Smith et al. 1991) and a high concentration response by *goosecoid* (*gsc*; Cho et al. 1991) expression. We observed a wave of *Xbra* gene activation moving out from the bead through the field of animal cap cells, in a way that increased with activin concentration and with time of exposure to a bead. This result is most easily interpreted as a differential effect of activin, each cell having a choice of at least three responses, namely, to a nil, low, or high concentration. We return to this point below.

EXPERIMENTAL PROCEDURE

Embryo injections. Embryos were microinjected at the two-cell stage, in 1 × MBS, 4% Ficoll, with either activin mRNA or RLDx, as described by Gurdon et al. (1994, 1996). For Figure 4, embryos were injected either ventrally with 500 pg of *Eomes* mRNA or equatorially with 1 ng of *Apod* mRNA.

Embryological manipulations. Animal caps were cut at stage 8 and conjugates made as described in the text and in Gurdon et al. (1994). Recombinant bovine activin βA (Genentech) beads (Affi-gel blue, 100–200 mesh, Bio-Rad) were prepared according to the method of Gurdon et al. (1994, 1996). Animal cap constructs were cultured at 23°C.

RNA expression constructs. Synthetic capped mRNA was synthesized as described previously (Lemaire et al. 1995) using a Megascript in vitro transcription kit (Ambion) and cap analog (New England Biolabs). *Xenopus* activin *βB* in pSR64T (Thomsen et al. 1990; a gift from Dr. D. Melton) was linearized with *Eco*RI and transcribed using SP3 RNA polymerase. *Eomes* in pBluescriptRN3P (Ryan et al. 1996) was linearized with *Sfi*I and transcribed with T3 RNA polymerase. *Apod* in pBluescriptRN3P was linearized with *Sfi*I and transcribed using T3 RNA polymerase.

RESULTS

Criteria for Establishing the Existence of Long Range Signaling and Morphogen Response

Cell division and cell movement. There are many examples in embryos of various species whose cells respond to a signal that originates several cell diameters away, suggesting long-range signaling. However, this situation could arise if a cell signals to its neighbor with which it is in direct contact and if the responding cell subsequently moves away from the signaling cell. A good example of extensive cell rearrangement and movement is that which takes place during the convergent extension of the notochord in vertebrate embryos during gastrulation, and therefore during signaling from the organizer region (Keller et al. 1992).

In the experimental system described here, cell division does not complicate the interpretation. Animal caps can be implanted with activin beads at any stage between 8 and 9.5. Within 2 hours of implantation, *Xbra* expression can be seen several cell diameters away from a bead. The cell number increase over a 2-hour period starting at stages 8, 9, and 9.5 is 23%, 19%, and 14%, respectively. Apart from the low cell division rate at these late blastula stages, there is little if any cell movement, as can be seen by interposing a continuous layer of fluorescein-labeled cells between activin beads and a rhodamine-labeled responding animal cap; the fluorescein-labeled cell layer near the bead remained intact, and *Xbra* expression was seen in the distant rhodamine-labeled cells (Gurdon et al. 1994). We can therefore conclude that, in this experimental system, the distant gene expression observed is clearly attributable to long-range signaling and not to cell movement.

A single cell can make different responses to the same morphogen. We referred above to the ripple of *Xbra* expression that moves away from a bead as the duration of bead exposure or the strength of activin on the bead is increased. We have also established that cell movement is not significant in blastula animal caps. To be sure that heterogeneity of the animal cap cell population does not account for the different gene responses observed, we need to know that most cells respond to any particular activin concentration, i.e., that there is not a mixed population of cells, some of which are predisposed to express *Xbra* and others to express the gene *goosecoid*. We have observed that some of the early expressing *Xenopus* genes produce transcripts soon after they start to be transcribed, and these are localized mainly in the nucleus, thereby making it easy to score each cell individually for expression of the transcribed gene.

Examples of such genes are *Xbrachyury* and *Xpo*, the latter being expressed in the posterior mesoderm (Sato and Sargent 1991). Both these genes are activated by a low concentration of activin (Gurdon et al. 1996), and in both cases a band two to four cells thick shows labeling by in situ hybridization in all cells (Fig. 1A,B). We have never seen labeled and unlabeled cells interspersed with each other in the same region. In the higher concentration range of activin, *gsc* and other genes, including *Eomes* and *Mix1* (Gurdon et al. 1996), are expressed uniformly, although individual nuclei are not stained with probes for these genes. If anything approaching 50% of cells had not been labeled, this would have been readily detectable. We therefore conclude that individual animal cap cells have at least three options, namely, a high-level response (such as *gsc*), a low-level response (*Xbra* or *Xpo*), and a nil response.

A direct or indirect response? If several gene expressions are induced by activin, this could result from sequential effects, such that one type of initial response leads directly to a later response, even though cells have made only one initial interpretation of activin concentration. The likelihood of such an effect is especially great in those instances in which responses are recorded a long time (12–24 hours) after the start of signaling. In the case of blastula animal caps, all the early gene responses re-

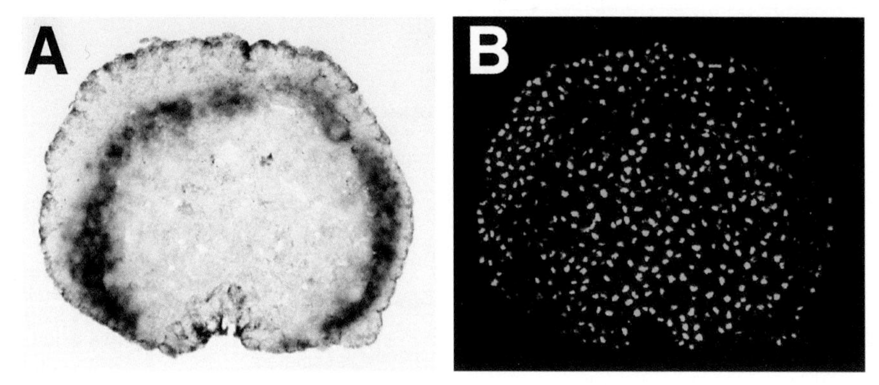

Figure 1. Cell heterogeneity does not account for differing gene responses to a morphogen. (*A*) A ripple of *Xbra* mRNA, seen by in situ hybridization, is 2–4 cells wide. All cells within the ripple express *Xbra*. A similar *Xbra* ripple is seen in the responding cap at all distances from the juxtaposed activin mRNA containing signaling animal cap, according to the concentration previously injected. (*B*) Hoechst staining of nuclei indicates individual cells.

ferred to here are direct, or "immediate early," in the sense that they take place in the presence of cycloheximide, and cannot therefore be a consequence of, or precedent to, other gene expression.

Mechanism of Gradient Perception

To begin to understand this, we need to know (1) the identity of the gradient molecules whose concentration cells can measure, (2) whether cells monitor concentration continuously or only at a particular time, and (3) whether each cell assesses concentration on its own or in cooperation with its neighbors.

The identity of distant signaling molecules. There are clear examples of long-range signaling in which the distant signaling is mediated by a different molecule from that emitted by the primary signaling cell. For example, this seems to be the case in Hedgehog-Dpp signaling in the wing disc in *Drosophila* (Hammerschmidt et al. 1997).

To identify distant signaling molecules, we need to distinguish two fundamentally different mechanisms of signal transmission, namely, by diffusion or relay. In a diffusion mechanism, the initial signaling molecule itself travels to distant cells where it is the signaling molecule, and the intervening cells make no active contribution. On the other hand, in a relay mechanism, each signaling molecule need travel only between one cell and its neighbor, and each intermediate cell in the pathway needs to send a signal, which may be different from that which it received. Although a relay mechanism has been proposed for signaling in *Xenopus* embryos (Reilly and Melton 1996), there seem to us to be difficulties with, and evidence against, this process as an explanation for long-range activin signaling. In the first place, there is the theoretical difficulty of explaining how a relay mechanism can achieve the consistent attenuation of signaling that is evident from our bead experiments referred to above. In addition, we have carried out the following experimental tests of relay processes.

One form of relay process may be described as wave propagation. Each relay member imparts not only a signal, but also a direction to its neighbor. The expanding ripples in a pond in which a stone has been dropped bear some resemblance to the ripple of *Xbra* expression in our activin bead conjugates. To ask whether activin signaling has directionality, like wave propagation, we carried out the experiment shown diagrammatically in Figure 2A. Animal caps were exposed to high-concentration activin beads for 1.5 hours, after which beads were removed and

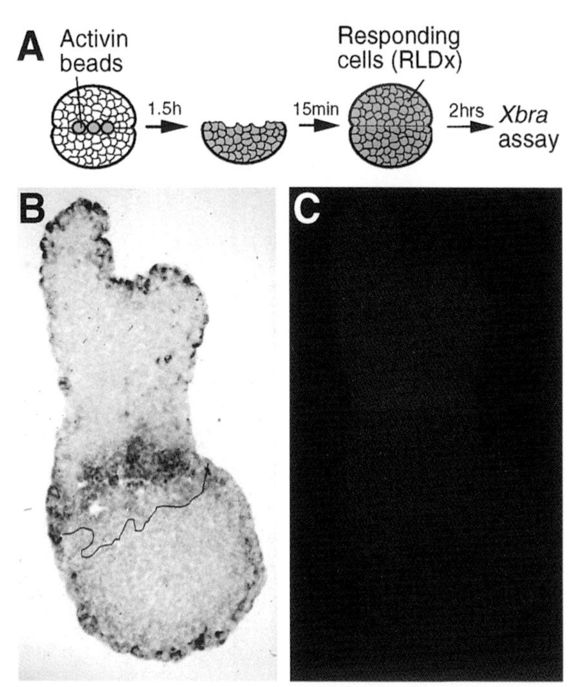

Figure 2. Signaling can progress in a backward as well as forward direction and does not have the properties of wave propagation. (*A*) Experimental design. The activin signal passes from beads into the animal caps. One of the activin-influenced animal caps (now shown in blue) is then placed next to a RLDx-labeled, nonactivin-treated animal cap and assayed, after a further 2 hr, for *Xbra*. (*B*) An *Xbra* band has been induced in the cells of the RLDx-labeled cap and at a distance from cells of the activin-treated signaling cap. The black line (in *B*) demarcates the boundary between the activin-treated and -untreated, RLDx-labeled tissue (as seen in *C*).

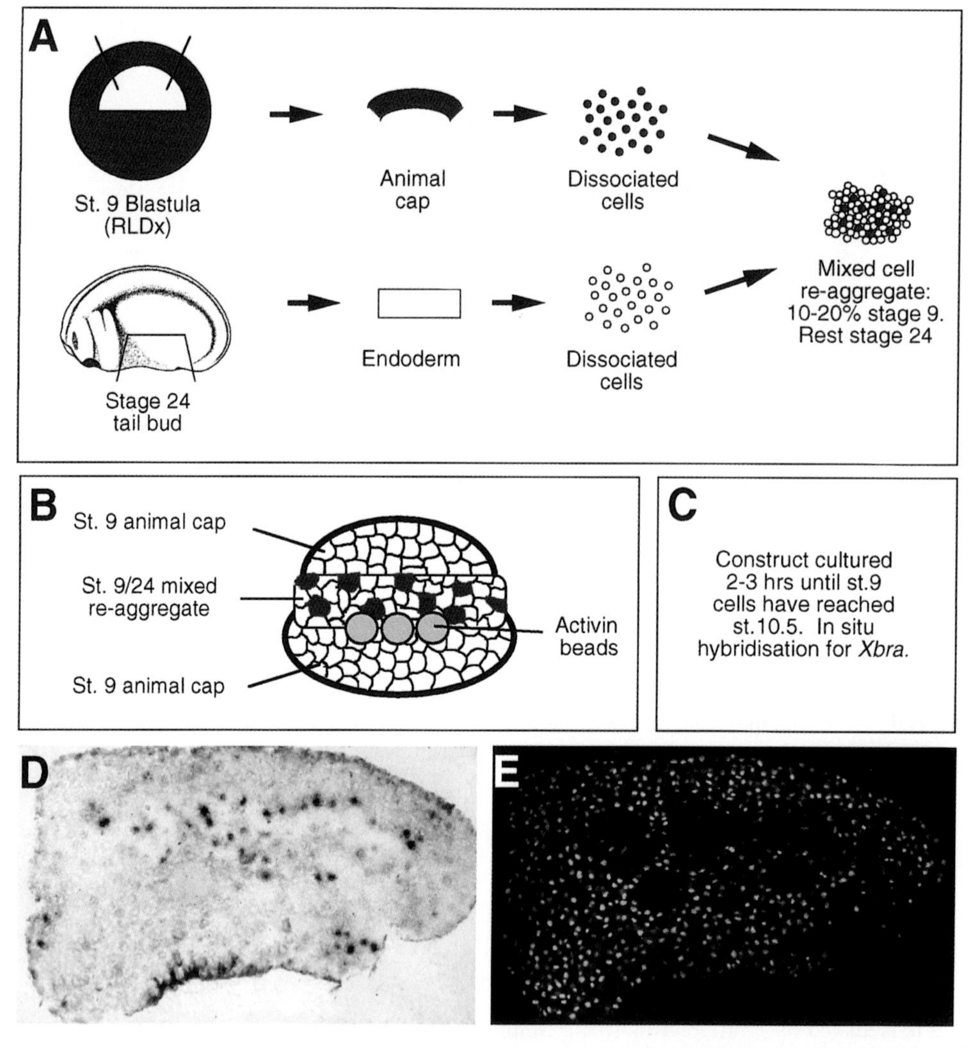

Figure 3. Local cell interactions are not required for cells to respond to a concentration gradient. (*A*) Experimental design. Animal cap cells, dissected from an RLDx-labeled, stage-9 blastula, are dissociated, then mixed and reaggregated with dissociated endoderm cells from a stage-24 tail bud embryo. (*B*) The reaggregated layer is sandwiched between activin-coated beads and an upper (protecting) animal cap. (*C*) Experimental procedure. (*D*) *Xbra* transcripts, although absent from the surrounding endoderm cells, are induced in the separated animal cap cells. (*E*) Hoechst staining indicates individual cell nuclei.

each cap placed next to an untreated lineage-marked cap. After a further 2 hours, in situ hybridization showed that *Xbra* had been induced as a broad band in the lineage-labeled untreated cap (Fig. 2B,C). The signal that had first moved into the primary cap causing *Xbra* expression had subsequently signaled in the reverse direction to induce *Xbra* in cells in the other cap. We conclude that there is no directionality to the signaling process, which does not therefore have the properties of a propagated wave. However, both relay and diffusion processes can have the nondirectional properties observed. To distinguish these, we have shown, using cells with dominant-negative receptors, that activin itself is needed in distant cells. Furthermore, the closely related but nonendogenous protein TGF-β1 can pass across cells that cannot respond to it and can then activate mesodermal genes in distant responsive tissue (McDowell et al. 1997). We have also shown that labeled activin leaves implanted beads, forming a gradient across the tissue and reaching distant responding cells. This series of experiments (McDowell et al. 1997) argues against any form of relay mechanism and provides direct support in favor of diffusion. Most importantly, they tell us that it is activin, rather than some other secondary signaling molecule, whose concentration cells detect.

The time of gradient perception. We have previously addressed the question of when cells sense the activin concentration around them by bead replacement experiments. Cells seem to monitor concentration continuously and by a ratchet mechanism in which they select genes for activation according to the highest concentration that they experience within their period of competence (Gurdon et al. 1995).

Is a community effect involved in gradient perception? We now turn to the question of whether cells sense a concentration gradient individually or whether some kind of interaction between neighboring cells is involved. A community effect by which like cells enhance each others' response to signaling factors has been shown to be required during gastrulation for cells to progress to muscle or notochord differentiation (Gurdon et al. 1993; Weston et al. 1994). It has also been suggested, from the culture of dissociated cells, that interactions between cells may be required for the concentration-dependent response to be achieved (Green et al. 1994; Wilson and Melton 1994). To test more clearly whether cells can respond to a concentration gradient independently of their neighbors, we tested the ability of animal cap cells surrounded by other nonresponsive cells to respond to an activin concentration (Fig. 3A–C). By fixing conjugates at an early stage after *Xbrachyury* expression starts, most transcripts are localized in the nucleus, and the response of single cells can be assessed. We find that single cells can activate *Xbra* to a normal extent even though surrounded by nonresponsive cells (Fig. 3D,E). The cells that express *Xbra* are in the region expected for a low concentration response to activin, which is in the middle distance from medium concentration beads. Cells nearer the bead, in the higher concentration region, did not express *Xbra*, as expected. We conclude that cells can sense their surrounding activin concentration independently of their neighbors and do not require the cooperation of other cells that respond in the same way. This conclusion is entirely consistent with the requirement for a community effect during gastrulation, and therefore after cells have made an initial response to the concentration gradient.

Interaction between High and Low Response Genes

The simplest explanation for how cells respond to low or high concentrations of the same morphogen (see Discussion below) does not readily account for why genes such as *Xbra* that respond to a low morphogen concentration, are not also expressed in a region of high morphogen concentration. This could be explained if expression of a gene activated by a high concentration inhibits the continuing transcription of a low-concentration-response gene. To test this idea, we have overexpressed the high-concentration-response gene *Eomes* in one region of the equator, by local mRNA injection at the two-cell stage. Subsequent in situ hybridization reveals localized inhibition of the usual equatorial band of *Xbra* RNA (Fig. 4A,B). Overexpression of *Apod* mRNA in the equatorial zone can also inhibit *Xbra* expression (Fig. 4C,D). Although *Xbra* is initially transcribed in a cycloheximide-independent way, such that it does not need to be activated by another previously newly transcribed gene, its continued transcription becomes increasingly cycloheximide-sensitive (Schulte-Merker and Smith 1995). We therefore suggest that genes such as *gsc, Eomes,* and *Apod,* which respond to high concentrations of activin, are able to inhibit the continuing expression of low-concentration-response genes such as *Xbra*.

DISCUSSION

We have used an experimental system to analyze the response to a morphogen gradient. We summarized above several reasons for believing that activin may be a natural signaling molecule with morphogen-like effects. Apart from this, the history of cell biology has shown that it is often productive to analyze an interesting mechanism with the most favorable experimental system. The principles that emerge are very likely to be relevant to any developmental process in which cells can make differential responses to secreted factor concentration. For example, there is now emerging evidence that the TGF-β molecule

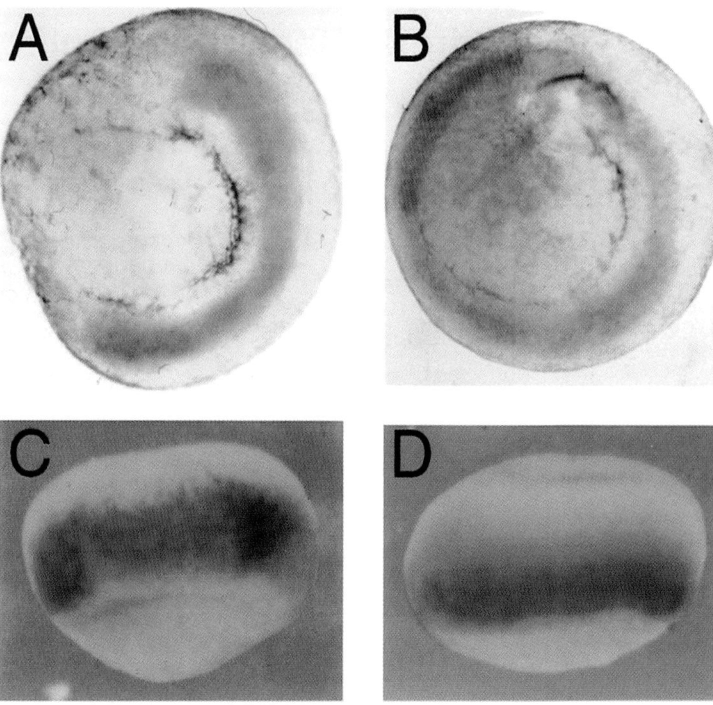

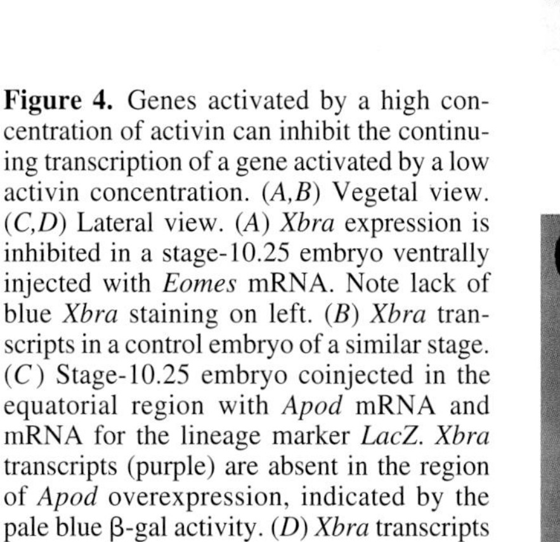

Figure 4. Genes activated by a high concentration of activin can inhibit the continuing transcription of a gene activated by a low activin concentration. (*A,B*) Vegetal view. (*C,D*) Lateral view. (*A*) *Xbra* expression is inhibited in a stage-10.25 embryo ventrally injected with *Eomes* mRNA. Note lack of blue *Xbra* staining on left. (*B*) *Xbra* transcripts in a control embryo of a similar stage. (*C*) Stage-10.25 embryo coinjected in the equatorial region with *Apod* mRNA and mRNA for the lineage marker *LacZ*. *Xbra* transcripts (purple) are absent in the region of *Apod* overexpression, indicated by the pale blue β-gal activity. (*D*) *Xbra* transcripts in a stage-10.25 control embryo.

BMP4 acts as a morphogen to pattern mesodermal and neural tissue during gastrulation (Dosch et al. 1997; Tonegawa et al. 1997). The basic mechanisms that we describe here in relation to activin are likely to be helpful in understanding the mechanism of action of other morphogen molecules.

Our current understanding of this experimental system is as follows. Activin protein is released from the implanted beads and makes a concentration gradient in 1–2 hours within the surrounding animal cap tissue and it does so by diffusion rather than by a relay mechanism. The concentration increases with time for a few hours, after which beads no longer release activin. Cells respond directly to activin throughout the concentration gradient and not to any other secondary signaling molecule. They assess continuously their surrounding concentration of activin and select genes for activation according to the highest concentration that they experience within their competent life. The simplest explanation for this ratchet effect is that the ligand activin has a high affinity, and therefore long occupancy, of its receptors; intracellular signaling would therefore continue, at least for about 2 hours, after a ligand has bound. In fact, stability could exist at any stage of the signaling process from ligand binding to gene transcription, including chromatin structure.

By what mechanism do cells sense the concentration of activin around them? There are two ideas. One is that there are two or more independent receptors, with independent pathways leading to the activation of different genes (Fig. 5A). This seems unlikely because the same receptors (type II and type I), and the same transduction pathway, Smad2, when overexpressed, can lead to the ex-

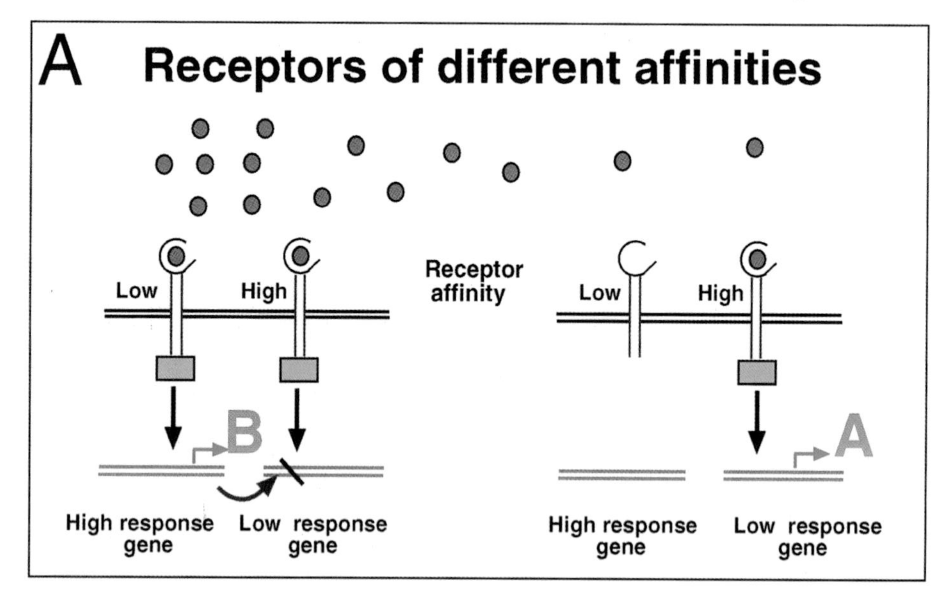

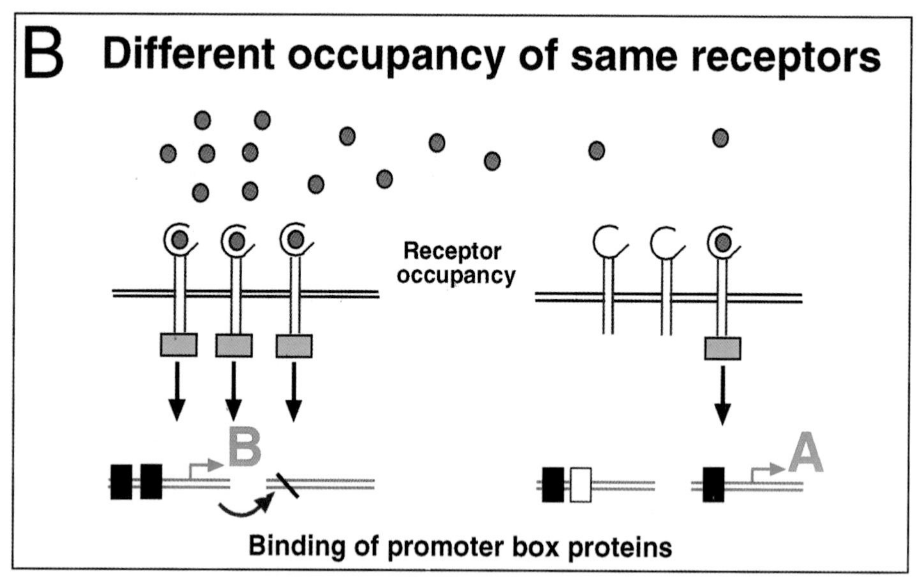

Figure 5. Hypothetical schemes to account for the concentration-dependent response of cells to activin. (*A*) Two types of receptors (of low or high affinity) and two transduction pathways lead to the activation of two different genes. (*B*) One type of receptor and one transduction pathway transmit different levels of receptor occupancy to more or less transcription factor phophorylation. The extent of proteins binding to promoters corresponds to the choice of gene activation. In each scheme, the transcription of a strong response gene inhibits transcription of a low concentration response gene (*left half of each diagram*).

pression of low as well as high response genes (Graff et al. 1996). The second idea is that the occupancy of receptors by activin is relayed to appropriate genes. Cells could recognize either the absolute number of occupied receptors or the ratio of occupied to unoccupied receptors; these values would correspond to increasing levels of intracellular kinase activity, which might in turn activate increasing amounts of a necessary transcription factor. Some gene promoters might require more of the active transcription factor than others, perhaps according to the number of the same promoter sites that need to be occupied (Fig. 5B).

Future progress in understanding the very fundamental problem of how cells sense concentration will require information on the duration and extent of receptor occupancy by ligands, as well as the identification of activin-response elements in the promoters of those genes that are activated as an immediate response to activin. Activin response elements have been identified with respect to *Mix2* (Huang et al. 1995; Chen et al. 1996; Vize 1996), XFD-1 (Kaufmann et al. 1996), *gsc* (Watabe 1995), and *HNF1α* (Weber et al. 1996), and they are being sought in *Eomes* (Ryan et al. 1996) and in *Apod* (Stennard et al. 1996).

ACKNOWLEDGMENTS

This work is supported by the Cancer Research Campaign. Personal support has been provided by NSERC of Canada (A.M.Z), the Wellcome Trust (N.McD.), and the Medical Research Council (D.J.C. and S.D.). We thank K. Butler, A. Mitchel, E. Tweed, R. Coulson, and M. Thoday for technical assistance. We also thank Genentech for activin protein.

REFERENCES

Asashima M., Nakano H., Uchiyama H., Sugino H., Nakamura T., Eto Y., Ejima E., Nishimatsu S.-I., Ueno N., and Kinoshita K. 1991. Presence of activin (erythroid differentation factor) in unfertilized eggs and blastulae of *Xenopus laevis*. *Proc. Natl. Acad. Sci.* **88:** 6511.

Chen X., Rubock M.J., and Whitman M. 1996. A transcriptional partner for MAD proteins in TGF-β signaling. *Nature* **383:** 691.

Cho K.W.Y., Blumberg B., Steinbeisser H., and De Robertis E.M. 1991. Molecular nature of Spemann's organizer: The role of the *Xenopus* homeobox gene *goosecoid*. *Cell* **67:** 1111.

Dosch R., Gawantka V., Delius H., Blumenstock C., and Niehrs C. 1997. BMP-4 acts as a morphogen in dorsoventral patterning in *Xenopus*. *Development* **124:** 2325.

Dyson S. and Gurdon J.B. 1997. Activin signaling has a necessary function in *Xenopus* early development. *Curr. Biol.* **7:** 81.

Graff J.M., Bansal A., and Melton D.A. 1996. *Xenopus* Mad proteins transduce distinct subsets of signals for the TGFβ superfamily. Cell **85:** 479.

Green J.B.A. and Smith J.C. 1990. Graded changes in dose of a *Xenopus* activin A homologue elicit stepwise transitions in embryonic cell fate. *Nature* **347:** 391.

Green J.B., New H.V., and Smith J.C. 1992. Response of embryonic *Xenopus* cells to activin and FGF are separated by multiple dose thresholds and correspond to distinct axes of the mesoderm. *Cell* **71:** 731.

Green J.B.A., Smith J.C., and Gerhart J.C. 1994. Slow emergence of a multithreshold response to activin requires cell-contact dependent sharpening but not prepattern. *Development* **120:** 2271.

Gurdon J.B., Mitchell A., and Mahony D. 1995. Direct and continuous assessment by cells of their position in a morphogen gradient. *Nature* **376:** 520.

Gurdon J.B., Mitchell A., and Ryan K. 1996. An experimental system for analysing response to a morphogen gradient. *Proc. Natl. Acad. Sci.* **93:** 9334.

Gurdon J.B., Harger P, Mitchell A., and Lemaire P. 1994. Activin signaling and response to a morphogen gradient. *Nature* **371:** 487.

Gurdon J.B., Tiller E., Roberts J., and Kato K. 1993. A community effect in muscle development. *Curr. Biol.* **3:** 1.

Hammerschmidt M, Brook A., and McMahon A.P. 1997. The world according to *hedgehog*. *Trends Genet.* **13:** 14.

Hemmati-Brivanlou A. and Melton D.A. 1992. A truncated activin receptor inhibits mesodermal induction and formation of axial structures in *Xenopus* embryos. *Nature* **359:** 609.

Huang H.-C., Murtaugh L.C., Vize P.D., and Whitman M. 1995. Identification of a potential regulator of early transcriptional responses to mesoderm inducers in the frog embryo. *EMBO. J.* **14:** 5965.

Keller R., Shih J., and Domingo C.. 1992. The patterning and functioning of protrusive activity during convergence and extension of the *Xenopus* organizer. *Development* (suppl.), p. 81.

Kaufmann E., Paul H., Friedle H., Metz A., Scheucher M., Clement J.H., and Knochel W. 1996. Antagonistic actions of activin A and BMP-2/4 control dorsal lip-specific activation of the early response gene *XFD-1′* in *Xenopus laevis* embryos. *EMBO J.* **15:** 6739.

Lawrence P.A. and Struhl G.. 1996. Morphogens, compartments, and pattern: Lessons from *Drosophila*. *Cell* **85:** 951.

Lemaire P., Garrett N., and Gurdon J.B. 1995. Expression cloning of *Siamois*, a *Xenopus* homeobox gene expressed in dorsal-vegetal cells of blastulae and able to induce a complete secondary axis. *Cell* **81:** 85.

Matzuk M.M., Kumar T.R., Vassalli A., Bickenbach J.R., Roop D.R., Jaenisch R., and Bradley A. 1995. Functional analysis of activins during mammalian development. *Nature* **374:** 354.

McDowell N., Zorn A.M., Crease J.D., and Gurdon J.B. 1997. Activin has direct long-range signalling activity and can form a concentration gradient by diffusion. *Curr. Biol.* **7:** 671.

Reilly K.M. and Melton D.A. 1996. Short-range signaling by candidate morphogens of the TGFβ family and evidence for a relay mechanism of induction. *Cell* **86:** 743.

Ryan K., Garrett N., Mitchell A., and Gurdon J.B. 1996. Eomesodermin, a key early gene in *Xenopus* mesoderm differentiation. *Cell* **87:** 989.

Sato S.M. and Sargent T.D. 1991. Localized and inducible expression of *Xenopus-posterior (Xpo)*, a novel gene active in early frog embryos, encoding a protein with a "CCHC" finger domain. *Development* **112:** 747.

Schulte-Merker S. and Smith J.C. 1995. Mesoderm formation in response to *Brachyury* requires FGF signaling. Curr. Biol. **5:** 62.

Smith J.C., Price B.M.J., Van Nimmen K., and Huylebroeck D. 1990. Identification of a potent *Xenopus* mesoderm-inducing factor as a homolog of activin A. *Nature* **345:** 729.

Smith J.C., Price B.M., Green J.B.A., Weigal D., and Herrmann B.G. 1991. Expression of a *Xenopus* homolog of *Brachyury* (T) is an immediate early response to mesoderm induction. *Cell* **67:** 79.

Stennard F., Carnac G., and Gurdon J.B. 1996. The *Xenopus* T-box gene, *Antipodean*, encodes a vegetally localized maternal mRNA and can trigger mesoderm formation. *Development* **122:** 4179.

Thomsen G., Woolf T., Whitman M., Sokol S., Vaughan J., Vale W., and Melton D.A. 1990. Activins are expressed early in *Xenopus* embryogenesis and can induce axial mesoderm and anterior structures. *Cell* **63:** 485.

Tiedemann H., Lottespeich F., Davids M., Knochel S., Hoppe P., and Tiedemann H. 1992. The vegetalizing factor: A member of the evolutionarily highly conserved activin family. *FEBS Lett.* **300:** 123.

Tonegawa A., Funayama N., Ueno N., and Takahashi Y. 1997. Mesodermal subdivision along the mediolateral axis in chicken controlled by different concentrations of BMP-4. *Development* **124:** 1975.

Turing A.M. 1952. The chemical basis of morphogenesis. *Philos. Trans. R. Soc. Lond. B Biol. Sci.* **237:** 37.

Vize P.D. 1996. DNA sequences mediating the transcriptional response of the *Mix.2* homeobox gene to mesoderm induction. *Dev. Biol.* **177:** 226.

Watabe T., Kim S., Candia A., Rothbacher U., Hashimoto C., Inoue K., and Cho K.W.Y. 1995. Molecular mechanisms of Spemann's organizer formation: Conserved growth factor synergy between *Xenopus* and mouse. *Genes Dev.* **9:** 3038.

Weber H., Holewa B., Jones E.A., and Ryfell G.U. 1996. Mesoderm and endoderm differentiation in animal cap explants: Identification of the HNF4-binding site as an activin A responsive element in the *Xenopus* HNF1α promoter. *Development* **122:** 1975.

Weston M.J.D., Kato K., and Gurdon J.B. 1994. A community effect is required for amphibian notochord differentiation. *Roux's Arch. Dev. Biol.* **203:** 250.

Wilson P.A. and Melton D.A. 1994. Mesodermal patterning by an inducer gradient depends on secondary cell-cell communication. *Curr. Biol.* **4:** 676.

Wittbrodt J. and Rosa F.M. 1994. Disruption of mesoderm and axis formation in fish by ectopic expression of activin variants: The role of maternal activin. *Genes Dev.* **8:** 1448.

Wolpert L. 1969. Positional information and the spatial pattern of cellular differentiation. *J. Theor. Biol.* **25:** 1.

———. 1989. Positional information revisited. *Development* (suppl.), p. 3.

Analysis of *Fgf8* Gene Function in Vertebrate Development

M. LEWANDOSKI,[1] E.N. MEYERS,[1,2] AND G.R. MARTIN[1]

[1] *Department of Anatomy and Program in Developmental Biology,* [2]*Department of Pediatrics, School of Medicine, University of California, San Francisco, California 94143-0452*

Intercellular interactions have a major role in regulating a wide variety of developmental processes. It is now evident that the molecules that mediate these interactions are members of a relatively small number of families of secreted "growth factors" or "ligands" that transmit signals by activating receptors on the cell surface of neighboring cells. The fibroblast growth factor (FGF) family of heparin-binding proteins is one such group of signaling molecules. Here, we briefly summarize current knowledge about the FGF ligands and their receptors and then focus attention on one member of this gene family, *Fgf8*, which is thought to have key roles in a variety of developmental processes, including limb and brain development. We then describe the genetic approach we have taken to explore *Fgf8* gene function in the mouse.

FGF LIGAND AND RECEPTOR GENES

In mammals, at least 14 different genes are classified as members of the FGF gene family because they contain a conserved sequence that encodes an approximately 120-amino-acid core sequence, which includes receptor-binding and heparin-binding domains (for review, see Basilico and Moscatelli 1992; Coulier et al. 1997; see also Smallwood et al. 1996). The orthologs of some of these genes have been identified in birds and amphibia (Isaacs et al. 1992; Tannahill et al. 1992; Zúñiga-Mejía-Borja et al. 1993; Niswander et al. 1994; Mahmood et al. 1995a), and recently, FGF family members have been identified in worms (Burdine et al. 1997) and flies (Sutherland et al. 1996). Although the first FGF proteins to be identified, FGF1 and FGF2 (previously known as acidic and basic FGF, respectively), were termed fibroblast growth factors because they had mitogenic effects on fibroblasts in vitro, it is now clear that FGFs can affect many different cell types and that they have a wide variety of biological activities, including stimulation or inhibition of cell proliferation, and regulation of cell survival and differentiation (for review, see Basilico and Moscatelli 1992; Baird 1994). Consistent with their proposed functions as intercellular signaling molecules, most FGFs studied in detail thus far and are secreted. However, this is not the case for FGF1 or FGF2, and the mechanism by which these proteins are released from cells has yet to be determined. Because FGFs appear to bind avidly to molecules such as heparan sulfate proteoglycans on the cell surface and in the extracellular matrix (Yayon et al. 1991), it is thought that they do not diffuse very far from the cells that produce them. Thus, they must act either as autocrine factors or as paracrine factors that influence nearby cells.

Most FGF ligand genes appear to have a relatively simple genomic organization: They contain three exons that encode a single gene product. However, in the case of both *Fgf2* (Florkiewicz and Sommer 1989; Prats et al. 1989) and *Fgf3* (Acland et al. 1990), proteins with amino-terminal extensions can be produced by initiation of protein translation at CUG codons upstream of the AUG codon normally used for translation initiation. Interestingly, these amino-terminally extended FGF2 and FGF3 protein isoforms appear to contain a nuclear localization signal (NLS), but the function of these nuclear protein isoforms is not known. In addition, there is evidence for some alternative splicing of FGF genes. *Fgf1* transcripts lacking exon 2 have been identified (Yu et al. 1992), and the chick *Fgf2* gene has been found to contain an alternative first exon, and therefore produces RNA isoforms that encode two proteins with unrelated amino termini (Zúñiga-Mejía-Borja et al. 1993). In contrast, *Fgf8*, which was originally identified as the gene encoding a secreted, androgen-induced growth factor (AIGF) that mediates the androgen-dependent growth of a mouse mammary tumor cell line, SC-3 (Tanaka et al. 1992), has a much more complex structure than any other FGF family member described thus far. Just upstream of the sequences that encode the conserved core is a genomic region containing at least three small exons, rather than the single exon found in other FGF family members. Differential exon usage and the differential use of splice donor and splice acceptor sites within these three exons together result in the production of at least seven different mRNAs, which encode a family of at least seven secreted FGF8 polypeptides, apparently differing only in a short domain that lies between the signal sequence and the start of the conserved FGF core (Crossley and Martin 1995; MacArthur et al. 1995b). As yet, the functional significance of this *Fgf8* isoform diversity is not known.

FGF signaling involves ligand binding to a high-affinity receptor-tyrosine kinase, which acts primarily through an evolutionarily conserved *ras*-dependent intracellular signal transduction pathway. In vertebrates, there are presently four genes known to code for a family of structurally related high-affinity FGF receptors (FGFRs), *Fgfr1*–*Fgfr4* (for review, see Johnson and Williams

1993; Wilkie et al. 1995). These FGF receptors contain an extracellular, ligand-binding domain that consists of three regions that each exhibits features of the immunoglobulin superfamily of structural modules (Ig-like loops), a transmembrane domain, and a split intracellular tyrosine kinase domain. A variety of receptor RNA isoforms can be produced by these genes through alternative splicing, including some that encode secreted proteins lacking the transmembrane and tyrosine kinase domains. Although the functional significance of most FGFR splice variants is not yet known, it is clear from in vitro studies that alternate splicing in the third Ig-like loop plays an important part in determining ligand binding specificities (for review, see Johnson and Williams 1993). However, it is also evident that individual FGFR proteins can bind multiple FGF ligands in vitro (Ornitz et al. 1996). Because FGF ligand binding is facilitated by molecules such as proteoglycans in the extracellular matrix, it is unknown to what extent these in vitro assays reflect the ligand binding specificity of the different FGFR proteins in vivo.

EVIDENCE FOR *Fgf8* GENE FUNCTION IN LIMB DEVELOPMENT

Much of what is known about the early development of the vertebrate limb has come from studies carried out in the chick embryo, which is easily manipulated in ovo. However, it is clear that the basic mechanisms of limb development have been evolutionarily conserved, and information obtained from studies in the chick can be generalized to other species (for review, see Tickle et al. 1975; Tabin 1991). Development of the vertebrate limb depends on the establishment and maintenance of discrete signaling centers within the limb bud: the apical ectodermal ridge (AER or ridge), a specialized ectoderm at the distal tip of the limb bud; the zone of polarizing activity (ZPA) in the mesoderm at the limb bud posterior margin; and the nonridge ectoderm of the limb bud (for review, see Hinchliffe and Johnson 1980; Johnson et al. 1994; Tickle and Eichele 1994; Martin 1995). The signals they produce act on mesodermal cells in the "progress zone" at the distal tip of the limb bud (Summerbell et al. 1973) or their descendants, which give rise to most of the mesenchymal elements of the limb. In turn, the progress zone produces signals that maintain the AER (for review, see Hinchliffe and Johnson 1980).

The functions of the AER, ZPA, and ectoderm were previously thought to be largely independent of one another, with the AER providing signals required for outgrowth along the proximal-distal (P/D) axis, the ZPA producing a "polarizing" signal that regulates patterning along the anteroposterior (A/P) axis, and the ectoderm supplying signals involved in patterning along the dorsoventral (D/V) axis. However, it is now clear that there are regulatory interactions among the different signaling centers and that their products work cooperatively to regulate and coordinate limb outgrowth and patterning along all three axes. For example, secreted signals from both the ridge and the dorsal ectoderm are required to maintain the activity of the ZPA, and the ZPA in turn influences gene expression in the ridge (for review, see Johnson et al. 1994; Tickle and Eichele 1994; Martin 1995).

Limb bud development can be considered as occurring in three phases: induction, initiation, and maintenance. During the induction phase, a signal(s) emanating from a tissue that lies along the main body axis acts to induce cells in the prospective limb territories to begin limb bud formation. At the stages when the limb is induced, competence to respond to the inducer appears to be widespread along much of the length of the embryo (Cohn et al. 1995), and thus induction of the limb, which develops from a small group of lateral plate mesoderm cells and overlying surface ectoderm, is likely to involve some mechanism for restricting the availability of the limb inducer. Some information is available on the tissue source of the limb inducer and the time of its action. Foil barrier (Stephens and McNulty 1981; Strecker and Stephens 1983) and tissue ablation (Geduspan and Solursh 1992) experiments in the chick suggest that the signals required for forelimb bud induction emanate from the intermediate mesoderm, which lies between the somites and the lateral plate mesoderm and develops into the embryonic kidney (mesonephros).

During the initiation phase, mesoderm cells in the prospective limb territory are stimulated to maintain a high rate of proliferation, whereas elsewhere along the length of the lateral plate mesoderm, the rate of proliferation decreases (Searls and Janners 1971). As the prospective limb bud mesoderm thickens, cells in its posterior region begin to express *Sonic hedgehog*, the molecule responsible for polarizing activity (Riddle et al. 1993), and develop into the ZPA. Shortly thereafter, cells in the overlying ectoderm form the AER.

During the maintenance phase, the limb elongates along the P/D axis. Differentiation becomes apparent as cells condense to form the primordia of individual skeletal elements. The primordia of the most proximal structure (e.g., the upper arm bone [humerus]) begins to differentiate first, followed at successively later times by the differentiation of the primordia of progressively more distal structures (e.g., forearm bones [radius and ulna], wrist bones, and then digits).

Evidence is mounting that several members of the FGF family have important roles in each of the three phases of limb development (for review, see Martin 1995; Cohn and Tickle 1996), and studies from our laboratory (Crossley et al. 1996b; Grieshammer et al. 1996) and other laboratories (Mahmood et al. 1995b; Vogel et al. 1996; Ohuchi et al. 1997) have suggested that *Fgf8* is one such family member (summarized in Fig. 1). *Fgf8* function during the induction phase is suggested by two observations. First, when a bead soaked in recombinant FGF8 (FGF8-bead) was inserted in the lateral plate mesoderm in the chick prospective interlimb region, an ectopic limb developed on the operated side. The FGF8-induced ectopic limbs were similar to those induced by other FGF family members (FGF1, FGF2, or FGF4), as described by Cohn et al. (1995). Thus, FGF8 is capable of inducing a limb. Second, in the chick, *Fgf8* is expressed in the intermediate mesoderm, the tissue hypothesized to be the

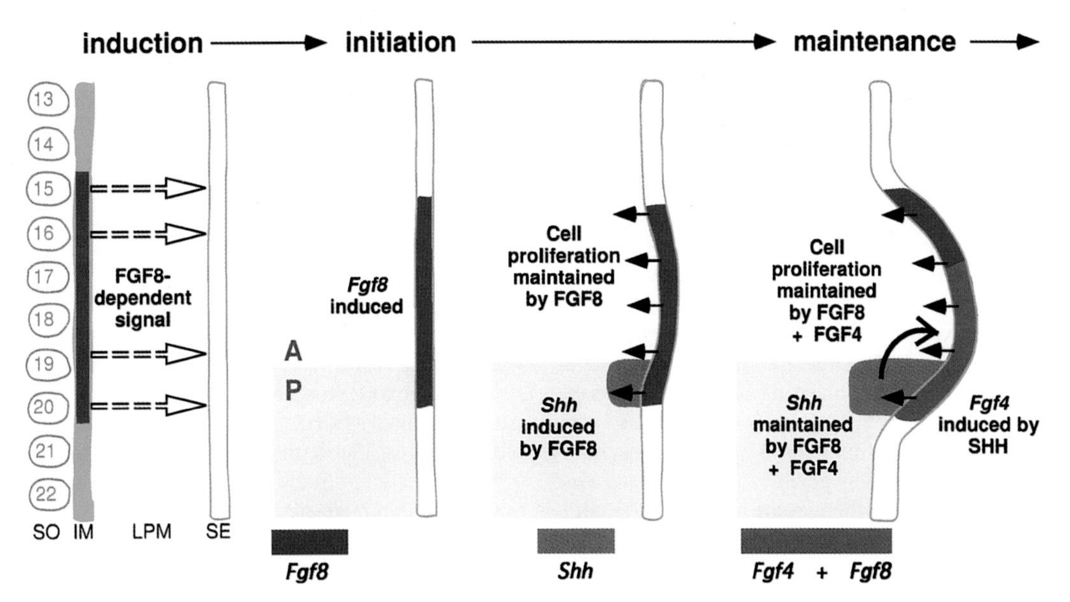

Figure 1. A model for the early stages of limb development illustrating proposed roles for FGF8. Limb induction is mediated by an FGF-dependent signal. FGF8 produced in the nephrogenic component of the intermediate mesoderm (IM) in the prospective forelimb territory (at the level of somites [so] 15–20 in the chick embryo) may provide this signal. During the initiation phase of limb bud formation, *Fgf8* is expressed in the overlying surface ectoderm (SE), and FGF8 protein secreted by the surface ectoderm, acting in conjunction with determinants of A/P positional information, induces expression of *Shh* in the lateral plate mesoderm (LPM). The A/P boundary at which *Shh* is induced is shown as being established during the initiation phase of limb bud development, but may be present earlier. In addition, FGF8 produced in the surface ectoderm may help to maintain proliferation of prospective limb bud cells in the LPM. SHH protein produced in the LPM directly or indirectly induces expression of *Fgf4* in the overlying ectoderm. During the maintenance phase, outgrowth and patterning of the limb bud mesenchyme depends on FGFs produced in the apical ectodermal ridge, which function to maintain *Shh* expression and stimulate cell proliferation in the underlying mesenchyme. The diagram does not illustrate the role(s) of D/V positional information in limb development. (Reprinted, with permission, from Crossley et al. 1996b [copyright Cell Press].)

source of the wing inducer, at the right stage to function as the wing inducer. At present, little is known about the source of the hindlimb inducer, but it is unlikely to be the intermediate mesoderm, because at the stages when the leg is presumably being induced, the intermediate mesoderm in the prospective hindlimb territory is still in the process of forming from the primitive streak. However, the observation that *Fgf8* is expressed at very high levels in the primitive streak at the time when hindlimb bud induction is likely to be occurring is consistent with the idea that FGF8 functions as a hindlimb inducer. Thus, FGF8 emanating from two different tissue sources may induce forelimbs and hindlimbs.

During the initiation phase of chick limb development, *Fgf8* is expressed in the surface ectoderm overlying the prospective limb-forming regions before the first morphological manifestation of limb bud outgrowth (Crossley et al. 1996b; Vogel et al. 1996). Interestingly, it appears that *Fgf8* expression in this domain may be induced by FGF10 produced in the underlying prospective limb bud mesoderm (Ohuchi et al. 1997). It has been suggested that *Fgf8* expression in the surface ectoderm of the nascent limb bud serves two functions. First, based on the finding that FGF8 can stimulate outgrowth of the lateral plate mesoderm, it was hypothesized that FGF8 produced in the surface ectoderm functions to stimulate the proliferation of the prospective limb bud mesoderm and thereby initiate limb bud formation. However, a careful analysis of *Fgf8* expression in the prospective wing bud of chick embryos homozygous for the *limbless* mutation, indicated that *Fgf8* is never expressed in that region of the mutant embryos (Grieshammer et al. 1996; Ros et al. 1996). Because early outgrowth of the mutant limb buds is indistinguishable from that of normal limb buds, these data provide evidence that FGF8 is not required for the initial phase of limb bud outgrowth, although they do not preclude the possibility that FGF8 does contribute some stimulatory activity, perhaps in conjunction with FGF10 (Ohuchi et al. 1997).

A second proposed function of FGF8 produced in the surface ectoderm during the initiation phase is as an inducer of *Shh* expression at the posterior margin of the limb bud. Several lines of evidence support this hypothesis, including our findings that *Fgf8* expression in the surface ectoderm precedes *Shh* expression in the mesoderm during development of both normal and ectopic limb buds (Crossley and Martin 1995) and that FGF8 protein can induce *Shh* expression in *limbless* limb buds, which do not normally express *Shh* or *Fgf8* (Grieshammer et al. 1996). Although these data strongly support the hypothesis that FGF8 is involved in the activation of *Shh* expression during normal limb development, the restriction of *Shh* expression to the posterior margin of each limb bud suggests that FGF8 alone is not sufficient. It appears that FGF8 must act in conjunction with determinants of A/P positional information, to promote *Shh* expression in its normal domain. As yet, little is known about how this positional information required for limb initiation is established in the lateral plate mesoderm. Once *Shh* is expressed, SHH protein produced in the posterior limb bud mesenchyme functions to induce *Fgf4* expression in the nascent AER (Laufer et al. 1994; Niswander et al. 1994).

In the maintenance phase, *Fgf8* is expressed throughout the AER (Crossley and Martin 1995), and *Fgf4* expression is restricted to the posterior half or two thirds of the AER (Niswander and Martin 1992). It is well established that signals from the apical ridge are required to maintain an active progress zone by stimulating cells in the progress zone to proliferate and remain undifferentiated and also to maintain polarizing activity (for review, see Tickle and Eichele 1994). Both FGF8 and FGF4 protein are capable of performing these two functions, but it remains to be determined by genetic experiments to what extent they each perform them in the normal limb bud.

EVIDENCE FOR *Fgf8* GENE FUNCTION IN BRAIN DEVELOPMENT

There is also evidence that *Fgf8* may have a key role in development of the brain. Although *Fgf8* expression is detected in several regions of the developing brain, one expression domain of particular interest is located at the junction of the mesencephalon (mes) and metencephalon (met) (prospective midbrain and anterior hindbrain, respectively). This region contains a morphological feature known as the isthmic constriction. Experimental studies in the chick have shown that a signaling center or "organizer" becomes localized at the mes/met junction early in the development of the neuraxis. This organizer is thought to be responsible for producing signals that act over a relatively long range to pattern tissue both rostral and caudal to it (for review, see Bally-Cuif and Wassef 1995; Puelles et al. 1996). When the region containing this "isthmic organizer" is transplanted to the caudal part of the diencephalon (prospective forebrain), it causes the surrounding tissue to develop into an ectopic midbrain that is the mirror image of the normal midbrain (Martinez et al. 1991; for review, see Puelles et al. 1996). Studies from our laboratory have demonstrated that an FGF8-bead has the same midbrain-inducing and polarizing effect as isthmic tissue (Crossley et al. 1996a), suggesting that FGF8 produced in the isthmic constriction is responsible for the midbrain-inducing activity of an isthmus graft in the chick.

To gain insight into the molecular mechanism by which FGF8 acts to induce an ectopic midbrain, we analyzed the expression following the insertion of an FGF8-bead in the caudal diencephalon of genes whose expression normally marks the isthmic region (Crossley et al. 1996a). These include *Wnt1,* which is expressed in a sharp transverse ring just rostral to the mes/met junction (McMahon et al. 1992), and *Fgf8*, which is normally expressed in a transverse ring of cells immediately caudal to the ring of *Wnt1*-expressing cells (Heikinheimo et al. 1994; Ohuchi et al. 1994; Crossley and Martin 1995; Mahmood et al. 1995b). An important function of WNT1 produced in this region is the maintenance of expression of the *Engrailed (En)* genes (Danielian and McMahon 1996). *En1* and *En2*, vertebrate homologs of the *Drosophila engrailed* gene, are broadly expressed in the mes/met and are required in the mouse from at least E9.5 for development of the midbrain and cerebellum (Wurst et al. 1994; Hanks et al. 1995).

We detected *Fgf8* RNA in the vicinity of an FGF8-bead within 24 hours of bead insertion, consistent with our observation that FGF8 protein induces *Fgf8* gene expression during limb development (Crossley et al. 1996b). We also found that expression of *En2* and *Wnt1* was detected in the vicinity of the FGF8-bead, whereas control beads soaked in phosphate-buffered saline had no effect on expression of these genes. More recently, we have obtained evidence that *En1* is also induced by FGF8 in the caudal diencephalon (P. Crossley et al., unpubl.). These data demonstrate that FGF8 protein is sufficient to induce a complete midbrain in competent neuroepithelium and suggest that FGF8 from the bead induces the ectopic structures by establishing an isthmus-like organizing center in the caudal diencephalon (Fig. 2).

Signals from this induced organizing region appear to have a long-range effect because they transform tissue at some distance from the bead to a midbrain fate. The reversed polarity of the ectopic midbrain indicates that tissue closest to the organizing center is transformed to a caudal fate, whereas tissue farther away is respecified to a more rostral fate. In some cases, the most rostral structure, the griseum tectalis, is absent from both the normal and ectopic midbrains. An analogous effect is seen in limb buds with polarizing regions on opposite sides, which develop a mirror-image digit pattern that sometimes lacks the most anterior digit (4-3-3-4 lacking digit 2) (Tickle et al. 1975). These effects presumably are due to overlap of patterning activities emanating from opposite ends of the responding tissue.

An important question is whether the ability of FGF8 to induce an ectopic midbrain reflects a normal function of FGF8 in the developing brain. There is evidence that *En* expression is induced in the neuroectoderm by a signal from the underlying mesoderm (Ang and Rossant 1993) and then is maintained by a *Wnt1*-dependent signal acting within the plane of the neuroepithelium (McMahon et al. 1992; Danielian and McMahon 1996). Our data demonstrate that FGF8 alone is sufficient to induce *En* expression, raising the possibility that during normal development, FGF8 induces *En* expression in the neuroepithelium. Because *Fgf8* is expressed in mesoderm (e.g., the prospective cardiogenic mesoderm) underlying the neural plate at a stage before *En* induction (Crossley and Martin 1995; Mahmood et al. 1995b), it is possible that FGF8 produced in the mesoderm functions as the normal inducer of *En* expression. Alternatively, *En* gene expression may be induced in the normal embryo by some other signaling molecule produced in the mesoderm, possibly another FGF family member, whose activity can be mimicked by FGF8 in the experimental assays.

It also remains to be determined what role the FGF8 produced in the isthmic organizer has in normal brain development. One possibility is that FGF8 is involved in the *Wnt1*-dependent maintenance of *En* gene expression.

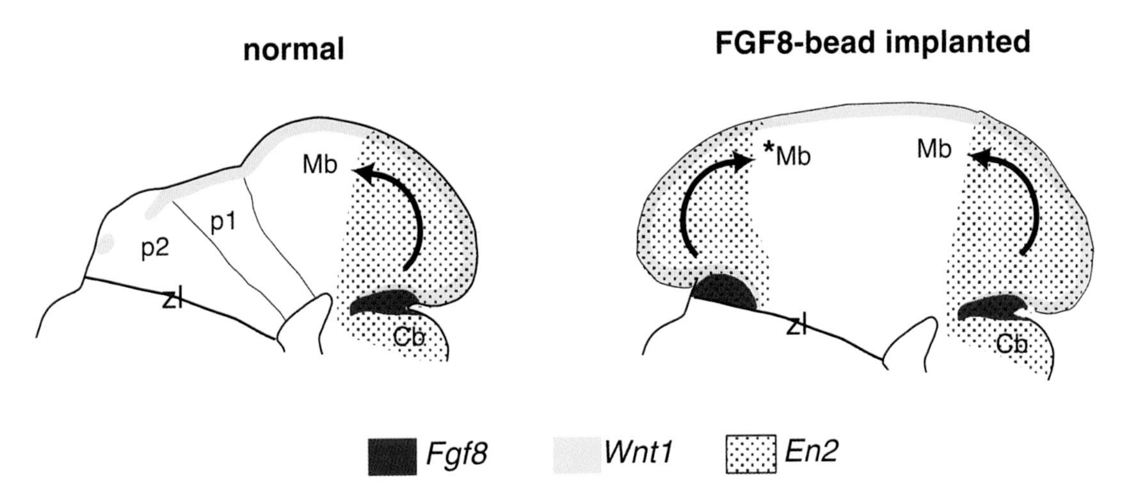

Figure 2. FGF8 induces ectopic midbrain development in the caudal diencephalon. Schematic diagrams illustrating domains of *Fgf8, Wnt1,* and *En* gene expression at stage 19 in the cerebellum (Cb), midbrain (Mb), and caudal diencephalon (prosomeres 1 and 2; p1, p2) of a normal chick embryo and one in which an FGF8-bead was implanted in p2 48 hr earlier. Note that in addition to its expression just rostral to the normal *Fgf8* expression domain, *Wnt1* is also expressed along the dorsal midline of the brain. The ectopic midbrain is indicated by an asterisk (*Mb). Arrows indicate the caudal to rostral polarity of midbrains. Note that inductive effects of the FGF8-bead do not appear to spread rostral of the zona limitans (zl). (Reprinted, with permission, from Crossley et al. 1996a [copyright Macmillan].)

This suggestion is based on the observation that the *Fgf8* expression domain is immediately caudal to the *Wnt1* expression domain in the isthmic region (Crossley and Martin 1995) and that the products of these two genes cooperate to accelerate mammary tumor formation in mice (MacArthur et al. 1995a). Interactions between WNT1 and FGF8 may be responsible not only for maintaining *En* expression, but perhaps also for other aspects of the organizing activity of the isthmic region during normal development.

RESULTS

Strategy for Determining *Fgf8* Gene Function in the Mouse Embryo

The experiments described above strongly suggest a role for *Fgf8* in limb and brain development, and data from expression studies suggest that *Fgf8* function might also be required for other morphogenetic processes including gastrulation, forebrain, and craniofacial development (Heikinheimo et al. 1994; Ohuchi et al. 1994; Crossley and Martin 1995; Mahmood et al. 1995b). However, such data do not provide conclusive evidence of *Fgf8* function in the normal embryo. To explore this issue, we sought to produce mice in which *Fgf8* gene function has been eliminated. This can be accomplished by the now-standard method of gene targeting to produce a loss-of-function allele in mice (Capecchi 1989). Although this is a remarkably powerful approach for studying gene function in vivo, it has several limitations. Perhaps the most serious is that it may be informative about the function of a particular gene only at the earliest time at which that gene is required for normal embryogenesis. If embryos affected by the lack of gene product at a particular stage do not survive to later stages, then the role the gene of interest plays at later stages of development cannot be studied. Studies of embryos deficient in *Fgf4* gene function provide a particularly relevant example of this problem: The mutant homozygotes die at the time of implantation (Feldman et al. 1995), precluding analysis of *Fgf4* function at later stages of development (e.g., during gastrulation or in the developing limb bud). We were concerned that *Fgf8* might likewise have an essential function in the early embryo and that the mutant homozygotes might therefore be uninformative with regard to *Fgf8* function in limb and brain development.

To circumvent this potential problem, we used an approach that takes advantage of methods for making gene alteration in mice conditional upon recombination mediated by a site-specific DNA recombinase such as Cre (for review, see Kilby et al. 1993). Cre recombinase-modifiable alleles contain two or more *loxP* sites, the 34-base-pair recognition sequence for Cre. In the progeny of a cross between mice carrying such an allele and transgenic mice carrying a *cre* recombinase gene, recombination of the *loxP* sites to a single site occurs in cells that express the recombinase, thereby deleting the gene sequences that lie between them. Recombinase expression in the germ line produces a modified gene (e.g., a knockout allele) that can be transmitted to progeny (Schwenk et al. 1995; Lewandoski et al. 1997), whereas recombinase expression in particular somatic tissues results in tissue-specific gene modification (Gu et al. 1994; Hennet et al. 1995; Tsien et al. 1996). The ability to carry out tissue-specific gene inactivation experiments is particularly important when homozygosity for a null allele of the gene results in death of the embryo before the specific tissue of interest is formed.

Production of Mouse Lines Required to Generate an *Fgf8* Null Allele

To produce an *Fgf8* allele that can be modified by Cre-mediated recombination, part of the strategy we used was to construct a targeting vector in which one *loxP* site was inserted in the intron upstream of exon 2 and another in the 3′-untranslated region (UTR) (Fig. 3A) (E.N. Meyers et al., in prep.). These two insertions served to "flox" (flank with *loxP* sites) the conserved *Fgf8* coding sequences in exons 2 and 3, which are presumably essential for *Fgf8* function. Therefore, Cre-mediated deletion of these sequences should convert the floxed *Fgf8* allele to a null allele. Thus, mice carrying the floxed allele should be useful, when mated to different *cre* transgenic lines, for generating a mouse line carrying an *Fgf8* null allele and also for performing Cre-mediated tissue-specific knockout experiments. The *Fgf8* targeting vector containing the two *loxP* sites as well as other modifications (not shown) was used to produce a mouse line carrying the targeted allele using standard gene targeting methods (Nagy and Rossant 1993). Mice heterozygous for this mutant allele were found to be phenotypically indistinguishable from their wild-type littermates.

To obtain mice carrying an *Fgf8* null allele, we proposed to mate the mice heterozygous for the floxed *Fgf8* gene to transgenic animals carrying a *cre* gene that is expressed in all cells of the early embryo. We sought to produce such a *cre* transgenic mouse line by zygote injection of a construct containing the *cre* recombinase gene, modified in two ways. First, we changed the sequences surrounding the ATG translation initiation codon to match those reported to be optimal for translation initiation in eukaryotic cells (Kozak 1991). Second, we introduced into the amino-terminal region of the *cre* open reading frame the coding sequence for the 7-amino-acid nuclear localization signal (NLS) of the large T-antigen of SV40 (Kalderon et al. 1984). This modified *cre* gene was then placed under the control of regulatory elements from the human β-*actin* gene, including its promoter, 5′ enhancer and intron, 3′-flanking UTR, and polyadenylation sequences (kindly provided by K. Sturm and R. Pedersen). Several independent lines of mice carrying this transgene were produced. One of these lines, designated β-*actin-cre,* was studied in detail as described below.

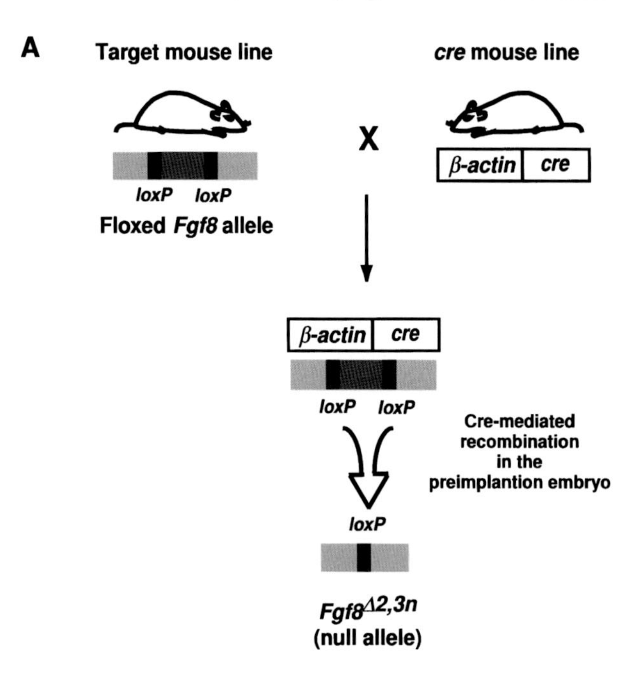

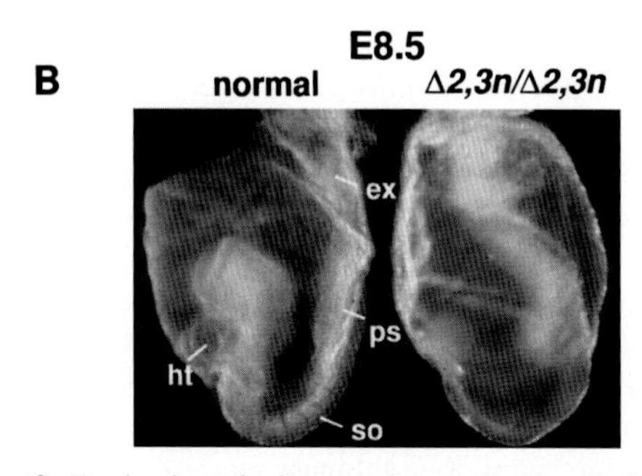

Figure 3. Production of mice carrying an *Fgf8* null allele and phenotype of null mutant homozygotes. (*A*) Schematic diagram illustrating the mating scheme used to produce a mouse line carrying $Fgf8^{\Delta 2,3n}$, a null allele of *Fgf8*. The target mouse line, in which *loxP* sites (*black bars*) flank the coding portions of *Fgf8* exons 2 and 3 (*dark gray box*), was mated with mice homozygous for the β-*actin-cre* gene. In all offspring that inherited both the floxed *Fgf8* gene and the β-*actin-cre* gene, Cre-mediated recombination occurred in the preimplantation embryo, resulting in the conversion of the floxed allele to a null allele ($Fgf8^{\Delta 2,3n}$). This allele was then passed to successive generations, establishing a line heterozygous for $Fgf8^{\Delta 2,3n}$, which could be used to produce mutant homozygotes. (*B*) At E8.5, a null mutant (*Δ2,3n/Δ2,3n*) homozygote is readily distinguishable from its normal (heterozygous or wild-type) littermate by its small size and lack of embryonic mesoderm-derived structures such as heart (ht) and somites (so). Note also the abnormal accumulation of cells in the primitive streak (ps) at the posterior end of the embryo and in the extraembryonic region (ex) of the mutant homozygote.

Analysis of Cre Function in β-*actin-cre* Mice

The goal of these studies was to determine whether the *cre* gene in β-*actin-cre* mice functions efficiently in the early embryo. The floxed "target" gene that served as the substrate for Cre-mediated recombination in this analysis was one for which we had previously developed a sensitive polymerase chain reaction (PCR)-based assay that could readily detect both the recombined and nonrecombined forms of the gene (Lewandoski et al. 1997). This PCR assay employs primers that hybridize to sequences flanking the *loxP* sites (Fig. 4A), and therefore yields amplification products of 1.75 kb and 0.25 kb when nonrecombined and recombined target genes, respectively, are present (Fig. 4B). When the substrate is wild-type mouse DNA, no PCR product is obtained because one of the primers hybridizes to sequences in a *neomycin-resistance* gene cassette present in the target gene but absent in wild-type DNA. Mice heterozygous for this target gene were mated to animals homozygous for the β-*actin-cre* gene. The progeny of this cross were collected at the blastocyst (64-cell) stage, and DNA was prepared from individual embryos for use as a PCR substrate. Blastocyst DNA from 51 progeny was analyzed; half the samples were expected to be $target^{+}/cre^{+}$, whereas the other half were expected to be $target^{-}/cre^{+}$. In 26 samples, the 0.25-kb amplification product was detected, whereas in the remaining 25

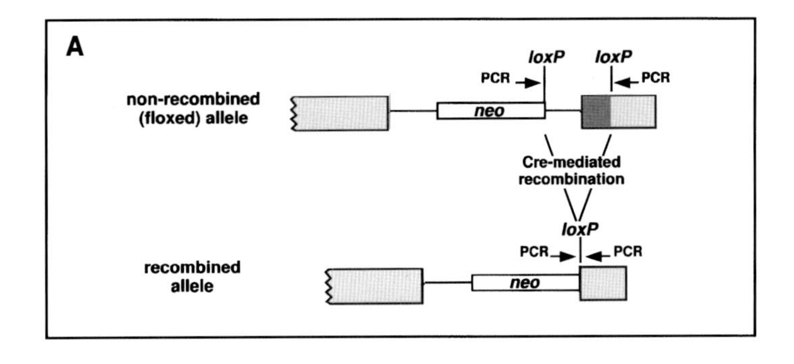

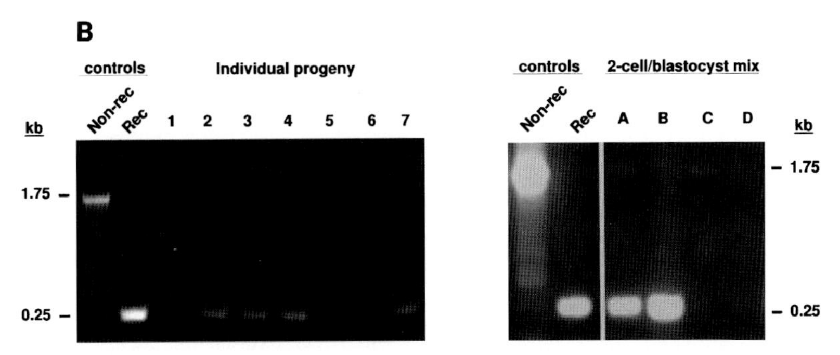

Figure 4. Assay for Cre-mediated DNA recombination in preimplantation embryos carrying the β-*actin-cre* gene. (*A*) Diagram of the target gene. The shaded boxes represent exons of the target gene *(Gbx2)* and the horizontal line represents intron DNA. Dark shading indicates exon DNA that will be deleted by Cre-mediated recombination. The white box represents a *neo* selection cassette, which was inserted into the intron. (*Top*) The floxed target gene, which contains *loxP* sites flanking parts of the intron and the 3′ exon. (*Bottom*) After Cre-mediated recombination, the DNA flanked by *loxP* sites is deleted and a single *loxP* site remains. The horizontal arrows represent primers used in the target gene PCR assay (5′-CTTGTGTAGCGCCAAGTG-3′ and 5′-GGTGCCTCTGAGTTCTGC-3′). (*B*) Assay for recombination status of the target gene in individual blastocysts from a cross between a mouse heterozygous for the target gene and a β-*actin-cre* homozygote. Control DNAs were isolated from mice carrying either the recombined or nonrecombined target gene, as indicated. (*Left panel*) DNA from individual blastocysts (lanes 1–7) was used as a substrate for the target gene PCR assay. In samples from all target$^+$ blastocysts, only the recombined form of the target gene was detectable. (*Right panel*) To confirm that the assay is sensitive enough to detect lack of recombination in a small number of cells in individual blastocysts, a two-cell embryo (either wild-type or carrying the target gene in its nonrecombined form) was mixed with a blastocyst (either wild-type or carrying the target gene in its recombined form), and DNA isolated from each embryo mixture was used as a substrate for the amplification reaction. The embryo mixtures were classified in four categories (1–4), as described in the text.

samples, no such amplification product was obtained. The 1.75-kb amplification product was not detected in any sample (Fig. 4B, left panel; data not shown).

The finding that the only amplification product detected represented the recombined target gene suggested that Cre-mediated recombination had occurred in every cell by the blastocyst stage of development. Alternatively, the assay may be too insensitive to detect the 1.75-kb amplification product representing the nonrecombined target gene. To determine if this was the case, we performed control experiments in which we sought to determine whether the 1.75-kb fragment would be amplified when the DNA substrate for the amplification reaction contained as few as two copies of the nonrecombined target gene mixed with DNA from a blastocyst carrying a recombined target gene. To obtain the desired mixture, we collected individual embryos at the two-cell stage from a cross between mice heterozygous for the nonrecombined target gene and wild-type mice. Half these embryos should carry the nonrecombined target gene. We then mixed a single two-cell embryo with an individual blastocyst from a cross between mice heterozygous for the recombined target gene and wild-type mice; half those blastocysts should carry the recombined target gene. Thus, we would expect that there would be four categories of mixtures, present at approximately equal frequency: (1) two-cell embryo carrying the nonrecombined target gene and blastocyst carrying the recombined target gene; (2) wild-type two-cell embryo (no target gene) and blastocyst carrying the recombined target gene; (3) two-cell embryo carrying the nonrecombined target gene and wild-type blastocyst (no target gene); and (4) wild-type two-cell and wild-type blastocyst (no target gene present). DNA from each of 39 such two-cell/blastocyst combinations was then used as a substrate in the PCR assay. As expected, the results could be classified in four categories: (1) Both the nonrecombined and recombined amplification products were detectable (7/39; 18%); (2) only the recombined amplification product was detectable (11/39; 28%); (3) only the nonrecombined amplification product was detectable (10/39; 26%); and (4) no amplification product was detected (11/39; 28%) (Fig. 4B, right panel; data not shown). These data demonstrate that the assay we employed is sufficiently sensitive to detect a very small number of cells carrying the nonrecombined allele in a blastocyst in which recombination has occurred in all of the remaining cells. Therefore, we conclude that the *cre* gene carried by our β-*actin-cre* mouse line functions efficiently in all cells of the embryo at or before the blastocyst stage.

Generation of an *Fgf8* Null Allele by Cre-mediated Recombination and Phenotype of Null Homozygotes

Mice heterozygous for the floxed *Fgf8* allele were then mated with β-*actin-cre* homozygotes. Half the progeny of this cross are expected to be target$^+$/cre$^+$ and half should be target$^-$/cre$^+$. Based on the data described above, we expected that recombination of the *Fgf8* floxed allele should occur by the blastocyst stage in all mice that inherited the target gene. PCR analysis indicated that only the recombined allele was detectable in DNA isolated from the tail of all target$^+$/cre$^+$ progeny assayed, suggesting that recombination of the $Fgf8^{neo}$ allele occurred in all cells of the offspring that inherited both the mutant *Fgf8* allele and *cre* (data not shown). These mice therefore carried a presumed null allele, $Fgf8^{\Delta 2,3n}$, which lacks the coding sequences in exons 2 and 3 (Fig. 3A). The $Fgf8^{\Delta 2,3n}$ and *cre* genes in these mice segregate independently, and a mouse line carrying $Fgf8^{\Delta 2,3n}$ but not *cre* was established. $Fgf8^{\Delta 2,3n}/+$ mice are phenotypically indistinguishable from their wild-type littermates, but can be identified by Southern blot analysis (not shown), and can be mated inter se to produce mutant homozygotes for analysis.

Because *Fgf8* is expressed in embryos before and during gastrulation (Ohuchi et al. 1994; Crossley and Martin 1995; Mahmood et al. 1995b), we examined the progeny of the intercross on embryonic day (E)7.5, at the end of the primitive streak stage of gastrulation, as well as on E8.5 and E9.5. Of the 56 embryos analyzed, 13 (23%) were genotyped as $Fgf8^{\Delta 2,3n}$ homozygotes, all of which were phenotypically abnormal. Mutant homozygotes were significantly smaller than their wild-type and heterozygous littermates and appeared to lack all embryonic mesoderm-derived structures, including the heart and somites (Fig. 3B, and data not shown). By E10.5, all mutant homozygotes were being resorbed. These data indicate that *Fgf8* function is essential for normal gastrulation.

DISCUSSION

We describe here the production of two mouse lines: One carries an *Fgf8* gene flanked by *loxP* sites, and therefore provides a valuable resource for studying *Fgf8* gene function; the other carries a modified *cre* gene expressed under the control of human β-*actin* regulatory sequences, and Cre protein functions at very high efficiency at preimplantation stages of development. When such β-*actin-cre* mice are mated to floxed target mice, recombination of the target gene in *cre*$^+$/target$^+$ progeny occurs in every cell by the 64-cell stage of embryogenesis. This mouse line is thus extremely useful as a "deletor" strain (Schwenk et al. 1995); i.e., a mouse line that produces the Cre necessary to delete floxed DNA fragments in the developing embryo.

The utility of such deletor mice for efficiently generating progeny that carry the recombined derivative of a floxed target gene is illustrated by the experiments in which we mated the β-*actin-cre* mice with animals carrying the floxed *Fgf8* gene. When tail DNA from the progeny of such a cross was assayed for recombination, only the recombined derivative of the target allele ($Fgf8^{\Delta 2,3n}$) was detected in the samples from *cre*$^+$/target$^+$ animals. Moreover, those *cre*$^+$/target$^+$ mice invariably transmitted the recombined target gene to their offspring, consistent with the conclusion that the β-*actin-cre* transgene is expressed very early in development and causes complete recombination in all cells of the embryo, including those that will establish the germ line. Similar results were obtained when the β-*actin-cre* mice were mated to numerous other floxed target mouse lines (data not shown), demonstrating that the β-*actin-cre* deletor mice are generally useful for establishing mouse lines that carry alleles in which floxed DNA segments have been deleted by Cre-mediated recombination.

The β-*actin-cre* mice described here are also especially useful for a second type of study, in which the goal is to determine the consequences of transgene expression when such expression causes lethality or some defect that precludes the establishment of a transgenic line. In that type of experiment, the target transgene could be one that causes ectopic expression of a normal gene, expression of a gain-of-function mutation, or cell ablation (via toxin-gene expression) in a particular tissue only after being activated by recombination. By mating mice carrying such a target gene to the β-*actin-cre* mice described here, one would obtain *cre*$^+$/target$^+$ animals in which every cell contains the target transgene in the recombined (i.e., activated) form, and as they develop they will display the mutant phenotype. In such studies, ubiquitous expression of *cre* in the early embryo, as occurs in our β-*actin-cre* mice, is particularly important because mosaicism with respect to recombination, due to lack of *cre* expression in all cells, would complicate the analysis of the mutant phenotype.

It should be noted that two other *cre* deletor mouse lines have been previously described. One such line contains a *cre* gene under the control of the human cytomegalovirus (CMV) minimal promoter (Schwenk et al. 1995), and in the other line, *cre* is expressed under the control of the adenovirus EIIa promoter (Lakso et al. 1996). Both lines appear to be very useful for producing animals that efficiently transmit the recombined form of a floxed target gene to their progeny. However, the extent to which these deletor lines are useful for target gene activation experiments is as yet unknown, because the stage and tissue specificity of *cre* expression in the early embryo has not been reported, and recombination may not occur in all cells of the embryo.

The availability of the mouse line carrying the recombined derivative of the floxed *Fgf8* gene, $Fgf8^{\Delta 2,3n}$, which we generated by mating mice carrying the floxed *Fgf8* allele with β-*actin-cre* mice, has enabled us to produce *Fgf8* null mutant homozygotes. Analysis of such embryos is still in progress, but the data thus far demonstrate that *Fgf8* has a key role in early embryogenesis. In the absence of *Fgf8* gene function, development is arrested during gastrulation and embryonic mesoderm fails to form. These results underscore the value of the approach we have taken to mutating the *Fgf8* gene. Had we simply produced mice carrying a deletion allele of *Fgf8*, then analysis of FGF8 function in processes such as limb and brain development would not have been possible because of the early embryonic lethality of the null mutant homozygotes. However, because the mice we produced carry a floxed *Fgf8* allele, they remain available for matings with different *cre* transgenic lines, such as those that express *cre* in the developing limb bud, in order to perform in tissue-specific knockout experiments. The results of such experiments should serve to resolve the question of whether *Fgf8* is required for normal development of the limbs and brain.

ACKNOWLEDGMENTS

This work was supported by grants from the National Institutes of Health to G.R.M. E.N.M. was supported by a grant from the Pediatric Scientist Development Program funded by the March of Dimes Birth Defects Foundation.

REFERENCES

Acland P., Dixon M., Peters G., and Dickson C. 1990. Subcellular fate of the Int-2 oncoprotein is determined by choice of initiation codon. *Nature* **343:** 662.

Ang S.-L. and Rossant J. 1993. Anterior mesendoderm induces mouse *Engrailed* genes in explant cultures. *Development* **118:** 139.

Baird A. 1994. Fibroblast growth factors: Activities and significance of non-neurotrophin neurotrophic growth factors. *Curr. Opin. Neurobiol.* **4:** 78.

Bally-Cuif L. and Wassef M. 1995. Determination events in the nervous system of the vertebrate embryo. *Curr. Biol.* **5:** 450.

Basilico C. and Moscatelli D. 1992. The FGF family of growth factors and oncogenes. *Adv. Cancer Res.* **59:** 115.

Burdine R., Chen E., Kwok S., and Stern M. 1997. *egl-17* encodes an invertebrate fibroblast growth factor family member required specifically for sex myoblast migration in *Caenorhabditis elegans*. *Proc. Natl. Acad. Sci.* **94:** 2433.

Capecchi M.R. 1989. Altering the genome by homologous recombination. *Science* **244:** 1288.

Cohn M. and Tickle C. 1996. Limbs: A model for pattern formation within the vertebrate body plan. *Trends Genet.* **12:** 253.

Cohn M.J., Izpisúa-Belmonte J.C., Abud H., Heath J.K., and Tickle C. 1995. Fibroblast growth factors induce additional limb development from the flank of chick embryos. *Cell* **80:** 739.

Coulier F., Pontarotti P., Roubin R., Hartung H., Goldfarb M., and Birnbaum D. 1997. Of worms and men: An evolutionary perspective on the fibroblast growth factor (FGF) and FGF receptor families. *J. Mol. Evol* **44:** 43.

Crossley P.H. and Martin G.R. 1995. The mouse *Fgf8* gene encodes a family of polypeptides and is expressed in regions that direct outgrowth and patterning in the developing embryo. *Development* **121:** 439.

Crossley P.H., Martinez S., and Martin G.R. 1996a. Midbrain development induced by FGF8 in the chick embryo. *Nature* **380:** 66.

Crossley P.H., Minowada G., MacArthur C.A., and Martin G.R. 1996b. Roles for FGF8 in the induction, initiation, and maintenance of chick limb development. *Cell* **84:** 127.

Danielian P. and McMahon A. 1996. *Engrailed-1* as a target of the *Wnt-1* signalling pathway in vertebrate midbrain development. *Nature* **383:** 332.

Feldman B., Poueymirou W., Papaioannou V.E., DeChiara T.M., and Goldfarb M. 1995. Requirement of FGF-4 for postimplantation mouse development. *Science* **267:** 246.

Florkiewicz R. and Sommer A. 1989. Human basic fibroblast growth factor gene encodes four polypeptides: Three initiate translation from non-AUG codons. *Proc. Natl. Acad. Sci.* **86:** 3978.

Geduspan J.S. and Solursh M. 1992. A growth-promoting influence from the mesonephros during limb outgrowth. *Dev. Biol.* **151:** 212.

Grieshammer U., Minowada G., Pisenti J.M., Abbott U.K., and Martin G.R. 1996. The chick *limbless* mutation causes abnormalities in limb bud dorsal-ventral patterning: Implications for the mechanism of apical ridge formation. *Development* **122:** 3851.

Gu H., Marth J.D., Orban P.C., Mossmann H., and Rajewsky K. 1994. Deletion of a DNA polymerase β gene segment in T cells using cell type-specific gene targeting. *Science* **265:** 103.

Hanks M., Wurst W., Anson-Cartwright L., Auerbach A.B., and Joyner A.L. 1995. Rescue of the *En-1* mutant phenotype by replacement of *En-1* with *En-2*. *Science* **269:** 679.

Heikinheimo M., Lawshé A., Shackleford G.M., Wilson D.B., and MacArthur C.A. 1994. *Fgf-8* expression in the post-gastrulation mouse suggests roles in the development of the face, limbs, and central nervous system. *Mech. Dev.* **48:** 129.

Hennet T., Hagen F.K., Tabak L.A., and Marth J.D. 1995. T-cell-specific deletion of a polypeptide N-acetylgalactosaminyl-transferase gene by site-directed recombination. *Proc. Natl. Acad. Sci.* **92:** 12070.

Hinchliffe J.R. and Johnson D.R. (1980). *The development of the vertebrate limb: An approach through experiment, genetics, and evolution.* Clarendon Press, Oxford, United Kingdom.

Isaacs H.V., Tannahill D., and Slack J.M.W. 1992. Expression of a novel FGF in the *Xenopus* embryo. A new candidate inducing factor for mesoderm formation and anteroposterior specification. *Development* **114:** 711.

Johnson D.E. and Williams L.T. 1993. Structural and functional diversity in the FGF receptor multigene family. *Adv. Cancer Res.* **60:** 1.

Johnson R.L., Riddle R.D., and Tabin C. 1994. Mechanisms of limb patterning. *Curr. Opin. Genet. Dev.* **4:** 535.

Kalderon D., Roberts B.L., Richardson W.D., and Smith A.E. 1984. A short amino acid sequence able to specify nuclear location. *Cell* **39:** 499.

Kilby N.J., Snaith M.R., and Murray J.A. 1993. Site-specific recombinases: Tools for genome engineering. *Trends Genet.* **9:** 413.

Kozak M. 1991. An analysis of vertebrate mRNA sequences: Intimations of translational control. *J. Cell Biol.* **115:** 887.

Lakso M., Pichel J.G., Gorman J.R., Sauer B., Okamoto Y., Lee E., Alts F.W., and Westphal H. 1996. Efficient *in vivo* manipulation of mouse genomic sequences at the zygote state. *Proc. Natl. Acad. Sci.* **93:** 5860.

Laufer E., Nelson C.E., Johnson R.I., Morgan B.A., and Tabin C. 1994. *Sonic hedgehog* and *Fgf-4* act through a signaling cascade and feedback loop to integrate growth and patterning of the developing limb bud. *Cell* **79:** 993.

Lewandoski M., Wassarman K.M., and Martin G.R. 1997. *Zp3-cre*, a transgenic mouse line for the activation or inactivation of *loxP*-flanked target genes specifically in the female germ line. *Curr. Biol.* **7:** 148.

MacArthur C.A., Shankar D.B., and Shackleford G.M. 1995a. *Fgf-8*, activated by proviral insertion, cooperates with the *Wnt-1* transgene in murine mammary tumorigenesis. *J. Virol.* **69:** 2501.

MacArthur C.A., Lawshé A., Xu J., Santos-Ocampo S., Heikinheimo M., Chellaiah A.T., and Ornitz D.M. 1995b. FGF-8 isoforms activate receptor splice forms that are expressed in mesenchymal regions of mouse development. *Development* **121:** 3603.

Mahmood R., Kiefer P., Guthrie S., Dickson C., and Mason I. 1995a. Multiple roles for FGF-3 during cranial neural development in the chicken. *Development* **121:** 1399.

Mahmood R., Bresnick J., Hornbruch A., Mahony C., Morton N., Colquhoun K., Martin P., Lumsden A., Dickson C., and Mason I. 1995b. A role for FGF-8 in the initiation and maintenance of vertebrate limb bud outgrowth. *Curr. Biol.* **5:** 797.

Martin G.R. 1995. Why thumbs are up (News and views). *Nature* **374:** 410.

Martinez S., Wassef M., and Alvarado-Mallart R.-M. 1991. Induction of a mesencephalic phenotype in the 2-day-old chick prosencephalon is preceded by the early expression of the homeobox gene *en. Neuron* **6:** 971.

McMahon A., Joyner A., Bradley A., and McMahon J. 1992. The midbrain-hindbrain phenotype of *Wnt-1/Wnt-1* mice results from stepwise deletion of *engrailed*-expressing cells by 9.5 days postcoitum. *Cell* **69:** 581.

Nagy A. and Rossant J. 1993. Production of completely ES cell-derived fetuses. In *Gene targeting: A practical approach* (ed. A. L. Joyner), p. 147. IRL Press, Oxford University Press, Oxford, United Kingdom.

Niswander L. and Martin G.R. 1992. *Fgf-4* expression during gastrulation, myogenesis, limb and tooth development in the mouse. *Development* **114:** 755.

Niswander L., Jeffrey S., Martin G., and Tickle C. 1994. A positive feedback loop coordinates growth and patterning in the vertebrate limb. *Nature* **371:** 609.

Ohuchi H., Yoshioka H., Tanaka A., Kawakami Y., Nohno T., and Noji S. 1994. Involvement of androgen-induced growth factor (FGF-8) gene in mouse embryogenesis and morphogenesis. *Biochem. Biophys. Res. Commun.* **204:** 882.

Ohuchi H., Nakagawa T., Yamamoto A., Araga A., Ohata T., Ishimaru Y., Yoshioka H., Kuwana T., Nohno T., Yamasaki M., Itoh N., and Noji S. 1997. The mesenchymal factor, FGF10, initiates and maintains the outgrowth of the chick limb bud through interaction with FGF8, an apical ectodermal factor. *Development* **124:** 2235.

Ornitz D., Xu J., Colvin J., McEwen D., MacArthur C., Coulier F., Gao G., and Goldfarb M. 1996. Receptor specificity of the fibroblast growth factor family. *J. Biol. Chem.* **271:** 15292.

Prats H., Kaghad M., Prats A., Klagsbrun M., Lelias J., Liauzun P., Chalon P., Tauber J., Amalric F., Smith J., and Caput D. 1989. High molecular mass forms of basic fibroblast growth factor are initiated by alternative CUG codons. *Proc. Natl. Acad. Sci.* **86:** 1836.

Puelles L., Marin F., Martinez de la Torre M., and Martinez S. 1996. The midbrain-hindbrain junction: A model system for brain regionalization through morphogenetic neuroepithelial interactions. In *Mammalian development* (ed. P. Lonai), pp. 173–197. Gordon and Breach/Harwood Academic, London.

Riddle R.D., Johnson R.L., Laufer E., and Tabin C. 1993. *Sonic hedgehog* mediates the polarizing activity of the ZPA. *Cell* **75:** 1401.

Ros M.A., López-Martínez A., Simandl B.K., Rodriguez C., Izpisúa-Belmonte J.-C., Dahn R., and Fallon J.F. 1996. The limb field mesoderm determines initial limb bud anteroposterior asymmetry and budding independent of *sonic hedgehog* or apical ectodermal gene expressions. *Development* **122:** 2319.

Schwenk F., Baron U., and Rajewsky K. 1995. A *cre*-transgenic mouse strain for the ubiquitous deletion of *loxP*-flanked gene segments including deletion in germ cells. *Nucleic Acids Res.* **23:** 5080.

Searls R.L. and Janners M.Y. 1971. The initiation of limb bud outgrowth in the embryonic chick. *Dev. Biol.* **24:** 198.

Smallwood P., Munoz-Sanjuan I., Tong P., Macke J., Hendry S., Gilbert D., Copeland N., Jenkins N., and Nathans J. 1996. Fibroblast growth factor (FGF) homologous factors: New members of the FGF family implicated in nervous system development. *Proc. Natl. Acad. Sci.* **93:** 9850.

Stephens T.D. and McNulty T.R. 1981. Evidence for a metameric pattern in the development of the chick humerus. *J. Embryol. Exp. Morphol.* **61:** 191.

Strecker T.R. and Stephens T.D. 1983. Peripheral nerves do not play a trophic role in limb skeletal morphogenesis. *Teratology* **27:** 159.

Summerbell D., Lewis J.H., and Wolpert L. 1973. Positional information in chick limb morphogenesis. *Nature* **224:** 492.

Sutherland D., Samakovlis C., and Krasnow M. 1996. *branchless* encodes a *Drosophila* FGF homolog that controls tracheal cell migration and the pattern of branching. *Cell* **87:** 1091.

Tabin C.J. 1991. Retinoids, homeoboxes, and growth factors: toward molecular models for limb development. *Cell* **66:** 199.

Tanaka A., Miyamoto K., Minamino N., Takeda M., Sato B., Matsuo H., and Matsumoto K. 1992. Cloning and characterization of an androgen-induced growth factor essential for the androgen-dependent growth of mouse mammary carcinoma cells. *Proc. Natl. Acad. Sci.* **89:** 8928.

Tannahill D., Isaacs H., Close M., Peters G., and Slack J. 1992. Developmental expression of the *Xenopus int-2* (FGF-3) gene: Activation by mesodermal and neural induction. *Development* **115:** 695.

Tickle C. and Eichele G. 1994. Vertebrate limb development. *Annu. Rev. Cell Biol.* **10:** 121.

Tickle C., Summerbell D., and Wolpert L. 1975. Positional signalling and specification of digits in chick limb morphogenesis. *Nature* **254:** 199.

Tsien J.Z., Chen D.F., Gerber D., Tom C., Mercer E.H., Anderson D.J., Mayford M., Kandel E.R., and Tonegawa S. 1996. Subregion- and cell type-restricted gene knockout in mouse brain. *Cell* **87:** 1317.

Vogel A., Rodriguez C., and Izpisúa-Belmonte J.-C. 1996. Involvement of FGF-8 in initiation, outgrowth and patterning of the vertebrate limb. *Development* **122:** 1737.

Wilkie A.O.M., Morriss-Kay G.M., Jones E.Y., and Heath J.K. 1995. Functions of fibroblast growth factors and their receptors. *Curr. Biol.* **5:** 500.

Wurst W., Auerbach A., and Joyner A. 1994. Multiple developmental defects in *Engrailed-1* mutant mice: An early mid-hindbrain deletion and patterning defects in forelimbs and sternum. *Development* **120:** 2065.

Yayon A., Klagsbrun M., Esko J., Leder P., and Ornitz D. 1991. Cell surface, heparin-like molecules are required for the binding of basic fibroblast growth factor to its high affinity receptor. *Cell* **64:** 841.

Yu Y., Kah H., Golden J., Migchielsen A., Goetzl E., and Turck C. 1992. An acidic fibroblast growth factor protein generated by alternative splicing acts like and antagonist. *J. Exp. Med.* **175:** 1073.

Zúñiga-Mejía-Borja A., Meijers C., and Zeller R. 1993. Expression of alternatively spliced bFGF first coding exons and antisense mRNAs during chicken embryogenesis. *Dev. Biol.* **157:** 110.

Patterning by Genes Expressed in Spemann's Organizer

E.M. De Robertis, S. Kim, L. Leyns, S. Piccolo, D. Bachiller, E. Agius, J.A. Belo, A. Yamamoto, A. Hainski-Brousseau, B. Brizuela, O. Wessely, B. Lu, and T. Bouwmeester
Howard Hughes Medical Institute, Department of Biological Chemistry, University of California, Los Angeles, California 90095-1662

Nearly three quarters of a century ago, Hans Spemann and Hilde Mangold carried out an experiment that has guided many of the efforts of amphibian embryologists since then (Spemann and Mangold 1924). They were studying the process of gastrulation, in which movements of sheets of cells lead to the formation of the three germ layers (ecto-, meso-, and endoderm) and to the establishment of the anteroposterior and dorsoventral axes of the embryo. When they transplanted the dorsal lip of the blastopore into the opposite (ventral) side of a host embryo, a new body axis, or Siamese twin, was obtained. The grafted tissue was able to organize, or recruit, cells from the host into the secondary axis. Two main activities were found: (1) the induction of a new central nervous system (CNS) and (2) the dorsalization of mesodermal tissues leading to the formation of somites in neighboring cells. The isolation of the cell-cell interaction signaling molecules that pattern the vertebrate body plan has been a Holy Grail for embryologists ever since.

With the advent of molecular cloning techniques, scientists searched for molecules specifically expressed in the dorsal lip or organizer region of the *Xenopus* embryo. The first one to be isolated was *goosecoid*, a homeodomain protein that was able to recapitulate in part Spemann's organizer phenomenon when injected into the ventral part of the embryo (Cho et al. 1991). A large number of organizer-specific molecules have been isolated since, consisting of transcription factors and of secreted proteins, as shown in Figure 1. The various functions of these molecules have been reviewed recently (Lemaire and Kodjabachian 1996; Steinbeisser 1996; Bouwmeester and Leyns 1997). Only 5 years since the isolation of *goosecoid*, the outline of the logic of how the organizer works at a mechanistic level is gradually becoming apparent. From a plethora of organizer-specific molecules, a biochemical pathway for organizer function seems to be emerging: (1) induction of dorsal mesoderm, (2) expression of organizer-specific homeobox genes such as *siamois, goosecoid,* and *Xlim-1*, (3) activation of downstream target genes encoding secreted factors such as *chordin, cerberus,* or *Frzb-1*, and (4) recruitment of neighboring cells into axial structures. For example, the homeobox gene *goosecoid* is an immediate-early response gene to induction by Activin (Cho et al. 1991). Activin mimics the induction of dorsal mesoderm. Microinjection of *goosecoid* mRNA leads to the formation of secondary axes and dorsalization of ventral mesoderm. *goosecoid* induces the expression of Chordin (Sasai et al. 1994), which is a secreted protein that antagonizes BMP-4. The induction of *chordin* can explain most of the phenotypic effects of *goosecoid* in *Xenopus*, in particular the recruitment of neighboring cells into twinned axes.

First, we review the results of a recent screen for dorsal-specific genes that led to the isolation of the novel secreted factors Cerberus and Frzb-1 (Bouwmeester et al. 1996). Second, we present the expression patterns of two other genes isolated during the screen which encode nuclear factors: *Sox-2* that serves as an excellent early panneural marker of the *Xenopus* CNS and developing peripheral nervous system (PNS), and a novel fork head gene, *Fkh-like*, that is an early neural marker with predilection for the floor plate region. Finally, we discuss observations concerning *Xenopus* Chordin, Noggin, and Frzb-1 which suggest that Spemann's organizer is a source of inhibitory secreted molecules that antagonize ventralizing signals by binding to them in the extracellular space.

EXPERIMENTAL PROCEDURES

Differential screening of an unamplified *Xenopus* dorsal lip library was performed on duplicate filters hybridized with dorsal marginal zone (DMZ) or ventral marginal zone (VMZ) probes subtracted with biotinylated ventral gastrula halves as described previously (Bouwmeester et al. 1996). After screening 80,000 plaques, 93 dorsal-enriched clones were isolated, of which one corresponded to a partial *Xenopus Sox-2*. Rescreening of the dorsal lip library resulted in 21 additional *Sox-2* clones, of which the longest (2.0 kb) was sequenced on both strands. The sequence of the *Sox-2* neural marker is accessible in the NCBI database (accession number AF005476). *Xenopus* Sox-2 is 91% identical to its avian homolog at the amino acid level. For antisense in situ hybridization probes, the full-length clone in Bluescript SK was linearized with *Eco*RI and transcribed with T7 RNA polymerase. For sense RNA, the plasmid can be linearized with *Xho*I and transcribed with T3 RNA polymerase. For reverse transcription-polymerase chain reaction (RT-PCR) studies, we find the following *Sox-2* primers useful: F, 5′-GAGGATGGACACTTATGCC-

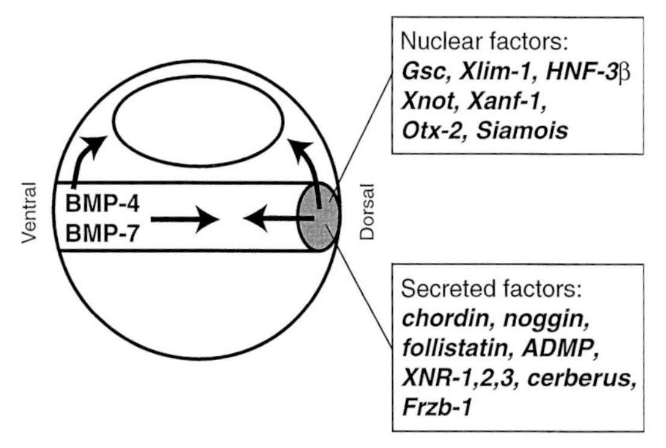

Figure 1. Molecules specifically expressed in the dorsal lip of the *Xenopus* blastopore that have been implicated in various aspects of axial patterning. The nuclear factors, mostly homeodomain proteins, would control the expression of secreted molecules that in turn influence the differentiation of neighboring cells by organizer tissue. For references concerning individual genes, see Lemaire and Kodjabachian (1996).

CAC; R, 5′-GGACATGCTGTAGGTAGGCGA, which yield a product of 213 base pairs.

Another dorsal cDNA obtained in this screen corresponded to *Fkh-like*. Rescreening yielded 28 additional clones. The sequence of the longest clone, as far as it is available at present, indicates that *Fkh-like* is a full-length cDNA distinct from all the murine fork head homologs found to date in the database (accession number pending). In situ hybridization probes can be prepared by linearizing the clone in Bluescript SK with *Eco*RI and transcribing with T7 RNA polymerase. *Fkh-like* primers found to be useful for RT-PCR studies are as follows: F, 5′-GCAGTGGGGAAGCACAAC; R, 5′-AGGTGAGGTAGCTGGGGA, yielding a product of 335 base pairs.

RESULTS

The results of the differential screen of Bouwmeester et al. (1996) are summarized in Table 1. The most abundant cDNA was *chordin*, with 70 independent isolates. Other previously known organizer-specific genes were isolated (*goosecoid*, *XFKH-1*, *Xnot-2*, and *Xlim-1*), whereas others were not (e.g., *noggin*, *XNR3*, and *Siamois*), indicating that the screen was not saturating. In addition, several novel genes were identified. We describe two of these new genes (*Sox-2* and *Fkh-like*) and together with the work on *cerberus* (Bouwmeester et al. 1996), *Frzb-1* (Leyns et al. 1997), and *Paraxial Protocadherin* (S.H. Kim et al., in prep.), this brings to a close the descriptive studies on this productive experiment.

Table 1. *Xenopus* Dorsal-specific cDNAs Isolated by Differential Screening by Bouwmeester et al. (1996)

Previously known genes	Gene product	No. of isolates
Chordin	secreted BMP-4-binding protein	70
Goosecoid	homeodomain protein	3
XFKH-1/HNF-3β	winged helix protein	2
Xnot-2	homeodomain protein	1
Xlim-1	homeodomain protein	1
New genes		
Cerberus	novel secreted protein	11
PAPC	paraxial protocadherin	2
Frzb-1	secreted frizzled-related protein	1
Sox-2	HMG-box factor	1
Fkh-like	novel winged helix factor	1

Cerberus

cerberus, with 11 isolates, was the second most abundant organizer-specific cDNA isolated in this experiment (Table 1). This gene has been studied in some detail in gain-of-function experiments and found to be able to induce small head-like structures when injected into ventral-vegetal blastomeres (Bouwmeester et al. 1996). Figure 2 shows the longest-surviving animal with a *cerberus*-induced secondary head, in which a prominent eye connected by an optic nerve to a small brain can be seen in the ventral abdomen. The tadpole was photographed at the hindlimb bud stage and died a few days after metamorphosis was complete.

In early embryos, *cerberus* is expressed in the leading edge of the involuting endomesoderm during gastrulation. This region gives rise to foregut, liver, pancreas, midgut, and, in more lateral areas, heart mesoderm (Bouwmeester et al. 1996). Only a single eye is formed in heads induced by *cerberus* mRNA; this is an indication that Cerberus inhibits prechordal plate formation, a structure that normally splits the eye field. The ectopic struc-

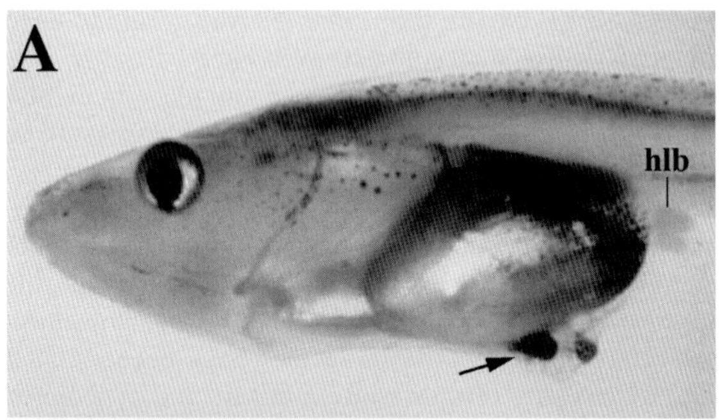

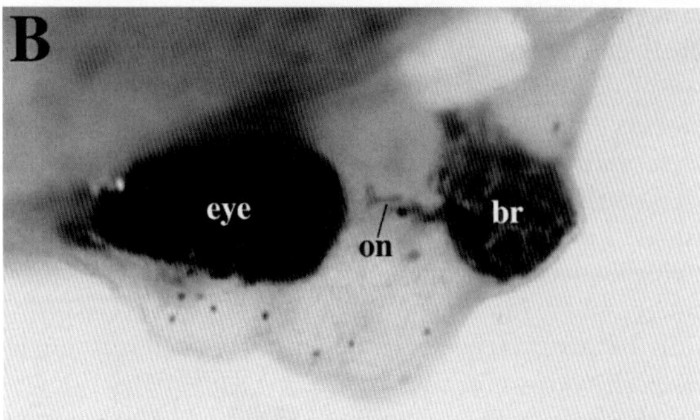

Figure 2. Cerberus induces head-like structures. (*A*) Tadpole injected with *cerberus* mRNA into a D4 blastomere at the 32-cell stage and grown for three months. The arrow indicates an ectopic eye formed in the ventral abdomen; hlb, hindlimb bud. (*B*) High-power view of the ectopic head shown in panel *A*. Note that the ectopic eye is connected via an optic nerve (*on*) to a small brain (*br*). The brain contains a few pigment cells in the surrounding meningeal membranes.

tures induced by *cerberus* include not only eye, forebrain and olfactory placodes, but also duplicated heart and livers. In animal cap explants, neural tissue expressing forebrain markers such as *Otx-2*, but not more posterior neural markers such as *engrailed*, *Krox-20,* or *Hoxb-9*, is induced by *cerberus* mRNA injection (Bouwmeester et al. 1996).

cerberus is also able to induce the pan-endodermal marker *endodermin* (Sasai et al. 1996) and the cardiac precursor marker *Nkx-2.5* (*tinman*) in animal caps. Because *cerberus* encodes a secreted product, the availability of Cerberus protein in soluble purified form could provide the opportunity to test its function in combination with other signaling molecules such as Chordin, Noggin, Frzb-1, and others in order to generate organs such as heart, liver, pancreas, and brain in vitro. If such studies could be extended to pluripotent mammalian cells, tissues might conceivably be prepared in vitro for replacement therapy purposes. In the meantime, the studies with *Xenopus cerberus*, together with parallel studies on the primitive endoderm of the early mouse embryo (Thomas and Beddington 1996; Varlet et al. 1997; for review, see Bouwmeester and Leyns 1997), have led to the unexpected realization that the vertebrate head may be induced by growth factors secreted by the anterior endoderm.

Frzb-1

Paraxial Protocadherin functions as a mesodermal mantle adhesion molecule and will be presented elsewhere (S.H. Kim et al., in prep.). *Frzb-1* encodes a secreted protein containing a cysteine-rich domain (CRD) similar to that of the frizzled family of transmembrane receptors. The CRD is a conserved extracellular domain that is sufficient to bind *Drosophila* Wingless to its receptor Dfz2 (Bhanot et al. 1996). There is a large family of frizzled receptors that have, in addition to the CRD, seven transmembrane domains characteristic of serpentine receptor proteins (Perrimon 1996; Wang et al. 1996). These receptors differ in their response to various Wnt proteins (He et al. 1997). In contrast, Frzb-1 lacks the transmembrane domains and is secreted into the culture medium by transfected human cells and is able to bind to the surface of cells expressing Wnt-1 tethered to their cell surface (Leyns et al. 1997). In the *Xenopus* gastrula, *Frzb-1* is expressed as a typical Spemann organizer component (Leyns et al. 1997; Wang et al. 1997). It is able to antagonize the activities of Xwnt-8, a secreted factor expressed in ventrolateral mesoderm, which was known to have ventralizing activities during gastrulation (Christian and Moon 1993). This conclusion, shown in diagrammatic form in Figure 3A, was reached from mRNA microinjection experiments in *Xenopus* (for review, see Moon et al. 1997).

Overexpression of *Frzb-1* leads to *Xenopus* embryos with shorter trunks and enlarged heads and blocks formation of trunk muscles marked by MyoD. At late gastrula, the amount of Spemann organizer in *Frzb-1*-injected embryos (marked by *chordin*) is greatly expanded at the expense of lateral mesoderm (Leyns et al. 1997). Taken together with work from other groups, these data support the working hypothesis that the normal function of Xwnt-8 during gastrulation is to promote the differentiation of trunk musculature and to limit the extent of tissues allocated to Spemann's organizer (Hoppler et al. 1996; Sokol 1996; Leyns et al. 1997; Moon et al. 1997; Wang et al. 1997).

Frzb-1 expression not only is confined to early embryonic development, but is also in fact one of the most abundant sequences containing a Fz-CRD to be found in searches of human EST databases of adult tissues. Several other secreted Frizzled-related proteins, or sFRPs, have been subsequently reported (Pfeffer et al. 1997; Rattner et al. 1997). In adult tissues, Frzb-1 and their relatives may serve additional functions that have not yet been revealed by the embryological assays in *Xenopus*. For example, they could serve as long-range transporters or facilitators of secretion of Wnt proteins, which are notoriously insoluble products. Another intriguing function could be to act as tumor suppressors, as it is well known that Wnts are potent oncogenes when overexpressed (Nusse and Varmus 1992; Moon et al. 1997). Frzb-1 protein has been independently isolated by entirely different methodologies, as part of protein fractions containing chondrogenic (Hoang et al. 1996) and angiogenic activity (Mayr et al. 1997). Interestingly, purified recombinant Frzb-1 lacked these activities, suggesting that they may be mediated by Wnt signaling molecules that copurify

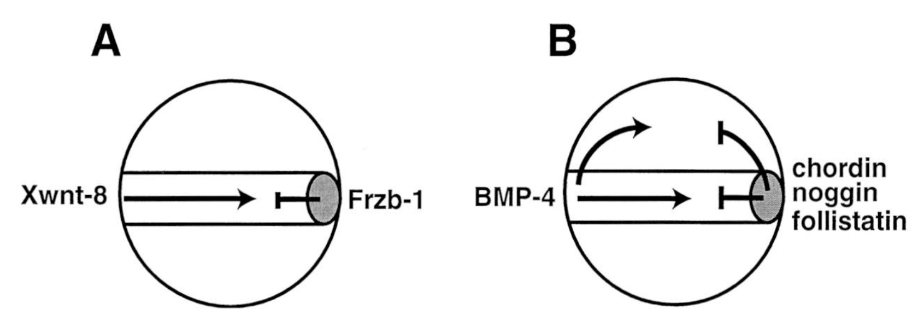

Figure 3. Spemann's organizer as a source of secreted factors that inhibit ventral signals provided by Xwnt-8 and BMP-4. (*A*) Function of Frzb-1 during *Xenopus* gastrulation. Frzb-1 is secreted by the Spemann organizer (*gray oval*) and antagonizes Xwnt-8, which is involved in patterning mesoderm in the ventrolateral marginal zone. (*B*) BMP-4 is secreted by the ventral side of the embryo and is antagonized by Chordin, Noggin, and Follistatin, which directly bind to it in the extracellular space of ectoderm and mesoderm. In this mechanistic view of Spemann's organizer, its function is to inhibit ventral signals.

with Frzb-1 in protein complexes. Thus, the availability of Frzbs (or sFRPs) may permit new methodologies for the purification and isolation of novel Wnt signaling molecules.

Sox-2 and *Fkh-like* Provide Early Neural *Xenopus* Markers

Useful bycatch of fishing expeditions such as the one described in Table 1 is the isolation of novel gene markers for embryological studies. We present two new early neural markers for *Xenopus*. Figure 4 shows the expression of *Sox-2*, an HMG-box transcription factor in *Xenopus*, a gene that had previously been isolated in mouse and chick and found to be ubiquitously expressed in the early CNS (Conlon et al. 1994; Collignon et al. 1996; Streit et al. 1997). Already at stage 10½ (early gastrula; Fig. 4A), a crescent-shaped region of ectoderm that will give rise to the neural plate can be observed. By mid-gastrula (stage 11), *Sox-2* expression is very strong and comprises most of the dorsal side of the embryo, starting several cell diameters above the blastopore (Fig. 4B). *Sox-2* is a pan-neural marker, including future neuronal, glial, and floor plate cells. By stage 13, early neurula, *Sox-2* expression demarcates the future neural plate, and it is our impression that this domain includes the future neural crest as well, given the width of the ectodermal expression (Fig. 4C). There is expression surrounding the blastopore at a distance; it remains to be determined whether the more ventral cells contribute to caudal neural tissue at later stages (Fig. 4C). During neurulation, neural precursors expressing the pan-neural *Sox-2* marker converge toward the midline, eventually forming the CNS and eyes (Fig. 4D–F).

A peculiarity of *Sox-2* is to mark all neural precursors as they are formed, in particular the forming placodes of the PNS of *Xenopus*. Figure 4, G and H, show the various placodes of the head region of *Xenopus* at the tailbud stage. The development of the post-otic posterior lateral line placode (Fig. 4G–I, arrowheads) illustrates the usefulness of this marker to study the development of the posterior migration of the sensory ridges that lead to the development of the lateral line organ. At later stages (3-day tadpole), an additional row of sensorial organs can be observed in the abdominal epidermis (Fig. 4I, arrows). The organogenesis of the lateral line organ has been studied in considerable detail in fish and in particular in axolotl (Northcutt 1992; Northcutt et al. 1994, 1995); the *Sox-2* marker could now permit studies of similar depth in the *Xenopus* PNS.

The final gene we present here is *Fkh-like*. Like *Sox-2*,

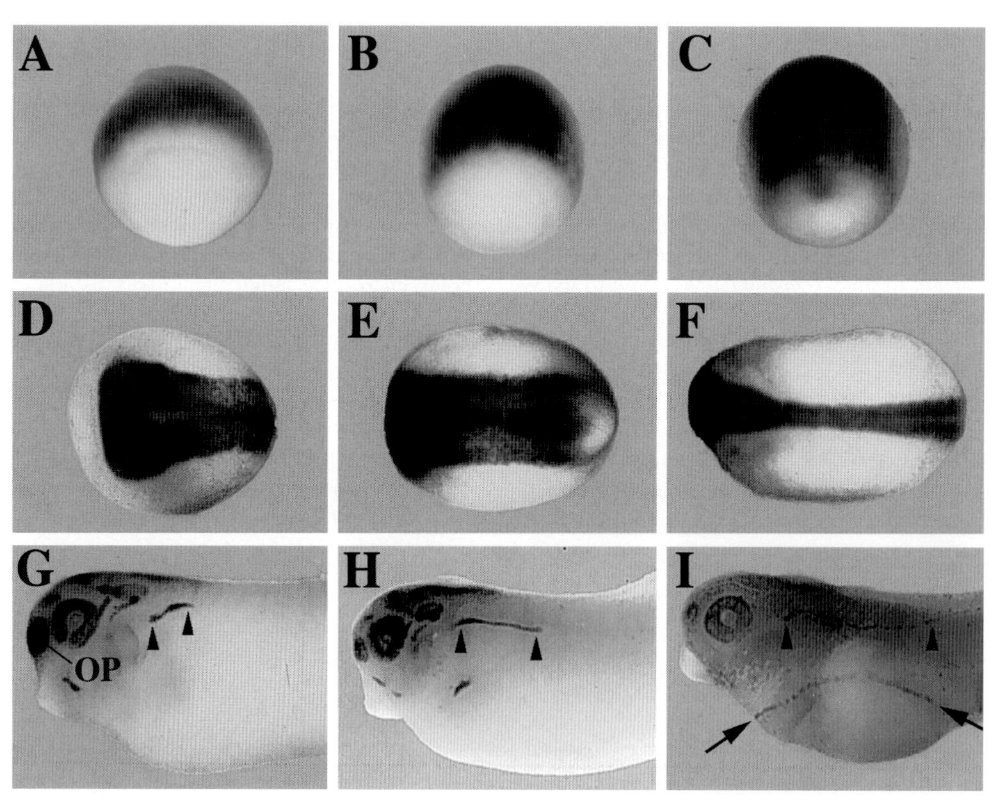

Figure 4. Expression pattern of *Sox-2*, an early pan-neural marker in *Xenopus* embryos. (*A*) Stage 10½, the future neural plate has a crescent shape. (*B*) Stage 11, expansion of the neural plate. (*C*) Stage 12, maximal width of the neural plate anlage which has a rectangular shape. (*D*) Stage 14, convergence toward the midline starts; the anterior neural plate is shown. (*E*) Stage 14, view including the posterior neural plate. (*F*) Stage 21, ectodermal convergence is completed as neural folds close. (*G*) Stage-33 tailbud embryo. Olfactory placode (OP) and many other sensory placodes are visible in the head region; the posterior lateral line placode is indicated by arrowheads. (*H*) Stage 36, the lateral line placode (arrowheads) elongates as it starts its posterior migration. (*I*) Stage 41, at 3 days of development, additional sensory organs, such as the row of spots indicated by arrows in the abdominal wall, become apparent. *Sox-2* is useful as an early CNS marker and as a PNS marker at later stages.

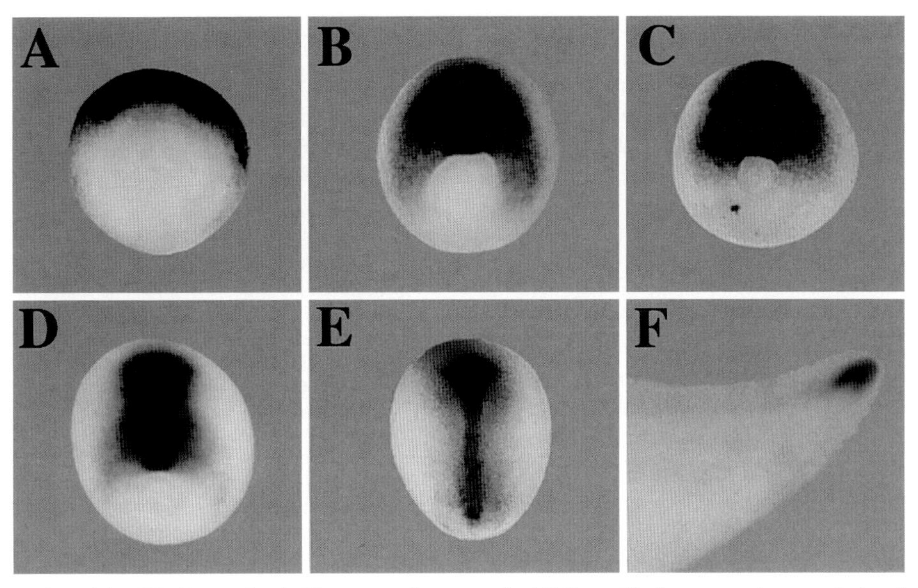

Figure 5. Expression of the *Fkh-like* marker in *Xenopus* embryogenesis. (*A*) Stage 10, the prospective neural plate forms a crescent on the dorsal side of the embryo. (*B*) Stage 11, expression in late gastrula ectoderm. (*C*) Stage 12, expression in the neural plate. (*D*) Stage 12½, expression in neural plate. Note that expression is narrower than that of *Sox-2* (Fig. 4) and that intense staining in the midline starts to appear. (*E*) Stage 14, expression is seen in the floor plate, including in its triangular-shaped expansion in the anterior. (*F*) Stage 32, expression in the tailbud. *Fkh-like* provides a marker for early neural plate and for floor plate.

it is a putative nuclear transcription factor that serves as a very early neural marker. *Fkh-like* expression becomes apparent in a dorsal crescent of neuroectoderm at late blastula, even before an external dorsal blastopore lip becomes evident (Fig. 5A). Together with the findings reported for *Sox-2*, this indicates that neural induction in *Xenopus* takes place earlier than previously thought. As gastrulation proceeds (Fig. 5B,C, stages 11 and 12) the size of the region expressing *Fkh-like* greatly increases along the length of the anteroposterior axis. It is possible, although not proven, that this elongation may correspond to convergence and extension movements in the neuroectodermal layer. Testing this hypothesis would require detailed lineage tracing studies which have not been performed. At the neurula stage, an expression domain emerges in the CNS midline, demarcating the floor plate from its posterior end at the blastopore to the anterior region in which the floor plate expands at its rostral tip (Fig. 5D,E). By the tailbud stage, *Fkh-like* is detected in the tailbud (Fig. 5F) and midbrain (not shown). In conclusion, *Fkh-like* provides a marker for neural induction whose expression is somewhat narrower than that of *Sox-2*. At the neurula stage, *Fkh-like* provides an excellent floor plate marker. It may be useful in the future to investigate whether *Fkh-like* regulates the expression of *sonic hedgehog*, or vice versa, in the neural midline.

DISCUSSION

The search for genes expressed in Spemann's organizer has led to the isolation of interesting new molecules. Some are transcriptional regulators, one is a cell adhesion molecule (*Paraxial Protocadherin*), and yet others are secreted signaling factors. We have presented here two transcriptional regulators: *Sox-2*, an HMG-box transcription factor and *Fkh-like*, a novel winged helix putative DNA-binding protein. The latter genes are not expressed in organizer tissue proper, but rather in the neuroectoderm induced by it and were isolated by virtue of their early expression on the dorsal side of the embryo. However, they provide useful markers that lead to the conclusion that neural induction starts much earlier than previously appreciated.

The Spemann organizer genes that encode secreted proteins are of considerable interest for the problem of how patterning is established in the embryo. Some, such as *cerberus*, presumably act through their own receptors and signaling pathways. Two main surprises, however, came from the long search for neural inducers and mesoderm dorsalizing factors in Spemann's organizer.

First, the same signals can induce neural tissue and dorsalize mesoderm (Fig. 3B) (for review, see De Robertis and Sasai 1996). This is true for *noggin* (Smith and Harland 1992; Lamb et al. 1993), *chordin* (Sasai et al. 1994, 1995), and *follistatin* (Hemmati-Brivanlou et al. 1994; Sasai et al. 1995) mRNAs microinjected into *Xenopus* embryos. This surprising observation has recently found loss-of-function support in the zebrafish ventralized mutant known as *dino*, more recently renamed *chordino*. In *chordino* mutants, the neural plate is greatly reduced and the amount of ventral mesoderm is expanded at the gastrula stage (Hammerschmidt et al. 1996a,b). The mutation results from a small deletion in the amino-terminal part of the zebrafish *chordin* gene, causing a frameshift (Schulte-Merker et al. 1997). This important finding establishes that mutation in a single gene can result in a loss of dorsal values in both ectoderm and mesoderm. This is precisely the type of patterning defect that could have been predicted from a partial loss of Spemann's organizer activity.

Second, proteins such as Chordin and Noggin do not appear to signal through their own receptor and signaling pathway but rather act by binding to and antagonizing the function of BMP-4 in the extracellular space. The dissociation constant (K_D) of Chordin for BMP-4 is 300 pM, i.e., the same as that of BMP-4 for its cognate receptors (Piccolo et al. 1996). Noggin, which shares no primary structural homology with Chordin, binds BMP-4 with a 15-fold higher affinity (Zimmerman et al. 1996). BMP-4 is expressed in most of the ectoderm and mesoderm of the *Xenopus* gastrula, except for the organizer proper (Fainsod et al. 1994). As shown in Figure 3B, organizer signals such as Chordin and Noggin act by neutralizing the action of ventral BMPs (for review, see Graff 1997; Sasai and De Robertis 1997). Follistatin may act in a similar way, for it is able to inhibit ventralizing signals by BMPs (Yamashita et al. 1995; Fainsod et al. 1997), although at lower affinities than its ability to block Activin. However, Follistatin may have some different functions as well, because Activin provides a dorsalizing (rather than ventralizing) signal, and Activin is not bound by Chordin or Noggin. As discussed above, the organizer also secretes Frzb-1, an extracellular antagonist of Xwnt-8 that as far as we know at present acts only in the mesodermal layer (Fig. 3A).

In conclusion, after three quarters of a century of studies on Spemann's organizer, a molecular mechanism is emerging for this embryological phenomenon. An important realization is that the organizer, a small group of embryonic cells with powerful inducing activities on the differentiation of neighboring cells, secretes proteins such as Chordin, Noggin, and Frzb-1 that bind to and inhibit ventral signals. In this view, patterning of the vertebrate embryo would result in part from antagonistic interactions mediated by protein binding in the extracellular space.

ACKNOWLEDGMENTS

We acknowledge fellowship support for S.P. from Comitato Promotore Telethon (Italy), for E.A. from Association pour le Recherche Contre le Cancer (France), for J.A.B. from Gulbenkian Foundation, for A.Y. from Nagasaki University, and for T.B. from EMBO and HFSPO. Long-term support from the National Institutes of Health for this research is gratefully acknowledged (R37 HD21502-11). E.D.R. is an investigator of the Howard Hughes Medical Institute.

REFERENCES

Bhanot P., Brink M., Samos C.H., Hsieh J.C., Wang Y., Macke J.P., Andrew D., Nathans J., and Nusse R. 1996. A new member of the *frizzled* family from *Drosophila* functions as a Wingless receptor. *Nature* **382:** 225.

Bouwmeester T. and Leyns L. 1997. Vertebrate head induction by anterior primitive endoderm. *BioEssays* (in press.)

Bouwmeester T., Kim S., Sasai Y., Lu B., and De Robertis E.M. 1996. Cerberus is a head-inducing secreted factor expressed in the anterior endoderm of Spemann's organizer. *Nature* **382:** 595.

Cho K.W.Y., Blumberg B., Steinbeisser H., and De Robertis E.M. 1991. Molecular nature of Spemann's organizer: The role of the *Xenopus* homeobox gene *goosecoid*. *Cell* **67:** 1111.

Christian J.L., and Moon R.T. 1993. Interactions between *Xwnt-8* and Spemann organizer signaling pathways generate dorsoventral pattern in the embryonic mesoderm of *Xenopus*. *Genes Dev.* **7:** 13.

Collignon J., Sockanathan S., Hacker A., Cohen-Tannoudji M., Norris D., Rastan S., Stevanovic M., Goodfellow P.N., and Lovell-Badge R. 1996. A comparison of the properties of *Sox-3* with *Sry* and two related genes, *Sox-1* and *Sox-2 Development* **122:** 509.

Conlon F.L., Lyons K.M., Takaesu N., Barth K.S., Kispert A., Herrmann B., and Robertson E.J. 1994. A primary requirement for *nodal* in the formation and maintenance of the primitive streak in the mouse. *Development* **120:** 1919.

De Robertis E.M. and Sasai Y. 1996. A common plan for dorsoventral patterning in Bilateria. *Nature* **380:** 37.

Fainsod A., Steinbeisser H., and De Robertis E.M. 1994. On the function of BMP-4 in patterning the marginal zone of the *Xenopus* embryo. *EMBO J.* **13:** 5015.

Fainsod A., Deissler K., Yelin R., Marom K., Epstein M., Pillemer G., Steinbeisser H., and Blum M. 1997. The dorsalizing and neural inducing gene follistatin is an antagonist of BMP-4. *Mech. Dev.* **63:** 39.

Graff J.M. 1997. Embryonic patterning: To BMP or not to BMP, that is the question. *Cell* **89:**171.

Hammerschmidt M., Serbedzija G.N. and McMahon A.P. 1996a. Genetic analysis of dorsoventral pattern formation in the zebrafish: Requirement of a BMP-like ventralizing activity and its dorsal repressor. *Genes Dev.* **10:** 2452.

Hammerschmidt M., Pelegri F., Mullins M.C., Kane D.A., van Eeden F.J.M., Granato M., Brandt M., Furutani-Seiki M., Haffter P., Heisenberg C.P., Jiang Y.J., Kelsh R.N., Odenthal J., Warga R.M., and Nússlein-Volhard C. 1996b. *dino* and *mercedes*, two genes regulating dorsal development in the zebrafish embryo. *Development* **123:** 95.

He X., Saint-Jeannet J.P., Wang Y., Nathans J., Dawid I., and Varmus H. 1997. A member of the Frizzled protein family mediating axis induction by Wnt-5A. *Science* **275:** 1652.

Hemmati-Brivanlou A., Kelly O.G., and Melton D.A. 1994. Follistatin, an antagonist of activin, is expressed in the Spemann organizer and displays direct neuralizing activity. *Cell* **77:** 283.

Hoang B., Moos M., Vukicevic S., and Luyten F.P. 1996. Primary structure and tissue distribution of FRZB, a novel protein related to *Drosophila* Frizzled, suggest a role in skeletal morphogenesis. *J. Biol. Chem.* **271:** 26131.

Hoppler S., Brown J.D., and Moon R.T. 1996. Expression of a dominant negative Wnt blocks induction of MyoD in *Xenopus* embryos. *Genes Dev.* **10:** 2805.

Lamb T.M., Knecht A.K., Smith W.C., Stachel S.E., Economides A.N., Stahl N., Yancopolous G.D., and Harland R.M. 1993. Neural induction by secreted polypeptide *noggin*. *Science* **262:** 713.

Lemaire P. and Kodjabachian L. 1996. The vertebrate organizer: Structure and molecules. *Trends Genet.* **12:** 489.

Leyns L., Bouwmeester T., Kim S.H., Piccolo S., and De Robertis E.M. 1997. Frzb-1 is a secreted antagonist of Wnt signaling expressed in the Spemann organizer. *Cell* **88:** 747.

Mayr T., Deutsch U., Kühl M., Drexler H.C.A., Lottspeich F., Deutzmann R., Wedlich D., and Risau W. 1997. Fritz: A secreted frizzled-related protein that inhibits Wnt activity. *Mech. Dev.* **63:** 109

Moon R.T., Brown J.D., Yang-Snyder J.A., and Miller J.R. 1997. Structurally related receptors and antagonists compete for secreted Wnt ligands. *Cell* **88:** 725.

Northcutt R.G. 1992. Distribution and innervation of lateral line organs in the axolotl. *J. Comp. Neurol.* **325:** 95.

Northcutt R.G., Brändle K., and Fritzsch B. 1995. Electroreceptors and mechanosensory lateral line organs arise from single placodes in axolotls. *Dev. Biol.* **168:** 358.

Northcutt R.G., Catania K.C., and Criley B.B. 1994. Development of lateral line organs in the axolotl. *J. Comp. Neurol.* **340:** 480.

Nusse R. and Varmus H.E. 1992. *Wnt* genes. *Cell* **69:** 1073.

Perrimon N. 1996. Serpentine proteins slither into the wingless and hedgehog fields. *Cell* **86:** 513.

Pfeffer P., De Robertis E.M., and Izpisúa-Belmonte J.C. 1997. *Crescent*, a novel chick gene encoding a Frizzled-like cysteine-rich domain, is expressed in anterior regions during early embryogenesis. *Int. J. Dev. Biol.* **41:** 449.

Piccolo S., Sasai Y., Lu B., and De Robertis E.M. 1996. Dorsoventral patterning in *Xenopus*: Inhibition of ventral signals by direct binding of Chordin to BMP-4. *Cell* **86:** 589.

Rattner A., Hsieh J.C., Smallwood P.M., Gilbert D.J., Copeland N.G., Jenkins N.A., and Nathans J. 1997. A family of secreted proteins contains homology to the cysteine-rich ligand-binding domain of frizzled receptors. *Proc. Natl. Acad. Sci.* **94:** 2859.

Sasai Y. and De Robertis E.M. 1997. Ectodermal patterning in vertebrate embryos. *Dev. Biol.* **182:** 5.

Sasai Y., Lu B., Piccolo S., and De Robertis E.M. 1996. Endoderm induction by the organizer secreted factors Chordin and Noggin in *Xenopus* animal caps. *EMBO J.* **15:** 4547.

Sasai Y., Lu B., Steinbeisser H., and De Robertis E.M. 1995. Regulation of neural induction by the *chd* and *BMP-4* antagonistic patterning signals in *Xenopus*. *Nature* **376:** 333.

Sasai Y., Lu B., Steinbeisser H., Geissert D., Gont L.K., and De Robertis E.M. 1994. *Xenopus chordin*: A novel dorsalizing factor activated by organizer-specific homeobox genes. *Cell* **79:** 779.

Schulte-Merker S., McMahon A.P., and Hammerschmidt M. 1997. The *chordino* gene is an essential component of the zebrafish organizer. *Nature* **387:** 862.

Smith W.C. and Harland R.M. 1992. Expression cloning of noggin, a new dorsalizing factor localized to the Spemann organizer in *Xenopus* embryos. *Cell* **70:** 829.

Sokol S.Y. 1996. Analysis of Dishevelled signaling pathways during *Xenopus* development. *Curr. Biol.* **6:** 1456.

Spemann H. and Mangold H. 1924. Über Induktion von Embryonalanlagen durch Implantation Artfremder Organisatoren. *Roux's Arch. Entw. Mech.* **100:** 599.

Steinbeisser H. 1996. The impact of Spemann's concepts on molecular embryology. *Int. J. Dev. Biol.* **40:** 63.

Streit A., Sockanathan S., Pérez L., Rex M., Scotting P.J., Sharpe P.T., Lovell-Badge R., and Stern C.D. 1997. Preventing the loss of competence for neural induction: HGF/SF, L5 and *Sox-2*. *Development* **124:** 1191.

Thomas P. and Beddington R. 1996. Anterior primitive endoderm may be responsible for patterning the anterior neural plate in the mouse embryo. *Curr. Biol.* **6:** 1487.

Varlet I., Collignon, J., and Robertson E.J. 1997. *noda* expression in the primitive endoderm is required for specification of the anterior axis during mouse gastrulation. *Development* **124:** 1033.

Wang S., Krinks M., Lin K., Luyten F.P., and Moos M. 1997. Frzb, a secreted protein expressed in the Spemann organizer, binds and inhibits Wnt-8. *Cell* **88:** 757.

Wang Y., Mack J.P., Abella B.S., Andreasson K., Worley P., Gilbert D.J., Copeland N.G., Jenkins N.A., and Nathans J. 1996. A large family of putative transmembrane receptors homologous to the product of the *Drosophila* tissue polarity gene *frizzled*. *J. Biol. Chem.* **271:** 4468.

Yamashita H., ten Dijke P., Huylebroeck D., Sampath T.K., Andries M., Smith J.C., Heldin C.H., and Miyazono K. 1995. Osteogenic protein-1 binds to activin type II receptors and induces certain activin-like effects. *J. Cell Biol.* **130:** 217.

Zimmerman L.B., De Jesús-Escobar J.M., and Harland R.M. 1996. The Spemann organizer signal noggin binds and inactivates bone morphogenetic protein-4. *Cell* **86:** 599.

Brainiac and Fringe Are Similar Pioneer Proteins That Impart Specificity to Notch Signaling during *Drosophila* Development

S. GOODE[1] and N. PERRIMON[2]

[1]*Department of Genetics, Harvard Medical School, and* [2]*Howard Hughes Medical Institute, Boston, Massachusetts 02115*

Notch (N) has been implicated in a plethora of signaling events in a variety of tissues throughout the *Drosophila* life cycle (for review, see Artavanis-Tsakonas et al. 1995). N receptors are found in animals spanning phylogeny, and mutations in mammalian N genes have been implicated in leukemia, breast cancer, stroke, and dementia (Ellisen et al. 1991; Robbins et al. 1992; Joutel et al. 1996). N receptors have a modular structure. The large extracellular domain consists of 34–36 tandem epidermal growth factor (EGF)-like repeats, at least two of which are known to be essential for binding to N ligands, and three extracellular cysteine-rich Notch/Lin-12 repeats of unknown function. The intracellular domain consists of six tandem Ankyrin repeats that mediate interactions with cytoplasmic proteins and are sufficient for induction of at least some N-mediated cell fate decisions (Lieber et al. 1993; Rebay et al. 1993; Struhl et al. 1993). Functionally, N receptors are involved in more than one kind of signaling event, including lateral specification of cell fates between groups of equivalent cells and induction of cell fates across fields of nonequivalent cells, as well as for the development and maintenance of sheets of polarized epithelial cells (Fig. 1) (Hartenstein et al. 1992; Artavanis-Tsakonas et al. 1995; Goode et al. 1996a). The involvement of N in many different types of interactions between cells and tissues in a multitude of contexts raises the issue of how specificity is generated from these receptors.

Two structurally similar N ligands, Serrate (Ser) and Delta (Dl), regulate N signaling events. Ser and Dl are members of a family of transmembrane molecules that comprise an amino-terminal, extracellular cysteine-rich DSL motif (named after family members *D*elta, *S*errate, and *L*ag-2), a variable number of extracellular EGF repeats, and a small intracellular domain of variable similarity. Abundant genetic evidence indicates that Dl and Ser interact with N, and Dl or Ser expressing cells have been shown to aggregate with N expressing cells. Overexpression of Dl or Ser during development causes phenotypes resembling gain of N function, whereas loss of Dl or Ser causes phenotypes resembling loss of N function. Combined with the finding that both Dl and Ser act non-cell-autonomously, these experiments demonstrate that these molecules act as N ligands (for review, see Artavanis-Tsakonas et al. 1995).

The differential expression of Dl and Ser suggests a means for producing specific N signals via differential, localized activation of N receptor, a hypothesis supported by their differential expression patterns (Fleming et al. 1990; Thomas et al. 1991). The best studied example is in the wing disk, in which Dl and Ser have complementary and distinct roles in defining the wing margin (Fig. 2) (Doherty et al. 1996). Ser is expressed in cells on the dorsal side of the margin and triggers N in ventral cells, whereas Dl is expressed in cells on the ventral side of the margin and triggers N in dorsal cells. N activation by either ligand induces expression of the margin-specific genes *wingless*, *cut*, and *vestigal* (Couso et al. 1995; Diaz-Benjumea and Cohen 1995; de Celis et al. 1996; Doherty et al. 1996; Neuman and Cohen 1996). In contrast, Ser specifically activates Dl expression in ventral cells, whereas Dl specifically activates Ser in dorsal cells, forming a positive feedback loop (Fig. 2) (Doherty et al. 1996; Panin et al. 1997). If Ser is expressed artificially in dorsal cells, or Dl in ventral cells, they cannot induce expression of the complementary ligand, indicating that Dl and Ser induce tissue-specific responses. Significantly, expression of constitutively active N, which is active independent of the N extracellular domain, can induce Ser in dorsal cells and Dl in ventral cells (Doherty et al. 1996; Panin et al. 1997). This striking result suggests that the tissue-specific effects of Dl and Ser result from direct or parallel modification of N signaling capacity.

In this paper, we summarize recent findings on a novel class of putative secreted factors, Fringe (Frg) and Brainiac (Brn), which may impart specificity to N signaling events by acting upstream or in parallel to N. Brn and Frg are involved in regulating N action in patterning fields of cells, essential for induction of cell fates and for the maintenance of epithelial cell polarity and differentiation, but apparently not for regulating lateral specification decisions. Thus, Brn and Frg appear to be essential for generating qualitatively distinct Notch signals.

Brn REQUIREMENT IN N SIGNALING EVENTS

Phenotypes associated with loss of *brn* and *N* function during oogenesis and early embryogenesis have revealed that Brn is crucial for a subset of processes in which N

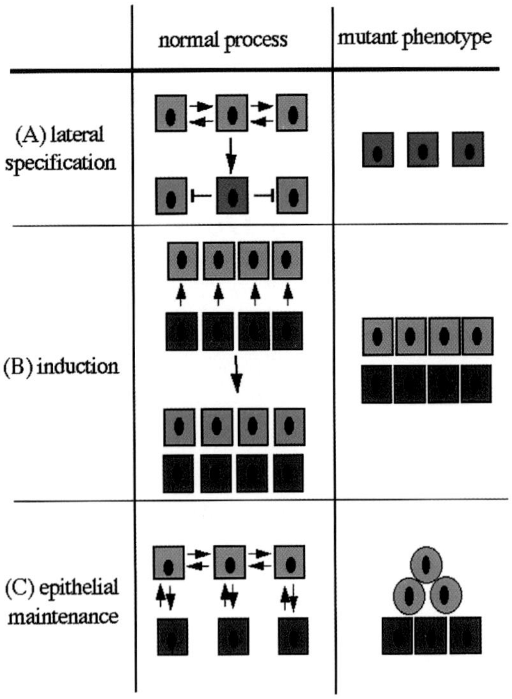

normal process | mutant phenotype
(A) lateral specification
(B) induction
(C) epithelial maintenance

Figure 1. Patterning processes regulated by N receptors. There are three types of patterning processes: lateral specification, induction, and epithelial maintenance, regulated by N receptors. Small arrows indicate D1 or Ser signals that impinge on N receptors. (*A*) Lateral specification. N signals act between individual cells of equal potential (*blue cells*). Through a presumed stochastic process, one cell adopts an alternative fate (*purple cell*). This cell then sends an inhibitory signal to adjacent cells, restricting them from adopting the same fate (*green cells*). In the absence of N (mutant phenotype), all cells adopt the same fate. (*B*) Induction. N signals act between fields of cells of nonequivalent potential. One block of cells (*red*) sends a signal to another (*blue*), switching their fate (*green cells*). Signals may also pass from the receiving cells back to the sending cells, changing the phenotype of the sending cells, as in the wing disk (Doherty et al. 1996; see text). In the absence of N signals (mutant phenotype), cells do not switch their fate. (*C*) Epithelial maintenance. N signals act between cells of nonequivalent potential and perhaps cells of equivalent potential. Without these signals (mutant phenotype), cells fail to develop and/or lose their polarized epithelial morphology and/or cannot complete the morphogenetic transitions essential for the development and maintenance of epithelial sheets. As during the process of induction, cells receiving N signals may send signals back to the sending cells, altering their morphology.

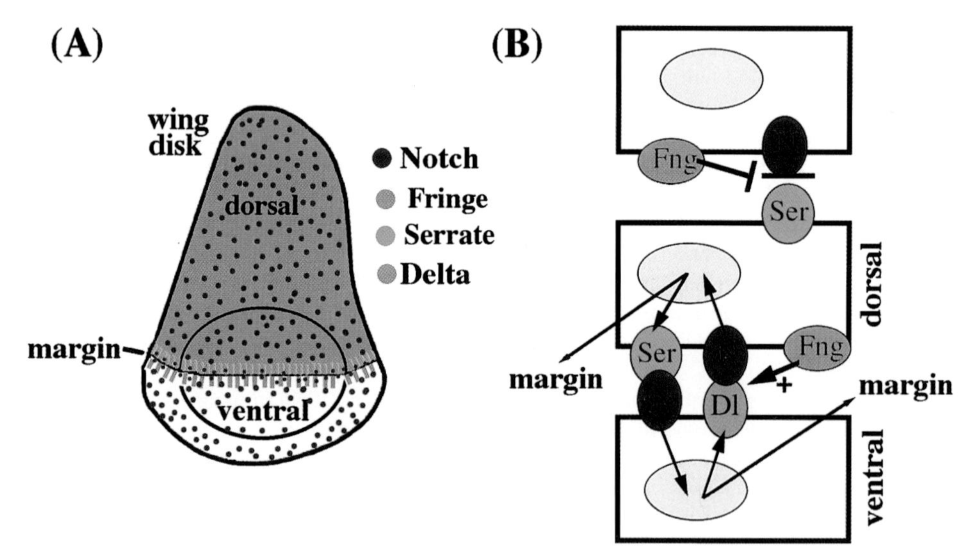

Figure 2. Patterning the wing margin. (*A*) Schematic of a developing wing disk, the tissue that differentiates into an adult wing. The margin is the interface at which dorsal and ventral compartments meet. The distribution of key signaling molecules that pattern the margin is indicated. Ser and Fng are expressed in the dorsal compartment, D1 is expressed in the ventral compartment, and N is expressed throughout the wing disk. Ser expression is induced, and wing margin is formed, wherever Frg expressing cells juxtapose Frg nonexpressing cells, at the margin in wild-type animals, or at ectopic sites under experimentally constructed conditions (Irvine and Wieschaus 1994; Kim et al. 1995). (Adapted from Doherty et al. 1996 and Panin et al. 1997.) (*B*) A model for Frg action. N activation by either D1 or Ser triggers spatially restricted expression of the margin organizing genes *wingless, vestigal*, and *cut* in both dorsal and ventral margin cells (not shown; Couso et al. 1995; Diaz-Benjumea and Cohen 1995; de Celis et al. 1996; Doherty et al. 1996; Neuman and Cohen 1996). Expression of D1 and Ser is maintained at the margin by a positive feedback loop, in which N activation by D1 in dorsal cells induces expression of Ser, and N activation by Ser in ventral cells induces D1 (Doherty et al. 1996; Panin et al. 1997). Frg acts to position this feedback loop apparently through a cell autonomous mechanism by which Frg potentiates activation of N by D1 (arrow), and blocks N activation by Ser (inhibitory symbol; described in detail in Panin et al. 1997). (Adapted from Doherty et al. 1996 and Panin et al. 1997.)

signaling is needed during early development. Brn, like N, is required for the segregation of neural precursor cells from epidermal precursor cells during early embryogenesis, as indicated by the *brn* maternal-effect "neurogenic" phenotype (Perrimon et al. 1989; Goode et al. 1992). This phenotype is similar to zygotic phenotypes associated with *N* mutant animals, as well as other "neurogenic" mutants (Lehmann et al. 1983; Goode et al. 1992). Although the *brn* and *N* embryonic neurogenic phenotypes have not been compared in detail, it is clear that the *brn* phenotype is not as severe as complete loss of *N* function (Goode et al. 1992). This consideration suggests that Brn may participate only in a subset of N signaling processes, and this idea has been substantiated by comparing *brn* and *N* phenotypes in other tissues.

For example, during pupal metamorphosis, N signaling is crucial for lateral specification of epidermal versus sensory organ precursor cell fates (see Fig. 1). Absence of N of D1, or most neurogenic genes within the pupal ectoderm, but not Brn, causes flies to have a bald phenotype because of hypertrophy of neuronal cells at the expense of epidermal cells (for review, see Posakony 1994). Likewise, during egg chamber morphogenesis, absence of N or D1 signals causes too many polar cells to accumulate at the expense of polar flanking cells, resulting from defective Notch signaling in the specification of polar versus polar-flanking cell fates (Fig. 2) (Ruohola et al. 1991). Loss of Brn has no consequence on these decisions (Fig. 2) (Goode et al. 1996a). Brn is involved in a second, apparently separate, N signaling event essential for main-

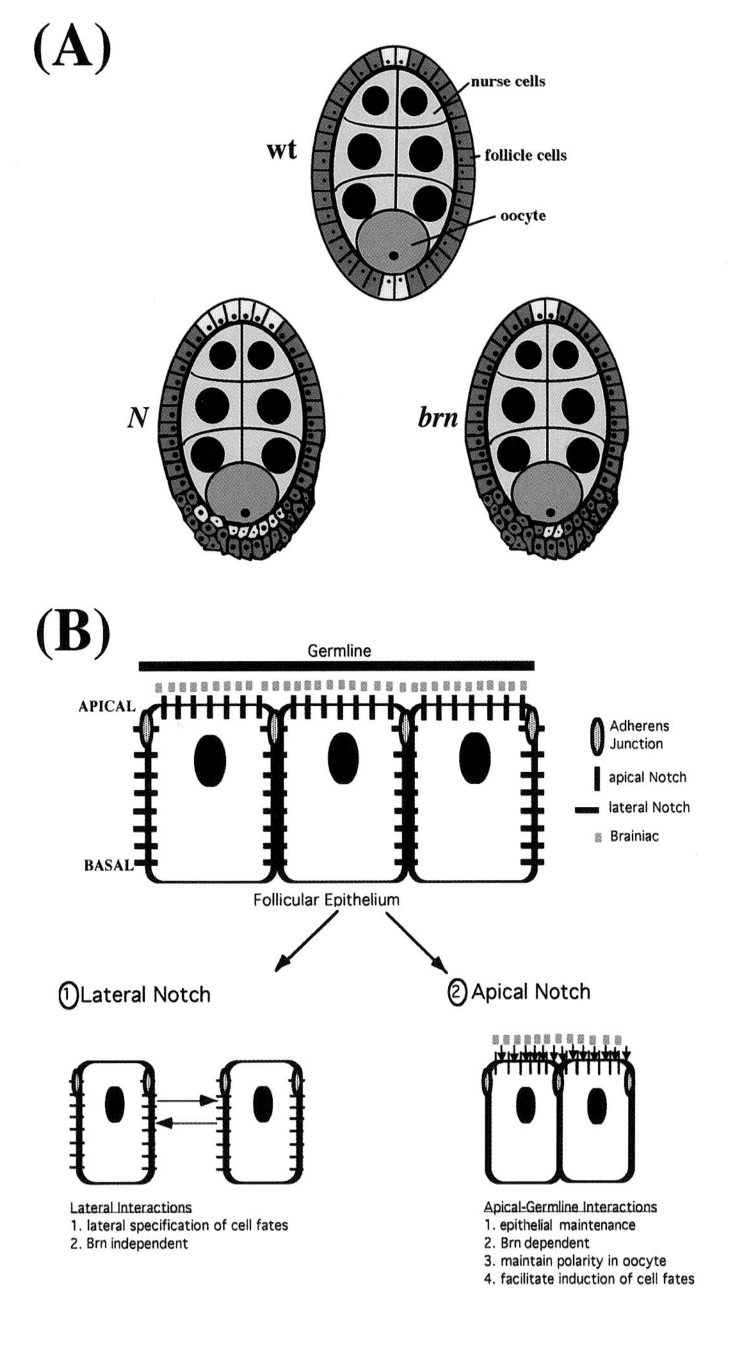

Figure 3. Roles of Brn and N in patterning the follicular epithelium. (*A*) Wild-type egg chambers comprise germ cells and follicle cells. There are two types of germ cells, the oocyte, and 15 nurse cells. Two polar follicle cells are found at the anterior and posterior poles of the egg chamber (*yellow cells*). N is expressed throughout the follicular epithelium. Brn is expressed in germ cells. In *N* mutant egg chambers, but not *brn* mutant egg chambers, too many polar cells segregate. This phenotype led to the model that N is required in a lateral specification process that ensures segregation of two polar cells from neighboring polar-flanking cells (see Fig. 1A; Ruohola et al. 1991). In *N* and *brn* mutant egg chambers, follicle cells lose polarity and accumulate in several layers specifically around the oocyte. This phenotype has led to the model that both N and Brn are required for maintaining the integrity of the follicular epithelium around the oocyte (Goode et al. 1996a). (*B*) A model to account for Brn and N action during oogenesis. The upper portion of the figure shows the distribution of N and presumptive distribution of Brn (Goode et al. 1996a,b). We propose that lateral N has distinct function(s) from apical N (below). (*1*) N expression in lateral membranes is likely to be required for mediating lateral specification of polar versus flanking cells. (*2*) N expression on the apical surface of follicle cells is likely to be essential for epithelial maintenance. The focus of Brn action between germ cells and follicle cells provides an explanation for the specificity of Brn's collaboration with N in epithelial maintenance. (Adapted from Goode et al. 1996.)

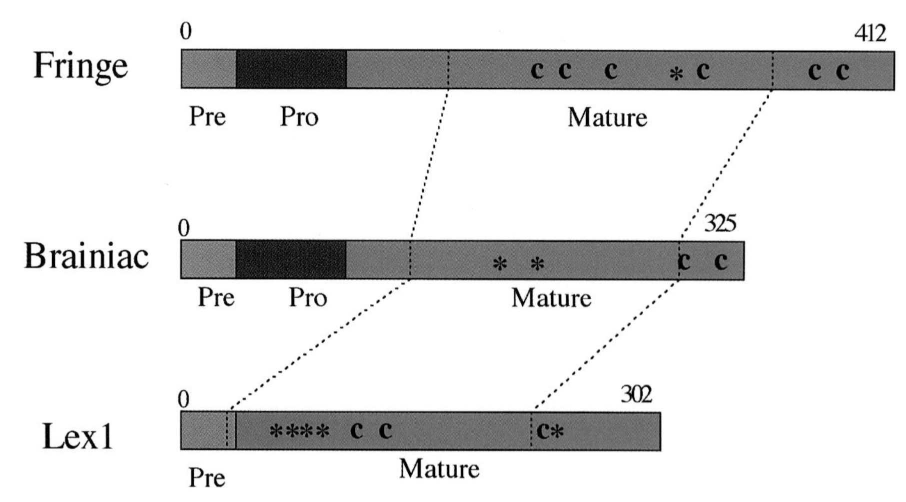

Figure 4. Structure of Frg, Brn, and Lex1 molecules. Frg and Brn share the same presumptive structure, a preregion (the signal peptide) suggestive of secretion, a proregion ending with a dibasic site for proteolytic processing, and a mature region. The central regions of Brn and Frg show limited similarity to Lex1 (*dotted lines*), a glycosyltransferase of the parasitic bacterium *Haemophilus influenzae* (Yuan et al. 1997). Unlike Frg and Brn, Lex1 apparently has no proregion. Neither Frg nor Brn has been shown to be a glycosyltransferase. The presumptive secreted nature of Frg has been shown in vivo (Panin et al. 1997), whereas a *Xenopus* Frg-like molecule, lunatic Frg, has been shown to undergo processing in a manner consistent with the suggested pre-pro structure (Wu et al. 1996). There is no direct evidence yet that Brn is a secreted molecule or that either Frg or Brn has glycosyltransferase activities. Putative Frg, Brn, and Lex1 glycosylation sites are shown (*). A striking difference between Frg and Brn and Lex1 structures is the number of cysteine residues (c). Frg contains six cysteine residues within the proregion of the molecule. Although these cysteines are not arranged in the cysteine knot pattern characteristic of classic cytokines, the pattern of cysteines is conserved in *Xenopus* and human Frg molecules (Wu et al. 1996), suggesting that they are likely to be crucial for determining Frg tertiary structure, perhaps by forming an atypical cysteine knot (Irvine and Wieschaus 1994). Although Brn has only two cysteine residues and Lex1 has only three cysteine residues, we do not believe that this makes it less likely that Brn serves as a signaling factor, because considerable heterogeneity has been described in chemokine cysteine residue patterns (Mackay 1997). Brn shares with Frg and Lex1 a predicted alternating arrangement of α-helices and β-strands suggestive of α/β folding within the central portion of each molecule (*dotted lines*; Yuan et al. 1997).

taining the apical-basal polarity of follicular epithelial cells (Goode et al. 1996a). In *brn* and *N* mutant animals, epithelial cells accumulate in multiple layers around the oocyte (Fig. 2) (Goode et al. 1996a).

How is Brn specificity in N signaling events during oogenesis achieved? Specificity may be accomplished, at least in part, by differential activation of N on the apical surface of follicle cells by Brn, expressed in germ cells (Fig. 3). N is expressed on both apical and lateral follicle cell surfaces. On the lateral surface of follicle cells, N can mediate lateral specification decisions, and on the apical surface, N can participate with germ line Brn in maintaining the follicular epithelium (Fig. 3) (Goode et al. 1996a). Because Brn is not expressed on lateral cell membranes, where the signals responsible for cell fate determination are generated, it apparently has no role in these decisions. How does Brn, a putative secreted factor (Fig. 4) with no similarity to membrane-spanning N ligands D1 and Ser, modulate N signaling? Currently, we can only speculate, but studies of Frg suggest that Brn may participate in N signaling by modulating N-ligand interactions.

Frg REQUIREMENT IN N SIGNALING EVENTS

Like Brn, Frg is essential for viability and is required in many different signaling events during adult metamorphosis to pattern eyes, wings, and legs (Irvine and Wieschaus 1994; Kim et al. 1995). Intensive focus has been placed on the role of Frg in wing development, where it is essential for positioning the D1-Ser/N feedback loop that establishes the wing blade (see Fig. 2) (see introduction; Irvine and Wieschaus 1994; Kim et al. 1995; Panin et al. 1997). Frg is expressed with Ser strictly in dorsal cells (see Fig. 2). Loss of Frg or Ser or D1 causes a "notched" wing phenotype similar to that found in hemizygous *N* mutant animals (Irvine and Wieschaus 1994; Kim et al. 1995). Mosaic analyses have shown that whenever Frg expressing and nonexpressing cells are juxtaposed, margin formation and wing growth occur in a pattern identical to expression of constitutively activated N signals (Irvine and Wieschaus 1994). Frg establishes a sharp wing margin boundary apparently by inhibiting the ability of N to respond to Ser in dorsal cells while potentiating the ability of N to respond to Dl in dorsal cells (see Fig. 2) (Panin et al. 1997). A similar mechanism for Frg action is likely to be crucial for patterning the apical ectodermal ridge, the apparent homologous organizer to the invertebrate wing margin in vertebrates limbs (Concepción et al. 1997; Laufer et al. 1997).

Loss of Frg activity in the wing and thorax apparently has no effect on N decisions in the lateral specification of thoracic or margin bristles, even though Frg signals appear to impinge on the field of cells in which these N decisions are taking place (Irvine and Wieshaus 1994). Thus, Fng is similar to Brn in that it appears to be specifically required in N inductive events that pattern nonequivalent fields of cells but is not required for lateral specification between individual cells within equivalence groups. As noted in the previous section, the specificity of Brn action may be accounted for by the fact that Brn signals are restricted to only one surface on cells in which N is expressed. A similar mode of specificity for Frg action has not been proposed and will have to await a detailed analysis of Frg and N localization and function in the imaginal wing epithelium.

Brn AND Frg SHOW LIMITED SIMILARITY AND BELONG TO A LARGE GENE FAMILY THAT MAY INCLUDE GLYCOSYLTRANSFERASES

Molecular characterization of Brn and Frg revealed that they share the same presumptive structure, a preregion (the signal peptide) suggestive of secretion, a proregion ending with a dibasic site for proteolytic processing, and a mature region that will be functionally active after cleavage from the precursor protein (see Fig. 2). Biochemical experiments have confirmed this structure for a *Xenopus* Frg homolog (Wu et al. 1996). The action of *Drosophila* Brn and Frg on cells adjacent to those in which they are expressed (described above) is consistent with their putative secreted structure.

Similarity between Brn and Frg pioneer proteins is fairly weak and is not detected using standard search algorithms, but emerges when conservation patterns gleaned by comparing Brn and Frg-like signaling molecules from *Caenorhabditis elegans* to humans are used to execute very sensitive motif and profile searches (Bork and Gibson 1996; Yuan et al. 1997). The Frg and Brn alignments predicted from these comparisons reveals an alternating arrangement of α-helices and β-strands suggestive of an α/β folding type for the central portion of each subfamily, similar to that of prokaryotic and eukaryotic glycosyltransferases (Yuan et al. 1997). Within a region spanning 150–180 amino acids, Brn and Frg show greatest similarity to Lex1 (Fig. 4), a glycosyltransferase found in the parasitic bacterium *Haemophilus influenzae* that is essential for the biosynthesis of its extracellular lipopolysaccharides (Yuan et al. 1997). The most conserved regions are also the hallmarks of the putative glycosyltransferase superfamily.

In contrast to Brn and Frg, glycosyltransferases do not appear to have a proregion that would serve as a target for proteolytic processing (Fig. 4). The proregion may have been added to Brn and Egh during evolution to impart an additional target for developmental regulation, sensitive to the action of proteolytic cascades. Brn or Frg may be glycosyltransferases that modulate N-ligand interactions by regulating accessibility of ligand to receptor or trigger N (or another receptor) activity by altering conformation of the receptor via carbohydrate modification, but there is currently no direct evidence that Brn or Frg have glycosyltransferase activity. Alternatively, Brn and Frg may be descendants of glycosyltransferases that have lost enzymatic activity and trigger receptor action in a manner similar to that of more classic cytokines.

Egh MAY FACILITATE Brn ACTIVITY

Although Brn is expressed in all germ cells throughout oogenesis, *brn* phenotypes are manifested specifically around the oocyte during mid-oogenesis, suggesting a high degree of spatiotemporal specificity to Brn signaling action (Goode et al. 1996a). Many hypotheses can be created to explain this anomaly. Differential expression of Egghead (Egh), a neurogenic mutant with phenotypes indistinguishable from Brn mutant phenotypes, offers a clue (Goode et al. 1996a). Egh becomes expressed in the oocyte precisely at the time and place that *brn* and *egh* epithelial defects become manifest (Fig. 5). Egh is a putative membrane protein of 457 amino acids, with a homo-

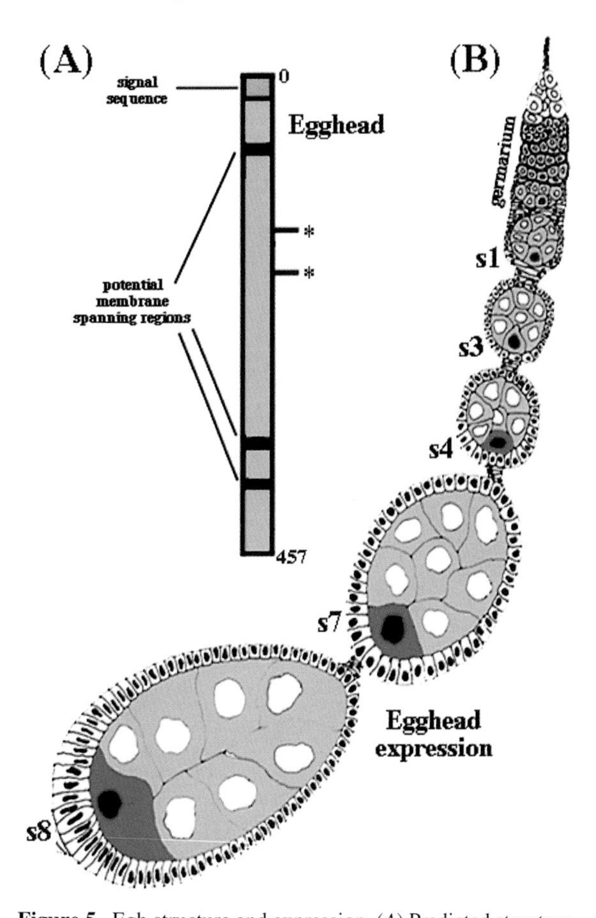

Figure 5. Egh structure and expression. (*A*) Predicted structure of the Egh protein (Goode et al. 1996a). Database searches indicate that Egh does not show similarity to proteins of known function, but Egh homologs have been identified in animals from *C. elegans* to humans. Egh has a putative signal sequence at the amino terminus and several putative transmembrane spanning regions. Egh thus appears to be a multipass membrane protein. Putative glycosylation sites (*) are consistent with this hypothesis. (*B*) Expression of Egh mRNA during oogenesis. Egh is expressed in all germ cells starting very early when the egg chamber is born in the germarium. At stage 4 (s4), Egh becomes differentially localized in the oocyte, corresponding to the time that epithelial defects first appear around the oocyte in animals harboring *brn* and *egh* mutant germ cells (Goode et al. 1996a).

log identified in *C. elegans* (Goode et al. 1996a). Egh is likely to participate directly in Brn signaling processes since *brn, egh* double-mutant animals do not display phenotypes any more severe than either mutant alone (Goode et al. 1996a). Egh may directly or indirectly bind to secreted Brn, limiting Brn diffusion, and/or increasing Brn activity, and/or modifying the structure of Brn. Additional factors similar to Egh have not been identified in the Frg pathway.

Brn AND Egh ARE REQUIRED FOR DORSAL VENTRAL PATTERNING

In addition to maintaining the apical-basal polarity of follicular epithelial cells, Brn and Egh are essential for dorsoventral (D/V) patterning during oogenesis. Brn and Egh are expressed in the oocyte during the time that D/V polarity of the egg is established, and eggs laid by both *brn* and *egh* mutant females have D/V defects, as indicated by a shift of morphological and molecular markers from dorsolateral to more dorsal positions on the egg shell and follicular epithelium (Goode et al. 1992, 1996a,b; Goode 1994). Temperature-shift experiments indicate that the requirement for Brn in D/V patterning is temporally separable from the requirement for Brn in maintaining the apical-basal polarity of the follicular epithelial cells (Goode et al. 1992). As described below, the involvement of Brn in D/V patterning is likely to occur via modulation of EGF receptor (Egfr) signals, the primary signals responsible for D/V patterning during oogenesis.

Grk, a transforming growth factor-α (TGF-α) homolog, is expressed on the dorsal side of the oocyte and triggers the *Drosophila* Egfr in dorsal follicle cells. In the absence of TGF-α or Egfr signals, all follicle cells assume a ventral cell fate, expression of dorsal follicle cell markers is completely abolished, and the oocyte also loses D/V polarity as indicated by the transformation of dorsal embryonic cell fates to more ventral cell fates in eggs laid by *grk* and Egfr mutant females (for review, see Ray and Schüpbach 1996). Brn phenotypes are subtle by comparison and have no consequence on D/V patterning in the early embryo.

To address the possibility that Brn acts by modulating Egfr signals, we asked whether TGF-α or Egfr signals are sensitive to reduction in Brn signals (Fig. 6). We approached this genetically, by looking for dominant and synergistic interactions between *brn* and *grk* or *Egfr*. We found defects in the D/V pattern of the eggshell in *brn*/+; *grk*/+ or *brn*/+; *Egfr*/+ doubly heterozygous animals and synergistic phenotypes in *brn*; *grk* and *brn*; *Egfr* and *N*; *grk* and *N*; *Egfr* double-mutant animals (Goode et al. 1992, 1996a,b; Goode 1994). These results strongly suggest that Brn signals significantly overlap with Grk signals. It will be of interest to examine further the interaction between Egfr and N signals in order to characterize at what levels these signaling pathways interact (Fig. 6).

SUMMARY AND FUTURE DIRECTIONS

In this review, we have made a simple comparison of functional parallels and molecular similarities between Brn and Frg. Analysis of Brn function in egg chambers suggests that Brn specificity in patterning fields of cells versus lateral specification is achieved by restricting Brn cooperation with N to the apical surface of follicle cells. Misexpression of Brn in the follicular epithelium, where it would have access to the N signaling process on the lateral surface of follicle cells, might be expected to interfere with N signaling processes, but this has not been

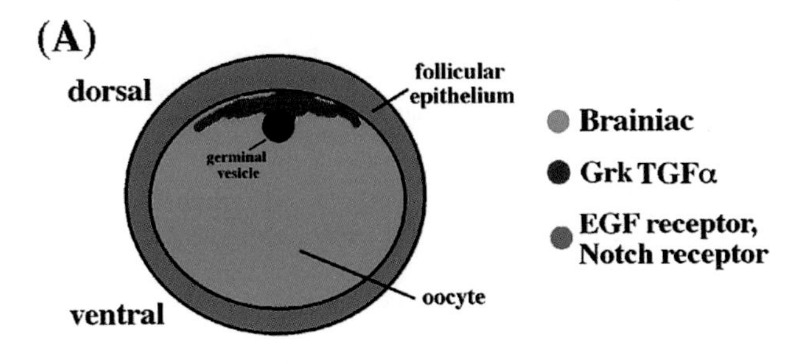

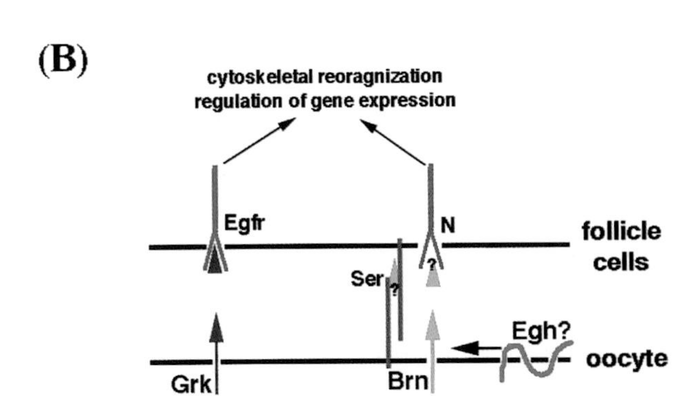

Figure 6. Expression and cooperativity of Brn and Grk in D/V patterning. (*A*) mRNA expression of Brn, Grk, Egfr, and protein expression of N in and around the oocyte from stages 8–9 of oogenesis (Xu et al. 1992; Goode et al. 1996b; Ray and Schüpbach 1996). Grk TGF-α is expressed on the dorsal side of the oocyte, where it triggers EGFr in dorsal follicle cells. Brn is expressed throughout the oocyte and may interact with the N signaling machinery in all follicle cells or with a yet unknown signaling system in a parallel pathway. (*B*) The relationship between Brn/N signals and Grk/Egfr signals is unclear, but an abundance of genetic evidence indicates that they interact at some level within follicle cells (see text; Goode et al. 1992, 1996a,b). Dl does not appear to be expressed in the oocyte or oocyte-associated follicle cells throughout the time that D/V pattern is established (Bender et al. 1993). The expression of Ser (?) has not been reported. It is not clear whether putative secreted Brn interacts with Ser, N, or some other extracellular factor(s) (?), but our current evidence indicates that Brn somehow regulates N signals. On the basis of results from genetic interaction experiments (see text), it seems likely that signals generated by N overlap with those generated by Egfr, at either a common cytoskeletal structure, a nuclear factor, or some component of the signal transduction machinery responsible for altering gene expression and rearranging the cytoskeleton.

tested. Furthermore, it will be of interest to test, as we propose for Brn, whether the specificity of Frg in patterning the wing disk relies on restricting Frg activity to a particular epithelial cell surface. We currently have no molecular knowledge of how Brn participates in the N signaling process. The apparent requirement of Frg to inhibit activation of N by Ser, and to potentiate activation of N by Dl, suggests avenues for investigation.

If conclusions drawn from analysis of Brn function during oogenesis can be applied to Brn function in early embryos, then we might expect to find that Brn is required for maintaining the epithelial integrity of the neurogenic ectoderm during neuroblast segregation but not for lateral specification between cells within the ectoderm (Goode et al. 1996a,b). Interestingly, N expression in mesodermal cells dramatically rescues the epidermal phenotype of *N* mutant embryos, which may imply that in addition to being required between "equivalent" ectodermal cells to ensure proper segregation of neural precursor cells, N acts between distinct tissue layers (Baker and Schubiger 1996). We find this hypothesis attractive because it suggests an alternative means by which Brn might be involved in regulating the segregation of neural precursor cells within the neurogenic ectoderm and is consistent with the tenet that Frg and Brn are specifically involved in N processes of induction and/or maintenance between nonequivalent fields of cells.

Brn and Frg are the first secreted factors to be implicated in N signaling processes. We would like to know whether Brn and Frg are diffusible factors and if their secretion is crucial for establishing their activity, perhaps by analyzing the functional consequences of expressing forms of the molecules that remain tethered to the membrane. Further understanding of Brn and Frg in N induction and maintenance processes will also depend on determining whether they act as glycosyltransferases or by a distinct mechanism to influence N-ligand interactions. It will also be important to understand how Brn and Frg differ in function, and to what degree, if any, they can substitute for each others' function. The proposal that N serves as a multifunctional receptor by using its multitude of EGF repeats to bind distinct ligands (Rebay et al. 1991) suggests the possibility that Brn and Frg might bind N either directly or through a distinct protein. Alternatively, Brn and Frg might modify the glycosylation state of N or another receptor to influence their ability to be activated by distinct ligands. Further analysis of Egh will be essential to demonstrate whether this protein is part of the Brn signaling process and whether Egh or an Egh homolog acts in the Frg signaling pathway.

ACKNOWLEDGMENTS

We thank Kevin Fitzgerald, Ben Vollrath, and Cliff Tabin for helpful comments. We thank Kenith Irvine for helpful discussion and Tony Mahowald for his interest and support. S.G. was funded by a National Institutes of Health postdoctoral fellowship (GM-17511-02) and is a Leukemia Society of America Special Fellow (3232-98). N.P. is an investigator of the Howard Hughes Medical Institute.

REFERENCES

Artavanis-Tsakonas S., Matsuno K., and Fortini M.E. 1995. Notch signaling. *Science* **268:** 225.

Baker R. and Schubiger G. 1996. Autonomous and nonautonomous Notch functions for embryonic muscle and epidermis development in *Drosophila. Development* **122:** 617.

Bender L.B., Kooh P.J., and Muskavitch M.A.T. 1993. Complex function and expression of *Delta* during *Drosophila* oogenesis. *Genetics* **133:** 967.

Bork P. and Gibson T. 1996. Applying motif and profile searches. *Methods Enzymol.* **266:** 162.

Concepción R.-E., Schwabe J.W.R., De La Pena J., Foys B., Eshelman B., and Izpisúa-Belmonte J.C. 1997. *Radical fringe* positions the apical ectodermal ridge at the dorsoventral boundary of the vertebrate limb. *Nature* **386:** 360.

Couso J.P., Knust E., and Martinez Arias A. 1995. *Serrate* and *wingless* cooperate to induce *vestigial* gene expression and wing formation in *Drosophila. Curr. Biol.* **5:** 1437.

de Celis J.F., Garcia-Bellido A., and Bray S.J. 1996. Activation and function of Notch at the dorsal-ventral boundary of the wing imaginal disc. *Development* **122:** 359.

Diaz-Benjumea F.J. and Cohen S.M. 1995. Serrate signals through Notch to establish a Wingless-dependent organizer at the dorsal/ventral compartment boundary of the *Drosophila* wing. *Development* **121:** 4215.

Doherty D., Feger G., Younger-Shepherd S., Jan L.Y., and Jan Y.N. 1996. Delta is a ventral to dorsal signal complementary to Serrate, another Notch ligand, in *Drosophila* wing formation. *Genes Dev.* **10:** 421.

Ellisen L.W., Bird J., West D.C., Soreng A.L., Reynolds T.C., Smith S.D., and Sklar J. 1991. *TAN-1*, the human homolog of the *Drosophila Notch* gene, is broken by chromosomal translocations in T lymphoblastic neoplasms. *Cell* **66:** 649.

Fleming R.J., Scottgale T.N., Diederich R.J., and Artavanis-Tsakonas S. 1990. The gene *Serrate* encodes a putative EGF-like transmembrane protein essential for proper ectodermal development in *Drosophila melanogaster. Genes. Dev.* **4:** 2188.

Goode S. 1994. "*brainiac* encodes a novel, putative secreted protein that cooperates with EGF RTK for the ontogenesis and polarization of the follicular epithelium of *Drosophila melanogaster.*" Ph. D. thesis. University of Chicago, Illinois.

Goode S., Wright D., and Mahowald A.P. 1992. The neurogenic locus *brainiac* cooperates with the *Drosophila* EGF receptor to establish the ovarian follicle and to determine its dorsal-ventral polarity. *Development* **116:** 177.

Goode S., Melnick M., Chou T.-B., and Perrimon N. 1996a. The neurogenic genes *egghead* and *brainiac* define a novel signaling pathway essential for epithelial morphogenesis during *Drosophila* oogenesis. *Development* **122:** 3863.

Goode S., Morgan M., Liang Y.-P., and Mahowald A.P. 1996b. *brainiac* encodes a novel, putative secreted protein that cooperates with *grk* TGFα in the genesis of the follicular epithelium. *Dev. Biol.* **178:** 35.

Hartenstein A.Y., Rugendorf A., Tepass U., and Hartenstein V. 1992. The function of the neurogenic genes during epithelial development in the *Drosophila* embryo. *Development* **116:** 1203–1220.

Irvine K.D. and Wieschaus E. 1994. *fringe*, a boundary-specific signaling molecule, mediates interactions between dorsal and ventral cells during *Drosophila* wing development. *Cell* **79:** 595.

Joutel A., Corpechot C., Ducros A., Vahedi K., Chabriat P., Mouton P., Alamowitch V., Domenga V., Cecillion M., Marechal E., Maciazek J., Vayssiere C., Cruaud C., Cabanis E.A., Ruchoux M.M., Weissenbach J., Bach J.F., Bousser M.G., and Tournier-Lasserve E. 1996. *Notch3* mutations in CADASIL, a hereditary adult-onset condition causing stroke and dementia. *Nature* **383:** 707.

Kim J., Irvine K.D., and Carroll S.B. 1995. Cell interactions and inductive signals at the dorsal/ventral boundary of the developing *Drosophila* wing. *Cell* **82:** 795.

Laufer E., Dahn R., Orozco O.E., Yeo C.-Y., Pisenti J., Henrique D., Abbot U.K., Fallon J.F., and Tabin C. 1997. Expression of *Radical fringe* in limb-bud ectoderm regulates apical ectodermal ridge formation. *Nature* **386:** 366.

Lehmann R., Jimeniz F., Dietrich U., and Campos-Ortega J.A. 1983. On the phenotype and development of mutants of early neurogenesis in *Drosophila melanogaster. Roux's Arch. Dev. Biol.* **192:** 62.

Lieber T., Kidd S., Alcamo E., Corbin V., and Young M.W. 1993. Antineurogenic phenotypes induced by truncated Notch proteins indicate a role in signal transduction and may point to a novel function for Notch in nuclei. *Genes Dev.* **7:** 1949.

Mackay, C.R. 1997. Chemokines: What chemokine is that? *Curr. Biol.* **7:** 384–386.

Neumann C.J. and Cohen S.M. 1996. A hierarchy of cross-regulation involving *Notch, wingless, vestigal*, and *cut* organizes the dorsal/ventral axis of the *Drosophila* wing. *Development* **122:** 3477.

Panin V.M., Papayannopoulos V., Wilson R., and Irvine K.D. 1997. Fringe modulates Notch-ligand interactions. *Nature* **387:** 908.

Perrimon N., Engstrom L., and Mahowald A.P. 1989. Zygotic lethals with specific maternal effect phenotypes in *Drosophila melanogaster*. I. Loci on the X chromosome. *Genetics* **121:** 333.

Posakony, J.W. 1994. Nature versus nurture: Asymmetric cell divisions in *Drosophila* bristle development. *Cell* **76:** 415.

Ray R.P. and Schüpbach T. 1996. Intercellular signaling and the polarization of body axes during *Drosophila* oogenesis. *Genes Dev.* **10:** 1711.

Rebay I., Fehon R.G., and Artivanis-Tsakonas S. 1993. Specific truncations of *Drosophila* Notch define dominant activated and dominant negative forms of the receptor. *Cell* **74:** 319.

Rebay I., Fleming R.J., Fehon R.G., Cherbas L., Cherbas P., and Artavanis-Tsakonas S. 1991. Specific EGF repeats of Notch mediate interactions with Delta and Serrate: Implications for Notch as a multifunctional receptor. *Cell* **67:** 687.

Robbins J., Blondel B.J., Gallahan D., and Callahan, R. 1992. Mouse mammary tumor gene *int-3:* A member of the *Notch* gene family transforms mammary epithelial cells. *J. Virol.* **66:** 2594.

Ruohola H., Bremer K.A., Baker D., Swedlow J.R., Jan L.Y., and Jan Y.N. 1991. Role of neurogenic genes in establishment of follicle cell fate and oocyte polarity during oogenesis in *Drosophila. Cell* **66:** 1.

Struhl G., Fitzgerald K., and Greenwald I. 1993. Intrinsic activity of the lin-12 and Notch intracellular domains in vivo. *Cell* **74:** 331.

Thomas U., Speicher S.A., and Knust E. 1991. The *Drosophila* gene *Serrate* encodes an EGF-like transmembrane protein with a complex expression pattern in embryos and wing discs. *Development* **111:** 749.

Wu J.Y., Wen L., Zhang W.-J., and Rao Y. 1996. The secreted product of *Xenopus* gene *lunatic Fringe*, a vertebrate signaling molecule. *Science* **273:** 355.

Xu T., Caron L.A., Fehon R.G., and Artavanis-Tsakonas S. 1992. The involvement of the *Notch* locus in *Drosophila* oogenesis. *Development* **115:** 913.

Yuan Y.P., Schultz J., Mlodzik M., and Bork P. 1997. Secreted Fringe-like signaling molecules may be glycosyltransferases. *Cell* **88:** 9.

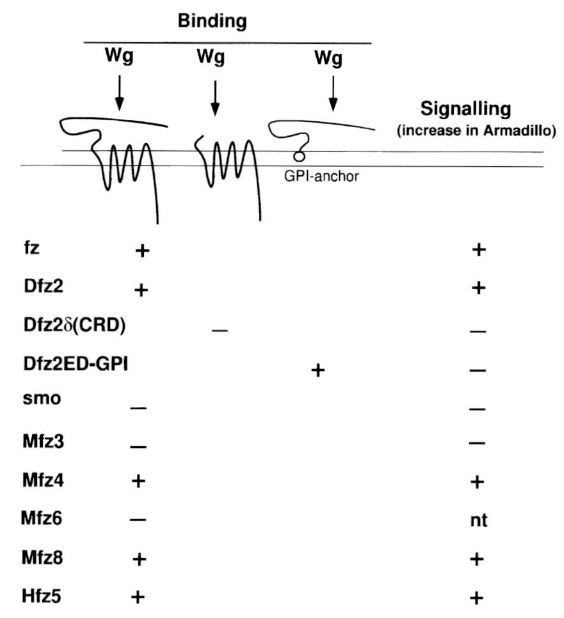

Figure 2. Wg protein binding to, and signal transduction by, several members of the frizzled gene family and mutant forms. Wg binds to full-length frizzled, Defrizzled-2 (D*fz*2), the GPI-linked extracellular domain of D*fz*2 (Bhanot et al. 1996), and several mammalian *fz* genes, as indicated. Binding to the full-length forms leads to signaling in S2 cells transfected with these Fz forms, as the levels of the Arm protein are increased by Wg in a receptor-dependent manner (*right side*). Fz variants that do not bind Wg (smoothened, D*fz*2 without the extracellular domain, and mouse Fz3 and mouse Fz6) do not elevate Arm either. The GPI-linked extracellular domain of D*fz*2 binds Wg but is unable to transduce the signal.

protein on their cell surface (Bhanot et al. 1996). Transfection of cells with D*fz*2 constructs lacking either the extracellular or intracellular domain of the protein demonstrated that the extracellular domain was required for binding (Fig. 2). Although a direct interaction between the Wg protein and D*fz*2 is still lacking, the data suggest that D*fz*2 can bind to and transduce the Wg signal. The Smo protein, when tested in the same assay, does not bind Wg, but of the identified mouse and human Fz proteins (Y. Wang et al. 1996, 1997b), several are also positive in the Wg binding assay described above (see Fig. 2; *smo* clone kindly provided by Marcel Van Den Heuvel and Phil Ingham). There is a correlation between the sequence distances between these various Fz family members and their capacity to bind Wg, because the receptors that are more distantly related to D*fz*2, including Smo, do not bind (Fig. 3). Presumably, these Fz family members bind other Wnt proteins.

We have also tested for signal transduction, by transfection of Fz constructs into S2 cells and measuring the Arm protein concentration before and after adding soluble Wg protein. Without exception, the Fz proteins that were able to bind Wg did also transduce the Wg signal to Arm (see Fig. 2). The original *fz* gene can confer Wg responsiveness to nonresponding cultured cells, as well as Wg binding. One possibility is that *fz* acts redundantly with D*fz*2 or other as yet unidentified frizzled proteins to transduce the *wg* signal in vivo. However, it does appear likely that some *Drosophila Wnt* gene is the ligand for the polarity function of *fz*.

Other Evidence for Wnt-frizzled Interactions

The results mentioned above, and the fact that there are no D*fz*2 mutants, make it uncertain whether D*fz*2 is the physiological *wg* receptor. We have obtained some evidence that D*fz*2 can act as a receptor for Wg by expressing the extracellular domain of the protein as a GPI-linked cell surface protein in *Drosophila* imaginal discs, using the GAL-4/UAS system. The wings of the resulting flies have marked defects in the margin (Fig. 4), known to be specified by Wg, and in other structures such as the eyes and the legs (data not shown). All the phenotypes observed are similar to loss of Wg function, which suggests that the extracellular domain binds Wg and inhibits its function.

The proposal of Fz molecules functioning as Wnt receptors was strengthened by recent work from nematodes and frogs. In the nematode *C. elegans*, there are two genes involved in asymmetric cell divisions of certain cell lineages. One of these genes (*lin-44*) encodes a *Wnt*

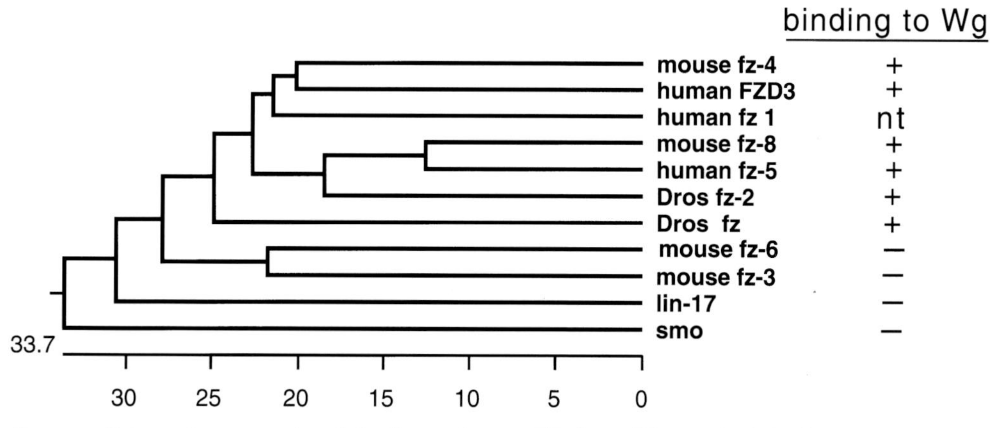

Figure 3. Sequence distances between various *fz* family members and binding to Wg protein. A dendrogram (produced by the DNA Star Software suite, Lasergene Inc.) shows the evolutionary distance between Fz protein sequences. Fz proteins closely related to D*fz*2 bind Wg, but more distantly related proteins (mfz3, mfz6, lin-17 and smo) do not.

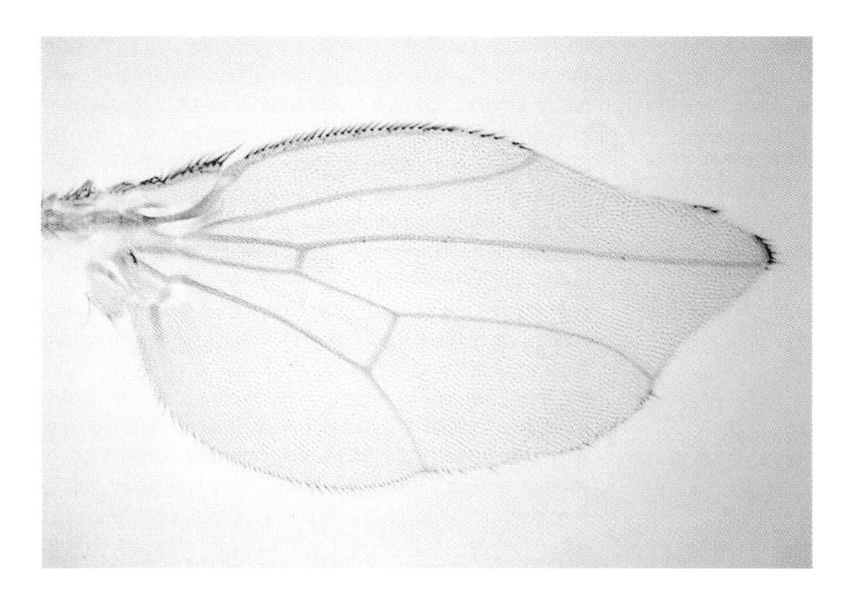

Figure 4. Effects of overexpression of the GPI-linked D*fz*2 extracellular domain in the *Drosophila* wing. The GPI-linked D*fz*2 extracellular domain was expressed using the Gal-4/UAS system. This results in loss of the wing margin, which is also a phenotype caused by loss of *wg* function (Couso et al. 1993).

gene (Herman et al. 1995), the other, *lin-17*, encodes a *frizzled* family member (Sawa et al. 1996). The phenotypes of the two mutants are similar but not identical, but the biochemical evidence summarized above suggests a ligand-receptor relationship for these genes as well.

In *Xenopus* embryos, injection of Wnt RNA has long been known to induce a secondary body axis. Some *Wnt* genes lack this activity; it has been suggested that they function through a different signaling mechanism (Christian et al. 1991). If these Wnts are coinjected with the appropriate frizzled gene's RNA, the axis duplication effect is restored (He et al. 1997). This suggests that the only difference between the two classes of frog *Wnt* genes may be their affinity for the endogenous Fz receptor. More recently, secreted forms of Fz proteins have been found in *Xenopus* and in other vertebrates (Finch et al. 1997; Leyns et al. 1997; Rattner et al. 1997; S.W. Wang et al. 1997). These proteins, called FRP or Frzb, are made by the Spemann organizer and counteract the ventralizing activity of Xwnt-8 during frog embryogenesis (Leyns et al. 1997; S.W. Wang et al. 1997).

Function of Dishevelled in Wg Signaling

On the basis of genetic analysis of the *wg* signal transduction pathway, *wg* activates *dsh*, which in turn inhibits *zw3*. *dsh* encodes a cytoplasmic protein (Dsh) with no known biochemical function and little homology with other proteins (Klingensmith et al. 1994; Theisen et al. 1994). Several *dsh* genes (Dvl-1, 2, and 3) have been cloned from the mouse by homology (Sussman et al. 1994). Sequence comparison of *dsh* genes reveals regions of high homology in the amino terminus and in the central domain, whereas the carboxyl terminus is divergent (Fig. 5). The central region of Dsh contains a PDZ domain (Fig. 5) (Ponting 1995). Structural analysis has shown that a carboxy-terminal four-residue motif (X-Thr/Ser-X-Val) binds to the PDZ domain of a number of

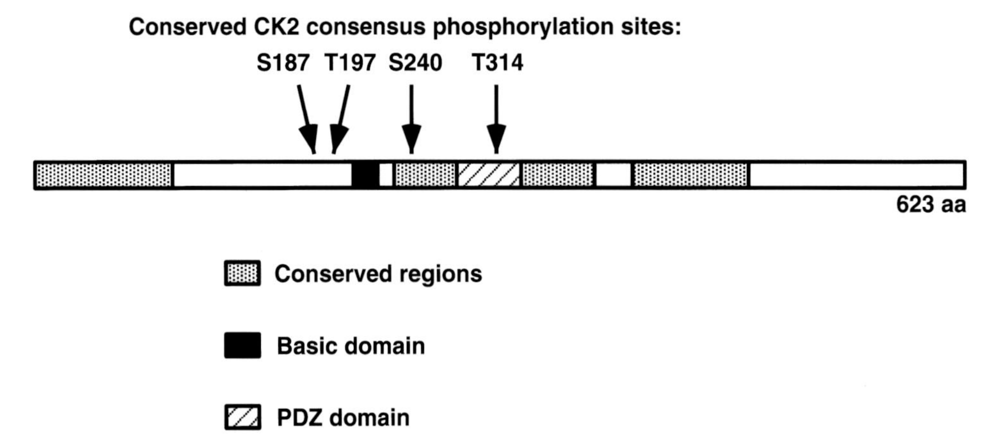

Figure 5. Schematic structure of the Dishevelled protein, with CK2 consensus substrate sites (Willert et al. 1997). Dsh has various highly conserved domains, including a DEP and a PDZ domain. CK2 sites are found at various positions, including one in the PDZ domain.

proteins (Doyle et al. 1996). However, no ligand for the PDZ domain of Dsh has yet been identified. Because several proteins have a similar carboxy-terminal motif as well, we have tested whether Dsh binds to D*fz*2, but these assays have failed to provide evidence that this is the case (data not shown).

The carboxyl side of Dsh contains a recently described domain referred to as the DEP domain (Ponting 1996). DEP domains are found in a variety of proteins many of which participate in G-protein signaling. Directly amino-terminal to the PDZ of Dsh is a conserved stretch of basic residues that, in many other proteins, serves as a nuclear localization signal (Fig. 5). However, cell biological studies have shown that Dsh is localized predominantly in the cytoplasm of the cell and not in the nucleus (Yanagawa et al. 1995). The functions of the PDZ, DEP, and basic domains in Dsh are unknown.

To test for possible biochemical functions of Dsh, we examined the protein in cells stimulated by Wg. Stimulation of the wing imaginal disc cell line clone 8 with Wg-conditioned medium leads to the accumulation of Arm and the hyperphosphorylation of Dsh (Yanagawa et al. 1995). This suggested that the hyperphosphorylated form of Dsh is the active form of Dsh. We were therefore interested to see that a kinase activity is associated with Dsh protein in cultured cells and in embryos and that this kinase phosphorylates *Dsh* in vitro. When Dsh protein is immunoprecipitated from those cells and the immune complex is subjected to an in vitro kinase reaction, it becomes phosphorylated (Willert et al. 1997). Phospho amino acid analysis of Dsh showed that most of the phosphorylation occurs on serine, some on threonine, and none on tyrosine.

We used the specific binding of the kinase to Dsh as a first step in the purification of this enzyme. A soluble protein extract from S2 *dsh* cells was applied to an anti-Dsh antibody affinity column and fractions were assayed for kinase activity toward a Dsh fusion protein. This resulted in two bands, of 38 and 28 kD, which were isolated and subjected to peptide sequencing. One tryptic peptide was sequenced, and the amino acid sequence revealed a perfect match with the α-subunit of *Drosophila melanogaster* caseine kinase 2 (CK2). We also found CK2 to coimmunoprecipitate with Dsh in lysates prepared from *Drosophila* embryos, demonstrating that CK2 is associated with Dsh in the *Drosophila* embryo (Willert et al. 1997). Dsh has a number of CK2 consensus phosphorylation sites (Fig. 5), and we are in the process of inactivating these sites to examine possible consequences for Dsh activity.

CK2 Is Associated with Dsh in Cells Overexpressing D*fz*2

Interestingly, we found a difference in CK2 binding to Dsh when we compared normal and D*fz*2-transfected S2 cells (Willert et al. 1997). In parental S2 cells, which do not express D*fz*2, Dsh migrates as a single band of 70 kD. Uninduced S2 D*fz*2 cells contain low, but detectable, levels of D*fz*2 protein and display a series of Dsh bands that migrate more slowly than Dsh in S2 cells. Clone-8 cells, which normally express D*fz*2, also contain a series of differently phosphorylated forms (data not shown). Unlike overexpression of Dsh, which leads to the accumulation of Arm protein, overexpression of D*fz*2 is not sufficient to increase Arm. Apparently, the phosphorylation of Dsh caused by D*fz*2 overexpression is not sufficient for the transduction of the signal to Arm. These findings point to complicated interactions between the various components of Wg signaling, to be addressed by further genetic and biochemical studies.

ACKNOWLEDGMENTS

These studies were supported by the Howard Hughes Medical Institute, of which R.N. is an investigator. Part of the work was also supported by a grant (DAMD17-94-J-4351) to R.N. from the USAMRMC.

REFERENCES

Alcedo J., Ayzenzon M., Vonohlen T., Noll M., and Hooper J.E. 1996. The *Drosophila smoothened* gene encodes a seven-pass membrane protein, a putative receptor for the Hedgehog signal. *Cell* **86:** 221.

Behrens J., Von Kries J.P., Kuhl M., Bruhn L., Wedlich D., Grosschedl R., and Birchmeier W. 1996. Functional interaction of β-catenin with the transcription factor LEF-1. *Nature* **382:** 638.

Bhanot P., Brink M., Harryman Samos C., Hsieh J.C., Wang Y.S., Macke J.P., Andrew D., Nathans J., and Nusse R. 1996. A new member of the frizzled family from *Drosophila* functions as a Wingless receptor. *Nature* **382:** 225.

Brunner E., Peter O., Schweizer L., and Basler K. 1997. *pangolin* encodes a Lef-1 homologue that acts downstream of Armadillo to transduce the Wingless signal in *Drosophila*. *Nature* **385:** 829.

Christian J.L., McMahon J.A., McMahon A.P. and Moon R.T. 1991. *Xwnt-8*, a *Xenopus Wnt-1/int-1*-related gene responsive to mesoderm-inducing growth factors, may play a role in ventral mesodermal patterning during embyrogenesis. *Development* **111:** 1045.

Clevers H.C. and Grosschedl R. 1996. Transcriptional control of lymphoid development: Lessons from gene targeting. *Immunol. Today* **17:** 336.

Cook D., Fry M.J., Hughes K., Sumathipala R., Woodgett J.R., and Dale T.C. 1996. Wingless inactivates glycogen synthase kinase-3 via an intracellular signalling pathway which involves a protein kinase c. *EMBO J.* **15:** 4526.

Couso J.P., Bate M., and Martínez Arias A. 1993. A *wingless*-dependent polar coordinate system in *Drosophila* imaginal discs. *Science* **259:** 484.

Doyle D.A., Lee A., Lewis J., Kim E., Sheng M., and MacKinnon R. 1996. Crystal structure of a complexed and peptide-free membrane protein-binding domain: Molecular basis of peptide recognition by PDZ. *Cell* **85:** 1067.

Finch P.W., He X., Kelley M.J., Uren A., Schaudies R.P., Popescu N.C., Rudikoff S., Aaronson S.A., Varmus H.E., and Rubin J.S. 1997. Purification and molecular cloning of a secreted, frizzled-related antagonist of wnt action. *Proc. Natl. Acad. Sci.* **94:** 6770.

He X., SaintJeannet J.P., Wang Y.S., Nathans J., Dawid I., and Varmus H. 1997. A member of the frizzled protein family mediating axis induction by Wnt-5A. *Science* **275:** 1652.

Herman M.A., Vassilieva L.L., Horvitz H.R., Shaw J.E., and Herman R.K. 1995. The *C. elegans* gene *lin-44*, which controls the polarity of certain asymmetric cell divisions, encodes a Wnt protein and acts cell nonautonomously. *Cell* **83:** 101.

Klingensmith J. and Nusse R. 1994. Signaling by *wingless* in *Drosophila. Dev. Biol.* **166:** 396.

Klingensmith J., Nusse R., and Perrimon N. 1994. The *Drosophila* segment polarity gene *dishevelled* encodes a novel protein required for response to the *wingless* signal. *Genes Dev.* **8:** 118.

Lawrence P.A., Sanson B., and Vincent J.P. 1996. Compartments, wingless and engrailed: Patterning the ventral epidermis of *Drosophila* embryos. *Development* **122:** 4095.

Leyns L., Bouwmeester T., Kim S.H., Piccolo S., and De Robertis E.M. 1997. Frzb-1 is a secreted antagonist of Wnt signaling expressed in the Spemann organizer. *Cell* **88:** 747.

Marigo V., Davey R.A., Zuo Y., Cunningham J.M., and Tabin C.J. 1996. Biochemical evidence that patched is the hedgehog receptor. *Nature* **384:** 176.

Molenaar M., Van de Wetering M., Oosterwegel M., Petersonmaduro J., Godsave S., Korinek V., Roose J., Destree O., and Clevers H. 1996. XTcf-3 transcription factor mediates β-catenin-induced axis formation in *Xenopus* embryos. *Cell* **86:** 391.

Neumann C.J. and Cohen S.M. 1997. Long-range action of Wingless organizes the dorsal-ventral axis of the *Drosophila* wing. *Development* **124:** 871.

Noordermeer J., Klingensmith J., Perrimon N., and Nusse R. 1994. *dishevelled* and *armadillo* act in the *wingless* signalling pathway in *Drosophila. Nature* **367:** 80.

Noordermeer J., Johnston P., Rijsewijk F., Nusse R., and Lawrence P. 1992. The consequences of ubiquitous expression of the *wingless* gene in the *Drosophila* embryo. *Development* **116:** 711.

Nusse R. 1997. A versatile transcriptional effector of wingless signaling. *Cell* **89:** 321.

Nusse R. and Varmus H.E. 1992. *Wnt* genes. *Cell* **69:** 1073.

Nüsslein-Volhard C. and Wieschaus E. 1980. Mutations affecting segment number and polarity in *Drosophila. Nature* **287:** 795.

Papkoff J. and Schryver B. 1990. Secreted int-1 protein is associated with the cell surface. *Mol. Cell. Biol.* **10:** 2723.

Peifer M., Sweeton D., Casey M., and Wieschaus E. 1994. *wingless* signal and zeste-white 3 kinase trigger opposing changes in the intracellular distribution of *armadillo. Development* **120:** 369.

Peifer M., Rauskolb C., Williams M., Riggleman B., and Wieschaus E. 1991. The segment polarity gene *armadillo* interacts with the *wingless* signaling pathway in both embryonic and adult pattern formation. *Development* **111:** 1029.

Ponting C.P. 1995. DHR domains in syntrophins, neuronal NO synthases and other intracellular proteins. *Trends Biochem. Sci.* **20:** 102.

———. 1996. Pleckstrin's repeat performance: A novel repeat in G-protein signaling? *Trends Biochem. Sci.* **21:** 245.

Rattner A., Hsieh J.C., Smallwood P.M., Gilbert D.J., Copeland N.G., Jenkins N.A., and Nathans J. 1997. A family of secreted proteins contains homology to the cysteine-rich ligand-binding domain of frizzled receptors. *Proc. Natl. Acad. Sci.* **94:** 2859.

Riggleman B., Schedl P., and Wieschaus E. 1990. Spatial expression of the *Drosophila* segment polarity gene *armadillo* is post-transcriptionally regulated by *wingless. Cell* **63:** 549.

Rubinfeld B., Albert I., Porfiri E., Fiol C., Munemitsu S., and Polakis P. 1996. Binding of GSK3β to the APC—β-catenin complex and regulation of complex assembly. *Science* **272:** 1023.

Sawa H., Lobel L., and Horvitz H.R. 1996. The *Caenorhabditis elegans* gene *lin-17*, which is required for certain asymmetric cell divisions, encodes a putative seven-transmembrane protein similar to the *Drosophila* Frizzled protein. *Genes Dev.* **10:** 2189.

Siegfried E., Wilder E.L., and Perrimon N. 1994. Components of wingless signalling in *Drosophila. Nature* **367:** 76.

Stone D.M., Hynes M., Armanini M., Swanson T.A., Gu Q.M., Johnson R.L., Scott M.P., Pennica D., Goddard A., Phillips H., Noll M., Hooper J.E., Desauvage F., and Rosenthal A. 1996. The tumour-suppressor gene patched encodes a candidate receptor for sonic hedgehog. *Nature* **384:** 129.

Sussman D.J., Klingensmith J., Salinas P., Adams P.S., Nusse R., and Perrimon N. 1994. Isolation and characterization of a mouse homolog of the *Drosophila* segment polarity gene *dishevelled. Dev. Biol.* **166:** 73.

Theisen H., Purcell J., Bennett M., Kansagara D., Syed A., and Marsh J. 1994. *dishevelled* is required during *wingless* signaling to establish both cell polarity and cell identity. *Development* **120:** 347.

Van Den Heuvel M. and Ingham P.W. 1996. *smoothened* encodes a receptor-like serpentine protein required for *hedgehog* signalling. *Nature* **382:** 547.

van de Wetering M., Cavallo R., Dooijes D., van Beest M., van Es J., Loureiro J., Ypma A., Hursh D., Jones T., Bejsovec A., Peifer M., Mortin M., and Clevers H. 1997. *armadillo* coactivates transcription driven by the product of the *Drosophila* segment polarity gene *dTCF. Cell* **88:** 789.

Van Leeuwen F., Harryman Samos C., and Nusse R. 1994. Biological activity of soluble wingless protein in cultured *Drosophila* imaginal disc cells. *Nature* **368:** 342.

Vinson C.R., and Adler P.N. 1987. Directional non-cell autonomy and the transmission of polarity information by the *frizzled* gene of *Drosophila. Nature* **329:** 549.

Vinson C.R., Conover S., and Adler P.N. 1989. A *Drosophila* tissue polarity locus encodes a protein containing seven potential transmembrane domains. *Nature* **338:** 263.

Wang S.W., Krinks M., Lin K.M., Luyten F.P., and Moos M. 1997. Frzb, a secreted protein expressed in the Spemann organizer, binds and inhibits Wnt-8. *Cell* **88:** 757.

Wang Y.-K., Harryman Samos C., People R., Perez-Jurado L., Nusse R., and Francke A. 1997. A novel human homologue of the *Drosophila frizzled wnt* receptor gene binds wingless protein and is in the Williams syndrome deletion at 17q11.13. *Hum. Mol. Genet.* **6:** 465.

Wang Y., Macke J., Abella B., Andreasson K., Worley P., Gilbert D., Copeland N., Jenkins N., and Nathans J. 1996. A large family of putative transmembrane receptors homologous to the product of the *Drosophila* tissue polarity gene *frizzled. J. Biol. Chem.* **271:** 4468.

Wieschaus E. and Riggleman R. 1987. Autonomous requirements for the segment polarity gene *armadillo* during *Drosophila* embryogenesis. *Cell* **49:** 177.

Willert K., Brink M., Wodarz A., Varmus H., and Nusse R. 1997. Casein kinase 2 associates with and phosphorylates dishevelled. *EMBO J.* **16:** 3089.

Yanagawa S., Van Leeuwen F., Wodarz A., Klingensmith J., and Nusse R. 1995. The Dishevelled protein is modified by Wingless signaling in *Drosophila. Genes Dev.* **9:** 1087.

Zecca M., Basler K., and Struhl G. 1996. Direct and long-range action of a *wingless* morphogen gradient. *Cell* **87:** 833.

Multiple Roles of Cholesterol in Hedgehog Protein Biogenesis and Signaling

P.A. Beachy,[1] M.K. Cooper,[1] K.E. Young,[1] D.P. von Kessler,[1] W.-J. Park,[1,3]
T.M. Tanaka Hall,[2] D.J. Leahy,[2] and J.A. Porter[1,4]
[1]*Department of Molecular Biology and Genetics,* [2]*Department of Biophysics and Biophysical Chemistry, Howard Hughes Medical Institute, Johns Hopkins University School of Medicine, Baltimore, Maryland 21205*

During the past decade, extracellular protein signals encoded by several gene families have emerged as central players in coordinating cell behavior and thus generating pattern during animal development. Members of the *hedgehog* (*hh*) gene family in particular are notable for their association with several well-studied patterning activities. In *Drosophila*, where *hh* was discovered and isolated, patterning functions include specification of positional identity within developing segments and appendages. In vertebrate embryos, the function of the *hh* family member *Sonic hedgehog* (*Shh*) is associated with the patterning influences of notochord and prechordal plate mesoderm on spinal cord and brain, as well as on other surrounding structures. *Shh* expressed in mesoderm at the posterior margin of the developing vertebrate limb bud also has a central role in controlling limb outgrowth and patterning. The patterning functions of *hh*-encoded proteins have been studied extensively (for a recent general review, see Hammerschmidt et al. 1997), and novel functions continue to emerge.

This paper presents a selective view of the Hh protein biogenesis and signaling pathways, with particular attention paid to the involvement of the abundant neutral lipid cholesterol. One role for cholesterol is as a covalent adduct for the biologically active form of the *hh*-encoded protein (Hh), which is formed as a product of an autoprocessing reaction that entails internal cleavage. Cholesterol attachment restricts the spatial deployment of the Hh signal, thus influencing the pattern of cellular responses in developing tissues. Here, we summarize our studies of the Hh autoprocessing reaction and of the role of cholesterol in this reaction. We also summarize more recent studies suggesting that in addition to its role in Hh signal production, cholesterol has an essential role in mediating the response to the Hh signal within target cells. This role is revealed by genetic or drug-induced perturbations of cholesterol homeostasis that render target tissues unresponsive to the Hh signal.

Present addresses: [3]Kwang-Ju Institute of Science and Technology, Department of Life Science, Kwangju, Korea 506-303; [4]Ontogeny Inc., 45 Moulton Street, Cambridge Massachusetts 02138.

EXPERIMENTAL PROCEDURES

Immunostaining and blotting of Hh protein. For immunostaining, stably transfected Schneider line 2 (S2) cultured cells harboring full-length Hh under metallothionein promoter control (Lee et al. 1994) were induced by adding $CuSO_4$ (0.5 mM) to the medium, incubated overnight, transferred to an eight-chamber slide (Nunc), and allowed to adhere for 1 hour. The cells were fixed with 4% paraformaldehyde phosphate-buffered saline (PBS) for 10 minutes at room temperature and washed several times with PBSS (PBS containing 0.1% saponin). The cells were treated as follows, with several PBSS washes at every reagent change: anti-N (1:100 dilution of anti-Hh-N described in Lee et al. 1994) for 1 hour and anti-rabbit-Texas Red (1:50 dilution, Jackson ImmunoResearch Laboratories) for 30 minutes. All incubations were performed at room temperature, and the antibodies were diluted in 1% bovine serum albumin (BSA)/PBSS. The cells were mounted with Vectashield (Vector Laboratories) and observed by confocal microscopy (Biorad).

For immunoblotting, medium and cells from cultures of stably transfected S2 cells expressing full-length or truncated Hh were suspended in SDS sample buffer, and equivalent fractions of the total culture were loaded onto a 12% polyacrylamide gel, electrophoresed, and transferred to nitrocellulose. Amino-terminal epitopes were detected with anti-Hh-N antibody (Lee et al. 1994), and bound antibody was detected with ECL (Amersham).

Jervine inhibition of cholesterol biosynthesis. Metabolic labeling and sterol analysis were carried out essentially as described previously (Rodriguez and Parks 1985; Popjak et al. 1989; Rilling et al. 1993), with minor modifications. Briefly, COS-7 cells were plated at approximately 35% confluence into two 60-mm dishes at 37°C in 4 ml each of Dulbecco's modified Eagle's medium (DMEM) supplemented with 10% fetal bovine serum (FBS). After 24 hours of growth, the medium in each dish was replaced with 2 ml of fresh medium with 10% FBS; [^{3}H]mevalonic acid (New England Nuclear NET 176) brought to a specific activity of 0.8 Ci/mmole in a 1% solution of BSA was added to this medium to a final concentration of 20 μM. At this time, one dish received 6 μl of a 4 mg/ml solution of jervine (a kind gift from W. Gaffield) in ethanol (final concentration 28 μM jervine), and the other received 6 μl of ethanol. After 24 hours of further incubation, cells were washed in PBS and extracted with methanol, and 1 M potassium hydroxide (KOH) was added to 10%. Following a 3-hour incubation at 60°C, the methanol/KOH mixture was extracted with diethyl ether and the extract was dried down, resuspended in isopropanol, and subjected to reverse-phase high-performance liquid chromatography (HPLC) analysis according to the method of Rodriguez and Parks (1985).

RESULTS AND DISCUSSION

Hedgehog Protein Autoprocessing

Figure 1 presents a view of Hh biosynthesis that although largely derived from studies of *Drosophila* Hh, likely applies to Hh proteins from all species. As suggested by genetic studies (Mohler 1988) and as predicted from sequence analysis (Lee et al. 1992; Mohler and Vani 1992; Tabata et al. 1992; Tashiro et al. 1993), the Hh protein enters the secretory pathway and is cleaved following a signal sequence located near the amino terminus (Lee et al. 1992). From earliest examination in in vitro translation experiments, the *Drosophila* Hh protein also revealed a propensity to undergo cleavage at another internal site (Lee et al. 1992); antibodies specifically directed against amino- or carboxy-terminal epitopes confirmed that the predominant forms of endogenous Hh protein correspond to the products of this internal cleavage (Lee et al. 1994). The internal cleavage depends on carboxy-terminal Hh sequences and can be observed with purified recombinant protein in vitro, thus indicating the operation of a self-directed processing activity (Lee et al. 1994).

Further in vitro analysis of this cleavage demonstrated that it occurs between the Gly and Cys residues within a conserved Gly-Cys-Phe tripeptide (Porter et al. 1995). This information permitted ectopic expression in transgenic *Drosophila* of constructs encoding either the amino-terminal or the carboxy-terminal cleavage products (Hh-N and Hh-C, respectively), and these studies demonstrated that biological signaling activity resides entirely within the precisely truncated Hh-N fragment (Porter et al. 1995). Similar transgenic experiments also demonstrated that mutations within Hh-C that interfere with processing but do not alter Hh-N sequences nevertheless block Hh function (Lee et al. 1994; Porter et al. 1995). Thus, whereas Hh-N suffices for signaling activity, Hh-C sequences are required to generate the active

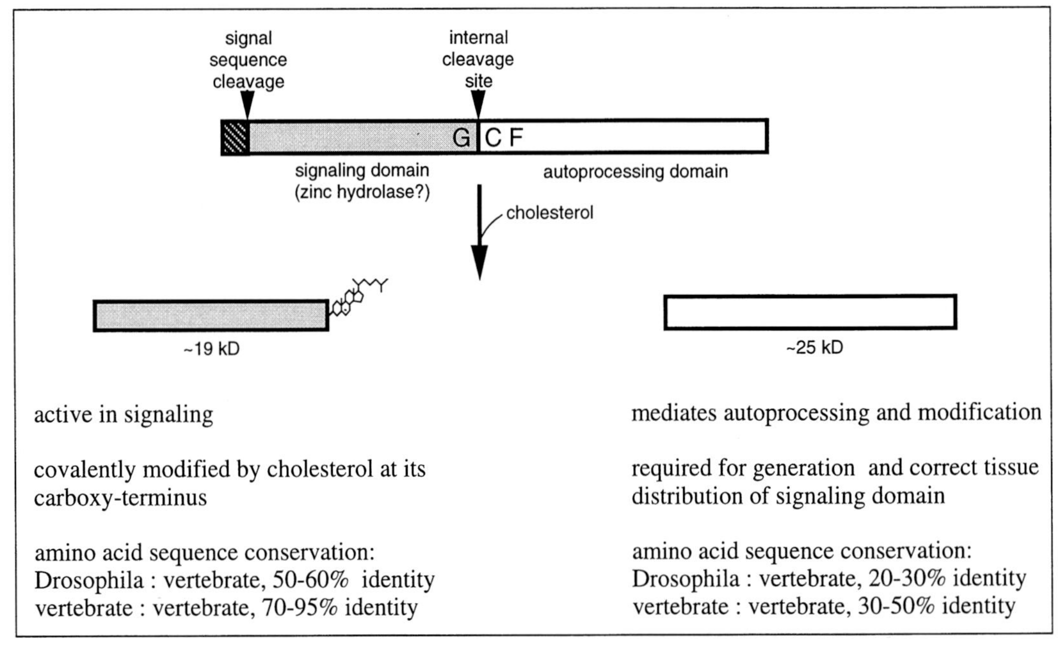

Figure 1. Autoprocessing of the Hedgehog protein precursor. Following signal sequence cleavage, the Hedgehog precursor (Hh) undergoes an autoprocessing reaction that entails cleavage between the Gly and Cys residues within a tripeptide, Gly-Cys-Phe, that is conserved among all Hh proteins; this cleavage is accompanied by attachment of cholesterol to the carboxyl terminus of the amino-terminal product. Whereas the amino-terminal domain is active in signaling, the carboxy-terminal domain mediates the autoprocessing reaction, and the resulting modification by cholesterol influences the tissue distribution of signaling activity. As indicated, amino acid sequence conservation among orthologs of the Hh family is greater within the amino-terminal domain as compared to the carboxy-terminal domain. Crystallographic analysis of the amino-terminal domain of Shh protein revealed striking similarity in folded structure of a portion of this domain to the catalytic domain of D,D carboxypeptidase, a zinc hydrolase from *Streptomyces* that acts on cell wall components (Dideberg et al. 1982; Hall et al. 1995; Murzin 1996); the significance of this similarity and the role of this putative hydrolase in Shh signaling are not known. Not shown in this figure, N-linked glycosylation of carboxy-terminal sequences has been reported within the carboxy-terminal domain of Shh (Bumcrot et al. 1995), but the glycosylation site is not uniformly conserved among Hh orthologs and its significance is unknown.

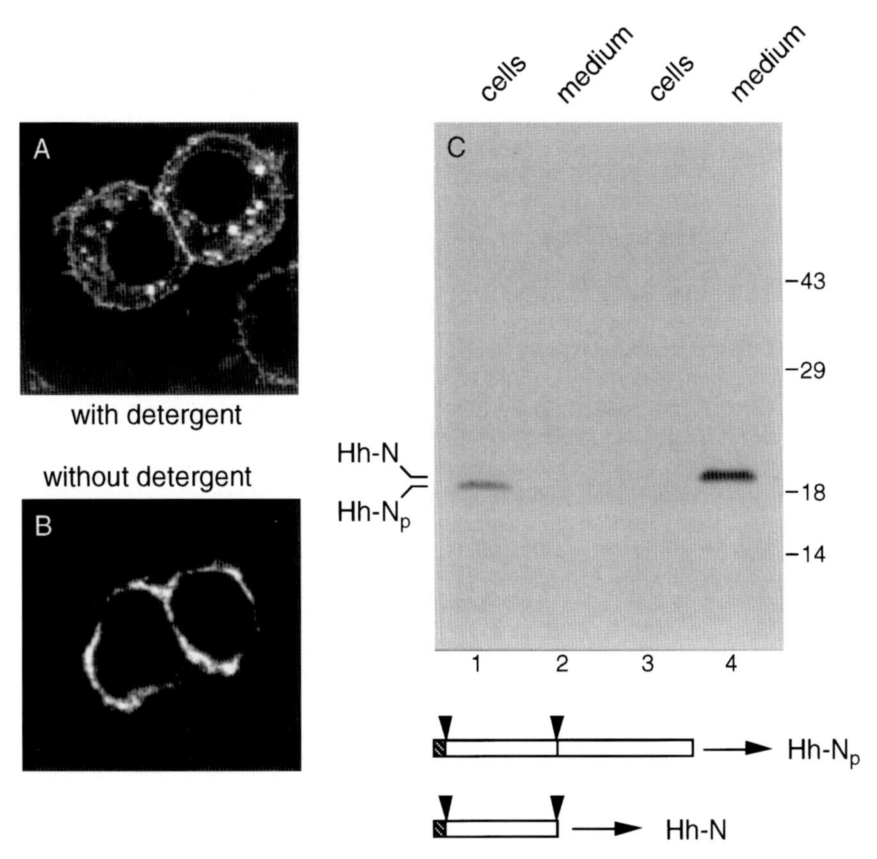

Figure 2. Cell surface association of the amino-terminal domain. Indirect immunofluorescence staining revealed a prominent cell surface association of amino-terminal epitopes (*A,B*) with the plasma membrane; this staining was observed in the absence of detergent permeabilization, indicating localization to the cell surface. The Western blot in *C* shows amino-terminal domain expression from full-length Hh protein (Hh-N_p; lanes *1,2*) or from a construct with a chain termination codon at the cleavage site (Hh-N; lanes *3,4*). Hh-N_p is retained by cells within the culture, whereas Hh-N is nearly quantitatively released into the medium. The greater mobility of Hh-N_p relative to Hh-N is characteristic of processed Hh amino-terminal domains.

amino-terminal signaling domain from precursor via autoprocessing (Fig. 1). Consistent with these conclusions, all molecularly characterized *hh* mutations in *Drosophila* either directly affect the Hh-N signaling domain or otherwise appear to block release of the signaling domain from precursor by affecting the Hh-C autoprocessing function (Porter et al. 1995).

Biological Role of Autoprocessing

Since the cleavage products of the Hh precursor are the predominant forms observed in vivo, the occurrence of the autoprocessing event appears not to be regulated. What then is the raison d'etre of autoprocessing? The answer to this question began to emerge from studies in cultured cells which demonstrated that processed amino-terminal domain protein generated from precursor remains tightly associated with the cell surface (Fig. 2A,B) (Lee et al. 1994). In contrast, protein expressed from a construct lacking the autoprocessing domain is almost quantitatively released into the culture medium (Fig. 2C) (Porter et al. 1995). Autoprocessing is thus associated with tethering of the amino-terminal signaling domain to the cell surface. Amino-terminal domain protein derived by processing of the Hh precursor is designated Hh-N_p, to distinguish it from amino-terminal domain derived from a truncated construct (Hh-N).

As would be expected for a potent secreted signal whose expression is spatially restricted within segments, autoprocessing and cell surface tethering have an important role in segmental patterning (Porter et al. 1996b). This role was revealed by comparing the effects of localized expression of Hh-N or Hh-N_p in transgenic *Drosophila* embryos: When activated at the normal sites of *hh* transcription, transgenes expressing Hh-N_p, which is also the form of the endogenous protein signal, showed no significant alterations in the normal segmental patterns of gene expression or cuticle formation. In contrast, similar localized expression of transgenic Hh-N caused disruption of segmental patterning equivalent to that caused by ubiquitous high-level expression of Hh protein. Autoprocessing thus appears to restrict the spatial distribution of Hh signaling activity, and immunofluorescence studies indeed confirm that Hh-N more readily diffuses from expressing cells to surrounding cells than does Hh-N_p (Porter et al. 1996b). These immunofluorescence studies also revealed a difference in subcellular localization within expressing cells, with Hh-N_p sequestered in large punctate structures in basolateral domains of epidermal cells. Hh-N protein from the truncated construct in contrast lacks this type of punctate localization and instead appears to be freely secreted to the apical surface of expressing cells within the epidermal epithelium (Porter et al. 1996b). The functional significance of this processing-dependent localization to the basolateral domain of expressing cells within the epidermal epithelium is not yet known.

The biological role of autoprocessing in vertebrates is particularly well illustrated by the role of *Shh* in neural tube patterning. Shh protein is processed similarly to the Hh protein (Chang et al. 1994; Bumcrot et al. 1995; Ekker et al. 1995; Lai et al. 1995; Porter et al. 1995; Roelink et al. 1995), and the normal cell surface association of the amino-terminal fragment also is processing-dependent (Bumcrot et al. 1995; Roelink et al. 1995; Porter et al. 1996a). Loss of *Shh* gene function in mouse embryos results in failure to differentiate floorplate cells and motor neurons (Chiang et al. 1996). Induction in naive neural plate explants of these and other ventral cell types by recombinant Shh-N protein occurs in a concentration-dependent manner, with motor neurons induced at low concentrations and floorplate cells induced at the expense of motor neuron fates at higher concentrations (Marti et al. 1995; Roelink et al. 1995; Ericson et al. 1997). In other explant experiments with embryonic tissues as inducers, structures such as the notochord can only induce floorplate cells in a contact-dependent manner (Placzek et al. 1993), whereas motor neuron induction does not require such contact (Yamada et al. 1993). The ability to circumvent contact dependence with high concentrations of soluble Shh-N protein suggests that one role for modification and surface association of the signaling domain is to generate large concentration differences between local and distant sites, with consequent sharp distinctions between the cell types induced. Consistent with this idea, the *Shh* signaling domain is found predominantly on the surface of notochord cells, and embryonic floorplate normally forms only in close proximity to the notochord.

The Autoprocessing Reaction

Given the striking differences in diffusibility and patterning activity of Hh-N and Hh-N_p, it was not surprising to find accompanying physical differences. As compared to Hh-N, Hh-N_p displays a slight difference in electrophoretic mobility (Fig. 2C), a dramatic increase in hydrophobic character, a greater mass associated with the carboxy-terminal fragment of cyanogen bromide (CNBr) digestion, and an insensitivity to digestion by carboxypeptidase (Porter et al. 1996b). These data together indicate that Hh-N_p carries a covalently attached lipophilic adduct at its carboxyl terminus whose addition depends on the autoprocessing activity of the carboxy-terminal domain. The presence of this adduct accounts for the tethering of Hh-N_p to the cell surface, since the lipid adduct would be expected to partition preferentially into the lipid bilayer.

Despite information about its mass and other properties, the identity of the lipid adduct could not be determined directly because quantities of purified Hh-N_p sufficient for chemical analysis proved to be difficult to obtain. Identification of the adduct therefore relied ultimately on a mechanistic understanding of the in vitro processing reaction and its use as an assay to identify a lipid capable of participating in the autoprocessing reaction. Initial insight into the autoprocessing reaction derived from the observation that the kinetics of cleavage in vitro were independent of starting protein concentration, indicating an intramolecular mechanism (Porter et al. 1995). From a limited number of proteins known to autoprocess by an intramolecular mechanism, a particularly strong analogy could be drawn to prohistidine decarboxylase (van Poelje and Snell 1990), which is capable of undergoing intramolecular cleavage with either a Cys or Ser residue at the position immediately following the scissile bond; Hh autocleavage also could be observed in vitro, albeit inefficiently, if a Ser residue replaced the normal Cys (Porter et al. 1996b). Contemporaneously with our studies of Hh autoprocessing, the self-splicing proteins have also emerged as intramolecular processing proteins with Ser or Cys residues at the site of cleavage (Xu and Perler 1996); we now know that the similarities between Hh and self-splicing proteins extend beyond mechanism to include sequence and structure (Hall et al. 1997; see below).

The feature common to all of these autoprocessing reactions is initiation by attack of a nucleophilic side chain upon the preceding carbonyl, with displacement of the peptide amine and formation of an ester or thioester intermediate. As seen in Figure 3A, this is the first step of the Hh autoprocessing reaction, with a labile thioester replacing the main chain peptide bond between amino- and carboxy-terminal domains (Porter et al. 1995, 1996b). The second step of the Hh autoprocessing reaction involves attack upon the same carbonyl by a second nucleophile, displacing the sulfur and severing the connection between Hh-N and Hh-C. The requirement for a second nucleophile in vitro can be met by a high concentration either of a thiol-containing molecule or of another small molecule with nucleophilic properties at neutral pH; these small nucleophiles can be shown to form covalent adducts to the amino-terminal product of the in vitro cleavage reaction (Porter et al. 1996b).

Of some interest in the in vitro studies of Hh autoprocessing was the use of Cys-initiated peptides as nucleophile in the second step: The initial linkage between the peptide and the amino-terminal product is a thioester, which can then rearrange to form an amide bond by reversal of the steps involved in thioester formation during the first part of the reaction (Porter et al. 1996b). The net effect of these reactions is the ligation of a Cys-initiated peptide at the site of cleavage and is analogous to the recent use of a chemically synthesized thioester intermediate for peptide ligation (Dawson et al. 1994).

Cholesterol Modification In Vitro and In Vivo

To account for the lipid modification in Hh-N_p, the in vivo reaction was presumed to occur with the participation of an endogenous lipid carrying the second nucleophilic moiety. This presumption led to use of the in vitro

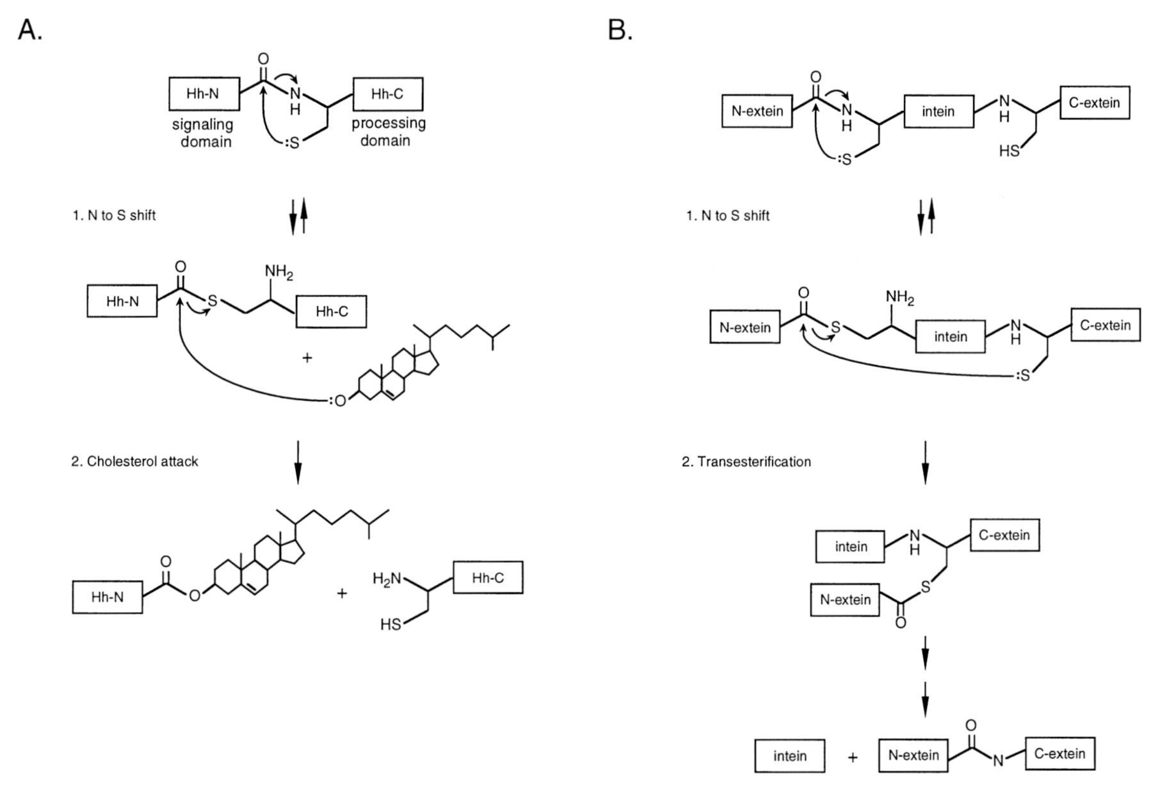

Figure 3. Hh autoprocessing and protein self-splicing are initiated by formation of a thioester intermediate. As described in the text, Hh autoprocessing (*A*) and protein self-splicing (*B*) are both initiated by formation of a thioester in place of a main-chain peptide. The reactions differ in the second step, which for Hh entails attack of the thioester intermediate by cholesterol. For self-splicing proteins, the second nucleophile is the side chain of the first residue in the carboxy-terminal extein (C extein); the resulting three-branched intermediate is then resolved to give rise to the free intein and ligated N and C exteins. For simplicity, proton transfers implicit in the activation of nucleophiles or of leaving groups are omitted from this scheme.

reaction as an assay that was applied to fractionated cell lipids, leading to the identification of cholesterol as a neutral lipid that at relatively low concentrations could supply the requirement for a nucleophile in the second step (Porter et al. 1996a). Cholesterol thus stimulates the in vitro autoprocessing reaction and forms a covalent linkage to the amino-terminal product of the cleavage reaction. This linkage is sensitive to base treatment, consistent with formation of an ester with the oxygen of the 3β hydroxyl of cholesterol. Confirming this role for cholesterol in vivo, [^{3}H]cholesterol was observed to label Hh-N_p or Shh-N_p expressed in *Drosophila* or in mammalian cultured cells, and this label could be removed by base treatment (Porter et al. 1996a; K.E. Young et al., unpubl.). The label hydrolyzed from Hh-N_p was further analyzed and shown to display chromatographic behavior identical to that of cholesterol, indicating that the in vivo adduct is cholesterol and not some other sterol derivative.

A somewhat surprising finding in the metabolic labeling experiments with mammalian cells was the apparent linkage of cholesterol to several other mammalian proteins. There is little evidence at present regarding the identity and function of these proteins or the mechanism of attachment of cholesterol. We have found that the cholesterol can be removed from these proteins by base treatment, suggestive of an ester linkage like that resulting from Hh autoprocessing (K.E. Young et al., unpubl.).

Thioesters as Intermediates in Protein Modification

The use of a Cys-derived thioester as an intermediate is a theme common to several other acyl transfers that result in covalent modifications of proteins. Following formation of the initial thioester in these systems, the acyl portion of the thioester (the acceptor, corresponding to Hh-N; see diagram in Fig. 4) can receive the final modification directly or alternatively may be transferred to other thiols in one or more subsequent steps before receiving the final modification (Fig. 4). The ubiquitin cascade represents such a reaction with multiple intermediates, whose role is to attach ubiquitin to proteins destined for degradation by the proteasome (Hochstrasser 1996). The acyl group for these thioesters is supplied by the carboxy-terminal Gly of ubiquitin, and the thiols come from Cys side chains in three distinct classes of enzymes. The first of these, E1, forms the initial thioester in an ATP-consuming reaction. Then, through trans(thio)esterification reactions, the ubiquitin forms thioesters sequentially with E2 and E3 enzymes, before final transfer to the ϵ amine of a Lys side chain. The protein receiving ubiquitin in the resulting amide linkage is thus marked for degradation.

The α-macroglobulin proteinase inhibitors and the C3, C4, and C5 complement proteins represent members of an ancient superfamily that use an intrachain thioester as

System	Acceptor (Acyl group)	Donor (Thiol)	Final Donor (Adduct)
Self-splicing proteins*	C-terminal residue of N extein, generally Gly	Thiol of N-terminal Cys in Hint module	N-terminal residue of downstream extein
Hedgehog*	C-terminal Gly of signaling domain	Thiol of N-terminal Cys in Hint module	Cholesterol (O of 3ß hydroxyl)
Novel *C. elegans* proteins*	C-terminal residue of N-terminal domain	Thiol of N-terminal Cys in Hint module	?
Peptide ligation (in vitro)	Synthetic peptide terminated by thioester	Thiol side chain of Cys-initiated peptide	S to N shift to form peptide bond
Ubiquitin	Carbonyl carbon of C-terminal Gly	E1 Activator (ATP consuming) E2 carrier, E3 ligase } sequential trans-thioesterification	Nucleophilic ε-NH_2 of lysine on protein targeted to proteasome
Complement C3,C4,C5	Amido carbon of Gln side chain	Intrachain Cys intramolecular thiol attack	Nucleophile from a protein on the surface of a cell targeted for lysis
α-Macroglobulin	Amido carbon of Gln side chain	Intrachain Cys intramolecular thiol attack	Nucleophilic group on the proteinase being inactivated

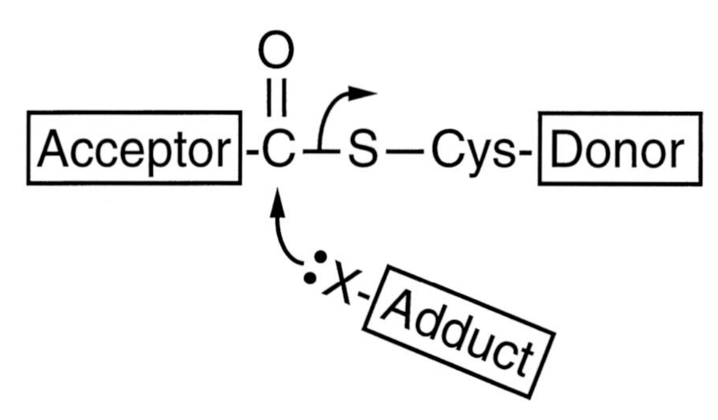

Figure 4. Thioester intermediates in protein modification. (*) Hint domain proteins (see also Figs. 3 and 5). Acyl group acceptors and thiol donors participating in the formation of thioester intermediates are indicated in the middle columns, and the final modifying adduct is listed in the rightmost column. See text.

a "spring-loaded" functionality that can be triggered for covalent attachment to target molecules (Chu and Pizzo 1994). The intrachain thioester is formed by thiol attack of a Cys side chain on the amido group of a Gln side chain. The final adducts in the case of the complement proteins are nucleophiles on the surface of cells to be targeted for lysis. In the case of α-macroglobulin, the final adduct is a nucleophile on a protease to be inactivated, which is targeted to α-macroglobulin through the presence of multiple cleavage sites for proteases of various specificities.

In the examples just discussed, the acyl group contributing to the thioester intermediate derives either from another protein or from an amino acid side chain. In contrast, the acyl group in the Hh thioester intermediate is linked to a main chain carbonyl, and the thioester therefore replaces an amide bond within the peptide backbone. Other proteins likely to utilize main chain ester or thioester intermediates in autoprocessing reactions include prohistidine decarboxylase and certain members of the Ntn (*N*-*t*erminal *n*ucleophile) hydrolase family that are processed by an intramolecular mechanism (Brannigan et al. 1995; Guan et al. 1996). The Ntn hydrolases are structurally related enzymes that are autoprocessed with internal cleavage, leaving the active-site nucleophile as the amino-terminal residue. The role of these reactions appears to be activation of a precursor protein and there is no net addition of a modifying adduct. Thus, although the prohistidine decarboxylase reaction was of heuristic value in understanding the mechanism of Hh autoprocessing, there is no evidence of any evolutionary relationship between Hh autoprocessing domains and either prohistidine decarboxylase or Ntn hydrolase proteins.

Ester Intermediates in Proteins Containing the Hint Domain

Hh proteins in contrast are evolutionarily related to two other groups of proteins, the self-splicing proteins and a group of novel nematode proteins containing Hh-C-like sequences. The self-splicing proteins undergo a reaction in which an internal portion of the protein, termed an intein, is excised and amino- and carboxy-terminal flanking regions, termed exteins, are ligated to form the mature protein (Perler et al. 1994). Inteins are found inserted into a wide variety of archaeal, bacterial, chloroplast, and yeast proteins. The intein portion mediates the protein-splicing reaction and typically also contains an endonuclease thought to act at the DNA level in mediating movement of intein-coding sequences. Similar to Hh autoprocessing, the protein-splicing reaction is initiated by intramolecular attack of a hydroxyl or thiol upon the preceding carbonyl, and the resulting ester or thioester intermediate replaces the peptide bond at the amino-terminal extein/intein boundary (Fig. 3) (Xu and Perler 1996). Unlike Hh proteins, the second nucleophilic attack in the protein self-splicing reaction involves the side chain of another Ser or Cys residue several hundred residues downstream. The resulting branched protein intermediate ultimately resolves into the ligated exteins and the free intein protein (Fig. 3).

Nematode proteins with Hh-C-like sequences were identified by searching for homology within the *Caenorhabditis elegans* genomic sequence database. At about 80% completion of the *C. elegans* genome, nine putative proteins that have homology with the Hh-C autoprocessing domain have been identified (Burglin 1996; Porter et al. 1996b; Hall et al. 1997; R. Mann et al., unpubl.). As in the Hh family, the Hh-C-like domain is located at the carboxyl terminus of these proteins and is preceded by an amino-terminal domain bearing a signal sequence. The amino-terminal domains of these nematode proteins, however, bear no sequence similarity to Hh-N. Instead, they resemble each other and can be divided into three families. The structures of these proteins suggest the possibility that they are secreted and undergo autoprocessing; a preliminary study of one family member in *Drosophila* cultured cells indeed demonstrates cleavage at the junction between amino- and carboxy-terminal domains (Porter et al. 1996b).

The level of amino acid sequence identity between these nematode proteins and Hh ranges from 24% to 32% in a region approximately corresponding to the amino-terminal two thirds of Hh-C. This same region of Hh-C also can be aligned with inteins, although alignment is complicated by the presence of sequences corresponding to the endonuclease (Dalgaard et al. 1997; Hall et al. 1997; Pietrokovski 1997). The level of amino acid identity between Hh-C and inteins with endonuclease sequences removed is approximately 10%, but most of the residues known to be essential for Hh-C processing activity are conserved.

A common evolutionary origin for these protein families is further indicated by a domain with a common fold that is present in the crystal structure of a portion of Hh-C and in the crystal structure of the 454-residue intein protein PI-SceI (Duan et al. 1997; Hall et al. 1997). Two additional domains not present in the Hh-C fragment are present in the intein structure: One of these is the endonuclease and the other is thought to aid in DNA binding. Remarkably, both of these additional domains are inserted into peripheral loops of the common domain, with little apparent effect upon its three-dimensional fold. The crystallized Hh fragment contains the amino-terminal 151 residues of Hh-C, of which the first 145 residues are well-ordered in the crystal structure; these residues correspond to the region conserved in the nematode proteins. This domain alone suffices for thioester formation, as indicated by the ability of an Hh protein truncated after this point to undergo cleavage in the presence of DTT (Hall et al. 1997), and this domain has been referred to as the Hint module (*H*edgehog, *int*ein).

Although the Hint module in Hh-C suffices for the first step of autoprocessing, at least some part of the 63 carboxy-terminal residues missing in the crystallized fragment is required for the second step of cholesterol addition (Hall et al. 1997). Because of its apparent role in sterol addition, this 63-residue region is referred to as SRR, for sterol recognition region. No clear alignment can be made between SRR and sequences within the nematode family, however, and sequences in these nematode proteins that extend carboxy-terminal to the Hint domain are tentatively designated ARR, for adduct recognition region. The differences in sequence between the SRR of Hh proteins and the ARR regions of nematode gene family members raise the possibility that molecules other than cholesterol may participate in the processing reaction and form novel protein-modifying adducts.

From these sequence and structure relationships, a plausible evolutionary history can be constructed in which all three protein groups diverged from an ancestral Hint domain (Fig. 5) (Hall et al. 1997). In one branch, the ancestral intein was formed by insertion of an endonuclease into a Hint domain and by adjustment (or preservation) of the chemistry to ensure that the second nucleophilic attack is made intramolecularly by the side chain of a downstream residue. In a second branch, Hh proteins were formed by association of a Hint domain with amino-terminal domains of the Hh and nematode proteins. The sequence of events leading to formation of these proteins is not known. One possibility is that the Hint and SRR modules may have been assembled into a cholesterol transfer unit prior to association with the Hh signaling domain; alternatively, the Hint module might have been inserted within a preassembled protein comprising a signaling domain and the SRR precursor. In the second scenario, the SRR precursor in the preassembled protein might have served some function related to sterol recognition, such as membrane association. Similarly, several scenarios are possible in assembly of the nematode proteins. The possibility also exists that additional proteins will be found in which the Hint module initiates novel splicing or transfer reactions.

Cholesterol Synthesis Inhibitors and Holoprosencephaly

One of the most striking aspects of the *Shh* loss-of-function phenotype in mice (Chiang et al. 1996) is its resemblance to holoprosencephaly (HPE), a term applied to a spectrum of human developmental malformations characterized by a loss of midline structures in the forebrain and face. In its most severe form, as seen in *Shh*$^{-/-}$ mice, HPE is associated with a cyclopic eye positioned beneath a proboscis consisting of fused nasal chambers (Cohen and Sulik 1992). Abnormal features of brain anatomy, for

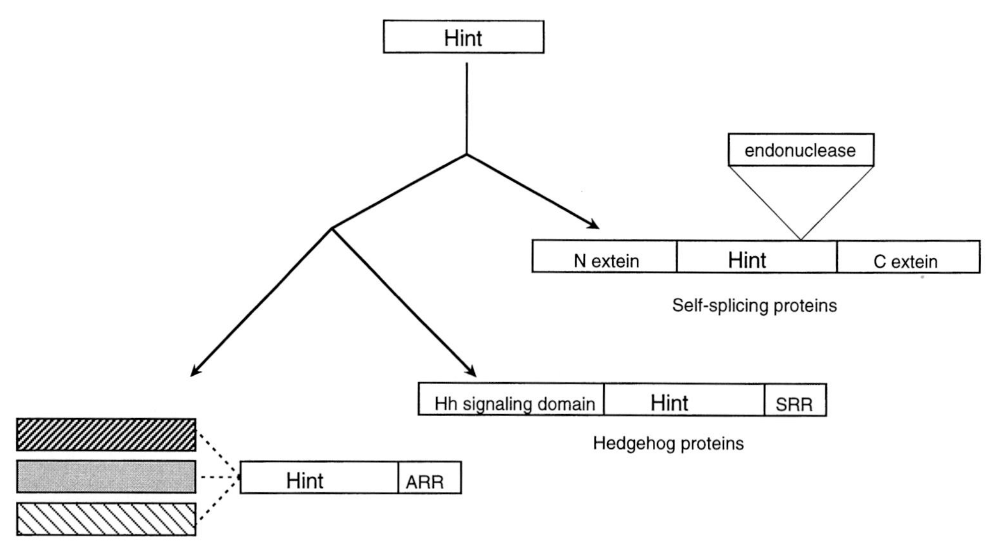

Figure 5. Evolutionary lineage of proteins containing Hint domains. Hint domains are found within three distinct protein families: the self-splicing proteins, the Hedgehog family of signaling proteins, and a novel family of *C. elegans* proteins of unknown function. In the self-splicing proteins, an endonuclease domain is inserted within a peripheral loop of the Hint domain (Duan et al. 1997; Hall et al. 1997). In Hh proteins, a sterol recognition region (SRR) appended to the carboxyl terminus of the Hint domain is required for cholesterol addition; in the absence of SRR sequences, only the first step of thioester formation occurs. The Hint domains of the *C. elegans* proteins are more closely related to those of the Hh family, but SRR sequences are replaced by other sequences of variable length, tentatively referred to as the adduct recognition region (ARR). Nine *C. elegans* proteins that contain Hint domains have been identified thus far (at ~80% completion of the *C. elegans* genomic sequence); this family can be subdivided further into three groups based on the presence of three unrelated types of sequences present within the amino-terminal domains of these proteins.

which the syndrome is named, include an absence of ventral forebrain structures and development of remaining forebrain structures as a single fused vesicle. Experimental manipulations of amphibian embryos carried out more than 60 years ago led to an understanding of cyclopia as a consequence of disrupting the influence normally exerted by prechordal plate mesoderm upon forebrain neuroepithelium (Mangold 1931; Adelmann 1936a,b). This influence is required for bilateral subdivision of the early eye field; in its absence, the eye field remains continuous across the midline, resulting in cyclopia and the loss of such ventral forebrain derivatives as the pituitary and the optic chiasm.

Shh expression in the prechordal mesoderm underlying the neural plate can first be detected in mid-streak mouse embryos (Echelard et al. 1993; Chang et al. 1994), a stage that coincides with or just precedes the requirement for prechordal plate signaling. All of these studies therefore are consistent with the view that Shh constitutes or contributes to the midline signal that passes from the prechordal plate mesoderm to forebrain neural plate, in a manner analogous to that in which Shh from the notochord induces regionalization and morphogenesis of the spinal cord. Recent studies indeed have demonstrated that an autosomal dominant form of human HPE is caused by mutations in the human *Shh* gene (Belloni et al. 1996; Roessler et al. 1996). The mutations described would be expected to cause a loss of Shh function, indicating that in contrast to the mouse *Shh* mutation, which is entirely recessive (Chiang et al. 1996), human *Shh* function is haploinsufficient. Consistent with this interpretation, the malformations associated with heterozygous human *Shh* mutations are variable, even among individuals carrying the same allele, and are far less severe than those in the homozygous mouse *Shh* mutation (Chiang et al. 1996).

Given this association of HPE with mutations in the *Shh* gene, a reasonable supposition would be that HPE could also be caused by other perturbations of the *Shh* signaling pathway. Of particular interest to us, in view of the role of cholesterol in Hh autoprocessing, were a series of observations published beginning more than 30 years ago which noted that HPE-like malformations can be induced by treatment of pregnant rats with the drugs Triparanol, AY 9944R, and BM 15.766 (Roux 1964, 1966; Roux et al. 1979; Dehart et al. 1997). These drugs inhibit enzymes of cholesterol biosynthesis and cause an abnormal accumulation of desmosterol (Triparanol) or of 7-dehydrocholesterol (AY 9944R and BM 15.766), which are the immediate precursors in alternate biosynthetic routes to cholesterol.

Genetic Perturbations of Cholesterol Synthesis and Transport

Further links between cholesterol and vertebrate embryonic development are provided by several mouse and human mutations affecting cholesterol synthesis or transport. Smith-Lemli-Opitz Syndrome (SLOS) is an autosomal recessive human genetic disease characterized by numerous developmental defects including microcephaly, pituitary agenesis, limb and genital abnormalities, and defects of the heart, kidneys, and pancreas (Opitz 1994; Tint et al. 1994; Salen et al. 1996). These patients lack the activity of 7-dehydrocholesterol reductase, the same enzyme inhibited by AY 9944R or BM 15.766, and as a consequence have abnormally low serum cholesterol levels and accumulate 7-dehydrocholesterol. Approximately 5% of SLOS patients display signs of HPE, with malformations that tend toward the milder end of the spectrum (Kelley et al. 1996). A possible explanation for the reduced severity of the defects as compared to those in the progeny of drug-treated rats is cholesterol supplementation from heterozygous mothers via placental exchange. Consistent with this idea, high dietary cholesterol can suppress the teratogenic effects of cholesterol synthesis inhibitors given to pregnant rats (Roux et al. 1979). Some of the developmental malformations in SLOS patients are likely to result from deficiencies in steroid hormone biosynthesis as well as from effects on other unknown targets.

Related developmental defects have also been described in mouse mutants lacking function of the endocytic receptor megalin or apolipoprotein B (Herz et al. 1997). The megalin protein, also referred to as gp330, is encoded by a member of the low-density lipoprotein (LDL) receptor gene family and is specifically expressed on the apical surfaces of embryonic neuroectoderm and neuroepithelium in the developing neural tube. Apolipoprotein B (apoB) is the major structural component of several lipoprotein particles that carry esterified cholesterol and other neutral lipids in the circulation. The defects in megalin-deficient mice include fusion of forebrain structures into a single vesicle, agenesis of the olfactory bulbs and pituitary, and absence of the corpus callosum, all malformations within the HPE sequence and therefore suggestive of a perturbation in the *Shh* signaling pathway (Willnow et al. 1996). The defects in mice lacking apoB function are more severe, appear less specific, and cause resorption of most homozygous embryos by 9.5 or 10.5 days of gestation (Farese et al. 1995; Huang et al. 1995). Mutations in both of these genes affect cholesterol transport, with the difference that the megalin effect may be restricted to cholesterol uptake in neural precursors, whereas the apoB defect would block all embryonic absorption of maternally derived cholesterol, normally transported via the yolk sac in the mouse (Farese et al. 1996). A hypomorphic mutation that produces a truncated but partly functional apoB protein is homozygous viable and does not consistently show developmental defects (Homanics et al. 1993). The reduced cholesterol levels in these mice, however, make them susceptible to teratogenesis with the compound BM 15.766, and the resulting defects, including HPE, are like those in treated rats (Lanoue et al. 1997). Normal mice are not susceptible to treatments with cholesterol synthesis inhibitors, possibly because their cholesterol levels are higher than those in rats.

Plant Teratogens as Cholesterol Synthesis Inhibitors

Another experimental model for HPE derives from the occurrence of epidemics of congenital craniofacial malformations among newborn lambs on sheep ranches in several National Forests of the western United States (Gaffield and Keeler 1996). The most dramatically affected lambs showed severe HPE, including true cyclopia and other craniofacial malformations characteristic of HPE. The occurrence of these defects was traced to grazing by pregnant ewes on the range plant *Veratrum californicum* (Binns et al. 1963). The compounds responsible were identified by Keeler and Binns (1968) as a family of steroidal alkaloids; the structures of two of these, cyclopamine and jervine, are shown as compared to cholesterol in Figure 6A.

Given the structural similarities of these compounds to cholesterol and the similar teratogenic effects of cholesterol synthesis inhibitors upon the offspring of pregnant rats, a reasonable mechanism to consider for the effects of these plant sterol derivatives was the inhibition of cholesterol biosynthesis. Accordingly, we tested COS7 cultured cells treated with jervine for defects in cholesterol biosynthesis by labeling with [^{3}H]mevalonic acid and then extracting and analyzing radioactively labeled, nonsaponifiable lipids. Figure 6B shows that treated cells synthesized reduced levels of cholesterol and accumulated increased levels of another sterol that we have provisionally identified as the cholesterol precursor, zymosterol. The natural product jervine at these concentrations thus inhibits cholesterol biosynthesis in cultured cells in much the same manner as the synthetic drugs discussed above, although the specific enzyme(s) affected appears to differ. Given the similarities in their teratogenic effects, this inhibition seems likely to underlie the teratogenic effects of both the synthetic and natural compounds.

Perturbations of Cholesterol Homeostasis Block the Response to Shh Signaling

As reviewed above, there is a striking correspondence between the developmental malformations in mouse and human *Shh* mutants and those caused by perturbations of cholesterol homeostasis. These malformations are caused by effects on either synthesis or transport of cholesterol; in the case of the hypomorphic *apob* allele combined with BM 15.766 treatment, effects on both synthesis and transport appear to synergize in generating severe HPE in mice, where neither effect alone suffices (Lanoue et al. 1997). The developmental malformations caused by these perturbations of cholesterol homeostasis strongly suggest that the *Shh* signaling pathway or its targets must somehow be affected. Our attention initially was drawn to these perturbations because of the role of cholesterol in Hh autoprocessing and the possibility that autoprocessing and hence signal production might be affected. But a second possibility is that instead of an effect on signal production, these perturbations of cholesterol homeostasis might interfere with the ability of target tissues to sense or transduce the Shh signal. To distinguish these two possibilities, we have examined the effects of these synthetic and plant-derived compounds on the autoprocessing reaction in vivo and in vitro and have also tested the ability of drug-treated neural plate explants to respond to recombinant Shh protein.

The autoprocessing reaction is not inhibited by these compounds in cultured cells, nor are cleavage and cholesterol modification inhibited in the in vitro reaction (M.K. Cooper et al., in prep.). The amino-terminal product of processing in drug-treated cultured cells displays a mobility suggestive of sterol modification. Since a number of cholesterol biosynthetic precursors are able to participate in the in vitro reaction, the adduct could be either cholesterol or one of the precursors whose accumulation is caused by drug treatment. The lack of any apparent effect on processing leaves open the second possibility that drug treatment affects the ability of target tissues to respond to the Shh signal. This possibility is supported by our observations with neural plate explants, in which target genes normally induced by recombinant Shh-N protein are unresponsive when the explants are treated with synthetic and plant-derived cholesterol synthesis inhibitors (M.K. Cooper et al., in prep.). An effect on target tissues also is consistent with the occurrence of HPE in megalin mutant mice and with the specific expression of megalin in embryonic neuroepithelium during the critical period of Shh signaling (Willnow et al. 1996). The megalin receptor is able to mediate LDL uptake (Stefansson et al. 1995), and no developmental defects are observed in mice or humans lacking function of the more generally expressed LDL receptor (Ishibashi et al. 1993). This suggests that megalin may function specifically to maintain cholesterol homeostasis in developing neuroepithelium, which is the target of Shh protein signaling during the HPE critical period. A link between cholesterol homeostasis and activity of the Hh pathway is interesting since most responses to Hh signaling in neuroepithelium or other developing tissues involve cell proliferation. This linkage might make sense from a regulatory point of view, given the importance of cholesterol as a membrane component in dividing cells.

Additional information consistent with a role for cholesterol in receiving the Shh signal derives from the recent isolation of the Niemann-Pick C (NP-C) disease gene from mice and humans (Carstea et al. 1997; Loftus et al. 1997). The NP-C protein contains 13–16 transmembrane domains, and two features of its sequence are notable in the context of Shh signaling (Carstea et al. 1997; Loftus et al. 1997). The first is that it resembles that of the Hh pathway protein Patched (Ptc) throughout most of its extent, and the second is that five of the NP-C transmembrane domains constitute an apparent sterol-sensing domain (SSD), which had not previously been noticed in Ptc.

SSD sequences are also present in HMG-CoA reductase and SCAP (SREBP cleavage activating protein; Hua et al. 1996), two proteins involved in maintenance of cholesterol homeostasis (Fig. 7). Although it is not yet known whether the SSD binds cholesterol directly or instead indirectly senses cholesterol-induced changes in membrane properties (see, e.g., Nezil and Bloom 1992), the SSD in

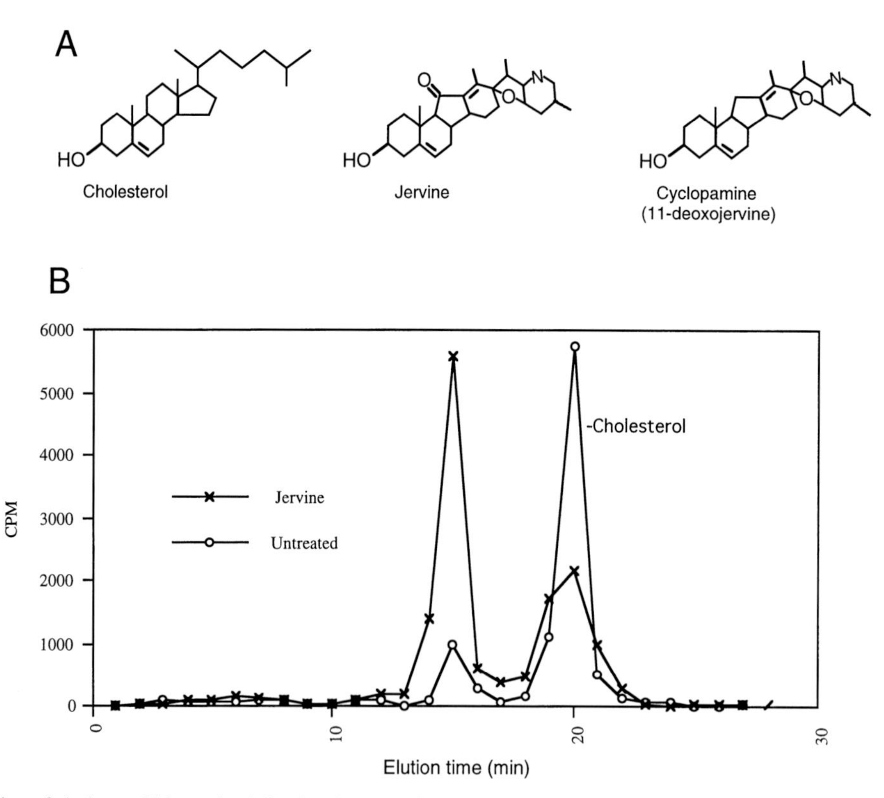

Figure 6. Inhibition of cholesterol biosynthesis by the plant steroidal alkaloid, jervine. Sterols were extracted and analyzed by HPLC from COS7 cells metabolically labeled with [^{3}H]mevalonic acid in the presence or absence of jervine, a teratogenic plant steroidal alkaloid. In the presence of 28 μM jervine, radioactively labeled cholesterol levels were reduced and another radioactively labeled sterol was found to accumulate. On the basis of its retention time in this reverse-phase HPLC method (Rodriguez and Parks 1985), this abnormal sterol is tentatively identified as zymosterol, an intermediate in the cholesterol biosynthetic pathway.

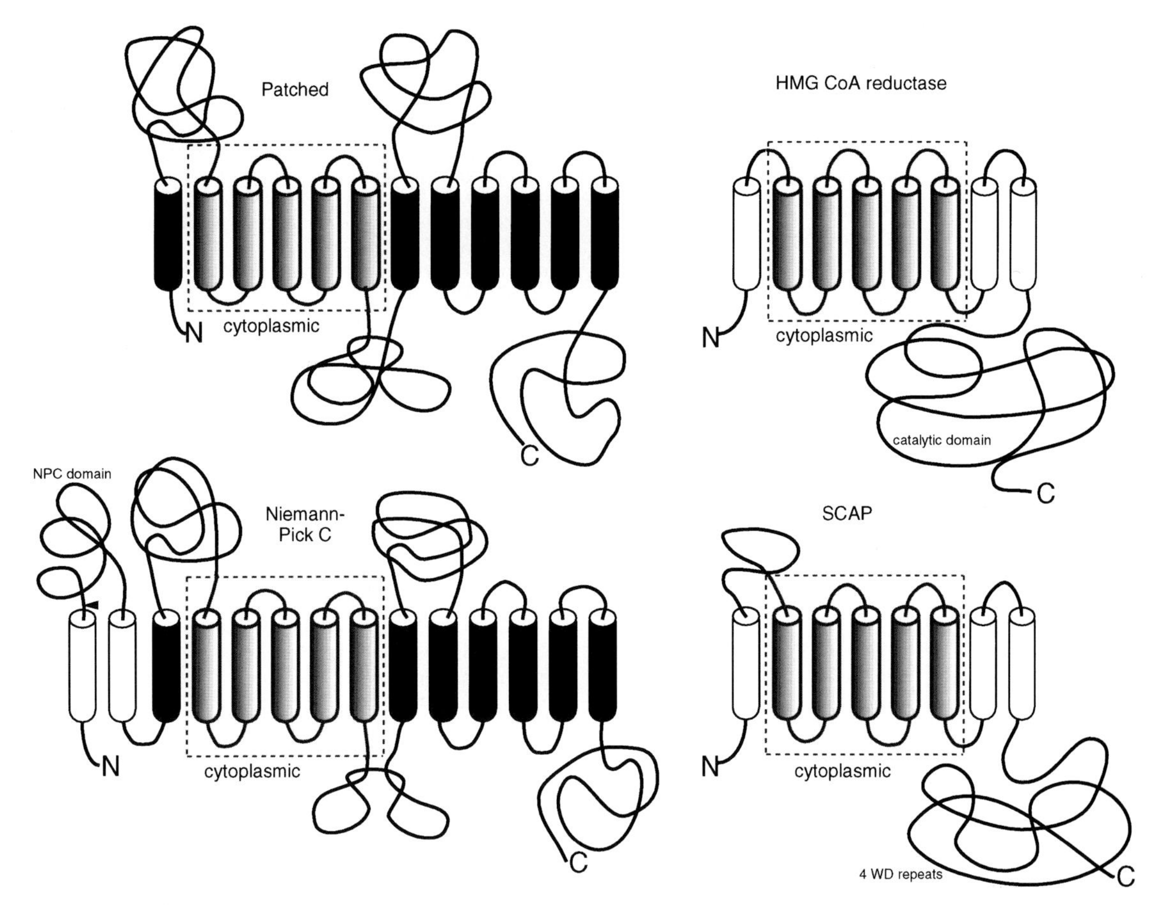

Figure 7. Proteins with sterol-sensing domains. Four proteins containing a sterol-sensing domain (SSD) are schematically depicted. The cylinders denote predicted transmembrane helices and the SSD of each protein is enclosed within the rectangle formed by the dashed lines. As indicated by the shading, the homology between Patched (Ptc) and the Niemann-Pick C disease protein (NP-C) extends beyond the SSD to include all 12 of the Ptc transmembrane domains. In the case of HMG CoA reductase, the topology of these transmembrane segments was experimentally determined (Olender and Simoni 1992; Roitelman et al. 1992). The topology of Ptc shown is as suggested by Goodrich et al. (1996), and that of SCAP as suggested by Brown and Goldstein (1997). The proposed topology of the NP-C protein is based on sequence analysis presented by Carstea et al. (1997) and the homology with Ptc; not shown are several transmembrane domains that are weakly predicted to exist in the human but not the mouse protein. The arrowhead after the first transmembrane domain of NP-C denotes a possible site of signal sequence cleavage as suggested by Carstea et al. (1997). The drawings only crudely approximate the extent of loops between transmembrane domains and are not intended to convey structural information. The labeling within these loops indicates the presence of the catalytic domain in carboxy-terminal portions of HMG CoA reductase, of four repeats of the WD protein-protein interaction domain in the carboxy-terminal portion of SCAP (Hua et al. 1996), and a region in the first loop of NP-C that is tightly conserved among all NP-C homologs from various species (Carstea et al. 1997; Loftus et al. 1997).

these proteins is required for a response to distinct levels of cellular cholesterol. In the case of HMG-CoA reductase, the rate-limiting enzyme in the cholesterol biosynthetic pathway, the SSD mediates decreased enzyme stability under conditions of cholesterol excess (Gil et al. 1985). The SCAP protein responds to sterol levels by modulating the cleavage and activation of SREBP (sterol response-element-binding protein), a transcription factor that controls expression of proteins involved in cholesterol synthesis and uptake (Brown and Goldstein 1997); in the absence of cleavage, SREBP remains anchored to the endoplasmic reticulum via two-transmembrane domains. The NP-C protein itself appears to have a role in cholesterol homeostasis by directly acting in or by regulating cholesterol transport, as indicated in NP-C cultured cells by a delayed response to challenge by LDL with a consequent accumulation of cholesterol in lysosomes.

The Ptc protein has been suggested to constitute or contribute to the Hh receptor mechanism, but its role is not that of a conventional receptor. In the absence of Ptc function, the Hh signaling pathway is constitutively activated (Ingham et al. 1991). The normal function of Ptc thus seems to be suppression of the pathway in target cells, and this suppression is alleviated by the presence of the Hh signal. In addition to this cell autonomous role in suppressing the pathway, the Ptc protein has a role in sequestration of the Hh signal within tissues, a non-cell-autonomous activity that affects the spatial extent of Hh signaling (Chen and Struhl 1996). These two activities can be genetically uncoupled, as demonstrated by a mutant protein that retains the sequestration function but, like other Ptc mutations, does not suppress the Hh pathway (Chen and Struhl 1996). Recent biochemical evidence suggests that Shh-N protein may interact directly with the mouse Ptc protein (Marigo et al. 1996; Stone et al. 1996), but assuming that a direct interaction can be confirmed with the use of purified components, it is unclear whether it would have a role in either, both, or neither of the Ptc functions described above.

These biochemical experiments were carried out with recombinant protein that lacked the cholesterol modification (Marigo et al. 1996; Stone et al. 1996). In light of the presence of an SSD within Ptc, it would be interesting to know whether the Hh cholesterol adduct influences the apparent affinity of the amino-terminal signaling domain for Ptc. Along these lines, it is interesting to note that *Drosophila* embryos expressing a truncated *hh* construct in the normal spatial pattern (Porter et al. 1996b) are similar to embryos that ectopically express full-length Hh and to embryos that lack Ptc function (Chang et al. 1994; Ingham 1993; Porter et al. 1996b): All three of these genotypes show spatially indiscriminate activation of the Hh pathway. One possible explanation of these similarities is that the cholesterol adduct not only increases the association of the Hh signaling domain with producing cells, but also contributes favorably to the interaction with Ptc. In the absence of the cholesterol adduct, the Hh-N protein would not be as effectively sequestered by Ptc, leading to ectopic Hh pathway activation.

An enhanced sequestration by Ptc of the cholesterol-modified signaling domain would represent a role distinct from that of mediating the response to cholesterol homeostasis within the target cell, although it is conceivable that the Ptc SSD might be involved in both. Given our incomplete knowledge of the mechanistic role of Ptc, all discussion as to the role of its SSD must be considered speculative. But the presence of an SSD in a protein with such a central role in regulating the Hh response is tantalizing as the possible link to the dual roles of cholesterol in limiting the spatial extent of Hh signaling and in facilitating transduction of the Hh signal within target cells.

ACKNOWLEDGMENTS

The authors thank D. Andrew and R. Mann for comments on the manuscript, and R. Mann and X. Wang for communication of unpublished data. P.A.B. and D.J.L. are investigators and M.K.C. is a physician postdoctoral fellow of the Howard Hughes Medical Institute.

REFERENCES

Adelmann H.B. 1936a. The problem of cyclopia (part I). *Q. Rev. Biol.* **11:** 116.

———. 1936b. The problem of cyclopia (part II). *Q. Rev. Biol.* **11:** 284.

Belloni E., Muenke M., Roessler E., Traverso G., Siegel-Bartelt J., Frumkin A., Mitchell H.F., Donis-Keller H., Helms C., Hing A.V., Heng H.H.Q., Koop B., Martindale D., Rommens J.M., Tsui L.-C., and Scherer S.W. 1996. Identification of *Sonic hedgehog* as a candidate gene responsible for holoprosencephaly. *Nat. Genet.* **14:** 353.

Binns W., James L.F., Shupe J.L., and Everett G. 1963. A congenital cyclopian-type malformation in lambs induced by maternal ingestion of a range plant, *Veratrum californicum*. *Am. J. Vet. Res.* **24:** 1164.

Brannigan J.A., Dodson G., Duggleby H.J., Moody P.C.E., Smith J.L., Tomchick D.R., and Murzin A.G. 1995. A protein catalytic framework with an N-terminal nucleophile is capable of self-activation. *Nature* **378:** 416.

Brown M.S. and Goldstein J.L. 1997. The SREBP pathway: Regulation of cholesterol metabolism by proteolysis of a membrane-bound transcription factor. *Cell* **89:** 331.

Bumcrot D.A., Takada R., and McMahon A.P. 1995. Proteolytic processing yields two secreted forms of *Sonic hedgehog*. *Mol. Cell. Biol.* **15:** 2294.

Burglin T.R. 1996. Warthog and Groundhog, novel families related to Hedgehog. *Curr. Biol.* **6:** 1047.

Carstea E.D., Morris J.A., Coleman K.G., Loftus S.K., Zhang D., Cummings C., Gu J., Rosenfeld M.A., Pavan W.J., Krizman D.B., Nagle J., Polymeropoulos M.H., Sturley S.L., Ioannou Y.A., Higgins M.E., Comly M., Cooney A., Brown A., Kaneski C.R., Blanchette-Mackie E.J., Dwyer N.K., Neufeld E.B., Chang T.-Y., Liscum L., Strauss J.F., III, Ohno K., Zeigler M., Carmi R., Sokol J., Markie D., O'Neill R.R., van Diggelen O.P., Elleder M., Patterson M.C., Brady R.O., Vanier M.T., Pentchev P.G., and Tagle D.A. 1997. Niemann-Pick C1 disease gene: Homology to mediators of cholesterol homeostasis. *Science* **277:** 228.

Chang D.T., Lopez A., von Kessler D.P., Chiang C., Simandl B.K., Zhao R., Seldin M.F., Fallon J.F., and Beachy P.A. 1994. Products, genetic linkage, and limb patterning activity of a murine *hedgehog* gene. *Development* **120:** 3339.

Chen Y. and Struhl G. 1996. Dual roles for patched in sequestering and transducing hedgehog. *Cell* **87:** 553.

Chiang C., Litingtung Y., Lee E., Young K., Corden J.L., Westphal H., and Beachy P. 1996. Cyclopia and defective axial patterning in mice lacking *Sonic hedgehog* gene function. *Nature* **383:** 407.

Chu C.T. and Pizzo S.V. 1994. α-2 macroglobulin, complement, and biological defense: Antigens, growth factors, microbial proteases, and receptor ligation. *Lab. Invest.* **71:** 792.

Cohen M.M. and Sulik K.K. 1992. Perspectives on holoprosencephaly. II. Central nervous system, craniofacial anatomy, syndrome commentary, diagnostic approach, and experimental studies. *J. Craniofacial Genet. Dev. Biol.* **12:** 196.

Dalgaard J.Z., Moser M.J., Hughey R., and Mian I.S. 1997. Statistical modeling, phylogenetic analysis and structure prediction of a protein splicing domain common to inteins and Hedgehog proteins. *J. Comput. Biol.* **4:** 193.

Dawson P.E., Muir T.W., Clark-Lewis I., and Kent S.B.H. 1994. Synthesis of proteins by native chemical ligation. *Science* **266:** 776.

Dehart D.B., Lanoue L., Tint G.S., and Sulik K.K. 1997. Pathogenesis of malformations in a rodent model for Smith-Lemli-Opitz syndrome. *Am. J. Med. Genet.* **68:** 328.

Dideberg O., Charlier P., Dive G., Joris B., Frére J.M., and Ghuysen J.M. 1982. Structure of a Zn^{2+}-containing D-alanyl-D-alanine-cleaving carboxypeptidase at 2.5 Å resolution. *Nature* **299:** 469.

Duan X., Gimble F.S., and Quiocho F.A. 1997. Crystal structure of PI-SceI, a homing endonuclease with protein splicing activity. *Cell* **89:** 555.

Echelard Y., Epstein D.J., St. Jacques B., Shen L., Mohler J., McMahon J.A., and McMahon A.P. 1993. Sonic hedgehog, a member of a family of putative signalling molecules, is implicated in the regulation of CNS polarity. *Cell* **75:** 1417.

Ekker S.C., Ungar A.R., Greenstein P., von Kessler D.P., Porter J.A., Moon R.T., and Beachy P.A. 1995. Patterning activities of vertebrate *hedgehog* proteins in the developing eye and brain. *Curr. Biol.* **5:** 944.

Ericson J., Rashbass P., Schedl A., Brenner-Morton S., Kawakami A., van Heyningen V., Jessell T.M., and Briscoe J. 1997. Pax6 controls progenitor cell identity and neuronal fate in response to graded Shh signaling. *Cell* **90:** 169.

Farese R.V., Ruland S.L., Flynn L.M., Stokowski R.P., and Young S.G. 1995. Knockout of the mouse apolipoprotein B gene results in embryonic lethality in homozygotes and protection against diet-induced hypercholesterolemia in heterozygotes. *Proc. Natl. Acad. Sci.* **92:** 1774.

Farese R.V., Cases S., Ruland S.L., Kayden H.J., Wong J.S., Young S.G., and Hamilton R.L. 1996. A novel function for apolipoprotein B: Lipoprotein synthesis in the yolk sac is critical for maternal-fetal lipid transport in mice. *J. Lipid Res.* **37:** 347.

Gaffield W. and Keeler R.F. 1996. Steroidal alkaloid teratogens: Molecular probes for investigation of craniofacial malformations. *J. Toxicol.* **15:** 303.

Gil G., Faust J.R., Chin D.J., Goldstein J.L., and Brown M.S. 1985. Membrane-bound domain of HMG CoA reductase is required for sterol-enhanced degradation of the enzyme. *Cell* **41:** 249.

Goodrich L.V., Johnson R.L., Milenkovic L., McMahon J.A., and Scott M.P. 1996. Conservation of the *hedgehog/patched* signaling pathway from flies to mice: Induction of a mouse *patched* gene by Hedgehog. *Genes Dev.* **10:** 301.

Guan C., Cui T., Rao V., Liao W., Benner J., Lin C.-L., and Comb D. 1996. Activation of glycosylasparaginase: Formation of active N-terminal threonine by intramolecular autoproteolysis. *J. Biol. Chem.* **271:** 1732.

Hall T.M.T., Porter J.A., Beachy P.A., and Leahy D.J. 1995. A potential catalytic site revealed by the 1.7-Å crystal structure of the amino-terminal signalling domain of Sonic hedgehog. *Nature* **378:** 212.

Hall T.M.T., Porter J.A., Young K.E., Koonin E.V., Beachy P.A., and Leahy D.J. 1997. Crystal structure of a Hedgehog autoprocessing domain: Conservation of structure, sequence, and reaction mechanism between Hedgehog and self-splicing proteins. *Cell.* **91:** 85.

Hammerschmidt M., Brook A., and McMahon A.P. 1997. The world according to hedgehog. *Trends Genet.* **13:** 14.

Herz J., Willnow T.E., and Farese R.V., Jr. 1997. Cholesterol, hedgehog and embryogenesis. *Nat. Genet.* **15:** 123.

Hochstrasser M. 1996. Ubiquitin-dependent protein degradation. *Annu. Rev. Genet.* **30:** 405.

Homanics G.E., Smith T.J., Zhang S.H., Lee D., Young S.G., and Maeda N. 1993. Targeted modification of the apolipoprotein B gene results in hypobetalipoproteinemia and developmental abnormalities in mice. *Proc. Natl. Acad. Sci.* **90:** 2389.

Hua X., Nohturfft A., Goldstein J.L., and Brown M.S. 1996. Sterol resistance in CHO cells traced to point mutation in SREBP cleavage-activating protein. *Cell* **87:** 415.

Huang L.-S., Voyiaziakis E., Markenson D.F., Sokol K.A., Hayek T., and Breslow J.L. 1995. apo B gene knockout in mice results in embryonic lethality in homozygotes and neural tube defects, male infertility, and reduced HDL cholesterol ester and apo A-1 transport rates in heterozygotes. *J. Clin. Invest.* **96:** 2152.

Ingham P.W. 1993. Localized *hedgehog* activity controls spatial limits of *wingless* transcription in the *Drosophila* embryo. *Nature* **366:** 560.

Ingham P.W., Taylor A.M., and Nakano Y. 1991. Role of the *Drosophila patched* gene in positional signalling. *Nature* **353:** 184.

Ishibashi S., Brown M.S., Goldstein J.L., Gerard R.D., Hammer R.E., and Herz J. 1993. Hypercholesterolemia in low density lipoprotein receptor knockout mice and its reversal by adenovirus-mediated gene delivery. *J. Clin. Invest.* **92:** 883.

Keeler R.F. and Binns W. 1968. Teratogenic compounds of *Veratrum californicum* (Durand). V. Comparison of cyclopian effects of steroidal alkaloids from the plant and structurally related compounds from other sources. *Teratology* **1:** 5.

Kelley R.I., Roessler E., Hennekam R.C.M., Feldman G.I., Kosaki K., Jones M.C., Palumbos J.C., and Muenke M. 1996. Holoprosencephaly in RSH/Smith-Lemli-Opitz syndrome: Does abnormal cholesterol metabolism affect the function of *Sonic hedgehog. Am. J. Med. Genet.* **66:** 478.

Lai C.-J., Ekker S.C., Beachy P.A., and Moon R.T. 1995. Patterning of the neural ectoderm of *Xenopus laevis* by the amino-terminal product of hedgehog autoproteolytic cleavage. *Development* **121:** 2349.

Lanoue L., Dehart D.B., Hinsdale M.E., Maeda N., Tint G.T., and Sulik K.K. 1997. Limb, genital, CNS and facial malformations result from gene/environment-induced cholesterol deficiency: Further evidence for a link to Sonic hedgehog. *Am. J. Med. Genet.* **73:** 24.

Lee J.J., von Kessler D.P., Parks S., and Beachy P.A. 1992. Secretion and localized transcription suggest a role in positional signaling for products of the segmentation gene *hedgehog. Cell* **71:** 33.

Lee J.J., Ekker S.C., von Kessler D.P., Porter J.A., Sun B.I., and Beachy P.A. 1994. Autoproteolysis in *hedgehog* protein biogenesis. *Science* **266:** 1528.

Loftus S.K., Morris J.A., Carstea E.D., Gu J.Z., Cummings C., Brown A., Ellison J., Ohno K., Rosenfeld M.A., Tagle D.A., Pentchev P.G., and Pavan W.J. 1997. Murine model of Niemann-Pick C disease: Mutation in a cholesterol homeostasis gene. *Science* **277:** 232.

Mangold O. 1931. Das determinationsproblem. III. Das wirbeltierauge in der entwicklung und regeneration. *Ergeb. Biol.* **7:** 193.

Marigo V., Davey R.A., Zuo Y., Cunningham J.M., and Tabin C.J. 1996. Biochemical evidence that Patched is the Hedgehog receptor. *Nature* **384:** 176.

Marti E., Bumcrot D.A., Takada R., and McMahon A.P. 1995. Requirement of 19K form of Sonic hedgehog for induction of distinct ventral cell types in CNS explants. *Nature* **375:** 322.

Mohler J. 1988. Requirements for *hedgehog*, a segmental polarity gene, in patterning larval and adult cuticle of *Drosophila. Genetics* **120:** 1061.

Mohler J. and Vani K. 1992. Molecular organization and embryonic expression of the *hedgehog* gene involved in cell-cell communication in segmental patterning of *Drosophila. Development* **115:** 957.

Murzin A.G. 1996. Structural classification of proteins: New superfamilies. *Curr. Opin. Struct. Biol.* **6:** 386.

Nezil F.A. and Bloom M. 1992. Combined influence of cholesterol and synthetic amphiphilic peptides upon bilayer thickness in model membranes. *Biophys. J.* **61:** 1176.

Olender E.H. and Simoni R.D. 1992. The intracellular targeting and membrane topology of 3-hydroxy-3-methylglutaryl-CoA reductase. *J. Biol. Chem.* **267:** 4223.

Opitz J.M. 1994. RSH/SLO ("Smith-Lemli-Opitz") syndrome: Historical, genetic, and developmental considerations. *Am. J. Med. Genet.* **50:** 344.

Perler F.B., Davis E.O., Dean G.E., Gimble F.S., Jack W.E., Neff N., Noren C.J., Thorner J., and Belfort M. 1994. Protein splicing elements: Inteins and exteins—A definition of terms and recommended nomenclature. *Nucleic Acids Res.* **22:** 1125.

Pietrokovski S. 1997. Modular organization of inteins and C-terminal autocatalytic domains. *Protein Sci.* (in press).

Placzek M., Jessell T.M., and Dodd J. 1993. Induction of floor plate differentiation by contact-dependent, homeogenetic signals. *Development* **117:** 205.

Popjak G., Meenan A., Parish E.J., and Nes W.D. 1989. Inhibition of cholesterol synthesis and cell growth by 24(R,S), 25-iminolanosterol and triparanol in cultured rat hepatoma cells. *J. Biol. Chem.* **264:** 630.

Porter J.A., Young K.E., and Beachy P.A. 1996a. Cholesterol modification of Hedgehog signaling proteins in animal development. *Science* **274:** 255.

Porter J.A., von Kessler D.P., Ekker S.C., Young K.E., Lee J.J., Moses K., and Beachy P.A. 1995. The product of *hedgehog* autoproteolytic cleavage active in local and long-range signalling. *Nature* **374:** 363.

Porter J.A., Ekker S.C., Park W.-J., von Kessler D.P., Young K.E., Chen C.-H., Ma Y., Woods A.S., Cotter R.J., Koonin E.V., and Beachy P.A. 1996b. Hedgehog patterning activity: Role of a lipophilic modification mediated by the carboxy-terminal autoprocessing domain. *Cell* **86:** 21.

Rilling H.S., Bruenger E., Leining L.M., Buss J.E., and Epstein W.W. 1993. Differential prenylation of proteins as a function of mevalonate concentration in CHO cells. *Arch. Biochem. Biophys.* **301:** 210.

Rodriguez R.J. and Parks L.W. 1985. High-performance liquid chromatography of sterols: Yeast sterols. *Methods Enzymol.* **111:** 37.

Roelink H., Porter J.A., Chiang C., Tanabe Y., Chang D.T., Beachy P.A., and Jessell T.M. 1995. Floor plate and motor neuron induction by different concentrations of the amino-terminal cleavage product of Sonic hedgehog autoproteolysis. *Cell* **81:** 445.

Roessler E., Belloni E., Gaudenz K., Jay P., Berta P., Scherer S.W., Tsui L.-C., and Muenke M. 1996. Mutations in *Sonic hedgehog* gene cause holoprosencephaly. *Nat. Genet.* **14:** 357.

Roitelman J., Olender E.H., Bar-Nun S., Dunn W.A., Jr., and Simoni R.D. 1992. Immunological evidence for eight spans in the membrane domain of 3-hydroxy-3-methylglutaryl coenzyme A reductase: Implications for enzyme degradation in the endoplasmic reticulum. *J. Cell Biol.* **117:** 959.

Roux, C. 1964. Action teratogene du triparanol chez l'animal. *Arch. Franc. Pediatr.* **21:** 451.

———. 1966. Action teratogene chez le rat d'un inhibiteur de la synthese du cholesterol, le AY9944. *C.R. Soc. Biol.* **160:** 1353.

Roux C., Horvath C., and Dupuis R. 1979. Teratogenic action and embryo lethality of AY 9944R: Prevention by a hypercholesterolemia-provoking diet. *Teratology* **19:** 35.

Salen G., Shefer S., Batta A.K., Tint G.S., Xu G., Honda A., Irons M., and Elias E.R. 1996. Abnormal cholesterol biosynthesis in the Smith-Lemli-Opitz syndrome. *J. Lipid Res.* **37:** 1169.

Stefansson S., Chappell D.A., Argraves K.M., Strickland D.K., and Argraves W.S. 1995. Glycoprotein 330/low density lipoprotein receptor-related protein-2 mediates endocytosis of low density lipoproteins via interaction with apolipoprotein B100. *J. Biol. Chem.* **270:** 19417.

Stone D.M., Hynes M., Armanini M., Swanson T.A., Gu Q., Johnson R.L., Scott M.P., Pennica D., Goddard A., Phillips H., Noll M., Hooper J.E., de Sauvage F., and Rosenthal A. 1996. The tumor-suppressor gene *patched* encodes a candidate receptor for Sonic hedgehog. *Nature* **384:** 129.

Tabata T., Eaton S., and Kornberg T.B. 1992. The *Drosophila hedgehog* gene is expressed specifically in posterior compartment cells and is a target of *engrailed* regulation. *Genes Dev.* **6:** 2635.

Tashiro S., Michiue T., Higashijima S., Zenno S., Ishimaru S., Takahashi F., Orihara M., Kojima T., and Saigo K. 1993. Structure and expression of *hedgehog*, a *Drosophila* segment-polarity gene required for cell-cell communication. *Gene* **124:** 183.

Tint G.S., Irons M., Elias E.R., Batta A.K., Frieden R., Chen T.S., and Salen G. 1994. Defective cholesterol biosynthesis associated with the Smith-Lemli-Opitz syndrome. *N. Engl. J. Med.* **330:** 107.

van Poelje P.D. and Snell E.E. 1990. Pyruvoyl-dependent enzymes. *Annu. Rev. Biochem.* **59:** 29.

Willnow T.E., Hilpert J., Armstrong S.A., Rohlman A., Hammer R.E., Burns D.K., and Herz J. 1996. Defective forebrain development in mice lacking gp330/megalin. *Proc. Natl. Acad. Sci.* **93:** 8460.

Xu M.-Q. and Perler F.B. 1996. The mechanism of protein splicing and its modulation by mutation. *EMBO J.* **15:** 5146.

Control of Cell Growth and Fate by *patched* Genes

R.L. JOHNSON AND M.P. SCOTT
Howard Hughes Medical Institute, Departments of Developmental Biology and Genetics, Stanford University School of Medicine, Stanford, California 94305-5427

When the Hedgehog (Hh) protein signals to a cell bearing the Patched (Ptc) receptor, events are triggered which can drastically affect the cell's fate and growth properties. Ptc exerts its profound effects by inhibiting the transcription of a number of different genes, such as members of the Wnt and transforming growth factor-β (TGF-β) families and the *ptc* gene itself. Hh opposes Ptc activity to induce gene transcription; cells receiving the Hh signal respond by inducing high levels of *ptc* (Fig. 1) and other transcriptional targets. The Hh/Ptc signaling system, originally identified in *Drosophila*, is now known to be broadly conserved in the animal kingdom (for review, see Hammerschmidt et al. 1997). The involvement of *ptc* mutations in different human tumor types demonstrates an important additional role for Hh and Ptc in the control of cell proliferation in adult tissues. Understanding how Hh and Ptc regulate gene expression will have substantial implications for comprehending animal development and tumorigenesis.

Ptc TRANSCRIPTION AS AN INDICATOR OF Hh SIGNALING DURING DEVELOPMENT

The outcome of Hh signaling is the transcriptional activation of specific target genes, and the target genes can vary with tissue and developmental stage. For example, in *Drosophila*, *wingless* (*wg*), a homolog of the vertebrate Wnt family (Rijsewijk et al. 1987), is activated by Hh in embryos and some imaginal discs (Hidalgo and Ingham 1990; Basler and Struhl 1994), whereas the TGF-β class gene *decapentaplegic* (*dpp*) (Padgett et al. 1987) is activated by Hh only in discs (Basler and Struhl 1994; Kojima et al. 1994; Tabata and Kornberg 1994; Felsenfeld and Kennison 1995; Ingham and Fietz 1995). Although the genes controlled by Hh vary from tissue to tissue, one common target of Hh regulation is the *ptc* gene itself (Hidalgo and Ingham 1990). When Ptc is inhibiting target gene transcription, it keeps its own transcript levels low. In the presence of Hh ligand, *ptc* transcripts are elevated (Hidalgo and Ingham 1990; Ingham et al. 1991). Expression of Hh in areas where it is normally not present results in increases in *ptc* levels (Basler and Struhl 1994; Kojima et al. 1994; Tabata and Kornberg 1994; Ingham and Fietz 1995). Hence, cells that receive the Hh signal respond by raising *ptc* expression.

Why should a ligand induce higher-level expression of a receptor while at the same time opposing the receptor's function? One possibility is that the buildup of Ptc leads to feedback damping of Hh signaling, limiting the duration of the effective signal. In keeping with this idea, increased production of Ptc overcomes normal Hh functions and turns off target genes where they should be activated (Schuske et al. 1994; Johnson et al. 1995). Another possibility is that the range of action of the Hh signal is restricted by binding and sequestration of Hh by Ptc, an idea supported by the increased range of Hh protein (Taylor et al. 1993) and Hh response (Chen and Struhl 1996) when cells lack Ptc. Whatever the function of Ptc induction, it is perhaps the most reliable reporter to date of Hh signal received.

High-level *ptc* transcription may provide a measure of how far Hh moves from its source. *ptc* induction appears to be due uniquely to Hh and is detected adjacent to all sources of Hh proteins reported to date. *ptc* expression patterns in *Drosophila* and vertebrates suggest that Hh can act over a broad range. In the fly embryo, Hh emanates from the posterior cells of each segment primordium and acts upon more anterior cells to maintain the expression of *wg* (Ingham et al. 1991). The range of normal Hh action in fly embryos appears to be about two cells, but expanded Hh expression readily induces *wg* transcription in additional anterior cells (Ingham 1993; Chang et al. 1994). In vertebrates, the Hh homolog Sonic Hedgehog (Shh), appears to have both short- and long-range actions. In the developing neural tube, Shh is first produced in the notochord to induce *ptc* in the adjacent floor plate. Here, Shh range is short, as indicated by *ptc* transcription (Goodrich et al. 1996; Hahn et al. 1996a; Marigo and Tabin 1996) and by the contact dependence of the notochord-derived signal (Placzek et al. 1990). Later, the floor plate itself becomes a source of Shh to induce motor neurons further away in the ventral lateral regions (Echelard et al. 1993; Roelink et al. 1994). At this time, *ptc* expression is absent from the floor plate but is highly elevated in the ventral neural tube where it appears as a gradient reaching far toward the dorsal neural tube (Goodrich et al. 1996; Hahn et al. 1996a; Marigo and Tabin 1996). The induction of *ptc* over a broad area of the neural tube suggests that Shh is acting over many cell diameters, a hypothesis supported by tissue culture experiments (Roelink et al. 1995; Ericson et al. 1996).

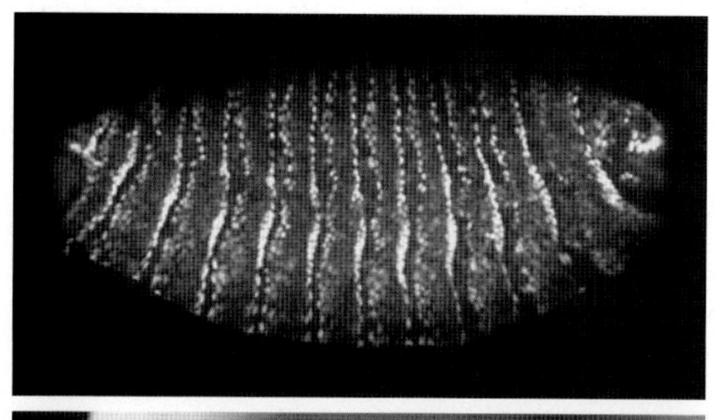

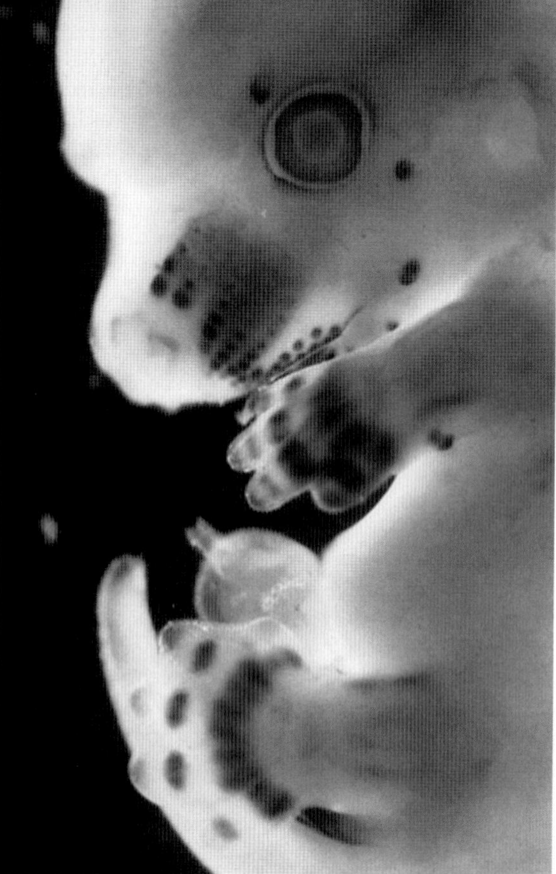

Figure 1. *ptc* expression in *Drosophila* and mouse. (*Top panel*) *ptc* LacZ expression in a *Drosophila* embryo. During late embryogenesis, *ptc* is expressed in all cells that lack the transcription factor, Engrailed. The bright pinstripes of *ptc* staining indicate cells where Ptc activity has been blocked by the adjacent sources of Hh expression. Low *ptc* staining in each segment indicates cells outside the range of Hh where Ptc is actively inhibiting its own transcription. (*Lower panel*) *ptc* mRNA expression in a mouse at about 13.5 days of development. High levels of *ptc* expression are detected in the developing limb bud and hair follicles adjacent to sources of Ihh and Shh, respectively. Elevated levels of *ptc* suggest that Ptc-negative autoregulation is inhibited by Hh family members.

ROLE OF PTC IN TUMOR SUPPRESSION

The first evidence implicating Ptc function in tumorigenesis was the discovery that individuals with the inherited disorder, basal cell nevus syndrome (BCNS; also known as Gorlin or nevoid basal cell carcinoma syndrome) carry mutations in *PATCHED* (*PTC*), a human *ptc* homolog (Hahn et al. 1996b; Johnson et al. 1996). The developmental defects in these individuals correlate well with the developing tissues known from mouse studies to employ Hh/Ptc signaling. For example, mouse *ptc* is expressed prominently in the developing limb, neural tube, and branchial arches (Goodrich et al. 1996; Hahn et al. 1996a), and BCNS patients can manifest polydactyly, spina bifida, and abnormal craniofacial development (Gorlin 1995; Bale 1997). BCNS patients also have a high incidence of certain tumors, most notably basal cell carcinoma (BCC) of the skin and medulloblastoma, a dangerous tumor most commonly observed in the cerebellum. Most BCNS patients develop large numbers of BCCs by age 30 (Kimonis et al. 1997), whereas about 3% acquire medulloblastoma as young children (Lacombe et al. 1990; Evans et al. 1991).

More often, BCCs and medulloblastomas occur sporadically in normal individuals. BCC is a very common cancer in people of northern European descent where the tumors arise in small numbers late in life (Miller 1991; Miller and Weinstock 1994). Both inherited and sporadic forms of BCC arise from mutations in a tumor suppressor locus on chromosome 9q (Gailani et al. 1992). BCNS individuals were predicted to inherit one normal and one defective copy of this gene, and BCCs were proposed to arise by mutation of both copies (Gailani et al. 1992; Chenevix-Trench et al. 1993; Bonifas et al. 1994; Compton et al. 1994; Goldstein et al. 1994a; Wicking et al. 1994). These predictions were borne out upon the identification of *PTC* mutations in BCNS patients and in sporadic BCCs (Hahn et al. 1996b; Johnson et al. 1996). In many sporadic and inherited BCCs, both alleles of *PTC* are mutated, suggesting that tumors arise from homozygous loss of *PTC* function (Gailani et al. 1996; Hahn et al. 1996b; Unden et al. 1996). In *Drosophila*, one indicator of Ptc activity is the negative regulation of its own transcription—the loss of Ptc function leads to the accumulation of high levels of *ptc* mRNA (Hidalgo and Ingham 1990). In human BCCs as well, transcripts of *PTC* accumulate to high levels, implying that Ptc function is greatly reduced in the tumor cells and that Ptc autoregulation is conserved between flies and vertebrates (Gailani et al. 1996). *PTC* mutations have been found in sporadic medulloblastomas, and, as in BCCs, both alleles appear to be damaged (Pietsch et al. 1997; Raffel et al. 1997; Xie et al. 1997). This suggests that loss of PTC function leads to the formation of several different types of tumors.

The tumor-suppressing properties of Ptc correlate with its role in *Drosophila* wing development and its inferred role in vertebrate limb formation. In both of these tissues, ectopic Hh (Basler and Struhl 1994; Kojima et al. 1994; Tabata and Kornberg 1994; Felsenfeld and Kennison 1995; Ingham and Fietz 1995) or Shh (Riddle et al. 1993; Chang et al. 1994) signaling can induce rapid and dramatic proliferation of cells. Correspondingly, the loss of Ptc function in the wing imaginal disc also leads to overgrowth (Capdevila et al. 1994). However, the absence of

Ptc activity does not always result in cell proliferation. In homozygous *Drosophila ptc*$^-$ embryos, cell fates and *wg* transcription are altered, but there is no excess growth (Hidalgo and Ingham 1990). Why cells respond differently to the Hh signal is not well understood. Possibly the cell's history or its neighbors influence the ways it interprets the signal. Understanding the link between cell differentiation and division, and how Hh influences these processes, will be critical for understanding how BCCs and medulloblastomas form.

SURPRISING COMPONENTS IN THE Hh SIGNALING PATHWAY

The membrane proteins, Ptc and Smoothened (Smo), are important in regulating the detection of Hh at the plasma membrane (Fig. 2). Ptc contains multiple transmembrane domains (Hooper and Scott 1989; Nakano et al. 1989) and constitutively inhibits activation of the pathway (Ingham et al. 1991). Ptc is thought to mediate its repressive effects by blocking the activity of Smo. Smo has seven potential transmembrane domains, is genetically downstream from Ptc, and is required to activate the pathway (Alcedo et al. 1996; van den Heuvel and Ingham 1996). Biochemical studies in vertebrate systems have shown that Shh binds to Ptc but not Smo (Marigo et al. 1996b; Stone et al. 1996). In addition, Ptc and Smo associate with one another in the membrane (Stone et al. 1996). These results suggest that in the absence of Hh, Ptc associates with and prevents Smo from activating the pathway. The binding of Hh to Ptc somehow blocks Ptc inhibition of Smo (although Ptc and Smo appear to remain associated) and permits activation of gene transcription.

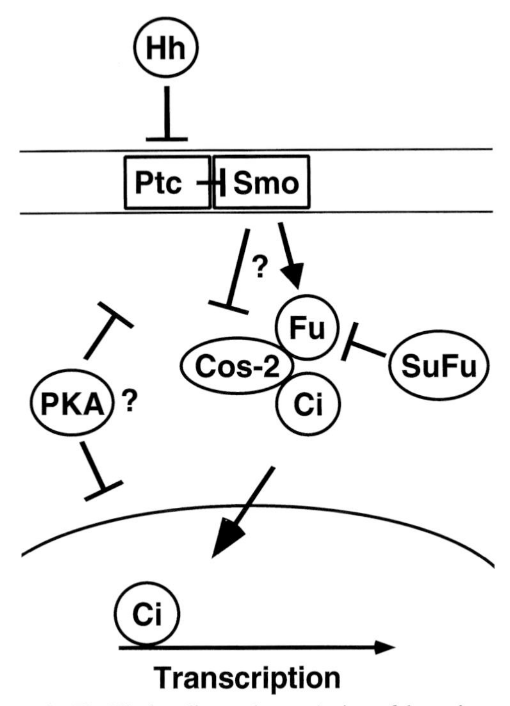

Figure 2. The Hh signaling pathway. A view of the pathway assembled from *Drosophila* and vertebrate studies. Hh is proposed to bind to Ptc and block its activity, thereby relieving Ptc inhibition of Smo. Smo transmits the signal to the cytosol in an unknown manner to either activate Fu or inhibit Cos2. Cos2, Fu, and Ci are part of a complex that regulates the entry of Ci into the nucleus. In the nucleus, Ci binds to upstream sequences of select genes to activate or repress transcription. PKA and SuFu block Hh signaling in an unknown manner.

Ptc, Hh, and Smo might have activities other than signaling in this pathway. In the embryonic neuroectoderm of *Drosophila*, Ptc inhibition of gene expression does not appear to be blocked by Hh but by an unidentified factor (Bhat 1996; Bhat and Schedl 1997). In the *Drosophila* epidermis, Hh has Ptc-independent influences on pattern (Bejsovec and Wieschaus 1993; Bokor and DiNardo 1996). Hence, Ptc may bind other ligands and Hh may have other receptors. Smo has an extracellular cysteine-rich domain that shares sequence similarity to the Wnt-binding region of the serpentine Frizzled receptors (Alcedo et al. 1996; van den Heuvel and Ingham 1996). This raises the possibility that Smo may bind a Wnt ligand and integrate Wnt and Hh signaling. Although Smo has some sequence hallmarks of a G-protein-coupled receptor (Alcedo et al. 1996; van den Heuvel and Ingham 1996), there is yet no biochemical or genetic evidence implicating heterotrimeric G-proteins in this pathway.

In *Drosophila*, a large cytoplasmic complex has been implicated in regulating the Hh signal to the nucleus (Aza-Blanc et al. 1997; Robbins et al. 1997; Sisson et al. 1997). Three proteins in this complex have been identified: the serine/threonine kinase, Fused (Fu) (Preat et al. 1990); the zinc finger protein, Cubitus interruptus (Ci) (Orenic et al. 1990); and the kinesin-related protein, Costal2 (Cos2) (Sisson et al. 1997). Fu and Ci are required for transmitting the Hh signal, whereas Cos2, like Ptc, negatively regulates the pathway (Forbes et al. 1993). Ci has homology with the Gli family of transcription factors in vertebrates (Orenic et al. 1990) and is proposed to control directly the transcription of Hh target genes (Alexandre et al. 1996; Dominguez et al. 1996; Aza-Blanc et al. 1997; Hepker et al. 1997; Von Ohlen et al. 1997). Cos2, which binds to microtubules in vitro, may inhibit Hh signaling by tethering the complex to the cytoskeleton and controlling the entry of Ci into the nucleus (Aza-Blanc et al. 1997; Robbins et al. 1997; Sisson et al. 1997). How this complex is regulated is not known, but phosphorylation may be important since Cos2 and Fu become phosphorylated in response to Hh (Therond et al. 1996a; Robbins et al. 1997).

Protein kinase A (PKA) (Jiang and Struhl 1995; Li et al. 1995; Pan and Rubin 1995) and Suppressor of fused (SuFu) (Preat et al. 1993; Pham et al. 1995) act in a manner similar to that of Ptc by being negative regulators of Hh signaling. Whereas *PKA* functions downstream from *smo* (van den Heuvel and Ingham 1996), it has not been placed in a linear pathway with Ptc because activated PKA cannot replace Ptc function (Li et al. 1995). Identifying the relevant targets of PKA phosphorylation will help determine if PKA is in a parallel pathway or different branch of the pathway. SuFu appears to be a nonessential novel protein, yet reduction of SuFu function in the absence of Fu can restore normal development (Preat et al. 1993; Pham et al. 1995). The loss of Fu kinase activity is apparently tolerable if SuFu function is also reduced. The effects of some Fu alleles are dramatically altered, and are in fact worsened, by reduced SuFu function. These alleles generally map to the potential regulatory domain of Fu, leaving the kinase domain possibly still active. Similarly, reduced Cos2 function suppresses or enhances Fu alleles with the same allele specificity as that of SuFu (Preat et al. 1993; Therond et al. 1996b). The molecular events underlying these remarkable genetic interactions are probably related to the Cos2/Fu/Ci protein complex but are not understood.

Most of the information about Hh signal transduction has come from studies in *Drosophila*, but the expected parallels in vertebrate development are emerging quickly. In mice, the Hh family members are *Shh*, *Desert hedgehog*, and *Indian hedgehog* (*Ihh*) (Echelard et al. 1993). The extent to which each protein sequence confers functional differences is unclear, but each is produced in different cells and has distinct developmental functions. Cells adjacent to the sources of all three Hh proteins activate *ptc* transcription, which suggests that Ptc may be used by all three Hh members. However, the identification of a second *ptc* gene in vertebrates (Concordet et al. 1996; Takabatake et al. 1997) could imply the specialization of different Ptc proteins for different Hh members or the existence of different signal transduction events. As in *Drosophila*, vertebrate PKA has been implicated in the repression of Shh target genes (Concordet et al. 1996; Epstein et al. 1996; Hammerschmidt et al. 1996) and Gli appears to be an activator of Shh targets (Marigo et al. 1996a; Sasaki et al. 1997). However, in contrast to flies, where Ci is not transcriptionally regulated by Hh (Motzny and Holmgren 1995; Slusarski et al. 1995), *Gli* transcription is induced by Shh (Marigo et al. 1996a; Sasaki et al. 1997). Vertebrate relatives of Fu, Cos2, and SuFu have not yet been identified.

Aside from *PTC*, can mutations in other genes in the Hh pathway also cause BCC? When Shh is overexpressed under the control of a keratin promoter in the epidermis of developing mice, tumors similar to BCCs are seen (Oro et al. 1997). Presumably, excess production of Shh binds to and inactivates Ptc, mimicking genetic loss of *PTC* in the skin. In addition, potential gain-of-function mutations in human *Shh* have been found in three types of sporadic tumors which raises the possibility that *Shh* is a oncogene (Oro et al. 1997). This evidence argues that activation of Shh target genes, whether by increasing Shh activity or decreasing Ptc function, causes BCC formation. Other components of the pathway, like *smo* or *Gli*, might also become oncogenic in some BCCs or medulloblastomas. Although *Gli* has not yet been implicated in these tumors, it has been found to be amplified in another brain tumor, glioblastoma (Kinzler et al. 1987)

CONSERVATION OF THE Ptc PROTEIN

Homologs of *ptc* have been identified in mice (Goodrich et al. 1996; Hahn et al. 1996a), humans (Hahn et al. 1996a; Johnson et al. 1996), chicks (Marigo et al. 1996c), zebrafish (Concordet et al. 1996), butterflies (L. Goodrich et al., unpubl.), and worms (Wilson et al. 1994). Hydropathy analysis of these homologs suggests that the topological structure of Ptc is well conserved, and Ptc may contain 12 potential transmembrane domains (Fig. 3)(Hooper and Scott 1989; Nakano et al. 1989; Goodrich et al. 1996). The transmembrane regions are arranged as a tandem array that is composed of one membrane-spanning domain separated from a cluster of five domains by a large hydrophilic loop. The large number of membrane-spanning regions and topologically duplicated structure suggest that Ptc may have transporter or channel function in addition to its proposed role as an Hh receptor and regulator of Smo. The highly conserved transmembrane domains may form an aqueous pore through which small molecules or ions pass or they may form a surface that contacts Smo. Unlike many transporters and channels, Ptc appears to lack sequence similarity between its first and second sets of six transmembrane domains. No sequence similarity has been detected between Ptc homologs and any member of the transporter or channel families, and an attempt to detect channel activity in *Xenopus* oocytes expressing Ptc had negative results (Marigo et al. 1996b).

An alignment of all seven Ptc protein sequences shows that about 30% of the residues are conserved over a span of about 1100 amino acids (Fig. 3). This conservation is rather evenly dispersed throughout the putative extracellular and transmembrane regions. In each set of six transmembrane domains, the first two domains have little conservation (16%), whereas the third, fourth, and sixth domains are much more conserved (47%). The first extracellular loop is composed of about 350 residues of which 25% are conserved including six invariant cysteines. The cysteine residues could potentially form disulfide linkages within the loop to form part of a interaction domain for binding to Hh proteins. In addition, an N-linked glycosylation site is also present in the first extracellular loop of five of the seven homologs. The second extracellular loop is smaller, consisting of about 250 amino acids of which 33% are conserved. This loop has no conserved cysteines but does contain an invariant N-linked glycosylation site. Ptc appears to be a glycoprotein since tunicamycin treatment of cells, which blocks N-linked glycosylation, results in reduced mobility of Ptc in polyacrylamide gels (Marigo et al. 1996b). The two conserved glycosylation sites present in the loops are the likely targets of this modification and suggest that the two hydrophilic loops are extracellular.

The amino-terminal, carboxyl-terminal, and central hydrophilic regions of the Ptc homologs are highly divergent in both length and amino acid composition and are proposed to be intracellular. For instance, the carboxyl terminus varies from 53 (zebrafish Ptc 1) to 256 amino acids in length (mouse Ptc), whereas the central cytosolic loop ranges from 70 (worm Ptc) to 159 amino acids (zebrafish Ptc 1). The conserved residues in these regions are located mostly within 20–30 amino acids of the putative transmembrane regions (Fig. 3).

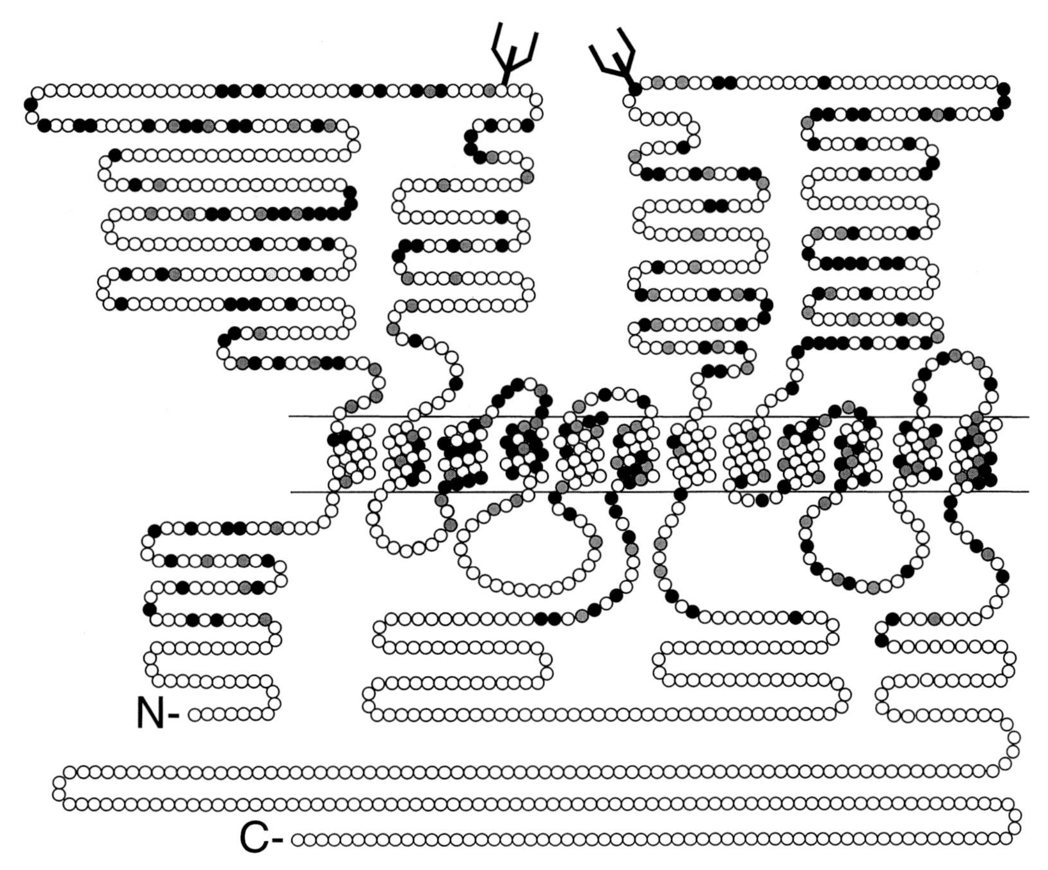

Figure 3. Conservation of the Ptc protein. A model of the mouse Ptc sequence is diagrammed indicating amino acid identity (*black*) and similarity (*gray*) between Ptc homologs from seven species (human, mouse, chick, zebrafish, worm, fly, and butterfly). Ptc is predicted to have 12 transmembrane domains and two extracellular loops, each containing a conserved N-linked glycosylation site.

HUMAN *ptc* MUTATIONS IN BCNS INDIVIDUALS AND BCCs

More than 70 mutations in *PTC* have been reported either in sporadic BCCs or in individuals with the BCNS (Chidambaram et al. 1996; Gailani et al. 1996; Hahn et al. 1996b; Johnson et al. 1996; Unden et al. 1996; Pietsch et al. 1997; Raffel et al. 1997; Wicking et al. 1997; Xie et al. 1997). *PTC* mutations map throughout the predicted protein sequence (Fig. 4) and do not appear to cluster in specific regions, in contrast to the *APC* tumor suppressor gene (Polakis 1995). Of the human *PTC* mutations, about three quarters are nonsense, frameshift, or splice site mutations predicted to result in truncation of *PTC* at positions from amino acid 85 to amino acid 1121. Two thirds of these truncation mutations are predicted to delete at least half of the protein. The most carboxy-terminal truncation mutation yet found involves two independently identified deletions that are predicted to truncate the protein just before transmembrane 11 at amino acid 1121 (Wicking et al. 1997; M. Aszterbaum and E. Epstein, pers. comm.). This mutation suggests that more than 10 of the 12 potential transmembrane domains are required for Ptc to function.

Missense mutations that result in amino acid changes, insertions, or in-frame deletions in *PTC* might indicate residues that are important for function. About half of these mutations map to the two putative extracellular loops of Ptc and may denote residues involved in contacting Hh or Smo. In addition, the loops might also be important in maintaining a stable protein conformation since a three-amino-acid duplication (PNI) at residue 815 in the second extracellular loop results in greatly reduced expression of the protein (Stone et al. 1996). Missense mutations have also been found in some of the putative transmembrane domains and cytosolic loops.

Only 9 of the 23 missense mutations alter highly conserved amino acids, so changes at less conserved positions in Ptc appear to be important. For instance, in two independent sporadic BCCs, an amino acid change and an in-frame deletion have been detected in the highly variable regions of the major cytosolic loop and carboxyl terminus, respectively. BCC formation appears to require complete loss of Ptc function, and in both of the BCCs, the second allele of *PTC* is deleted, so these mutations are unlikely to be benign polymorphisms (Gailani et al. 1996). Therefore, although much of the major cytosolic loop and of the carboxyl terminus is highly variable between species, their composition may be functionally important within a species.

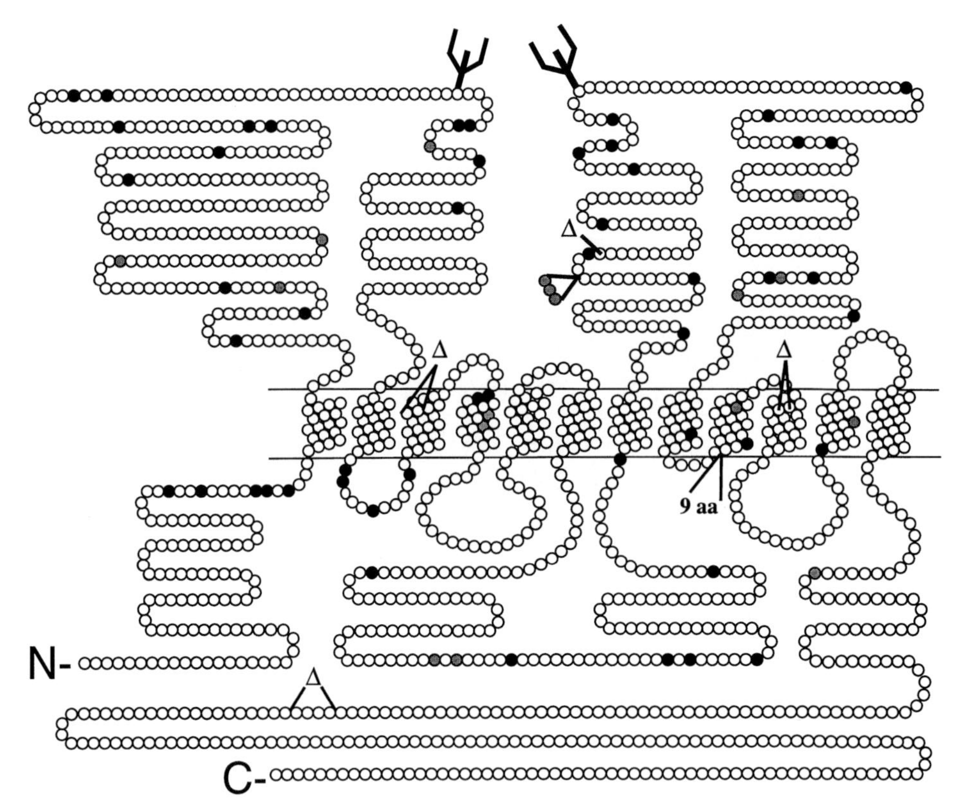

Figure 4. Locations of mutations in human *PTC*. The positions of mutations predicted to result in truncation (*black*), amino acid change (*gray*), and in-frame deletion (Δ) are shown for the human Ptc sequence. 9aa indicates a nine-amino-acid insertion.

The type and severity of developmental defects vary widely within and between BCNS families (Goldstein et al. 1994b; Gorlin 1995). The severity of BCNS phenotypes does not appear to correlate with the extent of predicted truncation of *PTC*, in contrast to findings for the *APC* gene (Polakis 1995). One study of 24 BCNS families found no correlation between the position of truncation mutations in *PTC* and the age of onset of BCCs or the number of defective features. Additionally, individuals with missense mutations did not appear to have milder symptoms (Wicking et al. 1997). This suggests that most mutations in PTC that give rise to identifiable BCNS may result in complete loss of function, whereas mutations that partially reduce *PTC* activity may be phenotypically normal. As yet, no correlation has been found between distinct developmental defects or tumor types in BCNS families and particular types of *PTC* mutations. On the contrary, two families that independently carry the same *PTC* mutation have distinct inherited developmental abnormalities; only one family carries heritable palate defects, whereas the second family does not (Wicking et al. 1997). These data and others (Chidambaram et al. 1996; Unden et al. 1996) suggest the existence of other genetic and environmental factors that influence the types and severity of phenotypes in BCNS individuals.

NPC, A GENE RELATED TO *ptc*, SUGGESTS NEW FUNCTIONS IN Hh SIGNALING

The recent identification of a *ptc*-related gene involved in a cholesterol trafficking disease suggests that Ptc and other components of the pathway may function in vesicle loading, movement, or targeting. Niemann-Pick type C (NPC) disease results from a defect in cholesterol homeostasis in which cells accumulate abnormal amounts of deesterifed cholesterol in the lysosomes and other intracellular compartments (Liscum and Faust 1987; Pentchev et al. 1987). The NPC gene encodes a large multiple transmembrane protein that has homology with Ptc in the hydrophobic regions. One region conserved between NPC and Ptc appears to be a sterol-sensing domain, based on similarity to the sterol-regulated proteins, HMG CoA reductase and SREBP cleavage-activating protein (Carstea et al. 1997; Loftus et al. 1997). This similarity implies that Ptc and NPC may detect and be regulated by sterols. This idea is strengthed by the finding that several conserved residues within the NPC homology regions of Ptc (including the putative sterol-sensing domain) are the locations of missense mutations in BCNS individuals (Fig. 5) (Chidambaram et al.1996; Wicking et al. 1997). The NPC protein also contains a presumed extracellular domain that is rich in cysteine residues and may, like Ptc, bind a ligand (Carstea et al. 1997; Loftus et al. 1997).

What new functions for Hh signaling does the NPC/Ptc homology imply? The presence of a sterol-sensing motif in Ptc suggests that Ptc may be regulated by sterol levels within the cell. Cholesterol or a derivative may regulate Ptc protein stability, subcellular localization, or its ability to signal. Moreover, the NPC homology with Ptc suggests that Ptc may function intracellularly in vesicular compartments. This suggests an elegant function for the kinesin-related protein, Cos2, for moving Ptc-containing vesicles within the cell (Sisson et al. 1997).

The unexpected linkage between Ptc and cholesterol is

intriguing in light of the fact that the amino-terminal half of Hh, a ligand for Ptc, is itself covalently attached to cholesterol and rendered lipophilic (Porter et al. 1996a,b). Perhaps Ptc binds both the Hh protein and its cholesterol moiety to sequester Hh within the receiving cell. This would require moving membrane-bound Hh between the lipid bilayers of the secreting and receiving cells, perhaps by exocytosis. Alternatively, the two connections between Hh/Ptc signaling and cholesterol may indicate separate aspects of the pathways. In this case, there would not be any transfer of Hh-cholesterol between cells, and sterol sensing by Ptc could involve cholesterol that is not linked to Hh.

NPC homologs exist in yeast and worms (Figs. 5 and 6) (Carstea et al. 1997; Loftus et al. 1997) and are similar to Ptc in the hydrophobic regions, including the sterol-sensing domain. A unique situation occurs in the worm, where at least 14 NPC/Ptc homologs have been identified and

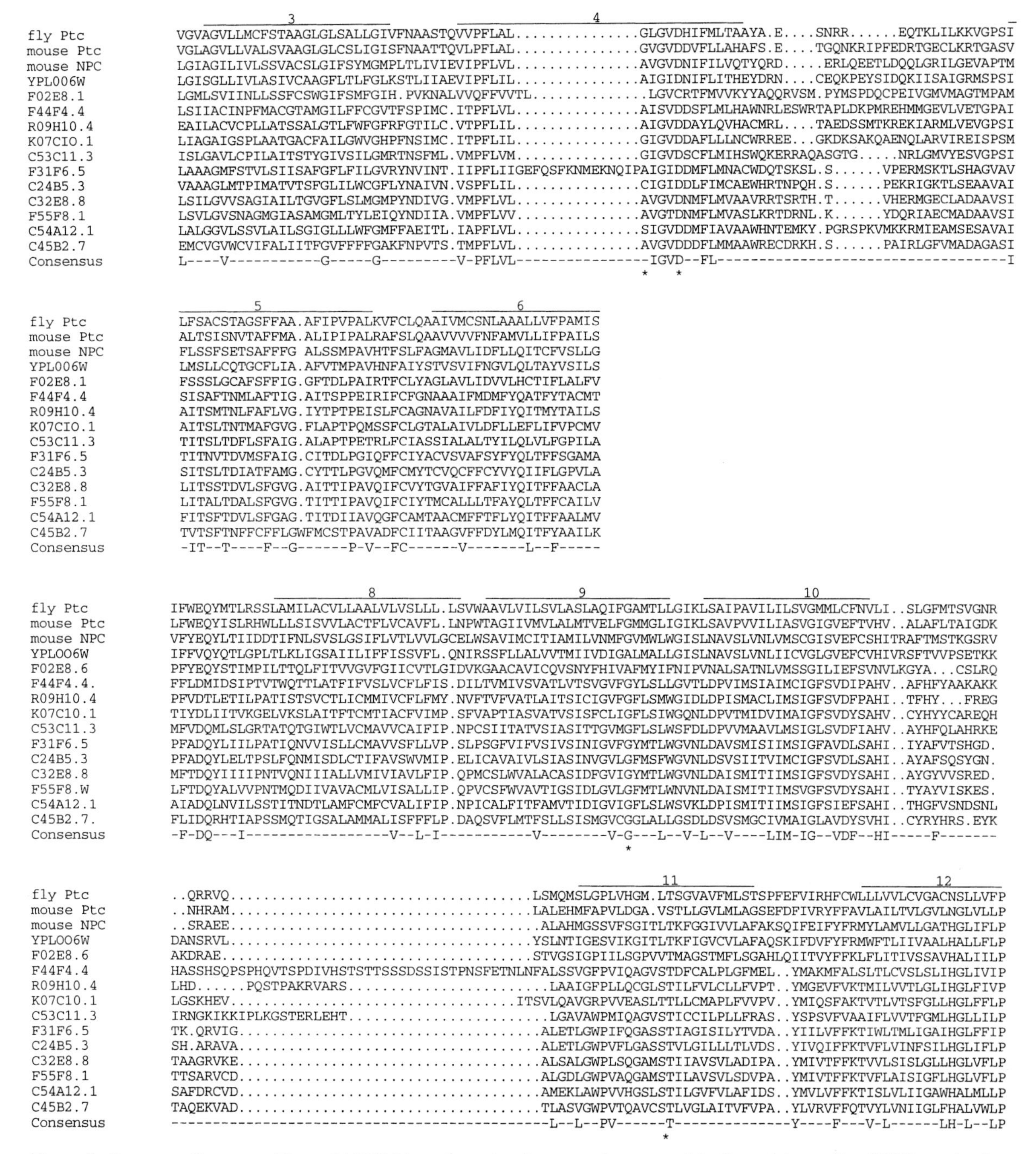

Figure 5. Sequence alignment of Ptc and 13 NPC homologs. An alignment of mouse and fly Ptc and the predicted NPC proteins from mouse, yeast, (YPL006), and worm (F02E8.1, F44F4.4, R09H10.4, K07C10.1, C53C11.3, F31F6.5, C24B5.3, C32E8.8, F55F8.1, C54A12.1, C45B2.7) was made using the GCG program, Pileup. Sequence similarity is detected in two separate regions of Ptc: putative transmembrane domains 3–6 (including the putative sterol sensing domain) and 8–12. Bars over sequence indicate potential membrane spanning segments of Ptc. Consensus sequence indicates similar or identical residues shared between 11 of the 15 sequences. Asterisks (*) indicate positions where missense mutations have been identified in *PTC* in BCNS patients or in BCCs.

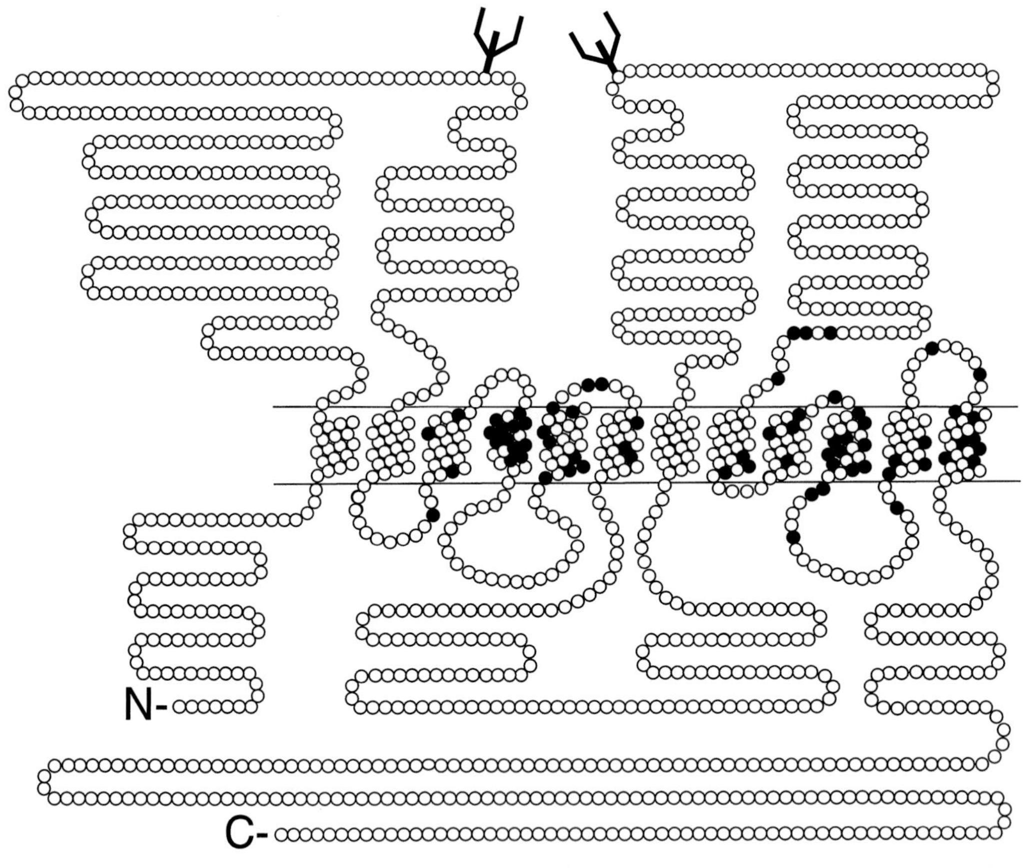

Figure 6. Conservation between Ptc and 13 NPC homologs. A model of the mouse Ptc sequence showing conserved residues with a family of NPC homologs. (*Closed circle*) Residues that are similar for at least 11 of 15 sequences.

appear to be transcribed since expressed sequence tags have been identified for most members (Wilson et al. 1994). The NPC/Ptc homologs might represent a family of receptors with functions, like Ptc, in cell-cell signaling. The conserved hydrophobic regions may form a common functional domain, whereas the unconserved hydrophilic loops might bind different ligands. Ligands for this group of proteins may be encoded by a family of putative secreted proteins in *C. elegans* that are related in sequence to the carboxy-terminal half of Hh but have divergent amino-terminal domains (Burglin 1996; Porter et al. 1996). One family member contains autocatalytic cleavage activity that releases the amino-terminal half in a manner like Hh (Porter et al. 1996b). These Hh-related genes might function as a group of novel signaling molecules that interact with the NPC/Ptc proteins to control diverse cell-cell signaling events. Alternatively, the worm NPC/Ptc homologs might have specialized functions in different cell types or within a cell for moving vesicles to different organelles.

PERSPECTIVES

The recent findings about the fundamental role of Hh signaling in vertebrate development and tumorigenesis have created much excitement and interest. However, many questions remain to be answered. How do two cells respond differently to the same Hh signal to induce the transcription of distinct sets of genes? How is the Hh signal transduced from the plasma membrane to the nucleus? The entire repertoire of pathway components has yet to be identified and the interactions among them understood. The similarities between NPC and Ptc suggest a function common to these proteins. Do the parallels between these two proteins extend to interacting components as well? For instance, is there a ligand and a kinesin motor involved in the NPC pathway? What role does cholesterol have in regulating Hh, Ptc, and other components of this pathway? Answering these questions is certain to reveal more surprises about this extraordinary pathway.

ACKNOWLEDGMENTS

We thank David Jung and Kaye Suyama for staining embryos and preparing images of gene expression. We also thank Drs. Anthony Oro, Ervin Epstein, and Stuart Kim for comments on the manuscript. R.L.J. is the recipient of a Cancer Research Fund of the Damon Runyon-Walter Winchell Foundation Fellowship and a Walter and Idun Berry Postdoctoral Fellowship. M.P.S. is an investigator of the Howard Hughes Medical Institute.

REFERENCES

Alcedo J., Ayzenzon M., Von Ohlen T., Noll M., and Hooper J.E. 1996. The *Drosophila smoothened* gene encodes a seven-pass membrane protein, a putative receptor for the hedgehog signal. *Cell* **86:** 221.

Alexandre C., Jacinto A., and Ingham P.W. 1996. Transcriptional activation of hedgehog target genes in *Drosophila* is mediated directly by the Cubitus interruptus protein, a member of the GLI family of zinc finger DNA-binding proteins. *Genes Dev.* **10:** 2003.

Altschul S.F., Gish W., Miller W., Myers E.W., and Lipman D.J. 1990. Basic local alignment search tool. *J. Mol. Biol.* **215:** 403.

Aza-Blanc P., Ramirez-Weber F.-A., Laget M.-P., Schwartz C., and Kornberg T.B. 1997. Proteolysis that is inhibited by Hedgehog targets Cubitus interruptus protein to the nucleus and converts it to a repressor. *Cell* **89:** 1043.

Bale A.E. 1997. Variable expressivity of patched mutations in flies and humans. *Am. J. Hum. Genet.* **60:** 10.

Basler K. and Struhl G. 1994. Compartment boundaries and the control of *Drosophila* limb pattern by hedgehog protein. *Nature* **368:** 208.

Bejsovec A. and Wieschaus E. 1993. Segment polarity gene interactions modulate epidermal patterning in *Drosophila* embryos. *Development* **119:** 501.

Bhat K.M. 1996. The patched signaling pathway mediates repression of gooseberry allowing neuroblast specification by wingless during *Drosophila* neurogenesis. *Development* **22:** 2921.

Bhat K.M. and Schedl P. 1997. Requirement for engrailed and invected genes reveals novel regulatory interactions between *engrailed/invected, patched, gooseberry* and *wingless* during *Drosophila* neurogenesis. *Development* **124:** 1675.

Bokor P. and DiNardo S. 1996. The roles of *hedgehog, wingless* and *lines* in patterning the dorsal epidermis in *Drosophila. Development* **122:** 1083.

Bonifas J.M., Bare J.W., Kerschmann R.L., Master S.P., and Epstein Jr., E. 1994. Parental origin of chromosome 9q22.3-q31 lost in basal cell carcinomas from basal cell nevus syndrome patients. *Hum. Mol. Genet.* **3:**447.

Burglin T.R. 1996. Warthog and groundhog, novel families related to hedgehog. *Curr. Biol.* **6:** 1047.

Capdevila J., Estrada M.P., Sanchez-Herrero E., and Guerrero I. 1994. The *Drosophila* segment polarity gene *patched* interacts with *ecapentaplegic* in wing development. *EMBO J.* **13:** 71.

Carstea E.D., Morris J.A., Coleman K.G., Loftus S.K., Zhang D., Cummings C., Gu J., Rosenfeld M.A., Pavan W.J., Krizman D.B., Nagle J., Polymeropoulos M.H., Sturley S.L., Ioannou Y.A., Higgins M.E., Comly M., Cooney A., Brown A., Kaneski C.R., Blanchette-Mackie E.J., Dwyer N.K., Neufeld E.B., Chang T.Y., Liscum L., Strauss J.F., Ohno K., Zeigler M., Carmi R., Sokol J., Markie D., O'Neill R.R., van Diggelen O.P., Elleder M., Patterson M.C., Brady R.O., Vanier M.T., Pentchev P.G., and Tagle D.A. 1997. Niemann-Pick C1 disease gene: Homology to mediators of cholesterol homeostasis. *Science* **277:** 228.

Chang D.T., Lopez A., von Kessler D.P., Chiang C., Simandl B.K., Zhao R., Seldin M.F., Fallon J.F., and Beachy P.A. 1994. Products, genetic linkage and limb patterning activity of a murine hedgehog gene. *Development* **120:** 3339.

Chen Y. and Struhl G. 1996. Dual roles for *patched* in sequestering and transducing Hedgehog. *Cell* **87:** 553.

Chenevix-Trench G., Wicking C., Berkman J., Sharpe H., Hockey A., Haan E., Oley C., Ravine D., Turner A., Goldgar D., Searle J., and Wainwright B. 1993. Further localization of the gene for nevoid basal cell carcinoma syndrome (NBCCS) in 15 Australasian families: Linkage and loss of heterozygosity. *Am. J. Hum. Genet.* **53:** 760.

Chidambaram A., Goldstein A.M., Gailani M.R., Gerrard B., Bale S.J., DiGiovanna J.J., Bale A.E., and Dean M. 1996. Mutations in the human homologue of the *Drosophila patched* gene in Caucasian and African-American nevoid basal cell carcinoma syndrome patients. *Cancer Res.* **56:** 4599.

Compton J.G., Goldstein A.M., Turner M., Bale A.E., Kearns K.S., McBride O.W., and Bale S.J. 1994. Fine mapping of the locus for nevoid basal cell carcinoma syndrome on chromosome 9q. *J. Invest. Dermatol.* **103:** 178.

Concordet J.P., Lewis K.E., Moore J.W., Goodrich L.V., Johnson R.L., Scott M.P., and Ingham P.W. 1996. Spatial regulation of a zebrafish patched homologue reflects the roles of sonic hedgehog and protein kinase A in neural tube and somite patterning. *Development* **122:** 2835.

Dominguez M., Brunner M., Hafen E., and Basler K. 1996. Sending and receiving the hedgehog signal: Control by the *Drosophila* Gli protein Cubitus interruptus (comments). *Science* **272:** 1621.

Echelard Y., Epstein D.J., St.-Jacques B., Shen L., Mohler J., McMahon J.A., and McMahon A.P. 1993. Sonic hedgehog, a member of a family of putative signaling molecules, is implicated in the regulation of CNS polarity. *Cell* **75:** 1417.

Epstein D.J., Marti E., Scott M.P., and McMahon A.P. 1996. Antagonizing cAMP-dependent protein kinase A in the dorsal CNS activates a conserved Sonic hedgehog signaling pathway. *Development* **122:** 2885.

Ericson J., Morton S., Kawakami A., Roelink H., and Jessell T.M. 1996. Two critical periods of Sonic Hedgehog signaling required for the specification of motor neuron identity. *Cell* **87:** 661.

Evans D.G., Farndon P.A., Burnell L.D., Gattamaneni H.R., and Birch J.M. 1991. The incidence of Gorlin syndrome in 173 consecutive cases of medulloblastoma. *Br. J. Cancer* **64:** 959.

Felsenfeld A.L. and Kennison J.A. 1995. Positional signaling by hedgehog in *Drosophila* imaginal disc development. *Development* **121:** 1.

Forbes A.J., Nakano Y., Taylor A.M., and Ingham P.W. 1993. Genetic analysis of hedgehog signalling in the *Drosophila* embryo. *Development Suppl.*, p. 115.

Gailani M.R., Stahle-Backdahl M., Leffell D.J., Glynn M., Zaphiropoulos P.G., Pressman C., Unden A.B., Dean M., Brash D.E., Bale A.E., and Toftgard R. 1996. The role of the human homologue of *Drosophila* patched in sporadic basal cell carcinomas. *Nat. Genet.* **14:** 78.

Gailani M.R., Bale S.J., Leffell D.J., DiGiovanna J.J., Peck G.L., Poliak S., Drum M.A., Pastakia B., McBride O.W., Kase R., Greene M., Mulvihill J.J., and Bale A.E. 1992. Developmental defects in Gorlin syndrome related to a putative tumor suppressor gene on chromosome 9. *Cell* **69:** 111.

Goldstein A.M., Stewart C., Bale A.E., Bale S.J., and Dean M. 1994a. Localization of the gene for the nevoid basal cell carcinoma syndrome. *Am. J. Hum. Genet.* **54:** 765.

Goldstein A.M., Pastakia B., DiGiovanna J.J., Poliak S., Santucci S., Kase R., Bale A.E., and Bale S.J. 1994b. Clinical findings in two African-American families with the nevoid basal cell carcinoma syndrome (NBCC). *Am. J. Med. Genet.* **50:** 272.

Goodrich L.V., Johnson R.L., Milenkovic L., McMahon J.A., and Scott M.P. 1996. Conservation of the hedgehog/patched signaling pathway from flies to mice: Induction of a mouse patched gene by Hedgehog. *Genes Dev.* **10:** 301.

Gorlin R.J. 1995. Nevoid basal cell carcinoma syndrome. *Dermatol. Clin.* **13:** 113.

Hahn H., Christiansen J., Wicking C., Zaphiropoulos P.G., Chidambaram A., Gerrard B., Vorechovsky I., Bale A.E., Toftgard R., Dean M., and Wainwright B. 1996a. A mammalian patched homolog is expressed in target tissues of sonic hedgehog and maps to a region associated with developmental abnormalities. *J. Biol. Chem.* **271:** 12125.

Hahn H., Wicking C., Zaphiropoulous P.G., Gailani M.R., Shanley S., Chidambaram A., Vorechovsky I., Holmberg E., Unden A.B., Gillies S., Negus K., Smyth I., Pressman C., Leffell D.J., Gerrard B., Goldstein A.M., Dean M., Toftgard R., Chenevix-Trench G., Wainwright B., and Bale A.E. 1996b. Mutations of the human homolog of *Drosophila* patched in the nevoid basal cell carcinoma syndrome. *Cell* **85:** 841.

Hammerschmidt M., Bitgood M.J., and McMahon A.P. 1996. Protein kinase A is a common negative regulator of Hedgehog signaling in the vertebrate embryo. *Genes Dev.* **10:** 647.

Hammerschmidt M., Brook A., and McMahon A.P. 1997. The world according to hedgehog. *Trends Genet.* **13:** 14.

Hepker J., Wang Q.T., Motzny C.K., Holmgren R., and Orenic T.V. 1997. *Drosophila cubitus interruptus* forms a negative feedback loop with *patched* and regulates expression of Hedgehog target genes. *Development* **124:** 549.

Hidalgo A. and Ingham P. 1990. Cell patterning in the *Drosophila* segment: Spatial regulation of the segment polarity gene *patched. Development* **110:** 291.

Hooper J.E. and Scott M.P. 1989. The *Drosophila patched* gene encodes a putative membrane protein required for segmental patterning. *Cell* **59:** 751.

Ingham P.W. 1993. Localized *hedgehog* activity controls spatial limits of *wingless* transcription in the *Drosophila* embryo. *Nature* **366:** 560

Ingham P.W. and Fietz M.J. 1995. Quantitative effects of *hedgehog* and *decapentaplegic* activity on the patterning of the *Drosophila* wing. *Curr. Biol.* **5:** 432.

Ingham P.W., Taylor A.M., and Nakano Y. 1991. Role of the *Drosophila patched* gene in positional signalling. *Nature* **353:** 184.

Jiang J. and Struhl G. 1995. Protein kinase A and hedgehog signaling in *Drosophila* limb development. *Cell* **80:** 563.

Johnson R.L., Grenier J.K., and Scott M.P. 1995. Patched overexpression alters wing disc size and pattern: Transcriptional and post-transcriptional effects on hedgehog targets. *Development* **121:** 4161.

Johnson R.L., Rothman A.L., Xie J., Goodrich L.V., Bare J.W., Bonifas J.M., Quinn A.G., Myers R.M., Cox D.R., Epstein Jr. E., and Scott M.P. 1996. Human homolog of *patched*, a candidate gene for the basal cell nevus syndrome. *Science* **272:** 1668.

Kimonis V.E., Goldstein A.M., Pastakia B., Yang M.L., Kase R., DiGiovanna J.J., Bale A.E., and Bale S.J. 1997. Clinical manifestations in 105 persons with nevoid basal cell carcinoma syndrome. *Am. J. Med. Genet.* **69:** 299.

Kinzler K.W., Bigner S.H., Bigner D.D., Trent J.M., Law M.L., O'Brien S.J., Wong A.J., and Vogelstein B. 1987. Identification of an amplified, highly expressed gene in a human glioma. *Science* **236:** 70.

Kojima T., Michiue T., Orihara M., and Saigo K. 1994. Induction of a mirror-image duplication of anterior wing structures by localized hedgehog expression in the anterior compartment of *Drosophila melanogaster* wing imaginal discs. *Gene* **148:** 211.

Lacombe D., Chateil J.F., Fontan D., and Battin J. 1990. Medulloblastoma in the nevoid basal-cell carcinoma syndrome: Case reports and review of the literature. *Genet. Couns.* **1:** 273.

Li W., Ohlmeyer J.T., Lane M.E., and Kalderon D. 1995. Function of protein kinase A in hedgehog signal transduction and *Drosophila* imaginal disc development. *Cell* **80:** 553.

Liscum L. and Faust J.R. 1987. Low density lipoprotein (LDL)-mediated suppression of cholesterol synthesis and LDL uptake is defective in Niemann-Pick type C fibroblasts. *J Biol. Chem.* **262:** 17002.

Loftus S.K., Morris J.A., Carstea E.D., Gu J.Z., Cummings C., Brown A., Ellison J., Ohno K., Rosenfeld M.A., Tagle D.A., Pentchev P.G., and Pavan W.J. 1997. Murine model of Niemann-Pick C disease: Mutation in a cholesterol homeostasis gene. *Science* **277:** 232.

Marigo V. and Tabin C.J. 1996. Regulation of patched by Sonic hedgehog in the developing neural tube. *Proc. Natl. Acad. Sci.* **93:** 9346.

Marigo V., Johnson R.L., Vortkamp A., and Tabin C.J. 1996a. Sonic hedgehog differentially regulates expression of GLI and GLI3 during limb development. *Dev. Biol.* **180:** 273.

Marigo V., Davey R.A., Zuo Y., Cunningham J.M., and Tabin C.J. 1996b. Biochemical evidence that patched is the Hedgehog receptor. *Nature* **384:** 176.

Marigo V., Scott M.P., Johnson R.L., Goodrich L.V., and Tabin C.J. 1996c. Conservation in hedgehog signaling: Induction of a chicken patched homolog by Sonic hedgehog in the developing limb. *Development* **122:**1225.

Miller D.L. and Weinstock M.A. 1994. Nonmelanoma skin cancer in the United States: Incidence. *J. Am. Acad. Dermatol.* **30:** 774

Miller S.J. 1991. Biology of basal cell carcinoma (part I). *J. Am. Acad. Dermatol.* **24:** 1.

Motzny C.K. and Holmgren R. 1995. The *Drosophila* cubitus interruptus protein and its role in the wingless and hedgehog signal transduction pathways. *Mech. Dev.* **52:** 137.

Nakano Y., Guerrero I., Hidalgo A., Taylor A., Whittle J.R., and Ingham P.W. 1989. A protein with several possible membrane-spanning domains encoded by the *Drosophila* segment polarity gene *patched. Nature* **341:** 508.

Orenic T.V., Slusarski D.C., Kroll K.L., and Holmgren R.A. 1990. Cloning and characterization of the segment polarity gene *cubitus interruptus Dominant* of *Drosophila. Genes Dev.* **4:** 1053.

Oro A.E., Higgins K.M., Hu Z., Bonifas J.M., Epstein Jr. E., and Scott M.P. 1997. Basal cell carcinomas in mice overexpressing Sonic hedgehog. *Science* **276:** 817.

Padgett R.W., St. Johnston R.D., and Gelbart W.M. 1987. A transcript from a *Drosophila* pattern gene predicts a protein homologous to the transforming growth factor-beta family. *Nature* **325:** 81

Pan D. and Rubin G.M. 1995. cAMP-dependent protein kinase and hedgehog act antagonistically in regulating decapentaplegic transcription in *Drosophila* imaginal discs. *Cell* **80:** 543.

Pham A., Therond P., Alves G., Tournier F.B., Busson D., Lamour-Isnard C., Bouchon B.L., Preat T., and Tricoire H. 1995. The Suppressor of fused gene encodes a novel PEST protein involved in *Drosophila* segment polarity establishment. *Genetics* **140:** 587

Pietsch T., Waha A., Koch A., Kraus J., Albrecht S., Tonn J., Sorensen N., Berthold F., Henk B., Schmandt N., Wolf H.K., von Deimling A., Wainwright B., Chenevix-Trench G., Wiestler O.D., and Wicking C. 1997. Medulloblastomas of the desmoplastic variant carry mutations of the human homologue of *Drosophila patched. Cancer Res.* **57:** 2085.

Pentchev P.G., Comly M.E., Kruth H.S., Tokoro T., Butler J., Sokol J., Filling-Katz M., Quirk J.M., Marshall D.C., Patel S., et al. 1987. Group C Niemann Pick disease: Faulty regulation of low-density lipoprotein uptake and cholesterol storage in cultered fibroblasts. *FASEB J.* **1:** 40.

Placzek M., Tessier-Lavigne M., Yamada T., Jessell T., and Dodd J. 1990. Mesodermal control of neural cell identity: Floor plate induction by the notochord. *Science* **250:** 985.

Polakis P. 1995. Mutations in the APC gene and their implications for protein structure and function. *Curr. Opin. Genet. Dev.* **5:** 66.

Porter J.A., Young K.E., and Beachy P.A. 1996a. Cholesterol modification of hedgehog signaling proteins in animal development. *Science* **274:** 255.

Porter J.A., Ekker S.C., Park W.J., von Kessler D.P., Young K.E., Chen C.H., Ma Y., Woods A.S., Cotter R.J., Koonin E.V., and Beachy P.A. 1996b. Hedgehog patterning activity: Role of a lipophilic modification mediated by the carboxy-terminal autoprocessing domain. *Cell* **86:** 21.

Preat T., Therond P., Limbourg-Bouchon B., Pham A., Tricoire H., Busson D., and Lamour-Isnard C. 1993. Segmental polarity in *Drosophila* melanogaster: Genetic dissection of fused in a Suppressor of fused background reveals interaction with costal-2. *Genetics* **135:** 1047.

Preat T., Therond P., Lamour-Isnard C., Limbourg-Bouchon B., Tricoire H., Erk I., Mariol M.C., and Busson D. 1990. A putative serine/threonine protein kinase encoded by the segment-polarity fused gene of *Drosophila. Nature* **347:** 87.

Raffel C., Jenkins R.B., Frederick L., Hebrink D., Alderete B., Fults D.W., and James C.D. 1997. Sporadic medulloblastomas contain PTCH mutations. *Cancer Res.* **57:** 842.

Riddle R.D., Johnson R.L., Laufer E., and Tabin C. 1993. Sonic hedgehog mediates the polarizing activity of the ZPA. *Cell* **75:** 1401.

Rijsewijk F., Schuermann M., Wagenaar E., Parren P., Weigel D., and Nusse R. 1987. The *Drosophila* homolog of the mouse mammary oncogene *int-1* is identical to the segment polarity gene *wingless*. *Cell* **50:** 649.

Robbins D.J., Nybakken K.E., Kobayashi R., Sisson J.C., Bishop J.M., and Therond P.P. 1997. Hedgehog elicits signal transduction by means of a large complex containing the kinesin-related protein Costal-2. *Cell* **90:** 225.

Roelink H., Porter J.A., Chiang C., Tanabe Y., Chang D.T., Beachy P.A., and Jessell T.M. 1995. Floor plate and motor neuron induction by different concentrations of the amino-terminal cleavage product of sonic hedgehog autoproteolysis. *Cell* **81:** 445.

Roelink H., Augsburger A., Heemskerk J., Korzh V., Norlin S., Ruiz i Altaba A., Tanabe Y., Placzek M., Edlund T., Jessell T.M., and Dodd J. 1994. Floor plate and motor neuron induction by *vhh-1*, a vertebrate homolog of *hedgehog* expressed by the notochord. *Cell* **76:** 761.

Sasaki H., Hui C., Nakafuku M., and Kondoh H. 1997. A binding site for Gli proteins is essential for HNF-3β floor plate enhancer activity in transgenics and can respond to Shh in vitro. *Development* **124:**1313.

Schuske K., Hooper J.E., and Scott M.P. 1994. *patched* overexpression causes loss of *wingless* expression in *Drosophila* embryos. *Dev. Biol.* **164:** 300.

Sisson J.C., Ho K.S., Suyama K., and Scott M.P. 1997. Costal-2, a novel kinesin-related protein in the Hedgehog signaling pathway. *Cell* **90:** 235.

Slusarski D.C., Motzny C.K., and Holmgren R. 1995. Mutations that alter the timing and pattern of cubitus interruptus gene expression in *Drosophila melanogaster*. *Genetics* **139:** 229.

Stone D.M., Hynes M., Armanini M., Swanson T.A., Gu Q., Johnson R.L., Scott M.P., Pennica D., Goddard A., Phillips H., Noll M., Hooper J.E., de Sauvage F., and Rosenthal A. 1996. The tumour-suppressor gene patched encodes a candidate receptor for Sonic hedgehog (comments). *Nature* **384:** 129.

Tabata T. and Kornberg T.B. 1994. Hedgehog is a signaling protein with a key role in patterning *Drosophila* imaginal discs. *Cell* **76:** 89

Takabatake T., Ogawa M., Takahashi T.C., Mizuno M., Okamoto M., and Takeshima K. 1997. Hedgehog and patched gene expression in adult ocular tissues. *FEBS Lett.* **410:** 485.

Taylor A.M., Nakano Y., Mohler J., and Ingham P.W. 1993. Contrasting distributions of patched and hedgehog proteins in the *Drosophila* embryo. *Mech. Dev.* **42:** 89

Therond P.P., Knight J.D., Kornberg T.B., and Bishop J.M. 1996a. Phosphorylation of the fused protein kinase in response to signaling from hedgehog. *Proc. Natl. Acad. Sci.* **93:** 4224.

Therond P., Alves G., Limbourg-Bouchon B., Tricoire H., Guillemet E., Brissard-Zahraoui J., Lamour-Isnard C., and Busson D. 1996b. Functional domains of fused, a serine-threonine kinase required for signaling in *Drosophila*. *Genetics* **142:** 1181.

Unden A.B., Holmberg E., Lundh-Rozell B., Stahle-Backdahl M., Zaphiropoulos P.G., Toftgard R., and Vorechovsky I. 1996. Mutations in the human homologue of *Drosophila patched* (*PTCH*) in basal cell carcinomas and the Gorlin syndrome: Different in vivo mechanisms of *PTCH* inactivation. *Cancer Res.* **56:** 4562.

van den Heuvel M. and Ingham P.W. 1996. smoothened encodes a receptor-like serpentine protein required for hedgehog signalling. *Nature* **382:** 547.

Von Ohlen T., Lessing D., Nusse R., and Hooper J.E. 1997. Hedgehog signaling regulates transcription through Cubitus interruptus, a sequence-specific DNA binding protein. *Proc. Natl. Acad. Sci.* **94:** 2404.

Wicking C., Berkman J., Wainwright B., and Chenevix-Trench G. 1994. Fine genetic mapping of the gene for nevoid basal cell carcinoma syndrome. *Genomics* **22:** 505.

Wicking C., Shanley S., Smyth I., Gillies S., Negus K., Graham S., Suthers G., Haites N., Edwards M., Wainwright B., and Chenevix-Trench G. 1997. Most germ-line mutations in the nevoid basal cell carcinoma syndrome lead to a premature termination of the PATCHED protein, and no genotype-phenotype correlations are evident. *Am. J. Hum. Genet.* **60:** 21.

Wilson R., Ainscough R., Anderson K., Baynes C., Berks M., Bonfield J., Burton J., Connell M., Copsey T., Cooper J., Coulson A., Craxton M., Dear S., Du Z., Durbin R., Favello A., Fraser A., Fulton L., Gardner A., Green P., Hawkins T., Hillier L., Jier M., Johnston L., Jones M., Kershaw J., Kirsten J., Laisster N., Latreille P., Lightning J., Lloyd C., Mortimore B., O'Callaghan M., Parsons J., Percy C., Rafken L., Roopra A., Saunders D., Shownkeen R., Sims M., Smaldon N., Smith A., Smith M., Sonnhammer E., Staden R., Sulston J., Thierry-Mieg J., Thomas K., Vaudin M., Vaughan K., Waterson R., Watson A., Weinstock L., Wilkinson-Sproat J., and Wohldman P. 1994. 2.2 Mb of contiguous nucleotide sequence from chromosome III of *C. elegans*. *Nature* **368:** 32

Xie J., Johnson R.L., Zhang X., Bare J.W., Waldman F.M., Cogen P.H., Menon A.G., Warren R.S., Chen L., Scott M.P., and Epstein E.H. 1997. The *PATCHED* gene in several types of sporadic extracutaneous tumors. *Cancer Res.* **57:** 2369.

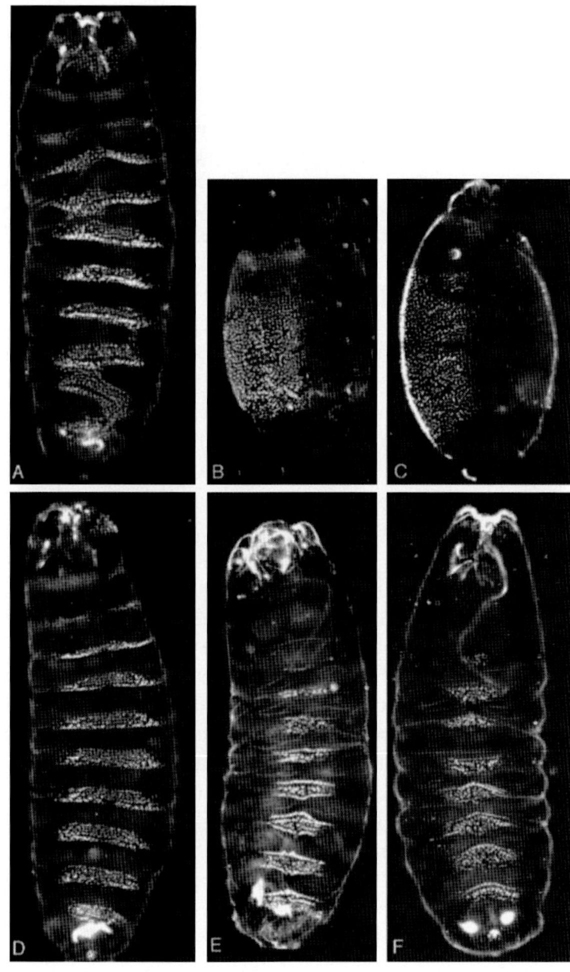

Figure 1. Effects of mutations in *hh*, *smo*, and *ptc* on embryonic patterning. Cuticular preparations of mutant *Drosophila* larvae viewed with dark field illumination. (*A*) *smo*^{3} *homozygote* derived from heterozygous *smo* mutant parents and raised at 25°C. The phenotype is highly variable, ranging from almost wild type to the occasional partial fusions of denticle belts, as shown here. (*B*) *smo*^{3} homozygote derived from a homozygous mutant female germ line clone and raised at 18°C. Note the complete elimination of segmentally reiterated denticle belts and reduction in overall size; the ventral surface of the cuticle is covered in a continuous lawn of denticles of similar size and randomized polarity. (*C*) *hh*^{13C} homozygote showing a phenotype very similar to that of the *smo*^{3} homozygote in *B*. (*D*) *smo*^{3} *ptc*^{G12} double homozygote derived from heterozygous parents and raised at 25°C. This animal shows an essentially wild-type phenotype, typified by the trapezoidal shape of each denticle belt (compare the ptcG12 homozygote in *E*). (*E*) *ptc*^{G12} homozygote; note the reduced numbers of denticle rows and the rectangular shape of each belt. (*F*) *GAL4* ^{e22c} *UAShh* heterozygote in which *hh* is expressed ubiquitously. Note the changes in denticle belt morphology similar to those seen in *ptc*^{G12} homozygotes (*E*).

exhibit an extreme phenotype indistinguishable from that of an amorphic *hh* homozygote (compare Fig. 1B,C).

smo Activity Is Required for Transduction of Hh But Not Wg

The similarity between the *smo* and *hh* loss-of-function phenotypes does not in itself establish that *smo* is required for *hh* signaling. One of the principal roles of *hh* activity in the developing embryo is maintenance of the spatially restricted transcription of the segment polarity gene *wingless*. The activity of the secreted Wg protein is essential for the patterning of each larval segment and loss of *wg* expression results in a phenotype similar to that caused by loss of *hh* (Baker 1988; Bejsovec and Martinez Arias 1991; Bejsovec and Wieschaus 1993). Like *hh* homozygotes, embryos devoid of both maternal and zygotic *smo* activity lack *wg* expression in each segment (van den Heuvel and Ingham 1996). However, because *wg* activity is also required to maintain its own expression (Hooper 1994), such an effect on *wg* transcription could be indicative of a role for *smo* in transduction of either signal. To discriminate between these two possible roles, we used an inducible *wg* transgene to ask whether *wg* activity can be transduced when expressed in an embryo lacking *smo* activity (van den Heuvel and Ingham 1996). The cuticular pattern of such embryos clearly reveals that *wg* signaling does not require *smo*, because the patterning of alternate segments is restored when *wg* is expressed in alternate segmental primordia (Fig. 2B). This indicates that the *smo* mutant phenotype is caused by the loss of transcriptional activation of *wg* by *hh* and not through a failure of *wg* signaling. This interpretation was confirmed by performing a similar experiment in which an *hh* transgene was expressed in alternate segments in a *smo* mutant embryo (van den Heuvel and Ingham 1996). In this case, there was no rescue of cuticle patterning in any of the segments (Fig. 2A), indicating that the transcriptional activation of *wg* by *hh* had not been restored.

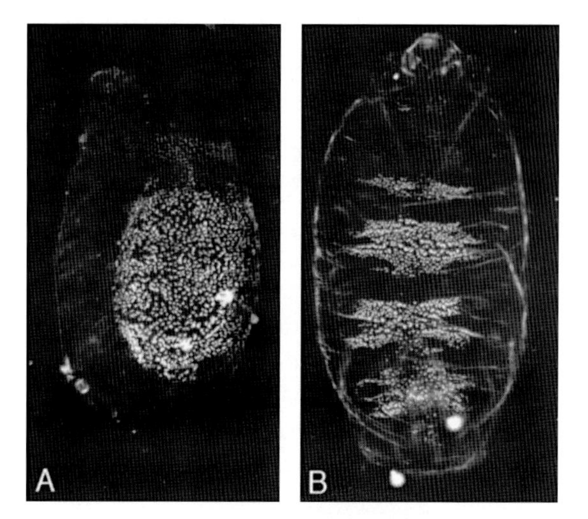

Figure 2. *smo* is required for *hedgehog* but not *wingless* signaling. (*A*) *smo*^{3} homozygote (raised at 18°C) derived from a homozygous mutant female germ line clone and expressing *hh* under the control of enhancers from the pair-rule gene *hairy* (*UAShh h-GAL4*). The phenotype is unmodified by the expression of *hh*. (*B*) *smo*^{3} homozygote (raised at 18°C) derived from a homozygous mutant female germ line clone and expressing *wg* under the control of enhancers from the pair-rule gene *hairy* (*UASwg h-GAL4*). Note the suppression of the *smo* phenotype, with recovery of naked cuticle in alternating segments, reflecting the pattern of expression of the *wg* transgene.

smo Acts Downstream from *ptc* to Transduce the Hh Signal

In contrast to the loss-of-function phenotype of *smo*, inactivation of *ptc* results in a larval cuticular phenotype very similar to that caused by the overexpression of *hh* (compare E and F in Fig. 1) (Ingham 1993; Tabata and Kornberg 1994). In both cases, this phenotype is due largely to the ectopic transcription of *wg*. Thus, the normal function of *ptc* is to repress expression of *hh* target genes, a repression that is overcome when cells are exposed to Hh protein (Ingham et al. 1991). Consistent with this, removal of *ptc* activity from *hh* homozygous mutants completely suppresses the *hh* phenotype (Ingham et al. 1991), as would be expected if *ptc* acts downstream from *hh*. Inactivation of *ptc* in embryos lacking all *smo* activity, in contrast, fails to suppress the *smo* phenotype (data not shown); thus, *smo* is essential for the activation of *wg* in the absence of *ptc*, indicating that *smo* must act downstream from *ptc* to transduce the *hh* signal (see Fig. 3).

In *smo ptc* double mutants raised at the semipermissive temperature (25°C), however, there is a mutual suppression of both mutant phenotypes, the larval cuticular pattern being restored almost to wild type (see Fig. 1D). Although such embryos still express *wg* ectopically, the ectopic expression of another segment polarity gene, *engrailed* (*en*), that typically occurs in *ptc* mutants in response to ectopic *wg* activity is largely suppressed (data not shown). We take this to indicate that the reduction in the levels of *smo* has a corresponding effect on the levels of ectopic *wg* transcription, which therefore fail to reach the threshold required for *en* activation.

smo Encodes a Serpentine Protein Highly Conserved in Evolution

The isolation of genomic clones from the *smo* locus allowed us to determine the complete sequence of the *smo* transcription unit (van den Heuvel and Ingham 1996) (GenBank Accession no: AFO30334). The gene encodes a 4.1-kb transcript containing five introns. The mature processed form of this transcript encodes a predicted protein of 115 kD, hydropathy analysis of which suggests it to have seven membrane-spanning domains (Alcedo et al. 1996; van den Heuvel and Ingham 1996). Searches of the databases revealed limited homology between Smo and the Frizzled family of seven transmembrane domain proteins, originally defined by the *Drosophila* tissue polarity gene *frizzled* (*fz*).

As a first step toward identifying a homolog of *smo* in vertebrates, we used a degenerate PCR-based strategy to amplify *smo*-related sequences from the silk-moth *Bombyx mori*. We designed degenerate primers (1 and 2; see Experimental Procedures) based on homology between the *Drosophila* Smo and Fz proteins and used these to amplify a fragment from cDNA prepared from moth eggs arrested postfertilization. Sequence analysis of the amplified fragment confirmed that it derives from a gene highly related to *Drosophila smo* (Fig. 4). A new set of degenerate primers (2 and 3; see Experimental Procedures) based

Inactive

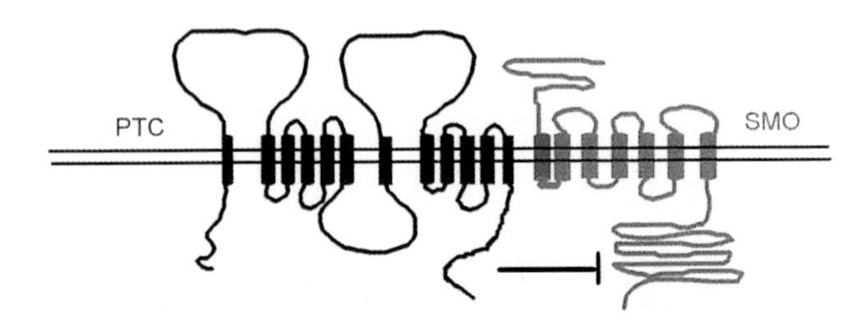

Active

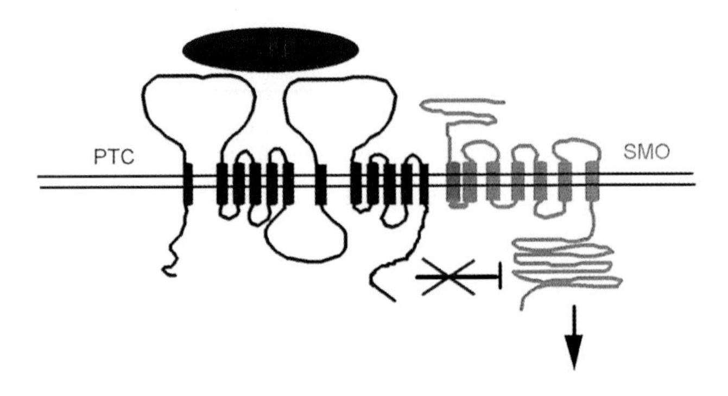

Figure 3. Model for the interaction between Hh, Ptc, and Smo proteins, based on genetic and biochemical data. The activity of Smo is normally inhibited by Ptc, most likely through a direct association between the two proteins. Binding of Hh to Ptc somehow antagonizes the inhibitory activity of the latter, releasing Smo from inhibition and thus activating intracellular components of the signaling pathway (*arrow*).

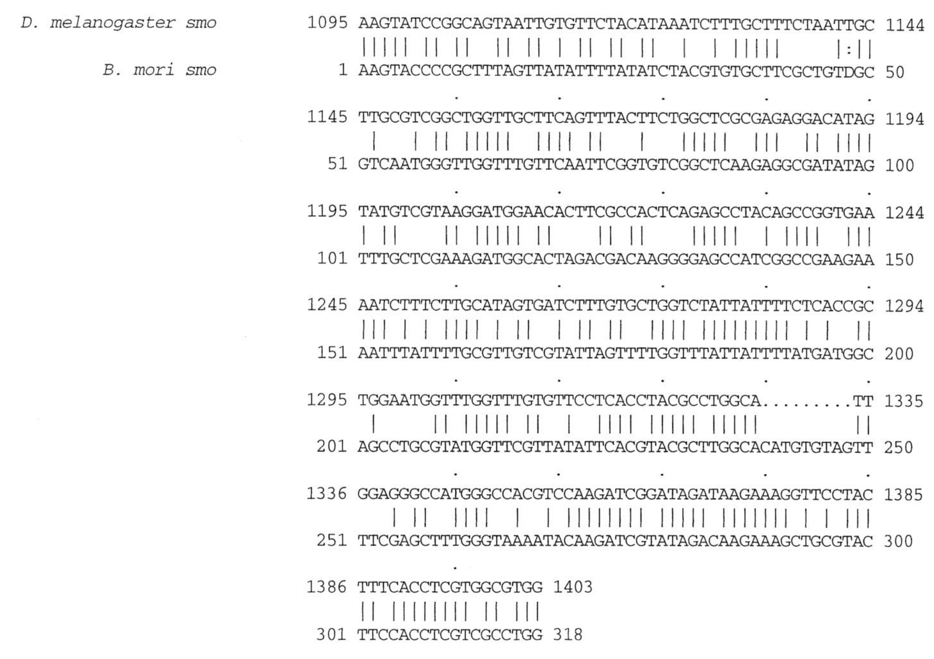

Figure 4. Sequence alignment of the *B. mori* PCR product amplified with primers 1 and 2 with the corresponding region of *D. melanogaster smo.*

on a three-way comparison between this moth fragment and the *Drosophila* Smo and Fz protein sequences was then synthesized and used in PCR amplification of chick cDNA. This yielded a fragment of the expected size, the sequence of which is consistent with it deriving from a chick *smo* homolog. A near full-length cDNA was assembled from partial clones isolated from stage-17 and stage-22 chick cDNA libraries screened with this PCR fragment. The sequence alignment of this cDNA (GenBank Accession no: AFO19977), which we designate *c-smo,* with human, rat, and *Drosophila smo* is shown in Figure 5.

Expression of *c-smo* during Embryogenesis

The expression pattern of c-*smo* during development was analyzed by in situ hybridization of antisense probes to chick embryos. In normal development, cells responding to signaling by different Hh proteins are characterized by their up-regulation of *ptc* transcription (Concordet et al. 1996; Goodrich et al. 1996; Marigo et al. 1996b). As expected for a gene required for the response to Hh signaling, c-*smo* is expressed in all such regions of high-level *ptc* expression, for example, in the ventral somites and the posterior limb mesenchyme (see Fig. 6). Expression of c-*smo* is not limited to these regions, however, but rather is more or less ubiquitous at all stages of development analyzed. Such ubiquitous expression is in line with the finding that most cells are capable of responding to Hh proteins when they are ectopically expressed. One notable exception to this rule appears to be the apical ectodermal ridge (AER) of the developing limb bud; this finding is in line with the previous observation that ectopic Shh activity cannot induce *ptc* expression in the AER (Marigo et al. 1996b), suggesting that these cells are incapable of responding directly to Shh signaling.

Human *Smo* Gene Is Located Close to *Sonic Hedgehog* on Chromosome 7

A human genomic clone, 244J22, containing sequences homologous to c-*smo* was obtained by screening a human P1-derived artificial chromosome (PAC) library (Ioannou et al. 1994) at low stringency with a radioactively labeled c-*smo* probe. DNA from the PAC clone isolated in this way was biotinylated and hybridized to metaphase spreads prepared from phytohemagglutinin-stimulated normal human lymphocytes using standard procedures (see Experimental Procedures). Forty spreads were analyzed, and signal was observed on both copies of chromosomes 7 band 7q32 (Fig. 7). No other consistent signal was observed.

DISCUSSION

Genetic analysis in *Drosophila* has identified the *ptc* and *smo* gene products as components of the *hh* signaling pathway. Both genes have been shown to encode large multipass membrane-spanning proteins; the predicted topology of Ptc is reminiscent of that of channels or transporter proteins, whereas Smo is clearly homologous to members of the Fz protein family, recently implicated as receptors for Wnt proteins. Notwithstanding this homology, the results of epistasis analysis indicate that Smo acts downstream from Ptc, suggesting that Smo does not itself interact directly with Hh. Consistent with this, vertebrate Hh proteins have been found to bind to vertebrate Ptc but not to vertebrate Smo (Stone et al. 1996). The sim-

```
        Human Smo   ~~~~~~~~~~ ~~~~~~~~~~ ~~~~~~~~~~ ~~~~~~~~~~ ~~~~~~~~~~ ~~~~~~~~~~ ~~~~~~~~~~ ~~~~~~~~Ma AarpaRGpEL p...lLglll
          Rat Smo   ~~~~~~~~~~ ~~~~~~~~~~ ~~~~~~~~~~ ~~~~~~~~~~ ~~~~~~~~~~ ~~~~~~~~~~ ~~~~~~~~~~ ~~~~~~~~Ma AgrpvRGpEL aprrlLqlll
        Chick Smo   ~~~~~~~~~~ ~~~~~~~~~~ ~~~~~~~~~~ ~~~~~~~~~~ ~~~~~~~~~~ ~~~~~~~~~~ ~~~~~~~~~~ ~~~~~~~~~~ ~~~~~~~~~~ ~~~~~~~~~~
D.melanogaster Smo  ~~~~~~~~~~ ~~~~~~~~~~ ~~~~~~~~~~ ~~~~~~~~~~ ~~~~~~~~~~ ~~~~~~~~~~ ~~~~~~~~~~ ~~~~~~~~~~ ~~~~~~~~~~ ~~~~~~~~~~
D.melanogaster  Fz  ~~~~~~~~~m wrqiLfilpt liqGvqrydq spldaspyyr sGgglmassG teldglPh.. hnRCEpITIs iCknIpYNMt impnLiGhtk qeEAgLEvhq
D.melanogaster Fz2  mrhnrlkvli lglvLlltsc radGplhsad hgmggmgmgg hGldaspapG ygvpaiPkdp nlRCEeITIp mCrgIgYNMt sfpnemnhEt qdEAgLEvhq

                    101                                                                                                200
        Human Smo   lllLgdpGrg AasSGnatgp gPRsAgGSAR RsAaVtgpp. pp.lSHCGRA ApCePL..Ry NVCLGSvLPY GATSTLLAgD SDSQEEAHGK LVLWSGLRNA
          Rat Smo   lvlLggrGrg AaLSGnvtgp gPRsAgGSAR RnAPVtspp. ppllSHCGRA AhCePL..Ry NVCLGSALPY GATtTLLAgD SDSQEEAHsK LVLWSGLRNA
         Chic Smo   ~~~~~~~~~~ ~~gpcwLwaL alglAlGprR cPAaplnas. aapperCrRp AaCerL..Rf gsCLGSALPY ahTSTLLAaD SgSQEEAHGK LlLWSGLRNA
D.melanogaster Smo  ttPasaqqsd veLepingtL nyRlyakkgR ddkPwfdgl. dsrhiqCvRr ArCyPtsnat NtCfGSkLPY .elSsLdltD fhtekElndK LndyyaLkhv
D.melanogaster  Fz  faPLvkIGcs ddLqlfLcsL yvpvc.tile RPiPpcRslC esArv.Cekl mktynfnWpe Nlecskfpvh Gged.Lcvae nt........ ..........
D.melanogaster Fz2  fwPLveIkcs pdLkffLcsm ytpicledyh kPlPVcRsvC erArSgCapi mqqysfeWpe rmacehlplh GdpdnLcmeq psytEagsGg ssggSGgsgs

                    201                                                                                                300
        Human Smo   PRCWAVIQPL LCAVYMPKCE ....NDrVEL PSRTLCQATR GPCAIVERER GWPDFLRCTP DrFPEGCtNE VQNIKFNSSG QCEvPLVRTD NPKSWYEDVE
          Rat Smo   PRCWAVIQPL LCAVYMPKCE ....NDrVEL PSRTLCQATR GPCAIVERER GWPDFLRCTP DhFPEGCpNE VQNIKFNSSG QCEAPLVRTD NPKSWYEDVE
        Chick Smo   PRCWdVIQPL LCAVYMPKCE ....dgqVEL PSqTLCQATR aPCtIVERER GWPDFLkCTP DrFPEGCpNE VQNIKFNSSG QCEAPLVRTy NPKSWYEDVE
D.melanogaster Smo  PkCWAaIQPf LCAVfkPKCE kingeDmVyL PSyemCriTm ePCrIlyntt ffPkFLRCne tlFPtkCtNg argmKFNgtG QClsPLVpTD tsaSyYpgiE
D.melanogaster  Fz  .......... .......... .tSSastaat PtR....... ...s.Vakvt trkhqtgves phrniGfvcp VQ.lKtplgm gyElkvggkD l.......h
D.melanogaster Fz2  gsgsggkrkq ggsgsggsga ggSSgststk PcRgrnsknc gnpq.gEkas GkecscsCrs pliflGkeql lQ.qqsqmpm mhhphhwymn ltvqriagVp

                                              TM1                                  TM2
                    301                                                                                                400
        Human Smo   GCGIQCQNPL FTEaEHqDMH SYIAAFGAVt GLcTLFTLAT FVaDWRNSNR YPAVILFYVN ACFFVGSIGW LA........ QFMDGARREI VCRADGTMRL
          Rat Smo   GCGIQCQNPL FTEaEHqDMH SYIAAFGAVt GLcTLFTLAT FVaDWRNSNR YPAVILFYVN ACFFVGSIGW LA........ QFMDGARREI VCRADGTMRf
        Chick Smo   GCGIQCQNPL FTEtEHreMH vYI.AFssVt iscTfFTLAT FVaDWRNSNR YPAVILFYVN ACFFVGSIGc vA........ QFMDGARdEI VCRADGTMRL
D.melanogaster Smo  vCGvrCkdPL yTddEHrqiH klIqwaGsic lLsnLFvvsT FfiDWkNaNk YPAVIvFYiN lCFliacvGW Ll........ QFtsGsRedI VCRkDGTlRh
D.melanogaster  Fz  dCGapChamf FpErErtvlr ywvgswaAVc vascLFTvlT FliD.ssfrR YPeraivfla vCylVvgcay vAglgagdsv screpfpppv klgrlqmMst
D.melanogaster Fz2  nCGIpCkgPf FsndEkdfag lwIAlwsglc fcsTLmTLtT FiiD.terfk YPerpivfls ACyFmvavGy Lsrnflqne. ........EI aCdglllres

                                          TM3                                  TM4
                    401                                                                                                500
        Human Smo   GEPTSNETLS CVIIFVIVYY aLMAGVVWFV VLTYAWHTSF KALGTTYQPl SGKTSYFHLl tWSLPFVLTV AILAVAQVDG DSVSGICFVG YKNYRYRAGF
          Rat Smo   GEPTSsETLS CVIIFVIVYY aLMAGVVWFV VLTYAWHTSF KALGTTYQPl SGKTSYFHLl tWSLPFVLTV AILAVAQVDG DSVSGICFVG YKNYRYRAGF
        Chick Smo   GEPTSNETLS CVIIFVIVYY sLMsGViWFV mLTYAWHTSF KALGTTYQPl lGKTSYFHLi tWSiPFVLTV AILAVAQVDG DSVSGICFVG YKNYRYRAGF
D.melanogaster Smo  sEPTagEnLS CivIFVlVYY fLtAGmVWFV fLTYAWH..w rAmGhvqdri dkKgSYFHLv aWSLPlVLTi ttmAfseVDG nSivGICFVG YiNhsmRAGl
D.melanogaster  Fz  itqghrqTtS CtvlFmalYf ccMAafaWws cLafAWfla. agLkwgheai enKshlFHLv aWavPalqTi svLAlAkVeG DilSGvCFVG qldthslgaF
D.melanogaster Fz2  stgph....S CtlvFlltYf fqMAssiWwV iLTftWfla. agLkwgneai tkhsqYFHLa aWliPtVqsV AvLllsaVDG DpilGICyVG nlNpdhlktF

                                  TM5                                           TM6
                    501                                                                                                600
        Human Smo   VLAPiGLVLI VGGYFLIRGV MTLFSIKS.. .NHPGLLSEK AASKINETML RLGIFGFLAF GFVLITFsCH FYdFFNQAEW ERSFRDYVLC QANVTIgLPT
          Rat Smo   VLAPiGLVLI VGGYFLIRGV MTLFSIKS.. .NHPGLLSEK AASKINETML RLGIFGFLAF GFVLITFsCH FYdFFNQAEW ERSFRDYVLC QANVTIgLPT
        Chick Smo   VLAPiGLVLI VGGYFLIRGV MTLFSIKS.. .NHPGLLSEK AASKINETML RLGIFGFLAF GFVfITFGCH FYdFFNQAEW ERSFReYVLC eANVTIatqT
D.melanogaster Smo  hLqPlcqVil sGGYFitRGm vmLFglKh.. .fandikSts AsnKIhliim RmGvcalLtl vFiLvaiaCH vteFrhadEW aqSFRqfiiC kissvfeeks
D.melanogaster  Fz  lilPlciyLs iGalFLlaGf isLFrIrtVm Ktd.....gK rtdKlerlML RiGfFsqLfi lpavqllGCl FYeyyNfdEW miqwhrdic. ...kpfsiPc
D.melanogaster Fz2  VLAPlfvyLv iGttFLmaGf vsLFrIrSVi KqqgGvgagv kAdKleklMi RiGIFsvLyt vpatIviGCy lYeaayfedW i......... ...kalacPc

                                                TM7
                    601                                                                                                700
        Human Smo   KqPIPdCEIK NRPSLLVEKI NLFAMFGTGI aMSTWVWTKA TLLIWRRTWC RLTGqSDDEP KRIKKSKMIA KAFSKRhELL QnPGqELSFS MHTVSHDGPV
          Rat Smo   KkPIPdCEIK NRPSLLVEKI NLFAMFGTGI aMSTWVWTKA TLLIWRRTWC RLTGhSDDEP KRIKKSKMIA KAFSKRrELL QnPGqELSFS MHTVSHDGPV
        Chick Smo   nkPIPeCEIK NRPSLLVEKI NLFAMFGTGI sMSTWVWTKA TLLIWkRTWC RLTGqSDDqP KRIKKSKMIA KAFSKRkELL rdPGrELSFS MHTVSHDGPV
D.melanogaster Smo  s.....CrIe NRPSvgVlql hLlclFssGI vMSTWcWTps sietWkRyir kkcGkevvEe vkmpKhKvIA qtwaKRKd.f edkGr.LSit lyn.tHtdPV
D.melanogaster  Fz  paarapgspe aRPifqifmv kylcsmlvGv tsSvWlyssk TmvsWRnfve RLqGkeprtr aqayv~~~~~ ~~~~~~~~~~ ~~~~~~~~~~ ~~~~~~~~~~
D.melanogaster Fz2  aqvkgpgk.. .kPlysVlml kyFmalavGI tsqvWiWsgk TLesWRRfWr RLlGapDrtg anqaliKqrp piphpyagsg mgmpvgsaag sllatpytqa

                    701                                                                                                800
        Human Smo   AGLAFDLNEP SA....DVSS AWAQHVTKMV ARRGAILPQD iSVTPVATPV PPEEQaNLWl VEAEiSPELQ KRlGRKK... KRRKRKKEVC PLaPpPELhp
          Rat Smo   AGLAFeLNEP SA....DVSS AWAQHVTKMV ARRGAILPQD VSVTPVATPV PPEEQaNLWl VEAEiSPELe KRlGRKK... KRRKRKKEVC PLgPaPELhh
        Chick Smo   AGLAFDiNEP SA....DVSS AWAQHVTKMV ARRGAILPQD VSVTPVATPV PPEErsNLWv VEAdvSPELQ KRsrKK.... rRkKkKeEVC ...Perragl
D.melanogaster Smo  .GLnFDvNdl nssetnDiSS tWAaylpqcV kRRmAltgaa tgnssshgPr knsldseisv svrhvSvEsr rnsvdsqvsv KiaemKtkVa srsrgkhggs
D.melanogaster  Fz  ~~~~~~~~~~ ~~~~~~~~~~ ~~~~~~~~~~ ~~~~~~~~~~ ~~~~~~~~~~ ~~~~~~~~~~ ~~~~~~~~~~ ~~~~~~~~~~ ~~~~~~~~~~ ~~~~~~~~~~
D.melanogaster Fz2  gGasvastsh hhlhhhvlkq paAsHV~~~~ ~~~~~~~~~~ ~~~~~~~~~~ ~~~~~~~~~~ ~~~~~~~~~~ ~~~~~~~~~~ ~~~~~~~~~~ ~~~~~~~~~~

                    801                                                                                                900
        Human Smo   pAPaPsT..i PRLPQLPRQK CLVAA..... .......... ...gAWGaGD sCRQG.AWTl VSNPFCPEPS PPQDPFLPsA p......... ..........
          Rat Smo   SAPvPaTSaV PRLPQLPRQK CLVAA..... .......... ...nAWGtGe pCRQG.AWTv VSNPFCPEPS PhQDPFLPgA S......... ..........
        Chick Smo   SvaplTpSsV PRLPrLPqQp CLVAiprhrg dtfiptvlpg lsngAgGlwD grRrahvphf itNPFCPEsg sPeDeenPgp SvghrQhNGg rrwPpeplpg
D.melanogaster Smo  SsnrrTqrrr dyiaaatgks srrresstsv esqvialkkt typnAshkvg vfahhsskkq hnytssmkrr tanagldPsi lneflQkNGd fifPflqnqd
D.melanogaster  Fz  ~~~~~~~~~~ ~~~~~~~~~~ ~~~~~~~~~~ ~~~~~~~~~~ ~~~~~~~~~~ ~~~~~~~~~~ ~~~~~~~~~~ ~~~~~~~~~~ ~~~~~~~~~~ ~~~~~~~~~~
D.melanogaster Fz2  ~~~~~~~~~~ ~~~~~~~~~~ ~~~~~~~~~~ ~~~~~~~~~~ ~~~~~~~~~~ ~~~~~~~~~~ ~~~~~~~~~~ ~~~~~~~~~~ ~~~~~~~~~~ ~~~~~~~~~~

                    901                                                                                               1000
        Human Smo   .aPVaWAhGR RqGLGPIHSR TNLMdtELmD ADSDF~~~~~ ~~~~~~~~~~ ~~~~~~~~~~ ~~~~~~~~~~ ~~~~~~~~~~ ~~~~~~~~~~ ~~~~~~~~~~
          Rat Smo   .aPrvWAqGR lqGLGsIHSR TNLMeAELLD ADSDF~~~~~ ~~~~~~~~~~ ~~~~~~~~~~ ~~~~~~~~~~ ~~~~~~~~~~ ~~~~~~~~~~ ~~~~~~~~~~
        Chick Smo   gsgVtrtrGR RaGLaPIHSR TNLvnAELLD ADlDF~~~~~ ~~~~~~~~~~ ~~~~~~~~~~ ~~~~~~~~~~ ~~~~~~~~~~ ~~~~~~~~~~ ~~~~~~~~~~
D.melanogaster Smo  mssssseedns Rasqkiqdln vvvkqqEise dDhDgikiee lpnskqvale nflknikksn esnsnrhsrn sarsqskksq krhlknpaad ldfrkdcvky
D.melanogaster  Fz  ~~~~~~~~~~ ~~~~~~~~~~ ~~~~~~~~~~ ~~~~~~~~~~ ~~~~~~~~~~ ~~~~~~~~~~ ~~~~~~~~~~ ~~~~~~~~~~ ~~~~~~~~~~ ~~~~~~~~~~
D.melanogaster Fz2  ~~~~~~~~~~ ~~~~~~~~~~ ~~~~~~~~~~ ~~~~~~~~~~ ~~~~~~~~~~ ~~~~~~~~~~ ~~~~~~~~~~ ~~~~~~~~~~ ~~~~~~~~~~ ~~~~~~~~~~

                    1001                                                                                              1100
        Human Smo   ~~~~~~~~~~ ~~~~~~~~~~ ~~~~~~~~~~ ~~~~~~~~~~ ~~~~~~~~~~ ~~~~~~~~~~ ~~~~~~~~~~ ~~~~~~~~~~ ~~~~~~~~~~ ~~~~~~~~~~
          Rat Smo   ~~~~~~~~~~ ~~~~~~~~~~ ~~~~~~~~~~ ~~~~~~~~~~ ~~~~~~~~~~ ~~~~~~~~~~ ~~~~~~~~~~ ~~~~~~~~~~ ~~~~~~~~~~ ~~~~~~~~~~
        Chick Smo   ~~~~~~~~~~ ~~~~~~~~~~ ~~~~~~~~~~ ~~~~~~~~~~ ~~~~~~~~~~ ~~~~~~~~~~ ~~~~~~~~~~ ~~~~~~~~~~ ~~~~~~~~~~ ~~~~~~~~~~
D.melanogaster Smo  rsndslscss eeldvaldvg sllnssfsgi smgkphsrns ktscdvgiqa npfelvpsyg edelqqamrl lnaasrqrte aanedfggte lqgllghshr
D.melanogaster  Fz  ~~~~~~~~~~ ~~~~~~~~~~ ~~~~~~~~~~ ~~~~~~~~~~ ~~~~~~~~~~ ~~~~~~~~~~ ~~~~~~~~~~ ~~~~~~~~~~ ~~~~~~~~~~ ~~~~~~~~~~
D.melanogaster Fz2  ~~~~~~~~~~ ~~~~~~~~~~ ~~~~~~~~~~ ~~~~~~~~~~ ~~~~~~~~~~ ~~~~~~~~~~ ~~~~~~~~~~ ~~~~~~~~~~ ~~~~~~~~~~ ~~~~~~~~~~

                    1101                  1122
        Human Smo   ~~~~~~~~~~ ~~~~~~~~~~ ~~
          Rat Smo   ~~~~~~~~~~ ~~~~~~~~~~ ~~
        Chick Smo   ~~~~~~~~~~ ~~~~~~~~~~ ~~
D.melanogaster Smo  hqreptfmse sdklkmlllp sq
D.melanogaster  Fz  ~~~~~~~~~~ ~~~~~~~~~~ ~~
D.melanogaster Fz2  ~~~~~~~~~~ ~~~~~~~~~~ ~~
```

Figure 5. Comparison of the predicted amino acid sequences of *Drosophila* Smo and Fz proteins with those of Smo from chick, rat, and human. (The latter two sequences are from Stone et al. 1996.) Conserved cysteine residues are shown in boldface. Putative membrane-spanning domains are underlined. X-Thr/Ser-X-Val motifs that could in principle interact with PDZ domains are boxed.

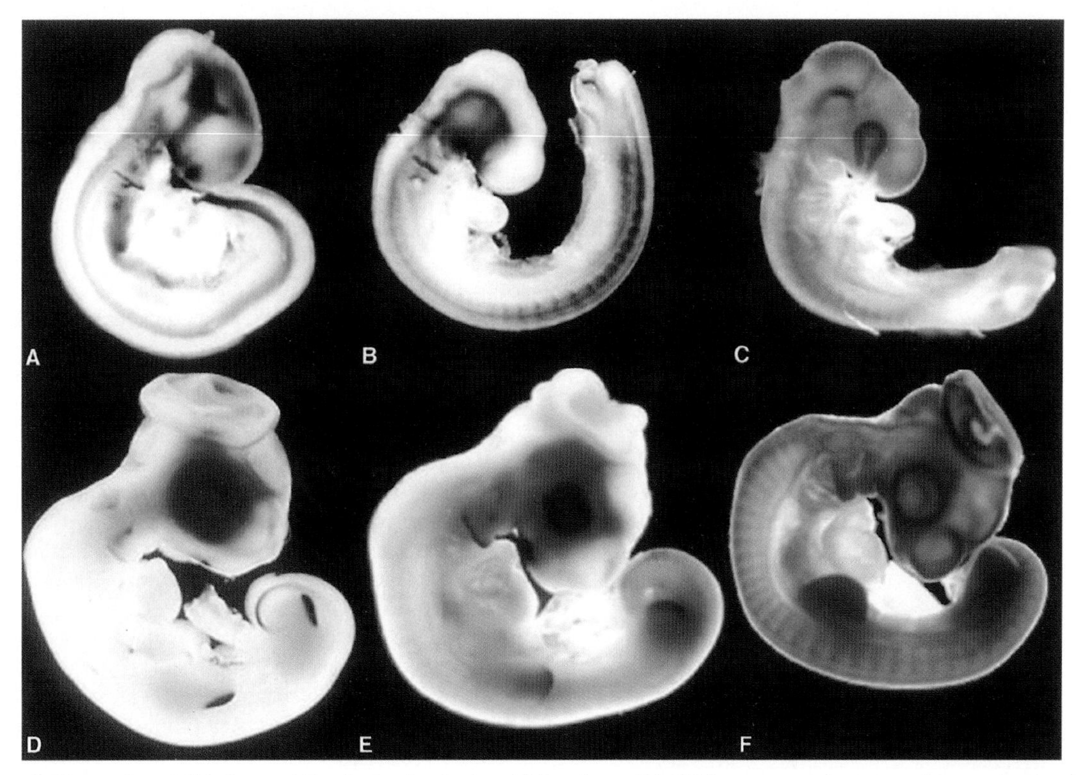

Figure 6. Expression of *Shh*, *Ptc*, and *Smo* in the developing chick embryo. (A) *Shh* is expressed throughout the notochord and ventral floor of the brain and neural tube of a stage-15 chick embryo. (B) At the same stage, *Ptc* expression is up-regulated in the ventral brain and neural tube and in the somites in response to *Shh* signaling. (C) In contrast to *Shh* and *Ptc*, *Smo* is essentially ubiquitously expressed at stage 15. By stage 22, *Shh* (D) and *Ptc* (E) are both expressed at high levels in the developing limb bud, whereas expression along the main body axis has largely dissipated; both genes are also expressed in the second branchial arch. *Smo* continues to be expressed ubiquitously (F), although transcript levels are noticeably higher in the limb buds and second branchial arch, where *Shh* and *Ptc* expression remains high.

plest explanation of these data is that Ptc and Smo form a receptor complex, Ptc binding the Hh ligand and Smo transducing its activity. Activation of Smo may occur through a conformational change or perhaps by the dissociation of the receptor complex, induced by Hh binding to Ptc. It is of course still possible that Smo interacts with some other unidentified ligand. Notably, most of the cysteine residues in the amino-terminal ligand-binding domain of Fz proteins (Bhanot et al. 1996) are conserved in Smo, suggesting that Smo itself might interact with a Wnt protein. Attempts to demonstrate such an interaction have, however, proved to be unsuccessful (Nusse et al., this volume), and there are no genetic data that would implicate any of the known *Drosophila* Wnt proteins in modulating the activity of Smo.

If the activity of *smo* is modulated exclusively in response to the repressive effects of *ptc*, it follows that removal of *ptc* should result in the constitutive activation of Smo in all cells that normally express Ptc protein. In this light, our finding that the cuticular pattern of embryos completely lacking Ptc activity can be restored almost to wild type by partial inactivation of Smo seems rather surprising. In such embryos, expression of *wg* is still activated ectopically yet, despite this, the patterning of cells in each segment in response to *wg* activity is essentially normal. One explanation for this may be that the levels of ectopic *wg* transcription induced by the mutant forms of the Smo protein are too low to have any detectable effect on the pattern. In any event, this result confirms the view that the positional specification of cells within each segment does not depend directly on their exposure to differing levels of Hh signaling because in the double mutant, all cells presumably experience the same, albeit reduced, level of *smo* activity.

The role of Smo in Hh signaling is well-established, but the mechanism by which it transduces the signal remains obscure. Although superficially similar to G-protein-coupled receptors, there is to date no evidence that either Smo or indeed any other member of the Fz protein family signals via heterotrimeric G-proteins. The intracellular carboxy-terminal tails of these proteins would seem to be likely candidates for domains involved in interaction with intracellular components of the pathway. In the case of the Wg pathway, the most receptor proximal component to be identified is the product of the *dishevelled* (*dsh*) gene (Klingensmith et al. 1994; Theisen et al. 1994). The Dsh protein contains a PDZ domain, which in other proteins has been shown to interact with the sequence X-Thr/Ser-X-Val (Doyle et al. 1996). Such a sequence occurs at the carboxyl terminus of Fz2, suggesting that it may interact directly with DSH; however, attempts to demonstrate such an interaction have so far proved negative (Nusse et al., this volume). Thus, whereas the presence of X-Thr/Ser-X-Val sequences in

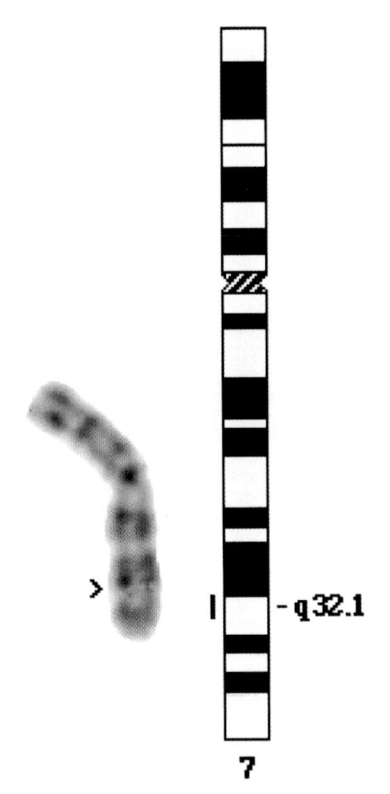

Figure 7. Cytogenetic localization of human *smo*. Partial image showing localization of PAC DNA (clone 244J22) to human chromosome band 7q32. FITC signal shown on G-banded chromosome and corresponding ideogram.

the carboxyl terminus of Smo might indicate an interaction with other PDZ-domain-containing proteins, there is no precedent for such an interaction. Moreover, it should be noted that such sequences are not conserved between *Drosophila* Smo and its vertebrate counterparts. Indeed, the carboxy-terminal portion of Smo proteins from different species is remarkable more for the extent to which it has diverged between *Drosophila* and vertebrates, rather than for any conserved motifs.

Signaling by *Sonic hedgehog* has a crucial role during the development of the vertebrate embryo, specifying ventral cell types in both the somites and the central nervous system (Johnson et al. 1994; Fan et al. 1995; Marti et al. 1995; Roelink et al. 1995; Chiang et al. 1996; Ericson et al. 1996). Not surprisingly, therefore, perturbations in the Hh pathway have been found to be associated with congenital defects in humans, with *Shh* and *Ptc* having been identified as the genes mutated in the inherited syndromes holoprosenchephaly type III (HPE3) (Belloni et al. 1996) and Gorlin's syndrome (Hahn et al. 1996; Johnson et al. 1996), respectively. Gorlin's syndrome is a dominant autosomal trait, causing a number of developmental abnormalities, including bifurcated ribs and facial anomalies, defects consistent with a derepression of *Shh* signaling. Affected individuals also suffer from a very high frequency of basal cell carcinomas (BCCs) because of somatic mutation of their second *Ptc* allele.

HPE3 patients, in contrast, present defects in midline neurogenesis that, in severe cases, may result in cyclopia. HPE3 has been shown by linkage study and mutational analysis to be associated both with translocations mapping centromeric to human *Shh* on 7q36 (Belloni et al. 1996) and with mutations within the *Shh* protein-coding region. In contrast to the mouse, where mutations in *Shh* appear to be strictly recessive (Chiang et al. 1996), most cases of HPE3 are dominant, indicating that in humans, the *Shh* locus, like *Ptc*, is haplo-insufficient. Although HPE3 is caused by a loss or reduction of Shh activity, recent studies have identified a specific class of *Shh* mutations, presumed to be gain-of-function alleles, associated with BCCs (Oro et al. 1997).

HPE3 is just one of a number of syndromes in humans associated with holoprosencephaly. Intriguingly, another of these, Smith-Lemli-Opitz syndrome (SLOS), has been mapped close to *Shh* at 7q32.1 (Wallace et al. 1994; Alley et al. 1995), precisely the position to which our FISH mapping has localized *Smo*. Given the similarity between the *hh* and *smo* phenotypes in *Drosophila*, this immediately raises the possibility that SLOS may be caused by mutations in *Smo*. Against this, however, a large body of evidence exists associating SLOS with a biochemical defect in cholesterol metabolism, resulting in raised serum levels of 7-dehydrocholesterol (Tint et al. 1994; Xu et al. 1995). On this basis, it has been proposed that the abnormal levels of serum 7-dehydrocholesterol and, subsequently, abnormally low levels of cholesterol in the serum are responsible for the observed SLOS symptoms, leading to the suggestion that the primary genetic lesion in SLOS occurs in the 7-dehydrocholesterol reductase gene (Tint et al. 1995). According to this view, the similarity between the SLOS and HPE3 syndromes could be accounted for by the fact that normal processing of the Shh protein requires cholesterol (Porter et al. 1996); thus, any reduction in cholesterol levels might be expected to result in a reduction in levels of active SHh protein. Although this interpretation seems plausible, the possibility that SLOS is in fact due to mutation of *Smo* deserves further investigation. The recent finding that the NPC1 protein, required for control of intracellular cholesterol levels, shows significant homology with Ptc (Carstea et al. 1997; Loftus et al. 1997) introduces a further unexpected link between Hh signaling and cholesterol; it may be that the observed cholesterol defects in SLOS patients and normal Smo function are not unrelated.

ACKNOWLEDGMENTS

We are grateful to John Moore for help with DNA sequencing and Ian Goldsmith for oligonucleotide synthesis. We thank the HGMP Resource Centre, Hinxton Hall, Cambridge, UK, for providing the PAC clones. This work was supported by the ICRF and by the Human Frontiers of Science Programme (HFSP).

REFERENCES

Alcedo J., Ayzenzon M., Vonohlen T., Noll M., and Hooper J.E. 1996. The *Drosophila smoothened* gene encodes a 7-pass membrane-protein, a putative receptor for the hedgehog signal. *Cell* **86:** 221.

Alley T., Gray B., Lee S., Scherer S., Tsui L., Tint G., Williams C., Zori R., and Wallace M. 1995. Identification of a yeast artificial chromosome clone spanning a translocation breakpoint at 7q32.1 in a Smith-Lemli-Opitz syndrome patient. *Am. J. Hum. Genet.* **56:** 1411.

Amaya E., Stein P., Musci T., and Kirschner M. 1993. FGF signaling in the early specification of mesoderm in *Xenopus. Development* **110:** 477.

Baker N. 1988. Embryonic and imaginal requirements for *wingless*, a segment polarity gene in *Drosophila. Dev. Biol.* **125:** 96.

Bejsovec A. and Martinez Arias A. 1991. Roles of *wingless* in patterning the larval epidermis of *Drosophila. Development* **113:** 471.

Bejsovec A. and Wieschaus E. 1993. Segment polarity gene interactions modulate epidermal patterning in *Drosophila* embryos. *Development* **119:** 501.

Belloni E., Muenke M, Roessler E., Traverso G., Siegel-Bartelt J., Frumkin A., Mitchell H., Donis-Keller H., Helms C., Hing A., Heng H., Koop B., Martindale D., Rommens J., Tsui L., and Scherer S. 1996. Identification of *Sonic hedgehog* as a candidate gene responsible for holoprosencephaly. *Nat. Genet.* **14:** 353.

Bhanot P., Brink M., Samos C.H., Hsieh J.-C., Wang Y., Macke J.P., Andrew D., Nathans J., and Nusse R. 1996. A new member of the *frizzled* family from *Drosophila* functions as a Wingless receptor. *Nature* **382:** 225.

Carstea E., Morris J., Coleman K., Loftus S., Zhang D., Cummings C., Gu J., Rosenfeld M., Pavan W., Krizman D., Nagle J., Polymeropoulos M., Sturley S., Ioannou Y., Higgins M., Comly M., Cooney A., Brown A., Kaneski C., Blanchette-Mackie E., Dwyer N., Neufeld E., Chang T.-Y., Liscum L., Strauss III J., Ohno K., Zeigler M., Carmi R., Sokol J., Markie D., ONeill R., van Diggelen O., Elleder M., Patterson M., Brady R., Vanier M., Pentchev P., and Tagle D. 1997. Niemann-Pick C1 disease gene: Homology to mediators of cholesterol homeostasis. *Science* **277:** 228.

Chiang C., Litingtung Y., Lee E., Young K.E., Corden J.L., Westphal H., and Beachy P.A. 1996. Cyclopia and defective axial patterning in mice lacking *Sonic hedgehog* gene function. *Nature* **383:** 407.

Concordet J.-P., Lewis K., Moore J., Goodrich L.V., Johnson R.L., Scott M.P., and Ingham P.W. 1996. Spatial regulation of a Zebrafish *patched* homologue reflects the roles of Sonic hedgehog and protein kinase A in neural tube and somite patterning. *Development* **122:** 2835.

Diaz-Benjumea F. and S. Cohen. 1995. Serrate signals through Notch to establish a *wingless*-dependent organizer at the dorsal/ventral compartment boundary of the *Drosophila* wing. *Development* **121:** 4215.

Doyle D., Lee A., Lewis J., Kim E., Sheng M., and MacKinnon R. 1996. Crystal structure of a complexed and peptide free membrane protein-binding domain: Molecular basis of peptide recognition by PDZ. *Cell* **85:** 1067.

Ericson J., Morton S., Kawakami A., Roelink H., and Jessell T.M. 1996. Two critical periods of Sonic hedgehog signaling required for the specification of motor neuron identity. *Cell* **87:** 661.

Fan C.-M., Porter J.A., Chiang C., Chang D.T., Beachy P.A., and Tessier-Lavigne M. 1995. Long-range induction of sclerotome by Sonic hedgehog: A direct role for the amino terminal product of autoproteolytic cleavage and modulation by the cyclic AMP signaling pathway. *Cell* **81:** 457.

Goodrich L.V., Johnson R.L., Milenkovic L., McMahon J.A., and Scott M.P. 1996. Conservation of the *hedgehog/patched* signaling pathway from flies to mice—Induction of a mouse *patched* gene by *hedgehog. Genes Dev.* **10:** 301.

Hahn H., Wicking C., Zaphiropoulos P.G., Gailani M.R., Shanley S., Chidambaram A., Vorechovsky I., Holmberg E., Unden A.B., Gillies S., Negus K., Smyth I., Pressman C., Leffell D.J., Gerrard B., Goldstein A.M., Dean M., Toftgard R., Chenevixtrench G., Wainwright B., and Bale AE. 1996. Mutations of the human homolog of *Drosophila patched* in the nevoid basal-cell carcinoma syndrome. *Cell* **85:** 841.

Henrique D., Adam J., Myat A., Chitnis A., Lewis J., and Ish-Horowicz D. 1995. Expression of a *Delta* homologue in prospective neurons in the chick. *Nature* **375:** 787.

Hogan B. 1996. Bone morphogenetic proteins in development. *Curr. Opin. Genet. Dev.* **6:** 432.

Hooper J. 1994. Distinct pathways for autocrine and paracrine Wingless signaling in *Drosophila* embryos. *Nature* **372:** 461.

Hooper J. and Scott M.P. 1989. The *Drosophila patched* gene encodes a putative membrane protein required for segmental patterning. *Cell* **59:** 751.

Ingham P.W. 1993. Localised *hedgehog* activity controls spatially restricted transcription of *wingless* in the *Drosophila* embryo. *Nature* **366:** 560.

———. 1995. Signaling by hedgehog family proteins in *Drosophila* and vertebrate development. *Curr. Opin. Genet. Dev.* **5:** 492.

———. 1996. Has the quest for a Wnt receptor finally frizzled out? *Trends Genet.* **12:** 382.

Ingham P.W. and Hidalgo A. 1993. Regulation of *wingless* transcription in the *Drosophila* embryo. *Development* **117:** 283.

Ingham P.W., Taylor A.M., and Nakano Y. 1991. Role of the *Drosophila patched* gene in positional signaling. *Nature* **353:** 184.

Ioannou P., Amemiya C., Garnes J., Kroisel P., Shizuya H., Chen C., Batzer M., and de Jong P. 1994. A new bacteriophage P1-derived vector for the propagation of large human DNA fragments. *Nat. Genet.* **6:** 84.

Johnson D. and Williams L. 1993. Structural and functional diversity in the FGF receptor multigene family. *Adv. Cancer Res.* **60:** 1.

Johnson R.L., Laufer E., Riddle R.D., and Tabin C. 1994. Ectopic expression of Sonic hedgehog alters dorsal-ventral patterning of somites. *Cell* **79:** 1165.

Johnson R.L., Rothman A.L., Xie J.W., Goodrich L.V., Bare J.W., Bonifas J.M., Quinn A.G., Myers R.M., Cox D.R., Epstein E.H., and Scott M.P. 1996. Human homolog of *patched*, a candidate gene for the basal-cell nevus syndrome. *Science* **272:** 1668.

Klingensmith J. and Nusse R. 1994. Signaling by *wingless* in *Drosophila. Dev. Biol.* **166:** 396.

Klingensmith J., Nusse R., and Perrimon N. 1994. The *Drosophila* segment polarity gene *dishevelled* encodes a novel protein required for response to the *wingless* signal. *Genes Dev.* **8:** 118.

Loftus S., Morris J., Carstea E., Gu J., Cummings C., Brown A., Elison J., Ohno K., MA Rosenfeld, Tagle D., Pentchev P., and Pavan W. 1997. Murine model of Niemann-Pick C disease: Mutation in a cholesterol homeostasis gene. *Science* **277:** 232.

MacNicol A., Muslin A., and Williams L. 1993. Raf-1 kinase is essential for early *Xenopus* development and mediates the induction of mesoderm by FGF. *Cell* **73:** 571.

Marigo V., Davey R., Zuo Y., Cunningham J., and Tabin C. 1996a. Biochemical-evidence that patched is the hedgehog receptor. *Nature* **384:** 176.

Marigo V., Scott M.P., Johnson R.L., Goodrich L.V., and Tabin C.J. 1996b. Conservation in *hedgehog* signaling: Induction of a chicken *patched* homolog by Sonic hedgehog in the developing limb. *Development* **122:** 1225.

Marti E., Bumcrot D.A., Takada R., and McMahon A.P. 1995. Requirement of 19kDalton form of Sonic hedgehog for induction of distinct ventral cell types in vertebrate CNS explants. *Nature* **375:** 322.

Massagué J. 1996. TGF-β signaling—Receptors, transducers, and mad proteins. *Cell* **85:** 947.

Massagué J., Hata A., and Liu F. 1997. TGF-β signaling through the Smad pathway. *Trends Cell Biol.* **7:** 187.

McMahon A. 1992. The *wnt* family of developmental regulators. *Trends Genet.* **8:** 236.

Mohler J. 1988. Requirements for *hedgehog*, a segment polarity gene, in patterning larval and adult cuticle of *Drosophila. Genetics* **120:** 1061.

Nakano Y., Guerrero I., Hidalgo A., Taylor A.M., Whittle R.R.S., and Ingham P.W. 1989. The *Drosophila* segment polarity gene *patched* encodes a protein with multiple potential membrane spanning domains. *Nature* **341:** 508.

Neumann C. and Cohen S. 1996. A hierarchy of cross-regulation involving *Notch*, *wingless*, *vestigial* and *cut* organizes the dorsal/ventral axis of the *Drosophila* wing. *Development* **122:** 3477.

Nüsslein-Volhard C. and Wieschaus E. 1980. Mutations affecting segment number and polarity in *Drosophila*. *Nature* **287:** 795.

Nüsslein-Volhard C., Wieschaus E., and Kluding H. 1984. Mutations affecting the pattern of the larval cuticle in *Drosophila melanogaster*. I. Zygotic loci on the second chromosome. *Roux's Arch. Dev. Biol.* **193:** 267.

Oro A., Higgins K., Hu Z., Bonifas J., Epstein E., and Scott M. 1997. Basal cell carcinomas in mice overexpressing Sonic hedgehog. *Science* **276:** 817.

Porter J., Young K., and Beachy P. 1996. Cholesterol modification of hedgehog signaling proteins in animal development. *Science* **274:** 255.

Riddle R., Johnson R.L., Laufer E., and Tabin C. 1993. *Sonic hedgehog* mediates the polarizing activity of the ZPA. *Cell* **75:** 1401.

Riddle R., Ensini M., Nelson C., Tsuchida T., Jessell R., and Tabin C. 1995. Induction of the LIM homeobox gene *Lmx1* by Wnt7a establishes dorsoventral pattern in the vertebrate limb. *Cell* **83:** 631.

Roelink H., Porter J., Chiang C., Tanabe Y., Chang D.T., Beachy P.A., and Jessell T.M. 1995. Floor plate and motor neuron induction by different concentrations of the amino terminal cleavage product of Sonic hedgehog autoproteolysis. *Cell* **81:** 445.

Sambrook J., Fritsch E.F., and Maniatis T. 1989. *Molecular cloning: A laboratory manual*, 2nd edition. Cold Spring Harbor Laboratory Press, Cold Spring Harbor, New York.

Stone D., Hynes M., Armanini M., Swanson T., Gu Q., Johnson R., Scott M., Pennica D., Goddard A., Phillips H., Noll M., Hooper J., Desauvage R., and Rosenthal A. 1996. The tumor-suppressor gene *patched* encodes a candidate receptor for Sonic hedgehog. *Nature* **384:** 129.

Sutherland D., Samakovlis C., and Krasnow M. 1996. *branchless* encodes a *Drosophila* FGF homolog that controls tracheal cell-migration and the pattern of branching. *Cell* **87:** 1091.

Tabata T. and Kornberg T.B. 1994. Hedgehog is a signaling protein with a key role in patterning *Drosophila* imaginal discs. *Cell* **76:** 89.

Theisen H., Purcell J., Bennett M., Kansagara D., Syed A., and Marsh J.L. 1994. *dishevelled* is required during *wingless* signaling to establish both cell polarity and cell identity. *Development* **120:** 347.

Tint G., Irons M., Elias E., Batta A., Frieden R., Chen T., and Salen G. 1994. Defective cholesterol biosynthesis associated with the Smith-Lemli-Opitz syndrome. *N. Engl. J. Med.* **330:** 107.

Tint G., Salen G., Batta A., Shefer S., Irons M., Elias E., Abuelo D., Johnson V., Lambert M., Lutz R., Schanen C., Morris C., Hoganson G., and Hughes-Benzie R. 1995. Correlation of severity and outcome with plasma sterol levels in variants of the Smith-Lemli-Opitz syndrome. *J. Pediatr.* **127:** 82.

van den Heuvel M. and Ingham P.W. 1996. *smoothened* encodes a serpentine protein required for *hedgehog* signaling. *Nature* **382:** 547.

Wallace M., Zori R., Alley R., Whidden E., Gray B., and Williams C. 1994. Smith-Lemli-Opitz syndrome in a female with a *de novo*, balanced translocation involving 7q32: Probable disruption of an SLOS gene. *Am. J. Med. Genet.* **50:** 368.

Whitman M. and Melton D. 1992. Involvement of p21^{ras} in *Xenopus* mesoderm induction. *Nature* **357:** 252.

Williams J., Paddock S., and Carroll S. 1993. Pattern-formation in a secondary field—A hierarchy of regulatory genes subdivides the developing *Drosophila* wing disc into discrete subregions. *Development* **117:** 571.

Xu G., Salen G., Shefer S., Ness G., Chen T., Zhao A., and Tint G. 1995. Reproducing abnormal cholesterol biosynthesis as seen in the Smith-Lemli-Opitz syndrome by inhibiting the conversion of 7-dehydrocholesterol to cholesterol in rats. *J. Clin. Invest.* **95:** 76.

Yamaguchi T. and Rossant J. 1995. Fibroblast growth-factors in mammalian development. *Curr. Opin. Genet. Dev.* **5:** 485.

WNT5 Is Required for Tail Formation in the Zebrafish Embryo

G.-J. Rauch,[1] M. Hammerschmidt,[2] P. Blader,[3] H.E. Schauerte,[1] U. Strähle,[3] P.W. Ingham,[4] A.P. McMahon,[5] and P. Haffter[1]

[1]*Max-Planck-Institut für Entwicklungsbiologie, 72076 Tübingen, Germany;* [2]*Spemann Laboratories, Max-Planck Institut für Immunbiologie, Stübeweg 51, 79108 Freiburg, Germany;* [3]*IGMBC, CNRS/INSERM/ULP, BP163, 67404 Illkirch, France;* [4]*The Krebs Institute, Sheffield S10 2TN, United Kingdom; and* [5]*Department of Molecular and Cellular Biology, Harvard University, The Biolabs, Cambridge, Massachusetts 02138*

The molecular analysis of genes identified by mutational screens in *Drosophila melanogaster* has identified several signal transduction pathways underlying a variety of developmental processes (Forbes et al. 1993). One of these is the *wingless* (*wg*) signaling pathway, which is essential for segmentation of the embryo and many other patterning events in *Drosophila* (Klingensmith and Nusse 1994). *wg* is a member of the *Wnt* family of genes that encode secreted glycoproteins that have been cloned from a variety of vertebrate and invertebrate species (Baker 1987; Rijsewijk et al. 1987; Nusse and Varmus 1992; Parr and McMahon 1994). The conservation of *Wnt* sequences and expression patterns among different vertebrates suggests that the activities and functions of these genes have been conserved during vertebrate evolution. As a consequence, it seems desirable to use all the advantages offered by the various vertebrate model systems to gain a full picture of the functions of different *Wnt* genes during vertebrate development.

Functional studies based on microinjection of synthetic *Wnt* mRNAs into *Xenopus laevis* embryos have grouped the *Wnt* genes into two distinct classes by virtue of the resultant phenotypes (Du et al. 1995). Members of the *Wnt1* class, when ectopically expressed in the ventral marginal zone of cleaving frog embryos, induce duplication of the embryonic axes (McMahon and Moon 1989; Du et al. 1995). In contrast, injection of members of the *Wnt5a* class does not induce complete axis duplication, but instead appears to decrease cell adhesion, thus perturbing morphogenetic movements during gastrulation (Moon et al. 1993; Du et al. 1995). Targeted mutants of several *Wnt* genes have been generated in the mouse, revealing their function in a variety of developmental processes including gastrulation, neural development, limb axis formation and kidney development (McMahon and Bradley 1990; Stark et al. 1994; Takada et al. 1994; Parr and McMahon 1995).

The zebrafish is becoming an increasingly popular model organism for studying pattern formation in the vertebrate embryo. The embryological and genetical properties of the zebrafish, mainly the optical clarity of the embryos and its high fecundity, have made it an ideal vertebrate to perform large-scale screens for mutations affecting early development (Streisinger et al. 1981; Kimmel 1989). Recent large-scale screens for mutants displaying a specific visible phenotype in the embryo or early larvae of the zebrafish have yielded more than 1850 mutations affecting early development (Driever et al. 1996; Haffter et al. 1996a). As judged from the variety of phenotypes displayed by these mutants, these genes seem to be involved in a large number of processes during morphogenesis, pattern formation, organogenesis, and cell differentiation. Nevertheless, many mutations affecting different genes cause similar or overlapping phenotypes, which allows them to be grouped into phenotypic classes. Some of the most striking examples for such classes are the dorsalized mutants, the *you*-type somite mutants, and the *accordion* group of motility mutants (Granato et al. 1996; Mullins et al. 1996; van Eeden et al. 1996). The similarities of the phenotypes within such a group suggest that the affected genes participate in a single developmental process and may encode components of a molecular signaling pathway.

Several members of the *Wnt* gene family have been cloned in the zebrafish (Krauss et al. 1992; Ungar et al. 1995; Blader et al. 1996). Although targeted gene-disruption technology is not yet available in the zebrafish, the ease of genetic linkage analysis makes it feasible to screen through a large number of mutants looking for mutations in candidate genes. This approach has already resulted in the molecular characterization of several mutations, including *no tail*, *floating head*, *no isthmus*, *nic1*, and *chordino* (Sepich and Westerfield 1993; Schulte-Merker et al. 1994, 1997; Talbot et al. 1995; Brand et al. 1996). Here, we describe the molecular characterization of mutations in the *pipetail* gene and show that it is identical to zebrafish *wnt5*.

MATERIALS AND METHODS

Fish stocks. Fish and fish stocks were kept under standard laboratory conditions as described by Mullins et al. (1994). The Tü (Tübingen) strain is the wild-type strain in which the *pipetail* mutations had been induced (Haffter et al. 1996a; Hammerschmidt et al. 1996a) and kept. The TL strain, which was used as a reference line, is homozygous for leo^{tl} and lof^{dt2} (Haffter et al. 1996b).

Whole-mount in situ hybridization. Whole-mount in situ hybridizations were carried out as described elsewhere (Schulte-Merker et al. 1992) using the modifications described in Hammerschmidt et al. (1996a).

Linkage analysis and PCR. The linkage analysis was carried out by polymerase chain reaction (PCR) using the following oligonucleotide primers flanking a polymorphic CA-repeat in the 3′ noncoding sequence of the *wnt5* mRNA: CGTGTGCCGTTGATAGTAATTTAA and GTTCTTAAGAGCTCTGACATCAG. PCR amplification was performed as follows: 1 μl of genomic fish DNA (100 ng/μl) was combined with 69 μl of ddH_2O, 16 μl of dNTPs (1.25 mM), 10 μl of 10× PCR buffer (Pharmacia), 1 μl of each primer (10 μM), and 2 μl of *Taq* polymerase (2 units/μl). The DNA was denatured for 2 minutes at 95°C, then amplified for 35 cycles (30 seconds at 94°C, 30 seconds at 60°C, and 1 minute at 73°C), followed by extension for 10 minutes at 73°C. The reaction (10 μl) was separated on a 2% agarose gel, a mix of regular agarose (QualexGold, AGS) and Metaphor (FMC) high-resolution agarose at a ratio of 1 : 1. The electrophoresis conditions were 5V/cm at 2 hours in TBE (18 mM Tris-borate at pH 8.0, 0.4 mM EDTA). The PCR amplification products were in the size range of approximately 300 bp, depending on the length of the CA-repeat.

DNA sequence analysis of the four alleles. For DNA sequencing of mutant *ppt* alleles, the four exons were amplified separately by PCR using genomic DNA from homozygous mutant embryos as templates. The following oligonucleotide primers corresponding to noncoding sequence flanking the exons were used for PCR amplification:

exon 1:
CAGACTCCGCGAGTGCCTCA
and GCCTCCACATTTGTTCTGAGCG
exon 2:
GATGATACCTAATAGATTTCCCAT
and TGTAAGGCTGGCAAATGGGTC
exon 3:
CATCCAAATCCAGAGCTGTTCA
and ACTTGCATGAACCCAAACTGGT
exon 4:
CTACACCATCAGTATATTTCACC
and CTATCAACGGCACACGTACAC.

The amplified PCR products were sequenced directly with the Thermo Sequenase fluorescent-labeled primer cycle sequencing with 7-deaza-dGTP kit (Amersham, Life Science) on an ALF™ Express sequencing machine (Pharmacia) using the following Cy5-labeled oligonucleotide primers:

exon 1:
CGCGCAGCTGCTGTTTTGAGG
and TTGTGCATCACTTCATCACACAC
exon 2:
CATCATTAGGGTAGCCTGATGA
and GGTGCTGCAGTGTTATCCAAAG
exon 3:
CACACAGCCCTAGTGGAATGTTT
and CCCAGTATGGCATCCACAGC
exon 4:
GTGCATGTTTGGAATCCCTGTT
and GTCCGTAAGGGTCCCGTTTC

All PCR amplifications and subsequent DNA sequencing reactions were performed in duplicate from separate isolates of *ppt* homozygous mutant embryos. Sequence analysis was done using the software package MacMolly® Tetra 1.2.3.

RESULTS

Expression of *wnt5* in Wild-type Embryos

The expression pattern of *wnt5* in the developing zebrafish embryo has been analyzed by in situ hybridization analysis (Fig. 1f,g) (Blader et al. 1996; the name of the gene has been changed from *ZfWnt5* to *wnt5* in keeping with standard zebrafish nomenclature). Analysis of early cleavage stage embryos indicates that zebrafish *wnt5* mRNA is maternally deposited in the egg. Subsequently, zygotic expression of *wnt5* is detected ubiquitously throughout gastrulation until early somitogenesis, when expression of *wnt5* becomes restricted to the developing tailbud (Fig. 1f). At 19 hours, strong expression is detected in the tailbud and in two to three stripes in the somitic mesoderm immediately rostral to the tailbud (Fig. 1g). The extent of the *wnt5* expression domain in the mesenchyme of the tailbud decreases gradually during somitogenesis, with expression no longer being detectable after somitogenesis is completed.

Anteriorly, *wnt5* expression is present in tissues ventrolateral to the head after 18 hours of development (Fig. 1g), which later extends posteriorly as far as the caudal end of the hindbrain. This position corresponds to cranial neural crest and paraxial mesoderm, which will later contribute to the pharyngeal arches. At 48 hours of development, *wnt5* expression can be seen in the mesenchyme of the pharyngeal arches but is not detected in mature chondrocytes of the differentiated jaw and gill arches.

In the pectoral fin field, *wnt5* is initially expressed uniformly, and as the fin bud increases in size, expression becomes restricted to the ventral half of the fin bud. By 36 hours, this expression is limited to the outermost mesenchymal cells of the fin fold. As with the jaw and gill arches, *wnt5* expression is excluded from differentiated chondrocytes of the developing pectoral fin.

Pipetail is Required for Tail Outgrowth and Jaw Formation

In a large-scale screen for mutations affecting morphogenesis and pattern formation during zebrafish embryogenesis, a number of mutants with abnormalities in gastrulation and tail formation were identified (Driever et al. 1996; Haffter et al. 1996a). Mutations in at least seven genes affect the establishment of the basic body plan, resulting in partial dorsalization or ventralization of the mutant embryos (Hammerschmidt et al. 1996b; Mullins et al. 1996; Solnica-Krezel et al. 1996). Other mutations cause unique phenotypes during specific aspects of early pro-

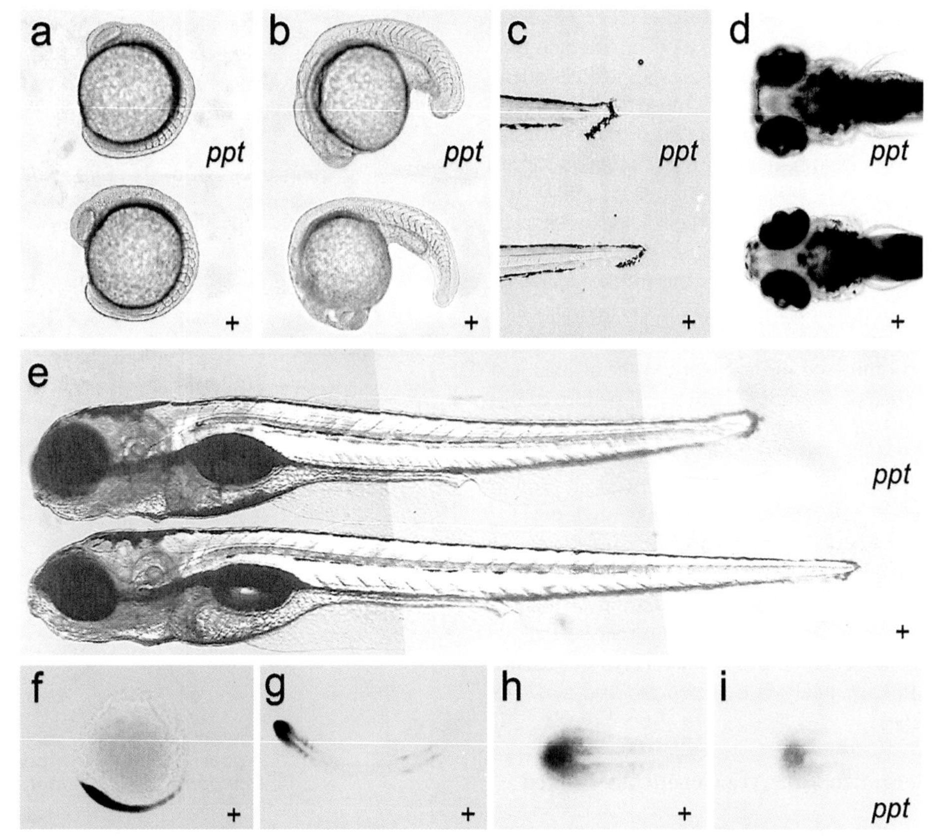

Figure 1. (*a–e*) Morphology of *pipetail* (*ppt*^{ta98} [*a–d*] and *ppt*^{th278} [*e*]) mutant embryos compared to wild type (+). (*a*) 12-somite stage, lateral view. *ppt* mutant embryos display a shortened body axis. (*b*) 24-somite stage, lateral view. The body axis of *ppt* mutant embryos is compressed. The tail fails to grow out and remains attached to the yolk tube. (*c*) Day 5, lateral view of the tail tip: The notochord and the neurocoel at the end of the tail are deformed. (*d*) Day 5, ventral view of the head: *ppt* mutant embryos lack tissue anterior to the eyes. (*e*) Day 5, lateral view of free-swimming larvae. *ppt* mutant larvae are about 80% of the wild-type length and display a deformed tip of tail and head skeleton. (*f–i*) *wnt5* in situ hybridization pattern on wild-type and *ppt* embryos. (*f*) Lateral view of 6-somite wild-type embryo. (*g*) Dorsal view of 18-hr wild-type embryo. (*h*) Dorsal view of 6-somite wild-type embryo. (*i*) Dorsal view of 6-somite *ppt*^{te1} mutant embryo with reduced *wnt5* expression compared to wild-type. (*a, b, c,* and *d* Reprinted, with permission, from Hammerschmidt et al. 1996a.)

cesses, such as epiboly (Kane et al. 1996). A number of additional genes were found to be required for other morphogenetic processes during gastrulation and tail formation (Hammerschmidt et al. 1996a; Solnica-Krezel et al. 1996). Among these, the phenotype of *pipetail* (*ppt*), described below, was suggestive of a defect in *wnt5* function.

Five alleles of *ppt* were isolated, *ppt*^{ta98}, *ppt*^{te1}, *ppt*^{tc271}, *ppt*^{ti265}, and *ppt*^{th278} (Hammerschmidt et al. 1996a). Homozygous *ppt* mutant embryos can first be recognized at the two-somite stage, by which they display a shortened body axis. This shortening is due to defects in morphogenesis after gastrulation: In wild-type embryos, the forming tailbud moves ventrally on the yolk sac, whereas in *ppt* mutants, the tailbud remains associated with the site of germ ring closure. At the 12-somite stage, *ppt* embryos are reduced in length and have an undulating notochord and somites that are reduced in width (Fig. 1a). At the 24-somite stage, the yolk tube is shorter and thicker, and the tail posterior to the yolk tube is several times shorter and remains attached to the yolk tube much longer than in wild-type siblings (Fig. 1b). By day 5 of development, *ppt* mutant larvae are approximately 80% of the wild-type length and display a deformed notochord and neurocoel at the tip of the tail (Fig. 1c,e).

ppt mutant larvae also display defects of the head skeleton (Piotrowski et al. 1996). This is most obvious by the lack of head tissue anterior to the eyes, which is a characteristic feature shared with the *hammerhead*-class of mutations affecting the head skeleton (Fig. 1d,e) (Piotrowski et al. 1996). The cartilaginous elements of the neurocranium and the pharyngeal skeleton are reduced in size and seem to be composed of much smaller but more numerous chondrocytes. Therefore, *ppt* seems to be required for proper outgrowth of the tail posteriorly and the head skeleton anteriorly.

Altered Gene Expression in *ppt* Mutants

To understand the *ppt* mutant phenotype better, we carried out in situ hybridization analysis with marker genes that are expressed in different regions of the tailbud and the tail (Hammerschmidt et al. 1996a). At early somite stages, the expression of *eve1* (Joly et al. 1993), *ntl* (Schulte-Merker et al. 1992), and *sna1* (Hammerschmidt and Nüsslein-Volhard 1993; Thisse et al. 1993), all of which are expressed in presumptive mesodermal cells of the tailbud, is normal, suggesting that the formation of the posterior mesoderm is unaffected in *ppt* mutants. Similarly, the expression of *pou2* (Takeda et al. 1994) and *cad1* (Joly et al. 1992) in posterior neurectoderm is normal. The expression of *ntl* and *myoD* (Weinberg et al. 1996) during somite stages reflects the trunk phenotype of *ppt* mutant embryos. *ntl* expression in the notochord of *ppt* embryos is undulated, and the somitic *myoD* expression is more compressed in the anteroposterior axis and broadened in posterior somites. The spatial pattern of expression of *wnt[a]*, a zebrafish PCR fragment that has high homology with the murine *Wnt3* and *Wnt3a* genes (Krauss et al. 1992), reflects the impaired ventral growth of the tailbud in *ppt*. However, expression levels of *wnt[a]* appear normal. In contrast, expression levels of *wnt5* were found to be reduced in ppt^{te1} (Fig. 1h,i), suggesting that *ppt* is involved in the regulation of the expression of *wnt5*, or, alternatively, *ppt* might be the zebrafish *wnt5* gene itself. To test the latter hypothesis, we performed a linkage analysis between *ppt* and *wnt5*, as described below.

ppt and Zebrafish *wnt5* Are Genetically Linked

To test whether *ppt* and *wnt5* are linked, we carried out a segregation analysis using a simple sequence length polymorphism (SSLP) linked to the *wnt5* gene (Fig. 2). Using PCR, we identified a length polymorphism in a CA-repeat in the 3′-untranslated region of the *wnt5* mRNA between the Tübingen strain, in which the ppt^{th278} mutation was originally induced, and the wild-type TL strain (Fig. 2a). In a linkage cross between a heterozygous ppt^{th278} carrier and a TL fish, we then followed the segregation of this SSLP in the F_2 progeny of F_1 pairs that were both heterozygous for ppt^{th278} and the SSLP. The F_2 offspring were sorted by mutant phenotype into pools of 50 embryos each and PCR was performed (Fig. 2c). By using this bulked segregant analysis, we found that only the fragment length linked to the ppt^{th278} allele segregated with the mutant pool of embryos, whereas the wild-type and the ppt^{th278}-associated fragments were found in the sibling pool at an expected ratio of 1 : 2 (mutant : wild-type) indicating that *wnt5* and *ppt* are genetically linked in the zebrafish genome.

Sequence Analysis of *ppt* Alleles

To characterize the molecular lesions underlying the *pipetail* mutant phenotype, we decided to sequence the entire coding region of the *wnt5* genes from four of the

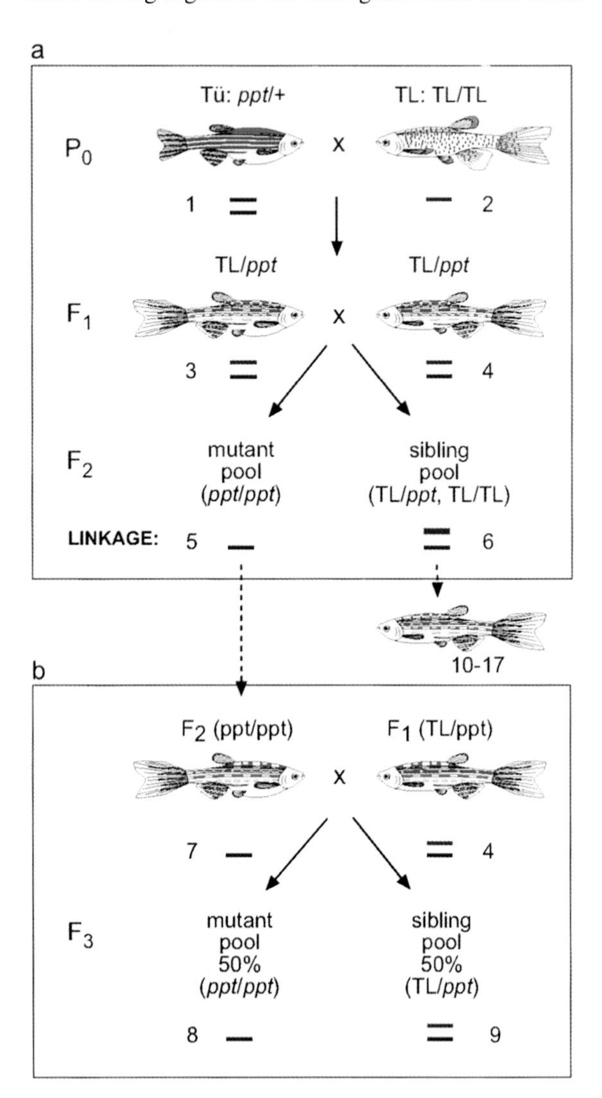

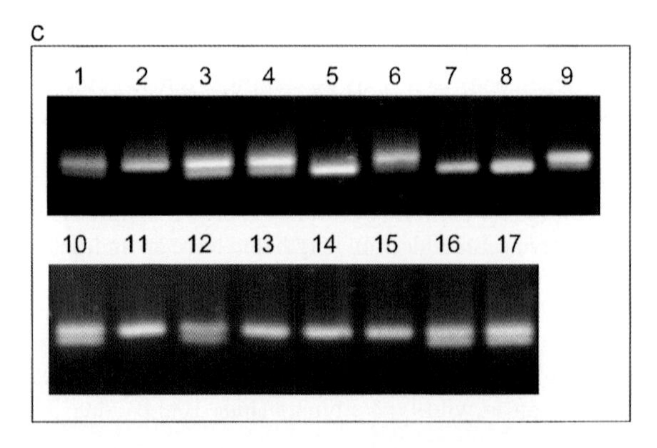

Figure 2. (*a*) Segregation analysis of *wnt5* with *ppt*. Fish and embryos used for PCR analysis were numbered with consecutive numbers (1–17). A heterozygous ppt^{th278} carrier (1) was crossed to a wild-type TL (2). Heterozygous carriers were identified among the F_1 progeny (3,4). These were crossed to each other and F_2 embryos were sorted into phenotypically mutant (5) and wild-type sibling (6) pools. (*b*) Embryos derived from females homozygous for ppt^{th278} do not display a maternal effect phenotype. Females homozygous for ppt^{th278} (7) were back-crossed to heterozygous F_1 males (4) and their offspring was sorted by phenotype into F_3 mutant (8) and sibling (9) pools of embryos. Individual F_2 siblings were also raised to adulthood and analyzed individually as a control (10–17). (*c*) PCR analysis of fish and embryos using an SSLP in the 3′-untranslated region of *wnt5*. The numbers of each reaction correspond to the consecutive numbers used for the fish and embryos above. The lower band is linked to the ppt^{th278} allele of *wnt5*; the upper band is linked to the TL allele.

ppt alleles (Fig. 3a; see Materials and Methods). In *ppt*^{ta98}, we identified a T to C change at position 968 of the open reading frame resulting in a cysteine to arginine change at the amino acid level. In *ppt*^{th278}, we identified a T to C change at position 1033, resulting in a cysteine to serine change at the amino acid level. Both cysteines affected in *ppt*^{ta98} and *ppt*^{th278}, respectively, belong to 24 cysteines that are absolutely conserved among all members of the Wnt5 subfamily of proteins (Fig. 3b) (Blader et al. 1996). In *ppt*^{ti265}, we identified an A to T change at position 1045 resulting in a premature TAG stop codon. The *ppt*^{te1} allele does not carry any mutation within the coding region of *wnt5*. However, because the amount of *wnt5* mRNA detected by whole-mount in situ hybridization is reduced in this allele, it is possible that it carries a mutation in the noncoding region that affects the stability or expression of the *wnt5* mRNA. Sequencing of the corresponding regions of the originally mutagenized founder fish from which *ppt*^{th278} and *ppt*^{ti265} were derived showed that the alterations in the *wnt5* coding regions of these alleles were not present before the mutagenesis and were probably induced by the treatment of the founder fish with ENU. Taken together, we conclude that *ppt* is due to mutations in the *wnt5* gene.

Embryos from Females Homozygous for *ppt*^{th278} Do Not Display a Maternal Effect Phenotype

In situ hybridization of early cleavage stage embryos indicated that *wnt5* is maternally deposited in the egg (Blader et al. 1996). To analyze the function of the maternally derived *wnt5* mRNA, we sorted *ppt*^{th278} homozygous mutant embryos for raising. Although most *ppt*^{th278} homozygous mutant embryos died before reaching feeding stages, we did obtain a few escapers that were subsequently raised and gave fertile homozygous adults (3/36). Two individual *ppt*^{th278} homozygous females were backcrossed to their *ppt*^{th278} heterozygous fathers. In both

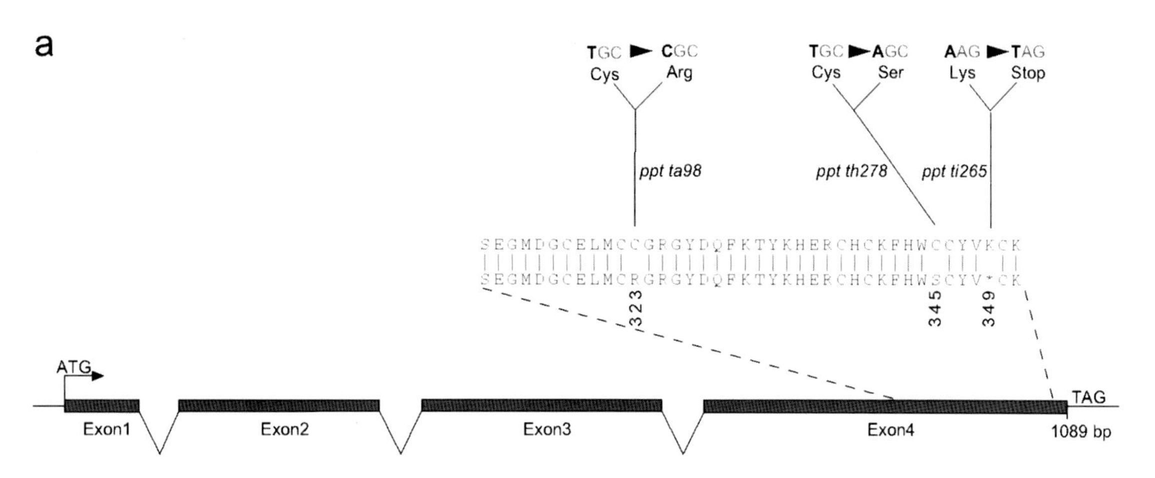

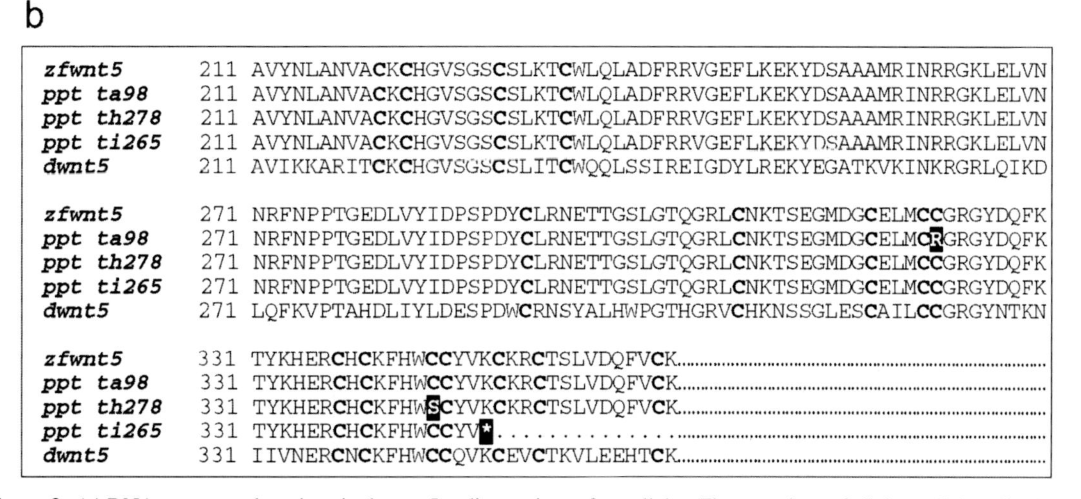

Figure 3. (*a*) DNA sequence alterations in the *wnt5* coding regions of *ppt* alleles. The mutations of all three alleles affecting the *wnt5* coding region fall within the fourth exon. (*b*) The *ppt*^{ta98} and *ppt*^{th278} mutations affect cysteines that are conserved among the Wnt5 subfamily of proteins, including the *wnt5* homolog of *Drosophila melanogaster* (*dwnt5*). The *ppt*^{ti265} mutation introduces a stop codon that truncates the Wnt5 protein by 15 amino acids.

cases, 50% of the progeny (225/453 and 228/453) displayed the *ppt* homozygous mutant phenotype, whereas the remaining 50% appeared wild type (see Fig. 2b). Because the ppt^{th278} homozygous females were F_2 individuals derived from the linkage cross that was used for the segregation analysis, we confirmed that both females were homozygous for $ppt2^{th278}$ using the SSLP linked to *wnt5*. We also genotyped their homozygous mutant offspring to show that they were homozygous for ppt^{th278} (Fig. 2c). Because we did not observe an increase in phenotypic strength or any additional phenotypes, it appears that the maternal expression of *wnt5* is not required for correct embryonic development.

DISCUSSION

This study describes the identification of mutations in the *wnt5* gene as the molecular cause for the phenotype of the zebrafish *pipetail* (*ppt*) mutation. We show that *wnt5* and *ppt* are genetically linked and that three out of four *ppt* alleles carry mutations in the *wnt5* coding sequence. Two alleles carry point mutations affecting highly conserved cysteine residues (Blader et al. 1996) that are predicted to impair Wnt5 function severely. A third mutation causes a stop codon, which truncates the carboxyl terminus of the Wnt5 protein by 14 amino acids. The fourth allele, ppt^{te1}, does not have a mutation in the coding sequence; instead, the expression or stability of the *wnt5* mRNA seems to be reduced relative to wild-type siblings, as shown by whole-mount in situ hybridization analysis, suggesting that the mutation resides in flanking noncoding sequences. We consider this evidence to be sufficient grounds for concluding that *ppt* is identical to the *wnt5* gene in the zebrafish.

The mutations in the three alleles that affect the *wnt5* coding sequence all fall within the fourth exon of the *wnt5* gene and therefore affect the carboxy-terminal part of the protein. Given that the fourth allele most likely results in lower-level expression of the wild-type protein, it is possible that none of the alleles that we have analyzed represent a completely amorphic condition of the gene. To determine the complete loss-of-function phenotype of *wnt5*, it would be desirable to recover alleles that truncate the protein in the amino-terminal portion or a deletion covering the *wnt5* gene.

Expression of the zebrafish *wnt5* gene correlates quite well with the *ppt* mutant phenotype. Zygotic expression of *wnt5* first becomes restricted to the developing tailbud at somite stages. This pattern of expression is in good agreement with the first visible phenotypes and the defects in tail outgrowth seen in *ppt* embryos. Similarly, expression of *wnt5* in cells giving rise to the head skeleton correlates with the impaired growth of the head skeleton observed in *ppt* larvae. The expression of *wnt5* in the mesenchyme of the pharyngeal skeleton until the appearance of mature chondrocytes is also consistent with the lack of normal cartilage differentiation in *ppt*.

Surprisingly, no obvious pectoral fin phenotype was observed in *ppt* mutant embryos. It is conceivable that the pectoral fin defect caused by the *ppt* mutations is too subtle to detect. Alternatively, this may reflect the presence of another member of the Wnt family that could compensate for the lack of *wnt5* function in the pectoral fin buds of *ppt* embryos.

Previous studies have suggested that the tail forms by a continuation of gastrulation in the tailbud and later in the tip of the tail (Gont et al. 1993; Tucker and Slack 1995). It was also suggested that tail formation depends on the same set of genes that are also required for gastrulation. Ectopic expression of different *Wnts* in *Xenopus* embryos have identified two distinct classes of *Wnt* genes (Du et al. 1995). The *Xwnt5a* class does not induce complete axis duplication but instead decreases cell adhesion and perturbs morphogenetic movements during gastrulation (Moon et al. 1993). The lack of normal tail outgrowth is therefore consistent with the proposed function of *Xwnt5* in morphogenetic movements. In *ppt* mutant embryos, however, gastrulation and the formation of the tailbud itself appear largely normal, suggesting either that *wnt5* is not required for gastrulation and specifically involved in the later tail forming process or that the lack of *wnt5* function during gastrulation can be compensated for by other *Wnt* family members.

The *ppt* mutant phenotype indicates that *wnt5* is required for very specific morphogenetic processes during tail formation, most likely in cells of the tailbud that are in close association with the yolk. These cells migrate ventrally on the yolk after germ ring closure, and finally detach from the yolk around the 15-somite stage. These processes most likely require changes in the intercellular organization of the cells that are involved in these interactions (cell-cell adhesion and/or cell-extracellular space interaction). Thus, one could imagine that Wnt5 regulates the expression of genes involved in cell-cell adhesion, as recently suggested for E-cadherin, which is required for epithelial organization of cells and which appears to be negatively regulated by the Wnt-signal transducers β-catenin and LEF-1 (Huber et al. 1996a,b). We are currently studying the expression of cadherins and other regulators of intercellular organization in wild-type and *ppt* mutant embryos in order to identify putative targets of the Wnt5 signal. Unfortunately, no other complementation groups with *ppt*-like phenotypes were isolated in the mutant screens, suggesting that mutations in the Wnt5 transducers and targets lead to no phenotype or to a different phenotype than mutations in *wnt5*, either because those transducers and targets are redundant or because they are also required for earlier or more general processes.

So far, only a single gene belonging to the *Wnt5* subfamily has been identified in the zebrafish, whereas two closely related genes, namely, *Wnt5a* and *Wnt5b*, have been described in the mouse (Gavin et al. 1990; Krauss et al. 1992; Blader et al. 1996). Interestingly, although the amino acid homology of the zebrafish *wnt5* gene is highest to the *Wnt5b* gene of the mouse, the expression pattern of *wnt5* in the zebrafish embryo strikingly resembles that of *Wnt5a* in the mouse (Parr et al. 1993). The consistency between the *wnt5* expression pattern and the *ppt* phenotype in the fish suggests that the mouse *Wnt5a* loss-of-function phenotype is likely to resemble the *ppt* phenotype and that the zebrafish *wnt5* and the mouse *Wnt5a* genes are probably true orthologs. With the absence of a second *wnt5* subfamily member to date, it is, however, conceivable that the function of the zebrafish *wnt5* gene is covered by two genes in the mouse.

To analyze the function of maternally derived *wnt5* mRNA, we raised ppt^{th278} homozygous females and analyzed their progeny for maternal effect phenotypes. As no increase in phenotypic strength or additional phenotypes were observed, we conclude that females homozygous for ppt^{th278} do not cause any obvious maternal effect phenotype in their offspring. We do not know, however, whether ppt^{th278} is a complete loss-of-function allele and cannot rule out the possibility that homozygous females for any of the other alleles would produce a maternal effect phenotype. An alternative explanation for the lack of a maternal effect phenotype is the possibility that another maternally derived Wnt could compensate for the lack of maternal *wnt5* mRNA in progeny of *ppt* homozygous females.

Intercellular communication is an important process during pattern formation in the vertebrate embryo. By genetic mapping, we are systematically screening the Tübingen stock collection for candidates that carry mutations in genes encoding signaling molecules such as the Wnts and the Hedgehogs. Once identified, these mutations will be powerful tools to determine the function of these signaling molecules during development of the vertebrate embryo.

SUMMARY

Intercellular signaling molecules, such as those encoded by the *Wnt* gene family, have a fundamental role in various aspects of pattern formation in the developing embryo. The zebrafish *wnt5* gene encodes a member of a subfamily of Wnt molecules thought to be involved in modulating cell behavior during vertebrate development. Here, we show that the zebrafish *pipetail* gene is identical to *wnt5*. The *pipetail* mutant phenotype is characterized by defects in tail formation and impaired maturation of the cells that contribute to cartilaginous elements of the head skeleton. This suggests a major role for *wnt5* in morphogenetic processes underlying tail outgrowth and cartilage differentiation in the head. To investigate the function of maternally derived *wnt5* mRNA, we generated females that were homozygous for *pipetail*. The lack of a maternal effect phenotype in the progeny of these females suggests that no obvious function for the maternal *wnt5* expression can be deduced.

ACKNOWLEDGMENTS

We are extremely grateful to Silke Geiger-Rudolph and Nadine Fischer for technical assistance and fish maintenance, to Simone Gränz for participation in the initial phases of the project, and to Hans-Georg Frohnhöfer from the Tübingen stock center for providing zebrafish strains. We also thank Christiane Nüsslein-Volhard for many helpful suggestions on the manuscript. We are grateful to the oligonucleotide synthesis and sequencing group of the IGBMC. P.B. was supported by a TMR fellowship from the European Community. U.S. is a recipient of a fellowship from the DFG. Part of this work was supported by the INSERM, CNRS, the Centre Hospitalier Universitaire Regional, ARC, and GREG.

REFERENCES

Baker N.E. 1987. Molecular cloning of sequences from *wingless*, a segment polarity gene in *Drosophila*. *EMBO J.* **6:** 1765.

Blader P., Strähle U., and Ingham P.W. 1996. Three *Wnt* genes expressed in a wide variety of tissues during development of the zebrafish, *Danio rerio*: Developmental and evolutionary perspectives. *Dev. Genes Evol.* **206:** 3.

Brand M., Heisenberg C.-P., Jiang Y.-J., Beuchle D., Furutani-Seiki M., Granato M., Haffter P., Hammerschmidt M., Kane D.A., Kelsh R.N., Mullins M.C., Odenthal J., van Eeden F.J.M., and Nüsslein-Volhard C. 1996. Mutations in zebrafish genes affecting the formation of the boundary between midbrain and hindbrain. *Development* **123:** 179.

Driever W., Solnica-Krezel L., Schier A.F., Neuhauss S.C.F., Malicki J., Stemple D.L., Stainier D.Y.R., Zwartkruis F., Abdelilah S., Rangini Z., Belak J., and Boggs C. 1996. A genetic screen for mutations affecting embryogenesis in zebrafish. *Development* **123:** 37.

Du S.J., Purcell S.M., Christian J.L., McGrew L.L., and Moon R.T. 1995. Identification of distinct classes of functional domains of *Wnt*s through expression of wild-type and chimeric proteins in *Xenopus* embryos. *Mol. Cell. Biol.* **15:** 2625.

Forbes A.J., Nakano Y., Taylor A.M., and Ingham P.W. 1993. Genetic analysis of *hedgehog* signalling in the *Drosophila* embryo. *Development* (suppl.), p. 115.

Gavin B.J., McMahon J.A., and McMahon A.P. 1990. Expression of multiple novel *Wnt-1/int-1*-related genes during fetal and adult mouse development. *Genes Dev.* **4:** 2319.

Gont L., Steinbeisser H., Blumberg B., and De Robertis E. 1993. Tail formation as a continuation of gastrulation: The multiple cell populations of the *Xenopus* tailbud derive from the late blastopore lip. *Development* **119:** 991.

Granato M., van Eeden F.J.M., Schach U., Trowe T., Brand M., Furutani-Seiki M., Haffter P., Hammerschmidt M., Heisenberg C.-P., Jiang Y.-J., Kane D.A., Kelsh R.N., Mullins M.C., Odenthal J., and Nüsslein-Volhard C. 1996. Genes controlling and mediating locomotion behaviour of the zebrafish embryo and larva. *Development* **123:** 399.

Haffter P., Granato M., Brand M., Mullins M.C., Hammerschmidt M., Kane D.A., Odenthal J., van Eeden F.J.M., Jiang Y.-J., Heisenberg C.-P., Kelsh R.N., Furutani-Seiki M., Vogelsang E., Beuchle D., Schach U., Fabian C., and Nüsslein-Volhard C. 1996a. The identification of genes with unique and essential functions in the development of the zebrafish, *Danio rerio*. *Development* **123:** 1.

Haffter P., Odenthal J., Mullins M.C., Lin S., Farrell M.J., Vogelsang E., Haas F., Brand M., van Eeden F.J.M., Furutani-Seiki M., Granato M., Hammerschmidt M., Heisenberg C.-P., Jiang Y.-J., Kane D.A., Kelsh R.N., Hopkins N., and Nüsslein-Volhard C. 1996b. Mutations affecting pigmentation and shape of the adult zebrafish. *Dev. Genes Evol.* **206:** 260.

Hammerschmidt M. and Nüsslein-Volhard C. 1993. The expression of a zebrafish gene homologous to *Drosophila*-snail suggests a conserved function in invertebrate and vertebrate gastrulation. *Development* **119:** 1107.

Hammerschmidt M., Pelegri F., Mullins M.C., Kane D.A., Brand M., van Eeden F.J.M., Furutani-Seiki M., Granato M., Haffter P., Heisenberg C.-P., Jiang Y.-J., Kelsh R.N., Odenthal J., Warga R.M., and Nüsslein-Volhard C. 1996a. Mutations affecting morphogenesis during gastrulation and tail formation in the zebrafish, *Danio rerio*. *Development* **123:** 143.

Hammerschmidt M., Pelegri F., Kane D.A., Mullins M.C., Serbedzija G.N., van Eeden F.J.M., Granato M., Brand M., Furutani-Seiki M., Haffter P., Heisenberg C.-P., Jiang Y.-J., Kelsh R.N., Odenthal J., Warga R.M., and Nüsslein-Volhard C. 1996b. Identification of *dino* and *mercedes*, two genes regulating dorsal development in the zebrafish embryo. *Development* **123:** 95.

Huber O., Bierkamp C., and Kemler R. 1996a. Cadherins and catenins in development. *Curr. Opin. Cell Biol.* **8:** 685.

Huber O., Korn R., McLaughlin J., Ohsugi M., Herrmann B.G., and Kemler R. 1996b. Nuclear localization of β-catenin by interaction with the transcription factor LEF-1. *Mech. Dev.* **59:** 3.

Joly J.S., Joly C., Schulte-Merker S., Boulekbache H., and Condamine H. 1993. The ventral and posterior expression of the zebrafish homeobox gene *eve1* is perturbed in dorsalized and mutant embryos. *Development* **119:** 1261.

Joly J.S., Maury M., Joly C., Duprey P., Boulekbache H., and Condamine H. 1992. Expression of a zebrafish caudal homeobox gene correlates with the establishment of posterior cell lineages at gastrulation. *Differentiation* **50:** 75.

Kane D.A., Hammerschmidt M., Mullins M.C., Maischein H.-M., Brand M., van Eeden F.J.M., Furutani-Seiki M., Granato M., Haffter P., Heisenberg C.-P., Jiang Y.-J., Kelsh R.N., Odenthal J., Warga R.M., and Nüsslein-Volhard C. 1996. The zebrafish epiboly mutants. *Development* **123:** 47.

Kimmel C.B. 1989. Genetics and early development of zebrafish. *Trends Genet.* **5:** 283.

Klingensmith J. and Nusse R. 1994. Signaling by *wingless* in *Drosophila*. *Dev. Biol.* **166:** 396.

Krauss S., Korzh V., Fjose A., and Johansen T. 1992. Expression of 4 zebrafish *wnt*-related genes during embryogenesis. *Development* **116:** 249.

McMahon A.P. and Bradley A. 1990. The *Wnt-1* (*int-1*) protooncogene is required for development of a large region of the mouse brain. *Cell* **62:** 1073.

McMahon A.P. and Moon R.T. 1989. Ectopic expression of the protooncogene *int-1* in *Xenopus* embryos leads to duplication of the embryonic axis. *Cell* **58:** 1075.

Moon R.T., Campbell R.M., Christian J.L., McGrew L.L., Shih J., and Fraser S. 1993. *Xwnt-5A*: A maternal *Wnt* that affects morphogenetic movements after overexpression in embryos of *Xenopus laevis*. *Development* **119:** 97.

Mullins M.C., Hammerschmidt M., Haffter P., and Nüsslein-Volhard C. 1994. Large-scale mutagenesis in the zebrafish: In search of genes controlling development in a vertebrate. *Curr. Biol.* **4:** 189.

Mullins M.C., Hammerschmidt M., Kane D.A., Odenthal J., Brand M., Furutani-Seiki M., Granato M., Haffter P., Heisenberg C.-P., Jiang Y.-J., Kelsh R.N., van Eeden F.J.M., and Nüsslein-Volhard C. 1996. Genes establishing dorsoventral pattern formation in the zebrafish embryo: The ventral determinants. *Development* **123:** 81.

Nusse R. and Varmus H.E. 1992. *Wnt* genes. *Cell* **69:** 1073.

Parr B.A. and McMahon A.P. 1994. *Wnt* genes and vertebrate development. *Curr. Opin. Genet. Dev.* **4:** 523.

———. 1995. Dorsalizing signal *Wnt-7a* required for normal polarity of D-V and A-P axes of mouse limb. *Nature* **374:** 350.

Parr B.A., Shea M.J., Vassileva G., and McMahon A.P. 1993. Mouse *Wnt* genes exhibit discrete domains of expression in the early CNS. *Development* **119:** 247.

Piotrowski T., Schilling T.F., Brand M., Jiang Y.-J., Heisenberg C.-P., Beuchle D., Grandel H., van Eeden F.J.M., Furutani-Seiki M., Granato M., Haffter P., Hammerschmidt M., Kane D.A., Kelsh R.N., Mullins M.C., Odenthal J., Warga R.M., and Nüsslein-Volhard C. 1996. Jaw and branchial arch mutants in zebrafish II: Anterior arches and cartilage differentiation. *Development* **123:** 345.

Rijsewijk F., Schuermann M., Wagenaar E., Parren P., Weigel D., and Nusse R. 1987. The *Drosophila* homolog of the mouse mammary oncogene *int-1* is identical to the segment polarity gene *wingless*. *Cell* **50:** 649.

Schulte-Merker S., Ho R.K., Herrmann B.G., and Nüsslein-Volhard C. 1992. The protein product of the zebrafish homologue of the mouse *T*-gene is expressed in nuclei of the germ ring and the notochord of the early embryo. *Development* **116:** 1021.

Schulte-Merker S., Lee K.J.L., McMahon A.P., and Hammerschmidt M. 1997. The zebrafish organizer requires *chordino*. *Nature* **387:** 862.

Schulte-Merker S., van Eeden F.J.M., Halpern M.E., Kimmel C.B., and Nüsslein-Volhard C. 1994. *no tail* (*ntl*) is the zebrafish homologue of the mouse *T* (*Brachyury*) gene. *Development* **120:** 1009.

Sepich D.S. and Westerfield M. 1993. Molecular characterization of zebrafish acetylcholine receptor mutation, *nic1*. *Soc. Neurosci. Abstr.* **19:** 1294.

Solnica-Krezel L., Stemple D.L., Mountcastle-Shah E., Rangini Z., Neuhauss S.C.F., Malicki J., Schier A.F., Stainier D.Y.R., Zwartkruis F., Abdelilah S., and Driever W. 1996. Mutations affecting cell fates and cellular rearrangements during gastrulation in zebrafish. *Development* **123:** 67.

Stark K., Vainio S., Vassileva G., and McMahon A.P. 1994. Epithelial transformation of metanephric mesenchyme in the developing kidney regulated by *Wnt-4*. *Nature* **372:** 679.

Streisinger G., Walker C., Dower N., Knauber D., and Singer F. 1981. Production of clones of homozygous diploid zebra fish (*Brachydanio rerio*). *Nature* **291:** 293.

Takada S., Stark K.L., Shea M.J., Vassileva G., McMahon J.A., and McMahon A.P. 1994. *Wnt-3a* regulates somite and tailbud formation in the mouse embryo. *Genes Dev.* **8:** 174.

Takeda H., Matsuzaki T., Oki T., Miyagawa T., and Amanuma H. 1994. A novel POU domain gene, zebrafish *pou2*—Expression and roles of two alternatively spliced twin products in early development. *Gene Dev.* **8:** 45.

Talbot W.S., Trevarrow B., Halpern M.E., Melby A.E., Farr G., Postlethwait J.H., Jowett T., Kimmel C.B., and Kimelman D. 1995. The organizer-specific homeobox gene *floating head* is essential for notochord development in the zebrafish. *Nature* **378:** 150.

Thisse C., Thisse B., Schilling T.F., and Postlethwait J.H. 1993. Structure of the zebrafish snail 1 gene and its expression in wild-type, spadetail and no-tail mutant embryos. *Development* **119:** 1203.

Tucker A.S. and Slack J.M.W. 1995. The *Xenopus laevis* tail-forming region. *Development* **121:** 249.

Ungar A.R., Kelly G.M., and Moon R.T. 1995. *Wnt4* affects morphogenesis when misexpressed in the zebrafish embryo. *Mech. Dev.* **52:** 153.

van Eeden F.J.M., Granato M., Schach U., Brand M., Furutani-Seiki M., Haffter P., Hammerschmidt M., Heisenberg C.-P., Jiang Y.-J., Kane D.A., Kelsh R.N., Mullins M.C., Odenthal J., Warga R.M., Allende M.L., Weinberg E.S., and Nüsslein-Volhard C. 1996. Mutations affecting somite formation and patterning in the zebrafish *Danio rerio*. *Development* **123:** 153.

Weinberg E.S., Allende M.L., Kelly C.L., Abdelhamid A., Andermann P., Doerre G., Grunwald D.J., and Riggleman B. 1996. Developmental regulation of zebrafish *myoD* in wild-type, *no tail,* and *spadetail* embryos. *Development* **122:** 271.

Genes That Control Organ Form: Lessons from Bone and Branching Morphogenesis

M.A. KRASNOW
Howard Hughes Medical Institute and Department of Biochemistry, Stanford University School of Medicine, Stanford, California 94305-5307

Ever since anatomical pioneers like Leonardo da Vinci began to describe the myriad shapes and structures of animal organs (Clayton 1992)—the ramifying network of epithelial tubes that compose the lung, the intricate shapes and variety of sizes of the bones, the complex wiring pattern of the nervous system—investigators have speculated on how such forms might arise. Although a quite detailed understanding of the morphogens and mechanisms that pattern the major body axes of *Drosophila melanogaster* has emerged over the past decade (Bate and Martinez-Arias 1993), the molecular mechanisms that control the three-dimensional forms of animal organs remain poorly understood. The extreme variety of structures would suggest that many different patterning mechanisms and molecules are necessary. Indeed, the intricate genetic programs that are being elucidated for each organ appear as different as their structures suggest, and each program is highly specialized. But do any principles of organogenesis emerge from the molecular and genetic studies?

Recent results indicate that for several well-studied organs there is a single gene or a small set of genes that specifies the basic form of the organ. These genes are expressed early in development in wonderfully complex patterns that prefigure the complex form the organs will take. They are not passive markers of organ structure, but rather are key morphogenetic regulators that drive the early developmental events that shape the organs. I introduce this emerging class of genes with an example from our own work on branching morphogenesis in *Drosophila* and follow it with examples from bone morphogenesis in mouse. I then describe how the elaborate expression patterns of these genes might arise by integration of earlier, more regular patterning information. This may define a common developmental strategy for generating the basic forms of organs.

FGFs AND PATTERNING OF AIRWAY BRANCHING

The tracheal system of *Drosophila* is a complex network of epithelial tubes that ramify throughout the body and transport oxygen to the internal tissues (for review, see Manning and Krasnow 1993). It arises during embryogenesis from 20 epithelial sacs, each of which undergoes sequential branch-forming events to generate an array of several hundred branches. The pattern of the 6 primary and approximately 20 secondary branches in each metamere is complex and asymmetric, yet it is similar from segment to segment and highly stereotyped between individuals. Surprisingly, this complex pattern of branching is largely controlled by a single gene called *branchless* (Sutherland et al. 1996), one of several identified genes required to initiate branching. Absence of any of these genes prevents branching and leaves the tracheal metameres as unbranched, amorphous sacs. Two of the genes have been analyzed in detail and encode components of a fibroblast growth factor (FGF) signaling pathway. The *breathless* gene encodes an FGF receptor that turns on in the tracheal cells as they form the epithelial sacs (Glazer and Shilo 1991; Klambt et al. 1992). The *branchless* gene encodes an FGF that turns on in the neighboring tissues just before branching begins (Sutherland et al. 1996).

The *branchless* FGF is expressed in a complex and dynamic pattern in the epidermal and mesodermal tissue surrounding each developing tracheal sac, at each position where a primary branch will bud (Fig. 1) (Sutherland et al. 1996). The secreted growth factor activates the *breathless* FGF receptor on tracheal cells and guides their migration as they grow out and form primary branches. It also induces the expression of secondary branch genes in cells at the ends of growing primary branches, thus specifying the positions where the next generation of branches will bud (Fig. 2). That expression of *branchless* is determinative of the branched structure has been shown by misexpressing the gene in new positions, and by eliminating normal patches of expression, which cause corresponding changes in the branching pattern (Sutherland et al. 1996; Vincent et al. 1997). Indeed, the expression pattern of this growth factor accounts for much of the complex branched structure of this organ, although there are also other regulators that modulate and refine the branching pattern (Sutherland et al. 1996; Hacohen et al. 1998).

BMPs AND PATTERNING OF BONES

The vertebrate skeleton is composed of an incredible variety of bones, each with a distinct size and shape tailored to its position and function in the skeleton. Bones

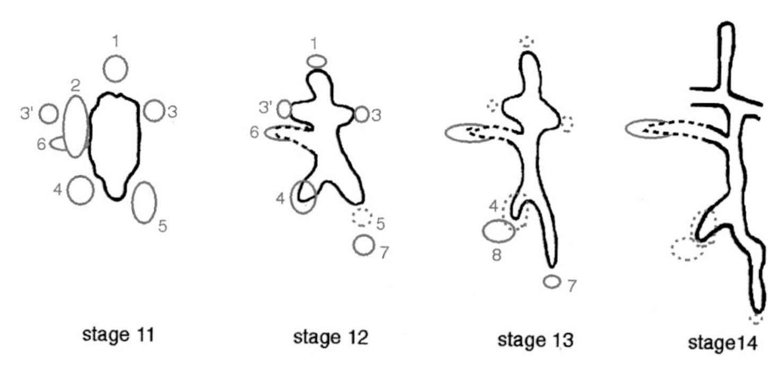

Figure 1. Expression of *branchless* FGF during embryonic tracheal development. The tracheal system arises from 20 epithelial sacs, one in each hemisegment. The formation of the six primary branches (labeled at embryonic stage 12) in a single sac (outlined in black) is shown during a 5-hr period of development. The regions of substantial *branchless* RNA expression are outlined in blue and labeled 1–8; region 3′ is from the neighboring metamere. Regions of weaker or variable expression are outlined by blue dots. (Reprinted, with permission, from Sutherland et al. 1996 [copyright Cell Press].)

develop from condensations of mesenchymal cells (for review, see Gruneberg 1963; Kingsley 1994; Erlebacher et al. 1995). These condensations take on the basic shape of the bone and then begin differentiating into cartilage and bone. Although the initial structures are subsequently modified by cell growth and death and by elaboration of extracellular matrix, the basic pattern of the bones is laid out in the mesenchymal condensations. Over the past few years, a set of TGF-β-like secreted signaling molecules called bone morphogenetic proteins (BMPs) have been shown to play a critical role in patterning of bones (for review, see Kingsley 1994; King et al. 1996). Such a role was suggested by their remarkable ability to induce ectopic bone formation at a site of implantation, an assay that allowed the identification of the founding members of the BMP family (Urist 1965; Urist and Strates 1971; Wozney et al. 1988). Their important roles in normal bone development were established more recently by genetic and molecular studies (Kingsley et al. 1992; King et al 1994; Storm et al. 1994). Loss-of-function mutations in the mouse *short ear* gene, which encodes the BMP5 protein, cause alterations in the size and shapes, or the complete absence, of many small bones and skeletal elements, including many in the external ear and specific ones in the rib cage and vertebral column. The *short ear* gene is expressed in a complex and dynamic pattern in the mesenchymal precursors of these skeletal structures, in a pattern that prefigures the mesenchymal condensations (King et al. 1994). The *short ear* mutations prevent formation or alter the structure of the mesenchymal condensations, demonstrating a requirement for BMP5 in the earliest morphological steps of bone formation. Because the striking expression pattern of BMP5 and other BMPs predicts the structure of individual bones, and because they can trigger ectopic bone formation in vivo, BMPs are proposed to be critical early determinants of bone form (Kingsley 1994).

Although the role of BMP5 in the development of specific bones has been established by these studies, its role in the formation of certain other bones is obscured by the overlapping expression patterns of other BMP family members. These apparently share many of the same functions as BMP5, because the BMP5 null mutations cause much less significant or no defects in these structures despite the fact that the gene is expressed early and in specific patterns in the precursors of these bones too. For example, several BMPs in addition to BMP5 are expressed in developing ribs and vertebrae, and much less severe defects are seen here compared to the external ear, where only BMP5 is expressed (King et al. 1996). Thus, whereas the structures of some skeletal elements rely critically on a single BMP, others may be specified by the combined activity of several BMPs expressed in overlapping patterns, a hypothesis that can be tested as mutations in other BMP genes become available.

There is evidence that one BMP family member functions in the opposite manner, as an inhibitor of bone formation that helps shape bone structure by restricting the

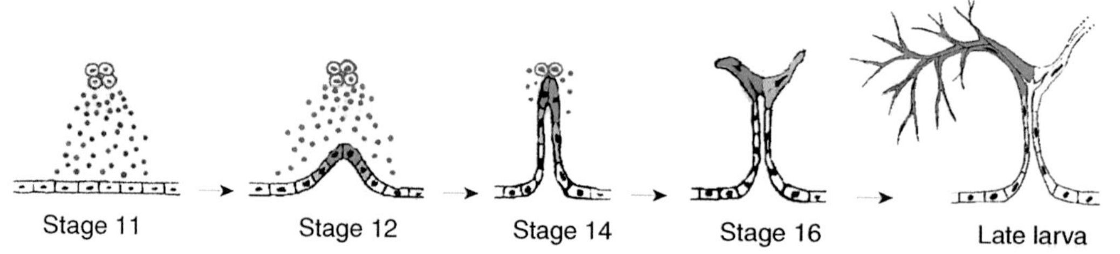

Figure 2. Model of *branchless* control of tracheal branching. A cluster of *branchless*-expressing cells (*blue circles*) is shown near the epithelium of an unbranched tracheal sac. The secreted *branchless* FGF (*blue dots*) causes budding (stage 12) and outgrowth (stage 14) of the epithelium to form a primary branch. It also induces expression of genes that promote the next generation of branching in cells at the end of the primary branch (*green fill*) which go on to form secondary (stage 16) and later terminal branches (larva).

positions where bone can arise. The mouse *brachypodism* gene, which encodes a BMP-like protein called GDF5, is expressed in the middle of many mesenchymal condensations, at sites where the condensations split in two to produce separate skeletal elements connected by synovial joints (Storm et al. 1994; Storm and Kingsley 1996). In *brachypodism* mutants, some of these joints do not form, and normally separate bones develop as a single composite structure (Gruneberg and Lee 1973; Storm and Kingsley 1996; Thomas et al. 1996).

A NEW CLASS OF ORGAN MORPHOGENESIS GENES

The examples described above illustrate how single genes expressed in complex patterns can dictate the general form of the *Drosophila* airways and a number of mouse bones. The structures of these organs are about as different as they come, and they develop in completely different ways, one by branching morphogenesis of an epithelium and the other by condensation of mesenchymal cells. In addition, the gene products that control these morphogenetic processes, although both secreted growth factors, are structurally unrelated and utilize different receptors and effector pathways. That individual genes have such profound effects on the forms of these disparate organs renders it likely that other organs are similarly shaped by a single gene or a small number of genes, and a good candidate for patterning of another organ is known (see below). I suggest the name "genomorphen" (geen′-o-mor′-fen) for this class of genes, to emphasize their key roles in generating organ form.

There are two defining characteristics of a genomorphen. First, the gene must be expressed in a spatial pattern that prefigures part or all of the form that the organ will take. It is the first manifestation, molecular or morphological, of this shape. Second, it must be a controller and not simply a marker of organ form. Eliminating the gene or expressing it in a new pattern must alter the structure in the predicted manner.

The nature of the gene product is left unspecified by this definition and can include transcription factors, RNA-binding proteins, or other types of developmental regulators, in addition to secreted growth factors as in the examples provided. The genes can also shape the organs in different ways. The simplest way is exemplified by the *short ear* BMP5 gene, which specifies the full shape the bone will initially take (Fig. 3a). Another way is to select only critical points that define the shape (Fig. 3b). *branchless* functions in this way, marking only the points toward which tracheal branches extend, like the stakes and guy wires that support a tent. A third way is for the gene to be expressed in a negative image of the ultimate structure, circumscribing the area in which the organ will form (Fig. 3b). The *brachypodism* gene displays aspects of such a negatively acting genomorphen in defining the ends of individual bones and the positions of the joints.

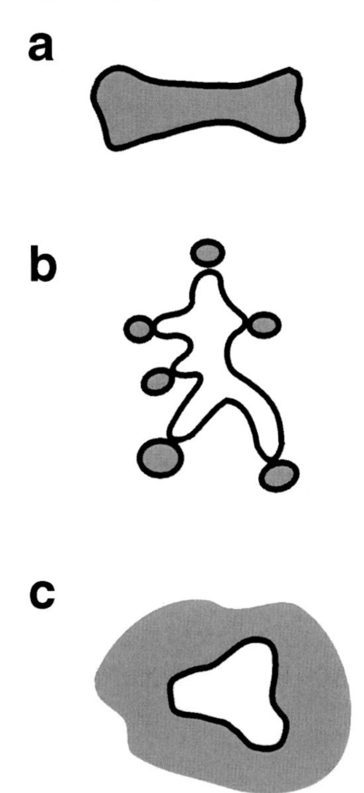

Figure 3. Models of how genomorphens can control organ form. Grey area shows the genomorphen expression domain. (*a*) The genomorphen is expressed in the full shape of the organ. (*b*) The genomorphen is expressed in specific positions that define the organ's shape. (*c*) The genomorphen is expressed in the region surrounding where the organ will form, generating an inverse image of the organ.

ORGAN MORPHOGENESIS GENES AS COMPLEX INTEGRATORS OF EARLIER SPATIAL INFORMATION

The most surprising aspect of the studies of the genomorphens *branchless* and *short ear* is how much spatial patterning information for the organs resides in the expression of individual genes. This raises the question of how their complex expression patterns are controlled. The characteristic segmental positions of *branchless* transcripts suggest that its expression is most likely controlled by the anteroposterior and dorsoventral gene regulatory hierarchies acting on the *cis*-regulatory regions of *branchless* (Sutherland et al. 1996). We have speculated that there might be different enhancer elements for each patch of *branchless* expression, as there are different enhancers for the different stripes of expression of the segmentation genes like *even-skipped* and *hairy* (Pankratz and Jäckle 1990). Each of the enhancers at the *branchless* gene would sense a distinct set of regulators differentially distributed along the anteroposterior and dorsoventral axes of the segment (Fig. 4). For example, expression patches 2 and 4 turn on just after and precisely in register with the stripe of expression of the *engrailed* segmentation gene (stripe I in Fig. 4a), so the *engrailed* gene product may be one of the transcription factors that binds and activates the enhancers that control these two patches (D. Sutherland and M.A. Krasnow, unpubl.). Each enhancer would also be responsive to factors differentially distributed along the dorsoventral axis. Consistent with this,

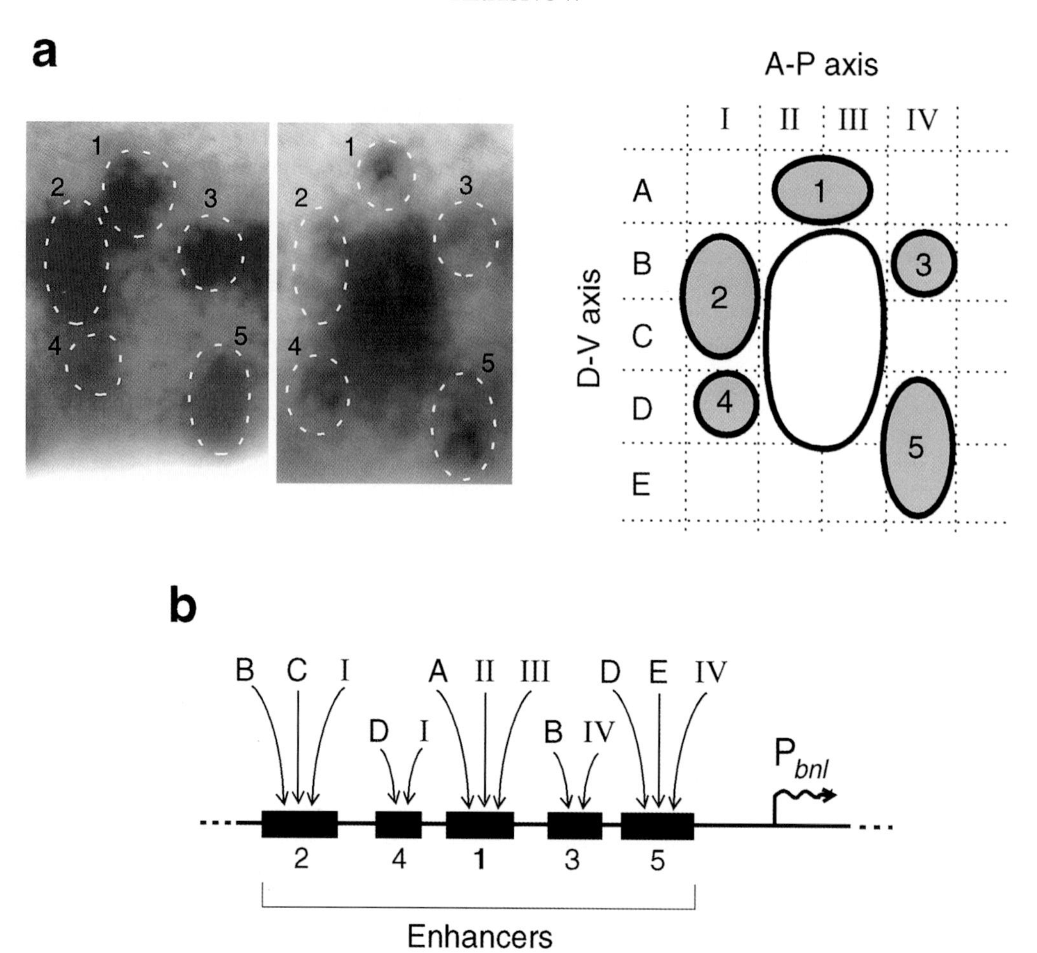

Figure 4. Hypothetical model of how the complex pattern of *branchless* expression might be controlled by different combinations of anteroposterior and dorsoventral patterning genes. (*a*) The 5 epidermal expression domains of *branchless* mRNA in a typical metamere at stage 11 are shown by the blue staining in the *left* and *center* panels. They flank the developing tracheal sac (brown staining, *center* panel). Superimposed on this in the diagram (*right* panel) are the anteroposterior stripes of expression of the segmentation gene *engrailed* (stripe I), wingless (stripe IV), and several other known or hypothetical anteroposterior and dorsoventral patterning genes whose approximate register with the *branchless* expression domains is shown. Expression of the anteroposterior and dorsoventral patterning genes intersect to form a grid-like pattern of positional values in each segment. These genes turn on before *branchless* expression begins, and the gene products are proposed to function directly or indirectly through the *cis*-regulatory elements of the *branchless* gene to activate its expression. (*b*) It is speculated that different regulatory regions of *branchless* contain different combinations of sites for factors differentially localized along the anteroposterior and dorsoventral axes, and thus control expression of *branchless* in the different expression domains. (Color panels reprinted with permission from Sutherland et al. 1996 [copyright *Cell* Press].)

the dorsal-most patch of *branchless* expression requires the DPP dorsal patterning pathway, but the more ventrally located patches do not (Vincent et al. 1997).

If this idea for the regulation of *branchless* expression proves true, it would show how the continuum of positional information provided by anteroposterior and dorsoventral morphogens like Bicoid protein and DPP flows first to downstream genes like *engrailed* that partition the animal into regular stripes representing different positional values along the major body axes, and then on to organ morphogenesis genes like *branchless*. In this scenario, the function of genomorphens is to integrate the regular, grid-like patterning information and transform it into complex three-dimensional patterns that represent the general forms of the organs. The proposed regulation of organ morphogenesis genes by axis patterning genes also serves another important function: It dictates the overall position in the grid that each organ will occupy and thus arranges the organs with respect to one another.

Although the upstream regulators that control expression of BMP genes during bone morphogenesis are unknown, the complex expression patterns of BMP genes imply that considerable patterning information is in place at the time they begin to be expressed. One can envision a scenario similar to the one postulated above for control of *branchless* expression, in which overlapping gradients of morphogens specify a grid-like pattern of spatial coordinates that is integrated in complex ways at the regulatory elements of BMP genes. Gradients of morphogens are present in developing structures like the limb bud before BMP expression and bone formation begin, and these could contribute to the regulation of BMP expression (for review, see Tickle and Eichele 1994; Kawakami et al. 1996). A major challenge for the bone morphogenesis field is to define the upstream regulatory pathways and the *cis*-regulatory elements that control BMP expression. Because individual BMP genes are expressed in unrelated patterns throughout the body, it is likely that they are controlled by a variety of pathways integrated through correspondingly complex control regions.

The modular nature of transcriptional control elements (Dynan 1989) is ideally suited to the integration function postulated for the organ morphogenesis genes. Genes can have many independent enhancers, each responding to a different set of upstream regulators and driving a different portion of the expression pattern. Expression domains controlled by different enhancers can lie next to one another or overlap to give rise to composite shapes of gene expression, as in the *fork head* gene that contributes to specification of the salivary gland primordium in *Drosophila* (Weigel et al. 1990; S. Beckendorf, pers. comm.). Expression domains can also be located far away from each other and even in different tissues, as are the different patches of expression of the *branchless* gene. If all the necessary regulatory elements cannot fit around a single gene, gene duplication events can provide extra copies of the gene around which additional regulatory elements can be arrayed. This may contribute to the large number of BMP genes (King et al. 1996).

IMPLICATIONS FOR EVOLUTION OF ORGAN STRUCTURE

The modular nature of transcriptional control regions makes it easy to alter the various domains of expression of a gene by adding or altering individual modules so that it responds differently to upstream regulators. Organ structures can thus readily evolve by mutation or recombination of regulatory elements of organ morphogenesis genes (Kingsley 1994), much like the pattern of tracheal branching was altered in transgenic embryos by altering the pattern of *branchless* expression (Sutherland et al. 1996). Organ structures can also evolve by altering the grid-like pattern of the upstream regulators, although this strategy would have more profound consequences because it would impact all structures dependent on the grid.

Related species often have different structures of their organs. This is true for bones in vertebrates (Romer 1967) as well as tracheal systems in insects (Keilin 1944). A stringent test of whether evolution has used the strategy of altering transcriptional control regions to specify these differences would be to experimentally transform the structure of an organ in one species into that of another species by replacing the genomorphen of the former with its cognate from the latter. If only the regulatory elements of the genomorphen have changed significantly during evolution and the grid has been maintained, this should transform the organ structure of the recipient into a structure resembling that of the gene donor.

ORGAN FORM IS ELABORATED BY SECONDARY MORPHOGENETIC PROGRAMS

Once the general form of an organ is specified by genomorphens in the manner outlined above, it can be modified in a variety of ways by activation of secondary morphogenetic programs within specific subdomains of the structure or even within individual cells (Follette and O'Farrell 1997). For example, localized expression of cell cycle regulators can cause growth of a particular part of an organ, or localized expression of a signaling molecule like *folded gastrulation* can lead to apical constriction of a group of cells to generate an invagination or crease at a particular position. Expression of subprogram regulators can rely on the same patterning cues postulated to control expression of the more global regulators of organ form, or they can utilize the more refined spatial information that arises after further integration of the grid-like patterning information has taken place.

ARE THERE GENOMORPHENS FOR OTHER ORGANS?

There is accumulating evidence that molecules and mechanisms formally similar to the ones that pattern tracheal branching in *Drosophila* may pattern early branching events in murine lung development. Branching of the bronchial tree appears to require FGF signaling, as branching is completely blocked by the expression of a dominant negative FGF receptor in the branching epithelium (Peters et al. 1994). Brigid Hogan (this volume) described the expression pattern of the *FGF10* gene in the mesenchyme that surrounds the developing bronchial tree. The transcripts were localized in discrete clusters of cells, near each position where a new branch was budding in the epithelium. This suggests that *FGF10* could be an important regulator of the structure of the developing lung, directing branch outgrowth much like patches of expression of the *branchless* FGF direct branching of the *Drosophila* airways. Ubiquitous expression of an FGF in the developing mouse lung severely disrupts the branching pattern, consistent with a requirement for localized FGF signaling in the process (Simonet et al. 1995). Proof of a controlling role for *FGF10* in branch patterning, however, will require the demonstration that elimination of *FGF10* expression or localized misexpression of the gene leads to the expected changes in branching.

Although single genomorphens may be found for other organs, it seems likely that the general form of many organs will be found to be controlled by several genomorphens, each contributing in important ways to the organ's shape. Already it appears that control of the shape of certain bones is distributed among several BMP genes. For some organs, the initial shape may be so extensively modified during development by morphogenetic subprograms that it may not be useful or possible to classify an earlier acting gene as a genomorphen. However, the flow of spatial information from regular patterns of early acting morphogens to complex patterns of expression of the downstream genes controlling organ morphogenesis is likely to be quite general.

ACKNOWLEDGMENTS

I thank the members of my laboratory and David Kingsley for stimulating discussions, and the members of my laboratory for their valuable comments on the manuscript. This work was supported in part by grants from the National Institutes of Health. M.A.K. is an associate investigator of the Howard Hughes Medical Institute.

REFERENCES

Bate M. and Martinez Arias A., Eds. 1993. *The Development of* Drosophila melanogaster, vol. 1. Cold Spring Harbor Laboratory Press, Cold Spring Harbor, New York.

Clayton M. 1992. *Leonardo da Vinci: The anatomy of man.* Little, Brown, and Company, Boston, Massachusetts.

Dynan W.S. 1989. Modularity in promoters and enhancers. *Cell* **58:** 1.

Erlebacher A., Filvaroff E.H., Gitelman S.E., and Derynck R. 1995. Toward a molecular understanding of skeletal development. *Cell* **80:** 371.

Follette P.J. and O'Farrell P.H. 1997. Connecting cell behavior to patterning: Lessons from the cell cycle. *Cell* **88:** 309.

Glazer L. and Shilo B.Z. 1991. The *Drosophila* FGF-R homolog is expressed in the embryonic tracheal system and appears to be required for directed tracheal cell extension. *Genes Dev.* **5:** 697.

Gruneberg H. 1963. *The pathology of development.* Blackwell Scientific, Oxford, United Kingdom.

Gruneberg H. and Lee A.J. 1973. The anatomy and develoment of brachypodism in the mouse. *J. Embryol. Exp. Morphol.* **30:** 119.

Hacohen N., Kramer S., Sutherland D., Hiromi Y., and Krasnow M.A. 1998. *sprouty* encodes a novel antagonist of FGF signaling that patterns apical branching of the *Drosophila* airways. *Cell* **92:** (in press).

Kawakami Y., Ishikawa T., Shimabara M., Tanda N., Enomoto-Iwamoto M., Iwamoto M., Kuwana T., Ueki A., Noji S., and Nohno T. 1996. BMP signaling during bone pattern determination in the developing limb. *Development* **122:** 3557.

Keilin D. 1944. Respiratory systems and respiratory adaptations in larvae and pupae of *Diptera. Parasitology* **36:** 1.

King J.A., Marker P.C., Seung K.J., and Kingsley D.M. 1994. BMP5 and the molecular, skeletal, and soft-tissue alterations in short ear mice. *Dev. Biol.* **166:** 112.

King J.A., Storm E.E., Marker P.C., Dileone R.J., and Kingsley D.M. 1996. The role of BMPs and GDFs in development of region-specific skeletal structures. *Ann. N. Y. Acad. Sci.* **785:** 70.

Kingsley D.M. 1994. What do BMPs do in mammals? Clues from the mouse short-ear mutation. *Trends Genet.* **10:** 16.

Kingsley D.M., Bland A.E., Grubber J.M., Marker P.C., Russell L.B., Copeland N.G., and Jenkins N.A. 1992. The mouse short ear skeletal morphogenesis locus is associated with defects in a bone morphogenetic member of the TGF-β superfamily. *Cell* **71:** 399.

Klambt C., Glazer L., and Shilo B.Z. 1992. *breathless*, a *Drosophila* FGF receptor homolog, is essential for migration of tracheal and specific midline glial cells. *Genes Dev.* **6:** 1668.

Manning G. and Krasnow M.A. 1993. Development of the *Drosophila* tracheal system. In *The Development of* Drosophila melanogaster (ed. M. Bate and A. Martinez Arias), p. 609. Cold Spring Harbor Laboratory Press, Cold Spring Harbor, New York.

Pankratz M.J. and Jäckle H. 1990. Making stripes in the *Drosophila* embryo. *Trends Genet.* **6:** 287.

Peters K., Werner S., Liao X., Wert S., Whitsett J., and Williams L. 1994. Targeted expression of a dominant negative FGF receptor blocks branching morphogenesis and epithelial differentiation of the mouse lung. *EMBO J.* **13:** 3296.

Romer A.S. 1967. Major steps in vertebrate evolution. *Science* **158:** 1629.

Simonet W.S., DeRose M.L., Bucay N., Nguyen H.Q., Wert S.E., Zhou L., Ulich T.R., Thomason A., Danilenko D.M., and Whitsett J.A. 1995. Pulmonary malformation in transgenic mice expressing human keratinocyte growth factor in the lung. *Proc. Natl. Acad. Sci.* **92:** 12461.

Storm E.E. and Kingsley D.M. 1996. Joint patterning defects caused by single and double mutations in members of the bone morphogenetic protein (BMP) family. *Development* **122:** 3969.

Storm E.E., Huynh T.V., Copeland N.G., Jenkins N.A., Kingsley D.M., and Lee S.J. 1994. Limb alterations in *brachypodism* mice due to mutations in a new member of the TGFβ-superfamily. *Nature* **368:** 639.

Sutherland D., Samakovlis C., and Krasnow M.A. 1996. *branchless* encodes a *Drosophila* FGF homolog that controls tracheal cell migration and the pattern of branching. *Cell* **87:** 1091.

Tickle C. and Eichele G. 1994. Vertebrate limb development. *Annu. Rev. Cell. Biol.* **10:** 121.

Thomas J.T., Lin K., Nandedkar M., Camargo M., Cervenka J., and Luyten F.P. 1996. A human chondrodysplasia due to a mutation in a TGF-β superfamily member. *Nat. Genet.* **12:** 315.

Urist M.R. 1965. Bone: Formation by autoinduction. *Science* **150:** 893.

Urist M.R. and Strates B.S. 1971. Bone morphogenetic protein. *J. Dent. Res.* **50:** 1392.

Vincent S., Ruberte E., Grieder N.C., Chen C.K., Haerry T., Schuh R., and Affolter M. 1997. DPP controls tracheal cell migration along the dorsoventral body axis of the *Drosophila* embryo. *Development* **124:** 2741.

Weigel D., Seifert E., Reuter D., and Jäckle H. 1990. Regulatory elements controlling expression of the *Drosophila* gene *fork head. EMBO J.* **9:** 1199.

Wozney J.M., Rosen V., Celeste A., Mitsock L.M., Whitters M.J., Kriz R.W., Hewick R.M., and Wang E.A. 1988. Novel regulators of bone formation: Molecular clones and activities. *Science* **242:** 1528.

Branching Morphogenesis in the *Drosophila* Tracheal System

B.-Z. SHILO, L. GABAY, L. GLAZER, M. REICHMAN-FRIED,[1] P. WAPPNER, R. WILK,[2] AND E. ZELZER

Department of Molecular Genetics, Weizmann Institute of Science, Rehovot 76100, Israel

Morphogenesis of branched tubular organs is an essential process in the development of all multicellular organisms. One aspect that is common to the formation of many tubular structures is the need to follow a highly stereotyped and elaborate process of migration, to generate the final architecture of the organ. Another feature shared by those tubular structures that arise simultaneously at different parts of the organism is the capacity to form a continuous network joining the different clusters.

Development of the vertebrate vascular system has been characterized in detail, and the strategy for forming these structures appears to be common to other tubular systems (for review, see Folkman and D'Amore 1996; Risau 1997). Initially, separate clusters of progenitor cells are induced to form angioblasts. This process requires activation of the fibroblast growth factor (FGF) receptors, and the vascular endothelial growth factor (VEGF) receptor Flk-1, both members of the receptor tyrosine kinase family. Migration of the angioblasts to form a continuous tubular network is also triggered by VEGF, mostly through another receptor for the same ligand, termed Flt-1. Concomitant with the guidance of stereotyped migration, VEGF also induces proliferation of the endothelial cells, thus generating the proper number of cells required to form the branched tubular network.

Cross-talk between the smooth muscle cells wrapping the vasculature and the endothelial cells is responsible for terminating endothelial cell divisions. The initial attraction of mesenchyme cells by the growing vascular system is also mediated by a receptor tyrosine kinase cascade. The Angiopoietin-1 ligand expressed by the mesenchyme triggers the Tie-2 receptor expressed on endothelial cells, thus activating, through an unknown mechanism, the production of attractive cues for the migration of mesenchyme cells (for review, see Folkman and D'Amore 1996).

Although development of the vasculature is highly stereotyped, the system retains its capacity to generate new vessels and collaterals in wound healing, or following pathological situations of ischemia or tumor growth. In these cases, the inductive cues for the system are not developmental. Rather, low oxygen levels at the affected tissue induce the same signals that were used during development. In hypoxic tissues, a transcription factor of the basic helix-loop-helix (bHLH)/PAS family termed HIF1α is stabilized. This protein functions as a heterodimer with another bHLH/PAS protein termed ARNT and induces cellular responses to hypoxia, including the transcriptional induction of key enzymes in glycolysis (Maxwell et al. 1993; Wang and Semenza 1993b; Wang et al. 1995). In parallel, the same factor is responsible for systemic responses such as induction of erythropoietin expression and the induction of VEGF transcription. In addition, another bHLH/PAS protein termed EPAS-1 is induced by hypoxia in endothelial cells, where it triggers the expression of glycolytic enzymes, as well as the Tie-2 receptor. EPAS-1 thus appears to initiate cross-talk between endothelial and mesenchyme cells (Tian et al. 1997).

The *Drosophila* tracheal system shares many common features with the vertebrate vascular system and is thus an attractive model organ to study the different phases of branching morphogenesis (for review, see Manning and Krasnow 1993). It is formed as ten tracheal placodes on each side of the embryo. Initially, each placode is composed of approximately 20 cells. Following two rounds of division, the final number of cells in the placode is approximately 80. No additional events of cell division take place, and the formation of the extended tubular network takes place from this finite cell number. Initially, all tracheal cells are identical to each other (Samakovlis et al. 1996a). The first specifications of tracheal subfates take place prior to cell migration. As migration proceeds, different cell types are generated. Most notably, the terminal tracheal cells are responsible for sending long extensions that will reach the cells in target tissues (Guillemin et al. 1996). The homotip cells, on the other hand, are necessary for forming the contacts between tubes in adjacent segments, to generate the continuous, interconnected tubular network (Samakovlis et al. 1996b; Tanaka-Matakatsu et al. 1996).

Formation of the tracheal system raises a host of questions. For example, how are tracheal cell fates induced?

Present addresses: [1]Department of CBRC, MGH-Harvard Medical School, Charlestown, Massachusetts 02129; [2]Department of Molecular and Medical Genetics, University of Toronto, Toronto, ON M56 1X8.

What guides migration of the tracheal cells? How is the correct number of cells in each tracheal branch determined? This paper presents data and models pertaining to these issues.

RESULTS AND DISCUSSION

Trachealess Induces Tracheal Cell Fates

Organogenesis in *Drosophila* and in other organisms does not take place at the onset of embryonic development. Rather, after the initial anteroposterior (A/P) and dorsoventral (D/V) coordinates have been specified on the ectoderm, it is possible to utilize them in order to define clusters of cells that will form the organ of choice. Furthermore, in cases where the precursors of a given organ undergo migration, preexisting coordinates on the ectoderm can be utilized as signposts for migration.

The concept of "master regulators" has been described for different organs, where initially, expression of a cardinal transcription factor is triggered in restricted sets of cells by the existing A/P and D/V grid. Expression of the factor can be maintained during organogenesis by an autoregulatory loop. A master regulator is both necessary and sufficient for development of a given organ, i.e., no development is observed in its absence, and ectopic organs can be induced simply by ectopic expression of the factor. Such observations define the position of the factor at the top of the regulatory hierarchy.

Formation of the tracheal placodes is induced in this manner by the Trachealess (Trh) protein, a bHLH/PAS transcription factor. In the absence of Trh, no placodes will develop. Conversely, ubiquitous expression of Trh induces the formation of extra tracheal placodes in segments that normally do not form placodes. Finally, Trh autoregulates its own transcription and continues to be expressed in the tracheal system throughout embryonic and larval development (Wilk et al. 1996).

Trh is likely to have a heterodimeric bHLH/PAS partner and possibly also associates with additional proteins that will contribute to the specificity of target-gene induction. Although not all components of the complex are known, it appears that some of them are also restricted to the placodes. This is corroborated by the observation that following ubiquitous expression of Trh in a *trachealess* mutant embryo, formation of tracheal placodes in the correct positions can be observed (Zelzer et al. 1997).

Trh is also expressed in the salivary ducts and the posterior spiracles. In *trachealess* mutant embryos, the salivary duct and posterior spiracle precursors remain on the ectoderm and fail to invaginate (Isaac and Andrew 1996). Thus, it appears to be required for the formation of additional tubular structures in the embryo. The implication is that there may be common target genes for Trh in all three tissues, which may be involved in processes of invagination and tube formation.

Target Specificity of bHLH/PAS Proteins

In addition to Trh, other bHLH/PAS proteins have been identified in *Drosophila*. Most notable is Single minded (Sim), which is both necessary and sufficient for induction of midline cell fates (Nambu et al. 1991). These cells comprise the central, ventralmost row of cells in the embryo that will give rise to glial and several neuronal lineages. *single minded* mutant embryos fail to develop a midline, and ectopic expression of Sim induces midline fates in the entire ventral ectoderm. The different and nonoverlapping target genes of Trh and Sim provided an opportunity to dissect the structural basis for target specificity of bHLH/PAS proteins.

In vertebrates, the bHLH/PAS proteins including HIF1α, EPAS1, and AhR appear to associate with another member of the bHLH/PAS family termed ARNT, to form a functional heterodimer (Hoffman et al. 1991). A similar DNA-binding site was identified for HIF1α/ARNT and for the *Drosophila* Sim complex (Wang and Semenza 1993a; Wharton et al. 1994).

The basic domains of Trh and Sim recognize the same half-site on the DNA. Replacement of the basic domain of Trh with that of Sim did not alter its capacity to induce ectopic tracheal structures (Zelzer et al. 1997). Identification of only a single *Drosophila* ARNT homolog suggests that it serves as a universal partner for Trh, Sim, and possibly other bHLH/PAS proteins (Zelzer et al. 1997). To determine if the binding sites of the Trh and Sim complexes are identical, the expression of a reporter linked to a multimer of the Sim or HIF1α-binding sites was monitored in embryos. The reporter is expressed only in the tissues in which Trh and Sim are normally expressed and functional, indicating that both heterodimers recognize the same DNA-binding site (Zelzer et al. 1997).

To identify the domain(s) responsible for target specificity, the PAS domain of Trh was exchanged with that of Sim. This chimeric protein lost the capacity to induce tracheal-target genes. Instead, it acquired the potential to induce midline genes, including *sim* itself, as well as a battery of Sim target genes (Zelzer et al. 1997). The PAS domain is thus responsible for conferring target specificity. How is this achieved?

We postulate that an additional *cis*-binding protein associates with the promoters of the tissue-specific target genes as well as with the heterodimeric complex. Together, these proteins form a functional complex. Thus, divergent binding sites for different *cis* elements combined with specific protein-protein interactions of these elements with the heterodimeric bHLH/PAS complex determine the repertoire of target genes for each complex (Fig. 1). By exchanging the PAS domain, the interactions with the *cis*-binding protein may have been altered.

A Common Evolutionary Origin for bHLH/PAS Proteins Inducing Hypoxic Responses?

The multiprotein complex of bHLH/PAS proteins requires the correct protein-DNA as well as protein-protein interactions for proper induction of target genes. Although the heterodimeric binding site appears to be universal for Trh, Sim, and HIF1α complexes, it was not possible to predict from the sequence of HIF1α whether it would be able to form protein-protein interactions that

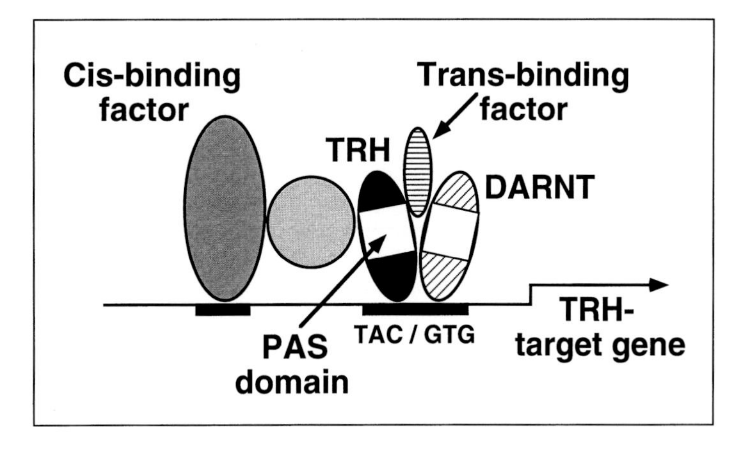

Figure 1. A model for the capacity of Trh to induce distinct target genes. In the tracheal cells, Trh and its partner, DARNT, are expressed and bind to the canonical binding site on the regulatory regions of tracheal as well as midline target genes. The binding is not sufficient, however, to induce expression of these genes. Transcription will be induced only for genes that also have a binding site for an additional, tracheal-specific *cis*-binding factor that is capable of forming (directly or indirectly) protein-protein interactions with the Trh complex. Activity of the complex also requires the association with a *trans*-acting factor. The PAS domain of Trh or Sim has a pivotal role in the complex, as it appears to mediate the interaction with the distinct *cis*- and *trans*-acting factors, as well as with DARNT.

would mimic those of Trh or Sim. Specifically, the PAS domain, which provides the nucleation site for these interactions, shows an equal degree of similarity to Trh and Sim.

Ectopic expression of human HIF1α in *Drosophila* embryos demonstrates that it is highly potent and capable of inducing Trh target genes and ectopic tracheal pits. In contrast, it suppresses the activity of endogenous Sim in the midline (Zelzer et al. 1997). Thus, HIF1α retained the capacity to interact with the *Drosophila* ARNT homolog, as well as with additional proteins in the transcription complex.

This observation may suggest the existence of a primordial bHLH/PAS complex that was dedicated to the response of the cells and the organism to hypoxic stress. The genes encoding this complex have subsequently duplicated in the evolving species and are capable of responding to hypoxic stress in different tissues. On the one hand, genes such as HIF1α function in all tissues of the organism. A putative bHLH/PAS protein in *Drosophila* that was found to be elevated in cells following hypoxia may also serve a similar function (Nagao et al. 1996). On the other hand, bHLH/PAS proteins such as Trh and EPAS-1 are required for the development of the tubular network that supplies oxygen to the target tissues.

Guided Migration of Tracheal Cells

Simultaneous migration of the tracheal cells to form the different tracheal branches in each segment is observed at stage 12. This migration is highly stereotyped and gives rise to the reproducible structure of the tracheal tree. All structures are formed only by cell migration and extension, without any further cell division. It was logical to assume that such a precise pattern cannot be determined by endogenous cues and must rely on signals emanating from the ectoderm, which is already patterned when migration ensues.

The first indication for exogenous cues that are required for migration was obtained when a *Drosophila* FGF receptor homolog was isolated and shown to be expressed in the tracheal tissue from the placode stage through all embryonic and larval phases (Glazer and Shilo 1991). Mutations in this gene do not disrupt the initial specification of the tracheal cells or the formation of the tracheal pit. However, no tracheal cell migration takes place in the absence of a functional receptor; hence, the gene was termed *breathless* (Klämbt et al. 1992). It is possible to inactivate the Breathless pathway after the initial tubular network has formed, by induction of a dominant-negative construct (Reichman-Fried and Shilo 1995). This protocol demonstrated that the receptor is essential not only for the initial migration of the tracheal cells, but also for the capacity of the terminal cells to extend long and branched processes during the larval phases.

Although the requirement for the Breathless protein indicated the involvement of a signaling system in cell migration, it did not necessarily prove that this receptor is responsible for receiving the cues that guide directionality. It was formally possible that the FGF receptor pathway has a more permissive role, and another guiding system (or systems which may be different for each of the branches) determines directionality. Isolation of the ligand for Breathless demonstrated conclusively that the receptor is providing the directional cues for migration (Sutherland et al. 1996). The ligand (termed Branchless) is an FGF-like molecule, and mutations in the gene give rise to a phenotype identical to that of *breathless*. Branchless is expressed in a dynamic pattern on the ectoderm and mesoderm, which prefigures the direction in which the branches will migrate. Furthermore, ectopic expression of Branchless was shown to target tracheal migration in ectopic directions.

Monitoring Activation of Breathless In Situ

Receptor tyrosine kinases activate a conserved cytoplasmic kinase signaling cascade. One of the last enzymes in this pathway is MAP kinase (ERK) that is activated by dual phosphorylation of threonine and tyrosine residues by MEK (Seger and Krebs 1995). The capacity to follow specifically the distribution of activated (diphospho) ERK should be extremely informative, as ERK provides a nodal point for different RTKs. In addition, since it is located at the bottom of the cascade, its activation should represent an amplification of the original signal elicited by the respective RTKs.

By utilizing a monoclonal antibody raised against diphospho-ERK (dp-ERK), it is indeed possible to follow the activation profiles of all RTKs in situ during *Drosophila* development (Gabay et al. 1997a). With respect to the activation profile of Breathless, a pattern consistent with the dynamic expression of Branchless is observed. Initially, at the pit stage, large patches of Branchless-expressing cells surround the trachea. Accordingly, large patches of tracheal cells display dp-ERK. This pattern is eliminated in *btl* mutant embryos. As the tracheal branches begin to migrate, the pattern of Branchless expression and accordingly the pattern of dp-ERK in the tracheal branches refine. Only a very small group of cells closest to the tip (and to the source of Branchless) display dp-ERK (Gabay et al. 1997b). Thus, the diffusion of Branchless is extremely limited. This restricted activation may indeed account for the capacity of different levels of Branchless activation to induce distinct cell fates (terminal vs. homotip cells) in the tracheal tips (Fig. 2).

Is Branchless the Only Guiding Cue?

The absence of any tracheal branch migration in *btl* or *bnl* mutant embryos demonstrates that the Branchless/Breathless system is essential for all aspects of migration. On the other hand, several results indicate that for some branches, additional redundant guiding cues may be present. This was first realized when a constitutively activated construct of Breathless was expressed in *btl* null mutant embryos (Reichman-Fried et al. 1994). In these embryos, an equal level of activated Breathless is present in all tracheal cells. If localized Breathless activation is the sole guiding cue, no migration should be observed. However, formation of a normal dorsal trunk and lateral branches was observed. Thus, for these branches, other guiding cues may provide directionality, in the presence of uniform Breathless activation. A complementary result was obtained when Branchless was expressed uniformly in all tracheal cells by the *btl-Gal4* driver, thus providing uniform activation of Breathless. Again, migration of the dorsal trunk and lateral branch cells was normal, whereas the other branches failed to form (Fig. 3).

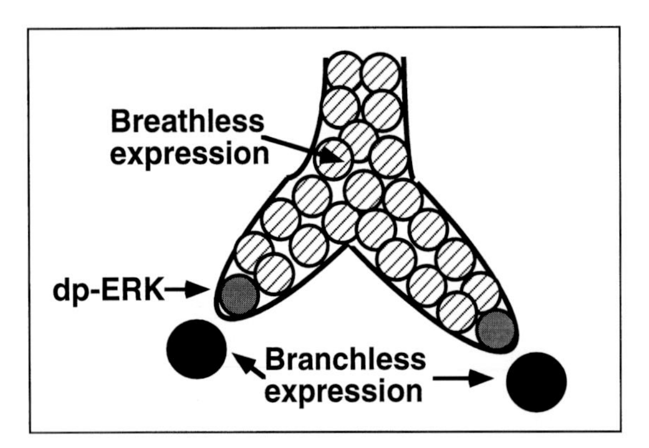

Figure 2. Localized activation of Breathless by Branchless in the tracheal branch tips. Breathless, an FGF receptor homolog, is expressed in all tracheal cells (*hatched*). Branchless, an FGF homolog that is a ligand for Breathless, is expressed in restricted clusters of ectodermal cells (*black*), predicting the direction of tracheal migration. By following the distribution of dp-ERK (*gray*), it is possible to identify the cells in which Breathless is activated. Only the leading tip cells closest to the ligand source show prominent dp-ERK.

The EGF Receptor and Dpp Signaling Pathways Determine Tracheal Branch Fates

In contrast to the development of the vascular system, where the generation of new vessels is accompanied by endothelial cell division, no cell divisions take place after formation of the tracheal pit. Thus, the final number of cells comprising each branch must be defined prior to migration. At the onset of tracheal migration, all branches emanating from the same pit initiate their migration simultaneously. The cell number allocated to each branch can be identified already from these early stages. Two signaling pathways are responsible for determination of tracheal branch fates.

An indication for activation of the EGF receptor (DER) pathway during tracheal development was presented by the prominent presence of dp-ERK in the tracheal placodes at stage 10, which is eliminated in mutants for DER or other elements in its signaling pathway (Wappner et al. 1997). Another indication was provided by the expression of Rhomboid in the tracheal placodes at this phase (Bier et al. 1990). Rhomboid is a novel multitransmembrane domain protein that appears to be essential for processing of the DER ligand Spitz. The expression of Rhomboid, which is very dynamic, marks the cells and tissue in which the DER signaling pathway will be induced, as can be ascertained from the corresponding pattern of dp-ERK (Gabay et al. 1997a).

The phenotypes of mutations in *DER* or in other components of the pathway, such as the ligand Spitz, or Rhomboid are similar. While some branches migrate normally (the dorsal branch and the lateral trunks), the dorsal trunk fails to migrate and the visceral branch is only rudimentary (Wappner et al. 1997).

A complementary phenotype was observed in mutations for the Dpp signaling pathway. Only mutations in some components of the pathway give an accurate reflection of the tracheal phenotype. Maternal components mask the phenotype of some genes, on one hand, whereas the early requirement for Dpp which precedes tracheal development may give rise to early general defects in ectodermal patterning, on the other. Most notably, Thick veins (Tkv), a type I TGF-β receptor is provided maternally during the early stages of embryogenesis. Prior to the tracheal placode stage, maternal *tkv* transcripts are eliminated. Prominent zygotic transcription of *tkv* is then detected in the tracheal placodes (Affolter et al. 1994). It is thus likely that a *tkv* mutant embryo would accurately reflect the loss of this signaling pathway in the trachea. In these mutants, the dorsal trunk and visceral branch form normally, whereas the dorsal and lateral branches are absent (Affolter et al. 1994; Vincent et al. 1997).

Since the DER and *tkv* mutant phenotypes are complementary, the prediction was that in a mutant embryo missing both signaling pathways, no tracheal migration

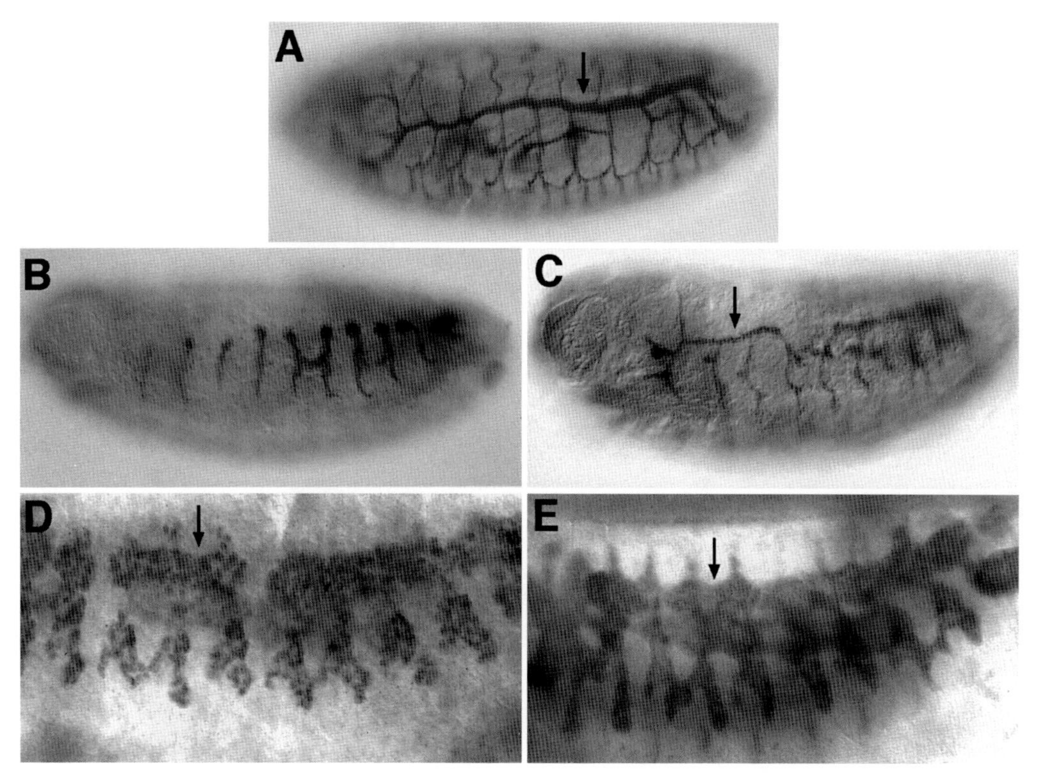

Figure 3. Directional migration of dorsal trunk cells in the absence of localized Breathless activation. (*A*) Wild-type embryonic tracheal tree after the completion of cell migration, visualized by tracheal-lumen antibody. Dorsal trunk is shown by arrow. (*B*) In *breathless* null mutant embryos, tracheal cell fates are specified, but the cells fail to migrate, remain in the tracheal pits, and secrete lumen components. (*C*) In *breathless* null embryos, migration of the dorsal trunk cells can be rescued by ubiquitous expression of the constitutively activated Torso^{D}/Breathless chimera. (*D*) In wild-type embryos in which Breathless has been activated in all tracheal cells (by *btl-Gal4/UAS-bnl*), migration of most branches is stalled, but the dorsal trunk cells migrate normally (*btl-Gal4* was kindly provided by S. Hayashi, and *UAS-bnl* by M. Krasnow). Tracheal cells are visualized by anti-Trh antibodies. (*E*) In these embryos, dp-ERK is detected in all tracheal cells (dp-ERK antibody was kindly provided by R. Seger).

would be observed. This was indeed the case (Wappner et al. 1997). We therefore suggest that activation of the DER and Dpp pathways is crucial for determination of tracheal branch identities, prior to the onset of migration. The branches affected by each pathway reflect the regions in which each pathway is activated. The DER pathway is induced more strongly in the central part of the placode, corresponding to the domain in which Rhomboid is expressed. Dpp, on the other hand, is expressed in two ectodermal stripes abutting the dorsal and ventral aspects of the placode, presumably giving rise to more prominent activation of the receptors in the dorsal and ventral aspects of the placode (Fig. 4).

Translation of Tracheal Cell Fates to Directed Migration

We assume that the roles of the DER and Dpp pathways are manifested directly in the tracheal cells, rather than on the surrounding ectodermal cells. For example, it was possible to rescue the DER tracheal phenotype by expressing the normal transcript only in the tracheal cells (Wappner et al. 1997). How are these fates then translated to the capacity of the tracheal cells to be recruited to one branch or another?

One possible model is that during branch migration, the cells within each branch are not identical to each other with respect to the chemotactic cues they encounter. This notion is corroborated by the restricted expression of the Breathless ligand Branchless and by the identification of dp-ERK only in the leading cell at the tip of each tracheal branch. The cells in a migrating branch may thus resemble train cars being pulled by a locomotive. If this is the case, the adhesive properties of the cells are crucial for the normal migration process.

It is possible that each signaling pathway triggers the expression of a different set of homophilic cell adhesion molecules. The molecules expressed by each cell will determine which cells it will adhere to and what leading cell it will follow. This notion is supported by the observation that migration of the leading cells can still be detected in double mutants for the DER and Dpp pathways (Wappner et al. 1997). In addition, mutations in several components of the adhesion complex such as D/E cadherin and Armadillo were shown to affect the formation of a specific tracheal branch, namely, the dorsal trunk (Uemura et al. 1996).

CONCLUDING REMARKS

The genetic and molecular dissection of tracheal development demonstrates several universal features and

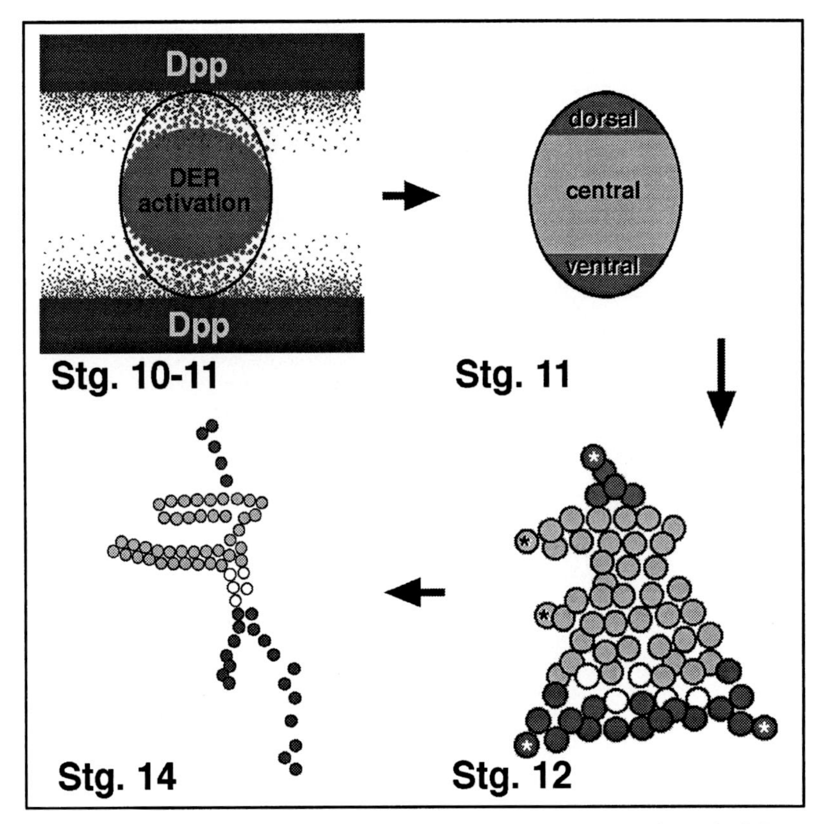

Figure 4. Model for patterning the tracheal cells by the DER and Dpp pathways. At stages 10–11, the DER pathway is activated in the tracheal placode. Due to Rhomboid expression in the central portion of the placode and diffusion of secreted Spitz, stronger activation levels are observed in the central part. On the other hand, Dpp is expressed on the ectoderm in two stripes abutting the tracheal placode. Thus, higher levels of Tkv/Put activation may be induced in the dorsal and ventral parts of the placode. As a result, two different cell populations are determined in the trachea: the DER-induced cells in the central part, and the Dpp-induced cells in the dorsal and ventral regions. Antagonistic interactions between the DER and Dpp pathways may also help to generate sharper borders between the different cell populations. When tracheal cell migration begins at stage 12, different tracheal cells will be recruited to different tracheal branches, according to the fate they have assumed. Since activation of Breathless by Branchless induces only the migration of the tip cells (marked by an asterisk), the fate of the trailing cells would determine which leading tip cell they will join.

highlights similarities to processes of vascularization. The initial determination of the tracheal precursors is achieved by localized expression of the Trh bHLH/PAS transcription factor, and possibly by additional elements in this complex. Once the tracheal cells are defined, the determination of specific branch fates is achieved by the DER and Dpp signaling pathways. Subsequently, the direction in which each branch will migrate is driven by localized expression of the Breathless ligand, Branchless. How receptor activation triggers directional migration within each cell is still an open question. Cell fates continue to be refined within each of the migrating branches by graded activation of Breathless.

Finally, an important set of interactions, yet to be elucidated, concerns the capacity of the terminal tracheal cells to branch and send long cytoplasmic processes toward target tissues. The elusive oxygen-sensing mechanism may be common to *Drosophila* and vertebrates, in view of the induction of a protein associating with the HIF1α DNA-binding site in *Drosophila* cells, following hypoxia. It will be interesting to determine if this process is driven by the hypoxic state of the target tissues, which induces the expression of diffusible ligands attracting the tracheal cells. An essential role of Breathless in this process was also demonstrated, suggesting that this putative ligand (possibly Branchless) may be required to activate Breathless.

ACKNOWLEDGMENTS

We thank current and past members of the Shilo lab for stimulating discussions, M. Krasnow, S. Hayashi, R. Seger, and M. Affolter for open exchange of information and reagents, and the Israel Academy of Sciences for support.

REFERENCES

Affolter M., Nellen D., Nussbaumer U., and Basler K. 1994. Multiple requirements for the receptor serine/threonine kinase *thick veins* reveal novel function of TGFβ homologs during *Drosophila* embryogenesis. *Development* **120:** 3105.

Bier E., Jan L. Y. and Jan Y. N. 1990. *rhomboid*, a gene required for dorsoventral axis establishment and peripheral nervous system development in *Drosophila melanogaster. Genes Dev.* **4:** 190.

Folkman J. and D'Amore P.A. 1996. Blood vessel formation: What is its molecular basis? *Cell* **87:** 1153.

Gabay L., Seger R., and Shilo B.-Z. 1997a. *In situ* activation pattern of the *Drosophila* EGF receptor pathway during development. *Science* **277:** 1103.

———. 1997b. MAP kinase in situ activation atlas during *Drosophila* embryogenesis. *Development* **124:** 3535.

Glazer L. and Shilo B.-Z. 1991. The *Drosophila* FGF-R homolog is expressed in the embryonic tracheal system and appears to be required for directed tracheal cell extension. *Genes Dev.* **5:** 697.

Guillemin K., Groppe J., Ducker K., Treisman R., Hafen E., Affolter M., and Krasnow M. 1996. The *pruned* gene encodes the *Drosophila* serum response factor and regulates cytoplasmic outgrowth during terminal branching of the tracheal system. *Development* **122:** 1353.

Hoffman E.C., Reyes H., Chu F.-F., Sander F., Conley L.H., Brooks B.A., and Hankinson O. 1991. Cloning of a factor required for activity of the Ah (Dioxin) Receptor. *Science* **252:** 954.

Isaac D.D. and Andrew D. 1996. Tubulogenesis in *Drosophila*: A requirement for the *trachealess* gene product. *Genes Dev.* **10:** 103.

Klämbt C., Glazer L., and Shilo B.-Z. 1992. *breathless,* a *Drosophila* FGF receptor homolog, is essential for migration of tracheal and specific midline glial cells. *Genes Dev.* **6:** 1668.

Manning G. and Krasnow M.A. 1993. Development of the *Drosophila* tracheal system. In *The development of* Drosophila melanogaster (Cold Spring Harbor Laboratory Press, (ed. M. Bate and A. Martinez Arias), p. 609. Cold Spring Harbor, New York.

Maxwell P.H., Pugh C.W., and Ratcliffe P.J. 1993. Inducible operation of the erythropoietin 3' enhancer in multiple cell lines: Evidence for a widespread oxygen-sensing mechanism. *Proc. Natl. Acad. Sci.* **90:** 2423.

Nagao M., Ebert B.L., Ratcliffe P.J., and Pugh C.W. 1996. *Drosophila melanogaster* SL2 cells contain a hypoxically inducible DNA binding complex which recognizes mammalian HIF-1 binding sites. *FEBS Lett.* **387:** 161.

Nambu J.R., Lewis J.O., Wharton J.K.A., and Crews S.T. 1991. The *Drosophila single-minded* gene encodes a helix-loop-helix protein that acts as a master regulator of CNS midline development. *Cell* **67:** 1157.

Reichman-Fried M. and Shilo B.-Z. 1995. *breathless*, a *Drosophila* FGF receptor homolog, is required for the onset of tracheal cell migration and tracheole formation. *Mech. Dev.* **52:** 265.

Reichman-Fried M., Dickson B., Hafen E., and Shilo B.-Z. 1994. Elucidation of the role of *breathless*, a *Drosophila* FGF receptor homolog, in tracheal cell migration. *Genes Dev.* **8:** 428.

Risau W. 1997. Mechanisms of angiogenesis. *Nature* **386:** 671.

Samakovlis C., Hacohen N., Manning G., Sutherland D.C., Guillemin K., and Krasnow M.A. 1996a. Development of the *Drosophila* tracheal system occurs by a series of morphologically distinct but genetically coupled branching events. *Development* **122:** 1395.

Samakovlis C., Manning G., Steneberg P., Hacohen N., Cantera R., and Krasnow M. 1996b. Genetic control of epithelial tube fusion during *Drosophila* tracheal development. *Development* **122:** 3531.

Seger R. and Krebs E.G. 1995. The MAPK signaling cascade. *FASEB J.* **9:** 726.

Sutherland D., Samakovlis C., and Krasnow M.A. 1996. *branchless* encodes a *Drosophila* FGF homolog that controls tracheal cell migration and the pattern of branching. *Cell* **87:** 1091.

Tanaka-Matakatsu M., Uemura T., Oda H., Takeichi M., and Hayashi, S. 1996. Cadherin-mediated cell adhesion and cell motility in *Drosophila* trachea. *Development* **122:** 3697.

Tian H., McKnight S.L., and Russell D.W. 1997. Endothelial PAS domain protein 1 (EPAS1), a transcription factor selectively expressed in endothelial cells. *Genes Dev.* **11:** 72.

Uemura T., Oda H., Kraut R., Hayashi S., Kataoka Y., and Takeichi M. 1996. Zygotic *Drosophila* E-cadherin expression is required for processes of dynamic epithelial cell rearrangement in the *Drosophila* embryo. *Genes Dev.* **10:** 659.

Vincent S., Ruberte E., Grieder N.C., Chen C.-K., Haerry T., Schuh R., and Affolter M. 1997. DPP controls tracheal cell migration along the dorsoventral body axis of the *Drosophila* embryo. *Development* **124:** 2741.

Wang G.L. and Semenza G.L. 1993a. Characterization of hypoxia-inducible factor 1 and regulation of DNA binding activity by hypoxia. *J. Biol. Chem.* **268:** 21513.

———. 1993b. General involvement of hypoxia-inducible factor 1 in transcriptional response to hypoxia. *Proc. Natl. Acad. Sci.* **90:** 4304.

Wang G.L., Jiang B.-H., Rue E.A., and Semenza G.L. 1995. Hypoxia-inducible factor 1 is a basic-helix-loop-helix-PAS heterodimer regulated by cellular O_2 tension. *Proc. Natl. Acad. Sci.* **92:** 5510.

Wappner P., Gabay L., and Shilo B.-Z. 1997. Interactions between the EGF receptor and Dpp pathways establish distinct cell fates in the tracheal placodes. *Development* (in press).

Wharton K.A., Franks R.G., Kasai Y., and Crews S.T. 1994. Control of CNS midline transcription by asymetric E-box-like elements: Similarity to xenobiotic responsive regulation. *Development* **120:** 3563.

Wilk R., Weizman I., and Shilo B.-Z. 1996. *trachealess* encodes a bHLH-PAS protein that is an inducer of tracheal cell fates in *Drosophila. Genes Dev.* **10:** 93.

Zelzer E., Wappner P., and Shilo B.-Z. 1997. The PAS domain confers target-gene specificity of *Drosophila* bHLH/PAS proteins. *Genes Dev.* **11:** 2079.

Branching Morphogenesis of the Lung: New Models for a Classical Problem

B.L.M. HOGAN,[1] J. GRINDLEY,[2] S. BELLUSCI,[2] N.R. DUNN,[2] H. EMOTO,[3] AND N. ITOH[3]

[1] *Howard Hughes Medical Institute and* [2] *Department of Cell Biology, Vanderbilt Medical Center, Nashville, Tennessee 37232-2175;* [3]*Department of Genetic Biochemistry, Faculty of Pharmaceutical Science, Kyoto University, Kyoto 606-01, Japan*

Branching morphogenesis is fundamental to the growth and development of many vertebrate organs, including the lung, pancreas, kidney, mammary gland, and prostate. In all cases, the organ primordium is a simple bud of relatively undifferentiated epithelial cells, surrounded by mesenchyme. As development proceeds, the bud grows and branches in a highly reproducible way, generating a characteristic three-dimensional, tree-like structure. Experiments in which the mesenchyme and epithelium of various branching organs are separated and then recombined in different ways support the idea that, in general, the mesenchyme determines the proliferation and branching pattern of the epithelium, but not its tissue-specific program of gene expression. However, the molecular mechanisms underlying the dynamic and reciprocal interactions between the mesenchyme and the epithelium are not well understood. Our goal is to identify the rules of intercellular behavior that are common to branching morphogenesis in all vertebrate organs and to investigate their molecular and genetic basis, using the embryonic mouse lung as a convenient model system.

DEVELOPMENT OF THE MOUSE LUNG

The lung primordium arises around E9.5 (25–27 somites) from the ventral foregut just anterior of the stomach. It consists of two parts: the future trachea and two endodermal buds that give rise to the left and right lobes of the distal lung (Fig. 1A) (Spooner and Wessells 1970). Both parts consist of an epithelial layer of endoderm surrounded by splanchnic mesoderm. The primary buds grow and branch in a specific pattern, but the trachea does not give rise to lateral buds in vivo, except under pathological conditions (Skandalakis et al. 1994). The branching pattern of the left and right buds is very different, reflecting an asymmetry that is presumably established, or even preexists, as soon as the buds form. In the mouse, the left bud generates a single lobe, whereas the right bud gives rise to four lobes. The branching pattern of each lobe can be traced easily up to about E12.5 and is highly reproducible from embryo to embryo.

There are two basic branching mechanisms used in the early, pseudoglandular stage of lung development (Ten Have-Opbroek 1981). The first is particularly important in establishing the length and overall shape of the main lobes. It involves a single bud growing out for an extended distance accompanied by the sprouting of secondary lateral buds at specific distances from the leading tip. In the second kind of branching, a terminal bud bifurcates into two branches, each of which continues to grow along a different axis, but not necessarily to the same extent. The two mechanisms—lateral sprouting and dichotomous branching—are shown in Figure 1B. Later in development, a process of septation, which divides the terminal saccules into multiple compartments, is important in generating the alveoli (Bostrom et al. 1996).

Evidence for Epithelial-Mesenchymal Interactions during Mouse Lung Development

More than 30 years ago, pioneering experiments were carried out in which small pieces of distal mesenchyme were grafted next to tracheal endoderm denuded of proximal mesoderm. This results in the formation of ectopic buds that grow, branch, and differentiate like normal pulmonary epithelium. Conversely, tracheal mesoderm grafted next to distal endoderm inhibits outgrowth and branching (Alescio and Cassini 1962; Spooner and Wessells 1970; Wessells 1970; Lawson 1983; Hilfer et al. 1985; Shannon 1994). Since the effect of the mesenchyme on epithelial morphogenesis still occurs in the presence of an intervening filter (Taderera 1967), it was concluded that the mesoderm secretes soluble factors that act on the endoderm and/or its basal lamina to bring about changes in cell behavior. We have therefore focused our attention on the expression by the embryonic lung of genes encoding families of highly conserved, secreted signaling molecules, known from genetic and biochemical studies in other systems to regulate cell growth and patterning.

Temporal and Spatial Patterns of Expression of Signaling Genes

A wide range of techniques, including immunohistochemistry and section and whole-mount in situ hy-

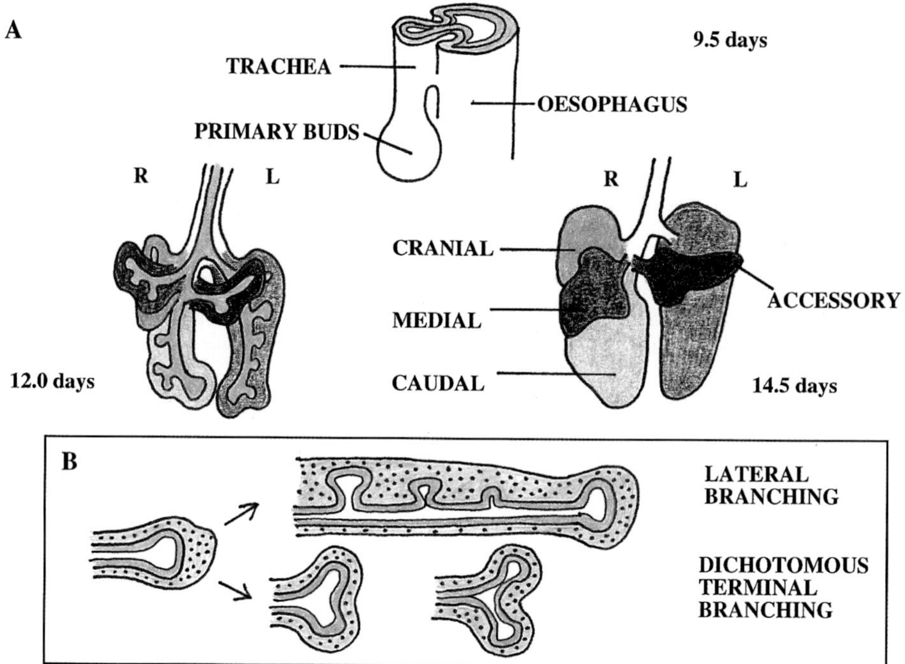

Figure 1. Schematic representation of mouse lung development. (*A*) About E9.5, the trachea and primary buds develop from the ventral foregut. The tracheal-oesophageal groove eventually closes and separates the alimentary and pulmonary systems. By E12.5, the precursors of the five primary lobes have been established, and the regular branching pattern of the epithelial endoderm can be clearly followed. Note the left-right (L-R) asymmetry of the lung. At E14.5, the five lobes (right cranial, medial, caudal, and accessory lobes and left lobe) are well developed. (*B*) Two different branching processes seen in early lung development, with endoderm in yellow and mesoderm in stippled gray. See text for details.

bridization, have been used to determine in the embryonic lung the temporal and spatial patterns of expression of signaling proteins and of their genes and receptors. These include members of the transforming growth factor-β (Tgf-β), bone morphogenetic protein (Bmp), epidermal growth factor (Egf), fibroblast growth factor (Fgf), Wnt, and Hedgehog (Hh) families (Heine et al. 1990; Fu et al. 1991; Warburton et al. 1992 and references therein). Several patterns of localized expression have emerged (Fig. 2). In particular, it is clear that high levels of RNA for some genes are tightly localized to the distal endoderm of both terminal and lateral buds; examples in this category include *Bmp4*, *Sonic hedgehog (Shh)* and *Wnt7b* (Bellusci et al. 1996; 1997; B. Hogan et al., unpubl.). In contrast, expression of other genes is restricted to the distal mesoderm immediately surrounding the buds; examples here include *Patched* (*Ptc,* which encodes a Shh receptor), *Wnt2,* and *Fgf10* (Bellusci et al. 1996, 1997 and in prep.). Other genes are expressed more widely, but nevertheless have spatial variations in transcript levels; the three *Gli* genes, *Gli, Gli2,* and *Gli3*, fall into this category, as discussed below (Grindley et al. 1997). Finally, the expression of *Bmp4* in the mesoderm, although lower than in the endoderm, is highly dynamic; between E11 and E12, it is clearly expressed in both the tracheal and distal mesoderm, but only in the ventral domain. At E12.5 and E14.5, it is expressed in the mesoderm around the primary and secondary bronchi and in the more proximal regions of the five lobes (Fig. 3). By birth, there are high levels of *Bmp4* expression throughout the lung (data not shown).

These patterns of gene expression have suggested various models for the flow of information between the epithelium and mesenchyme during branching morphogenesis. In particular, they suggest that the distal end buds are acting as signaling centers, with factors secreted by the endoderm affecting the adjacent mesoderm, and vice versa. As discussed below, supporting evidence has come from the study of transgenic embryos in which genes are misexpressed throughout the distal endoderm and of mutant embryos in which specific genes have been inactivated either by spontaneous mutations or by homologous recombination in embryonic stem cells. Finally, the function of various gene products has been tested in vitro, taking advantage of the ease with which embryonic mouse lungs can be cultured under conditions in which branching continues.

ROLE OF SONIC HEDGEHOG IN LUNG DEVELOPMENT

As shown in Figure 2B, *Shh* is expressed throughout the lung endoderm, but at very high levels in the distal end buds. Northern analysis shows that the level of endogenous *Shh* transcripts declines toward birth, as lung growth slows. We have therefore used regulatory regions of the gene encoding Surfactant Protein-C (SP-C) (Wert et al. 1993) to drive high levels of *Shh* in transgenic em-

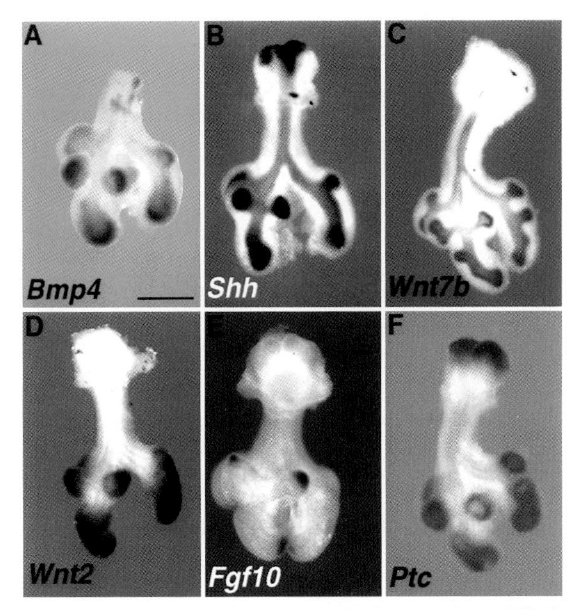

Figure 2. Gene expression in the E11.5 mouse lung. Wholemount in situ hybridization with digoxygenin-labeled antisense riboprobes was used to localize transcripts for Bmp4 (*A*) (Bellusci et al. 1996), Shh (*B*) (Bellusci et al. 1997), Wnt7b (*C*), Wnt2 (*D*) (Bellusci et al. 1996), Fgf10 (*E*), and Ptc (F) (Bellusci et al. 1997) in lungs from E11.5 embryos. Note the reproducible pattern of branching between different lungs.

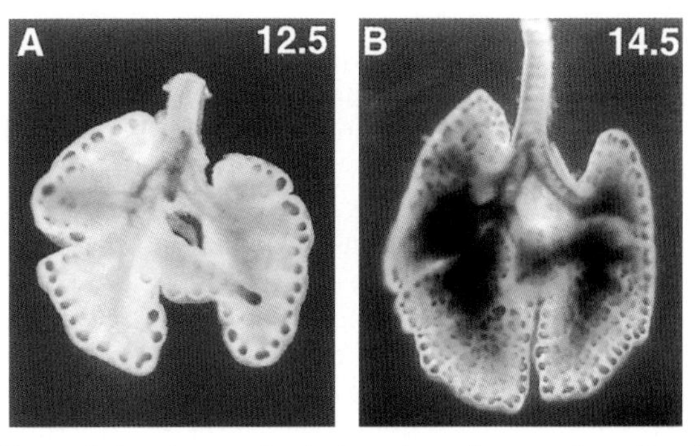

Figure 3. *Bmp4* expression in the embryonic mouse lung. *Bmp4* expression was revealed by X-gal staining of lungs taken at E12.5 (*A*) and E14.5 (*B*) from embryos heterozygous for a $Bmp4^{lacZ}$ allele. In this allele, the bacterial *lacZ* gene has been inserted in-frame into the first protein-coding exon of the endogenous *Bmp4* gene (N.R. Dunn., unpubl.). In all tissues studied, *lacZ* expression faithfully follows endogenous *Bmp4* expression, as judged from in situ hybridization. Note high-level *Bmp4* expression in the endoderm of distal end buds at both stages. At E14.5, *Bmp4* expression is also seen in the mesoderm associated with the formation of cartilage rings in the trachea and primary bronchi.

bryos. SP-C is expressed exclusively in the lung endoderm from the time of its first formation, with increasing efficiency as development proceeds. SP-C-*Shh* transgenic lungs are very abnormal, and by E17.5, they show excessive mesoderm accumulation, accompanied by a higher than normal rate of mesodermal cell proliferation (Bellusci et al. 1997). This suggests that Shh protein secreted by the distal endoderm stimulates, directly or indirectly, cell proliferation in the mesoderm. This conclusion is compatible with the observation that *Ptc*, which encodes a receptor for Shh, is predominantly expressed in the mesenchyme immediately adjacent to the endoderm. Moreover, the level of *Ptc* RNA is up-regulated in the mesoderm of SP-C-Shh transgenic lungs (Bellusci et al. 1997; Grindley et al. 1997). This finding is consistent with the up-regulation of *ptc* in *Drosophila* cells responding to Hh signaling and with the increase in *ptc* expression seen in other vertebrate systems where *Shh* is ectopically expressed (Goodrich et al. 1996; Marigo et al. 1996b). Finally, embryos lacking a functional *Shh* gene that survive to E18 have severely abnormal lungs, with only a few distal buds (C. Chiang and Y. Litingtung, pers. comm.).

All these lines of evidence support the idea that Shh acts on the mesoderm, but, at present, we cannot rule out the possibility that Shh also acts directly on the lung endoderm, since *Ptc* is expressed at low levels in this cell type and there is some increase in proliferation of the endoderm in SPC-*Shh* lungs (Bellusci et al. 1997). Alternatively, or in addition, the effect of Shh on the endoderm may be indirect, as a result of changes in the mesoderm. There has been one report of a large lung cyst in a patient with nevoid basal cell carcinoma syndrome, a condition associated with mutations in the human *PTC* gene (Totten 1980; Johnson et al. 1997).

EVIDENCE FOR A ROLE FOR *GLI* GENES

In *Drosophila*, the gene *cubitus interruptus* (*ci*) encodes a zinc finger transcription factor that positively regulates *ptc* expression in a number of tissues in response to Hh signaling (Alexandre et al. 1996). In cells responding to Hh, Ci is activated and translocated to the nucleus from the cytoplasm (Aza-Blanc et al. 1997), and the level of Ci protein is also up-regulated, possibly at the level of translation (Johnson et al. 1995; Motzny and Holmgren 1995). In vertebrates, there are three *ci*-related genes, *Gli*, *Gli2*, and *Gli3*; they are widely transcribed in the mouse embryo, often in domains adjacent to cells in which one of the three hedgehog genes (*Shh*, *Ihh*, and *Dhh*) is expressed (Hui et al. 1994). In the chick wing bud, ectopic Shh induces *Gli* transcription but represses that of *Gli3*, implying a complex interrelation between the two gene families (Marigo et al. 1996a). In situ hybridization studies have shown that all three *Gli* genes are expressed in the distal mesoderm of the lung, in somewhat different but overlapping patterns (Grindley et al. 1997). For example, transcripts for *Gli* appear to be higher immediately adjacent to the endoderm of lateral and terminal buds than those of *Gli2* and *Gli3*. Several lines of evidence suggest that this expression of *Gli* genes has functional significance in lung morphogenesis. For example, *Gli* (but not *Gli2* or *Gli3*) is up-regulated in the lungs of SP-C-*Shh* transgenic embryos. In addition, the lungs of $Gli3^{Xt}/Gli3^{Xt}$ embryos, homozygous for a deletion of the *Gli3* gene, are smaller than wild type and have abnormalities in length and shape, more severe on some genetic backgrounds than others (Grindley et al. 1997). Finally, a range of pulmonary defects have been reported in humans heterozygous for frameshift mutations in the *GLI3* gene associated with Pallister-Hall syndrome (OMIM #146510) (Kang et al. 1997). It seems likely that the truncated mutant Gli3 protein acts as a dominant-negative inhibitor of other Gli proteins.

In conclusion, the evidence that Shh secreted by the distal lung endoderm acts on the adjacent mesoderm through a signaling pathway involving Ptc and one or more Gli transcription factor(s) is quite persuasive. However, much more needs to be learnt about the downstream effector genes and how mesodermal cell proliferation is up-regulated in response to Shh.

ROLE OF FGF10 IN LUNG DEVELOPMENT

In 1994, new light was shed on the mechanism of lung branching morphogenesis by experiments in which a dominant-negative form of Fgf receptor 2 splice variant IIIb (FgfR2IIIb) was misexpressed in the distal endoderm under the control of the SP-C enhancer/promoter (Peters et al. 1994). Although the trachea and two primary bronchi were intact, no lateral branches were present, demonstrating that at least one Fgf ligand is required for normal lung development. This result posed the question of the nature and localization of the Fgf ligand(s) blocked by the mutant receptor. Of the different Fgf ligands known at that time, Fgf7 acts through FgfR2 with high efficiency (Ornitz et al. 1996), so attention naturally focused on this candidate. The *Fgf7* gene is indeed expressed in embryonic lung mesoderm, but only very low levels of transcript are present at early stages, when branching is well under way (Mason et al. 1994; Post et al. 1996). Moreover, mice in which both copies of the *Fgf7* gene have been inactivated by homologous recombination have no obvious abnormalities in pulmonary development (Guo et al. 1996). Therefore, the finding of a new *Fgf* gene expressed at high levels in the mesoderm of the rat embryo lung was of particular interest (Yamasaki et al. 1996). Indeed, as shown in Figure 2, the mouse *Fgf10* gene fulfills at least two of the criteria required for a gene involved in branching morphogenesis: It is expressed at high levels in specific regions from the earliest stages of development, and the temporal and spatial pattern of expression is compatible with a specific role in the induction and outgrowth of primary and lateral buds.

A potential role of Fgf10 both in inducing lateral branches and in directing bud outgrowth is illustrated in Figure 4. This shows, by whole-mount in situ hybridization, expression in the right caudal lobe at intervals between E11.5 and E12.5. Throughout this time, expression is maintained at high levels in the mesoderm of the distal tip, consistent with the protein playing a part in the extended outgrowth of the primary branch. Expression is also seen in the mesoderm adjacent to the first lateral bud to appear (bud 1). It then extends caudally to anticipate the outgrowth of the next lateral buds (buds 2, 3, and 4). Buds grow toward the domain of highest *Fgf10* expression, and as they do so, the level of *Fgf10* transcripts increases. However, it is important to note that a zone of mesodermal cells that does *not* express high levels of *Fgf10* lies immediately adjacent to the distal bud endoderm. This suggests that some factor made by the endoderm locally inhibits Fgf10 expression. Having the highest level of *Fgf10* production set at a distance from the distal tip may help to entice forward outgrowth of the bud, possibly along a gradient of ligand.

The Action of Recombinant Fgf10 on Lung Buds in Culture

In debating models for the role of Fgf10 in lung morphogenesis, it is important to know whether its primary effect is on the endoderm or mesoderm, and what responses it can elicit in target tissues. Studies have shown that three of the four known Fgf receptor genes, including *FgfR2*, are expressed in the lung endoderm, but not in the mesoderm. In addition, endoderm denuded of mesoderm is able to respond to both Fgf1 and Fgf7 when cultured in Matrigel (a mixture of basal lamina components such as laminin, entactin, type IV collagen, and heparan sulfate proteoglycan) (Nogawa and Ito 1995; Bellusci et al. 1997; Cardoso et al. 1997). This shows that the ligands can act directly on the endoderm, without a requirement for the mesoderm. However, in this assay, the two proteins elicit rather different responses, even when added at the same concentrations (e.g., 250 ng/ml). Both proteins promote cell survival and proliferation, but Fgf1 elicits the formation of multiple buds within the first 24 hours of culture, whereas Fgf7 promotes the formation of an expanded spheroid of endoderm. We have found that the response of isolated endoderm to recombinant human Fgf10 is initially very similar to that of Fgf7, rather than Fgf1. This is perhaps to be expected, given that the amino acid sequence of Fgf10 is more similar to that of Fgf7 in the conserved region than Fgf1. However, in the case of Fgf10, the spheroids give rise to multiple buds after 4–60 hours (Bellusci et al. 1997).

Similar experiments were carried out with distal end buds isolated from E11.5 lungs. The buds, consisting of small pieces of distal endoderm surrounded by mesoderm, were placed in Matrigel culture and treated with varying concentrations of Fgf10 (30–500 ng/ml). Initially, the endoderm expands, forming a single, spheroid-like structure, and then between 48 and 72 hours, this gives rise to multiple buds (Fig. 4). However, a very different response is seen when the distal buds are cultured in a matrix of type I collagen. In this case, the endoderm expands, but never gives rise to secondary buds. Absence of bud outgrowth in type I collagen is also seen with Fgf1 and Fgf7. This experiment demonstrates that the morphogenetic response of endoderm to Fgf10 varies depending on the extracellular matrix to which the endoderm is exposed. This has important implications for models of branching morphogenesis, as discussed below.

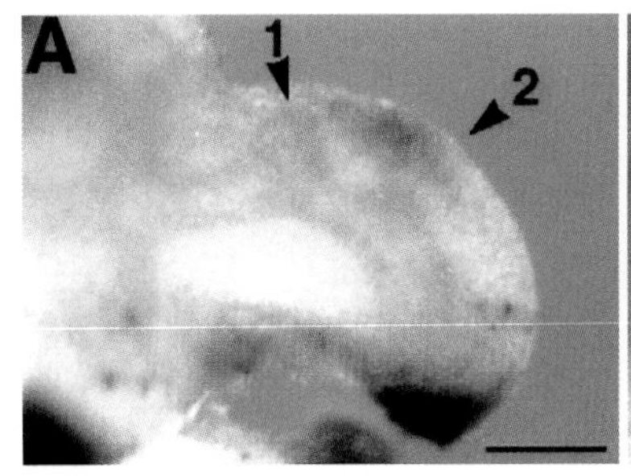

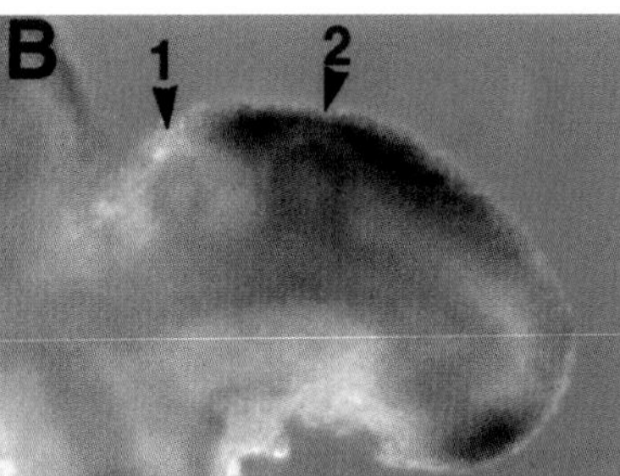

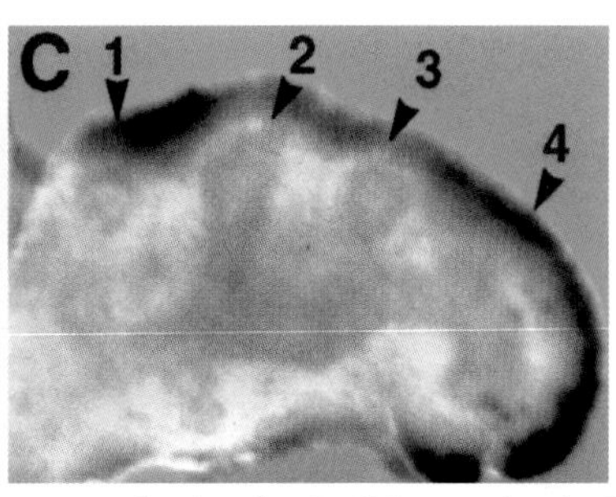

Figure 4. Fgf10 expression during lateral branching. Whole-mount in situ hybridization was used to localize *Fgf10* transcripts in the right caudal lobe during the process of primary branch outgrowth and lateral branchin. See text for details.

A Conserved Role for Fgfs in Branching Morphogenesis?

We have found that *Fgf10* is expressed at high levels in the distal mesoderm, in close proximity to outgrowing primary and secondary endodermal buds. Moreover, buds appear to grow toward domains of high *Fgf10* expression. We do not yet know whether Fgf10 is sufficient to elicit lateral bud formation, or only necessary; the observation that lateral buds form at very specific distances from the terminal bud suggests that additional factors are involved in controlling bud spacing, possibly by inhibiting new bud formation close to the tip. Although many more details need to be worked out concerning Fgf10 function, our findings suggest remarkable parallels between branching morphogenesis in the vertebrate lung and the insect tracheal system. In each abdominal segment of the *Drosophila* larva, a small epidermal placode gives rise to a cluster of postmitotic tracheal cells. These migrate, elongate, and branch to form a tree-like system of primary, secondary, and terminal tubes. Genetic screens have identified a number of genes required for and regulating distinct stages in the morphogenesis of the tracheal system; among these are *branchless (bnl)* and *breathless (btl)*, encoding an Fgf ligand and Fgf receptor, respectively (Klambt et al. 1992; Lee et al. 1996; Sutherland et al. 1996). *btl* is expressed in the tracheal cells, whereas *bnl* is expressed in small clusters of epidermal or mesodermal cells, providing an external source of Fgf ligand toward which the tracheal cells extend. Ectopic expression of *bnl* results in the misdirection of tracheal cells. These findings raise the exciting possibility that other genes regulating *Drosophila* tracheal development have also been conserved during evolution and that their homologs have a role in vertebrate lung morphogenesis and in the growth and patterning of many other branches organ systems. Among these genes might be *sprouty*, discussed in the article by Krasnow (this volume), which encodes an inhibitor of Fgf activity on tracheal cells.

TOWARD AN INTEGRATED MODEL FOR BRANCHING MORPHOGENESIS

Previous models of branching morphogenesis in organs such as the lung have tended to focus on the assembly and degradation of the extracellular matrix as major determinants of branching pattern. This is perhaps best illustrated by the "matrix remodeling hypothesis" of Bernfield (1981). According to this model, the proximal epithelium or branch stalk is constrained by a well-organized, sheath-like basal lamina and extracellular matrix that includes type I collagen fibrils (Wessells 1970). In contrast, the rapid growth and expansion of the distal tip, coupled with the matrix degrading activity of the adjacent mesenchyme, lead to disruption of the basal lamina, which cannot be assembled fast enough to keep pace with proliferation. This results in the local collapse of the epithelial sheet and the formation of a cleft in which assembly of an intact basal lamina can be restored and growth restrained. Dichotomous branching then occurs as a result of the higher proliferation and expansion of epithelial cells lateral to the cleft. The sprouting of lateral buds is less well served by this model, but could be seen to involve the localized degradation of basal lamina and extracellular matrix by the mesoderm around the proximal epithelium, allowing a bud to grow out through the site of disruption. Numerous lines of evidence point to extracellular matrix and cell-matrix receptors such as integrins, playing important parts in the branching of organs (including the lung) (Chen and Little 1987; Kreidberg et al. 1996; Wu and Santoro et al. 1996; Kadoya et al. 1997). However, the matrix remodeling hypothesis does not take into account the more recent evidence, reviewed briefly here, that secreted signaling molecules also have a role in regulating the epithelial-mesenchymal interactions that underlie the branching process. An alternative model is therefore needed that integrates both approaches to the problem. A tentative example is shown in Figure 5, in which branching is dividing into three phases: bud outgrowth, tip arrest, and lateral branching.

Bud Outgrowth

From the very early stages of bud outgrowth, the distal tip epithelium secretes Shh (Bellusci et al. 1996). This stimulates adjacent mesodermal cell proliferation by a mechanism that involves *Ptc* and a *Gli* gene, but not apparently a direct effect on *Fgf10* expression (Bellusci et al. 1977; Grindley et al. 1997). We propose that some other factor, for example, a Wnt (Wnt7b is suggested in Fig. 5), secreted by the distal endoderm up-regulates

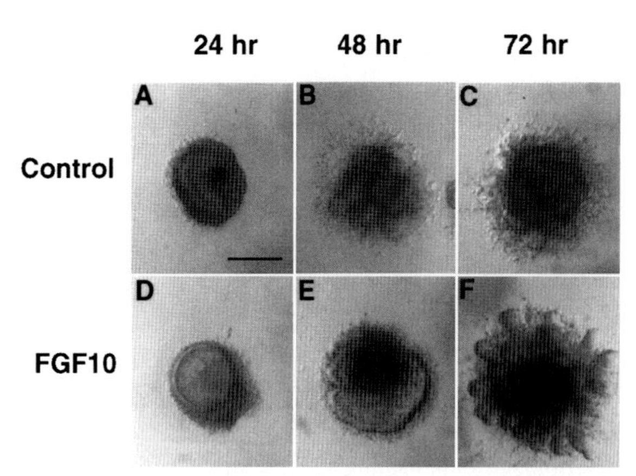

Figure 5. Effect of Fgf10 on lung cells in culture. Distal end buds (endoderm plus mesoderm) were isolated from E11.5 lungs and cultured in Matrigel either alone (*A–C*) or in the presence of 250 ng/ml human recombinant Fgf10 expressed from a baculovirus vector in insect cells (*D–F*). Note the initial swelling of the endoderm in response to Fgf10, followed after 48 hr by extensive budding.

Fgf10 expression, possibly by a concentration-dependent mechanism, with low concentrations being most active. Fgf10, in turn, acts on the distal endoderm to stimulate cell proliferation and also possibly to reorganize cell-cell associations, promote fluid secretion into the lumen, and induce cell migration. All these result in the expansion of the epithelial layer and its outgrowth. Subsequently, a self-propelling feedback mechanism is set up that drives forward movement of the branch toward a localized source of Fgf10, unless interrupted by some kind of "brake."

Tip Arrest

A key feature of the embryonic mouse lung is the high level of expression of *Bmp4* in the endoderm of distal buds. Overexpression of *Bmp4* in transgenic lungs under the control of SP-C regulatory elements leads to an inhibition of endodermal cell proliferation and disruption of branching morphogenesis (Bellusci et al. 1996). One possibility is that the expression of *Bmp4* in the epithelium increases as distal outgrowth proceeds and that there is a gradient of Bmp4 within the bud, with the highest levels at the tip. Bmp4 may inhibit or counteract the effect of Fgf10 on the epithelium, so that Fgf10 acts with greatest efficiency on cells lateral to the tip. Tip arrest may occur when the concentration of Bmp4 has built up to such a level that the effect of Fgf10 on the most terminal cells is completely blocked. Obviously, the time when this occurs could be influenced by factors such as noggin and chordin, proteins known to bind and inactivate Bmp4. Other factors that bind and influence the activity of Fgfs, such as heparin sulfate proteoglycans, may also regulate the process of bud outgrowth.

Dichotomous Branching

Where do extracellular matrix molecules fit into the picture? Is there an alternative to the scenario proposed

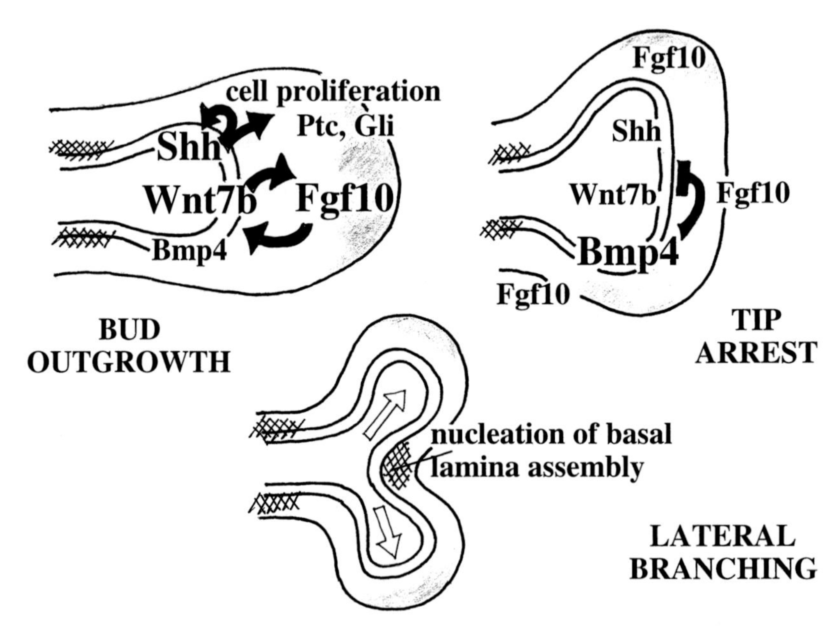

Figure 6. Tentative model for the interaction between signaling molecules and matrix assembly in branching morphogenesis. During bud outgrowth, Shh produced at high levels by the distal tip endoderm stimulates cell proliferation in the mesoderm by a mechanism involving *Ptc* and *Gli* genes. Shh may also act directly within the endoderm. Fgf10 made in the mesoderm (gray shading) stimulates cell proliferation and forward outgrowth of the distal endoderm. Wnt7b or some other factor made by the endoderm may positively regulate Fgf10 expression. As outgrowth progresses, the level of Bmp4 in the distal tip increases and reaches a level where the effect of Fgf10 is inhibited. Fgf10 expression also shifts more laterally where it stimulates outgrowth of new buds (*open arrows*). High levels of Bmp4 may also activate the nucleation and assembly of an organized basal lamina underneath the most distal endoderm (*cross-hatching*). Incorporation of type I collagen into this matrix both distally and proximally may help to inhibit branching.

by Bernfield (1981) in his original model that nevertheless incorporates the absence of an intact basal lamina around the distal tip endoderm, but a well-organized lamina around the branch stalk? First, we have shown that budding in response to Fgfs is inhibited by type I collagen. This supports the idea that a sleeve of collagen fibrils around the proximal region of the branch stalk inhibits the sprouting of lateral buds and serves as a physical constraint promoting forward expansion. On the other hand, Matrigel components are permissive for Fgf-induced budding. One possibility is that, in vivo, high concentrations of Bmp4 at the distal tip promote the expression in the epithelium of specific integrins that "nucleate" the attachment of epithelial cells to matrix proteins such as laminin made by the mesenchyme. These epithelial cells then change their shape and initiate the local assembly and deposition of an organized basal lamina, leading to the formation of a cleft. This model differs from that of Bernfield's in that local basal lamina assembly precedes cleft formation, rather than vice versa. In isolated endoderm in culture, formation of these nucleation sites would be stochastic, since there is no regular pattern to the budding. In vivo, they would be determined by the integrated activity and spatial localization of many different signaling molecules, ligand-binding proteins, and receptors.

This tentative model is proposed as a framework on which to design future experiments. The goal is to bring together what we learn about intercellular signaling pathways, cell-cell and cell-matrix interactions, and proliferation in the embryonic lung into a comprehensive cellular and molecular description of branching morphogenesis, and to compare this with models for morphogenesis of other organs, such as the kidney and pancreas.

ACKNOWLEDGMENTS

This work was supported by National Institutes of Health grant HD-28955, by a grant-in-aid for scientific research from the Ministry of Education, Science, and Culture, Japan, and by National Institutes of Health predoctoral training grant 5T32CA09385 to N.R.D. B.L.M.H. is an investigator of the Howard Hughes Medical Institute.

REFERENCES

Alescio T. and Cassini A. 1962. Induction in vitro of tracheal buds by pulmonary mesenchyme grafted on tracheal epithelium. *J. Exp. Zool.* **150:** 83.

Alexandre C., Jacinto A., and Ingham P.W. 1996. Transcriptional activation of *hedgehog* target genes in *Drosophila* is mediated directly by the Cubitus interruptus protein, a member of the GLI family of zinc finger DNA-binding proteins. *Genes Dev.* **10:** 2003.

Aza-Blanc P., Ramirez-Weber F.-A., Laget M.-P., Schwartz C., and Kornberg T.B. 1997. Proteolysis that is inhibited by *hedgehog* targets cubitus interruptus protein to the nucleus and converts it to a repressor. *Cell* **89:** 1043.

Bellusci S., Grindley J., Emoto H., Itoh N., and Hogan B.L.M. 1997. Fibroblast growth factor 10 (FGF10) and branching morphogenesis in the embrhonic mouse lung. *Development* (in press).

Bellusci S., Henderson R., Winnier G., Oikawa T., and Hogan B.L.M. 1996. Evidence from normal expression and targeted misexpression that bone morphogenetic protein-4 (Bmp4) plays a role in mouse embryonic lung morphogenesis. *Development* **122:** 1693.

Bellusci S., Furuta Y., Rush M.G., Henderson R., Winnier G., and. Hogan B.L.M. 1997. Involvement of Sonic hedgehog (Shh) in mouse embryonic lung growth and morphogenesis. *Development* **124:** 53.

Bernfield M.R. 1981. Organization and remodeling of the extracellular matrix in morphogenesis. In *Morphogenesis and pattern formation* (ed. T.G. Connelly), p. 139. Raven Press, New York.

Bostrom H., Willetts K., Pekny M., Leveen P., Lindahl P., Hedstrand H., Pekna M., Hellstrom M., Gebre-Medhin S., Schalling M., Nilsson M., Kurland S., Tornell J., Heath J.K., and Betsholtz C. 1996. PDGF-α signaling is a critical event in lung alveolar myofibroblast development and alveogenesis. *Cell* **85:** 863.

Cardoso W.V., Itoh A., Nogawa H., Mason I., and Brody J.S. 1997. FGF-1 and FGF-7 induce distinct patterns of growth and differentiation in embryonic lung epithelium. *Dev. Dyn.* **208:** 398.

Chen J.-M. and Little C.D. 1987. Cellular events associated with lung branching morphogenesis including the deposition of collagen type IV. *Dev. Bio.* **120:** 311.

Fu Y.-M., Spirito P., Yu Z.-X., Biro S., Sasse J., Lei J., Ferrans V.J., Epstein S.E., and Casscells W. 1991. Acidic fibroblast growth factor in the developing embryo. *J. Cell Biol.* **114:** 1261.

Goodrich L.V., Johnson R.L., Milenkovic L., McMahon J.A., and Scott M.P. 1996. Conservation of the *hedgehog/patched* signaling pathway from flies to mice: Induction of a mouse *patched* gene by Hedgehog. *Genes Dev.* **10:** 301.

Grindley J.C., Bellusci S., Perkins D., and Hogan B.L.M. 1997. Evidence for the involvement of the Gli gene family in embryonic mouse lung development. *Dev. Biol.* **188:** 337.

Guo, L., Degenstein L., and Fuchs E. 1996. Keratinocyte growth factor is required for hair development but not for wound healing. *Genes Dev.* **10:** 165.

Heine U.I., Munoz E.F., Flanders K.C., Roberts A.B., and Sporn M.B. 1990. Colocalization of TGF-β 1 and collagen I and III, fibronectin and glycosaminoglycans during lung branching morphogenesis. *Development* **109:** 29.

Hilfer S.R., Rayner R.M., and Brown J.W. 1985.Mesenchymal control of branching pattern in the fetal mouse lung. *Tissue Cell* **17:** 523.

Hui C.-C., Slusarski D., Platt K.A., Holmgreen R., and Joyner A.L. 1994. Expression of three mouse homologs of the *Drosophila* segment polarity gene *cubitus interruptus*, *Gli*, *Gli-2* and *Gli-3*, in ectoderm-and mesoderm-derived tissues suggests multiple roles during postimplantation development. *Dev. Biol.* **162:** 402.

Johnson R.L., Grenier J.K., and Scott M.P. 1995. Patched overexpression alters wing disc size and pattern: Transcriptional and post-transcriptional effects on hedgehog targets. *Development* **121:** 4161.

Johnson R.L., Rothman A.L., Xie J., Goodrich L.V., Bare J.W., Bonifas J.M., Quinn A.G., Myers R.M., Cox D.R., Epstein E.H., and Scott M.P. 1996. Human homolog of *Patched*, a candidate gene for the basal cell nevus syndrome. *Science* **272**:1668.

Kadoya Y., Salmivirta K., Talts J.F., Kadoya K., Mayer U., Timpl R., and Ekblom P. 1997. Importance of nidogen binding to laminin γ1 for branching epithelial morphogenesis of the submandibular gland. *Development* **124:** 683.

Kang S., Graham J.M., Olney A.H., and Bieseker L.G. 1997. GLI3 frameshift mutations cause autosomal dominant Pallister-Hall syndrome. *Nat. Genet.* **15:** 266.

Klambt C., Glazer L., and Shilo B.-Z. 1992. *breathless*, a *Drosophila* FGF receptor homolog, is essential for migration of trachael and specific midline glial cells. *Genes Dev.* **6:** 1668.

Kreidberg J., Donovan M., Goldstein S., Rennke H., Shepherd

K., Jones R., and Jaenisch R. 1996. α 3 β1 integrin has a crucial role in kidney and lung organogenesis. *Development* **122:** 3537.

Lawson K.A. 1983. Stage specificity in the mesenchyme requirement of rodent lung epithelium in vitro: A matter of growth control? *J. Embryol. Exp. Morph.* **74:** 183.

Lee T., Hacohen N., Krasnow M., and Montell D.J. 1996. Regulated *breathless* receptor tyrosine kinase activity required to pattern cell migration and branching in the *Drosophila* tracheal system. *Genes Dev.* **10:** 2912.

Marigo V., Johnson R., Vortkamp A., and Tabin C. 1996a. Sonic hedgehog differentially regulates expression of GLI and GLI3 during limb development. *Dev. Biol.* **180:** 273.

Marigo V., Scott M.P., Johnson R.L., Goodrich L.V., and Tabin C.J. 1996b. Conservation in *hedgehog* signaling: Induction of a chick *patched* homolog by *Sonic hedgehog* in the developing limb. *Development* **122:** 1225.

Mason I.J., Fuller-Pace F., Smith R., and Dickson C. 1994. FGF-7 (keratinocyte growth factor) expression during mouse development suggests roles in myogenesis, forebrain regionalisation and epithelial-mesenchymal interactions. *Mech. Dev.* **45:** 15.

Motzny C.K. and Holmgren R. 1995. The *Drosophila* cubitus interruptus protein and its role in the *wingless* and *hedgehog* signal transduction pathways. *Mech. Dev.* **52:** 137.

Nogawa H. and Ito T. 1995. Branching morphogenesis of embryonic mouse lung epithelium in mesenchyme-free culture. *Development* **121:** 1015.

Ornitz D.M., Xu J., Colvin J.S., McEwen D.G., MacArthur C.A., Coulier F., Gao G., and Goldfarb M. 1996. Receptor specificity of the fibroblast growth factor family. *J. Biol. Chem.* **271:** 15292.

Peters K., Werner S., Liao X., Wert S., Whitsett J., and Williams L. 1994. Targeted expression of a dominant negative FGF receptor blocks branching morphogenesis and epithelial differentiation of the mouse lung. *EMBO J.* **13:** 3296.

Post M., Souza P., Liu J., Tseu T., Wang J., Kuliszewski M., and Tanswell A.K. 1996. Keratinocyte growth factor and its receptor are involved in reglating early lung branching. *Development* **122:** 3107.

Shannon J.M. 1994. Induction of alveolar type II cell differentiation in fetal trachael epithelium by grafted distal lung mesenchyme. *Dev. Biol.* **166:** 600.

Skandalakis J.E., Gray S.W., and Symbas P. 1994. The trachea and the lungs. In *Embryology for surgeons* (ed. J.E. Skandalakis and S.W. Gray), p. 414. Williams and Wilkins, Baltimore, Maryland.

Spooner B.S. and Wessells N.K. 1970. Mammalian lung development: Interactions in primordium formation and bronchial morphogenesis. *J. Exp. Zool.* **175:** 445.

Sutherland D., Samakovlis C., and Krasnow M. 1996. *branchless* encodes a *Drosophila* FGF homolog that controls tracheal cell migration and the pattern of branching. *Cell* **87:** 1091.

Taderera J.V. 1967. Control of lung differentiation in vitro. *Dev. Biol.* **16:** 489.

Ten Have-Opbroek A.A.W. 1981. The development of the lung in mammals: An analysis of concepts and findings. *Am. J. Anat.* **162:** 201.

Totten J.R. 1980 The multiple nevoid basal cell carcinoma syndrome. *Cancer* **46**: 1456.

Warburton D., Seth R., Shum L., Horcher P.G., Hall F.L., Werb Z., and Slavkin H.C. 1992. Epigenetic role of epidermal growth factor expression and signalling in embryonic mouse lung morphogenesis. *Dev. Biol.* **149:** 123.

Wert S.E., Glasser S.W., Korfhagen T.R., and Whitsett J.A. 1993. Transcriptional elements from the human SP-C gene direct expression in the primordial respiratory epithelium of transgenic mice. *Dev. Biol.* **156:** 426.

Wessells N.K. 1970. Mammalian lung development: Interactions in formation and morphogenesis of tracheal buds. *J. Exp. Zool.* **175:** 455.

Wu J.E. and Santoro S.A. 1996. Differential expression of integrin α subunits supports distinct roles during lung branching morphogenesis. *Dev. Dyn.* **206**: 169.

Yamasaki M., Miyake A., Tagashira S., and Itoh N. 1996. Structure and expression of the rat mRNA encoding a novel member of the fibroblast growth factor family. *J. Biol. Chem.* **271:** 15918.

The Enamel Knot: A Putative Signaling Center Regulating Tooth Development

I. THESLEFF AND J. JERNVALL
Institute of Biotechnology, FIN-00014 University of Helsinki, Finland

Since the discovery by Spemann and Mangold (1924) of the organizing activity of the blastopore lip, a number of other organizing tissues have been identified in vertebrate embryos. The activity of the blastopore lip was found by transplanting pieces of embryonic tissues to heterotopic locations, and similar methodology was used thereafter by experimental embryologists who revealed organizing activity in other tissues. For instance, the notochord was shown to regulate the formation of somites and the patterning of the neural tube (Watterson et al. 1954), and in the limb bud, the zone of polarizing activity (ZPA) was discovered, which regulated anteroposterior (A/P) patterning of the digits (Tickle et al. 1975). More recently, genetic experiments in *Drosophila* have demonstrated the existence of a number of organizing tissues in the fly as well (Serrano and O'Farrell 1997). An increasing number of signaling tissues are being uncovered, notably in the mouse, with molecular approaches by localizing signal molecules and by creating transgenic mice. For instance, in the limbs, A/P, proximodistal (P/D), and dorsoventral (D/V) pattern is now known to be regulated by an interplay of signals from three different signaling tissues, the apical ectodermal ridge (AER), the ZPA, and the dorsal ectoderm (Tickle 1995).

One of the most intriguing discoveries in the field has been how conserved, or similar, the molecular basis of the organizing activities really are. It has become clear, and the evidence is constantly increasing, that the same signal molecules are used for patterning in a wide variety of tissues and across different animal groups. A remarkable breakthrough was the demonstration that Sonic hedgehog (Shh) mimicks the organizing activity in the notochord and ZPA, thus revealing the common molecular basis in the patterning of the neural tube and the limbs (Echelard et al. 1993; Riddle et al. 1993). Subsequently, in addition to Shh, signals belonging to the wnt/wg, FGF, and BMP/DPP families have been shown to mediate the patterning effects in practically all signaling tissues (Liem et al. 1995; Parr and McMahon 1995; Crossley et al. 1996). In this paper, we present our recent observations suggesting that tooth shape may be regulated by signaling or organizing centers present in transient epithelial structures, called the enamel knots.

ANATOMY OF TOOTH MORPHOGENESIS

Teeth are epithelial appendages found in the oral region of vertebrates. They are typical examples of organs that develop from epithelial and mesenchymal tissue components and in which sequential and reciprocal interactions between the two tissues constitute a central regulatory mechanism guiding morphogenesis. Teeth develop from stomodeal or pharyngeal epithelium and the underlying neural-crest-derived mesenchymal cells; their early evolution is believed to be associated with the appearance of the neural crest (Smith and Hall 1993). The initial morphological features of tooth morphogenesis closely resemble those of skin derivatives such as hair, feathers, and scales (for review, see Thesleff et al. 1995).

The first morphological sign of tooth development is a thickening of the oral epithelium, which subsequently buds to the underlying mesenchyme, and the mesenchymal cells condense around the bud. In mammals, which have evolved the highest diversity of tooth shapes, the epithelium undergoes folding morphogenesis that results in the establishment of the form of the tooth crown during the following cap and bell stages (Fig. 1). The mesenchyme, which becomes surrounded by the dental epithelium during the cap stage, forms the dental papilla, giving rise to tooth pulp and the odontoblasts. The more peripheral cells of the condensed dental mesenchyme form the dental follicle and extend around the epithelial dental component, the enamel organ. The terminal differentiation of the dentine-forming odontoblasts and the enamel-forming ameloblasts takes place during the bell stage at the interface of the epithelium and mesenchyme. Dentin resembles bone in many respects, whereas enamel matrix is composed of unique proteins that direct the mineralization of the enamel into the hardest tissue in the body. After crown morphogenesis, the roots of the teeth develop, and subsequently, the teeth erupt into the oral cavity (Fig. 1) (for more details, see Ten Cate 1994). Hence, tooth crown is fully established before its function

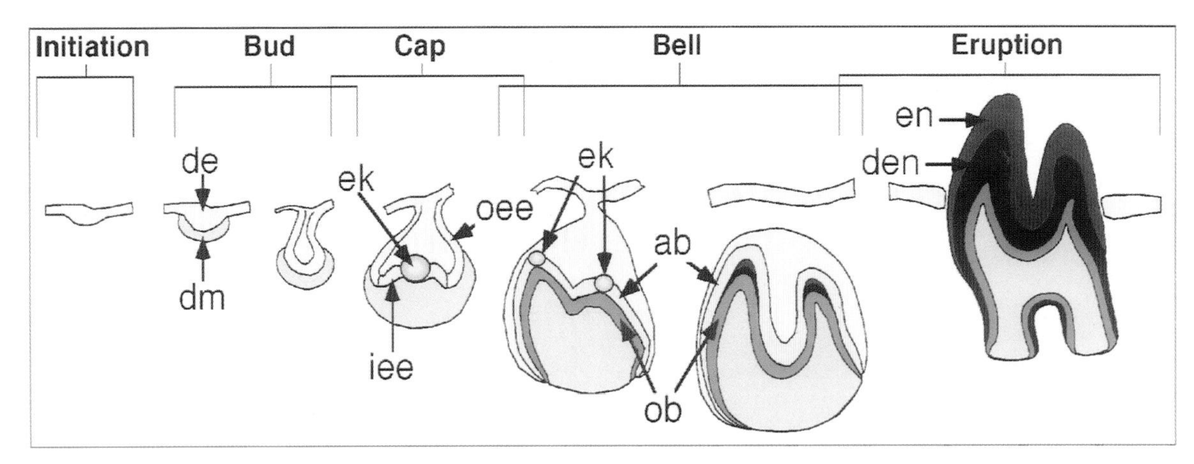

Figure 1. Transverse representation of the main stages of tooth development. The inner enamel epithelium will give rise to the enamel-forming ameloblasts and the underlying mesenchyme will differentiate into the dentine-forming odontoblasts. Tooth shape is a result of the unequal growth of the inner enamel epithelium. (ab) Ameloblasts; (de) dental epithelium; (den) dentine; (dm) dental mesenchyme; (en) enamel; (ek) enamel knot; (iee) inner enamel epithelium; (ob) odontoblasts; (oee) outer enamel epithelium.

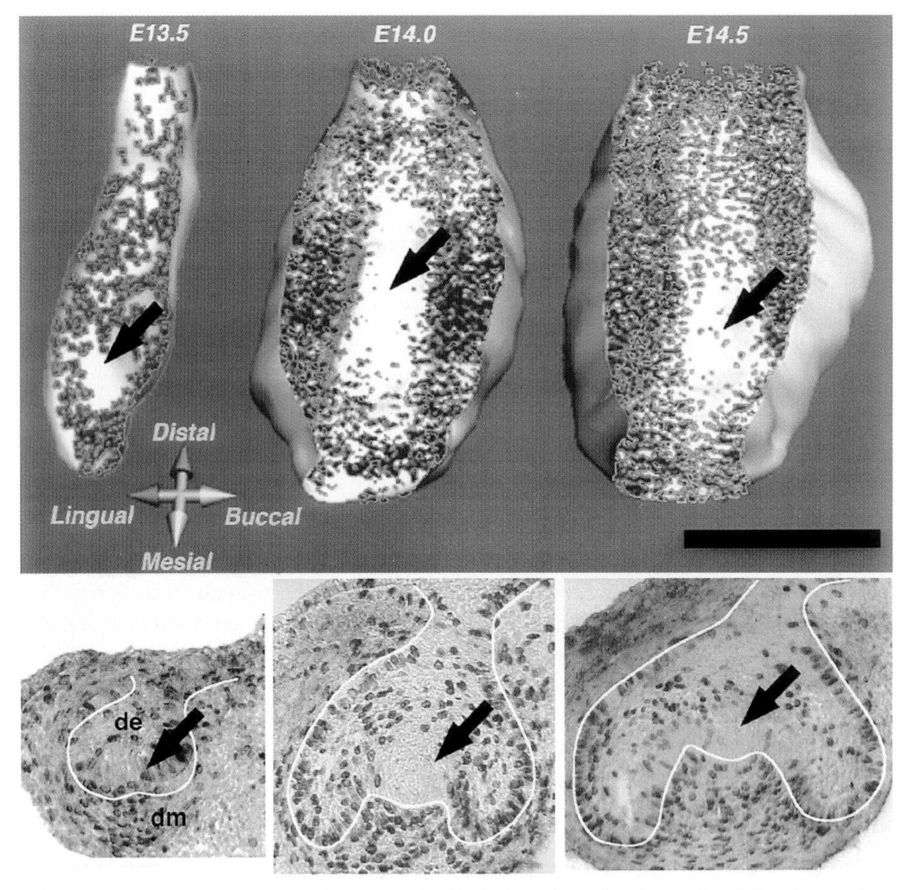

Figure 2. Three-dimensional reconstructions and transverse histological sections showing cell proliferation, as detected by BrdU labeling, around the mouse molar enamel knot (E13.5–14.5). The enamel knot (*arrows*) begins to form in the mesial part of the tooth germ and progresses distally (E13.5–14). The removal of the knot starts from the distal end of the tooth germ (E14.5). Bar, 100 μm. (For methods, see Jernvall et al. 1994; Vaahtokari et al. 1996b.)

(eruption and occlusion), and thus external factors (e.g., nutrition and physical activity) influence very little developing tooth shape.

LIFE HISTORY OF THE PRIMARY ENAMEL KNOT

The enamel knots were observed more than a century ago and thereafter described in many histological studies (see, e.g., Ohlin 1896; Ahrens 1913; Gaunt 1955; Butler 1956). They were described as cell clusters in the dental epithelium located in the center of cap-stage tooth germs, and in some studies, it was speculated that they function in regulating the folding of the dental epithelium and thus the morphogenesis of tooth cusps (see Butler 1956). However, more recently it was thought that they functioned as a reservoir for cells to be incorporated into the growing epithelium. Enamel knots were found in the incisors and molars of several mammalian species, including the mouse, and more recently they were described also in crocodilians (Westergaard and Ferguson 1987). However, the enamel knots have been largely neglected during the last decades.

We "rediscovered" the enamel knot when analyzing cell proliferation patterns in the first mandibular molar tooth germs in mouse embryos. Cells incorporating bromodeoxyuridine (BrdU) were localized by immunohistochemistry from serial sections, and the distributions were analyzed from three-dimensional computer-based reconstructions. A striking lack of cell proliferation was evident in the enamel knot cells, whereas intense cell division took place at all other locations (Fig. 2) (Jernvall et al. 1994).

Subsequent analysis has shown that the enamel knots appear at late bud stage just before morphogenesis of the bud into a cap. This transition from the bud to cap stage involves the invagination of the undersurface of the epithelial bud and the formation of the dental papilla in the adjacent mesenchyme. A few cells at the tips of the tooth buds stop proliferating, as seen by the lack of BrdU incorporation (Fig. 2) (Jernvall 1995). The enamel knot grows as the bud transforms to a cap, at which stage it is readily visible in histological sections as a tightly packed epithelial cell aggregate. It is a torpedo-shaped structure, the development of which starts in the mesial part of the tooth germ and progresses distally (Jernvall et al. 1994).

Interestingly, the enamel knot cells undergo apoptosis (Vaahtokari et al. 1996a). Apoptotic cells, as detected by Tunel staining, are abundant in the cap-stage enamel knot. The disappearance starts 24 hours after the first appearance of the knot and, interestingly, it starts in its distal end (Jernvall et al. 1997). The removal therefore occurs in the opposite direction from which it is formed. Consequently, the last traces of the enamel knot can be seen at the end of the cap stage.

An indication that the enamel knot represents a specific cell lineage with a unique fate is the expression of the cyclin-dependent kinase inhibitor *p21*. Expression of *p21* precedes slightly the exit of the first enamel knot cells from the cell cycle at late bud stage and thereafter correlates closely with the nondividing enamel knot cells and disappears with the enamel knot (Jernvall et al. 1997). p21 is known to inhibit cell division at the G_1/S transition, and it is associated with terminal differentiation of many cell types (Parker et al. 1995; Harper and Elledge 1996). Interestingly, *p21* is also expressed in the AER in the limb buds (Harper and Elledge 1996). It is possible that p21 is associated with apoptosis of the enamel knot cells, as it does in some other cell types, but this is not known.

SIGNALING ACTIVITIES IN THE ENAMEL KNOT

The first indication that the enamel knot may have a signaling function came from in situ hybridization studies in which *Fgf-4* gene expression was localized in cap-stage tooth germs exclusively in the enamel knot (Fig. 3) (Niswander and Martin 1992; Jernvall et al. 1994). Subsequently, we detected the expression of several other signals in the enamel knot. These include *Bmp-2, -4,* and *-7,* and *Shh* (Vaahtokari et al. 1996b), as well as *Fgf-9* (Kettunen and Thesleff 1998).

The analysis of three-dimensional reconstructions of serial sections after in situ hybridization showed that the different signals are expressed in nested patterns and that there is variation in the onset and termination of expression (Fig. 3). *Fgf-4* expression correlates well with the nonmitotic cells, although the expression domain is a few cell layers more restricted and starts slightly later (Jernvall et al. 1994). *Shh, Bmp-2,* and *Bmp-7* are present in budding epithelium before the exit of the distal tip cells from the cell cycle. *Bmp-4,* on the other hand, begins to be expressed during advanced cap stage, starting from the distal region of the knot (Jernvall et al. 1997).

The expression of several important morphogenetic signals specifically in the enamel knot indicates to us that it can be considered a signaling center. However, so far, we have no direct evidence for an organizing capacity of the enamel knot in tooth formation. We have attempted to analyze the function of the enamel knot by transplantation experiments, but they have not been informative because of technical problems in incorporation of the isolated epithelial enamel knot to heterotopic locations in tooth epithelium. An organizing role of the enamel knot, however, is supported by the fact that expression of the same signals is accumulated in well-known organizing centers in the embryo. *Shh* is expressed in the developing limb in the ZPA, which regulates A/P patterning (Riddle et al. 1993), and *Fgf-4* is expressed in the AER, which controls P/D growth in the limb and interacts with the ZPA (Niswander and Martin 1992). In addition, both ZPA and AER express the same *Bmps* as the enamel knot (Tickle 1995). The notochord, which regulates patterning of the neural tube and somites, also expresses *Shh,* and the ectoderm overlying the neural tube, which participates in this patterning, expresses *Bmp-4* and *Bmp-7* (Echelard et al. 1993; Fan and Tessier-Lavigne 1994; Liem et al. 1995).

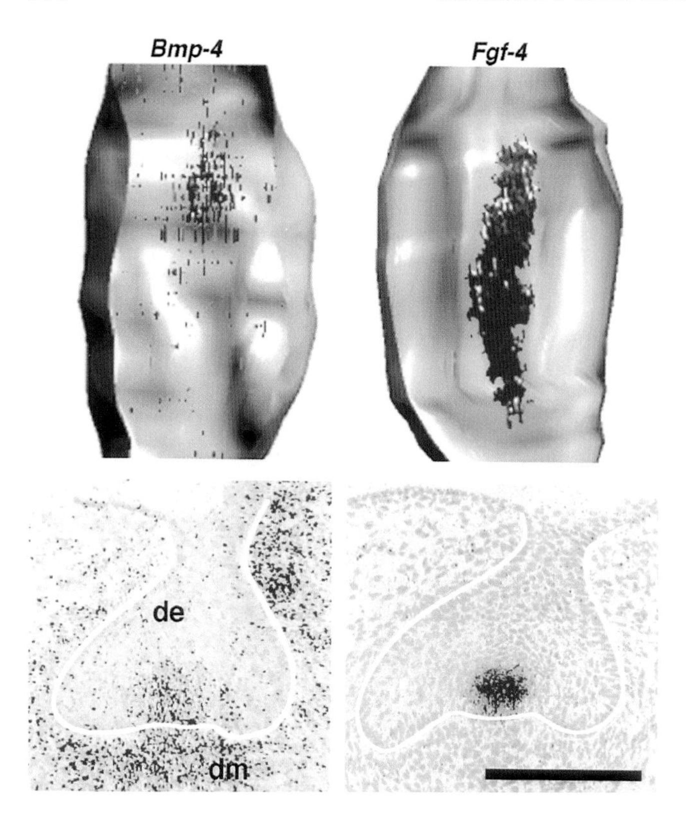

Figure 3. Three-dimensional reconstructions and transverse histological sections showing expression domains of *Bmp-4* and *Fgf-4* in E14 mouse tooth germ. The three-dimensional reconstructions show the tooth germ from above (mesial side toward the bottom). Note how *Bmp-4* transcripts are present only in the distal end of the enamel knot, whereas *Fgf-4* transcripts extend the full length (cf. Fig. 2). Bar, 100 μm. (For methods, see Jernvall et al. 1994; Vaahtokari et al. 1996b.)

ROLES OF THE DIFFERENT ENAMEL KNOT SIGNALS

It is noteworthy that all of the signals that we have localized in the enamel knot, except *Fgf-4,* are expressed also at other locations during various phases of tooth development. Hence, their functions in tooth development are not restricted to the enamel knot. This is not surprising and fits with the current concept on the signaling cascades and networks. It has become apparent that the same signals, in addition to being used in many independent morphogenetic processes, are also used in successive stages of a given morphogenetic process (Davidson 1993).

When discussing the modes of action of the various signals expressed in the enamel knot, the extensive growth that takes place in the tooth germ at the time of enamel knot activity should be borne in mind. In general, morphogenesis is accompanied by rapid growth (Serrano and O'Farrell 1997). This is the case also in the developing tooth, particularly at the time of the development of tooth form. The increase in the size of the mouse molar tooth germ from the bud to the advanced bell stage when the shape of the crown is established is about fourfold. Hence, tooth morphogenesis occurs in the context of rapid growth (except in the enamel knots and the sheets of terminally differentiated odontoblasts and ameloblasts that have left the cell cycle), and locally regulated differences in the rates of cell proliferation may provide a central mechanism in the regulation of tooth morphology.

It follows that the signals expressed in the enamel knots presumably affect both pattern formation and growth. In *Drosophila* and in vertebrates, many of the signals that direct pattern formation also regulate growth. Shh/hh and BMP/Dpp have well-established roles in patterning in the flies and vertebrates, and in addition, they regulate growth (Serrano and O'Farrell 1997). Dpp mediates the Shh signal in patterning of the *Drosophila* limb and eye. Different concentrations of hh specify distinct cell types in *Drosophila* (Heemskerk and DiNardo 1994) and there is similar evidence for the Shh in notochord (Ericson et al. 1996). In addition, hh stimulates proliferation in follicle somatic cells in flies and Dpp stimulates proliferation in wing discs (Serrano and O'Farrell 1997). Fibroblast growth factors (FGFs) also appear to link pattern and growth. FGF stimulates proliferation in a wide variety of cells, including mesenchymal cells under the AER in the limb buds. In the limbs, the signals for the P/D growth and A/P patterning are coupled, and FGFs act in regulatory feedback loops with Shh, and they also regulate homeobox genes, including *Msx* and *Evx* genes in the limb mesenchyme. In addition, *Hox* genes have been suggested to regulate rates of cell proliferation (Dollé et al. 1993).

FGF-4 and FGF-9 in the Enamel Knot

Several of the FGF family members have been associated with tooth morphogenesis and, of these, *Fgf-4* and *Fgf-9* are expressed in the enamel knots (Jernvall et al. 1994; Kettunen and Thesleff 1998). FGF-4 and FGF-9 may regulate the growth of the cusps by stimulation of cell proliferation (Fig. 4). This function can be particularly attributed to FGF-4 because it shows restricted expression also to the secondary enamel knots (see below).

This function would be analogous to the limb where FGFs promote the P/D growth. FGFs may act on both epithelial and mesenchymal cells, as their respective tyrosine kinase receptors are present in both epithelial and mesenchymal tissues in the tooth (Peters et al. 1992; P. Kettunen et al., unpubl.), and FGF-4 has been shown to stimulate the proliferation of both cell types in cap-stage teeth (Jernvall et al. 1994).

FGFs may also have regulatory functions in patterning and cell differentiation. FGF-4 and FGF-9 up-regulate *Msx-1* expression in the dental mesenchyme when applied with heparin acrylic beads in vitro (Chen et al. 1996; Kettunen and Thesleff 1998). Another putative function for FGFs is prevention of untimely apoptosis in the enamel knot cells. FGF-4 prevents apoptosis in both the dental mesenchyme and epithelium (Vaahtokari et al. 1996a; Jernvall et al. 1997).

The biological effects of various FGFs are largely similar in different in vitro assays. Also in dental tissues, FGF-4 and FGF-9 affect cell proliferation and *Msx1* gene expression similarly (Kettunen and Thesleff 1998). Hence, it is possible that there is functional redundancy between FGF-4 and FGF-9 in the enamel knot. On the other hand, FGF-9 may have functions not shared by FGF-4 in the dental epithelium at bell stage. Its expression, like that of *Shh* as well as *Bmp-2* and *Bmp-7,* spreads from the primary enamel knot and subsequently covers the coronal part of the dental epithelium, including the secondary enamel knots (see below). There is an apparent correlation between this expression and the differentiation of the underlying mesenchymal cells into odontoblasts.

Sonic Hedgehog and BMP-2, BMP-4, and BMP-7 in the Enamel Knot

Shh is expressed in dental epithelium at several stages starting in the early epithelial thickenings and then reappearing in the enamel knot; expression subsequently spreads along the dental epithelium, correlating closely with *Fgf-9* as well as *Bmp-2* and *Bmp-7* (Bitgood and McMahon 1995; Iseki et al. 1996; Vaahtokari et al. 1996b). *Patched*, the Shh receptor, is expressed in the dental papilla mesenchyme, suggesting that the mesenchyme may be the target for the action of Shh (I. Thesleff and J. Jernvall, unpubl.). In *Drosophila*, hedgehog acts as a morphogen, and different concentrations specify distinct cell types (Heemskerk and DiNardo 1994).

BMPs constitute one of the signal families that has multiple signaling functions throughout the animal kingdom (Hogan 1996). The *Bmps* shift between the epithelial and mesenchymal tissues, and they apparently signal in both directions between the epithelium and mesenchyme. In addition, they are associated with the differentiation and secretory activities of odontoblasts and ameloblasts (Begue-Kirn et al. 1992; Åberg et al. 1997). Several of the serine-threonine kinase receptors for BMPs are expressed during morphogenesis, but the patterns do not appear to be as clearly spatiotemporally restricted as those of the ligands (Verschueren et al. 1995; T. Åberg et al., unpubl.). The expression patterns of *Bmps* are largely overlapping, suggesting that there may be functional redundancy between the coexpressed genes. This possibility has been supported by many in vitro studies. In our studies, BMP-2 and BMP-4 had similar effects

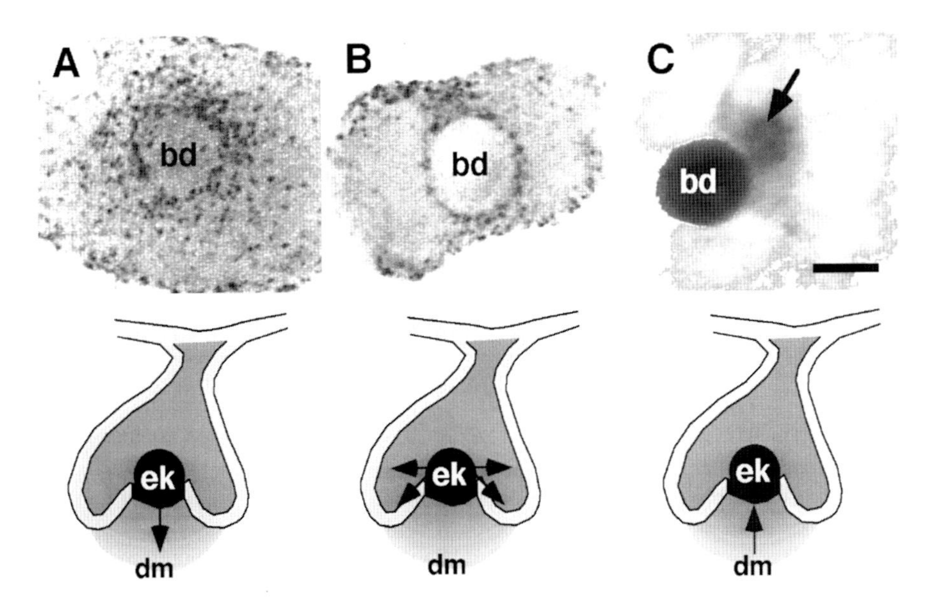

Figure 4. Stimulation of cell proliferation by FGF-4 in isolated dental mesenchyme (*A*) and epithelium (*B*), and the stimulation of *p21* expression in isolated dental epithelium by BMP-4 (*C*). The effect of FGF-4-releasing beads on tooth epithelium and mesenchyme in vitro is detected using BrdU incorporation, and the effect of BMP-4-releasing beads on *p21* expression is detected using whole-mount in situ hybridization. The illustrations below show the inferred signal direction in vivo. Bar, 100 μm. (ek) Enamel knot; (bd) bead. (For methods, see Jernvall et al. 1994, 1997.)

on dental epithelium and mesenchyme (Vainio et al. 1993).

BMP-4 appears to have different functions as compared with BMP-2 and BMP-7. Its expression in tooth development correlates with the odontogenic potential, first in the epithelium and thereafter in mesenchyme. There is evidence that it acts as an epithelial signal inducing the dental mesenchyme and thereafter as a mesenchymal signal inducing the enamel knot (Fig. 4) (Vainio et al. 1993; Jernvall et al. 1997). In the enamel knot, *Bmp-4* is expressed later than the other *Bmps*, and its expression pattern shows striking correlation with apoptosis (see Fig. 3) (Jernvall et al. 1997). As BMP-4 regulates programmed cell death during embryonic development in the rhombomeres and in the interdigital mesenchyme (Graham et al. 1994; Zou and Niswander 1996), it is possible that this signal may have a similar function in the enamel knot (Jernvall et al. 1997).

Dpp, the *Drosophila* homolog of BMP-2 and BMP-4, mediates some of the effects of hh in the patterning of the wing and eye; in the vertebrate limb bud, BMP-2 mediates the action of Shh in the ZPA. The close correlation between *Shh* and *Bmp-2* and *Bmp-7* in the dental epithelium, including the enamel knots, supports similar relationships between Shh and BMPs (Åberg et al. 1997). Complex regulatory networks between *Shh, Bmp*s, and *Fgf*s exist within and between various organizing tissues; it is conceivable that these three families of signals interact also in the enamel knot. Their analysis in the tooth is difficult because all signals are expressed in the same signaling center, unlike the situation in the patterning of the limb and neural tube. *Shh* expression and the expression of *Bmp*s have been localized to budding epithelium during hair and feather development and in the lungs (Nohno et al. 1995; Tingberreth and Chuong 1996; Bellusci et al. 1997), suggesting conserved actions during epithelial appendage development. However, no morphologically distinct epithelial structures resembling the enamel knot have been described in these epithelial appendages.

PRIMARY AND SECONDARY ENAMEL KNOTS

After the apoptotic removal of the enamel knot during late cap stage, the morphogenesis proceeds to the bell stage when the cusps develop (see Fig. 1). Interestingly, in the molar teeth of the mouse, new enamel knots appear at the sites of tooth cusp formation (Fig. 5). To differentiate the two types of enamel knots, we call them primary and secondary enamel knots. Like the primary enamel knots, the secondary enamel knots comprise packed epithelial cells that do not incorporate BrdU and that express locally *Fgf-4* (Jernvall et al. 1994). However, they do not show similar restricted expression of the other signaling molecules present in the primary enamel knot. *Shh, Bmp-2, Bmp-7,* and *Fgf-9* expression starts to spread from the primary enaml knot before its disappearance to the neighboring dental epithelium; subsequently, their expression covers the entire coronal area of the inner dental epithelium, including the secondary enamel knots. Hence, the gene expression patterns suggest that the primary and secondary enamel knots have some common functions but that they also are functionally different.

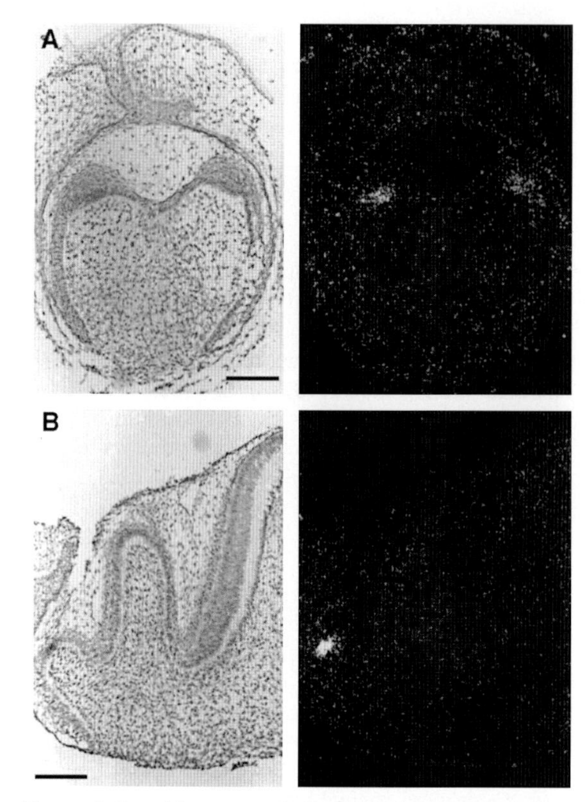

Figure 5. Day-18 mouse embryonic first lower molar (*A*), and 7-day postnatal second lower molar (*B*) showing *Fgf-4* expression (dark field images on the right) in the secondary enamel knots. Bar, 100 μm.

On the basis of the molecular differences between primary and secondary enamel knots, we have proposed that in multicusped teeth, the primary enamel knot regulates the spatial and temporal activation of the secondary enamel knots (Vaahtokari et al. 1996b). The timing of initiation of cusp formation determines its relative height as compared with other cusps and therefore a cusp that starts to form first will become the highest, and the subsequently forming cusps will end up being progressively shallower (see, e.g., Jernvall 1995). Hence, by determining both spatial and temporal activation of the secondary enamel knots, the primary enamel knot would be involved in the regulation of tooth shape (Fig. 6). The secondary enamel knots (which express only *Fgf-4* in a restricted manner) would in turn promote the growth of individual tooth cusps.

Apparently, the primary enamel knot is present in all types of teeth, including homodont teeth, as shown in crocodilians (Westergaard and Ferguson 1987) and different tooth families in mammals. For example, the cells of the primary enamel knot in mouse incisor do not proliferate and they express *Fgf-4* (Jernvall et al. 1994), as well as other signal molecules (T. Åberg et al., unpubl.). The incisors, however, do not have multiple cusps like the molars, and no secondary enamel knots appear in the incisor tooth germs.

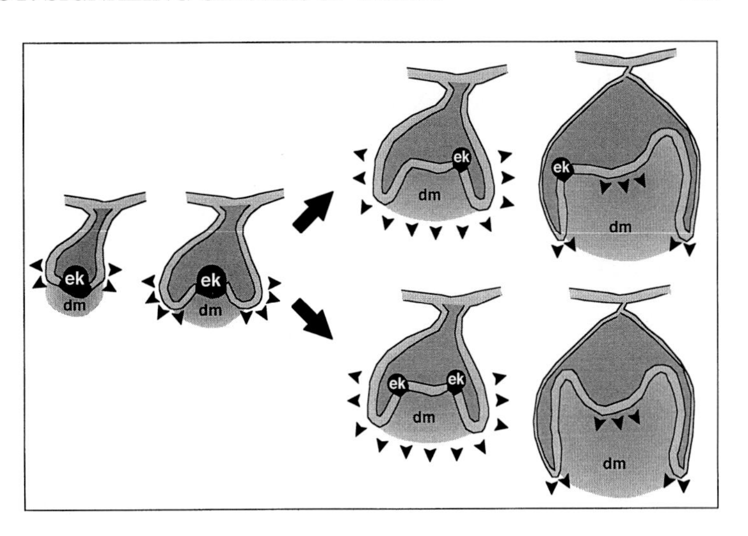

Figure 6. Generalized illustration (transverse sections) of the possible role of enamel knot in the development of tooth crown form. Nonproliferative enamel knot (ek) shown black. Arrowheads show the direction of growth. At cap stage, tooth epithelium grows lateral to the primary enamel knot and around dental mesenchyme (dm). The final development of individual cusps in multicusped teeth corresponds to the secondary enamel knots at the cusp tips and unequal growth of the inner enamel epithelium (epithelium facing the dm). If there is a temporal gap between the initiation of secondary enamel knots, the resulting cusps are of unequal height (*top*). If the secondary enamel knots appear at the same time, the cusps are equal in height.

EPITHELIAL-MESENCHYMAL INTERACTIONS AND THE INDUCTION OF THE ENAMEL KNOT

Tooth morphogenesis, like that of all other epithelial appendages, is regulated by a sequence of reciprocal interactions between the epithelial and mesenchymal tissue components. This has been demonstrated in classic tissue recombination experiments. The presumptive dental epithelium governs tooth morphogenesis before the bud stage, but this potential to instruct tooth morphogenesis shifts to the dental mesenchyme during budding of the epithelium (Mina and Kollar 1987; Lumsden 1988). Subsequently, during cap and bell stage, the mesenchymal dental papilla can regulate development when combined with nondental epithelium. The mesenchyme also regulates tooth shape development, i.e., a molar tooth develops when mouse molar mesenchyme is cultured with mouse incisor epithelium, and vice versa (Kollar and Baird 1969, 1970). Furthermore, the dental mesenchyme is able to instruct the differentiation of ameloblasts and enamel secretion in nondental epithelium (Kollar and Baird 1970). Hence, in addition to morphogenesis, cell differentiation is regulated by epithelial-mesenchymal interactions.

Because the primary enamel knots appear during late bud stage, when the instructive signaling capacity resides in the dental mesenchyme, it is reasonable to assume that the dental mesenchyme is involved in the regulation of enamel knot formation. Our recent experiments have given experimental support to this assumption. We have shown in organ culture studies that the expression of two markers of the enamel knot, the cyclin-dependent kinase inhibitor *p21* and *Msx-2,* is regulated in isolated dental epithelium by dental mesenchyme and that this inductive capacity is mimicked by beads releasing BMP-4 protein (see Fig. 4C) (Jernvall et al. 1997).

Other aspects of BMP-4 also have been associated with the epithelial-mesenchymal interactions in early tooth development. *Bmp-4* expression shifts from epithelium to mesenchyme during budding and it mimics the effect of the early dental epithelium on mesenchyme by inducing the expression of the homeobox genes *Msx-1* and *Msx-2* (Vainio et al. 1993). Furthermore, *Bmp-4* is expressed intensely in the dental mesenchyme during the bud stage and might be involved in the induction of enamel knot formation (Jernvall et al. 1997). Evidence for this also comes from studies on tissues of *Msx-1*-deficient mouse embryos. Tooth development is arrested at late bud stage when *Msx-1* is knocked out in transgenic mice (Satokata and Maas 1994). In mutant teeth, *Bmp-4* expression in mesenchyme is not up-regulated as in wild-type animals, indicating that Msx-1 acts also upstream of BMP signaling (Chen et al. 1996). When the mutant teeth were exposed to BMP-4 protein in culture, their phenotype was partly restored and some of the teeth reached the cap stage. Hence, in these cultured teeth, BMP-4 apparently stimulated the formation of an enamel knot.

In addition to *Msx-1, Lef-1*, an HMG domain transcription factor associated with Wnt signaling, is absolutely required for tooth morphogenesis, as transgenic null mutant mice lack teeth (Satokata and Maas 1994; van Genderen et al. 1994). In addition, in *Lef-1* knockout mice, tooth morphogenesis is arrested at the bud stage. Tissue recombination experiments between mutant and wild-type tissues demonstrated that Lef-1 was needed early in the dental epithelium and that it presumably regulates the expression of a signal acting reciprocally on the dental mesenchyme (Kratochwil et al. 1996). Hence, the arrest in the development in both cases of knockout teeth may be associated with the lack of the formation of the enamel knot (Jernvall et al. 1997). Interestingly, vestigial teeth present in the diastema region in mouse embryos are also arrested at the bud stage (Turesková et al. 1995). Taken together, the induction of the enamel knot may be a key event in tooth morphogenesis, and, because both Lef-1 and Msx-1 are needed for its induction, it is possible that proper enamel knot formation and maintenance require the coordination of several signaling pathways.

REGULATION OF TOOTH SHAPE AND THE INDUCTION OF THE SECONDARY ENAMEL KNOTS

Mammalian teeth are sequentially arranged structures, and the different tooth groups (incisors, canines, premolars, and molars) show characteristic differences in morphology as do individual tooth types (e.g., molars) among different species. It is interesting that there is currently a dearth of theories to explain the developmental mechanisms behind the species-specific tooth shape (e.g., molar tooth of mice and humans), whereas theories to explain regulation of tooth identity are plentiful (Jernvall 1995). Theories to explain the segmental patterning and shape differences between the tooth groups can be broadly divided into two groups. The field, or gradient, theories assume that concentrations of chemical morphogens regulate different morphogenesis of initially identical primordia (see Lumsden 1979; Butler 1995 and references therein). According to the second group of models, the so-called clone models, the stem cells giving rise to different classes of teeth differ from each other initially (Osborn 1973, 1978). Recent studies on neural crest have given support to the clone models.

Studies on neural crest cell migration have shown that the tooth-bearing parts of the mandibular and maxillary processes are colonized by neural crest cells from the midbrain region (Imai et al. 1996; Köntges and Lumsden 1996). Furthermore, the final position of the cells in the maxillary and mandibular processes is determined by the original position of the cells in the neural crest as well as by the time the cells leave the crest (Imai et al. 1996). The neural crest cells have been shown to maintain their patterns of *Hox* gene expression during migration; hence, it is possible that this array of expressed genes, which has been called the *Hox* code (Hunt et al. 1991), determines their identity also in tooth development. This kind of combinatorial code has been suggested to comprise in developing teeth several homeobox-containing transcription factors, including *Dlx* genes in the jaw mesenchyme, which would specify the identity of teeth (incisor, canine, premolar, or molar; see Sharpe 1995; Weiss et al. 1995). However, it is also possible that the dental identity of the migrating cells is further regulated by the environment in which they migrate. The cells have been shown to migrate as a sheet along a subectodermal pathway as far as the distal aspect of the branchial arches (Trainor and Tam 1995).

One gene has been actually identified that shows tooth-type-specific expression. The homeobox gene, *Barx1*, is present in mouse molar mesenchyme and absent from incisors during morphogenesis (Tissier-Seta et al. 1995). Evidence for the "homeobox code" model has also recently been obtained from studies of tooth development in transgenic mice with deficient function of both *Dlx-1* and *Dlx-2* genes (Qiu et al. 1997). The development of maxillary molars is arrested already before bud stage, whereas mandibular molars and incisors develop normally. This implies that development of teeth in different regions of the jaws involves independent genetic control, but the *Dlx* genes apparently do not account (at least they are not the only determining genes) for the difference between the incisor and molar morphology.

Although the sequential and reciprocal epithelial mesenchymal interactions have been analyzed extensively and resolved at the molecular level in tooth development (Thesleff and Nieminen 1996), the specific functions of the different tissues in the determination of individual tooth shapes are still unclear. The mechanism whereby the site of the primary enamel knot in the buccolingual and A/P dimensions is determined is not understood, but recent evidence from *Drosophila* and chick organizing centers may provide clues for the spatial regulation of the enamel knots.

In *Drosophila*, genetic experiments have revealed that organizing centers are formed at boundaries of territories that express different regulatory genes (Serrano and O'Farrell 1997). Of these, Fringe and engrailed were recently shown to be involved in the regulation of AER formation in the chick limb buds, thus suggesting that the mechanism of the determination of organizing tissues is conserved between flies and vertebrates (Rodriguez-Esteban et al. 1997). It is also possible that the location of the enamel knot in the tip of the epithelial tooth bud is determined by similar mechanisms. In fact, *Bmp-4* expression in the mesenchyme and *Msx-2* expression in the epithelium show clear buccal localization with a boundary in the area of the enamel knot (MacKenzie et al. 1992; Åberg et al. 1997). It will be interesting to see which other genes show lingual or buccal expression and whether similar signaling interactions are involved in the determination of the site of the enamel knot, as they are in other organizing tissues.

How then are the sites of the secondary enamel knots determined in the multicusped teeth? It is possible that they also form at boundaries of gene expression. The elucidation of this will require careful analysis of gene expression patterns. We do not know if our hypothesis of the role of the primary enamel knot in the regulation of the patterning of the secondary knot holds true. This theory appears to be in contradiction with the prevailing concept that the mesenchymal dental papilla determines the shape of teeth. The tissue recombination studies have clearly shown that the dental papilla mesenchyme has an instructive role in regulating the shape of teeth. During cap stage, the molar dental papilla directs molar tooth development when associated with incisor enamel organ epithelium, and vice versa (Kollar and Baird 1969, 1970). However, because these experiments were done using tissues from the same species (mouse), it is not possible to identify whether the mesenchyme is instructing the tooth type only (incisor/molar) or also the individual species-specific tooth cusp patterns. Therefore, it is possible that in these recombinants, the dental papilla induces in the heterotypic epithelium the formation of a primary enamel knot that is tooth-type-specific and that subsequently directs the formation of secondary enamel knots and cuspal pattern inherent to the dental epithelium. Alternatively, it is possible that the mesenchymal dental papilla alone, without epithelial influence, is capable of determining the temporospatial patterns of secondary enamel knots.

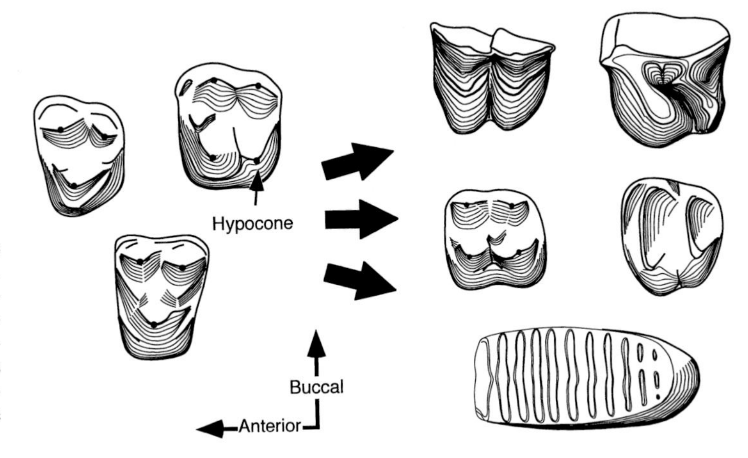

Figure 7. Upper molars of dentally primitive mammals (*left*) and advanced mammals (*right*). Acquisition of the fourth cusp, the hypocone, enabled the evolution of complicated crown patterns composed of cusps and crests. The advanced molars are of a deer and a rhinoceros (*top*), of a pig and a tapir (*middle*), and of a mammoth (*bottom*). Examples are not to the same scale.

Hence, it still remains to be demonstrated how the mesenchymal dental papilla and the epithelial enamel knot interact in the regulation of the secondary enamel knots and subsequently the cusp pattern of multicusped teeth.

EVOLUTIONARY IMPLICATIONS OF THE ENAMEL KNOTS

The basic developmental anatomy of tooth morphogenesis has been conserved to a high degree, and the morphological steps of early tooth development are very similar in all vertebrates. There are, however, a variety of modifications in the ways in which the dentitions are organized and in which they function in different animals. The numbers of teeth vary, and they come in a variety of forms that serve special functions in catching, chopping, and chewing food. In addition to the number of teeth, there are major differences in the shapes of teeth between different mammalian species. In fact, the evolution of mammals has been best documented in the teeth. It is obvious that because of the good fossil record, teeth offer special advantages compared with other organs in evolutionary studies.

The shapes of teeth reflect the diets of the animals (see, e.g., Fortelius 1985; Jernvall 1995). In particular, evolutionary changes in the size and location of the cusps and crests in the premolars and molars are correlated with changes in the diet of mammals (Fig. 7). For example, the last-forming cusp of the upper molars, the so-called hypocone, has evolved independently more than 20 times (Hunter and Jernvall 1995). As the hypocone effectively doubles the occlusal area devoted to crushing food, its evolution is related to increase in consumption of plants. Thus, the hypocone is observed to have evolved independently at the bases of mammalian radiations leading to specialized herbivory. Moreover, a recent study showed that a concomitant increase in the number of crests (shearing blades) on molars with hypocones has happened independently among northern continents (Jernvall et al. 1996). A high number of crests is correlated with consumption of fibrous plants, and the acquisition of crests is thus related to the shift toward specialized herbivory, ultimately driven by climatic cooling and drying.

Although the evolution of mammalian molar patterns can be causally linked to ecology, the generation of this diversity obviously stems from alterations in the developmental program regulating tooth morphogenesis. Comparative studies may provide new insights into the underlying developmental mechanisms that have allowed the high evolutionary diversity of mammalian molar tooth shapes, while at the same time keeping tooth shapes highly heritable. Comparative molecular embryology can be used to elucidate the molecular basis of morphological diversity. Examples of such approaches are comparisons of the activities in regulatory genes and morphological variation in arthropod appendages and in the vertebrate axial skeleton (Duboule 1994; Panganiban et al. 1995). Teeth provide unique models for such comparative studies because the macroevolution of teeth is well documented and because they offer many examples of morphological change. Therefore, by using suitable extant model animals with different tooth shapes, we can begin to understand the parts played by the various genes that have been associated with the enamel knot and, presumably, tooth shape.

ACKNOWLEDGMENTS

This review is based largely on our studies supported by the Finnish Academy.

REFERENCES

Åberg T., Wozney J., and Thesleff I. 1997. Expression patterns of BMPs in the developing mouse tooth suggests roles in morphogenesis and cell differentiation. *Dev. Dyn.* (in press).

Ahrens K. 1913. Die Entwicklung der menschlichen Zähne. *Arb. Anat. Inst. Wiesbaden* **48:** 169.

Begue-Kirn C., Smith A.J., Ruch J.V., Wozney J.M., Purchio A. Hartmann D., and Lesot H. 1992. Effects of dentin proteins, transforming growth factor β1 (TGFβ1) and bone morphogenetic protein 2 (BMP2) on the differentiation of odontoblast in vitro. *Int. J. Dev. Biol.* **36:** 491.

Bellusci S., Furuta Y., Rush M.G., Henderson R., Winnier G., and Hogan B.L.M. 1997. Involvement of Sonic hedgehog (*Shh*) in mouse embryonic lung growth and morphogenesis. *Development* **124:** 53.

Bitgood M.J. and McMahon A.P. 1995. *Hedgehog* and *Bmp* genes are coexpressed at many diverse sites of cell-cell interaction in the mouse embryo. *Dev. Biol.* **172:** 126.

Butler P.M. 1956. The ontogeny of molar pattern. *Biol. Rev.* **31:** 30.

———. 1995. Ontogenetic aspects of dental evolution. *Int. J. Dev. Biol.* **39:** 25.

Chen Y., Bei M., Woo I., Satokata I., and Maas R. 1996. *Msx1* controls inductive signaling in mammalian tooth morphogenesis. *Development* **122:** 3035.

Crossley P.H., Minowada G., MacArthur C.A., and Martin G.R. 1996. Roles for FGF8 in the induction, initiation, and maintenance of chick limb development. *Cell* **84:** 127.

Davidson E.H. 1993. Later embryogenesis: Regulatory circuitry in morphogenetic fields. *Development* **118:** 665.

Dollé P., Dierich A., LeMeur M., Schimmang T., Schuhbaur B., Chambon P., and Duboule D. 1993. Disruption of the *Hoxd-13* gene induces localized heterochrony leading to mice with neotenic limbs. *Cell* **75:** 431.

Duboule D. 1994. Temporal colinearity and the phylotypic progression: A basis for the stability of a vertebrate Bauplan and the evolution of morphologies through heterochrony. *Development* (suppl.), p. 135.

Echelard Y., Epstein D.J., St-Jacques B., Shen L., Mohler J., McMahon J.A., and McMahon A.P. 1993. Sonic hedgehog, a member of a family of putative signaling molecules, is implicated in the regulation of CNS polarity. *Cell* **75:** 1417.

Ericson J., Morton S., Kawakami A., Roelink H., and Jessell T.M. 1996. Two critical periods of sonic hedgehog signaling required for the specification of motor neuron identity. *Cell* **87:** 661.

Fan C.M. and Tessier-Lavigne M. 1994. Patterning of mammalian somites by surface ectoderm and notochord: Evidence for sclerotome induction by a hedgehog homolog. *Cell* **79:** 1175.

Fortelius M. 1985. Ungulate cheek teeth: Developmental, functional, and evolutionary interrelations. *Acta Zool. Fenn.* **180:** 1.

Gaunt W.A. 1955. The development of the molar pattern of the mouse (*Mus musculus*). *Acta Anat.* **24:** 249.

Graham A., Francis-West P., Brickell P., and Lumsden A. 1994. The signalling molecule BMP4 mediates apoptosis in the rhombencephalic neural crest. *Nature* **372:** 684.

Harper J.W. and Elledge S.J. 1996. Cdk inhibitors in developmental cancer. *Curr. Opin. Genet. Dev.* **6:** 56.

Heemskerk J. and DiNardo S. 1994. *Drosophila hedgehog* acts as morphogen in cellular patterning. *Cell* **76:** 449.

Hogan B.L. 1996. Bone morphogenetic proteins in development. *Curr. Opin. Genet. Dev.* **6:** 432.

Hunt P., Gulisano M., Cook M., Sham M.H., Faiella A., Wilkinson D., Boncinelli E., and Krumlauf R. 1991. A distinct Hox code for the branchial region of the vertebrate head. *Nature* **353:** 861.

Hunter J.P. and Jernvall J. 1995. The hypocone as a key innovation in mammalian evolution. *Proc. Natl. Acad. Sci.* **92:** 10718.

Imai H., Osumi-Yamashita N., Ninomiya Y., and Eto K. 1996. Contribution of early-emigrating midbrain crest cells to the dental mesenchyme of mandibular molar teeth in rat embryos. *Dev. Biol.* **176:** 151.

Iseki S., Araga A., Ohuchi H., Nohno T., Yoshioka H., Hayashi F., and Noji S. 1996. Sonic hedgehog is expressed in epithelial cells during development of whisker, hair, and tooth. *Biochem. Biophys. Res. Commun.* **218:** 688.

Jernvall J. 1995. Mammalian molar cusp patterns: Developmental mechanisms of diversity. *Acta Zool. Fenn.* **198:** 1.

Jernvall J., Hunter J.P., and Fortelius M. 1996. Molar tooth diversity, disparity, and ecology in Cenozoic ungulate radiations. *Science* **274:** 1489.

Jernvall J., Åberg T., Kettunen P., Keränen S., and Thesleff I. 1997. The life history of an embryonic signaling center: BMP-4 induces p21 and is associated with apoptosis in the mouse tooth enamel knot. *Development* (in press).

Jernvall J., Kettunen P., Karavanova I., Martin L.B., and Thesleff I. 1994. Evidence for the role of the enamel knot as a control center in mammalian tooth cusp formation: Non-dividing cells express growth stimulating *Fgf-4 gene. Int. J. Dev. Biol.* **38:** 463.

Kettunen P. and Thesleff I. 1998. Expression and function of FGFs-4, -8 and -9 suggest functional redundancy and repetitive use as epithelial signals during tooth morphogenesis. *Dev. Dyn.* (in press).

Kollar E.J. and Baird G.R. 1969. The influence of the dental papilla on the development of tooth shape in embryonic mouse tooth germs. *J. Embryol. Exp. Morphol.* **21:** 131.

———. 1970. Tissue interactions in embryonic mouse tooth germs. II. The inductive role of the dental papilla. *J. Embryol. Exp. Morphol.* **24:** 173.

Kratochwil K., Dull M., Fariñas I., Galceran J., and Grosschedl R. 1996. *Lef1* expression is activated by BMP-4 and regulates inductive tissue interactions in tooth and hair development. *Genes Dev.* **10:** 1382.

Köntges G. and Lumsden A. 1996. Rhombencephalic neural crest segmentation is preserved throughout craniofacial ontogeny. *Development* **122:** 3229.

Liem K.F., Jr., Tremml G., Roelink H., and Jessell T.M. 1995. Dorsal differentiation of neural plate cells induced by BMP-mediated signals from epidermal ectoderm. *Cell* **82**: 969.

Lumsden A.G. 1979. Pattern formation in the molar denition of the mouse. *J. Biol. Buccale* **7:** 77.

———. 1988. Spatial organization of the epithelium and the role of neural crest cells in the initiation of the mammalian tooth germ. *Development* (suppl.) **103:** 155.

MacKenzie A., Ferguson M.W.J., and Sharpe P.T. 1992. Expression patterns of the homeobox gene, *Hox-8,* in the mouse embryo suggest a role in specifying tooth initiation and shape. *Development* **115:** 403.

Mina M. and Kollar E.J. 1987. The induction of odontogenesis in non-dental mesenchyme combined with early murine mandibular arch epithelium. *Arch. Oral. Biol.* **32:** 123.

Niswander L. and Martin G.R. 1992. *Fgf-4* expression during gastrulation, myogenesis, limb and tooth development in the mouse. *Development* **114:** 755.

Nohno T., Ishikawa T., Saito T., Hosokawa K., Noji S., Wolsing D.H., and Rosenbaum J.S. 1995. Identification of a human type II receptor for bone morphogenetic protein-4 that forms differential heteromeric complexes with bone morphogenetic protein type I receptors. *J. Biol. Chem.* **270:** 22522.

Ohlin A. 1896. Om tandutvecklingen hos Hyperoodon. *K. Sven. Vetenskapsakad. Handl.* (suppl. 4) **22:** 1.

Osborn J.W. 1973. The evolution of denitions. *Am. Sci.* **61:** 548.

———. 1978. Morphogenetic gradients: Fields versus clones. In *Development, function and evolution of teeth* (ed. P.M. Butler et al.), p. 171. Academic Press, New York.

Panganiban G., Sebring A., Nagy L., and Carroll S. 1995. The development of crustacean limbs and the evolution of arthropods (comments). *Science* **270:** 1363.

Parker S.B., Eichele G., Zhang P., Rawls A., Sands A., Bradley A., Olson E.N., Harper J.W., and Elledge S.J. 1995. p53-Independent expression of p21^{Cip1} in muscle and other terminally differentiating cells. *Science* **267:** 1024.

Parr B.A. and McMahon A.P. 1995. Dorsalizing signal *Wnt-7a* required for normal polarity of D-V and A-P axes of mouse limb. *Nature* **374:** 350.

Peters K.G., Werner S., Chen G., and Williams L.T. 1992. Two FGF receptor genes are differentially expressed in epithelial and mesenchymal tissues during limb formation and organogenesis in the mouse. *Development* **114:** 233.

Qiu M., Bulfone A., Ghattas I., Meneses J.J., Sharpe PT., Presley R., Pedersen R.A., and Rubenstein J.L.R. 1997. Role of Dlx-1 and -2 in proximodistal skeletal elements derived from the first and second arches. *Dev. Biol.* **185:** 165.

Riddle R.D., Johnson R.L., Laufer E., and Tabin C. 1993. Sonic hedgehog mediates the polarizing activity of the ZPA. *Cell* **75:** 1401.

Rodriguez-Esteban C., Schwabe J.W.R., De La Pena J., Foys B., Eshelman B., and Izpisúa-Belmonte J.C. 1997. Radical fringe positions the apical ectodermal ridge at the dorsoventral boundary of the vertebrate limb. *Nature* **386:** 360.

Satokata I. and Maas R. 1994. *Msx1* deficient mice exhibit cleft palate and abnormalities of craniofacial and tooth development. *Nat. Genet.* **6:** 348.

Serrano N. and O'Farrell H.O. 1997. Limb morphogenesis: Connections between patterning and growth. *Curr. Biol.* **7:** R186.

Sharpe P.T. 1995. Homeobox genes and orofacial development. *Connect. Tissue Res.* **32:** 17.

Smith M.M. and Hall B.K. 1993. A developmental model for evolution of the vertebrate exoskeleton and teeth: The role of cranial and trunk neural crest. *Evol. Biol.* **27:** 387.

Spemann H. and Mangold H. 1924. Uber Induktion von Embryonanlagen durch Implantation artfremder Organisatoren. *Roux's Arch Entw. Mech.* **100:** 599.

Ten Cate A.R. 1994. *Oral histology.* Mosby-Year Book, St. Louis, Missouri.

Thesleff I. and Nieminen P. 1996. Tooth morphogenesis and cell differentiation. *Curr. Opin. Cell. Biol.* **8:** 844.

Thesleff I., Vaahtokari A., and Partanen A.M. 1995. Regulation of organogenesis. Common molecular mechanisms regulating the development of teeth and other organs. *Int. J. Dev. Biol.* **39:** 35.

Tickle C. 1995. Vertebrate limb development. *Curr. Opin. Genet. Dev.* **5:** 478.

Tickle C., Summerhill D., and Wolpert L. 1975. Positional signalling and specification of digits in chick limb morphogenesis. *Nature* **254:** 199.

Tingberreth S.A. and Chuong C.M. 1996. Sonic hedgehog in feather morphogenesis—Induction of mesenchymal condensation and association with cell death. *Dev. Dyn.* **207:** 157.

Tissier-Seta J.P., Mucchielli M.L., Mark M., Mattei M.G., Goridis C., and Brunet J.F. 1995. *Barxl,* a new mouse homeodomain transcription factor expressed in cranio-facial ectomesenchyme and the stomach. *Mech. Dev.* **51:** 3.

Trainor P.A. and Tam P.P. 1995. Cranial paraxial mesoderm and neural crest cells of the mouse embryo: Co-distribution in the craniofacial mesenchyme but distinct segregation in branchial arches. *Development* **121:** 2569.

Turecková J., Sahlberg C., Åberg T., Ruch J.V., Thesleff I., and Peterkova R. 1995. Comparison of expression of the *msx-1, msx-2, bmp-2* and *bmp-4* genes in the mouse upper diastemal and molar tooth primordia. *Int. J. Dev. Biol.* **39:** 459.

Vaahtokari A., Åberg T. and Thesleff I. 1996a. Apoptosis in the developing tooth—Association with an embryonic signaling center and suppression by EGF and FGF-4. *Development* **122:** 121.

Vaahtokari A., Åberg T., Jernvall J., Keränen S., and Thesleff I. 1996b. The enamel knot as a signaling center in the developing mouse tooth. *Mech. Dev.* **54:** 39.

Vainio S., Karavanova I., Jowett A., and Thesleff I. 1993. Identification of BMP-4 as a signal mediating secondary induction between epithelial and mesenchymal tissues during early tooth development. *Cell* **75:** 45.

van Genderen C., Okamura R.M., Farinas I., Quo R.G., Parslow T.G., Bruhn L., and Grosschedl R. 1994. Development of several organs that require inductive epithelial-mesenchymal interactions is impaired in LEF-1-deficient mice. *Genes Dev.* **8:** 2691.

Verschueren K., Dewulf N., Goumans M.J., Lonnoy O., Feijen A., Grimsby S., Vande Spiegle K., ten Dijke P., Moren A., Vanscheeuwijck P., Heldin C.-H., Miyazono K., Mummery C., Van Den Eijnden-Van Raaij J., and Huylebroeck D. 1995. Expression of type I and type IB receptors for activin in midgestation mouse embryos suggests distinct functions in organogenesis. *Mech. Dev.* **52:** 109.

Watterson R.L., Fowler I., and Fowler B.J. 1954. The role of the neural tube and notochord in development of the axial skeleton of the chick. *Am. J. Anat.* **95:** 337.

Weiss K.M., Ruddle F.H., and Bollekens J. 1995. *Dlx* and other homeobox genes in the morphological development of the dentition. *Connect. Tissue Res.* **32:** 35.

Westergaard B. and Ferguson M.W.J. 1987. Development of denition in *Alligator mississippiensis.* Later development in the lower jaws of embryos, hatchlings and young juveniles. *J. Zool.* **212:** 191.

Zou H. and Niswander L. 1996. Requirement for BMP signaling in interdigital apoptosis and scale formation. *Science* **272:** 738.

BMP Signaling and Vertebrate Limb Development

H. ZOU,[1] K.-M. CHOE,[2] Y. LU,[2] J. MASSAGUÉ,[1,3] AND L. NISWANDER[2]
[1]*Cell Biology and* [2]*Molecular Biology Programs and* [3]*Howard Hughes Medical Institute, Memorial Sloan-Kettering Cancer Center, New York, New York 10021*

Bone morphogenetic protein (BMP) signaling is involved in a great number of developmental processes in the vertebrate embryo. These roles are beginning to be revealed by genetic and experimental approaches. Gene targeting has been used to generate loss-of-function alleles of a few *Bmp* genes (see Hogan et al.; Varlet et al.; both this volume). Targeted mutagenesis has also been used to inactivate the gene *Noggin* which sequesters BMP ligand. In the absence of *Noggin*, BMP signaling is effectively enhanced (R. Harland, pers. comm.). A mutagenesis screen in zebrafish has uncovered a mutation in the *Chordin* gene, another secreted factor that binds to BMPs and inhibits their function (see DeRobertis et al., this volume). Our studies have investigated the role of BMP signaling by application of recombinant BMP or by using mutant forms of the BMP receptors (BMPR) to generate loss-of-function (dominant-negative) and gain-of-function (constitutively active) mutations to alter BMP signaling in the developing chick limb.

RESULTS AND DISCUSSION

Bmp Genes Are Downstream Targets of Sonic Hedgehog Signaling

It has been suggested that *Bmp* genes act downstream from *Sonic Hedgehog* (*Shh*) to pattern the vertebrate limb along the anteroposterior (A/P) axis. This is based largely on two sets of data: expression patterns in the vertebrate limb and functional studies in *Drosophila*. In the chick limb, *Bmp2* and *Bmp7* expression overlaps the *Shh* domain in the posterior mesenchyme. *Bmp4* is detected in the posterior mesenchyme but is also expressed at high levels in the anterior mesenchyme, and at lower levels in the distal tip mesenchyme. In addition, all three genes are expressed in the apical ectodermal ridge (Francis et al. 1994; Francis-West et al. 1995). It should be noted that *Bmp* gene expression differs in the developing chick and mouse limb. For instance, chick *Bmp7* expression in later-stage limbs is more similar to that observed for mouse *Bmp4* (for an overview of mouse expression, see Hogan 1996). Although the expression of individual genes is not necessarily correlative, the overall *Bmp* expression pattern is similar.

Application of *Shh*-expressing cells to the anterior chick limb bud induces the expression of *Bmp2* and *Bmp7* (Francis et al. 1994; Laufer et al. 1994; Kawakami et al. 1996; normal expression of *Bmp4* in the anterior mesenchyme precludes similar analysis) and ultimately results in mirror-image duplication of the distal skeletal elements (alteration from normal digit 2-3-4 pattern to 4-3-2-2-3-4 pattern; Riddle et al. 1993). Rather similarly, in the *Drosophila* wing and leg imaginal discs, Hedgehog induces the expression of the *Bmp2/Bmp4* homolog, *Decapentaplegic (Dpp)*. Furthermore, ectopic expression of either *Hedgehog* or *Dpp* is sufficient to repattern the imaginal disc along the A/P axis (Basler and Struhl 1994; Capdevila and Guerrero 1994; Tabata and Kornberg 1994). Returning to vertebrates, application of cells stably transfected with *Bmp2* (replication-defective virus) to the anterior chick limb bud can affect A/P patterning in that an additional anterior digit (digit 2) is often formed (Duprez et al. 1996a). However, the *Bmp2*-expressing cells are not capable of reproducing all of the patterning effects produced in response to SHH. Taken together, these pieces of data have led to the suggestion that BMPs may have an active role in A/P patterning of the vertebrate limb.

Exploring the Role of BMPs in A/P Limb Patterning

Recombinant BMPs. To explore further the role of BMPs in A/P patterning, we have applied recombinant BMP to the anterior limb bud. However, in contrast to the results obtained using stably transfected cells expressing *Bmp2* (Duprez et al. 1996a), we found that recombinant BMP2 does not cause digit duplication. Instead, this treatment results in extensive cell death such that often the anterior skeletal elements are missing (Fig. 1A, loss of radius or radius and digit 2). Similar results have recently been published by Macias et al. (1997). We observe a dramatic effect on cell death following application of recombinant BMP2, BMP4, or BMP7 homodimeric protein or BMP2/6 heterodimeric protein (proteins kindly provided by the Genetics Institute). The cell death phenotype is seen over a wide range of protein concentrations in which the bead is soaked (40 μg/ml to 2 mg/ml). Lower protein concentrations result in normal skeletal patterns.

In accordance with the results produced by application of exogenous BMP, we have also observed that a constitutively active mutant form of BMPR-IB (caBMPR-IB) also promotes cell death in the limb (Zou et al. 1997). Widespread caBMPR-IB misexpression results in exten-

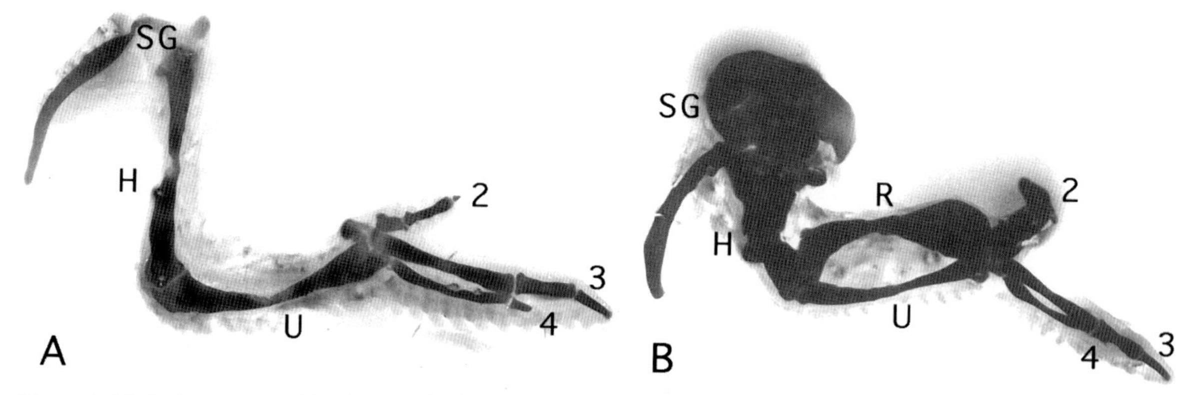

Figure 1. Limb phenotype resulting from application of recombinant BMP2 homodimeric protein (*A*) or a pellet of chick embryo fibroblasts expressing *Bmp2*/RCAS (*B*) under the anterior apical ridge of a stage-20 limb bud. In *A*, the radius is absent as a consequence of extensive cell death (not shown) elicited by BMP. In *B*, the anterior skeletal elements (shoulder girdle, humerus, radius, and digit 2) are much thicker than normal. (H) Humerus; (R) radius; (SG) shoulder girdle; (U) ulna; (2, 3, or 4) digit 2, 3, or 4, respectively.

sive cell death such that often the limb is truncated. Conversely, a dominant-negative BMPR-IB (dnBMPR-IB) can inhibit apoptosis in the limb (Zou and Niswander 1996). Thus, it appears that at least one function of BMP signaling is to regulate the cell death program in the chick embryo.

The different results produced by BMP-expressing cells versus recombinant BMP suggest that the cells expressing *Bmp* may also produce other factors, potentially including survival factors. When recombinant protein is used, an effect on limb patterning presumably would not be revealed if the cells were dead. Therefore, we sought to test whether we could prevent the cell death and thereby reveal a potential role of BMPs in limb patterning. To begin to test this idea, we applied a BMP-soaked bead and a fibroblast growth factor (FGF)-soaked bead to the anterior limb bud, as previous studies indicated that FGF could counteract the negative effects of BMP on cell proliferation and death (Niswander and Martin 1993). We found that FGF could rescue the cell death phenotype produced by recombinant BMP (data not shown). However, in most cases, the skeleton was normal. In a few cases, an extra anterior digit 2 was formed, but this phenotype is observed following application of an FGF bead alone (data not shown; Riley et al. 1993). Therefore, our results to date using recombinant BMP have not been able to recapitulate the digit-patterning effects of application of BMP-expressing cells to the anterior limb bud. This again suggests that the stably transfected cell lines may produce other factors that work in conjunction with BMP to polarize the limb. As a final note with regard to different activities of recombinant protein versus transfected cells, application of cells that express *Bmp2* from a replication-competent retrovirus (BMP2/RCAS) does not affect cell death or digit patterning but instead stimulates the formation of ectopic cartilage (Fig. 1B) (Duprez et al. 1996b).

This leads us back to the apparent paradox that in the *Drosophila* wing disc, DPP can mimic all of the effects of Hedgehog signaling. However, at least on the basis of experiments carried out to date, BMPs are not capable of fully substituting for SHH to give complete A/P pattern duplications of the vertebrate limb. This returns us to the likely possibility that in vertebrates, SHH may activate other genes that are necessary to function in conjunction with BMP in the regulation of patterning. Additional studies are required to determine if this is the case and, if so, to identify these factors.

If BMP signaling is at least a component of A/P pattern formation, then one can envision other ways in which to block BMP signaling and perhaps reveal its role in patterning. BMP signaling can be blocked in at least two ways by use of (1) as a dominant-negative form of the BMP receptor (dnBMPR; see below) or (2) Noggin which binds BMP and prevents BMP from interacting with its receptor(s) (Zimmerman et al. 1996). We have tried both of these methods (Zou et al. 1997; S. Pizette and L. Niswander, unpubl.). However, in each situation, cartilage formation is blocked and hence no skeleton forms. In the absence of skeleton, it is very difficult to determine whether a change in limb patterning has occurred, as molecular markers of downstream patterning events have not been identified. Therefore, these blocking experiments have also not been informative as to the role of BMPs in limb patterning (for additional thoughts as to the relationship between patterning and initial cartilage formation, see section on BMP receptor mutations).

Heterodimeric BMP proteins. The data nevertheless still suggest that BMPs have at least some role in patterning the limb. An interesting possibility is that limb patterning may be controlled by different combinations of BMP homodimeric and heterodimeric proteins. BMPs are synthesized as large precursors that undergo dimerization, processing, and secretion to yield an active dimeric protein. *Bmp* genes are expressed in dynamic patterns such that at different times, cells within different regions of the limb may produce a variety of BMP dimeric species. These might then be involved in differentially regulating pattern along the A/P axis. It is not yet clear whether BMP heterodimers may elicit different signaling events and consequently different downstream activities.

However, this is a distinct possibility. It has been shown that BMP receptors form a heteromeric complex (two or more type I receptors with two or more type II receptors; Weis-Garcia and Massagué 1996). Moreover, our studies described below and in Zou et al. (1997) indicate that different BMPRs mediate distinct outcomes, presumably by triggering different intracellular pathways. Therefore, it is possible that different combinations of ligands (homodimeric or heterodimeric) could activate different BMPR-mediated signal transduction pathways.

One method to test this hypothesis would be to apply heterodimeric BMP protein to the anterior limb bud to determine whether it may cause a response different from that noted with homodimeric protein. Our studies to date using putative BMP heterodimeric protein (kindly provided by Ali Hemmati-Brivanlou and Atsushi Suzuki) have not resulted in gross morphological changes (limbs appear normal in terms of patterning and cell death). However, it is likely that the concentration of BMP used may be too low to elicit an effect in the chick limb bud assay. Therefore, to date, these studies have not been able to determine whether BMP homodimeric and heterodimeric proteins have different activities in the limb, potentially in contributing to A/P patterning.

BMP receptor mutations. Another means we have used to explore the role of BMP signaling in chick limb development is to study the expression of the two known type I BMPRs and to create point mutations in these BMPRs to either generate dominant-negative BMPRs that cannot respond to BMP signaling or constitutively active BMPRs to elicit signal transduction in the absence of ligand. The mutant receptors were then introduced into the chick embryo by retroviral-mediated gene transfer using a replication-competent retrovirus.

RNA in situ hybridization analyses indicate that *BmpR-IB* is strongly expressed in precondensing mesenchyme prefiguring the cartilagenous condensations (Kawakami et al. 1996; Zou et al. 1997). Introduction of the dnBMPR-IB mutation dramatically inhibits cartilage formation both in vitro and in vivo. Embryonic limb mesenchyme cells can be cultured, and over a period of 4–5 days, they spontaneously differentiate into cartilage. Infection of the cultures with dnBMPR-IB results in complete inhibition of cartilage formation (Kawakami et al. 1996; Zou et al. 1997). In vivo, dnBMPR-IB infection of the developing limb results in digit truncation and reduction or absence of other skeletal elements (Zou et al. 1997). These results indicate that BMP-mediated signaling is necessary for cartilage formation.

This role has also been explored using the opposite mutation, the activated BMPR-IB. Both our in vitro and in vivo studies demonstrate that BMPR-IB activity is sufficient to elicit cartilage formation. Infection of in vitro cultures causes a large increase in the number and size of chondrogenic nodules, whereas in vivo infection results in dramatic expansion of the chondrogenic regions such that the limb is largely composed of cartilage with very little muscle or soft tissue (Zou et al. 1997).

Taken together, these studies indicate that BMP signaling through BMPR-IB is important for formation of the cartilage primordium. With this in mind, we return to the original question of the role of BMPs in A/P patterning. How is it that cells acquire positional information and turn it into the pattern we see, which in the limb is the skeleton? One hypothesis we propose is that BMPs are targets of SHH that could serve to link patterning signals to downstream effectors that translate the positional information into the formation of the skeletal primordium. In this model, SHH induces *Bmp* expression. We suggest that this activation may occur in a temporally and spatially specific manner such that different cells may produce different combinations of BMP homodimeric and heterodimeric proteins. BMP signaling would then activate BMPR-IB, leading to the initial formation of the cartilage elements. It is not clear how the posteriorly restricted, and potentially graded, SHH signal ultimately results in a pattern of periodic cartilage condensations/noncondensations. However, it is worth noting that mesenchyme condensations that form the cartilage primordium do not occur all at one time but instead occur in a progressive wave starting from the posterior side and proceeding toward the anterior. In addition, it is thought that the condensations themselves produce a short-range inhibitory signal that prevents condensation of nearby cells (see models in Hinchliffe and Johnson 1980). Cells farther away from the inhibitory signal can undergo condensation. Therefore, a combination of longer-range positive signals and short-range negative signals could be involved in transmitting the patterning information for condensation across the A/P axis of the limb. We propose that BMPs could be an important link between patterning and cartilage formation.

ACKNOWLEDGMENTS

We gratefully acknowledge Atsushi Suzuki and Ali Hemmati-Brivanlou (Rockefeller University, New York) for providing BMP heterodimeric protein, Genetics Institute (Cambridge, Massachusetts) for BMP homodimeric protein, and Delphine Duprez for BMP2/RCAS virus DNA.

REFERENCES

Basler K. and Struhl G. 1994. Compartment boundaries and the control of *Drosophila* limb pattern by *hedgehog* protein. *Nature* **368:** 208.

Capdevila J. and Guerrero I. 1994. Targeted expression of the signaling molecule decapentaplegic induces pattern duplications and growth alterations in *Drosophila* wings. *EMBO J.* **13:** 4459.

Duprez D.M., Kostakopoulou K., Francis-West P.H., Tickle C., and Brickell P.M. 1996a. Activation of expression of FGF-4 and HoxD gene expression by BMP-2 expressing cells in the developing chick limb. *Development* **122:** 1821.

Duprez D., Bell E.J., Richardson M.K., Archer C.W., Wolpert L., Brickell P.M., and Francis-West P. H. 1996b. Overexpression of BMP-2 and BMP-4 alters the size and shape of developing skeletal elements in the chick limb. *Mech. Dev.* **57:** 145.

Francis P.H., Richardson M.K., Brickell P.M., and Tickle C. 1994. Bone morphogenetic proteins and a signalling pathway that controls patterning in the developing chick limb. *Development* **120:** 209.

Francis-West P.H., Robertson K., Ede D.A., Rodriguez C.,

Izpisúa-Belmonte J.-C., Houston B., Burt D.W., Gribbin C., Brickell P.M., and Tickle C. 1995. Expression of genes encoding bone morphogenetic proteins and Sonic Hedgehog in talpid (ta3) limb buds: Their relationships in the signalling cascade involved in limb patterning. *Dev. Dyn.* **203:** 187.

Hinchliffe J.R. and Johnson D.R. 1980. *The development of the vertebrate limb: An approach through experiment, genetics, and evolution.* Clarendon Press, Oxford, United Kingdom.

Hogan B.L.M. 1996. Bone morphogenetic proteins: Multifunctional regulators of vertebrate development. *Genes Dev.* **10:** 1580.

Kawakami Y., Ishikawa T., Shimabara M., Tanda N., Enomoto-Iwamoto M., Iwamoto M., Kuwana T., Ueki A., Noji S., and Nohno T. 1996. BMP signaling during bone pattern determination in the developing limb. *Development* **122:** 3557.

Laufer E., Nelson C., Johnson R.L., Morgan B.A., and Tabin C. 1994. *Sonic hedgehog* and *Fgf-4* act through a signaling cascade and feedback loop to integrate growth and patterning of the developing limb bud. *Cell* **79:** 993.

Macias D., Ganan Y., Sampath T.K., Peidra M.E., Ros M.A., and Hurle J.M. 1997. Role of BMP-2 and OP-1 (BMP-7) in programmed cell death and skeletogenesis during chick limb development. *Development* **124:** 1109.

Niswander L. and Martin G.R. 1993. FGF-4 and BMP-2 have opposite effects on limb growth. *Nature* **361:** 68.

Riddle R.D., Johnson R.L., Laufer E., and Tabin C. 1993. *Sonic hedgehog* mediates the polarizing activity of the ZPA. *Cell* **75:** 1401.

Riley B.B., Savage M.P., Simandl B.K., Olwin B.B., and Fallon J.F. 1993. Retroviral expression of FGF-2 (bFGF) affects patterning in chick limb bud. *Development* **118:** 95.

Tabata T. and Kornberg T. 1994. Hedgehog is a signaling protein with a key role in patterning *Drosophila* imaginal discs. *Cell* **76:** 89.

Weis-Garcia F. and Massagué J. 1996. Complementation between kinase-defective and activation-defective TGF-β receptors reveals a novel form of receptor cooperativity essential for signaling. *EMBO J.* **15:** 276.

Zimmerman L.B., Jesús-Escobar J.M., and Harland R.M. 1996. The Spemann organizer signal noggin binds and inactivates bone morphogenetic protein 4. *Cell* **86:** 599.

Zou H. and Niswander L. 1996. Requirement for BMP signaling in interdigital apoptosis and scale formation. *Science* **272:** 738.

Zou H., Wieser R., Massagué J., and Niswander L. 1997. Distinct roles of type I bone morphogenetic protein receptors in the formation and differentiation of cartilage. *Genes Dev.* **11:** 2191.

Hox Genes and Mammalian Development

M.R. Capecchi
Howard Hughes Medical Institute, Department of Human Genetics, University of Utah School of Medicine, Salt Lake City, Utah 84112

Phylogenetic relationships among all animal and plant species are characterized by the themes of unity and diversity, an observation brilliantly synthesized and explicated by Darwin in his *Origin of the Species.* However, only following the last decade and a half of intense molecular and genetic analyses, principally of three species, *Caenorhabditis elegans, Drosophila melanogaster,* and *Mus musculus,* has it become apparent that the foundation for these two themes is deeply rooted in the molecular circuits used to guide the development of all species. Unexpectedly, parallel molecular circuits have been discovered that guide the formation of the basic body plan, the eyes, the heart, and neural circuits, to name just a few systems, in both vertebrates and invertebrates. From such analyses, it has become apparent that structures previously viewed as being disparate, such as the invertebrate and vertebrate eye or the invertebrate trachea and vertebrate lungs, may have common origins. More importantly, such discoveries reemphasize that investigators of invertebrate and vertebrate development can communicate with one another through a common language. Whatever is learned about the molecular circuits guiding development in one species is likely to have direct relevance to revealing similar circuits in other species. Finally, these recent discoveries of developmental parallelism underscore the wisdom of using model organisms, such as bacteria, yeast, *C. elegans, Drosophila,* and the mouse, to study all biological phenomena, human or nonhuman.

In no genetic system have the above principles been more clearly illustrated than with the *Hox* complexes (McGinnis et al. 1984b; Scott and Weiner 1984). The genes within these complexes encode transcription factors of the *Antennapedia homeodomain* class and may be used to establish the body plans of all metazoans of the animal kingdom (Carrasco et al. 1984; McGinnis et al. 1984a; Slack et al. 1993). It was the discovery of the structural and functional similarities between the *Drosophila* and mouse *Hox* complexes that sparked the current revolution in our ability to appreciate the common threads that weave through the molecular fabric that guides development in all species. Figure 1 shows a comparison of the *Hox* complexes from *Drosophila* and the mouse. *Drosophila* has 8 *Hox* genes present in two subcomplexes, the *Antennapedia* and *Bithorax complexes*, whereas the mouse has 39 genes distributed on four linkage groups designated Hox A, B, C, and D. Humans and mice, and perhaps all mammals, have the same set of 39 *Hox* genes. The mammalian organization is believed to have arisen early in vertebrate phylogeny by quadruplication of an ancestral complex common to vertebrates and invertebrates (Pendleton et al. 1993; Holland and Garcia-Fernandez 1996). From Figure 1 it is apparent that the order of *Hox* genes on the chromosomes of these two species has not altered significantly since the lineages of insects and vertebrates diverged approximately 530 million years ago. On the basis of DNA sequence similarities and the position of the genes on their respective chromosomes, individual members of the four mouse linkage groups have been classified into 13 paralogous families (Scott 1992). Members of a paralogous family share both DNA sequence similarities and similarities in their patterns of expression. As first demonstrated by Ed Lewis for the *Drosophila Bithorax complex*, the position of a homeotic gene within a complex correlates with the expression of that gene along the embryo's anteroposterior axis, a phenomenon termed spatial colinearity (Lewis 1978; Duboule and Dollé 1989; Graham et al. 1989). Furthermore, in vertebrates, there is also a temporal colinearity (Duboule 1994) in which the position of a gene on the chromosome correlates with the time that the gene is activated in the embryo. Thus, a 3′ *Hox* gene is activated prior to and in a more anterior region of the mouse embryo than its 5′ neighbors.

So far, I have stressed the unity among *Hox* complexes in different species. It is equally important to understand how modulations of these molecular circuits can be responsible for generating the enormous diversity of body plans that enrich our planet. Modifications of this system could occur in a number of different ways, including, for example, simple changes in the number and type of *Hox* genes retained within the complex. In this context, it is interesting that the puffer fish, *Fugu rubripes*, has recently been shown to retain only 31 *Hox* genes, having lost, relative to the mammalian complexes, 5 genes from the Hox D complex alone (Aparicio et al. 1997). Other possible variations that could contribute to the formation of vastly different body plans are changes in *Hox* gene expression patterns, changes in the transcriptional cofactors that interact with *Hox* genes, and qualitative and quantitative changes in the target genes controlled by these transcriptional regulators.

The two homeotic complexes that have been most extensively analyzed are those belonging to *Drosophila* and

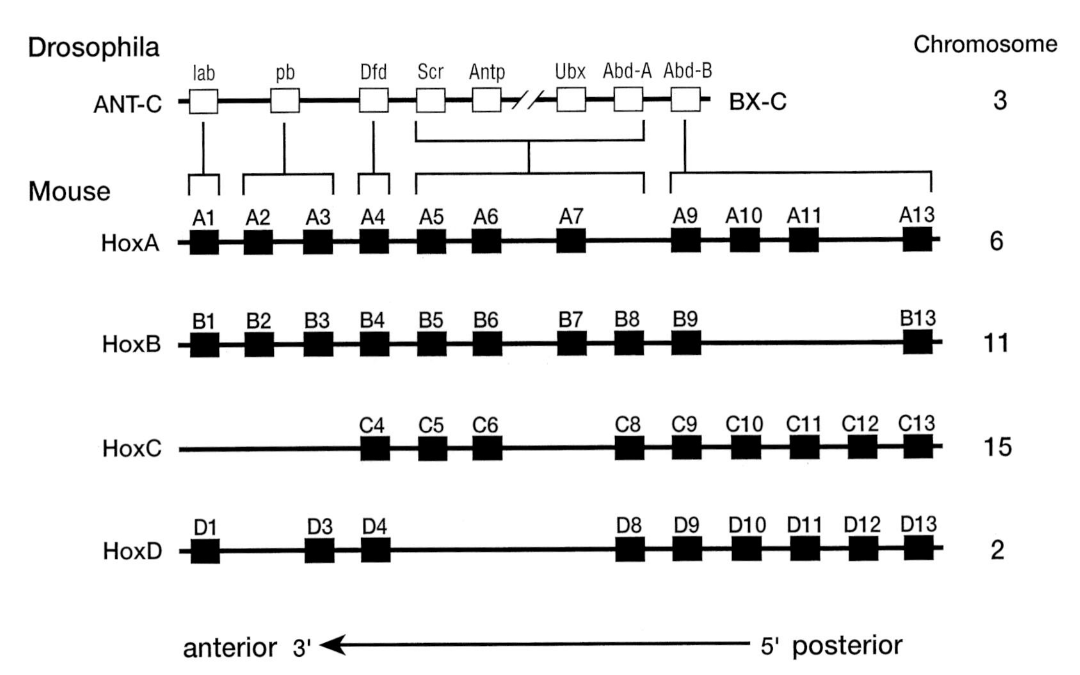

Figure 1. Comparison of the mouse *Hox* complex with the *Drosophila Homeotic Complex* (*HomC*). The *Drosophila HomC* is depicted at the top, and the four mouse linkage groups Hox A, B, C, and D are depicted below. (*Open and closed boxes*) Genes belonging to the *Drosophila HomC* and mouse *Hox* complex, respectively (Scott 1992). The relationships based on DNA and protein sequence similarities, between the eight *Drosophila HomC* genes and the mammalian *Hox* genes, are denoted by solid lines. Thus, the *Drosophila* labial (*lab*) gene is most closely related to the mouse *Hoxa1* gene and its paralogous family members *Hoxb1, Hoxd1,* and so on. Mice and humans have the identical network of *Hox* genes. All of the mammalian *Hox* genes are transcribed from the same DNA strand. The polarity of transcription is indicated at the bottom of the figure. In the mouse, a 3′ *Hox* gene is expressed before and in a more anterior region of the embryo than its 5′ neighbors. (*A, B*, reprinted, with permission, from Chisaka and Capecchi 1991; *C–F*, reprinted, with permission, from Condie and Capecchi 1994 [copyright Macmillan].)

the mouse. The *Drosophila HomC* genes are used to establish the identity of parasegments through specification of cell identity (Akam 1987). A mutation in *HomC* genes does not change the number of parasegments in the *Drosophila* embryo, but rather the identity of the parasegment. The situation in the mouse is much more complex. First of all, quadruplication, of perhaps the entire genome, appears to have preceded many of the innovations and elaborations that characterize more complex vertebrates, such as an internal skeletal system, cranial ganglia, expansion of the brain with the concomitant remodeling of the entire head, the acquisition of teeth, and so on. In fact, expansion of this gene complex may have contributed directly to the progression from invertebrates to vertebrates by supplying the complexity in this genetic network required to accommodate the development of the more complex vertebrate body plan. In this context, it is important to determine whether paralogous *Hox* genes provide merely redundant functions and thus serve only to increase the fidelity of the system or whether, following the expansion of the complex, novel regulatory mechanisms arose, utilizing multiple *Hox* genes, that directly contributed to the explosion of innovations that characterize vertebrates.

To address such issues, we have initiated a systematic genetic analysis of the mouse *Hox* complex. A critical comparison of *Hox* gene function in *Drosophila* and the mouse may shed light on these broader issues and perhaps reveal whether combinations of *Hox* genes are used in novel ways in vertebrates to modulate the amplitudes, times, and sites of expression of downstream target genes. The first phase of this project was to generate loss-of-function mutations in the individual *Hox* genes. This phase is nearing completion. We have generated mice with targeted disruptions in 37 of the 39 *Hox* genes. These mice have been used not only to define the individual functions of *Hox* genes during development, but also to explore the interactions among *Hox* genes by introduction of multiple mutations into the same mouse. From this analysis, an understanding of how multiple *Hox* genes cooperate to direct the formation of the myriad of structures under their guidance is emerging. It is anticipated that, in mammals, *Hox* genes will be found to function not as individual entities, but rather as members of a highly integrated network with paralogous genes, adjacent genes in the same linkage group, and even nonparalogous genes in separate linkage groups, interacting positively, negatively, and in parallel with each other to orchestrate the morphological regionalization of the embryo.

In this paper, I describe three stories involving interactions among *Hox* genes in development of the mouse. Each story has been chosen to illustrate particular features of mammalian *Hox* gene function that contrast with the more familiar roles of the *Homeotic* complex in *Drosophila* development. The first story involves the interactions among group-3 paralogous genes in forming the cervical vertebrae, the second involves potential roles of *Hox* genes in the formation and/or maintenance of the segmental paradigm in the hindbrain, and the third involves the role of *Hox* genes in forming the limbs.

INTERACTIONS AMONG *Hoxa3, Hoxb3,* AND *Hoxd3*

The mutant phenotypes resulting from targeted disruption of *Hoxa3, Hoxb3,* and *Hoxd3* are distinct (Chisaka and Capecchi 1991; Condie and Capecchi 1993; N.R. Manley and M.R. Capecchi, in prep). Disruption of *Hoxa3* causes defects in tissues derived from mesenchymal neural crest, whereas *Hoxd3* mutant mice show abnormalities in somitic mesoderm-derived structures. Thus, *Hoxa3* mutant homozygotes are athymic, aparathyroid, have reduced thyroid tissue, and malformations in throat cartilages, whereas *Hoxd3* mutant mice show transformations of the cervical vertebrae, C1 and C2, which acquire characteristics associated with more anterior structures (Fig. 2). In contrast, mice homozygous for a *Hoxb3* mutation have milder defects at lower penetrance both in somitic-mesoderm-derived structures and in neural-crest-derived tissues. Although *Hoxb3* mutant homozygotes also show malformation in the first and second cervical vertebrae, the set of defects is distinct from those observed in *Hoxd3* mutant mice (Fig. 2).

Mice mutant for either *Hoxa3* or *Hoxd3* alone do not show overlapping defects, suggesting that although these two paralogous genes operate in the same region of the embryo, they function in tissues of separate embryonic origins. However, analysis of mice mutant for both genes reveals a more complex picture. In the double mutant, it is clear that the presence of *Hoxd3* mutant alleles exacerbates the *Hoxa3* defects and, conversely, that the *Hoxa3* mutation enhances the *Hoxd3* mutant phenotype (Fig. 3) (Condie and Capecchi 1994). Furthermore, the degree of exacerbation is proportional to the total number of disrupted alleles carried by the mutant mice; i.e., mice heterozygous for the *Hoxa3* mutation and homozygous for the *Hoxd3* mutation, or vice versa, show intermediate mutant phenotypes compared with mice homozygous for a mutation in either *Hoxa3* or *Hoxd3* alone, and the double mutants. These results suggested that both *Hoxa3* and *Hoxd3* are functional in neural-crest-derived and somitic-mesoderm-derived tissues but that the role of *Hoxa3* in forming the atlas and axis only becomes apparent in the absence of *Hoxd3* function. Similarly, the role of *Hoxd3* in the development of neural-crest-derived tissues was only evident in *Hoxa3* mutant homozygotes. In addition, the interactions between the *Hox* genes are very sensitive to the concentration of Hox protein present within these cells, since intermediate genotypes showed intermediate phenotypes.

Analysis of *Hoxa3/Hoxb3* and *Hoxb3/Hoxd3* double mutants extended this concept by showing that not only do these paralogous genes operate in multiple tissues in the same region of the embryo, but they also often appear to be performing equivalent functions within these tissues. This is particularly evident in the formation of the cervical vertebrae, where *Hoxa3/Hoxd3* and *Hoxb3/Hoxd3* double mutants have indistinguishable defects, the deletion of the entire atlas (Fig. 4). Similar examples of equivalent functions among these paralogous genes are apparent in other tissues (N.R. Manley and M.R. Capecchi, in prep).

The observation that the contributions of *Hoxa3* and *Hoxb3* to the formation of the atlas and axis appears to be equivalent in *Hoxd3* mutant homozygotes suggests that the identity of the *Hox* gene operating in the developing tissue is not as critical a factor as the number of *Hox*

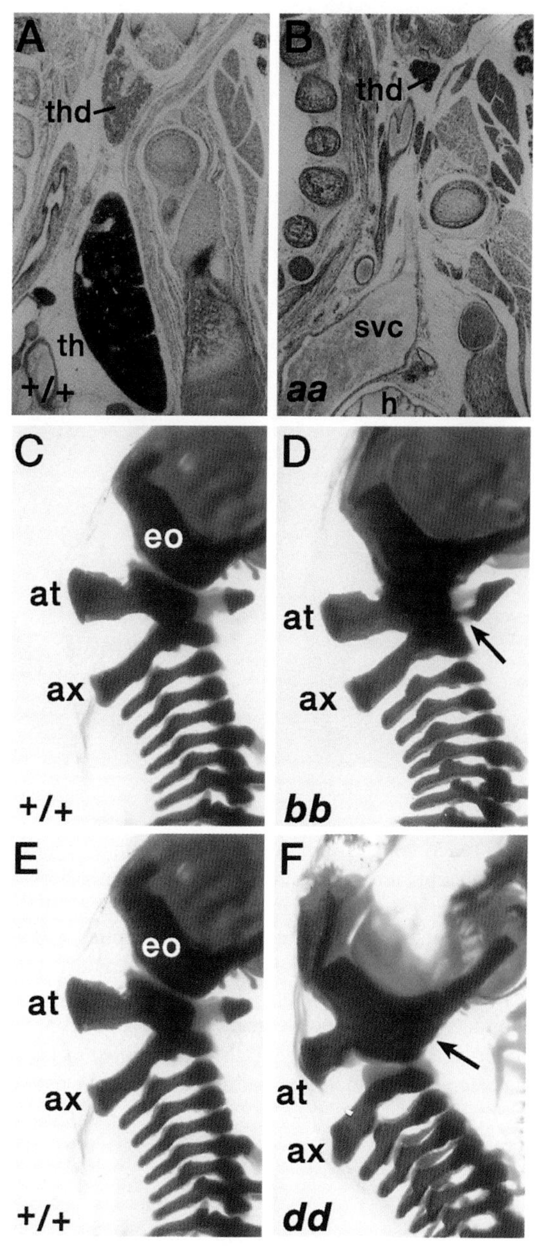

Figure 2. Defects observed in *Hoxa3, Hoxb3,* and *Hoxd3* mutant homozygotes. Disruption of *Hoxa3* results primarily in abnormalities of tissues derived from mesenchymal neural crest. Thus, *Hoxa3* mutant homozygotes are athymic (*B*). *Hoxb3* mutant homozygotes show milder and lower penetrant defects in tissues derived from either neural crest cells or somitic mesoderm. (*D*) Ossified fusion between the anterior arch of the atlas and the dens which is specific to *Hoxb3* mutant homozygotes. *Hoxd3* mutant homozygotes have defects in somitic mesoderm-derived structures where the first and second cervical vertebrae acquire shape characteristics normally associated with their immediate anterior structures. These transformations are shown in panel *F*. (+/+) Normal, wild-type controls; (aa, bb, and dd) *Hoxa3, Hoxb3,* and *Hoxd3* mutant homozygotes; (th, thd, svc, and h) thymus, thyroid, superior vena cava, and heart, respectively; (at and ax) atlas and axis, respectively. (Reprinted, with permission, from Condie and Capecchi 1994 [copyright Macmillan].)

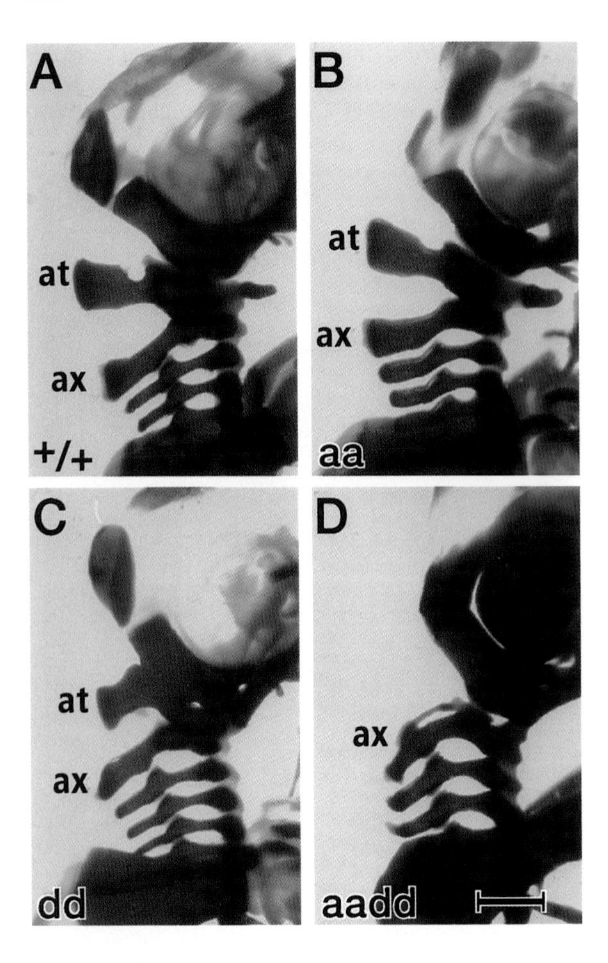

Figure 3. Comparison of skeletal defects in mice homozygous for the *Hoxa3* or *Hoxd3* mutation with mice homozygous for both mutations. Lateral views of cleared skeletal preparations of newborn mice. (*A*) Wild-type newborn, the first cervical vertebra or atlas (at) and the second cervical vertebra or axis (ax) are labeled. (*B*) *Hoxa3* mutant newborn. The skeletons of *Hoxa3* mutant homozygotes are indistinguishable from wild type. (*C*) *Hoxd3* mutant newborn displaying the characteristic homeotic transformation of the atlas and axis. (*D*) *Hoxa3/Hoxd3* double-mutant homozygotes. In these mutant newborns, the complete atlas is deleted and the axis is transformed in shape to resemble a more typical cervical vertebra, C3-C5. +/+, aa, dd, and aadd denote wild type, *Hoxa3* mutant homozygotes, *Hoxd3* mutant homozygotes, and *Hoxa3/Hoxd3* double-mutant homozygotes, respectively.

genes functioning within this tissue. In further support of this concept, mice that are heterozygous for both the *Hoxa3* and *Hoxb3* mutations and homozygous for the *Hoxd3* mutation have the same mutant phenotype as *Hoxa3/Hoxd3* or *Hoxb3/Hoxd3* double-mutant mice. Thus, the perspective changes from a qualitative one to a quantitative one. It is not critical which *Hox* genes are operating in a given tissue, but rather the number of paralogous *Hox* genes functional within that tissue. A molecular interpretation of this phenomenon would suggest that individual *Hox* genes may not be individually responsible for implementing unique developmental programs (as is observed in *Drosophila*), but rather that multiple *Hox* genes physically function together to mediate a program by controlling common target genes through common *cis* elements, and that for proper development, it is the stoichiometry of paralogous *Hox* genes operating within a developing tissue that is critical.

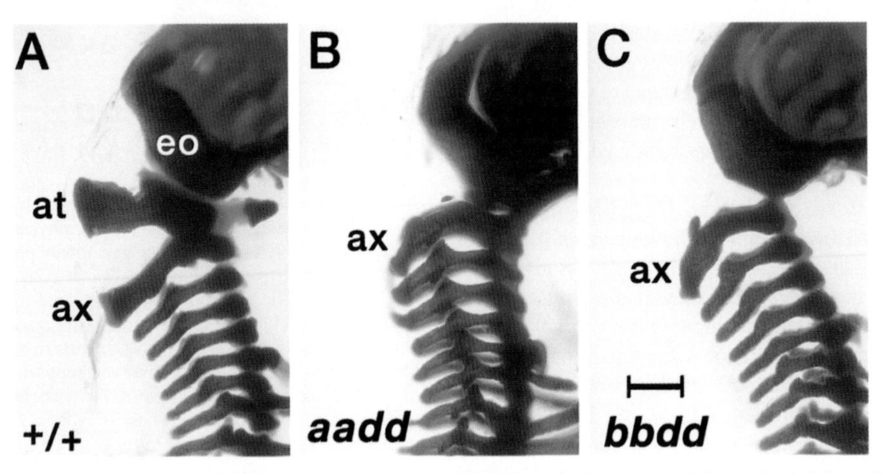

Figure 4. *Hoxa3/Hoxd3* and *Hoxb3/Hoxd3* double mutants show identical cervical vertebral defects, the deletion of the entire atlas. This suggests that in *Hoxd3* mutant homozygotes, *Hoxa3* and *Hoxb3* have equivalent functions. +/+, aadd, and bbdd denote wild type, *Hoxa3/Hoxd3* double-mutant homozygotes, and *Hoxb3/Hoxd3* double-mutant hymozygotes, respectively.

In the absence of *Hoxd3*, the roles of *Hoxa3* and *Hoxb3* in forming the cervical vertebrae appear to be equivalent. Nevertheless, *Hoxd3* is more important than its paralogs in mediating the formation of the vertebrae. Similarly, *Hoxa3* predominates over *Hoxb3* and *Hoxd3* in directing the formation of neural-crest-derived structures. These rankings of normal functional roles might derive from either qualitative or quantitative differences. A qualitative difference would postulate that Hoxd3 and Hoxa3 proteins physically interact with different transcriptional cofactors, conferring on one Hox protein specificity for directing somitic mesodermal function and on the other mesenchymal neural crest function. Alternatively, a quantitative explanation would be based either on the production of more Hoxd3 or Hoxa3 protein in somitic mesoderm or neural crest cells, respectively, or on a higher affinity of Hoxd3 and Hoxa3 proteins for common vertebral or neural crest tissue target gene *cis* elements, respectively.

In summary, we have demonstrated extensive genetic interactions among all three group-3 paralogous *Hox* genes in forming multiple tissues in the throat region. The concentration of Hox proteins operating in this region must be very tightly regulated since twofold differences in the concentration of these proteins show marked differences in mutant phenotypes. In many of the genetic interactions exhibited by these paralogous genes, the identity of the mutated gene appears to be less critical than the total number of normal alleles that remain functional within the affected region.

ROLES OF *Hoxa1* AND *Hoxb1* IN HINDBRAIN DEVELOPMENT

The mouse hindbrain is a particularly attractive target for molecular genetic analysis. Although it is an enormously complex structure controlling numerous autonomic and voluntary functions, its complexity is generated during development by a rather simple and commonly used paradigm: the generation and diversification of repeated units. Early in development, the mouse hindbrain anlage is transiently subdivided along its rostrocaudal axis into eight metameric segments called rhombomeres (Vaage 1969; Lumsden and Keynes 1989). Rhombomeres can function as compartments that limit cell movement and thereby create centers capable of independent development and diversification through localized gene activity and cell interactions (Fraser et al. 1990). However, it is also apparent that through intercompartmental communication, these units can function collectively to form a scaffold upon which a coherent neural network is built (Glover and Petursdottir 1991; Clarke and Lumsden 1993).

Mouse hindbrain development has become particularly amenable to molecular genetic analysis following identification of members of the *Hox* complex as major components of the molecular network that specifies cell identity within rhombomeres. I will also argue that *Hox* genes are involved in the establishment and/or maintenance of hindbrain segmentation itself (Chisaka et al. 1992; Carpenter et al. 1993; Dollé et al. 1993; Mark et al. 1993). This is a major departure from what is observed in *Drosophila* development, where the *gap, pair rule,* and *segment polarity* genes are used to establish and maintain segmentation, whereas the *Hom C* genes are subsequently used to determine the parasegmental identities (Akam 1987).

Suspicion that *Hox* genes were involved in hindbrain development arose from an examination of their expression patterns (for review, see Krumlauf 1994). The expression boundaries of the rostral (3′) set of *Hox A* genes are illustrated in Figure 5. *Hox* gene expression commences early in embryogenesis, typically during the formation of the primitive streak (mouse gestation day 7.5, E7.5). From this posterior position, expression moves rostrally to a specific anterior boundary, characteristic for each *Hox* gene, thus forming a nested set of transcripts extending along the anteroposterior axis. From Figure 5, it is evident that *Hox* gene expression respects rhombomere boundaries. However, the anterior limits of *Hox* gene expression are established prior to the formation of rhombomere boundaries. Thus, the timing of their ex-

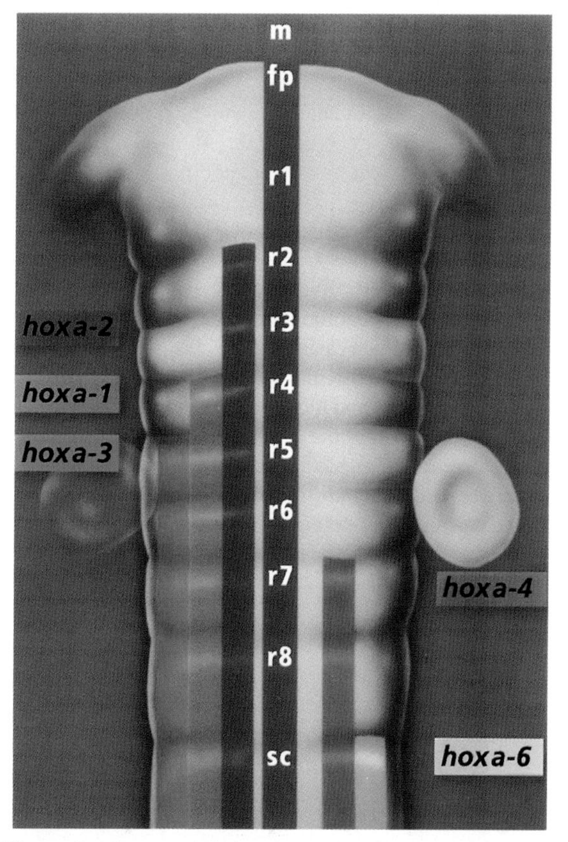

Figure 5. The rostral limits of expression in the murine hindbrain for the 3′ members of the Hox A linkage group. The hindbrain anlage is transiently subdivided into eight segments called rhombomeres (r). The rhombomere boundaries are depicted by darker shading. The otocyst is positioned lateral to the hindbrain adjacent to rhombomeres 5 and 6. The position of major neural crest production in the hindbrain is depicted by red shading. Note that the anterior boundaries of expression of these *Hox* genes respect rhombomere boundaries. (mfp) Midline floor plate; (sc) spinal cord.

pression is such that *Hox* genes could be involved in the establishment and/or maintenance of rhombomeres, as well as in the specification of cell identities within rhombomeres. What is not evident from Figure 5 is that *Hox* gene expression is very dynamic, changing rapidly during development. For example, expression of *Hoxa1* reaches the presumptive boundary between rhombomere 3 (r3) and rhombomere 4 (r4) by 8 days of gestation and promptly recedes caudally. By E8.5, *Hoxa1* expression is no longer detectable within the hindbrain. In addition, the level of *Hox* gene expression is not uniform from one rhombomere to another. Each rhombomere is characterized by the expression of a unique set of highly expressed *Hox* genes, suggesting that combinations of Hox proteins may be used to specify cell identities.

Targeted disruption of *Hoxa1* in mice results in severe defects in the formation of the hindbrain and associated cranial ganglia and nerves (Chisaka et al. 1992; Carpenter et al. 1993; Dollé et al. 1993; Mark et al. 1993). From the analysis of molecular markers expressed within specific rhombomeres of *Hoxa1* mutants, it is apparent that defects extend from r3 through r8, but, most importantly, r5 is absent. Thus, disruption of this gene affects not only the identity of cells within rhombomeres, but also the integrity of segmentation itself. Whether this is a result of a defect in the specification or in the outgrowth and/or maintenance of specific rhombomeres remains to be determined. Whatever the mechanism, the outcome is that a mutation in a *Hox* gene alters the number of hindbrain rhombomeres present in the mutant embryos.

In contrast, disruption of the paralogous *Hox* gene, *Hoxb1*, does not alter the number or pattern of rhombomeres, but rather the properties of neurons within r4 (Goddard et al. 1996; Studer et al. 1996). Specifically, there is a loss of the motor component of the VIIth nerve, which normally innervates the muscles of facial expression (Fig. 6). Normally, the cell bodies of these neurons arise medially in r4 and then migrate caudally into r5. On reaching the r6 boundary, these cell bodies abruptly change course and migrate laterally to their final destination in the pons. In *Hoxb1* mutant homozygotes, this population of migrating neurons is not present, resulting in paralysis and subsequent degeneration of the muscles of facial expression. Instead, there is a more lateral population of cell bodies in r4, the identity of which needs further characterization. By these criteria, *Hoxb1* function is more similar to that of a *Drosophila Hox* gene than is *Hoxa1*. However, as with the group-3 paralogous *Hox* genes described above, in *Hoxa1/Hoxb1* double mutants both *Hoxa1* and *Hoxb1* mutant phenotypes are exacerbated, demonstrating that both genes are involved in rhombomere specification and/or maintenance, as well as in decisions involving cell indentities (M. Rossel and M.R. Capecchi, unpubl.).

In the hindbrain, each rhombomere is characterized by the expression of specific sets of *Hox* genes. This relationship further supports the hypothesis that *Hox* genes are involved in the establishment and/or maintenance of rhombomere compartments. Such an intimate relationship is not observed between *Hox* gene expression patterns and somitic mesoderm compartments. Thus, *Hox* genes do not appear to be involved directly in the establishment and/or maintenance of compartments within the somitic-derived cell lineages. However, it is interesting that in sclerotome-derived tissues, the boundaries of *Hox* gene expression do respect the boundaries between different classes of prevertebrae (i.e., the transitions between the anlage that will form the cervical and thoracic vertebrae, thoracic and lumbar vertebrae, and so on; Burke et al. 1995). The latter observation suggests a role for *Hox* genes in mediating distinction between different vertebral classes.

HOX GENES AND LIMB DEVELOPMENT

During development, nested sets of *Hox* gene transcripts are observed not only along the major anteroposterior (A/P) axis of the embryo, but also in subcomponents of the embryo such as the gut, gonadal tissues, and the limbs (Dollé et al. 1989, 1991; Izpisúa-Belmonte et al. 1991; Yokouchi et al. 1991, 1995; Haack and Gruss 1993; Roberts et al. 1995). The nested set of transcripts in limbs suggested a role for *Hox* genes in limb patterning. This role was confirmed by the demonstration that loss-

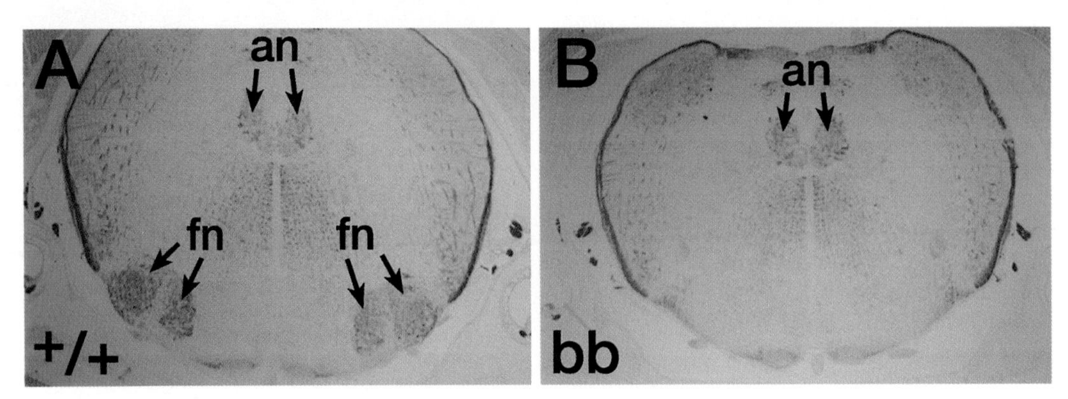

Figure 6. Absence of the facial motor nucleus in *Hoxb1* mutant homozygous mice. Transverse, 10 μm paraffin sections of wild-type (*A*) and *Hoxb1* mutant homozygous (*B*) newborn mice. These sections were subjected to immunohistochemistry with the 2H3 antibody against neurofilament protein. Arrow indicates the abducens nuclei (an) and the facial motor nuclei (fn). (Reprinted, with permission, from Goddard et al. 1996.)

of-function mutations in 5′ *Hox* genes cause limb malformations (Dollé et al. 1993; Small and Potter 1993; Davis and Capecchi 1994, 1996; Davis et al. 1995; Favier et al. 1995, 1996; Fromental-Ramain et al. 1996; van der Hoeven et al. 1996; Zákány and Duboule 1996). However, the malformations are not readily interpretable in terms of simple patterning paradigms. For example, the set of defects does not exhibit a polarity with respect to the A/P axis. Thus, *Hox* genes do not appear to be used to interpret, in any simple way, a gradient of a morphogen, such as sonic hedgehog (*shh*), whose concentration varies along the A/P axis (Riddle et al. 1993). Instead, these *Hox* mutants exhibit delays in the timing of ossification of autopodal bones. In addition, reduction, complete absence, or improper segmentations occur, and localized reductions in growth of numerous limb bones are apparent. A correlation is also seen between the bones of the autopod that are most affected by the 5′ *Hox* mutations and the order in which these bones are made during normal development (Shubin and Alberch 1986), with the most affected bones being those that are made last. The latter result suggests that one common consequence of 5′ *Hox* gene mutations may be the reduction in the number of prechondrogenic precursor cells available for limb formation. This would result in the last bony components made being the ones most severely affected, since those cartilaginous elements would be left competing for the few remaining precursor cells.

An additional factor that contributes to the complexity of interpreting the phenotype in terms of patterning defects arises from the fact that the *Hox* genes do not affect formation of the limb at a single point in development but rather at multiple points. *Hox* genes are functional within the limb buds not only during the initial phases of limb development, when the prechondrogenic mesenchymal condensations are accruing, but also at later stages when the outgrowth of the long bones is taking place. Thus, *Hox* gene expression is detectable in the growth plates of all the long bones. The final phenotypic outcome of a mutation in these *Hox* genes results from a summation of defects accumulated at multiple steps in bone development.

Nevertheless, combinations of *Hox* gene mutations that separately affect limb development have dramatically demonstrated that *Hox* genes do have a role in proximodistal patterning of the limb. Disruptions of either *Hoxa11* or *Hoxd11* alone result in rather mild defects in the formation of the radius and ulna (Small and Potter 1993; Davis and Capecchi 1994). However, in *Hoxa11/Hoxd11* double mutants, only a vestige of the radius and ulna remains (Fig. 7) (Davis et al. 1995). The initial bifurcation of the radius and ulna anlage is still detected in these double mutants (Fig. 8), but the enormous subsequent outgrowth of these long bones does not occur. The apparently normal initiation of prechondrogenic condensations for radius and ulna in *Hoxa11/Hoxd11* double mutants argues either that *Hox11* genes are not involved in the initial specification of these bones or that the presence of functional *Hoxc11* genes in these mutant mice is sufficient to initiate this process. This latter hypothesis can be tested by generation of mice mutant for all three

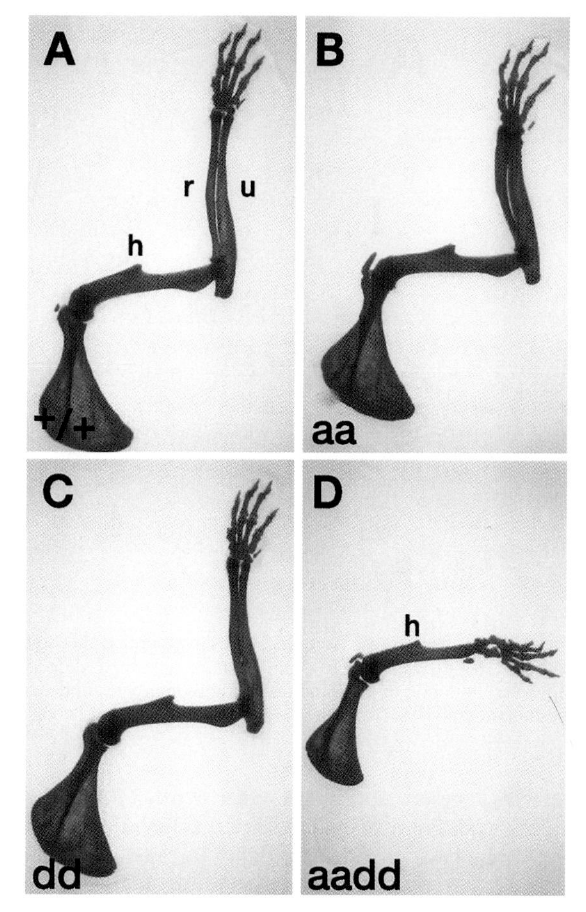

Figure 7. The loss of the radius and ulna in *Hoxa11/Hoxd11* double-mutant mice. Dorsal view of the right forelimbs from adult mouse skeleton preparations. (*A*) Wild-type mouse (+/+) with a normal humerus (h), radius (r), and ulna (u). (*B*) *Hoxa11* mutant homozygote (aa) with a slightly thicker and shorter zeugopod. (*C*) *Hoxd11* mutant homozygote (dd) with subtle defects in the zeugopod and autopod. (*D*) *Hoxa11/Hoxd11* double-mutant homozygote (aadd) showing an unexpected dramatic reduction of the radius and ulna. The forepaw is also malformed and rotated 90° compared with wild type. (Reprinted, with permission, from Davis et al. 1995 [copyright Macmillan].)

group-11 paralogous genes. Technically, this is a difficult experiment because loss-of-function mutations in each of the group-11 paralogous genes affect the fertility of the mutant mice.

To test for genetic interactions between adjacent genes on the same linkage group, a series of *trans*-heterozygotes was constructed, in which one chromosome has a mutation in one *Hox* gene and the other chromosome has a mutation in the adjacent *Hox* gene (i.e., *Hoxd11*$^{+/-}$/*Hoxd12*$^{-/+}$). Mice heterozygous for a mutation at only a single *Hox* locus usually show mild defects with low penetrance in the formation of the limb. In each case, the *trans*-heterozygote shows a much more severe and more highly penetrant set of defects (Davis and Capecchi 1996). The severity of the mutant phenotypes in these *trans*-heterozygotes increases from (*Hoxd11/Hoxd12*) through (*Hoxd11/Hoxd13*) to (*Hoxd-12/Hoxd-13*) *trans*-heterozygotes. The pattern of observed

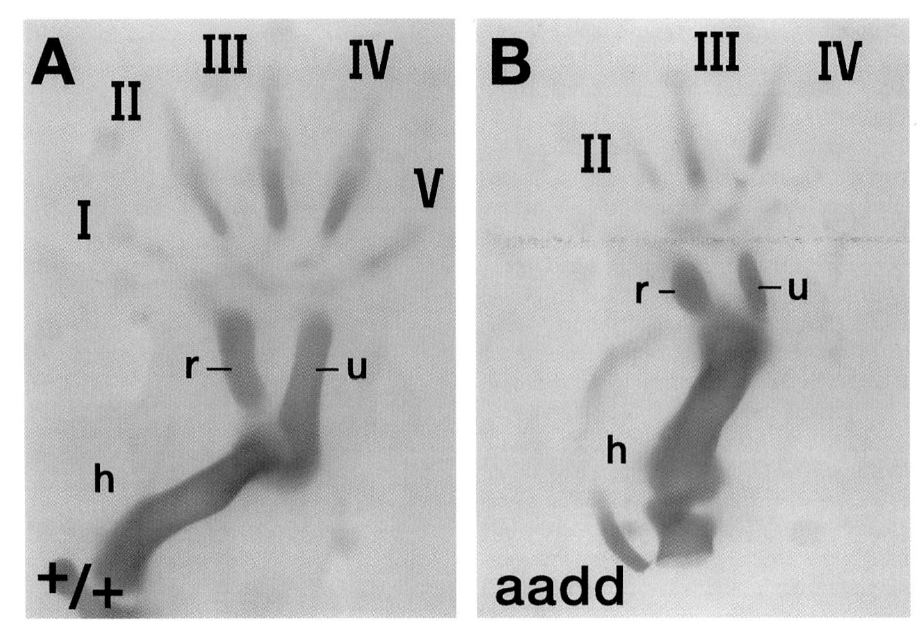

Figure 8. Cartilage condensations in the developing right forelimb of wild-type (*A*) and *Hoxa11/Hoxd11* double-mutant homozygous embryos (*B*) at E13.5. In the double-mutant zeugopod, defects are already evident with the reduction of the radius and ulna anlage. However, initiation of the bifurcation of mesenchyme condensations that will give rise to the radius and ulna has occurred in the double-mutant embryos. (Reprinted, with permission, from Davis et al. 1995 [copyright Macmillan].)

defects suggests a dominant role for *Hoxd13* in forming the autopod. From an evolutionary standpoint, this hierarchy may have provided flexibility for modulating the autopod. A limb can be divided into three zones. The most proximal is the stylopod containing one large support bone. This is followed by the middle zeugopod with two similar bones that usually function as a hinge to allow rotation. Finally, at the most distal end is the autopod with several carpals, metacarpals, and phalangeal bones. The most distal end of the limb holds the greatest potential for exploitation by natural selection. Thus, it is advantageous that the last *Hox* genes (paralogous group-13 members) involved in producing the last structures, the autopod, have the greatest sensitivity to potential modulations that have been randomly accrued during evolution. Thus, changes in the expression pattern of the group-13 genes could dramatically alter the form of the autopod. This concept may underlie the well-documented paleological observation of proximal stability/distal variability in tetrapod limb structures (Hinchliffe 1991).

In summary, we have examined the interactions of *Hox* genes in forming cervical vertebrae, hindbrain, and limbs. In each case, it is apparent that individual *Hox* genes are performing individual functions but that more profound roles are apparent when they act in combination with other *Hox* genes. The observed interactions suggest that multiple *Hox* genes function in concert to regulate overlapping sets of target genes. This suggestion is particularly strong in the interactions observed among the group-3 paralogous genes in formation of the cervical vertebrae and among *Hoxd11, Hoxd12,* and *Hoxd13* in formation of the autopod. In each case, the cumulative effect of combining multiple mutations is the deletion of structures, resulting from either lack of specification or lack of proliferation of the precursor cells needed to form the structures. Similarly, the combination of *Hoxa1* and *Hoxb1* mutations results in more extensive deletions of anterior structures than is apparent in mice homozygous for either individual mutation. All of the results, both of single and combined mutations, are compatible with a role of *Hox* genes in the early regionalization of the embryo. In the absence of *Hox* gene function, formation of the axes and germ cell layers of the embryo still occurs. At this point, the *Hox* genes are activated to initiate the formation of multiple tissues and structures specific to each region of the embryo by conferring positional value along the major axes of the embryo. Perhaps the most primitive function of *Hox* genes is their innate ability, through their chromosomal organization, to convert a series of temporal signals into morphological direction, a conversion of time's arrow into a spatial vector.

REFERENCES

Akam M.E. 1987. The molecular basis for metameric pattern in the *Drosophila* embryo. *Development* **101:** 1.

Aparicio S., Hawker K., Cottage A., Mikawa Y., Zuo L., Venkatesh B., Chen E., Krumlauf R., and Brenner S. 1997. Organization of the *Fugu rubripes Hox* cluster: Evidence for continuing evolution of vertebrate *Hox* complexes. *Nat. Genet.* **16:** 79.

Burke A.C., Nelson C.E., Morgan B.A., and Tabin C. 1995. *Hox* genes and the evolution of vertebrate axial morphology. *Development* **121:** 333.

Carpenter E.M., Goddard J.M., Chisaka O., Manley N.R., and Capecchi M.R. 1993. Loss of *Hox-A1* (*Hox-1.6*) function results in the reorganization of the murine hindbrain. *Development* **118:** 1063.

Carrasco A.E., McGinnis W., Gehring W.J., and De Robertis E.M. 1984. Cloning of an *X. laevis* gene expressed during early embryogenesis coding for a peptide region homologous to *Drosophila* homeotic genes. *Cell* **37:** 409.

Chisaka O. and Capecchi M.R. 1991. Regionally restricted developmental defects resulting from targeted disruption of the mouse homeobox gene *Hox-1.5*. *Nature* **350:** 473.

Chisaka O., Musci T.S., and Capecchi M.R. 1992. Developmental defects of the ear, cranial nerves and hindbrain resulting from targeted disruption of the mouse homeobox gene *Hox-1.6*. *Nature* **355:** 516.

Clarke J.D.W. and Lumsden A. 1993. Segmental repetition of neuronal phenotype sets in the chick embryo hindbrain. *Development* **118:** 151.

Condie B.G. and Capecchi M.R. 1993. Mice homozygous for a targeted disruption of *Hoxd-3* (*Hox-4.1*) exhibit anterior transformations of the first and second cervical vertebrae, the atlas and the axis. *Development* **119:** 579.

———. 1994. Mice with targeted disruptions in the paralogous genes *Hoxa-3* and *Hoxd-3* reveal synergistic interactions. *Nature* **370:** 304.

Davis A.P. and Capecchi M.R. 1994. Axial homeosis and appendicular skeleton defects in mice with targeted disruption of *Hoxd-11*. *Development* **120:** 2187.

———. 1996. A mutational analysis of the 5′ *Hox D* genes: Dissection of genetic interactions during limb development in the mouse. *Development* **122:** 1175.

Davis A.P., Witte D.P., Hsieh-Li H.M., Potter S.S., and Capecchi M.R. 1995. Absence of radius and ulna in mice lacking *Hoxa-11* and *Hoxd-11*. *Nature* **375:** 791.

Dollé P., Izpisúa-Belmonte J.-C., Brown J.M., Tickle C., and Duboule D. 1991. *Hox-4* genes and the morphogenesis of mammalian genitalia. *Genes Dev.* **5:** 1767.

Dollé P., Izpisúa-Belmonte J.-C., Falkenstein H., Renucci A., and Duboule D. 1989. Coordinate expression of the murine *Hox-5* complex homeobox-containing genes during limb pattern formation. *Nature* **342:** 767.

Dollé P., Dierich A., LeMeur M., Schimmang T., Schuhbaur B., Chambon P., and Duboule D. 1993. Disruption of the *Hoxd-13* gene induces localized heterochrony leading to mice with neotenic limbs. *Cell* **75:** 431.

Duboule D. 1994. Temporal colinearity and the phylotypic progression: A basis for the stability of a vertebrate Bauplan and the evolution of morphologies through heterochrony. *Development* (suppl.), p. 135.

Duboule D. and Dollé P. 1989. The structural and functional organization of the murine *Hox* gene family resembles that of *Drosophila* homeotic genes. *EMBO J.* **8:** 1497.

Favier B., LeMeur M., Chambon P., and Dollé P. 1995. Axial skeletal homeosis and forelimb malformations in *Hoxd-11* mutant mice. *Proc. Natl. Acad. Sci.* **92:** 310.

Favier B., Rijli F.M., Fromental-Ramain C., Fraulob V., Chambon P., and Dollé P. 1996. Functional cooperation between the non-paralogous genes *Hoxa-10* and *Hoxd-11* in the developing forelimb and axial skeleton. *Development* **122:** 449.

Fraser S., Keynes R., and Lumsden A. 1990. Segmentation in the chick embryo hindbrain is defined by cell lineage restrictions. *Nature* **344:** 431.

Fromental-Ramain C., Warot X., Lakkaraju S., Favier B., Haack H., Birling C., Dierich A., Dollé P., and Chambon P. 1996. Specific and redundant functions of the paralogous *Hoxa-9* and *Hoxd-9* genes in forelimb and axial skeleton patterning. *Development* **122:** 461.

Glover J.C. and Petursdottir G. 1991. Regional specificity of developing reticulospinal, vestibulospinal and vestibulo-ocular projections in the chicken embryo. *J. Neurobiol.* **22:** 353.

Goddard J.M., Rossel M., Manley N.R., and Capecchi M.R. 1996. Mice with targeted disruption of *Hoxb-1* fail to form the motor nucleus of the VIIth nerve. *Development* **122:** 3217.

Graham A., Papalopulu N., and Krumlauf R. 1989. The murine and *Drosophila* homeobox gene complexes have common features of organization and expression. *Cell* **57:** 367.

Haack H. and Gruss P. 1993. The establishment of murine *Hox-1* expression domains during patterning of the limb. *Dev. Biol.* **157:** 410.

Hinchliffe R. 1991. Developmental approaches to the problem of transformation of limb structure in evolution. In *Developmental patterning of the vertebrate limb* (ed. J.R. Hinchliffe et al.), p. 313. Plenum Press, New York.

Holland P.W.H. and Garcia-Fernandez J. 1996. *Hox* genes and chordate evolution. *Dev. Biol.* **173:** 382.

Izpisúa-Belmonte J.-C., Falkenstein H., Dollé P., Renucci A., and Duboule D. 1991. Murine genes related to the *Drosophila AbdB* homeotic gene are sequentially expressed during development of the posterior part of the body. *EMBO J.* **10:** 2279.

Krumlauf R. 1994. *Hox* genes in vertebrate development. *Cell* **78:** 191.

Lewis E.B. 1978. A gene complex controlling segmentation in *Drosophila*. *Nature* **276:** 565.

Lumsden A. and Keynes R. 1989. Segmental patterns of neuronal development in the chick hindbrain. *Nature* **337:** 424.

Mark M., Lufkin T., Vonesch J.-L., Ruberte E., Olivo J.-C., Dollé P., Gorry P., Lumsden A., and Chambon P. 1993. Two rhombomeres are altered in *Hoxa-1* mutant mice. *Development* **119:** 319.

McGinnis W., Garber R.L., Wirz J., Kuroiwa A., and Gehring W.J. 1984a. A homologous protein-coding sequence in *Drosophila* homeotic genes and its conservation in other metazoans. *Cell* **37:** 403.

McGinnis W., Levine M.D., Hafen E., Kuroiwa A., and Gehring W.J. 1984b. A conserved DNA sequence in homeotic genes of the *Drosophila Antennapedia* and *bithorax* complexes. *Nature* **308:** 428.

Pendleton J.W., Nagai B.K., Murtha M.T., and Ruddle F.H. 1993. Expansion of the *Hox* gene family and the evolution of chordates. *Proc. Natl. Acad. Sci.* **90:** 6300.

Riddle R.D., Johnson R.L., Laufer E., and Tabin C. 1993. *Sonic hedgehog* mediates the polarizing activity of the ZPA. *Cell* **75:** 1401.

Roberts D.J., Johnson R.L., Burke A.C., Nelson C.E., Morgan B.A., and Tabin C. 1995. Sonic hedgehog is an endodermal signal inducing *Bmp-4* and *Hox* genes during induction and regionalization of the chick hindgut. *Development* **121:** 3163.

Scott M.P. 1992. Vertebrate homeobox gene nomenclature. *Cell* **71:** 551.

Scott M.P. and Weiner A.J. 1984. Structural relationships among genes that control development: Sequence homology between the *Antennapedia, Ultrabithorax* and *fushi tarazu* loci of *Drosophila*. *Proc. Natl. Acad. Sci.* **81:** 4115.

Shubin N.H. and Alberch P. 1986. A morphogenetic approach to the origin and basic organization of the tetrapod limb. *Evol. Biol.* **20:** 319.

Slack J.M.W., Holland P.W.H., and Graham C.F. 1993. The zootype and the phylotypic stage. *Nature* **361:** 490.

Small K.M. and Potter S.S. 1993. Homeotic transformations and limb defects in *HoxA-11* mutant mice. *Genes Dev.* **7:** 2318.

Studer M., Lumsden A., Ariza-McNaughton L., Bradley A., and Krumlauf R. 1996. Altered segmental identity and abnormal migration of motor neurons in mice lacking *Hoxb-1*. *Nature* **384:** 630.

Vaage S. 1969. The segmentation of the primitive neural tube in chick embryos (*Gallus domesticus*). *Adv. Anat. Embryol. Cell Biol* **41:** 1.

van der Hoeven F., Zákány J., and Duboule D. 1996. Gene transpositions in the *HoxD* complex reveal a hierarchy of regulatory controls. *Cell* **85:** 1025.

Yokouchi Y., Sasaki H., and Kuroiwa A. 1991. Homeobox gene expression correlated with the bifurcation process of limb cartilage development. *Nature* **353:** 443.

Yokouchi Y., Nakazato S., Yamamoto M., Goto Y., Kameda T., Iba H., and Kuroiwa A. 1995. Misexpression of *Hoxa-13* induces cartilage homeotic transformation and changes cell adhesiveness in chick limb buds. *Genes Dev.* **9:** 2509.

Zákány J. and Duboule D. 1996. Synpolydactyly in mice with a targeted deficiency in the *HoxD* complex. *Nature* **384:** 69.

Intercompartmental Signaling and the Regulation of *vestigial* Expression at the Dorsoventral Boundary of the Developing *Drosophila* Wing

J. KIM, J. MAGEE, and S.B. CARROLL
Howard Hughes Medical Institute, Laboratory of Molecular Biology, University of Wisconsin, Madison, Wisconsin 53706

Appendage development is a dynamic process of growth and patterning. Signals from cells of the embryonic body wall initiate formation of limb primordia at discrete positions and establish each secondary patterning field in which the growth and differentiation of the limb are regulated. At a molecular level, growth and patterning are controlled through the precise temporal and spatial expression of successive organizing signals which, in turn, activate tissue-specific effector genes. One of the principal current challenges is to understand the signaling pathways and transcriptional regulatory mechanisms through which key appendage-patterning genes are controlled.

In the *Drosophila* embryo, the formation of the wing primordia is promoted by the decapentaplegic (Dpp) and wingless (Wg) signals and repressed by the spitz (Spi) ligand (Cohen et al. 1993; Goto and Hayashi 1997) and by selected *Hox* genes in segments in which wings will not form (Vachon et al. 1992; Carroll et al. 1995). In response to these signals, cells in the embryonic body wall segregate to form the imaginal disc that will give rise to the wing and adjacent adult body wall. The cells of growing larval wing imaginal discs are subdivided into four distinct populations, the anterior (A), posterior (P), dorsal (D), and ventral (V) compartments (Garcia-Bellido1975). The lineage restriction and identity of each compartment are determined by the activity of regulatory genes called selector genes (Garcia-Bellido 1975; Blair 1995; Lawrence and Struhl 1996). The selector gene *engrailed* is expressed in the posterior compartment and controls *hedgehog* (Hh) expression (Morata and Lawrence 1975; Lawrence and Morata 1976; Sanicola et al. 1995) by repressing *Cubitus interruptus* (Ci) expression (Eaton and Kornberg 1990) which in turn represses *hh* in the anterior compartment (Orenic et al. 1990; Domínguez et al. 1996). Another selector gene, the LIM-homeodomain protein *apterous* (Ap), is responsible for the dorsoventral (D/V) subdivision of the wing field (Cohen et al. 1992; Dias-Benjumea and Cohen 1993; Williams et al. 1993). Apterous-expressing cells in the dorsal compartment activate the *fringe* (*fng*) gene, which encodes a putative extracellular protein (Irvine and Wieschaus 1994). When *fng*-expressing dorsal cells are juxtaposed with ventral cells that do not express *fng*, the Serrate (Ser) protein, a ligand for the Notch (N) receptor (Spreicher et al. 1994), is upregulated in dorsal boundary cells (Kim et al. 1995).

Short-range interactions between anteroposterior (A/P) compartments through the Hh signaling pathway (Basler and Struhl 1994; Dias-Benjumea et al. 1994; Tabata and Kornberg 1994; Ingham and Feitz 1995; Zecca et al. 1995) and D/V compartments through the N signaling pathway (Couso et al. 1995; Dias-Benjumea and Cohen 1995; Kim et al. 1995; Rulifson and Blair 1995; Doherty et al. 1996; Neumann and Cohen 1996; de Celis et al. 1996a,b) induce the expression of two long-range signals, Dpp and Wg, that organize wing pattern (Lecuit et al. 1996; Nellen et al. 1996; Zecca et al. 1996; Neumann and Cohen 1997). Thus, cells along the compartment boundaries serve as discrete signaling sources that regulate the expression of genes throughout the wing field (Brook et al. 1996; Kim et al. 1996; Lawrence and Struhl 1996). One of these genes, *vestigial* (*vg*), is expressed in the presumptive wing blade under the control of two separate enhancers, the boundary and quadrant enhancers (Kim et al. 1996). The boundary enhancer drives *vg* expression in a stripe that bisects the wing pouch along the D/V boundary and marks the presumptive adult wing margin (Williams et al. 1994). The quadrant enhancer drives *vg* expression in the rest of the wing pouch (Kim et al. 1996). The boundary enhancer is activated in parallel to the *wingless* gene in boundary cells via a pathway involving the membrane receptor N and the transcription factor Suppressor of Hairless (Su[H]) (Kim et al. 1995, 1996). The quadrant enhancer is regulated by Dpp, as well as by a signal produced by the D/V boundary cells via N signaling (Kim et al. 1996, 1997).

The insect wing requires symmetrical growth and patterning around the D/V boundary for the elaboration of its normal morphology. Therefore, growth of the future dorsal and ventral surfaces of the wing must be coordinated in order for these surfaces to later fuse and to form the wing blade. To achieve this symmetry, patterning genes such as *vg* and *wg* must be activated on both sides of the D/V compartment border, and thus the N pathway must also be activated symmetrically. This process is initiated by the *ap* selector gene via the *fng* cell recognition molecule and Ser ligand produced by dorsal cells (Kim et al. 1995). The Ser signal is received by *N* in ventral cells and transduced via the Su(H) transcription factor to activate D/V boundary-specific enhancers such as the *vg* boundary enhancer (Kim et al. 1996), the *wg* wing margin enhancer (Neumann and Cohen 1996), and genes of the *E(spl)* complex (Bailey and Posakony 1995; Lecourtois and Schweisguth 1995). However, only ventral cells are competent to receive the Ser signal (Couso et al. 1995; Kim et al. 1995). This implies the presence of a ventral to dorsal signaling molecule(s) that mirrors the Ser dorsal signal (Kim et al. 1995). Two possible mechanisms could account for the reciprocal circuitry at the D/V boundary. If the pathways are acting in parallel, the dorsal to ventral and ventral to dorsal signals may act simultaneously and independently. Alternatively, a linear pathway could involve a primary signal produced in one compartment that in turn triggers the production of the secondary signal in the opposite compartment. This secondary signal would then signal back to the first compartment to complete the pathway.

In this work, we show that reciprocal signaling at the D/V boundary is a sequential pathway that initiates with dorsal to ventral signaling, rather than two parallel signaling events. We also present evidence that other factors, in addition to Su(H), must exist to confer tissue-specific expression of *vg* and to regulate the spatial and temporal features of *vg* expression within the wing pouch itself.

METHODS

Drosophila *stocks and antibodies.* *Su(H)^{SF8}* mutant flies were obtained from Francois Schweisguth (Institute Jacques Monod, CNRS-Universite Paris VII, France). The *w^{1118}; P[mini-w$^+$: hs-M]36F, P[ry$^+$: hs-neo: FRT] 40A* fly stock was obtained from the Bloomington *Drosophila* Stock Center. Mouse anti-human c-Myc hybridoma line 1-9E10.2 was obtained from the American Type Culture Collection. The hybridoma culture supernatant was used at a dilution of 1 : 5. Mouse anti-human c-Myc hybridoma line 1-9E10.2 was obtained from the American Type Culture Collection. The hybridoma culture supernatant was used at a dilution of 1:5. Rabbit anti-Vg, rat anti-gal (Williams et al. 1993, 1994), and rat anti-Ap (Kim et al. 1995) antibodies were generated against purified proteins. Rabbit anti-Wg antibody 65 was obtained from R. Nusse (Stanford University). Secondary and tertiary antibodies conjugated with the fluorochromes FITC, LRSC, or Cy5 were purchased from Jackson Immuno Research Lab. Inc. (West Grove, Pennsylvania).

Generation of* Su(H)^{SF8} *mosaic clones. The mitotic recombination was induced by irradiating the *Su(H)^{SF8}/P[mini-w$^+$, hs-M]36F, P[ry$^+$;hs-neo;FRT] 40A; P[mini-w$^+$, vg D/V enhancer-LacZ]* larvae with 4000 Rads of γ-ray 42–50 hours after egg laying (for the clones crossing the D/V boundary) or 60–65 hours after egg laying (for the clones not crossing the D/V boundary). The larvae were heat-shocked to induce the Myc epitope tag for 90 minutes at 37°C, dissected, fixed, and stained with anti-Myc antibody in combination with anti-β-galactosidase, anti-Vg, and/or anti-Wg antibody.

Immunohistochemistry. Wing imaginal discs were dissected, fixed, and stained with antibodies as described previously (Kim et al. 1995). The following experimental steps were performed at 4°C. Imaginal disc complexes dissected in phosphate-buffered saline (pH 7.0) were fixed for 30 minutes in fixative [0.1 M PIPES [pH 6.9], 150 mM NaCl, 1 mM EGTA [pH 6.9], 1% Triton X-100, 2% formaldehyde). The fixed disc complexes were incubated for 4 hours in Block Buffer (50 mM Tris [pH 6.8], 150 mM NaCl, 0.5% NP-40, 5 mg/ml BSA). The disc complexes were incubated with primary antibodies overnight in Wash Buffer (50 mM Tris [pH 6.8], 150 mM NaCl, 0.5% NP-40, 1 mg/ml BSA). The disc complexes were washed four times for 20 minutes each with the Wash Buffer and incubated with secondary antibodies for 3 hours. The disc complexes were soaked overnight in Mounting Solution (50 mM Tris [pH 8.8], 150 mM NaCl, 30% glycerol [v/v]). The wing discs were dissected and mounted in the Mounting Solution. When an FITC-conjugated reagent was used, 0.5 mg/ml of *p*-phenylethylene diamine (Sigma) was added to the Mounting Solution. Confocal images were collected with a Bio-Rad MRC1024 laser scanning confocal microscope (Bio-Rad Microscience Division, Hercules, California) and assembled with Adobe Photoshop 3.0 (Adobe Systems, Mountain View, California).

Reporter gene construction. Deletion constructs were made by polymerase chain reaction (PCR) from the wild-type enhancers. Primers used in the PCRs correspond to the coordinates given in the construct name in Figure 4. Nucleotide numbers were assigned as described by Williams et al. (1994). When constructs contained internal deletions, each half of the construct was PCR-amplified separately and assembled in a cloning vector. Constructs Δ80-98 and Δ110–197 were assembled directly in the Hsp70-Casper-LacZ shuttle vector, and they contain no linker to maintain spacing. Constructs Δ197–246, Δ241–282, Δ281–322, and Δ321–361 were assembled in a pBC-SK vector (Stratagene) with a 40-base-pair linker to maintain spacing. These constructs were subsequently cloned into Hsp70-Casper-LacZ (Nelson and Laughon 1993). All constructs were sequenced and injected into yw embryos. For each construct, at least two transformant lines were examined for reporter gene expression. β-galactosidase expression was analyzed immunohistochemically with a rat anti-β-galactosidase primary antibody and an LRSC-conjugated donkey anti-rat secondary antibody (Jackson).

RESULTS

Ser Determines vg Expression Domains in the Wing Pouch

The boundary enhancer expression pattern consists of three separable domains. In the second instar wing disc, the enhancer is expressed in a broad stripe across the D/V

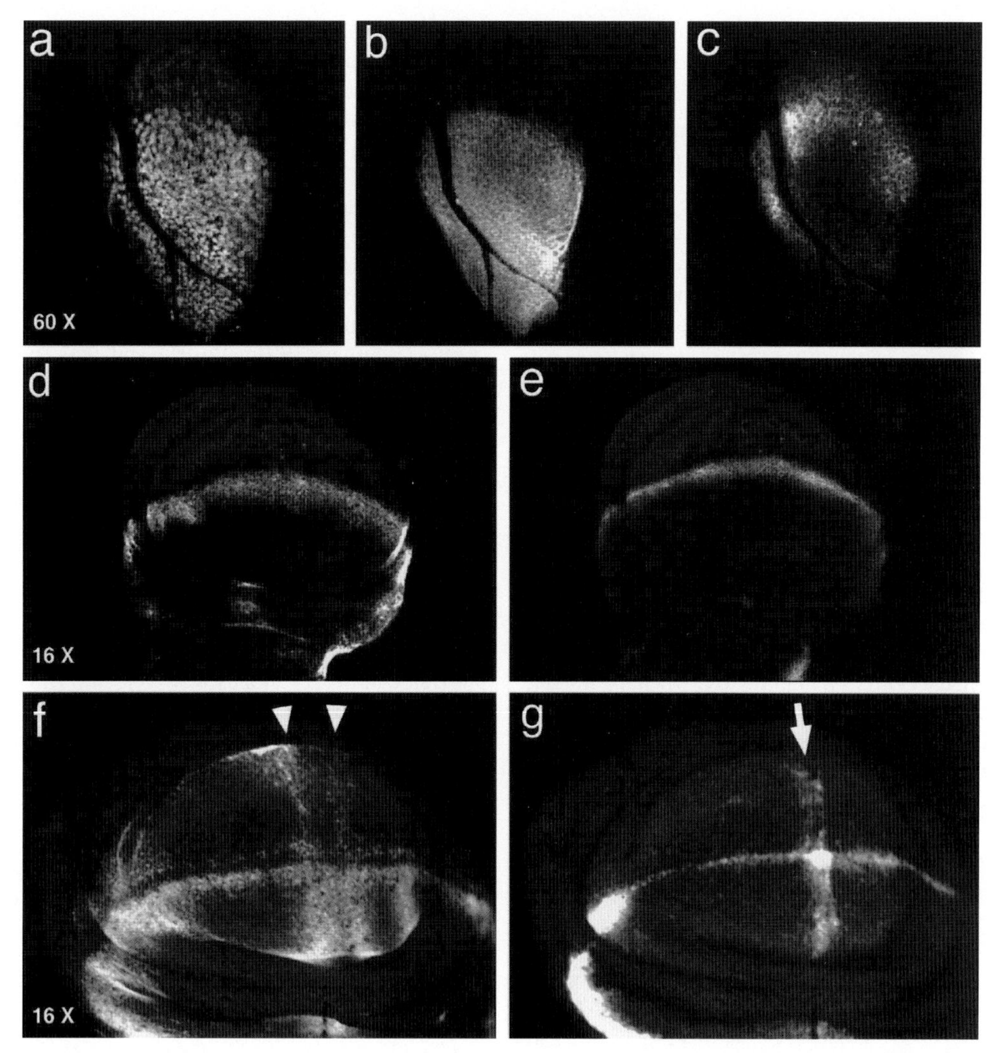

Figure 1. Spatiotemporal regulation of the *vg* boundary enhancer correlates with Ser expression domains. Wing imaginal discs were dissected from larvae carrying two copies of a β-galactosidase transgene driven by the *vg* boundary enhancer. In all panels, the anterior is to the left, and ventral is at the top. Relative magnification is indicated in the lower left panel. (*a–c*) A mid late second instar wing imaginal disc triple-stained with anti-ap (*a*), anti-Ser (*b*), and anti-β-galactosidase antibodies (*c*). At this stage, Ser is uniformly expressed in the dorsal compartment (*b*) where ap protein is expressed (*a*) and the *vg* boundary enhancer is activated in a broad, graded pattern along the D/V boundary (*c*). (*d,e*) A mid-third instar wing imaginal disc double-stained with anti-Ser (*d*) and anti-β-galactosidase antibodies. During the early to mid third instar stage, Ser expression recedes to the perimeter of the dorsal wing pouch, leaving a stripe near the D/V boundary (*d*), and the *vg* boundary enhancer expression is refined to a thin stripe that straddles the D/V boundary symmetrically (*e*). (*f,g*) A late third-instar wing imaginal disc double-stained with anti-Ser (*f*) and anti-β-galactosidase (*g*) antibodies. Ser is expressed in the pre-vein regions and the edge cells on both sides of the D/V compartment boundary (*f*) in late third instar. At this stage, the *vg* boundary enhancer is also activated in the intervein region (*g*, arrow) which is flanked by two longitudinal Ser stripes in the pre-vein 3 and pre-vein 4 (*f*, arrowheads).

boundary (Fig. 1c) that is refined by the mid third instar to a thin stripe that straddles the D/V boundary symmetrically (Fig. 1e). The enhancer is also expressed in two stripes in the notum of the disc and in the late third instar; the enhancer is expressed in a third domain along the A/P boundary of the wing pouch (Fig. 1g). The D/V boundary and notal patterns are clearly the product of distinct regulatory inputs because an enhancer carrying a mutant Su(H)-binding site fails to express a *lacZ* reporter gene at the D/V boundary, whereas the notal pattern remains unaffected (Kim et al. 1996). The relationship between the regulation of the A/P pattern and the other expression domains is not as clear, however.

Here, we examined the relationship between the expression of the Ser ligand and vg boundary enhancer, and we suggest that the A/P stripe is also N-dependent. Others have shown that in the late third instar, Fng expression expands into the ventral compartment. At this point, the Ser expression domain expands into the ventral compartment, forming a stripe along the A/P boundary within the wing pouch (Fig. 1f). The Ser stripe coincides with the vg A/P stripe, suggesting that the A/P stripe is N-regulated (compare f and g in Fig. 1). This hypothesis is further supported by the fact that mutation of the Su(H)-binding site in the enhancer abolishes the A/P stripe in addition to the D/V stripe. Thus, all wing pouch expression of the boundary enhancer appears to be regulated by the N/Su(H) pathway.

Su(H) Has Differential Roles in the Dorsal and Ventral Boundary Cells

We have previously shown that activation of the *vg* boundary enhancer at the D/V boundary ultimately requires the direct binding of the Su(H) protein to the enhancer in response to activation of the Notch pathway (Kim et al. 1996). However, the role of the ventral boundary cells in activating the *vg* boundary enhancer in dorsal cells and the role of N signaling in the Ser-producing dorsal cells are not clear.

To examine the role of N signaling in dorsal and ventral boundary cells separately, we selectively removed Su(H) activity in each individual compartment. *Su(H)* null mutant clones generated in the late second to early third instar stage obey compartmental restrictions and stay within the compartment from which their parental cells originated. Thus, late clones tend to smoothly abut cells in the neighboring compartment without crossing the D/V boundary (Fig. 2). Importantly, when the mutant cell patches are induced after the D/V boundary is established, ventral *Su(H)*$^-$ clones that touch the D/V boundary abolish *vg* boundary enhancer activity and *wg* expression on both sides of the D/V boundary (Fig. 2a,c). In contrast, dorsal clones that touch the D/V boundary lose boundary enhancer and *wg* expression but do not affect ventral boundary cells (Fig. 2b,d).

These results reveal three features of gene regulation and signaling at the D/V boundary. First, Su(H) function is necessary for the expression of *vg* and *wg* within cells on both sides of the D/V boundary (Kim et al. 1996). Second, Su(H) is required within ventral cells not only for receiving Ser signals from the dorsal cells, but also for the signaling to dorsal cells to express *vg* and *wg*. Since Su(H) protein is a transcription factor (Fortini and Artavanis-Tsakonas 1994; Bailey and Posakony 1995; Lecourtois and Schweisguth 1995) and itself cannot signal to other cells, we deduce that the Su(H) protein in the ventral boundary cells is involved in the production of an unknown ventral signal that signals back to the dorsal boundary cells. Third, *vg* and *wg* expression in dorsal cells also requires the N-Su(H) signaling cascade.

Su(H) Represses the vg Quadrant Enhancer at the D/V Boundary

Although *Su(H)*$^-$ clones that cross the D/V boundary abolish *vg* boundary enhancer activity (Fig. 3b), vg protein itself is still expressed (Fig. 3c). Moreover, Su(H)$^-$ clones, unlike *vg* null mutant clones, can survive in the wing pouch. Since *vg* expression in the pouch is regulated

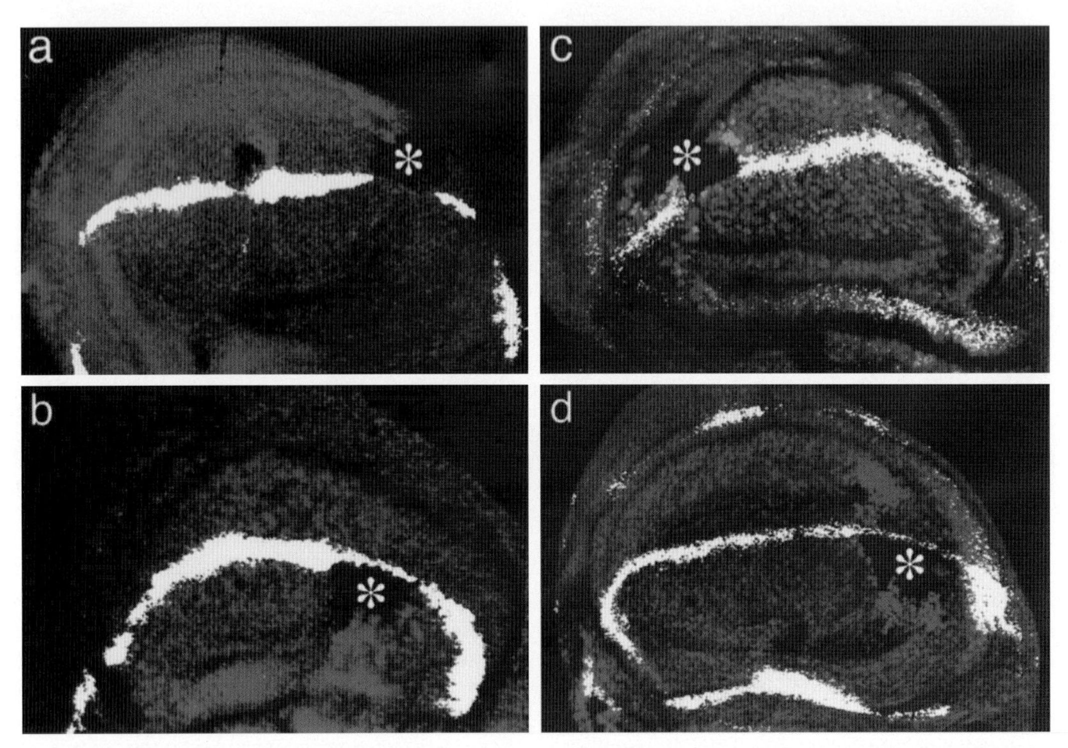

Figure 2. Asymmetric requirement for Su(H) function in the ventral and dorsal boundary cells for *vg* D/V boundary enhancer and *wg* expression. The Su(H)$^-$ null mutant cell patches (blank areas marked with asterisks) were generated by inducing mitotic recombination in the *Su(H)* null heterozygote larvae harboring two copies of the *vg* boundary enhancer-*lacZ* transgene in which the wild-type second chromosome is marked with a Myc-epitope tag. In all panels, the expression of Myc epitope tag is shown in gray. The expression of the *vg* D/V boundary enhancer (*a* and *b*) and wg protein (*c* and *d*) are shown in white. The anterior is to the left, and ventral is at the top. (*a,c*) The ventral Su(H) null mutant cell patch touching the D/V boundary from the ventral side causes nonautonomous loss of the *vg* D/V enhancer (*a*) and *wg* (*c*) expression in the adjacent dorsal boundary cells as well as autonomous loss in the mutant ventral boundary cells. (*b,d*) In contrast, the dorsal Su(H)-null clones touching the D/V boundary from the dorsal side cause autonomous loss of the *vg* D/V boundary enhancer (*b*) and *wg* (*d*) expression.

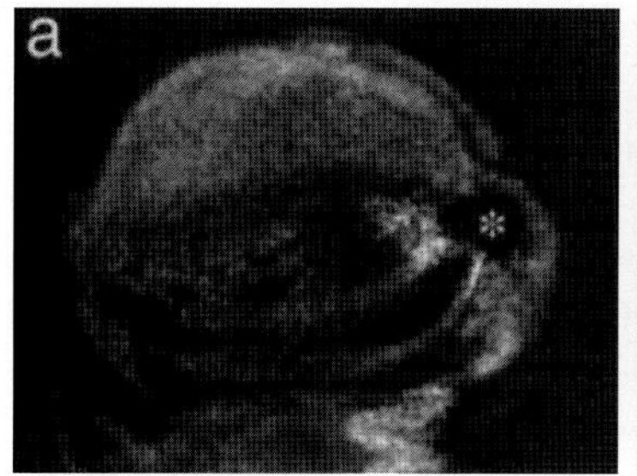

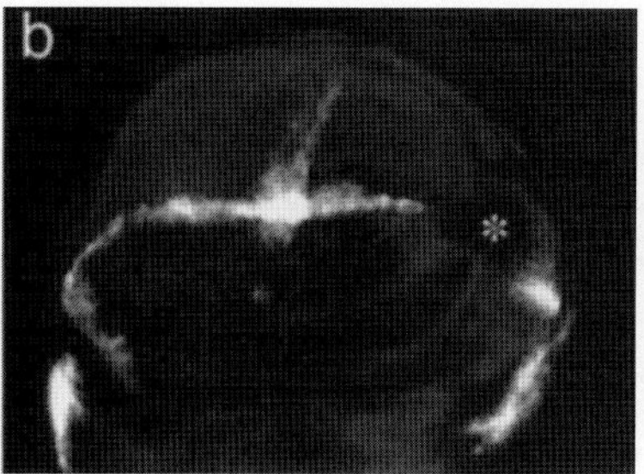

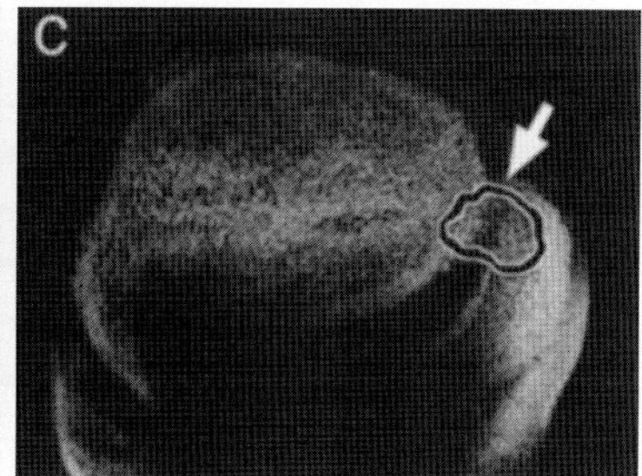

Figure 3. Loss of Su(H) function causes derepression of the quadrant enhancer along the D/V boundary. The *Su(H)* null clones were generated in larvae harboring two copies of the *vg* boundary enhancer-*lacZ* transgene. The wild-type second chromosome was marked with a Myc-epitope tag. The imaginal discs were dissected and triple-stained with anti-Myc (*a,b*; shown in gray), anti-β-galactosidase (*b*; shown in white), and anti-vg (*c*) antibodies. In all panels, the anterior is to the left, and ventral is at the top. Su(H) null mutant cells at the D/V boundary (blank area marked with asterisks in *a* and *b*, and circled with a bold line *c*) still express vg proteins (*c*, arrow) even though the *vg* boundary enhancer is not activated (*b*).

by two enhancers, and the boundary enhancer is inactive in $Su(H)^-$ clones, we conclude that the quadrant enhancer is now active in these cells and compensates for the loss of the *vg* boundary enhancer activity. This suggests that the repression of the *vg* quadrant enhancer along the D/V boundary is Su(H)-dependent. Since there are no Su(H)-binding sites in the *vg* quadrant enhancer (Kim et al. 1996), this repression is probably indirect and mediated by another Su(H) target that prevents the quadrant enhancer from being activated along the D/V boundary. The *E(spl)m8* repressor protein is activated by *Su(H)* at the D/V boundary (Bailey and Posakony 1995), and there is an E(spl)-binding site in the quadrant enhancer (Kim et al. 1996). We propose that *E(spl)m8* is the best candidate for this repressor.

Quantitative Inputs to the Boundary Enhancer

The N/Su(H) pathway is involved in many well-studied developmental processes such as the development of photoreceptor cells in the eye and the sensory organs of the body wall and wing margin (Artavanis-Tsakonas et al. 1995). The N/Su(H)-regulated boundary enhancer, however, is activated only in the wing and haltere discs (Williams et al. 1993). Clearly, other regulatory inputs must coincide with N/Su(H) signaling to distinguish *vg* expression from other N-target genes. In addition, there may be factors that control the temporal or quantitative expression of the enhancer. To explain these inputs, we have dissected the boundary enhancer by sequentially deleting segments and assaying enhancer function when fused to a *lacZ* reporter gene (Fig. 4a).

The *D. melanogaster* boundary enhancer is a 750-bp element that is highly conserved with a homologous element in *Drosophila virilis* (Fig. 4a), a species approximately 50 million years diverged from *D. melanogaster* (Williams et al. 1994). It has been shown that the first 360 bp of the element are sufficient to control the minimal pattern of expression, although the level of activity was significantly reduced. Therefore, we designed our deletion analysis to target the 360-base minimal element in the context of the entire 750-bp enhancer (Fig. 4a). The Su(H) site, located within bases 99–109, was left intact in each construct.

None of the deletions eliminated enhancer function altogether. However, four deletions reduced *lacZ* expression at the D/V boundary (Fig. 4). Similar results were seen for the notal expression domain. Of the four deletions that affected reporter gene levels at the D/V boundary, the Δ0–80 construct shows an almost complete loss of function in the mid third instar (Fig. 4b). However, unlike the Su(H) site knockout, the Δ0–80 construct appears to be expressed at normal levels in the second instar (not shown) and weaker levels in the early third instar (Fig. 4c). Since the Δ0–40 deletion had no effect, we deduced that the essential sequences lie between base pairs 40 and 80 in the enhancer. An interesting deviation in the spatial pattern of expression was seen in the Δ110–197 deletion. When this unconserved region is deleted, the enhancer displays reduced levels of expression along the D/V boundary and in the notum. However, the pattern is unaffected at the interface between the A/P and D/V boundaries (Fig. 4e).

These data show that the Su(H) transcription factor is the primary determinant of the qualitative pattern of enhancer activation, whereas several quantitative inputs mediate the level of enhancer activity. Moreover, these quantitative inputs appear to vary both spatially and temporally.

Three deletions caused ectopic *lacZ* expression in the wing disc or elsewhere in the body (Δ80–98, Δ197–246, and Δ241–282; Fig. 4d,f–h). The Δ80–98 deletion causes ectopic expression of the element within the hinge regions of the wing itself (Fig. 4d). The Δ197–246 and Δ241–282 deletions generate similar patterns of ectopic expression (Fig. 4f–h and not shown). In both constructs, reporter gene expression was observed in the most proximal regions of the leg and in the brain (Fig. 4f, lower right, and data not shown). In Δ241–282, expression was detected in the eye behind the morphogenetic furrow as well (Fig. 4h). These patterns may or may not be N-mediated. The ectopic leg pattern does not correlate with Ser expression that arises in the leg in the concentric folds (J. True and S. Carroll unpubl.). However, the eye and brain expression patterns may correlate with N-mediated lateral inhibition activity. These data suggest a role for re-

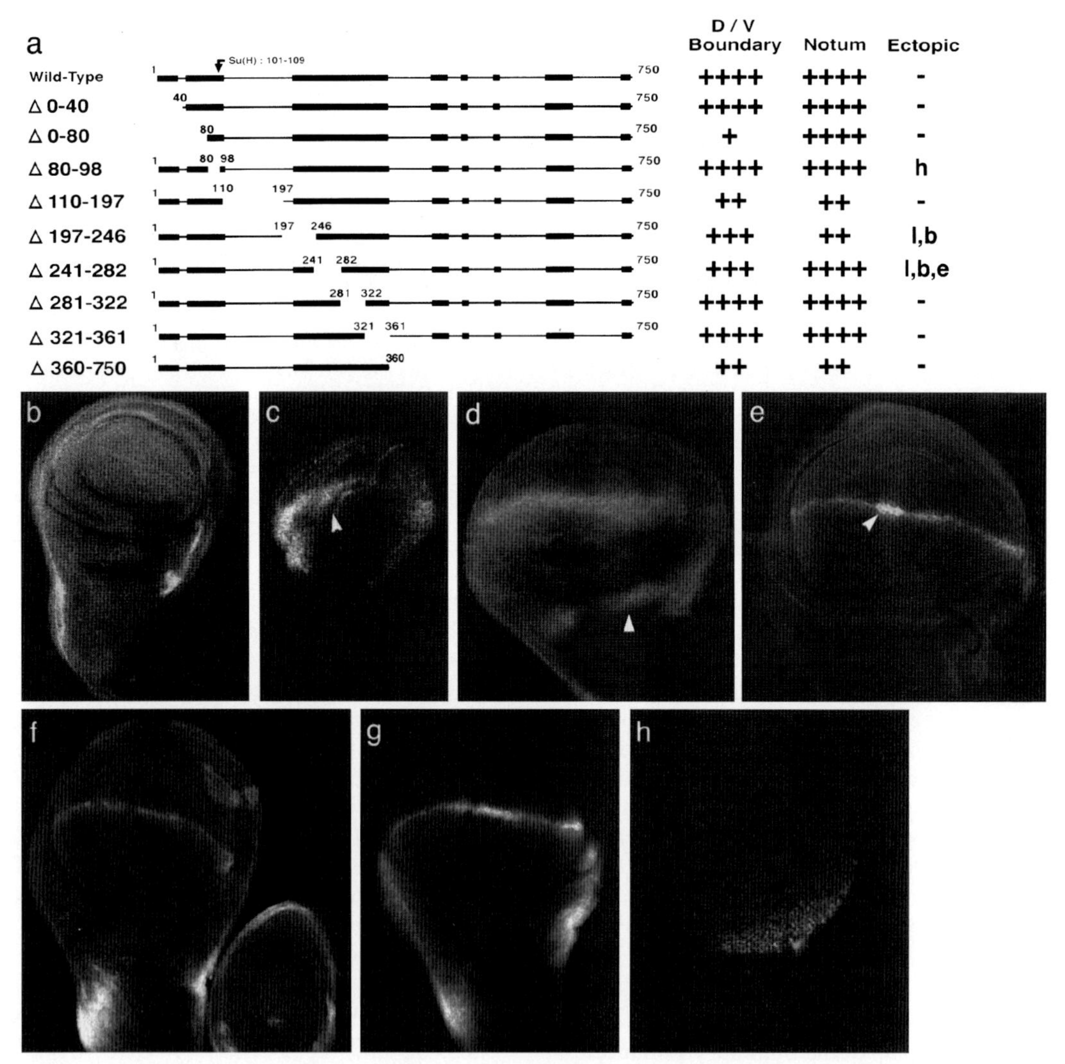

Figure 4. Deletion analysis of the *vg* boundary enhancer. (*a*) Each deletion construct is shown along with a summary of its effect on expression of the *lacZ* reporter gene on in specific regions of the wing disc. The "ectopic" column summarizes expression in the hinge of the wing disc (h), the leg (l), the brain (b), or the eye (e). Conserved regions of the enhancer are marked by bold boxes. (*b,c*) The Δ0–80 deletion almost completely abolishes enhancer function during the mid third instar (*b*), although expression is detectable in the early third instar (*c*, arrowhead). (*d*) The Δ80–98 deletion induces ectopic *lacZ* expression in the hinge region of the wing disc (arrowhead). (*e*) The Δ110–197 deletion reduces notal and boundary expression, except at the interface of the D/V and A/P boundaries (arrowhead). (*f*) The Δ197–246 deletion reduces *lacZ* expression at the D/V boundary and in part of the notum (*left*). Ectopic *lacZ* expression is shown in the proximal regions of the leg imaginal disc (*lower right*). (*g,h*) The Δ241–282 deletion reduces D/V boundary expression (*g*) while inducing ectopic enhancer activity in the eye behind the morphogenetic furrow (*h*) and in the leg and brain (not shown).

pressor factors, in addition to the quantitative activators that regulate the boundary enhancer.

DISCUSSION

Intercompartmental Signaling at the D/V Boundary

Drosophila limb fields are specified by localized signals in the embryonic body wall and then segregate as imaginal discs in the larval stages. An early step toward regulating the growth and patterning of this field is the partitioning of the fields into distinct compartments (Garcia-Bellido 1975; Morata and Lawrence 1975). One of the major functions of compartmentalization appears to be the specification of compartment boundary cells as signaling sources which organize the growth and patterning of the fields (for review, see Blair 1995; Lawrence and Struhl 1996).

The A/P compartment boundary is inherited from embryonic ectoderm (Wilkins and Gubb 1991), and the signaling between these two compartments is unidirectional (Basler and Struhl 1994; Tabata and Kornberg 1994; Zecca et al. 1995). In contrast, the D/V compartment boundary is established in the larval imaginal discs by localized expression of the selector gene *apterous* (Dias-Benjumea and Cohen 1993). Moreover, the expression of effector genes, as well as growth and patterning along the D/V boundary, is symmetrical and requires reciprocal signaling between the two compartments (Williams et al. 1994; Kim et al. 1995).

In this work, we have shown that the reciprocal signaling between the D/V compartments is a circuit of sequential signaling events, rather than two independent, simultaneous events (Fig. 5). We found that the synthesis or release of a signal from ventral cells to the dorsal boundary cells is dependent on reception of the dorsal signal, Ser (see Fig. 2). Because *vg* and *wg* expression in the dorsal boundary cells also depends on Su(H) activity, the ventral signal is likely to be a N ligand or a modifier of the interaction between N and its ligands. The ventral signal is most likely not wg as ventral *wg*$^{-}$ clones that touch the D/V boundary do not cause the loss of *wg* expression in the adjacent dorsal cells (Rulifson and Blair 1995). *Delta* (*Dl*) has been proposed as the ventral signal (Doherty et al. 1996; Panin et al. 1997). *Dl* is expressed in both sides of the D/V boundary (de Celis et al. 1996a) and can induce wing overgrowth in both compartments when expressed ectopically (Doherty et al. 1996), although the greatest effect is in the dorsal compartment (Panin et al. 1997). However, *Dl* null clones that cross or touch the D/V boundary from the ventral side do not abolish the *wg* expression at the dorsal boundary cells as would be expected if Dl was the exclusive ventral signal (Micchelli et al. 1997).

Dorsal cells, which express *fng*, are normally insensitive to the Ser signal unless they are adjacent to cells not expressing *fng* (Kim et al. 1995; Panin et al. 1997). Thus, it is possible that the ventral signal modifies N to make it sensitive to Ser, even in the presence of Fng. This modification could include posttranslational modification of N itself or modification of N interactions with other membrane proteins. Interestingly, the *Abruptex(Ax)* mutations of *N* make the domains of *wg* and *E(Spl)* expression expand asymmetrically toward the dorsal compartment where Ser is present (de Celis et al. 1996a). This implies that *Ax* mutations modify N so that it is able to respond to Ser, even in the presence of the Fng protein. It is also possible that the putatative ventral signaling molecule forms a complex with N or N-interacting protein(s) to modulate N function by protein-protein interaction(s). The N ligand that activates *wg* and the *vg* boundary enhancer in dorsal boundary cells may indeed be Ser (Fig. 5) (Couso et al. 1995; Kim et al. 1995).

The Structure of the Boundary Enhancer: The Unit of *cis*-Regulatory Element Function

The *cis*-regulatory regions of genes in higher organisms often contain multiple discrete enhancer elements that can act as independent functional units. One of the features of this mode of *cis*-element organization is that each individual enhancer can regulate gene activity in a temporal-, tissue-, or cell-specific manner. For example, the *vg* gene is activated in D/V boundary cells in response to D/V intercompartmental signaling through N, whereas

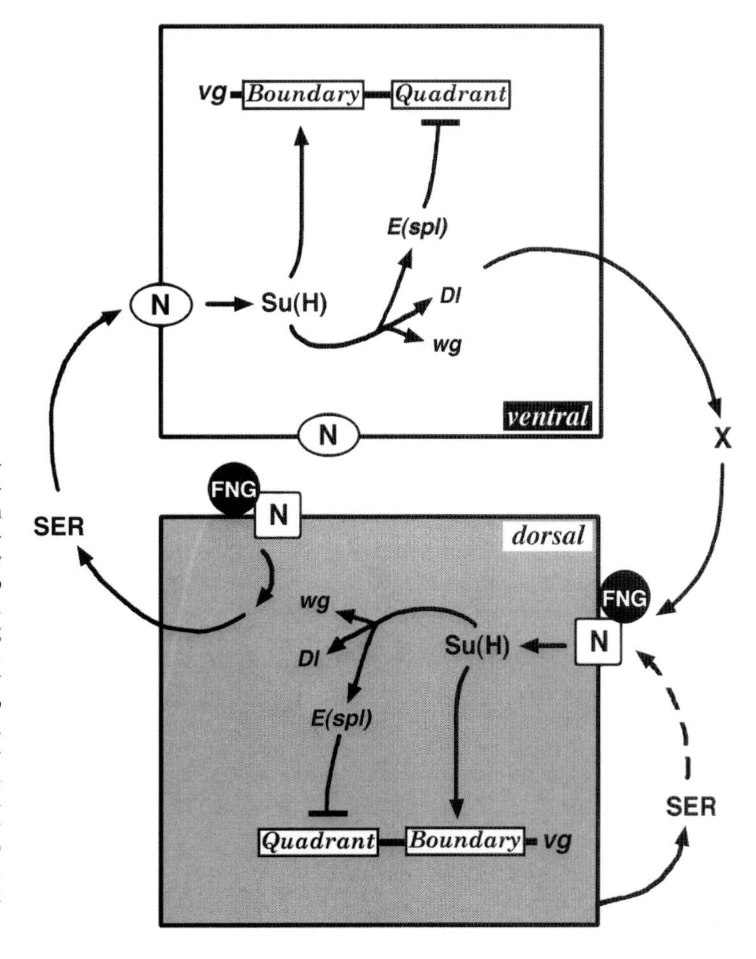

Figure 5. Schematic illustration of the signaling events between the dorsal and ventral boundary cells that result in the symmetrical activation of *vg*, *wg*, *Dl*, and *E(spl)* along the D/V boundary. The dorsal boundary cell expressing *fng* (*bottom*) produces Ser when they are adjacent to a ventral cell (*top*) which is not expressing *fng*. The Ser signal is transmitted to the neighboring ventral cells and received by the N receptors. Activation of the N receptors in ventral cells by the Ser signal causes the Su(H) protein to translocate into the nucleus and activate the *vg*, *wg*, *Dl*, *E(spl)*, and unknown target genes including an as yet unidentified secondary signal X. The signal X is transmitted back to the dorsal boundary cells and sensitizes the dorsal N-signaling pathway, which is normally insensitive to their own Ser ligand, to activate *vg*, *wg*, *E(spl)*, and other D/V boundary genes, again through the nuclear translocation of the Su(H) proteins.

vg expression in wing pouch cells is activated by Dpp and a signal emanating from the D/V boundary cells. The differential regulation of the two *vg* enhancers within the D/V boundary cells illustrates another functional advantage to organizing *cis*-regulatory elements as discrete units. In the D/V boundary cells, the *vg* boundary enhancer is activated and the *vg* quadrant enhancer is repressed (Fig. 3) (Kim et al. 1996). By separating these two enhancers, repression of the *vg* quadrant enhancer does not affect the *vg* boundary enhancer and consequently does not inactivate the *vg* gene. Thus, the boundary cells are able to produce vg independently of the graded expression that Dpp or the D/V organizing signal might confer.

Optimal enhancer activity typically depends on the composite activities of several classes of *trans*-regulatory factors. These include general activators and repressors, signal-dependent activators and repressors, tissue-specific activators and repressors (Arnone and Davidson 1997), and architectural proteins (Werner and Burley 1997). Our deletion analysis shows that the *vg* boundary enhancer also utilizes multiple inputs. Some portions of the enhancer sequences are necessary for the proper level of activation, whereas others are required for the tissue-specific restriction of enhancer activity (Fig. 4). A single qualitative input, the N/Su(H) pathway, activates the enhancer, although an unknown factor must bind the 40–80-nucleotide region to maintain transcription. Hypothetically, the multiple regulatory inputs may be priming the enhancer to respond or stabilizing the response to the single qualitative signaling cue.

Tissue-specific control of the *vg* boundary enhancer also appears to be the result of collaboration among multiple *cis*-regulatory sites. Specific deletions yielded ectopic enhancer activity in specific regions of the larval body, yet none of the deletions enabled the enhancer to respond globally to N signaling without any tissue restrictions. Therefore, the enhancer must carry different repressor elements, each capable of down-regulating *vg* in a specific tissue, or other tissues lack the *trans*-acting cofactors required for N-dependent activation of the boundary enhancer. Identification of the transcription factors that mediate the tissue specificity of the *vg* enhancers will provide crucial insights into the general mechanisms by which the same set of signals activate different sets of target genes in different tissues and thereby control tissue-specific patterns of morphogenesis.

ACKNOWLEDGMENTS

We thank Kathy Vorwerk for technical assistance and Jamie Wilson for help with manuscript preparation. This work was supported by the Howard Hughes Medical Institute.

REFERENCES

Arnone M. and Davidson E. 1997. The hardwiring of development: Organization and function of genomic regulatory systems. *Development* **124:** 1851.

Artavanis-Tsakonas S., Matsuno K., and Fortini M. 1995. Notch signaling. *Science* **268:** 225.

Bailey A. and Posakony J.W. 1995. Suppressor of hairless directly activates transcription of enhancer of split complex genes in response to Notch receptor activity. *Genes Dev.* **9:** 2609.

Basler K. and Struhl G. 1994. Compartment boundaries and the control of *Drosophila* limb pattern by hedgehog protein. *Nature* **368:** 208.

Blair S. 1995. Compartments and appendage development in *Drosophila. BioEssays* **17:** 299.

Brook W., Diaz-Benjumea F., and Cohen S. 1996. Organizing spatial pattern in limb development. *Annu. Rev. Cell Dev. Biol.* **12:** 161.

Carroll S., Weatherbee S., and Langeland J. 1995. Homeotic genes and the regulation and evolution of insect wing number. *Nature* **375:** 58.

Cohen B., Simcox A.A., and Cohen S.M. 1993. Allocation of the thoracic imaginal primordia in the *Drosophila* embryo. *Development* **117:** 597.

Cohen B., McGriffin M.E., Pfeifle C., Segal D., and Cohen S.M. 1992. *apterous*, a gene required for imaginal disc development in *Drosophila* encodes a member of the LIM family of developmental regulatory proteins. *Genes Dev.* **6:** 715.

Couso J., Knust E., and Martinez Arias A. 1995. *Serrate* and *wingless* cooperate to induce *vestigial* gene expression and wing formation in *Drosophila. Curr. Biol.* **5:** 1437.

de Celis J., Garcia-Bellido A., and Bray S. 1996a. Activation and function of *Notch* at the dorsal-ventral boundary of the wing imaginal disc. *Development* **122:** 359.

de Celis J.F., de Celis J., Ligoxygakis P., Preiss A., Delidakis C., and Bray S. 1996b. Functional relationships between *Notch, Su(H)*, and the bHLH genes of the *E(spl)* complex: The *E(spl)* genes mediate only a subset of *Notch* activities during imaginal development. *Development* **122:** 2719.

Dias-Benjumea F. and Cohen S. 1993. Interaction between dorsal and ventral cells in the imaginal disc directs wing development in *Drosophila. Cell* **75:** 741.

———. 1995. Serrate signals through Notch to establish a Wingless-dependent organizer at the dorsal/ventral compartment boundary of the *Drosophila* wing. *Development* **121:** 4215.

Dias-Benjumea F., Cohen B., and Cohen S. 1994. Cell interaction between compartments establishes the proximal-distal axis of *Drosophila* legs. *Nature* **372:** 175.

Doherty D., Feger G., Younger-Shephard S., Jan L., and Jan Y. 1996. Delta is a ventral to dorsal signal complementary to Serrate, another Notch ligand, in *Drosophila* wing formation. *Genes Dev.* **10:** 421.

Domínguez M., Brunner M., Hafen E., and Basler K. 1996. Sending and receiving the hedgehog signal: Control by the *Drosophila* gli protein Cubitus interruptus. *Science* **272:** 1621.

Eaton S. and Kornberg T. 1990. Repression of *Ci-D* in posterior compartments of *Drosophila* by *engrailed. Genes Dev.* **4:** 1068.

Fortini M. and Artavanis-Tsakonas S. 1994. The suppressor of hairless protein participates in notch receptor signaling. *Cell* **79:** 273.

Garcia-Bellido, A. 1975. Genetic control of wing disc development in *Drosophila. Ciba Found. Symp.* **29:** 161.

Goto S. and Hayashi S. 1997. Specification of the embryonic limb primordium by graded activity of Decapentaplegic. *Development* **124:** 125.

Ingham P. and Feitz M. 1995. Quantitative effects of *hedgehog* and *decapentaplegic* activity on the patterning of the *Drosophila* wing. *Curr. Biol.* **5:** 432.

Irvine K. and Wieschaus E. 1994. *fringe*, a boundary-specific signaling molecule, mediates interactions between dorsal and ventral cells during *Drosophila* wing development. *Cell* **79:** 595.

Kim J., Irvine K., and Carroll S. 1995. Cell recognition, signal induction, and symmetrical gene activation at the dorsal-ventral boundary of the developing *Drosophila* wing. *Cell* **82:** 795.

Kim J., Johnson K., Chen H.J., Carroll S.B., and Laughon A. 1997. MAD binds to DNA and directly mediates activation of vestigial by DPP. *Nature* **388:** 304.

Kim J., Sebring A., Esch J., Kraus M., Vorwerk K., Magee J., and Carroll S. 1996. Integration of positional signals and regulation of wing formation and identity by *Drosophila vestigial* gene. *Nature* **382:** 133.

Lawrence P. and Morata G. 1976. Compartments in the wing of *Drosophila*: A study of the *engrailed* gene. *Dev. Biol.* **50:** 321.

Lawrence P. and Struhl G. 1996. Morphogens, compartments, and pattern: Lessons from *Drosophila*? *Cell* **85:** 951.

Lecourtois M. and Schweisguth F. 1995. The neurogenic suppressor of hairless DNA-binding protein mediates the transcriptional activation of the enhancer of split complex genes triggered by Notch signaling. *Genes Dev.* **9:** 2598.

Lecuit T., Brook W., Ng M., Calleja M., Sun H., and Cohen S. 1996. Two distinct mechanisms for long-range patterning by Decapentaplegic in the *Drosophila* wing. *Nature* **381:** 387.

Micchelli C., Rulifson E., and Blair S. 1997. The function and regulation of *cut* expression on the wing margin of *Drosophila*: Notch, Wingless and a dominant negative role for Delta and Serrate. *Development* **124:** 1485.

Morata G. and Lawrence P. 1975. Control of compartment development by the *engrailed* gene in *Drosophila. Nature* **255:** 615.

Nellen D., Burke R., Struhl G., and Basler K. 1996. Direct and long-range actions of a Dpp morphogen gradient. *Cell* **85:** 357.

Nelson H. and Laughon A. 1993. The DNA binding specificity of the *Drosophila* fushi tarazu protein: A possible role for DNA bending in homeodomain recognition. *Roux's Arch. Dev. Biol.* **202:** 341.

Neumann C. and Cohen S. 1996. A hierarchy of cross-regulation involving *Notch*, *wingless*, *vestigial* and *cut* organizes the dorsal/ventral axis of the *Drosophila* wing. *Development* **122:** 3477.

———. 1997. Long-range action of Wingless organizes the dorsal-ventral axis of the *Drosophila* wing. *Development* **124:** 871.

Orenic T., Slusarski D., Kroll K., and Holmgren R. 1990. Cloning and characterization of the segment polarity gene cubitus interruptus Dominant of *Drosophila. Genes Dev.* **4:** 1053.

Panin V., Papayannopoulos V., Wilson R., and Irvine K. 1997. Fringe modulates Notch-ligand interactions. *Nature* **387:** 908.

Rulifson E. and Blair S. 1995. *Notch* regulates *wingless* expression and is not required for reception of the paracrine *wingless* signal during wing margin neurogenesis in *Drosophila. Development* **121:** 2813.

Sanicola M., Sekelshy J., Elson S., and Gelbart W. 1995. Drawing a stripe in *Drosophila* imaginal discs: Negative regulation of *decapentaplegic* and *patched* expression. *Genetics* **139:** 745.

Spreicher S., Thomas U., Hinz U., and Knust E. 1994. The *Serrate* locus of *Drosophila* and its role in morphogenesis or imaginal discs: Control of cell proliferation. *Development* **120:** 535.

Tabata T. and Kornberg T. 1994. Hedgehog is a signalling protein with a key role in patterning *Drosophila* imaginal discs. *Cell* **76:** 89.

Vachon G., Cohen B., Pfeifle C., McGuffin M., Botas J., and Cohen S. 1992. Homeotic genes of the Bithorax complex repress limb development in the abdomen of the *Drosophila* embryo through the target gene. *Cell* **71:** 437.

Werner M. and Burley S. 1997. Architectural transcription factors: Proteins that remodel DNA. *Cell* **88:** 733.

Wilkins A. and Gubb D. 1991. Pattern formation in the embryo and imaginal discs of *Drosophila*: What are the links? *Dev. Biol.* **145:** 1.

Williams J.A., Paddock S.W., and Carroll S.B. 1993. Pattern formation in a secondary field: A hierarchy of regulatory genes subdivides the developing *Drosophila* wing disc into discrete sub-regions. *Development* **117:** 571.

Williams J., Paddock S., Vorwerk K., and Carroll S. 1994. Organization of wing formation and induction of a wing-patterning gene at a compartment boundary. *Nature* **368:** 299.

Zecca M., Basler K., and Struhl G. 1995. Sequential organizing activities of engrailed, hedgehog and decapentaplegic in the *Drosophila* wing. *Development* **121:** 2265.

———. 1996. Direct and long-range action of a wingless morphogen gradient. *Cell* **87:** 833.

The Dance of the *Hox* Genes: Patterning the Anteroposterior Body Axis of *Caenorhabditis elegans*

C.J. Kenyon, J. Austin, M. Costa, D.W. Cowing, J.M. Harris, L. Honigberg, C.P. Hunter, J.N. Maloof, M.M. Muller-Immerglück, S.J. Salser, D.A. Waring, B.B. Wang, and L.A. Wrischnik

Department of Biochemistry and Biophysics, University of California, San Francisco, California 94143-0554

Hox mutations are fascinating. Like magic, they can turn antennae into legs or create extra wings. What makes these genes so talented? How can they make such high-level decisions? Are there simple rules that can explain the effects they have on the development of individual cells? Do the genes act multiple times during the development of a tissue to micromanage individual cell fate decisions, or can they act relatively early to initiate developmental programs that run independently of their further input?

When the functions of *Drosophila Hox* genes were first analyzed, it seemed reasonable to suppose that they would be expressed uniformly within their functional domains and act within these regions in combination with cell-type and position-specific regulators to select between alternative developmental pathways. However, when investigators in many different fields began to visualize Hox gene products during development, it became apparent that the patterns of expression of these genes were not always uniform, but often appeared heterogeneous at the level of single cells. Moreover, the expression patterns often appeared to change over time. These and other unexpected observations, in both arthropods and vertebrates, made this simple view of *Hox* gene function questionable (for excellent reviews of *Hox* gene expression and function in arthropods and vertebrates, see Lawrence 1992; Krumlauf 1994; Lawrence and Morata 1994).

To ascertain the functions of *Hox* genes at the single-cell level, it is advantageous to be able to determine exactly which cells express which *Hox* genes during development. In addition, it is helpful to be able to correlate a particular expression pattern with the subsequent development of specific cells and their descendants, and to investigate the causality of the expression patterns by altering *Hox* gene expression within individual cells. In *C. elegans*, such high-resolution analysis is possible because the fate of each cell and its descendants is known and because it is possible to manipulate gene expression in individual cells (for review, see Riddle et al. 1997). With this organism, it is possible to ask exactly which cells express a *Hox* gene during development and what role this expression plays in the development of that cell and its descendants.

Using these approaches, we have investigated the roles of the *C. elegans Hox* genes, and we have found that the patterning strategies that utilize *Hox* genes are remarkably diverse. *Hox* genes are employed in different ways in different tissues; they can even be employed differently in different cells within the same tissue. In some cases, *Hox* gene expression patterns are relatively uniform, and complexity is generated by combinatorial interactions between multiple Hox proteins or between Hox proteins and other regulatory factors. In other cases, complex patterns of cell types and structures arise because the pattern of *Hox* gene expression itself is complex. In some instances, a *Hox* gene acts multiple times within a lineage to sequentially program many different cell fate decisions, whereas in other cases, a *Hox* gene appears to act briefly at the beginning of a lineage to initiate a complex sequence of events that unfolds without its further participation. In this paper, we describe studies that we have carried out to investigate the roles of the *C. elegans Hox* genes in the lateral and ventral epidermis of the animal and in the peripheral nervous system. In addition, we describe the roles that *Hox* genes play in the migrations of cells that traverse their domains of function.

A different but equally intriguing aspect of *Hox* gene function is their positional specificity. *Hox* genes control the development of specific body regions, and this is because each *Hox* gene is expressed in a position-specific fashion. How is this pattern of expression set up? In *Drosophila*, where this question has been investigated in detail, it is quite clear that the position of a cell (actually a nucleus) within the early syncytial embryo determines which *Hox* gene it ultimately expresses (for review, see Lawrence 1992). However, the strategy for positioning *Hox* gene expression in the leech seems to be quite different from the strategy used in *Drosophila* (Nardelli-Haefliger et al. 1994). In the leech, certain tissue anlage are formed by parallel bandlets of cells, each of which is formed by the sequential budding of large precursor cells called teloblasts. If a teloblast is temporarily prevented from undergoing cell division, its descendants eventually occupy anteroposterior (A/P) positions that are out of phase with the rest of the animal. Nevertheless, these cells, which never reach their normal A/P positions, express the same *Hox* gene that they would during normal development. This finding indicates that in the leech, the spatial pattern of *Hox* gene expression is not dependent on a cell's position, but instead is established by a mechanism that couples the timing of *Hox* gene expression to the rate of teloblast cell division. The mode of early development of *C. elegans* is not the same as that of either *Drosophila* or the leech, in the sense that cell fate determination in *C. elegans* appears to occur by mechanisms that program specific blastomeres to generate complex lineages (for review, see Schnabel and Priess 1997). Therefore, we were very interested in investigating whether or not cell-extrinsic, A/P positional factors would set up *Hox* gene expression patterns in this organism. This question was particularly interesting because cells that come to occupy the same A/P region of the larvae, and thus the same *Hox* domain, do not all originate from the same early blastomeres, or even from different blastomeres located in the same body region. Instead,

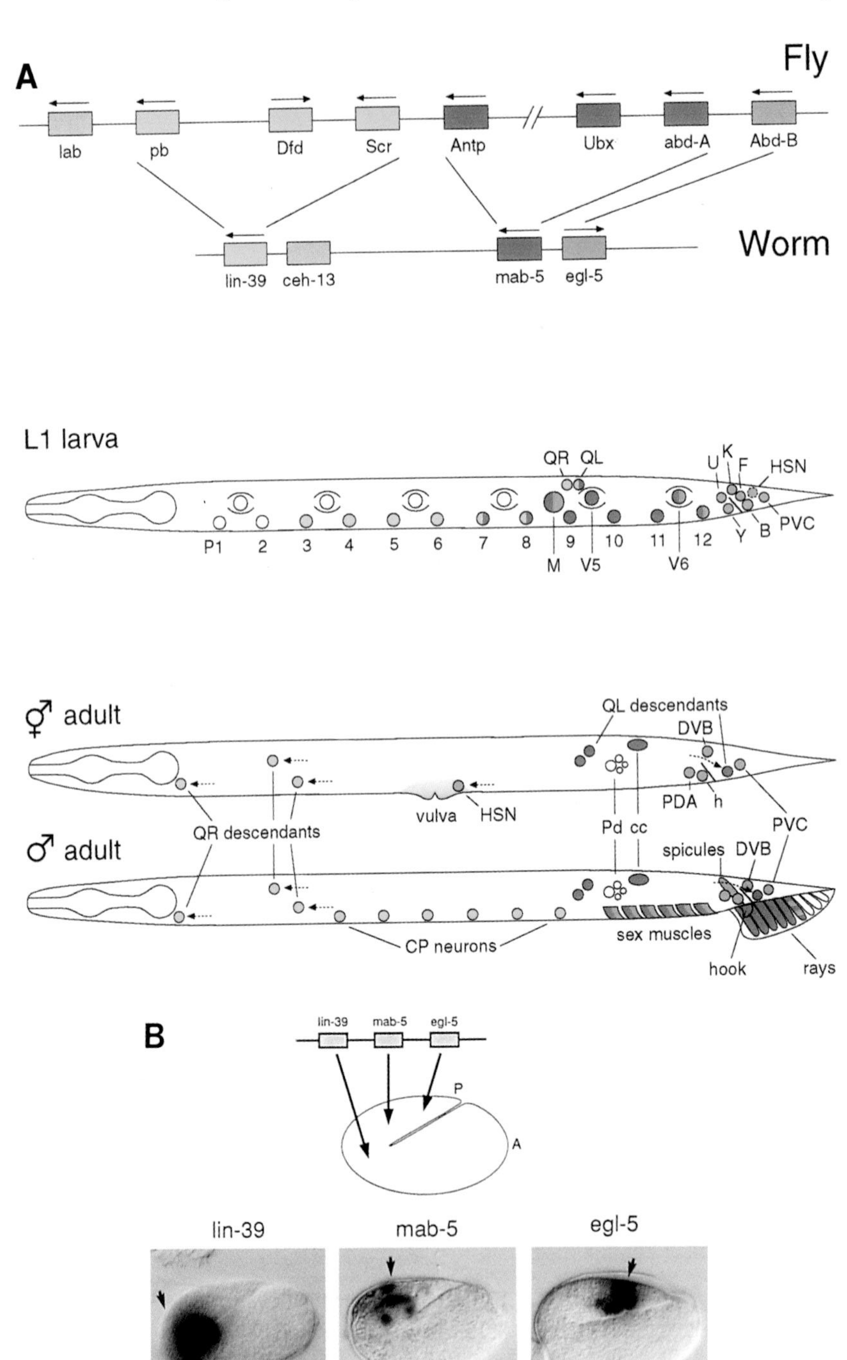

Figure 1. The *C. elegans Hox* genes. (*A*) The *C. elegans Hox* complex is shown at the top. Cells whose fates are affected by each of the (color-coded) *Hox* genes are shown in the newly hatched larva and in the adult. (*B*) Position-specific pattern of *Hox* gene expression in the embryo. (Adapted, with permission, from Wang et al. 1993 [copyright Cell Press].)

they arise from a variety of early blastomeres located in different A/P positions in the embryo, much as the different branches of Congress (Senate, House) are each made up of individuals drawn from all 50 states. Our findings are not easily accommodated to the simple model that a cell expresses its *Hox* gene solely because it finds itself within the range of a cell-extrinsic positional signal that marks a specific A/P body region. (The type of signal we are talking about here would be a cell-extrinsic version of *Drosophila bicoid*, which assigns anterior-specific developmental fates to the nuclei that it contacts.) Although our findings do not indicate how the pattern of *Hox* gene expression is actually established, they raise the possibility that *Hox* gene expression is regulated, at least in part, by segregation of regulatory information into particular cells during the divisions of the early blastomeres, presumably guided by lineage history and extrinsic polarity information that orients the segregation apparatus along the A/P axis. These studies are discussed in the final section of this paper.

OVERVIEW OF THE *C. ELEGANS HOX* GENES

C. elegans contains a small Hox complex consisting of four *Hox* genes, *ceh-13,* a *labial* homolog; *lin-39,* a homolog of *Sex combs reduced (Scr), Deformed (Dfd),* and *proboscipedia (pb); mab-5,* a homolog of *Antennapedia (Antp), abdominal A (abd-A)* and *Ultrabithorax (Ubx)*; and *egl-5,* a homolog of *Abdominal-B (Abd-B)* (Fig. 1A, B) (Costa et al. 1988; Burglin et al. 1991; Kenyon and Wang 1991; Clark et al. 1993; Wang et al. 1993). The genomic organization of these genes shares some, but not all, features characteristic of *Hox* clusters in other organisms: The genes *ceh-13* and *lin-39* are tightly linked, as are the genes *mab-5* and *egl-5.* These two pairs are separated by a stretch of several hundred kilobases that contains several unrelated genes (Sulston et al. 1992; Wang et al. 1993). In addition, the orientations of *ceh-13* (the *labial* homolog) and *lin-39* (the *Scr/Dfd/pb* homolog) are inverted relative to their homologs in other organisms (Sulston et al. 1992).

The functions of three of these four genes are known: Like *Hox* genes in other organisms, each of these influences primarily the development of a specific A/P body region (Fig. 1A, B). *lin-39* specifies patterns in the central body region (Clark et al. 1993; Wang et al. 1993), *mab-5* specifies pattern in the posterior body region (Kenyon 1986), and *egl-5* specifies cell fates in the tail (Chisholm 1991). In addition, like *Hox* genes in other organisms, these genes are expressed in a position-specific fashion within their domains of function (Fig. 1B) (Costa et al. 1988; Cowing and Kenyon 1992; Wang et al. 1993). The level of amino acid identity between the homeoboxes of these genes and their homologs ranges from 53% to 77% (Costa et al. 1988; Burglin et al. 1991; Clark et al. 1993; Wang et al. 1993), which is lower than that seen for many other organisms. However, not only are their functions similar to those of classic *Hox* genes, but their activities can also be partially mimicked by their closest *Drosophila* homologs (Hunter and Kenyon 1995). Specifically, *Drosophila Antp* can promote the development of several body structures normally specified by its *C. elegans* homolog *mab-5,* and *Drosophila Scr* can promote the development of several body structures normally specified by *lin-39.* The degree to which the specificity of these *Hox* genes has been conserved is particularly striking for one set of cells, the migratory Q cells (described below). *mab-5* and its homolog *Antp* cause the descendants of these cells to migrate toward the tail, whereas *lin-39* and its homolog *Scr* cause the same cells to migrate toward the head. Together these findings demonstrate that these *C. elegans* genes are real *Hox* genes and imply that they were derived from a common precursor of the arthropod and vertebrate Hox complexes (see Kenyon and Wang 1991; Kenyon 1994).

Unlike arthropods and vertebrates, which require *Hox* gene function during embryogenesis, *C. elegans* mutants carrying single, double, or even triple *Hox* mutations are viable and fertile (Kenyon 1986; Chisholm 1991; Clark et al. 1993; Wang et al. 1993; Wrischnik and Kenyon 1997). Thus, *C. elegans* generates all of its essential tissues, undergoes all of its early morphogenesis, and hatches into a small, active, feeding worm in the absence of *Hox* gene function (although we note that the function of the *labial* homolog *ceh-13* is not known and could have an early developmental role). This surprising finding can be rationalized by the fact that the dramatic differences in body pattern that come to distinguish different body regions develop largely during postembryonic development; this regional diversification does require *Hox* gene function (Fig. 1A). Our efforts to investigate the role of *Hox* gene function in several of these tissues are described in the next sections.

HOX GENES AND THE GENERATION OF A/P PATTERN IN THE LATERAL EPIDERMIS

The lateral epidermis generates a beautiful pattern in the male, consisting of lateral stripes of alae (cuticular structures) that run along the sides of the animal, and finger-like sensory rays that radiate out into the copulatory fan of the tail (not shown). These structures are derived from the lineages of the lateral epidermal seam cells V(1–6), which are located in the body (Fig. 2A), and T, which is located in the tail. The seam cells go through repeated rounds of stem cell division during postembryonic development, generating copies of themselves as well as other types of epidermal cells and neuroblasts. The anterior V cells, V(1–4), undergo a simple sequence of cell divisions and generate only epidermal cells, whereas the posterior V5 and V6 cells generate, in the male, more complex lineages consisting of additional proliferative cell divisions as well as divisions that generate two types of neuroblasts, the V5-derived postdeirid neuroblast and the V5- and V6-derived sensory rays (see Fig. 2A).

The *Hox* gene *mab-5* plays an important role in the generation of this A/P pattern among the lateral epidermal lineages (Kenyon 1986). In *mab-5*(−) mutants, the sensory rays normally produced by V5 and V6 are not made, and instead, alae extend into the tail. These pattern

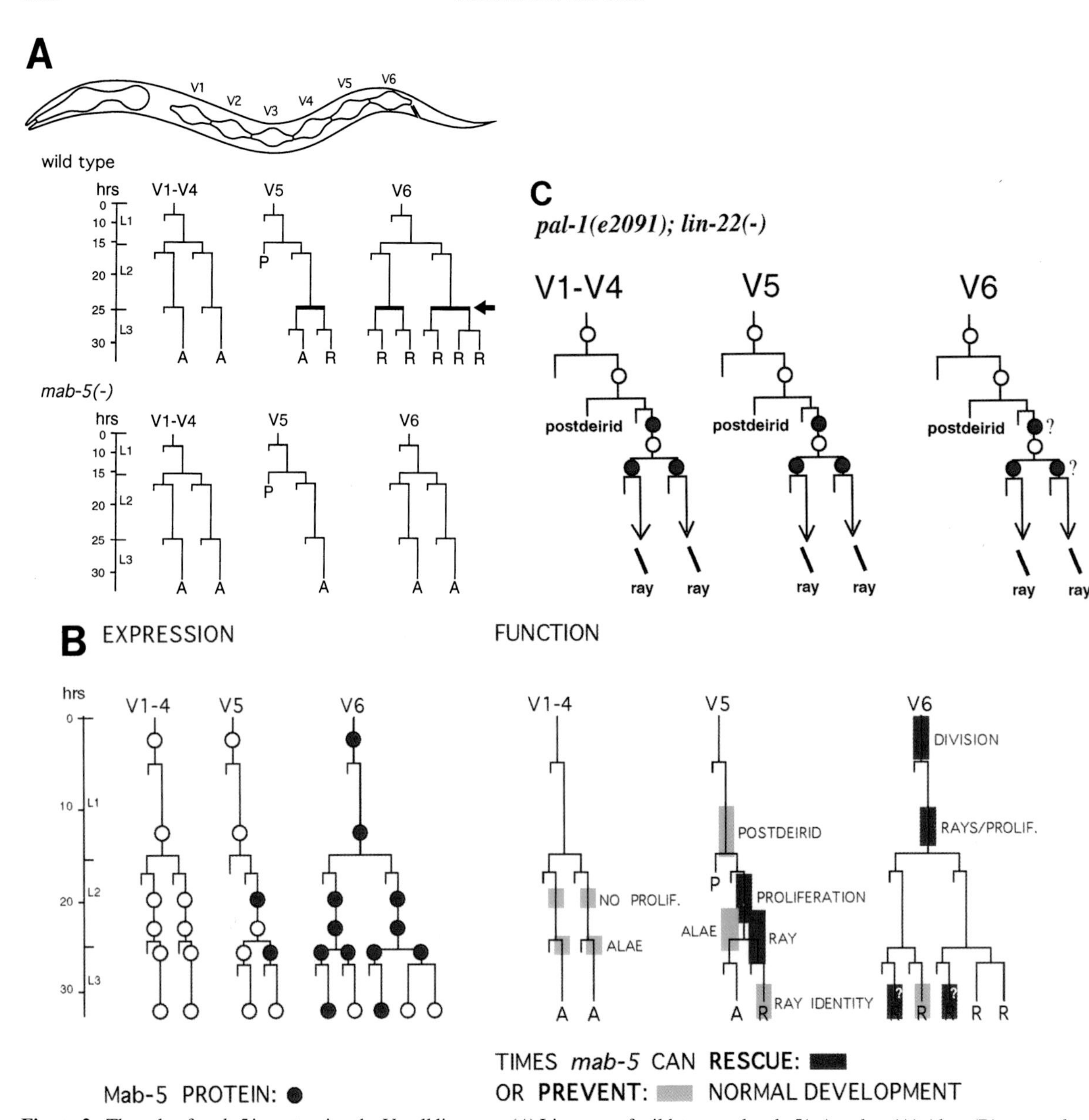

Figure 2. The role of *mab-5* in patterning the V-cell lineages. (*A*) Lineages of wild-type and mab-5(−) males. (A) Alae; (R) ray. *mab-5* is required for the proliferative cell division (*bold line*) and for production of rays. (*B*) Pattern of *mab-5* expression (assayed using anti-MAB-5 antisera) in the V-cell lineages, and the consequences of changing the expression pattern. (*C*) V5-like cell lineages generated when both *pal-1* and *lin-22* activities have been removed. The postulated pattern of *mab-5* expression is inferred from the lineage; it has not been determined directly. (*A* and *B*, reprinted, with permission, from Salser and Kenyon 1996.)

transformations are the consequence of underlying cell lineage transformations (Fig. 2A). In *mab-5* mutants, the V6 cell generates a lineage similar to those of V(1–4), as does V5, with the exception that the V5-derived postdeirid is still produced. This implies that *mab-5* is needed for the proliferative stem cell divisions in the V5 and V6 lineages, and also for the generation of the rays.

The complexity of these lineages made them an attractive experimental system to investigate the roles that *Hox* genes play over time within specific lineages. In particular, it was possible to ask whether the pattern of expression of a single *Hox* gene was simple or complex, and if it was complex, whether the complexity was required for proper development. Using anti-MAB-5 antisera to visualize MAB-5 protein, and a heat-shock-*mab-5* (hs-*mab-5*) fusion to alter *mab-5* expression, we found that in the V5 lineage, the expression pattern of *mab-5* was highly complex and dynamic, and furthermore, that all aspects of its complexity were critical for formation of the normal pattern (Salser and Kenyon 1996). As diagrammed in Figure 2B, *mab-5* is OFF in the early part of the V5 lineage, and this OFF state is necessary to allow the postdeirid to form. *mab-5* then switches ON prior to the proliferative cell division, and it is required for this division to take place. *mab-5* is then kept OFF in one daughter cell and this OFF state is necessary to prevent that branch of the lineage from generating a ray. In contrast, *mab-5* is switched back ON within the sister lineage, and this ON state is necessary for that branch of the lineage to produce a ray. Finally, after all the cells have been generated, *mab-5* switches OFF in the ray cell group. In fact, the pattern of *mab-5* expression changes to an alternating ON/OFF spatial pattern within the row of developing sensory rays generated by V5 and V6. This alternating pattern was found to be required for the different rays to each adopt their particular identities. If *mab-5* was expressed uniformly at this time, then nearby rays fused together, suggesting that cell-surface distinctions that allow them to develop autonomously may have been abolished. Neighboring rays have also been found to fuse with one another when the dosage of *Hox* genes is altered (Chow and Emmons 1994).

Surprisingly, the role of *mab-5* in V5's neighbor V6 is very different (Fig. 2B) (Salser and Kenyon 1996). In the V6 cell, which generates a complex lineage similar to that of V5, *mab-5* is expressed uniformly as the cells undergo their divisions. Moreover, in this lineage, *mab-5* does not micromanage individual cell fate decisions, since the entire wild-type lineage can be rescued by a single, transient pulse of heat-shock *mab-5* delivered soon after hatching. Thus, in this lineage, *mab-5* initiates a complex developmental program, but it is not required to assign individual cell fates (until the very end, when its expression changes and it influences ray morphogenesis, as described above; see Fig. 2B). This is a nice example of a case in which testing the relevance of expression patterns was very informative. *mab-5* is required for this lineage, and it is expressed throughout this lineage, but, since it is only required transiently, it probably does not directly regulate the genes required for the execution of this lineage, or the differentiation of the cells it generates. A similar example of *mab-5's* acting early and transiently within a lineage occurs in that of the ventral ectoblast P12, where its early activity causes a cell to undergo programmed cell death at a later time (Salser et al. 1993).

The V cells can be divided into three groups with respect to *mab-5* expression and function: In the anterior V(1–4) cells, *mab-5* is OFF; in the V5 cell, *mab-5* is switched ON and OFF repeatedly, and in the V6 lineage, *mab-5* is simply ON. Moreover, these three sets of cells respond differently to pulses of heat-shock MAB-5 in a *mab-5*(−) background. As mentioned above, V6 responds strongly to an early, transient pulse of MAB-5, and V5 requires timed pulses that coincide with individual cell fate decisions. V(1–4) are able to generate posterior-specific lineages and structures in response to *mab-5* activity, but they are much less responsive than V5. What causes these three groups of cells to express *mab-5* differently and to respond differently to pulses of heat-shock *mab-5*? For the V6 lineage, at least part of the answer is the caudal homolog *pal-1*, which activates *mab-5* expression in V6 in the embryo (Waring and Kenyon 1990, 1991; C. P. Hunter et al., in prep). Interestingly, *pal-1* is also required for *mab-5*(−) V6 cells to respond to early pulses of hs-MAB-5 (Salser and Kenyon 1996). In a *pal-1(e2091)* mutant, in which PAL-1 protein appears to be absent in V6 (C. P. Hunter et al., in prep.), V6 generates a V(1–4)-like lineage and also responds to pulses of hs-*mab-5* the same way that V(1–4) respond (Salser and Kenyon 1996). Thus, *pal-1* activity is needed both to turn on *mab-5* expression and to do something else that makes the V6 cell very sensitive to *mab-5* activity. A different gene, the *hairy* homolog *lin-22* (Wrischnik and Kenyon 1997), gives the V(1–4) lineage their distinctive features. This gene prevents V(1–4) from generating lineages similar to those of V5 (in *lin-22* mutants, the V5 lineage also generates an additional ray, and so do the transformed V(1–4) lineages) (Fixsen 1985; Wrischnik and Kenyon 1997). If *lin-22* activity is removed in a *pal-1(e2091)* mutant, then V6, too, generates a V5-like lineage (Waring and Kenyon 1990). Thus, in this double mutant, all the V cells generate similar V5-like lineages (Fig. 2C). Not surprisingly, the *mab-5* expression pattern correlates with the lineage transformations seen in *lin-22* mutants (Wrischnik and Kenyon 1997). In the normal V5 lineage, *mab-5* is switched ON and OFF repeatedly, and the same is true of the V(1–4) lineages in *lin-22* mutants (and presumably of V6 in *pal-1; lin-22* double mutants, although this has not been examined). So far, no mutations have been found that transform the V5 lineage into either the V(1–4) or the V6 lineages. Thus, surprisingly, the V5 lineage, with its dynamic pattern of *mab-5* expression, may represent the ground state of this tissue.

The fact that a *Hox* gene can be switched on and off repeatedly within a lineage to regulate cell division, neuroblast formation, and morphogenesis independently is fascinating and raises the question of how this precise and dynamic expression pattern is regulated. This is still a mystery. However, we know how one aspect of this expression pattern is regulated: that is, the repression of *mab-5* expression in the early portion of the V5 lineage. If the neighbors of V5 are ablated with a laser microbeam, V5 fails to generate a postdeirid (Sulston and White 1980; Waring et al. 1992). We have found that the intercellular signaling required for postdeirid formation takes place between the daughters of the V cells, the Vn.p cells (Austin and Kenyon 1994). The Vn.p cells, which become seam cells like their parents, extend thin processes across their sister cells, which adopt another epidermal fate (Fig. 3). The extending seam cell processes ultimately contact one another and stop growing. We found that once these processes had contacted one another, it was no longer possible to affect postdeirid development by ablating neighboring cells. Moreover, if the time of contact between the Vn.p cells was delayed experimentally, the time of the signaling event was delayed until the new time of contact (Austin and Kenyon 1994). Thus, it seems that the critical signaling event occurs when the Vn.p cells touch one another.

How does signaling between the Vn.p cells permit V5 to make a postdeirid? We have found that this signaling affects postdeirid formation by preventing cells in the V5 lineage from expressing *mab-5* precociously (Austin and Kenyon 1994; C. P. Hunter et al., in prep.). Moreover, it does so by preventing the activity of a Wnt signaling pathway that would otherwise turn *mab-5* expression ON (C. P. Hunter et al., in prep.). It will be very interesting to learn how contact between the Vn.p processes inhibits the Wnt signaling pathway.

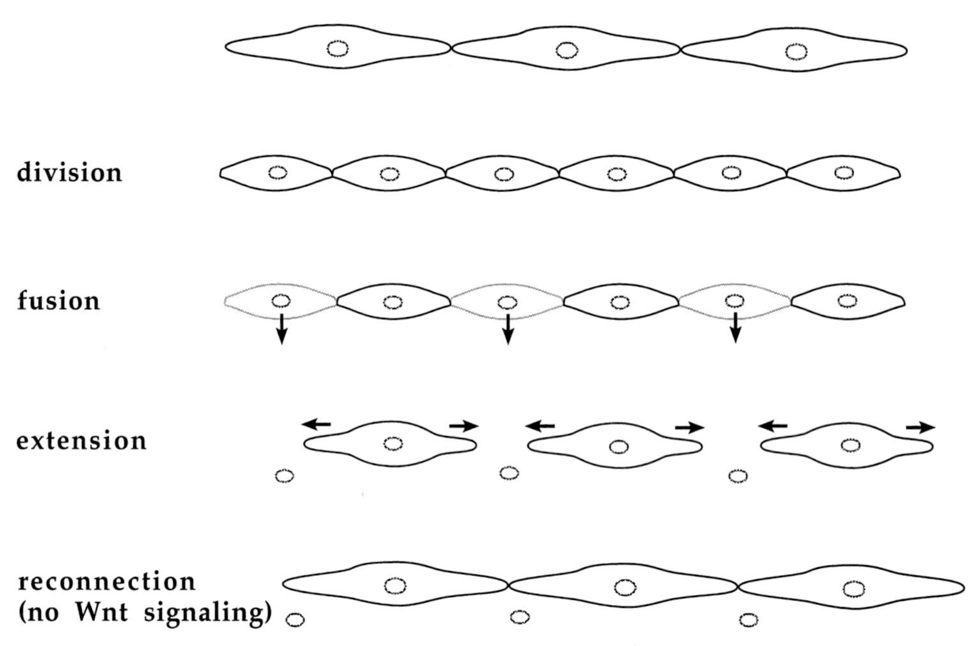

Figure 3. The seam cell cycle of division and extension.

REGULATION AND FUNCTION OF *HOX* GENES IN THE VENTRAL ECTODERM

The ventral ectodermal precursor cells P(1–12) generate lineages that are quite different from those of the V cells (Fig. 4). These cells, which are located in a row along the ventral surface of the animal, each divide soon after hatching to generate a Pn.p epidermal daughter and a Pn.a neuroblast (Fig. 4A,B). At a later stage in postembryonic development, centrally located Pn.p cells in the hermaphrodite generate the vulva, and posteriorly located Pn.p cells in the male generate specialized copulatory structures. Each of the Pn.a neuroblasts divides several times in a stereotypical sequence to produce five daughter cells that either undergo programmed cell death, differentiate into motoneurons or reproductive neurons, or become neuroblasts. In the absence of the *Hox* genes, each P cell would ultimately generate a similar complement of six descendants, and cells that occupied analogous positions in the different lineages (here called homologs) would adopt the same fates. The *Hox* genes act on this underlying repetitive, metameric, pattern to allow certain cells in these lineages to adopt fates that differ from their homologs in other P-cell lineages (Fig. 4C,D) (Kenyon 1986; Clark et al. 1993; Salser et al. 1993; Wang et al. 1993). In the P-cell lineages, the patterns of *Hox* gene expression are not particularly complex; instead, for the most part, the *Hox* genes are expressed uniformly throughout their domains of function (Salser and Kenyon 1996). In these cell lineages, local complexity arises because cells that occupy different positions in the P-cell lineages have different developmental potentials, and large-scale complexity arises, at least in part, because of the way that the *Hox* genes interact with one another and with other factors (for review, see Salser and Kenyon 1994).

Two *Hox* Genes Act Combinatorially to Regulate Pn.p Cell Fusion

The first cell fate decision taken by the epidermal Pn.p cells is whether or not to fuse with the large epidermal syncytium hyp7, which, at hatching, covers the dorsal surface of the animal. In hermaphrodites, only the central P(3–8).p cells remain unfused; these mononucleate cells subsequently become the vulval equivalence group, whose members generate the vulva. The ability of the P(3–8).p cells to remain unfused is due to *lin*-39 (Clark et al. 1993; Wang et al. 1993), which is expressed in P(3–8).p (Wang et al. 1993; Maloof and Kenyon, 1998). Thus, the activity of this *Hox* gene establishes the vulval equivalence group. In males, the fusion pattern is more complex (Fig. 3B). The posterior P(9–11).p cells do not fuse; instead, they form an equivalence group that generates male mating structures (P12.p adopts a different fate). In addition, the central P(3–6).p cells remain unfused. The intervening P(7,8).p cells do undergo cell fusion. This complex pattern of fusion results from combinatorial interactions between the *Hox* genes *lin-39* and *mab-5,* which have overlapping domains of function (Kenyon 1986; Clark et al. 1993; Wang et al. 1993). As in hermaphrodites, *lin-39* is expressed in the male P(3–8).p cells, and *mab-5* is expressed in an overlapping domain that spans P(7–11).p (Fig. 4C,D). The expression of either gene alone prevents cell fusion, whereas the expression of both genes, like the expression of neither gene, permits the cells to fuse (Clark et al. 1993; Salser et al. 1993; Wang et al. 1993). The mechanism by which this combinatorial control occurs is not known; however, it is unlikely that *lin-39* and *mab-5* act stoichiometrically to cancel out one other's activities, since changing the relative levels of the two Hox proteins within these cells does not affect the combinatorial code (Salser et al. 1993). Recently, we have identified a gene specifically required for this combinatorial activity, which may lead to molecular insights in the future (S. Alper and C. Kenyon, unpubl.).

lin-39 and *mab-5* Interact in Different Ways to Generate Boundaries at Different A/P Positions in Different Sets of P-derived Homologs

One of the most curious features of *mab-5* activity in the male P lineages is that the anterior boundary of its functional domain is located in different A/P positions depending on the P-derived homologs in question. For the Pn.p cells, it is located the furthest anteriorly, between the P6 and P7 metamers as just described; for the Pn.ap cells, it is located two P-cell units posteriorly, between P8 and P9, and for the Pn.aaap cells, it is located two more P-cell units posteriorly, between P10 and P11 (Fig. 4D). How can the boundaries of a single *Hox* gene be so different for different homologs within a single kind of cell lineage? In principle, this complexity could have arisen from a complex pattern of *mab-5* expression, a graded pattern of *mab-5* expression (assuming that different P homologs had different thresholds for *mab-5*), or a simple, uniform pattern of *mab-5* expression, assuming that other factors created the different boundaries. Thus, learning the pattern of expression of *mab*-5 was crucial. *mab-5* was found to be expressed in all of the descendants of the posterior P cells: it was ON within the P(7–11) lineages, and OFF in more anterior lineages (expression of *mab-5* in the P12 lineage was turned off by *egl-5* soon after hatching (Salser et al. 1993). As described above, the anterior boundary of *mab-5* function within the Pn.p cells was determined by the anterior extent of *mab-5* expression. However, in the Pn.aap homologs, the anterior boundary coincided with the posterior border of *lin-39* expression, located two units posteriorly to the boundary of *mab-5* expression. This was because where both *lin-39* and *mab-5* were expressed, *lin-39* prevented *mab-5* from functioning. (Note that this is a different type of interaction from the *one* that occurs in the Pn.p cells, in which the same two Hox proteins act combinatorially to specify a new fate). Finally, the anterior boundary of *mab-5* function in the Pn.aaap cells was shifted so far posterior because another inferred (but unidentified) gene activity restricted *mab-5* function to just those two lineages (Salser et al. 1993). Even if *mab-5* was expressed at high uniform levels using a hs-*mab-5* construct, *mab-5* was only able to affect the fates of these two posterior Pn.aaap homologs. Thus, in the P cells, the patterning machinery does not create complexity by generating a complex *Hox* gene expression pattern the way it does in the V5 lineage. Instead, in these lineages the patterns of *Hox* gene expression are for the most

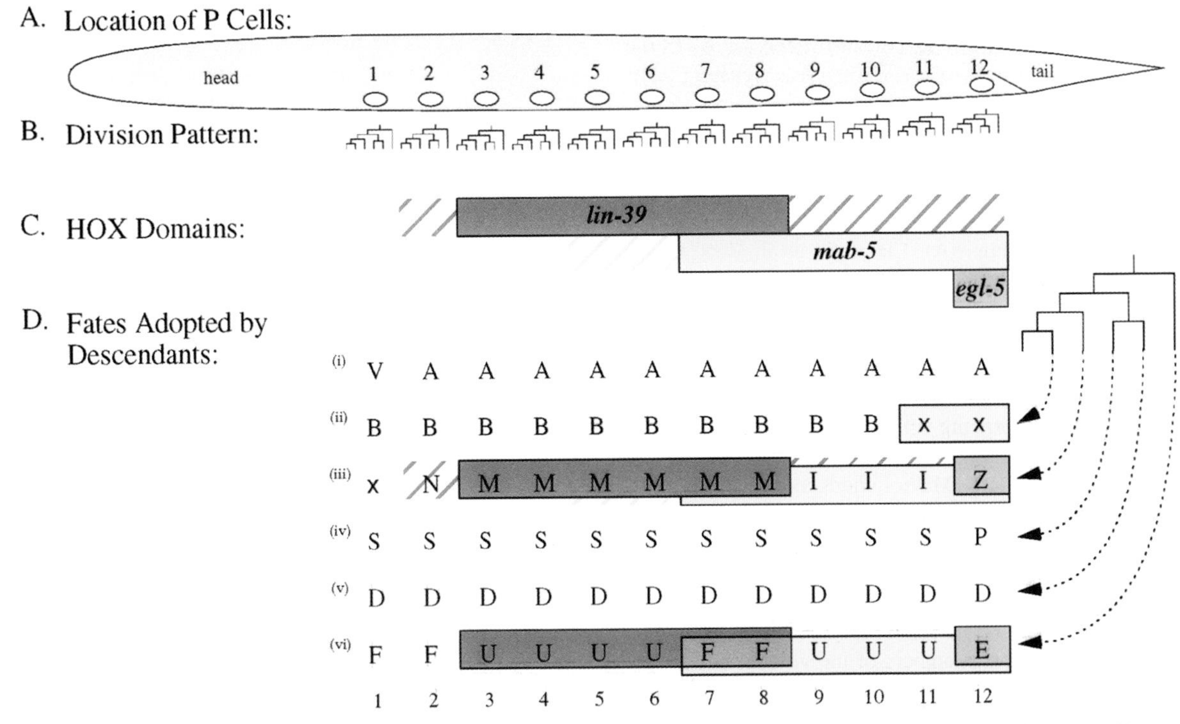

Figure 4. Generation of A/P pattern within the male P-cell lineages by the *Hox* genes. (*A, B*) Lineages generated by the P cells. (*C*) Domains of *Hox* gene function, color coded. The distribution of MAB-5 protein and *lin-39-LacZ* expression coincide with the indicated (*filled*) domains. (*D*) Fates of P-cell descendants in the different P lineages. Letters represent the fate of each homolog descended from each of the P cells in the wild-type male. A, B, S, D, N, P, and V stand for specific neural cell fates. M indicates a division that produces two neurons, one of which is serotonergic; I indicates a division that produces two neurons, neither of which produces serotonin. E indicates a division that produces one cell death and one rectal epithelial cell. (U) Unfused; (F) fuses with hyp7. Cell fates that are color coded require the function of the indicated *Hox* gene. The orange color represents the new fate specified by the combination of *lin-39* and *mab-5*. Cross-hatching represents body regions in which *mab-5* or *lin-39* influences certain cell fates in a variable fashion (described in Salser et al. 1993). (Reprinted, with permission, from Salser and Kenyon 1994.)

part uniform and static; complex spatial patterns of cell types are generated by allowing these Hox proteins to interact with one another with other proteins at the post-translational level in diverse, cell-type-specific ways.

Hox Genes Are Required for the Ras Pathway to Trigger Cell Division during Vulval Development and Also to Select between Alternative Organ Fates

The most intensively studied descendants of the P cells are the members of the vulval equivalence group, which generate the hermaphrodite vulva (for reviews, see Horvitz and Sternberg 1991; Kenyon 1995, Greenwald 1997). The *C. elegans* vulva is induced by a cell in the underlying gonadal tissue called the anchor cell, which activates a classic Ras/MAP kinase pathway in the Pn.p cells. This pathway, together with a lateral signaling pathway, causes P(5–7).p to generate vulva-specific cell lineages. Because the precursors of the vulva undergo cell fusion in *lin-39* mutants, it was not clear whether *lin-39* might have a role in vulva development itself. However, when we found that LIN-39 protein was present in the vulva cells, we decided to address this question directly (Maloof and Kenyon 1997). To do this, we analyzed *lin-39* mutants that had been given an early pulse of hs-LIN-39 to prevent the members of the vulval equivalence group from fusing with hyp7 soon after their birth. Surprisingly, we found that the initiation of vulval development requires not only the activity of the Ras and lateral signaling pathways, but also the *Hox* gene *lin-39.* In the absence of *lin-39* activity, even if the Ras signaling pathway is activated, vulval cell lineages are not initiated. Genetic and molecular epistasis experiments indicate that both *lin-39* and the *ras* pathway are absolutely required for vulval induction: Constitutively activating either one is not sufficient to trigger vulval cell divisions unless the other is present (Fig. 5A). Clandinin et al. (1997) have reached similar conclusions using a weak, temperature-sensitive allele of *lin-39* to rescue the early *lin-39* function. They have also found that *mab-5* has the opposite role of preventing the P(7,8).p cells from responding efficiently to Ras signaling.

In addition to allowing vulval induction to proceed, *lin-39* activity plays an important role in determining the outcome of *ras* signaling (Maloof and Kenyon 1997). If *lin-39* instead of *mab-5* is expressed in posterior Pn.p cells (which would normally generate male mating structures, such as the copulatory hook, under the auspices of *mab-5* activity), the posterior cells generate vulva-like lineages. Conversely, if *mab-5* instead of *lin-39* is expressed in the central body region, members of the vulval equivalence group generate posterior-specific Pn.p lineages and a structure that resembles a hook. One plausible model to explain how *Hox* genes and the Ras pathway together generate the vulva is the following (Fig. 5B): Activation of the Ras pathway, by inactivating the ETS-like transcription factor *lin-1,* renders a set of genes competent for expression but does not actually activate their expression. For these genes to be transcribed, another transcriptional regulator, a *Hox* gene, is required. In the central body region, where *lin-39* is expressed, *lin-39* selects and activates only a subset of genes rendered competent for expression by Ras activation. This specific subset of genes specifies vulval cell lineages. There is reason to think that this model may be at least partially correct. This is because one such downstream gene that has these properties has been identified, and that is *lin-39* itself. Expression of *lin-39* is up-regulated by the *ras* pathway, provided that basal levels of *lin-39* are present. Thus, we imagine that Ras activation may remove LIN-1 protein from *lin-39*-regulatory sequences, allowing LIN-39 protein to up-regulate its own expression. High levels of LIN-39 protein would then efficiently activate other genes that have been rendered competent for expression by the inactivation of LIN-1.

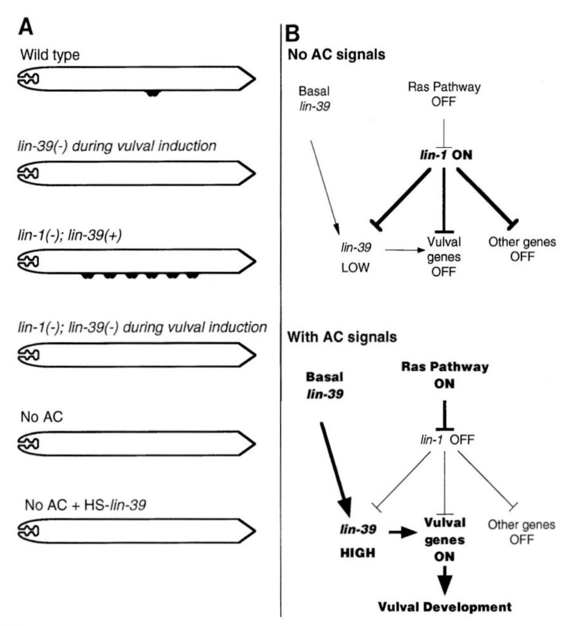

Figure 5. Two roles for *lin-39* in vulval induction. (*A*) In the wild type, Ras activation by the anchor cell (AC) triggers the central Pn.p cells to generate the vulva. This occurs because the Ras pathway inactivates LIN-1. In *lin-1*(−) mutants, all members of the vulval equivalence group generate vulval lineages, as long as *lin-39* is present. All *lin-39*(−) animals were treated with hs-*lin-39* early in development to prevent the Pn.p cells from fusing with hyp7. (*B*) Model for *lin-39* function. *lin-39* plays two roles in vulval induction: It acts in parallel with the Ras pathway to allow vulval cell divisions to take place, and it confers specificity on the Ras pathway.

HOX GENES AND THE REGULATION OF CELL MIGRATION

Hox genes are best known for their effects on patterns that develop within stationary fields of cells; however, they also influence a number of cell migrations. In *mab-5* mutants, the male sex myoblasts migrate toward the anterior instead of the posterior (Kenyon 1986), and in *lin-39* mutants, Pn.a cells located in the central body region, which normally stay put, migrate toward the anterior (Salser et al. 1993). Axon guidance as well as cell migration can be influenced by Hox mutations; for example, ectopic expression of *mab-5* can redirect the axons of certain serotonergic neurons from the anterior toward the posterior of the animal (unpublished data).

The *Hox* genes *mab-5* and *lin-39* play a particularly interesting role in regulating the migrations of neuroblasts in the QL and QR lineages (Chalfie et al. 1983; Kenyon 1986; Salser and Kenyon 1992; Clark et al. 1993; Wang et al. 1993). The Q cells are bilateral homologs that originate within the epidermal epithelium and then delaminate and migrate along the anteroposterior body axis. The cells execute stereotypical sequences of cell division as they migrate, and ultimately generate several types of peripheral neurons. Their migrations are particularly intriguing because they are left-right asymmetric: QL and its descendants migrate posteriorly, whereas QR and its descendants migrate anteriorly (Fig. 6A). *lin-39* and *mab-5* play related, but opposite, roles in programming the migrations of cells in the Q lineage (Fig. 6). The initial phase of Q-cell migration occurs independently of either *Hox* gene (although it is variably abnormal in the double mutant; Wang et al. 1993). At this time, QR migrates anteriorly and QL migrates posteriorly. The subsequent migrations of QL's descendants require *mab-5* gene activity. Just as QL completes its short posteriorward migration, a Wnt signaling pathway causes QL to switch on *mab-5* expression (Salser and Kenyon 1992; Harris et al. 1996; J. Whangbo et al., unpubl.). (Curiously, the Wnt pathway does not activate *mab-5* in QR, which never expresses *mab-5*.) *mab-5* activity is subsequently required for the descendants of QL to execute their normal posterior migrations instead of migrating anteriorly like the descendants of QR (Fig. 6B,C) (Chalfie et al. 1983; Salser and Kenyon 1992).

The anterior migrations of QR's descendants also require *Hox* gene function; in this case *lin-39* (Fig. 6E) (Clark et al. 1993; Wang et al. 1993). In *lin-39* null mutants, the descendants of QR still migrate toward the anterior, but they stop migrating prematurely at variable positions along the A/P axis. Both genes appear to have a specific role in cell migration but do not appear to be required for other aspects of Q-cell development, such as the pattern of cell divisions it undergoes or the types of neurons it produces.

Both *lin-39* and *mab-5* function cell-autonomously within the Q descendants to influence cell migration. Genetic mosaics in which cells in the QR lineage were mutant for *lin-39* but all surrounding cells were wild type exhibited the mutant migration phenotype (Clark et al. 1993). Mosaic analysis, and the *mab-5* expression pattern, was also consistent with a cell-autonomous role for *mab-5* in regulating QL migrations (Kenyon 1986; Salser and Kenyon 1992). More compellingly, laser activation of hs-*mab-5* specifically within a single Q descendant migrating toward the head is sufficient to cause that cell to reverse direction and migrate toward the tail (Harris et al. 1996).

The *lin-39* and *mab-5* proteins have the ability to act in a competitive manner with one another. Ectopic expression of *mab-5* causes descendants of both QL and QR to migrate posteriorly (Salser and Kenyon 1992), and ectopic expression of *lin-39* causes descendants of both QL and QR to migrate anteriorly (Hunter and Kenyon 1995), suggesting that the relative levels of either the Hox proteins or their downstream targets determine their direction of migration. It is obviously crucial to learn the identities of the downstream migration genes that determine the migratory properties of these genes. One candidate for a downstream target of *mab-5* is the gene *vab-8*. This gene is partially required for the posterior migration of the QL descendants, and it is required for many other

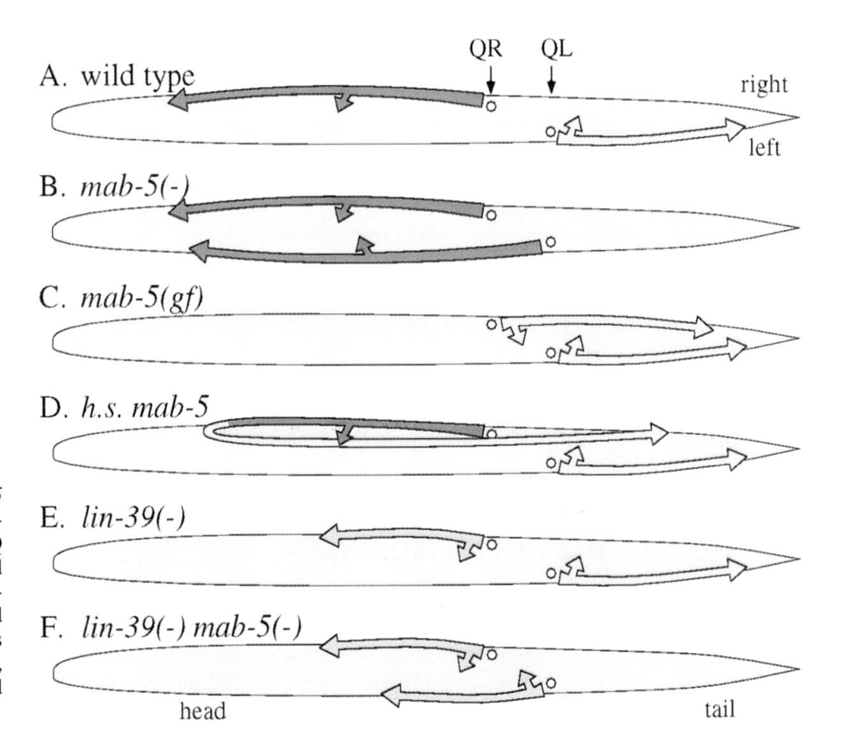

Figure 6. The roles of *lin-39* and *mab-5* in the migrations of cells in the Q lineages. The initial migrations of the Q cells (not shown) shift QR anteriorly and QL posteriorly. The short arrow represents the migrations of the Q.p cells, and the long arrow represents the migrations of cells in the Q.a lineage. (Reprinted, with permission, from Salser and Kenyon 1994.)

posterior migrations in the animal (Wightman et al. 1996).

Little is known about the extracellular information that presumably tells the cells in the Q lineages which direction to migrate and where to stop. Q cells migrating anteriorly near the head will respond to pulses of hs-*mab-5* by turning around and migrating posteriorly, and Q cells migrating posteriorly near the tail will reverse direction and migrate anteriorly if they run out of *mab-5* (Salser and Kenyon 1992; Harris et al. 1996). These findings indicate that the information these cells require to migrate in either direction is accessible all along the body axis. However, little is known about the mechanism of guidance along the A/P axis; in fact, it is not even clear whether the Q cells are specified to migrate in certain *directions,* or whether instead they are programmed to migrate to specific *positions*, which just happen to be located anterior or posterior to themselves. *mab-5* is not absolutely necessary for cells in the Q lineage to migrate toward the posterior, since other mutations that appear to affect this guidance system can allow cells in the Q lineage to migrate posteriorly even in the absence of *mab-5* (Harris et al. 1996). Clearly, this is a first-rate mystery that will be great fun to solve.

SUMMARY OF *HOX* GENE FUNCTION

Together these studies indicate that *Hox* genes do not function in one single, predictable manner during development. These genes have a complex and dynamic expression pattern in some tissues, but a simple expression pattern in others, and they have a wide variety of functions at the cellular level. They can be cell fate micromanagers, or they can act transiently to initiate entire complex lineages. In other organisms as well, it is becoming clear that *Hox* gene functions are quite complex and diverse. For example, in *Drosophila, Hox* genes have also been shown to play more than one role in a single cell lineage (Castelli-Gair et al. 1994).

Assuming that the expression and function of *Hox* genes in the common ancestor was simple, the complexity that we observe in modern organisms implies that these genes have tremendous intrinsic plasticity and flexibility that allows them to be incorporated into many different strategies for regulating cell fate determination and generating body pattern. The diversity of their functions in generating pattern within their primary domains is truly remarkable; nevertheless, it was their positional specificity, and not the details of their activities, that first made them so enchanting. How *C. elegans* achieves this positional specificity is discussed next.

ESTABLISHING THE PATTERN OF *HOX* GENE EXPRESSION DURING EMBRYOGENESIS

Since the positional specificity of *Hox* gene expression and function is their most striking and memorable feature, we have asked how the expression of these genes is targeted to specific body regions in *C. elegans* (Fig. 7a, b). The simplest hypothesis would be that position-specific expression patterns are set up by cell-extrinsic positional signals located along the A/P body axis. The best way to test whether a cell's position dictates its *Hox* gene expression pattern is to transplant that cell to a new position. Unfortunately, this type of experiment is not currently possible in *C. elegans,* so we were forced to use less direct methods to address this important issue. We carried out two experiments (Cowing and Kenyon 1996). First, we blocked the migration of a mesodermal cell called M, which originates in the anterior body region and then migrates into the posterior where it switches on *mab-5*, and found, to our astonishment, that we did not disrupt its ability to express its normal posterior-specific *Hox* gene (Fig. 7c). Thus, this cell does not have to reside in its normal A/P position to express its normal *Hox* gene. Second, we effectively altered the positions of early blastomeres by inducing them to undergo homeotic transformations that turned them into copies of blastomeres located in other regions of the body. Previously, it had been shown that if blastomeres producing inductive signals (EMS and P_2) are ablated, one can cause the embryo to generate multiple copies of a blastomere that normally produces two *mab-5*-expressing cells, the V6L (left) and V6R (right) cells, but no other cells that would normally express *mab-5* (Fig. 7d). As a consequence, the embryos produce multiple cells that are lineally equivalent to V6. We carried out similar experiments, and asked which, if any, cells in these disturbed embryos might express *mab-5*. There were many possible outcomes of this experiment. If a source of positional information had been ablated or transformed away, we might expect no (or all) cells to express *mab-5*. If positional signals had been redistributed, we would have expected a somewhat random pattern of *mab-5* expression. Instead, we found something quite striking and unexpected: We found that almost all of the *mab-5*-expressing cells turned out to be lineally equivalent to the V6 cells, irrespective of their location in the animal. Likewise, with only a very few exceptions, cells that would not be predicted by lineage to express *mab-5* did not (Fig. 7e).

These findings are not at all what the A/P positional model (i.e., "extracellular bicoid") would predict. However, we emphasize that these experiments do not rule out the possibility that A/P positional information contributes to *Hox* gene expression in *C. elegans*. It could act on other cells that we did not examine, or, at least in the case of the M cell, it could act redundantly with other mechanisms to ensure that the Hox expression pattern is reproducible. However, the experiments that we carried out raise new questions about how lineage-associated mechanisms could contribute to Hox gene regulation. First, can these findings be completely explained by postulating the existence of local signaling centers? One could imagine that such local signaling centers exist, and that they were transformed along with the V6 equivalents in these experiments (see Schnabel and Schnabel 1997). This seems unlikely, because most of the normal neighbors of V6 were either ablated or transformed away by these ablations, and because these embryos were highly disorganized. Thus, if signaling centers did exist, they would have to be close lineal relatives of V6. However, even if

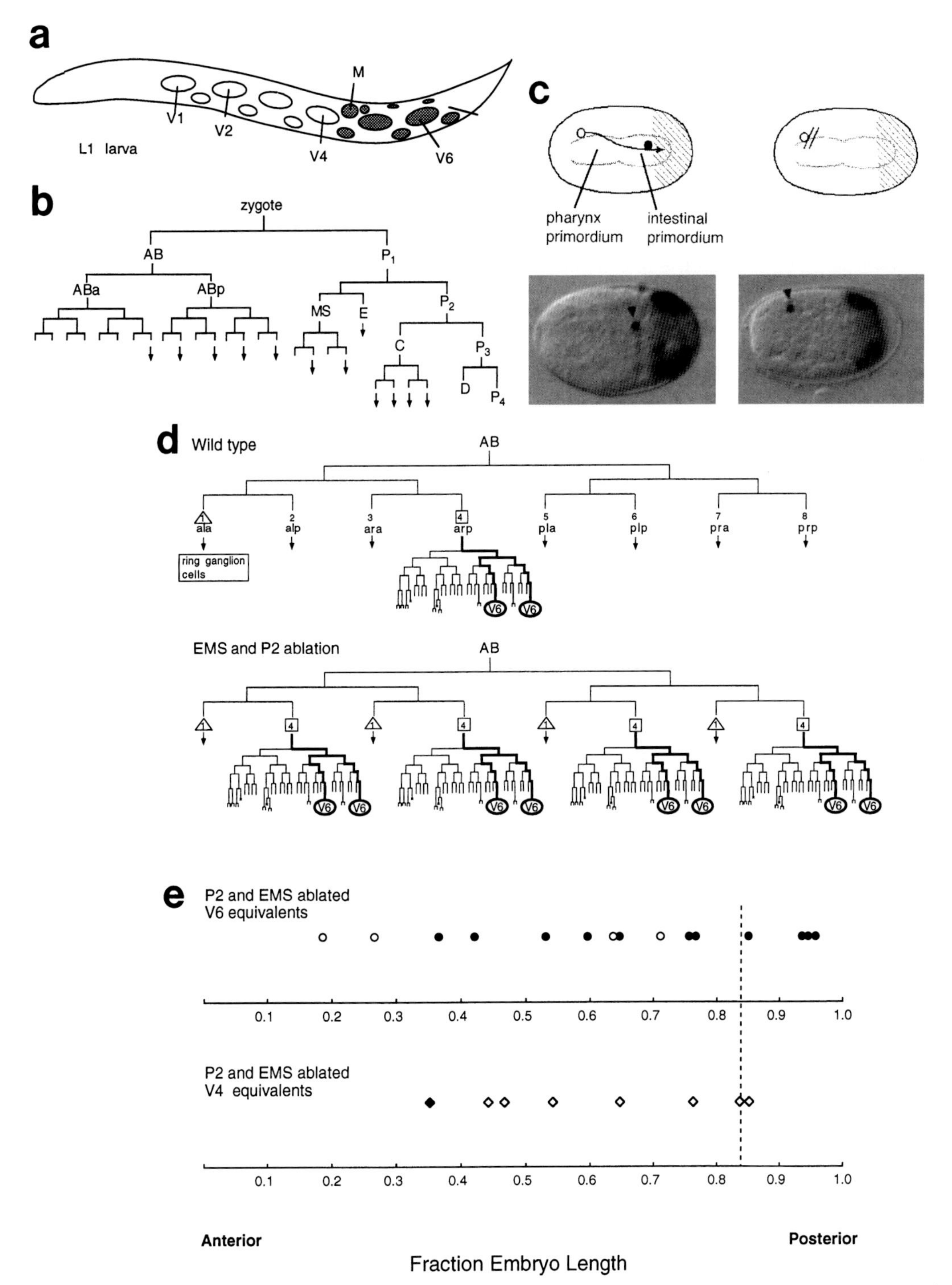

Figure 7. Does position along the A/P body axis govern *mab-5* expression? (*a*) Cells that express *mab-5* at hatching. (*b*) Origins of *mab-5*-expressing cells in the lineage (these cells are located in the posterior body region). (*c*) (*Left panels*) The M cell expresses *mab-5* as it nears the posterior body region. (This expression begins before M enters the heart of the *mab-5* domain, which originally made us think it was particularly sensitive to low levels of a *mab-5*-inducing positional signal.) When this migration is blocked using cytochalasin or colchicine (*right panels*), the cell expresses *mab-5* even though it is located far anterior to its normal site of *mab-5* activation (*lower right*). (*d*) Lineage transformations caused by ablating the signaling cells EMS and P_2 in the early embryo. Often the transformations seen are incomplete. (*e*) Positions of the V6 equivalents following ablation of EMS and P_2. Filled circles represent cells that express *mab-5*. Note that although more posterior cells than anterior cells express *mab-5*, the majority of *mab-5*-expressing cells were located anterior to the normal site of *mab-5* expression (*dashed vertical line*). Note also that the V4 cells, which are sisters of V6 and do not express *mab-5* during normal development, do not express *mab-5* in these embryos either. These cells make particularly convincing internal controls. In this study, the great majority of the *mab-5*-expressing cells we could identify were V6 equivalents; the only exceptions were the one V4 depicted here, and a head epidermal cell. (Adapted, with permission, from Cowing and Kenyon 1996 [copyright Macmillan].)

Note: This study has been challenged by Schnabel and Schnabel (1997). We take this opportunity to note that their quantitative analysis of both our M and V6 experiments was incorrect (as a careful reading of our tables and figure legends indicates) and that their example actually supports rather than contradicts our interpretation, for the reasons explained in the text and in Cowing and Kenyon (1996).

we assume that local signaling between close lineal relatives takes place (which is certainly possible), it is nearly impossible to explain why such hypothetical signaling cells would not have induced *mab-5* expression in a little rosette of surrounding cells instead of only in the cells that were lineally equivalent to V6. Thus, the inescapable (or at least hard-to-escape) conclusion of these experiments is that cell lineage plays an important role in establishing the early pattern of *mab-5* expression, at least in some cells. Either lineal mechanisms directly activate *mab-5*, or else, if local cell-cell signaling triggers *mab-5* expression, lineal mechanisms give particular cells the unique ability to respond to local *mab-5*-inducing signals (see Cowing and Kenyon 1996). We know that the *C. elegans* caudal homolog *pal-1* turns on *mab-5* expression in V6 (Salser and Kenyon 1996; C. P. Hunter et al., in prep.). This means that by investigating in detail how *pal-1* activity is targeted to this cell we may be able to learn more about the general mechanisms by which *Hox* gene expression patterns are established in the embryo.

If the entire *Hox* expression pattern is set up by lineal mechanisms triggered by inductive signals that specify fates at the level of individual blastomeres, this would be truly mind-boggling, since there are so many cells with such diverse lineal origins that come to express the same *Hox* gene (Fig. 7b). First, how could such a mechanism actually operate, and second, how could it have evolved?

One possible mechanism we have suggested (Kenyon 1994) is that a tissue polarity system governs the segregation of developmental potential, including the potential to express *Hox* genes, within the lineage. In *Drosophila*, it is known that gradients of Wnt signals can cause responding cells to express particular genes (see Lawrence and Struhl 1996; Lecuit and Cohen 1997), and at least within developing segments, these include *Hox* genes (for review, see Lawrence 1992). It is also clear from studies of tissue polarity in the wing that similar signaling systems can orient the cytoskeleton and in this way influence the direction of the hairs produced by individual cells (Park et al. 1994). Given these findings, it seems possible that in *C. elegans* a similar type of signaling pathway could orient a segregation apparatus that segregates determinants to particular cells at cell division. In fact, Wnt signaling systems have been found to govern the polarity of asymmetric cell divisions during both embryonic and postembryonic development (Herman et al. 1995; Rocheleau et al. 1997; Thorpe et al. 1997; Harris et al. 1996). This segregation model also has the advantage of potentially simplifying the patterning problem, since many of the cells that express *mab-5* occupy relatively *posterior* positions within their respective lineages. Perhaps determinants (qualitative or quantitative in nature) that activate specific *Hox* genes have different potentials to segregate anteriorly versus posteriorly within a lineage.

The only key difference between this hypothetical system and those described in *Drosophila* is the coupling of the cell polarity system to a segregation apparatus. Thus, it seems plausible that the two could have evolved from a common signaling system originally (in fact, it seems likely that both types of systems would operate in certain contexts in both organisms).

These findings also prompt the question of how the ancestral metazoan set up its *Hox* gene expression pattern. Which came first: lineal, positional, or temporal control? The simplest model we can think of is that in the original ancestor, position, lineage, and temporal control were all correlated with one another. This could happen if segments were added by repeated divisions of a stem cell precursor, one at a time. In *Drosophila*, the relative importance of positional mechanisms would have increased; in the leech, the relative importance of temporal controls would have increased, and in *C. elegans*, the relative importance of lineal controls would have increased.

Finally, it is interesting to note that in *C. elegans* embryos, "homeotic selector genes," which are best known for the large-scale transformations they cause in *Drosophila* embryos, are not the *Hox* genes! The *C. elegans Hox* genes behave like homeotic selector genes during postembryonic development, but not in the early embryo. Instead, in the early embryo, the functional equivalents of selector genes are the growing number of developmental genes in which mutations trigger large-scale homeotic transformations that convert one blastomere into another, as assayed by the lineages (that is, the pattern of divisions and differentiated cell types) they generate (for review, see Schnabel and Priess 1997). These cell fate specification genes create differences not between cells located in different A/P positions per se, but instead between different blastomeres, such as the different descendants of the AB cell at the 8-AB cell stage. At this stage in development, *Hox* genes behave simply like cell-type-specific markers. Only later, after hatching, do they behave as homeotic selector genes. Perhaps the larval stage of *C. elegans* is the most ancient and most closely resembles the common ancestor.

REFERENCES

Austin J. and Kenyon C. 1994. Cell contact regulates neuroblast formation in the *Caenorhabditis elegans* lateral epidermis. *Development* **120:** 313.

Burglin T.R., Ruvkun G., Coulson A., Hawkins N.C., McGhee J.D., Schaller D., Wittman C., Muller F., and Waterston R.H. 1991. Nematode homeobox cluster. *Nature* **351:** 703.

Castelli-Gair J., Greig S., Micklem G., and Akam M. 1994. Dissecting the temporal requirements for homeotic gene function. *Development* **120:** 1983.

Chalfie M., Thomson J.N., and Sulston J.E. 1983. Induction of neuronal branching in *C. elegans. Science* **263:** 803.

Chisholm A. 1991. Control of cell fate in the tail region of *C. elegans* by the gene *egl-5. Development* **111:** 921.

Chow K.L. and Emmons S.W. 1994. HOM-C/Hox genes and four interacting loci determine the morphogenetic properties of single cells in the nematode male tail. *Development* **120:** 2579.

Clandinin T.R., Katz W.S., and Sternberg P.W. 1997. *Caenorhabditis elegans* HOM-C genes regulate the response of vulval precursor cells to inductive signal. *Dev. Biol.* **182:** 150.

Clark S.G., Chisholm A.D., and Horvitz H.R. 1993. Control of cell fates in the central body region of *C. elegans* by the homeobox gene *lin-39. Cell* **74:** 43.

Costa M., Weir M., Coulson A., Sulston J., and Kenyon C. 1988. Posterior pattern formation in *C. elegans* involves position-specific expression of a gene containing a homeobox. *Cell* **55:** 747.

Cowing D.W. and Kenyon C. 1992. Expression of the homeotic

gene *mab-5* during *Caenorhabditis elegans* embryogenesis. *Development* **116:** 481.
———. 1996. Correct Hox gene expression established independently of position in *Caenorhabditis elegans. Nature* **382:** 353.
Fixsen W.D. 1985. "The genetic control of hypodermal cell lineages during nematode development." Ph.D. thesis, Massachusetts Institute of Technology, Cambridge, Massachusetts.
Greenwald I. 1997. Development of the vulva. In C. elegans *II* (ed. D.L. Riddle et al.), p. 519. Cold Spring Harbor Laboratory Press, Cold Spring Harbor, New York.
Harris J., Honigberg L., Robinson N., and Kenyon C. 1996. Neuronal cell migration in *C. elegans*: Regulation of *Hox* gene expression and cell position. *Development* **122:** 3117.
Herman M.A., Vassilieva L.L., Horvitz H.R., Shaw J.E., and Herman R.K. 1995. The *C. elegans* gene *lin-44*, which controls the polarity of certain asymmetric cell divisions, encodes a Wnt protein and acts cell nonautonomously. *Cell* **83:** 101.
Horvitz H.R. and Sternberg P.W. 1991. Multiple intercellular signaling systems control the development of the *Caenorhabditis elegans* vulva. *Nature* **351:** 535.
Hunter C.P. and Kenyon C. 1995. Specification of anteroposterior cell fates in *Caenorhabditis elegans* by *Drosophila* Hox proteins. *Nature* **377:** 229.
Kenyon C. 1986. A gene involved in the development of the posterior body region of *C. elegans. Cell* **46:** 477.
———. 1994. If birds can fly, why can't we? Homeotic genes and evolution. *Cell* **78:** 175.
———. 1995. A perfect vulva every time: Gradients and signaling cascades in *C. elegans. Cell* **82:** 171.
Kenyon C. and Wang B. 1991. A cluster of *Antennapedia*-class homeobox genes in a non-segmented animal. *Science* **253:** 516.
Krumlauf R. 1994. Hox genes in vertebrate development. *Cell* **78:** 191.
Lawrence P.A. 1992. *The making of a fly.* Blackwell Scientific, London.
Lawrence P.A. and Morata G. 1994. Homeobox genes: Their function in *Drosophila* segmentation and pattern formation. *Cell* **78:** 181.
Lawrence P.A. and Struhl G. 1996. Morphogens, compartments, and pattern: Lessons from *Drosophila*? *Cell* **85:** 951.
Lecuit T. and Cohen S.M. 1997. Proximal-distal axis formation in the *Drosophila* leg. *Nature* **388:** 139.
Maloof J.M. and Kenyon C.J. 1997. The Hox gene *lin-39* is required during *C. elegans* vulval induction to select the outcome of Ras signaling. *Development* (in press).
Nardelli-Haefliger D., Bruce A.E., and Shankland M. 1994. An axial domain of HOM/Hox gene expression is formed by morphogenetic alignment of independently specified cell lineages in the leech *Helobdella. Development* **120:** 1839.
Park W.J., Liu J., and Adler P.N. 1994. Frizzled gene expression and development of tissue polarity in the *Drosophila* wing. *Dev. Genet.* **15:** 383.
Riddle D.L., Blumenthal T., Meyer B.J., and Priess J.R., Eds. 1997. C. elegans *II.* Cold Spring Harbor Laboratory Press, Cold Spring Harbor, New York.
Rocheleau C.E., Downs W.D., Lin R., Wittmann C., Bei Y., Cha Y., Ali M., Priess J.R., and Mello C.C. 1997. Wnt signaling and an APC-related gene specify endoderm in early *C. elegans* embryos. *Cell* **90:** 707.
Salser S.J. and Kenyon C. 1992. Activation of a *C. elegans Antennapedia* homolog in migrating cells controls their direction of migration. *Nature* **355:** 255.
———. 1994. Patterning *C. elegans*: Homeotic cluster genes, cell fates and cell migrations. *Trends Genet.* **10:** 159.
———. 1996. A *C. elegans* Hox gene switches on, off, on and off again to regulate proliferation, differentiation and morphogenesis. *Development* **122:** 1651.
Salser S.J., Loer C.M., and Kenyon C. 1993. Multiple HOM-C gene interactions specify cell fates in the nematode central nervous system. *Genes Dev.* **7:** 1714.
Schnabel R. and Priess J.R. 1997. Specification of cell fates in the early embryo. In C. elegans *II* (D.L. Riddle et al.), p. 361. Cold Spring Harbor Laboratory Press, Cold Spring Harbor, N.Y.
Schnabel R. and Schnabel H. 1997. Hox genes misled by local environments. *Nature* **385:** 588.
Sulston J.E. and White J.G. 1980. Regulation and cell autonomy during postembryonic development of *Caenorhabditis elegans. Dev. Biol.* **78:** 577.
Sulston J., Du Z., Thomas K., Wilson R., Hillier L., Staden R., Halloran N., Green P., Thierry-Mieg J., Qiu L., Dear S., Coulson A., Craxton M., Durbin R., Berks M., Metzstein M., Hawkins T., Ainscough R., and Waterston R. 1992. The *C. elegans* genome sequencing project: A beginning. *Nature* **356:** 37.
Thorpe C.J., Schlesinger A., Carter J.C., and Bowerman B. 1997. Wnt signaling polarizes an early *C. elegans* blastomere to distinguish endoderm from mesoderm. *Cell* **90:** 695.
Wang B.B., Muller-Immergluck M.M., Austin J., Robinson N.T., Chisholm A., and Kenyon C. 1993. A homeotic gene-cluster patterns the anteroposterior body axis of *C. elegans. Cell* **74:** 29.
Waring D.A. and Kenyon C. 1990. Selective silencing of cell communication influences anteroposterior pattern formation in *C. elegans. Cell* **60:** 123.
———. 1991. Regulation of cellular responsiveness to inductive signals in the developing *C. elegans* nervous system. *Nature* **350:** 712.
Waring D.A., Wrischnik L., and Kenyon C. 1992. Cell signals allow the expression of a preexistent neural pattern in *C. elegans. Development* **116:** 457.
Wightman B., Clark S.G., Taskar A.M., Forrester W.C., Maricq A.M., Bargmann C.I., and Garriga G. 1996. The *C. elegans* gene *vab-8* guides posteriorly directed axon outgrowth and cell migration. *Development* **122:** 671.
Wrischnik L.A. and Kenyon C.J. 1997. The role of *lin-22,* a *hairy/Enhancer of split* homolog, in patterning the peripheral nervous system of *C. elegans. Development* **124:** 2875.

The Regulation of Enhancer-Promoter Interactions in the *Drosophila* Embryo

J. Zhou, H.N. Cai,[1] S. Ohtsuki, and M. Levine
Department of Molecular and Cell Biology, Division of Genetics, University of California, Berkeley, California 94720

We have been interested in characterizing the *cis* regulatory machinery that is responsible for converting relatively few transcription factors into a diverse array of gene expression patterns in the early *Drosophila* embryo. Initially, our studies emphasized the importance of *cis* regulatory modules, or complex enhancers, in directing basic stripes and bands of gene expression during embryogenesis. Such enhancers convert the crude maternal Dorsal gradient into a series of thresholds of gene activity, which subdivide the dorsoventral (D/V) axis of the embryo into mesoderm, neurogenic ectoderm, and dorsal ectoderm. Given the importance of complex enhancers in directing basic patterns of gene expression in the early embryo, we have become interested in the regulation of enhancer-promoter interactions within complex genetic loci. Recent studies suggest that promoter selection depends on a combination of insulator DNAs and the intrinsic activities of different core promoter elements.

THE DORSAL GRADIENT

The Dorsal protein is initially distributed throughout the cytoplasm of oocytes and early embryos, but shortly after fertilization, it is subject to a regulated nuclear transport process (Roth et al. 1989; Rushlow et al. 1989; Steward 1989; for review, see Wasserman 1993; Steward and Govind 1993; Belvin et al. 1995). Dorsal protein enters nuclei in ventral regions but remains in the cytoplasm in dorsal regions. In lateral regions, just a fraction of the protein is released from the cytoplasm to nuclei. The broad Dorsal nuclear gradient differentially regulates a number of target genes in a concentration-dependent fashion. For example, peak levels of the gradient activate the target gene *snail*, which is expressed in ventral regions of the embryo where it helps define the limits of the invaginating mesoderm (Fig. 1A) (Alberga et al. 1991; Kosman et al. 1991; Leptin 1991; Ip et al. 1992). Low levels of Dorsal are insufficient to activate *snail*, but succeed in triggering the expression of another target gene, *short gastrulation* (*sog*), which is expressed in broad lateral stripes that help define the limits of the neurogenic ectoderm (Fig. 1B) (Francois et al. 1994; Holley et al. 1995). The Dorsal gradient also functions as a transcriptional repressor that restricts the expression of certain genes, such as *zen* and *dpp*, to dorsal regions that form derivatives of the dorsal ectoderm (Fig. 1C) (Huang et al. 1993; Jiang et al. 1993; Kirov et al. 1993).

DUPLICATION OF THE DORSOVENTRAL AXIS

Dorsal nuclear transport is initiated by the local activation of the Toll transmembrane receptor in ventral regions of early embryos (Shelton and Wassserman 1993; Morisato and Anderson 1994; Schneider et al. 1994; Bergmann et al. 1996; Reach et al. 1996). A mutant form of Toll, $Toll^{10b}$, contains a single-amino-acid substitution that results in the full constitutive activation of the receptor in a ligand-independent fashion (Erdelyi and Szabad 1989; Schneider et al. 1991). The mutant $Toll^{10b}$ mRNA was expressed in anterior regions of transgenic embryos using the 3′-untranslated region (UTR) localization sequence from the *bicoid* gene (Huang et al. 1997). The $Toll^{10b}$-bcd chimeric mRNA appears to be tightly localized to the anterior pole of early embryos. Nonetheless, this localized RNA triggers a broad D/V patterning response along the anteroposterior (A/P) axis of the embryo.

Dorsal is distributed in a broad A/P nuclear gradient in transgenic embryos. Most or all of the protein enters nuclei in the anterior third of the embryo. Just a fraction of the protein is released from the cytoplasm in central regions; little or no protein enters nuclei in the posterior third of the embryo (Huang et al. 1997). This ectopic A/P Dorsal gradient establishes the full repertory of D/V patterning responses along the length of the embryo. For example, *snail* is activated by high concentrations of the Dorsal gradient in the anterior third of the embryo (Fig. 1D; compare with 1A). Low levels of the ectopic gradient activate *sog* in central regions (Fig. 1E; compare with 1B), and *zen* expression is restricted to the posterior third of transgenic embryos (Fig. 1F; compare with 1C). These studies indicate that Dorsal target genes can be differentially activated in the absence of an extracellular Spatzle (Spz) gradient. An implication of these studies is that Toll signaling components that function downstream from Spz might be able to diffuse and influence the overall slope of the normal D/V Dorsal gradient.

[1]*Present address*: Department of Cellular Biology, University of Georgia, Athens, Georgia 30602.

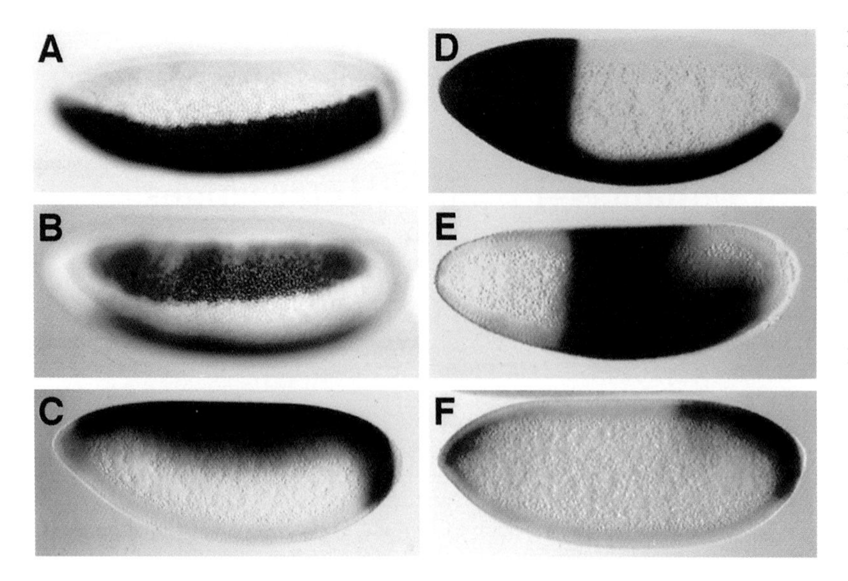

Figure 1. Misexpression of dorsal target genes along the A/P axis. *Toll*^{10b} transgenic embryos are undergoing cellularization and are oriented with dorsal up and anterior to the left. The embryos were hybridized with various digoxigenin-labeled antisense RNA probes. (*A*) *snail* expression pattern in a wild-type embryo. Staining is restricted to ventral regions that will form the mesoderm. (*B*) *sog* expression pattern in a wild-type embryo. Staining is detected in two broad lateral stripes that encompass the entire presumptive neurogenic ectoderm. Only one of the stripes can be seen in this ventrolateral view. The other stripe is out of the plane of focus. (*C*) *zen* expression pattern in a wild-type embryo. Staining is detected in dorsal regions, and at the anterior and posterior poles. The endogenous Dl nuclear gradient represses *zen* expression in ventral and lateral regions. The Torso RTK pathway blocks the ability of Dorsal to function as a repressor at the poles (Rusch and Levine 1994). (*D*) *snail* expression pattern in an embryo derived from a transgenic female. Ectopic staining is observed in the anterior third of the embryo. Because the *Toll*^{10b} transgene is expressed in a wild-type strain, the normal, endogenous staining pattern is also observed in the ventral mesoderm. (*E*) *sog* expression pattern. Ectopic staining is detected in a broad domain in central regions of the embryo. The endogenous lateral stripes are also observed. Both the ectopic and endogenous staining patterns are excluded from the anterior third of the embryo by the Snail repressor (see *A*). (*F*) *zen* expression pattern. Staining is detected at the anterior and posterior poles, and in a dorsal patch in the presumptive abdomen. The latter site of expression corresponds to the only region that lacks both the endogenous and ectopic Dorsal nuclear gradients (Fig. 2C and data not shown). Staining at the poles coincides with regions where the Dl repressor is masked by the Torso RTK pathway.

COMPLEX ENHANCERS

In general, localized patterns of gene expression depend on complex enhancers that are typically 300–900 bp in length. For example, *zen* is regulated by a 600-bp *cis* regulatory module that contains binding sites for Dorsal, corepressors, and one or more ubiquitious activators (Jiang et al. 1993; Kirov et al. 1993). In principle, the activators mediate expression throughout the embryo, but the pattern is kept off in ventral and lateral regions by the Dorsal gradient (see Fig. 1C). Dorsal is inherently a tran-

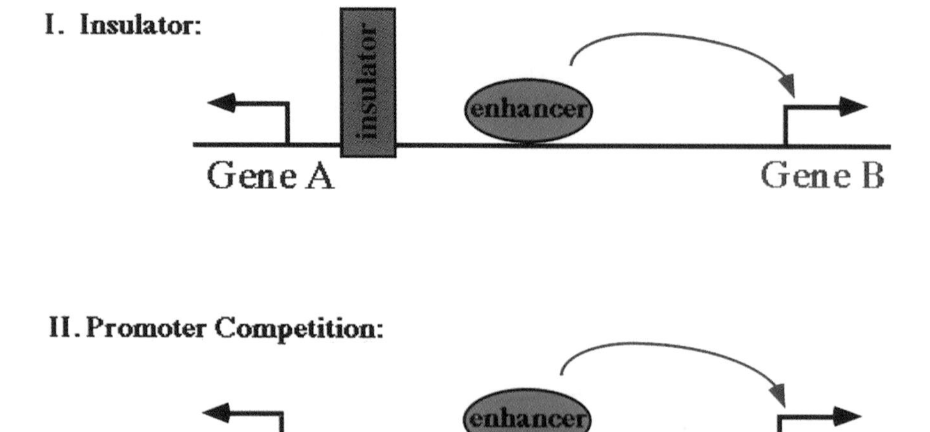

Figure 2. Regulation of enhancer-promoter interactions. The diagram shows a shared enhancer located between two linked genes, *A* and *B*. At least two mechanisms can be envisioned for the specific interaction of the enhancer with gene *B*. In principle, an insulator DNA can block interactions of the enhancer with gene *A* and not affect enhancer-gene *B* interactions. Alternatively, it is possible that the enhancer can interact with both genes *A* and *B*, but it prefers the core promoter sequence in gene *B*. As a result, the enhancer tends to occupy promoter *B*, thereby precluding interactions with gene *A*.

scriptional activator, but it functions as a silencer in the context of the *zen* promoter due to neighboring corepressor elements.

Given the importance of enhancers in directing basic patterns of gene expression, we have become interested in the regulation of enhancer-promoter interactions within complex genetic loci, such as the *Antennapedia* and *Bithorax complexes* (Lewis 1978; Kaufman et al. 1980). At least two mechanisms can be envisioned for restricting the interaction of a shared enhancer with an inappropriate target promoter (Fig. 2). In principle, an insulator DNA can block the interaction of the enhancer with gene *A*, but not affect interactions with gene *B*. A second potential mechanism is promoter competition. According to this scenario, the enhancer can interact with both genes, but prefers the core promoter sequence present in gene *B*. The interaction of the enhancer with gene *B* precludes interactions with promoter *A*. This type of mechanism was first described in the chicken globin gene complex (Choi and Engel 1988; Gallarda et al. 1989; Foley and Engel 1992; Foley et al. 1994; Mason et al. 1996), and we have recently obtained evidence that a similar mechanism might be used in the *Drosophila* embryo.

INSULATOR DNAs

The *Abdominal-B* (*Abd-B*) gene of the Bithorax complex (BX-C) is regulated by an extended *cis* regulatory region that encompasses approximately 100 kb of DNA (Lewis 1978; Sanchez-Herrero et al. 1985; Lewis et al. 1995; Martin et al. 1995). This region contains a series of tissue-specific enhancers, including IAB5, which is located approximately 57 kb downstream from the *Abd-B* transcription start site (Karch et al. 1985; Celniker et al. 1989, 1990; DeLorenzi and Bienz 1990; Boulet et al. 1991; Sanchez-Herrero 1991; Busturia and Bienz 1993). It has been proposed that the *Abd-B* regulatory region is punctuated by boundary elements, or insulator DNAs, that organize the different enhancers into separate chromatin loop domains (Fig. 3A) (Gyurkovics et al. 1990;

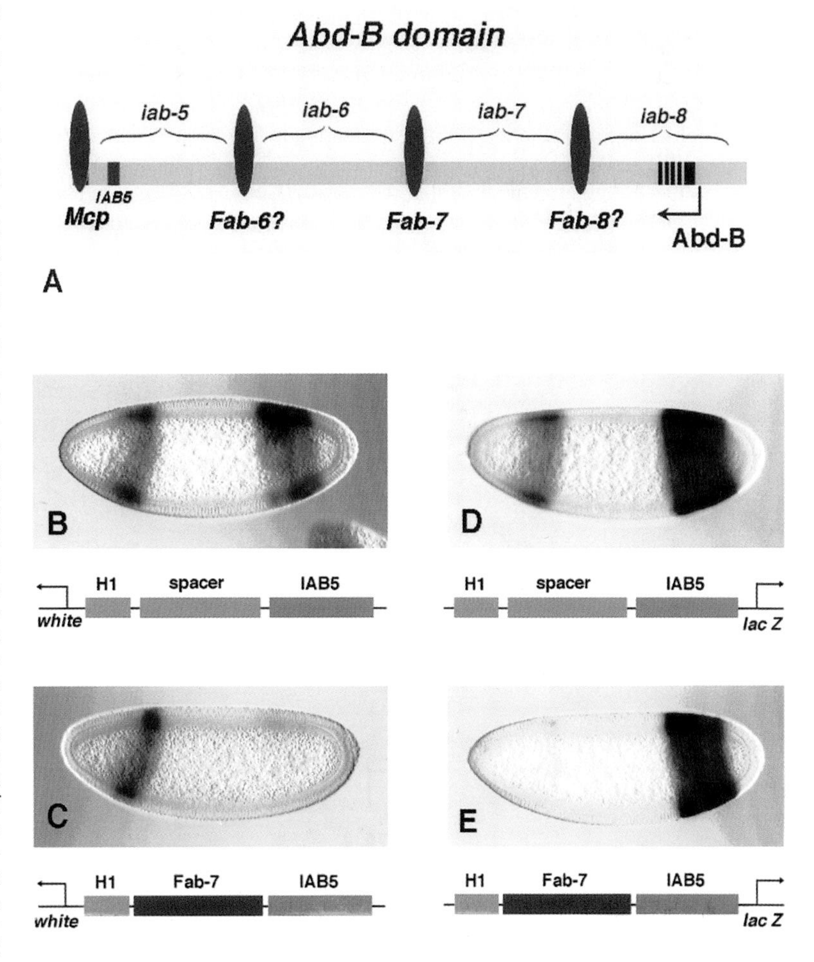

Figure 3. Insulator DNAs in the *Abd-B* domain of the BX-C. (*A*) The leftward arrow indicates the position of the *Abd-B* transcription start site. The 1-kb IAB5 enhancer is located in a distal region of *iab-5*, approximately 57 kb downstream from the *Abd-B* transcription start site. The Fab-7 insulator is located between *iab-6* and *iab-7*, approximately 35 kb from the *Abd-B* promoter. The Mcp insulator is located between *iab-4* and *iab-5*, about 64 kb from the *Abd-B* promoter (Lewis et al. 1995; Martin et al. 1995). It has been proposed that the *Abd-B cis* regulatory region contains two additional insulators, Fab-6 and Fab-8 (Karch et al. 1994). (*B–E*) Transgenic embryos were hybridized with either a digoxygenin-labeled *white* or *lacZ* antisense RNA probe. Hybridization signals were visualized by histochemical staining using alkaline phosphatase. All embryos are undergoing cellularization, about 150 min postfertilization; they are oriented with anterior to the left and dorsal up. (*B*) *white* expression pattern directed by the fusion promoter summarized in the diagram below the embryo. The synthetic gene complex contains two divergently transcribed genes, the leftward *white* gene and a rightward *lacZ* reporter gene (driven by a minimal *Tp* promoter sequence). Two different enhancers were placed between the *white* and *lacZ* transcription units; a 228-bp enhancer that directs expression of *hairy* stripe 1 (H1) as well as the 1-kb IAB5 enhancer from the *iab-5 cis* regulatory region of *Abd-B*. The two enhancers are separated by a 1-kb spacer sequence. An additive *white* expression pattern is observed consisting of a head stripe (directed by the H1 enhancer) and a band of staining in the presumptive abdomen (directed by IAB5). (*C*) Same as *A* except that the spacer sequence was replaced with the 1.2-kb Fab-7 element. The distal IAB5 enhancer is attenuated, so that there is only weak expression of the *white* reporter gene in the abdomen. In contrast, the proximal H1 enhancer is unaffected and directs a strong head stripe. (*D*) Expression of the rightward *lacZ* gene directed by the same fusion promoter shown in *A*. As for *white*, the fusion promoter directs an additive staining pattern consisting of a head stripe and abdominal band. (*E*) Same as *C* except that the spacer sequence was replaced with the Fab-7 element. The distal H1 enhancer is selectively blocked, so that there is only residual staining in head regions. The IAB5 enhancer continues to direct strong expression in the presumptive abdomen.

Galloni et al. 1993; Karch et al. 1994). We have employed a "stripe expression assay" in transgenic embryos to show that the *cis* regulatory region contains at least three different insulator DNAs, Fab-7, Mcp, and a newly identified sequence, Fab-8 (Fig. 3A) (Gyurkovics et al. 1990; Galloni et al. 1993; Karch et al. 1994; Hagstorm et al. 1996; Zhou et al 1996; J. Zhou and M. Levine, unpubl.). Each of these insulators is less than 1 kb in length and selectively blocks the interactions of distal, not proximal, enhancers with a target promoter (Fig. 3B–E).

Some of our studies involved the use of a fusion promoter that contains the *hairy* H1 head stripe enhancer (Howard and Struhl 1990; Riddihough and Ish-Horowicz 1991) and the IAB5 enhancer. This H1-IAB5 promoter was placed between two divergently transcribed reporter genes, *white* to the left and *lacZ* to the right (Fig. 3B–E). When a neutral spacer was placed between the two enhancers, both enhancers interacted with the *white* and *lacZ* genes to direct composite patterns of gene expression (Fig. 3B,C). Both *white* and *lacZ* are expressed in an anterior stripe and a posterior band within the presumptive abdomen. However, very different staining patterns are observed when the spacer sequence is replaced with the Fab-7 insulator DNA. The *white* reporter gene is expressed exclusively within the limits of the H1 *hairy* stripe in anterior regions, whereas the distal IAB5 enhancer is blocked so that there is little or no staining in the abdomen (Fig. 3D). Conversely, *lacZ* staining is restricted to the abdomen, indicating that the distal H1 enhancer is blocked (Fig. 3E). Similar results were obtained with the Mcp and Fab-8 insulators. These results demonstrate that individual insulator DNAs block the interactions of a distal enhancer with a target promoter. Future studies will determine whether combinations of insulators can mediate the formation of loop domains and possibly facilitate long-range enhancer-promoter interactions.

PROMOTER COMPETITION

The AE1 enhancer is located in the intergenic region between *Sex combs reduced* (*Scr*) and *fushi tarazu* (*ftz*) within the *Antennapedia* gene complex (ANT-C; Kuroiwa et al. 1985; Ingham 1988; Gindhart et al. 1995; Gorman and Kaufman 1995). *Scr* is a homeotic selector gene that controls the morphogenesis of posterior head structures and anterior thorax, whereas *ftz* is a pair-rule gene that initiates the segmentation pattern. AE1 is an auto-regulatory element that maintains the seven-stripe *ftz* pattern during gastrulation and germ band elongation (Pick et al. 1990; Schier and Gehring 1992). It specifically interacts with the *ftz* promoter but does not activate *Scr*, even though AE1 is evenly spaced between the two promoters (Gindhart et al. 1995; Gorman and Kaufman 1995; also see summary in Fig. 4). We have obtained evidence that promoter competition helps exclude interactions between AE1 and *Scr*, as discussed below. The two genes contain diverse core promoter sequences. *Scr* lacks a TATA box but contains an initiator element (DPE; see diagram in Fig. 4). Conversely, *ftz* contains a TATA sequence but lacks an initiator. A number of experiments were conducted to determine whether these different core elements contribute to the observed regulatory specficity exhibited by the AE1 enhancer.

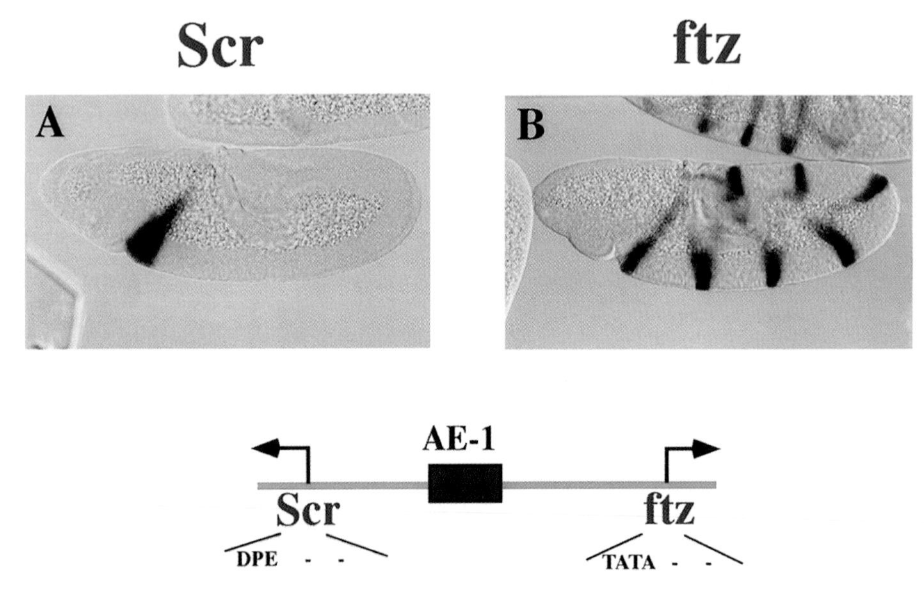

Figure 4. Regulatory specficity in the *Scr-ftz* interval of the ANT-C. Embryos were hybridized with either a dioxigenin-labeled *Scr* or *ftz* antisense RNA probe and visualized via histochemical staining. The embryos are undergoing the rapid phase of germ band elongation (4–5 hr postfertilization) and are oriented with anterior to the left and dorsal up. *Scr* is expressed within the primordia of parasegment 2 (PS2), which gives rise to regions of the labial and prothoracic segments (*A*). *ftz* is expressed in a series of pair-rule stripes (*B*). The diagram below the embryos shows the location of the AE1 enhancer within the *Scr-ftz* interval. AE1 specifically interacts with the *ftz* promoter to maintain the seven-stripe pattern. It does not activate the linked *Scr* gene, which is regulated by separate enhancers within the ANT-C. *ftz* and *Scr* contain distinct core promoter sequences (see diagram). *ftz* contains a TATA sequence, but lacks initiator elements. In contrast, *Scr* lacks TATA but contains a close match to the downstream promoter element (DPE), which is often associated with initiator-class promoters (see Burke and Kadonaga 1996).

AE1 was challenged with different classes of core promoter sequences. In the first such experiment, the divergently transcribed *white* and *lacZ* reporter genes were placed under the control of the *miniwhite* promoter and the core promoter from the *P*-element transposase gene (*Tp*; Kaufman et al. 1989). AE1 activates both promoters so that *white* and *lacZ* are expressed in a series of stripes in response to the endogenous ftz activator. Both the *white* and *Tp* promoters lack a TATA box but contain strong initiator elements (Pirrotta et al. 1985; Kaufman et al. 1989).

The substitution of the *Tp* promoter with a core *eve* promoter sequence results in a substantial reduction in the expression of the linked *white* gene. AE1 interacts with *eve* to direct *lacZ* expression in a series of stripes. However, *white* expression is virtually silent. This result suggests that AE1/*eve* interactions preclude activation of the *white* gene. In the absence of the preferred *eve* promoter, AE1 is able to activate the linked *white* gene in a series of stripes. The *eve* promoter is similar to *ftz* in that it contains an optimal TATA box but lacks an initiator element (Frash et al. 1987; Lo and Smale 1996). These results suggest that promoter competition might regulate enhancer-promoter interactions within the ANT-C. AE1-*ftz* interactions might preclude AE1-Scr interactions.

We propose that different core promoters possess distinct regulatory activities. It is conceivable that the TFIID complex adopts different conformations on these distinct types of promoters. The ftz activator might make more efficient contact with the basal targets arranged on TATA. Perhaps these targets are somewhat sequestered in the alternate conformation formed on initiator-class core promoters.

ACKNOWLEDGMENT

This work was funded by grants from the National Institutes of Health (GM-46638 and GM-34431).

REFERENCES

Alberga A., Boulay J.L., Kempe E., Dennefield C., and Haenlin M. 1991. The *snail* gene required for mesoderm formation in *Drosophila* is expressed dynamically in derivatives of all three germ layers. *Development* **111:** 983.

Belvin M.P., Jin Y., and Anderson K.V. 1995. Cactus protein degradation mediates *Drosophila* dorsal-ventral signaling. *Genes Dev.* **9:** 783.

Bergmann A., Stein D., Geisler R., Hagenmaier S., Schmid B., Fernandez N., Schnell B., and Nüsslein-Volhard C. 1996. A gradient of cytoplasmic Cactus degradation establishes the nuclear localization gradient of the Dorsal morphogen in *Drosophila. Mech. Dev.* **60:** 109.

Boulet A.M., Lloyd A., and Sakonju S. 1991. Molecular definition of the morphogenetic and regulatory functions of the *cis*-regulatory elements of the *Drosophila Abd-B* homeotic gene. *Development* **111:** 393.

Burke T.W. and Kadonaga J.T. 1996. *Drosophila* TFIID binds to a conserved downstream basal promoter element that is present in many TATA-box-deficient promoters. *Genes Dev.* **10:** 711.

Busturia A. and Bienz M. 1993. Silencers in *Abdominal-B*, a homeotic *Drosophila* gene. *EMBO J.* **12:** 1415.

Celniker S.E., Keelan D.J., and Lewis E.B. 1989. The molecular genetics of the bithorax complex of *Drosophila*—Characterization of the *Abdominal-B* domain. *Genes Dev.* **3:** 1424.

Celniker S.E., Sharma S., Keelan D., and Lewis E.B. 1990. The molecular genetics of the bithorax complex of *Drosophila: cis*-regulation in the *Abdominal-B* domain. *EMBO J.* **9:** 4277.

Choi O.R. and Engel J.D. 1988. Developmental regulation of β-globin gene switching. *Cell* **55:** 17.

DeLorenzi M. and Bienz M. 1990. Expression of Abdominal-B homeoproteins in *Drosophila* embryos. *Development* **108:** 323.

Erdelyi M. and Szabad J. 1989. Isolation and characterization of dominant female sterile mutations of *Drosophila melanogaster.* I. Mutations on the third chromosome. *Genetics* **122:** 111.

Foley K.P. and Engel J.D. 1992. Individual stage selector element mutations lead to reciprocal changes in β- vs. ε-globin gene transcription: Genetic confirmation of promoter competition during globin gene switching. *Genes Dev.* **6:** 730.

Foley K.P., Pruzina S., Winick J.D., Engel J.D., Grosveld F., and Fraser P. 1994. The chicken β/ε-globin enhancer directs autonomously regulated, high-level expression of the chicken ε-globin gene in transgenic mice. *Proc. Natl. Acad. Sci.* **91:** 7252.

Francois V., Solloway M., O'Neill J.W., Emery J., and Bier E. 1994. Dorsal-ventral patterning of the *Drosophila* embryo depends on a putative negative growth factor encoded by the *short gastrulation* gene. *Genes Dev.* **8:** 2602.

Frasch M., Hoey T., Rushlow C., Doyle H., and Levine M. 1987. Characterization and localization of the Even-skipped protein of *Drosophila. EMBO J.* **6:** 749.

Gallarda J.L., Foley K.P., Yang Z.Y., and Engel J.D. 1989. The β-globin stage selector element factor is erythroid-specific promoter/enhancer binding protein NF-E4. *Genes Dev.* **3:** 1845.

Galloni M., Gyurkovics H., Schedl P., and Karch F. 1993. The *bluetail* transposon: Evidence for independent *cis*-regulatory domains and domain boundaries in the bithorax complex. *EMBO J.* **12:** 1087.

Gindhart J.G., Jr., King A.N., and Kaufman T.C. 1995. Characterization of the *cis*-regulatory region of the *Drosophila* homeotic gene *Sex combs reduced. Genetics* **139:** 781.

Gorman M.J. and Kaufman T.C. 1995 Genetic analysis of embryonic *cis*-acting regulatory elements of the *Drosophila* homeotic gene *Sex combs reduced. Genetics* **140:** 557.

Gyurkovics H., Gausz J., Kummer J., and Karch F. 1990. A new homeotic mutation in the *Drosophila* bithorax complex removes a boundary separating two domains of regulation. *EMBO J.* **9:** 2579.

Hagstorm K., Muller M., and Schedl P. 1996. Fab-7 functions as a chromatin domain boundary to ensure proper segment specification by the *Drosophila* bithorax complex. *Genes Dev.* **10:** 3202.

Holley S.A., Jackson P.D., Sasai Y., De Robertis E.M., Hoffman F.M., and Ferguson E.L. 1995. A conserved system for dorsal-ventral patterning in insects and vertebrates involving *sog* and *chordin. Nature* **376:** 249.

Howard K.R. and Struhl G. 1990. Decoding postitional information—Regulation of the pair-rule gene *hairy. Development* **110:** 1223.

Huang A.M., Rusch J., and Levine M. 1997. An anteroposterior Dorsal gradient in the *Drosophila* embryo. *Genes Dev.* **11:** 1987.

Huang J.D., Schwyter D.H., Shirokawa J.M., and Courey A.J. 1993. The interplay between multiple enhancer and silencer elements defines the pattern of *decapentaplegic* expression. *Genes Dev.* **7:** 694.

Ingham P.W. 1988. The molecular genetics of embryonic pattern formation in *Drosophila. Nature* **335:** 25.

Ip Y.T., Park R.E., Kosman D., Yazdanbakhsh K., and Levine M. 1992. Dorsal-Twist interactions establish *snail* expression in the presumptive mesoderm of the *Drosophila* embryo. *Genes Dev.* **6:** 1518.

Jiang J., Cai H., Zhou Q., and Levine M. 1993. Conversion of a Dorsal-dependent silencer into an enhancer: Evidence for Dorsal corepressors. *EMBO J.* **12:** 3201.
Karch F., Galloni M., Sipos L., Gausz J., Gyurkovics H., and Schedl P. 1994. Mcp and Fab-7: Molecular analysis of putative boundaries of *cis*-regulatory domains in the bithorax complex of *Drosophila melanogaster. Nucleic Acids Res.* **22:** 3138.
Karch F., Weiffenbach B., Peifer M., Bender W., Duncan I., Celniker S., Crosby M., and Lewis E.B. 1985. The abdominal region of the bithorax complex. *Cell* **43:** 81.
Kaufman P.D., Doll R.F., and Rio D.C. 1989. *Drosophila P* element transposase recognizes internal *P* element DNA sequences. *Cell* **59:** 359.
Kaufman T.C., Lewis R., and Wakimoto B. 1980. Cytogenetic analysis of chromosome 3 in *Drosophila melanogaster:* The homeotic gene complex in polytene chromosomal interval 84A,B. *Genetics* **94:** 115.
Kirov N., Zhelnin L., Shah J., and Rushlow C. 1993. Conversion of a silencer into an enhancer: Evidence for a co-repressor in Dorsal-mediated repression in *Drosophila. EMBO J.* **12:** 3193.
Kosman D., Ip Y.T., Levine M., and Arora K. 1991. Establishment of the mesoderm-neuroectoderm boundary in the *Drosophila* embryo. *Science* **254:** 118.
Kuroiwa A., Kloter U., Baumgartner P., and Gehring W.J. 1985. Cloning of the homeotic *Sex combs reduced* gene on *Drosophila* and *in situ* localization of its transcripts. *EMBO J.* **4:** 3757.
Leptin M. 1991. Twist and Snail as positive and negative regulators during *Drosophila* mesoderm development. *Genes Dev.* **5:** 1568.
Lewis E.B. 1978. A gene complex controlling segmentation in *Drosophila. Nature* **276:** 565.
Lewis E.B., Knafels J.D., Mathog D.R., and Celniker S.E. 1995. Sequence analysis of the cis-regulatory regions of the bithorax complex of *Drosophila. Proc. Natl. Acad. Sci.* **92:** 8403.
Lo K. and Smale S.T. 1996. Generality of a functional initiator consensus sequence. *Gene* **182:** 13.
Martin C.H., Mayeda C.A., Davis C.A., Ericsson C.L., Knafels J.D., Mathog D.R., Celniker S.E., Lewis E.B., and Palazzolo M.J. 1995. Complete sequence of the bithorax complex of *Drosophila. Proc. Natl. Acad. Sci.* **92:** 8398.
Mason M.M., Grasso J.A., Gavrilova O., and Reitman M. 1996. Identification of functional elements of the chicken ϵ-globin promoter involved in stage-specific interaction with the β/ϵ enhancer. *J. Biol. Chem.* **271:** 25459.
Morisato D. and Anderson K.V. 1994. The *spatzle* gene encodes a component of the extracellular signaling pathway establishing the dorsal-ventral pattern of the *Drosophila* embryo. *Cell* **76:** 677.
Pick L., Schier A., Affolter M., Schmidt-Glenewinkel T., and Gehring W.J. 1990. Analysis of the *ftz* upstream element: Germ layer-specific enhancers are independently autoregulated. *Genes Dev.* **4:** 1224.
Pirrotta V., Steller H., and Bozzetti M.P. 1985. Multiple upstream regulatory elements control the expression of the *Drosophila white* gene. *EMBO J.* **4:** 3501.
Riddihough G. and Ish-Horowicz D. 1991. Individual stripe regulatory elements in the *Drosophila hairy* promoter respond to maternal, gap, and pair-rule genes. *Genes Dev.* **5:** 840.
Reach M., Galindo R.L., Towb P., Allen J.L., Karin M., and Wasserman S.A. 1996. A gradient of Cactus protein degradation establishes dorsoventral polarity in the *Drosophila* embryo. *Dev. Biol.* **180:** 353.
Roth S., Stein D., and Nüsslein-Volhard C. 1989. A gradient of nuclear localization of the Dorsal protein determines dorsoventral pattern in the *Drosophila* embryo. *Cell* **59:** 1189.
Rusch J. and Levine M. 1994. Regulation of Dorsal morphogen by the Toll and Torso signaling pathways: A receptor tyrosine kinase selectively masks transcription repression. *Genes Dev.* **8:** 1247.
Rushlow C.A., Han K., Manley J.L., and Levine M. 1989. The graded distribution of the Dorsal morphogen is initiated by selective nuclear transport in *Drosophila. Cell* **59:** 1165.
Sanchez-Herrero E. 1991. Control of the expression of the bithorax complex *abdominal-A* and *abdominal-B* by *cis*-regulatory regions in *Drosophila* embryos. *Development* **111:** 437.
Sanchez-Herrero E., Vernos I., Marco R., and Morata G. 1985. Genetic organization of the *Drosophila* bithorax complex. *Nature* **313:** 108.
Schier A.F. and Gehring W.J. 1992. Direct homeodomain-DNA interaction in the autoregulation of the *fushi tarazu* gene. *Nature* **356:** 804.
Schneider D.S., Hudson K.L., Lin T.Y., and Anderson K.V. 1991. Dominant and recessive mutations define functional domains of Toll, a transmembrane protein required for dorsal-ventral polarity in the *Drosophila* embryo. *Genes Dev.* **5:** 797.
Schneider D.S., Jin Y., Morisato D., and Anderson K.V. 1994. A processed form of Spatzle protein defines dorsal-ventral polarity in the *Drosophila* embryo. *Development* **120:** 1243.
Shelton C.A. and Wasserman S.A. 1993. *pelle* encodes a protein kinase required to establish dorsoventral polarity in the *Drosophila* embryo. *Cell* **72:** 515.
Steward R. 1989. Relocalization of the Dorsal protein from the cytoplasm to the nucleus correlates with its function. *Cell* **59:** 1179.
Steward R. and Govind S. 1993. Dorsal-ventral polarity in the *Drosophila* embryo. *Curr. Opin. Genet. Dev.* **3:** 556.
Wasserman S.A. 1993. A conserved signal transduction pathway regulating the activity of the Rel-like proteins dorsal and NF-κB. *Mol. Biol. Cell* **4:** 767.
Zhou J., Barolo S., Szymanksy P., and Levine M. 1996. The Fab7 element of the bithorax gene complex attenuates enhancer-promoter interactions in the *Drosophila* embryo. *Genes Dev.* **10:** 3195.

Cross-Regulatory Interactions between *Hox* Genes and the Control of Segmental Expression in the Vertebrate Central Nervous System

S. Nonchev, M. Maconochie, A. Gould, A. Morrison, and R. Krumlauf
Division of Developmental Neurobiology, National Institute for Medical Research, The Ridgeway, Mill Hill, London NW7 1AA, United Kingdom

In animals, many morphological differences along the anteroposterior (A/P) body axis are controlled by a conserved set of transcription factors encoded by the *Hox* gene family. For example, the vertebrate hindbrain is segmentally organized and develops from a series of lineage-restricted units termed rhombomeres (r) (for review, see Lumsden and Krumlauf 1996). Within this metameric plan, *Hox* genes display rhombomere-specific and nested domains of expression (for review, see Keynes and Krumlauf 1994). By analogy to their counterparts in *Drosophila melanogaster*, the homeotic selector genes of the ANT-C and BX-C complexes (Lewis 1978; Kaufman et al. 1980; Struhl 1984; Tiong et al. 1987), the vertebrate *Hox* complexes were postulated to regulate regional identity. In agreement with this hypothesis, mutational analyses and ectopic expression studies in vertebrates (for review, see McGinnis and Krumlauf 1992; Krumlauf 1993b) have clearly demonstrated that the *Hox/HOM-C* genes are key regulators of patterning in the vertebrate hindbrain. Ectopic expression of *Hoxa1* (Zhang et al 1994; Alexandre et al. 1996) or induction of *Hoxa1* and *Hoxb1* by retinoids (Marshall et al. 1992; Kessel 1993; Hill et al. 1995) in mouse and fish embryos led to an r2 to r4 transformation. Conversely, loss-of-function experiments of mouse *Hoxb1* (Goddard et al. 1996; Studer et al. 1996) and *Hoxa1* (Carpenter et al. 1993; Dollé et al. 1993; Mark et al. 1993) led to anterior shifts in rhombomere identity or rhombomere deletions, respectively.

A hallmark of the *Hox* loci is their organization into complexes that display the property of colinearity, in which the position of a gene within a complex correlates with its relative A/P expression domain (for review, see Duboule 1992; McGinnis and Krumlauf 1992; Krumlauf 1994; Carroll 1995). Mutational analysis reveals that these restricted A/P domains also correspond to functional domains (Akam et al. 1988; Krumlauf 1994), indicating that the *Hox* clusters contain within their genomic structure regulatory information that is translated into A/P patterning in the embryo. The reasons for this colinear relationship between organization, spatial expression, and function are not clear but are believed to involve conserved *Hox* regulatory mechanisms.

In most vertebrates, there are four *Hox* complexes located on separate chromosomes, which are related to one another by duplication and divergence from a common ancestor (Boncinelli et al. 1989; Kappen et al. 1989; Duboule 1992; McGinnis and Krumlauf 1992; Garcia-Fernandez and Holland 1994). Until recently, it was thought that all vertebrates had four complexes with exactly the same complement of *Hox* genes; however, cloning and analysis of the *Hox* complexes from the pufferfish *Fugu rubripes* have illustrated that there is a continuing evolution of *Hox* genes from the early vertebrate ancestor (Aparicio et al. 1997). Several genes are missing and new members are present in the four pufferfish complexes.

Considerable emphasis has been placed on the degree of similarity between the arthropod and vertebrate *Hox* complexes, but there are significant differences. The *Drosophila Hox* cluster is naturally divided into two complexes (Lewis 1978; Kaufman et al. 1980), and experimentally, the BX-C itself can be split into two with relatively minor phenotypic consequences (Struhl 1984; Tiong et al. 1987). Furthermore, the vertebrate complexes are small (~120 kb), with all of the genes transcribed in the same direction, and there are no non-*Hox* loci in the clusters (McGinnis and Krumlauf 1992). However, the ANT-C contains several non-*Hox* loci, the *Deformed* gene is transcribed in a direction opposite to all other ANT-C *Hox* genes (Kaufman et al. 1990; McGinnis and Krumlauf 1992), and the complexes are large, with the *Drosophila Antennapedia* (*Antp*) gene nearly as large as *HoxB* itself (Scott et al. 1983; Schneuwly et al. 1986; Duboule and Dollé 1989). Thus, it appears that the vertebrate genes have remained closer to an idealized ancestral colinear organization than have their *Drosophila* counterparts (McGinnis and Krumlauf 1992; Garcia-Fernandez and Holland 1994; Carroll 1995). This suggests that there are strong evolutionary constraints keeping vertebrate *Hox* genes colinear and tightly clustered.

Despite the important role that *Hox* genes have in regulating morphogenesis, there is not much information available on direct target genes through which the *Hox* genes act. Major aspects of the patterning pathway downstream from *Hox* genes have not been revealed by genetic screens (Nüsslein-Volhard and Wieschaus 1980), presumably due to the multiple roles of such targets in many tissues and processes. Insight into the nature of target sites for *Hox* action has come from analysis of the *Hox* genes themselves. In *Drosophila,* the *Hox* genes are known to cross and auto-regulate each other by direct and indirect mechanisms. With respect to direct regulation, this often results in the down-regulation of anterior genes through negative cross-regulation by more posteriorly expressed loci. The *Drosophila* midgut represents a good example of indirect auto- and cross-regulation between homeotic genes (Bienz 1994).

It is not known if a similar system operates in vertebrates and in general the contribution of cross-regulatory interactions in maintaining the A/P boundaries and characteristic nested domains of expression. Auto-regulation appears to be important as analysis of the *Hoxb1* gene in mice has demonstrated that there is a highly conserved and direct auto-regulatory loop involved in maintaining segment-specific expression in the developing hindbrain (Popperl et al. 1995). Hoxb1 in combination with a Pbx protein, a vertebrate equivalent of the extradentical (exd) homeodomain protein (Rauskolb et al. 1993), as a cofactor binds to its own 5′ enhancer and positively regulates rhombomere 4 expression (Popperl et al. 1995; Studer et al. 1996). These bipartite *Hox/Pbx*-binding motifs appear to be generally important, as the vertebrate motifs when placed in the *Drosophila* germ line function in a *Hox*- and *exd*-dependent manner and show specific interactions in vitro that reflect their in vivo selectivity (Chambers 1994; Chan and Mann 1996; Chan et al. 1996, 1997).

To investigate whether cross-regulatory interactions are fundamentally important as a general mechanism in the regulation of vertebrate *Hox* complexes, we have employed genetic approaches in mice. Previously, using *Hox/lacZ* reporter genes in transgenic mice, we have found evolutionarily conserved enhancers that reconstruct segmental patterns of expression of *Hoxb4* (Whiting et al. 1991; Aparicio et al. 1995; Morrison et al. 1995) and *Hoxb2* (Sham et al. 1993; Vesque et al. 1996) in the hindbrain. In this paper, analysis of these enhancers reveals that positive and direct cross-regulatory interactions are integrally involved in regulating these patterns of segmental expression. These findings demonstrate the fundamental role that cross-regulation has in vertebrate *Hox* regulation and point to important and general mechanistic differences between regulation of the arthropod and vertebrate complexes.

METHODS

DNA constructs. Reporter constructs with *Hoxb2* and *Hoxb4* genomic regions were in the BGZ40 *lacZ* reporter vector (Yee and Rigby 1993). Constructs for the ectopic expression of *Hoxa1* and *Hoxb1* using the human β-actin promoter/enhancer have been described previously (Zhang et al. 1994; Popperl et al. 1995). Similar ectopic vectors for *Hoxb-3*, *Hoxb-4*, *Hoxb-5*, *Hoxb-9*, and *Hoxd-4* cDNAs were polymerase chain reaction (PCR)-amplified from sequenced cDNA clones with an optimized Kozak translational start site (Kozak 1989) and inserted into β-actin or *Wnt1* vectors (Gould et al. 1997). The *Hoxb2* and *Drosophila labial* ectopic expression constructs were prepared from a full-length *Hoxb2* cDNA (M.K. Maconochie, unpubl.) or an end-filled *Eco*RI-*Hin*dIII fragment from pLabial *Ssp*I (Chouinard and Kaufman 1991) into the same β-actin expression vector, respectively. Enhancer mutations (deletions and substitutions) were generated by site-directed mutagenesis and confirmed by sequencing.

Generation and analysis of transgenic mice. Generation of transgenic embryos by microinjection and *lacZ* reporter activity was assayed as described previously (Whiting et al. 1991). In the *Hoxb2 trans*-activation experiments, males homozygous for the *Hoxb2* r4 enhancer were mated to superovulated F_1 hybrid females to collect fertilized eggs for the subsequent microinjection of various β-actin/cDNA ectopic expression constructs. F_0 embryos were then harvested to assay for changes in *lacZ* expression. The *Hoxb4 trans*-activation assays were performed by pronuclear injection of β-actin constructs into fertilized eggs derived from a stable line carrying the CR3 r6/7 enhancer from *Hoxb4* (construct 8 in Gould et al. 1997). The same line was crossed to strains of mice carrying replacement alleles at the *Hoxb4* and *Hoxd4* loci (Ramirez-Solis et al. 1993; Horan et al. 1995).

RESULTS

Identification of a *Hox/Pbx* Site in the *Hoxb2* r4 Enhancer

In the developing mouse hindbrain, the *Hoxb2* gene is initially expressed in a uniform manner to an anterior limit at the future rhombomere (r) 2/3 boundary (8.0 dpc), but within 12 hours (8.5 dpc), the gene is up-regulated specifically in the three segments r3, r4, and r5 (Sham et al. 1993). In transgenic analysis, a 2.1-kb fragment 5′ of the mouse *Hoxb2* gene behaved as an enhancer directing reporter expression at high levels in r3, r4, and r5 in a manner analogous to that of the endogenous *Hoxb2* gene (Sham et al. 1993; Vesque et al. 1996). Here, we were interested in using this enhancer to determine the molecular mechanisms governing r4-restricted expression of *Hoxb2*, as previously we demonstrated that elevated expression in r3 and r5 is independently controlled by the zinc finger transcription factor Krox20 (Sham et al. 1993; Vesque et al. 1996). Transgenic approaches with *lacZ* reporter genes in mice were used to map the regions involved in r4 restricted expression.

A 1.4-kb subfragment of the full enhancer diagrammed in Figure 1A, which deleted the Krox20-binding sites necessary for r3/r5 expression, mediated reporter staining in r4 and neural crest in the second branchial arch (Fig. 2A). Progressive deletions further narrowed the *cis*-acting regions required and identified a 181-bp *Stu*I fragment (Fig. 1B) also able to confer expression of the reporter specifically in r4 and associated crest, in a manner similar to that of the 1.4-kb subfragment (data not shown; Maconochie et al. 1997). Before performing further dele-

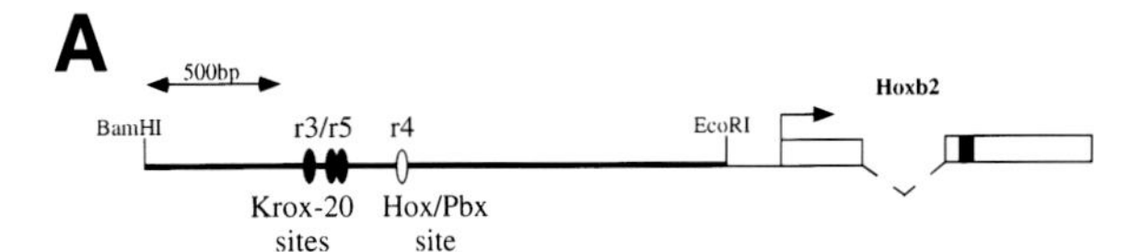

B

StuI
AGGCCTTTTTAAGGGATATGCAATTTAGGTTGTTCCCTCT

GCTTTCCCAAAGAGCCAAATTCTTTGATGCAATCGGAGGG

AGCTGTCAGGGGGCTAAGATTGATCGCCTCATCTCCTCTT

GGTCCCTCTGACCCACTTATTTAAAAATCTTCCCCCAGCC

GACTTTTCATTTTCAGCTTTGAGGCCT
StuI

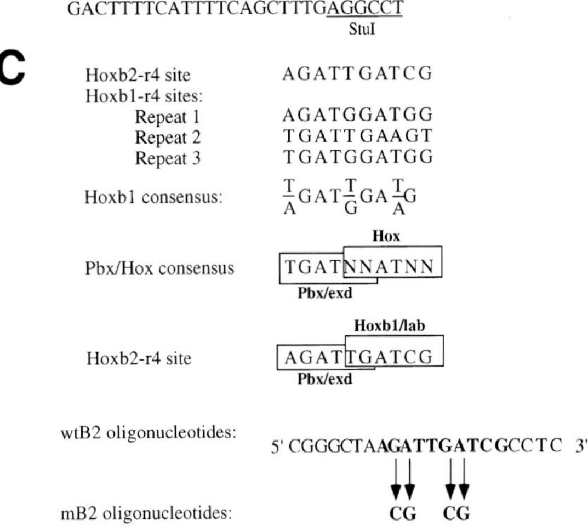

Figure 1. Identification of a Hoxb2 r4 enhancer motif. (*A*) A restriction map of *Hoxb2* 5′-flanking regions where the three solid filled ellipses represent the position of the Krox20-binding sites which regulate r3 and r5 expression of *Hoxb2* (Sham et al. 1993), and the open ellipse indicates the Hox/Pbx site identified in panels *B* and *C*. (*B*) Sequence of the 181-bp r4 *Stu*I enhancer. (*C*) Alignment of the bipartite Hox/Pbx site in the *Hoxb2* enhancer with the three repeats from a *Hoxb1* r4 enhancer and a Hox/Pbx consensus sequence. The boxes in *B* and *C* indicate the Hox- and Pbx-binding sites and the bipartite consensus motif. Sequence of wild-type (wtB2) and mutant (mB2) oligonucleotides used for in vitro EMSA and transgenic analysis. (Reprinted, in part, from Maconochie et al. 1997.)

tion analysis, we examined the sequence (Fig. 1B) as a first step toward identifying upstream factors. Although database comparisons with binding sites for known transcription factors generated no obvious candidates for r4 activity, we noted the presence of a single motif (boxed in Fig. 1B) highly related to a repeated motif we previously identified in the *Hoxb1* locus (Popperl et al. 1995).

An alignment of the site upstream of *Hoxb2* with the three sites from the *Hoxb1* r4 enhancer (Fig. 1C) indicates that this motif is related to a bipartite Pbx/Hox consensus site (Chan and Mann 1996). These sites represent separate binding sites for both Hox and Pbx proteins which overlap by 2 bp, with a Pbx site on the 5′ side of the consensus and Hox sites on the 3′ side (Fig. 1C). The part of the *Hoxb2* motif corresponding to the Pbx-binding site (5′-AGATTG-3′) closely matches the consensus Pbx/Exd-binding site from the *Hoxb1* elements. However, the Hox-binding site (5′-TGATCG-3′) in the bipartite motif differs from all of those in *Hoxb1* at position 5, showing that the *Hoxb2* motif is not identical to any of the motifs in *Hoxb1*.

The *Hoxb2* Bipartite Pbx/Hox Site Is Necessary and Sufficient for r4 Activity

To address the in vivo significance of the *Hoxb2* motif, we analyzed the function of this site in the context of the 2.1-kb *Bam*HI-*Eco*R enhancer (Fig. 1A), which contains control elements necessary for up-regulation of *Hoxb2* in r3, r4, and r5 (Fig. 2C) (Sham et al. 1993). A 7-

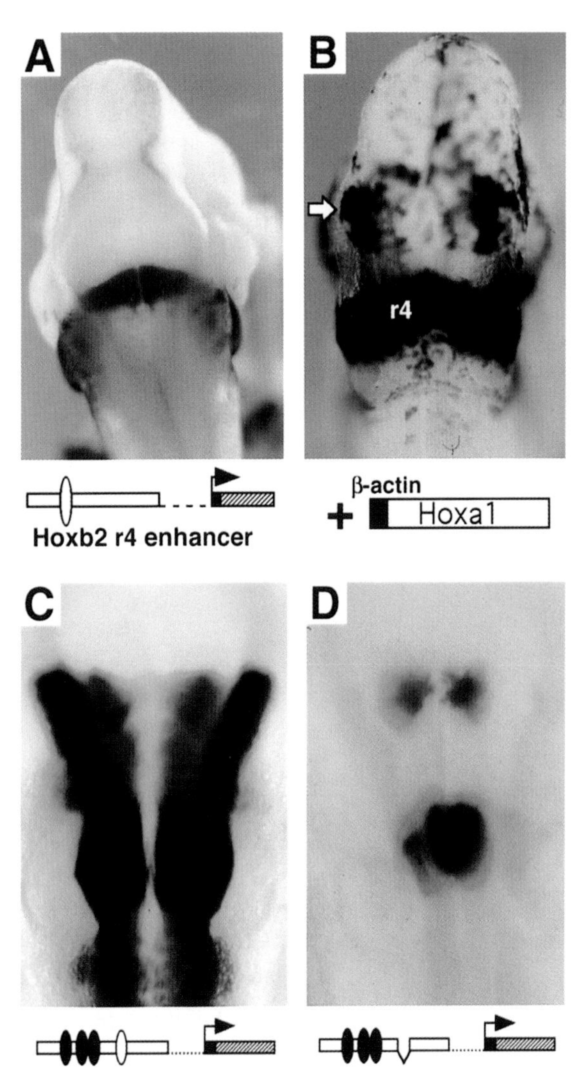

Figure 2. Mapping and selective *trans*-activation of the *Hoxb2* r4 enhancer by *labial*-related genes. (*A*) Reporter staining in a control transgenic embryo carrying a 1.4-kb r4 element. (*B*) Ectopic expression of *Hoxa1 trans*-activates the *Hoxb2* enhancer in regions anterior to r4 (see open arrowheads). (*C*) Expression of the wild-type 2.1-kb *Bam*HI-*Eco*RI enhancer in r3, r4, and r5. (*D*) Specific loss of r4 expression if the Hox/Pbx site is deleted in the 2.1-kb enhancer. Below each panel is indicated the construct tested or gene used for *trans*-activation. All embryos are assayed at 9.5 dpc. (Reprinted, in part, from Maconochie et al. 1997.)

bp deletion in the core of the *Hoxb2* motif led to a loss of expression of the *lacZ* reporter specifically in r4, whereas the r3 and r5 domains of expression regulated by the *Krox20* sites were unaffected (Fig. 2D). In addition, a construct carrying a 4-bp mutation in the core of the *Hoxb2* motif also leads to the loss of expression specifically in r4 (data not shown; Maconochie et al. 1997).

The sequence of the *Hoxb2* motif is different from those found in *Hoxb1* (Fig. 1C) and there is only one copy in the enhancer. The three Hoxb1/Pbx-binding sites in the 5′ region of *Hoxb1* are not functionally equivalent in vivo and in vitro (Popperl et al. 1995); therefore, we examined whether the *Hoxb2* motif was sufficient to direct r4 expression. Three copies of a double-stranded 21-bp oligonucleotide (wtB2, Fig. 1D) linked to a *lacZ* reporter gene was sufficient to confer r4-restricted expression in transgenic mouse embryos. Staining was strong ventrally in r4, with a sharp anterior boundary at the junction between r3 and r4, and there was some midline staining in r5 and r6 similar to that observed with the full 181-bp enhancer (data not shown; Maconochie et al. 1997). These experiments indicate that the bipartite Hox/Pbx motif from *Hoxb2* is both necessary and sufficient for r4 expression.

Ectopic Expression of *Hoxb2* Does Not Activate the r4 Enhancer But Does Result in Reprogramming of the Axial Skeleton

The sequence variations in the *Hoxb2* motif compared with those in *Hoxb1* might reflect a different in vivo specificity or preference for a *Hox* partner. We wondered whether this motif was a *Hoxb2* response element mediating an auto-regulatory influence of *Hoxb2* itself. To test this idea, we examined the ability of ectopic *Hoxb2* expression to *trans*-activate expression from the *Hoxb2* r4 reporter in vivo. In the background of a transgenic line of mice carrying the 1.4-kb fragment from *Hoxb2*, which directs r4-specific expression (Fig. 2A), we generated widespread ectopic expression of *Hoxb2* using a human β-actin promoter/enhancer vector. Although this construct produces functional protein, we never detected *trans*-activation of the r4-*lacZ* reporter (data not shown; Maconochie et al. 1997), demonstrating that *Hoxb2* is unable to interact with this site in vivo under the conditions of this assay.

Despite the lack of *trans*-activation, we observed highly penetrant skeletal phenotypes induced by the ectopic *Hoxb2* expression (Fig. 3). *Hoxb2* is normally expressed in paraxial mesoderm in the first few cranial

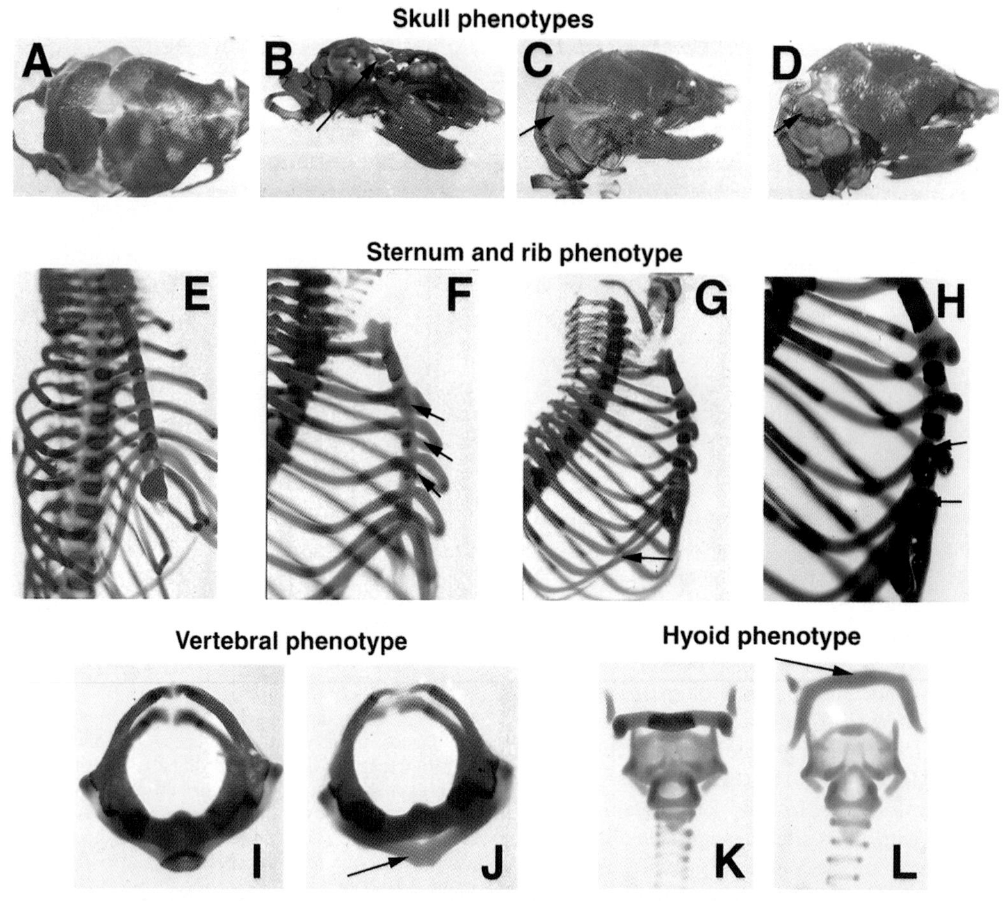

Figure 3. Skeletal abnormalities induced by ectopic expression of *Hoxb2*. All cleared skeletons from embryos at 18.5 dpc were stained with Alcian blue (cartilage) and Alizarin red (ossified bone). (*A,E,I,K*) Wild-type skeletal elements. (*B–D*) Skull phenotypes in three separate *Hoxb2* transgenics. (*F–H*) Sternal and rib abnormalities in transgenic skeletons. (*J,L*) Vertebral and hyoid phenotypes induced by *Hoxb2*. Arrows indicate the affected structures in each panel.

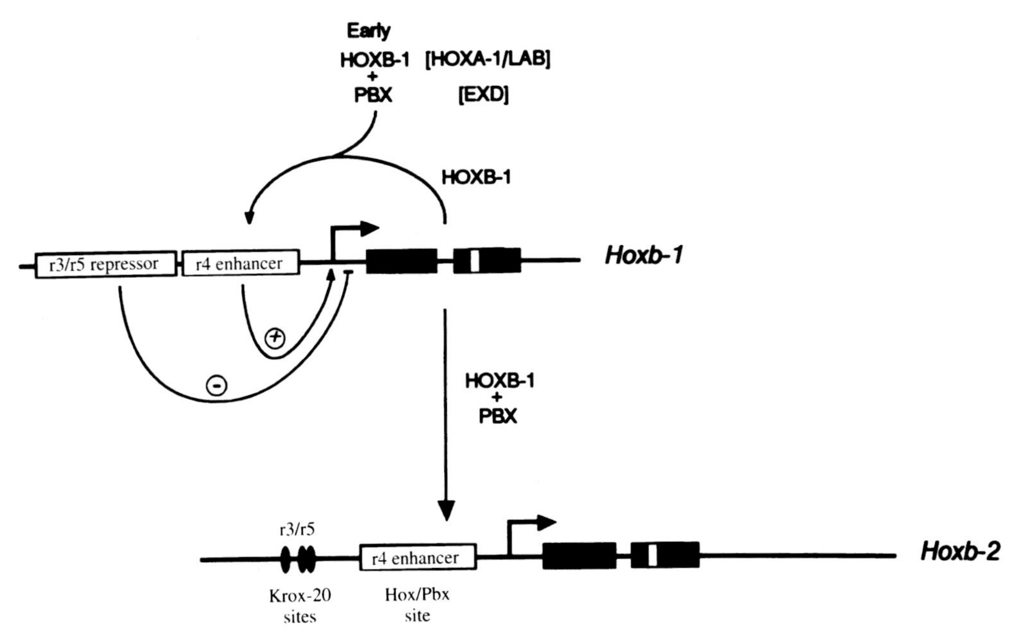

Figure 4. Model for direct auto-regulation of Hoxb1 and cross-regulation of Hoxb2 in r4 by labial-related proteins with Pbx/exd proteins as cofactors.

somitomeres and at high levels in second arch and more posterior cranial neural crest (Hunt et al. 1991a). However, the β-actin vector stimulates expression in more anterior domains. At 11.5 dpc, 4/15 β-actin/*Hoxb2* transgenic embryos displayed an obvious exencephaly, and at 18.5 dpc, 4/4 displayed skeletal abnormalities of variable severity. In the skull, we observed in one case complete absence of the supraoccipital, parietal, and exoccipital bones (Fig. 3B). The basioccipital bone is normal, and the tympanic bone and the otic capsule are reduced. Two of the other transgenic embryos lack the exoccipital bones and display poor ossification of the supraoccipital (Fig. 3C). In the fourth embryo, part of the parietal and the whole of the supraoccipital bone are missing, whereas a tiny ossified remnant of the exoccipital bone is still visible (Fig. 3D).

All four embryos displayed abnormalities of the sternebrae and the intersternaebral discs. In one case only, one out of the five intersternaebral discs was properly ossified (Fig. 3F). The remaining four discs as well as the xiphoid process show a number of foci where small groups of cells form areas of reduced ossification in the cartilage. In the three other embryonic skeletons, most of the anterior intersternaebral discs and the xiphoid process seem to be normally ossified, and only intersternaebral discs situated immediately above the xiphoid process appear to be delayed in their ossification (Fig. 3G,H). In two of these embryos, the sternebrae themselves present abnormal thickening. At least three out of six sternebrae in these embryos form cartilagineous outgrowths and bulges either protruding and sticking out of the sternum or curving the outgrowth toward the lower sternebrum. The attachment of the ribs to the vertebral column seem to be normal, but in two of the skeletons, additional pairs of ribs were attached to the xiphoid process (Fig. 3G).

In the vertebrae, the only abnormality observed in two of the four transgenic skeletons was the lack of formation and/or ossification of the anterior arch of the atlas (arcus anterior atlantis, or AAA) (Fig. 3I). The other elements of the vertebral column in the affected embryos look normal. In one of them, the whole sternum is also poorly ossified as described above; however, the other three out of five intersternebral discs are normally formed and ossified. This shows that the AAA can be altered independent of the sternum. In addition, in one case, the shape and the ossification of the hyoid process are severely disrupted (Fig. 3L). Although in the wild-type embryos, the central portion of the hyoid cartilage is completely ossified and ossification foci appear on the sides (Fig. 3K), in the transgenic littermate, the cartilage displays outgrowth in posterior direction, and there are no signs of ossification centers. All of the phenotypes noted above correlate with new sites and levels of *Hoxb2* expression and are concentrated in regions more anterior than its normal domain of expression, suggesting that there has been a posterior transformation of skeletal components.

The *Hoxb2* Enhancer Mediates a Differential *Hox* Response In Vivo

Rhombomere 4 expression of *Hoxb2* might arise through direct auto- or cross-regulatory mechanisms involving Hox and Pbx proteins acting through the *Hoxb2* bipartite motif. The *Hoxb2* enhancer did not respond in an auto-regulatory manner in vivo, but it might be capable of mediating a response to other *Hox* genes. Therefore, we used the human β-actin promoter/enhancer vector to generate widespread expression in vivo of a number of *Hox* genes, to address whether the *Hoxb2* r4 enhancer was capable of responding to any ectopically expressed *Hox* genes in transgenic mice. Since only members of paralogous groups 1 and 2 are normally expressed in the region of r4 (Krumlauf 1993a) and we eliminated the ability of Hoxb2 to induce expression, we initially focused on *Hoxb1* and *Hoxa1*.

In the background of a transgenic line containing the 1.4-kb *Hoxb2* r4 enhancer, ectopic expression of *Hoxa1* activates the *lacZ* transgene in regions anterior to the normal r4 domain compared with control transgenic embryos (Fig. 2A,B). Furthermore, ectopic expression of *Hoxb1* and the *Drosophila labial* gene also *trans*-activates the *lacZ* reporter in a manner similar to that of *Hoxa1* (data not shown; Maconochie et al. 1997). These findings show that labial-related proteins in general have conserved the ability to stimulate this enhancer in vivo (Fig. 4). We have also tested the ability of Hoxb1 and exd to bind cooperatively to this *Hoxb2* motif and find that it displays a degree of cooperativity in vitro similar to that seen with the related *Hoxb1* bipartite motifs (Popperl et al. 1995; Maconochie et al. 1997). Therefore, in vitro labial-related proteins in conjunction with exd/Pbx proteins as cofactors display a strong preference for sites defined by the Hoxb1 and Hoxb2 r4 enhancers (Chan and Mann 1996; Chan et al. 1996, 1997; Maconochie et al. 1997).

Since the *Hoxb2* motif is able to discriminate between Hoxb2 and group-1 Hox proteins in vivo, we wanted to determine if this differential response is unique to *Hoxb2*. Therefore, we tested *Hoxb4* and found that it also failed to activate the reporter (data not shown; Maconochie et al. 1997), although it is capable of activating other target sequences in vivo (Gould et al. 1997). Hence, the *Hoxb2* enhancer mediates a selective response in vivo, which displays a preference for labial-related proteins.

A

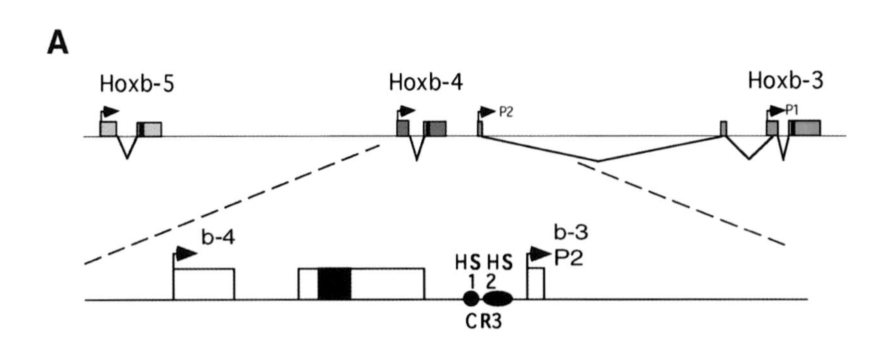

B

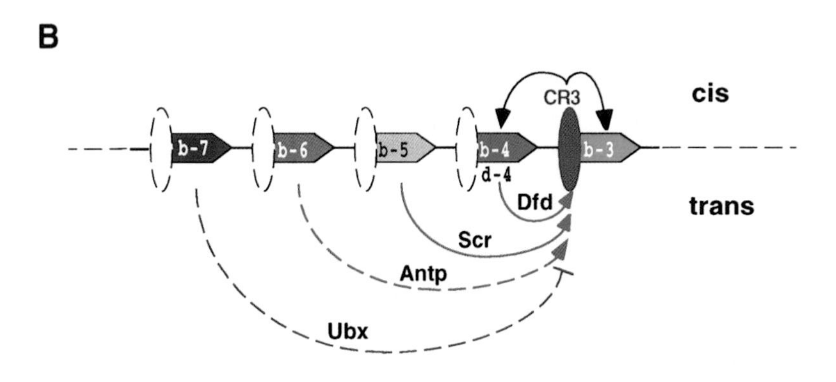

C

```
                 HS1                        HS2
         ----ggcc-------------cgggcc---
mouse    GAGAATTATAC-AGAA-AACCATTAATCACT
chicken  GAGAATTATAC-AGAA-AATCATTAATCACT
fugu     GAGAATTATACTAGACCAATCATTAATCACG

cons     GAGAATTATAC-AGA--AATCATTAATCAC-

HOX/PBC                      TC
                             CAATTAATCA
                             DFD    EXD
```

Figure 5. Shared elements in a Hoxb3/Hoxb4 enhancer involve cross-regulation through a bipartite Hox/Pbx motif. (*A*) Diagram of the Hoxb5-Hoxb3 genomic region at top and expanded below the position of conserved region 3 (CR3) and the two homeodomain-binding sites HS1 and HS2. (*B*) Summary of the cross-regulatory interactions in mouse and *Drosophila* embryos involving CR3. *Hox* genes are represented by arrow-shaped boxes with CR3 (*closed oval*) and other potential CR3-like elements (*dashed ovals*) indicated. The regulatory interactions of CR3, in *cis* (above dashed line) and in *trans* (below dashed line) are shown: Arrowheads and flat bars indicate positive and negative auto/cross-regulations, respectively. Dashed arrows indicate speculative vertebrate cross-regulations inferred from *Drosophila* CR3 experiments. (*C*) Sequence conservation of TAAT motifs and alignment of HS1/2 region of CR3 from three vertebrates. Point mutations in HS1 and HS2 used in EMSA and transgenic constructs are indicated above the mouse sequence. cons indicates those residues conserved between vertebrates and HOX/PBC shows the consensus for DFD and EXD binding aligned to HS2. (Reprinted, in part, from Gould et al. 1997.)

In Vivo Evidence That labial and exd Proteins Are Required for Activity of the *Hoxb2* Motif

Further evidence that labial-related proteins are able to interact on the *Hoxb2* motif in vivo comes from the ability of the *Hoxb2* motif to direct expression in *Drosophila* embryos. A *lacZ* reporter gene, *3XB2-lacZ*, which has three copies of the *Hoxb2* motif upstream of a minimal promoter driving *lacZ*, is expressed in a *labial*-like pattern in *Drosophila* embryos, and expression is dependent on both endogenous labial and exd (Maconochie et al. 1997).

In vertebrates, *Hoxb1* is the best candidate responsible for the cross-regulation of *Hoxb2* in r4, as it is the only member of paralogous group 1 expressed at a high level in r4 when this up-regulation occurs (Hunt et al. 1991b; Murphy and Hill 1991; Krumlauf 1993a). We have recently generated a null mutation in the *Hoxb1* gene (Studer et al. 1996) and have examined *Hoxb2* expression in these mutants to address the issue of whether *Hoxb1* is required for *Hoxb2* up-regulation. In homozygous *Hoxb1* mutant embryos, high-level up-regulation of endogenous *Hoxb2* in r4 is never observed and remains very weak (data not shown; Maconochie et al. 1997). This confirms that the r4-restricted domain of *Hoxb2* in the mouse hindbrain is dependent on a cross-regulatory interaction by the *Hoxb1* gene in this segment. Taken together, these data lead us to conclude that *Hoxb2* is a direct target of the *Hoxb1* gene and further underline the importance of cross-regulation in the vertebrate Hox complexes. Figure 4 illustrates the auto- and cross-regulatory roles *Hoxb1* has in directly regulating the segmental identity of r4.

A Shared Enhancer between *Hoxb4* and *Hoxb3* Is Required for r6/7 Expression

To broaden the analysis of cross-regulation, we have also examined the mechanisms controlling *Hoxb4* and *Hoxb3* expression in the hindbrain. *Hoxb3* and *Hoxb4* have domains of expression in the hindbrain that partially overlap and map to the r6/7 junction (Wilkinson et al. 1989; Sham et al. 1992; Gould et al. 1997). A neural enhancer 3′ of the *Hoxb4* gene (region A) directs this r6/7 pattern of expression (Whiting et al. 1991), and this enhancer is highly conserved in the *Hoxb4* gene of other vertebrates (Aparicio et al. 1995; Morrison et al. 1995, 1997). This neural r6/7 enhancer, although 3′ of *Hoxb4,* is directly adjacent to a distal promoter of the *Hoxb3* gene (Sham et al. 1992), and we have recently found that this *cis*-element is shared by the *Hoxb4* and the *Hoxb3* genes (Gould et al. 1997).

The overlapping domains of *Hoxb3* and *Hoxb4*, along with their proximity in the *Hoxb* complex, suggest that a common mechanism might be used to regulate both genes. We used a transgenic approach, linking a *lacZ* reporter gene to genomic fragments to examine elements within the Hoxb4 r6/7 neural enhancer (region A). Previous sequence comparisons of region A from pufferfish, chick, and mouse identified a conserved region (CR3) potentially involved in the r6/7 regulation of *Hoxb4* (Aparicio et al. 1995; Morrison et al. 1995). To investigate the function of CR3 in mouse region A independent of the *Hoxb3* P2 promoter that it contains (Fig. 5A), we have tested the activities of various deletions on a neutral minimal β-globin promoter (Yee and Rigby 1993). CR3 alone is capable of directing a neural pattern with a sharp anterior boundary at r6/7 and strong staining in the posterior spinal-cord-like region A (Fig. 6G) (Whiting et al. 1991). Furthermore, a deletion of CR3 from region A specifically abolished anterior staining up to the r6/7 boundary but gave strong expression posteriorly (data not shown; Gould et al. 1997). These experiments show that CR3 is necessary and sufficient for an r6/7 boundary of expression from *Hoxb4* and *Hoxb3* and that other neural elements within region A are responsible for the more posterior staining.

CR3 Depends on Group-4 *Hox* Genes

We wanted to address the nature of the upstream components controlling CR3. Hoxb4 protein is expressed in the neural plate as early as 8.0 dpc but CR3 is first active only at 9.25 dpc (data not shown). Hence, CR3 appears to be involved in the maintenance and not the establishment of the *Hoxb4* neural domain. As it is known that *Drosophila Deformed* (*Dfd*), an ortholog of *Hoxb4*, maintains its expression, in part, by auto-regulation (Zeng et al. 1994), we tested the ability of CR3 to function as a *Hoxb4* response element. The endogenous *Dfd* gene can be autoactivated by a ubiquitous pulse of DFD protein but only in restricted ectopic locations (Kuziora and McGinnis 1988). We used a similar transgenic assay described above for *Hoxb2*, using a β-actin promoter to drive *Hoxb4* expression throughout the mouse embryo to test for CR3 *trans*-activation (Fig. 6C,E). Ectopic Hoxb4 is sufficient to induce reporter expression mediated by CR3 but only within the confines of the central nervous system (CNS) and only anterior to the r6/7 boundary. *Trans*-activation by *Hoxb4* is particularly strong in ventrolateral regions of the midbrain compared with control embryos (Fig. 6C,E).

Since CR3 responds to ectopic *Hoxb4,* we tested whether it was dependent on endogenous *Hoxb4* activity. Previously, in embryos homozygous for a targeted replacement allele of *Hoxb4*, the anterior boundary of expression in the CNS was found to be diffuse and not as sharply defined as in wild-type embryos (Ramirez-Solis et al. 1993). Surprisingly in mouse embryos homozygous for this same mutation, we find that CR3 activity appears normal (data not shown). One explanation for the inconsistency between our reporter staining and the previous in situ expression (Ramirez-Solis et al. 1993) is that the transcriptional profile of the *Hoxb4* promoter in the replacement allele has been disrupted in *cis* by the nearby insertion of the *gpt* and *neo* transcription units.

It is known that the paralogue of *Hoxb4, Hoxd4*, also has an anterior expression boundary at r6/7 (Hunt et al. 1991b), and functional compensation between group-4 genes might account for the expression from CR3 in the *Hoxb4* mutants. In direct support of this idea, ectopic expression of *Hoxd4* using the β-actin vector is able to acti-

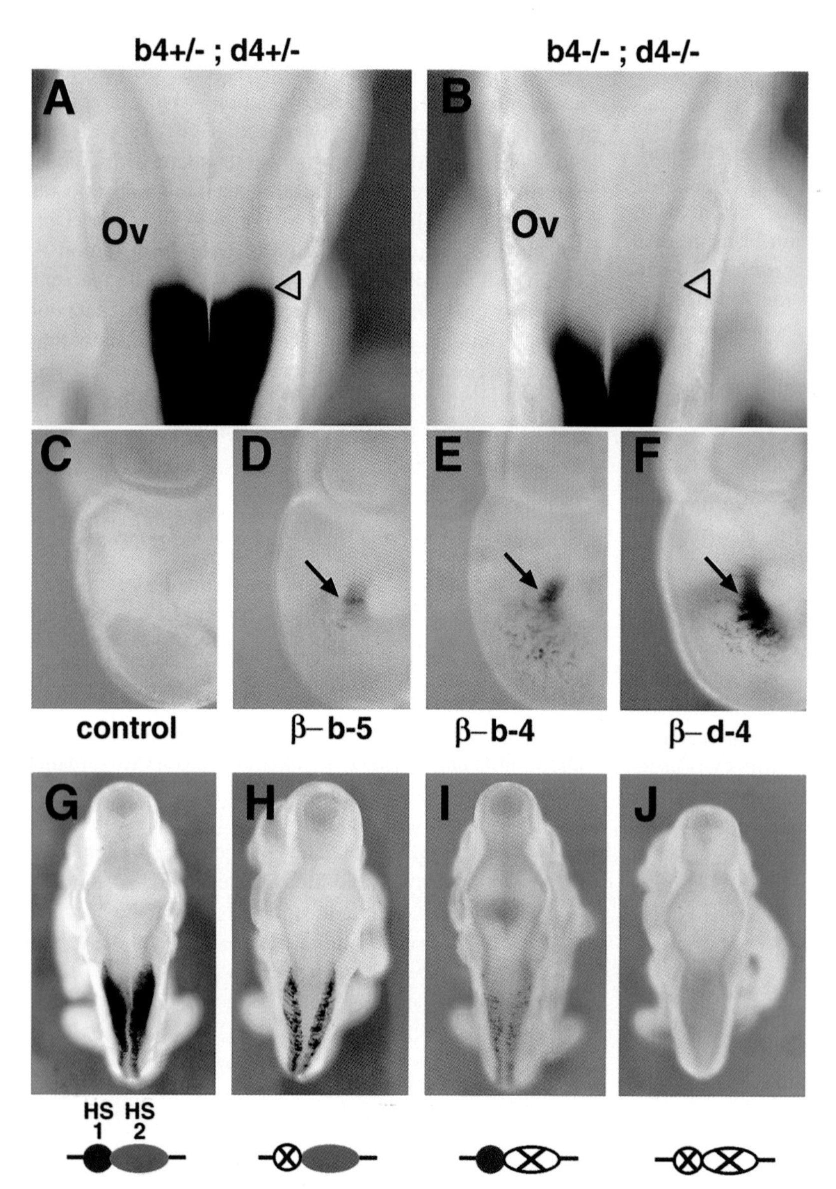

Figure 6. CR3 is a Hox-selective response element in vivo. (*A*) A line carrying CR3/*lacZ* is expressed at the r6/7 boundary in Hoxb4 and Hoxd4 compound heterozygotes. (*B*) The expression mediated by CR3 shifts posteriorly in double homozygous *Hoxb4* and *Hoxd4* mutants. Hence, expression of the reporter requires both of these proteins in vivo. (*C*) Wild-type midbrain of line carrying CR3/*lacZ* reporter. (*D–F*) Induction of the CR3 reporter in embryos where ectopic expression of Hoxb5 (*D*), Hoxb4 (*E*), and Hoxd4 (*F*) was generated using a β-actin vector. (*G–J*) HS1 and HS2 are essential components of CR3. (*G*) Dorsal views of 10.5-dpc embryos transgenic for-wild type-CR3 reporter construct, and altered versions containing mutation in HS1 (*H*), HS2 (*I*), or HS1 and HS2 (*J*). Construct variants are indicated below each panel. (Reprinted, in part, from Gould et al. 1997.)

vate CR3 expression (Fig. 6F). Furthermore, when mutations in both *Hoxb4* and *Hoxd4* (Horan et al. 1995) were combined, a dramatic loss of the CR3-mediated r6/7 boundary of expression was observed in double homozygous embryos (Fig. 6B). In *Hoxd4* homozygous mutants alone or in compound heterozygotes between *Hoxb4* and *Hoxd4,* there was no change in reporter expression (Fig. 6A), indicating that one or more wild-type alleles at either the *Hoxb4* or *Hoxd4* loci are sufficient to restore the wild-type r6/7 expression boundary. Thus, the anterior r6/7 expression mediated by CR3 can be specified by either gene, illustrating an important role for auto/cross-regulatory interactions between these two paralogues.

Selective Response to *Hox* Genes

In a manner similar to that used for the *Hoxb2* motif, we wanted to test the selectivity of the CR3 response to *Hox* genes. We note that in the double *Hoxb4/Hoxd4* homozygotes, CR3 expression is not abolished completely, rather the rostral limit is shifted posteriorly (Fig. 6B). Hence, other *Hox* genes might be involved in controlling the more posterior neural expression of CR3. Therefore, we again ectopically expressed a number of *Hox* genes to test their ability to *trans*-activate the CR3 enhancer. No changes in reporter expression were observed when *Hoxb1*, *Hoxa1*, *Hoxb3*, and *Hoxb9* were used (see Methods). In contrast, ectopic *Hoxb5* did induce expression in midbrain regions in a manner similar to that of both *Hoxb4* and *Hoxd4* (Fig. 6F). From the six *Hox* genes tested, we conclude that members of group 4 and 5 can *trans*-activate CR3, whereas groups 1, 3, and 9 will not. Together, these loss and gain of *Hox* function experiments indicate that CR3 functions as a target for the action of multiple *Hox* genes. The group-4 genes (*Hoxb4* and *Hoxd4*) are involved in setting the r6/7 boundary of CR3, whereas more 5′ genes, such as *Hoxb5*, contribute to the activation of CR3 in more caudal regions of the

CNS. We have also examined the ability of CR3 to direct expression in *Drosophila* embryos. Again, in both gain- and loss-of-function homeotic backgrounds, a dependence and preference for group-4–7 *HOMC* genes were observed in accord with the vertebrate results (Gould et al. 1997).

Is CR3 a Direct Target of *Hox* Genes?

To investigate in more detail if *Hox* responsiveness of CR3 is direct, deletions of CR3 were analyzed in transgenic mice to map the elements involved. Minimal overlaps of expressing constructs (data not shown) defined a critical role for a 61-bp region within the enhancer. Inspection of the sequence in this region between species indicates the presence of two closely spaced and highly conserved TAAT motifs (HS1 and HS2 sites, Fig. 5A,C). These motifs form a core recognition sequence for many Hox proteins, including those of group 4 (Regulski et al. 1991; Popperl and Featherstone 1992; Mann and Chan 1996).

To first test the role of these two sites, we examined the in vitro binding ability of Hoxb4. Specific binding of Hoxb4 to oligonucleotides containing HS2, or both HS1 and HS2, was efficiently competed by wild type but not TAAT mutant versions of the sites (data not shown; Gould et al. 1997). To test the function of the HS1 and HS2 sites in vivo, the same series of point mutations that eliminated in vitro binding were introduced into the CR3 enhancer and examined for activity in transgenic mice. Mutation of HS1 gave a moderate reduction in expression (Fig. 6H) compared with wild-type CR3 (Fig. 6G), whereas mutation of HS2 resulted in a more severe reduction in transgene expression (Fig. 6I). Mutations in both sites abolished all CR3-dependent expression in mouse embryos (Fig. 6J). These same mutations in HS1 and HS2 also abolished expression from CR3 in *Drosophila* embryos (data not shown).

These results show that the same two sites are critical for enhancer activity in both mouse and *Drosophila* embryos. It is important to note that in both species, mutation in HS1 and HS2 abolishes all of the CR3-mediated expression and not just those domains dependent on group-4 *Hox* genes. Since CR3 also responds to group-5–7 members, it appears that HS1 and HS2 are capable of mediating all of the in vivo *Hox* regulatory inputs that we have observed. Together with the in vitro-binding data, this strongly suggests that CR3 is a direct target for control by multiple *Hox* genes.

DISCUSSION

The studies above on *Hoxb2* and *Hoxb3/Hoxb4* serve to highlight the general importance of cross-regulatory interactions in controlling expression of *Hoxb* genes in the vertebrate hindbrain. Unlike the case in *Drosophila,* these interactions in the mouse are positive rather than negative, which may be important in the vertebrate complexes for maintaining appropriate nested patterns of gene expression. The fact that the same control elements are shared between *Hoxb3* and *Hoxb4* has important implications for maintaining the *Hox* complexes themselves (Gould et al. 1997). It is possible that although many of the regulatory studies in transgenic mice show that *Hox* genes can function outside of the endogenous complex, removing these genes or regions might also alter control regions needed by adjacent genes in the complex and hence be required for proper regulation of multiple genes. This would provide a basis for keeping genes clustered to maintain their appropriate expression patterns.

In this study, as in the previous case of auto-regulation of *Hoxb1*, the motifs involved in the cross-regulation and Hox response contain bipartite *Hox/Pbx*-binding sites. In vivo, these sites show a high degree of specificity and selectivity. The motifs from Hoxb1 and Hoxb2 alone will generate r4-restricted expression, whereas the CR3 motifs produce a completely different pattern of expression even though they are highly related (Chan et al. 1997). In vitro, these sites display rather promiscuous binding, suggesting that other factors in addition to Hox and Pbx/exd proteins are involved in directing restricted expression and selectivity in vivo.

Perhaps the best illustration of the importance of auto- and cross-regulation comes from the interaction of *Hoxb1* and *Hoxb2* in r4. This study in conjunction with our previous results (Popperl et al. 1995) allows us to propose a model (Fig. 4) whereby early expression of *Hoxb1* or *Hoxa1* is triggered through retinoid signaling in the mouse embryos. The vertebrate-labial-related proteins in combination with Pbx proteins as cofactors then bind directly to the Hox/Pbx motif 5′ of *Hoxb1* and stimulate a direct auto-regulatory loop. A repressor serves to limit the activity of this auto-regulatory enhancer to r4 by blocking expression in r3 and r5 (Studer et al. 1994). In addition, in *trans,* the Hoxb1 and Pbx proteins also directly bind to the *Hox/Pbx* motif 5′ of *Hoxb2* and stimulate r4 expression of that gene. This demonstrates the key role that Hoxb1 has in maintaining the regulatory cascade that controls r4 identity (Studer et al. 1996).

ACKNOWLEDGMENTS

We thank Z. Webster for animal husbandry; J. Sharpe for help with Photoshop; G. Horan and R. Behringer for *Hoxd-4* mutants; W. McGinnis for *Dfd* mutants and antibodies, M. Bienz, G. Struhl, E. Wieschaus, and the Bloomington Stock Center for fly stocks; and R. White for collaborations on the fly experiments. A.G was a Beit Memorial Fellow, A.M. and M.M. were MRC Training Fellows. S.N. was supported by an European Commission Biotechnology grant (BIO CT 930060) and all work was supported by the Medical Research Council.

REFERENCES

Akam M., Dawson I. and Tear G. 1988. Homeotic genes and the control of segment diversity. *Development* **104:** 123.

Alexandre D., Clarke J.D., Oxtoby E., Yan Y.-L., Jowett T., and Holder N. 1996. Ectopic expression of *Hoxa-1* in the zebrafish alters the fate of the mandibular arch neural crest and

phenocopies a retinoic acid-induced phenotype. *Development* **122:** 735.
Aparicio S., Hawker K., Cottage A., Mikawa Y., Zuo L., Chen E., Krumlauf R., and Brenner S. 1997. Organisation of the *Fugu rubripes Hox* clusters, evidence for continuing evolution of vertebrate *Hox* complexes. *Nat. Genet.* **16:** 79.
Aparicio S., Morrison A., Gould A., Gilthorpe J., Chaudhuri C., Rigby P.W.J., Krumlauf R., and Brenner S. 1995. Detecting conserved regulatory elements with the model genome of the Japanese puffer fish *Fugu rubripes. Proc. Natl. Acad. Sci.* **92:** 1684.
Bienz M. 1994. Homeotic genes and positional signalling in the *Drosophila* viscera. *Trends Genet.* **10:** 22.
Boncinelli E., Acampora D., Pannese M., D'Esposito M., Somma R., Gaudino G., Stornaiuolo A., Cafiero M., Faiella A., and Simeone A. 1989. Organization of human class I homeobox genes. *Genome* **31:** 745.
Carpenter E.M., Goddard J.M., Chisaka O., Manley N.R., and Capecchi M.R. 1993. Loss of *Hoxa-1* (*Hox-1.6)* function results in the reorganization of the murine hindbrain. *Development* **118:** 1063.
Carroll S.B. 1995. Homeotic genes and the evolution of arthropods and chordates. *Nature* **376:** 479.
Chambers C.A. 1994. TKO'ed: Lox, stock and barrel. *BioEssays* **16:** 865.
Chan S.-K. and Mann R. 1996. A structural model for a homoetic protein-extradenticle-DNA complex accounts for the choice of HOX protein in the heterodimer. *Proc. Natl. Acad. Sci.* **93:** 5223.
Chan S.-K., Popperl H., Krumlauf R., and Mann R.S. 1996. An extradenticle-induced conformational change in a Hox protein overcomes an inhibitory function of the conserved hexapeptide motif. *EMBO J.* **15:** 2476.
Chan S.-K., Ryoo H.-D., Gould A., Krumlauf R., and Mann R. 1997. Switching the in vivo specificity of a minimal *HOX*-responsive element. *Development* **124:** 2007.
Chouinard S., and Kaufman T.C. 1991. Control of expression of the homeotic labial (lab) locus of *Drosophila melanogaster*: Evidence for both positive and negative autogenous regulation. *Development* **113:** 1267.
Dollé P., Lufkin T., Krumlauf R., Mark M., Duboule D., and Chambon P. 1993. Local alterations of *Krox-20* and *Hox* gene expression in the hindbrain of *Hoxa-1*(*Hox-1.6*) homozygote null mutant embryos. *Proc. Natl. Acad. Sci.* **90:** 7666.
Duboule D. 1992. The vertebrate limb: A model system to study the Hox/HOM gene network during development and evolution. *BioEssays* **14:** 375.
Duboule D. and Dollé P. 1989. The structural and functional organization of the murine HOX gene family resembles that of *Drosophila* homeotic genes. *EMBO J.* **8:** 1497.
Garcia-Fernandez J. and Holland P.W.H. 1994. Archetypal organisation of the *amphioxus Hox* gene cluster. *Nature* **370:** 563.
Goddard J., Rossel M., Manley N., and Capecchi M. 1996. Mice with targeted disruption of *Hoxb1* fail to form the motor nucleus of the VIIth nerve. *Development* **122:** 3217.
Gould A., Morrison A., Sproat G., White R., and Krumlauf R. 1997. Positive cross-regulation and enhancer sharing: Two mechanisms for specifying overlapping *Hox* expression patterns. *Genes Dev.* **11:** 900.
Hill J., Clarke J.D.W., Vargesson N., Jowett T., and Holder N. 1995. Exogenous retinoic acid causes specific alterations in the development of the midbrain and hindbrain of the zebrafish embryo including positional respecification of the Mauthner neuron. *Mech. Dev.* **50:** 3.
Horan G., Kovacs E., Behringer R., and Featherstone M. 1995. Mutations in paralogous *Hox* genes result in overlapping homeotic transformations of the axial skeleton: Evidence for unique and redundant function. *Dev. Biol.* **169:** 359.
Hunt P., Wilkinson D., and Krumlauf R. 1991a. Patterning the vertebrate head: Murine *Hox2* genes mark distinct subpopulations of premigratory and migrating neural crest. *Development* **112:** 43.
Hunt P., Gulisano M., Cook M., Sham M., Faiella A., Wilkinson D., Boncinelli E., and Krumlauf R. 1991b. A distinct *Hox* code for the branchial region of the head. *Nature* **353:** 861.
Kappen C., Schugart K., and Ruddle F. 1989. Two steps in the evolution of *Antennapedia*-class vertebrate homeobox genes. *Proc. Natl. Acad. Sci. USA* **86:** 5459.
Kaufman T.C., Lewis R., and Wakimoto B. 1980. Cytogenetic analysis of chromosome 3 in *Drosophila melanogaster*: The homeotic gene complex in polytene chromosomal interval 84A,B. *Genetics* **94:** 115.
Kaufman T., Seeger M., and Olsen G. 1990. Molecular and genetic organisation of the *Antennapedia* gene complex of *Drosophila melanogaster. Adv. Genet.* **27:** 309.
Kessel M. 1993. Reversal of axonal pathways from rhombomere 3 correlates with extra *Hox* expression domains. *Neuron* **10:** 379.
Keynes R. and Krumlauf R. 1994. *Hox* genes and regionalization of the nervous system. *Annu. Rev. Neurosci.* **17:** 109.
Kozak M. 1989. The scanning model for translation: An update. *J. Cell Biol.* **108:** 229.
Krumlauf R. 1993a. *Hox* genes and pattern formation in the branchial region of the vertebrate head. *Trends Genet.* **9:** 106.
———. 1993b. Mouse *Hox* genetic functions. *Curr. Opin. Genet. Dev.* **3:** 621.
———. 1994. *Hox* genes in vertebrate development. *Cell* **78:** 191.
Kuziora M.A. and McGinnis W. 1988. Autoregulation of a *Drosophila* homeotic selector gene. *Cell* **55:** 477.
Lewis E. 1978. A gene complex controlling segmentation in *Drosophila. Nature* **276:** 565.
Lumsden A. and Krumlauf R. 1996. Patterning the vertebrate neuraxis. *Science* **274:** 1109.
Maconochie M., Nonchev S., Studer M., Chan S.-K., Popperl H., Sham M.-H., Mann R., and Krumlauf R. 1997. Cross-regulation in the mouse *HoxB* complex: The expression of *Hoxb2* in rhombomere 4 is regulated by *Hoxb1. Genes Dev.* **11:** 1885.
Mann R. and Chan S.-K. 1996. Extra specificity from extradenticle: The partnership between HOX and PBX/EXD homeodomain proteins. *Trends Genet.* **12:** 258.
Mark M., Lufkin T., Vonesch J.-L., Ruberte E., Olivo J.-C., Dollé P., Gorry P., Lumsden A., and Chambon P. 1993. Two rhombomeres are altered in *Hoxa-1* mutant mice. *Development* **119:** 319.
Marshall H., Nonchev S., Sham M.H., Muchamore I., Lumsden A., and Krumlauf R. 1992. Retinoic acid alters hindbrain *Hox* code and induces transformation of rhombomeres 2/3 into a 4/5 identity. *Nature* **360:** 737.
McGinnis W. and Krumlauf R. 1992. Homeobox genes and axial patterning. *Cell* **68:** 283.
Morrison A., Ariza-McNaughton L., Gould A., Featherstone M., and Krumlauf R. 1997. *HOXD4* and regulation of the group 4 paralog genes. *Development* **124:** 3135.
Morrison A., Chaudhuri C., Ariza-McNaughton L., Muchamore I., Kuroiwa A., and Krumlauf R. 1995. Comparative analysis of chicken *Hoxb-4* regulation in transgenic mice. *Mech. Dev.* **53:** 47.
Murphy P. and Hill R.E. 1991. Expression of the mouse labial-like homeobox-containing genes, *Hox 2.9* and *Hox 1.6,* during segmentation of the hindbrain. *Development* **111:** 61.
Nüsslein-Volhard C. and Wieschaus E. 1980. Mutations affecting segment number and polarity in *Drosophila. Nature* **287:** 795.
Popperl H. and Featherstone M.S. 1992. An autoregulatory element of the murine *Hox-4.2* gene. *EMBO J.* **11:** 3673.
Popperl H., Bienz M., Studer M., Chan S.K., Aparicio S., Brenner S., Mann R.S., and Krumlauf R. 1995. Segmental expression of *Hoxb-1* is controlled by a highly conserved autoregulatory loop dependent upon *exd/Pbx. Cell* **81:** 1031.
Ramirez-Solis R., Zheng H., Whiting J., Krumlauf R., and Bradley A. 1993. *Hox-b4* (*Hox-2.6*) mutant mice show homeotic transformation of cervical vertebra and defects in the closure of the sternal rudiments. *Cell* **73:** 279.

Rauskolb C., Peifer M., and Wieschaus E. 1993. *extradenticle,* a regulator of homeotic gene activity, is a homolog of the homeobox-containing human proto-oncogene *pbx1. Cell* **74:** 1101.
Regulski M., Dessain S., McGinnis N., and McGinnis W. 1991. High-affinity binding sites for the Deformed protein are required for the function of an autoregulatory enhancer of the Deformed gene. *Genes Dev.* **5:** 278.
Schneuwly S., Kuroiwa A., Baumgartner P., and Gehring W. 1986. Structural organization and sequence of the homeotic gene *Antennapedia* of *Drosophila melanogaster. EMBO J.* **5:** 733.
Scott M., Weiner A., Hazelrigg T., Polisky B., Pirotta V., Scalenghe F., and Kaufman M. 1983. The molecular organization of the *Antennapedia* locus of *Drosophila. Cell* **35:** 763.
Sham M.-H., Hunt P., Nonchev S., Papalopulu N., Graham A., Boncinelli E., and Krumlauf R. 1992. Analysis of the murine *Hox-2.7* gene: Conserved alternative transcripts with differential distributions in the nervous system and the potential for shared regulatory regions. *EMBO J.* **11:** 1825.
Sham M.-H., Vesque C., Nonchev S., Marshall H., Frain M., Das Gupta R., Whiting J., Wilkinson D., Charnay P., and Krumlauf R. 1993. The zinc finger gene *Krox-20* regulates *Hoxb-2* (*Hox2.8*) during hindbrain segmentation. *Cell* **72:** 183.
Struhl G. 1984. Splitting the bithorax complex of *Drosophila. Nature* **308:** 454.
Studer M., Lumsden A., Ariza-McNaughton L., Bradley A., and Krumlauf R. 1996. Altered segmental identity and abnormal migration of motor neurons in mice lacking *Hoxb-1. Nature* **384:** 630.
Studer M., Popperl H., Marshall H., Kuroiwa A., and Krumlauf R. 1994. Role of a conserved retinoic acid response element in rhombomere restriction of *Hoxb-1. Science* **265:** 1728.
Tiong S.Y., Whittle J.R., and Gribbin M.C. 1987. Chromosomal continuity in the abdominal region of the bithorax complex of *Drosophila* is not essential for its contribution to metameric identity. *Development* **101:** 135.
Vesque C., Maconochie M., Nonchev S., Ariza-McNaughton L., Kuroiwa A., Charnay P., and Krumlauf R. 1996. *Hoxb-2* transcriptional activation by *Krox-20* in vertebrate hindbrain requires an evolutionary conserved cis-acting element in addition to the *Krox-20* site. *EMBO J.* **15:** 5383.
Whiting J., Marshall H., Cook M., Krumlauf R., Rigby P.W.J., Stott D., and Allemann R.K. 1991. Multiple spatially specific enhancers are required to reconstruct the pattern of *Hox-2.6* gene expression. *Genes Dev.* **5:** 2048.
Wilkinson D.G., Bhatt S., Cook M., Boncinelli E., and Krumlauf R. 1989. Segmental expression of *Hox-2* homeobox-containing genes in the developing mouse hindbrain. *Nature* **341:** 405.
Yee S.-P. and Rigby P.W.J. 1993. The regulation of *myogenin* gene expression during the embryonic development of the mouse. *Genes Dev.* **7:** 1277.
Zeng C., Pinsonneault J., Gellon G., McGinnis N., and McGinnis W. 1994. Deformed protein binding sites and cofactor binding sites are required for the function of a small segment-specific regulatory element in *Drosophila* embryos. *EMBO J.* **13:** 2362.
Zhang M., Kim H.-J., Marshall H., Gendron-Maguire M., Lucas A.D., Baron A., Gudas L.J., Gridley T., Krumlauf R., and Grippo J.F. 1994. Ectopic *Hoxa-1* induces rhombomere transformation in mouse hindbrain. *Development* **120:** 2431.

Role of the Brn-3 Family of POU-domain Genes in the Development of the Auditory/Vestibular, Somatosensory, and Visual Systems

M. XIANG,[1] L. GAN,[2] D. LI,[3] L. ZHOU,[4,6] Z.-Y. CHEN,[1] D. WAGNER,[2] B.W. O'MALLEY, JR.,[3] W. KLEIN,[2] J. NATHANS[4,5,6]

[1]Center for Advanced Biotechnology and Medicine, Department of Pediatrics, UMDNJ-Robert Wood Johnson Medical School, Piscataway, New Jersey 08854; [2]Department of Biochemistry and Molecular Biology, University of Texas M.D. Anderson Cancer Center, Houston, Texas 77030; Departments of [3]Otolaryngology, Head and Neck Surgery, [4]Molecular Biology and Genetics, [5]Neuroscience, and [5]Ophthalmology, [6]Howard Hughes Medical Institute, Johns Hopkins University School of Medicine, Baltimore, Maryland 21205

Understanding the genetic regulatory networks that specify neuronal identity is one of the central challenges in developmental neurobiology. A number of transcription factors have been implicated in decisions related to neuronal versus nonneuronal cell fate, regional specification in the nervous system, or determination of the terminally differentiated phenotype. For example, basic helix-loop-helix (bHLH) factors such as neuroD and the achaete-scute family control neural versus ectodermal cell fates (Jan and Jan 1993; Lee et al. 1995), and *Hox* genes control regional specification along the neuraxis (Keynes and Krumlauf 1994). Among the transcriptional regulators that control neuronal development in both vertebrates and invertebrates are members of the POU-domain family. This family was initially defined by the mammalian pituitary-specific transcription factor Pit-1/GHF-1, the octamer-binding proteins Oct-1 and Oct-2, and the *Caenorhabditis elegans* gene *unc-86* (Herr et al. 1988). The POU-domain functions as a bipartite DNA-binding domain that contains an approximately 70-amino-acid POU-specific domain and an approximately 60-amino-acid POU-homeodomain, joined by a variable linker. Many members of this gene family have distinctive patterns of expression in the developing and adult nervous systems, consistent with a role for these factors in neural development (Wegner et al. 1993).

Genetic studies in mice and humans show that many POU-domain genes function in the terminal stages of nervous system development (Fig. 1). *SCIP/Tst-1/Oct-6* controls the differentiation of Schwann cells (Weinstein et al. 1995; Bermingham et al. 1996; Jaegle et al. 1996); *Pit-1/GHF-1* is required for the normal development of the anterior pituitary (Li et al. 1990); *Brn-4/RHS2/POU3F4* is required for the normal development of the inner and middle ear (Bitner-Glindzicz et al. 1995; de Kok et al. 1995); and *Brn-2* is required for the specification of subsets of neurons in the hypothalamus (Nakai et al. 1995; Schonemann et al. 1995).

The POU-domain family has been divided into six classes on the basis of primary sequence similarities in the POU-domain (Wegner et al. 1993). The class IV POU-domain group is defined by the *unc-86* gene (Finney et al. 1988), the *Drosophila I-POU* gene (Treacy et al. 1991, 1992), and the three vertebrate *Brn-3* genes (Gerrero et al. 1993; Theil et al. 1993; Xiang et al. 1993, 1995; Turner et al. 1994). The Unc-86 protein is found exclusively within a subset of neurons and neuroblasts, and *unc-86* loss-of-function mutations affect some of these cells by causing a daughter cell to assume the fate of its mother or by altering cell phenotypes postmitotically (Chalfie et al. 1981; Desai et al. 1988; Finney and Ruvkin 1990). The *Drosophila I-POU* gene encodes two isoforms that are generated by alternative splicing (Treacy et al. 1991, 1992; Turner 1996). In mammals, there are three highly homologous class IV POU-domain genes, *Brn-3a*, *Brn-3b*, and *Brn-3c* (also referred to as *Brn-3.0*, *Brn-3.2*, and *Brn-3.1*, respectively). Each *Brn-3* gene is expressed in a distinct pattern in the developing and adult brainstem, retina, inner ear, and dorsal root and trigeminal ganglia (Gerrero et al. 1993; Ninkina et al. 1993; Xiang et al. 1993, 1995, 1997; Turner et al. 1994; Fedtsova and Turner 1996). In this paper, we review our work on the *Brn-3* family and the role of these genes in sensory system development (Xiang et al. 1993, 1995, 1996, 1997; Gan et al. 1996).

METHODS

For detailed descriptions of the experimental methods, see Xiang et al. (1993, 1995, 1996, 1997) and Gan et al. (1996).

RESULTS

Identification and Characterization of the *Brn-3* Subfamily of POU-domain Genes

The first member of the *Brn-3* family was identified by He et al. (1989) as a polymerase chain reaction (PCR) product encoding a novel POU-domain. By degenerate PCR and low-stringency DNA hybridization, we and others subsequently identified three genes with POU-domains that are identical or highly homologous to that re-

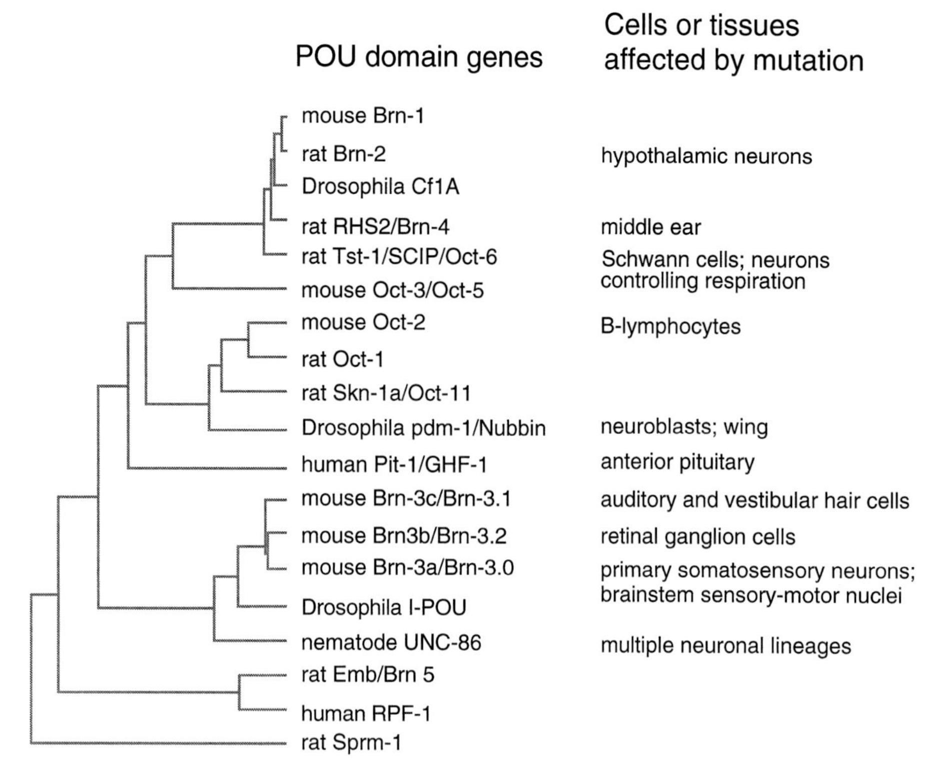

Figure 1. Dendrogram of POU-domains (*left*) and the phenotypes associated with mutation of the corresponding genes (*right*). The dendrogram was constructed by aligning POU-domain amino acid sequences and determining percent amino acid identity beginning with the first amino acid of the POU-specific domain (in Oct-1: EEPS...) and ending with the 58th amino acid of the POU-homeodomain (in Oct-1: ...KEKR) as defined in Klemm et al. (1994). Loss-of-function phenotypes are described in the following references: Brn-2 (Nakai et al. 1995; Schonemann et al. 1995), RHS2/Brn-4 (de Kok et al. 1995), Tst-1/SCIP/Oct-6 (Weinstein et al. 1995; Bermingham et al. 1996; Jaegle et al. 1996), Oct-2 (Corcoran et al. 1993), pdm-1/Nubbin (Ng et al. 1995; Yeo et al. 1995), Pit-1/GHF-1 (Li et al. 1990), Brn-3a (McEvilly et al. 1996; Xiang et al. 1996;), Brn-3b (Erkman et al. 1996; Gan et al. 1996), Brn-3c (Erkman et al. 1996; Xiang et al. 1997), and Unc-86 (Chalfie et al. 1981; Desai et al. 1988; Finney and Ruvkin 1990).

ported by He et al. (1989). These genes, identified in humans, mice, and rats, are refered to as *Brn-3a*, *Brn-3b*, and *Brn-3c*. They reside on distinct autosomes and encode proteins of 423, 410, and 338 amino acids in length, respectively. The intron-exon structures of the three *Brn-3* genes are identical, consisting of two coding exons interrupted by a small intron. In pairwise comparisons, the Brn-3 proteins share approximately 95% amino acid identity within the POU-domain, and approximately 70% identity outside of the POU-domain. A comparison of POU-domain sequences shows that the *Brn-3* family is most closely related to the *Drosophila I-POU* and the *C. elegans unc-86* genes (Fig. 1).

We have localized the expression of *Brn-3a*, *Brn-3b*, and *Brn-3c* by RNA-blot hybridization, RT-PCR, and immunostaining with affinity-purified antibodies raised against the most divergent regions of the Brn-3 proteins. Each of the anti-Brn-3 antibodies specifically recognizes the Brn-3 family member against which it was raised and does not cross-react with other Brn-3 proteins as determined by (1) Western blotting against recombinant fusion proteins produced in *Escherichia coli*, (2) the distinctive patterns of tissue staining obtained with each antibody, and (3) the selective elimination of immunostaining in mice lacking the corresponding gene (Xiang et al. 1995, 1996, 1997; Gan et al. 1996). During development and in the adult, expression of the *Brn-3* family is found principally within the central nervous system (CNS). Each *Brn-3* gene is expressed in a subset of retinal ganglion cells, somatosensory neurons in the trigeminal and dorsal root ganglia, and scattered cells within a small number of brainstem nuclei. *Brn-3a* and *Brn-3b* are expressed in many neurons within the spiral and vestibular ganglia, and *Brn-3c* is expressed in auditory and vestibular hair cells (Fig. 2). Other regions of the adult CNS, including the cortex and cerebellum, do not express the *Brn-3* genes. Figure 3 shows Brn-3a, Brn-3b, and Brn-3c immunolocalization in the adult mouse retina (Fig. 3A–F) and Brn-3b immunolocalization in the developing mouse retina (Fig. 3G–J). In all mammalian retinas tested to date (mouse, rabbit, cat, and macaque), Brn-3 immunoreactivity is confined to ganglion cells. In cat and macaque retinas, the expression pattern of individual Brn-3 family members correlates with previously defined morphologic and functional classes of retinal ganglion cells.

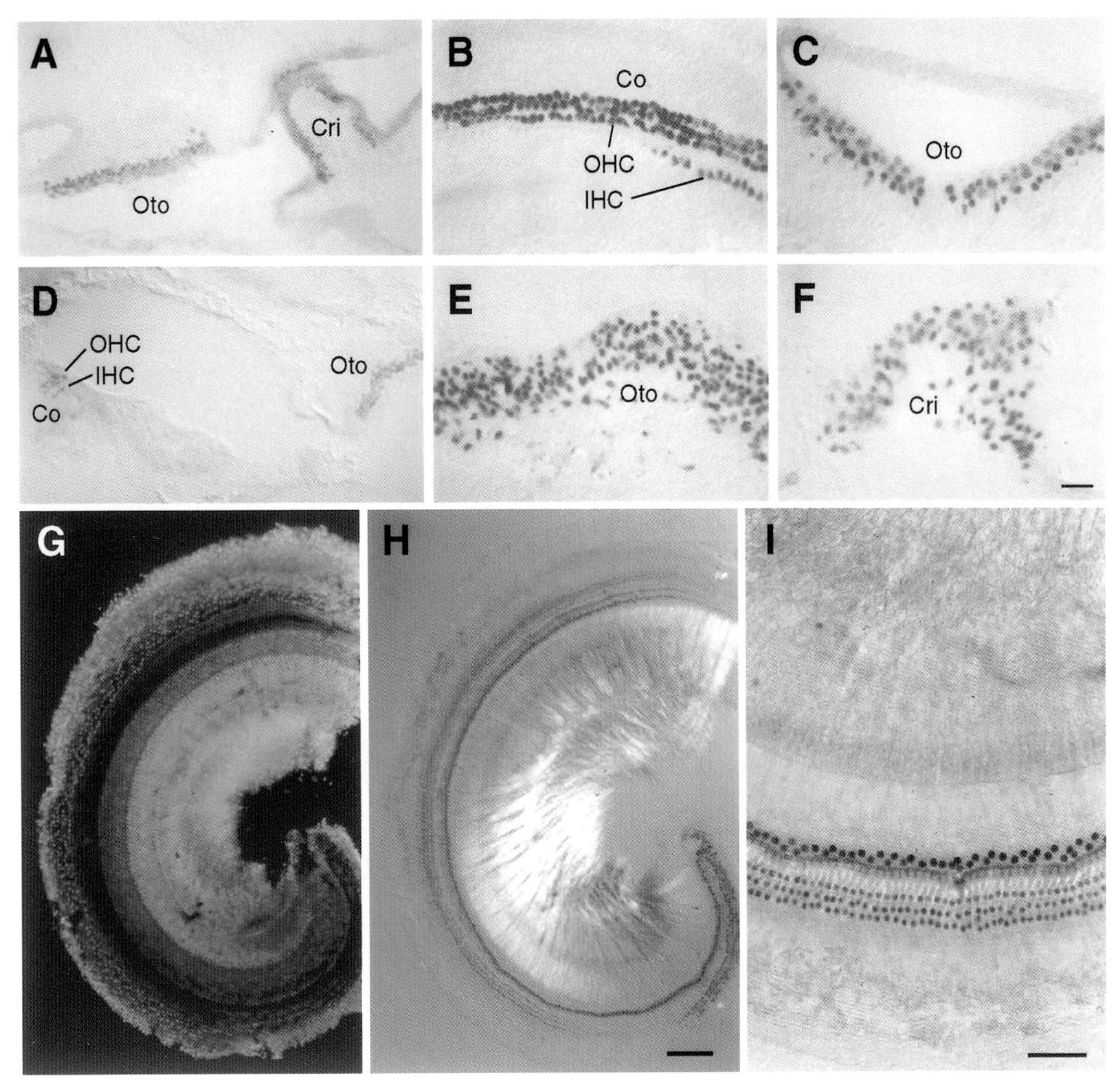

Figure 2. Brn-3c in the developing and adult mouse inner ear. Immunostaining of sections of the inner ear with anti-Brn-3c antibodies at e17.5 (*A–C*) and P5 (*D–F*). Staining is present in the nuclei of developing hair cells at all stages shown and is absent from other cell types. (*G–I*) Brn-3c in the adult organ of Corti. One turn of a wild-type organ of Corti adjacent to the apex is shown. The cochlea was immunostained as a whole mount with affinity-purified anti-Brn-3c antibodies, and the dissected organ of Corti was then incubated with DAPI. (*G*) DAPI staining reveals nuclei of supporting cells that are not immunostained. (*H,I*) Anti-Brn-3c immunoreactivity is found exclusively within the single row of inner hair cell nuclei and the three rows of outer hair cell nuclei. All hair cell nuclei are immunostained. Note that the precipitate formed by reaction of 3-amino-9-ethylcarbozole, the immunoperoxidase substrate, quenches DAPI fluorescence. (Cri) Crista; (Co) cochlea; (Oto) otolith organ; (IHC) inner hair cells; (OHC) outer hair cells. Bar in F: (*A*) 50 μm; (*B,C,E,F*) 25 μm; (*D*) 100 μm. Bar in H**:** (*G,H*) 100 μm. Bar in I**:** 50 μm. (Reprinted, with permission, from Xiang et al. 1997.)

DNA-binding Properties of Brn-3 Proteins

An optimal DNA-binding site for the Brn-3 proteins was identified using iterative cycles of in vitro binding and PCR amplification (the selex method; Thiesen and Bach 1990). In these experiments, we selected from a pool of random sequences that subset which bound to a fusion protein containing the POU domain of Brn-3b fused to glutathione *S*-transferase (GST). Following several cycles of enrichment, the selected double-stranded DNA segments were cloned, sequenced, and individually tested for their ability to bind the GST-POU domain fusion protein. Of 33 cloned segments, representing 30 different DNA sequences, 31 were found to bind the Brn-3b POU-domain and each of the 31 segments contained the consensus (A/G)TTAATGAG(C/T) or a close derivative of it. (In the original description of this experiment [Xiang et al. 1995], we refer to the complementary strand of this sequence.) To test the DNA-binding properties of the intact Brn-3b protein, whole-cell protein extracts were

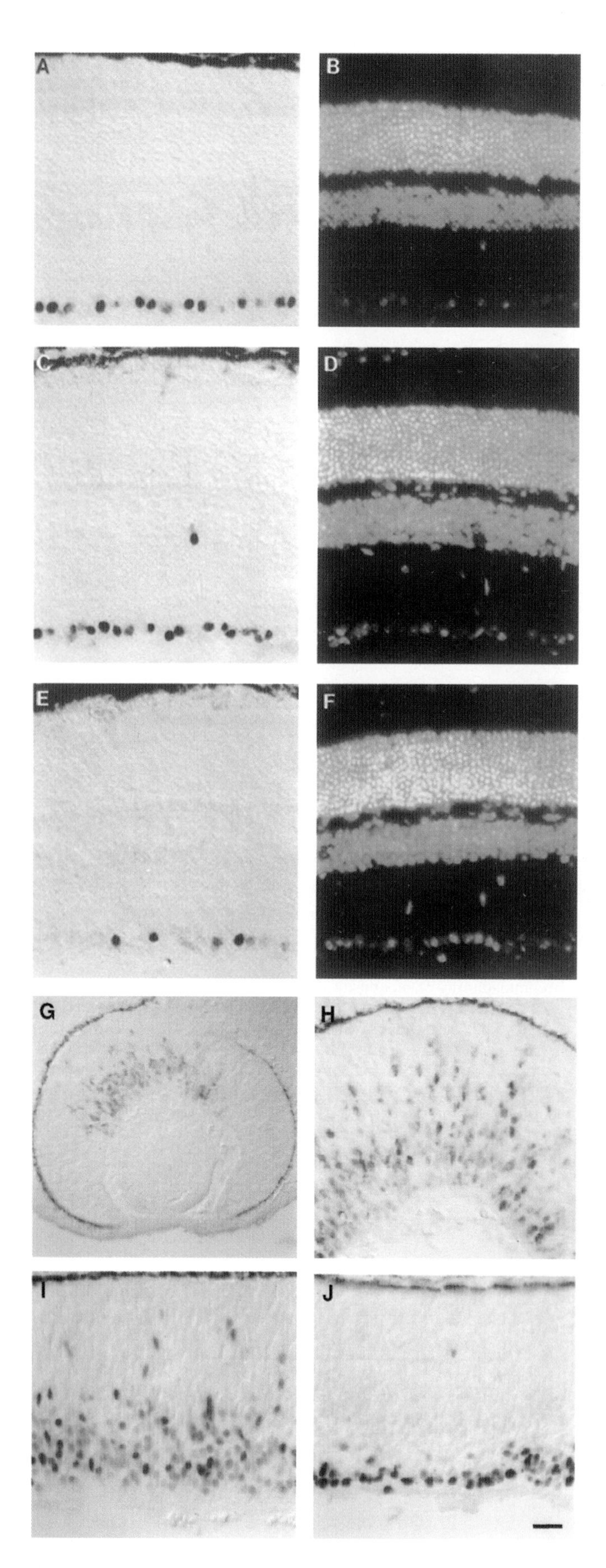

Figure 3. Anti-Brn-3a, -Brn-3b, and -Brn-3c immunoreactivity in the mouse retina. To localize Brn-3 proteins and visualize nuclei simultaneously, mouse retina sections were double-stained with anti-Brn-3a (*A*) and DAPI (*B*), anti-Brn-3b (*C*) and DAPI (*D*), or anti-Brn-3c (*E*) and DAPI (*F*). Note that the purple HRP product of the immunostaining reaction partially quenches DAPI fluorescence when both are present in the same nucleus. The three layers of nuclei are (from top to bottom): outer nuclear layer, inner nuclear layer, and ganglion cell layer. (*G–J*) Brn-3b appears early in ganglion cell development. Sections through the developing mouse retina at e12.5 (*G*), e13.5 (*H*), e16.5 (*I*), and P1.5 (*J*) immunostained with anti-Brn-3b antibodies. Brn-3b is initially present in both the developing ganglion cell layer and the proximal part of the mitotic zone. As development proceeds, Brn-3b is increasingly confined to cells within the developing ganglion cell layer. The localized expression of *Brn-3b* within the central retina at e12.5 reflects the earlier development of this region. In all micrographs of cross sections, the inner retina is at the bottom. Bar in J: (*H–J*) 25 μm; (*G*) 50 μm. (Reprinted, with permission, from Gan et al. 1996.)

prepared from untransfected human embryonic kidney cells (293) or from 293 cells that had been transiently transfected with a *Brn-3b* expression construct or an *Oct-2* expression construct (Fig. 4A). Brn-3b produced in 293 cells showed strong binding to the selected Brn-3 consensus site (C) and no detectable binding to a canonical octamer site (Oct; Singh et al. 1986). In contrast, endogenous Oct-1 from 293 cells and expressed Oct-2 bound well to the octamer site but poorly or not at all to the selected Brn-3 consensus site.

The regions of the selected Brn-3 consensus site that are essential for binding were delineated using a series of derivative sites containing base substitutions in or around the consensus region (M1-M6). These segments were tested for binding by full-length Brn-3b or GST-POU-domain fusion proteins derived from Brn-3a, Brn-3b, Brn-3c, and Oct-1 (Fig. 4B,C and data not shown). Two other segments of identical length, containing either a canonical octamer site or an Unc-86 site (M7; Xue et al. 1992), were also tested. The selected Brn-3 consensus site binds to full-length Brn-3b and to each of the Brn-3 GST-POU-domain fusion proteins with high specificity, and mutating pairs of nucleotides within the selected consensus sequence TTAATGAG dramatically reduces DNA binding (mutant sites M2-M5). Surprisingly, the GST-POU-domain fusions derived from Brn-3a and Brn-3c bind the Unc-86 consensus site (M7) poorly, and the isolated Brn-3b POU-domain binds this site with a reduced affinity relative to the selected Brn-3 consensus site (Xiang et al. 1995). However, full-length Brn-3b binds to the Unc-86-binding site and to the selected Brn-3 consensus site with similar affinities.

These in-vitro-binding experiments show that the Brn-3 POU-domains can bind to a selected consensus site (TTAATGAG) distinct from the site shown previously to bind Unc-86 (CAT(N)$_3$TAAT) and that full-length Brn-3b can bind to both sites. The numerous differences between the two sites suggest that the Brn-3 POU-domains may contact the selected DNA site by rearranging the positions and/or orientations of the POU-specific domain and the POU homeodomain relative to that observed in the Oct-1-DNA crystal structure (Klemm et al. 1994), as postulated by Li et al. (1993) for the interaction between Brn-2 and its DNA target in the promotor of the corticotropin-releasing hormone gene (Fig. 4D). Unlike most other POU-domain proteins described thus far, the Brn-3

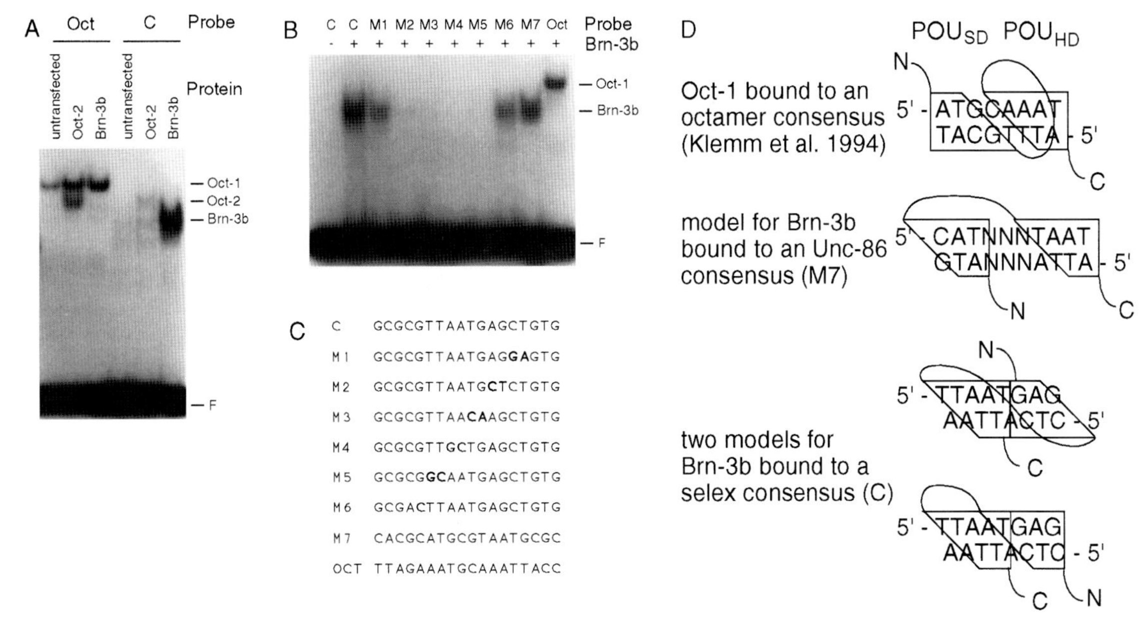

Figure 4. DNA-binding specificity of Brn-3b. (*A*) Comparison of the binding of Brn-3b and Oct-2 to the selected Brn-3 consensus site (*C*) and the canonical octamer site (Oct). Whole-cell protein extracts from 293 cells (293), and 293 cells transiently transfected with a Brn-3b expression construct (Brn-3b) or an Oct-2 expression construct (Oct-2) were utilized for gel mobility shift assays. (*B*) DNA-binding affinity of Brn-3b for the selected Brn-3 consensus site (*C*), mutant sites (M1-6), the Unc-86-binding site (M7), and the octamer site (Oct). Whole-cell protein extracts from 293 cells transiently transfected with the Brn-3b expression construct were used for gel mobility shift assays. Mobility-shifted complexes with Brn-3b, Oct-1, and Oct-2 are indicated. Note that all of the 293 cell extracts contain endogenous Oct-1. (*F*) Free probe. (*C*) Nucleotide sequences of one strand of the double-stranded DNA segments used for mobility shift assays in *A* and *B*. M1-M6 each differ by two nucleotides from the consensus obtained by the selex method. (In Xiang et al. [1995], the opposite strands of sequences C and M1-M6 were shown.) (*D*) Possible geometries of Brn-3 POU-domain-DNA interactions, based on the protein-DNA contacts observed in the Oct-1-DNA crystal structure (Klemm et al. 1994). The amino-terminal POU-specific domain (POU$_{SD}$) and the carboxy-terminal POU-homeodomain (POU$_{HD}$) are represented by trapezoids connected by a spacer; the labels POU$_{SD}$ and POU$_{HD}$ at the top of the figure refer to the Oct-1-DNA complex. In the model of Brn-3b binding to the Unc-86 consensus site (M7), the POU$_{SD}$ is positioned to the left of the POU$_{HD}$, and in the two models of Brn-3b binding to the selex consensus site (C), the POU$_{SD}$ is to the right of the POU$_{HD}$. (N) Amino terminus; (C) carboxyl terminus. (Reprinted, with permission, from Xiang et al. 1995.)

proteins bind poorly to the canonical octamer site, and, conversely, Oct-1 and Oct-2 bind poorly to the selected Brn-3 consensus site. These observations suggest that the Brn-3 proteins act on target genes in vivo that are distinct from those of previously described POU-domain proteins.

Brn-3a Is Required for the Development of Primary Somatosensory Neurons and Select Brainstem Sensory and Motor Nuclei

To determine the role of the *Brn-3* family in vivo, mice lacking either *Brn-3a*, *Brn-3b*, or *Brn-3c* were generated by homologous recombination in embryonic stem cells. The phenotypes associated with the three targeted deletions are described in Gan et al. (1996) and Xiang et al. (1996, 1997) and are summarized below. All of the phenotypes that we have been able to identify are recessive with complete penetrance. Similar results have been obtained by Erkman et al. (1996) and McEvilly et al. (1996).

Brn-3a (–/–) mice exhibit grossly normal prenatal development but do not survive beyond 24 hours after birth. *Brn-3a* (–/–) neonates display two overt behavioral defects: They lack a suckle reflex (and therefore do not nurse), and they lack the coordinated limb and trunk movements required to right themselves. When a *Brn-3a* (–/–) neonate is placed on its back or side, it typically stretches its limbs ineffectually and displays extended postures not observed in the wild type (Fig. 5A). Anatomically and histologically, *Brn-3a* (–/–) mice at e20 and P0 show a twofold decrease in the volume of the trigeminal ganglion and a severalfold decrease in the density of trigeminal neurons relative to the wild type (Fig. 5B–E). Although there is little or no difference between *Brn-3a* (+/+) and (–/–) animals in the appearance or number of dorsal root ganglion cells, Brn-3b immunoreactive cells are nearly absent in the trigeminal and dorsal root ganglia of *Brn-3a* (–/–) mice, and Brn-3c immunoreactive cells are diminished in number by five- to tenfold (Fig. 5F–I). Three sites of *Brn-3a* expression in the brainstem are also affected in *Brn-3a* (–/–) mice: The medial habenula and the caudal region of the inferior olivary nucleus show a modest diminution in cell number, and the large neurons of the red nucleus, which are clearly seen in wild-type mice as two symmetric clusters, are not seen in *Brn-3a* (–/–) mice (Fig. 5J–O). Other parts of the CNS, including those regions of the brainstem and retina that normally express *Brn-3a*, are not detectably affected.

The behavioral defects seen in *Brn-3a* (–/–) mice could arise from somatosensory dysfunction, motor dysfunction, or a combination of the two. Although the observed anatomic defects do not unequivocally define the sites responsible for each behavioral defect, they suggest plausible correlations between the two. In particular, a decrease in the number of trigeminal ganglion neurons could produce sensory defects in the face and mouth that impair the suckling response. With respect to coordination of limb and trunk movement, the observed phenotype could arise from defects in one or more intrinsic spinal reflexes or in any of the pathways that integrate ascending information from the dorsal root ganglia or descending information from the motor cortex or cerebellum. As the red nucleus conveys information from the cerebellum and the cerebral cortex to the spinal cord via the rubro-spinal tract, and the inferior olivary nucleus integrates sensory information and sends its output to cerebellar Purkinje cells, defects in these two nuclei could plausibly impair limb and trunk coordination or posture (Paxinos 1995). Functional alterations in the dorsal root ganglia related to the loss of *Brn-3b* and *Brn-3c* expression may also be relevant.

Brn-3b Is Required for Retinal Ganglion Cell Development

Brn-3b (–/–) and (+/–) mice are indistinguishable from their wild-type littermates in viability, growth rate, fertility, gait, and response to handling and loud sounds. These characteristics suggest that in *Brn-3b* (–/–) and (+/–) mice, the somatosensory, motor, and auditory/vestibular systems are grossly intact. Visual system function has not been assessed behaviorally, but it is likely to be at least partially functional as determined by recording light responses from retinal ganglion and/or amacrine cells with an array electrode (E. Soucy et al., unpubl.).

The only gross anatomic defects in *Brn-3b* (–/–) mice are associated with the eye: *Brn-3b* (–/–) mice show, on average, a fivefold reduction in the cross-sectional area of the optic nerve and a corresponding reduction in ganglion cell number. Whereas the overall structure of the *Brn-3b* (–/–) retina resembles that of the wild type, retinas from *Brn-3b* (–/–) mice are on average 20% thinner, due primarily to a decrease in the thickness of the inner plexiform, ganglion cell, and nerve fiber layers (Fig. 6A–D). The number of nuclei in the ganglion cell layer, the inner nuclear layer, and the outer nuclear layer is reduced by 30%, 15%, and 10%, respectively (Fig. 6A–F).

In the ganglion cell layer of both mouse and rat retinas, displaced amacrine cells and ganglion cells are equally abundant. To assess independently these cell types, retinas were immunostained for Thy-1, which is present in the axons and dendrites of all ganglion cells, for phosphorylated neurofilaments (monoclonal antibody SMI-32), which is found predominantly in the axons of large ganglion cells, and for the amacrine cell markers tyrosine hydroxylase and glutamic acid decarboxylase. As seen in Figure 6, G–J, *Brn-3b* (–/–) mice show a large reduction in both anti-Thy-1 and SMI-32 immunoreactive processes, suggesting a significant loss of retinal ganglion cells. Ganglion cells expressing *Brn-3a* and *Brn-3c* are reduced in number approximately sixfold (Fig. 6K,L,O,P). In contrast, the density of dopaminergic and GABAergic amacrine cells does not differ between *Brn-3b* (+/+) and (–/–) retinas. Other parts of the CNS, including those regions where *Brn-3b* is expressed, appear to be unaffected.

In the mouse, [^{3}H]thymidine-labeling studies show that between embryonic days 11 and 18 (e11 and e18), most

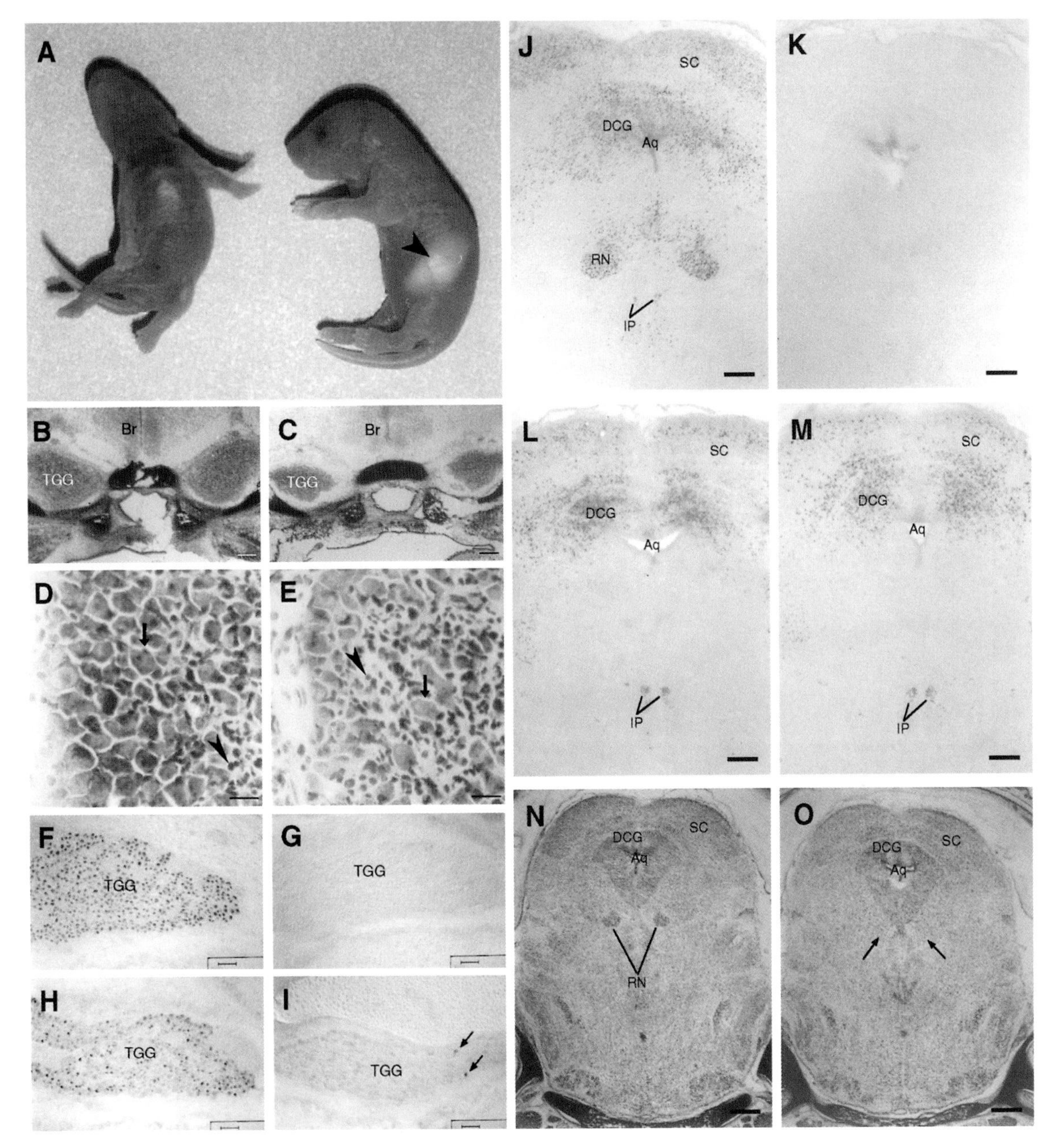

Figure 5. Phenotypes of *Brn-3a* (*–/–*) mice. (*A*) Behavioral differences between *Brn-3a* (+/+) and *Brn-3a* (*–/–*) siblings at P0. The *Brn-3a* (+/+) mouse (*right*) has nursed as seen by the presence of milk in its stomach; the *Brn-3a* (*–/–*) mouse (*left*) has not. Both animals were placed on their backs and photographed over the ensuing several minutes. The *Brn-3a* (+/+) mouse uses a forelimb to right itself; the *Brn-3a* (*–/–*) animal extends its limbs, head, and trunk but fails to right itself. Cresyl violet staining (*B–E,N,O*) and immunostaining with anti-Brn-3a antibodies (*F,G,J,K*) or anti-Brn-3b antibodies (*H,I,L,M*) of sections from e20 *Brn-3a* (+/+) mice (*B,D,F,H,J,L,N*; i.e., the left member of each pair of matched panels) or *Brn-3a* (*–/–*) mice (*C,E,G,I,K,M,O*; i.e., the right member of each pair of matched panels). (*B–I*) Trigeminal ganglia in *Brn-3a* (*–/–*) mice at e20. (*B–E*) Cresyl violet staining shows a twofold reduction in size of the *Brn-3a* (*–/–*) trigeminal ganglia (*C*) relative to the *Brn-3a* (+/+) trigeminal ganglion (*B*) and a selective loss of large cells (compare *D* and *E*; large vertical arrows, large cells; arrowheads, small cells). In *Brn-3a* (+/+) trigeminal ganglia, Brn-3a immunoreactivity is present in most neurons (*F*); in *Brn-3a* (*–/–*) trigeminal ganglia, Brn-3a immunoreactivity is absent as expected (*G*). In *Brn-3a* (*–/–*) trigeminal ganglia, the density of Brn-3b immunoreactive cells is greatly reduced (compare *H* and *I*; rare Brn-3b immunoreactive cells are indicated by small arrows). (*J–O*) The midbrain at the level of the red nucleus at e20. In *Brn-3a* (+/+) mice, Brn-3a immunoreactive neurons are present in the superior colliculus, dorsal central grey, red nucleus, and the interpeduncular nucleus (*J*); in the *Brn-3a* (*–/–*) midbrain, Brn-3a immunoreactivity is absent as expected (*K*). Brn-3b immunoreactive neurons are present in the superior colliculus, dorsal central grey, and the interpeduncular nucleus regardless of *Brn-3a* genotype (*L,M*). By cresyl violet staining, the large neurons of the red nucleus are readily visualized in *Brn-3a* (+/+) mice (*N*) but are absent in *Brn-3a* (*–/–*) mice (*O*; arrows indicate the expected location of the red nucleus). (TGG) Trigeminal ganglion; (Br) brain; (Aq) aquaduct; (DCG) dorsal central grey; (IP) interpeduncular nucleus; (RN) red nucleus; (SC) superior colliculus. Bars**:** (*B,C*) 200 μm; (*D,E*) 25 μm; (*F–I*) 50 μm; (*J–M*) 200 μm; (*N,O*) 400 μm. (Reprinted, with permission, from Xiang et al. 1996.)

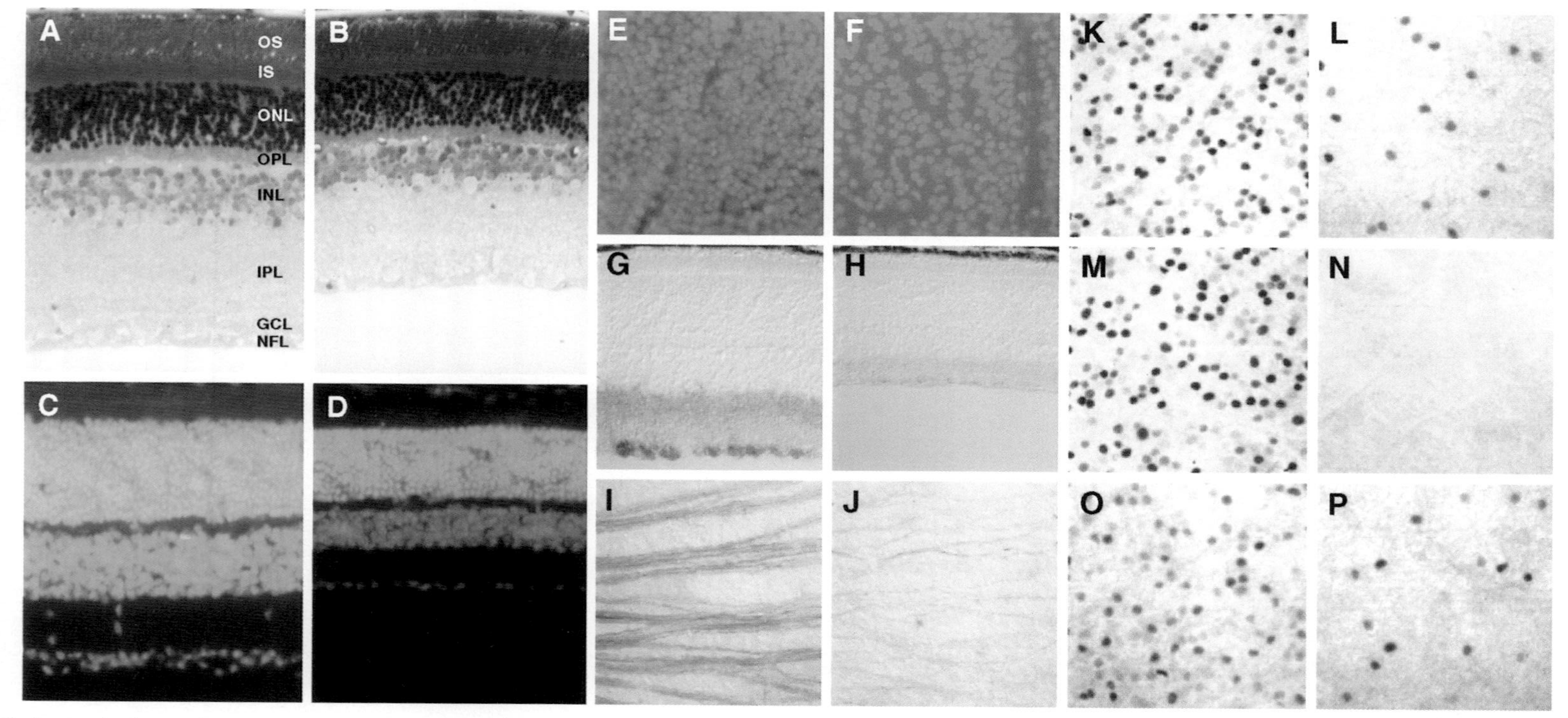

Figure 6. **R**eduction in the number of retinal ganglion cells in *Brn-3b* (–/–) mice. For each pair of micrographs, a *Brn-3b* (+/+) retina is on the left (*A,C,E,G,I,K,M,O*) and a *Brn-3b* (–/–) retina is on the right (*B,DF,H,J,L,N,P*). The paired samples were obtained from adult littermates. In all micrographs of cross sections, the inner retina is at the bottom. (*A,B*) 1-μm plastic sections of retina stained with toluidine blue. (*C,D*) DAPI staining to visualize nuclei in 10-μm sections of retina. (*E,F*) Retinal whole mounts in which nuclei are visualized by staining with SYTOX; the region shown is in the vicinity of the optic disc and the plane of focus is in the ganglion cell layer. (*G,H*) 10-μm sections immunostained with anti-Thy-1 monoclonal antibody, a marker for ganglion cells. (*I,J*) Retinal whole mounts immunostained with mAb SMI-32, a marker for large ganglion cells and their axons. (*K–P*) Retinal whole mounts immunostained with affinity-purified anti-Brn3a (*K,L*), anti-Brn-3b (M,N), or anti-Brn-3c (*O,P*). In *Brn-3b* (–/–) mice, the number of Brn-3a and Brn-3c immunoreactive nuclei is reduced approximately fivefold; Brn-3b immunoreactivity is absent as expected. As described previously (Xiang et al. 1993 1995), the abundance of the Brn-3 proteins shows a characteristic heterogeneity among different cells. (OS) Outer segments; (IS) inner segments; (ONL) outer nuclear layer; (OPL) outer plexiform layer; (INL) inner nuclear layer; (IPL) inner plexiform layer; (GCL) ganglion cell layer; (NFL) nerve fiber layer. Bars, 25 μm. Bars in *B* and *O* refer to *A–D* and *E–P*, respectively. (Reprinted, with permission, from Gan et al. 1996.)

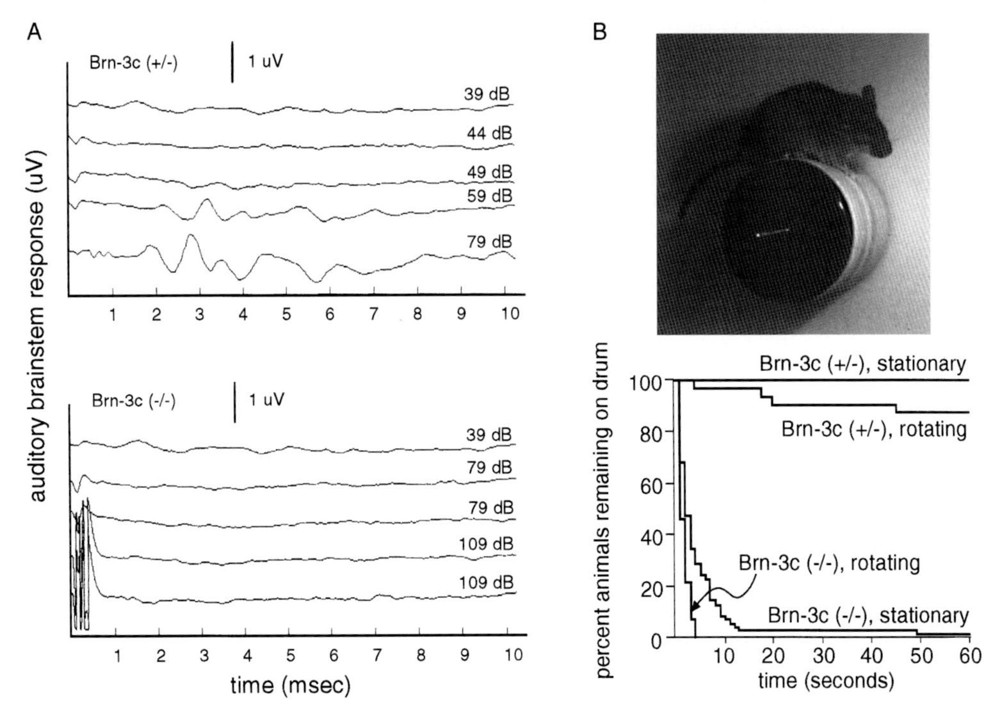

Figure 7. Physiologic defects in *Brn-3c* (–/–) mice. (*A*) Auditory brainstem responses in *Brn-3c* (+/–) and *Brn-3c* (–/–) mice. Representative auditory brainstem responses to a click stimulus are shown at different stimulus intensities. *Brn-3c* (+/–) mice show a threshold between 49 and 59 dB SPL with robust responses at and above 59 dB SPL. *Brn-3c* (–/–) mice show no response at any stimulus level, including 109 dB SPL, the highest level shown here. The traces at 109 dB SPL show a stimulus artifact in the first 1 msec. Traces were produced by averaging more than 1000 responses. (*B*) Balancing defects in *Brn-3c* (–/–) mice. (*Top)* A *Brn-3c* (+/–) mouse is shown balancing on the drum while it is held stationary. (*Bottom*) Percent of animals remaining on the drum during the 60 sec following their placement on it. Each of seven animals of the indicated genotypes was tested in ten trials in which the drum was stationary and in ten trials in which the drum was rotating at 7 rpm, a total of 70 trials for each genotype and test condition. (Reprinted, with permission, from Xiang et al. 1997.)

ganglion cell precursors become postmitotic and migrate from the proliferative zone in the outer retina to the future ganglion cell layer at the inner surface of the retina (Sidman 1961). As seen in Figure 3, G–J, *Brn-3b* expression commences in presumptive ganglion cells during their migration from the proliferative zone; *Brn-3a* and *Brn-3c* expression begins 1-2 days later (M. Xiang, unpubl.). In the *Brn-3b* (–/–) retina, a significant decrease in cell number or an increase in cell death in the ganglion cell layer is not seen during the prenatal period. However, a fivefold decrease in the number of cells expressing *Brn-3a* and *Brn-3c* is observed begining at e13.5, the earliest times at which Brn-3a and Brn-3c immunoreactivity can be detected. Recently, experiments in which a β-galactosidase reporter has been targeted to the deleted *Brn-3b*-coding region indicate that in the *Brn-3b* (–/–) retina, those cells that were destined to express *Brn-3b* persist into early postnatal life (L. Gan and W. Klein, unpubl.). Thus, the reduction in the number of ganglion cells expressing *Brn-3a* and *Brn-3c* in the embryonic retina reflects a block in the expression of the *Brn-3a* and *Brn-3c* genes, indicative of an early developmental defect in these cells. The ultimate fate of these aberrant cells remains to be determined.

Brn-3c Is Required for Auditory and Vestibular Hair Cell Development

Brn-3c (–/–) mice have normal viability, but are10–20% smaller than wild type, have low fertility, and spend much of their time running in circles, a behavior that has been described for a number of mouse lines with inner ear defects (Fuller and Wimer 1966). *Brn-3c* (–/–) mice also lack a startle response to sound and show no auditory brainstem response, even at stimulus levels 60 dB higher than the threshold for wild-type mice (Fig. 7A). To test vestibular function, mice were placed individually on a horizontal drum, and the time elapsed until they fell from the drum was recorded (Fig. 7B). *Brn-3c* (+/+) and (+/–) animals rarely fell from the drum, even when it was slowly rotated. In contrast, *Brn-3c* (–/–) mice exhibit extremely poor balance, typically falling from a stationary drum within 10 seconds. When the drum was slowly rotated, forcing the *Brn-3c* (–/–) animals to walk, none remained on the drum after 5 seconds. As a second test of vestibular function, we monitored the ability of mice to remain upright and swim effectively in a tub of water. In this test, *Brn-3c* (+/+) and (+/–) mice remained upright and swam, whereas *Brn-3c* (–/–) mice tumbled about ineffectively.

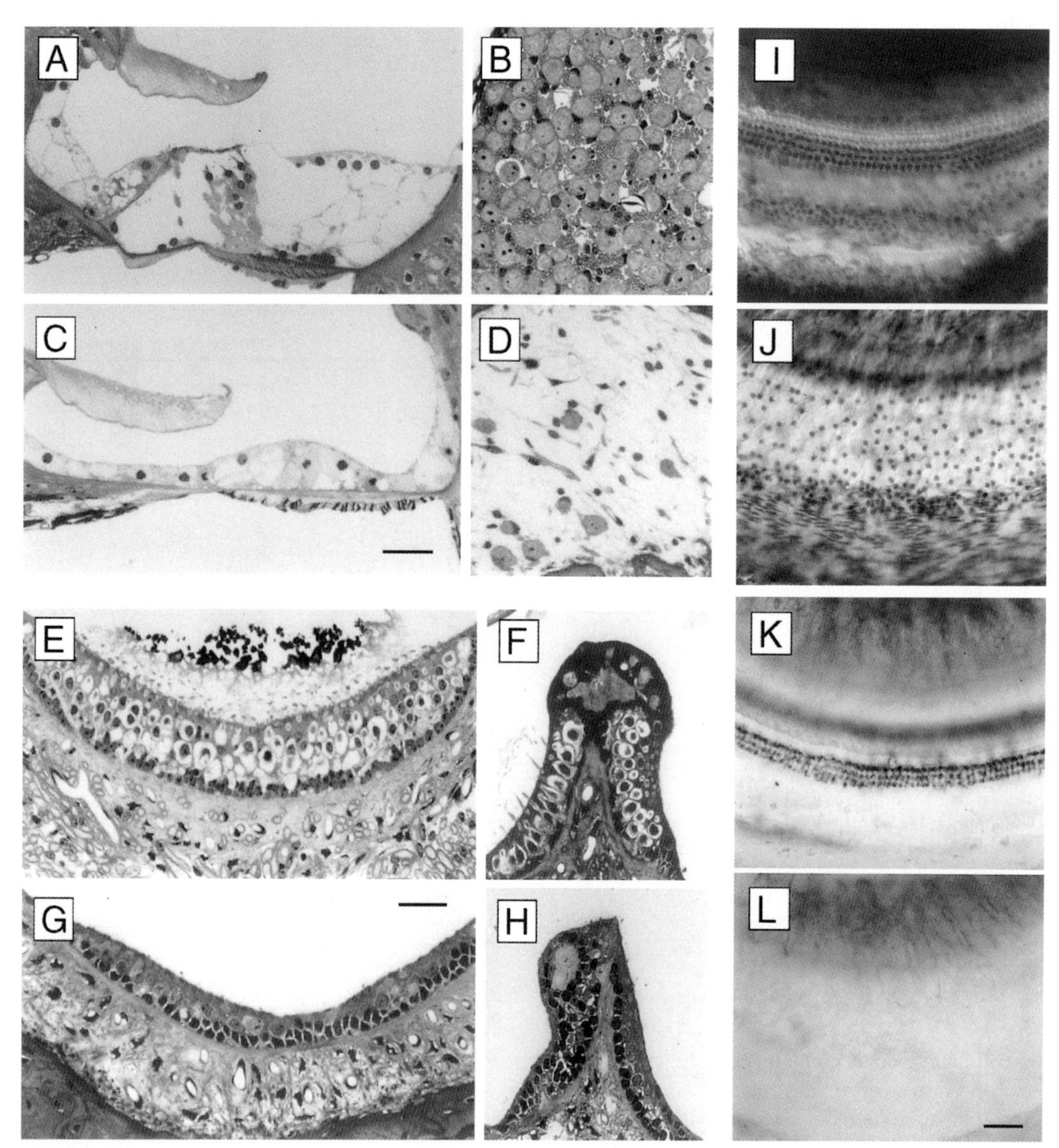

Figure 8. Anatomic defects in *Brn-3c* (–/–) mice. (*A–D*) Absence of hair cells in the organ of Corti and defects in the spiral ganglion in *Brn3c* (–/–) mice. (*A*) The organ of Corti from a *Brn-3c* (+/–) mouse. The single inner hair cell, the three outer hair cells, and the three underlying Deiter's cells are clearly visible. (*B*) Part of a spiral ganglion from a *Brn-3c* (+/–) mouse showing a dense packing of myelinated axons and neuronal cell bodies. (*C*) The organ of Corti from a *Brn-3c* (–/–) mouse. Inner and outer hair cells are missing and the epithelium beneath the tectorial membrane contains only supporting cells. (*D*) The spiral ganglion from a *Brn-3c* (–/–) mouse contains fewer than 10% as many myelinated axons and neuronal cell bodies as the spiral ganglia of *Brn-3c* (+/+) or (+/–) mice. (*E–H*) Absence of hair cells in the otolith organs and cristae of *Brn-3c* (–/–) mice. (*E*) An otolith organ from a *Brn-3c* (+/–) mouse. Type I hair cells, the most abundant hair cell class, have a lightly stained nerve chalice surrounding a darkly stained cell body. Ciliary bundles can be seen protruding into the densely stained otoliths at the top. A single layer of supporting cells lies beneath the layer of hair cells. (*F*) The crista from a *Brn-3c* (+/+) mouse. Type I hair cells are abundant and their ciliary bundles are seen at the left side of the ampullary crest where the section is nearly perpendicular to the apical surface. (*G,H*) An otolith organ (*G*) and a crista (*H*) from a *Brn-3c* (–/–) mouse are devoid of hair cells. Presumptive supporting cells are seen within the epithelium. The density of axon bundles beneath the otolith organ is greatly reduced in the *Brn-3c* (–/–) animal. Tissues were embedded in Spurr's resin and 1-μm sections were stained with methylene blue. (*I–L*) Whole-mount preparations of the organ of Corti in *Brn-3c* (+/+), (+/–), and (–/–) mice. (*I*) The organ of Corti from a *Brn-3c* (+/–) mouse stained with cresyl violet shows the wild-type arrangement of a single row of inner hair cells (out of focal plan) and three rows of outer hair cells (in the focal plane). (*K*) The organ of Corti from a *Brn-3c* (+/+) mouse stained histochemically for acetylcholine esterase reveals efferent synapses on the three rows of outer hair cells. The organ of Corti from a *Brn-3c* (–/–) mouse lacks identifiable hair cells (*J*) and cholinergic innervation (*L*). Bars**:** (*A–H*) 25 μm; (*I–L*) 50 μm. (Reprinted, with permission, from Xiang et al. 1977.)

Anatomically and histologically, *Brn-3c* (–/–) mice show a rapid and progressive loss of auditory and vestibular hair cells during late gestation and early postnatal life. Spiral and vestibular ganglion neurons degenerate over the ensuing weeks, but the middle ear and the overall architecture of the inner ear are unaffected. By adulthood, auditory and vestibular hair cells are completely absent and their associated ganglia contain few neuronal cell bodies or fibers (Fig. 8). Since *Brn-3c* is not expressed in supporting cells or in the spiral or vestibular ganglia during development, loss of spiral and vestibular ganglion neurons is likely to be secondary to the loss of hair cells. *Brn-3c* (–/–) mice show no differences in the size, number, or arrangement of neurons in the retina, dorsal root and trigeminal ganglia, and midbrain.

DISCUSSION

The high degree of sequence similarity between the POU-domains of *unc-86* and the *Brn-3* family, together with the conserved *Brn-3* gene structures, suggests that the *Brn-3* genes arose by duplication and divergence from an ancestral *unc-86*-like gene. As discussed below, the *unc-86* and *Brn-3* genes also bear intriguing functional similarities as seen in their expression in subsets of neurons and in the phenotypic consequences of their mutation.

In *C. elegans*, *unc-86* is expressed exclusively in neurons and neuroblasts. Of the 302 neurons in the adult, 57 express *unc-86*, including sensory neurons, motor neurons, and interneurons, and the *unc-86* phenotype includes defects in mechanosensation, chemotaxis, and egg laying (Hodgkin et al. 1979; Chalfie et al. 1981; Finney and Ruvkin 1990). Analysis of *unc-86* mutants at the single-cell level reveals defects both during and following the period of cell proliferation leading to subtle changes in neuronal phenotype, to neuronal loss, and to the generation of supernumerary neurons. Thus, *unc-86* appears to be involved in a variety of developmental decisions that differ depending on the cellular context.

In mammals, each member of the *Brn-3* family is expressed in a small subset of neurons in the brainstem and in the auditory/vestibular, somatosensory, and visual systems. Expression in each sensory system is confined to cells close to or at the site of sensory transduction: cochlear and vestibular hair cells and their associated ganglia, primary somatosensory neurons, and retinal ganglion cells. Interestingly, the phenotypes associated with mutation in each *Brn-3* gene reveal a greater degree of functional specialization than the partially overlapping zones of expression would suggest: The only anatomic defect thus far identified in *Brn-3b* (–/–) mice is a decrease of 70% in retinal ganglion cell number; and the only defects apparent in *Brn-3c* (–/–) mice are a loss of vestibular and cochlear hair cells and a secondary degeneration of their associated sensory ganglia. *Brn-3a* (–/–) mice show a decrease in the number of neurons in the trigeminal ganglia and in select brainstem nuclei but do not show any abnormalities in the retina. However, the early lethality of *Brn-3a* (–/–) animals means that a requirement for *Brn-3a* in the postnatal survival of other neuronal populations would have been missed in the analyses performed thus far.

The functional diversification of *Brn-3* family members indicates that the expansion and evolution of specialized sensory systems in more complex organisms have been accompanied by a parallel expansion and evolution of genetic regulatory proteins. These observations suggest that despite the great differences in transduction mechanism, sensory organ structure, and central information processing between the auditory/vestibular, somatosensory, and visual systems, there may be fundamental homologies in the genetic regulatory events that control their development.

ACKNOWLEDGMENTS

This work was supported by the National Eye Institute and the National Institute of Child Health and Development (NIH), the Robert A. Welch Foundation, the Retina Foundation, and the Howard Hughes Medical Institute.

REFERENCES

Bermingham J.R., Scherer S.S., O'Connell S., Arroyo E., Kalla K.A., Powell F.L., and Rosenfeld M.G. 1996. Tst-1/Oct-6/SCIP regulates a unique step in peripheral myelination and is required for normal respiration. *Genes Dev.* **10:** 1751.

Bitner-Glindzicz M., Turnpenny P., Hoglund P., Kaariainen H., Sankila E.M., van der Maarel S.M., de Kok Y.J.M., Ropers H.-H., Cremers F.P.M., Pembrey M., and Malcolm S. 1995. Further mutations in *Brain-4* (*POU3F4*) clarify the phenotype in the X-linked mixed deafness, DFN3. *Hum. Mol. Genet.* **4:** 1467.

Chalfie M., Horvitz H.R., and Sulston J.E. 1981. Mutations that lead to reiterations in the cell lineages of *C. elegans*. *Cell* **24:** 59.

Corcoran L.M., Karvelas M., Nossal G.J.V., Ye Z.-S., Jacks T., and Baltimore D. 1993. *Oct-2*, although not required for early B-cell development, is critical for later B-cell maturation and for postnatal survival. *Genes Dev.* **7:** 570.

de Kok Y.J.M., van der Maarel S.M., Bitner-Glindzicz M., Huber I., Monaco A., Malsolm S., Pembrey M.E., Ropers H.-H., and Cremers F.P.M. 1995. Association between X-linked mixed deafness and mutations in the POU domain gene *POU3F4*. *Science* **267:** 685.

Desai C., Garriga G., McIntyre S.L., and Horvitz H.R. 1988. A genetic pathway for the development of the *Caenorhabditis elegans* HSN motor neurons. *Nature* **336:** 638.

Erkman L., McEvilly R.J., Luo L., Ryan A.K., Hooshmand F., O'Connell S.M., Keithley E.M., Rapaport D.H., Ryan A.F., and Rosenfeld M.G. 1996. Role of transcription factors Brn-3.1 and Brn-3.2 in auditory and visual system development. *Nature* **381:** 603.

Fedtsova N.G. and Turner E.E. 1995. Brn-3.0 expression identifies early post-mitotic CNS neurons and sensory neural precursors. *Mech. Dev.* **53:** 291.

Finney M. and Ruvkin G. 1990. The *unc-86* gene product couples cell lineage and cell identity in *C. elegans*. *Cell* **63:** 895.

Finney M., Ruvkin G., and Horvitz H.R. 1988. The *C. elegans* cell lineage and differentiation gene *unc-86* encodes a protein with a homeodomain and extended similarity to transcription factors. *Cell* **56:** 757.

Fuller J.L. and Wimer R.E. 1966. Neural, sensory, and motor functions. In *Biology of the laboratory mouse* (ed. E.L. Green), p. 609. McGraw Hill, New York.

Gan L., Xiang M., Zhou L., Wagner D.S., Klein W.H., and Nathans J. 1996. The POU domain factor Brn-3b is required for the development of a large set of retinal ganglion cells. *Proc. Natl. Acad. Sci.* **93:** 3920.

Gerrero M.R., McEvilly R., Turner E., Lin C.R., O'Connell S., Jenne K.J., Hobbs M.V., and Rosenfeld M.G. 1993. Brn-3.0: A POU-domain protein expressed in the sensory, immune, and endocrine systems that functions on elements distinct from known octamer motifs. *Proc. Natl. Acad. Sci.* **90:** 10841.

He X., Treacy M.N., Simmons D.M., Ingraham H.A., Swanson

L.W., and Rosenfeld M.G. 1989. Expression of a large family of POU-domain regulatory genes in mammalian brain development. *Nature* **340:** 35.

Herr W., Sturm R.A., Clerc R.G., Corcoran L.M., Baltimore D., Sharp P.A., Ingraham H.A., Rosenfeld M.G., Finney M., Ruvkin G., and Horvitz H.R. 1988. The POU domain: A large conserved region in the mammalian *Pit-1*, *Oct-1*, *Oct-2* and *Caenorhabditis elegans unc-86* gene products. *Genes Dev.* **2:** 1513.

Hodgkin J., Horvitz H.R., and Brenner S. 1979. Nondisjunction mutants of the nematode *Caenorhabditis elegans. Genetics* **91:** 67.

Jaegle M., Mandemakers W., Broos L., Zwart R., Karis A., Visser P., Grosveld F., and Meijer D. 1996. The POU factor *Oct-6* and Schwann cell differentiation. *Science* **273:** 507.

Jan Y.N. and Jan L. 1993. HLH proteins, fly neurogenesis, and vertebrate myogenesis. *Cell* **75:**827.

Keynes R. and Krumlauf R. 1994. Hox genes and regionalization of the nervous system. *Annu. Rev. Neurosci.* **17:** 109.

Klemm J.D., Rould M.A., Aurora R., Herr W., and Pabo C. 1994. Crystal structure of the *Oct-1* POU domain bound to an octamer site: DNA recognition with tethered DNA-binding modules. *Cell* **77:** 21.

Lee J.E., Hollenberg S.M., Snider L., Turner D.L., Lipnick N., and Weintraub H. 1995. Conversion of *Xenopus* ectoderm into neurons by NeuroD, a basic helix-loop-helix protein. *Science* **268:** 836.

Li S., He X., Gerrero M.R., Mok M., Aggarwal A., and Rosenfeld M.G. 1993. Spacing and orientation of bipartite DNA-binding motifs as potential functional determinants for POU domain factors. *Genes Dev.* **7:** 2483.

Li S., Crenshaw E.B., Rawson E.J., Simmons D.M., Swanson L.W., and Rosenfeld M.G. 1990. Dwarf locus mutants lacking three pituitary cell types result from mutations in the POU-domain gene *Pit-1. Nature* **347:** 528.

McEvilly R.J., Erkman L., Luo L., Sawchenko P.E., Ryan A.F., and Rosenfeld M.G. 1996. Requirement for Brn-3.0 in differentiation and survival of sensory and motor neurons. *Nature* **384:** 574.

Nakai S., Kawano H., Yudate T., Nishi M., Kuno J., Nagata A., Jishage K., Hamada H., Fujii H., Kawamura K., Shiba K., and Noda T. 1995. The POU domain transcription factor Brn-2 is required for the determination of specific neuronal lineages in the hypothalamus of the mouse. *Genes Dev.* **9:** 3109.

Ng M., Diaz-Benjumea F.J., and Cohen S.M. 1995. *nubbin* encodes a POU-domain protein required for proximal-distal patterning in the *Drosophila* wing. *Development* **121:** 589.

Ninkina N.N., Stevens G.E.M., Wood J.N., and Richardson W.D. 1993. A novel Brn3-like POU transcription factor expressed in subsets of rat sensory and spinal cord neurons. *Nucleic Acids Res.* **21:** 3175.

Paxinos G. 1995. *The rat nervous system.* Academic Press, San Diego.

Schonemann M.D., Ryan A.K., McEvilly R.J., O'Connell S.M., Arias C.A., Kalla K.A., Li P., Sawchenko P.E., and Rosenfeld M.G. 1995. Development and survival of the endocrine hypothalamus and posterior pituitary gland requires neuronal POU domain factor Brn-2. *Genes Dev.* **9:** 3122.

Sidman R.L. 1961. Histogenesis of mouse retina studied with thymidine-3H. in *The structure of the eye* (ed. G. Smelser), p. 487. Academic Press, New York.

Singh H., Sen R., Baltimore D., and Sharp P.A. 1986. A nuclear factor that binds to a conserved sequence motif in transcriptional control elements of immunoglobulin genes. *Nature* **319:** 154.

Theil T., McLean-Hunter S., Zornig M., and Moroy T. 1993. Mouse BRN-3 family of POU transcription factors: A new aminoterminal domain is crucial for the oncogenic activity of BRN-3A. *Nucleic Acids Res.* **21:** 5921.

Thiesen H.-J. and Bach C. 1990. Target detection assay (TDA): A versatile procedure to determine DNA binding sites as demonstrated on Sp1 protein. *Nucleic Acids Res.* **18:** 3203.

Treacy M., He X., and Rosenfeld M.G. 1991. I-POU: A POU domain protein that inhibits neuron-specific gene activation. *Nature* **350:** 577.

Treacy M.N., Neilson L.I., Turner E.E., He X., and Rosenfeld M.G. 1992. Twin of I-POU: A two amino acid difference distinguishes an activator from an inhibitor of transcription. *Cell* **68:** 491.

Turner E.E. 1996. Similar DNA recognition properties of alternatively spliced *Drosophila* POU factors. *Proc. Natl. Acad. Sci.* **93:** 15097.

Turner E.E., Jenne K.J., and Rosenfeld M.G. 1994. Brn-3.2: A Brn-3-related transcription factor with distinctive central nervous system expression and regulation by retinoic acid. *Neuron* **12:** 205.

Wegner M., Drolet D.W., and Rosenfeld M.G. 1993. POU-domain proteins: Structure and function of developmental regulators. *Curr. Opin. Cell Biol.* **5:** 488.

Weinstein D.E., Burrola P.G., and Lemke G. 1995. Premature Schwann cell differentiation and hypermyelination in mice expressing a targeted antagonist of the POU transcription factor SCIP. *Mol. Cell. Neurosci.* **6:** 212.

Xiang M., Zhou L., Peng Y.W., Eddy R.L., Shows T.B., and Nathans J. 1993. Brn-3b: A POU domain gene expressed in a subset of retinal ganglion cells. *Neuron* **11:** 689.

Xiang M., Gan L., Zhou L., Klein W.H., and Nathans J. 1996. Targeted deletion of the mouse POU domain gene *Brn-3a* causes a selective loss of neurons in the brainstem and trigeminal ganglion, uncoordinated limb movement, and impaired suckling. *Proc. Natl. Acad. Sci.* **93:** 11950.

Xiang M., Gan L., Li D., Chen Z.-Y., Zhou L., O'Malley B.W., Klein W., and Nathans J. 1997. Essential role of POU-domain factor Brn-3c in auditory and vestibular hair cell development. *Proc. Natl. Acad. Sci.* **94:** 9445.

Xiang M., Zhou L., Macke J., Yoshioka T., Hendry S.H.C., Eddy R.L., Shows T.B., and Nathans J. 1995. The Brn-3 family of POU-domain factors: Primary structure, binding specificity, and expression in subsets of retinal ganglion cells and somatosensory neurons. *J. Neurosci.* **15:** 4762.

Xue D., Finney M., Ruvkun G., and Chalfie M. 1992. Regulation of the *mec-3* gene by *C. elegans* homeoproteins UNC-86 and MEC-3. *EMBO J.* **11:** 4969.

Yeo S.L., Lloyd A., Kozak K., Dinh A., Dick T., Yand X., Sakonju S., and Chia W. 1995. On the functional overlap between two *Drosophila* POU homeodomain genes and the cell fate specification of a CNS neural precursor. *Genes Dev.* **9:** 1223.

Upstream and Downstream from *Brachyury*, a Gene Required for Vertebrate Mesoderm Formation

J.C. Smith,[1] N.A. Armes,[1] F.L. Conlon,[1] M. Tada,[1] M. Umbhauer,[2] and K.M. Weston[3]

[1]Division of Developmental Biology, National Institute for Medical Research, London NW7 1AA, United Kingdom; [2]Centre National de la Recherche Scientifique, Université Pierre et Marie Curie (Paris 6), Laboratoire de Biologie Experimentale, 75005 Paris, France; and [3]Institute of Cancer Research, Chester Beatty Laboratories, London SW3 6JB, United Kingdom

Brachyury (or *T*) is an interesting gene for researchers studying mesoderm formation in the early vertebrate embryo. One reason is that the gene is expressed in mouse, frog, fish, and chick embryos in the presumptive mesoderm (Herrmann et al. 1990; Wilkinson et al. 1990; Smith et al. 1991; Schulte-Merker et al. 1992; Kispert et al. 1995b; Knezevic et al. 1997). Initially, expression occurs in all, or virtually all, the nascent primitive streak, germ ring, or marginal zone. As gastrulation proceeds, however, transcripts are lost from prospective lateral and ventral mesoderm but persist in the notochord and tailbud. Second, lack of *Brachyury* function results in loss of the tissues in which the gene is expressed at the highest levels for the longest time. Thus, the mouse *T* mutant lacks mesoderm posterior to somite 7 and fails to form a properly differentiated notochord (Herrmann 1991, 1995). Similarly, the zebrafish *no tail* mutant, which also lacks *Brachyury* function, has no tail and fails to form a notochord (Halpern et al. 1993; Schulte-Merker et al. 1994). In *Xenopus*, overexpression of a dominant-negative *Brachyury* construct produces a similar phenotype (Conlon et al. 1996).

Finally, *Brachyury* is of interest because misexpression of the gene in prospective ectoderm of *Xenopus* embryos causes ectopic mesoderm to form in a dose-dependent fashion (Cunliffe and Smith 1992; O'Reilly et al. 1995). Low levels of *Xenopus Brachyury* (*Xbra*) result in the formation of ventral mesodermal cell types such as smooth muscle, whereas higher levels result in the formation of skeletal muscle. Interestingly, even the highest levels of *Xbra* expression do not result in the formation of notochord, but this tissue does form if *Xbra* is coexpressed with the secreted protein noggin (Smith and Harland 1992; Cunliffe and Smith 1994), which inhibits BMP signaling (Zimmerman et al. 1996), or with the transcription factor *Pintallavis*, a homolog of HNF-3β (Ruiz i Altaba and Jessell 1992; O'Reilly et al. 1995).

These observations mark *Brachyury* as a key gene in vertebrate mesoderm formation, and it therefore becomes important to understand how expression of the gene is controlled and how it goes on to exert its effects. In this paper, we describe our efforts to address these questions. At present, our knowledge is limited, but techniques are becoming available that should speed progress in the next few years.

EXPERIMENTAL PROCEDURES

***Xenopus* embryos, microinjection, and dissection.** *Xenopus* embryos were obtained by in vitro fertilization (Smith and Slack 1983). They were maintained in 10% normal amphibian medium (NAM; Slack 1984) and staged according to the method of Nieuwkoop and Faber (1975). *Xenopus* embryos at the one- to two-cell stage or at the 32-cell stage were injected with RNA dissolved in 10 nl or 1 nl water, respectively, as described by Smith (1993). For animal cap assays, embryos were transferred to 75% NAM at the desired stages, and animal caps were dissected and cultured in 75% NAM or 75% NAM containing 0.1% bovine serum albumin when mesoderm-inducing factors were in the culture medium. *Xenopus* basic fibroblast growth factor (bFGF) was prepared by J. Green using an expression plasmid provided by D. Kimelman and M. Kirschner. A crude preparation of recombinant human activin A was prepared from the conditioned medium of COS cells transfected with a human inhibin βA cDNA. The cells were a gift from Dr. G. Wong (Genetics Institute Inc., Cambridge, Massachusetts). A unit of activin activity is defined by Cooke et al. (1987). Dexamethasone (Sigma) was dissolved in ethanol to a concentration of 2 mM and then diluted to a final concentration of 1 μM in 75% NAM for animal caps or in 10% NAM for whole embryos.

In vitro transcription. RNA for microinjection was synthesized according to the method of Smith (1993). Constructs were v-*ras*, N17-*ras*, C4*raf*, MEK1$^{S217E/S221/E}$, Xp42^{D324N}, Xp42$^{K57R/D324N}$, X17C, all described by Umbhauer et al. (1995). Xbra is described by Cunliffe and Smith (1992), Pintallavis is described by O'Reilly et al. (1995), Xbra-En^{R} is described by Conlon et al. (1996), and Xbra-GR and Xbra-GR-HA are described by Tada et al. (1997).

RNA isolation and RNase protection assays. RNA isolation and RNase protection analysis were carried out as described by Jones et al. (1995) using RNase T1 alone for all samples. Probes included cardiac actin (Mohun et al. 1984), EF-1α (Sargent and Bennett 1990), ornithine decarboxylase (ODC) (Isaacs et al. 1992), *goosecoid* (Cho et al. 1991), *Xbra* (Smith et al. 1991), *Pintallavis* (Ruiz i Altaba and Jessell 1992; O'Reilly et al. 1995), and *Xhox3* (Ruiz i Altaba and Melton 1989).

Histology and immunocytochemistry. For histological analysis, specimens were fixed, sectioned, and stained as described by Smith (1993). Whole-mount immunocytochemistry with monoclonal antibody 12/101 (Kintner and Brockes 1984), specific for muscle, was performed as described by Smith (1993).

Yeast transcription assays and transposition mutagenesis. $GAL4^{DBD}$-Xbra and $GAL4^{DBD}$-Ntl were generated by mutating the initiating methionine of Xbra (Smith et al. 1991) and Ntl (Schulte-Merker et al. 1992, 1994) to an *Eco*RI site by polymerase chain reaction (PCR) mutagenesis. The corresponding fragments were cloned in-frame to the *Eco*RI site of the GAL4 DNA-binding domain ($GAL4^{DBD}$) contained in the yeast vector pGBT9. Fusions were confirmed by sequencing. Transposon TnXR, a 4888-bp derivative of Tn*1000* (Morgan et al. 1996), was introduced into the $GAL4^{DBD}$-Xbra and $GAL4^{DBD}$-Ntl plasmids (Sedgwick and Morgan 1994). Targeted constructs were identified by restriction mapping and PCR analysis using primers specific for GAL4 and TnXR. Exact positions of the insertions were determined by transposon-primed nucleotide sequencing into flanking DNA. 5′ and internal deletions were created by conventional techniques. Transcription activation was assessed by the expression of $GAL4^{UAS}$-*lacZ* and $GAL4^{Uas}$-Ura reporters in *Saccharomyces cerevisiae* strains YN190 and Y166. At least three transformations were performed for each deletion, with a minimum of 250 colonies scored per transformation.

Transient transfection analyses. Expression plasmids encoding full-length Xbra and Ntl were derived from the corresponding pGBT9 constructs (see above). Xbra and Ntl deletion constructs were generated by PCR or conventional techniques. The $Xbra-En^{R}$ construct was generated by fusing the Xbra DNA-binding domain (amino acids 1–232) to a fragment encoding amino acids 2–298 of the *Drosophila* engrailed protein. This region of engrailed was derived from plasmid MEnT (Badiani et al. 1994), and the Myc tag (Evan et al. 1985) of MEnT was included in $Xbra-En^{R}$. All constructs were cloned in-frame into the *Eco*RI site in the plasmid MLVplink (Dalton and Treisman 1992). Fusions were confirmed by sequencing, and in vitro translation of all constructs gave proteins of the correct size (not shown). The chloramphenicol acetyltransferase (CAT) reporter plasmid pBLCAT2 (Luckow and Schutz 1987) was modified such that two copies of an oligonucleotide containing the Brachyury-binding site (Kispert and Herrmann 1993) were inserted into the *Sal*I site upstream of the promoter region. Lipofections were carried out according to the method of Marais et al. (1995). Cells were cultured for 48 hours, and extracts were then assayed for CAT activity. MLV *lacZ* was cotransfected as a control for transfection efficiency (Hill et al. 1993). All constructs were translated in a reticulocyte lysate system and shown by bandshift assays to bind to a Brachyury-binding site with equivalent affinity.

RESULTS

Signal Transduction Pathways and Activation of *Brachyury* Expression

The mesoderm of the amphibian embryo, and perhaps that of all vertebrates, is formed through an inductive interaction in which a signal from the vegetal hemisphere of the embryo acts on overlying equatorial cells (Nieuwkoop 1969; Sive 1993; Slack 1994; Smith 1995). Candidates for the mesoderm-inducing signal(s) include members of the transforming growth factor-β (TGF-β) and FGF families, particularly activin (Asashima et al. 1990; Smith et al. 1990; Thomsen et al. 1990; Dyson and Gurdon 1996) and Vg1 (Rebagliati et al. 1985; Dale et al. 1993; Thomsen and Melton 1993; Kessler and Melton 1995) from the former group, and FGF-2 (Kimelman and Kirschner 1987; Slack et al. 1987, 1989; Amaya et al. 1991, 1993; Song and Slack 1994) from the latter.

Both activin and FGF-2 can activate expression of *Xbra* in prospective ectodermal tissue in an immediate-early fashion (Smith et al. 1991), and recently, the signal transduction pathways employed by these factors have been studied. In the case of activin, overexpression of constitutively active type I activin receptors, particularly ALK-4, strongly activates expression of *Xbra* in animal pole tissue (Jones et al. 1996; Armes and Smith 1997); the same is true of members of the Smad family (Baker and Harland 1996; Graff et al. 1996). It is not yet known, however, whether Smad proteins bind directly to *Xbra* regulatory sequences in the same way that *Drosophila* Mad binds to a *vestigial* enhancer (Kim et al. 1997).

One important signaling pathway from the FGF receptor tyrosine kinase involves $p21^{ras}$ and components of the MAP kinase pathway (Marshall 1995); work in our laboratory (Umbhauer et al. 1995) and elsewhere (Gotoh et al. 1995; LaBonne et al. 1995) has demonstrated that overexpression of active forms of $p21^{ras}$, or MEK1, or MAP kinase itself is sufficient to activate expression of *Xbra* and to cause the formation of mesodermal cell types (Fig. 1). Furthermore, inhibition of MAP kinase signaling prevents mesoderm formation in response to FGF and to upstream components of the MAP kinase pathway. For example, a dominant-negative Raf construct inhibits the effects of FGF-2 and of v-*ras*, but has no effect on induction by constitutively active MEK1 (Fig. 1b) (Umbhauer et al. 1995). The effects of MEK1 are, however, inhibited by the MAP kinase phosphatase X17C (Umbhauer et al. 1995). Although these experiments point to the MAP kinase pathway being critical in the activation of *Xbra* expression by FGF-2, nothing is known,

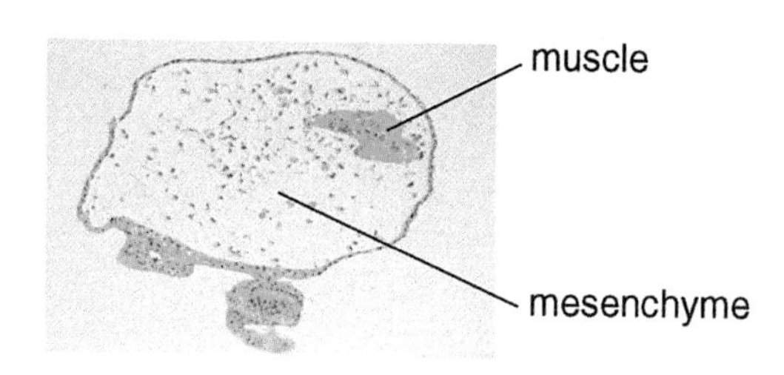

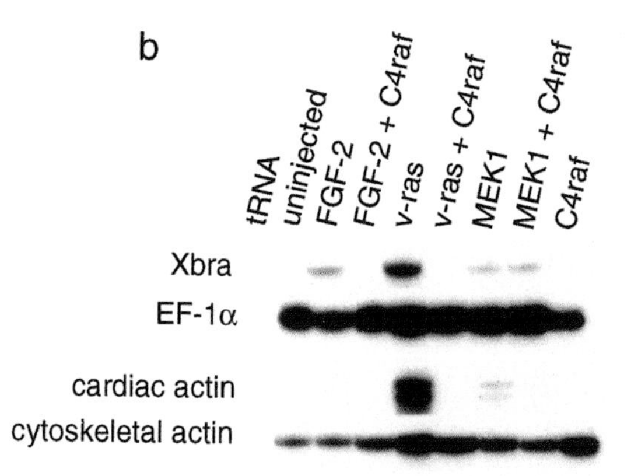

Figure 1. (*a*) Expression of MEK1$^{S217E/S221/E}$ in *Xenopus* animal pole tissue causes formation of mesodermal cell types, including mesenchyme and muscle. (*b*) A dominant-negative Raf construct (C4*raf*) inhibits induction of *Xbra* by FGF-2 and by v-*ras*, but has no effect on induction by constitutively active MEK1. C4*raf* also inhibits muscle formation in response to v-*ras* and constitutively active MEK1.

at present, as to how activated MAP kinase exerts its effects on the *Xbra* promoter.

Maintenance of *Xbra* Expression by an Indirect Autocatalytic Loop Involving eFGF

The above experiments address the activation of *Xbra* expression by the mesoderm-inducing factors activin and FGF-2, but they do not address the maintenance of *Xbra* expression. The importance of this point is made by the expression pattern of *Xbra* in the intact embryo. Initially, expression occurs throughout most of the marginal zone. During gastrulation, however, newly involuted cells rapidly lose *Xbra* transcripts, with the exception of those cells destined to form the notochord, where expression persists until at least the end of the gastrula stage. These observations therefore raise the question of how *Xbra* expression is down-regulated in newly involuted cells, and how it is maintained in the notochord.

One of the first clues as to how *Xbra* expression might be maintained came from experiments in which prospective mesoderm of the *Xenopus* embryo was cultured either as an intact piece of tissue or as dispersed cells. If cultured as an intact piece of tissue, *Xbra* expression persisted; if cultured as dispersed cells, however, expression declined precipitously, but this decline could be prevented if the cells were cultured in the presence of FGF (Isaacs et al. 1994; Schulte-Merker and Smith 1995). These results, together with the observations that ectopic expression of *Xbra* in prospective ectodermal tissue causes activation of eFGF expression (Isaacs et al. 1994; Schulte-Merker and Smith 1995), that *Xbra* and eFGF are coexpressed (Isaacs et al. 1995), and that inhibition of FGF function causes a down-regulation of *Xbra* expression (Amaya et al. 1993; Isaacs et al. 1994; Kroll and Amaya 1996) led to the idea that *Xbra* and eFGF might be components of an indirect autocatalytic loop, in which *Xbra* induces expression of eFGF and eFGF maintains expression of *Xbra* (Fig. 2). The importance of this loop is illustrated by the fact that the ability of *Xbra* to induce mesoderm in presumptive ectodermal tissue is inhibited if FGF signaling in those cells is blocked by overexpression of ΔFGFr, a truncated FGF receptor (Schulte-Merker and Smith 1995).

The existence of this indirect autocatalytic loop may provide an explanation for the "community effect," in which it is observed that muscle differentiation, for example, will not occur unless there is a threshold number of similar cells all bent on activating muscle-specific genes (Gurdon 1988). Thus, if there are only a few cells expressing *Xbra*, the extracellular concentration of eFGF might be too low for expression of *Xbra* to be maintained and for muscle differentiation to occur. If there are many *Xbra*-expressing cells, the extracellular concentration of eFGF would be high enough to maintain *Xbra* expression and to promote subsequent muscle differentiation. A similar indirect autoregulatory loop occurs in *Drosophila*, in which control of expression of *Ultrabithorax* is at least partly indirect and mediated by the extracellular signaling molecules wingless and decapentaplegic (Thuringer and Bienz 1993; Thuringer et al. 1993).

More recent experiments suggest that the FGF-*Brachyury* autoregulatory loop operates predominantly in the notochord. For example, if *Xbra* function is blocked using the Xbra-En^R construct (see below), expression of the gene is down-regulated in the notochord but persists

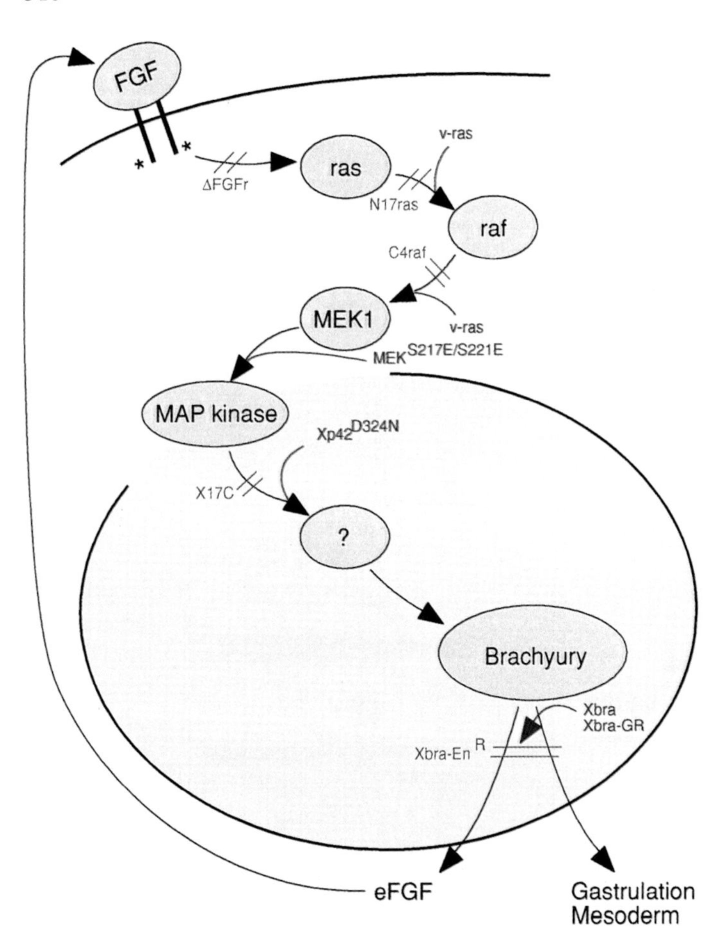

Figure 2. *Xbra*-eFGF autocatalytic loop and the reagents used to study it. Molecules indicated in red (ΔFGFr, N17*ras*, C4*raf*, X17C, Xbra-EnR) all inhibit a component of the loop; molecules in blue (v-*ras*, v-*raf*, $MEK1^{S217E/S221/E}$, $Xp42^{D324N}$, Xbra, Xbra-GR) represent constitutively active forms of MAP kinase pathway components or of Brachyury. The protein represented by "?" is a presumed target of MAP kinase activity.

in the marginal zone (Conlon et al. 1996). The autoregulatory loop therefore cannot be required for *Xbra* expression in the marginal zone. Similarly, expression of zebrafish *Brachyury, ntl,* declines in the notochord of *ntl* mutant embryos but persists in the germ ring (Schulte-Merker et al. 1994). The final piece of evidence comes from experiments involving chimeric mouse embryos (Schmidt et al. 1997). When *Brachyury*-mutant ES cells carrying a *Brachyury*-promoter *lacZ* construct (Clements et al. 1996) are introduced into wild-type embryos, *lacZ* activity is detected in the primitive streak in these cells even in the absence of functional Brachyury protein, and even at a distance of five to eight cell diameters from wild-type cells that might be providing FGF to maintain *Brachyury* expression. Thus, it seems unlikely that expression of *Brachyury* in the primitive streak, germ ring, or marginal zone of the embryo requires an autocatalytic loop, but, as in *Xenopus* and zebrafish, it remains possible that such a loop does function in the notochord (Schmidt et al. 1997).

Xbra Is an Activator of Transcription

Initial analysis of the protein encoded by *Brachyury* revealed only weak homology with the transcription factors MyoD and Rel (Willison 1990), but as *Brachyury*-like genes became identified in other species, it soon became clear that the amino-terminal half of Brachyury defines a conserved domain that became known as the T-box (Bollag et al. 1994; Herrmann 1995; Kavka and Green 1997; Smith 1997). Immunocytochemical analysis demonstrated that Brachyury is a nuclear protein (Schulte-Merker et al. 1992; Cunliffe and Smith 1994; Kispert and Herrmann 1994), and binding-site selection experiments established that the T-box binds a consensus 20-bp internally palindromic sequence T(G/C)ACACCTAGGTGTGAAATT (Kispert et al. 1995a).

Kispert et al. (1995a) and Conlon et al. (1996) went on to show that Brachyury is capable of activating transcription, and both groups used deletion analysis to map the regions of the protein responsible for this effect. Study of mouse Brachyury revealed two transcription activation domains and two repressor domains, all confined to the carboxy-terminal half of the protein (Kispert et al. 1995a). Interestingly, similar experiments using *Xenopus* and zebrafish Brachyury (Fig. 3) identified only a single activation domain (Conlon et al. 1996). The significance of this difference between the species is unclear, but together the experiments suggest strongly that the main role of Brachyury in the vertebrate embryo is to activate the transcription of downstream genes.

Transcription Activation by Xbra Is Essential for Xbra Function

The conclusion that the main role of Brachyury is to activate transcription was tested by replacing the transcription activation domain of Xbra with the repressor domain of the *Drosophila* engrailed protein (Jaynes and O'Farrell 1991; Han and Manley 1993; Badiani et al. 1994; Conlon et al. 1996). Transient transfection experiments confirmed that this Xbra-En^R construct efficiently inhibits transcription activation caused by Xbra (Fig. 4), and when RNA encoding Xbra-En^R was microinjected into *Xenopus* or zebrafish embryos, this resulted in the formation of embryos lacking a tail (Fig. 5) and, in many cases, a notochord (not shown). These embryos therefore resembled mouse and zebrafish embryos lacking Brachyury function and led to the conclusion that the main role of Brachyury is indeed to activate transcription.

Hormone-inducible Xbra

If the role of Brachyury is to activate transcription, what are the genes whose expression it induces? This is an important question, and one that applies to all transcription factors expressed in the early embryo; but for all that, little is known about transcription factor targets in early development. We plan to address this question using an approach that makes use of hormone-inducible *Xbra* constructs.

Work in cell culture has led to the identification of many "immediate-early" genes whose transcription is induced by serum or defined growth factors such as

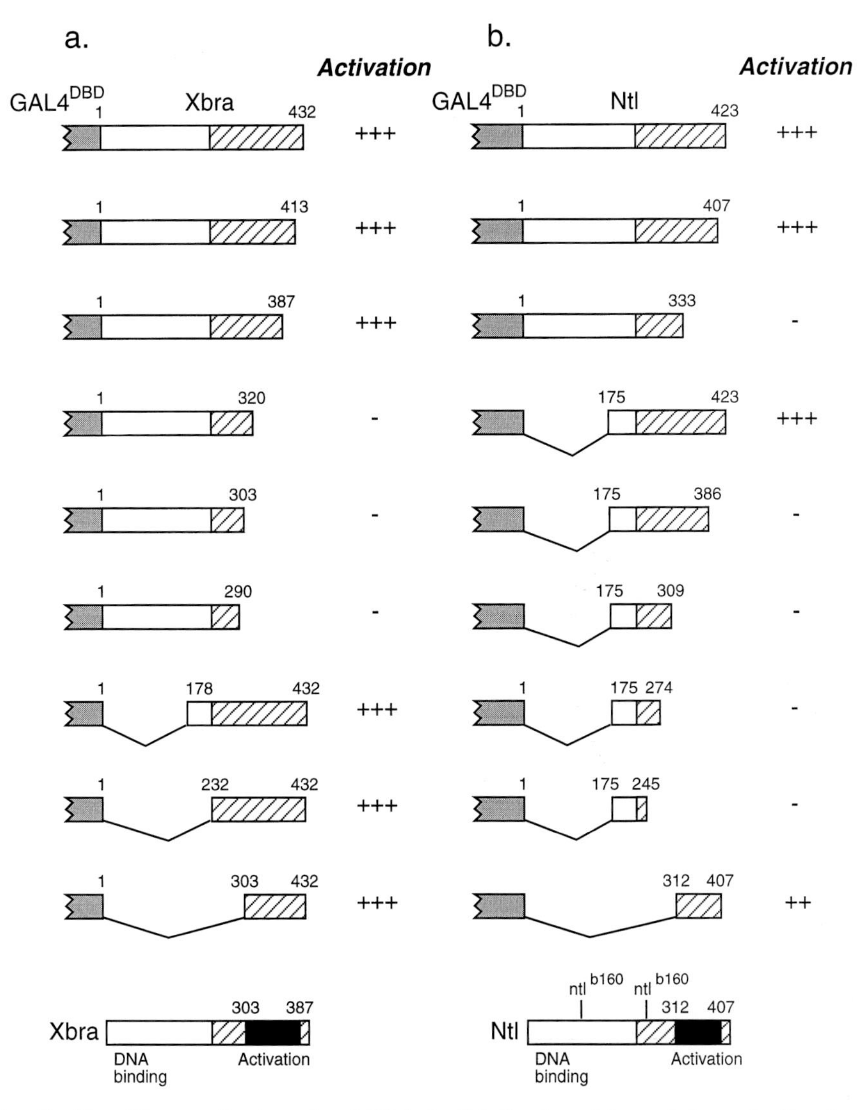

Figure 3. Mapping the Xbra (*Xenopus* Brachyury) and Ntl (zebrafish Brachyury) transcription activation domains. The GAL4 DNA-binding domain was fused to a 3′ deletion series of Xbra or Ntl generated by transposition mutagenesis or to 5′ truncations and deletions generated by conventional techniques. Transcription activation was demonstrated by induction of expression of a *$GAL4^{UAS}$-lacZ* reporter gene. (+++) Strong activation; (++) weaker activation; (−) no activation. (*a*) Results obtained with Xbra; (*b*) results obtained with Ntl. Interruptions of coding sequence due to the *ntl^{b160}* and *ntl^{b195}* mutations are indicated. In both *a* and *b*, the DNA-binding domain is shown unshaded.

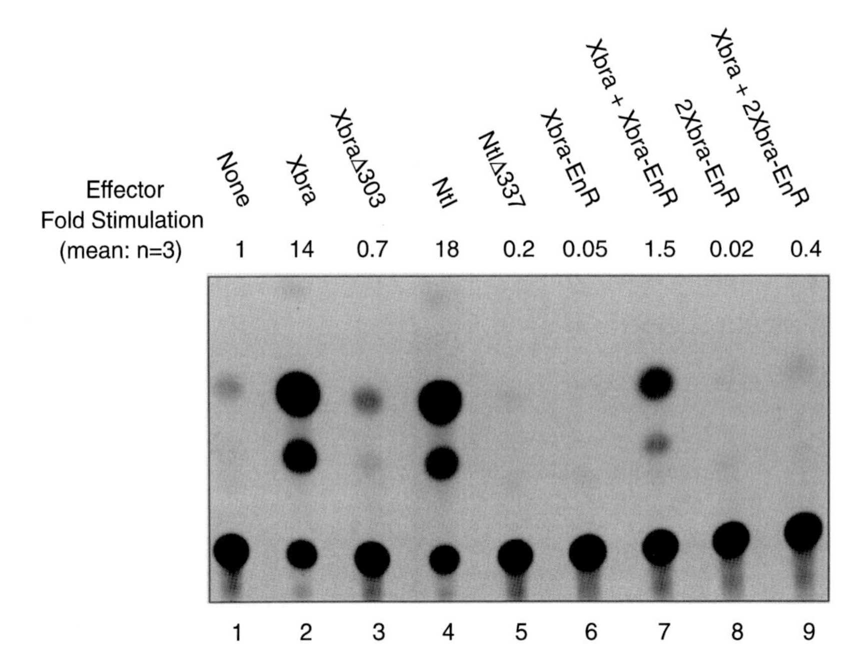

Figure 4. Transcription activation by Xbra and Ntl and repression of activation by a dominant-interfering Xbra-En^R construct. NIH-3T3 cells were lipofected with the indicated effector plasmids together with a CAT reporter plasmid carrying two copies of the Brachyury-binding site (Kispert and Herrmann 1993) upstream of a minimal promoter. Plasmids encoding wild-type Xbra (lane *2*) or Ntl (lane *4*) cause activation of transcription, whereas truncated versions of the two proteins have no effect (lanes *3* and *5*). A plasmid encoding Xbra-En^R does not activate transcription (lanes *6* and *8*), and it inhibits activation caused by Xbra in a dose-dependent manner, with complete inhibition achieved with a twofold excess of Xbra-En^R over Xbra (lanes *7* and *9*).

platelet-derived growth factor (Cochran et al. 1983). The identification of these genes has taken advantage of the fact that it is possible to treat cells with the factor in question for a short defined time and then to carry out a differential screen of some sort. It is not straightforward to perform this kind of experiment with *Xenopus* transcription factors because injected RNA is translated virtually immediately (Snape and Smith 1996) and it is therefore not possible to treat cells for the required short, defined time. To overcome this difficulty, we have developed a hormone-inducible version of Xbra, in which the open reading frame of Xbra is fused to the ligand-binding domain of the glucocorticoid receptor (GR). A similar approach has been used by Kolm and Sive (1995) and Gammill and Sive (1997) in their analyses of MyoD and Otx2. The rationale of this approach is that in the absence of ligand, the Xbra-GR fusion protein is sequestered by the heat shock apparatus of the cell and rendered nonfunctional. When dexamethasone is added, however, the fusion protein is liberated and is able to exert its effects. In principle, therefore, it should be possible to inject *Xenopus* eggs with RNA encoding Xbra-GR, to dissect animal caps at the late blastula stage, and then to treat them with dexamethasone for the desired short, defined time.

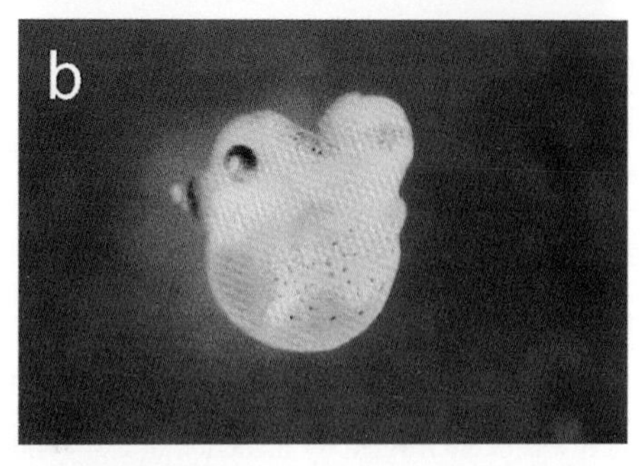

Figure 5. The phenotype of *Xenopus* embryos injected with RNA encoding Xbra-En. (*a*) Control embryo; (*b*) embryo expressing Xbra-En. Note that the head of the embryo is normal but that it lacks a tail.

Indirect Autoinduction of *Xbra*

We have tested this approach by asking whether the autoinduction of *Xbra* (Rao 1994) is direct or indirect; i.e., whether *Xbra* is a direct target of itself. Preliminary experiments demonstrated that an Xbra-GR construct (Fig. 6) induces mesoderm in animal cap explants in a hormone-dependent manner. RNA encoding Xbra-GR was injected into fertilized *Xenopus* eggs, and animal pole regions were dissected at the late blastula stage and left untreated or were treated with dexamethasone. Mesoderm

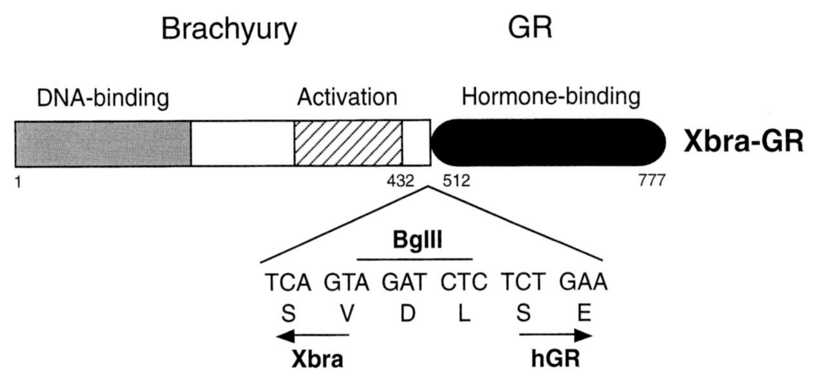

Figure 6. Xbra-GR. The hormone-binding domain of the hGR amino acids 512–777 is fused to the carboxyl terminus of Xbra (1–432) to generate Xbra-GR. This fusion replaces the stop codon of Xbra with two amino acids (DL) generated by the insertion of a *Bgl*II site. The DNA-binding and activation domains of Xbra are shown by hatched and shaded boxes, respectively (see Kispert and Herrmann 1993; Conlon et al. 1996). The hormone-binding domain of hGR is shown by a solid box.

formation was assessed by morphological, histological, and molecular criteria, and the results indicate clearly that mesoderm is only formed in response to Xbra-GR if dexamethasone is present (Tada et al. 1997). Further experiments (Fig. 7) demonstrated that animal pole cells are competent to respond to *Xbra* until late gastrula stages, beyond the time at which they can respond to the mesoderm-inducing factors FGF-2 and activin (Tada et al. 1997).

Four independent experiments then addressed the question of whether the autoinduction of *Xbra* is direct or indirect. In the first, the time required for autoinduction of

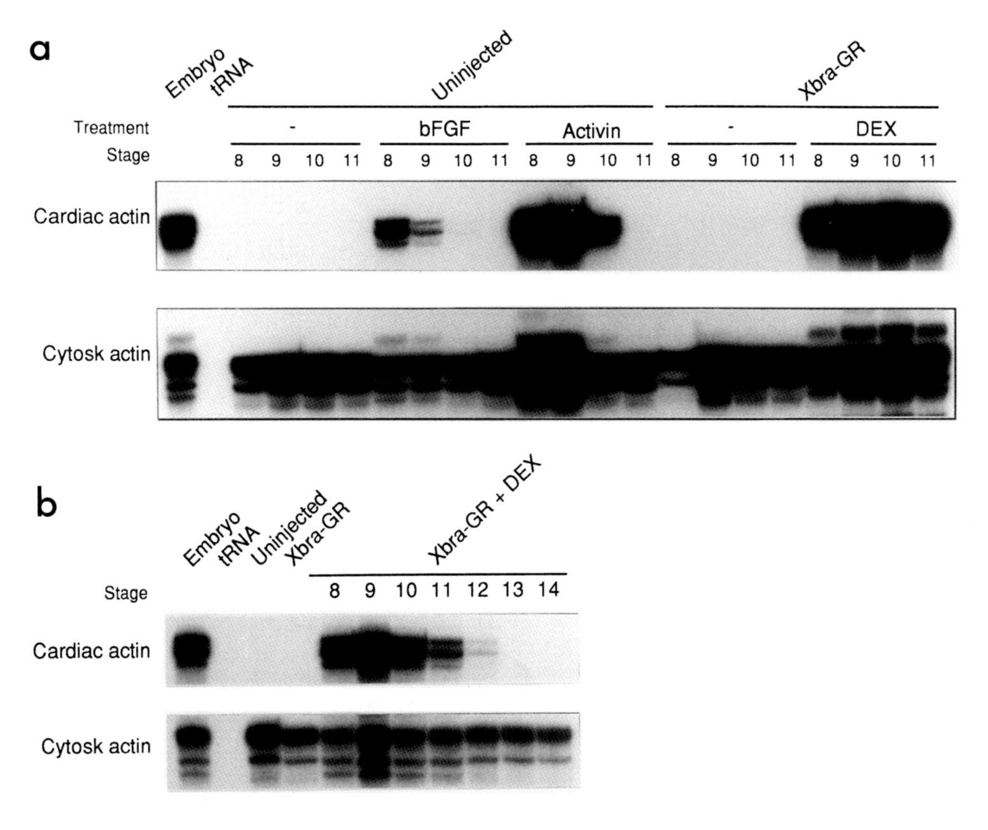

Figure 7. Competence of animal caps to respond to mesoderm inducers. (*a*) Embryos at the one-cell stage were left uninjected or injected with 100 pg of RNA encoding Xbra-GR. Animal caps were dissected at stage 8, 9, 10, or 11, then treated with 100 ng/ml bFGF, 8 units/ml activin, or 10^{-6} M DEX, and cultured to stage 23, when they were analyzed for expression of actin genes by RNase protection. Competence in response to either bFGF or activin is gradually lost by the early gastrula stage, whereas the response to Xbra persists. (*b*) Competence of animal caps to respond to Xbra. Embryos at the one-cell stage were injected with 100 pg of RNA encoding Xbra-GR. Animal caps were dissected at stage 8, treated with 10^{-6} M DEX at the indicated stages, and cultured to stage 23, when they were analyzed for expression of actin genes by RNase protection.

Xbra was compared with the time required for induction of *Xbra* by FGF-2 (which activates *Xbra* in an immediate-early fashion). Induction of *Xbra* in animal pole explants was first detectable 2 hours after exposure to FGF, whereas in animal caps derived from embryos injected with RNA encoding Xbra-GR, *Xbra* transcripts were not observed until 3 hours after addition of dexamethasone (Tada et al. 1997). This indicates that the autoinduction of *Xbra* is slow compared with induction by FGF-2. In the second experiment, animal pole explants derived from embryos injected with Xbra-GR RNA were treated with dexamethasone in the continuous presence or in the absence of cycloheximide. Dexamethasone-induced *Xbra* expression was dramatically reduced by cycloheximide (Tada et al. 1997), suggesting that autoinduction of *Xbra* requires new protein synthesis, and therefore is unlikely to be direct. The cycloheximide experiment was not conclusive, however, because *Xbra* induction by FGF was also slightly reduced, presumably because of inhibition of cell division of animal pole blastomeres by cycloheximide. In the third experiment, animal pole blastomeres were dissociated before being treated with dexamethasone; if *Xbra* autoinduction is direct, this treatment should not interfere with the process. We observe, however, that dexamethasone-induced *Xbra* expression was not seen in dissociated animal cap cells, whereas *Xbra* expression was able to be activated by FGF under the same conditions. Taken together with the result of the time course experiment, these last two results suggest that *Xbra* autoinduction requires protein synthesis and intercellular signals.

Finally, as discussed above, several lines of evidence suggest that *Xbra* and eFGF are involved in an autoregulatory loop in which *Xbra* induces eFGF expression, and eFGF is required for maintenance of *Xbra* expression (Isaacs et al. 1994; Schulte-Merker and Smith 1995). To examine the possibility that autoinduction of *Xbra* involves FGF signaling, RNA encoding a truncated FGF receptor was injected into *Xenopus* embryos along with Xbra-GR RNA. These experiments revealed that *Xbra* autoinduction does require FGF signaling. Analysis of *Xbra* expression at different times after addition of dexamethasone reveals that the truncated FGF receptor blocks the initial induction of *Xbra* and that the lack of expression is not due to inhibition of maintenance of expression (data not shown).

A Screen for Xbra Targets?

The above experiments suggest that one approach that may prove useful in isolating direct targets of *Xbra* will indeed be to inject *Xenopus* eggs with RNA encoding Xbra-GR, to dissect animal caps at the late blastula stage, and then to treat with dexamethasone for a short time. It should then be possible to isolate putative Xbra targets by differential screening and to test whether or not they are direct by studying the time course of activation, and by use of cycloheximide, dispersed cells, and truncated FGF receptors. Work of this sort is in progress.

DISCUSSION

The experiments described in this paper summarize the present state of knowledge concerning the transcriptional activation of *Xenopus Brachyury,* the function of *Brachyury* during development, and the identification of putative transcriptional targets. Even from the brief account presented here, it is obvious that there is much we do not know about *Brachyury*, and it is also obvious that things are not going to get any simpler. For example, it has recently become clear that *Brachyury* is a member of a family of proteins all of which share the T-box, the domain comprising the amino-terminal half of Brachyury itself. In *Xenopus,* there are two other T-box-containing proteins that, like *Xbra*, are expressed in the mesoderm. These are *eomesodermin* (Ryan et al. 1996) and another gene, related to mouse Tbx6 (Chapman et al. 1996), which has been isolated by no fewer than four groups and given four names: *Antipodean, VegT, Xombi*, and *Brat* (Lustig et al. 1996; Stennard et al. 1996; Zhang and King 1996; Horb and Thomsen 1997). It will be intriguing to discover whether these different transcription factors preferentially bind different sequences, to what extent they exert different effects, and to what extent their functions are redundant. Whatever the answers, it is clear that the T-box-containing genes will tell us a great deal about mesoderm formation in the early vertebrate embryo.

ACKNOWLEDGMENTS

This work is supported by the Medical Research Council. J.C.S. is an International Scholar of the Howard Hughes Medical Institute.

REFERENCES

Amaya E., Musci T.J., and Kirschner M.W. 1991. Expression of a dominant negative mutant of the FGF receptor disrupts mesoderm formation in *Xenopus* embryos. *Cell* **66:** 257.

———. 1993. FGF signaling in the early specification of mesoderm in *Xenopus. Development* **118:** 477.

Armes N.A. and J.C. Smith. 1997. The ALK-2 and ALK-4 activin receptors transduce distinct mesoderm-inducing signals during early *Xenopus* development but do not co-operate to establish thresholds. *Development* **124:** 3797.

Asashima M., Nakano H., Shimada K., Kinoshita K., Ishii K., Shiba H., and Ueno N. 1990. Mesodermal induction in early amphibian embryos by activin A (erythroid differentiation factor). *Roux's Arch. Dev. Biol.* **198:** 330.

Badiani P., Corbella P., Kioussis D., Marvel J., and Weston K. 1994. Dominant interfering alleles define a role for c-*myb* in T-cell development. *Genes Dev.* **8:** 770.

Baker J.C. and Harland R.M. 1996. A novel mesoderm inducer, Madr2, functions in the activin signal transduction pathway. *Genes Dev.* **10:** 1880.

Bollag R.J., Siegfried Z., Cebra-Thomas J.A., Garvey N., Davison E.M., and Silver L.M. 1994. An ancient family of embryonically expressed mouse genes sharing a conserved protein motif with the T locus. *Nat. Gene.* **7:** 383.

Chapman D.L., Agulnik I., Hancock S., Silver L.M., and Papaioannou V.E. 1996. *Tbx6*, a mouse T-box gene implicated in paraxial mesoderm formation at gastrulation. *Dev. Biol.* **180:** 534.

Cho K.W.Y., Blumberg B., Steinbeisser H., and De Robertis E.M. 1991. Molecular nature of Spemann's organizer: The role of the *Xenopus* homeobox gene *goosecoid*. *Cell* **67:** 1111.

Clements D., Taylor H.C., Herrmann B.G., and Stott D. 1996. Distinct regulatory control of the *Brachyury* gene in axial and non-axial mesoderm suggests separation of mesodermal lineages early in mouse gastrulation. *Mech. Dev.* **56:** 139.

Cochran B.H., Reffel A.C., and Stiles C.D. 1983. Molecular cloning of gene sequences regulated by platelet-derived growth factor. *Cell* **33:** 939.

Conlon F.L., Sedgwick S.G., Weston K.M., and Smith J.C. 1996. Inhibition of *Xbra* transcription activation causes defects in mesodermal patterning and reveals autoregulation of *Xbra* in dorsal mesoderm. *Development* **122:** 2427.

Cooke J., Smith J.C., Smith E.J., and Yagoob M. 1987. The organization of mesodermal pattern in *Xenopus laevis:* Experiments using a *Xenopus* mesoderm-inducing factor. *Development* **101:** 893.

Cunliffe V. and Smith J.C. 1992. Ectopic mesoderm formation in *Xenopus* embryos caused by widespread expression of a *Brachyury* homologue. *Nature* **358:** 427.

———. 1994. Specification of mesodermal pattern in *Xenopus laevis* by interactions between *Brachyury, noggin* and *Xwnt-8*. *EMBO J.* **13:** 349.

Dale L., Matthews G., and Colman A. 1993. Secretion and mesoderm-inducing activity of the TGF-β related domain of *Xenopus* Vg1. *EMBO J.* **12:** 4471.

Dalton S. and Treisman R. 1992. Characterization of SAP-1, a protein recruited by serum response factor to the c-*fos* serum response element. *Cell* **68:** 597.

Dyson S. and Gurdon J.B. 1996. Activin signalling has a necessary function in *Xenopus* early development. *Curr. Biol.* **7:** 81.

Evan G.I., Lewis G.K., Ramsay G., and Bishop J.M. 1985. Isolation of monoclonal antibodies specific for the human c-*myc* oncogene product. *Mol. Cell. Biol.* **75:** 3610.

Gammill L. and Sive H.L. 1997. Identification of *otx2* target genes and restrictions in ectodermal competence during *Xenopus* cement gland formation. *Development* **124:** 471.

Gotoh Y., Masuyama N., Suzuki A., Ueno N., and Nishida E. 1995. Involvement of the MAP kinase cascade in *Xenopus* mesoderm induction. *EMBO J.* **14:** 2491.

Graff J.M., Bansal A., and Melton D.A. 1996. *Xenopus* Mad proteins transduce distinct subsets of signals for the TGFβ superfamily. *Cell* **85:** 479.

Gurdon J.B. 1988. A community effect in animal development. *Nature* **336:** 772.

Halpern M.E., Ho R.K., Walker C., and Kimmel C.B. 1993. Induction of muscle pioneers and floor plate is distinguished by the zebrafish *no tail* mutation. *Cell* **75:** 99.

Han K. and Manley J.L. 1993. Functional domains of the *Drosophila* Engrailed protein. *EMBO J.* **12:** 2723.

Herrmann B.G. 1991. Expression pattern of the *Brachyury* gene in whole-mount T^{Wis}/T^{Wis} mutant embryos. *Development* **113:** 913.

———. 1995. The mouse *Brachyury* (*T*) gene. *Semin. Dev. Biol.* **6:** 385.

Herrmann B.G., Labeit S., Poustka A., King T.R., and Lehrach H. 1990. Cloning of the *T* gene required in mesoderm formation in the mouse. *Nature* **343:** 617.

Hill C.S., Marais R., John S., Wynne J., Dalton S., and Treisman R. 1993. Functional analysis of growth factor-responsive transcription factor complex. *Cell* **73:** 395.

Horb M.E. and Thomsen G.H. 1997. A vegetally-localized T-box transcription factor in *Xenopus* eggs specifies mesoderm and endoderm and is essential for embryonic mesoderm formation. *Development* **124:** 1689.

Isaacs H.V., Pownall M.E., and Slack J.M.W. 1994. eFGF regulates *Xbra* expression during *Xenopus* gastrulation. *EMBO J.* **13:** 4469.

———. 1995. eFGF is expressed in the dorsal mid-line of *Xenopus laevis*. *Int. J. Dev. Biol.* **39:** 575.

Isaacs H.V., Tannahill D., and Slack J.M.W. 1992. Expression of a novel FGF in the *Xenopus* embryo. A new candidate inducing factor for mesoderm formation and anteroposterior specification. *Development* **114:** 711.

Jaynes J.B. and O'Farrell P.H. 1991. Active repression of transcription by the Engrailed homeodomain protein. *EMBO J.* **10:** 1427.

Jones C.M., Armes N., and Smith J.C. 1996. Signalling by TGF-β family members: Short-range effects of Xnr-2 and BMP-4 contrast with the long-range effects of activin. *Curr. Biol.* **6:** 1468.

Jones C.M., Kuehn M.R., Hogan B.L.M., Smith J.C., and Wright C.V.E. 1995. Nodal-related signals induce axial mesoderm and dorsalize mesoderm during gastrulation. *Development* **121:** 3651.

Kavka A.I. and Green J.B.A. 1997. Tales of tails: *Brachyury* and the T-box genes. *Biochim. Biophys. Acta* (in press).

Kessler D.S. and Melton D.A. 1995. Induction of dorsal mesoderm by soluble, mature Vg1 protein. *Development* **121:** 2155.

Kim J., Johnson K., Chen H.J., Carroll S., and Laughon A. 1997. *Drosophila* Mad binds to DNA and directly mediates activation of *vestigial* by Decapentaplegic. *Nature* **388:** 304.

Kimelman D. and Kirschner M. 1987. Synergistic induction of mesoderm by FGF and TGF-β and the identification of an mRNA coding for FGF in the early *Xenopus* embryo. *Cell* **51:** 869.

Kintner C.R. and Brockes J.P. 1984. Monoclonal antibodies recognise blastemal cells derived from differentiating muscle in newt limb regeneration. *Nature* **308:** 67.

Kispert A, and Herrmann B.G. 1993. The *Brachyury* gene encodes a novel DNA binding protein. *EMBO J.* **12:** 3211.

———. 1994. Immunohistochemical analysis of the *Brachyury* protein in wild-type and mutant mouse embryos. *Dev. Biol.* **161:** 179.

Kispert A., Korschorz B., and Herrmann B.G. 1995a. The T protein encoded by *Brachyury* is a tissue-specific transcription factor. *EMBO J.* **14:** 4763.

Kispert A., Ortner H., Cooke J., and Herrmann B.G. 1995b. The chick *Brachyury* gene: Developmental expression pattern and response to axial induction by localized activin. *Dev. Biol.* **168:** 406.

Knezevic V., De Santo R., and Mackem S. 1997. Two novel chick T-box genes related to mouse *Brachyury* are expressed in different, non-overlapping mesodermal domians during gastrulation. *Development* **124:** 411.

Kolm P. and Sive H.L. 1995. Efficient hormone-inducible protein function in *Xenopus laevis*. *Dev. Biol.* **171:** 262.

Kroll K.L. and Amaya E. 1996. Transgenic *Xenopus* embryos from sperm nuclear transplantations reveal FGF signalling requirements during gastrulation. *Development* **122:** 3173.

LaBonne C., Burke B., and Whitman M. 1995. Role of MAP kinase in mesoderm induction and axial patterning during *Xenopus* development. *Development* **121:** 1475.

Luckow B. and Schutz G. 1987. CAT constructions with multiple unique restriction sites for the functional analysis of eukaryotic promoters and regulatory elements. *Nucleic Acids Res.* **15:** 5490.

Lustig K.D., Kroll K.L., Sun E.E., and Kirschner M.W. 1996. Expression cloning of a *Xenopus T*-related gene (*Xombi*) involved in mesodermal patterning and blastopore lip formation. *Development* **122:** 4001.

Marais R.M., Light Y., Paterson H.F., and Marshall C.J. 1995. Ras recruits Raf-1 to the plasma membrane for activation by tyrosine phosphorylation. *EMBO J.* **14:** 3136.

Marshall C.J. 1995. Specificity of receptor tyrosine kinase signalling: Transient versus sustained extracellular signal-regulated kinase activation. *Cell* **80:** 179.

Mohun T.J., Brennan S., Dathan N., Fairman S., and Gurdon J.B. 1984. Cell type-specific activation of actin genes in the early amphibian embryo. *Nature* **311:** 716.

Morgan B.A., Conlon F.L., Manzanares M., Millar J.B.A., Kanuga N., Sharpe J., Krumlauf R., Smith J.C., and Sedgwick S.G. 1996. Transposon tools for recombinant DNA manipulation: Characterization of transcriptional regulators from yeast, *Xenopus* and mouse. *Proc. Natl. Acad. Sci.* **93:** 2801.

Nieuwkoop P.D. 1969. The formation of mesoderm in Urodelean amphibians. I. Induction by the endoderm. *Wilhelm Roux Arch. Entwicklungsmech. Org.* **162:** 341.

Nieuwkoop P.D. and Faber J. 1975. Normal table of *Xenopus laevis* (Daudin). North-Holland, Amsterdam.

O'Reilly M.-A.J., Smith J.C., and Cunliffe V. 1995. Patterning of the mesoderm in *Xenopus:* Dose-dependent and synergistic effects of *Brachyury* and *Pintallavis. Development* **121:** 1351.

Rao Y. 1994. Conversion of a mesodermalizing molecule, the *Xenopus Brachyury* gene, into a neuralizing factor. *Genes Dev.* **8:** 939.

Rebagliati M.R., Weeks D.L., Harvey R.P., and Melton D.A. 1985. Identification and cloning of localized maternal RNAs from *Xenopus* eggs. *Cell* **42:** 769.

Ruiz i Altaba A. and Jessel T.M. 1992. *Pintallavis*, a gene expressed in the organizer and midline cells of frog embryos: Involvement in the development of the neural axis. *Development* **116:** 81.

Ruiz i Altaba A. and Melton D.A. 1989. Bimodal and graded expression of the *Xenopus* homeobox gene *Xhox3* during embryonic development. *Development* **106:** 173.

Ryan K., Garrett N., Mitchell A., and Gurdon J.B. 1996. Eomesodermin, a key early gene in *Xenopus* mesoderm differentiation. *Cell* **87:** 989.

Sargent M.G. and Bennett M.F. 1990. Identification in *Xenopus* of a structural homologue of the *Drosophila* gene *Snail. Development* **109:** 967.

Schmidt C., Wilson V., Stott D., and Beddington R.S.P. 1997. *T* promoter activity in the absence of functional T protein during axis formation and elongation in the mouse. *Dev. Biol.* (in press).

Schulte-Merker S. and Smith J.C. 1995. Mesoderm formation in response to *Brachyury* requires FGF signalling. *Curr. Biol.* **5:** 62.

Schulte-Merker S., Ho R.K., Herrmann B.G., and Nüsslein-Volhard C. 1992. The protein product of the zebrafish homologue of the mouse *T* gene is expressed in nuclei of the germ ring and the notochord of the early embryo. *Development* **116:** 1021.

Schulte-Merker S., van Eeden F.M., Halpern M.E., Kimmel C.B., and Nüsslein-Volhard C. 1994. *no tail* (*ntl*) is the zebrafish homologue of the mouse *T* (*Brachyury*) gene. *Development* **120:** 1009.

Sedgwick S.G. and Morgan C.B.A. 1994. Locating, DNA sequencing, and disrupting yeast genes using tagged Tn1000. *Methods Mol. Genet.* **3:** 131.

Sive H.L. 1993. The frog prince-ss: A molecular formula for dorsoventral patterning in *Xenopus. Genes Dev.* **7:** 1.

Slack J.M.W. 1984. Regional biosynthetic markers in the early amphibian embryo. *J. Embryol. Exp. Morphol.* **80:** 289.

———. 1994. Inducing factors in *Xenopus* early embryos. *Curr. Biol.* **4:** 116–126.

Slack J.M., Darlington B.G., Heath J.K., and Godsave S.F. 1987. Mesoderm induction in early *Xenopus* embryos by heparin-binding growth factors. *Nature* **326:** 197.

Slack J.M., Darlington B.G., Gillespie L.L., Godsave S.F., Isaacs H.V., and Paterno G.D. 1989. The role of fibroblast growth factor in early *Xenopus* development. *Development* (suppl.) **107:** 141.

Smith J.C. 1993. Purifying and assaying mesoderm-inducing factors from vertebrate embryos. In *Cellular interactions in development*—A practical approach (ed. D. Hartley), p. 181. Oxford University Press, United Kingdom.

———. 1995. Mesoderm-inducing factors and mesodermal patterning. *Curr. Opin. Cell Biol.* **7:** 856.

———. 1997. *Brachyury* and the T-box genes. *Curr. Opin. Genet Dev.* **7:** 474.

Smith J.C. and Slack J.M.W. 1983. Dorsalization and neural induction: Properties of the organizer in *Xenopus laevis. J. Embryol. Exp. Morphol.* **78:** 299.

Smith J.C., Price B.M.J., Van Nimmen K., and Huylebroeck D. 1990. Identification of a potent *Xenopus* mesoderm-inducing factor as a homologue of activin A. *Nature* **345:** 732.

Smith J.C., Price B.M.J., Green J.B.A., Weigel D., and Herrmann B.G. 1991. Expression of a *Xenopus* homolog of *Brachyury* (*T*) is an immediate-early response to mesoderm induction. *Cell* **67:** 79.

Smith W.C. and Harland R.M. 1992. Expression cloning of noggin, a new dorsalizing factor localized to the Spemann organizer in *Xenopus* embryos. *Cell* **70:** 829.

Snape A.M. and Smith J.C. 1996. Regulation of embryonic cell division by a *Xenopus* gastrula-specific protein kinase. *EMBO J.* **15:** 4556.

Song J. and Slack J.M.W. 1994. Spatial and temporal expression of basic fibroblast growth factor (FGF-2) mRNA and protein in early *Xenopus* development. *Mech. Dev.* **48:** 141.

Stennard F., Carnao G., and Gurdon J.B. 1996. The *Xenopus* T-box gene, *Antipodean*, encodes a vegetally localised maternal mRNA and can trigger mesoderm formation. *Development* **122:** 4179.

Tada M., O'Reilly M.-A.J., and Smith J.C. 1997. Analysis of competence and of *Brachyury* autoinduction by use of hormone-inducible *Xbra. Development* **124:** 2225.

Thomsen G.H. and Melton D.A. 1993. Processed Vg1 protein is an axial mesoderm inducer in *Xenopus. Cell* **74:** 433.

Thomsen G., Woolf T., Whitman M., Sokol S., Vaughan J., Vale W., and Melton D.A. 1990. Activins are expressed early in *Xenopus* embryogenesis and can induce axial mesoderm and anterior structures. *Cell* **63:** 485.

Thuringer F. and Bienz M. 1993. Indirect autoregulation of a homeotic *Drosophila* gene mediated by extracellular signaling. *Proc. Natl. Acad. Sci.* **90:** 3899.

Thuringer F., Cohen S.M., and Bienz M. 1993. Dissection of an indirect autoregulatory response of a homeotic *Drosophila* gene. *EMBO J.* **12:** 2419.

Umbhauer M., Marshall C.J., Mason C.S., Old R.W., and Smith J.C. 1995. Mesoderm induction in *Xenopus* caused by activation of MAP kinase. *Nature* **376:** 58.

Wilkinson D.G., Bhatt S., and Herrmann B.G. 1990. Expression pattern of the mouse *T* gene and its role in mesoderm formation. *Nature* **343:** 657.

Willison K. 1990. The mouse *Brachyury* gene and mesoderm formation. *Trends Genet.* **6:** 104.

Zhang J. and King M.L. 1996. *Xenopus* VegT RNA is localized to the vegetal cortex during oogenesis and encodes a novel T-box transcription factor involved in mesodermal patterning. *Development* **122:** 4119.

Zimmerman L.B., De Jesus Escobar J.M., and Harland R.M. 1996. The Spermann organizer signal noggin binds and inactivates bone morphogenetic protein 4. *Cell* **86:** 599.

Signal Transduction Downstream from RAS in *Drosophila*

G.M. RUBIN, H.C. CHANG, F. KARIM, T. LAVERTY, N.R. MICHAUD,[1] D.K. MORRISON,[1] I. REBAY, A. TANG, M. THERRIEN, AND D.A. WASSARMAN

Howard Hughes Medical Institute and Department of Molecular and Cell Biology, University of California, Berkeley, California 94720-3200; [1]Molecular Mechanisms of Carcinogenesis Laboratory, National Cancer Institute, Frederick Cancer Research and Development Center, Frederick, Maryland 21701

RAS1 activation has a pivotal role in signal transmission from the sevenless (Sev) receptor tyrosine kinase during *Drosophila* eye development, and activating mutations in *Ras1* can bypass the requirement for Sev activation (Simon et al. 1991; Fortini et al. 1992). Since the *Drosophila* compound eye is dispensable for viability and fertility, most deleterious effects associated with the constitutively active *Ras1* gene, whose widespread expression would be lethal, can be avoided by expressing it under the control of the sev-enhancer/promoter. The *sev-Ras1*^{V12} transgene triggers the presumptive R7 cell to adopt a photoreceptor fate and transforms many of the nonneuronal cone cells into supernumerary R7 cells. The production of ectopic R7 cells by *sev-Ras1*^{V12} disrupts the exterior eye morphology and causes it to become rough in appearance. We used the sev-Ras1^{V12} rough eye phenotype to screen for dominant suppressors and enhancers. The premise of this dominant suppressor/enhancer screen is that, in the sensitized background of *sev-Ras1*^{V12}, a twofold reduction in the dose of a downstream gene (i.e., by mutating one copy of two present in the diploid genome) will alter signaling efficiency and thereby visibly modify the rough eye phenotype. This and related screens illustrate how genetics in *Drosophila*, combined with biochemical and molecular analyses, can be used to place individual genes within pathways and networks involved in a complex cell fate decision.

Ras-mediated signal transduction pathways function in the control of both cell proliferation and cellular differentiation (for review, see Lowy and Willumsen 1993; Moodie and Wolfman 1994). Recent genetic and biochemical studies have demonstrated that Ras GTPase activity is tightly controlled by growth factor activation of receptor tyrosine kinases (RTKs; for review, see Schlessinger 1993; van der Geer et al. 1994). Elucidation of the steps by which a signal is transmitted from an RTK to Ras, and then from Ras to the nucleus, is critical to understanding Ras' diverse regulatory roles.

The developing *Drosophila* retina provides an experimental system where the molecular mechanisms of short-range inductive interactions can be studied at the level of individual cells (for review, see Wolff and Ready 1993). The adult eye is made up of a simple array of approximately 800 20-cell units, called ommatidia. Each ommatidium contains 8 photoreceptor cells, R1–R8, as well as 4 lens-secreting cone cells and 8 other accessory cells. The fates of the cells that make up the ommatidia appear to be governed by the specific combination of signals received by each cell from its immediate neighbors. One of our long-term goals has been to gain an understanding of how such signals are generated, sensed, and responded to by cells.

The R7 photoreceptor is the last of the 8 photoreceptors to be recruited to the developing ommatidium. The presumptive R7 cell appears to face a simple choice between two alternative cell fates: It will develop into an R7 photoreceptor if it receives a signal that is initiated by activation of the sevenless protein tyrosine kinase receptor; otherwise it will adopt a nonneuronal cone cell fate (for review, see Zipursky and Rubin 1994). Neuronal differentiation of the other photoreceptors also requires RAS1 signaling, although the precise sequence of events has been less extensively characterized then for the R7 cell.

THE GENETIC SCREEN

Genes encoding general factors involved in Ras1 signal transduction are expected to function during multiple signaling events at various stages of development. Therefore, mutations in these genes are expected to be homozygous lethal and would not be recovered in a conventional F2 eye phenotype screen. To circumvent this and identify new components of the Ras1 signal transduction pathway, we (Karim et al. 1996) carried out an F1 screen for dominant suppressors and enhancers of an activated Ras1 allele (*Ras1*^{V12}). This activated Ras1 allele was introduced into flies by P-element-mediated transformation and expressed under control of the sev-enhancer/promoter (*sev-Ras1*^{V12}). *sev-Ras1*^{V12} transforms nonneuronal cone cells into supernumerary R7 photoreceptor cells, even in the absence of Sev activity (Fortini et al. 1992). These extra R7 cells disrupt the regular ommatidial array and cause a rough eye phenotype. The degree of eye roughness is dose-sensitive; i.e., flies carrying two copies of the *sev-Ras1*^{V12} transgene have significantly rougher eyes than flies that carry only one copy.

The correlation between the degree of eye roughness

and the strength of RAS1 signaling suggests that a twofold reduction in the dose of a downstream gene (by mutating one copy of the two present in the diploid genome) might alter the signaling strength and modify the rough eye phenotype. For example, mutating one copy of a positively acting downstream gene should reduce signaling strength sufficiently to suppress the sev-Ras1^{V12} rough eye, whereas mutating one copy of a negatively acting downstream gene is expected to increase signaling strength and therefore enhance the sev-Ras1^{V12} rough eye. The ability to detect a phenotype with these mutations as heterozygotes is critical, since many of these genes will be required at earlier stages of development, and the loss of both copies (in a homozygous mutant animal) is expected to cause lethality.

On the basis of these premises, we screened approximately 250,000 ethylmethanesulfonate (EMS)-mutagenized flies and approximately 600,000 X-ray-mutagenized flies. In total, the 282 dominant suppressors and 577 dominant enhancers we isolated fell into about 30 lethal complementation groups. Several suppressors and a large number of enhancers did not fall into any lethal complementation group. Interestingly, some of these are homozygous viable with no apparent phenotype, yet they interact genetically with several components of the Ras signaling pathway in a number of tissues, suggesting that these mutant genes may have broad roles in signaling, even though they are themselves dispensable for viability.

There are many advantages to this type of dominant suppressor/enhancer screen. First, large numbers of mutagenized genomes can be screened, since one is scoring the phenotype of F_1 individuals. Second, because many of these dominant suppressors and enhancers are homozygous cell-lethal or homozygous-viable with no phenotype, they would not have been identified in screens based on recessive eye phenotypes, either in the whole organism or in homozygous somatic clones. There are also certain limitations to this approach. There are several reasons why we might not expect to isolate mutations in all critical signaling genes. As noted earlier, the premise of the screen is that a 50% reduction in the dose of a critical gene might increase or decrease signaling efficiency enough to alter the rough eye phenotype. However, if a particular signaling protein is in vast excess, then a greater than 50% reduction may be required to visibly alter the rough eye phenotype. In this case, only dominant-negative alleles, which are relatively rare, would modify the rough eye sufficiently to be detected. An example of this is provided by 14-3-3, in which we isolated only two mutations as compared to over 100 alleles of the gene encoding MAPK.

The overall effectiveness of this approach is demonstrated by the fact that we isolated mutations in over a dozen critical signaling genes, in addition to those in the conserved RAS1/MAPK cascade (Fig. 1). Some of these genes appear to be general factors that function in multiple Ras1 signaling events, whereas others may be cell-specific. We present several examples to illustrate the range of further genetic, molecular, and biochemical analyses of these loci that are now in progress.

SOME EXAMPLES

14-3-3e Positively Regulates Ras Signaling

We (Chang and Rubin 1997) isolated two mutations in the gene encoding a *Drosophila* 14-3-3e as suppressors of the RAS1^{V12} rough eye phenotype. By comparing the properties of these mutations to a null allele, which we generated by imprecise excision of a P-element inserted within the gene, we were able to establish that the mutations we isolated in the screen were dominant-negative. Homozygotes for the null allele show suppression of the activated RAS1 phenotype, but heterozygotes do not. Removal of 14-3-3e function impairs RAS1 signaling efficiency, indicating that 14-3-3e functions positively in RAS1-mediated signaling. We found that 14-3-3e appears to function in multiple RTK pathways, suggesting that 14-3-3e is a general component of the RAS1 signaling cascade. Sequence analysis of the two dominant negative alleles we isolated, as well as one allele from a screen carried out for suppressors of an activated Raf allele (Dickson et al. 1996), defined two regions of 14-3-3e that participate in RAS1 signaling. We also found that the 14-3-3 e and z, two 14-3-3 protein family members, are partially redundant for RAS1 signaling in photoreceptor formation and in animal viability. Our genetic data suggest that 14-3-3e functions downstream from RAF, consistent with its proposed role as a modulator of activity or specificity of RAF kinase (Morrison 1994), or as a scaffold for the assembly of signaling complexes (Brasselman and McCormick 1995; Muslin et al. 1996).

KSR, a New Component of the Ras Signaling Cascade

Mutations in the *ksr* gene were isolated as suppressors of activated RAS1 (Therrien et al. 1995). We found that

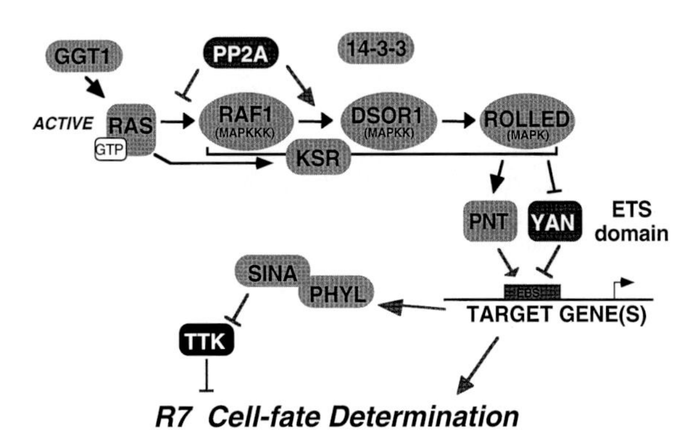

Figure 1. Summary of the signal transduction pathway downstream from RAS1 that triggers neural development in the R7 photoreceptor as revealed by the genetic analysis described in the text. RAS activates the RAF/MAPKK/MAPK cascade of protein kinases, which leads to modification of the activity of the Ets-domain containing transcription factors Yan and Pointed (Pnt) by direct phosphorylation. KSR regulates signal transduction through the RAF/MAPKK/MAPK cascade. Transcription of Phyl is induced, which, acting together with Sina, leads to degradation of the transcriptional repressor Ttk^{88}. The phosphatase PP2A has been shown to regulate the activity of the pathway (Wassarman et al. 1995) and GGT1 is required for the prenylation of the RAS protein (Therrien et al. 1995). See text for additional references and explanation.

ksr mutations also suppressed activated Torso receptor, but had no effect on the phenotype produced by activated RAF. Given its expression in multiple cell types throughout embryogenesis, its requirement for cell proliferation and its involvement in two independent RTK pathways, KSR is likely to be a general component of RTK-dependent signal transduction pathways. A genetic screen in *Caenorhabditis elegans* designed to isolate mutations that suppress a multivulva phenotype caused by an activating Ras mutation allowed two groups to independently identify a gene, *ksr-1*, that encodes a kinase similar to KSR (Kornfeld et al. 1995; Sundaram and Han 1995). The identification of a KSR family member as a positively acting component of Ras-dependent pathways in both *Drosophila* and *C. elegans*, as well as our isolation of mammalian KSR homologs, argues very strongly (see Sidow and Thomas 1994) in favor of a general and evolutionarily conserved role for this class of kinases in RAS signaling.

To further address the role of KSR in RAS signaling, we (Therrien et al. 1996) initiated a functional characterization of a murine KSR homolog (mKSR1). We found that mKSR1 cooperates with RAS to enhance its biological effects and that the mKSR1 function itself is regulated by RAS. In addition, analysis of a series of mKSR1 deletion mutants revealed unexpected behaviors for both the amino-terminal and the carboxy-terminal domains. Expression of the amino-terminal domain cooperated with oncogenic RAS, although less efficiently than full-length mKSR1, to promote *Xenopus* oocyte maturation and cellular transformation. In contrast, expression of the isolated carboxy-terminal kinase domain efficiently blocked RAS-mediated signaling and prevented MEK1 and MAPK activation but did not inhibit RAF-1 activation. Finally, we demonstrated an in vivo association between mKSR1 and RAF-1 that is RAS-dependent and occurs at the plasma membrane. Together, these findings suggest that mKSR1 is a component of the MAPK cascade whose function is to regulate signal propagation between RAF-1, MEK1, and MAPK.

Although several models can be envisioned to explain our findings, one model, which accounts for the behavior of both the mKSR1 amino-terminal and carboxy-terminal domains, is that mKSR1 may function to regulate the formation and subsequent disruption of a RAF-1/MEK1/MAPK complex. The formation of such a complex would be required to avoid inappropriate signaling transmission due to promiscuous enzyme/substrate interactions. The occurrence of such complexes has been demonstrated in *Saccharomyces cerevisae*, where the Ste5 protein has been shown to function as a scaffolding molecule bridging the three kinases comprising a MAPK module (Choi et al. 1994). By analogy, mKSR1 alone or together with RAF-1 may form a complex that facilitates signal propagation among RAF-1, MEK1, and MAPK. On the basis of this model, the amino-terminal domain of mKSR1 would serve to stabilize a RAF-1/MEK1/MAPK complex and perhaps provide a docking site for additional molecules required for the activation of this module. In concert, activation of the kinase domain of mKSR1 would generate the signal necessary for the disruption of the complex, thereby allowing MAPK to relocalize with its substrates. How the activity of mKSR1 is regulated is unknown, but an attractive possibility is that the signal activating the mKSR1 kinase domain may be provided by the newly activated MAPK itself. Evidence supporting this possibility comes from the observation that three consensus MAPK phosphorylation sites are located within the mKSR1 amino-terminal domain and that this domain is an excellent substrate for MAPK in vitro. In addition, we detected a strong interaction between the amino-terminal domain of Dm KSR and *Drosophila* MAPK using a yeast two-hybrid interaction assay. Therefore, once activated, MAPK could phosphorylate the amino-terminal domain of mKSR1, relieving its negative regulatory effect on the kinase domain. This step in turn would activate the catalytic function of the kinase domain, resulting in the disruption of the RAF-1/MEK1/MAPK complex. If this model or a variation of it is valid, then separation of the amino- and carboxy-terminal domains would uncouple the normal regulation of mKSR thereby allowing the kinase domain to phosphorylate its target at an inappropriate time. Therefore, the dominant-negative effect of isolated carboxy-terminal kinase domain could be explained by the constitutive activity of the kinase domain disrupting or preventing the formation of the complex. To test this model, we are attempting to purify and to identify the components of the RAS/RAF/KSR complex. Among its constituents should be the physiological substrate of KSR. We have also carried out additional genetic screens in *Drosophila* to isolate enhancers and suppressors of the rough eye phenotype produced by expression of the KSR kinase domain under the control of the sevenless promoter. Molecular and phenotypic analyses of the genes identified by these mutations are under way.

Genes Acting Downstream from MAPK

Cell fate determination is a complex process, governed by the combinatorial effects of positive and negative signaling events. Ultimately, downstream transcription factors — whose abundance and activities are modulated by these signals — mediate the specific patterns of gene expression that lead to a particular cell fate. Several nuclear components that act downstream from MAPK in the presumptive R7 cell have been identified (for review, see Dickson 1995). Efforts to identify direct targets of MAPK have implicated several nuclear transcription factors. In vertebrate systems, activated MAPK translocates from the cytoplasm to the nucleus where it phosphorylates several targets. In *Drosophila*, recent work has revealed that activated MAPK modulates the activities of two Ets-related proteins, Pointed and Yan (Brunner et al. 1994; O'Neill et al. 1994). Genetically, *pointed* appears to be a positive regulator of photoreceptor development, whereas *yan* appears to be a negative regulator, acting as an antagonist of the RAS1 proneural signal (Lai and Rubin 1992; O'Neill et al. 1994). Experiments assaying the activities of these proteins in transfected S2 cells have shown that Pointed is a transcriptional activator whose activity is stimulated by the RAS1/MAPK pathway, whereas Yan functions as a transcriptional repressor negatively regulated by the RAS1/MAPK pathway (O'Neill et al. 1994). Yan contains eight putative MAPK phosphorylation consensus sequences (Clark-Lewis et al. 1991), whereas Pointed contains one such sequence (Lai and Rubin 1992; Klambt 1993); both proteins can be directly phosphorylated in vitro by MAPK (Brunner et al. 1994).

As a negative regulator of a signal transduction pathway, Yan offers a unique opportunity to investigate an intriguing aspect of development, namely, how cells are inhibited from responding inappropriately to the complex array of external signals to which they are exposed. Antibodies raised against Yan show that it is expressed prominently in basally located nuclei of undifferentiated cells in the larval imaginal eye disc (Lai and Rubin 1992). However, as cells begin to differentiate and their nuclei migrate to a more apical position within the disc epithelium, Yan expression is abruptly down-regulated (Lai and Rubin 1992).

On the basis of these observations, it has been proposed that Yan performs its role as a negative regulator of photoreceptor development by maintaining cells in an undifferentiated state (Lai and Rubin 1992). According to this model, the Yan-mediated block to differentiation must be removed in order for a cell to respond to specific developmental cues. To test the model, we (Rebay and Rubin 1995) used in vitro mutagenesis to alter the eight MAPK phosphorylation consensus sequences in Yan and then examined the effects of overexpressing these mutants both in vivo and in cell culture assays. Overexpression of a constitutively activated form of Yan (Yan-Act), in which all eight MAPK sites have been mutated to a nonphosphorylatable form, blocks differentiation of multiple cell types during development. These results indicated that phosphorylation of Yan by MAPK affects the stability and subcellular localization of the protein, resulting in rapid down-regulation of Yan activity, and that Yan not only functions as a negative regulator in R7 photoreceptor development, but also acts as a much more general inhibitor of cell fate specification at multiple points during *Drosophila* development.

The expression of Yan-Act results in a dominant rough eye phenotype and has provided the starting point for a genetic interaction screen designed to identify additional downstream components of the RAS1 signaling pathway. Approximately 190,000 flies were screened (~80,000 with EMS and 110,000 with X-rays), yielding more than 300 modifiers, both enhancers and suppressers that fell into 22 complementation groups. The genes corresponding to several of these groups were quickly identified by crossing to other elements in the pathway that we knew interacted with Yan-Act. Thus, two enhancer groups proved to be *rolled/MAPK* and *pointed*, giving us a good indication that the screen was working properly in the sense that some of the expected genes were isolated. Next, a series of genetic tests were designed to help determine which groups are likely to represent signaling components of interest. For example, the modifiers of Yan-Act were crossed to a gain-of-function *yan* allele isolated in our RAS1 modifier screen (Karim et al. 1996). In contrast to Yan-Act which is an engineered mutant in which all eight MAPK consensus sequences have been mutated, the gain-of-function allele encodes a truncated protein that has lost the carboxy-terminal PEST-rich region, including the last two MAPK sites, but leaves the first six intact. Phenotypically, in the eye, the two mutations behave the same. Thus, if the modifiers interact similarly with both, they are more likely to encode relevant components of the pathway. On the basis of this and other tests, the number of groups warranting further analysis was reduced to 11 from the original 22. Of these 11, 6 correspond to known genes, leaving 5 potential new elements of the pathway to pursue in the future. Phenotypic and molecular characterization of these genes is currently under way.

In addition to these transcription factors that have their activities modified by direct MAPK phosphorylation, three other nuclear components of the pathway have been identified—Phyl, Sina, and Ttk^{88} (Carthew and Rubin 1990; Xiong and Montell 1993; Chang et al. 1995; Dickson et al. 1995). The *phyl* gene encodes a nuclear protein (Phyl) that is required for R1, R6, and R7 cell fate determination, as well as for proper development of the embryonic peripheral nervous system (PNS) (Chang et al. 1995; Dickson et al. 1995). Ectopic expression of Phyl during eye development can recruit the cone cell precursors to an R7 cell fate and induce expression of a neuronal marker in the pigment cells. Transcription of the *phyl* gene is up-regulated by RAS1/MAPK signaling during photoreceptor determination, suggesting that *phyl* is one of the earliest trancriptional targets of this signaling pathway (Chang et al. 1995; Dickson et al. 1995). Phyl contains no protein domains of known function.

The *sina* gene encodes a ring-finger-containing nuclear protein (Sina) which is required for R7 cell fate specification, as well as for the proper development of other sensory structures (Carthew and Rubin 1990). Ring finger domains are thought to mediate protein-protein interactions (Saurin et al. 1996). Sina has mammalian counterparts that are greater than 80% identical in amino acid sequence and both human and mouse *sina* genes have been shown to be induced during apoptosis (Amson et al. 1996; Nemani et al. 1996).

The *tramtrack* (*ttk*) gene encodes two alternatively spliced zinc finger transcription factors first identified by their ability to repress transcription of segmentation genes during embryogenesis (Harrison and Travers 1990; Brown et al. 1991; Read and Manley 1992; Read et al. 1992; Brown and Wu 1993). The role of the *ttk* gene in cell fate determination has been most closely examined in the developing eye and PNS. In both cases, Ttk acts to repress specific neuronal fates (Xiong and Montell 1993; Guo et al. 1995; Lai et al. 1996). In the eye, loss of function of the 88-kD gene product (Ttk^{88}) results in supernumerary R7 cell formation (Xiong and Montell 1993).

Genetic experiments indicate that Phyl, Sina, and Ttk^{88} are the most downstream known components of the signaling pathway that specifies the R7 cell fate. Mutation of *sina* blocks the ability of activated RAS1 (Fortini et al. 1992), activated MAPK (Brunner et al. 1994), ectopically expressed Phyl (Chang et al. 1995; Dickson et al. 1995), or loss of the repressor Yan (Lai and Rubin 1992) to induce R7 cell formation. Indeed, the only known case where R7 cells can be formed in a *sina* mutant background is when Ttk^{88} has also been mutationally inactivated (Lai et al. 1996; Yamamoto et al. 1996), indicating that Ttk acts downstream from Sina or in a parallel pathway.

We (Tang et al. 1997) investigated the functional relationship among the Sina, Phyl, and Ttk^{88} proteins. Ttk^{88} is a repressor of photoreceptor cell fate determination. We found that induction of Phyl expression leads to down-regulation of Ttk^{88} by a mechanism that requires Sina. These three proteins appear to physically interact with each other, and Sina genetically and physically interacts with UbcD1, a component of the ubiquitin-dependent protein degradation pathway. The genetic and physical interactions detected between Sina and UBCD1 raise the possibility that Sina may target Ttk^{88} to the ubiquitin-dependent proteolytic pathway. Our data support biochemical functions for the Sina and Phyl proteins and support a model for a novel mechanism of signaling by the RAS1/MAPK cascade—transcriptional induction of a protein that then targets a specific transcriptional repressor for degradation.

Our observation that development of all photoreceptors can be blocked by Ttk^{88} expression indicates that Ttk^{88} acts as a repressor of neuronal development. Ttk^{88} does not block the formation of nonneuronal cell types, however, in contrast to the transcription factor Yan, which acts as a general repressor of differentiation in the developing eye. Thus, whereas Yan can be thought of as controlling the general decision to differentiate, Ttk^{88} appears to act at a later decision point to block neuronal differentiation. The repressing effects of both Ttk^{88} and Yan are alleviated by down-regulating the levels of these proteins in response to RAS/MAPK signaling. However, unlike Yan, the mechanism we propose for Ttk^{88} does not involve its direct phosphorylation by MAPK.

CONCLUDING REMARKS

Screens for second-site mutations that modify the phenotype of an existing mutation have proven to be a useful method for discovering additional components of genetic pathways in organisms as diverse as bacteria, yeast, nematodes, and fruit flies. We have described studies of the signal transduction pathway that acts downstream from RAS activation as an example of how genetics in *Drosophila* can be used to place individual genes within pathways and networks. These studies have revealed the outlines of the pathway from RAS1 activation at the membrane to changes in gene expression in the nucleus.

The approach we have used of looking for dominant suppressors and enhancers in a genetic background that has been engineered to render it sensitive to small perturbations was modeled after an earlier genetic screen aimed at identifying components of the signal transduction pathway that acts between the transmembrane tyrosine kinase receptor Sevenless and RAS1 (Simon et al. 1991). This approach has proven applicable to a wide variety of processes and is particularly powerful when combined with biochemical and other functional analyses which, given the high evolutionary conservation of these pathways, can be performed in a variety of organisms and cell culture systems. *Drosophila* has proven particularly well-suited for this type of analysis. Among the metazoan organisms, *Drosophila* offers the most sophisticated experimental genetic system. It is also the closest of the commonly used invertebrate model systems to humans as estimated by conservation of gene sequence and gene function.

ACKNOWLEDGMENTS

We thank our colleagues in the Rubin and Morrison laboratories for many helpful discussions and technical assistance during the course of the work reviewed here. This work was supported in part by the Howard Hughes Medical Institute and the National Institutes of Health (G.M.R), the National Cancer Institute, Department of Health and Human Services, under contract with ABL (N.R.M. and D.K.M.), and the National Cancer Institute of Canada and the Medical Research Council of Canada (M.T.). A.T. was supported by an National Institutes of Health postoctoral fellowship, D.A.W. was a Helen Hay Whitney postdoctoral fellow, F.K. was a senior fellow of the American Cancer Society, California Division, and I.R. was a Burroughs-Welcome fellow.

REFERENCES

Amson R.B., Nemani M., Roperch J.P., Israeli D., Bougueleret L., Le Gall I., Medhioub M., Linares-Cruz G., Lethrosne F., Pasturaud P., Piouffre L., Prieur S., Susini L., Alvaro V., Millaseau P., Guidicelli C., Bui H., Massart C., Cazes L., Dufour F., Bruzzoni-Giovanelli H., Owadi H., Hennion C., Charpak G., Dausset J., Calvo F., Oren M., Cohen D., and Tellerman A. 1996. Isolation of 10 differentially expressed cDNAs in p53-induced apoptosis: Activation of the vertebrate homologue of the *Drosophila seven in absentia* gene. *Proc. Natl. Acad. Sci.* **93:** 3953.

Brasselman S. and McCormick F. 1995. BCR and RAF form a complex in vivo via 14–3-3 proteins. *EMBO J.* **14:** 4839.

Brown J.L. and Wu C. 1993. Repression of *Drosophila* pair-rule segmentation genes by ectopic expression of *tramtrack. Development* **117:** 45.

Brown J.L., Sonoda S., Ueda H., Scott M. P., and Wu C. 1991. Repression of the *Drosophila fushi tarazu* (*ftz*) segmentation gene. *EMBO J.* **10:** 665.

Brunner D., Ducker K., Oellers N., Hafen E., Scholz H., and Klambt C. 1994. The ETS domain protein Pointed-P2 is a target of MAP kinase in the Sevenless signal transduction pathway. *Nature* **370:** 386.

Carthew R.W. and Rubin G.M. 1990. seven in absentia: A gene required for specification of R7 cell fate in the *Drosophila* eye. *Cell* **63:** 561.

Chang H.C. and Rubin G.M. 1997. 14–3-3e positively regulates Ras-mediated signaling in *Drosophila. Genes Dev.* **11:** 1132.

Chang H.C., Solomon N.M., Wassarman D.A., Karim F.D., Therrien M., Rubin G.M. and Wolff T. 1995. *phyllopod* functions in the fate determination of a subset of photoreceptors in *Drosophila. Cell* **80:** 463.

Choi K.Y., Satterberg B., Lyons D.M., and Elion E.A. 1994. Ste5 tethers multiple protein kinases in the MAP kinase cascade required for mating in *S. cerevisiae. Cell* **78:** 499.

Clark-Lewis I., Sanghera J.S., and Pelech S.L. 1991. Definition of a consensus sequence for peptide substrate recognition by p44mpk, the meiosis activated myelin basic protein kinase. *J. Biol. Chem.* **266:** 15180.

Dickson B. 1995. Nuclear factors in sevenless signaling. *Trends Genet.* **11:** 106.

Dickson, B.J., Dominquez M., van der Straten A., and Hafen E. 1995. Control of *Drosophila* photoreceptor cell fates by Phyllopod, a novel nuclear protein acting downstream of the Raf kinase. *Cell* **80:** 453.

Dickson B.J., van der Straten A., Dominquez M., and Hafen E. 1996. Mutations modulating Raf signaling in *Drosophila* eye development. *Genetics* **142:** 163.

Fortini M.E., Simon M.A., and Rubin G.M. 1992. Signaling by the sevenless protein tyrosine kinase is mimicked by Ras1 activation. *Nature* **355:** 559.

Guo M., Bier E., Jan L.Y., and Jan Y.N. 1995. *tramtrack* acts downstream of numb to specify distinct daughter cell fates during asymmetric cell divisions in the *Drosophila* PNS. *Neuron* **14:** 913.

Harrison S.D. and Travers A.A. 1990. The tramtrack gene encodes a *Drosophila* finger protein that interacts with the *ftz* transcriptional regulatory region and shows a novel embryonic expression pattern. *EMBO J.* **9:** 207.

Karim F.D., Chang H.C., Therrien M., Wassarman D.A., Laverty T., and Rubin G.M. 1996. A screen for genes that function downstream of Ras1 during *Drosophila* eye development. *Genetics* **143:** 315.

Klambt, C. 1993. The *Drosophila* gene *pointed* encodes two ETS-like proteins which are involved in the development of the midline glial cells. *Development* **117:** 163.

Kornfeld K., Hom D.B., and Horvitz H.R. 1995. The *ksr-1* gene encodes a novel protein kinase involved in Ras-mediated signaling in *C. elegans*. *Cell* **83:** 903.

Lai Z.-C. and Rubin G.M. 1992. Negative control of photoreceptor development in *Drosophila* by the product of the *yan* gene, an ETS domain protein. *Cell* **70:** 609.

Lai Z.-C., Harrison S.D., Karim F., Li Y., and Rubin G.M. 1996. Loss of *tramtrack* gene activity results in ectopic R7 cell formation, even in a *sina* mutant background. *Proc. Natl. Acad. Sci.* **93:** 5025.

Lowy D.R. and Willumsen B.M. 1993. Function and regulation of Ras. *Annu. Rev. Biochem.* **62:** 851.

Moodie S.A. and Wolfman A. 1994. The 3Rs of life: Ras, Raf and growth regulation. *Trends Genet.* **10:** 44.

Morrison D. 1994. Modulators of signaling proteins? *Science* **266:** 56.

Muslin A.J., Tanner J.W., Allen P.M., and Shaw A.S. 1996. Interaction of 14-3-3 with signal proteins is mediated by the recognition of phosphoserine. *Cell* **84:** 889.

Nemani M., Linares-Cruz G., Bruzzoni-Giovanelli H., Roperch J. P., Tuynder M., Bougueleret L., Cherif D., Medhioub M., Pasturaud P., Alvaro V., Sarkissan H.D., Cazes L., Le Paslier D., Le Gall I., Israeli D., Dausset J., Sigaux F., Chumakov I., Oren M., Calvo F., Amson R.B., Cohen D., and Tellerman A. 1996. Activation of the human homologue of the *Drosophila sina* gene in apoptosis and tumor suppression. *Proc. Natl. Acad. Sci.* **93:** 9039.

O'Neill E.M., Rebay I., Tjian R., and Rubin G.M. 1994. The activities of two Ets-related transcription factors required for *Drosophila* eye development are modulated by the Ras/MAPK pathway. *Cell* **78:** 137.

Read D. and Manley J.L. 1992. Alternatively spliced transcripts of the *Drosophila tramtrack* gene encode zinc finger proteins with distinct DNA binding specificities. *EMBO J.* **11:** 1035.

Read D., Levine M., and Manley J.L. 1992. Ectopic expression of the *Drosophila tramtrack* gene results in multiple embryonic defects, including repression of *even-skipped* and *fushi tarazu*. *Mech. Dev.* **38:** 183.

Rebay I. and Rubin G.M. 1995. Yan functions as a general inhibitor of differentiation and is negatively regulated by activation of the Ras1/MAPK pathway. *Cell* **81:** 857.

Saurin A.J., Borden K.L.B., Boddy M.N., and Freemont P.S. 1996. Does this have a familiar ring? *Trends Biochem. Sci.* **21:** 208.

Schlessinger J. 1993. How receptor tyrosine kinases activate Ras. *Trends Biochem. Sci.* **18:** 273.

Sidow A. and Thomas W.K. 1994. A molecular evolutionary framework for eukaryotic model organisms. *Curr. Biol.* **4:** 596.

Simon M.A., Bowtell D.D.L., Dodson G.S., Laverty T.R., and Rubin G.M. 1991. Ras1 and a putative guanine nucleotide exchange factor perform crucial steps in signaling by the sevenless protein tyrosine kinase. *Cell* **67:** 701.

Sundaram M. and Han M. 1995. The *C. elegans ksr-1* gene encodes a novel Raf-related kinase involved in Ras-mediated signal transduction. *Cell* **83:** 889.

Tang A.H., Neufeld T.P., Kwan E., and Rubin G.M. 1997. PHYL acts to down-regulate TTK^{88}, a transcriptional repressor of neuronal cell fates, by a SINA-dependent mechanism. *Cell* **90:** 459.

Therrien M., Michaud N.R., Rubin G.M., and Morrison D.K. 1996. KSR modulates signal propagation within the MAPK cascade. *Genes Dev.* **10:** 2684.

Therrien M., Chang H.C., Solomon N.M., Karim F.D., Wassarman D.A., and Rubin G.M. 1995. Ksr, a novel protein kinase required for Ras signal transduction. *Cell* **83:** 879.

van der Geer P., Hunter T., and Lindberg R.A. 1994. Receptor protein-tyrosine kinases and their signal transduction pathways. *Annu. Rev. Cell Biol.* **10:** 251.

Wassarman D.A., Solomon N.M., Chang H.C., Karim F.D., Therrien M., and Rubin G.M. 1995. Protein phosphatase 2A postively and negatively regulates Ras-mediated photoreceptor development in *Drosophila*. *Genes Dev.* **10:** 272.

Wolff T. and Ready D. 1993. Pattern formation in the *Drosophila* retina. In *The development of* Drosophila melanogaster (ed. M. Bate and A. Martinez Arias), p. 1277. Cold Spring Harbor Laboratory Press, Cold Spring Harbor, New York.

Xiong W.C. and Montell C. 1993. Tramtrack is a transcriptional repressor required for cell fate determination in the *Drosophila* eye. *Genes Dev.* **7:** 1085.

Yamamoto D., Nihonmatsu I., Matsuo T., Miyamoto H., Kondo S., Hirata K., and Ikegami Y. 1996. Genetic interactions of *pokkuri* with *seven in absentia*, *tramtrack* and downstream components of the sevenless pathway in R7 photoreceptor induction in *Drosophila melanogaster*. *Roux's Arch. Dev. Biol.* **205:** 215.

Zipursky S.L. and Rubin G.M. 1994. Determination of neuronal cell fate: Lessons from the R7 neuron of *Drosophila*. *Annu. Rev. Neurosci.* **17:** 373.

Mutations That Perturb Vulval Invagination in *C. elegans*

T. HERMAN AND H.R. HORVITZ
Howard Hughes Medical Institute, Department of Biology, Massachusetts Institute of Technology, Cambridge, Massachusetts 02139

During development, animals undergo dramatic changes in patterning as a result of the movement and shaping of epithelial cell layers (Bard 1990). Sheets of epithelial cells fold, branch, spread, and detach from or fuse with one another to generate the three-dimensional topology of an embryo. One example of such a morphogenetic process is epithelial invagination, which generates tubular structures from flat epithelia. To invaginate, a flat sheet of epithelial cells bends and moves inward, its apical surface facing the concave side of the depression. During sea urchin and *Drosophila* gastrulation, an epithelium bends inward about a point to form the digestive tract. The result is a tube that is perpendicular to the original sheet (Sweeton et al. 1991; for review, see Davidson et al. 1995). In contrast, vertebrate neurulation, which forms the spinal cord and brain, is initiated by the invagination of a sheet that bends along a line and results in a tube that is parallel to the original sheet (for review, see Schoenwolf and Smith 1990).

Many models have been proposed for the mechanistic basis of epithelial invagination (for reviews, see Ettensohn 1985; Fristrom 1988; Davidson et al. 1995). In general, these models postulate a mechanical force that decreases the area of the apical surface of the epithelial sheet relative to the area of its basal surface, causing an inward bend. The models differ in which cellular process initially provides this mechanical force.

A simple cytoskeleton-based model is that invagination results from individual changes in cell shape. Apically localized actin filaments contract and thereby decrease the apical surface area of each epithelial cell in the sheet, resulting in inward bending. This type of mechanism has been proposed for *Drosophila* gastrulation (Sweeton et al. 1991; Costa et al. 1994) and for neurulation (Clausi and Brodland 1993). A more complex cytoskeleton-based model involves cell crawling: Cells in the invaginating region pull themselves toward the center of the region by extending their apices along the extracellular matrix (Burke et al. 1991). If the volume and basal surface areas of each cell remain constant, this apical extension decreases the apical surface area, again resulting in bending of the sheet. This type of mechanism has been proposed for sea urchin gastrulation (Burke et al. 1991).

Changes in cell-cell adhesion also could initiate epithelial invagination (Gustafson and Wolpert 1963). If the adhesion between adjacent cells increases, it would be energetically favorable for the contact between them to increase in extent, causing the cells to increase in height and become thin and columnar. However, if the basal side of each cell remains strongly adherent to a basal lamina, forcing the basal surface area to remain constant, the increase in cell height would decrease only the apical surface area, and the epithelial sheet would bend inward. There are also more complex models of invagination based on local differences in cell-cell adhesivity within a sheet (Nardi 1981; Mittenthal and Mazo 1983). Although adhesion-based models explain the columnarization of epithelial cells that is nearly always observed during invagination (Fristrom 1988), they seem to be less popular than cytoskeleton-based models.

A third type of model, which has been proposed for sea urchin gastrulation, is based on changes in the composition of the extracellular matrix (Lane et al. 1993). According to this model, the apical surface of the epithelial sheet in the region where invagination is to occur secretes a proteoglycan, which forms a hygroscopic layer of matrix between the epithelium and the original extracellular matrix. The inner matrix layer swells as it becomes hydrated, increasing in surface area, whereas the outer layer remains the same. This difference in surface areas results in the bending inward of the outer matrix layer and a concomitant bending inward of the underlying epithelial sheet.

Although each of these models proposes that a single cellular process provides the mechanical force required for epithelial invagination, each requires the coordinated regulation of other cellular processes. For instance, if the cells in a sheet adhere too greatly to the apical extracellular matrix, they will not be able to constrict their apical surfaces appropriately. Similarly, if the cells in a sheet have too rigid a cytoskeleton, an increase in cell-cell adhesion will not cause an increase in cell height. Thus, disruption of a cellular process that is not the primary initiator of epithelial invagination may affect the invagination process. Furthermore, these models are not mutually exclusive: The mechanical force required for epithelial invagination might be provided simultaneously by several cellular processes, with partially or fully redundant abilities to initiate invagination. Finally, different instances of epithelial invagination may have different mechanistic bases.

To identify molecules involved in epithelial invagination, we have been analyzing an example of this process in the nematode *Caenorhabditis elegans*. During the larval development of the *C. elegans* hermaphrodite, a specialized group of outer epithelial cells invaginates to form the vulva, a tube that connects the outer epithelium to the uterine epithelium and allows outward passage of eggs and inward passage of male sperm. Many basic problems of developmental biology have been studied by analyzing the development of the *C. elegans* vulva, including the generation of cell diversity and the determination of cell type, intercellular signaling and signal transduction, and the control of developmental timing (Ferguson et al. 1987; Clark et al. 1993; Labouesse et al. 1994; Euling and Ambros 1996; for review, see Horvitz and Sternberg 1991 Kornfeld 1997). We have begun a genetic analysis of vulval invagination.

THE VULVA IS FORMED BY THE DESCENDANTS OF THREE EPITHELIAL CELLS

C. elegans larvae and adults are enclosed by a single layer of epithelial cells interconnected by desmosomes. The apical surface of this epithelium faces outward and secretes the cuticle, a multilayered extracellular structure consisting in large part of collagens. The basal surface contacts a basal lamina (White 1988).

The vulva is formed by the descendants of three epithelial cells, P5.p, P6.p, and P7.p (Fig. 1). During the third larval (L3) stage, these vulval precursor cells lie in an outer epithelial sheet, and their basal sides abut a basement membrane that lies between them and the developing gonad. A specialized cell in the developing gonad, the anchor cell, overlies P6.p. During the L3 stage, the anchor cell signals the vulval precursor cells to divide. The middle cell, P6.p, gives rise to a symmetric lineage termed "primary" and generates four binucleate descendants (two E cells and two F cells). P5.p and P7.p, the two cells flanking P6.p, give rise to asymmetric mirror-image lineages termed "secondary;" each generates two binucleate (A and C) and three mononucleate (B1, B2, and D) descendants (for review, see Greenwald 1997).

Just before and during the final round of vulval cell divisions, the epithelium thickens slightly in the region at which invagination will occur. As these final divisions are completed, the central vulval cells, consisting of all of the primary descendants and some of the secondary descendants, detach from the rigid cuticle, and the epithelial layer buckles inward, resulting in a fluid-filled invagination space between the vulva and the cuticle. During the fourth larval stage (L4), each of the 14 vulval cells becomes U-shaped, extending lateral arms toward its mirror-image equivalent (J. White, pers. comm.). These cells fuse pairwise, resulting in seven toroidal cells stacked

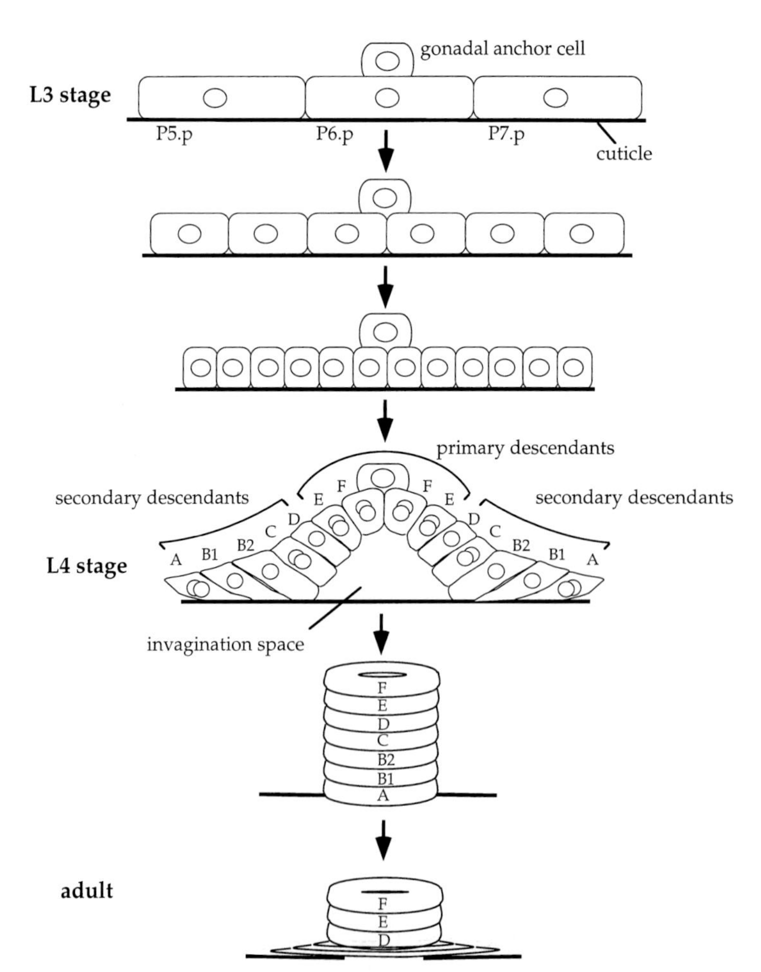

Figure 1. Schematic of *C. elegans* vulval development. The descendants of three epithelial cells, P5.p, P6.p, and P7.p, invaginate and fuse pairwise to form a stack of toroids. The inner hole of the toroids forms a passage between the gonad and the outside. Cell nuclei are indicated by small ovals but are not indicated in the stacks of toroids. The cuticle and cell bodies are indicated in each diagram. Toroids B1 and B2 are fused in the adult. See text for details.

one on top of another and connected by desmosomes (Podbilewicz and White 1994; Newman and Sternberg 1996). The vulval cell at the top of the stack forms an attachment with the uterine epithelium, whereas the vulval cell at the bottom remains attached to the outer epithelium (Newman et al. 1996). The inner holes of these vulval toroids define the tube through which eggs and sperm pass in the adult.

During the second half of the L4 stage, the inner and outer diameters of the vulval toroids decrease. At the molt between the L4 and adult stages, the vulva everts, causing the outer toroids to become considerably stretched (White 1988; J. White, pers. comm.).

THE ANCHOR CELL IS REQUIRED FOR WILD-TYPE VULVAL INVAGINATION

Several cells are in direct contact with the vulval cells during vulval morphogenesis and are thus candidates for affecting vulval invagination. The function of some of these cells in invagination has been tested by eliminating or displacing them and observing the consequences on vulval development.

The anchor cell is required not only to signal the vulval precursor cells to divide, but also to organize the invagination of the P6.p descendants. This conclusion is based on experiments in which a laser microbeam was used to ablate the anchor cell just after it had signaled P5.p, P6.p, and P7.p to divide (Kimble 1981) and from observations of *dig-1* mutant hermaphrodites, in which the entire gonad, including the anchor cell, detaches from the basement membrane adjoining the vulval cells and becomes mislocalized (Thomas et al. 1990). In both cases, the vulval precursors often undergo a completely wild-type pattern of cell divisions, but, in the absence of direct contact with the anchor cell, their descendants invaginate abnormally: The primary descendants do not remain symmetrically arranged and fail to form toroids, and no inner invagination space is formed between them. The secondary descendants appear to invaginate correctly. Within the gonad, the anchor cell is not only necessary, but also sufficient to organize the primary cells: The rest of the gonad can be ablated without affecting the initial stages of vulval invagination (Kimble 1981), although the absence of the uterine uv1 cells prevents the attachment of the developing vulva to the uterus (Newman et al. 1996).

Although the vulval muscles form specific attachments to the vulval toroids and the outer epithelium during the L4 stage, these muscles are not required for vulval invagination. *sem-4* mutant hermaphrodites, which entirely lack vulval muscles (Basson and Horvitz 1996), appear to have wild-type invaginations and can receive sperm through their adult vulvae (Desai et al. 1988; T. Herman et al., in prep.).

Throughout their development, the vulval cells are surrounded by hyp7, the main body epithelium. The function of hyp7 in vulval invagination cannot easily be tested, since ablation of hyp7 is lethal to the animal (Sulston et al. 1983).

MUTATIONS IN EIGHT GENES AFFECT VULVAL INVAGINATION

To identify genes involved in vulval invagination, we sought mutations that affected the invagination of the vulva but did not alter vulval cell lineages. Although one might expect that a hermaphrodite abnormal in vulval invagination would be unable to lay eggs as an adult, previous screens for fertile mutants with defects in laying eggs (Trent et al. 1983) had not identified any that met our criteria. We therefore performed a genetic screen for mutants with abnormal vulval invaginations in a way that allowed the recovery of mutations that additionally caused sterility or maternal-effect lethality (T. Herman et al., in prep.). Single F_1 progeny of hermaphrodites mutagenized with ethylmethanesulfonate (EMS) were placed on petri plates, and their L4 progeny were examined first with a dissecting microscope for abnormal vulval invaginations and then with Nomarski differential interference contrast microscopy to determine whether their vulval cell lineages were normal. Mutations that caused sterility were recovered in heterozygous siblings from the same F_2 brood. In a screen of 12,000 mutagenized haploid genomes, we identified 25 mutations that define eight genes required for wild-type vulval invagination. We named these genes *sqv-1* to *sqv-8,* for *sq*uashed *v*ulva. None of the *sqv* mutations affects the number or lineage of the vulval cells. Nearly all cause sterility.

All 25 mutations isolated result in the same abnormal vulval phenotype (Fig. 2). In the *sqv* mutants, from the time the final round of vulval cell divisions begins, there is a reduction in the size of the space that separates the anterior and posterior halves of the vulva (T. Herman et al., in prep.). Both primary and secondary descendants are affected: *lin-12(lf) sqv-3* double mutants, in which the vulval precursor cells generate primary but no secondary lineages (Greenwald et al. 1983), have decreased invaginations when compared with *lin-12(lf)* single mutants.

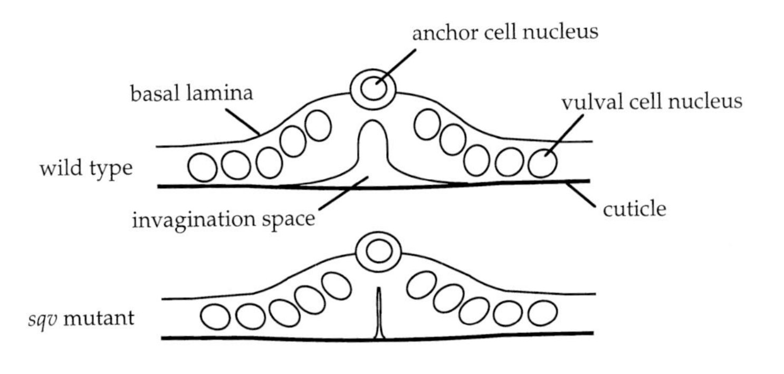

Figure 2. Schematic of wild-type and *sqv* mutant vulvae at the early L4 stage, as they appear in their central focal plane when viewed with Nomarski optics. Cell nuclei, the basal lamina, and the separation between the vulval cells and the cuticle are visible. A *sqv* mutant has a considerably reduced vulval invagination space compared with the wild type.

Similarly, *lin-12(gf) sqv-3* double mutants, in which the vulval precursor cells generate secondary but no primary lineages (Greenwald et al. 1983), have decreased invaginations when compared with *lin-12(gf)* single mutants (T. Herman et al., in prep.). This reduction in the invagination space, and later in the vulval tube, remains evident throughout the L4 stage until adulthood, when both the mutant and wild-type vulval tubes are closed and therefore indistinguishable.

Despite being abnormal in invagination, *sqv* mutant vulval cells are wild-type in lineage, movement toward the anchor cell, and detachment from the cuticle, and mutant vulval nuclei migrate to and occupy wild-type positions throughout the L4 and adult stages.

THE DECREASED INVAGINATION SPACE IN *sqv-3* MUTANTS IS A CONSEQUENCE OF ELONGATED VULVAL CELLS

To examine the ultrastructural basis of the Sqv vulval defect, we analyzed electron micrographs of serial sections of wild-type and *sqv-3* mutant vulvae at the early L4 stage (T. Herman et al., in prep.). We detected two main differences. First, at least some *sqv-3* mutant vulval cells appear to be more elongated than their wild-type equivalents. The central vulval cells, in particular, appear to extend too far into the region that is normally occupied by the invagination space.

Second, the material within the vulval invagination space is more electron-dense and more uniform in *sqv-3* animals than that in the wild type. This electron-dense material may be qualitatively identical to that in the wild type but may look different because it is compressed into a smaller volume. Alternatively, the difference in appearance may reflect a difference in the composition of the material in the mutant invagination space. Electron microscopic studies confirm that the *sqv-3* mutant vulval cell bodies are arranged in grossly wild-type positions relative to one another and that the appropriate cells have detached from the cuticle.

sqv VULVAL CELLS FORM A PARTIALLY FUNCTIONAL TUBE

Despite the abnormal appearance of the *sqv* vulval invagination, *sqv* mutant vulval cells form a vulval tube. The antibody MH27, which recognizes desmosomal connections between epithelial cells (Waterston 1988), stains the wild-type mid-L4 vulva in a pattern of rings, the points of contact between the vulval toroids. A *sqv* mutant vulva stained with MH27 has the same pattern of rings as a wild-type vulva, suggesting that the vulval toroids have formed correctly (T. Herman et al., in prep.).

Furthermore, *sqv* mutant adults are able to lay eggs, indicating the presence of a functional vulval tube; nonetheless, these animals retain an abnormally large number of eggs in their uterus (T. Herman et al., in prep.). This defect in egg laying is likely to be a consequence of the vulval abnormality, since the other components of the egg-laying system, the vulval muscles and the HSN neurons (Trent et al. 1983), are present in *sqv* mutant animals, although their adult differentiation and function have not been tested (T. Herman et al., in prep.).

SEVERAL MODELS COULD EXPLAIN THE Sqv VULVAL PHENOTYPE

The *sqv* mutations result in an abnormal elongation of the central vulval cells but do not prevent the bending inward of the vulval epithelium or the subsequent formation of the stack of toroidal vulval cells. One simple explanation for the Sqv phenotype can be derived from the adhesion-based model of invagination discussed above (Gustafson and Wolpert 1963). During normal vulval invagination, the adhesiveness between adjacent vulval cells may increase, resulting in an increased contact between them and their visible increase in cell height (Fig. 3). The Sqv phenotype could result from a greater than normal increase in the adhesiveness between adjacent vulval cells. Such a greater adhesiveness could abnormally increase the extent of contact between the vulval cells, resulting in an elongation of *sqv* mutant vulval cells compared with wild-type vulval cells. The sheet of *sqv* mutant vulval cells would still bend inward, since this greater elongation would still decrease the apical surface area of the sheet without affecting its basal surface area.

Alternatively, the *sqv* mutations may affect a cellular process not normally involved in vulval invagination. For instance, the *sqv* mutations might cause an abnormal accumulation of adhesive material in the extracellular matrix adjoining the vulval cells, causing them to adhere to it abnormally as they invaginate and thus to become ab-

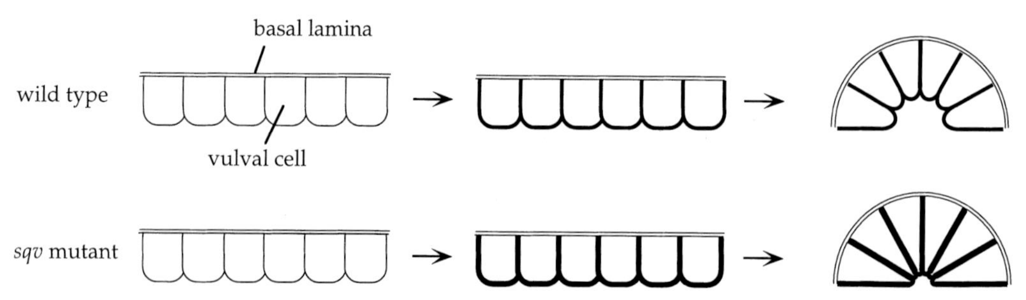

Figure 3. One model of the Sqv phenotype. The extent of contact between vulval cells is determined by the balance between their adhesiveness and their rigidity. During normal vulval invagination, an increase in adhesiveness occurs and drives an increase in the extent of contact between vulval cells. Because the vulval cell surface area in contact with the basal lamina is maintained, the resulting increase in cell height causes apical narrowing and invagination. The Sqv phenotype results from an abnormally great increase in adhesiveness. In this figure, the thickness of the cell membranes indicates their degree of adhesiveness.

normally elongated, or they might cause a decrease in the hydrostatic pressure within the invagination space relative to the internal hydrostatic pressure of the worm, resulting in a collapse of the vulval invagination space; such a collapse can result from a punctured cuticle (P. Sternberg, pers. comm.).

We have begun to clone and molecularly analyze the *sqv* genes. The molecular identities of the SQV-3, SQV-7, and SQV-8 proteins suggest that they may be required for the synthesis of a carbohydrate moiety.

SQV-3 IS SIMILAR TO A FAMILY OF GLYCOSYLTRANSFERASES

The SQV-3 protein has sequence similarity to a protein family that includes two proteins with glycosyltransferase activity: mammalian β(1,4)-galactosyltransferase (GalT) and *Lymnaea stagnalis* β(1,4)-*N*-acetylglucosaminyltransferase (T. Herman and H.R. Horvitz, in prep.). Other members of this family include a predicted *C. elegans* protein and several human proteins, all with unknown biochemical activity. SQV-3 has not been demonstrated to have glycosyltransferase activity, but the amino acid residues of GalT that are implicated in substrate binding (Aoki et al. 1990) are conserved in SQV-3.

Most glycosyltransferases are transmembrane proteins located in the endoplasmic reticulum (ER) or Golgi membrane, where they catalyze the sequential addition of sugars to glycoproteins and/or glycolipids (Kleene and Berger, 1993). Each glycosyltransferase removes a donor sugar from a nucleotide-sugar molecule and adds it to an acceptor sugar or sugars on a glycoconjugate. GalT catalyzes the addition of galactose (Gal) from UDP-Gal to *N*-acetylglucosamine (GlcNAc) residues.

SQV-8 IS SIMILAR TO A GLUCURONYLTRANSFERASE INVOLVED IN THE SYNTHESIS OF THE HNK-1 EPITOPE ON GLYCOPROTEINS

The SQV-8 protein has sequence similarity to a second glycosyltransferase, glucuronyltransferase (GlcAT-P), which transfers glucuronic acid (GlcA) to Gal on proteins but not lipids (Terayama et al. 1997; T. Herman and H.R. Horvitz, in prep.). In vertebrates, this reaction is the final glycosyltransferase reaction in the assembly of the carbohydrate epitope recognized by the HNK-1 monoclonal antibody. The HNK-1 epitope was originally identified on human natural killer cells (Abo and Balch 1981) and has subsequently been found on a large number of neural adhesion molecules in the vertebrate nervous system (for review, see Schachner and Martini 1995). The carbohydrate structure of glycolipids and a glycoprotein recognized by the HNK-1 antibody share the terminal trisaccharide SO_4-3-GlcA-β(1,3)-Gal-β(1,4)-GlcNAc, which is therefore likely to be the HNK-1 epitope itself. SQV-3 and SQV-8, if they have the biochemical activities predicted from their sequence similarities, would therefore be the two glycosyltransferases required for the formation of this structure: SQV-3 would add Gal to GlcNAc, and SQV-8 would add GlcA to Gal (Fig. 4). The HNK-1 antibody cross-reacts with invertebrates (Dennis et al. 1988; Bajt et al. 1990), including *C. elegans* (K. Nomura, pers. comm. cited in Terayama et al. 1997).

SQV-7 IS SIMILAR TO A PROTEIN REQUIRED FOR TRANSPORT OF A NUCLEOTIDE-SUGAR

The nucleotide-sugar substrates used by glycosyltransferases are synthesized in the cytoplasm and transported

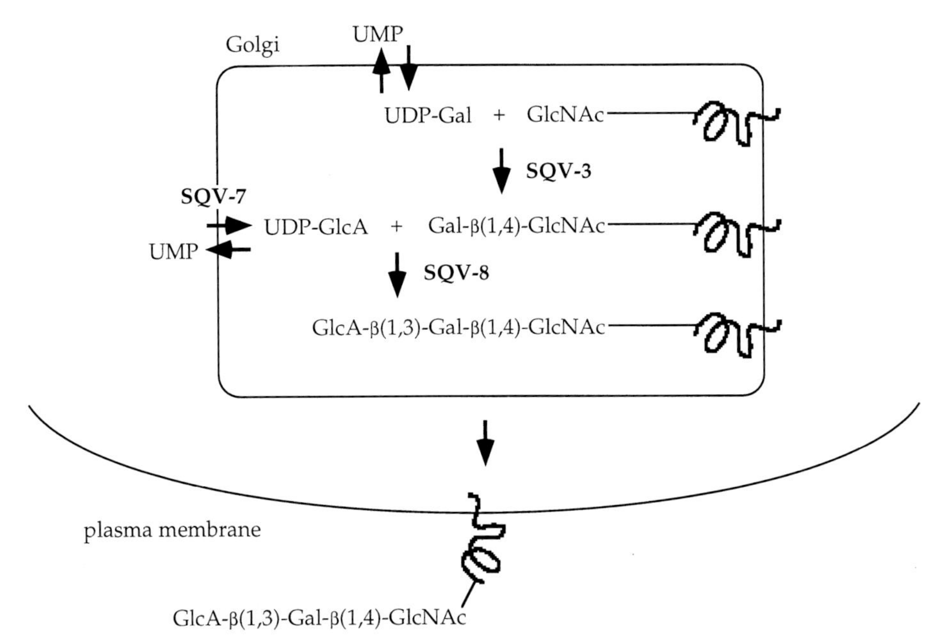

Figure 4. Model: SQV-3, SQV-7, and SQV-8 catalyze the formation of an oligosaccharide present on one or more cell surface glycoproteins. Part or all of this oligosaccharide is depicted as an unsulfated version of the HNK-1 epitope, but the biochemical activities of SQV-3, SQV-7, and SQV-8 have not been defined, and the addition of other sugars and/or sulfate groups to the putative GlcA-β(1,3)-Gal-β(1,4)-GlcNAc oligosaccharide has not been investigated. See text for details.

into the ER and Golgi compartments by specific antiports (Abeijon et al. 1997). In exchange, these antiports export the nucleotide-monophosphate products of the glycosyltransferase reactions.

The SQV-7 protein has sequence similarity to a family of proteins implicated in transport across membranes and is most closely related to the *Leishmania donovani* protein LPG2, which is required for assembly of a specific galactose-mannose repeat on glycoconjugates (Ma et al. 1997; T. Herman and H.R. Horvitz, in prep.). Microsomal vesicles derived from an *lpg2⁻* deletion mutant are defective in the uptake of the nucleotide-sugar donor GDP-mannose. Because LPG2 is predicted to contain multiple transmembrane domains and has sequence similarity to members of a large family of potential transporters, LPG2 is proposed to be the antiport for GDP-mannose. There are so far two human proteins that are very similar in sequence to SQV-7/LPG2, suggesting that higher eukaryotes may have multiple such transporters. Since only yeast and protozoa are known to transport GDP-mannose into the Golgi (Abeijon et al. 1997), it seems likely that some of these family members, including SQV-7, have diverged to transport other nucleotide-sugars.

We hypothesize that the SQV-7 antiport may directly provide either the SQV-3 or SQV-8 glycosyltransferase with its nucleotide-sugar donor (Fig. 4). A human UDP-Gal antiport has been cloned and is not similar to SQV-7/LPG2 (Miura et al. 1996). We therefore propose that SQV-7 acts as a GlcA antiport. It is, however, possible that SQV-7 provides SQV-3 with its substrate, either UDP-Gal or another nucleotide-sugar. It is also possible that one or more additional glycosyltransferases are involved in vulval invagination and that SQV-7 instead provides one of these enzymes with its nucleotide-sugar substrate.

SUGAR MODIFICATIONS IN MULTICELLULAR ORGANISMS MAY BE REQUIRED PRIMARILY FOR CELL-CELL INTERACTIONS

In general, defects in glycosylation have little or no affect on the viability of cell lines; even mammalian cell lines that lack nearly all glycosylation are able to grow normally (Krieger et al. 1989; Stanley and Ioffe 1995). In contrast, targeted gene disruptions of glycosyltransferases in mice suggest that multicellular organisms require complex sugar modifications to progress normally through development. For example, mice lacking *N*-acetylglucosaminyltransferase I, which is required early in the formation of N-linked carbohydrates, die at mid gestation (Ioffe and Stanley 1994). As might be expected, disruption of glycosyltransferases that act later in the glycosylation pathway and hence affect a smaller number of glycoconjugates causes less severe and more specific developmental defects. For example, mice lacking fucosyltransferase VII, which is required for the final step in the formation of the selectin ligands sLex and 6′-sulfo sLex, are viable but are specifically defective in selectin-mediated adhesion between endothelial cells and leukocytes (Maly et al. 1996). There are no vertebrate mutants that lack GlcAT-P glycosyltransferase, and so the function of the HNK-1 epitope has not been tested genetically.

Experiments in which the glycosylation sites on particular proteins have been eliminated suggest that sugar modifications can be required for the folding, stability, trafficking, and/or activity of secreted and cell-surface proteins and can mediate cell-cell recognition, cell-cell adhesion, and axonal guidance (for review, see Varki 1993). Several sugar modifications have been implicated in directly modulating cell-cell adhesion. One example is that of the regulation of cell-cell adhesion by polysialic acid moieties on the neural adhesion molecule N-CAM. Polysialylation decreases N-CAM homophilic binding and can also decrease cell-cell adhesion mediated by other molecules, perhaps by physically increasing the intercellular space and thereby preventing their interaction (for review, see Rutishauser 1996).

ABNORMAL GLYCOSYLATION OF ADHESION MOLECULES COULD CAUSE THE Sqv PHENOTYPE

Each of the models presented above to explain the Sqv vulval defect could be consistent with the loss of a carbohydrate modification. The first model was based on the degree of adhesiveness between *sqv* mutant vulval cells being abnormally high. SQV-3-, SQV-8-, and SQV-7-dependent glycosylation of one or more substrates might prevent excessive cell-cell adhesion in a way analogous to the way that polysialic acid decreases the cell-cell adhesion mediated by N-CAM.

The two other models mentioned above would also be consistent with abnormalities caused by alterations in glycoconjugates. One model postulated an abnormally adhesive extracellular matrix and the other an abnormally leaky cuticle; both the extracellular matrix and the cuticle contain glycoproteins and carbohydrates, and the absence of SQV-3-, SQV-8-, or SQV-7-dependent glycosylation might affect the properties of one or both of these structures.

sqv GENES ARE REQUIRED FOR EARLY DEVELOPMENT

In addition to the vulval defect discussed above, *sqv* mutations also cause hermaphrodites to be self-sterile (T. Herman et al., in prep.). This sterility is likely to be caused by a defect in *sqv* mutant oocytes, since *sqv* mutant hermaphrodites have grossly normal gonads with both oocytes and sperm, and sperm from *sqv* mutant hermaphrodites can produce viable cross-progeny in artificial insemination experiments.

The most severe terminal phenotype of progeny from *sqv-1* to *sqv-7* hermaphrodites is arrest before the first cell division (T. Herman et al., in prep.). The arrested one-cell eggs maintain their shape when laid, suggesting that they have eggshells and hence have been fertilized by sperm (Kemphues and Strome 1997), but unlike eggs with normal eggshells, the arrested eggs are permeable to hypochlorite. Glycosylation may therefore be involved in oocyte or early embryonic development. Since the arrest caused by strong mutations in *sqv-1* to *sqv-7* occurs well before the first events of epithelial morphogenesis, we have not ascertained whether *sqv-1* to *sqv-7* are required for these later events as well.

sqv-8 MAY AFFECT EPITHELIAL DEVELOPMENT MORE SPECIFICALLY THAN DO *sqv-1* TO *sqv-7*

Progeny of *sqv-8* hermaphrodites arrest development much later than the one-cell stage, at which the progeny of *sqv-1* to *sqv-7* hermaphrodites are blocked (T. Herman et al., in prep.). The self-progeny of hermaphrodites homozygous for strong alleles of *sqv-8* arrest about halfway through embryogenesis just as the outer epithelium would normally enclose the embryo. After this enclosure, the outer epithelial cells normally constrict in a roughly cylindrical pattern, squeezing the ovoid embryo into the elongated shape of a worm (Priess and Hirsh 1986). Progeny of *sqv-8* mutant hermaphrodites are often defective in epithelial enclosure and fail to elongate. Maternal *sqv-8* product, and hence SQV-8-modified glycoproteins, may therefore be involved in these epithelial morphogenetic events in addition to vulval morphogenesis.

Genetic and molecular evidence suggests that this failure in embryonic epithelial enclosure is the null phenotype of *sqv-8*. For example, one of the strong *sqv-8* alleles is a nonsense mutation predicted to eliminate the carboxy-terminal 78% of the protein (T. Herman and H.R. Horvitz, in prep.). *sqv-8* therefore appears to have later and presumably more specific effects on development than does *sqv-3*. This conclusion is consistent with our predicted molecular pathway, in which the SQV-8 glucuronyltransferase acts after the SQV-3 galactosyltransferase and modifies only a subset of proteins modified by SQV-3. The SQV-7 antiport may supply SQV-3 with its substrate or may be required to supply substrate to SQV-8 and to one or more additional glycosyltransferases that are less developmentally specific than SQV-8.

ADDITIONAL GENES ARE PROBABLY INVOLVED IN VULVAL INVAGINATION

All of the mutations isolated in our screen cause the same abnormal vulval phenotype, a reduced invagination space. Interestingly, none blocked invagination per se. It is possible that mutations in other genes involved in vulval invagination cause an earlier zygotic lethality, have only a maternal effect on the vulva, also cause vulval lineage defects, or are functionally redundant so that they disrupt only one of multiple processes with overlapping functions in invagination. Such genes would not have been identified in our screen, and we plan to seek them in other ways.

sqv MUTATIONS MAY GENETICALLY DEFINE CONSERVED COMPONENTS OF A GLYCOSYLATION PATHWAY

SQV-3, SQV-7, and SQV-8 are similar to components in a pathway for the synthesis of a carbohydrate moiety in vertebrates, the HNK-1 epitope, which has been implicated in adhesive interactions. It seems likely that SQV-3, SQV-7, and SQV-8 are required in *C. elegans* for the synthesis of a carbohydrate moiety and that the remaining *sqv* genes will prove to define additional conserved components required for assembly of this carbohydrate. One possible explanation for the Sqv vulval phenotype is that the *sqv* genes are normally involved in mediating cell-cell adhesion during vulval invagination. If so, a similar regulation of adhesion by glycoconjugates might therefore be involved in epithelial invagination in other species. Whether or not the *sqv* genes are normally involved in vulval invagination, these genes seem likely to define components of a glycosylation pathway conserved among organisms as diverse as nematodes and mammals.

ACKNOWLEDGMENTS

We thank Mark Alkema, Mark Metzstein, Rajesh Ranganathan, Gillian Stanfield, and Jeffrey Thomas for critically reading this manuscript. This work was supported by U.S. Public Health Service research grant GM-24663. H.R.H. is an investigator of the Howard Hughes Medical Institute.

REFERENCES

Abeijon C., Mandon E.C., and Hirschberg C.B. 1997. Transporters of nucleotide sugars, nucleotide sulfate and ATP in the Golgi apparatus. *Trends Biochem. Sci.* **22:** 203.

Abo T. and Balch C.M. 1981. A differentiation antigen of human NK and K cells identified by a monoclonal antibody (HNK-1). *J. Immunol.* **127:** 1024.

Aoki D., Appert H.E., Johnson D., Wong S.S., and Fukuda M.N. 1990. Analysis of the substrate binding sites of human galactosyltransferase by protein engineering. *EMBO J.* **9:** 3171.

Bajt M.L., Schmitz B., Schachner M., and Zipser B. 1990. Carbohydrate epitopes involved in neural cell recognition are conserved between vertebrates and leech. *J. Neurosci. Res.* **27:** 276.

Bard J. 1990. *Morphogenesis: The cellular and molecular basis of developmental anatomy.* Cambridge University Press, United Kingdom.

Basson M. and Horvitz H.R. 1996. The *Caenorhabditis elegans* gene *sem-4* controls neuronal and mesodermal cell development and encodes a zinc finger protein. *Genes Dev.* **10:** 1953.

Burke R.D., Myers R.L., Sexton T.L., and Jackson C. 1991. Cell movements during the initial phase of gastrulation in the sea urchin embryo. *Dev. Biol.* **146:** 542.

Clark S.G., Chisholm A.D., and Horvitz H.R. 1993. Control of cell fates in the central body region of *C. elegans* by the homeobox gene *lin-39. Cell* **74:** 43.

Clausi D.A. and Brodland G.W. 1993. Mechanical evaluation of theories of neurulation using computer simulations. *Development* **118:** 1013.

Costa M., Wilson E.T., and Wieschaus E. 1994. A putative cell signal encoded by the *folded gastrulation* gene coordinates cell shape changes during *Drosophila* gastrulation. *Cell* **76:** 1075.

Davidson L.A., Koehl M.A., Keller R., and Oster G.F. 1995. How do sea urchins invaginate? Using biomechanics to distinguish between mechanisms of primary invagination. *Development* **121:** 2005.

Dennis R.D., Antonicek H., Wiegandt H., and Schachner M. 1988. Detection of the L2/HNK-1 epitope on glycoproteins and acidic glycolipids of the insect *Calliphora vicina. J. Neurochem.* **51:** 1490.

Desai C., Garriga G., McIntire S.L., and Horvitz H.R. 1988. A genetic pathway for the development of the *Caenorhabditis elegans* HSN motor neurons. *Nature* **336:** 638.

Ettensohn C.A. 1985. Mechanisms of epithelial invagination. *Q. Rev. Biol.***60:** 289.

Euling S. and Ambros V. 1996. Heterochronic genes control cell cycle progress and developmental competence of *C. elegans* vulva precursor cells. *Cell* **84:** 667.

Ferguson E.L., Sternberg P.W., and Horvitz H.R. 1987. A genetic pathway for the specification of the vulval cell lineages of *Caenorhabditis elegans. Nature* **326:** 259.

Fristrom D. 1988. The cellular basis of epithelial morphogenesis. A review. *Tissue Cell* **20:** 645.

Greenwald I. 1997. Development of the vulva. In C. elegans *II* (ed. D.L. Riddle et al.), p.519. Cold Spring Harbor Laboratory Press, Cold Spring Harbor, New York.

Greenwald I.S., Sternberg P.W., and Horvitz H.R. 1983. The *lin-12* locus specifies cell fates in *C. elegans. Cell* **34:** 435.

Gustafson T. and Wolpert L. 1963. The cellular basis of morphogenesis and sea urchin development. *Int. Rev. Cytol.* **15:** 139.

Horvitz H.R. and Sternberg P.W. 1991. Multiple intercellular signaling systems control the development of the *Caenorhabditis elegans* vulva. *Nature* **351:** 535.

Ioffe E. and Stanley P. 1994. Mice lacking N-acetylglucosaminyltransferase I activity die at mid-gestation, revealing an essential role for complex or hybrid N-linked carbohydrates. *Proc. Natl. Acad. Sci.* **91:** 728.

Kemphues K.J. and Strome S. 1997. Fertilization and establishment of polarity in the embryo. In C. elegans *II* (ed. D.L. Riddle, et al.), p. 335. Cold Spring Harbor Laboratory Press, Cold Spring Harbor, New York.

Kimble J. 1981. Alterations in cell lineage following laser ablation of cells in the somatic gonad of *Caenorhabditis elegans. Dev. Biol.* **87:** 286.

Kleene R. and Berger E.G. 1993. The molecular and cell biology of glycosyltransferases. *Biochim. Biophys. Acta* **1154:** 283.

Kornfeld K. 1997. Vulval development in *Caenorhabditis elegans. Trends Genet.* **13:** 55.

Krieger M., Reddy P., Kozarsky K., Kingsley D., Hobbie L., and Penman M. 1989. Analysis of the synthesis, intracellular sorting, and function of glycoproteins using a mammalian cell mutant with reversible glycosylation defects. *Methods Cell Biol.* **32:** 57.

Labouesse M., Sookhareea S., and Horvitz H.R. 1994. The *Caenorhabditis elegans* gene *lin-26* is required to specify the fates of hypodermal cells and encodes a presumptive zinc-finger transcription factor. *Development* **120:** 2359.

Lane M.C., Koehl M.A., Wilt F., and Keller R. 1993. A role for regulated secretion of extracellular matrix during epithelial invagination in the sea urchin. *Development* **117:** 1049.

Ma D., Russell D.G., Beverley S.M., and Turco S.J. 1997. Golgi GDP-mannose uptake requires *Leishmania LPG2*. A member of a eukaryotic family of putative nucleotide-sugar transporters. *J. Biol. Chem.* **272:** 3799.

Maly P., Thall A., Petryniak B., Rogers C.E., Smith P.L., Marks R.M., Kelly R.J., Gersten K.M., Cheng G., Saunders T.L., Camper S.A., Camphausen R.T., Sullivan F.X., Isogai Y., Hindsgaul O., von Andrian U.H., and Lowe J.B. 1996. The α-(1,3)fucosyltransferase Fuc-TVII controls leukocyte trafficking through an essential role in L-, E-, and P-selectin ligand biosynthesis. *Cell* **86:** 643.

Mittenthal J.E. and Mazo R.M. 1983. A model for cell shape generation by strain and cell-cell adhesion in the epithelium of an arthropod leg segment. *J. Theor. Biol.* **100:** 443.

Miura N., Ishida N., Hoshino M., Yamauchi M., Hara T., Ayusawa D., and Kawakita M. 1996. Human UDP-galactose translocator: Molecular cloning of a complementary DNA that complements the genetic defect of a mutant cell line deficient in UDP-galactose translocator. *J. Biochem.* **120:** 236.

Nardi J.B. 1981. Induction of invagination in insect epithelium: Paradigm for embryonic invagination. *Science* **214:** 564.

Newman A.P. and Sternberg P.W. 1996. Coordinated morphogenesis of epithelia during development of the *Caenorhabditis elegans* uterine-vulval connection. *Proc. Natl. Acad. Sci.* **93:** 9329.

Newman A.P., White J.G., and Sternberg P.W. 1996. Morphogenesis of the *C. elegans* hermaphrodite uterus. *Development* **122:** 3617.

Podbilewicz B. and White J.G. 1994. Cell fusions in the developing epithelia of *C. elegans. Dev. Biol.* **161:** 408.

Priess J.R. and Hirsh D.I. 1986. *Caenorhabditis elegans* morphogenesis: The role of the cytoskeleton in elongation of the embryo. *Dev. Biol.* **117:** 156.

Rutishauser U. 1996. Polysialic acid and the regulation of cell interactions. *Curr. Opin. Cell Biol.* **8:** 679.

Schachner M. and Martini R. 1995. Glycans and the modulation of neural-recognition molecule function. *Trends Neurosci.* **18:** 183.

Schoenwolf G. and Smith J.L. 1990. Mechanisms of neurulation: Traditional viewpoint and recent advances. *Development* **109:** 243.

Stanley P. and Ioffe E. 1995. Glycosyltransferase mutants: Key to new insights in glycobiology. *FASEB J.* **9:** 1436.

Sulston J.E., Schierenberg E., White J.G., and Thomson J.N. 1983. The embryonic cell lineage of the nematode *Caenorhabditis elegans. Dev. Biol.* **100:** 64.

Sweeton D., Parks S., Costa M., and Wieschaus E. 1991. Gastrulation in *Drosophila*: The formation of the ventral furrow and posterior midgut invaginations. *Development* **112:** 775.

Terayama K., Oka S., Seiki T., Miki Y., Nakamura A., Kozutsumi Y., Takio K., and Kawasaki T. 1997. Cloning and functional expression of a novel glucuronyltransferase involved in the biosynthesis of the carbohydrate epitope HNK-1. *Proc. Natl. Acad. Sci.* **94:** 6093.

Thomas J.H., Stern M.J., and Horvitz H.R. 1990. Cell interactions coordinate the development of the *C. elegans* egg-laying system. *Cell* **62:** 1041.

Trent C., Tsung N., and Horvitz H.R. 1983. Egg-laying defective mutants of the nematode *C. elegans. Genetics* **104:** 619.

Varki A. 1993. Biological roles of oligosaccharides: All of the theories are correct. *Glycobiology* **3:** 97.

Waterston R.H. 1988. Muscle. In *The Nematode* Caenorhabditis elegans (ed. W.B. Wood and the community of *C. elegans* researchers), p. 281. Cold Spring Harbor Laboratory, Cold Spring Harbor, New York.

White J. 1988. The anatomy. In *The Nematode* Caenorhabditis elegans (ed. W.B. Wood and the community of *C. elegans* researchers), p. 81. Cold Spring Harbor Laboratory, Cold Spring Harbor, New York.

Genetics and the Evolution of Plant Form: An Example from Maize

J. DOEBLEY AND R.-L. WANG
Department of Plant Biology, University of Minnesota, St. Paul, Minnesota 55108

The evolution of form has been studied using two distinct approaches: comparative and genetic. The comparative approach examines the differences in shape among organisms and infers from these differences the underlying processes responsible for their evolution. In plants, this approach has been pioneered by morphologists, paleontologists, and systematists. This approach achieves greater power by extending observations to the early stages of ontogeny, allowing one to observe directly how alterations in ontogeny produce different adult forms (see, e.g., Tucker 1984). During the past decade, the comparative approach has been extended to the molecular level in animal systems, revealing conservation or change in the pattern of expression for genes involved in relevant developmental pathways (Raff 1996). Remarkably, comparative analysis of gene expression has not been widely applied in plants (but see Sinha and Kellogg 1996). The power of the comparative approach is that it enables one to study macroevolutionary changes that distinguish the higher-level taxonomic groups.

A complementary approach looks at the evolution of form on the genetic level. Through the study of the patterns of inheritance within and among species, the genetic changes involved in the evolution of new forms can be directly observed. One might assay the amount of genetic variation for a trait within species or populations to learn the extent of the raw material available for selection to act upon (see, e.g., MacKay and Langley 1990) or study the inheritance of fixed differences between related species or subspecies (see, e.g., Bradshaw et al. 1995). Genetic analyses of evolutionary change have been pioneered by population and quantitative geneticists (Falconer and MacKay 1996). The genetic approach is restricted to the analysis of recently evolved traits that distinguish or vary within cross-compatible taxa, and as such is anchored within the realm of microevolution. The power of this approach is that it enables one to identify the specific genes responsible for evolutionary change. This opens the window for addressing a variety of questions: (1) How many genes are typically involved in the evolution of a new trait, (2) what types of genes are typically involved (e.g., regulatory or target genes), and (3) what is the nature of the changes in these genes?

Our group has been studying the morphological evolution of cultivated maize (*Zea mays* ssp. *mays*) from its wild ancestor, teosinte (*Zea mays* ssp. *parviglumis*) using a genetic approach. As a recent progenitor-derivative "species-pair," maize and teosinte are fully cross-compatible. Both have ten gametic chromosomes, and gene arrangement on these is collinear. Nevertheless, as a result of human selection for higher yield and greater harvestability, maize and teosinte differ radically in plant and inflorescence architecture (Fig. 1). Thus, maize and teosinte represent an ideal pair for the application of a genetic analysis of morphological evolution. Their utility is further enhanced in that the many tools of maize genetics developed over the preceding century readily transport to teosinte. In this paper, we review our efforts to elucidate the genetic basis of the morphological evolution of maize from teosinte. The focus is on the evolution of overall differences in plant architecture. We summarize what has been learned about the numbers of genes responsible for the differences in plant architecture, the molecular cloning of one of these genes, *teosinte branched1* (*tb1*), and our current model for how changes in *tb1* generated the differences in plant form between maize and teosinte.

THE PLANT ARCHITECTURE OF MAIZE AND TEOSINTE

Teosinte plants have main stalks that typically bear an elongate lateral branch at most nodes (Fig. 1a,b). The lateral branches and the main stalk are both composed of a series of nodes and elongated (15 cm or more long) internodes with a leaf attached at each node. The number of internodes (or leaves) in each lateral branch is roughly equivalent to the number in the main stalk above the point of attachment of the branch. Thus, a branch attached on the third node below the main tassel will be composed of about three internodes, whereas one attached at the sixth node below the tassel will have about six internodes. The leaves along the lateral branches are fully formed and composed of two parts—a sheath that clasps around the stem and a blade that extends away from the plant. The leaves on the branches are arranged in an alternate phyllotaxy; i.e., each leaf is borne on the opposite side of the stem relative to the leaves at the nodes above and below. Both the main stem and the primary lateral branches of the teosinte plant are tipped by male inflorescences (tas-

Figure 1. (*a*) Teosinte (*Zea mays* ssp. *mexicana*) plant and (*b*) its axillary branch with terminal tassel (T) and silks (S) emerging from teosinte ears hidden within the leaf sheaths. (*c*) Maize plant and (*d*) ear shoot. (*e*) *tb1-ref* mutant maize plant and (*f*) axillary branches that have terminal male inflorescences and lack ears.

sels), whereas the slender female inflorescences (ears) of teosinte are borne on secondary branches in the axils of the leaves along the primary branches. Each of these female inflorescences is surrounded by a single, bladeless leaf or husk. The ears occur in clusters of one to five (or more) at each node along the branch.

The construction of the lateral branches of the maize plant is strikingly different from that of teosinte (Fig. 1c,d). Maize typically produces branches at only two or three of the nodes along the main stem. At most other nodes, axillary buds are present but they are arrested early in development. Each of the branches that are produced is composed of nodes and short internodes, averaging about 1 cm in length. Unlike teosinte, the number of internodes in the lateral branch is greater than the number in the main stalk above the point of attachment of the branch. For maize line W22, a branch attached at the fifth node below the main tassel will be composed of about 12 internodes and have 12 husk leaves. The leaves (husks) along the lateral branch are composed largely of sheath with only a small (if any) blade attached to it. The husks are arranged in a spiral phyllotaxy along the branch, rather than the alternate phyllotaxy for leaves on the main stem of the plant. Secondary branches are normally absent (aborted). Finally, the lateral branch is terminated by a female inflorescence or ear, which is tightly enclosed within the spirally arranged husks because of the failure of the internodes of the branch to fully elongate.

The divergent morphologies of maize and teosinte lateral branches reveal the complex set of evolutionary changes involved in the origin of maize plant architecture. These include (1) arresting the full elongation of the internodes, (2) arresting the growth of the leaf blade, (3) arresting the outgrowth of some axillary buds, (4) suppressing the outgrowth of secondary branches, (5) increasing the number of leaves (husks) formed on the branch, (6) changing leaf (husk) phyllotaxy from alternate to spiral, and (7) replacing the tassel (male inflorescence) at the tip of the lateral branch with an ear (female inflorescence). Noticeably, most of these changes involve arresting the growth of particular organs. This suggests that a common developmental process (repression of organ growth) expressed at distinct points in space or time of organogenesis produced this suite of correlated changes.

A QUANTITATIVE GENETIC APPROACH

To investigate the inheritance of the differences in plant architecture between maize and teosinte, our group constructed and analyzed two maize-teosinte F_2 populations (Doebley and Stec 1993). Examination of the individual F_2 plants revealed that the traits involved did not segregate in a discrete Mendelian manner, but rather demonstrated continuous variation. For example, the branch length, the length of the blade on the sheath, and the number of visible branches on a plant all showed continuous or quantitative variation. Even the sex of the inflorescence terminating the lateral branches varied from purely female through various degrees of mixed-sex to purely male.

The inheritance of such quantitative traits is most powerfully analyzed by quantitative trait locus or QTL mapping (Tanksley 1993). This method relies upon genetic linkage analysis and the availability of marker loci about every 20 cM throughout the genome. Essentially, one regresses the phenotypic values for each trait onto the marker locus genotypes and asks whether there is a statistically significant association between them. Where such associations are found, one has evidence for a QTL. By QTL-mapping, one can estimate the number of loci controlling a trait, the chromosomal locations of these loci, the proportion of the phenotypic variance for which each locus accounts, and the mode of gene action of each locus (i.e., dominant, additive, or overdominant). QTL mapping has limits and biases. QTL of very small effects would require impractically large populations to detect. Multiple, tightly linked QTL cannot be discerned from a single QTL. QTL linked in repulsion phase may not be detected. Estimates of the effects and positions of QTL can be biased, especially when analyzing small populations (Beavis 1994). These limitations notwithstanding, QTL mapping offers a powerful and proven means of dissecting the inheritance of complex traits. Moreover, the reliability of the results can be assessed by repeated and independent analyses (Doebley and Stec 1993).

To investigate the inheritance of plant architecture in maize and teosinte, our group has applied QTL mapping to a variety of traits (Doebley and Stec 1993; J. Doebley, unpubl.). These include the number of tillers, the length of the internodes in the lateral branches, the number of ears along the lateral branch (prolificacy), the length of the blade of leaves on the lateral branch, and the sex of the inflorescence terminating the lateral branch. The results of these analyses for one F_2 population are presented graphically in Figure 2. These analyses reveal several features about the inheritance of plant architecture. First, the traits shows multigenic rather than monofactorial inheritance. Second, although inheritance is multigenic, there is a considerable range of effects among QTL with some accounting for nearly 40% of the phenotypic variance. Third, there are QTL for multiple traits on chromosome arms 1L, 3L, 4S, 5S, and 6L. This coincidence of QTL location suggests that the individual QTL have pleiotropic effects on multiple traits. Finally, for four of the five traits, there is a QTL of large effect on the long arm of chromosome 1. From these analyses, one can infer that the evolution of maize plant architecture from that of teosinte involved changes in a relatively small number of effective genes including one of large effect on the long arm of chromosome 1.

FROM QTL TO GENE

A goal of our genetic dissection of the morphological evolution of maize was to go beyond the rather amorphous concept of QTL and precisely identify the underlying genes. Our best chance at realizing this goal was with QTL of large effect whose phenotypes are most stably expressed. Our chances would be enhanced if the QTL in question were associated with a known major mutant of

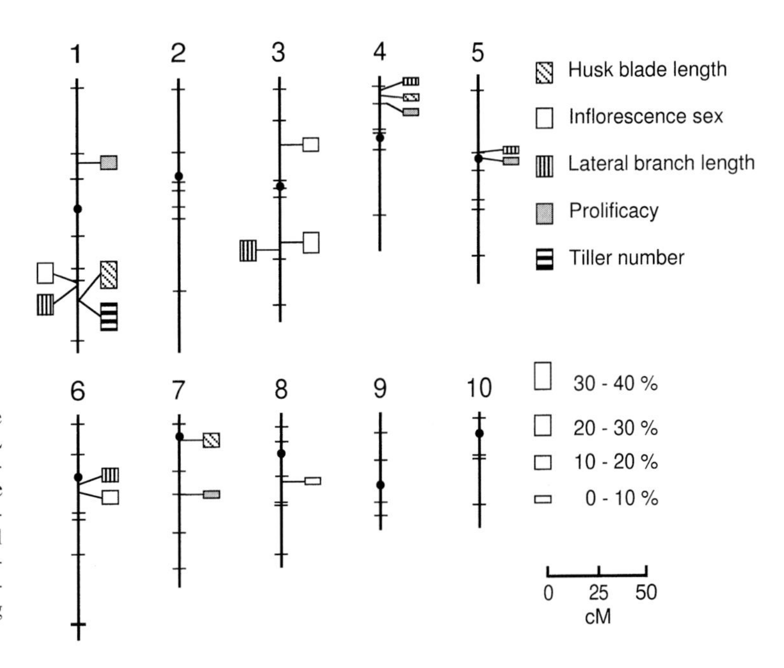

Figure 2. Map of the ten maize-teosinte chromosomes showing the positions of QTL for traits related to plant architecture (Doebley and Stec 1993). Cross-markings on the chromosomes indicate the positions of molecular markers used in the analysis and small black dots indicate the positions of the centromeres. For each chromosome, the cytologically defined short arm is above and the long arm is below.

maize such that the mutant locus could be used as a tool for the analysis and cloning of the QTL. Attempting to associate a QTL with a known major mutant has been referred to as the "candidate locus" approach.

Examination of the genetic map for maize revealed a promising candidate locus, *teosinte branched1* (*tb1*), for the QTL of large effect on chromosome arm 1L. *tb1* is a plant architecture mutant of maize that, as its name suggests, makes the maize plant resemble a teosinte plant. Like teosinte, plants homozygous for the reference allele (*tb1-ref*) have long lateral branches tipped by tassels at some upper nodes of the main culm (Fig. 1e,f). *tb1-ref* plants also have many tillers at the basal nodes. In W22 genetic background, the uppermost lateral branch of *tb1-ref* plants has about 5 or 6 leaves with mixed alternate-spiral phyllotaxy rather than the 12 spirally arranged husks of wild-type W22 plants. These differences in leaf/husk number and phyllotaxy also parallel differences between maize and teosinte as described above. *tb1-ref* plants differ from teosinte in that they do not form normal ears, their secondary branches typically bearing only sterile, tassel-like inforescences where teosinte bears its ears (cf. Fig. 1b,f). The inability of *tb1-ref* plants to form ears indicates that TB1 function is necessary for normal ear development. Since teosinte produces ears, it seems likely that teosinte has a functional copy of *tb1*, but one that is acting differently from its maize counterpart.

The coincident positions of *tb1* and our QTL, although suggestive, do not prove that they are the same. To establish this link more definitively, we first employed a standard genetic complementation test (Doebley et al. 1995). This test was possible because the teosinte allele of our QTL had a phenotype like that of *tb1*, and both *tb1* and our teosinte QTL were fully recessive for the conversion of the ear into a tassel. Results of the complementation test were definitive, showing that the teosinte allele of our QTL does not complement *tb1*. This result enabled us to conclude that *tb1* was our QTL.

As an additional test of the correspondence between *tb1* and QTL-1L, we have recently fine-mapped the *tb1* region of the chromosome arm 1L (R.-L. Wang and J. Doebley, unpubl.). This experiment employed a series of recombinant nearly isogenic lines, each possessing a different portion of the teosinte chromosome segment surrounding *tb1* in the genetic background of a maize inbred line (W22). These lines were created using molecular markers within the region to identify crossovers between markers. One of these markers was *tb1* itself. These lines were grown in a randomized design and their phenotypes evaluated. Statistical analysis of the phenotypic data using stepwise multiple regression enabled us to map the QTL to specific segments (or molecular markers) within the region. By measuring multiple traits, we could also test whether pleiotropy or linkage of multiple QTL accounted for the different phenotypic effects.

Results of the fine-scale mapping confirmed that our QTL maps to *tb1* (Fig. 3). Each trait has a large effect that maps to the *tb1* marker. For some traits, an additional QTL of smaller effect also maps within the segment. The fact that the principal effect on each trait mapped to *tb1* supports the interpretation that these different phenotypic effects represent the pleiotropic effects of a single gene (*tb1*) rather than linkage of multiple genes each controlling a different trait.

THE *teosinte branched1* GENE

With the results of the QTL mapping and complementation test in hand, our next step was to clone *tb1*. Direct cloning of a QTL is impractical in maize; however, cloning of recessive mutants by transposon mutagenesis is a straightforward if tedious procedure. Accordingly, we used the *Mutator* transposon system of maize and cloned *tb1* (Doebley et al. 1997). An active *Mu* stock was crossed to a stock carrying *tb1-ref*, and approximately 26,000 hybrids were grown and examined. Among these, three new *tb1* mutants were observed, and molecular analyses demonstrated that each contained a *Mu* insertion within *tb1*.

Genbank searches with the *tb1* sequence revealed that *tb1* is a member of a novel group of genes whose molecular function is unclear. It has homology with *cycloidea*

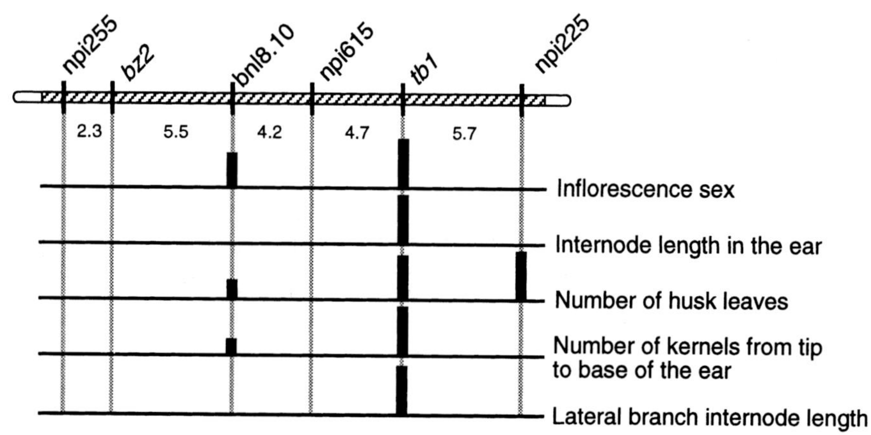

Figure 3. Map of a portion of the long arm of chromosome 1 surrounding *tb1*. Molecular marker names appear above the chromosome and map distances in centiMorgans appear between markers below it. Black columns show the positions of QTL for traits related to plant architecture. Column heights for traits with two QTL are scaled in relation to the magnitudes of the effects of the QTL (R.-L. Wang and J. Doebley, unpubl.).

(*cyc*) of snapdragon (*Antirrhinum majus*) and several expressed sequence tags (ESTs) of *Arabidopsis* and maize (Luo et al. 1996; Doebley et al. 1997). The sequence homology among these genes is restricted to two domains of the protein (Fig. 4a). The ATM (*Antirrhinum*-teosinte-maize) domain is approximately 60 amino acids in length, occurs in all sequences, and contains a putative bipartite nuclear localization signal (NLS) in *tb1* and *cyc*. The presence of this conserved NLS in *tb1* and *cyc* suggests that these genes have a role in the regulation of transcription (Luo et al. 1996). The shorter, KARE domain is composed of a conserved mix of basic (arginine and lysine) and neutral or acidic (arginine and glycine) amino acids. The core of this domain has alternating basic and nonbasic residues.

Phylogenetic analysis of the ATM domain indicates that *tb1* and *cyc* are closer to each other than to other members of this gene family (Fig. 4b). The expected number of amino acid differences per site is 0.43 between *cyc* and *tb1*, but averages 0.67 between these two genes and the ESTs. Thus, *tb1* and *cyc* may be truly homologous as opposed to merely being members of the same gene family. Sequence variation within the KARE domain is concordant with this interpretation since *tb1* and *cyc* are more similar to one another than to the ESTs (Fig. 4a).

At first glance, the phenotypes of *tb1* and *cyc* appear unrelated, *tb1* controlling apical dominance and *cyc* controlling floral symmetry. Nevertheless, there are similarities which suggest that the underlying gene function may be conserved. First, both mutants affect axillary structures, either flowers (*cyc*) or branches (*tb1*). Second, both genes have been proposed to function as repressors of organ growth (Doebley et al. 1995; Luo et al. 1996). Third, both genes affect the growth of petals and stamens. CYC arrests the growth of both the dorsal petal and stamen of the snapdragon flower. In the ear where *tb1* is expressed, growth of the lodicules (petals) and stamens is arrested.

The pattern of expression of *tb1* as revealed by Northern blot analysis provides some insights into the potential function of the protein it encodes. *tb1* is expressed in immature internodes of the axillary branch, in the ear primordium at the tip of the axillary branch, and in the immature husk leaves that surround the branch (Doebley et al. 1997). The husk and internodes are both reduced in size in wild-type maize as compared to *tb1* mutant plants. This suggests that TB1 functions to arrest or repress the growth of these organs. The ear is a complex structure and Northern analysis does not reveal where in the ear *tb1* is expressed. One possibility is that *tb1* is expressed in flowers (stamens and petals) as is its homolog *cyc* (Luo et al. 1996). Here, TB1 would arrest stamen growth, and thereby allow pistil development and ear formation. With

a

ATM Domain

```
tb1           AAARKDRHSKICTAGGMRDRRMRLSLDVARKFFALQDMLGFDKASKTVQWLLNTSKSAIQEI
cyc           QAVKKDRHSKIYTSQGPRDRRVRLSIGIARKFFDLQEMLGFDKPSKTLDWLLTKSKTAIKEL
AT 75H5T7     STGRKDRHSKVCTAKGPRDRRVRLSAPTAIQFYDVQDRLGFDRPSKAVDWLITKAKSAIDDL
AT 161L4T7    ASGGKDRHSKVLTSKGPRDRRVRLSVSTALQFYDLQDRLGYDQPSKAVEWLIKAAEDSISEL
AT 163F23T7   xxxxxxxxxxxxxxxxxxxxxxxxxVSTALQFYDLQDRLGYDQPSKAVEWLIKAAEDSISEL
ZM php20581   ASxGKDRHSNVYTAKGIRDRRVRLSVATAxxxYDLQDRxGYDQPSKAIEWLIMVAAEAIDKL
```

KARE Domain

```
tb1           KETRAKARERARERTKEK
cyc           KESRAKARARARERTKEK
AT 161L4T7    SELRDKARERARERTAKE
AT 163F23T7   SELRDKARERARERTAKE
```

b

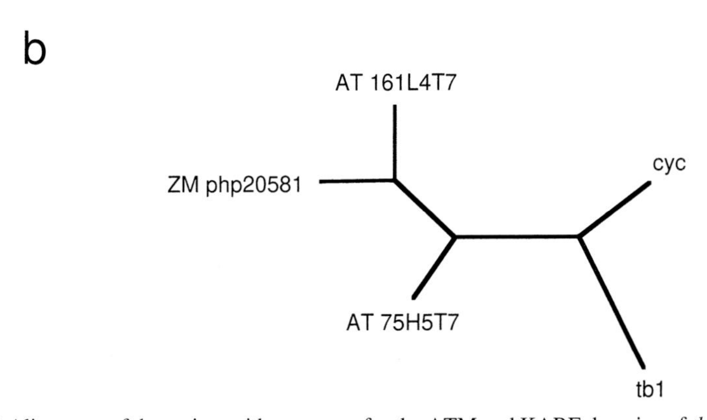

Figure 4. (*a*) Alignment of the amino acid sequences for the ATM and KARE domains of *tb1* and *cyc* proteins. Underlining indicates the position of a putative bipartite nuclear localization signal found in *tb1*, *cyc*, and one of the ESTs. (*b*) Unrooted neighbor-joining phylogenetic tree for the SAT domain.

loss of *tb1* expression in stamen primordia as occurs with *tb1* mutant alleles, stamen growth is derepressed, and we hypothesize that the growing stamens would produce a secondary signal that blocks pistil development (cf. Dellaporta and Calderon-Urrea 1994).

A MODEL FOR MAIZE EVOLUTION

A complete understanding of the role of *tb1* in maize evolution would allow us to describe specific nucleotide sequence differences between the maize and teosinte alleles, to specify how these differences affect gene expression or function, and to explain how such functional differences alter the course of ontogeny to give rise to the very different adult morphologies of these plants. This is a goal that we have yet to attain. Nevertheless, we have a sufficient number of pieces of the puzzle to propose a reasonable, if still incomplete, model for the role of *tb1* in maize evolution.

As discussed above, the proposed function of TB1 is a repressor of organ growth. This is based on the correlation between domains of *tb1* expression and a reduction in organ size. Repressor function could also explain the role of *tb1* in ear formation if *tb1* represses stamen growth and thereby blocks a signal originating in the stamens that represses pistil growth. Two aspects of the *tb1-ref* phenotype are not readily explained by a repressor model. First, the number of leaves/husks (or internodes) in the axillary branches (ear shoots) is larger in wild-type maize than in *tb1-ref* plants. Thus, rather than repressing leaf initiation, *tb1* appears to promote it. Second, leaf phyllotaxy in wild-type maize ear shoots is spiral, whereas it is at least partially alternate in *tb1-ref* axillary branches. How *tb1* controls these differences must be explained as well.

Assuming TB1 functions as a repressor of organ growth, most differences in plant form between teosinte and maize can be readily explained by changes in the spatial pattern and/or level of *tb1* expression. In teosinte, *tb1* should be off or expressed at low levels in the primordia that form the primary branches (Fig. 5). This would enable the growth of these primordia into fully elongated branches. *tb1* should also be off (or at low levels) in the inflorescence primordia terminating the primary branch and its stamen primordia, so that these stamens would not be repressed and a tassel rather than an ear would be formed. In teosinte, *tb1* would be expressed more strongly in the primordia that form the secondary branches, governing their conversion into short ear shoots with a surrounding bladeless husk leaf. With this starting point, the evolution of maize would require an increase in *tb1* expression in the primary axillary branch primordium and its terminal inflorescence primordium so that they form short ear shoots rather than elongated, tassel tipped branches. Similarly, an increase in the level of expression in the secondary branch primordia could cause full abortion of these branches as is found in most varieties of maize.

Given this model, what would one expect upon loss of TB1 function in *tb1* mutant maize? First, the primordia that form the primary branches would be derepressed and develop into elongated branches (Fig. 5). Second, with loss of TB1 function in the inflorescence primordium, stamens would be derepressed, promoting tassel rather than ear formation. Third, secondary branches would also be derepressed, but they could not form ears as in teosinte since they lack TB1 function needed for stamen repression and ear formation. Instead, the sterile tassel-like inflorescences are formed on secondary branches. Finally, since TB1 is required for repression of all axillary buds, *tb1* mutant plants would exhibit the unrestrained outgrowth of all axillary branches, producing an extremely bushy plant.

There is presently some data to support the above model (Doebley et al. 1997). First, comparison of the maize and teosinte alleles for the level of *tb1* message accumulation indicated that the maize allele is expressed at about twice the level of the teosinte allele in the inflorescence primordia terminating the primary lateral branches.

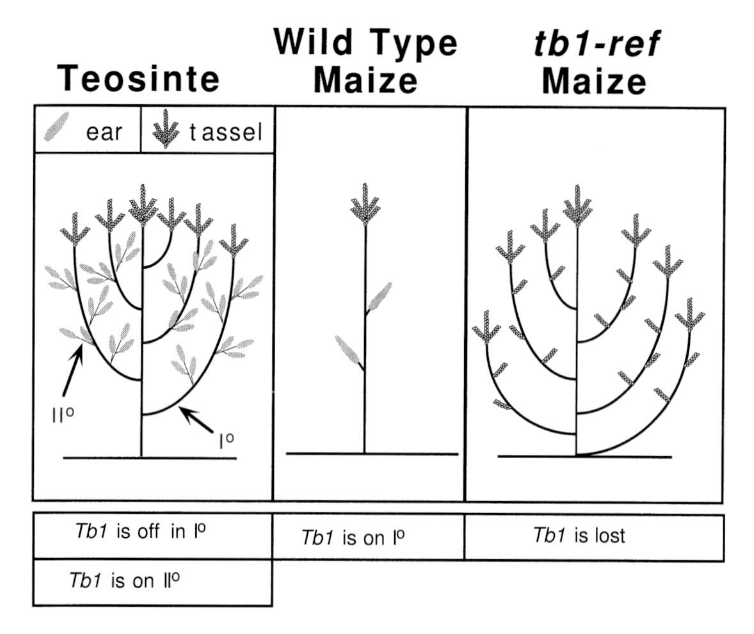

Figure 5. Diagrammatic representation of a model for the role of *tb1* in the evolution of maize plant architecture. In teosinte, *tb1* is off (or below a critical threshold) in the primary (I°) branch primordia so that they develop into long branches tipped by tassels, but on in secondary (II°) branch primordia so that they develop into short branches tipped by ears. In wild-type maize, *tb1* is on in the primary branch primordia so that they develop into short branches tipped by ears. In *tb1-ref* mutant maize, normal TB1 function is lost entirely so that primary branch primordia develop into long branches tipped by tassels and secondary branch primordia produce tassels rather than ears since TB1 is required for normal ear formation.

Second, there are no fixed amino acid difference between maize and teosinte, suggesting that a change in the function of TB1 has not occurred. If the function of the protein has not changed, then it is more likely that changes in expression underlie the evolutionary differences between maize and teosinte.

CONCLUSION

Our research has taken a genetic approach to understanding the events involved in the evolution of maize from its wild ancestor, teosinte. QTL mapping suggests that a relatively small set of genes was responsible for the evolution of the overall differences in plant form between maize and teosinte. Moreover, one of these QTL has a rather large effect and principally controls the differences between maize and teosinte plant form. Using the power of genetics, we were able to establish that the QTL of largest effect corresponds to a known maize mutant, *tb1*. The identification of *tb1* as our QTL was a crucial step in our research since it enabled us to move beyond the QTL, which is little more than a statistical rendering of a gene, and begin to analyze the actual gene controlling the traits of interest. With the gene in hand, many new avenues of investigation have been opened. A knowledge of the gene and its putative function enabled us to propose a specific model for the developmental and molecular events that underlie the differences in adult morphology between maize and teosinte. We have been able to partially test this model and further tests are in progress. Ultimately, it will be possible to learn how specific differences between the maize and teosinte alleles of *tb1* alter gene expression/function which in turn alters the course of development.

The morphological evolution of maize from teosinte is a remarkable example of how human selection can drive evolutionary change over the relatively short periods of time required for plant domestication. Could the maize-teosinte example also be informative as to how plant form has evolved among natural species over longer evolutionary periods? For example, among the grasses, there is considerable diversity in branching habit, differential organ abortion, and the sexuality of inflorescences, the same traits that differentiate maize and teosinte and are under the control of *tb1*. Although clearly no single magic gene or even class of genes generated this diversity, *tb1*-like genes can be considered reasonable candidates for the causal agents that produced at least some of the diversity in plant form that exists in nature. This possibility could be examined by a comparative analysis of *tb1* expression patterns among diverse species. If *tb1* is shown to have been involved in plant evolution more broadly, it would support our view that the domestication of maize provides an appropriate model for the morphological evolution of plants in general.

ACKNOWLEDGMENTS

The research presented was supported by grants from the National Science Foundation (MCB-9513573) and USDA (93-37300-8773). We thank Korise Rasmusson for comments on this manuscript and the organizers of the symposium for the invitation to participate.

REFERENCES

Beavis W. 1994. The power and deceit of QTL experiments: Lessons from comparative QTL studies. In *Proceedings of the 49th Annual Corn and Sorghum Research Conference*, p. 250. American Seed Trade Association, Washington, D.C.

Bradshaw H.D., Wilbert S.M., Otto K.G., and Schemske D.W. 1995. Genetic mapping of floral traits associated with reproductive isolation in monkey flowers (*Mimulus*). *Nature* **376:** 762.

Dellaporta S. and Calderon-Urrea A. 1994. The sex determination process in maize. *Science* **266:** 1501.

Doebley J. and Stec A. 1993. Inheritance of the morphological differences between maize and teosinte: Comparison of results for two F_2 populations. *Genetics* **134**: 559.

Doebley J., Stec A., and Gustus C. 1995. *Teosinte branched1* and the origin of maize: Evidence for epistasis and the evolution of dominance. *Genetics* **141:** 333.

Doebley J., Stec A., and Hubbard L. 1997. The evolution of apical dominance in maize. *Nature* **386:** 485.

Falconer D.S. and MacKay T.F.C. 1996. *Introduction to quantitative genetics*. Longman Press, Edinburgh, Scotland.

Luo D., Carpenter R., Vincent C., and Coen E. 1996. Origin of floral asymmetry in *Antirrhinum*. *Nature* **383:** 794.

MacKay T.F.C. and Langley C.H. 1990. Molecular and phenotypic variation in the *achaete-scute* region of *Drosophila melanogaster*. *Nature* **348:** 64.

Raff R.A. 1996. *The shape of life: Gene, development, and the evolution of animal form.* University of Chicago Press, Illinois.

Sinha N. and Kellogg E.A. 1996. Parallelism and diversity in multiple origins of C4 photosynthesis in the grass family. *Am. J. Bot.* **83:** 1458.

Tanksley S.D. 1993. Mapping polygenes. *Annu. Rev. Genet.* **27:** 205.

Tucker S.C. 1984. Origins of symmetry in flowers. In *Contemporary problems in plant anatomy* (ed. R.A. White and W.C. Dickison), p. 351. Academic Press, New York.

Control of Cell Division Patterns in Developing Shoots and Flowers of *Arabidopsis thaliana*

E.M. MEYEROWITZ
Division of Biology, California Institute of Technology, Pasadena, California 91125

Development of flowering plants is different from that of the animal model systems. In flowering plants such as *Arabidopsis,* embryogenesis serves primarily to establish the shoot and root apical meristems, which later develop into the mature plant. Most morphogenesis and pattern formation thus occur postembryonically. The shoot apical meristem (SAM), which is a small collection of undifferentiated cells, first forms and then arrests during embryogenesis; after seed germination, it activates and becomes the source of the cells that will later make up the entire above-ground part of the plant (Steeves and Sussex 1989; Meyerowitz 1997). The morphogenetic activities of the SAM after seed germination consist of a small number of stereotyped programs, with the environment playing an important part in which programs are selected. This allows the plant to respond continually to changes in its environment by altering its growth and development, rather than responding as do most animals by activities of brain and muscles. The fundamental programs followed by the *Arabidopsis* SAM, in addition to maintenance (which allows the meristem to serve as a continuing population of stem cells), are either production on its flanks of leaves with axillary secondary SAMs or production of flowers (Fig. 1A,B,C). A secondary SAM can behave as the primary SAM, making third-order meristems or flowers, or it can be held in a state of developmental arrest. These simple sets of meristematic choices thus provide a tool kit for plant architecture, and the choice made at any time by each meristem sums to the total form of the plant.

Little is known about the control of pattern formation in SAMs and their derivatives, but one thing is clear: It depends very directly on control of planes and numbers of cell divisions. This can be inferred from the small number of processes available for plant morphogenesis. There is no cell migration in meristems, nor any slippage of cells relative to one another, as the cells are encased in a cellulose-based wall. Although plants use programmed cell death for many things, it does not appear that they use it to regulate the number of cells in meristems. Instead, it is used to make dead structures, such as xylem or autumn leaves, or as a local response to pathogens. Without migration, slippage, or cell death to regulate cell number and cell position in meristems, only highly regulated cell division is left.

Beyond inference, there is much direct evidence of tight control of planes and numbers of cell divisions in SAMs. Even in the embryo, an *Arabidopsis* SAM has three distinct cell layers, which remain clonally distinct through the life of the plant, by continued anticlinal division in the outer two layers (Fig. 1D). The L1 layer is on the surface and is ancestral to the epidermal cell layer of the shoots, leaves, and flowers. The L2 layer is directly beneath the L1 layer; its derivatives are the subepidermal cells of stems, leaves, and floral organs. In addition to the anticlinal divisions that maintain this layer, the development of leaves and floral organs involves regulated periclinal divisions of L2 cells, so that in a mature organ, L2 derivatives can provide several layers of cells. Among the L2 derivatives are the germ cells, found in pollen grains and ovules. The L3, or corpus, is not a single cell layer, but a collection of cells in which divisions occur in all planes. The corpus derivatives include the pith and vasculature of the stem and the most central cells of leaves and floral organs.

In addition to this layering, which is well demonstrated in many plants by experiments with genetic mosaics (Tilney-Bassett 1986), observations of cell division patterns and cellular morphology in SAMs show that the meristem is divided in a different way, into domains of cell division activity that cut across the clonal boundaries (see Fig. 1C). The central zone is at the meristematic tip and consists of slowly dividing, relatively inactive cells. Surrounding the central zone is the peripheral zone, consisting of metabolically active and rapidly dividing cells. It is cell divisions in the peripheral zone that establish the leaf/shoot or floral units that develop on the meristematic flanks. The core of the meristem is the rib meristem, which has its own cell division rate and pattern, making long lines of cells that are the lineages contributing to stem growth (Steeves and Sussex 1989; Meyerowitz 1997).

The maintenance of meristem structure and shape throughout the life of a plant and the formation of structures with characteristic positions, shapes, and sizes, such as leaves and floral organs, also demonstrate a very tight and regulated control of cell division (and cell elongation) patterns. Furthermore, the coordinated growth of the clonal layers and demonstrations that reduction in cell division rates in one layer are accommodated by increased cellular proliferation in other layers (Tilney-Bassett 1986) indicate that cells in different meristematic regions communicate cell division information to each other. We

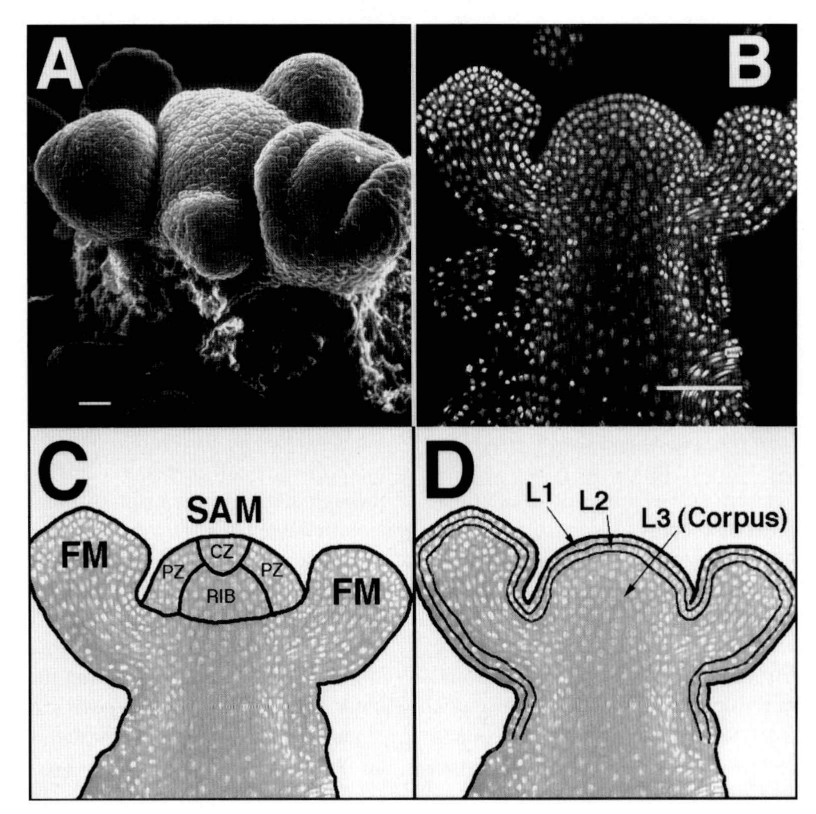

Figure 1. Shoot apical meristem structure. (*A*) Scanning electron micrograph of an *Arabidopsis thaliana* (Landsberg *erecta* wild-type) SAM, surrounded by secondary meristems (which will develop into flowers). Bar, 10 μm. (*B*) Laser confocal microscope optical section through an apex similar to that in *A*, but from ecotype Ws-0. Nuclei were stained with propidium iodide. Bar, 50 μm. (*C*) Same apex shown in *B*, but with different meristems and meristematic regions labeled. (SAM) Shoot apical meristem; (FM) floral meristem; (CZ) central zone; (PZ) peripheral zone; (RIB) rib meristem zone. (*D*) The same apex as in *B* and *C*, this time labeled to show the clonal layers.

do not know how meristem cells send or receive cell division information. Finding out will illuminate the basic processes of plant development and also allow experimental control of plant form.

My laboratory has taken a genetic approach toward finding how meristematic cells communicate cell division information, by inducing mutations and collecting mutant lines in which the usual regulated pattern of cell divisions in meristems has become abnormal. One sensitive indicator of changes in meristematic cell division pattern is change in floral organ number. *Arabidopsis* flowers derive from floral meristems, which are products of SAMs but are not themselves like shoot apical meristems, in that they make a different set of organs and are determinate in their growth pattern (Smyth et al. 1990). Nonetheless, many genes that regulate patterns and numbers of cell divisions in SAMs also appear to be important in cell division control in floral meristems; in floral meristems, extra cells result in extra floral organs, whereas reduced cell number in floral meristems results in reduced numbers or absence of floral organs (Koornneef et al. 1983; Clark et al. 1993, 1995, 1996; Laux et al. 1996). Changes in the numbers of floral organs are easily detected in mutant screens and lend themselves to quantitation by counting of organs. They thus serve as a convenient phenotype by which to find mutations affecting the regulation of meristematic cell divisions. Although it is equally easy to screen for mutations with reduced or extra floral organs, we have concentrated on extra organ mutants because these cannot be explained by the trivial possibility that the plants are simply unable to make cellular components such as cell wall, or are improperly nourished.

RESULTS AND DISCUSSION

The first extra-organ mutations that we studied in detail were recessive and semidominant alleles of *CLAVATA1* (*CLV1*, Leyser and Furner 1992; Clark et al. 1993; Crone and Lord 1993). Loss-of-function mutations in *CLV1* cause progressive enlargement of the shoot apical meristem during plant growth. In mature *clv1* embryos, the SAM is slightly larger than normal; it becomes progressively larger as postgermination growth occurs. This results in increasing thickness of the main stem with time, sometimes proceeding to grossly abnormal, fasciated forms, and loss of the usual phyllotactic pattern of leaves and secondary meristems. Secondary and higher-order SAMs are similarly affected, and floral meristems are larger than normal (because they have extra cells) from early stages. Wild-type floral meristems at a stage before the primordia of the inner floral organs have formed (stage 3; Smyth et al. 1990) are flattened domes, with a diameter of approximately 50 μm and a height of approximately 12 μm. Floral meristems of homozygotes for a strong *clv1* mutant allele like *clv1-4* average about 70 μm in diameter and 40 μm in height, with the same cell size as in wild type. The consequence of this for flowers is that each flower has many extra organs. A wild-type *Arabidopsis* flower has four sepals, four petals, six stamens, and a compound ovary made of two fused carpels. Flowers of a line homozygous for *clv1-4* may average five or six sepals, five petals, nine or ten stamens, and have ovaries of four to six carpels. *clv1* mutations also cause continued cell division in the center of developing flowers, beyond the stage when such cell division would ordinarily stop. This causes extra ovaries to form within the fourth whorl ovary, in a Russian-doll type of arrangement. This slight loss of determinacy is greatly exaggerated in double-mutant combinations with genes that are required for floral determinacy, such as *AGAMOUS* (*AG*; Yanofsky et al. 1990). *ag-2 clv1-4* double-mutant floral meristems, for example, can fasciate and grow to enormous sizes, all the while making endless whorls of floral organs (Clark et al. 1993; Meyerowitz 1997).

The phenotypes of *clv1* mutants point to several possible models for the action of the gene. One prominent possibility is that the gene is involved in the cell-cell signaling that regulates the number of cell divisions in shoot and floral meristems and that the wild-type function of the gene is to act in the reception and interpretation of cell division signals. Loss-of-function alleles cause extra cell division, so the normal function of the gene would be to repress excess meristematic cell divisions.

To clarify the biochemical function of *CLV1*, we cloned the gene by chromosome walking (Clark et al. 1997). It codes for an apparent transmembrane receptor kinase. The encoded protein has at its amino-terminal end a potential signal peptide, followed by a putative extracellular domain of 21 complete leucine-rich repeats. This is followed by what is likely to be a transmembrane domain. The putative intracellular domain has all of the residues found conserved among serine/threonine protein kinases. Consistent with this, the putative kinase domain, as expressed in an *Escherichia coli* protein expression system, shows protein kinase activity, autophosphorylating on serine (Williams et al. 1997).

As leucine-rich repeats are well-characterized protein-binding motifs (Buchanan and Gay 1996), it would appear that the CLAVATA1 protein binds an extracellular protein or peptide ligand, with binding activating a Ser/Thr protein kinase, and through a signal transduction cascade, repressing certain meristematic cell divisions. In situ hybridization with an antisense probe complementary to a non-cross-hybridizing portion of the *CLV1* RNA shows that the RNA is present in both shoot and floral meristems. Expression is seen at least as early as mature seeds in the shoot apical meristem (H. Sakai and E.M. Meyerowitz, unpubl.), and in postgermination plants, the shoot apical meristem shows *CLV1* RNA in a region that approximately corresponds to the rib meristem (Clark et al. 1997). If it is true that *CLV1* acts in the cells where its RNA is expressed, then the *CLV1* mutant phenotype is in part nonautonomous: Loss-of-function mutations in *CLV1* cause enlargement of the entire shoot apical meristem, including all three clonal layers. In situ hybridization does not detect *CLV1* RNA in the L1, and probably not in the L2 layer, although in *clv1* mutants, the layered nature of the meristem is not disrupted. Thus, excess cell division in the deeper cells of the meristem seems to activate comparable excess division in the L1 and L2 layers, indicating that there is communication of cell division information between meristematic cells. A reasonable speculation would be that CLV1 is involved in such communication, as a receptor in rib meristem cells.

One can then ask what might be some of the other components of the meristematic communication system. The products of other genes with mutant phenotypes similar to those of *clv1* are clear candidates. We have studied one such gene in some detail; it is called *CLAVATA3* (Fig. 2) (Alvarez and Smyth 1994; Clark et al. 1995). The mutant phenotypes of the weaker *clv3-1* and stronger *clv3-2* alleles are identical to those of *clv1* alleles—extra cells in the SAM of the mature embryo, increasing cell number and size of the SAM throughout the life of the plant, stem thickening and occasional stem fasciation, and floral meristems with extra cells, leading to flowers with excess numbers of organs, and nested ovaries. *CLV1* maps to the first chromosome of *Arabidopsis*, whereas *CLV3* maps to chromosome 2, so the mutations in *clv1* and *clv3* are clearly not allelic. Nonetheless, doubly heterozygous plants (*clv1/+*; *clv3/+*) show a mutant phenotype of somewhat enlarged SAM and extra floral organs. This nonallelic noncomplementation is one indicator that the products of the two genes may act in the same or in closely related pathways: Reduction in the level of one protein sensitizes the plant to a reduction in the level of the other (Clark et al. 1995). Another indicator that *CLV3* may act in the pathway defined by *CLV1* is the phenotype of plants homozygous for loss-of-function alleles in both genes. Double homozygotes (*clv1/clv1*; *clv3/clv3*) have the same phenotype as either single homozygote; formally, that is, *clv1* is epistatic to *clv3*, and *clv3* is epistatic to *clv1*. This is what would be expected if recessive alleles of each eliminate a signal transduction pathway, and the proteins coded by each gene act in different steps in the same pathway (Clark et al. 1995).

From these data, it is not possible to assign a role to *CLV3* in the pathway, although an appealing possibility is that *CLV3* codes for the *CLV1* ligand (Fig. 3). It could just as easily be true though that *CLV3* codes for a cytoplasmic or nuclear protein that responds to *CLV1* activation. We are in the process of trying to obtain molecular clones of *CLV3*, but there is as yet no clue to the nature of the encoded protein.

Given that *CLV1* and *CLV3* seem to be acting in the same pathway, can any additional gene products that act in this pathway be identified? One prominent possibility for an additional component is KAPP, a kinase-associated protein phosphatase identified by Walker and his colleagues (Stone et al. 1994). KAPP associates with the kinase domain of RLK5, an *Arabidopsis* leucine-rich re-

Figure 2. Wild-type and *clavata3-2* mutant inflorescence apices. (*A*) Wild-type inflorescence apex viewed from above. The spiral phyllotactic pattern of secondary floral meristems is evident, with younger meristems closer to the center. (*B*) *clavata3-2* mutant inflorescence apex, at somewhat lower magnification than *A*. The developing flowers surround a clearly visible shoot apical meristem, which is extremely enlarged relative to wild type. Extra floral organs are evident in the more mature flowers, especially extra sepals and carpels.

peat receptor kinase that resembles the CLV1 protein but that is of unknown function in the plant. KAPP also interacts with the kinase domain of the CLV1 protein, as shown by the in vitro association of the kinase interaction domain of KAPP with the CLAVATA1 kinase domain produced in and purified from an *E. coli* expression system (Williams et al. 1997). Furthermore, KAPP acts as a phosphatase that removes phosphate from the phosphoserine of the CLV1 kinase domain. If phosphorylated CLV1 protein is the active form (as indicated by the loss-of-function phenotype of kinase domain amino acid substitutions; Clark et al. 1997), and KAPP dephosphorylates CLV1 in vivo, then a plant overexpressing KAPP should show a *clv1* mutant phenotype (Fig. 3). We have overexpressed KAPP from a constitutive promoter, and the transgenic plants do indeed have a weak *clv1* phenotype, with increased carpel number in the flowers (Williams et al. 1997).

One additional gene also may code for a further component of the CLV1 pathway, the product of the *Arabidopsis* gene *WUSCHEL* (*WUS*; Laux et al. 1996). *wus* loss-of-function mutations have a phenotype opposite that of *clv1* mutants. Rather than an increase in meristem

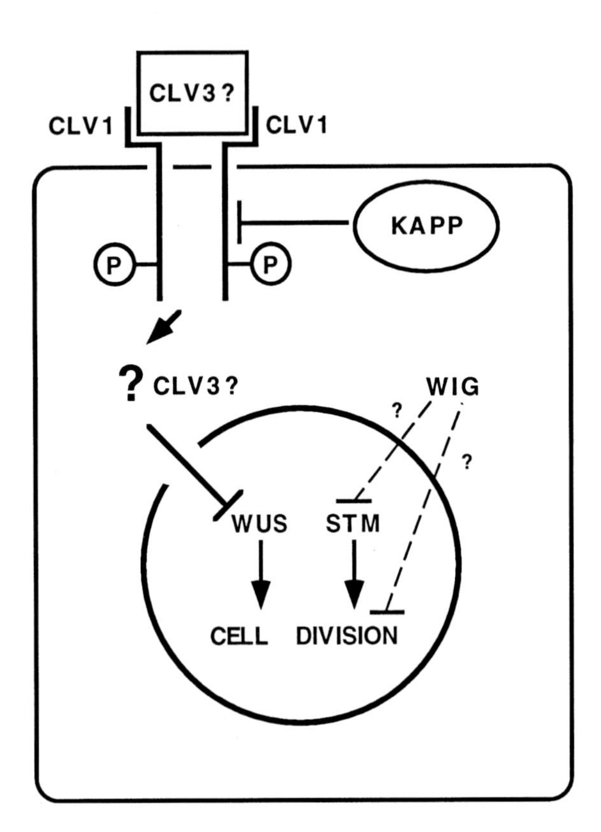

Figure 3. One possible model for the CLAVATA1 pathway and for the parallel pathway for cell division control that involves STM and WIG. CLV1 is a transmembrane receptor kinase, phosphorylated on serine. CLV3 could be the ligand or it could be an element that acts downstream from CLV1—it is shown in both positions, followed by a question mark. KAPP is a phosphatase that dephosphorylates CLV1; because CLV1 acts as a repressor of cell division, KAPP is thus formally an activator of cell division. WUS is the most downstream element in this hypothetical CLV1 pathway; because its loss-of-function phenotype is reduced cell division, it is shown as repressed by active CLV1, and as an activator of cell division. STM, a homeodomain protein, acts in the nucleus as an activator of cell division, but in a pathway not under the direct control of CLV1 (as indicated by the double-mutant experiments described in the text). WIG is a repressor of cell division that acts independently of CLV1/CLV3/WUS, perhaps through repression of STM or perhaps independently of STM.

size and cell number, there is a decrease. In *wus* embryos, the SAM is not recognizable. After germination, the region of the meristem is flattened rather than domed as in wild type, and only after weeks are any leaves formed. Following this late leaf formation, the apex stops growing; much later in the plant's life, adventitious meristems apparently form, as ectopic leaves and shoots emerge from various regions. Just as in the primary growth region, these adventitious meristems then stop their growth. On rare occasion, inflorescence meristems and a small number of flowers are formed; the flowers lack central organs. This again is opposite of the *clv1* mutant phenotype, which includes extra central organs and extra whorls of carpels. Although several interpretations for the wild-type action of WUS are possible, one is that WUS is required for there to be sufficient cell division in the SAM and in floral meristems, and therefore that WUS acts oppositely of CLV1 in shoot and floral meristems. If WUS acts downstream from CLV1, and CLV1 either represses *WUS* gene expression or decreases WUS protein activity (perhaps via a kinase cascade), one would expect *wus* mutations to be epistatic to *clv1* mutations. They are (Laux et al. 1996). WUS may therefore fit into the CLV1 pathway as shown in Figure 3, as a protein negatively regulated by CLV1 when CLV1 is in the activated state.

There is another *Arabidopsis* gene with a mutant phenotype very similar or identical to that of *wus*, called *SHOOT MERISTEMLESS*, or *STM* (Barton and Poethig 1993; Clark et al. 1996; Long et al. 1996; Endrizzi et al. 1996). Just as for *WUS, STM* acts opposite to the *CLV* genes: The wild-type function of the *CLV* genes is to repress cell division in meristems, whereas *STM* and *WUS* serve to activate meristematic cell division and are necessary for SAM initiation as well as maintenance. That *STM* is required for both processes is indicated by the meristem initiation phenotype of strong mutant alleles (*stm-1*, Barton and Poethig 1993) and the maintenance phenotype of weak alleles (*stm-2*, Clark et al. 1996). The STM protein is a member of the homeobox family, and *STM* RNA is found in all layers of *Arabidopsis* SAMs (Long et al. 1996). The interactions of *stm* mutations with *clv1* and *clv3* mutations are not at all the same as those of *wus* and the *clv* alleles. *clv1* and *clv3* mutations partially suppress *stm-1* and *stm-2* phenotypes and are capable of acting as dominant suppressors, despite their generally recessive character. In addition, although *stm* mutations are recessive, *stm* serves as a dominant suppressor of *clv* homozygous phenotypes. *clv stm* double mutants are intermediate between the singly homozygous plants, having more meristematic growth than *stm* alone, but less than in *clv* homozygotes. Thus, unlike *wus* mutations, *stm* mutations are not epistatic to *clv1* or *clv3* (Clark et al. 1996). One explanation for this is that STM and the CLV proteins act in separate pathways that lead to the same endpoint, a cell division decision. The expression patterns indicate that they could be acting in this fashion in the same cells, although the broader expression domain of STM indicates that it may also have activities in cells that do not express CLV1. One interpretation of the action of STM in cells that also express CLV1 is shown in Figure 3.

Given that STM seems to act in a pathway parallel (although perhaps cross-regulating) to that of CLV1, CLV3, KAPP, and WUS in the rib meristem cells, one can ask if there are any mutations among those collected with excess-cell phenotypes that might act in this parallel path. The one such gene that has been analyzed sufficiently to tell is *WIGGUM*, the mutant phenotype of which is a modest increase in floral organ number, correlated with increased cell number in early floral meristems (Running 1997). In this respect, *wig* mutants resemble *clv1* and *clv3* mutants. *wig* mutations do not act similarly to the *clv* mutants in tests of genetic interaction, however. *wig clv1* and *wig clv3* double mutants have a phenotype much more extreme than *wig* or *clv* single mutants, with massive overgrowth of the SAM and extreme disruption of floral meristems, including sometimes complete loss of determinacy (Running 1997). The SAM phenotype of the double mutants is striking, as the SAM produces undifferentiated, callus-like tissue and can achieve a diameter of more than 1 cm by the end of the plant's life. This is approximately 100 times the diameter of a wild-type SAM, indicating as much as a million-fold increase in volume in the double-mutant SAMs, as compared with wild-type. One interpretation of these results is that WIG acts to repress meristematic cell division in a separate but parallel and partly redundant pathway to CLV1 and CLV3. Eliminating any one of the pathways causes a modest increase in SAM cell number, whereas eliminating both causes SAM cell division to be completely unregulated. This interpretation is shown in Figure 3. It is not yet known if WIG acts through STM (i.e., if *stm* mutations are epistatic to *wig* mutations); thus, WIG is shown as acting independently of STM, although future experiments may indicate that *WIG* and *STM* define, together, a single pathway of cell division control. Furthermore, because the product of the *WIG* gene is unknown, it is not known if WIG is a cell surface receptor, ligand, element in a signal transduction pathway, or nuclear effector of cell division. Indeed, WIG could act in a different population of cells than CLV1, with loss of cell division control in two adjacent, communicating populations of cells causing a complete loss of cell division repression in the meristem.

All of the known mutant phenotypes and genetic interactions of *CLV1, CLV3, STM, WIG*, and *WUS*, and the known molecular properties and overexpression phenotype of KAPP, can be fit into the speculative model for cell division control shown in Figure 3. This model, even if true, would hold only for the cells in which CLV1, KAPP, STM, and so forth are known to be active. The only cells in which the CLV1, KAPP, and STM RNAs are found together are in those roughly coincident with the rib meristem of the SAM and a comparable region of developing floral meristems. Cells, for example, in the peripheral zone of the shoot meristem, which do not have detectable CLV1 RNA, would have to be controlled by a different set of gene products. It could be though that each meristematic zone has its own equivalent of CLV1 acting as a receptor for cell division information from nearby cells and also its own mechanism for producing a ligand for the receptors found in adjacent cells. If so, then variants of the pathway shown in Figure 3 might exist in each population of dividing cells in plants, and each population would be controlled by different ligands secreted by different sets of neighbors. Such a mechanism may account for the overall coordination of the cell divisions that occur in meristems and thus may account for the ability of meristems to maintain populations of stem cells, and at the same time produce lateral organs and additional meristems, all without substantially changing their shape, size, and clonal layering.

One scheme by which each set of meristematic cells might respond to different ligands would be for each to have a cell surface receptor like CLV1, but with an altered extracellular domain. It is already known that there is a leucine-rich repeat receptor kinase family in plants: In *Arabidopsis,* the first member of the family found was TMK1 (Chang et al. 1992), followed by RLK5 (Walker 1993) and TMKL1 (Valon et al. 1993). Mutants are not known for any of these genes, nor has any study of their expression patterns been done, so their functions in plants are unknown. Additional LRR-kinase genes in *Arabidopsis* include *ERECTA* (Torii et al. 1996), which has a mutant phenotype of shorter stems and fruits than wild type. Whether this is due to reduced cell division or to a reduction in cell elongation has not been established, but it is possible that ER, like CLV1, acts in the regulation (although positive, not negative, regulation) of cell division in particular plant regions. There are numerous additional LRRs and kinase domains related to CLV1 in the *Arabidopsis*-expressed sequence tag database. Few of the entered sequences are long enough to include the whole protein-coding region of the sequenced cDNA, so it is not known how many of these sequences code for LRR transmembrane receptor kinases, and how many for other LRR proteins or for different types of kinases. Nonetheless, the LRR receptor kinase family in *Arabidopsis* has at least four well-characterized members, and there could be dozens more.

To get a better idea of the number and expression pattern of *CLV1*-related genes in the *Arabidopsis* genome, we have used the part of the *CLV1* gene that codes for the kinase domain as a labeled probe for screens of an *Arabidopsis* genomic bacteriophage λ library (R.W. Williams and E.M. Meyerowitz, unpubl.). Genes for several additional and previously unknown LRR kinases were found, and preliminary in situ hybridization results indicate that they have a variety of expression patterns, with some showing expression in specific subdomains of the SAM and some expressed in different domains of the floral meristem. It is thus true that there are numerous LRR kinases in *Arabidopsis* and that at least some of them are expressed in SAMs and floral meristems, as if they might indeed serve as sensors to help cells in different meristematic regions assess their environments. It may be that each functional and clonal region of shoot and floral meristems has its own set of kinases, and secretes its own set of ligands when dividing or elongating, and that by this sort of intercommunication, the coordinated cell division activities of the meristems are controlled.

ACKNOWLEDGMENTS

I thank Bobby Williams for careful reading of the manuscript, and Mark Running for providing photographs for Figures 1 and 2. My laboratory's work on meristem cell division control is funded by U.S. National Science Foundation grant MCB-9603821 and a Strategic Research Fund grant from Zeneca Agrochemicals.

REFERENCES

Alvarez J. and Smyth D.R. 1994. Flower development in *clavata3*, a mutation that produces enlarged floral meristems. In Arabidopsis: *An atlas of morphology and development* (ed. J. Bowman), p. 254. Springer-Verlag, New York.

Barton M.K and Poethig R.S. 1993. Formation of the shoot apical meristem in *Arabidopsis thaliana*—An analysis of development in the wild-type and in the *SHOOT MERISTEMLESS* mutant. *Development* **119:** 823.

Buchanan S.G.S.C. and Gay N.J. 1996. Structural and functional diversity in the leucine-rich repeat family of proteins. *Prog. Biophys. Mol. Biol.* **65:** 1.

Chang C., Schaller G.E., Patterson S.E., Kwok S., Meyerowitz E.M., and Bleecker A.B. 1992. The TMK1 gene from *Arabidopsis* codes for a protein with structural and biochemical characteristics of a receptor protein kinase. *Plant Cell* **4:** 1263.

Clark S.E., Running M.P., and Meyerowitz E.M. 1993. *CLAVATA1*, a regulator of meristem and flower development in *Arabidopsis. Development* **119:** 397.

———. 1995. *CLAVATA3* is a specific regulator of shoot and floral meristem development affecting the same processes as *CLAVATA1. Development* **121:** 2057.

Clark S.E., Jacobsen S.E., Levin J.Z., and Meyerowitz E.M. 1996. The *CLAVATA* and *SHOOT MERISTEMLESS* loci competitively regulate meristem activity in *Arabidopsis. Development* **122:** 1567.

Clark S.E., Williams R.W., and Meyerowitz E.M. 1997. The *CLAVATA1* gene encodes a putative receptor-kinase that controls shoot and floral meristem size in *Arabidopsis. Cell* **89:** 575.

Crone W. and Lord E.M. 1993. Flower development in the organ number mutant *clavata1-1* of *Arabidopsis thaliana* (Brassicaceae). *Am. J. Bot.* **80:** 1419.

Endrizzi K., Moussian B., Haecker A., Levin J.Z., and Laux T. 1996. The *SHOOT MERISTEMLESS* gene is required for maintenance of undifferentiated cells in *Arabidopsis* shoot and floral meristems and acts at a different regulatory level than the meristem genes *WUSCHEL* and *ZWILLE. Plant J.* **10:** 967.

Koornneef M., Van Eden J., Hanhart C.J., Stam P., Braaksma F.J., and Feenstra W.J. 1983. Linkage map of *Arabidopsis thaliana. J. Hered.* **74:** 265.

Laux T, Mayer K.F.X., Berger J., and Jürgens G. 1996. The *WUSCHEL* gene is required for shoot and floral meristem integrity in *Arabidopsis thaliana. Development* **122:** 87.

Leyser H.M.O. and Furner I.J. 1992. Characterization of three shoot apical meristem mutants of *Arabidopsis thaliana. Development* **116:** 397.

Long J.A., Moan E.I., Medford J.I., and Barton M.K. 1996. A member of the KNOTTED class of homeodomain proteins encoded by the *STM* gene of *Arabidopsis. Nature* **379:** 66.

Meyerowitz E.M. 1997. Genetic control of cell division patterns in developing plants. *Cell* **88:** 299.

Running M. 1997. "Molecular genetics of floral patterning in *Arabidopsis thaliana*." Ph.D. thesis. California Institute of Technology, Pasadena.

Smyth D.R., Bowman J.L., and Meyerowitz E.M. 1990. Early flower development in *Arabidopsis. Plant Cell* **2:** 755.

Steeves T.M and Sussex I.M. 1989. *Patterns in plant development*, 2nd edition. Cambridge University Press, United Kingdom.

Stone J.M., Collinge M.A., Smith R.D., Horn M.A., and Walker J.C. 1994. Interaction of a protein phosphatase with an *Arabidopsis* serine-threonine receptor kinase. *Science* **266:** 793.

Tilney-Bassett R.A.E. 1986. *Plant chimeras*. E. Arnold, London.

Torii K.U., Mitsukawas N., Oosumi T., Matsuura Y., Yokoyama R., Whittier R., and Komeda Y. 1996. The *Arabidopsis ERECTA* gene encodes a putative receptor protein kinase with extracellular leucine-rich repeats. *Plant Cell* **8:** 735.

Valon C., Smalle J., Goodman H.M., and Giraudat J. 1993. Characterization of an *Arabidopsis thaliana* gene (TMKL1) encoding a putative transmembrane protein with an unusual kinase-like domain. *Plant Mol. Biol.* **23:** 415.

Walker J.C. 1993. Receptor-like protein kinase genes of *Arabidopsis thaliana. Plant J.* **3:** 451.

Williams R.W., Wilson J.M., and Meyerowitz E.M. 1997. A possible role for kinase-associated protein phosphatase in the *Arabidopsis* CLAVATA1 signaling pathway. *Proc. Natl. Acad. Sci.* **94:** 10467.

Yanofsky M.F., Ma H., Bowman J.L., Drews G.N., Feldmann K.A., and Meyerowitz E.M. 1990. The protein encoded by the *Arabidopsis* homeotic gene *AGAMOUS* resembles transcription factors. *Nature* **346:** 35.

Pancreas Development in the Chick Embryo

S.K. KIM, M. HEBROK, AND D.A. MELTON
Department of Molecular and Cellular Biology and Howard Hughes Medical Institute, Harvard University, Cambridge, Massachusetts 02138

Early development of endoderm and adjacent mesenchyme into the vertebrate respiratory and digestive tracts requires mechanisms that ensure a stereotyped and reproducible pattern of organogenesis along the three embryonic axes. Mechanisms, for example, must ensure that the pancreas forms caudal and dorsal to the stomach and that the trachea and lungs develop ventral to the esophagus. After selection of organ position, usually marked by an epithelial evagination, epithelial-mesenchymal cell interactions appear to be at least one important mechanism for controlling organ morphogenesis and cytodifferentiation (for reviews, see Mizuno and Yasugi 1990; Birchmeier and Birchmeier 1993). Although much has been learned about these later cell interactions, little is known about the mechanisms that initially pattern vertebrate endoderm and mesenchyme along the embryonic axes, resulting in gastrointestinal and respiratory tract order. We have focused on the first steps of pancreas formation in chick embryos which provide many experimental advantages (Bronner-Fraser 1996), particularly with respect to studying endodermal patterning and development.

During embryogenesis, the first morphological sign of vetebrate pancreas development is an epithelial bud forming at the dorsal side of the duodenum. In chick embryos, the dorsal bud fuses with two ventral buds that form on the opposite side of the gut tube, whereas in rodents and humans, only one ventral bud forms and fuses with the dorsal bud. The mature vertebrate pancreas has both endocrine and exocrine functions.

Exocrine tissue forms from multiple acinar glands that store and release digestive enzymes into the intestinal lumen. The endocrine cells of the pancreas are clustered in islets of Langerhans and secrete metabolic regulatory hormones including insulin, glucagon, somatostatin, and pancreatic polypeptide directly into the bloodstream (for review, see Slack 1995). Phenotypic similarities between pancreatic endocrine cells and cells known to be derived from the neural crest had suggested that pancreatic endocrine cells might derive from the neural crest (for review, see Le Douarin 1988). However, chick-quail chimera studies (Andrew 1976; Fontaine and Le Douarin 1977) do not support a neural crest origin for endocrine cells. Instead, evidence suggests that endocrine cells originate from the endoderm (Gu and Sarvetnick 1993; Gittes et al. 1996). The functions of two transcription factors necessary for endocrine cell development have been recently reported. Characterization of *pax-4* (Sosa-Pineda et al. 1997) homozygous mutant mice suggests that *pax-4* activity is necessary for differentiation of insulin-producing β-cells and somatostatin-producing δ-cells. *pax-6* homozygous mutants lack glucagon-producing α-cells (St. Onge et al. 1997). Mice lacking both *pax-4* and *pax-6* activity fail to develop any mature endocrine cells, suggesting that both genes are required for pancreatic endocrine cell differentiation (St. Onge et al. 1997).

Studies focused on the interactions between pancreatic endoderm and mesenchyme have shown that permissive signals from pancreatic mesenchyme to endoderm are necessary for pancreas development in mice (Golosow and Grobstein 1962; Wessels and Cohen 1967; Spooner et al. 1970) and chicks (Dieterlen-Lièvre 1970). The roles of two transcription factors crucial to later epithelial-mesenchymal interactions and endocrine cell differentiation in the pancreas have been reported. The homeodomain-containing factor PDX1 is expressed initially in the dorsal and ventral pancreatic endoderm evaginations and later in stomach and duodenum (Jonsson et al. 1994; Offield et al. 1996). Mice with homozygous null mutations in *pdx1* are born apancreatic (Jonsson et al. 1994; Offield et al. 1996) but do initiate early pancreatic bud formation with dorsal buds that express glucagon and insulin. These results argue that *pdx1* does not initially specify endoderm to a pancreatic fate. Recombination studies with mesenchyme and epithelium from normal and *pdx1*-mutant embryos suggest instead that PDX1 function is required to make pancreatic epithelium competent to respond to mesenchymal signals (Ahlgren et al. 1996). Recent studies show that ISL1 function is required for dorsal, but not ventral, pancreatic mesenchyme formation and all pancreatic islet cell differentiation (Ahlgren et al. 1997). In embryos with homozygous null mutations in *Isl1*, lack of mesenchyme interferes with exocrine pancreas development.

In mice (Wessels and Cohen 1967; Spooner et al. 1970) and chicks (Le Douarin and Bussonnet 1966; Dieterlen-Lièvre 1970), recombination of heterologous lung or stomach mesenchyme with pancreatic endoderm in vitro results in pancreatic differentiation. The chick dorsal and ventral pancreas buds are histologically distinct before their fusion, and Dieterlen-Lièvre (1970) has shown that recombination of dorsal mesenchyme with ventral endoderm results in pancreas tissue with a typical ventral histology. Together, these results suggest that the competence to form pancreas and the pattern of cytodifferentiation within the pancreas are intrinsic to the endoderm. However, it is not yet understood which tissues or possible inductive signals are important for endodermal commitment to a pancreatic fate.

In this paper, we describe studies of early morphogenesis and gene expression in the chick pancreas. We suggest that there may be parallels between patterning of pancreatic endoderm and neuroectoderm and describe methods to study the developmental steps leading to pancreas formation.

METHODS

Chick embryo growth and dissection methods. All experiments were performed on white Leghorn chick embryos from SPAFAS (Preston, Connecticut). Eggs were incubated at 38°C and staged according to the method of Hamburger and Hamilton (1951) (referred to afterward as "HH stage"). Endoderm including the anlage of the dorsal pancreatic bud was isolated using needles and forceps before visible morphogenesis at stage 15. The position of the dorsal pancreatic anlage was determined from the endodermal mapping studies of Le Douarin (1964) and Rosenquist (1971) and studies of chick pancreas formation by Dieterlen-Lièvre (1965, 1970). Tissue recombinants were grown in collagen-matrix gels as described previously (Dickinson et al. 1995), except 10× M199 medium at pH 4.0 (GIBCO/BRL) was substituted for 10× Dulbecco's modified Eagle's medium (DMEM).

RNA preparation and reverse transcription–polymerase chain reaction (RT-PCR). Dissected embryonic pancreas rudiments were dissolved in Trizol (GIBCO/BRL), and total RNA was prepared according to methods from the manufacturer. RT-PCR was performed as described in Wilson and Melton (1994). Cloning of chick *pdx1* and carboxypeptidase A cDNA is described elsewhere (Kim et al. 1997).

Histological analysis, microscopy, and photography. Paraffin sections (6 μm) were stained by hematoxylin-eosin (H & E; Sechrist and Marcelle 1996) by in situ hybridization (Henry et al. 1996) using sense and antisense digoxigenin probes (Riddle et al. 1993) or by immunoperoxidase techniques. Slides were photographed on a Zeiss Axiophot, and photos were scanned into Adobe Photoshop 3.0 for printing.

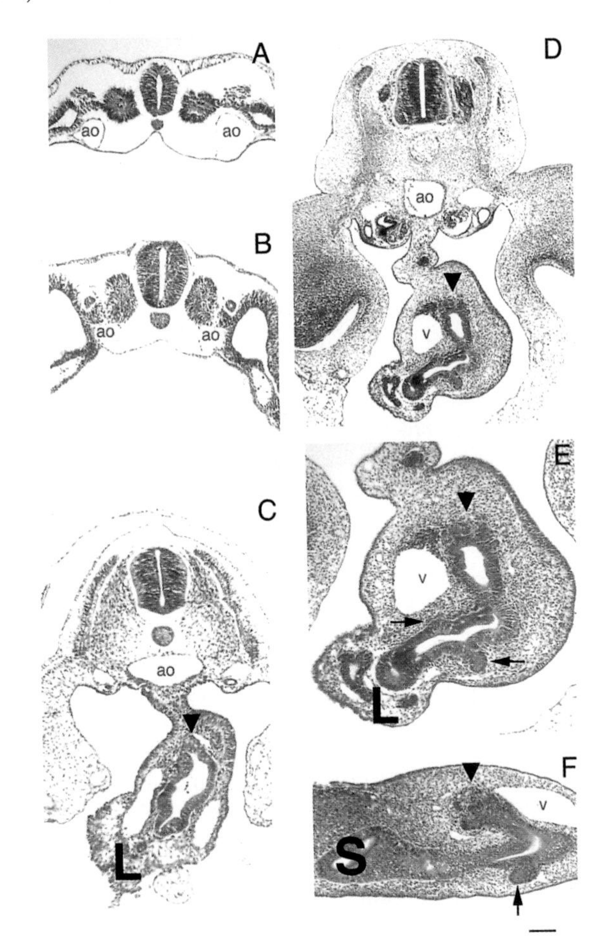

Figure 1. Pancreatic morphogenesis in chick. Transverse sections through foregut at the level of the pancreatic anlage were stained with hematoxylin-eosin. Dorsal side is toward the page top. (*A*) At stage 10, pancreatic endoderm is one-cell-layer thick and contacts notochord in the midline. The paired dorsal aortas (ao) are indicated. (*B*) At stage 13, midline endoderm is adjacent to, but no longer in contact with, the notochord. Splanchnic mesoderm lines more lateral endoderm. (*C*) At stage 15, the aortas have fused in the midline, separating notochord from pancreatic endoderm. Columnar dorsal endoderm has formed a discernible pancreas bud (arrowhead). Early liver tissue (L) is detected ventrally at this rostrocaudal level. (*D*) By stage 21–22, the dorsal pancreatic bud is prominent (arrowhead) adjacent to the right omphalomesenteric vein (v). (*E*) 2× magnification of the pancreas shown in *D* revealing details of the dorsal bud, including small apical cell cluster at the endoderm/mesenchyme junction (arrowhead), and evidence of ventral bud formation (arrows). Liver (L) tissue that has also grown ventrally is now characteristically on the right side of the embryonic gut, as a result of normal gut rotation. (*F*) Sagittal section of dissected stage 21 embryonic foregut showing dorsal pancreas endoderm with apical cell clusters (arrowhead), the left ventral bud (arrow), and their relationship to the stomach (S) and omphalomesenteric vein (v). Anterior is to the left. Bars: (*A,B*) 200 μm; (*C,D,F*) 150 μm; (*E*) 75 μm.

RESULTS AND DISCUSSION

Pancreatic Morphogenesis in Chick Embryos

The pancreas forms from the fusion of distinct dorsal and ventral buds. Before morphogenesis, the endoderm at the 10-somite stage (HH stage 10) destined to line the dorsal pancreas bud is a single layer of squamous epithelial cells (Fig. 1A) that touch the notochord. The ventral bud derives from endoderm lateral and rostral to the axial midline (see Rosenquist 1971); thus, the ventral bud originates from endoderm that does not touch notochord. Early studies by Le Douarin (1964) suggest that the pancreatic anlage is adjacent to somites 7 through 15 along the rostrocaudal axis. Laterally, the prepancreatic endoderm is adjacent to structures including the somitic mesoderm, endothelium of the paired dorsal aortas, and splanchnic mesoderm.

By 19 somites (HH stage 13), the prepancreatic endoderm remains near the notochord but is no longer in direct contact (Fig. 1B). At this stage, this region of the foregut has not yet formed a closed tube. The dorsal aortas at this rostrocaudal level have moved medially compared with earlier stages. The somites are also displaced more dorsally relative to the endoderm. Splanchnic mesoderm does not yet cover midline endoderm.

The first indication of dorsal bud morphogenesis in chicks occurs by the 26-somite stage (HH stage 15), the same somite age at which the dorsal bud is first seen in mice (Wessels and Cohen 1967). By this stage at the level of the pancreas, midline fusion of the paired dorsal aortas has completely separated endoderm from the notochord and the foregut has formed a closed tube. The pancreatic endoderm, now columnar in shape, has formed a dorsal bud (Fig. 1C) and is lined dorsally and laterally by mesenchyme that bulges slightly to the left. Ventrally, cords of vascularizing liver tissue are evident. During the next day, the dorsal bud endoderm proliferates (Romanoff 1960), and small apical cell clusters appear at the interface between endoderm and dorsal mesenchyme. Glucagon is first detectable by immunohistochemistry at stage 17 (day 3 of incubation), and insulin is detectable several hours later by stage 19 (Beaupain and Dieterlen-Lièvre 1974; Dieterlen-Lièvre and Beaupain 1974; Kim et al. 1997). These endocrine hormones are expressed in the small cell clusters at the endodermal apex of the dorsal bud.

By stage 21–22, the right ventral pancreatic bud and slightly larger left ventral bud are first detected (Fig. 1E,F). The right omphalomesenteric vein is adjacent to the dorsal bud. The dorsal pancreatic bud later gives rise to the splenic lobe or "tail" of the mature pancreas, whereas the ventral lobes give rise to the pancreatic "head." In the chick, insulin-secreting islets are present thoughout the pancreas, but the dorsally derived tail is richer in large glucagon-secreting islets than the ventrally derived pancreatic head. The exocrine pancreas begins its visible differentiation as early as the eighth day (Potvin and Arun 1927 cited in Romanoff 1960), but expression of carboxypeptidase A can be detected by immunohistochemistry by 4 days of incubation (HH stage 22) in the dorsal bud (Kim et al. 1997). Recently, we have shown that the sequence of in ovo morphologic development of the pancreas described above can be reproduced in chick embryos grown to HH stage 22 in vitro on albumin-agar plugs.

In vitro tissue culture and recombination experiments suggest that the dorsal pancreatic endoderm in mice is committed to a pancreatic fate by as early as the 10–12-somite stage (Wessels and Cohen 1967). Similar experiments in chicks using molecular markers suggest that dorsal endoderm acquires the ability to form pancreas as early as the 13-somite stage (Dieterlen-Lièvre and Beaupain 1974; Sumiya and Mizuno 1987). We are currently studying expression of pancreas genes in isolated chick prepancreatic endoderm grown in tissue culture (Dickinson et al. 1995) to determine the stage at which endoderm commitment to a pancreatic fate occurs.

Pancreatic Gene Expression

Expression patterns of genes shown to be involved in pancreas and endoderm development have been studied mainly in mice and rats (see references below). The similarities in the timing and morphogenesis between chick and rodent pancreas development suggested that expression patterns of genes important in endoderm and pancreas development might also be similar. Figure 2 shows expression patterns of several chick genes in the pancreas and adjacent tissues at stage 19–20 when a dorsal bud is clearly present.

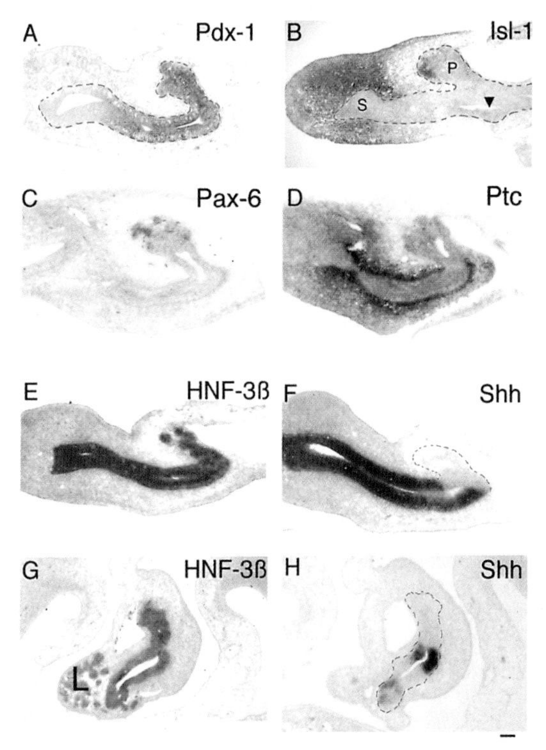

Figure 2. Pancreatic gene expression patterns revealed by in situ hybridization of sagittal and transverse sections from dissected stage-20 chick foregut. Dorsal is toward the page top and anterior is to the left in sagittal sections. (*A*) Pdx-1; red dashes outline endoderm including stomach, duodenum, and dorsal pancreas bud. (*B*) Isl-1; red dashes outline endoderm of stomach (S), dorsal pancreas (P), and duodenum. Arrowhead indicates Isl-1 expression in the left ventral pancreatic evagination. (*C*) Pax-6; expression detected at the apical portion of the dorsal pancreas bud. (*D*) Ptc; low-level expression detected in endoderm and at high levels in mesenchyme adjacent to endoderm expressing SHH. Note absence of ptc staining in mesenchyme surrounding the dorsal pancreas bud. (*E*) HNF-3β; sagittal section showing endoderm-specific expression including the dorsal pancreatic endoderm. (*F*) Shh; note expression in stomach and duodenal endoderm but no detectable expression in the pancreas bud. (*G*) HNF-3β; transverse section revealing HNF-3β expression in dorsal and ventral pancreas endoderm, duodenum, and liver (L). (*H*) Shh; transverse section demonstrating absence of Shh expression in the dorsal and presumptive ventral pancreas endoderm. Duodenal and ventral biliary duct endoderm do express Shh. Bars: (*A–F*) 75 μm; (*G,H*) 60 μm.

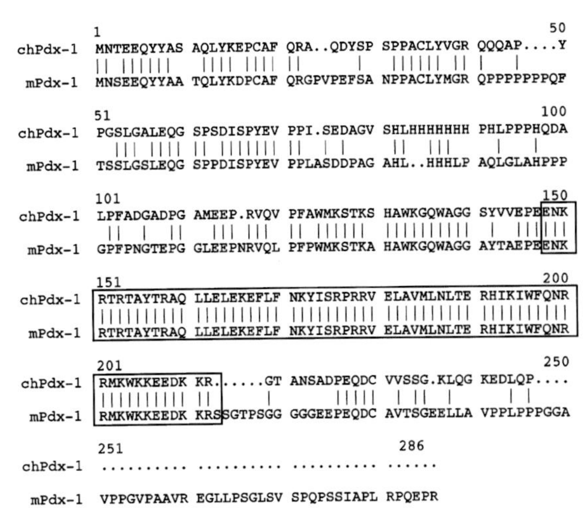

Figure 3. Alignment of deduced amino acid sequence of chick and mouse (Ohlsson et al. 1993) PDX-1. The vertical dashes represent identities. The homeodomain is boxed with a solid

pdx-1 is a homeodomain-containing transcription factor expressed in mouse in endoderm of the pancreatic anlage and later in the endoderm of duodenum and stomach antrum (Ohlsson et al. 1993; Jonssen et al. 1994; Offield et al. 1996). We cloned the chick cDNA homologous to *pdx-1* from a chick pancreatic cDNA library (Fig. 3). In situ hybridization with a chick specific *pdx-1* probe at stage 19 (equivalent to mice at embryonic day 10.5) reveals expression in the dorsal pancreatic bud, duodenum, and antrum of the stomach (Fig. 2).

Isl-1 is a LIM-homeodomain protein whose function is required for motor neuron development (Ericson et al. 1992; Pfaff et al. 1996) and differentiation of all pancreatic islet cells (Ahlgren et al. 1997). *Isl-1* in chicks is expressed in mesenchymal cells near the dorsal pancreatic evagination and proximal stomach; *Isl-1* mesenchymal expression extending to the stomach is more extensive than described in mice (Ahlgren et al. 1997). Expression is clearly detected in the dorsal pancreatic epithelium, and faint staining is also consistently detected in the epithelium of the presumptive left ventral pancreatic bud at this stage (Fig. 2B, arrowhead). In contrast, little to no staining of the mesenchyme adjacent to the ventral pancreatic buds is detected. Thus, similar to expression in mouse (Ahlgren et al. 1997), the pancreatic mesenchyme in chick appears to differentially express *Isl-1* along the dorsoventral axis.

Pax-6, a member of the Pax gene family, is expressed in the developing eye, nose, central nervous system, and pancreas (Goulding et al. 1993; Turque et al. 1994) and is necessary for differentiation of pancreatic α cells (St. Onge et al. 1997). Early in chick development, *pax-6* expression is observed in a small patch of foregut (Li et al. 1994). Later, we detect *pax-6*-expressing cells in clusters at the apex of the dorsal bud at stage 19 (Fig. 2). This *pax-6* expression in a subset of cells of the chick pancreatic epithelium that express insulin and glucagon (Dieterlen-Lièvre and Beaupain 1974; Kim et al. 1997) is also observed in mice (St. Onge et al. 1997).

The winged helix transcription factor *HNF-3β* is expressed throughout the embryonic gut including the pancreatic bud in mice (Ahlgren et al. 1996) and chicks (Fig. 2). At the same developmental stage, the secreted protein sonic hedgehog (Shh), a candidate effector of embryonic intestinal polarity, is also expressed generally in the embryonic endoderm, except in endoderm of the mouse dorsal and ventral pancreatic buds (Ahlgren et al. 1997). Examination of in situ hybridization patterns in the stage-19 embryonic chick pancreas also reveals a striking absence of *shh* staining in the dorsal bud endoderm and the endoderm from which the paired ventral buds evaginate (Fig. 2). Patched (Ptc) is a candidate transmembrane receptor of Shh, whose transcription is induced in tissues adjacent to sources of Shh (Goodrich et al. 1996; Marigo et al. 1996). The absence of *ptc* in pancreatic mesenchyme adjacent to the dorsal pancreatic

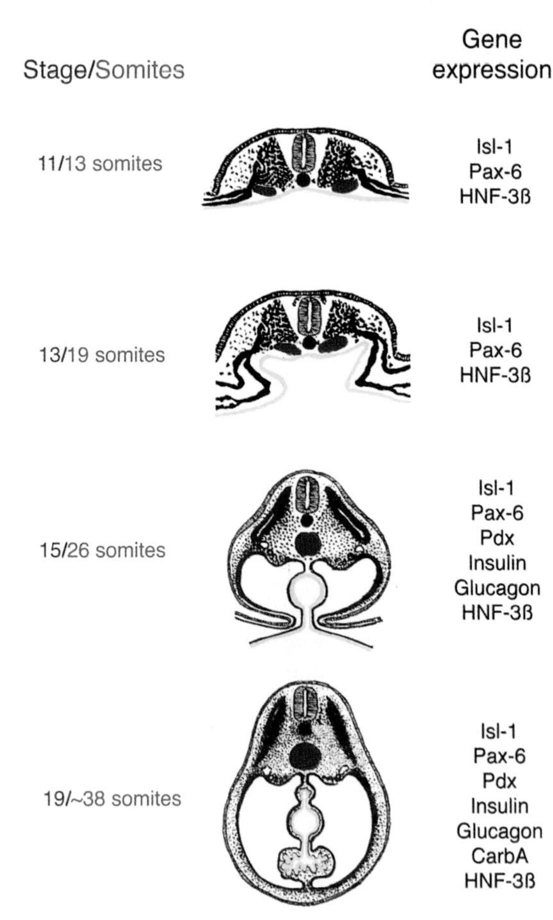

Figure 4. Summary schematic of dorsal pancreas morphogenesis in transverse view and gene expression during 2 days of development. Dorsal is toward the page top. Endoderm in yellow, aorta(s) in red, and notochord labeled blue. (Schematic drawings modified from Patten and Carlson 1974.) "Gene expression" refers to detectable transcription of genes in the dorsal pancreas at the indicated stages by either RT-PCR or in situ hybridization analysis. In the transverse view at stage 19 (~38 somites), the duodenum is shown with the pancreatic bud forming dorsally, opposite the ventral liver bud.

bud, compared with high levels of *ptc* expression in stomach and duodenal mesenchyme (Fig. 2), is consistent with the lack of expression of *shh* in pancreatic endoderm.

In addition to spatial patterns of gene expression, we have examined the temporal expression patterns of *Isl-1*, *pdx-1, pax-6, HNF-3*β, glucagon, insulin, and carboxypeptidase A by in situ hybridization and RT-PCR of staged embryonic pancreases. As summarized in Figure 4, we first detect *Isl-1* and *pax-6* transcription in endoderm from the dorsal pancreatic anlage before foregut closure. By 26 somites, when the dorsal pancreas bud is first forming, we detect *pdx-1*, glucagon, and low levels of insulin. By the third day of incubation, when the ventral buds are forming, high levels of insulin and glucagon and initial low levels of carboxypeptidase A are detected in the dorsal bud (Kim et al. 1997).

Parallels between Patterning Neuroectoderm and Endoderm

For both neuroectoderm and somitic mesoderm, patterning is partly controlled by inductive interactions with the notochord, an axial mesoderm derivative and known source of intercellular signals (see Pourquié et al. 1993; for review, see Placzek and Furley 1996). Several observations suggest that notochord signals could also influence endoderm and dorsal pancreatic development: (1) Specific developmental abnormalities in human vertebra, ascribed to defects of notochord development, have been correlated with congenital gastrointestinal defects (Elliott et al. 1970). (2) The endoderm of the dorsal pancreas is in direct contact with notochord when commitment to a pancreatic fate occurs (Wessels and Cohen 1967; Pictet et al. 1972), in contrast to all other endodermally derived organs that form from lateral or ventral endoderm evaginations (Remak 1854 cited in Dieterlen-Lièvre 1965). (3) Many genes expressed in developing neuroectoderm are also expressed in the developing pancreas (Fig. 5) (Rudnick et al. 1994; Turque et al. 1994; Ahlgren et al. 1997; Kim et al. 1997), and patterning of some of these genes including *pax-6* (Goulding et al. 1993), *Isl-1* (Ericson et al. 1992; Roelink et al. 1994), *nkx2.2* (Barth and Wilson 1995), *HNF-3*β (Ruiz i Altaba et al. 1995), and *shh* (see Placzek and Furley 1996) can be influenced by signals from notochord. (4) Four of these genes, *Isl-1*, *pax-6*, *shh*, and *ptc*, appear to be expressed in restricted domains along the dorsoventral axis in both neural tube (Yamada et al. 1991; Goulding et al. 1993) and pancreatic mesoendoderm (Fig. 2) (Ahlgren et al. 1997; this paper). Alternately, the expression pattern of these genes in the pancreas may reflect mechanisms that restrict mesoendodermal gene expression along the rostrocaudal axis of the gastrointestinal tract (Yokouchi et al. 1995; for review, see Bienz 1994). We are currently using the methods described here to analyze the influence of notochord and other mesoderm on morphogenesis of the chick dorsal and ventral pancreas during the time when pancreas fate is specified.

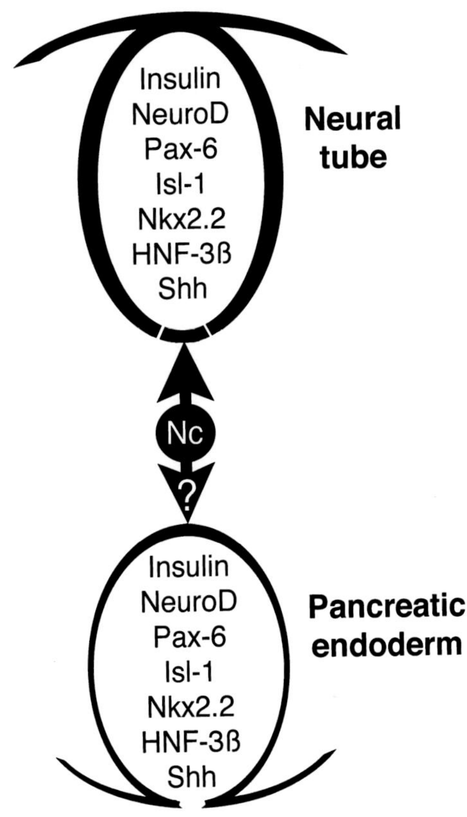

Figure 5. Summary of data from many researchers (see text) demonstrating expression of the indicated genes in developing neuroectoderm and the pancreas, which is derived from the endoderm of the foregut. The closed neural tube is patterned, in part, by contact-dependent and diffusible signals from the notochord (black arrow). Ventral floor plate is demarcated by white lines. It is not yet known whether structures like the notochord can pattern endoderm (indicated by the question mark). See text for details and references.

ACKNOWLEDGMENTS

We thank Mark Solloway and Drs. Tom Schultheiss, Mary Dickinson, Andrea Münsterberg, Cliff Tabin, Andrew McMahon, and Elizabeth Robertson for materials and helpful discussions. We also thank M. Buszczak, G. Chen, and Q. Wu for excellent and cheerful technical assistance. This work was supported by a Deutsche Forschungsgemeinschaft and a Howard Hughes Medical Institute postdoctoral fellowship to M.H. and a HHMI physician postdoctoral fellowship to S.K.K. D.A.M. is an investigator of the HHMI.

REFERENCES

Ahlgren U., Jonsson J., and Edlund H. 1996. The morphogenesis of the pancreatic mesenchyme is uncoupled from that of the pancreatic epithelium in IPF1/PDX1-deficient mice. *Development* **122:** 1409.

Ahlgren U., Pfaff S.L., Jessell T.M., Edlund T., and Edlund H. 1997. Independent requirement for ISL1 in formation of pancreatic mesenchyme and islet cells. *Nature* **385:** 257.

Andrew A. 1976. An experimental investigation into the possible neural crest origin of pancreatic APUD (islet) cells. *J. Embryol. Exp. Morphol.* **35:** 577.

Barth K.A. and Wilson S.N. 1995. Expression of zebrafish *nkx2.2* is influenced by sonic hedgehog/vertebrate hedgehog-1 and demarcates a zone of neuronal differentiation in the embryonic forebrain. *Development* **121:** 1755.

Beaupain D. and Dieterlen-Lièvre F. 1974. Étude immunocytologique de la différenciation du pancréas endocrine chez l'embryon de poulet. II. Glucagon. *Gen. Comp. Endocrinol.* **23:** 421.

Bienz M. 1994. Homeotic genes and positional signaling in the *Drosophila* viscera. *Trend. Genet.* **10:** 22.

Birchmeier C. and Birchmeier W. 1993. Molecular aspects of mesenchymal-epithelial interactions. *Annu. Rev. Cell Biol.* **9:** 511.

Bronner-Fraser M. 1996. Methods in avian embryology. *Methods Cell Biol.* vol. 51.

Dickinson M.E., Selleck M.A., McMahon A.P., and Bronner-Fraser M. 1995. Dorsalization of the neural tube by the nonneural ectoderm. *Development* **121:** 2099.

Dieterlen-Lièvre F. 1965. Étude morphologique et expérimentale de la différenciation du pancréas chez l'embryo de poulet. *Bull. Biol. Fr. Belg.* **99:** 3.

———. 1970. Tissus exocrine et endocrine du pancréas chez l'embryon de poulet: Origine et interactions tissulaires dans la différenciation. *Dev. Biol.* **22:** 138.

Dieterlen-Lièvre F. and Beaupain D. 1974. Étude immunocytologique de la différenciation du pancréas endocrine chez l'embryon de poulet. I. Îlots à insuline. *Gen. Comp. Endocrinol.* **22:** 62.

Elliott G.B., Tredwell S.J., and Elliott K.A. 1970. The notochord as an abnormal organizer in production of congenital intestinal defects. *Am. J. Roentgenol. Radium Ther. Nucl. Med.* **110:** 628.

Ericson J., Thor S., Edlund T., Jessell T.M., and Yamada T. 1992. Early stages of motor neuron differentiation revealed by expression of homeobox gene *Islet-1*. *Science* **256:** 1555.

Fontaine J. and Le Douarin N. 1977. Analysis of endoderm formation in the avian blastoderm by the use of quail-chick chimaeras. The problem of the neuroectodermal origin of the cells of the APUD series. *J. Embryol. Exp. Morphol.* **170:** 209.

Gittes G.K., Galante P.E., Hanahan D., Rutter W.J., and Debas H.T. 1996. Lineage-specific morphogenesis in the developing pancreas: Role of mesenchymal factors. *Development* **122:** 439.

Golosow N. and Grobstein C. 1962. Epitheliomesenchymal interaction in pancreatic morphogenesis. *Dev. Biol.* **4:** 242.

Goodrich L.V., Johnson R.L., Milenkovic L., McMahon J.A., and Scott M.P. 1996. Conservation of the *hedgehog/patched* signaling pathway from flies to mice: Induction of a mouse *patched* gene by Hedgehog. *Genes Dev.* **10:** 801.

Goulding M.D., Lumsden A., and Gruss P. 1993. Signals from the notochord and floor plate regulate the region-specific expression of two Pax genes in the developing spinal cord. *Development* **117:** 1001.

Gu D. and Sarvetnick N. 1993. Epithelial cell proliferation and islet neogenesis in IFN-γ transgenic mice. *Development* **118:** 33.

Hamburger V. and Hamilton H.L. 1951. A series of normal stages in the development of the chick embryo. *J. Morphol.* **88:** 49.

Henry G.L., Brivanlou I.H., Kessler D.S., Hemmati-Brivanlou A., and Melton D.A. 1996. TGF-β signals and a prepattern in *Xenopus laevis* endodermal development. *Development* **122:** 1007.

Jonsson J., Carlsson L., Edlund T., and Edlund H. 1994. Insulin-promoter-factor 1 is required for pancreas development in mice. *Nature* **371:** 606.

Kim S.K., Hebrok M., and Melton D.A. 1997. Notochord to endoderm signaling is necessary for pancreas development. *Development* **124:** (in press).

Le Douarin N. 1964. Étude expérimentale de l'organogenèse du tube digestif et du foie chez l'embryon de poulet. *Bull. Biol. Fr. Belg.* **98:** 544.

———. 1988. On the origin of pancreatic endoderm cells. *Cell* **53:** 169.

Le Douarin N. and Bussonnet C. 1966. Determination precoce et role inducteur de l'endoderme pharyngien chez l'embryon de poulet. *C.R. Acad. Sci.* **263D:** 1241.

Li H.-S., Yang J.-M., Jacobson R.D., Pasko D., and Sundin O. 1994. *Pax6* is first expressed in a region of ectoderm anterior to the early neural plate: Implications for stepwise determination of the lens. *Dev. Biol.* **162:** 181.

Marigo V., Scott M.P., Johnson R.L., Goodrich L.V., and Tabin C.J. 1996. Conservation in hedgehog signaling: Induction of a chicken patched homologue by Sonic hedgehog in the developing limb. *Development* **122:** 1225.

Mizuno T. and Yasugi S. 1990. Susceptibility of epithelia to directive influence of mesenchyme during organogenesis: Uncoupling of morphogenesis and cytodifferentiation. *Cell Differ. Dev.* **31:** 151.

Offield M.F., Jetton T.L., Labosky P.A., Ray M., Stein R., Magnuson M.A., Hogan B.L.M., and Wright C.V.E 1996. PDX-1 is required for pancreatic outgrowth and differentiation of the rostral duodenum. *Development* **122:** 1983.

Ohlsson H., Karlsson K., and Edlund T. 1993. IPF1, a homeodomain-containing transactivator of the insulin gene. *EMBO J.* **12:** 4251.

Patten B.M. and Carlson B.M., Eds. 1974. Structure of chicks from 50 to 60 hours of incubation. *Foundations of embryology*, 3rd edition, p. 236. McGraw-Hill, New York.

Pfaff S.L., Mendelsohn M., Stewart C.L., Edlund T., and Jessell T.M. 1996. Requirement for LIM homeobox gene *Isl-1* in motor neuron generation reveals a motor neuron-dependent step in interneuron differentiation. *Cell* **84:** 309.

Pictet R., Clark W.R., Williams R.H., and Rutter W.J. 1972. An ultrastructural analysis of the developing embryonic pancreas. *Dev. Biol.* **29:** 436.

Placzek M. and Furley A. 1996. Neural development: Patterning cascades in the neural tube. *Curr. Biol.* **6:** 526.

Pourquié O., Coltey M., Teillet M.A., Ordahl C., and Le Douarin N.M. 1993. Control of dorso-ventral patterning of somite derivatives by notochord and floorplate. *Proc. Natl. Acad. Sci.* **90:** 5242.

Riddle R.D., Johnson R.L., Laufer E., and Tabin C. 1993. *Sonic hedgehog* mediates the polarizing activity of the ZPA. *Cell* **75:** 1401.

Roelink H., Augsburger A., Heemskerk J., Korzh V., Norlin S., Ruiz A. Altaba i, Tanabe Y., Placzek M., Edlund T., Jessell T.M., and Dodd J. 1994. Floor plate and motor neuron induction by *vhh-1*, a vertebrate homolog of *hedgehog* expressed by the notochord. *Cell* **76:** 761.

Romanoff A.L., Ed. 1960. The digestive system. In *The avian embryo*, p. 526. Macmillan, New York.

Rosenquist G.C. 1971. The location of the pregut endoderm in the chick embryo at the primitive streak stage as determined by radioautographic mapping. *Dev. Biol.* **26:** 323.

Rudnick A., Ling T.Y., Odagiri H., Rutter W.J., and German M.S. 1994. Pancreatic β cells express a diverse set of homeobox genes. *Proc. Natl. Acad. Sci.* **91:** 12203.

Ruiz i Altaba A., Placzek M., Baldassare M., Dodd J., and Jessell T.M. 1995. Early steps of notochord and floor plate development in the chick embryo defined by normal and induced expression of HNF-3β. *Dev. Biol.* **170:** 299.

Sechrist J. and Marcelle C. 1996. Cell division and differentiation in avian embryos: Techniques for study of early neurogenesis and myogenesis. In *Methods in avian embryology* (ed. M. Bronner-Fraser), p. 301. Academic Press, San Diego, California.

Slack J.M.W. 1995. Developmental biology of the pancreas. *Development* **121:** 1569.

Sosa-Pineda B., Chowdhury K., Torres M., Oliver G., and Gruss P. 1997. The *Pax4* gene is essential for differentiation of insulin-producing β-cells in the mammalian pancreas. *Nature* **386:** 399.

Spooner B.S., Walther B.T., and Rutter W.J. 1970. The development of the dorsal and ventral mammalian pancreas in vivo and in vitro. *J. Cell Biol.* **47:** 235.

St.-Onge L., Sosa-Pineda B., Chowdhury K., Mansouri A., and Gruss P. 1997. *Pax6* is required for differentiation of glucagon-producing α-cells in mouse pancreas. *Nature* **387:** 406.

Sumiya M. and Mizuno T. 1987. Étude immunohistologique de l'expression de glucagon dans l'endoderme pancréatique dorsal chez l'embryon précoce de poulet. *C.R. Seances Soc. Biol. Fil.* **181:** 718.

Turque N., Plaza S., Radvanyi F., Carriere C., and Saule S. 1994. *Pax-QNR/Pax6*, a paired box- and homeobox-containing gene expressed in neurons, is also expressed in pancreatic endocrine cells. *Mol. Endocrinol.* **8:** 929.

Wessells N.K. and Cohen J.H. 1967. Early pancreas organogenesis: Morphogenesis, tissue interactions, and mass effects. *Dev. Biol.* **15:** 237.

Wilson P.A. and Melton D.A. 1994. Mesodermal patterning by an inducer gradient depends on secondary cell-cell communication. *Curr. Biol.* **4:** 676.

Yamada T., Placzek M., Tanaka M., Dodd J., and Jessel T.M. 1991. Control of cell pattern in the developing nervous system: Polarizing activity of the floor plate and notochord. *Cell* **64:** 635.

Yokouchi Y., Sakiyama J.-I., and Kuroiwa A. 1995. Coordinated expression of *Abd-B* subfamily genes of the *HoxA* cluster in the developing digestive tract of chick embryo. *Dev. Biol.* **169:** 76.

The Specification of Muscle in *Drosophila*

M.K. Baylies,[1] M. Bate,[2] and M. Ruiz Gomez[2]
[1]*Molecular Biology Program, Memorial Sloan-Kettering Cancer Institute, New York, New York 10021;*
[2]*Department of Zoology, Cambridge CB2 3EJ, England*

During embryogenesis in *Drosophila,* a complex pattern of somatic muscle fibers forms on the developing body wall (Bate 1993). This intricate arrangement of muscles provides us with a unique opportunity to study the machinery of myogenesis, from the initial segregation of the myogenic lineage in the early mesoderm to the final specification and differentiation of individual muscle fibers within the pattern. Here, we focus on the mechanisms that underlie the segregation of the myogenic lineage in *Drosophila* and the deployment of these cells to form the segmental muscle pattern. In particular, we consider the role of the basic helix-loop-helix (bHLH) protein Twist (Thisse et al. 1988) in regulating entry into somatic myogenesis and the function of specific cells in the myogenic lineage in seeding the formation of individual muscles.

INTRINSIC AND EXTRINSIC FACTORS IN MESODERMAL PATTERNING

The mesoderm in *Drosophila* forms as a mid ventral population of cells in the blastoderm stage embryo that invaginates at gastrulation and migrates dorsally, coating the inner face of the ectoderm (Fig. 1A) (Leptin and Grunewald 1990). A number of different tissues will be formed from this invaginated population of mesodermal cells, including the somatic muscles, the fat body, the heart, and the visceral muscles that line the gut. Transplantation experiments (Beer et al. 1987) have shown that, as the mesoderm invaginates, its cells have not yet been assigned to any of these different developmental pathways. However, shortly thereafter, patterns of gene expression (Azpiazu and Frasch 1993; Azpiazu et al. 1996) and the appearance of morphologically distinct groups of cells within the mesoderm (Dunin Borkowski et al. 1995) indicate that it has been subdivided into sets of cells from which the different derivatives will develop. The progenitor populations are organized into metamerically repeated groups along the body axis, such that every segment contributes cells toward the formation of the heart, the visceral muscles, the fat body, and the body wall muscles (Bate 1993; Dunin Borkowski et al. 1995). Recent experiments show that these assignments result from the activity of regulatory factors that are intrinsic and extrinsic to the mesoderm (Bate and Baylies 1996). In particular, patterns of pair rule gene expression in the blastoderm are maintained in the mesodermal cells as they invaginate at gastrulation, establishing intrinsic differences between mesodermal cells that are reiterated along the anteroposterior axis (Bate 1993; Azpiazu et al. 1996). At the same time, inductive cues from the ectoderm subdivide the mesoderm into sectors—for example, expression of the transforming growth factor-β (TGF-β) family member, Decapentaplegic (Dpp), by cells of the dorsal ectoderm regulates gene expression in the underlying mesoderm and leads to the formation of distinct dorsal and ventral populations of mesodermal cells (Frasch 1995; Staehling-Hampton et al. 1994). For example, a combination of intrinsic and extrinsic factors of this kind establishes the progenitors of the visceral musculature in the anterior, dorsal quadrant of each segment of the mesoderm. This assignment is signaled by the local expression in these cells of the homeobox-containing gene *bagpipe* (Azpiazu and Frasch 1993) which is limited to dorsal cells by a requirement for Dpp (Frasch 1995; Staehling-Hampton et al. 1994) and anterior cells by a requirement for the pair rule gene, *even-skipped* (Azpiazu et al. 1996).

THE ROLE OF TWIST IN MESODERMAL SUBDIVISION AND MYOGENESIS

A key element in executing the subdivision of the mesoderm and in regulating entry into the somatic muscle lineage is the bHLH protein Twist (Thisse et al. 1988). *twist* was originally identified as a gene required for gastrulation, and it is indeed expressed in the mesoderm prior to and during gastrulation (Fig. 1A). Initially, Twist is regulated by locally high intranuclear levels of maternally derived Dorsal protein, and in the absence of Twist, ventral cells fail to invaginate and many mesoderm-specific genes are not expressed (Simpson 1983; Roth et al. 1989; Leptin and Grunewald 1990). In wild-type embryos, *twist* continues to be expressed in the mesoderm after gastrulation in a complex pattern (Fig. 1B) that declines radically only when the terminal differentiation of somatic muscles begins. Interestingly, at this stage, Twist is maintained in a small number of cells that includes the persistent myoblast population from which the adult muscles will be formed at a later stage of development (Fig. 1C). These cells remain undifferentiated, retain Twist, and divide during larval life to form a pool of myoblasts (Bate et al. 1991). Just like the larval myoblasts in the embryo, these cells lose Twist as they differentiate to form

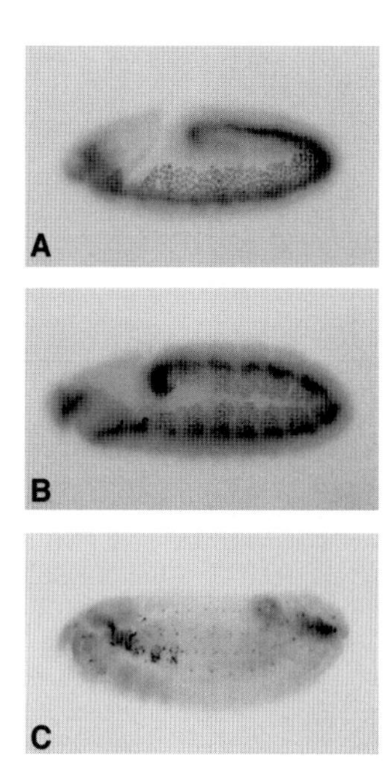

Figure 1. Wild-type patterns of Twist expression after mesoderm specification. (*A*) Embryo immediately after gastrulation showing high levels of Twist expression throughout the mesoderm. (*B*) Stage-10 embryo showing the modulated pattern of Twist expression, with low levels of Twist in the anterior of the segment and high levels in the posterior of the segment. (*C*) Stage-15 embryo showing persistent Twist expression in the progenitors of the adult muscles.

the adult muscles (Currie and Bate 1991). These distinctive patterns of expression were the first indication that Twist might have a significant role in mesodermal differentiation, other than its initial function in gastrulation and that it might be a regulatory protein in the pathway leading to somatic myogenesis.

One possibility that was excluded at an early stage of investigation was that Twist might act to repress somatic myogenic differentiation (Baylies and Bate 1996). Such a role would fit with the declining levels of Twist late in embryogenesis as the larval muscles differentiate and its persistent expression in the nondifferentiating proliferative population of adult myoblasts. However, by using the Gal4 targeted expression system (Brand and Perrimon 1993) and a late mesodermal enhancer, it was possible to maintain high levels of Twist throughout the mesoderm just prior to, during, and after the normal period of myogenesis was completed. This altered expression revealed that somatic muscle differentiation occurs normally and on schedule despite the sustained presence of Twist in the differentiating muscle cells (Baylies and Bate 1996).

In wild-type embryos, the early postgastrulation expression of *twist* is modulated in a metameric fashion that subdivides each mesodermal segment (Fig. 1B) into an anterior domain of low *twist* expression and a posterior domain of high expression (Dunin Borkowski et al. 1995; Baylies and Bate 1996). This pattern corresponds to the subdivision of the mesoderm in each segment into progenitor populations of different tissues. The progenitors of the visceral mesoderm are drawn from the dorsal part of the anterior domain of low *twist* expression in each segment, whereas the muscles of the body wall arise from a population of cells that is restricted to the posterior domain of relatively high *twist* expression (Dunin Borkowski et al. 1995). This coincidence of Twist expression with the apparent subdivision of the mesoderm after gastrulation suggested that Twist might have a decisive role in assigning mesodermal cells to their different developmental pathways and in particular that high levels of Twist might be a prerequisite for entry into the myogenic lineage.

We tested this hypothesis by using the *twist* promoter linked to Gal4 to drive Twist expression in the mesoderm (Baylies and Bate 1996). Although this may seem at first sight a curious experiment, it depends on the delay inherent in the Gal4 system for its effect. During normal development, the modulated pattern of Twist develops as cells in the anterior part of the mesodermal segment lose the high levels of Twist characteristic of all mesodermal cells at gastrulation. Thus, in wild-type embryos, cells that have expressed Twist at high levels lose Twist expression and form a domain of low Twist expression. However, because *twist*-Gal4 driving UAS-*twist* acts with a delay, in embryos carrying these constructs, high levels of Twist are maintained in all mesodermal cells during early embryogenesis.

The effects of this misregulation on the development of the mesoderm are striking: The differentiation of somatic muscles that are normally derived from domains of high Twist expression in the early embryo is unaffected; however, tissues formed from domains of lower Twist expression in normal embryos (e.g., the visceral mesoderm, the heart) are reduced and are partially replaced by ectopic somatic muscles (Fig. 2, compare A and B, D and E, G and H). Moreover, Twist is capable of transforming not only visceral muscle and heart, but also ectoderm into somatic muscle. Misexpression of high levels of Twist in the early ectoderm of the *Drosophila* embryo eliminates all ectodermal tissues, such as the nervous system and epidermis, transforming those cells into a mesodermal cell type that resembles somatic muscle (Baylies and Bate 1996). Further studies of embryos in which dorsoventral polarity is altered such that all cells in the embryo adopt a ventral fate corroborate this result: All cells now express high levels of Twist (Leptin et al. 1992); these cells adopt a somatic muscle-like fate and essentially no ectodermal tissues are found (M.K. Baylies and M. Bate, unpubl.).

Additional experiments using a temperature-sensitive combination of *twist* alleles show that if Twist function is reduced after gastrulation (i.e., levels of expression characteristic of mesoderm that does not form body wall muscles in normal embryos), the somatic muscles are missing or disturbed, whereas heart and visceral mesoderm are relatively unaffected (Fig. 2C, F, I). Thus, it appears that, in the normal development of the mesoderm, the level of Twist is a critical determinant of cell fate. High levels of Twist propel cells into the somatic muscle-forming pathway and the early subdivision of the mesoderm into domains of high and low Twist expression is a decisive event, defining distinct populations of cells which will contribute to different mesodermal derivatives.

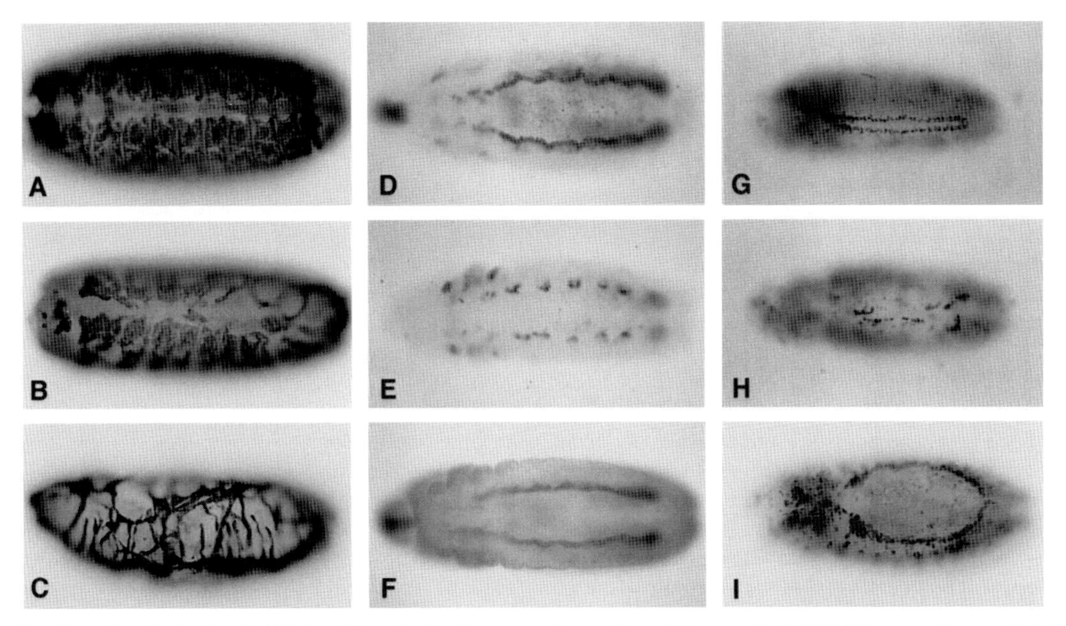

Figure 2. Alterations in the level of Twist affect the development of mesodermal tissues. (*A,D,G*) Wild-type embryos. (*B,E,H*) Embryos in which Twist levels have been elevated using two copies of both *twist*-GAL4 and UAS-*twist* constructs. (*C,F,I*) Embryos that carry a *twist* temperature-sensitive allele combination and have been shifted to the nonpermissive temperature after gastrulation (Baylies and Bate 1996). (*A,B,C*) Compare what happens to the development of the somatic muscles when levels of Twist are increased (*B*) or decreased (*C*). Dorsal views (*A* and *B*) of the somatic muscles show that myogenesis proceeds normally despite increased levels of Twist. Myogenesis is abnormal when Twist levels are decreased (lateral view, *C*) (anti-myosin). (*D,E,F*) Compare development of the visceral mesoderm progenitors (ventral views, anti-fasciclin III). Increased Twist leads to the reduction of the progenitors, whereas decreased Twist has no effect. (*G,H,I*) Contrast the effects of altered Twist levels on the pericardial and cardial cells of the heart (dorsal views). Increased levels of Twist (*H*) lead to loss of heart cells, whereas decreased levels have no effect (*I*).

A question that is currently under investigation is how *twist* is capable of effecting its different roles of regulating the behavior of mesodermal cells as they gastrulate, subdividing these cells into different progenitor populations and acting in the precursors of the adult muscles. Twist, a bHLH transcriptional regulator, is capable of activating a number of genes essential to somatic muscle differentiation. For example, Twist is required for the activation of DMEF2, a MADS box transcriptional activator necessary for myoblast fusion, construction of the contractile apparatus, proper attachment of the muscle to the epidermis, and assembly of the neuromuscular synapse (Lilly et al. 1995; Prokop et al. 1996). Unlike that of Twist, ectopic expression of DMEF2 in the ectoderm fails to convert ectoderm fully into muscle, despite the activation of genes such as myosin and tropomyosin in these cells (Beatty 1997; Lin et al. 1997). Hence, to ensure the production of somatic muscle, Twist must participate with DMEF2 in the execution of the somatic muscle program either directly or through the coordinated activation of other downstream genes. Experiments in vertebrate cell culture have also proposed that bHLH and MADS domain proteins may synergize to activate transcription of muscle-specific targets (Molkentin et al. 1995).

In addition, our experiments indicate that Twist has a repressive role, inhibiting the execution of the visceral muscle or heart fate. We suggest that Twist may accomplish this either by the direct activation of a downstream repressor or through alteration in activity of the Twist protein itself. Although protein activity can be regulated in a number of ways, such as modification through phosphorylation, we propose a role for potential dimerization partners/cofactors in modifying Twist activity. The HLH domain of Twist and other members of this class of transcriptional regulators mediate dimerization. It has been well documented that choice of dimerization partner can alter either the DNA target or the ability to activate the target gene (Jones 1990; Garrell and Campuzano 1991). Our preliminary genetic and biochemical results indicate that Twist can function as a homodimer in vivo and in vitro to regulate downstream genes. However, our genetic results also suggest that there may be a role for Twist as a heterodimer with other HLH proteins. We postulate that the choice of dimerization partner may alter Twist activity, leading to changes in target selection or type of regulation in vivo. We are currently testing genetically and biochemically a number of potential dimerization partners for Twist. The mechanism by which Twist activity is modulated by partner choice may have its parallel in vertebrates: Although no activator function of Twist homodimers has been reported, work from vertebrate cell culture indicates that Twist, as a heterodimer, can act as a repressor (Spicer et al. 1997).

From Myogenic Lineage to Muscle Pattern: The Role of Muscle Founder Cells

The issue of how muscles acquire their distinctive characteristics in the context of a general pathway of myogenic differentiation is common to all organisms. Indeed, the function of the neuromuscular system depends on the formation of a particular pattern of muscles, each with special mechanical functions and an appropriate type of innervation. In *Drosophila,* and it is likely in vertebrate embryos as well, segmental differences between muscles are regulated autonomously by patterns of homeotic gene expression in the mesoderm (Greig and Akam 1993; Michelson 1994). However, the formation of 30 different muscle fibers within a segment requires a different kind of control. So far as is known, the larval muscles in *Drosophila* are, biochemically and physiologically, a uniform set of contractile elements (Bate 1993; Bernstein et al. 1993). However, each muscle is a unique component of the neuromuscular system. In this sense, the functionally important distinctions between the muscles depend on their different positions, their anchorage sites in the epidermis, their size, and, most importantly, their specific connections with innervating motoneurons. The way in which these distinctive properties are acquired is not well understood, but they certainly depend on, for example, the expression by individual muscles of cell adhesion molecules such as Connectin (Nose et al. 1992) and Fasciclin III (Patel et al. 1987) that are important determinants of innervation (Nose et al. 1994; Chiba et al. 1995).

Like syncytial muscle fibers in other organisms, the muscles in *Drosophila* are formed by myoblast fusion: Groups of neighboring cells are recruited at specific locations in the muscle-forming mesoderm and each group combines to form the precursor of an individual muscle fiber (Bate 1990). The key to understanding how muscle characteristics are specified during this process of recruitment and fusion is the finding that the muscle-forming cells themselves are of two kinds: a majority of fusion-competent myoblasts and a minority (albeit a very significant one) of founder cells, each of which seeds the formation of a different muscle (Fig. 3). The accumulating evidence for the view that founder cells are crucial determinants of the muscle pattern is of various kinds: simple examination of the fusion process, the observation that expression of genes characteristic of muscle subsets or of individual muscles occurs in single myoblasts at specific locations before fusion begins, and the fact that neighboring myoblasts are recruited to these expression patterns as they fuse with founder cells to form muscle precursors (Dohrmann et al. 1990). However, the decisive finding is that in mutants where myoblast fusion fails to occur, the myoblasts are demonstrably of two kinds (Fig. 3). In such mutants, the majority of myoblasts remain rounded and undifferentiated, whereas the putative founder myoblasts manifest their characteristic patterns of gene expression and differentiate to form mononucleate muscle-like fibers (Rushton et al. 1995). These tiny fibers have the attachment sites, innervation, and patterns of gene expression typical of the large, syncytial fibers that would form at the same locations in a wild-type embryo. Thus, the myogenic population is subdivided into two classes: founder cells that initiate the formation of

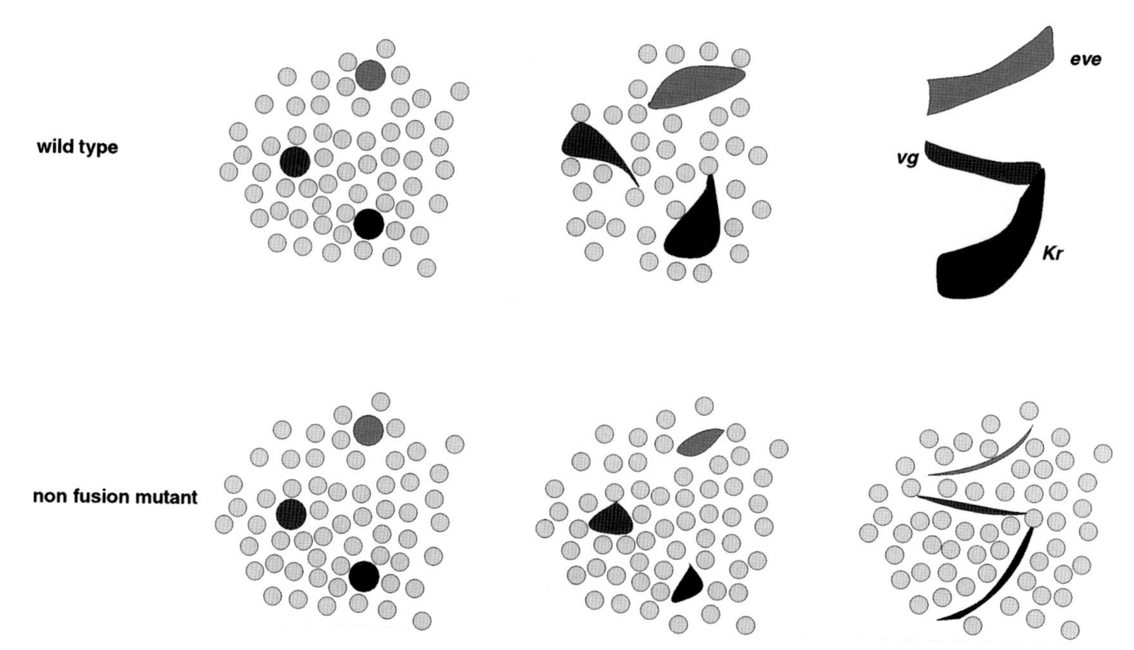

Figure 3. Schematic representation of the behavior of the myogenic populations in wild-type and in mutant embryos lacking myoblast fusion. Myogenic cells fall into two classes: a majority of fusion competent myoblasts (*small circles*) and a minority of muscle founder cells (*larger circles*) that express marker genes characteristic of subsets of muscles (*even-skipped* expression is represented by light gray, *vestigial* expression by dark gray, and *Krüppel* expression by black). During normal development, founder cells fuse with neighboring fusion competent myoblasts and recruit them to these patterns of gene expression. Thus, the final pattern of muscles depends on the segregation of uniquely specified muscle founders. This is confirmed by the different behavior of the two populations in mutant embryos where fusion is blocked. In such embryos, only the founder cells express specific muscle marker genes. They can differentiate into stretched cells that have the characteristics of individual muscles in the final pattern; they find their normal sites of insertion in the epidermis and receive the correct innervation. On the other hand, the fusion competent cells remain as undifferentiated rounded myoblasts.

specific muscles and fusion competent myoblasts that remain as undifferentiated, rounded cells, unless they are recruited by fusion with a founder cell to form part of a syncytial muscle precursor (Fig. 3).

The identification of founder cells and the part they play in myogenesis illuminate the issue of muscle patterning by focusing attention on a limited number of myoblasts that have the capacity to initiate muscle formation. To understand how the pattern forms, we have to understand how these cells are generated at particular locations in the mesoderm and how genes expressed in these cells can condition the general pathway of myogenesis, leading to the formation of individual muscles with unique characteristics.

We now know that there are very close similarities between the process of specifying single cells in the mesoderm during myogenesis and the specification of cells in the ectoderm during neurogenesis (Carmena et al. 1995). Founder cells are generated as pairs from cells known as muscle progenitors which are in turn selected from clusters of cells in the somatic mesoderm. These cell clusters express the proneural gene *lethal of scute (l'sc)* and progenitors are selected from them by a process of lateral inhibition depending on the activation of the Notch signaling pathway (Carmena et al. 1995). By analogy with the proneural clusters of the peripheral nervous system (PNS) (Ghysen and Dambly-Chaudière 1988), it is likely that these "promuscle" clusters are the endpoint for a process of integrating positional signals. This then leads to the formation of a uniquely specified progenitor at that location in the mesoderm. This process of specification will dictate the characteristics of muscle founder cells and muscles formed at this position, just as positional differences between neurons are likely to depend on the sense organ precursor from which they are derived (Merritt and Whitington 1995). Indeed, the data suggest that two signals involved in the patterning of the ectoderm, namely,Wingless and Decapentaplegic, provide the positional information necessary for founder specification in the mesoderm along the anteroposterior and dorsoventral axes, respectively (Baylies et al. 1995; M.K. Baylies, unpubl.; A. Michelson et al., pers. comm.).

To illustrate how the segregation of muscle progenitors and muscle founders leads to the generation of a pattern of different muscle fibers, we focus on one part of the myogenic lineage where progenitors and founders, their patterns of gene expression, and the muscles they give rise to are particularly well defined. The cells concerned are shown in Figure 4. During embryonic stage 11 (Campos-Ortega and Hartenstein 1985), a ventral cluster of *l'sc*-expressing cells gives rise in sequence to two progenitor cells (Fig. 5) (Carmena et al. 1995). Both of these cells express the homeobox-containing gene *S59* (Dohrmann et al. 1990), and the more dorsal progenitor coexpresses the segmentation gene *Krüppel* (*Kr*) (Ruiz Gomez et al. 1997). The more dorsal progenitor divides first giving rise to the founders of muscles VA1 and VA2 (muscle nomenclature according to Bate 1993), followed by the more ventral progenitor that produces the VA3 founder and the precursor of a ventral adult muscle (VaP) (Fig. 5). Initially, both the VA1 and VA2 founders coexpress *Kr* and *S59*, but as development proceeds, the expression of both genes disappears from VA1. *Kr* expression is lost first, soon after the VA1 founder is formed, and *S59* disappears once VA1 has fused with neighboring cells to form a recognizable precursor. In contrast, the sibling founder cell (VA2) maintains *S59* and *Kr* expression and gives rise to a muscle precursor which expresses both genes. *S59* continues to be expressed in the differentiated VA2 muscle, but *Kr* expression is lost (Fig. 5). The VA1 muscle is formed close to VA2, but the two fibers have different orientations and VA1 inserts more anteriorly in the epidermis (Fig. 4). Thus, a unique progenitor in which *Kr* and *S59* are coexpressed gives rise to two muscles with distinct patterns of gene expression and final morphology (Fig. 4). The founder of VA3 and the adult muscle precursor both express S59 when they are produced by division of the more ventral of the two progenitors. This expression is then lost, and a clear distinction between the two cells develops as Twist expression

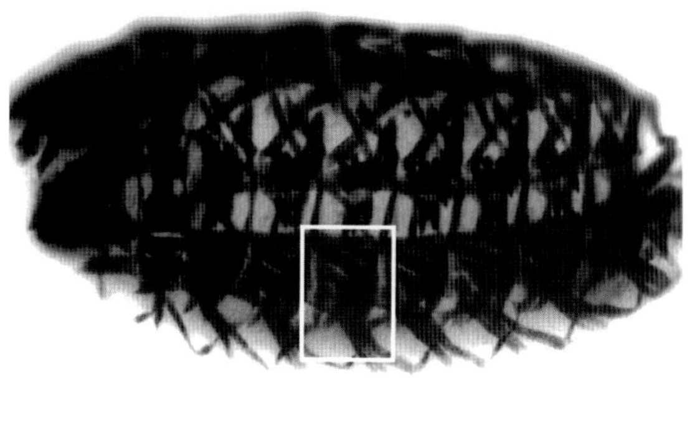

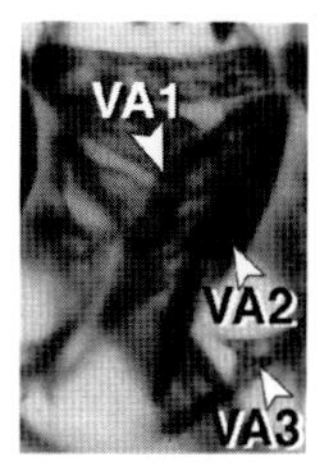

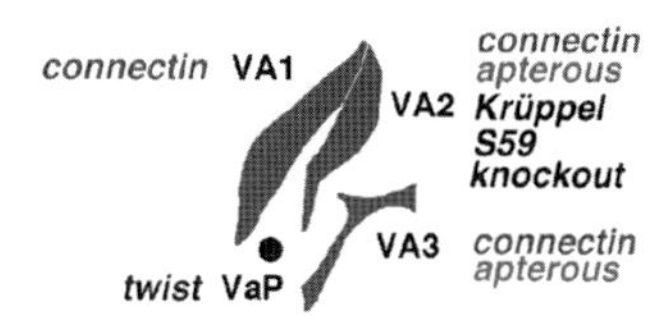

Figure 4. Muscles VA1–3 and the ventral adult muscle precursor. (*Upper panel*) Regular pattern of muscles in a late embryo stained with an antibody against muscle myosin. The white frame encloses the ventral sector of an abdominal segment in which VA1–3 are formed. A higher magnification view of this region (*middle panel*) shows the characteristic shapes and positions of these muscles. The lineages of these three muscles together with the ventral adult precursor (VaP) are known (see text). The relative positions of the three muscles and VaP are shown in the *bottom panel*, together with their distinctive patterns of gene expression. VA1–3 express the cell adhesion molecule *connectin* (Nose et al. 1992); VA2 and 3 express the transcription factor, *apterous*. VA2 additionally expresses *Krüppel* and *S59*, together with *knockout* (Hartmann et al. 1997), a gene required for normal innervation. VaP is characterized by the maintained expression of *twist*.

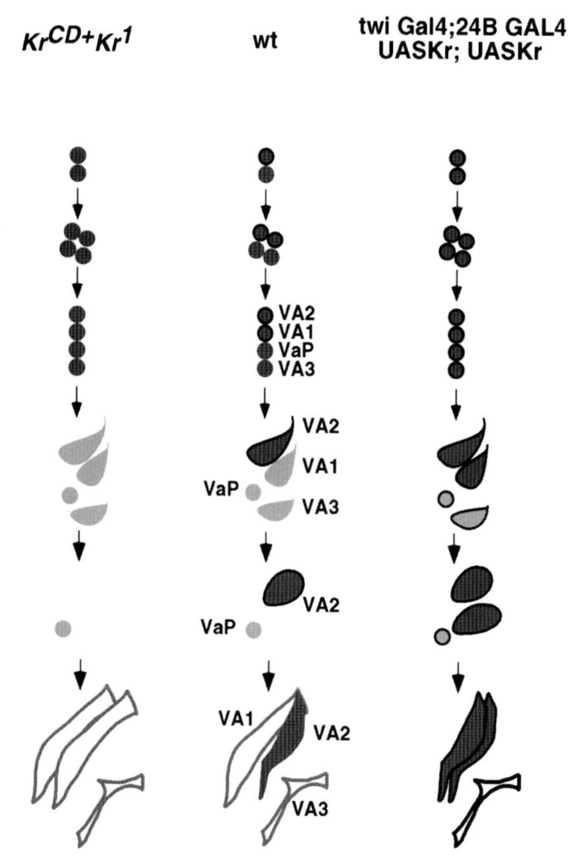

Figure 5. Diagram showing the consequences of lack of function and excess of function of *Kr* in the development of muscles VA1–3 and the ventral adult muscle precursor (VaP). Light and dark shading indicate levels of S59; *Kr* expression is represented by a black outline. Only the cells that can be identified by *S59* or Myosin (*bottom line*) expression are represented. During normal development (*middle column*), two *S59*-positive progenitors give rise to the three *S59*-positive founders that will seed the formation of muscles VA1–3 and VaP. Only the more dorsal progenitor and its two founders will express *Kr*. *Kr* is lost in the VA1 founder and *S59* decays in precursors VA1 and VA3 and in VaP, whereas both *S59* and *Kr* are maintained in the VA2 precursor. In the absence of *Kr* (*left-hand column*), the segregation of *S59*-positive progenitors and founders is not affected. However, *S59* expression declines in all the precursors by stage 13, indicating that the maintenance, but not the initiation, of *S59* expression in VA2 is dependent on *Kr*. Under these conditions, VA2 is transformed toward its non-*S59*-expressing sibling, VA1. When *Kr* is ectopically provided in the whole mesoderm (*right-hand column*), the segregation of *S59*-expressing cells is unaffected, confirming that *Kr* is unable to initiate *S59* expression. Ectopic *Kr* has no effect on *S59* expression in those cells where it is not normally expressed (VA3 and VaP). However, it can maintain *S59* in the VA1 precursor and muscle, and this is now transformed toward the *S59*-expressing VA2 fate.

is lost from one cell (VA3) and maintained (or possibly enhanced) in the other (VaP) (Fig. 4).

Since sibling founder cells give rise to muscle precursors with distinctly different patterns of gene expression, and these in turn give rise to muscles with different characteristics, it seems highly likely that it is the regulated expression of transcription factors such as *S59* and others (Fig. 6) that conditions the development of some or all of the characteristics of individual muscles. Although this suggestion has been made many times, there has been no clear-cut case where altering patterns of gene expression in muscle precursors lead to predictable changes in muscle characteristics. In the case of the *apterous*-expressing founders and precursors, it was possible to show that loss of function could eliminate some of the *apterous*-expressing muscles (Fig. 6) and that ectopic *apterous* expression sometimes leads to a duplication of some of the normally *apterous*-expressing muscles (Bourgouin et al. 1992). Similar results were obtained after overexpression of *nautilus* in the mesoderm (Keller et al. 1997). Although these results are interesting, they do not specifically address the important question of whether altering patterns of gene expression in muscle precursors can lead to predictable changes in the differentiation of muscles. In this view of myogenesis, switching patterns of gene expression from those characteristic of one precursor to those typical of another should lead to a corresponding transformation of the individual muscle phenotype.

We chose to work with the VA1 and VA2 founders (Figs. 4 and 5) to test the idea that it is the differential expression of *Kr* and *S59* in these cells that endows them with their unique characteristics (Ruiz Gomez et al. 1997). Because of the early requirement for *Kr* in establishing the embryonic body plan, we used a *Kr* construct $Kr^{CD+}Kr^{1}$ that rescues the early *Kr* phenotype by promoting *Kr* expression in the central body region of the early embryo (Romani et al. 1996). At later stages, however, there is no *Kr* expression in the CNS or the mesoderm. We took advantage of this fact to look at the effects of loss of *Kr* function on the differentiation of *Kr*-expressing muscles. In such embryos, the expression of *S59* is initiated normally in the VA1/2 progenitor but is no longer retained in the VA2 precursor (Fig. 5). When we examine the muscle that now forms at the VA2 position, it is transformed and has the characteristics of VA1 so that two muscles with the orientations and insertion sites of VA1 are now present (Fig. 5). These findings suggest that *Kr* is required to maintain *S59* expression in the VA2 precursor and that loss of *Kr* from VA2 alters its developmental fate into that of a muscle that expresses neither *Kr* nor *S59* (i.e., VA1). If this is so, then the distinction between the alternative fates of the two sibling cells (VA1 and VA2) depends on the restriction of *Kr* expression to one of them. We tested this by analyzing the effects of maintained *Kr* expression on the fate of the VA1 and VA2 founders. When the Gal4 system is used to express *Kr* throughout the mesoderm (Brand and Perrimon 1993) and therefore in both VA1 and VA2, *S59* expression is initiated normally but is then sustained in the VA1 precursor from which it would normally be lost. Most importantly, the opposite transformation now occurs and two muscles with the orientation and insertion sites of VA2 are formed (Fig. 5).

These experiments not only show that local expression of factors such as Kr in the myogenic lineage can regulate the diversification of muscles, but also provide some insight into the way in which such factors may interact with the myogenic pathway in general. One possibility is that

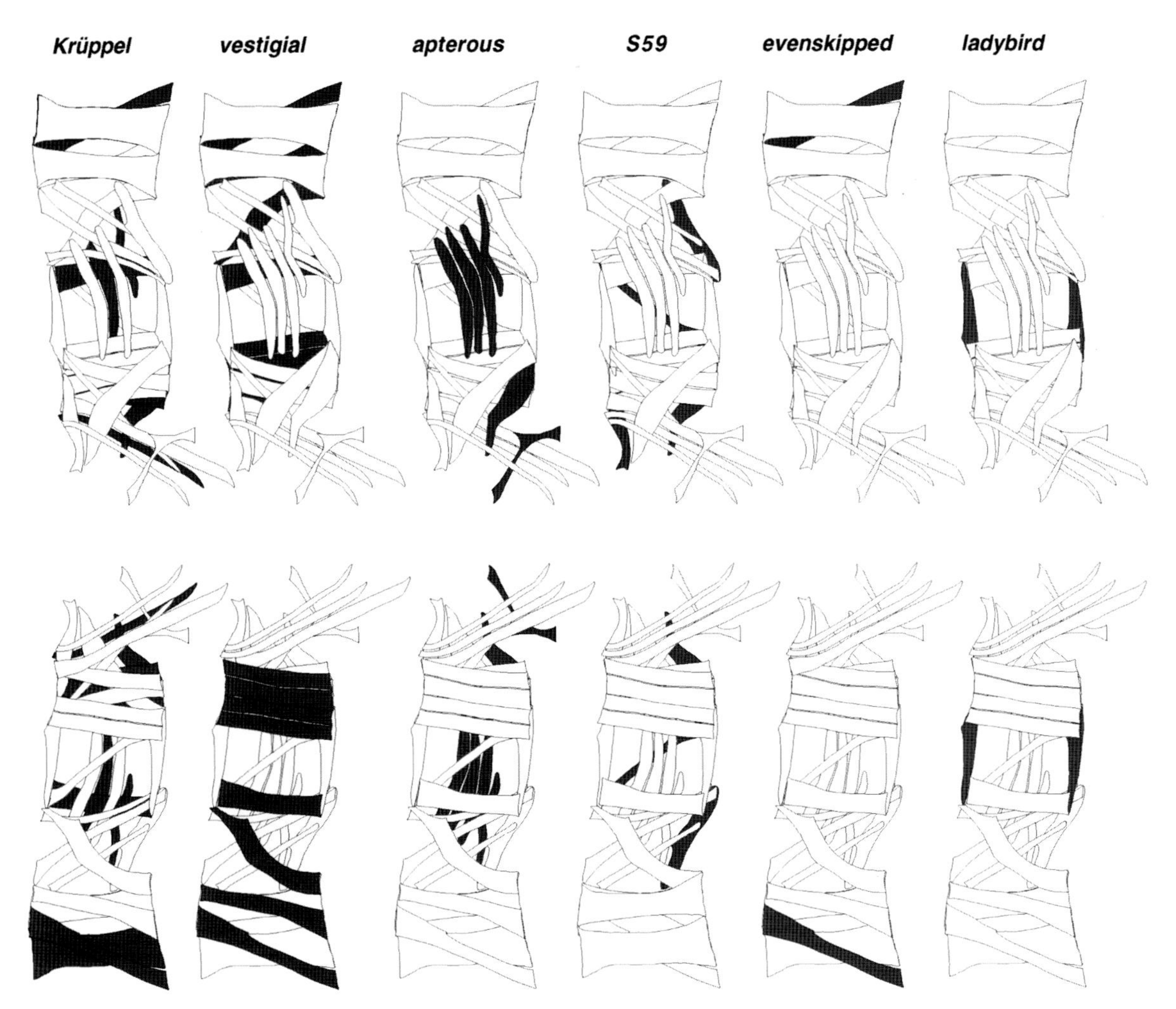

Figure 6. Expression of founder cell marker genes in specific muscle subsets.The diagram shows external (*upper line*) and internal (*lower line*) views of the abdominal muscle pattern. In each case, the muscles that express the particular transcription factor are colored in gray below the corresponding name. Note that subsets of muscles expressing different marker genes are partially overlapping so that muscles differ in the specific combinations of genes they express. For references, see *Krüppel* (Ruiz Gomez et al. 1997); *vestigial* (Bate and Rushton 1993); *apterous* (Bourgouin et al. 1992); *S59* (Dohrmann et al. 1990); *even-skipped* (Frasch et al. 1987); *ladybird* (Jagla et al. 1997).

the expression of genes such as *Kr* and *S59* is necessary for the formation of individual muscles. However, our experiments show that loss of *Kr* does not prevent muscle differentiation but alters the specific characteristics of individual muscles in which it is normally expressed. Thus, *Kr* acts in concert with the myogenic pathway to define specific muscle properties; it does not control myogenesis itself. Our evidence suggests that *Kr* exerts its effect by regulating the expression of genes such as *S59* in specific muscle precursors (Ruiz Gomez et al. 1997).

CONCLUSIONS

The analogy between neurogenesis and myogenesis that these findings suggest should not be overemphasized. The significant observation is probably that in myogenesis as in neurogenesis, the specification of individual cells depends on the prior specification of a spaced pattern of progenitor cells within a cell layer. Given this requirement, it is less surprising that the same machinery should be used in the mesoderm and the ectoderm to single out these cells. On the other hand, it is also clear that genes encoding bHLH proteins such as Twist and the genes of the Achaete-Scute Complex have an important role in defining the lineages from which these progenitor patterns will arise. It is likely that in the mesoderm, as in the ectoderm, these genes and their products act to integrate the positional cues that underlie the initial partitioning of embryonic cell populations as well as the final specification of individual progenitors.

ACKNOWLEDGMENTS

We are grateful to Helen Skaer and Ruben Artero for helpful comments on the manuscript and to Alfonso Martinez Arias and other colleagues for their help throughout this work. We would also like to thank Maria Leptin, Siegfried Roth, and Dan Kiehart for antibodies. The research reported here is supported by grants from The Society of Memorial Sloan-Kettering Cancer Center to M.K.B., National Institutes of Health Cancer Center Support grant NCI-P30-CA-08748 to Memorial Sloan-Kettering Cancer Center and from the Wellcome Trust to M.B.

REFERENCES

Azpiazu N. and Frasch M. 1993. *tinman* and *bagpipe*: Two homeo box genes that determine cell fate in the dorsal mesoderm of *Drosophila*. *Genes Dev.* **7:** 1325.

Azpiazu N., Lawrence P.A., Vincent J.-P., and Frasch M. 1996. Segmentation and specification of the *Drosophila* mesoderm. *Genes Dev.* **10:** 3183.

Bate M. 1990. The embryonic development of larval muscles in *Drosophila*. *Development* **110:** 791.

——— 1993. The mesoderm and its derivatives. In *The development of* Drosophila melanogaster (ed. M. Bate and A. Martinez-Arias), vol. 2, p. 1013. Cold Spring Harbor Laboratory Press, Cold Spring Harbor, New York.

Bate M. and Baylies M.K. 1996. Intrinsic and extrinsic determinants of mesodermal differentiation in *Drosophila*. *Semin. Cell Dev. Biol.* **7:** 103.

Bate M. and Rushton E. 1993. Myogenesis and muscle patterning in *Drosophila*. *C.R. Acad. Sci.* **316:** 1055.

Bate M., Rushton E., and Currie D.A. 1991. Cells with persistent *twist* expression are the embryonic precursors of adult muscles in *Drosophila*. *Development* **113:** 79.

Baylies M.K. and Bate M. 1996. *twist*: A myogenic switch in *Drosophila*. *Science* **272:** 1481.

Baylies M.K., Martinez-Arias A., and Bate M. 1995. *wingless* is required for the formation of a subset of muscle founder cells during *Drosophila* embryogenesis. *Development* **121:** 3829.

Beatty K. 1997. "Characterization of a novel transcription factor, DMEF2, expressed in the mesoderm of *Drosophila melanogaster.*" Ph. D. thesis, Cambridge University, United Kingdom.

Beer J., Technau G., and Campos-Ortega J.A. 1987. Lineage analysis of transplanted individual cells in embryos of *Drosophila melanogaster*. IV. Commitment and proliferative capabilities of individual mesodermal cells. *Roux's Arch. Dev. Biol.* **196:** 222.

Bernstein S.I., O'Donnell P.T., and Cripps R.M. 1993. Molecular genetic analysis of muscle development, structure and function in *Drosophila*. *Int. Rev. Cytol.* **143:** 63.

Bourgouin C., Lundgren S.E., and Thomas J.B. 1992. *apterous* is a *Drosophila* LIM domain gene required for the development of a subset of embryonic muscles. *Neuron* **9:** 549.

Brand A.H. and Perrimon N. 1993. Targeted gene expression as a means of altering cell fates and generating dominant phenotypes. *Development* **118:** 401.

Campos-Ortega J.A. and Hartenstein V. 1985. The embryonic development of *Drosophila melanogaster*. Springer Verlag, Berlin.

Carmena A., Bate M., and Jiménez F. 1995. *lethal of scute*, a proneural gene, participates in the specification of muscle progenitors during *Drosophila* embryogenesis. *Genes Dev.* **9:** 2373.

Chiba A., Snow P., Keshishian H., and Hotta Y. 1995. Fasciclin III as a synaptic target recognition molecule in *Drosophila*. *Nature* **374:** 166.

Currie D.A. and Bate M. 1991. The development of adult abdominal muscles in *Drosophila*: Myoblasts express *twist* and are associated with nerves. *Development* **113:** 91.

Dohrmann C., Azpiazu N., and Frasch M. 1990. A new *Drosophila* homeobox gene is expressed in mesodermal precursor cells of distinct muscles during embryogenesis. *Genes Dev.* **4:** 2098.

Dunin Borkowski O.M., Brown N.H., and Bate M. 1995. Anterior-posterior subdivision and the diversification of the mesoderm in *Drosophila*. *Development* **121:** 4183.

Frasch M. 1995. Induction of visceral and cardiac mesoderm by ectodermal Dpp in the early *Drosophila* embryo. *Nature* **374:** 464.

Frasch M., Hoey T., Rushlow C., Doyle H.J., and Levine M. 1987. Characterization and localization of the *even-skipped* protein of *Drosophila*. *EMBO J.* **6:** 749.

Garrell J. and Campuzano S. 1991. The helix-loop-helix domain: A common motif for bristles, muscles, and sex. *BioEssays* **13**: 493.

Ghysen A. and Dambly-Chaudière C. 1988. From DNA to form: The *achaete-scute* complex. *Genes Dev.* **2:** 495.

Greig S. and Akam M.E. 1993. Homeotic genes autonomously specify one aspect of pattern in the *Drosophila* mesoderm. *Nature* **362:** 630.

Hartmann C., Landgraf M., Bate M., and Jäckle H. 1997. *Krüppel* dependent *knockout* activity is required for proper innervation of a specific set of *Drosophila* larval muscles. *EMBO J.* (in press).

Jagla K., Jagla T., Heitzler P., Dretzen G., Bellard F., and Bellard M. 1997. *ladybird*, a tandem of homeobox genes that maintain late *wingless* expression in terminal and dorsal epidermis of the *Drosophila* embryo. *Development* **124:** 91.

Jones N. 1990. Transcriptional regulation by dimerization: Two sides to an incestuous relationship. *Cell* **61:** 9.

Keller C.A., Erickson M.S., and Abmayr S.M. 1997. Misexpression of *nautilus* induces myogenesis in cardioblasts and alters the pattern of somatic muscles fibers. *Dev. Biol.* **181:** 197.

Leptin M. and Grunewald B. 1990. Cell shape changes during gastrulation in *Drosophila*. *Development* **110:** 73.

Leptin M., Casal J., Grunewald B., and Reuter R. 1992. Mechanisms of early *Drosophila* mesoderm formation. *Development Suppl.*, p. 23.

Lilly B., Zhao B., Ranganayakulu G., Paterson B.M., Schulz R.A., and Olson E. 1995. Requirements of MADS domain transcription factor D-MEF2 for muscle formation in *Drosophila*. *Science* **267:** 688.

Lin M.H., Bour B., Abmayr S., and Storti R. 1997. Ectopic expression of MEF2 in the epidermis induces epidermal expression of muscle genes and abnormal muscle development in *Drosophila*. *Dev. Biol.* **182:** 240.

Merritt D. J. and Whitington P.M. 1995. Central projections of sensory neurons in the *Drosophila* embryo correlate with sensory modality, soma position and proneural gene function. *J. Neurosci.* **15:** 1755.

Michelson A.M. 1994. Muscle pattern diversification in *Drosophila* is determined by the autonomous function of homeotic genes in the embryonic mesoderm. *Development* **120:** 755.

Molkentin J., Black B., Martin J., and Olson E. 1995. Cooperative activation of muscle gene expression by MEF2 and myogenic bHLH proteins. *Cell* **83:** 1125.

Nose A., Mahajan V.B., and Goodman C.S. 1992. Connectin: A homophilic cell adhesion molecule expressed on a subset of muscles and the motoneurons that innervate them in *Drosophila*. *Cell* **70:** 553.

Nose A., Takeichi M., and Goodman C.S. 1994. Ectopic expression of connectin reveals a repulsive function during growth cone guidance and synapse formation. *Neuron* **13:** 525.

Patel N.H., Snow P.M., and Goodman C.S. 1987. Characterization and cloning of Fasciclin III: A glycoprotein expressed on a subset of neurons and axon pathways in *Drosophila*. *Cell* **48:** 975.

Prokop A., Landgraf M., Rushton E., Broadie K., and Bate M. 1996. Presynaptic development at the *Drosophila* neuromuscular junction: The assembly and localisation of presynaptic active zones. *Neuron* **17:** 617.

Romani S., Jimenez F., Hoch M., Patel N.H., Taubert H., and Jäckle H. 1996. *Krüppel*, a *Drosophila* segmentation gene, participates in the specification of neurons and glial cells. *Mech. Dev.* **60:** 95.

Roth S., Stein D., and Nüsslein-Volhard C. 1989. A gradient of the Dorsal protein determines dorsoventral pattern in the *Drosophila* embryo. *Cell* **59:** 1129.

Ruiz Gomez M., Hartmann C., Romani S., Jäckle H., and Bate M. 1997 Specific muscle identities are regulated by Krüppel during *Drosophila* embryogenesis. *Development* (in press).

Rushton E., Drysdale R., Abmayr S.M., Michelson A.M., and Bate M. 1995. Mutations in a novel gene, *myoblast city*, provide evidence in support of the founder cell hypothesis for *Drosophila* muscle development. *Development* **121:** 1979.

Simpson P. 1983. Maternal-zygotic gene interactions during the formation of the dorso-ventral pattern in *Drosophila* embryos. *Genetics* **105:** 615.

Spicer D., Rhee J., Cheung W., and Lassar A. 1997 Inhibition of myogenic bHLH and MEF2 transcription factors by the bHLH protein Twist. *Science* **272:** 1476.

Staehling-Hampton K., Hoffmann F.M., Baylies M.K., Rushton E., and Bate M. 1994. *dpp* induces mesodermal gene expression in *Drosophila*. *Nature* **372:** 22.

Thisse B., Stoetzel C., Gorostiza-Thisse C., and Perrin-Schmitt F. 1988. Sequence of the *twist* gene and nuclear localization of its protein in endomesodermal cells of early *Drosophila* embryos. *EMBO J.* **7:** 2175.

Homeobox Genes and Heart Development

C. Biben, S. Palmer, D.A. Elliott, and R.P. Harvey
Development and Neurobiology Group, The Walter and Eliza Hall Institute of Medical Research, Royal Melbourne Hospital, Victoria 3050 Australia

The heart is the first organ to form and function within the mammalian embryo. Heart development begins during gastrulation when cells fated to the cardiac lineages migrate to the anterior and anterior-lateral aspects of the embryo and acquire their committed state (Rawles 1943; Rosenquist 1966; DeRuiter et al. 1992). From this point on, a series of complex morphogenic events determines the structure of the linear heart tube and then the four-chambered organ seen in adults. These processes must occur both with precision and in harmony with an increasing functional output, serving the demands of the rapidly growing embryo. The complexity of heart development and its apparent sensitivity to perturbation is highlighted by the incidence of cardiac structural abnormalities in humans, about 1 in a 100 for live borns and up to 1 in 10 for stillborns. The development of the heart is now the subject of intensive investigation, and the first clues to genetic control of its cell lineages and morphogenesis have come to light, in part through analogies drawn between heart formation in the *Drosophila* and vertebrate embryo systems (Harvey 1996; Bodmer et al. 1997).

One of the key questions in the cardiac field is when and how mesodermal cells become committed to a cardiac fate. Studies in frog, chicken, and mouse embryos have shown that commitment occurs during gastrulation, after ingression of progenitor cells through the primitive streak (Jacobson and Sater 1988; Gonzalez-Sanchez and Bader 1990; Montgomery et al. 1994; Tam et al. 1997). Formation of cardiac muscle from its progenitor population and further morphogenesis of the heart tube are believed to depend on inductive interactions with endoderm (Jacobson and Sater 1988; Sugi and Lough 1994; Nascone and Mercola 1995; Schultheiss et al. 1995). Recently, members of the bone morphogenic protein (BMP) family have been shown to be involved in the initial steps of cardiac induction (Schultheiss et al. 1997). These signaling factors can induce markers of cardiac muscle and/or a beating phenotype in mesodermal explants that never form heart, but only in the continued presence of anterior endoderm. Thus, at least one other critical endodermal factor remains to be identified.

Downstream from these inductive events, cardiac transcription factors are thought to control genetic programs specific for the various heart lineages. The skeletal myogenic transcription factors related to *myoD* are not expressed in cardiac muscle, despite the fact that cardiac and skeletal muscles express numerous myofilament and metabolic genes in common (Lyons 1994). Thus, the genetic programs driving cardiac and skeletal myogenesis are distinct to some extent, although their common origins are reflected by shared expression of the MEF2 transcriptional regulators, members of the MADS family (Edmondson et al. 1994).

The *Drosophila* embryo has been instrumental in alerting us to some of the key transcription factors for vertebrate heart development. The *NK*-class homeobox gene *tinman* is expressed in heart and gut muscle lineages during early fly development (Bodmer 1993), and in *tin* mutants, these muscles do not form and their progenitors cannot be detected (Bodmer 1993). Thus, *tinman* is a key patterning gene for the mesoderm, and its continued expression in heart muscle cells suggests a later role in heart differentiation or morphogenesis. In mice, screens based on the *tinman* homeodomain sequence led to the identification of *Nkx2-5* (also called *Csx*) which, like *tinman* itself, is expressed in early cardiac mesoderm and in the heart muscle lineage throughout life (Komuro and Izumo 1993; Lints et al. 1993). Seven members of this homeobox gene subclass have now been isolated from mammalian genomes (*Nkx2-1* to *Nkx2-6*; *Nkx2.8*), with an additional member present in zebrafish (*nkx2.7*) (Harvey 1996; Boettger and Kessel 1997; Brand et al. 1997). However, *Nkx2-5* is the only member to be expressed in the hearts of all four vertebrate embryo models (mouse, chick, *Xenopus*, and zebrafish) from the cardiac progenitor stage onward. This pattern suggests that there is an early and key role for *Nkx2-5* in the cardiac regulatory hierarchy and that genetic pathways underlying cardiac development have been conserved in evolution.

The function of *Nkx2-5* is being studied on a number of levels. Its role as a positive transcription factor in myogenic differentiation has been suggested for several genes (Lyons et al. 1995; Chen et al. 1996; Durocher et al. 1996; Gajewski et al. 1997; Zou et al. 1997). Furthermore, the knockout phenotype suggests that *Nkx2-5* has a role, direct or indirect, in cardiac morphogenesis (Lyons et al. 1995). The overall picture is that myogenic differentiation of the heart and its complex morphogenesis are not separable processes and that a single regulatory gene can direct aspects of both.

In *Nkx2-5* mutants, morphogenesis of the heart beyond the linear tube stage is blocked (Lyons et al. 1995). The key process affected in the mutants would appear to be looping morphogenesis, the profound rightward bending of the heart tube that is an essential part of chamber, valve, and vessel development. Cardiac looping is a complex and critical event, as evidenced by the spectrum of life-threatening cardiac structural abnormalities that can be attributed to defects in the process (Burn 1991; Bowers et al. 1996). Looping is also the first gross morphological event that betrays developmental asymmetry in the left-right body axis, occurring just before other asymmetries appear associated with embryonic turning and visceral organogenesis. Various types of experiments have suggested that laterality information in the embryo influences the direction of heart looping and if that information is confused or eliminated, hearts loop with random situs (Boterenbrood and Nieuwkoop 1973; Layton et al. 1980; Levin et al. 1995; Yost 1995; Hyatt et al. 1996).

Recent reports have described several genes in the embryonic laterality pathway which show asymmetrical expression in the trunk region of the embryo (Levin et al. 1995; Collignon et al. 1996; Lowe et al. 1996; Meno et al. 1996; Isaac et al. 1997). Other genes are expressed asymmetrically in the heart itself (Tsuda et al. 1996; Isaac et al. 1997; Smith et al. 1997). Perturbation of trunk-expressed genes demonstrates that laterality information in the embryo develops in or around the node and primitive streak, before also becoming manifest in lateral plate mesoderm (LPM) just before the onset of cardiac looping (Levin et al. 1995; Isaac et al. 1997). Genes expressed exclusively in left or right LPM have been detected. *nodal* and *lefty*, encoding signaling factors of the transforming growth factor-β (TGF-β) superfamily, are expressed in left-sided LPM up to and including the caudal region of the heart, but not within the ventricular region that undergoes rightward looping (Levin et al. 1995; Collignon et al. 1996; Lowe et al. 1996; Meno et al. 1996). In the chick, the transcription factor gene *cSnR* is expressed in right-sided LPM in a similar fashion (Isaac et al. 1997). Manipulation of *nodal* and *cSnR* expression affects both the direction and timing of cardiac looping, suggesting that development of asymmetry in the heart depends on information from both left and right sides of the embryo (Collignon et al. 1996; Lowe et al. 1996; Isaac et al. 1997).

The molecular processes underlying cardiac looping are no doubt extremely complex. How the heart receives and interprets laterality information is not at all clear, nor is the role of transcription factors such as *Nkx2-5*. In this paper, we review our current understanding of the role of *Nkx2-5* in cardiac commitment, myogenesis, and morphogenesis. We present two new genes that appear to be under *Nkx2-5* control: those encoding atrial natriuretic factor (ANF) and SM-22. The diverse ways in which *Nkx2-5* controls the spatial patterns of target genes are discussed, as well as how these phenomena might relate to patterning mechanisms in the primitive heart tube.

EXPERIMENTAL PROCEDURES

In situ hybridization. *Nkx2-5*$^{-/-}$ embryos from heterozygous matings were genotyped using a polymerase chain reaction (PCR) assay performed on yolk sac DNA as described recently (Biben and Harvey 1997). After dissection, embryos from timed matings were fixed in 4% paraformaldehyde overnight. Noon of the day of plugging was taken as 0.5 day postcoitum (E0.5). In situ hybridization using digoxygenin-labeled cRNA probes was performed as described by Biben and Harvey (1997). Templates for myosin light chains, MLC1A and MLC2V, eHand and dHand, and SM-22 were prepared as described previously (Lyons et al. 1995; Li et al. 1996; Biben and Harvey 1997). The rat ANF template was provided by J. Molkentin and E. Olson. After linearization of the template with *Xho*I, a 134-nucleotide probe was synthesized with T7 RNA polymerase. Hybridization for this rat probe was performed at 65°C.

cDNA synthesis and hybridization analysis. Hearts from *Nkx2-5*$^{-/-}$ embryos and wild-type siblings at E8.5 were dissected and RNA was extracted using the guanidinium isothiocyanate technique (Chomzynski and Saachi 1987). DNA was extracted from remaining embryonic tissue for genotype analysis (Lyons et al. 1995) with a common forward primer (5′-cagtggagctggacaaagcc-3′) used in combination with either a mutant-allele-specific primer (5′-aacttcctgactaggggagg-3′) or a wild-type-specific primer (5’-gaagctccagagtctggtcc-3′). cDNA was produced from extracted RNA using the protocol described in the CapFinder PCR cDNA Library Construction Kit (Clontech Laboratories). Synthesized cDNA was electrophoresed on 1.5% agarose gels and blotted onto Hybond N+ (Amersham) for hybridization in RapidHyb (Amersham) with *SM-22, ANF, MLC2V, MLC1A*, and *MLC3F* gene-specific probes. In each case, probes hybridized to two bands with the predicted size of each full-length cDNA corresponding to the upper band. The identity of the lower band is uncertain, but it may represent single-stranded DNA.

RESULTS

The *Nkx2-5* Knockout Phenotype

Nkx2-5 has been inactivated in the mouse using a gene-targeting strategy: A PGK promoter-neomycin resistance cassette was cloned into the third helix of the homeodomain, an insertion predicted to disrupt Nkx2-5 protein:DNA interaction (Lyons et al. 1995). The *Nkx2-5* mutation leads to a recessive lethal phenotype with full penetrance. Homozygous mutant embryos have a grossly abnormal heart and die around E9.5, presumably from cardiac insufficiency. Mutant hearts develop to the linear tube stage and contract at about the same frequency as wild-type or heterozygous hearts, but then fail to undergo looping morphogenesis and remain largely in a linear conformation. Much of the downstream morphogenesis associated with looping is also blocked: Trabeculation is poor or absent, and the atrioventricular canal and endocardial cushions do not form.

This targeted mutation of *Nkx2-5* demonstrates an essential role for the gene in cardiac development but clearly does not lead to a *tinman*-like phenotype. This could be due to functional compensation by other members of the gene family, and indeed, a new *NK-2* homeobox gene related to *Nkx2-5* and expressed in the early chicken heart has been identified recently (Boettger and Kessel 1997; Brand et al. 1997).

Nkx2-5 and Cardiac Myogenesis

Hearts in *Nkx2-5* knockout embryos undergo rhythmic contractions, suggesting that the cardiocyte myofilament has formed normally, as confirmed by transmission electron microscopy (Lyons et al. 1995). Nevertheless, semiquantitative comparative RT-PCR analysis of myofilament gene expression and in situ hybridization unravel defects in some myogenic pathways. First, the gene encoding MLC2V, a ventricle-specific regulatory light chain and one of the first myofilament genes to be expressed in a regionally restricted fashion in the embryonic heart, is all but abolished in the mutant background. Sometimes, a small patch of positive cells on the dorsal side of the atrioventricular chamber/outflow tract junction can be detected in mutants.

Recently, in collaboration with K. Chien and colleagues, we have found that the gene encoding a novel nuclear protein, CARP (for cardiac ankyrin repeat protein), is also severely down-regulated in *Nkx2-5* mutant hearts (Zou et al. 1997). CARP was isolated via its association with the transcription factor YB-1, implicated in positive regulation of the *MLC2V* promoter. However, in cell culture assays, CARP appears to have negative effects on *MLC2V* transcription, indicating either that a key positive factor is missing from the complex or that a totally separate YB-1 partner positively regulates the *MLC2V* gene. The direct involvement of Nkx2-5 in *MLC2V* regulation is still an open question: A small region of the *MLC2V* promoter conferring most of the temporal and spatial information for *MLC2V* regulation in transgenic mice is still expressed correctly and robustly in the *Nkx2-5* mutant background. One interpretation is that an uncharacterized Nkx2-5-dependent element lies outside of this proximal promoter region.

New Genes under *Nkx2-5* Control

We have recently identified two additional genes that may fall under *Nkx2-5* control: those encoding ANF, a vasoactive hormone controlling fluid and sodium balance in blood (Vesely 1995), and SM-22, a calponin-related protein thought to regulate myofilament activity (Schmidt et al. 1995). In the adult, *ANF* is normally expressed only in the atrial chambers, but in early development, expression is complex and dynamic (Zeller et al. 1987). In the looping heart tube, expression, initially restricted to the left ventricle, is then detected in right ventricle and atrial chambers. In the ventricles, expression is initially in both trabeculae and the compact layer of the myocardium, but quickly becomes restricted to just the trabecular component, and this mode persists until just after birth. In contrast, *SM-22* is expressed across the whole heart, although by late embryogenesis, cardiac expression is down-regulated, remaining only in smooth muscle cells (Li et al. 1996).

We first detected abnormalities in ANF and SM-22 expression in *Nkx2-5* mutants by Southern analysis of amplified cDNA populations derived from dissected wild-type and mutant hearts (Fig. 1). This method presents a rapid and convenient way to score for genes potentially affected in a mutant context, although since the assay is PCR-based, the results can only be considered as semiquantitative. Nevertheless, probes for genes known to be expressed in mutant embryos at normal levels (*MLC1A*, *MLC3F*), or not at all (*MLC2V*), show the expected results. In this assay, *ANF* mRNA was totally abolished in the *Nkx2-5* mutant background, a finding supported by in situ hybridization to wild-type and mutant embryos (Fig. 2A). Transfection studies in neonatal cardiocytes have shown that the *ANF* promoter is regulated directly by Nkx2-5 (Durocher et al. 1996), and our studies on *Nkx2-5$^{-/-}$* hearts support these results. However, the complex and dynamic pattern of ANF expression in embryos suggests the involvement of other spatially restricted regulators.

SM-22 is down-regulated a fewfold, as judged from Southern analysis (Fig. 1). In situ hybridization (Fig. 2B,C) demonstrates that its expression is spatially disrupted in a dramatic way in *Nkx2-5* mutant hearts. Although the gene is expressed across the whole heart in wild-type embryos, it is only expressed in the sinus venosa and perhaps atrial cells of the mutants. Expression in the ventricle and outflow tract is abolished (see Discussion).

Nkx2-5 and Cardiac Morphogenesis

Nkx2-5 mutant hearts fail to undergo looping morphogenesis. Two other transcription factor genes have also been implicated in cardiac looping, those encoding the basic helix-loop-helix (bHLH) transcription factors eHand/Hxt/Th1 and dHand/Hed/Th2 (Cross et al. 1995; Cserjesi et al. 1995; Hollenberg et al. 1995). *eHand* was isolated in yeast two-hybrid and expression screens designed to detect dimerization partners of the ubiquitous class-A bHLH proteins E12, E47, and daughterless. *dHand* was subsequently isolated by homology screening. For *eHand*, a role in trophoblast differentiation has been suggested (Cross et al. 1995).

The *Hand* genes are expressed in several tissues during development, including extraembryonic membranes for *eHand*, decidua for *dHand*, and heart, lateral plate mesoderm, and neural crest derivatives in common (Cserjesi et al. 1995; Srivastava et al. 1997). In chick hearts, both *Hand* genes are expressed across the myocardium, and inhibition of their expression using antisense oligonucleotides leads to an arrest of heart development at early looping stages (Srivastava et al. 1995).

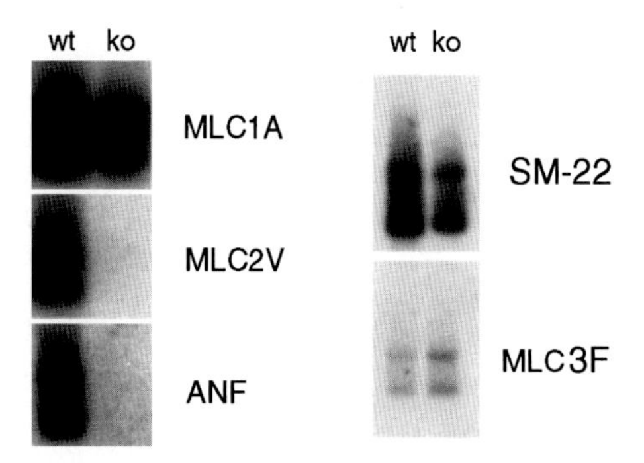

Figure 1. Hybridization of PCR-amplified cDNAs from *Nkx2-5$^{-/-}$* and wild-type hearts with probes for *MLC1A, 2V, 3F, ANF,* and *SM-22.* Expression of some genes is unaffected by the mutation (*MLC1A, MLC3F*), whereas others are down-regulated (*SM-22*) or not expressed at all (*MLC2V, ANF*).

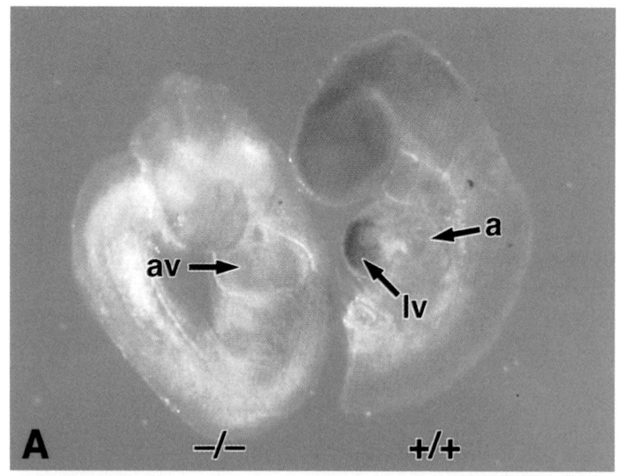

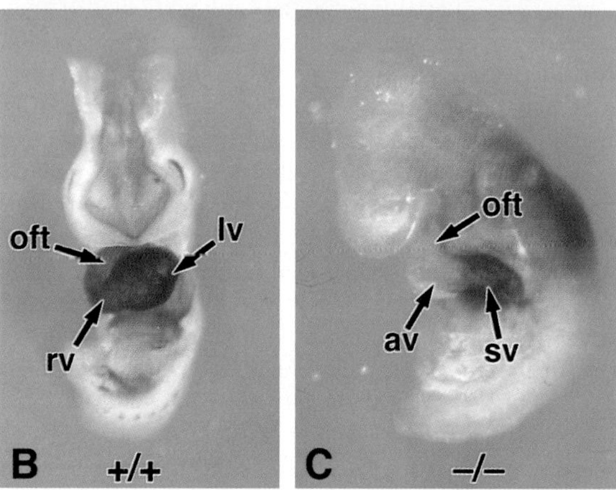

Figure 2. (*A*) Wild-type (*right*) and *Nkx2-5$^{-/-}$* (*left*) embryos (E9.0–9.5) hybridized with a probe for *ANF*. Expression on the outer side of the left ventricle seen in wild-type embryos is absent in *Nkx2-5$^{-/-}$* hearts. (*B*) Wild-type embryo (E9.5) hybridized with a probe for *SM-22*. The gene is expressed throughout the heart tube. (*C*) *Nkx2-5$^{-/-}$* embryo (E9.5) hybridized with a probe for *SM-22*. Note the absence of the anterior expression domain (ventricles and outflow tract). (a) Atrial chamber; (av) atrioventricular chamber; (lv) left ventricle; (oft) outflow tract; (rv) right ventricle; (sv) sinus venosa.

To investigate the suggested role of *Hand* genes in cardiac looping, we have examined expression throughout early stages of mouse heart development in normal embryos and in those carrying mutations affecting looping morphogenesis (Biben and Harvey 1997). In contrast to the situation in the chick, mouse *Hand* genes are expressed in a complex and dynamic fashion from early stages of cardiac development, with the patterns only partially overlapping. *dHand* transcripts can be seen in all myocytes of the cardiac crescent and forming heart tube, before the expression pattern evolves to become dominant in the outer curvature (right side) of the right ventricle. The prominent domain of *eHand* expression is at first caudal and ventral, before becoming dominant in the outer curvature of the left ventricle (Fig. 3) and later in the right side of the right ventricle. These provocative patterns highlight the complex molecular events occurring in the looping heart tube and the possibility of a role for *Hand* genes in both ventricular chamber specification and cardiac morphogenesis.

Careful examination of the *eHand* expression pattern suggests that it is modified by the embryonic left/right (L/R) patterning system (Biben and Harvey 1997). *eHand* expression is initially bilateral in its caudal domain, and then coincident with looping, the pattern evolves to the left side of the left ventricle. Formally, this pattern could be generated by torsion alone. In some embryos, however, at the earliest stages of heart looping, we find enhanced expression on the left side of the bilateral domain, suggesting influence from the L/R patterning system. Additional evidence for this comes from examination of atrial expression. As the bilateral expression in forming atrial chambers and sinus venosa is fading, there is transient unilateral expression in left atrium. These modifications suggest that *eHand* is responsive to L/R signaling, with a possible role in guiding or interpreting aspects of the dextral cardiac loop (see Discussion).

In *Nkx2-5$^{-/-}$* embryos, we find that *eHand* expression in the heart is severely down-regulated. Both left- and right-sided expressions appear to be abolished, with only low-level bilateral expression remaining in the caudal region of the abnormal heart. *eHand* expression at several sites where *Nkx2-5* is not normally expressed is unaffected. Observations suggest that loss of *eHand* expression in the mutant background is not due to indirect effects on gene expression in a failing heart, implying a genuine regulatory link between *Nkx2-5* and *eHand* in a pathway essential for chamber specification and/or cardiac looping.

DISCUSSION

The *Nkx2-5* gene, isolated by virtue of its structural homology with *Drosophila tinman*, is expressed in the myocardial lineage from the early paired progenitor stage in a number of vertebrates (Harvey 1996). *tinman* is considered a commitment gene because *tinman* mutant flies do not form cardiac progenitor cells, nor those of the gut (Bodmer 1993). Evidence that *Nkx2-5* and *tinman* are functionally homologous is emerging from rescue experiments in flies: *Nkx2-5* can substitute for *tinman* in gut muscle development (Bodmer et al. 1997), and, under some circumstances, also for heart development (G. Ran-

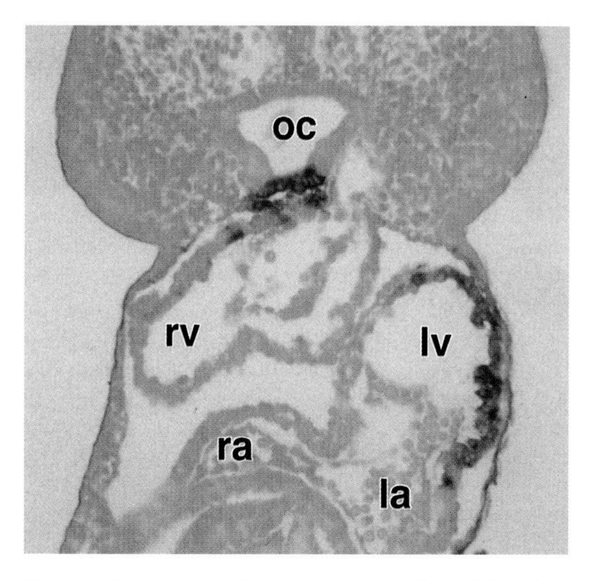

Figure 3. Section through the heart region of an E8.5 embryo hybridized with an *eHand* probe using the whole-mount protocol. Note expression in the outer region of the left ventricle, pericardium, and oral cavity ectoderm. (la) Left atrium; (lv) left ventricle; (oc) oral cavity; (ra) right atrium; (rv) right ventricle.

ganayakula and E. Olson, unpubl.). Thus, it seems likely that *Nkx2-5* could function as a commitment gene for heart development in vertebrates. The fact that *Nkx2-5* knockout mice do not show a *tinman*-like heart phenotype (no lineage detectable) may turn out to be due to functional compensation by another heart-expressed *NK-2* homeobox gene (Lyons et al. 1995; Boettger and Kessel 1997; Brand et al. 1997).

Nkx2-5 is required for some aspects of cardiac myogenesis. Whereas most myofilament genes are expressed normally in the absence of *Nkx2-5*, *MLC2V*, *CARP*, and *SM-22* have been found to be dysregulated (Lyons et al. 1995; Zou et al. 1997; this paper). However, although in vitro studies implicate *Nkx2-5* in direct control of the *alpha-cardiac actin* gene through collaboration with serum response factor (Chen et al. 1996), this gene is expressed normally in mutant hearts (Lyons et al. 1995). Why only a limited subset of muscle genes require Nkx2-5 is still obscure, although partial functional compensation is a plausible explanation.

The most challenging aspect of our studies on the *Nkx2-5* gene relates to spatial patterning within the heart. Genes that fall under *Nkx2-5* control are spatially dysregulated in a variety of ways in the mutant background (Fig. 4). *MLC2V* and *ANF* expression is all but abolished (Lyons et al. 1995; this paper); *CARP* expression is severely down-regulated—abolished in the outflow tract, but retained at a low level in the atrioventricular chamber (Zou et al. 1997). *eHand* expression does not develop on the left side of the heart, although low-level bilateral expression, reminiscent of the early phase of *eHand* expression, remains in the caudal domain (Biben and Harvey 1997). *SM-22* expression is abolished in the atrioventricular chamber, but robust expression remains in sinus venosa.

Is Ventricular Identity Specified in *Nkx2-5* Mutant Hearts?

Atrial and ventricular chambers arise as consecutive segments along the anteroposterior (A/P) axis of the heart tube (Stalsberg and DeHaan 1969). In the chick, atrial and ventricular identities are specified by mid gastrulation, and retinoic acid can respecify ventricular identity to that of atria (Yutzey et al. 1995), as it can the A/P identity of other tissues such as hindbrain and paraxial mesoderm (Kessel and Gruss 1991; Guthrie 1996). In zebrafish embryos, atrial and ventricular cells have separate clonal origins as early as the midblastula stage (Stainier et al. 1993). A/P patterning therefore appears to be established in some crude form early in heart development, possibly before the onset of myogenic commitment. However, the pattern must be refined considerably as development proceeds, since most "chamber-specific" myofilament genes are initially expressed across the whole heart, before becoming restricted at variably later times (Lyons 1994).

Less is known about L/R patterning in the heart and its relationship to A/P pattern. The connection between both patterning information systems in the heart becomes manifest during looping, where two different processes can be seen to occur: One is a physical relocation of ventricular chamber primordia, initially A/P neighbors, in the L/R axis, and the other is an imposition of L/R values upon the atria. Fate-mapping studies in the chick have demonstrated that progenitors of the left and right ventricles arise with an A/P neighbor relationship, left ventricular cells lying just caudal to those of the right (Stalsberg and DeHaan 1969). In contrast, the chicken fate map suggests that left atrium develops from heart progenitors that arise on the left side of the embryo, and right atrium from the right (Stalsberg and DeHaan 1969). Unlike ventricles, therefore, the atria appear to have an L/R neighbor relationship at the outset, and laterality information comes to bear on their individual identities without rearrangement.

Nkx2-5 mutation specifically disrupts ventricular expression of at least five genes: *CARP*, *ANF*, *eHand*, *SM-22*, and *MLC2V* (see Fig. 4) (Lyons et al. 1995; Biben and Harvey 1997; this paper). A plausible hypothesis is that the primary defect in *Nkx2-5*$^{-/-}$ hearts is loss of ventricular identity. Perhaps the open chamber in these hearts has atrial or a confused identity. Few chamber-specific markers are expressed early enough to be useful in addressing this issue. However, two markers, *myosin heavy chain-β* and *cyclin D2*, are expressed in mutants in a way that suggests ventricular and atrial identities have been specified to some degree (Lyons et al. 1995).

The left ventricular expression of *eHand* and the vasoactive hormone gene *ANF* does not develop in *Nkx2-5* mutants. Moreover, a novel gene recently isolated in this laboratory, *chisel* (S. Palmer, pers. comm.), and mainly expressed in the ventricles, is also down-regulated in the mutant heart, further suggesting that ventricular development is affected in the mutant. Indirect evidence from expression of an MLC2V-LacZ transgene supports this observation (Ross et al. 1996). The transgene carries 250 bp of *MLC2V* proximal regulatory sequences and is ex-

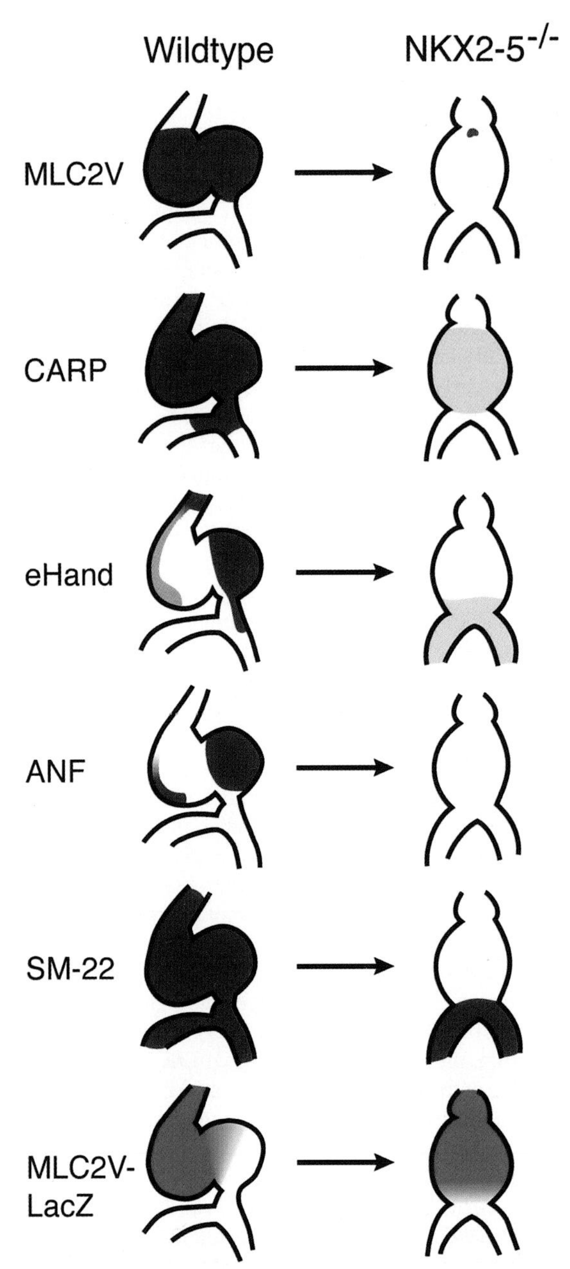

Figure 4. Summary of perturbations to gene expression seen in *Nkx2-5*$^{-/-}$ mutant hearts: Expression patterns of *ANF, SM-22, MLC2V, eHand, CARP,* and the MLC2V-LacZ transgene in hearts of E9.0 wild-type and mutant embryos are represented.

pressed predominantly in right ventricle and outflow tract. Unlike the endogenous gene, expression does not occur significantly in left ventricle and, in particular, is excluded from the *eHand* expression domain within that chamber. The transgene is also robustly expressed in the *Nkx2-5* mutant hearts. This is a surprising result, since the endogenous gene is down-regulated in these hearts and is most likely a consequence of the artificial situation created by using a truncated transgene. Interestingly, in *Nkx2-5* mutant hearts, LacZ staining occupies most of the open ventricular region (see Fig. 4). A possible interpretation is that the abnormal mutant chamber has *right* ventricular identity and that left ventricular cells have been lost. Thus, the morphogenetic defect in *Nkx2-5* mutant hearts could relate, in part, to the loss of the *eHand*-positive left ventricular cells. This possibility is highlighted by studies on *dHand* knockout mice (Srivastava et al. 1997). These mice have an abnormal heart with a very rudimentary dextral loop and hypoplastic ventricles. Marker expression and heart anatomy indicate specific loss of the right ventricle, where *dHand* expression during looping is normally highest (Biben and Harvey 1997). Thus, *dHand* may be essential for expansion or survival of right ventricular myocytes. *eHand* could perform the same function in left ventricle, although confirmation of this must await the heart phenotype in *eHand* knockout mice.

What Is the Role of *Nkx2-5*?

As discussed above, the spatial patterns of genes that fall under *Nkx2-5* control are disturbed in diverse and interesting ways in the mutant background (Fig. 4). How this relates to patterning sytems in the heart is uncertain, since there is currently no consistent modality to the way *Nkx2-5*-dependent genes are affected. Gene expression is variously and differentially altered in ventricles, outflow tract, sinus venosa, or across the whole heart, a situation also seen in *dHand* and *MEF2C* knockout hearts (Srivastava et al. 1997; Lin et al. 1997).

These observations may mean that we simply do not have enough data to see an inherent logic: Acquisition of complexity in the vertebrate heart during its evolution could have occurred entrained to a still unknown linear patterning system, similar to a Hox code. Alternatively, the addition of chambers, valves, and septa may have occurred by chance adaptions in expression or function of available morphoregulatory molecules, the selectional advantages given to improvements in heart performance during evolution being enormous. It is interesting that for a number of heart-expressed genes whose regulation has been examined by transgenic techniques, expression in different heart chambers seems to require different *cis*-regulatory regions (Kelly et al. 1995; Moessler et al. 1996; Ross et al. 1996). Although we clearly need more information, the heart appears to be a highly modular structure, with myogenic regulation in different modules requiring distinct *cis* elements and different combinations of transcription factors, Nkx2-5, MEF2C, dHand, and eHand potentially among them. It may be erroneous to assume that the underlying logic of heart patterning is neat. Further insights into this issue will come from analysis of the transcription factors involved in chamber-specific gene regulation.

What Is the Relationship between *eHand* and *Nkx2-5*?

Nkx2-5 mutant hearts do not show the definitive expression pattern of *eHand*. There are several plausible explanations for this, although we tend to favor the hypothesis of a direct regulatory relationship between *Nkx2-5* and *eHand* for the following reasons: (1) Mutant hearts are to some degree asymmetrical in their caudal region, indicating that they are not blind to laterality information (Biben and Harvey 1997; data not shown). Moreover, the cranial expression domain of *eHand* in the heart, which does not seem to be related to establishment of laterality, is also abolished in the mutant context. Therefore, an impairment of transmission of laterality information is unlikely to account for the loss of the definitive expression domain of *eHand.* (2) Ectopic expression of *eHand* is noticeable in the posterior part of the pharyngeal floor in *Nkx2-5* mutant embryos (Biben and Harvey 1997), where *Nkx2-5* was also expressed at an earlier time. This is suggestive that *Nkx2-5* and *eHand* interact directly, with the outcome dependent on the cellular context.

What Is Downstream from *eHand*?

The *eHand*-expressing cells may be missing in *Nkx2-5* mutant hearts. If *Nkx2-5* is a direct regulator of *eHand*, this would suggest that these cells do not survive, or do not proliferate, or adopt a right ventricle identity by default. Little information is available on *eHand* function. The early expression pattern of this gene within trophoblast and in trophoblastic cell lines suggests that it behaves more as a differentiation factor than a proliferative one (Cross et al. 1995). The situation could, of course, be different in the heart. Since *dHand* knockout mice lack a right ventricle (Srivastava et al. 1997), both *Hand* genes could perform similar functions in distinct chambers of the heart. The right ventricle also appears to be absent in an insertional mouse mutation, *hdf* (heart defect) (Yamamura et al. 1997).

Using whole-mount in situ hybridization, we have noted that the *ANF* pattern in the early looping heart is reminiscent of that of *eHand*, expression occurring first on the outer curvature of the left ventricle, followed by the outer curvature (right side) of the right ventricle. *Trans*-activation studies show that Nkx2-5 protein can directly activate the *ANF* promoter (Durocher et al. 1996). However, *ANF* expression during the early phases of heart development is complex and dynamic, suggesting involvement of regulators other than Nkx2-5 to set the spatial pattern (Zeller et al. 1987; Lyons 1994; this paper). Thus, ANF expression may utilize *eHand* for spatial regulation, and indeed, the proximal 700 bp of the ANF promoter contains seven E-box motifs (Genbank: K02062), possible sites of eHand binding. Experiments are under way to determine their significance in ANF spatial regulation.

What Is the Relationship between *Hand* Gene Expression and Laterality Information?

As discussed above, the *dHand* and *eHand* expression patterns evolve to the right and left sides of the heart, respectively (Biben and Harvey 1997; Srivastava et al. 1997). We have observed features of the *eHand* pattern that suggest it is modified by the L/R patterning system (Biben and Harvey 1997). Most importantly, *eHand* expression is seen transiently in the developing left atrium. As noted above, atria have an L/R relationship at the outset (Stalsberg and DeHaan 1969), and left atrial expression should reflect a genuine L/R difference. Although these observations are provocative, it is currently unclear to what extent L/R influences affect the final pattern, or indeed downstream function. The complex dynamics of *Hand* gene expression in mice stand in stark contrast to the more global patterns seen in chick hearts (Srivastava et al. 1995). Looping is an ancient and conserved outcome of the laterality pathway in vertebrates, and one would expect that expression of key regulators would be conserved between species. However, this is clearly not the case for all genes in the laterality pathway (Levin et al. 1995; Collignon et al. 1996), and indeed, a left-dominant pattern of *XeHand* expression in the *Xenopus* heart has also been noted (D. Sparrow and T. Mohun, pers. comm.). Thus, although the *Hand* genes clearly have key roles in patterning the heart, their exact roles in looping and to what extent their expression is influenced by laterality in different species await further investigation.

ACKNOWLEDGMENTS

This work was supported by funds from the National Health and Medical Research Council of Australia, the National Heart Foundation of Australia (G96M 4675), and the Human Frontiers Science Program (RG-308/95). S.P. holds a Wellcome Trust postdoctoral fellowship.

REFERENCES

Biben C. and Harvey R.P. 1997. Homeodomain factor *Nkx2-5* controls left-right asymmetric expression of bHLH *eHand* during murine heart development. *Genes Dev.* ***11:*** 1357.

Bodmer R. 1993. The gene *tinman* is required for specification of the heart and visceral muscles in *Drosophila. Development* ***118:*** 719.

Bodmer R., Golden K., Lockwood W.B., Ocorr K.A., Park M., Su M.-T., and Venkatesh T.V. 1997. Heart development in *Drosophila. Adv. Dev. Biochem.* ***5:*** 201.

Boettger T. and Kessel T. 1997. The chick *Nkx2-8* gene: A novel member of the *NK-2* family. *Dev. Genes Evol.* ***207:*** 65.

Boterenbrood E.C. and Nieuwkoop P.D. 1973. The formation of the mesoderm in urodelan amphibians. 5. Its regional induction by the endoderm. *Wilhelm Roux's Arch. Dev. Biol.* ***173:*** 319.

Bowers P.N., Brueckner M., and Yost H.J. 1996. Laterality disturbances. *Prog. Pediatr. Cardiol.* ***6:*** 53.

Brand T., Andree B., Schneider A., and Arnold H.-H. 1997. Chicken *Nkx2.8*, a novel homeobox gene expressed during early heart and foregut development. *Mech. Dev.* ***65:*** 53.

Burn J. 1991. Disturbance of morphological laterality in humans. In *Biological asymmetry and handedness* (ed. G.R. Bock and J. Marsh), p. 282. Wiley, New York.

Chen C.Y., Croissant J., Majesky M., Topouzis S., McQuinn T., Frankovsky M.J., and Schwartz R.J. 1996. Activation of the cardiac α-actin promoter depends upon serum response factor, Tinman homologue, *Nkx-2.5*, and intact serum response elements. *Dev. Genet.* ***19:*** 119.

Chomzynski P. and Saachi N. 1987. Single step method of RNA isolation by acid guanidinium thiocyanate-phenol-chloroform extraction. *Anal. Biochem.* ***162:*** 156.

Collignon J., Varlet I., and Robertson E.J. 1996. Relationship between asymmetric *nodal* expression and the direction of

embryonic turning. *Nature* **381:** 155.
Cross J.C., Flannery M.L., Blanar M.A., Steingrimsson E., Jenkins N.A., Copeland N.G., Rutter W.J., and Werb Z. 1995. *Htx* encodes a basic helix-loop-helix transcription factor that regulates trophoblast cell development. *Development* **121:** 2513.
Cserjesi P., Brown B., Lyons G.E., and Olson E.N. 1995. Expression of the novel basic helix-loop-helix gene *eHAND* in neural crest derivatives and extraembryonic membranes during mouse development. *Dev. Biol.* **170:** 664.
DeRuiter M.C., Poelmann R.E., VanderPlas-de Vries I., Mentink M.M.T., and Gittenberger-de Groot A.C. 1992. The development of the myocardium and endocardium in mouse embryos. *Anat. Embryol.* **185:** 461.
Durocher D., Chen C.-Y., Ardati A., Schwartz R.J., and Nemer M. 1996. The atrial natriuretic factor promoter is a downstream target for *Nkx-2.5* in the myocardium. *Mol. Cell. Biol.* **16:** 4648.
Edmondson D.G., Lyons G.E., Martin J., and Olson E.N. 1994. Mef2 gene expression marks the cardiac and skeletal muscle lineages during mouse embryogenesis. *Development* **120:** 1251.
Gajewski K., Kim Y., Lee Y.M., Olson E.N., and Schulz R.A. 1997. D-mef2 is a target for Tinman activation during Drosophila heart. *EMBO J.* **16:** 515.
Gonzalez-Sanchez A. and Bader D. 1990. *In vitro* analysis of cardiac progenitor cell differentiation. *Dev. Biol.* **139:** 197.
Guthrie S. 1996. Patterning the hindbrain. *Curr. Opin. Neurobiol.* **6:** 41.
Harvey, R.P. 1996. *NK-2* homeobox genes and heart development. *Dev. Biol.* **178:** 203.
Hollenberg S.M., Sternglanz R., Cheng P.F., and Weintraub H. 1995. Identification of a new family of tissue-specific basic helix-loop-helix proteins with a two hybrid system. *Mol. Cell. Biol.* **15:** 3813.
Hyatt B.A., Lohr J.L., and Yost H.J. 1996. Initiation of vertebrate left-right axis formation by maternal Vg1. *Nature* **384:** 62.
Isaac A., Sargent M.G., and Cooke J. 1997. Control of vertebrate left-right asymmetry by a *Snail*-related zinc finger gene. *Science* **275:** 1301.
Jacobson A.G. and Sater A.K. 1988. Features of embryonic induction. *Development* **104:** 341.
Kelly R., Alonso S., Tajbakhsh S., Cossu G., and Buckingham M.E. 1995. Myosin light chain 3F regulatory sequences confer regionalized cardiac and skeletal muscle expression in transgenic mice. *J. Cell Biol.* **129:** 383.
Kessel M. and Gruss P. 1991. Homeotic transformations of murine vertebrae and concomitant alteration of *Hox* codes induced by retinoic acid. *Cell* **67:** 1.
Komuro I. and Izumo S. 1993. *Csx*: A murine homeobox-containing gene specifically expressed in the developing heart. *Proc. Natl. Acad. Sci.* **90:** 8145.
Layton W.M., Manasek M.D., and Manasek D.M.D. 1980. Cardiac looping in early *iv/iv* mouse embryos. In *Etiology and morphogenesis of congenital heart disease* (ed. R. van Pragh), p. 109. Futura, Mount Kisco, New York.
Levin M., Johnson R.L., Stern C.D., Kuehn M., and Tabin C. 1995. A molecular pathway determining left-right asymmetry in chick embryogenesis. *Cell* **82:** 803.
Li L., Miano J.M., Cserjesi P., andOlson E.N. 1996. SM-22α, a marker of adult smooth muscle, is expressed in multiple myogenic lineages during embryogenesis. *Circ. Res.* **78:** 188.
Lin Q., Schwarz J., Bucana C., and Olson E.N. 1997. Control of mouse cardiac morphogenesis and myogenesis by transcription factor MEF2C. *Science* **276:** 1404.
Lints T.J., Parsons L.M., Hartley L., Lyons I., and Harvey R.P. 1993. *Nkx-2.5*: A novel murine homeobox gene expressed in early heart progenitor cells and their myogenic descendants. *Development* **119:** 419.
Lowe L.A., Supp D.M., Sampath K., Yokoyama T., Wright C.V.E., Potter S.S., Overbeek P., and Kuehn M.R. 1996. Conserved left-right asymmetry of *nodal* expression and alterations in murine *situs inversus*. *Nature* **381:** 158.
Lyons G.E. 1994. In situ analysis of the cardiac muscle gene program during embryogenesis. *Trends Cardiovasc. Med.* **4:** 70.
Lyons I., Parsons L.M., Hartley L., Li R., Andrews J.E., Robb L., and Harvey R.P. 1995. Myogenic and morphogenetic defects in the heart tubes of murine embryos lacking the homeobox gene *Nkx2-5*. *Genes Dev.* **9:** 1654.
Meno C., Saijoh Y., Fujii H., Ikeda M., Yokoyama T., Yokoyama M., Toyada Y., and Hamada H. 1996. Left-right asymmetric expression of the TGFβ-family member *lefty* in mouse embryos. *Nature* **381:** 151.
Moessler H., Mericskay M., Li Z., Nagl S., Paulin D., and Small J.V. 1996. The *SM-22* promoter directs tissue-specific expression in arterial but not in venous or visceral smooth muscle cells in transgenic mice. *Development* **122:** 2415.
Montgomery M.O., Litvin J., Gonzalez-Sanchez A., and Bader D. 1994. Staging of commitment and differentiation of avian cardiac myocytes. *Dev. Biol.* **164:** 63.
Nascone N. and Mercola M. 1995. An inductive role for the endoderm in *Xenopus* cardiogenesis. *Development* **121:** 515.
Rawles, M.E. 1943. The heart forming area of the chick blastoderm. *Physiol. Zool.* **16:** 22.
Rosenquist G.C. 1966. A radioautographic study of labeled grafts in the chick blastoderm. Development from primitive streak stages to stage 12. *Carnegie Inst. Wash. Contrib. Embryol.* **38:** 71.
Ross R.S., Navankasattusas S., Harvey R.P., and Chien K.R. 1996. An HF-1a/HF-1b/MEF-2 combinatorial element confers cardiac ventricular specificity and establishes an anterior-posterior gradient of expression via an *Nkx2-5* independent pathway. *Development* **122:** 1799.
Schmidt U.S., Troschka M., and Pfitzer G. 1995. The variable coupling between force and myosin light chain phosphorylation in Triton-skinned chicken gizzard fibre bundles: Role of myosin light chain phosphatase. *Pflueg. Arch. Eur. J. Physiol.* **429:** 708.
Schultheiss T.M., Burch J.B.E., and Lassar A.B. 1997. A role for bone morphogenetic proteins in the induction of cardiac myogenesis. *Genes Dev.* **11:** 451.
Schultheiss T.M., Xydas S., and Lassar A.B. 1995. Induction of avian cardiac myogenesis by anterior endoderm. *Development* **121:** 4203.
Smith S.M., Dickman E.D., Thompson R.P., Sinning A.R., Wunsch A.M., and Markwald R.R. 1997. Retinoic acid directs cardiac laterality and the expression of early markers of precardiac asymmetry. *Dev. Biol.* **182:** 162.
Srivastava, D., Cserjesi P., and Olson E.N. 1995. A Subclass of bHLH proteins required for cardiac morphogenesis. *Science* **270:** 1995.
Srivastava D., Thomas T., Lin Q., Brown D., and Olson E.N. 1997. Regulation of cardiac mesodermal and neural crest development by the bHLH transcription factor, *dHAND*. *Nat. Genet.* **16:** 154.
Stainier D.Y.R., Lee R.K., and Fishman M.C. 1993. Cardiovascular development in the zebrafish. I. Myocardial fate map and heart tube formation. *Development* **119:** 31.
Stalsberg H. and R.L. DeHaan. 1969. The precardiac areas and formation of the tubular heart in the chick. *Dev. Biol.* **19:** 128.
Sugi Y. and Lough J. 1994. Anterior endoderm is a specific effector of terminal cardiac myocyte differentiation in cells from the embryonic heart forming region. *Dev. Dyn.* **200:** 155.
Tam P.P.L., Parameswaran M., Kinder S.J., and Weinberger R.P. 1997. The allocation of epiblast cells to the embryonic heart and other mesodermal lineages: The role of ingression and tissue movement during gastrulation. *Development* **124:** 1631.
Tsuda T., Philp N., Zile M.H., and Linask K.K. 1996. Left-right asymmetric localization of flectin in the extracellular matrix during heart looping. *Dev. Biol.* **173:** 39.
Vesely D.L. 1995. Atrial natriuretic hormones originating from the N-terminus of the atrial natriuretic factor prohormone. *Clin. Exp. Pharmaco. Physiol.* **22:** 108.
Yamamura H., Zhang M., Markwald R.R., and Mjaatvedt C.H. 1997. A heart segmental defect in the anterior-posterior axis

of a transgenic mutant mouse. *Dev. Biol.* ***186:*** 58.
Yost H.J. 1995. Vertebrate left-right development. *Cell* ***82:*** 689.
Yutzey K.E., Gannon M., and Bader D. 1995. Diversification of cardiomyogenic cell lineages *in vitro*. *Dev. Biol.* ***170:*** 531.
Zeller R., Bloch K.D., Williams R.S., Arceci R.J., and Seidma C.E. 1987. Localized expression of the atrial natriuretic factor gene during cardiac embryogenesis. *Genes Dev.* ***1:*** 693.
Zou Y., Evans S., Chen J., Kou H.-C., Harvey R.P., and Chien K.R. 1997. CARP, a cardiac ankyrin repeat protein, is downstream in the *Nkx2-5* homeobox gene pathway. *Development* ***124:*** 793.

A Transcriptional Pathway for Cardiac Development

Q. Lin,[1] D. Srivastava,[1,2] and E.N. Olson[1]

[1]*Department of Molecular Biology and Oncology and* [2]*Pediatrics, Division of Cardiology, The University of Texas Southwestern Medical Center at Dallas, Dallas, Texas 75235-9148*

The heart is one of the most fascinating yet least understood organs in the body. Despite elegant anatomical descriptions of cardiac development in a variety of vertebrate species, little is known of the molecular pathways that underlie the formation of the mature multichambered heart (Olson and Srivastava 1996). The transcriptional mechanisms that control cardiomyocyte gene expression are also only beginning to be elucidated.

Heart formation during vertebrate embryogenesis begins when a bilaterally symmetric population of mesodermal cells in the anterior lateral plate of the gastrulating embryo become committed to a cardiogenic fate in response to inductive signals that eminate from the adjacent endoderm (Fig. 1) (DeHaan 1965; Nascone and Mercola 1996). These cardiogenic precursor cells migrate to the ventral midline to form a linear cardiac tube that initiates rhythmic contractions at about embryonic day 8.0 (E8) in the mouse. Embryologic studies have demonstrated that precursors of the different regions of the mature heart can be fate-mapped to specific segments along the anteroposterior (A/P) axis of the linear cardiac tube (Stanier and Fishman 1992; Yutzey and Bader 1995). From anterior to posterior, these segments give rise to the conotruncus (or outflow tract), the right and left ventricles, the atria, and the inflow tract, respectively. In all vertebrate species, the heart tube undergoes rightward looping (Brown and Wolpert 1990; Yost 1995), which converts A/P patterning to left-right patterning, such that the more anterior region of the linear heart tube becomes displaced to the right and the more posterior region to the left. Cardiac looping is essential for alignment of the inflow and outflow tracts and for orienting the atrial and ventricular chambers. Following looping, the individual chambers differentiate, the cardiac septa and cushions are formed, and trabeculation occurs.

Recent studies have begun to reveal the components of a left-right signaling system that control the direction of cardiac looping, as well as the orientation of other visceral organs (Levin 1997). The asymmetric expression of the transforming growth factor-β (TGF-β)-like growth factors Nodal and Lefty in the lateral plate mesoderm on the left side of the embryo results in looping of the heart to the opposite side (Collignon et al. 1996; Lowe et al. 1996; Meno et al. 1996). In the chick, sonic hedgehog controls the asymmetric expression of Nodal (Levin et al. 1995). There is also evidence that the TGF-β -like protein Vg1 is required for establishment of embryonic left-right asymmetry (Hyatt et al. 1996). In addition, the zinc finger transcription factor SnR-1, which is related to *Drosophila* Snail, has been shown to be expressed specifically in the right lateral plate mesoderm of the chick embryo and to be required for rightward looping of the heart (Isaac et al. 1997). It is unclear, however, how left-right signals are interpreted by the linear heart tube or how the looping process is controlled.

The establishment of asymmetry in the heart has important consequences for cardiac function. The right and left ventricles, for example, perform different functions

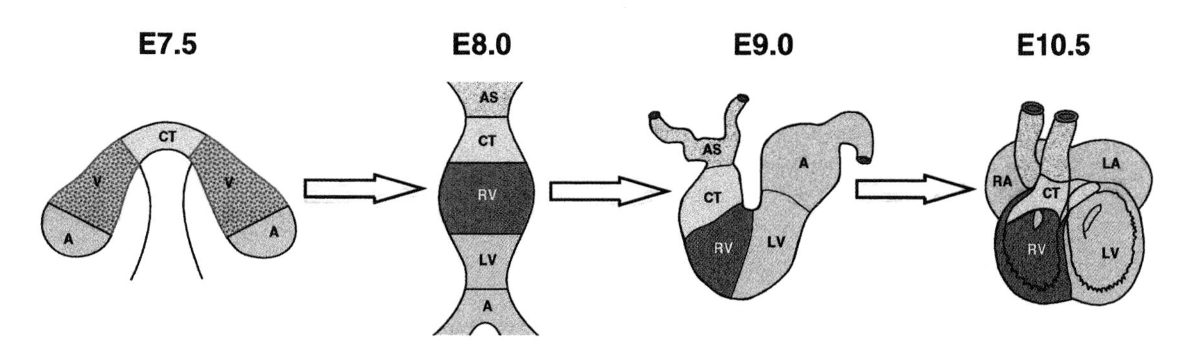

Figure 1. Schematic diagram of cardiac development. Cardiogenic precursors from the cardiac crescent (*left*) migrate to the ventral midline of the embryo to form the linear heart tube, which undergoes rightward looping and eventual formation of the mature multichambered heart. Different populations of cardiogenic precursors that are fated to form the aortic sac (AS), conotruncus (CT), right ventricle (RV), left ventricle (LV), and atria (A) are shown.

and exhibit distinct patterns of gene expression that account for their unique contractile and electrophysiological properties. The transcription factors that might confer left-right identities on cardiomyocytes have not been identified.

Neural crest cells also have an important role in formation of the heart and its vascular connections (Kirby and Waldo 1995). Following cardiac looping, neural crest cells populate the anterior region of the heart, where they contribute to the aortic arch arteries and outflow tract. Neural crest ablation experiments in avian embryos have demonstrated the essential role for these cells in formation of the heart and the catastrophic consequences of neural crest defects on heart development.

To begin to dissect the molecular mechanisms that control myogenesis and morphogenesis in the early heart, we have focused on the functions of several transcription factors implicated in regulation of cardiac muscle genes. These studies have revealed several important steps in the cardiogenic pathway, and they suggest the existence of an evolutionarily conserved pathway for the initial steps in cardiac development.

REGULATION OF CARDIAC MESODERMAL AND NEURAL CREST DEVELOPMENT BY THE HAND FAMILY OF BHLH TRANSCRIPTION FACTORS

The formation of skeletal muscle is dependent on the MyoD family of basic helix-loop-helix (bHLH) transcription factors, which act at multiple points in the skeletal muscle lineage to control myoblast determination and differentiation (Olson and Klein 1994). These myogenic factors are not expressed in the heart. However, two related bHLH proteins, called dHAND and eHAND (also known as Hed/Thing-2 and Hxt/Thing-1, respectively), are expressed during the early stages of cardiogenesis (Cserjesi et al. 1995; Hollenberg et al. 1995; Cross et al. 1995; Srivastava et al. 1995). In the mouse, dHAND and eHAND are expressed in the cardiac crescent. dHAND is then expressed throughout the linear heart tube before becoming restricted to the right ventricular region (Fig. 2). In contrast, eHAND expression becomes localized to two specific segments of the linear heart tube: the anterior-most segment that gives rise to the conotruncus and the more posterior region fated to form the left ventricular region (Srivastava et al. 1997; C. Biben and R. Harvey, pers. comm.). The *HAND* genes are the earliest known markers for the future right and left ventricular segments of the heart tube. Following formation of the ventricular chambers, dHAND and eHAND expression is down-regulated.

In addition to their expression in the developing heart, dHAND and eHAND are expressed in neural crest cells that populate the heart, as well as in the lateral mesoderm. dHAND and eHAND are also expressed in the deciduum and extraembryonic membranes, respectively. We have focused our attention on their roles in the heart.

We have inactivated the mouse *dHAND* gene by homologous recombination (Srivastava et al. 1997). Whereas mice heterozygous for the targeted mutation are normal, homozygous mutant embryos died between E10.5 and E11.0 and displayed severe defects in the mesodermal and neural-crest-derived components of the heart. Homozygous mutant embryos appeared to be normal until about E9.0, when they began to show delayed growth. The linear heart tube of the mutant then underwent incorrect looping, and the region of the heart tube that would normally give rise to the right ventricle does not develop. As a result, the heart is displaced to the left, and the ventricular region is reduced in size (Figs. 3 and 4). These morphologic defects are localized to regions of the developing heart tube in which dHAND is normally expressed, and they suggest that dHAND is essential for right ventricular development. Consistent with this conclusion is the observation that eHAND continues to be expressed in the left-sided region of the heart tube of *dHAND* mutant embryos.

At present, we can only speculate as to how dHAND controls the development of the future right ventricle. A possibility we favor is that dHAND is required within the heart tube for expansion of the population of cardiogenic cells destined to form the right ventricular region. Alternatively, dHAND might be required for the specification of this population of cells. It remains to be determined whether the cells that would have contributed to the right ventricle become misspecified or deleted in the *dHAND* mutant.

In contrast to their mutually exclusive expression patterns in the heart tube of the mouse, dHAND and eHAND are expressed homogeneously throughout the heart tube of the chick (Srivastava et al. 1995). Their overlapping expression in the chick may explain the finding that both genes needed to be inactivated by antisense oligonucleotides to observe cardiac defects in chick embryos. It will be of particular interest to determine whether eHAND has a unique role in left ventricular development in the mouse. Consistent with this notion, mice lacking the cardiac homeobox gene *Nkx-2.5* do not express eHAND in the heart (C. Biben and R. Harvey, per. comm.) and exhibit defects in looping morphogenesis (Lyons et al. 1995). Thus, eHAND may lie downstream from Nkx-2.5 in a pathway for left ventricular development.

The aortic arch arteries, which are normally populated by neural crest cells, also fail to form in *dHAND* mutant embryos (Srivastava et al. 1997). dHAND is expressed in this neural crest cell population coincident with their aggregation at their destination within the branchial arches (Srivastava et al. 1995). This suggests that dHAND controls the differentiation or organization of neural crest cells into vessels, rather than their initial emigration from the neural folds.

REGULATION OF CARDIAC MYOGENESIS AND MORPHOGENESIS BY THE MEF2 FAMILY OF TRANSCRIPTION FACTORS

The majority of cardiac- and skeletal-muscle-specific genes contain binding sites for members of the myocyte

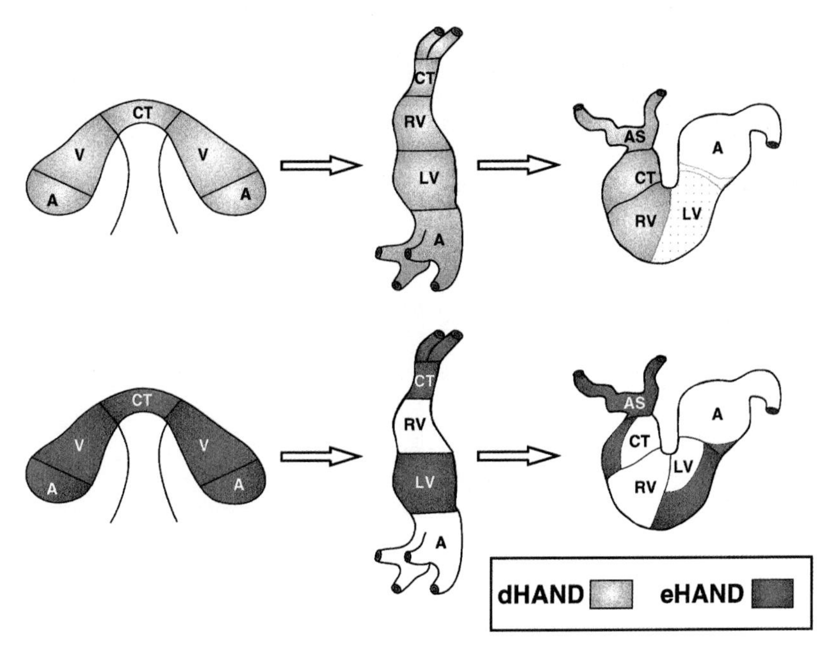

Figure 2. Summary of dHAND and eHAND expression patterns during cardiac development. The patterns of expression of dHAND and eHAND during mouse cardiogenesis are shown.

enhancer factor-2 (MEF2) family of transcription factors in their control regions (Gossett et al. 1989; Pollock and Treisman 1991; Olson et al. 1995). There are four *MEF2* genes in vertebrates, referred to as *MEF2A, -B, -C*, and *-D*, which are expressed in precursors of cardiac, skeletal, and smooth muscle cells during early embryogenesis, but become more widely expressed later in pre- and postnatal development (Edmondson et al. 1994; Molkentin et al. 1996b). The MEF2 proteins share high homology within a MADS (MCM1, Agamous, Deficiens, Serum response factor)-box at their amino terminus and an adjacent motif known as the MEF2 domain. Together, these regions, which encompass 86 amino acids, mediate dimerization of MEF2 monomers and binding to the DNA consensus sequence $CTA(T/A)_4TAG/A$ (Pollock and Treisman 1991; Molkentin et al. 1996a).

In *Drosophila*, a single *MEF2* gene, called *D-mef2*, is expressed in developing cardiac, skeletal, and visceral muscle cells (Lilly et al. 1994; Nguyen et al. 1994). Loss-of-function mutations have shown that D-MEF2 is required for differentiation of all muscle cell types in the embryo (Bour et al. 1995; Lilly et al. 1995; Ranganayakulu et al. 1995). This suggests that there is a commonality among different myogenic regulatory programs and that MEF2 is a central component of the programs that control differentiation of all muscle cell types. In skeletal muscle cells, members of the MEF2 family have been shown to interact with members of the MyoD family of bHLH proteins to control the expression of skeletal-muscle-specific genes (Molkentin et al. 1995; Molkentin and Olson 1996). The interactions between these two classes of transcription factors are mediated by their DNA-binding domains such that the proteins can associate with each other and with their respective DNA-binding sites simultaneously. In addition, either factor can bind DNA and recruit the other factor to a muscle regulatory region through direct protein-protein interactions without the necessity for both factors to bind DNA. Athough biochemical and genetic evidence strongly supports the conclusion that MEF2 and myogenic bHLH factors interact to establish a combinatorial code for skeletal muscle gene expression, it is unclear exactly how these proteins cooperate to orchestrate the entire program for skeletal myogenesis. Perhaps these proteins together create a unique protein conformation that is uniquely able to activate skeletal muscle genes, but not other genes that contain binding sites for these factors in their control regions.

On the basis of the cooperative interactions between MEF2 and myogenic bHLH factors in the skeletal muscle lineage and the requirement of D-MEF2 for differentiation of all muscle cell types in *Drosophila*, we have proposed that members of the MEF2 family may control different sets of muscle genes by acting combinatorially with other transcription factors unique to each muscle cell lineage (Molkentin et al. 1995; Molkentin and Olson 1996). This type of combinatorial control of gene expression is reminiscent of the mechanism whereby the MADS-box protein MCM1 interacts with different cofactors to regulate different sets of target genes in yeast (Shore and Sharrocks 1995).

During vertebrate embryogenesis, MEF2C is the first member of the MEF2 family to be expressed (Edmondson et al. 1994). MEF2C transcripts are detected initially in the precardiac mesoderm of the mouse at E7.5 at about the same time as Nkx-2.5 is expressed (Lints et al. 1993).

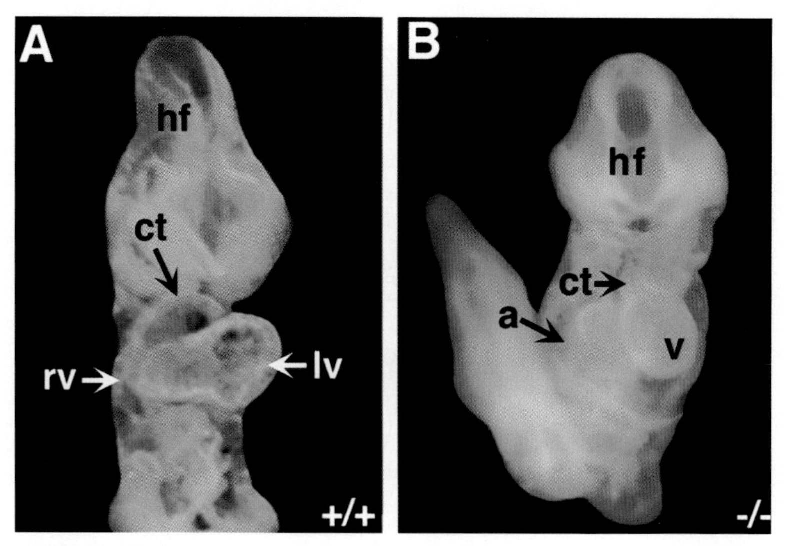

Figure 3. Defects in heart formation in *dHAND* mutant embryos. Frontal views of wild-type (*A*) and *dHAND* mutant (*B*) embryos are shown. The right ventricular region is absent in the mutant. (A) Atrium; (ct) conotruncus; (hf) headfold; (lv) left ventricle; (rv) right ventricle; (v) ventricular chamber.

Thereafter, MEF2C transcripts continue to be expressed throughout the developing heart tube, as well as in skeletal and smooth muscle cells.

To begin to determine the functions of *MEF2C* during mouse development, we inactivated the *MEF2C* gene by introducing a mutation that deleted the region encoding the MADS and MEF2 domains (Lin et al. 1997). Mice heterozygous for this mutation were phenotypically normal, whereas *MEF2C*-null mice died at E9.5 and showed severe cardiac defects. At E8.0, the linear heart tube stage, wild-type and mutant embryos were indistinguishable. However, at the time of cardiac looping, mutant embryos began to exhibit a delay in growth and the heart tube exhibited morphologic abnormalities (Fig. 5). Instead of undergoing rightward looping, the hearts of the mutants remained essentially linear, the ventricular chamber appeared hypoplastic, and the atrial chamber was enlarged. Morphologically, it appeared that the future right ventricular region of the heart tube was absent in the mutant.

dHAND transcripts were expressed normally in the linear heart tube in *MEF2C* mutant embryos, but expression was down-regulated at the time of looping, concomitant with the failure of the right ventricular region to form. eHAND was expressed in the mutant heart, but expression was contiguous throughout the heart tube, without the gap in expression that normally characterizes the future right ventricular region. The distortion of eHAND expression and the down-regulation of dHAND expression in the heart tube provide molecular evidence in support of the conclusion that the right ventricle is absent in the mutant and the left ventricular segment is hypoplastic.

The hearts of the mutants also exhibited abnormal contractions. In wild-type embryos, the heart tube exhibits a rhythmic beat that becomes sequential after looping with strong atrial contractions preceding ventricular contraction. In mutants, the atrial chamber exhibited weak contractions, and the hypoplastic ventricular chamber appeared to vibrate only in response to atrial contractions; there was no evidence of independent ventricular contractions. In *MEF2C* mutants, the hearts lacked trabeculation within the ventricular wall, the cardiomyocytes and endocardial cells appeared to be disorganized, and the cardiac jelly, which forms a layer between the myocardial wall and the endocardium, was absent.

In light of the requisite role of MEF2 for expression of cardiac muscle contractile protein genes in *Drosophila*, we anticipated that *MEF2C* mutant embryos would display abnormalities in cardiac gene expression. Indeed, numerous cardiac muscle genes, including *myosin heavy chain*, *atrial natriuretic factor*, and *myosin light chain* (*MLC*)-*1A* failed to be up-regulated in the mutant heart. In contrast, other cardiac structural genes, such as *MLC2V* and *MLC2A*, were expressed normally in the hearts of the mutants. These genes are known to contain MEF2-binding sites in their control regions that are essential for expression in cardiomyocytes (Navankasattusas et al. 1992). That they are expressed in the *MEF2C* mutant suggests that another member of the MEF2 family may compensate to support their expression. MEF2B would be a likely candidate to fulfill such a role because it is coexpressed with MEF2C during the early stages of cardiogenesis (Molkentin et al. 1996b) and it was up-regulated approximately sixfold in the mutant hearts (Lin et al. 1997). In this regard, we have generated *MEF2B* null mice and have thus far not observed any phenotype. This suggests that all of the functions of MEF2B can be performed by another member of the family, presumably MEF2C. Generation of *MEF2B/MEF2C* double mutants is under way and should allow these possibilities to be tested.

How does the absence of MEF2C, which is normally expressed homogeneously throughout the heart tube, result in the specific absence of the right ventricular segment of the heart? We favor a model in which MEF2C regulates the expression or activity of a regionally re-

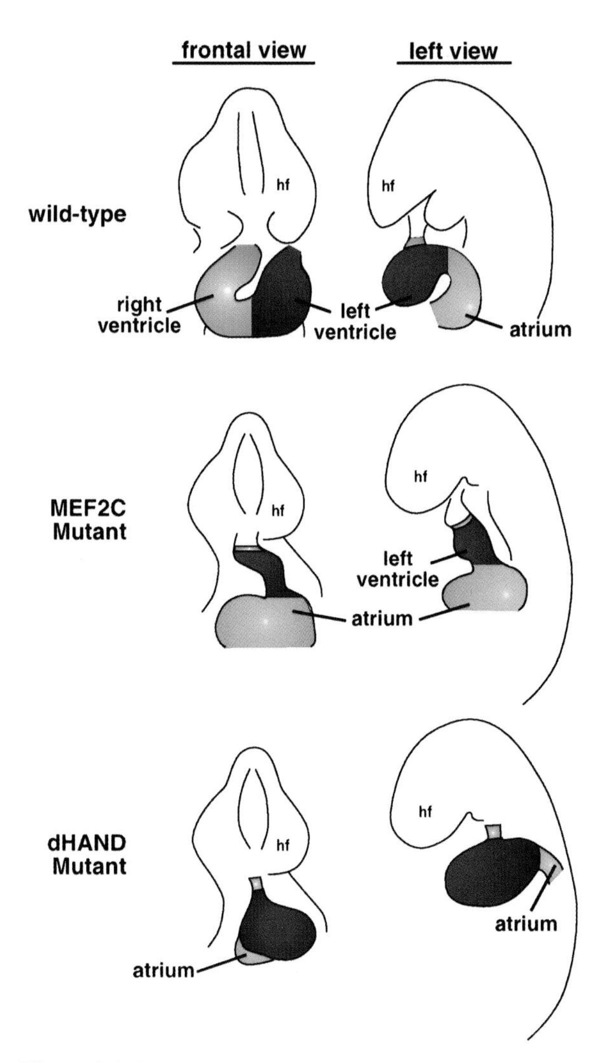

Figure 4. Schematic diagrams of cardiac defects in mutant embryos. Diagrams of frontal and left views of wild-type and mutant embryos are shown. In both the *MEF2C* and the *dHAND* mutant, the right ventricular chamber is hypoplastic and the heart tube fails to loop properly. The atrium is located behind the left ventricular chamber in the *dHAND* mutant as in the wild type. (Hf) Head fold.

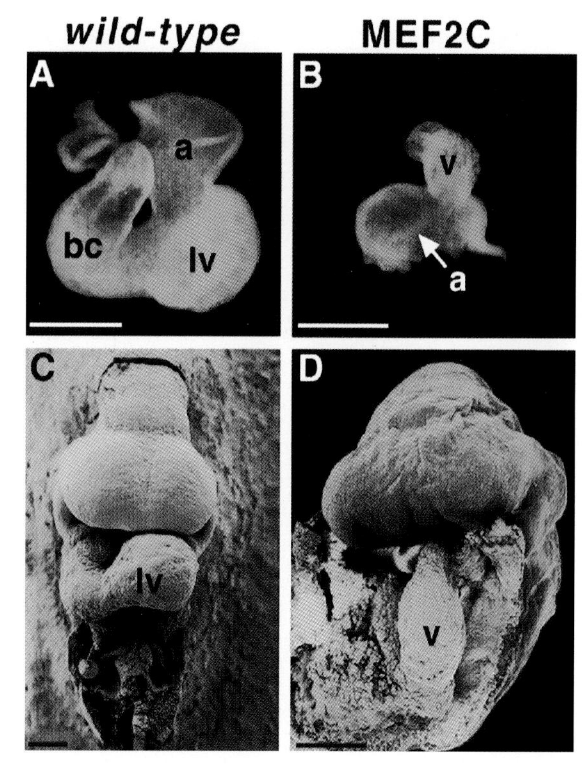

Figure 5. Defects in heart formation in *MEF2C* mutant embryos. Hearts of wild-type (*A*) and MEF2C mutant (*B*) embryos and scanning electron micrographs of wild-type (*C*) and mutant (*D*) embryos at E9.0 are shown. In *A* and *B*, the hearts were dissected from the embryos to reveal their morphologies clearly. (A) Atrium; (bc) bulbus cordis; (lv) left ventricle; (v) ventricular chamber. Bars, 2 μm.

stricted regulatory factor required for right ventricular development. dHAND is an obvious candidate for such a coregulator of right ventricular development. Indeed, although the cardiovascular defects in *MEF2C* mutant embryos are more severe than those of *dHAND* mutants, there are some striking similarities in the right ventricular defects which suggest that these genes may cooperate to control development of this region of the heart in a manner analogous to the combinatorial roles of MEF2 and myogenic bHLH factors in skeletal muscle development. The possibility that MEF2C might participate in a right ventricular regulatory program is supported by studies which showed that the *desmin* and *MLC2V* gene promoters, which contain essential MEF2-binding sites, direct transgene expression specifically in the right ventricle (Kuisk et al. 1996; Ross et al. 1996).

AN EVOLUTIONARILY CONSERVED MOLECULAR PATHWAY FOR HEART DEVELOPMENT

The phenotypes of mouse mutants deficient in cardiogenic transcription factors have made it possible to begin to envision cardiogenesis within the framework of a multistep transcriptional pathway (Fig. 6). Similarities in the expression patterns and putative functions of many cardiogenic transcription factors also suggest that many of the early events in the pathway for cardiac development are evolutionarily ancient and conserved.

In *Drosophila*, the NK-type homeobox gene product Tinman is expressed in early cardiogenic precursors and is required for formation of the heart-like structure known as the dorsal vessel (Bodmer et al. 1990; Bodmer 1993; Azpiazu and Frasch 1993). At least four *tinman*-related genes, including *Nkx-2.5*, have been identified in vertebrates, each of which is expressed throughout the developing heart (for review, see Harvey 1996). Tinman directly activates the transcription of *D-mef2* in the heart, and *D-mef2* is then required to activate cardiac contractile protein genes (Gajewski et al. 1997). In vertebrate embryos, members of the MEF2 family are expressed concomitant with the cardiac NK-type homeobox genes, consistent with the notion that the regulatory relationship between these two classes of cardiogenic genes has been

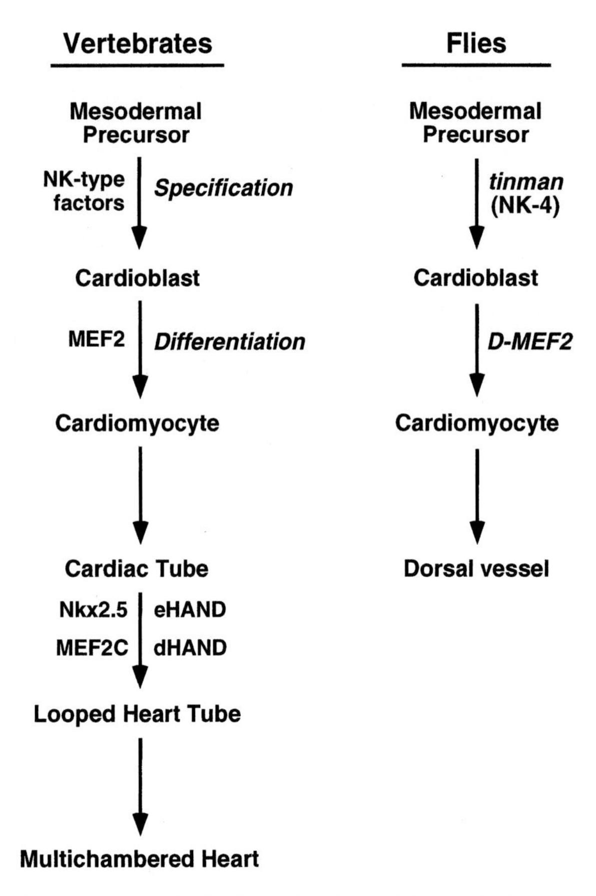

Figure 6. A model for heart development in vertebrates and flies. Steps in cardiac development are shown with genes that have been implicated in each step. The pathways from mesodermal precursor to formation of the linear heart tube are similar in vertebrates and flies, but the vertebrate pathway continues with the formation of the looped heart tube and ultimately the mature four-chambered heart.

conserved. However, whether NK-type homeobox gene products specify the cardiogenic lineage and directly activate the *MEF2* gene in the vertebrate heart, as has been shown to occur in *Drosophila*, remains to be determined.

The phenotype of *MEF2C* mutant embryos demonstrates that this member of the MEF2 family is essential for normal cardiomyocyte differentiation (Lin et al. 1997). However, in contrast to the single MEF2 factor in fruit flies, which appears to regulate all contractile protein genes, MEF2C regulates only a subset of these genes in the mouse. Presumably another member of the MEF2 family supports the expression of those cardiac genes that are MEF2C-independent.

In vertebrates and fruit flies, differentiated cardiomyocytes become organized into a linear heart tube. At that point, vertebrate cardiogenesis continues as the heart tube undergoes looping and chamber specification, whereas the fly heart does not develop beyond the linear heart tube stage. Intriguingly, it appears that certain of the genes that have evolutionarily conserved roles in the early steps of cardiogenesis are reemployed at later steps in cardiogenesis in vertebrates. *Nkx-2.5*, for example, would be expected to have a role in specification of cardioblasts that is likely to be obscured due to functional redundancy with other members of this family, but it also has a role in cardiac looping (Lyons et al. 1995). Similarly, our results show that MEF2C is required for activation of the complete set of cardiac contractile protein genes during cardiomyocyte differentiation in the mouse, which is similar to its function in differentiation of cardioblasts in *Drosophila* (Bour et al. 1995; Lilly et al. 1995; Ranganayakulu et al. 1995). However, MEF2C also has a role in cardiac looping and in chamber maturation, which are unique to vertebrate cardiogenesis. Thus, although *Drosophila* has been and will continue to be a powerful system for uncovering fundamental mechanisms of cardiogenesis, without complications from genetic redundancy of multigene families, there are many aspects of cardiac morphogenesis that are unique to vertebrates.

Now that several important transcription factors that control key steps in this pathway have been identified and their loss-of-function phenotypes are being defined in the mouse, the next challenge will be to identify the target genes for these factors, to define their combinatorial interactions, and to identify the upstream activators of these genes. Finally, given the fact that cardiac defects are the most frequently occurring congenital abnormalities in humans, being observed in an estimated 1% of newborns (Hoffman 1995), this knowledge will hopefully contribute to the eventual genetic diagnosis and treatment of these diseases.

ACKNOWLEDGMENTS

This work was supported by grants from the National Institutes of Health, the American Heart Association, and the Human Frontiers Science Program to E.N.O. and from the National Institutes of Health and American Heart Association to D.S.

REFERENCES

Azpiazu N. and Frasch M. 1993. *Tinman* and *bagpipe*: Two homeobox genes that determine cell fates in the dorsal mesoderm of *Drosophila. Genes Dev.* **7:** 1325.

Bodmer R. 1993. The gene *tinman* is required for specification of the heart and visceral muscles in *Drosophila. Development* **118:** 719.

Bodmer R., Jan L.Y., and Jan Y.N. 1990. A new homeobox containing gene, *msh-2*, is transiently expressed early during mesoderm formation in *Drosophila. Development* **110:** 661.

Bour B.A., O'Brien M.A., Lockwood W.L., Goldstein E.S., Bodmer R. Taghert P.H., Amayr S.M., and Nguyen H.T. 1995. *Drosophila* MEF2, a transcription factor that is essential for myogenesis. *Genes Dev.* **9:** 730.

Brown N.A. and Wolpert L. 1990. The development of handedness in left/right asymmetry. *Development* **109:** 1.

Collignon J., Varlet I., and Robertson E.J. 1996. Relationship between asymmetric *nodal* expression and the direction of embryonic turning. *Nature* **381:**155.

Cross J.C., Flannery, M.L., Blanar M.A., Steingrimsson E., Jenkins N.A., Copeland N.G., Rutter W.J., and Werb Z. 1995. Hxt encodes a basic helix-loop-helix transcription factor that regulates trophoblast cell development. *Development* **121:** 2513.

Cserjesi P., Brown D., Lyons G.E., and Olson E.N. 1995. Expression of the novel basic helix-loop-helix gene *eHAND* in neural crest derivatives and extraembryonic membrances during mouse development. *Dev. Biol.* **170:** 664.

DeHaan R.L. 1965. Morphogenesis of vertebrate heart. In *Oncogenesis* (ed. R.L. DeHann and Y.H. Ursprung), p. 377. Holt, Rinehart & Winston, New York.

Edmondson D.G., Lyons G.E., Martin J.F., and Olson E.N. 1994. Mef2 gene expression marks the cardiac and skeletal muscle lineages during mouse embryogenesis. *Development* **120:**1251.

Gajewski K., Kim Y., Lee Y.M., Olson E.N., and Schulz R.A. 1997. D-mef2: A target for Tinman activation during *Drosophila* heart development. *EMBO J.* **16:** 515.

Gossett L.A., Kelvin D.J., Sternberg E.A., and Olson E.N. 1989. A new myocyte-specific enhancer-binding factor that recognizes a conserved element associated with multiple muscle-specific genes. *Mol. Cell. Biol.* **9:** 5022.

Harvey R.P. 1996. NK-2 homeobox genes and heart development. *Dev Biol* **178:** 203.

Hoffman, J.I. 1995. Incidence of congenital heart disease. II. Prenatal incidence. *Pediatr. Cardiol.* **16:** 155.

Hollenberg S.M., Sternglanz R., Cheng P.F., and Weintraub H. 1995. Identification of a new family of tissue-specific basic helix-loop-helix proteins with a two-hybrid system. *Mol. Cell. Biol.* **15:** 3813.

Hyatt B.A., Lohr J.L., and Yost H.J. 1996. Initiation of vertebrate left-right axis formation by maternal Vg1. *Nature* **384:** 62.

Issac A., Sargent M.G., and Cooke J. 1997. Control of vertebrate left-right asymmetry by a *Snail*-related zinc finger gene. *Science* **275:** 1301.

Kirby M.L. and Waldo K.L. 1995. Role of neural crest in congenital heart disease. *Circ. Res.* **77:** 211.

Kuisk I.R., Li H., Tran D., and Capetanaki Y.A. 1996. A single MEF2 site governs desmin transcription in both heart and skeletal muscle during mouse embryogenesis. *Dev. Biol.* **174:** 1.

Levin M. 1997. Left-right asymmetry in vertebrate embryogenesis. *BioEssays* **19:** 287.

Levin M., Johnson R.L., Stern C.D., Kuehn M., and Tabin C. 1995. A molecular pathway determining left-right asymmetry in chick embryogenesis. *Cell* **82:** 803.

Lilly B., Galewsky S., Firulli A.B., Schulz R.A., and Olson E.N. 1994. D-MEF2: A MADS box transcription factor expressed in the differentiating mesoderm and muscle cell lineages during *Drosophila* embryogenesis. *Proc. Natl. Acad. Sci.* **91:** 5662.

Lilly B., Zhao B., Ranganayakulu G., Paterson B.M., Schulz R.A., and Olson E.N. 1995. Requirement of MADS domain transcription factor D-MEF2 for muscle formation in *Drosophila. Science* **267:** 688.

Lin Q., Schwartz J.A., and Olson E.N. 1997. Control of cardiac morphogenesis and myogenesis by the myogenic transcription factor MEF2C. *Science* **276:**1404.

Lints T.J., Parsons L.M., Hartley L., Lyons I., and Harvey R.P. 1993. *Nkx-2.5*: A novel murine homeobox gene expressed in early heart progenitor cells and their myogenic descendants. *Development* **119:** 419.

Lowe L.A., Supp D.M., Sampath K., Yokoyama T., Wright C.V., Potter S.S., Overbeek P., and Kuehn M.R. 1996. Conserved left-right asymmetry of nodal expression and alterations in murine situs inversus. *Nature* **381:** 158.

Lyons I., Parsons L.M., Hartley L., Li R., Andrews J.E., Robb L. and Harvey R.P. 1995. Myogenic and morphogenetic defects in the heart tubes of murine embryos lacking the homeobox gene *Nkx2-5. Genes Dev.* **9:** 1654.

Meno C., Saijoh Y., Fujii H., Ikeda M., Yokoyama T., Yokoyama, Y. Toyoda, and Hamada H. 1996. Left-right asymmetric expression of the TGFβ-family member *lefty* in mouse embryos. *Nature* **381:** 151

Molkentin J.D. and Olson E.N. 1996. Combinational control of muscle development by bHLH and MADS-box transcription factors. *Proc. Natl. Acad. Sci.* **93:** 9366.

Molkentin J.D., Black B.L., Martin J.F., and Olson E.N. 1995. Cooperative activation of muscle gene expression by MEF2 and myogenic bHLH proteins. *Cell* **83:** 1125.

———. 1996a. Mutational analysis of the DNA binding, dimerization, and transcriptional activation of MEF2C. *Mol. Cell. Biol.* **16:** 2627.

Molkentin J.D., Firulli A., Black B., Lyons G., Edmondson D., Hustad C.M., Copeland N., Jenkins N., and Olson E.N. 1996a. MEF2B is a potent transactivator expressed in early myogenic lineages. *Mol Cell. Biol.* **16:** 3814.

Navankasattusas S., Zhu H., Garcia A.V., Evans S.M., and Chien K.R. 1992. A ubiquitous factor (HF-1a) and a distinct muscle factor (HF-1b/MEF-2) form an E-box-indpendent pathway for cardiac muscle gene expression. *Mol. Cell. Biol.* **12:** 1469.

Nascone N. and Mercola M. 1996. Endoderm and cardiogenesis. *Trends Cardiovasc. Med.* **6:** 211.

Nguyen H.T., Bodmer R., Abmayr S.M., McDermott J.C., and Spoerel N.A. 1994. D-MEF2: A *Drosophila* mesoderm-specific MADS box-containing gene with a biphasic expression profile during embryogenesis. *Proc. Natl. Acad. Sci.* **91:** 7520.

Olson E.N. and Klein W.H. 1994. bHLH factors in muscle development: Dead lines and commitments, what to leave in and what to leave out. *Genes Dev.* **8:** 1.

Olson E.N. and Srivastava D. 1996. Molecular pathways controlling heart development. *Science* **272:** 671.

Olson E.N., Perry M., and Schulz R.A. 1995. Regulation of muscle differentiation by the MEF2 family of MADS box transcription factors. *Dev. Biol.* **172:** 2.

Pollock R. and Treisman R. 1991. Human SRF-related proteins: DNA-binding properties and potential regulatory targets. *Genes Dev.* **5:** 2327.

Ranganayakulu G., Zhao B., Dokidis A., Molkentin J.D., Olson E.N. and Schulz R.A. 1995. A series of mutations in the D-Mef2 transcription factor reveals multiple functions in larval and adult myogenesis in *Drosophila. Dev. Biol.* **171:** 169.

Ross R.S., Navankasattusas S., Harvey R.P., and Chien X.R. 1996. An HF-1a/HF-1b/MEF2 combinatorial element confers cardiac ventricular specificity and establishes anterior-posterior gradient of expression an *Nkx 2.5* independent pathway. *Development* **122:** 1799.

Shore P. and Sharrocks D. 1995. The MADS-box family of transcription factors. *Eur J. Biochem.* **229:** 1.

Srivastava D., Cserjesi P., and Olson E.N. 1995. A subclass of bHLH proteins required for cardiogenesis. *Science* **270:** 1995.

Srivastava D., Thomas T., Lin Q., and Olson E.N. 1997. Regulation of cardiac mesodermal and neural crest development by the cardiac bHLH transcription factor, dHAND. *Nat. Genet.* **16:** 154.

Stanier D.Y.R. and Fishman M.C. 1992. Patterning the zebrafish heart tube: Acquisition of anteroposterior polarity. *Dev. Biol.* **153:** 91.

Yost H.J. 1995. Vertebrate left-right development. *Cell* **82:** 689

Yutzey K.E. and Bader D. 1995. Diversification of cardiomyogenic cell lineages during early heart development. *Circ. Res.* **77:** 216

Induction of Chick Cardiac Myogenesis by Bone Morphogenetic Proteins

T.M. SCHULTHEISS AND A.B. LASSAR
Department of Biological Chemistry and Molecular Pharmacology, Harvard Medical School, Boston, Massachusetts 02115

In birds and mammals, the heart is the first embryonic tissue to differentiate. The early timing of heart development, together with the experimental accessibility of the avian embryo, has made avian heart development a model system for the study of cell fate determination during development. Extensive fate-mapping studies have shown that precardiac cells invaginate through the midportion of the primitive during early gastrulation (stage 3) and then migrate anteriorly through the lateral mesoderm (Rosenquist 1966; Rosenquist and DeHaan 1966; Garcia-Martinez and Schoenwolf 1993). The bilateral heart primordia reach the anterior ventral midline by stage 7, where they begin terminal differentiation (Han et al. 1992) and subsequently fusion into a single heart tube. Explant studies suggest that precardiac cells become specified to a cardiac fate shortly after gastrulation, during stages 4 to 5 (Rawles 1943; Gonzalez-Sanchez and Bader 1990; Holtzer et al. 1990; Antin et al. 1994); by stage 6, the majority of precardiac cells are committed to cardiac differentiation by a number of criteria (Orts-Llorca and Collado 1969; Chacko and Joseph 1974; Holtzer et al. 1990; Montgomery et al. 1994).

Experiments using amphibian and avian experiments have investigated the tissue interactions involved in specifying the heart primordia. In chick, we have recently found that the anterior endoderm, over which the precardiac mesoderm migrates after gastrulation, has cardiac inducing properties. Signals from the anterior endoderm can induce the formation of cardiac muscle in posterior primitive streak tissue, whose normal fate is to form blood and extraembryonic membranes (Schultheiss et al. 1995). Data from amphibians have implicated the pharyngeal endoderm (Bacon 1945; Jacobson 1960, 1961; Jacobson and Duncan 1968; Jacobson and Sater 1988) and the deep endoderm (Nascone and Mercola 1995) in cardiac specification and have also pointed to a role for the organizer in specification of cardiac lineages (Sater and Jacobson 1990; Nascone and Mercola 1995). However, until recently, the molecular mechanisms of vertebrate cardiac induction have been obscure.

In *Drosophila*, the homeobox-containing gene *tinman* is expressed in the heart and is required for heart formation (Bodmer et al. 1990; Bodmer 1993). Expression of *tinman* in the *Drosophila* heart is partially under the control of the transforming growth factor-β (TGF-β) family signaling molecule decapentaplegic (dpp) (Frasch 1995). In the chick and other vertebrates, a *tinman* homolog, *Nkx-2.5*, is expressed in the heart and heart precursors (Komuro and Izumo 1993; Lints et al. 1993; Tonissen et al. 1994; Schultheiss et al. 1995; Lee et al. 1996). Recently, we have shown that the dpp homologs bone morphogenetic proteins 2 and 4 (BMP-2 and BMP-4) have a role in cardiac specification. BMPs are expressed in tissues adjacent to the heart primordia as the heart is being specified and are capable of inducing heart formation from certain tissues that are not normally cardiogenic (Schultheiss et al. 1997). Here, we present additional data concerning the induction of heart muscle by BMPs, and we argue that BMPs act combinatorially with a currently unknown factor(s) in the anterior endoderm to induce heart formation in the anterior lateral (i.e., ventral) region of the avian embryo.

METHODS

The materials and methods used in these studies are described in Schultheiss et al. (1997).

RESULTS

Expression of Molecular Markers of Cardiogenesis in the Avian Embryo

Classical fate-mapping studies have traced the migration pathway of cardiac precursors in the chick embryo (Rosenquist and DeHaan 1966; Garcia-Martinez and Schoenwolf 1993). Precardiac cells leave the mid primitive streak during early gastrulation and enter the mesodermal layer, through which they migrate anteriorly and laterally. At stage 6, they form a bilateral crescent extending from the anterior border of the mesoderm to a region slightly posterior to Hensen's node (Fig. 1A). Subsequently, as the foregut forms, the lateral heart precursors are brought together on the ventral surface of the forming gut, and the bilateral heart primordia fuse in the ventral midline (Fig. 1B,C).

Whole-mount in situ hybridization was performed to follow the expression of genes encoding several proteins that have been implicated in heart development, including the transcription factors Nkx-2.5 (Schultheiss et al. 1995) and GATA-4 (a zinc finger transcription factor im-

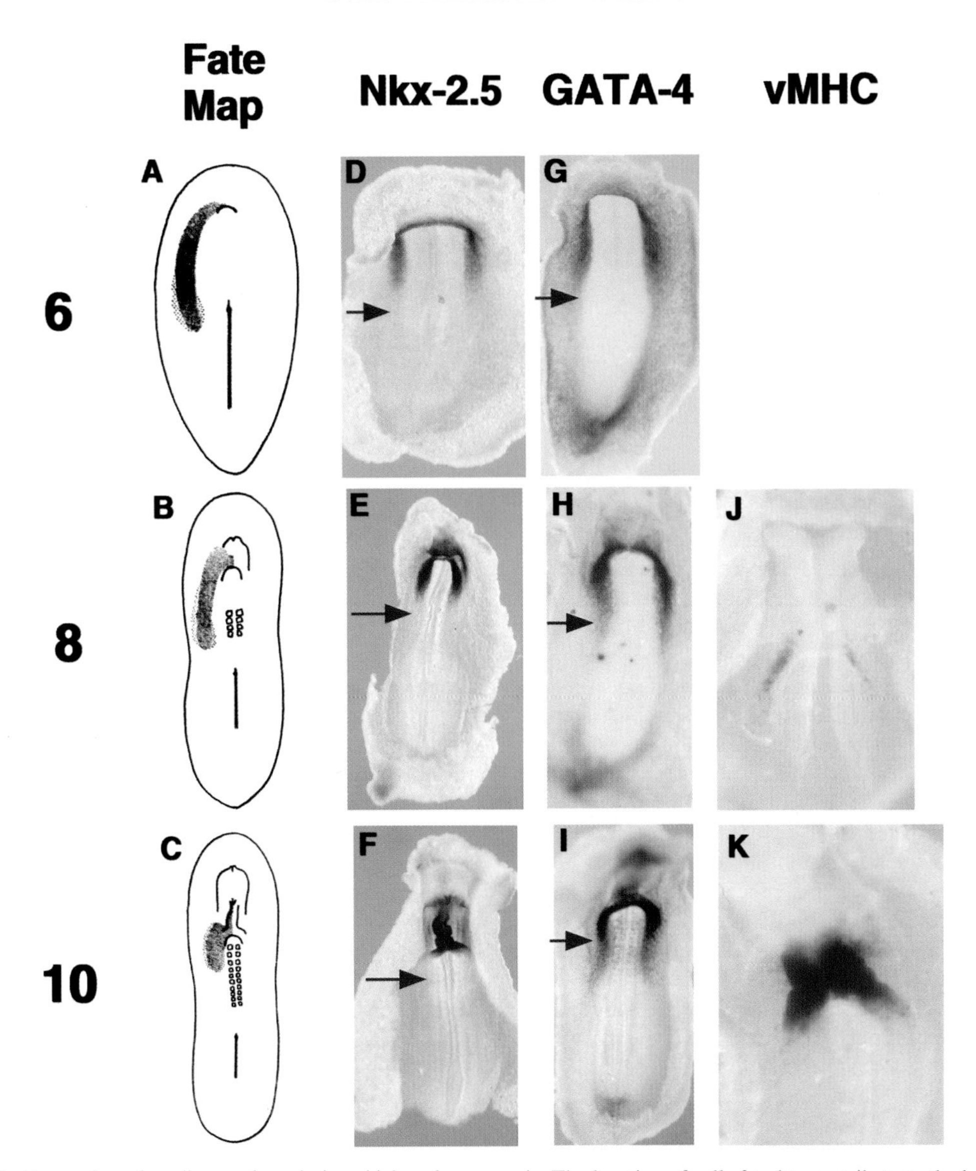

Figure 1. Expression of cardiac markers during chick embryogenesis. The location of cells fated to contribute to the heart at stages 6, 8, and 10, as determined by autoradiographic mapping, is shown by the shaded regions in panels *A–C* (adapted from Rosenquist and DeHaan 1966); only one side of the fate map is shown, although the cardiac primordia are bilateral. The cardiac fate map is compared with the expression patterns of Nkx-2.5 (*D–F*), GATA-4 (*G–I*), and vMHC (*J–K*), as determined by whole-mount in situ hybridization. The red arrows in *D–I* indicate the approximate posterior border of cells fated to contribute to the heart.

plicated in the expression of a number of cardiac genes (Laverriere et al. 1994) and the myofibrillar protein, ventricular myosin heavy chain (vMHC) (Bisaha and Bader 1991). Figure 1 compares the expression of these genes in relation to the location of the cardiac precursors as determined by fate-mapping studies. At stage 6, Nkx-2.5 is confined to the anterior region of the embryo, where its expression pattern overlaps the cardiac fate map (Fig. 1D). At later stages, Nkx-2.5 continues to be associated with the cardiac fate map (Fig. 1E,F); at these stages, Nkx-2.5 expression is not confined to the mesodermal heart tissues but is also expressed in the endoderm and ectoderm of the pharyngeal region (Schultheiss et al. 1995).

GATA-4 is also expressed in the cardiac region in the anterior part of the embryo at stage 6 (Fig. 1G), although its expression domain extends more laterally than the cardiac fate map (see also Schultheiss et al. 1997), and it is also expressed in the posterior region of the embryo. At stage 8 (Fig. 1H), the GATA-4 expression domain appears to include the entire cardiac fate map, in a pattern similar to that of CNkx-2.5, although GATA-4 expression at this stage is confined to the mesoderm (data not shown). As development proceeds, and the anterior intestinal portal moves posteriorly, the expression domain of GATA-4 also moves posteriorly, so that by stage 10, the posterior-most cells expressing GATA-4 no longer contribute to the heart, but rather to the lateral plate of the body wall (compare C and I in Fig. 1). Thus, whereas CNkx-2.5-expressing mesodermal cells contribute primarily to the heart, GATA-expressing cells contribute to multiple cell types, which share the property of being located in the lateral mesoderm during their development.

vMHC expression is not detectable at stage 6 (data not shown; Bisaha and Bader 1991). It begins to be expressed about stage 8+ in the forming heart tubes (Fig. 1J), and thereafter is strongly expressed in the myocardium of the heart (Fig. 1K). It thus serves as a marker of terminal cardiac differentiation.

Induction of Cardiogenesis by BMP-2 In Vivo

Because in *Drosophila* the gene *tinman* is regulated by the signaling molecule dpp (Frasch 1995), we set out to determine whether the *tinman* homolog Nkx-2.5 was regulated in a similar fashion in vertebrates. In the following account, we focus on BMP-2, which is the closest known homolog relative to *Drosophila* dpp (Hogan 1996). Information on the related molecules BMP-4 and BMP-7 is presented elsewhere (Schultheiss et al. 1997).

At stage 6, BMP-2 is expressed in the anterior and lateral regions of the embryo in a pattern that overlaps to a remarkable degree the expression domain of Nkx-2.5 (compare A and B in Fig. 2). It should be noted that the BMP-2 and Nkx-2.5 expression domains are not identical. The lateral domain of BMP-2 expression extends more posteriorly than that of Nkx-2.5, and BMP-2, unlike Nkx-2.5, is expressed in the posterior primitive streak. Nevertheless, in the anterior regions of the embryo, the expression domains are quite similar, as evaluated by whole-mount in situ hybridization. On sections through the precardiac region, it is evident that Nkx-2.5 is expressed in the mesoderm (where the heart precursors are located), whereas BMP-2 is expressed in the underlying endoderm, which will contribute to the ventral foregut (Fig. 2C,D).

The overlapping expression domains of BMP-2 and Nkx-2.5 raised the possibility that BMP-2 signaling might regulate Nkx-2.5 expression. To examine this possibility, beads soaked in BMP-2 were placed into ectopic locations in chick embryos at stages 3 to 4. After a period of further development, the embryos were examined for expression of Nkx-2.5. The anterior paraxial mesoderm normally gives rise primarily to skeletal muscle of the head and does not express cardiac markers such as Nkx-2.5. BMP-2-soaked beads which were implanted into the anterior paraxial mesoderm induced strong ectopic expression of CNkx-2.5 in a localized region around the bead (Fig. 3A–D). Interestingly, only a limited region of the embryo was competent to express ectopic CNkx-2.5 in response to BMP-2 (Fig. 3E,F). This competent region had the same anterior-posterior boundaries as the normal CNkx-2.5 expression domain but extended into medial regions of the anterior domain of the embryo (Fig. 3G). Ectopic GATA-4 was also induced by BMP-2 beads but to a much weaker extent than Nkx-2.5 (Schultheiss et al. 1995). Ectopic BMP-2 did not induce ectopic vMHC, even when embryos were incubated through stage 14 (data not shown).

Induction of Cardiogenesis by BMP Signaling In Vitro

To examine BMP regulation of cardiogenesis in more detail, an in vitro tissue culture system was devised. When placed in culture, stage-6 anteromedial (A/M) mesoendoderm (whose normal fate is to form skeletal muscle and connective tissue of the head) does not express cardiac markers (Fig. 4, lanes 2,4,6), whereas anterolateral (A/L) mesoendoderm from the same embryo, which lies within the cardiac fate map, differentiates into beating heart tissue in vitro (Fig. 4, lane 1). If the A/M tissue is treated with soluble recombinant BMP-2, it is induced to differentiate into beating tissue which expresses all of the cardiac markers that were examined (Fig. 4, lanes 3,5). Thus, BMP-2 can induce cardiac myogenesis in regions of the embryo whose normal fate is not cardiogenic.

Interestingly, BMP-2 could only induce a complete cardiogenic response in the A/M mesoendoderm if that tissue was separated from the axial structures which nor-

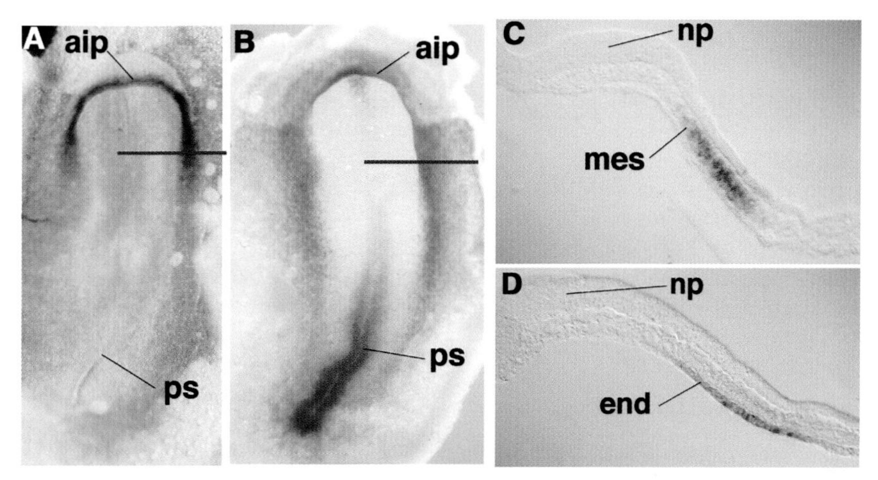

Figure 2. Comparison of expression of Nkx-2.5 and BMP-2. Whole-mount in situ hybridization was performed on stage-6 embryos for Nkx-2.5 (*A,C*) and BMP-2 (*B,D*). (*C* and *D*) Sections through the embryos in *A* and *B*, respectively, at the approximate axial levels indicated by the red lines. In the area through which the sections were taken, Nkx-2.5 is expressed in the mesoderm, whereas BMP-2 is expressed in the endoderm. (aip) Anterior intestinal portal; (end) endoderm; (mes) mesoderm; (np) neural plate; (ps) primitive streak. (Adapted from Schultheiss et al. 1997.)

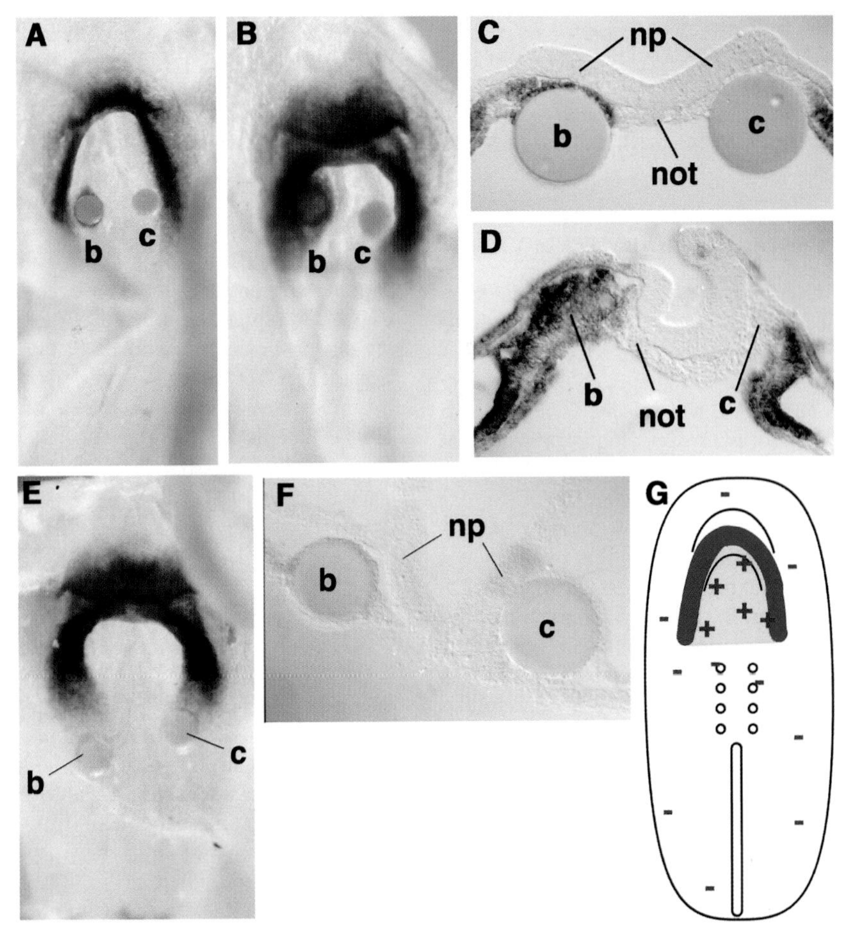

Figure 3. Induction of Nkx-2.5 by beads carrying recombinant BMP-2. Heparin-acrylamide beads soaked in human recombinant BMP-2 (*b*) or control beads (*c*) were placed in stage-3–4 chick embryos, and the embryos were allowed to develop to stage 8–10, following which they were fixed and analyzed for Nkx-2.5 expression by whole-mount in situ hybridization. (*A–D*) BMP-2 beads induced ectopic Nkx-2.5 expression in the area around the bead. (*C*) Section through the embryo in *A*; (*D*) section through the embryo in *B* at an axial level just posterior to the beads. Note that in *D*, the BMP-2 bead not only induced ectopic Nkx-2.5 expression, but also caused a truncation of the neural tube on the side of the bead. (*E–F*) BMP-2 bead placed more posteriorly than in *A–D* does not induce ectopic Nkx-2.5. (*F*) Section through the embryo in *E*. (*G*) Diagram of the stage-8 chick embryo, with blue indicating the normal expression domain of Nkx-2.5, and yellow indicating the region of the embryo that is competent to express ectopic Nkx-2.5 in response to ectopic BMP-2. The red "+" and "–" indicate whether or not Nkx-2.5 was induced when BMP-2 beads were placed in the indicated locations. A total of 99 embryos were analyzed. (not) Notochord; (np) neural plate. (Adapted from Schultheiss et al. 1997.)

mally surround it. If the A/M mesoendoderm was cultured in combination with the surrounding neural plate and notochord, then BMP-2 could only induce Nkx-2.5 and, to a lesser degree, GATA-4, whereas vMHC was not induced (Fig. 4, lanes 7,8). Since this more limited induction of cardiac markers was also seen when BMP-2 beads were implanted in the A/M region in vivo (Fig. 3), it suggests that the axial tissues may limit the cardiogenic response of the A/M mesoderm to BMP signaling in vivo.

BMP Signaling Is Required for Cardiogenesis

The previous experiments show that BMP signaling can promote cardiogenesis in tissues from the anterior embryo that normally form more dorsal derivatives such as head muscle and suggest that BMP expression in the lateral regions of early embryos has a role in specifying heart tissues in vivo. To determine whether BMP signaling is *required* for cardiac differentiation, we made use of noggin, a secreted molecule that has been shown to bind BMP-2 and BMP-4 directly, thereby antagonizing BMP activity (Smith and Harland 1992; Smith et al. 1993; Hogan 1996; Zimmerman et al. 1996). As shown in Figure 5, noggin completely inhibited cardiogenesis when added to cultures of stage-4 A/L mesoendoderm. When administered to stage-5 A/L mesoendoderm, noggin blocked cardiogenesis only partially, whereas stage-6 mesoendoderm was resistant to the effects of noggin. These data imply that prior to stage 5, precardiac cells require BMP signaling in order to undergo cardiac differentiation but that BMP signals are only required transiently for cardiac myocyte specification.

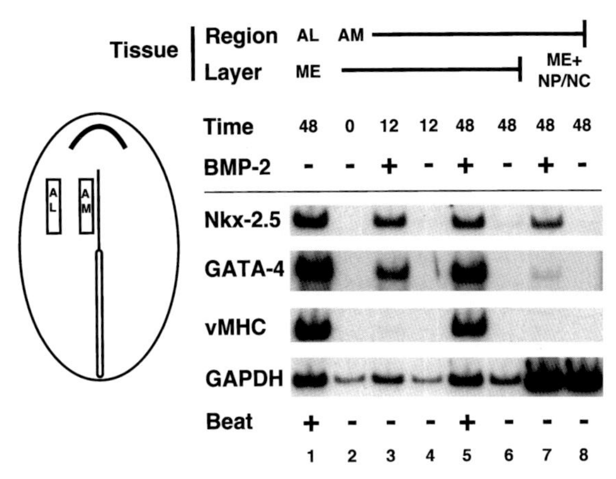

Figure 4. Induction of cardiogenesis by BMP-2 in explant culture. The indicated tissues were dissected from stage-6 embryos and placed in culture for the indicated lengths of time, in the presence or absence of 200 ng/ml of BMP-2, and analyzed by RT-PCR for the indicated genes. Cultures were also analyzed for the presence of rhythmic beating. In the presence but not the absence of BMP-2, the A/M mesoendoderm differentiated into beating heart tissue which expressed all the cardiac markers examined. See text for further description. (AL) Anterior lateral; (AM) anterior medial; (ME) mesoendoderm; (NP/NC) neural plate + notochord. (From Schultheiss et al. 1997.)

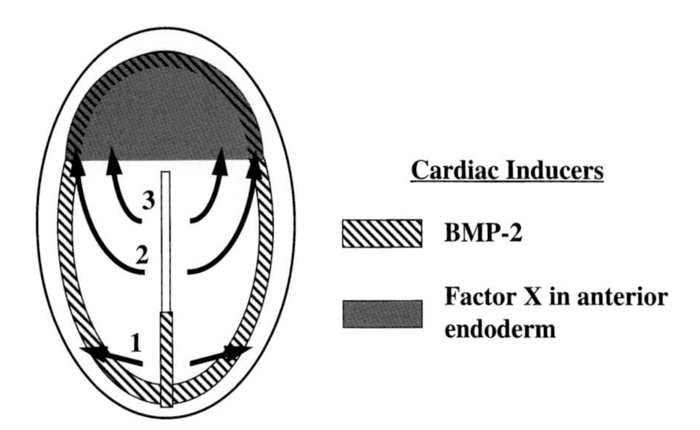

Figure 6. Model of cardiogenesis. Two cardiac-inducing activities have been identified in the chick embryo. One is BMP activity (only BMP-2 is illustrated here, but several other BMPs have similar patterns of expression), and the other is a currently unknown factor present in the anterior endoderm (Factor X). Cardiogenesis occurs where the inducing activities are both expressed, in the A/L region of the embryo. The numbered arrows indicate migration pathways of cells at various levels of the primitive streak. Only cells that follow pathway 2 encounter both inducing activities and differentiate as heart. Cells following pathway 1 are exposed to BMPs but not to Factor X. They differentiate as blood and extraembryonic membranes but can differentiate into heart if exposed to anterior endoderm (Schultheiss et al. 1995). Cells following pathway 3 encounter Factor X but not BMP signaling. They develop into skeletal muscle and connective tissue of the head but can form heart if exposed to BMP signals (Schultheiss et al. 1997; this paper).

DISCUSSION

The data presented here, together with our recently published study (Schultheiss et al. 1997), suggest that BMP signaling has a role in the induction of chick cardiogenesis. BMP signals, on their own, are not sufficient to induce cardiogenesis. Rather, BMP signals appear to determine which cells, from within a cardiogenic field in the anterior mesoderm, will actually differentiate into heart. In earlier studies, we have found that chick anterior endoderm has a cardiac-inducing activity distinct from BMP signals, since it can induce cardiogenesis from tissues that are expressing high levels of multiple BMPs (Schultheiss et al. 1995). It is significant that the mesoderm that overlies this anterior endoderm coincides well with the cardiac field in the anterior mesoderm which is competent to undergo cardiogenesis in response to BMP signaling. We are thus led to a two-step model of cardiogenesis in which the anterior endoderm induces a cardiogenic field in the anterior mesoderm, and BMP signaling selects which subdomain within this field will differentiate into heart. Another way of stating the same concept is that cardiogenesis in the chick occurs where anterior endoderm and BMP signals overlap, namely, in the A/L mesoderm (Fig. 6).

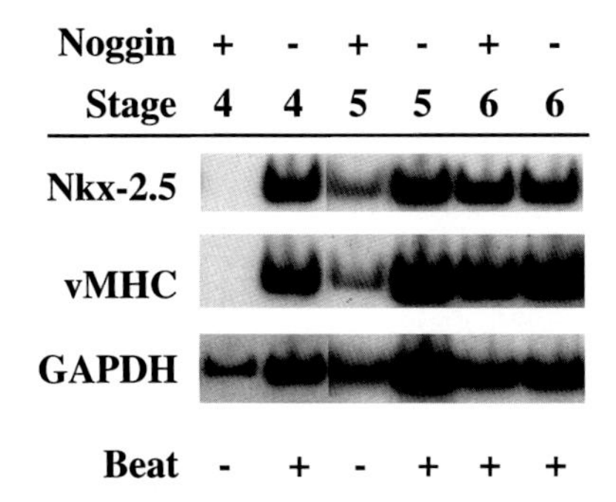

Figure 5. Inhibition of cardiogenesis by the BMP antagonist noggin. A/L mesoendoderm was dissected from embryos of the indicated stages, and cultured for 48 hr in the presence or absence of culture supernatant conditioned by a noggin-expressing cell line. Cultures were evaluated by RT-PCR for expression of the indicated genes and were examined for the presence of beating tissue.

The data presented here suggest that the requirement for BMP signaling is a transient one and is no longer required after stage 6, since the differentiation of stage 6 or older precardiac mesoendoderm is resistant to the administration of the BMP antagonist noggin. The chemicals bromodeoxyuridine (BrdU) and 12-O-tetradecanoylphorbal-13-acetate (TPA) have also been found to inhibit cardiogenesis, and precardiac tissues become resistant to these agents by stage 6 (Chacko and Joseph 1974; Montgomery et al. 1994). The mechanisms of action of noggin, TPA, and BrdU are very different. However, the similar windows of susceptibility to these agents suggest that by stage 6, the precardiac tissues have attained a high degree of commitment to the cardiac fate. At this stage, the precardiac cells have begun to express the homeobox gene *Nkx-2.5* at high levels (Schultheiss et al. 1995) and within several hours begin the expression of genes coding for myofibrillar proteins (Han et al. 1992).

In *Drosophila*, the gene *tinman* is expressed in the heart and is regulated in part by the TGF-β signaling molecule dpp (Bodmer et al. 1990; Bodmer 1993; Frasch 1995). Several groups have shown that the *tinman*-related gene *Nkx-2.5* is expressed in the vertebrate heart (Komuro and Izumo 1993; Lints et al. 1993; Tonissen et al. 1994; Schultheiss et al. 1995; Lee et al. 1996). Here, and in a recent publication (Schultheiss et al. 1997), we show that the BMPs, which are closely related to *Drosophila* dpp, play a part in regulating Nkx-2.5 and heart differentiation in the chick. These results point to a new level of conservation in the cardiogenic program between insects and vertebrates. It is not clear, at this point, whether other cardiac inductive mechanisms are conserved between these two groups of organisms. In *Drosophila*, the signaling molecule wingless cooperates with dpp to induce cardiogenesis (Wu et al. 1995), whereas in many vertebrates, a factor in the anterior endoderm (that cannot be mimicked by BMP signals) has an important role in cardiac induction (Jacobson 1960, 1961; Jacobson and Duncan 1968; Sugi and Lough 1994; Nascone and Mercola 1995; Schultheiss et al. 1995; Wu et al. 1995). In the future, it will be of interest to determine whether wingless/Wnt signaling is a component of this anterior endodermal activity or whether the anterior endodermal-inducing factor represents a component of cardiac regulation that is unique to vertebrates.

ACKNOWLEDGMENTS

The authors thank Maureen O'Brien, Steve Xydas, and Tak Chun for technical assistance; Randy Johnson, Tom Jessell, and Katherine Yutzey for providing probes for in situ hybridization; the Genetics Institute for providing recombinant BMP-2, BMP-4, and BMP-7; Karen Symes for providing the activin-containing COS cell supernatant; and Richard Harland for providing CHO cells expressing *Xenopus* noggin. This work was supported by a grant to A.B.L. from The Council for Tobacco Research U.S.A. This work was done during the tenure of an established investigatorship from the American Heart Association to A.B.L. T.M.S. is a Howard Hughes Medical Institute Physician postdoctoral fellow.

REFERENCES

Antin P.B., Taylor R.G., and Yatskievych T. 1994. Precardiac mesoderm is specified during gastrulation in quail. *Dev. Dyn.* **200:** 144.

Bacon R.L. 1945. Self-differentiation and induction in the heart of Amblystoma. *J. Exp. Zool.* **98:** 87.

Bisaha J.G. and Bader D. 1991. Identification and characterization of a ventricular-specific avian myosin heavy chain, VMHC1: Expression in differentiating cardiac and skeletal muscle. *Dev. Biol.* **148:** 355.

Bodmer R. 1993. The gene *tinman* is required for specification of the heart and visceral muscles in *Drosophila. Development* **118:** 719.

Bodmer R., Jan L.Y., and Jan Y.N. 1990. A new homeobox-containing gene, *msh-2*, is transiently expressed early during mesoderm formation in *Drosophila. Development* **110:** 661.

Chacko S. and Joseph X. 1974. The effect of 5-bromodeoxyuridine (BrdU) on cardiac muscle differentiation. *Dev. Biol.* **40:** 340.

Frasch M. 1995. Induction of visceral and cardiac mesoderm by ectodermal Dpp in the early *Drosophila* embryo. *Nature* **374:** 464.

Garcia-Martinez V. and Schoenwolf G.C. 1993. Primitive streak origin of the cardiovascular system in avian embryos. *Dev. Biol.* **159:** 706.

Gonzalez-Sanchez A. and Bader D. 1990. In vitro analysis of cardiac progenitor cell differentiation. *Dev. Biol.* **139:** 197.

Han Y., Dennis J.E., Cohen-Gould L., Bader D.M., and Fischman D.A. 1992. Expression of sarcomeric myosin in the presumptive myocardium of chicken embryos occurs within six hours of myocyte commitment. *Dev. Dyn.* **193:** 257.

Hogan B.L.M. 1996. Bone morphogenetic proteins in development. *Curr. Opin. Genet. Dev.* **6:** 432.

Holtzer H., Schultheiss T., Dilullo C., Choi J., Costa M., Lu M., and Holtzer S. 1990. Autonomous expression of the differentiation programs of cells in the cardiac and skeletal myogenic lineages. *Ann. N.Y. Acad. Sci.* **599:** 158.

Jacobson A.G. 1960. Influences of ectoderm and endoderm on heart differentiation in the newt. *Dev. Biol.* **2:** 138.

———. 1961. Heart determination in the newt. *J. Exp. Zool.* **146:** 139.

Jacobson A.G. and Duncan J.T. 1968. Heart induction in salamanders. *J. Exp. Zool.* **167:** 79.

Jacobson A.G. and Sater A.K. 1988. Features of embryonic induction. *Development* **104:** 341.

Komuro I. and Izumo S. 1993. Csx: A murine homeobox-containing gene specifically expressed in the developing heart. *Proc. Natl. Acad. Sci.* **90:** 8145.

Laverriere A.C., MacNeill C., Mueller C., Poelmann R.E., Burch J.B.E., and Evans T. 1994. GATA-4/5/6, a subfamily of three transcription factors transcribed in developing heart and gut. *J. Biol. Chem.* **269:** 23177.

Lee K.-H., Xu Q., and Breitbart R.E. 1996. A new *tinman*-related gene, *nkx-2.7*, anticipates the expression of *nkx-2.5* and *nkx-2.3* in zebrafish heart and pharyngeal endoderm. *Dev. Biol.* **180:** 722.

Lints T.J., Parsons L.M., Hartley L., Lyons I., and Harvey R.P. 1993. *Nkx-2.5:* A novel murine homeobox gene expressed in early heart progenitor cells and their myogenic descendants. *Development* **119:** 419.

Montgomery M.O., Litvin J., Gonzalez-Sanchez A., and Bader D. 1994. Staging of commitment and differentiation of avian cardiac myocytes. *Dev. Biol.* **164:** 63.

Nascone N. and Mercola M. 1995. An inductive role for the endoderm in *Xenopus* cardiogenesis. *Development* **121:** 515.

Orts-Llorca F. and Collado J.J. 1969. The development of heterologous grafts of the cardiac area (labelled with thymidine-^{3}H) to the caudal area of the chick blastoderm. *Dev. Biol.* **19:** 213.

Rawles M.E. 1943. The heart-forming areas of the early chick blastoderm. *Physiol. Zool.* **41:** 22.

Rosenquist G.C. 1966. A radioautographic study of labelled grafts in the chick blastoderm. *Carnegie Inst. Wash. Contrib. Embryol.* **38:** 71.

Rosenquist G.C. and DeHaan R.L. 1966. Migration of precardiac cells in the chick embryo: A radioautographic study. *Carnegie Inst. Wash. Contrib. Embryol.* **38:** 111.

Sater A.K. and Jacobson A.G. 1990. The role of the dorsal lip in the induction of heart mesoderm in *Xenopus laevis. Development* **108:** 461.

Schultheiss T.M., Burch J.B.E., and Lassar A.B. 1997. A role for bone morphogenetic proteins in the induction of cardiac myogenesis. *Genes Dev.* **11:** 451.

Schultheiss T.M., Xydas S., and Lassar A.B. 1995. Induction of avian cardiac myogenesis by anterior endoderm. *Development* **121:** 4203.

Smith W.C. and Harland R.M. 1992. Expression cloning of noggin, a new dorsalizing factor localized to the Spemann organizer in *Xenopus* embryos. *Cell* **70:** 829.

Smith W.C., Knecht A.K., Wu M., and Harland R.M. 1993. Secreted noggin protein mimics the Spemann organizer in dorsalizing *Xenopus* mesoderm. *Nature* **361:** 547.

Sugi Y. and Lough J. 1994. Anterior endoderm is a specific effector of terminal cardiac myocyte differentiation of cells from the embryonic heart forming region. *Dev. Dyn.* **200:** 155.

Tonissen K.F., Drysdale T.A., Lints T.J., Harvey R.P., and Krieg P.A. 1994. *XNkx-2.5*, a *Xenopus* gene related to *Nkx-2.5* and *tinman:* Evidence for a conserved role in cardiac development. *Dev. Biol.* **162:** 325.

Wu X., Golden K., and Bodmer R. 1995. Heart development in *Drosophila* requires the segment polarity gene *wingless. Dev. Biol.* **169:** 619.

Zimmerman L.B., Jesus-Escobar J.D., and Harland R.M. 1996. The Spemann organizer signal noggin binds and inactivates bone morphogenetic protein-4. *Cell* **86:** 599.

Expression of *Wnt* and *Frizzled* Genes during Chick Limb Bud Development

M. KENGAKU, V. TWOMBLY, AND C. TABIN
Department of Genetics, Harvard Medical School, Boston, Massachusetts 02115

Inductive cell-cell interactions are essential for the patterning of all multicellular organisms. Recent studies have shown that a number of signaling molecules involved in such interactions are remarkably conserved between species. An important family of such conserved signals encoded by *Wnt* genes have key roles in regulating patterning of both vertebrates and invertebrates.

The *Wnt* gene family encodes structurally related, secreted glycoproteins that comprise at least 16 mouse and 3 *Drosophila* members. Extensive studies have revealed that *Wnt* genes are expressed in restricted spatiotemporal patterns during embryonic development and have specific roles in a wide range of early morphogenic events. For instance, vertebrate *Wnt1* and *Wnt3a* exhibit similar expression patterns in the dorsal midline of developing neural tube (Parr et al. 1993). It has been shown that the products of both genes have the ability to induce undifferentiated somites to form myotome, suggesting that they are functionally redundant in somite differentiation (Münsterberg et al. 1995). However, targeted disruption of each gene elicits distinct phenotypes: Null alleles of *Wnt1* result in the loss of midbrain and adjacent components of cerebellum in mouse embryos (McMahon and Bradley 1990; McMahon et al. 1992; Fritzsch et al. 1995), whereas disruption of *Wnt3a* causes truncation of the posterior body axis caudal to the forelimb, although anterior neural tissue including midbrain develops normally (Takada et al. 1994; Greco et al. 1996). Another family member, *Wnt4*, is expressed in metanephric mesenchyme cells and its derivatives. Mice lacking Wnt4 activity fail to form the mesenchymal cell clusters that generate epithelial tubules, indicating that Wnt4 acts as an inducer of the mesenchymal to epithelial transition during nephron development (Stark et al. 1994). *Wnt4* is also expressed in a broad dorsal domain of the developing spinal cord; however, disruption of *Wnt4* does not cause any severe defect in neural development. These data suggest that Wnt family proteins are mediated by distinct signaling pathways to induce a diverse array of important functions during animal development.

The *Drosophila* homolog of *Wnt1* is the segment polarity gene *Wingless*. *Wingless* is required for patterning in each embryonic parasegment, as well as in patterning the imaginal discs. It has been demonstrated that Wingless protein is secreted and acts over several cell diameters, inducing the expression and activation of intracellular signaling molecules in a dose-dependent manner in responding cells (van den Heuvel et al. 1989; Gonzalez et al. 1991; Zecca et al. 1996). On the basis of studies of genetic interactions in *Drosophila*, many of the intracellular signaling molecules in the Wingless pathway were identified (for review, see Peifer 1995). Subsequent work suggested that their vertebrate counterparts have similar roles in transducing the Wnt1 signal (for review, see Moon et al. 1997). However, the receptor molecule that directly binds to Wnt/wingless, mediating the downstream cascades, remained unknown for a long time.

Recently, the *Drosophila frizzled 2* gene has emerged as a strong candidate for the Wingless receptor (Bhanot et al. 1996). *Drosophila Frizzled 2* (*Dfz2*) encodes a putative seven-transmembrane protein with a large extracellular cysteine-rich domain (CRD). Wingless protein binds to the cells transfected with *Dfz2* and increases Wingless signaling in those cells, as assayed by the level of the Armadillo protein. In vertebrates, there is a large family of genes related to *Dfz2*, with at least eight members in mouse, and more than ten in zebrafish (Chan et al. 1992; Wang et al. 1996). All of the vertebrate *Frizzled* genes share the structure of an extracellular CRD followed by seven transmembrane regions. An intriguing possibility is that distinct classes of Wnts might have different affinities for specific Frizzled receptors. There is some evidence in support of this idea. For example, it has been shown that some, but not all, of the members of the mammalian Frizzled family bind Wingless protein (Bhanot et al. 1996). A possible functional significance of this differential binding was indicated by a misexpression study in *Xenopus* embryos which showed that rat *Rfz1* is able to specifically recruit Xwnt8, but not the functionally distinct Xwnt5a, to the plasma membrane. Treatment with Xwnt8 but not Xwnt5a induces a duplication in the embryonic axis. On the other hand, *Xwnt5a* elicits duplications similar to those of *Xwnt8* when coinjected with human *Hfz5* (He et al. 1997). These data raise the interesting possibility that the vertebrate Frizzled members mediate distinct Wnt pathways to regulate divergent developmental processes.

The development of well-patterned vertebrate limbs has provided an important model system for understanding vertebrate morphogenesis and the signaling factors that guide it. In the early mouse limb buds, at least seven members of the *Wnt* family are expressed in differential patterns (Gavin et al. 1990; Parr et al. 1993). Among these, the functions of *Wnt7a* and *Wnt5a* have been best studied thus far. *Wnt7a* is expressed only in dorsal ectoderm and is implicated in the dorsoventral (D/V) patterning of the developing limb. Wnt7a emanating from the dorsal limb ectoderm induces *Lmx1* expression in the underlying dorsal mesenchyme, which in turn specifies dorsal fate in the limb distal mesoderm (Riddle et al. 1995; Vogel et al. 1995). Loss of *Wnt7a* function in the mouse results in a reciprocal phenotype, ventralization of distal limb structures (Parr and McMahon 1995). A second gene, *Wnt5a*, is expressed both in ectoderm and in mesenchyme, forming a proximodistal gradient with the highest level of expression distally. Targeted deletion of *Wnt5a* has revealed that *Wnt5a* is required for long bones in the limb to reach their normal length (A. McMahon, pers. comm.) Another family member, *Wnt11*, has a differential pattern of expression in both mesoderm and ectoderm including the apical ectodermal ridge (AER) in later stages of development (Kispert et al. 1996). Four other *Wnt* genes including *Wnt3, Wnt4, Wnt6*, and *Wnt7b* are expressed fairly uniformly throughout the limb ectoderm (Parr et al. 1993), suggesting that they have roles different from those of *Wnt7a* and *Wnt5a*.

To explore whether signaling by different Wnt ligands is mediated by different Frizzled family members in the limb, we cloned three members of the *Frizzled* family expressed in early chick limb buds. In situ hybridization showed that these *Frizzled* genes are expressed in overlapping but distinct patterns during limb development. We have also performed an analysis of *Wnt* gene expression in chick limb buds compared to published patterns in the mouse. Surprisingly, expression patterns of *Wnt* genes in the limb bud differ between mouse and chick. The spatial and temporal patterns of *Frizzled* and *Wnt* gene expression suggest that they may cooperatively play divergent parts in chick limb development.

METHODS

Cloning of chick* Frizzled *homologs. Poly(A)$^+$ selected RNA from stage-20–22 chick limb buds were primed with either oligo(dT) or random hexamers and reverse-transcribed with SuperScript II (GIBCO/BRL) for 1 hour at 37°C. Degenerate polymerase chain reaction (PCR) primers with flanking *Bam*HI and *Hin*dIII restriction sites were designed for the three amino acid sequences, PERPII, WWVIL, and TWFLAA, and used for PCR under the following conditions: 1x (94°C, 30 sec), 30x (94°C, 30 sec; 50°C, 2 min; 72°C, 1 min), 1x (72°C, 5 min). PCR products were cleaved by *Bam*HI and *Hin*dIII and subcloned into pBluescript (Stratagene) and individually sequenced.

About 10^6 colonies of a random-primed pBluescript KS(+) stage-22 limb bud cDNA library were transferred to nylon filters (Colony/Plaque Screen, New England Nuclear) and regrown on ampicillin plates. The filters were hybridized in 30% formamide, 10% dextran sulfate, 2x SSC, and 0.5% SDS at 4°C with a mixture of the three ^{32}P-labeled PCR products. Several positive colonies were isolated and sequenced. Two clones were found to cover the entire coding regions of *Chfz1* and *Chfz7*, respectively, and others were overlapping partial cDNAs for *Chfz1, Chfz2, and Chfz7.*

In situ hybridizations. Whole-mount in situ hybridizations were performed as described by Riddle et al. (1993). Section in situ hybridizations were performed as described by Marigo et al. (1996), except that riboprobes were made with [^{33}P]UTP.

RESULTS

Cloning of Chick *Frizzled* Family Members Expressed in Early Limb Bud

To identify *Frizzled* family genes expressed in the developing chick limb bud, we performed RT-PCR with degenerate PCR primers. Alignment of the rat, mouse, and human sequences revealed three stretches of amino acid identity that were used to design degenerate PCR primers (Chan et al. 1992; Wang et al. 1996). Using poly(A)$^+$ RNA from stage-20–22 chick limb buds, we performed RT-PCR and obtained partial cDNA products corresponding to multiple *Frizzled* family members. A mixture of these cloned PCR products were used in a low-stringency screen of HH stage-20–22 embryonic chick limb bud library. Three independent cDNA clones were isolated with apparently complete coding regions and named *Chfz1, Chfz2,* and *Chfz7* based on sequence identity to rat and mouse orthologs. We did not obtain clones for other two PCR products, one of which was identical to *Chfz16* (Wang et al. 1996) and another is likely a novel member of *Frizzled* family.

Chfz1, Chfz2, and *Chfz7* each share high sequence identity to their mammalian homologs, greater than 80% at the amino acid level (Fig. 1). In the coding region of the *Chfz2* clone, there is an additional 20-bp nucleotide stretch that has no homology with *Mfz2*, which causes a frameshift. The *Chfz2* clone we obtained is therefore possibly a splice variant or perhaps a partially processed precursor of the mature mRNA. Other than this 20-bp insert, the mouse and chick *Frizzled* genes have similar structures. Following the variant putative signal sequence, they have a long extracellular loop with a highly conserved cysteine-rich domain (CRD) of 120 amino acids. Between the CRD and transmembrane regions, there is a

Figure 1. Comparison of the deduced amino acid sequences of murine and chick Frizzled 1 (*a*) and Frizzled 7 (*b*). Positions of identity with reference to chick sequences are indicated by periods. Gaps are introduced to allow alignment of the sequences. Lines above the sequences indicate the locations of predicted transmembrane domains. Asterisks mark conserved cysteine residues in putative extracellular loops.

a

```
Chfz1   M A E R R G P A G G - - - - - - - - - - - - - - - - G S G E V G G G R R A G G D 24
rat Fz1 . . . E A V . S E S R A A G R P S L E L C A V A L P . R R . E V . H Q D T A . H 40

Chfz1   R C P R R P P - - - A L P L L L L W - - - - - - - - - - - - - - - A A A L P A G 47
rat Fz1 . R . . A H S R C W . R G . . . . . . L L E A P L L L G V R A Q P . G Q V S G P 80

Chfz1   G Q - - - P A A Q P A A - - - - - L S E R G I S I P D H G Y C Q P I S I P L C T 79
rat Fz1 . . Q R P . P P . . Q Q G G Q Q Y N G . . . . . . . . . . . . . . . . . . . . . 120

Chfz1   D I A Y N Q T I M P N L L G H T N Q E D A G L E V H Q F Y P L V K V Q C S A E L 119
rat Fz1 . . . . . . . . . . . . . . . . . . . . . . . . . . . . . . . . . . . . . . . . 160

Chfz1   K F F L C S M Y A P V C T V L E Q A L P P C R S L C E R A R Q G C E A L M N K F 159
rat Fz1 . . . . . . . . . . . . . . . . . . . . . . . . . . . . . - . . . . . . . . . . 199

Chfz1   G F Q W P D T L R C E K F P V H G A G E L C V G Q N A S E R G T P T P A L R P E 199
rat Fz1 . . . . . . . . K . . . . . . . . . . . . . . . . . T . D K . . . . . S . L . . 239

Chfz1   S W T S N P H R G G G A - - - - - - G G S G P G E A R G R F S C P R A L K V P S 233
rat Fz1 F . . . . . Q H . . . G Y R G G Y P . . A . . V . - . . K . . . . . . . R . . . 278

Chfz1   Y L N Y R F L G E K D C G A P C E P G R L Y G L M Y F G P E E L R F S R T W I G 273
rat Fz1 . . . . H . . . . . . . . . . . . . T K V . . . . . . . . . . . . . . . . . . . 318

Chfz1   I W S V L C C A S T L F T V L T Y L V D M K R F S Y P E R P I I F L S G C Y T A 313
rat Fz1 . . . . . . . . . . . . . . . . . . . . . R . . . . . . . . . . . . . . . . . . 358

Chfz1   V A V A Y I A G F L L E E R V V C N E R F A E D G S R T V A Q G T K R E G C T I 353
rat Fz1 . . . . . . . . . . . . D . . . . . D K . . . . . A . . . . . . . . K . . . . . 398

Chfz1   L F M M L Y F F G M A S S I W W V I L S L T W F L A A G M K W G H E A I E A N S 393
rat Fz1 . . . . . . . . S . . . . . . . . . . . . . . . . . . . . . . . . . . . . . . . 438

Chfz1   Q Y F H L A A W A V P A I K T I T I L A L G Q V D G D V L S G V C F V G I N N V 433
rat Fz1 . . . . . . . . . . . . . . . . . . . . . . . . . . . . . . . . . . . . L . . . 478

Chfz1   D A L R G F V L A P L F V Y L F I G T S F L L A G F V S L F R I R T I M K H D G 473
rat Fz1 . . . . . . . . . . . . . . . . . . . . . . . . . . . . . . . . . . . . . . . . 518

Chfz1   T K T E K L E K L M V R I G I F S V L Y T V P A T I V I A C Y F Y E Q A F R E Q 513
rat Fz1 . . . . . . . . . . . . . . V . . . . . . . . . . . . . . . . . . . . . . . D . 558

Chfz1   W E R S W V T Q S C K S Y A I P C P N N H S S - - - - H H P P M S P D F T V F M 549
rat Fz1 . . . . . . A . . . . . . . . . . . H L Q G G G G V P P . . . . . . . . . . . . 598

Chfz1   I K Y L M T L I V G I T S G F W I W S G K T L N S W R K F Y T R L T N S K Q G E 589
rat Fz1 . . . . . . . . . . . . . . . . . . . . . . . . . . . . . . . . . . . . . . . . 638

Chfz1   T T V                                                                           593
rat Fz1 . . .                                                                           641
```

b

```
Chfz7     M R P A A G E A G A G L R W L G L A A L L A A L L G T - - - - P C A A A H H E D 36
mouse Fz7 . . G P G T A . S H S P - - . . . C . . V L . . . . A L P T D T R . Q P Y . G E 38

Chfz7     K A I S V P D H G F C Q P I S I P L C T D I A Y N Q T I L P N L L G H T N Q E D 76
mouse Fz7 . G . . . . . . . . . . . . . . . . . . . . . . . . . . . . . . . . . . . . . . 78

Chfz7     A G L E V H Q F Y P L V K V Q C S A E L K F F L C S M Y A P V C T V L E Q A I P 116
mouse Fz7 . . . . . . . . . . . . . . . . . P . . R . . . . . . . . . . . . . . D . . . . 118

Chfz7     P C R S L C E R A R Q G C E A L M N K F G F Q W P E R L R C E N F P V H G A G E 156
mouse Fz7 . . . . . . . . . . . . . . . . . . . . . . . . . . . . . . . . . . . . . . . . 158

Chfz7     I C V G Q N T S D A P P G P G G A G G R G A T A Q P T A G Y L P D L - L T P P Q 195
mouse Fz7 . . . . . . . . . - - - . S . . . . . - S P . . Y . . . P . . . . P P F . A M S 194

Chfz7     P A A G - - - - - F S F S C P R Q L K V P P Y L G Y R F L G E R D C G A P C E P 230
mouse Fz7 . S D . R G R L S . P . . . . . . . . . . . . . . . . . . . . . . . . . . . . . 234

Chfz7     G R P N G L M Y F K E A E V R F A R L W V G V W S V L C C A S T L F T V L T Y L 270
mouse Fz7 . . A . . . . . . . . E . R . . . . . . . . . . . . . S . . . . . . . . . . . . 274

Chfz7     V D M R R F S Y P E R P I I F L S G C Y F M V A V A Y A A G F L L E E R V V C L 310
mouse Fz7 . . . . . . . . . . . . . . . . . . . . . . . . . . H V . . . . . . D . A . . V 314

Chfz7     E R F S E D G Y R T V A Q G T K K E G C T I L F M I L Y F F G M A S S I W W V I 350
mouse Fz7 . . . . D . . . . . . . . . . . . . . . . . . . . V . . . . . . . . . . . . . . 354

Chfz7     L S L T W F L A A G M K W G H E A I E A N S Q Y F H L A A W A V P A V K T I T I 390
mouse Fz7 . . . . . . . . . . . . . . . . . . . . . . . . . . . . . . . . . . . . . . . . 394

Chfz7     L A M G Q V D G D V L S G V C Y V G I Y S V D S L R G F V L A P L F V Y L F I G 430
mouse Fz7 . . . . . . . . . L . . . . . . . . L S . . . A . . . . . . . . . . . . . . . . 434

Chfz7     T S F L L A G F V S L F R I R T I M K H D G T K T E K L E K L M V R I G V F S V 470
mouse Fz7 . . . . . . . . . . . . . . . . . . . . . . . . . . . . . . . . . . . . . . . . 474

Chfz7     L Y T V P A T I V V A C Y F Y E Q A F R S T W E K T W L L Q T C K T Y A V P C P 510
mouse Fz7 . . . . . . . . . L . . . . . . . . . . E H . . R . . . . . . . . S . . . . . . 514

Chfz7     S - H F A P M S P D F T V F M I K Y L M T M I V G I T T G F W I W S G K T L Q S 549
mouse Fz7 P R . . S . . . . . . . . . . . . . . . . . . . . . . . . . . . . . . . . . . . 554

Chfz7     W R R F Y H R L S T G S K G E T A V                                             568
mouse Fz7 . . . . . . . . . H S . . . . . . .                                             572

Decoration 'Decoration #1': Hide (as '.') residues that match Chfz7 exactly.
```

Figure 1. (*See facing page for legend.*)

divergent stretch of hydrophilic amino acids that may act as an extended linker (Wang et al. 1996). The carboxy-terminal half of the proteins are remarkably well-conserved, especially the intracellular loops of hydrophilic residues and the flanking seven-transmembrane segments with 20–25 hydrophobic amino acids. The extracellular domains between the transmembrane stretches are relatively variable; however, all of the cysteine residues, which possibly form disulfide bonds, are conserved. The carboxyl termini of both *Chfz1* and *Chfz7* end with the sequence T-X-V. This motif is known as the PDZ-binding domain, which was shown to interact with proteins containing PDZ domains. The PDZ-binding domains vary between Frizzled family members but are conserved between orthologs in different species.

Expression Patterns of Chick *Frizzled* Genes during Chick Embryogenesis

In an attempt to clarify the function of *Frizzled* genes in chick development, we analyzed the expression patterns of *Chfz1, Chfz2,* and *Chfz7* transcripts in embryos at various stages of development by whole-mount and section in situ hybridization.

At stage 12, the earliest stage we analyzed, expression of three *Frizzled* genes is observed in the developing somites (Fig. 2a,b,c). *Chfz1* and *Chfz2* show similar expression pattern at this stage: Both are expressed in the more mature somites in the rostral half of the embryo but not in newly forming somites. *Chfz7* is expressed in all somites along the rostral-caudal axis as well as in presegmental mesenchyme caudal to the developing somites. Expression of all three *Frizzled* genes is also observed in the developing retina (data not shown). At stage 15, expression of three *Frizzled* genes is detected in all the developing somites (Fig. 2d,e,f). Sections reveal that the *Frizzled* genes are expressed uniformly in the caudal somites at this stage (Fig. 2g,h,i). Rostrally, expression becomes progressively restricted to the dorsal-most region making the future dermomyotome (data not shown). We detected all three *Frizzled* genes in the branchial arches. *Chfz7* is also expressed in the ventricular zone of the neural tube by stage 15.

As the limb buds emerge at stage 17, expression of three *Frizzled* genes is detected relatively uniformly in ectoderm and mesenchyme of the limb primordia (data not shown). However, as the limb buds grow, *Frizzled* genes show distinct expression patterns (Figs. 3 and 4). From stage 20 through stage 23, we observed two distinct domains of *Chfz1* expression: elevated expression on the ventral side of both ectoderm and mesenchyme cells, and intense expression in thin stripes of ectodermal cells

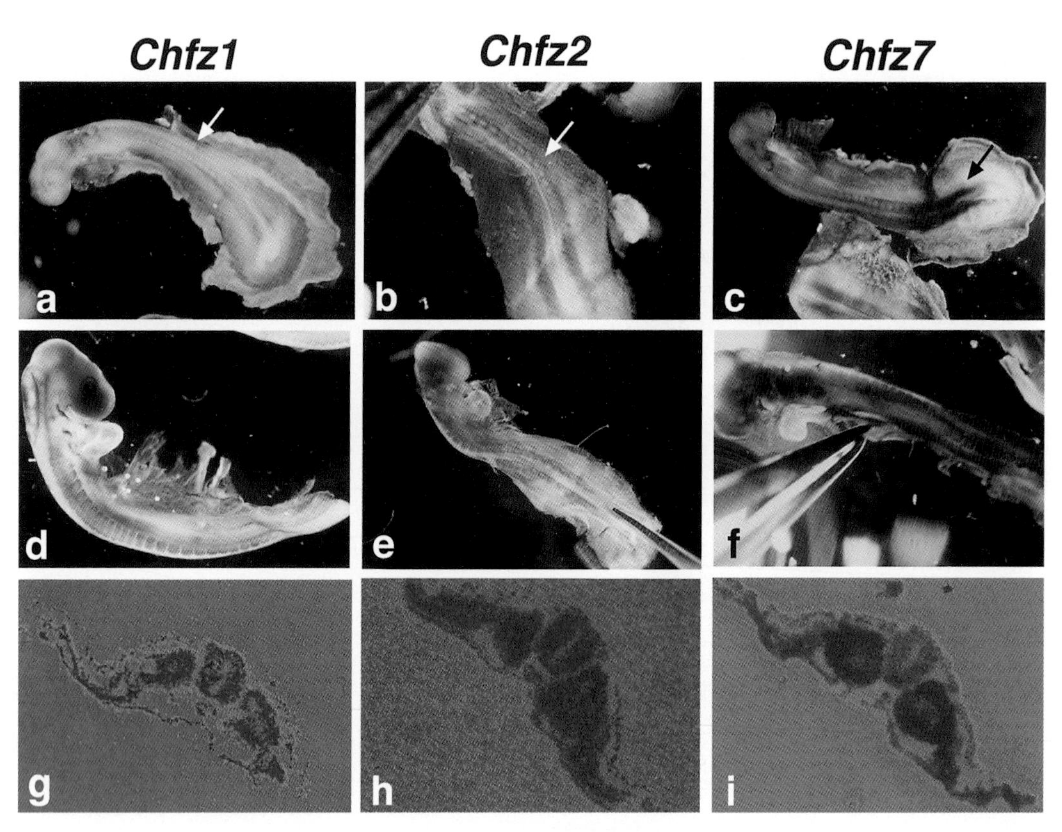

Figure 2. Expression of chick *Frizzled* genes in embryos at somite stages. Embryos were hybridized with probes for *Chfz1* (*a,d,g*), *Chfz2* (*b,e,h*), and *Chfz7* (*c,f,i*). (*a–f*) Whole-mount in situ hybridizations of stage-12 (*a–c*) and stage-15 (*d–f*) embryos. (*g–i*) Transverse sections of stage-15 embryos. *Chfz1* and *Chfz2* are expressed in the more mature somites at the rostral end at stage 12 and are expressed in all the developing somites at stage 15. Arrows in *a* and *b* indicate the caudal-most somites expressing *Chfz1* and *Chfz2*, respectively. *Chfz7* is expressed in all the developing somites and in paraxial mesoderm at stage 12 (arrow in *c*) and stage 15.

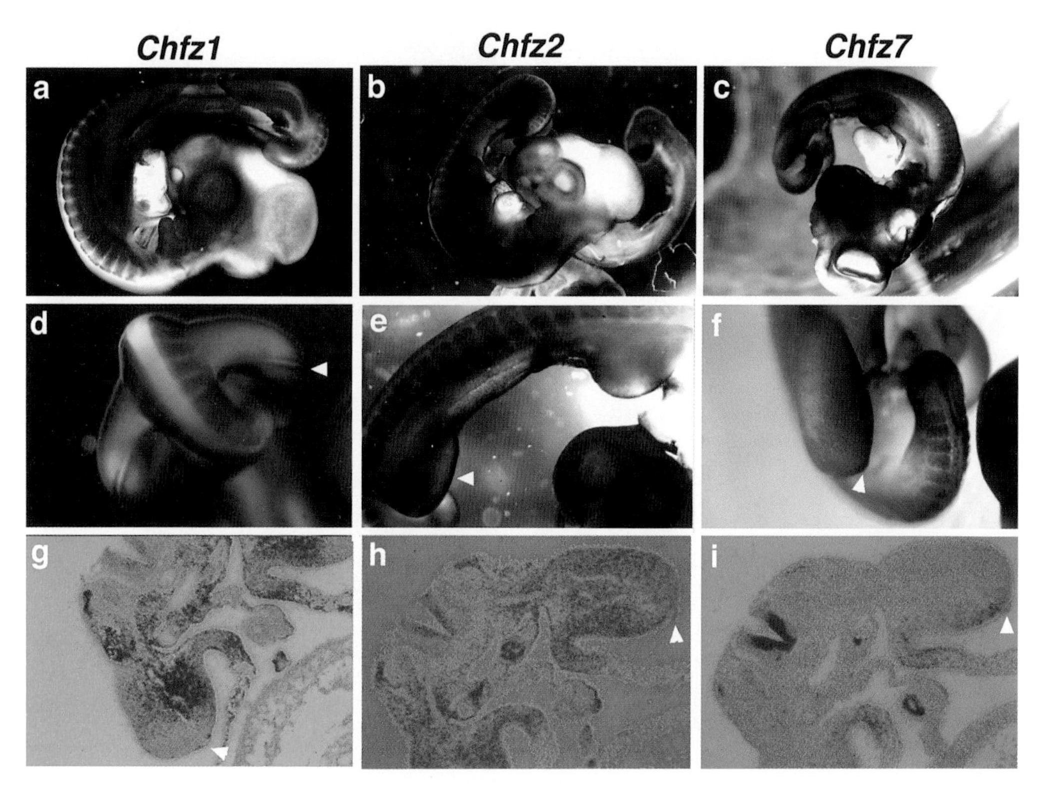

Figure 3. Expression of chick *Frizzled* genes in stage-20 embryos. (*a–c*) Lateral view of whole-mount in situ hybridizations and (*d–f*) higher magnification view of the limb buds. (*g–i*) Transverse sections at forelimb level. *Chfz1* is strongly expressed on the proximoventral side of the limb buds, in ectodermal cells surrounding the AER, and in the dermomyotomes (*a,d,g*). *Chfz2* is expressed fairly uniformly in the limb buds except the AER (*b,e,h*). *Chfz7* is strongly expressed in the progress zone of the limb buds, the dermomyotomes, and the ventricular zone of the neural tube (*c,f,i*). Arrowheads point to the AER.

flanking the AER. Expression of *Chfz1* adjacent to the AER persists until at least stage 29. The ventral mesenchyme cells expressing *Chfz1* are later confined to the condensing cartilaginous rudiments. Although the expression of *Chfz1* in cartilage is still stronger in the ventral half at stage 26, it becomes uniform along the dorsoventral axis in the perichondrium at stage 30. *Chfz1* is also strongly expressed in the dermomyotomes and migrating sclerotomal cells forming vertebrae from stage 20 to 26.

From stage 20 to 26, *Chfz2* is expressed fairly uniformly in the limb mesenchyme and ectoderm except for the AER. Little or no *Chfz2* expression is detected in AER at any stage of development. When the bone rudiments are detectable in limbs at stage 30, *Chfz2* expression appears to cease in the cartilaginous elements. However, a high level of *Chfz2* expression is still seen in mesoderm surrounding the cartilage.

By stage 22, *Chfz7* shows differential expression along the proximodistal axis, with the highest level in the distalmost mesoderm underlying AER. This domain is called progress zone, which undergoes active proliferation in response to signal(s) from AER. High level of expression of *Chfz7* is still observed in the most distal mesoderm of the autopod at stage 30. There is little tissue specificity in *Chfz7* expression, and low level of expression is observed in all the mesoderm derivatives in the limb buds. From stage 20 through 30, *Chfz7* is also strongly expressed in the ventricular zone of the neural tube from the forebrain to the spinal cord. *Chfz7* is also strongly expressed in the dermomyotomes and the tail buds. At these stages, all *Frizzled* genes are expressed in differentiating mesonephric ducts.

Expression of *Wnt* Genes during Chick Limb Development

Frizzled genes show dynamic expression patterns in chick limb bud development. To elucidate their functions, it is important to clarify the ligand-receptor relationships of Wnts and Frizzled. In mouse embryos, at least seven *Wnt* genes are known to be expressed in limb buds. However, it has been shown that some *Wnt* homologs are expressed deferentially in the mouse and the chick nervous systems (Hollyday et al. 1995). Therefore, we next analyzed the expression of several *Wnt* genes in developing chick limb buds.

The detailed expression patterns of *Wnt5a* and *Wnt7a* in the chick limb bud have been reported previously (Dealy et al. 1993). Besides them, we observed expression of at least five other *Wnt* genes in the developing chick limb bud.

The expression of *Wnt11* in the chick is markedly different from that of its mouse ortholog. As previously re-

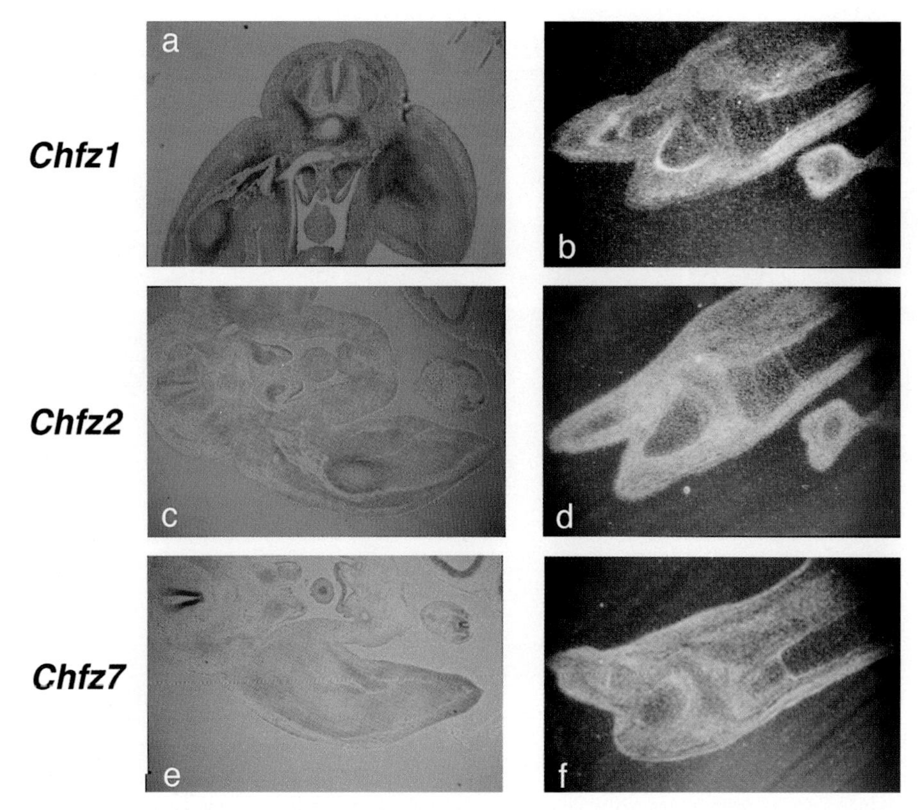

Figure 4. Expression of chick *Frizzled* genes at chondrogenic stages. (*a,c,e*) Transverse sections of stage-26 embryos. (*b,d,e*) Longitudinal sections of stage-30 limb buds. *Chfz1* is expressed in the cartilaginous rudiments at stage 26 which is later confined to the perichondria at stage 30 (*a,b*). *Chfz2* is uniformly expressed in the limb buds at stage 26, and is later excluded in the cartilaginous rudiments at stage 30 (*c,d*). *Chfz7* shows graded expression pattern with the highest level in the distal mesodermal tissues at both stage 26 and stage 30 (*e,f*).

ported (Tanda et al. 1995), *Wnt11* expression in the chick limb buds is first detected in mesenchyme underneath the dorsal ectoderm at stage 23. Later expression is also observed in ventral mesenchyme adjacent to ectoderm. At stage 30, *Wnt11* is expressed in differentiating tendons and in feather buds. Unlike mice, expression is observed neither in the AER nor in the cartilaginous rudiments (data not shown).

More surprisingly, *Wnt3a* is strongly expressed in the limb buds from stage 17 to 26, despite the fact that mouse *Wnt3a* is neither expressed in the limb buds nor required for limb development. *Wnt3a* expression is localized in ectoderm corresponding to overt AER formation. It is first detected in the ectoderm of the presumptive limb field at stage 15 and later confined to the mature AER, where it is maintained until at least stage 26 (Fig. 5).

Three other *Wnt* genes also expressed in the chick AER show patterns similar to those of their mouse homologs: *Wnt10b*, which is specific for the AER; *Wnt6*, which is in both the AER and the rest of the limb ectoderm; and *Wnt5b*, which is in the AER and adjacent mesenchyme (data not shown).

Expression of *Chfz1* and *Chfz7* suggests that they might be involved in AER formation and/or function. To assess the possibility that *Wnt* genes are involved in AER function, we examined the expressions of the chick *Wnt* genes in the limb buds of *eudiplopodia* mutant chick embryos (Fig. 6). In the *eudiplopodia* embryo, functional AERs are ectopically formed in dorsal ectoderm, which give rise to extra digits (Rosenblatt et al. 1959). Five of the *Wnt* genes tested are expressed in the ectopic *eudiplopodia* AERs as well as in the normal AERs. *Wnt3a* and *Wnt10b* are specifically expressed in both the normal and ectopic AERs, suggesting that they may have some roles in AER function. *Wnt5b* is strongly expressed in both normal and ectopic AERs and faintly expressed in mesenchyme adjacent to entire limb ectoderm. *Wnt5a* shows a graded expression pattern along the distal-proximal axis with the strongest point in the normal and the ectopic AERs. *Chfz7* also shows a similar expression pattern as *Wnt5a* (data not shown). *Wnt7a* is expressed throughout dorsal ectoderm but excluded from the ectopic mutant AERs as reported previously (Laufer et al. 1997). *Wnt6* is expressed throughout trunk and limb ectoderm, including the AER in both wild-type and mutant embryos (data not shown).

DISCUSSION

Molecular Identity of *Frizzled* Genes Expressed in Chick Limb Buds

We cloned cDNAs for the chick orthologs of mammalian *Frizzled 1*, *Frizzled 2*, and *Frizzled 7*. Amino acid sequences are well-conserved, especially in presumptive

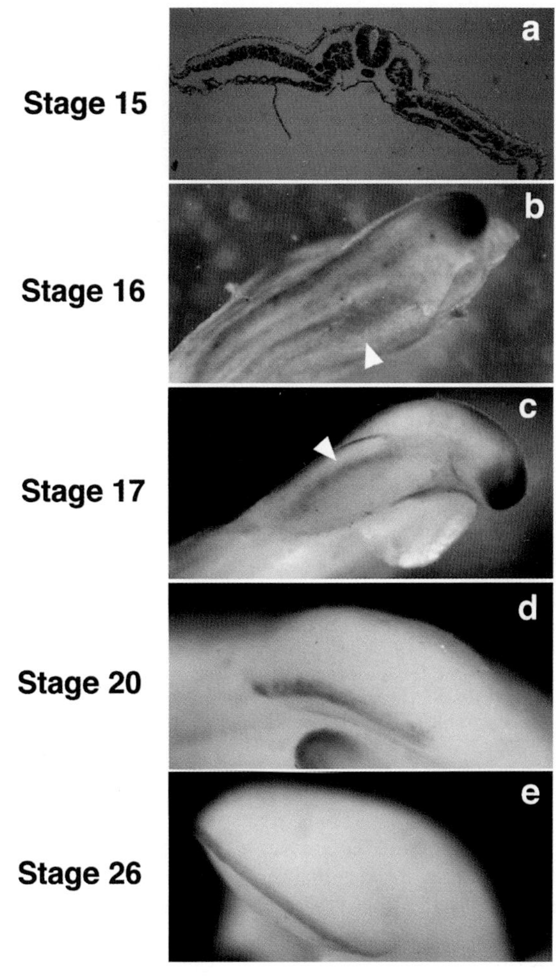

Figure 5. Expression of *Wnt3a* during chick limb bud development. *Wnt3a* expression is detected before AER formation in presumptive limb ectoderm by stage 15 (*a*). As limb outgrowth commences, this expression is maintained and enhanced in the presumptive AER field (*arrowheads*) at stages 16 (*b*) and 17 (*c*). After morphological AER formation, expression is detected in primarily the AER and persists until at least stage 26 (*d,e*).

functional motifs, between chick and mammalian orthologs.

High sequence identity in the CRD domain supports the idea that this domain is important for binding to the ligand. Despite the fact that the extracellular loops between the seven-transmembrane domains are relatively diverged, all of the cysteine residues in these loops are conserved. This might suggest that these extracellular loops are also involved in protein folding to form the ligand-binding site.

The PDZ-binding domain at the carboxyl terminus of each *Frizzled* member is also conserved between respective mouse and chick orthologs. Dishevelled, the earliest-acting intracellular component of Wnt signaling identified so far, contains a PDZ domain. However, a recent biochemical analysis of the specificity of PDZ-binding domains suggested that none of the PDZ-binding domains of the Frizzled family proteins are likely to bind to the PDZ domain of Dishevelled (Sokol 1996; Songyang et al. 1997). The amino acid sequences of the PDZ-binding motifs of the Frizzled family are closely related to that of the Shaker-type K^+ channel. The PDZ-binding motif of the Shaker channel has been shown to bind to a synaptic protein PSD-95, which is thought to localize the Shaker channel at appropriate synaptic sites (Kim et al. 1995, 1997). It has been shown that *Caenorhabditis elegans frizzled, lin-17,* regulates cell polarity during embryogenesis, suggesting that *lin-17* activity is asymmetric in the cell (Sawa et al. 1996). *Drosophila Frizzled* is also implicated in cell polarities of many tissues, such as ommatidia, hairs, and bristles (Gubb 1993). Taken together, it is possible that vertebrate Frizzled proteins bind to a PSD-95-like protein by their PDZ-binding domains and make clusters at specific sites within cells. The function of the PDZ-binding domains of Frizzled is an interesting aspect for future study.

Expression Pattern of Chick *Frizzled* Genes during Development

The temporal and spatial profiles of chick *frizzled* gene expression often prefigure overt patterning or differentiation of the tissues. It is noteworthy that in many cases, the expression domains of *frizzled* genes were observed in embryonic tissues where members of the *Wnt* family are expressed and are known to regulate morphogenesis. Nevertheless, expression of each *frizzled* gene does not coincide with the expression of specific members of the *Wnt* family. This suggests that there may be overlapping specificities in the ligand-receptor relationships of Wnt and Frizzled family proteins.

Somites. One primary region of chick *Frizzled* expression is the developing somites. At stage 12, expression of *Chfz1* and *Chfz2* was detected in the more mature somites at the rostral end of body axis. *Chfz7* expression precedes *Chfz1* and *Chfz2* expression in somitogenesis and is detected in all of the developing somites as well as in paraxial mesoderm caudal to segmenting somites. *Chfz7* is thus possibly required for segmentation and/or early differentiation of somites. As the somites mature and become compartmentalized into sclerotome, myotome, and dermomyotome, expression of these three *Frizzled* genes becomes strong in the dermomyotomes. It has been shown that several *Wnt* family members expressed in the dorsal midline of neural tube (i.e., *Wnt1*, *Wnt3*, *Wnt3a*, and *Wnt4*) have a strong ability to induce differentiation of dermomyotome from undifferentiated somites (Münsterberg et al. 1995). The expression data may indicate that the dorsal somitic cells expressing *Frizzled* genes receive Wnt signals from the adjacent neural tube and are thereby induced to differentiate into dermomyotome. The overlapping expression pattern of three *Frizzled* genes in the dermomyotome suggests a possible functional redundancy of *Frizzled* family members in dermomyotome development.

Chfz1 is also strongly expressed in the sclerotome and demarcates condensing vertebrae at stage 25. High level of expression of *Chfz1* is also observed in perichondrium of long bones in the limb buds at stage 26 and stage 30.

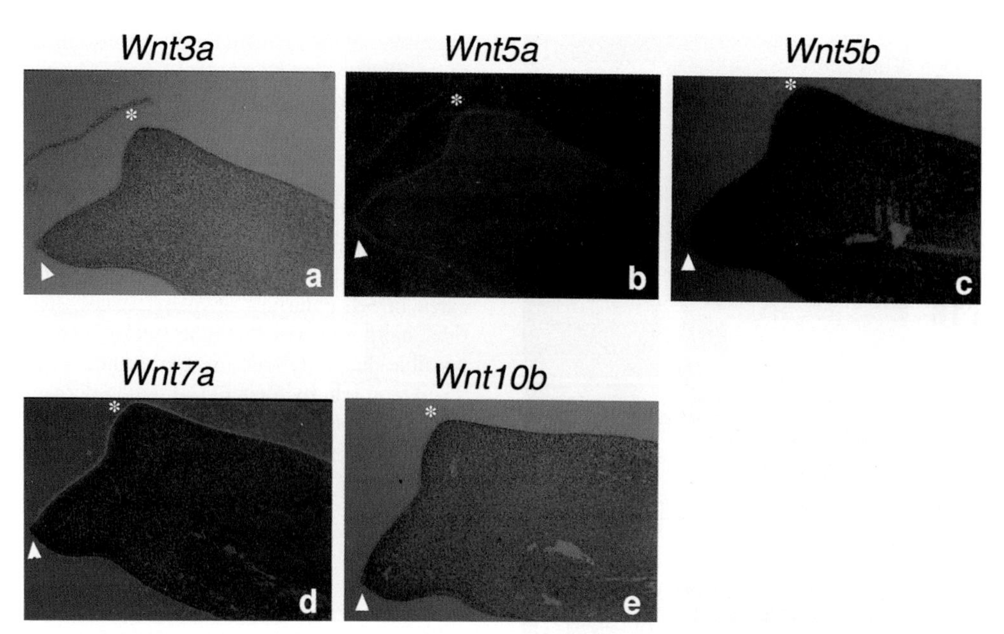

Figure 6. Expression of *Wnt* genes in the *eudiplopodia* mutant limb buds. Adjacent transverse sections at the forelimb level were hybridized with probes for *Wnt3a* (*a*), *Wnt5a* (*b*), *Wnt5b* (*c*), *Wnt7a* (*d*), or *Wnt10b* (*e*). *Wnt3a* and *Wnt10b* are predominantly expressed in normal and mutant AERs. *Wnt5a* is expressed in a graded manner with the strongest level in the normal (*arrowheads*) and the ectopic AERs (*asterisks*). *Wnt5b* is strongly expressed in normal and mutant AERs and is faintly expressed in the surface mesenchymal cells all over the limb bud. *Wnt7a* is expressed in the dorsal ectoderm on either side of the ectopic AER.

These data suggest a possibility that *Chfz1* and Wnt signaling is involved in aspects of endochondral ossification.

Limb buds. Chick *frizzled* genes show overlapping but distinct expression patterns in the developing limb buds. They are first detected in the mesenchyme of the presumptive limb fields and presage overt limb bud formation. As the limb bud grows, *Chfz1* expression in mesenchyme becomes elevated on the ventral side. As mentioned above, the D/V polarity of the limb mesenchyme is regulated by *Wnt7a* expression in the dorsal ectoderm. Preliminary results indicate that the expression of *Chfz1* is down-regulated by misexpression of *Wnt7a* in the ventral ectoderm (data not shown). The D/V gradient in *Chfz1* expression in the limb mesenchyme suggests that *Chfz1* may contribute to patterning along this axis. *Chfz1* is also strongly expressed in ectoderm outlining the AER after its appearance. It may therefore be regulated by factors provided by the AER and may act to maintain the distinct edge of the AER structure. Expression of *Chfz7* is markedly distinct from that of *Chfz1* but also related to the AER. Strong expression of *Chfz7* is observed in the distal mesenchyme called the progress zone. The cells in the progress zone receive signals from the AER, which induce proliferation and lead to the formation of the complete set of the long bones along the proximal distal axis of the limb. Consistent with the idea that *Chfz7* receives a signal from the AER, we observed that some *Wnt* genes are expressed in the AER. *Wnt3a* expression is observed in the presumptive AER field at stage 15 and persists in the AER until at least stage 26.

In *Drosophila*, *Wingless* is expressed in the margin cells of the wing imaginal disc and regulates the proximodistal patterning of the wing. There are remarkable parallels between mechanisms of the formation of the *Drosophila* wing margin and the AER of the vertebrate limb bud. It has been recently demonstrated that the formation of the AER is regulated by the homolog of *Drosophila Fringe,* which is implicated in the determination of the wing margin of the imaginal discs (Laufer et al. 1997; Rodriguez-Esteban et al. 1997). Both the AER and the wing margin are formed at the D/V border between *Fringe*-expressing cells and nonexpressing cells. In the *Drosophila* wing disc, *Fringe* induces Notch signaling at the margin, which in turn regulates the expression of *Wingless* in the margin cells (Panin et al. 1997). Ectopic expression of an activated form of *Notch* produces limb buds with ectopic AERs, which is strikingly similar to the mutant phenotype by *Fringe* misexpression (J.C. Izpisúa-Belmonte, pers. comm.). The conserved *Fringe/Notch* genetic cascade suggests that Wnt/Frizzled signaling may also have a role in the formation and/or function of the AER, analogous to that of Wingless in the *Drosophila* wing.

On the basis of the *Frizzled* and *Wnt* gene expression patterns presented here, the functions and the ligand/receptor relationships are likely to be complex. Nüsse and his colleagues have reported functional analysis of *Drosophila frizzled 2* using a dominant-negative mutant clone. Combinatorial analysis of misexpression of *Wnt* and dominant-negative *Frizzled* mutants will be helpful to investigate the signaling pathways and the functions of *Wnt* and *Frizzled* during chick development.

ACKNOWLEDGMENTS

We thank Andy McMahon and Yingzi Yang for sharing data prior to publication and for providing *Wnt* probes. This work was supported by a grant from the American Cancer Society to C.T. and by a fellowship from Japan Society for the Promotion of Science to M.K.

REFERENCES

Bhanot P., Brink M., Samos C.H., Hsieh J.C., Wang Y., Macke J.P., Andrew D., Nathans J., and Nusse R. 1996. A new member of the *frizzled* family from *Drosophila* functions as a Wingless receptor. *Nature* **382:** 225.

Chan S.D., Karpf D.B., Fowlkes M.E., Hooks M., Bradley M.S., Vuong V., Bambino T., Liu M.Y., Arnaud C.D., Strewler G.J., and Nissenson R.A. 1992. Two homologs of the *Drosophila* polarity gene *frizzled* (*fz*) are widely expressed in mammalian tissues. *J. Biol. Chem.* **267:** 25202.

Dealy C.N., Roth A., Ferrari D., Brown A.M., and Kosher R.A. 1993. *Wnt-5a* and *Wnt-7a* are expressed in the developing chick limb bud in a manner suggesting roles in pattern formation along the proximodistal and dorsoventral axes. *Mech. Dev.* **43:** 175.

Fritzsch B., Nichols D.H., Echelard Y., and McMahon A.P. 1995. Development of midbrain and anterior hindbrain ocular motoneurons in normal and *Wnt-1* knockout mice. *J. Neurobiol.* **27:** 457.

Gavin B.J., McMahon J.A., and McMahon A.P. 1990. Expression of multiple novel *Wnt-1/int-1*-related genes during fetal and adult mouse development. *Genes Dev.* **4:** 2319.

Gonzalez F., Swales L., Bejsovec A., Skaer H., and Martinez Arias A. 1991. Secretion and movement of wingless protein in the epidermis of the *Drosophila* embryo. *Mech. Dev.* **35:** 43.

Greco T.L., Takada S., Newhouse M.M., McMahon J.A., McMahon A.P., and Camper S.A. 1996. Analysis of the vestigial tail mutation demonstrates that *Wnt-3a* gene dosage regulates mouse axial development. *Genes Dev.* **10:** 313.

Gubb D. 1993. Genes controlling cellular polarity in *Drosophila*. *Development* (suppl.), p. 269.

He X., Saint-Jeannet J.P., Wang Y., Nathans J., Dawid I., and Varmus H. 1997. A member of the Frizzled protein family mediating axis induction by *Wnt-5A*. *Science* **275:** 1652.

Hollyday M., McMahon J.A., and McMahon A.P. 1995. *Wnt* expression patterns in chick embryo nervous system. *Mech. Dev.* **52:** 9.

Kim E., Niethammer M., Rothschild A., Jan Y.N., and Sheng M. 1995. Clustering of Shaker-type K^+ channels by interaction with a family of membrane-associated guanylate kinases. *Nature* **378:** 85.

Kim E., Naisbitt S., Hsueh Y.P., Rao A., Rothschild A., Craig A.M., and Sheng M. 1997. GKAP, a novel synaptic protein that interacts with the guanylate kinase- like domain of the PSD-95/SAP90 family of channel clustering molecules. *J. Cell Biol.* **136:** 669.

Kispert A., Vainio S., Shen L., Rowitch D.H., and McMahon A.P. 1996. Proteoglycans are required for maintenance of *Wnt-11* expression in the ureter tips. *Development* **122:** 3627.

Laufer E., Dahn R., Orozco O.E., Yeo C.Y., Pisenti J., Henrique D., Abbott U.K., Fallon J.F., and Tabin C. 1997. Expression of Radical fringe in limb-bud ectoderm regulates apical ectodermal ridge formation (comments). *Nature* **386:** 366.

Marigo V., Scott M.P., Johnson R.L., Goodrich L.V., and Tabin C.J. 1996. Conservation in hedgehog signaling: Induction of a chicken patched homolog by Sonic hedgehog in the developing limb. *Development* **122:** 1225.

McMahon A.P. and Bradley A. 1990. The *Wnt-1* (*int-1*) proto-oncogene is required for development of a large region of the mouse brain. *Cell* **62:** 1073.

McMahon A.P., Joyner A.L., Bradley A., and McMahon J.A. 1992. The midbrain-hindbrain phenotype of *Wnt-1-/Wnt-1-* mice results from stepwise deletion of engrailed-expressing cells by 9.5 days postcoitum. *Cell* **69:** 581.

Moon R., Brown J., and Torres M. 1997. *Wnts* modulate cell fate and behavior during vertebrate development. *Trends Genet.* **13:** 157.

Münsterberg A.E., Kitajewski J., Bumcrot D.A., McMahon A.P., and Lassar A.B. 1995. Combinatorial signaling by Sonic hedgehog and *Wnt* family members induces myogenic bHLH gene expression in the somite. *Genes Dev.* **9:** 2911.

Panin V.M., Papayannopoulos V., Wilson R., and Irvine K.D. 1997. Fringe modulates Notch-ligand interactions. *Nature* **387:** 908.

Parr B.A. and McMahon A.P. 1995. Dorsalizing signal *Wnt-7a* required for normal polarity of D-V and A-P axes of mouse limb. *Nature* **374:** 350.

Parr B.A., Shea M.J., Vassileva G., and McMahon A.P. 1993. Mouse *Wnt* genes exhibit discrete domains of expression in the early embryonic CNS and limb buds. *Development* **119:** 247.

Peifer M. 1995. Cell adhesion and signal transduction: The *Armadillo* connection. *Trends Cell Biol.* **5:** 224.

Riddle R.D., Johnson R.L., Laufer E., and Tabin C. 1993. Sonic hedgehog mediates the polarizing activity of the ZPA. *Cell* **75:** 1401.

Riddle R.D., Ensini M., Nelson C., Tsuchida T., Jessell T.M., and Tabin C. 1995. Induction of the LIM homeobox gene *Lmx1* by WNT7a establishes dorsoventral pattern in the vertebrate limb. *Cell* **83:** 631.

Rodriguez-Esteban C., Schwabe J.W., De La Pena J., Foys B., Eshelman B., and Izpisúa-Belmonte J.C. 1997. Radical fringe positions the apical ectodermal ridge at the dorsoventral boundary of the vertebrate limb (comments). *Nature* **386:** 360.

Rosenblatt L., Kreutziger G., and Taylor L. 1959. Eudiplopodia. *Poult. Sci.* **38:** 1242.

Sawa H., Lobel L., and Horvitz H.R. 1996. The *Caenorhabditis elegans* gene *lin-17*, which is required for certain asymmetric cell divisions, encodes a putative seven-transmembrane protein similar to the *Drosophila* frizzled protein. *Genes Dev.* **10:** 2189.

Sokol S.Y. 1996. Analysis of Dishevelled signalling pathways during *Xenopus* development. *Curr. Biol.* **6:** 1456.

Songyang Z., Fanning A.S., Fu C., Xu J., Marfatia S.M., Chishti A.H., Crompton A., Chan A.C., Anderson J.M., and Cantley L.C. 1997. Recognition of unique carboxyl-terminal motifs by distinct PDZ domains. *Science* **275:** 73.

Stark K., Vainio S., Vassileva G., and McMahon, A.P. 1994. Epithelial transformation of metanephric mesenchyme in the developing kidney regulated by *Wnt*. *Nature* **372:** 679.

Takada S., Stark K.L., Shea M.J., Vassileva G., McMahon J.A., and McMahon A.P. 1994. *Wnt-3a* regulates somite and tailbud formation in the mouse embryo. *Genes Dev.* **8:** 174.

Tanda N., Ohuchi H., Yoshioka H., Noji S., and Nohno T. 1995. A chicken *Wnt* gene, *Wnt-11*, is involved in dermal development. *Biochem. Biophys. Res. Commun.* **211:** 123.

van den Heuvel M., Nusse R., Johnston P., and Lawrence P.A. 1989. Distribution of the *wingless* gene product in *Drosophila* embryos: A protein involved in cell-cell communication. *Cell* **59:** 739.

Vogel A., Rodriguez C., Warnken W., and Izpisúa-Belmonte J.C. 1995. Dorsal cell fate specified by chick *Lmx1* during vertebrate limb development (erratum in *Nature* [1996] **379:** 848). *Nature* **378:** 716.

Wang Y., Macke J.P., Abella B.S., Andreasson K., Worley P., Gilbert D.J., Copeland N.G., Jenkins N.A., and Nathans J. 1996. A large family of putative transmembrane receptors homologous to the product of the *Drosophila* tissue polarity gene frizzled. *J. Biol. Chem.* **271:** 4468.

Zecca M., Basler K., and Struhl G. 1996. Direct and long-range action of a wingless morphogen gradient. *Cell* **87:** 833.

Outgrowth and Patterning of the Vertebrate Limb

J.W.R. Schwabe, C. Rodriguez-Esteban, J. De La Peña, A.T. Tavares, J.K. Ng, E.M. Banayo, B. Foys, B. Eshelman, J. Magallon, R. Tam, and J.C. Izpisúa-Belmonte
Salk Institute, Gene Expression Laboratory, La Jolla, California 92037-1099

Understanding the mechanism by which "patterns" are established, propagated, and elaborated during embryogenesis is one of the major goals of our study of development. Although there are many different systems in which patterning can be studied, arguably one of the most approachable vertebrate model systems is that of the developing chick limb, since these limbs can be readily visualized and manipulated without compromising the viability of the embryo.

Early experiments focused primarily on direct physical manipulations of developing limb tissues. These experiments established the existence of three primary "organizer centers" that appear to have a crucial role in limb patterning. These are the cells of the ectoderm covering the limb bud, the apical ectodermal ridge (AER), a special stratified epithelium at the dorsoventral (D/V) boundary of the limb bud, and a group of cells in the posterior mesoderm termed the polarizing region or ZPA (Fig. 1a,m). In addition to these three organizing centers, the distal mesoderm underlying the AER (termed the progress zone) plays an important regulatory part in controlling limb development (for review, see Tickle and Eichele 1994).

With the availability of an increasing repertoire of molecular techniques, it has become possible to identify and characterize the function of genes that are expressed in these organizer centers and to direct the outgrowth and patterning of the limb. In this paper, we first present the expression domains of genes known to be involved in early patterning of the limb and relate these to the organizer centers that were deduced as a result of the classical manipulation experiments. We discuss the different approaches available to unravel the molecular functions of these diverse genes and present some of the more informative experiments toward this goal. Interestingly, the phenotypes observed as a result of such genetic manipulations often recapitulate the phenotypes that result from tissue transplantations, suggesting that we are starting to uncover the molecular basis of these experiments and indeed the molecular processes that pattern the developing limb.

RESULTS AND DISCUSSION

Spatiotemporal Patterns of Gene Expression

A brief survey of the expression patterns of the genes seen in Figure 1 immediately reveals that although at a histological level, the early limb bud appears as a uniform ball of mesodermal cells surrounded by an ectodermal jacket, it is in fact a highly patterned structure. Indeed, even before it is possible to discern a limb bud, some of the genes described in Figure 1 show restricted domains of expression within the ectodermal and mesodermal tissues of the limb field that is fated to form the limb bud some hours later.

For instance, mRNA transcripts for *Wnt-7a* and *En-1* appear at around stage-14 HH (Hamburger and Hamilton 1951) in the presumptive dorsal and ventral ectoderm, respectively (Fig. 1c,e) (Riddle et al. 1995; Vogel et al. 1995). This first hint of D/V asymmetry is followed by the appearance at stage 15 of *Lmx-1* (Fig. 1d) and *R-fng* (Fig. 1g) transcripts in the presumptive dorsal mesoderm (*Lmx-1*) and ectoderm (*R-fng*) (Laufer et al. 1997; Rodriguez-Esteban et al. 1997). At approximately the same stage, other genes including *Lhx2* and *Msx-1* are expressed with no D/V asymmetry in the incipient limb bud mesoderm (Hill et al. 1989; Robert et al. 1989, 1991; Davidson et al. 1991; Suzuki et al. 1991; Muneoka and Sassoon 1992; C. Rodriguez-Esteban et al., in prep.). As the limb bud becomes discernible (stage 17), the expression of *R-fng* becomes concentrated to the most distal ectoderm (AER). *Lhx2* (Fig. 1f) and *Msx-1* (Fig. 1k) expression is localized to the proliferating mesoderm at the distal end of the bud in the progress zone. At the same time, transcripts of *Fgf-8, Serrate, Notch*, and *Dll* also reach high levels in the forming AER (Fig. 1h,i,j,l) (Dollé et al. 1992; Myat et al. 1996; Vogel et al. 1996; Laufer et al. 1997; C. Rodriguez-Esteban et al., in prep.). Asymmetry of genes expressed along the anteroposterior (A/P) axis (right half of Fig. 1) becomes apparent at stages 16–17 with the appearance of the nested expression pattern of the *Hox* genes (illustrated by *Hoxd* 9,11,13 in Fig. 1r,q,p) (Izpisúa-Belmonte et al. 1991). This is closely followed by the appearance of *Shh, Bmp-2*, and *Ptc* (Fig. 1s,o,t) (Riddle et al. 1993; Francis-West et al. 1994; Laufer et al. 1994; Marigo et al. 1996) in the posterior mesoderm and *Fgf-4* (Fig. 1u) (Laufer et al. 1994; Niswander et al. 1994) in the posterior region of the AER.

Both the spatial and temporal expression patterns of these genes provide some clues as to how limb development proceeds. First, the expression domains of many of these genes overlap, in a general way, with the organizer regions as determined by classical manipulation experiments. This confirms the spatial definition of these orga-

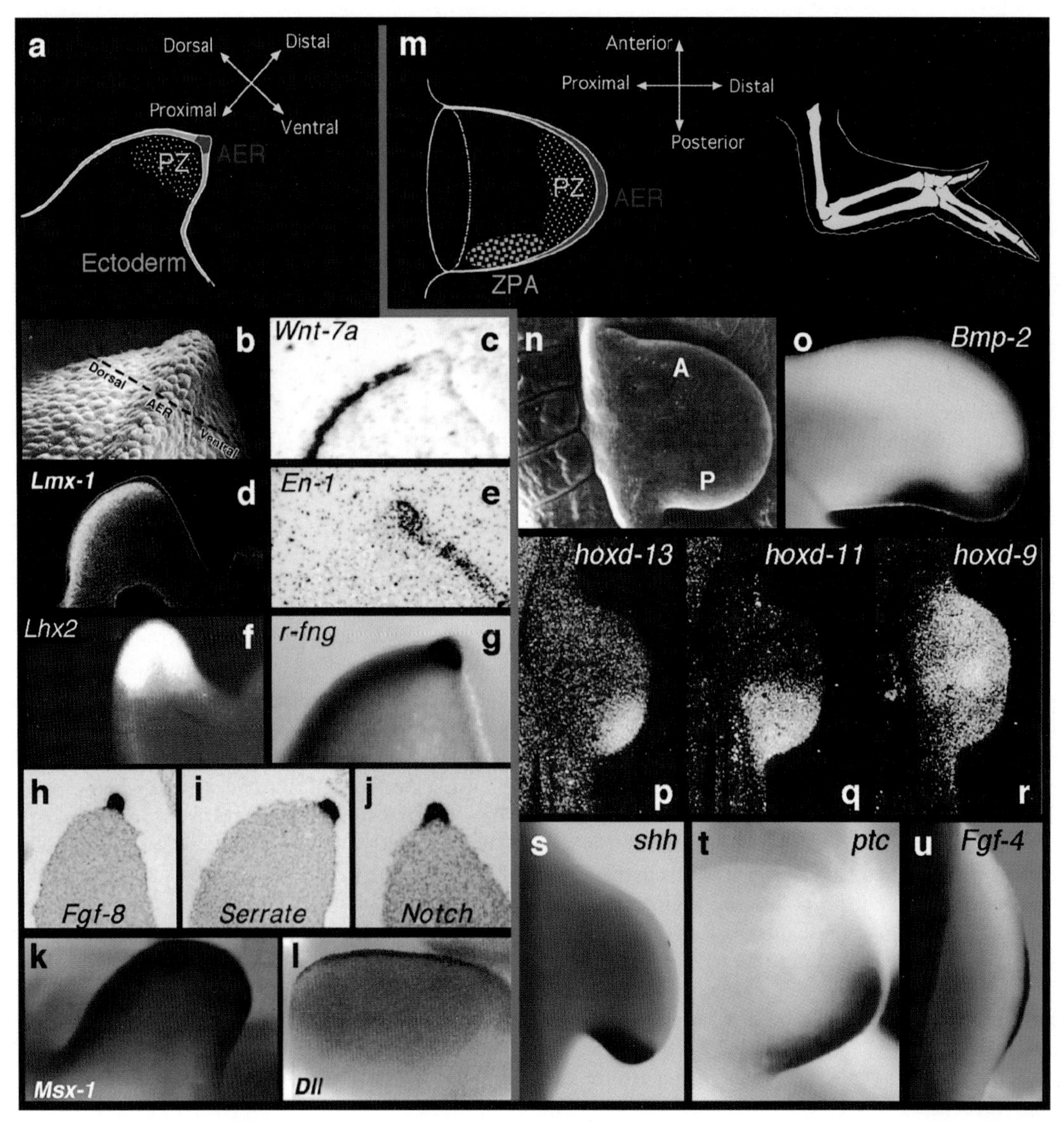

Figure 1. (*a,m*) Schematic representation showing the organizer centers (ectoderm, apical ectodermal ridge-AER, progress zone-PZ, and zone of polarizing activity-ZPA) and axes of the chick limb bud. (*b,n*) Scanning electron micrographs showing D/V and A/P views of the developing limb bud at stage 22. (*c–l, o–u*) In situ hybridizations (either whole mount or sectioned chicks) illustrating the expression domains of various genes involved in the outgrowth and patterning of the vertebrate limb.

nizer centers and suggests that some of these genes (which range from proteins of unknown function to transcription factors, enzymes, surface signals, and their receptors) may have a role in establishing these organizer centers and/or directing the fate of these or neighboring cells during the embryogenesis of the limb. Through looking in detail at the expression domains of these genes, we can define more precisely the spatial and temporal properties of these organizer centers. For instance, careful analysis of the location of *Fgf-8* transcripts suggests that the AER is not the uniform tissue that it appears, since there is a single line of cells at the boundary between the dorsal and ventral halves of the AER in which *Fgf-8* transcripts cannot be detected (our unpublished results). This raises the question of whether there may be some functional difference between the ventral and dorsal cells of the AER. This is supported by the finding that the selector gene, *En-1*, is expressed only in the ventral half of the AER. In addition to this careful spatial analysis of expression boundaries, examination of the time at which genes are expressed is also very informative and, as we will discuss below, must be taken into account when interpreting functional experiments. Even though analyses of spatiotemporal expression patterns can be very useful, it is evident that a deeper understanding of limb development necessitates the use of functional and ultimately biochemical experiments. Perhaps the most straightforward functional approach has been to extend the analysis of expression patterns to naturally occurring mutant organisms in which limb development has been impaired. Such studies have provided critical data concerning the ways in which limb development can be perturbed. However, there are as yet too few limb mutants in vertebrates for this to take us very far toward identifying genetic hierarchies. Consequently, there has been considerable impetus to devise other approaches that allow specific up- and down-regulation of genes in specific cells at specific times.

Approaches to Determining Gene Function

Gene knockouts. One of the most generally accepted approaches to studying the biological role of genes is to create mice in which the gene of interest has been ablated. The phenotypes of mice lacking *Wnt-7a, En-1, Hox*, and *Shh* genes have been particularly interesting to investigators studying limb development. Of these, perhaps the knockouts of *Wnt-7a* and *En-1* are the simplest to understand. Mice lacking *Wnt-7a* exhibit biventral limb phenotypes (Parr and McMahon 1995) and those lacking *En-1* show bidorsal phenotypes (Loomis et al. 1996). It is striking that these phenotypes are consistent with the normal expression of these genes in the dorsal and ventral limb ectoderm, respectively. The phenotype of mice lacking *Shh* could not easily have been predicted on the basis of its expression pattern (the posterior limb mesoderm) or from tissue deletion experiments. These mice exhibit arrested limb outgrowth, apparently as a consequence of a degenerate AER (Chiang et al. 1996). Although the resulting limbs are severely truncated, a proximal bone is normally present and exhibits seemingly normal A/P patterning, suggesting that the role of *Shh* is related to the growth and proliferation of the limb.

The phenotypes of mice lacking individual *Hox* genes have been more difficult to interpret. The overall view obtained from studying many different knockouts, as well as the phenotypes of mice lacking more than one *Hox* gene, suggests that these genes are used in a combinatorial fashion to direct local patterning within the developing limb (see van der Hoeven 1996 and references therein).

Gene misexpression. While knockout technology has in general been applied to ablating gene function in mice, complementary techniques have been applied to study limb development in chicks. Through using retroviral technology, genes can be overexpressed either within their normal domain of expression or ectopically in other areas of the limb (see, e.g., Rodriguez-Esteban et al. 1997). Indeed, for some genes (i.e., those that are normally secreted), it can even be sufficient to simply apply purified protein (see, e.g., Cohn et al. 1995). More recently, retroviral technology has been extended to include dominant-negative experiments in which transcription factor function is specifically down-regulated by generating chimeric genes with the repressor domain from the *Drosophila eng* gene (C. Rodriguez-Esteban et al., in prep.).

Figure 2 illustrates some of the phenotypes that have been seen in these kinds of experiments. *Lmx-1* is a gene normally expressed in the dorsal mesoderm. When this gene is ectopically expressed on the ventral side of the limb, it induces the formation of dorsal structures (Fig.

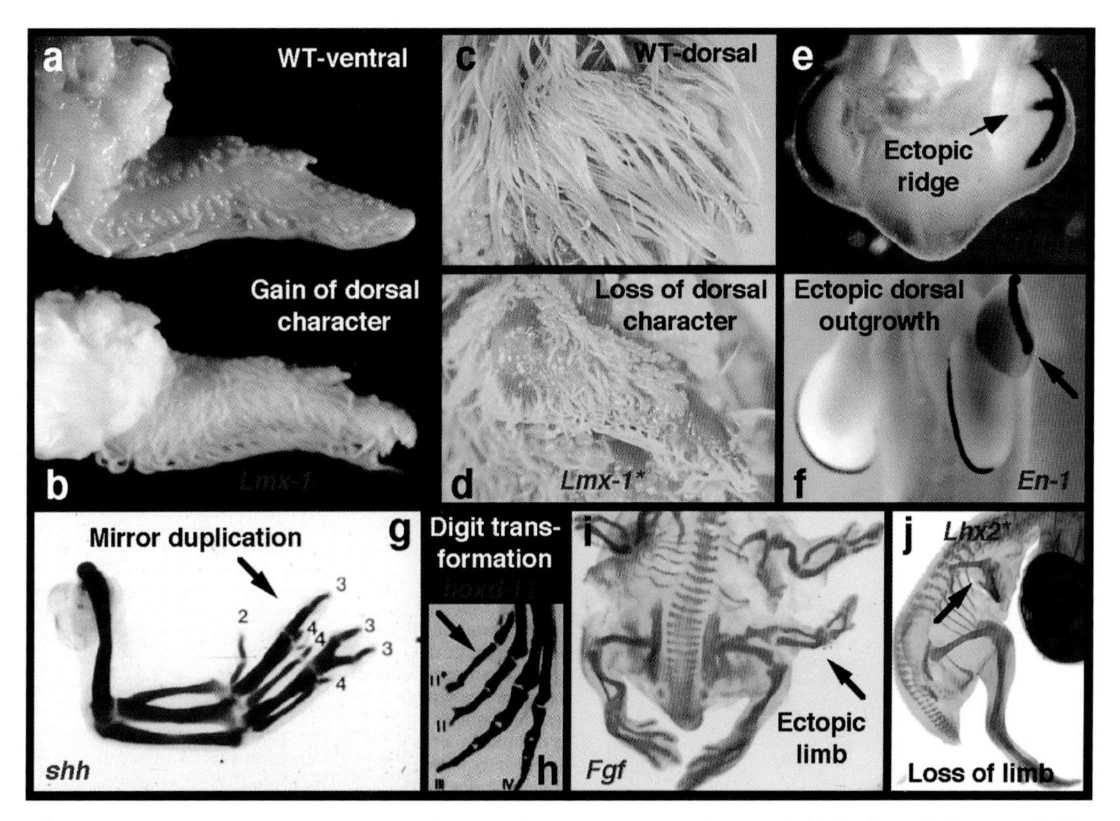

Figure 2. A summary of the phenotypes obtained after ectopic expression (either by retroviral infection or by beads soaked in protein) of various genes involved in the outgrowth and patterning of the vertebrate limb.

2a,b) (Riddle et al. 1995; Vogel et al. 1995). This phenotype is complementary to the loss of dorsal character seen when *Lmx-1* activity is down-regulated on the dorsal side through the expression of dominant-negative chimeras with the *eng* repressor (Fig. 2c,d) (C. Rodriguez-Esteban et al., in prep.). The *Wnt-7a* knockout and misexpression studies of *Lmx-1* suggest that these genes are primarily involved in dorsal specification.

Misexpression of the ventral gene, *En-1*, on the dorsal side of the limb induces two phenotypes: (1) down-regulation of *Wnt-7a* and *Lmx-1* expression and (2) disturbed AER formation that in some cases results in arrested limb outgrowth and in others leads to the formation of additional AERs on the dorsal side of the limb bud (Fig. 2f) (Laufer et al. 1997; Rodriguez-Esteban et al. 1997). The first phenotype is consistent with the observation that the knockout results in bidorsal limbs. The second suggests a role for *En-1* in proximodistal outgrowth and seems to be a result of *En-1* repressing *R-fng* expression on the dorsal side. In complementary experiments when *R-fng* is misexpressed on the ventral side of the limb bud, ectopic AERs are formed in many embryos, but unlike the ectopic dorsal AERs seen when *En-1* is misexpressed, the *R-fng*-induced AERs are located on the ventral side of the limb bud (Fig. 2e). In a number of embryos, misexpression of *R-fng* resulted in embryos lacking limbs. Altogether, this seems to be a result of excess *R-fng* over the whole of the limb bud. This phenotype resembles the striking loss of limbs seen when *Lhx2* activity (a gene shown to induce *R-fng* expression) is down-regulated through misexpression of an "*eng* repression domain/*Lhx2*" chimera (Fig. 2j) (C. Rodriguez-Esteban et al., in prep.).

When genes that are asymmetrically distributed along the A/P axis are misexpressed, the observed phenotypic variations are largely restricted to the distal end of the limb. In the case of *Shh*, misexpression in the anterior of the limb bud results in mirror image duplications of the distal limb elements (Fig. 2g). This seems to be a result of *Shh* inducing the *Hox* genes, among others, at the anterior margin of the limb bud (Riddle et al. 1993). It remains to be proven, however, that this is the role of *Shh* during normal limb development since the limb buds of mice lacking *Shh* appear to exhibit normal A/P patterning.

Misexpression of *Hoxd-11* (Fig. 2h) results in the transformation of the most anterior digit into a more posterior structure (digit I to II) (Morgan et al. 1992). The misexpression of other *Hox* genes also results in perturbed bone formation (Nelson et al. 1996).

Perhaps one of the most dramatic results of the misexpression technology can be seen in experiments in which beads soaked in Fgf protein are implanted in the flank of the embryo at stage 15–16. In these embryos, the Fgf protein appears to initiate de novo the limb developmental program resulting in an entire additional limb (Fig. 2i) (Cohn et al. 1995; Crossley et al. 1996; Vogel et al. 1996).

Toward Molecular Mechanisms

It is clear that studies of vertebrate limb development have gone from the simple morphological description of developing limbs to the physical manipulation of limb outgrowth through tissue deletion and transplantation experiments. Now we are beginning to discern a molecular description of limb development through the analysis of patterns of gene expression and, furthermore, to perform molecular manipulation of the developing limb by what might be termed molecular deletion and transplantation experiments. So our position now is that we have identified and partially characterized around two dozen genes that seem to be involved in establishing pattern along the three axes of the developing limb. Although it is evident that there are genetic hierarchies among these genes, it is striking that as yet very few, if any, of these genes have been shown to interact directly at a biochemical level with any other. This strongly suggests that there are many

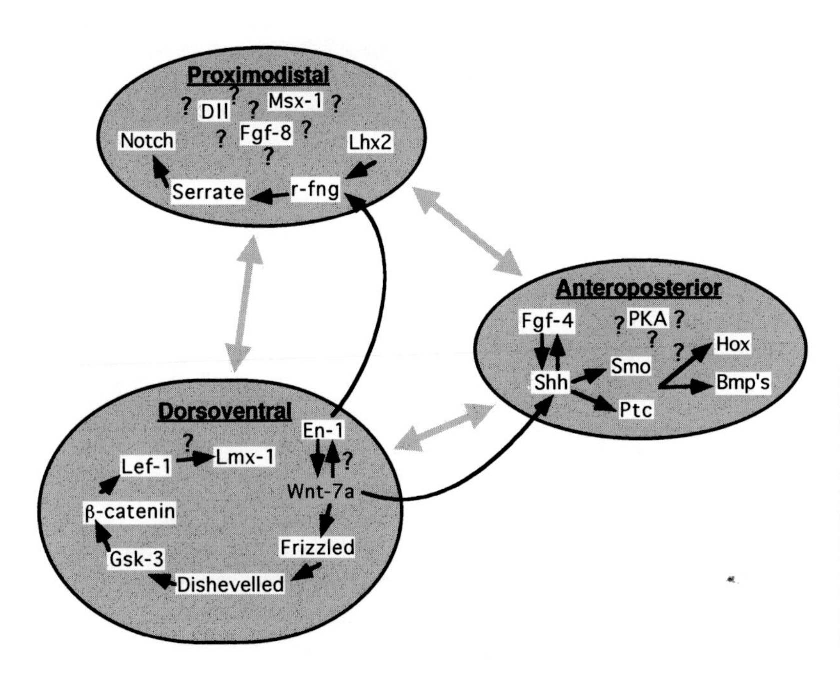

Figure 3. A schematic view of our current understanding of the relationships between genes and axes that pattern the vertebrate limb. Black arrows indicate established hierarchies among the genes illustrated. The *Wnt-7a* to *Lef-1* pathway has yet to be demonstrated in the limb. Question marks illustrate uncertainty. Note that there is communication between the genes of the three axes, being this the reason for the gray double-headed arrows.

more genes to be identified. No doubt, quite a few of these will be (or have already been) identified in other species or in different parts of the embryo, and more will follow from genetic, molecular, and biochemical studies. Figure 3 illustrates in a very schematic fashion our current understanding of the relationships between genes and axes that pattern the vertebrate limb. (Note, however, that the *Wnt-7a* to *Lef-1* pathway has yet to be demonstrated in the vertebrate limb.) The task ahead will be not only to identify new genes, but also to understand at a biochemical level how they operate and interact together. No doubt, there will be many surprises as we unravel all the molecular pathways that are used to build a limb. Right now, however, it appears that a major goal will be to explain how the molecules that establish and pattern the three different axes communicate and why it is that perturbations of gene activity in any one of these axes frequently disturbs the correct patterning of the other axes. As the details emerge, it is likely that we will be able not only to apply our understanding to other developmental processes, but also to gain a deeper understanding of the evolutionary relationships between the widely diverse limbs that can be seen across the different existing phyla.

ACKNOWLEDGMENTS

We thank Lorraine Hooks for her careful assembling of the manuscript and Tina De La Peña for expert technical assistance. This work was supported by grants from the Human Frontier Science Program Organization, National Institutes of Health, National Science Foundation, and the G. Harold and Leila Y. Mathers Charitable Foundation.

REFERENCES

Chiang C., Litingtung Y., Lee E., Young K.E., Corden J.L., Westphal H., and Beachy P.A. 1996. Cyclopia and defective axial patterning in mice lacking *Sonic hedgehog* gene function. *Nature* **383:** 407.

Cohn M., Izpisúa-Belmonte J.C., Abud H., Heath J.K., and Tickle C. 1995. FGF-2 application can induce additional limb formation from the flank of chick embryos. *Cell* **80:** 739.

Crossley P.H., Minowada G., MacArthur C.A., and Martin G.R. 1996. Roles for FGF-8 in the induction, initiation and maintenance of chick limb development. *Cell* **84:** 127.

Davidson D.R., Crawley A., Hill R.E., and Tickle C. 1991. Position-dependent expression of two related homeobox genes in developing vertebrate limbs. *Nature* **352:** 429.

Dollé P., Price M., and Duboule D. 1992. Expression of the mouse *Dlx-1* homeobox gene during facial, ocular and limb development. *Differentiation* **49:** 93.

Francis-West P.H., Richardson M.K., Brickell P.M. and Tickle C. 1994. Bone morphogenetic proteins and a signalling pathway that controls patterning in the developing chick limb. *Development* **120:** 209.

Hamburger V. and Hamilton H. 1951. A series of normal stages in the development of the chick embryo. *J. Morphol.* **88:** 49.

Hill R.E., Jones P.F., Rees A.R., Sime C.M., Justice M.J., Copeland N.G., Jenkins N.A., Graham E., and Davidson D.R. 1989. A new family of mouse homeobox-containing genes: Molecular structure, chromosomal location, and developmental expression of *Hox 7.1. Genes Dev.* **3:** 26.

Izpisúa-Belmonte J.C., Tickle C., Dollé P., Wolpert L., and Duboule D. 1991. Expression of the homeobox *Hox-4* genes and the specification of position in chick wing development. *Nature* **350:** 585.

Laufer E., Nelson C.E., Johnson R.L., Morgan B.A., and Tabin C. 1994. *Sonic hedgehog* and *Fgf-4* act through a signaling cascade and feedback loop to integrate growth and patterning of the developing limb bud. *Cell* **79:** 993.

Laufer E., Dahn R., Orozco O.E., Yeo C.-Y., Pisenti J., Abbott U.K., Fallon J.F., and Tabin C. 1997. The *Radical fringe* expression boundary in the limb bud ectoderm regulates AER formation. *Nature* **386:** 366.

Loomis C.A., Harris E., Michaud J., Wurst W., Hanks M., and Joyner A.L. 1996. The *Engrailed-1* gene and ventral limb patterning. *Nature* **382:** 360.

Marigo V., Scott M.P., Johnson R.L., Goodrich L.V., and Tabin C.J. 1996. Conservation in *Hedgehog* signaling: Induction of a chicken patched homologue by *Sonic hedgehog* in the developing limb. *Development* **122:** 1225.

Morgan B.A., Izpisúa-Belmonte J.C., Duboule D., and Tabin C.J. 1992. Targeted misexpression of *Hox-4.6* in the avian limb bud causes apparent homeotic transformations. *Nature* **358:** 236.

Muneoka K. and Sassoon D. 1992. Molecular aspects of regeneration in developing vertebrate limbs. *Dev. Biol.* **152:** 37.

Myat A., Henrique D., Ish-Horowicz D., and Lewis J. 1996. A chick homologue of *serrate* and its relationship with *notch* and *delta* homologues during central neurogenesis. *Dev. Biol.* **174:** 233.

Nelson C.E., Morgan B.A., Burke A.C., Laufer E., DiMambro E., Murtaugh L.C., Gonzales E., Tessarollo L., Parada L.F., and Tabin C. 1996. Analysis of *Hox* gene expression in the chick limb bud. *Development* **122:** 1449.

Niswander L., Jeffrey S., Martin G.R., and Tickle C. 1994. A positive feedback loop coordinates growth and patterning in the vertebrate limb. *Nature* **371:** 609.

Parr B.A. and McMahon A.P. 1995. Dorsalizing signal *Wnt-7a* required for normal polarity of D–V and A–P axes of mouse limb. *Nature* **374:** 350.

Riddle R.D., Johnson R.L., Laufer E., and Tabin C. 1993. *Sonic hedgehog* mediates the polarizing activity of the ZPA. *Cell* **75:** 1401.

Riddle R., Ensini M., Nelson C., Tsuchida T., Jessell T., and Tabin C. 1995. Induction of the LIM homeobox gene *Lmx-1* by *Wnt-7a* establishes dorsoventral pattern in the vertebrate limb. *Cell* **83:** 631.

Robert B., Lyons G., Simandl B.-K., Kuroiwa A., and Buckingham M. 1991. The apical ectodermal ridge regulates *Hox-7* and *Hox-8* gene expression in developing limb buds. *Genes Dev.* **5:** 2363.

Robert B., Sassoon D., Jacq B., Gehring W., and Buckingham M. 1989. Expression of a novel *Hox* gene, *Hox-7*, during mouse embryogenesis, is associated with morphogenetic phenomena. *EMBO J.* **8:** 91.

Rodriguez-Esteban C., Schwabe J.W., De La Peña J., Foys B., Eshelman B., and Izpisúa-Belmonte J.C. 1997. *Radical fringe* positions the apical ectodermal ridge at the dorsoventral boundary of the vertebrate limb. *Nature* **386:** 360.

Suzuki H.R., Pandanilam B.J., Vitale E., Ramirez F., and Solursh M. 1991. Repeating developmental expression of G-*Hox 7*, a novel homeobox-containing gene in the chicken. *Dev. Biol.* **148:** 375.

Tickle C. and Eichele G. 1994. Vertebrate limb development. *Annu. Rev. Cell Biol.* **10:** 121.

van der Hoeven F. 1996. Gene transposition in the *HoxD* complex reveal a hierarchy of regulatory controls. *Cell* **85:** 1025.

Vogel A., Rodriguez C., and Izpisúa-Belmonte J.C. 1996. Involvement of FGF-8 in initiation, outgrowth and patterning of the vertebrate limb. *Development* **122:** 1737.

Vogel A., Rodriguez C., Warnken W., and Izpisúa-Belmonte J.C. 1995. Dorsal cell fate specified by chick *Lmx1* during vertebrate limb development. *Nature* **378:** 716.

Retrovirus-mediated Insertional Mutagenesis in Zebrafish and Identification of a Molecular Marker for Embryonic Germ Cells

A. AMSTERDAM, C. YOON, M. ALLENDE, T. BECKER, K. KAWAKAMI, S. BURGESS, N. GAIANO, AND N. HOPKINS
Center for Cancer Research, Biology Department, Massachusetts Institute of Technology, Cambridge, Massachusetts 02139

We are interested in identifying genetic pathways for developmental processes and simple behaviors in the zebrafish. A particularly powerful approach to the identification of genes that specify developmental processes has been the application of large-scale Mendelian genetics followed by molecular cloning of mutant genes. Although this approach has long been considered impractical in vertebrate animals, some years ago George Streisinger challenged this notion by proposing that the zebrafish might be a suitable vertebrate for the "forward genetic" approach (Streisinger et al. 1981). Zebrafish possess the two traits required for the successful application of large-scale Mendelian genetics to the study of early development: First, zebrafish can be bred, raised, and maintained in large numbers in the laboratory, and second, fish embryos are numerous, transparent, and develop outside the mother so that developmental defects can readily be seen. It was shown some years ago that important developmental mutations could be identified in the zebrafish, and this year several labs reported the results of large-scale mutagenesis screens aimed at identifying all the embryonic lethal and visible phenotypes of the fish (Grunwald et al. 1985; Kimmel et al. 1989; Mullins et al. 1994; Driever et al. 1996; Haffter et al. 1996; Wiley 1996). These screens employed ethylnitrosourea (ENU) to induce mutations (Grunwald and Streisinger 1992; Mullins et al. 1994).

The large-scale chemical mutagenesis screens in fish reveal that there are roughly 2400 genes which when mutated are essential or produce a visible embryonic phenotype, with most being embryonic lethals (Haffter et al. 1996). The screens did not achieve saturation; rather, it is estimated that mutations in approximately half the embryonic lethal or visible genes have been seen to date. The mutants display a remarkable range of phenotypes, with different mutations affecting the development of virtually every embryonic organ system, as well as simple behaviors of the embryo, including motility (Westerfield et al. 1990; Wiley 1996). Small-scale chemical mutagenesis screens have shown that mutations affecting embryonic vision and smell can also be identified (Clark 1981; Brockerhoff et al. 1995; K. Whitlock and M. Westerfield, pers. comm.).

Although chemical mutagenesis screens in zebrafish are invaluable for displaying the full range of mutant phenotypes that can be obtained for particular developmental pathways, they suffer from the fact that it is difficult to clone the mutated genes. Currently this must be done by laborious and costly positional cloning or by the candidate gene approach. Chemically induced mutants are usually single-base changes, making positional cloning time-consuming even in invertebrate organisms such as flies or worms. The problem is more acute in the fish, whose genome is estimated to be about half to two-thirds the size of the mouse genome. In some organisms, a powerful complement to chemical mutagenesis screens has been insertional mutagenesis or gene tagging. Recognizing the special benefit such a method would have in the zebrafish, we set out to develop insertional mutagenesis for this organism. This is a significant technical challenge because the number of insertions needed to induce mutations in a substantial fraction of the genes is expected to be large. If the fish genome is approximately 1.5×10^9 bp, and if one were to insert exogenous fragments of DNA into the genome at 10-kb intervals (which might be sufficient to mutate every gene), one would need 150,000 insertions to hit every gene once. Because in practice insertions would not be evenly spaced, several times this number of insertions would be needed to achieve saturation.

We have developed a method that will allow us to generate several hundred thousand insertions of exogenous DNA in the fish germ line with about 6 months of work (Gaiano et al. 1996a). We infected blastula-stage embryos with a retroviral vector (Lin et al. 1994; Gaiano et al. 1996a). Furthermore, we carried out a small pilot screen and determined that proviral insertions are mutagenic (Allende et al. 1996; Gaiano et al. 1996b). Mutant genes proved even easier to clone than we had anticipated. In this paper, we review the results of our pilot insertional mutagenesis screen and discuss plans to perform a large-scale insertional mutagenesis screen.

Given our interest in manipulating the fish germ line, we have long been interested in identifying primordial germ cells of zebrafish. In some fish species, primordial germ cells have been identified as early as somite stages by using morphological criteria and working backward from stages where germ cells can be identified with certainty within the embryonic gonad. To locate and track primordial germ cells in zebrafish, we sought a molecular marker for these cells. We cloned the probable zebrafish homolog of the *Drosophila vasa* gene, which is a germ-line marker in flies, and used this sequence for whole mount in situ hybridization in fish embryos. Our results indicate that the zebrafish *vasa* homolog is a marker for primordial germ cells and suggest that germ plasm is set aside as early as the four-cell stage by a novel mechanism involving RNA localization to specific regions of the cleavage membranes (Yoon et al. 1997). Besides their intrinsic biological interest, these results suggest new ways in which one might be able to purify or enrich for the primordial germ cells of the zebrafish.

RESULTS

Retroviral Vectors as Insertional Mutagens in Zebrafish

Mouse retroviral vectors can be used to generate hundreds of thousands of insertions in the zebrafish germ line. Mouse retroviruses have restricted host ranges, which in many cases are determined by their envelope glycoproteins. These proteins must interact specifically with receptors on the cell surface to allow the virus to productively infect the cell. The host range of mouse retroviruses can be broadened by co-infecting cells with a mouse retrovirus plus vesicular stomatitis virus (VSV), a lytic virus that belongs to the Rhabdovirus family and has an extremely broad host range (Zavada 1972). The expanded host range of mouse retroviruses obtained by co-infection with VSV results from the incorporation into retrovirus particles of a single VSV-coded gene product, the G envelope glycoprotein (Rose and Bergmann 1982; Emi et al. 1991). A mouse retrovirus with a VSV-coded envelope glycoprotein in place of its own glycoprotein is called a pseudotype. Although the potential usefulness of VSV pseudotype viruses was recognized immediately, it took many years of effort before these viruses could be grown free of contaminating VSV and could be grown to high titers. Effort in achieving high titers of pure pseudotype viruses was driven by the potential usefulness of these viruses for human gene therapy and was ultimately accomplished by Friedmann's lab at the University of California at San Diego (Burns et al. 1993; Hopkins 1993). As expected, high-titer VSV pseudotype viruses can infect cells of virtually all species that have been tested, and they are currently favored retrovirus vectors for use in human gene therapy.

A critical question in using VSV pseudotypes as mutagens for zebrafish was whether the virus could be delivered to the fish germ line. Almost nothing was known about the location of the zebrafish primordial germ cells when we began these experiments. Because early fish development is so rapid, proceeding from a fertilized egg to a moving embryo within 24 hours, whereas mouse retroviral infections were thought to require at least 5–6 hours before proviral DNA is synthesized and integrated into the host genome, it was uncertain whether the virus would synthesize and integrate provirus in time to productively infect a significant fraction of the cells destined to become the germ line. We found that virus injected into blastula stage embryos can infect the fish germ line (Lin et al. 1994). Although this unpredictable result was a critically important step in developing a methodology, in initial experiments only 15% of injected eggs grew to be founder fish that transmitted a retrovirus provirus to their F_1. Furthermore, most founders transmitted only a single insertion to only about 2–4% of their F_1 progeny. At this frequency it would not be feasible to generate large numbers of proviral insertions in the fish germ line.

We prepared higher-titer stocks of VSV-pseudotyped mouse retroviral vectors than those used in our initial experiments (Gaiano et al. 1996a). Using our most recent stocks of virus, the percent of transgenic founder fish has risen from 15% to 100% (Fig. 1). Furthermore, each founder fish (of 200 tested) now transmits an average of 12 insertions to its F_1 progeny, with each insertion being inherited in a few percent of the F_1 fish. Most transgenic F_1 fish inherit a single provirus, but some inherit 2–5 insertions (see Fig. 1). We attribute the improvement in transgenic frequency to the higher titer of virus used in the later experiments.

At the frequency shown in Figure 1, and given the large number of eggs that can be injected in a day and survive to sexual maturity, we estimate that our lab could generate founder fish harboring about 250,000 insertions in about 6 months. This requires an efficient fish facility and a team of 3–4 people who perform injections 5 days a week.

Proviral insertions are mutagenic. Once we had established that we could generate several hundred thousand insertions in the fish germ line, we needed to determine whether the proviral insertions were mutagenic and, if so, at what frequency. This question is related to the question of whether the sites of integration of the provirus are random. Many studies have addressed the question of whether murine retroviral provirus integration sites are random. Although the answer remains somewhat controversial, the results indicate that mouse retroviral proviruses integrate into an extremely large number of sites (Chen 1994; Chowdhury et al. 1997). Some experiments suggest that integration is essentially random in most mouse cell types, whereas others suggest that mouse proviruses are preferentially integrated into active genes, possibly with a preference for the 5′ ends of the genes.

If the size of the fish genome and the number of embryonic lethal genes were known with certainty, and if proviral insertions were random, one should be able to estimate the mutagenic frequency of insertions making an assumption only about the target size of an average gene. For example, assuming a target size of 10,000 bp/gene, if there are 2400 embryonic lethals and the genome is 1.5×10^9 bp, then if insertions were evenly spaced, one would expect a mutagenic frequency of one insertional mutation per 60 insertions. In practice, however, both the numbers and assumptions are sufficiently inaccurate that such calculations are unreliable. In mice, retroviral integrations are mutagenic and the frequency has been estimated to be

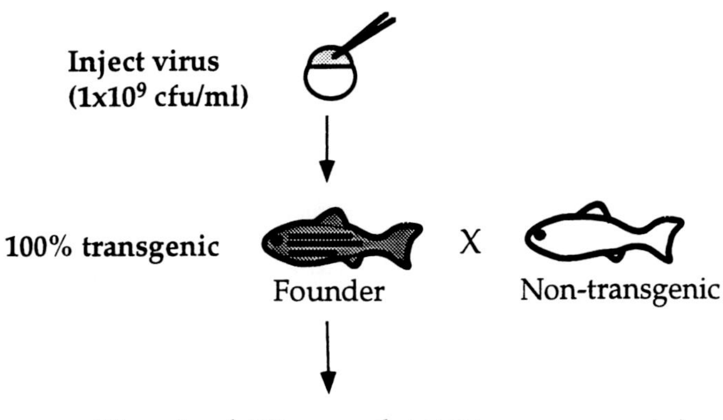

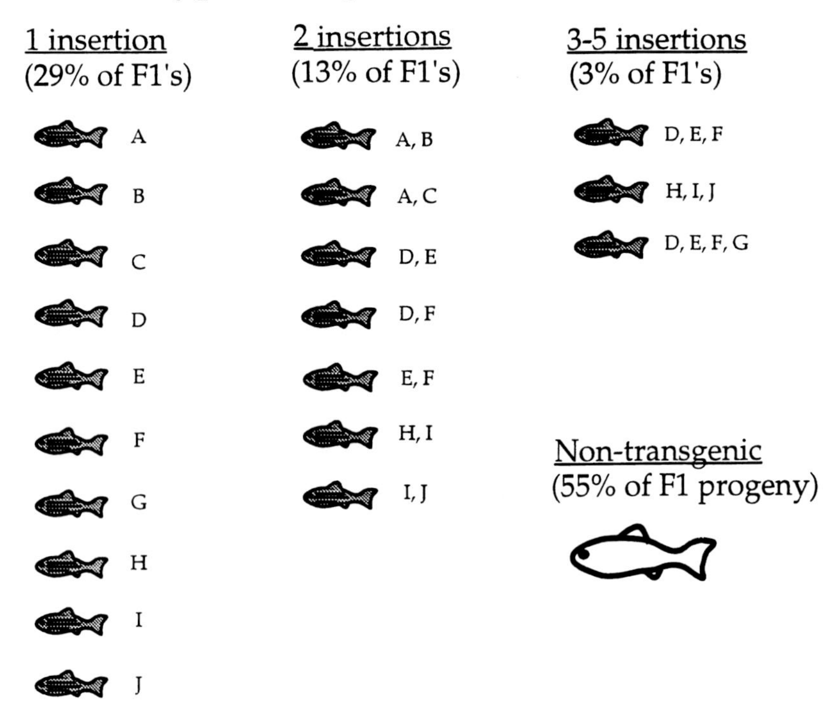

Figure 1. Percentage of transgenic F_1 lines obtained from a typical founder fish. Every blastula-stage embryo injected with high-titer virus grows to be a founder fish that transmits proviral insertions to its F_1 progeny. The percentage of transgenic F_1 progeny and the number of insertions inherited in each F_1 are shown for a typical founder. Data averaged from 200 founders whose progeny were analyzed by Southern blotting of fin-clip DNAs of about 100 F_1 progeny.

1 lethal mutation induced for every 20 proviral insertions, a frequency that seems surprisingly high (Jaenisch 1988).

To determine the mutagenic frequency of proviral insertions in fish, we performed a pilot insertional mutagenesis screen. F_1 fish inheriting the same single insertion were crossed to one another, or F_1 fish with a single insertion were outcrossed, and transgenic F_2 fish were identified and mated to one another. Progeny were examined for embryonic mutant phenotypes in 25% of the embryos. We identified 3 insertional mutants among 217 insertions screened, for a calculated frequency of 1 mutation per 70 insertions (Allende et al. 1996; Gaiano et al. 1996b). Subsequently, we isolated 3 additional embryonic lethal mutations and 1 dominant adult phenotype (see below). The frequency has remained approximately the same. Currently we estimate that about 1/70 to 1/100 insertions induces an embryonic lethal or visible mutation.

Evidence that the mutants we isolated were induced by insertions is based on several criteria, but the primary evidence is linkage data (Allende et al. 1996; Gaiano et al. 1996b). Using inverse polymerase chain reaction (PCR) we clone a fragment of DNA adjacent to the proviral insertion in each mutant line, identify a single-copy probe from this fragment, and then determine a restriction enzyme digest that allows us to distinguish the chromosome bearing the proviral insertion from its homolog lacking the insertion. We then mate fish heterozygous for the insertion in question. DNA extracted from individual mutant and phenotypically wild-type embryos is analyzed by Southern blot. If all mutant embryos are homozygous for the proviral insertion, and no wild-type embryos are homozygous, the insertion must be linked to the mutation. If enough embryos are analyzed to show that the insertion and the mutation are very tightly linked, it is highly likely that the insertion caused the mutation. Additional evidence that our mutations are caused by proviral insertions has come from cloning genes into which the proviruses have integrated. Before describing the genes, however, we first consider the phenotypes of the mutants we obtained.

Phenotypes of insertional mutants. An important question is whether the phenotypes of insertional mutants display the same spectrum as that seen in the chemical mutagenesis screens. In chemical screens, mutants were classified into two broad categories, specific and nonspecific, with the latter being further subdivided into mutations whose most striking defect is extensive apoptosis or necrosis in the CNS, and mutants with multiple defects and frequently displaying edemas and retarded growth (Mullins et al. 1994; Driever et al. 1996; Haffter et al. 1996). Nonspecific mutants comprised 70% of the total, with about 20% having extensive cell death in the CNS and 50% having multiple defects. About 25–30% of the mutants had specific defects, with one or a few organ systems affected.

As shown in Table 1, our seven insertional mutants can be assigned to the categories defined by the chemical mutagenesis screens, with four mutants belonging to the nonspecific phenotype group, and three to the category of specific phenotypes. One mutant, *dead eye* (*dye*), displays extensive cell death in the brain and neural tube beginning at day 1 of development (Allende et al. 1996). Three mutants, *no arches* (*nar*), *pescadillo* (*pes*), and *80A*, display multiple defects and exhibit edema at day 5 (Gaiano et al. 1996b). Two of our embryonic lethal inser-

Table 1. Retrovirus-induced Insertional Mutants in Zebrafish

	Mutant	Phenotype	Candidate gene	References
Nonspecific phenotypes	*no arches (nar)*	embryonic lethal, onset at day 3; lacks branchial arches, other defects	homolog of *Drosophila clipper*, a ribonuclease (Bai and Tolias 1996); no mutants known in other species; transcript maternally supplied in flies and fish, expressed in embryo but low in adult except in ovary	Gaiano et al. (1996b)
	pescadillo (pes)	embryonic lethal, onset at day 3; defective in growth of embryonic liver, gut, fins, arches, and brain, but not several other organs	homolog of ESTs; no known function; maternally supplied; expressed in embryonic sites that fail to grow in mutants, expression low in adult except in ovary; may belong to BRCT superfamily which includes human breast cancer gene BRCA-1 (S. Altschul and E. Koonin, pers. comm.)	Allende et al. (1996)
	dead eye (dye)	embryonic lethal, onset at day 2; extensive apoptosis in eyes, brain, and neural tube	homolog described in frog (Hudson et al. 1996); possible nuclear pore component, weak homology to yeast NIC96 (nucleoporin interacting component 96) (Grandi et al. 1993); yeast loss of function is lethal; in fish RNA is maternally supplied and gene is expressed in CNS in embryos	Allende et al. (1996)
	80A	embryonic lethal, onset at day 2; multiple defects	not yet cloned	Gaiano et al. (1996b)
Specific phenotypes	*399*	post-embryonic lethal; small eyes at day 5; dies at about 2 weeks of age; defective in the photoreceptor cell layer of the eye	transcription factor NRF-1; unique DNA-binding motif; studied biochemically in human, rat, chicken, and sea urchin (Evans and Scarpulla 1990; Calzone et al. 1991; Gomez-Cuadrato et al. 1995); in flies a mutant in homologous gene is *erect wings* (DeSimone et al. 1995)	T. Becker et al. (unpubl.)
	D1	homozygous viable; dominant adult pigmentation phenotype; onset at 3 weeks; body stripe pattern breaks up into spots	homolog of ESTs; protein of unknown function with similarity to WD-repeat proteins	K. Kawakami et al. (unpubl.)
	891	embryonic lethal, onset at day 1; yolk darkens beginning at the yolk extension, after which the embryo dies	not yet cloned	M.L. Allende et al. (unpubl.)

tional mutants have specific developmental defects. Of the latter, *399* displays a small-eye phenotype at day 5 but otherwise appears normal, although the mutant embryos die at about 2 weeks of age (T. Becker et al., unpubl.). The phenotype of mutant *891* is yolk degeneration beginning from the yolk extension at day 1 of development and proceeding until the entire yolk changes color and recedes, after which the embryo rapidly declines (M. L. Allende and N. Hopkins, unpubl.). Our seventh insertional mutant displays a dominant adult phenotype and is homozygous viable (K. Kawakami and N. Hopkins, unpubl.). In this mutant, the stripes on the body of fish heterozygous or homozygous for the insertion break up into blobs. Adult phenotypes affecting pigmentation, body shape, and fins were also seen in the chemical mutagenesis screens and are assigned to the phenotypically specific class of mutants. Although the number of mutants we have isolated is still very small, the fact that they are representative of the different types seen in chemical mutagenesis screens is reassuring.

Molecular cloning of genes disrupted by proviral insertions. If our putative insertional mutants are indeed caused by a proviral integration, we would predict that each insertion that is linked to a mutant phenotype should lie in a gene and disrupt its expression. Furthermore, the disrupted genes should be expressed in wild-type fish at a time and place such that perturbation of gene expression could explain the mutant phenotype seen in animals homozygous for the corresponding insertion. To determine if this is the case, we have attempted to clone genes in the vicinity of the proviral insertions that are linked to mutant phenotypes. This proved to be easier than we had anticipated, and to date we have identified candidate genes for five of the seven mutants (Allende et al. 1996; Gaiano et al. 1996b; see below).

To identify a gene in the vicinity of the provirus, we use inverse PCR to clone host DNA flanking the provirus. The host DNA sequence is then used either for additional rounds of inverse PCR, or as a probe to clone a larger genomic fragment from a library. We then sequence host DNA adjacent to the proviral insertion and use the sequence to search the database of gene sequences. To date, fish genes have been identified because of the large number of cDNA and genomic sequences from other organisms that are in the database, and because of the high conservation of amino acid sequences between homologous genes from different organisms. As shown in Figure 2, in four of five cases, we found the mutagenic provirus to lie within the first exon or intron of a gene, whereas in the fifth case, the provirus lies in a large (~25 kb) intron.

The first four examples in Figure 2 correspond to four recessive embryonic lethal mutations. In these cases, the gene that has been disrupted by proviral integration was found to be expressed in phenotypically wild-type embryos whereas mutant embryos homozygous for the proviral insertions were found to lack detectable gene expression as determined by Northern blot, reverse transcriptase (RT) PCR, and/or whole-mount in situ hybridization. This result supports the conclusion that disruption of these genes is responsible for the mutant phenotypes observed. The fifth case represents the proviral insertion linked to the dominant, homozygous-viable mutation. In this case, the provirus lies in a large intron, and RNA expression from the gene is not abolished.

Further evidence that disruption of the genes we have identified is responsible for the mutant phenotypes observed comes from the finding that, in cases where whole-mount in situ hybridization has been performed on embryos, in wild-type embryos the genes are expressed in tissues that are affected in mutant embryos. This is true of *pescadillo*, *dead eye*, and mutant *399*.

Definitive proof that the genes we identified are responsible for the mutant phenotypes we have observed would require rescue of the mutant phenotype or recreation of the mutant phenotype by knockout technology. However, the latter technology is not yet available in the fish, and the former is still a demanding experiment rather than a routine procedure (Stuart et al. 1990; Amsterdam et al. 1995). Nevertheless, for the embryonic lethal mutations in which gene expression is profoundly affected, the evidence described above strongly suggests that disruption of the candidate genes we identified is likely to be responsible for the mutant phenotypes observed. The case of D1, the dominant adult phenotype, is less certain, however. In this case, expression of the candidate gene, which is normally expressed in embryos and adult fish, is not abolished and we have not been able to detect aberrant expression of this gene in heterozygous or homozygous mutant fish. It is possible that proviral integration results in aberrant expression of the gene at a low level in some tissues, and this might not have been detected by experiments to date.

Thus far we have failed to identify a gene in the vicinity of the proviral insertion linked to mutation *80A*, although we have sequenced DNA flanking this provirus. It is possible that this provirus has integrated in a gene that has no homolog in the database yet, or a gene that is sufficiently different in amino acid sequence from its homologs that we have not recognized it. Alternatively, the provirus may lie at a considerable distance from the gene it disrupts or from exons of a gene, and we may not yet have sequenced enough DNA to find the gene. Attempts to clone the gene presumably mutated by the insertion linked to mutation *891* are under way.

Feasibility and design of a large-scale insertional mutagenesis screen in zebrafish. Our results show that it is relatively easy to generate a large number of insertions in the fish germ line, that insertions are mutagenic, and that it is easy to clone genes disrupted by the insertions and presumably responsible for the mutant phenotypes observed. The results also show, however, that with the viral vector used, the frequency of insertional mutagenesis is low: About 1 in 70–100 insertions is mutagenic. Given these findings, is a large-scale insertional mutagenesis screen feasible in the fish using this technology? Recent findings in our lab suggest that the answer is yes,

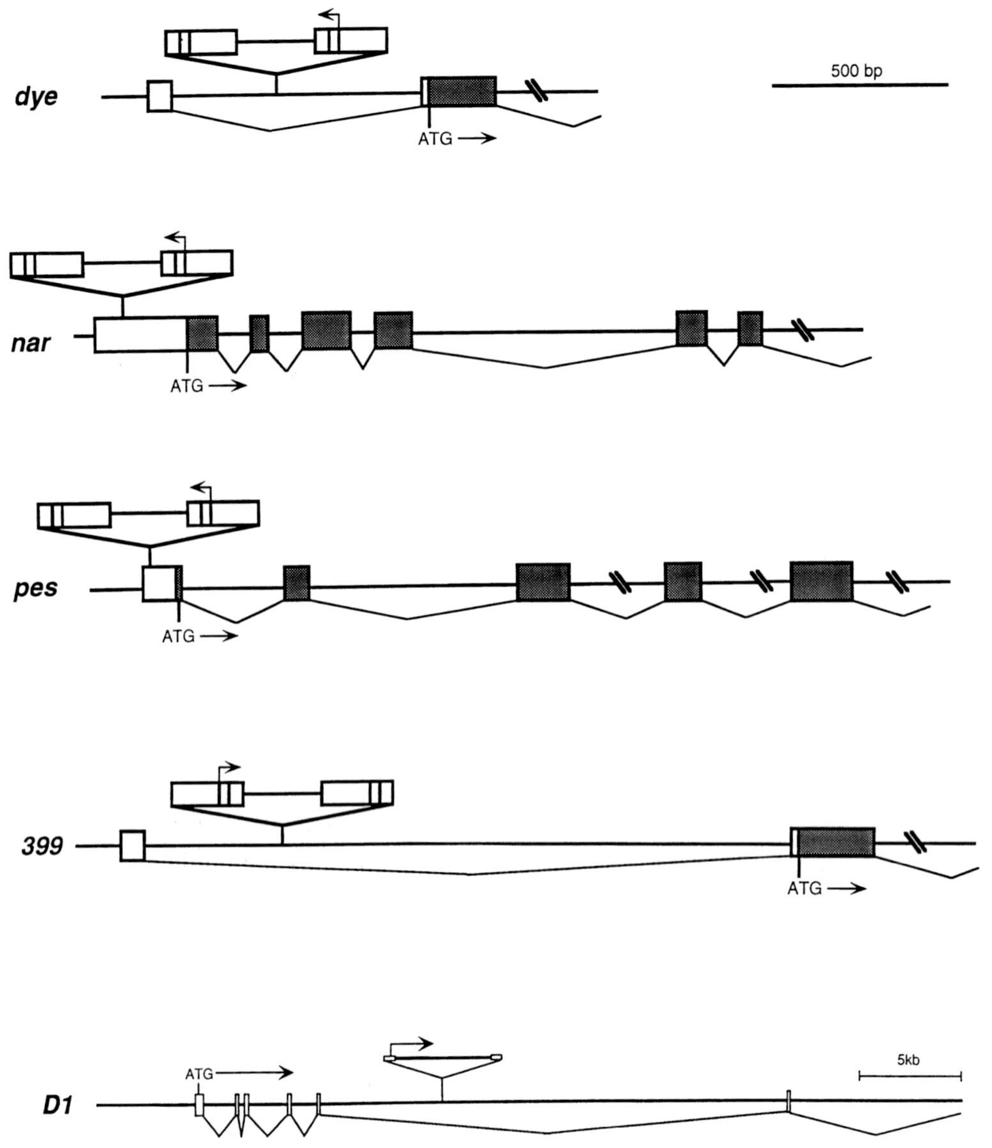

Figure 2. Genes disrupted by proviral integrations in five insertional mutants. Diagrams show the position of proviral insertions relative to genes they disrupt. The first four diagrams represent the positions of proviruses that are genetically linked to recessive embryonic lethal mutations. The fifth diagram represents the position of a provirus that is linked to a dominant adult phenotype. The genes are drawn to scale, the proviruses are not. Note that the fifth diagram is drawn to a different scale from the first four. In the case of the first four examples, the proviral insertions abolish or alter detectable transcript from the genes, whereas in the fifth case alteration in gene expression has not yet been detected (see text).

and that the number of mutants that can be obtained justifies this arduous undertaking.

In our pilot screen, we identified insertional mutants in most cases by following the protocol shown in part A of Figure 3. F_1 fish with single insertions were identified by Southern analysis of fin clip DNAs and were then outcrossed. Transgenic F_2 fish were identified among the progeny and crossed to one another, and F_3 embryos were examined to determine if 25% displayed a mutant phenotype. This strategy makes it easy to immediately clone host DNA flanking a proviral insertion, since there is only one insertion per fish. This strategy means, however, that each fish tank can accommodate fish harboring just one insertion.

The number of insertions that can be screened in a given number of fish tanks can be increased substantially using the scheme shown in Figure 3B. In this scheme, F_1 fish that harbor several insertions are identified and crossed with each other to generate F_2 families. Then, just as in a chemical mutagenesis screen, pairs are mated blindly from the F_2 family. Every insertion present in the F_1 parents will be transmitted to 50% of the F_2 fish. If six successful pair matings of F_2 fish are obtained, every insertion has an 85% chance of coming to homozygosity among the six pair matings. Using the scheme in Figure 3B rather than A, the number of mutants that can be obtained from the same size fish facility increases by about fivefold.

We have found that when founder fish are mated to one another, about 20% or more of their F_1 offspring harbor 3–5 insertions per fish (A. Amsterdam et al., unpubl.). It is these fish that must be identified to carry out the protocol shown in Figure 3B. This is done by Southern analysis of fin clip DNAs of a large number of F_1 fish. Although laborious, this task is justified by the greater number of mutants that can be obtained.

On the basis of the frequencies obtained in our pilot screen, we estimate that we could generate about 800–1100 insertional embryonic lethal mutations over a period of 3 years. If insertions are random, this should correspond to about 600–800 genes, or 25–33% of the total number of 2400 embryonic lethals estimated from chemical screens.

A Molecular Marker for Embryonic Germ Cells in Zebrafish

A zebrafish homolog of the* Drosophila *gene* vasa *identifies the putative primordial germ cells of the zebrafish. Genetic studies by Walker and Streisinger (1983) suggested that during early blastula stages about 5 cells in the embryo are destined to give rise to the fish germ line. Studies in our own lab by Lin et al. (1992) demonstrated that cells that lie within the blastoderm at early blastula stages can be transplanted between embryos and contribute to the germ line. Neither of these studies answered the question of when or how germ cells acquire their identity, and other than these two findings, nothing was known about the number or location of primordial germ cells (PGCs) in the zebrafish. In other fish species, however, PGCs have been identified as early as somite stages (Hamaguchi 1982; Timmermans and Taverne 1989). This was achieved using morphological criteria by beginning at a stage of embryogenesis when germ cells can be clearly identified within an embryonic gonad, then tracing these cells in earlier and earlier embryos on the basis of their morphology.

There are several reasons why we wished to develop a molecular marker for embryonic germ cells in the fish. First, by knowing the location and time of determination of the germ cells we might be able to better deliver virus to the cells and thus increase our frequency of insertional mutagenesis. Second, it might be possible to culture PGCs and return them to the embryo, a method that has been achieved in mice.

In the hope of identifying a molecular marker expressed specifically in germ cells, we cloned the probable zebrafish homolog of the *Drosophila* gene *vasa* (Schupbach and Wieschaus 1986; Yoon et al. 1997). *vasa* is a member of the DEAD box family and encodes an RNA helicase whose function in germ cell determination is not yet understood (Hay et al. 1988; Lasko and Ashburner 1988). In flies, *vasa* transcripts are maternally supplied but translated only in the pole cells, and zygotic transcription of the gene in embryos is limited to pole cells. The fish and fly genes are about 41% identical at the amino acid level (Yoon et al. 1997).

To determine if the zebrafish *vas* gene is a marker for germ cells, we performed whole mount in situ hybridization on embryos at different stages of development (Yoon et al. 1997; see below). As shown in Figure 4, at early blastula stages exactly four cells contain significant levels of *vas* transcript. Evidence that these are likely to be the primordial germ cells is based on the appearance of

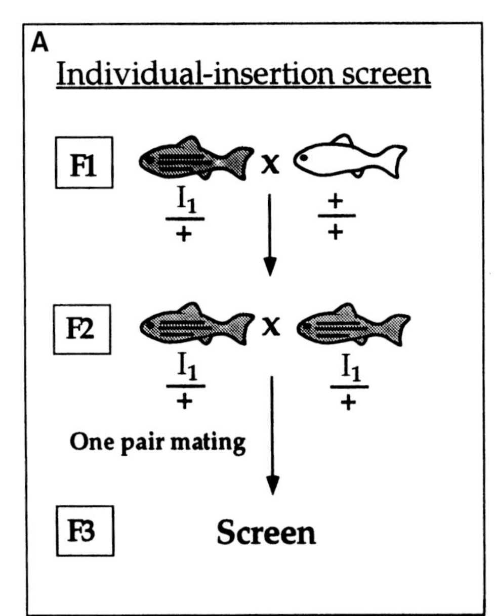

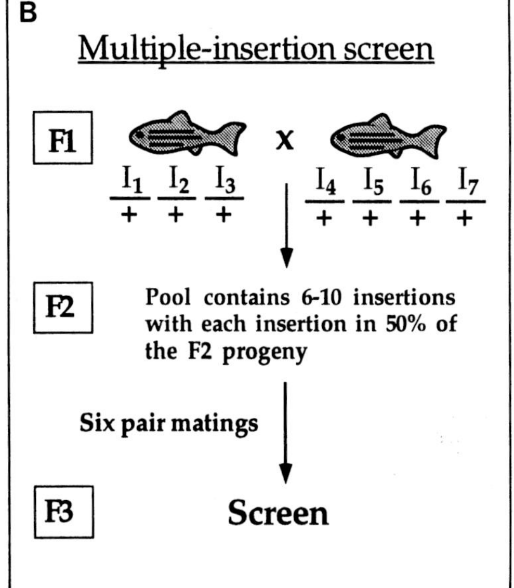

Figure 3. Two schemes for performing insertional mutagenesis screens. (*A*) Scheme for breeding single insertions to homozygosity that was used in a small pilot screen in our lab. (*B*) Scheme for inbreeding multiple insertions simultaneously. The second scheme makes it possible to screen a much larger number of proviral insertions with a given number of fish tanks.

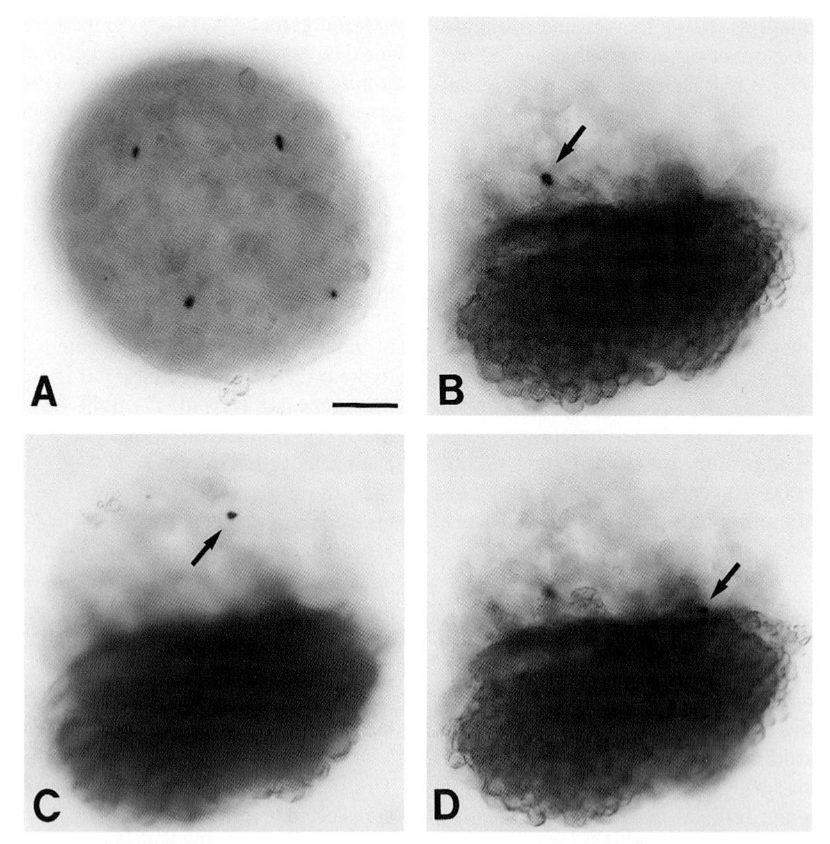

Figure 4. Zebrafish *vas* homolog expression at the 1000-cell stage. Whole mount RNA in situ hybridization using a digoxigenin-labeled *vas* cDNA fragment-riboprobe. The same 1000-cell-stage embryo is shown in all panels. (*A*) The embryo viewed from the top. (*B–D*) Side views in different focal planes. (*A*) *vas* transcript is detected in 4 cells at the 1000-cell stage. (*B–D*) Viewing the same embryo from the side reveals that the *vas*-expressing cells can be located at different heights above the yolk. The *vas*-expressing cells are indicated with an arrow. Bar, 100 μm.

whole mount in situ hybridizations to embryos of increasing age. During somitogenesis, the position of *vas*-hybridizing cells corresponds to that seen in other fish species (Fig. 5). Signal was detected up to 10 days postfertilization, by which time an embryonic gonad is clearly visible (Fig. 5). Only cells within this structure, and having the appearance of germ cells in other fish species, expressed detectable transcript. Thus, although our evidence is based entirely on static pictures, it appears that *vas* is expressed zygotically in the PGCs of the zebrafish.

vas *transcripts are maternally supplied and localize along the cleavage planes of 2- and 4-cell stage embryos.* Embryos at the 1000–2000-cell stages have exactly four *vas*-expressing cells detectable by whole-mount in situ hybridization, and when the embryo is viewed from above, the four cells are invariably seen to lie far apart and as if at the corners of a square (Yoon et al. 1997). Viewed from the side, it appears that the cells can lie at different heights above the yolk (Fig. 4). The results of whole-mount in situ hybridization at earlier stages of embryogenesis suggests a possible explanation for the presence of four *vas*-hybridizing cells at the 1000-cell stage. When whole-mount in situ hybridization was performed on fertilized eggs at the 1-cell stage, no clear signal was observed. However, from the 2–cell stage on, a striking localization of *vas* transcript was seen. As shown in Figure 6, at the 2-cell stage *vas* RNA is seen to

Figure 5. Evidence that *vas* RNA is expressed in the zebrafish primordial germ cells. Embryos and larvae were hybridized with a *vas* riboprobe. In *A–C*, specimens are oriented such that anterior is to the left. (*A, B*) Dorsal views of embryos at 100% epiboly (*A*) and 8-somite-stage (*B*) (flattened, yolk removed) that were hybridized with *vas* (*arrow*, purple stain) and *MyoD* (Weinberg et al. 1996) (*arrowhead*, red stain). As seen in *A*, during epiboly, *vas*-expressing cells (*arrow*) migrate toward the dorsal side of the embryo. By early somitogenesis (*B*), the *vas*-expressing cells (*arrow*) have clustered together into two groups to the left and right of the midline, adjacent to the anterior somites (*arrowhead*, red stain). The position of these cells is similar to the position of the primordial germ cells of other fish at comparable stage (see text for details). (*C*) Side view of a 3-day larva. *vas*-expressing cells (*arrow*) extend posteriorly in two bilateral rows during late embryonic and early larval development. (*D, E*) Transverse sections (2 μm) of 10-day larvae were prepared following in situ hybridization. The *vas*-expressing cells are located in the gonad, as determined by its position relative to the other organs (*D*). A higher-magnification view of a representative section through the gonad (*E*) shows that *vas* RNA is present in the cytoplasm. The *vas*-expressing cells have greatly increased in number by this stage of development. (g) Gut; (m) muscle; (n) notochord; (nt) neural tube; (p) pancreas; (s) swim bladder. Bars: *A–C*, 100 μm; *D, E*, 10 μm.

localize to two discrete regions along the cleavage plane. At the 4-cell stage there are four such aggregates. At the 8-cell stage, in most embryos there are still only four aggregates of *vas* RNA, and they still lie along the cleavage planes, although in some embryos additional weaker signal is also seen along additional cleavage planes. By the 16- and 32-cell stages (Fig. 6), instead of the linear aggregates, we see four subcellular blobs of *vas* RNA, spaced far apart in a square pattern. The blobs persist as cleavages continue. By the dome stage, two cell divisions after mid-blastula transition, *vas* transcripts appear to fill the cytoplasm of the cells that are detected by whole mount in situ hybridization (Fig. 6). These results suggest that at mid-blastula transition, which occurs at the 1000-cell stage, the four cells that receive the *vas*-containing blobs may begin to express *vas* zygotically and divide (Kane and Kimmel 1993).

The fact that *vas* RNA is detected in whole mount in situ hybridizations before the mid-blastula transition suggests that the RNA is maternally supplied. Northern blots

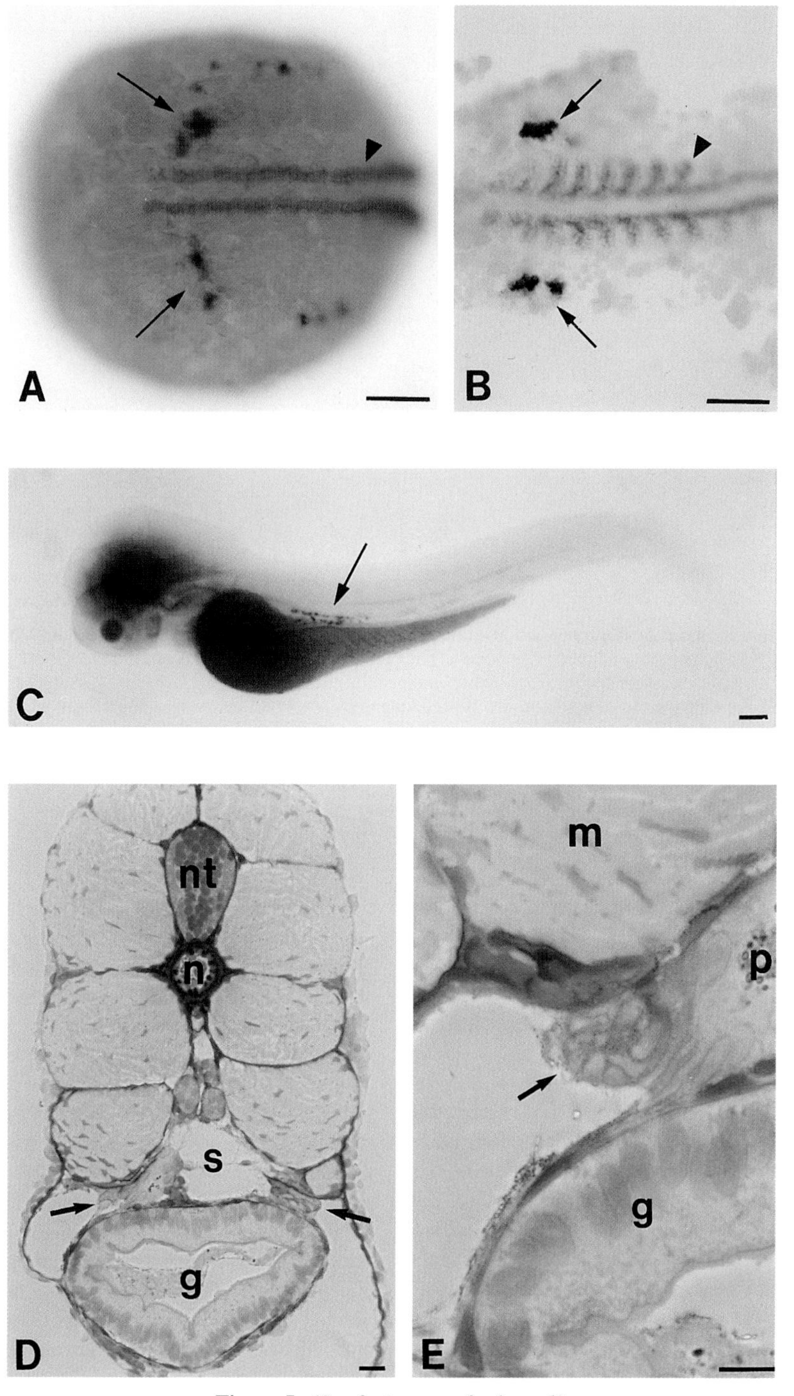

Figure 5. (*See facing page for legend.*)

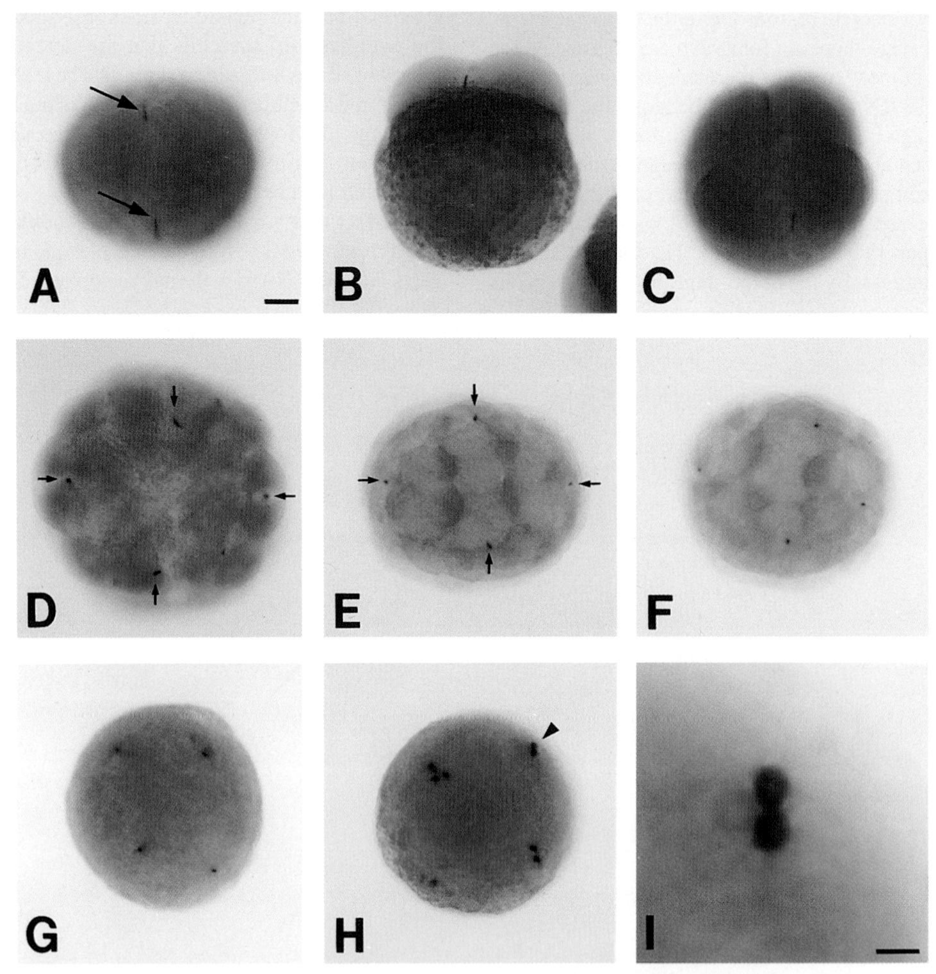

Figure 6. *vas* RNA expression during cleavage and blastula stages. Embryos were hybridized with a *vas* cDNA fragment riboprobe. All panels show top views of embryos, except *B* (side view). The cells expressing *vas* RNA are indicated with an arrow. *vas* transcript is detected as stripes along the cleavage planes (*arrows*) of embryos at the 2-cell stage (*A, B*) and 4-cell stage (*C*). *vas* RNA is condensed into four subcellular clumps sometime from the 16-cell stage (*D*) to the 32-cell stage (*E, F*; 2 different embryos) and remains in this configuration through the 2000-cell stage (*G*). By the 4000-cell stage (dome stage), there are more *vas*-expressing cells, suggesting that cells that express the gene may have divided (*H*). The pair of cells marked with an arrowhead in *H* are shown at higher magnification in *I*. *vas* RNA is no longer subcellularly localized and appears to fill the cytoplasm by the dome stage. Bars: *A–H*, 100 μm; *I*, 20 μm.

of RNA prepared from newly fertilized eggs reveal that *vas* transcript is present in these eggs, again suggesting that the RNA is maternally supplied. In Northern blots in which lanes are loaded with equal amounts of total RNA, *vas* is detected from 0 to 6 hours after fertilization, but not at later time points (Yoon et al. 1997).

Although Northern blots and in situ hybridization protocols both suggest that *vas* transcripts are maternally supplied, the two techniques yield apparently contradictory results in two respects: Northern blotting detects *vas* transcript at the 1-cell stage, but clear signal was not detected at this stage by whole-mount in situ hybridization; Conversely, whole mount in situ hybridization reveals *vas* expression in the putative PGCs up to 10 days postfertilization, but Northern blots failed to reveal transcripts later than about 6 hours after fertilization. We think these discrepancies might be explained as follows. (1) Why does in situ hybridization not reveal *vas* RNA at the 1-cell stage? It may be that the RNA is diffusely located at this stage and hence remains below the level of detection with the protocol used. Support for this statement is the fact that when unfertilized eggs were held for about 2.5 hours before being fixed and analyzed by whole mount in situ hybridization, small blobs of *vas*-hybridizing material were detected, but these were not arranged on cleavage planes as they are when embryos cleave normally (Fig. 7). This suggests that in order to become detectable by our whole mount in situ protocol, maternally supplied *vas* RNA must first aggregate. (2) Why aren't *vas* transcripts detected in Northern blots of total embryo RNA later than about 6 hours postfertilization? This discrepancy might be due to the fact that the percentage of cells in the embryo that are expressing *vas* RNA declines rapidly as the embryo develops and thus falls below the level of detection of Northern blotting when equal amounts of RNA are loaded.

In summary, although our data consist primarily of static pictures of whole mount in situ hybridizations, it is

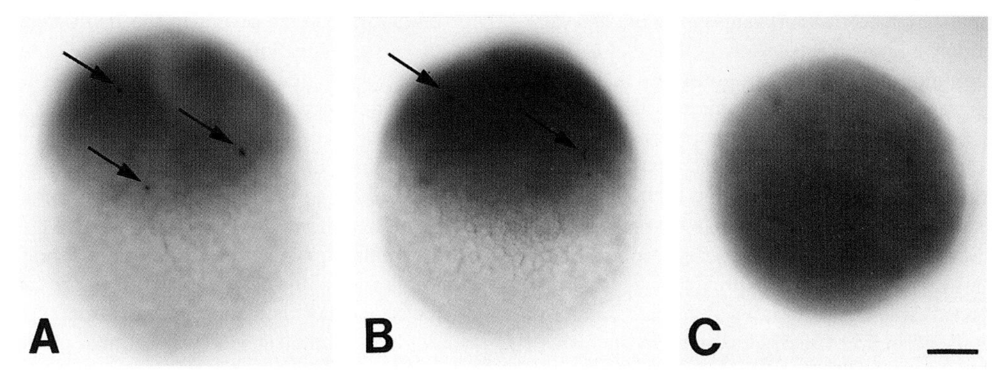

Figure 7. *vas* RNA expression in 2.5-hr unfertilized eggs. Unfertilized eggs were incubated at 28.5°C for 2.5 hr before fixation, followed by whole mount RNA in situ hybridization. In contrast to the lack of detectable expression in 1-cell-stage embryos (data not shown), we detect small clumps of *vas* RNA in 2.5-hr unfertilized eggs. These clumps appear to vary in number and location within the cytoplasm. Three different eggs are shown for comparison. (*A,B*) Side views; (*C*) top view. Bar, 100 μm.

tempting to impose the following interpretation on these pictures (see Fig. 8): *vas* transcripts are maternally supplied. At the first two cleavages, *vas* RNA aggregates along specific regions of the cleavage planes, forming part of a putative zebrafish germ plasm. By the next cleavages, most of the material needed to form aggregates has been used up so that new ones do not form. (It should be noted that we do not know what fraction of ma-

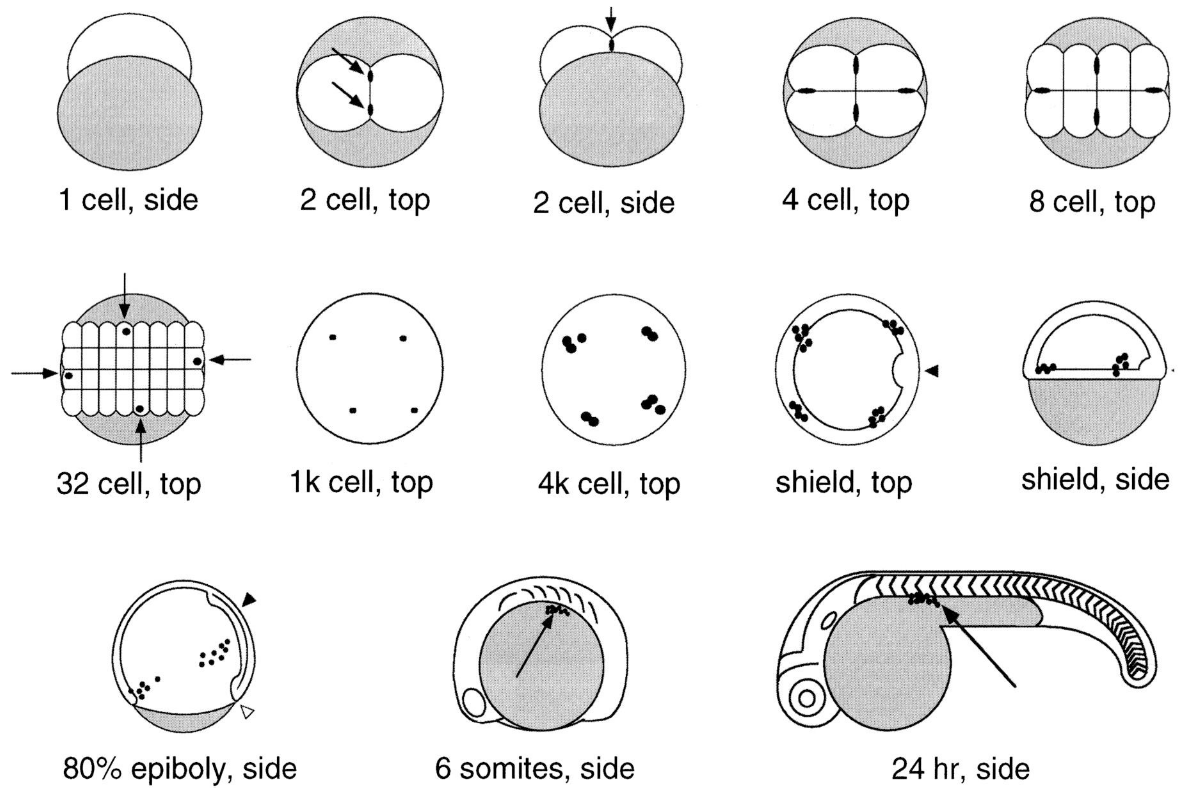

Figure 8. Summary of zebrafish germ line development during embryogenesis. Schematic drawings of *vas* RNA expression in zebrafish embryos as determined by whole mount in situ hybridization (Figs. 4–6) (Yoon et al. 1997). The data suggest that *vas* RNA is expressed in the zebrafish primordial germ cells (PGCs). The stages and views are indicated below each drawing and the yolk is shaded. Arrows point to the *vas*-expressing cells in selected stages. In shield-stage embryos, the arrowhead indicates the dorsal shield. In the 80% epiboly embryo, the black arrowhead indicates the developing notochord and the white arrowhead indicates the germ ring. We interpret the whole mount in situ hybridization results as follows: *vas* RNA is detected along the first two cleavage planes (*arrows*), then condenses into four subcellular clumps by the 32-cell stage (*arrows*). These four clumps are segregated to four cells through the 1000-cell stage. By the 4000-cell stage, the cells that inherited *vas* RNA, the PGCs, have begun to divide and *vas* RNA is located throughout the cytoplasm. Mitoses continue during early gastrulation to generate about 25–30 PGCs. *vas* RNA detected up to the 1000-cell stage is presumably maternal RNA, and after that it is presumed to be derived by zygotic transcription. During epiboly until early somitogenesis, the PGCs migrate toward the dorsal side of the embryo, forming two clusters of cells to the right and left of the notochord, adjacent to the third to fifth somite. They remain in this position through early larval stages, extending posteriorly for a variable distance to form two bilateral rows of cells in the gonadal anlagen and later resume mitosis (not shown).

ternally supplied *vas* RNA aggregates along the cleavage planes, nor do we know if it is *vas* transcript that is limiting in determining the number of aggregates that form.) The four aggregates of *vas* RNA condense into blobs that are considerably smaller in diameter than the cells at this stage. As cleavage continues, the blobs persist while the cells around them are becoming smaller. At the 1000-cell stage, cells that inherit the blobs begin to express *vas* zygotically. Shortly after this, *vas* transcripts appear to fill the cytoplasm of the cells that express the gene. These cells are presumably the primordial germ cells and will give rise to the germ line. Clearly, many experiments will be needed to determine if this interpretation is correct.

DISCUSSION

Insertional Mutagenesis

We developed a method of insertional mutagenesis that should allow us to isolate a significant fraction of the genes essential for the embryonic development of the zebrafish over the next few years. We estimate that the fraction of embryonic lethal or visible mutations we could isolate in 3 years is about 25% of total, assuming there are roughly 2400 such genes. Is this fraction sufficient to make the screen valuable, and is it possible that improvements in the methodology could increase the number of mutants we could isolate?

In invertebrate model organisms it is possible to perform Mendelian genetic screens on a large enough scale to identify *all* the genes that are essential for the process of interest, the so-called "saturation mutagenesis screen," and this has proved to be an important aspect of the approach. Achieving saturation in a mutagenesis screen in zebrafish is extraordinarily difficult technically, even using chemical mutagens, and in fact has not been achieved in screens to date. It is not yet clear whether the results of achieving saturation will be as profound in fish as they have been in invertebrate animals, because the genetic redundancy seen in vertebrates may preclude the identification of key genes in pathways even when saturation is achieved. Despite this possible limitation, it must still be acknowledged that saturation remains the goal in genetic screens, including in the zebrafish.

Several considerations lead us to conclude that an insertional mutagenesis screen using our current methodology is a desirable pursuit even though the frequency of mutagenesis is considerably lower than that of chemical mutagenesis. First, because the chemical mutagenesis screens have been performed, our insertional mutants form part of a larger picture. Our ability to interpret the phenotypes of the tiny group of mutants we have obtained to date is clearly enhanced by the findings of the chemical mutagenesis screens. Second, although we will obtain a smaller percentage of all the essential genes than was obtained in the chemical screens, as we have shown, the mutated genes are readily clonable, and this is a tremendous advantage. As we discovered in our small pilot screen, many of the genes we will find will be novel in that they will not previously have been associated with phenotypes in any organism. Many of the genes essential for embryonic development in the fish could provide a novel entry to studying important developmental processes in vertebrates. The ease of cloning the genes for insertional mutants means that one requires far less certainty that a phenotype is "interesting" before embarking on the cloning, and this should help to eliminate the bias which is likely to favor mutants that we can already understand. Insertional mutagenesis may also prove a powerful stimulus for behavioral screens in zebrafish. In behavioral screens, the desired phenotypes are usually more difficult to assay than morphological defects, and thus cloning the mutated genes by positional cloning is even more difficult when the mutations have been induced by chemicals. The greater ease of tracking mutants and cloning mutated genes using insertional mutagenesis may make such screens more manageable, even given the lower frequency of mutagenesis.

Although we plan to apply our current methodology on a large scale, we are also interested in trying to improve retrovirus-mediated insertional mutagenesis. Even small improvements in the efficiency of some of the steps would lead to a substantial decrease in the amount of work needed to generate mutants. Some possibilities for improving the method are the following.

1. *Higher-titer viruses.* Considerable effort is required to generate the F_1 fish with multiple insertions that are needed for the screen (see Fig. 3B). These F_1 fish are obtained from crosses of two founder fish, each harboring about 12 insertions in their germ line (see Fig. 1). At present, about 20% of the F_1 fish from such crosses have 3–5 insertions and thus are useful for the screen. It is possible that founders generated from injection of higher-titer virus stocks would yield a higher frequency of multiple-insertion F_1 fish.
2. *More mutagenic viruses.* In mice, about 1 in 20 proviral insertions results in an embryonic lethal or visible mutation, whereas in our studies the number is only about 1 in 70–100 insertions (Jaenisch 1988; Allende et al. 1996; Gaiano et al. 1996b). There are several possible explanations: Mice may have more embryonic lethal genes than fish, mouse retroviral vectors may integrate preferentially into genes in mouse cells but not in fish cells, or viruses may be more mutagenic in mice than in fish. We identified one mutant, designated D1, in which the provirus that is genetically linked to the mutation lies in a large intron of a gene whose disruption may explain the mutant phenotype. However, in this case, we have not yet detected an alteration in transcripts from the gene. This observation suggests that proviral insertion into a gene does not necessarily affect its expression dramatically, if at all, a result that has been seen in mice as well. It may be possible to generate a virus that is more mutagenic such that integration into large introns invariably disrupts transcription. Alteration of the splicing and termination signals within the virus might help in this respect. If so, this would increase the target size for generating mutagenic proviral integrations and hence increase the mutagenic frequency.
3. *Gene traps.* If the retroviral vector used for the muta-

genesis carried a reporter gene that was activated upon integration into an active gene, then F_1 fish with integrations into genes could be detected as heterozygotes (Friedrich and Soriano 1991; Chen 1994). The ideal reporter would be GFP, since this would make it possible to isolate live F_1 embryos with proviral insertions in genes expressed during embryogenesis. If one could pre-select all such F_1 fish for inbreeding, one could isolate a much larger number of embryonic mutants, conceivably even achieving saturation of certain phenotypes, assuming proviruses can integrate into all possible chromosomal sites.

4. *Haploid screens*. For phenotypes that can be screened in haploid embryos, for example, morphological defects arising in the first 1–2 days of development, or possibly, defects detectable at later times by whole mount in situ hybridization, antibody staining, or other procedures, it might be possible to screen a larger number of insertions than by performing a diploid screen as shown in Figure 3B (Westerfield 1995). However, the greater effort needed to generate haploid embryos as opposed to simply letting fish mate must be taken into consideration in making this decision.

Finally, it is possible, of course, that other insertional elements will become available for mutagenesis in fish and that these will complement or supplant the method described here. For example, Tc-1-related elements have been studied extensively for this purpose by Hackett and his collaborators with promising results (Izsvak et al. 1995).

The Zebrafish Germ Line

We have presented evidence that the zebrafish homolog of the *Drosophila* gene *vasa* is a useful marker for the putative primordial germ cells of the fish. Our evidence that the gene is expressed specifically in embryonic, or primordial, germ cells (PGCs) rests on the fact that only a small population of cells in the developing embryo and larva express the gene, and that by 10 days of age the *vas*-expressing cells are located in a structure which, in appearance and location, would seem clearly to be the embryonic gonad. Many more experiments will be needed to confirm that the transcript we detect during embryogenesis is restricted to germ plasm and then to primordial germ cells, but the whole mount in situ hybridization analysis performed to date on embryos of closely spaced ages presents a compelling picture to support this interpretation (see Figs. 4–8).

Assuming that our simple interpretation is correct, several interesting areas for future study emerge from these observations. First, the ability to mark PGCs suggests ways in which one might be able to label this population of cells, purify them, and possibly develop conditions to selectively culture them. If successful, genetic manipulation of the cells in vitro might be possible. One should also be able to use a *vas* probe in a genetic screen to detect defects in germ cell formation and migration in the fish, an approach that has been productive in *Drosophila* (Lehmann, this volume). Another area of research suggested by these findings is to determine the basis for the interesting localization of *vas* transcript during early cleavage stages. What does the RNA attach to? Is *vas* RNA part of a zebrafish germ plasm, and if so, what other components are present in this putative germ plasm?

Finally, visualization of the PGCs of the fish is helpful in thinking about how to better manipulate the germ line in vivo. Specifically, when and where might viruses or other insertional mutagens be injected to maximize the number of insertions in the DNA of these cells? The presence of about 20–30 primordial germ cells at the time the cells begin to migrate during gastrulation, and the number of proviral integrations we observe in a single founder fish by our current method of infection, suggest that proviral integrations probably occur during this time period, before the number of cells increases by further cell divisions. Earlier infection might favor larger clone sizes for each integration event, and later infection might favor an even larger number of different integration events per founder.

Note Added in Proof

Recently, M. Allende in collaboration with the laboratory of S. Lin cloned a gene that is disrupted by provirus insertion in mutant 891 (unpublished).

ACKNOWLEDGMENTS

Developing a method of insertional mutagenesis was supported by grants from the National Science Foundation, the Human Frontiers Science Program, the National Institutes of Health, and Amgen, as well as several private sources. The pilot screen was supported by Amgen. The work on the zebrafish germ line was supported by the National Science Foundation, the Human Frontiers Science Program, and Amgen. Both projects were also supported by a core grant from the National Cancer Institute to the Center for Cancer Research. We also acknowledge postdoctoral fellowships from the Damon Runyon-Walter Winchell Foundation, the National Institutes of Health, the Yamada Science Foundation, and the Toyobo Biotechnology Foundation, and predoctoral stipends from the National Science Foundation and the Howard Hughes Medical Institute. We thank Bob Bosselman and Arthur Merrill for their encouragement and support.

REFERENCES

Allende M.L., Amsterdam A., Becker T., Kawakami K., Gaiano N., and Hopkins N. 1996. Insertional mutagenesis in zebrafish identifies two novel genes, *pescadillo* and *dead eye*, essential for embryonic development. *Genes Dev.* **10**: 3141.

Amsterdam A., Lin S., and Hopkins N. 1995. The *Aequorea victoria* green fluorescent protein can be used as a reporter in live zebrafish embryos. *Dev. Biol.* **171**: 123.

Bai C. and Tolias P.P. 1996. Cleavage of RNA hairpins mediated by a developmentally regulated CCCH zinc-finger protein. *Mol. Cell. Biol.* **16:** 6661.

Brockerhoff S.E., Hurley J.B., Janssen-Bienhold V., Neuhauss S.C.F., Driever W., and Dowling J.E. 1995. A behavioral screen for isolating zebrafish mutants with visual system defects. *Proc. Natl. Acad. Sci.* **92**:10545.

Burns J.C., Friedmann T., Driever W., Burrascano M., and Yee J.K. 1993. Vesicular stomatitis virus G glycoprotein pseudotyped retroviral vectors: Concentration to very high titer and efficient gene transfer into mammalian and non-mammalian cells. *Proc. Natl. Acad. Sci.* **90**: 8033.

Calzone F.J., Hoog C., Teplow D.B., Cutting A.E., Zeller R.W., Britten R.J., and Davidson E.H. 1991. Gene regulatory factors for the sea urchin embryo. I. Purification by affinity chromatography and cloning of P3A2, a novel DNA-binding protein. *Development* **112**: 335.

Chen J., Degregori J., Hicks G., Roshon M., Shi E.-G., Scherer S., and Ruley E. 1994. Gene trap retroviruses. *Methods Mol. Genet.* **4:** 123.

Chowdhury K., Bonaldo P., Torres M., Stoykova A., and Gruss P. 1997. Evidence for the stochastic integration of gene trap vectors into the mouse genome. *Nucleic Acids Res.* **25**:1531.

Clark D.T. 1981. "Visual responses in the developing zebrafish (*Brachydanio rerio*)." Ph.D. thesis, University of Oregon, Eugene.

DeSimone S., Coelho C., Roy S., VijayRaghavan K., and White K. 1995. ERECT WING, the *Drosophila* member of a family of DNA-binding proteins is required in imaginal myoblasts for flight muscle development. *Development* **121**: 31.

Driever W., Solnica-Krezel L., Schier A.F., Neuhauss S.C.F., Malicki J., Stemple D.L., Stainier D.Y.R., Zwartkruis F., Abdeliah S., Rangini Z., Belak J., and Boggs C. 1996. A genetic screen for mutations affecting embryogenesis in zebrafish. *Development* **123**: 37.

Emi N., Friedmann T., and Yee J.-K. 1991. Pseudotype formation of murine leukemia virus with the G protein of vesicular stomatitis virus. *J. Virol.* **65**: 1202.

Evans M.J. and Scarpulla R.C. 1990. NRF-1: A *trans*-activator of nuclear-encoded respiratory genes in animal cells. *Genes Dev.* **4**: 1023.

Friedrich G. and Soriano P. 1991. Promoter traps in embryonic stem cells: A genetic screen to identify and mutate developmental genes in mice. *Genes Dev.* **5**: 1513.

Gaiano N., Allende M., Amsterdam A., Kawakami K., and Hopkins N. 1996a. Highly efficient germ-line transmission of proviral insertions in zebrafish. *Proc. Natl. Acad. Sci.* **93**: 7777.

Gaiano N., Amsterdam A., Kawakami K., Allende M., Becker T., and Hopkins N. 1996b. Insertional mutagenesis and rapid cloning of essential genes in zebrafish. *Nature* **383**: 829.

Gomez-Cuadrado A., Martin M., Noel M., and Ruiz-Carrillo A. 1995. Initation binding receptor, a factor that binds to the transcription initiation site of the histone *h5* gene, is a glycosylated member of a family of cell growth regulators. *Mol. Cell. Biol.* **15**: 6670.

Grandi P., Doye V., and Hunt E.C. 1993. Purification of NSP1 reveals complex formation with "GLFG" nucleoporins and a novel nuclear pore protein NIC96. *EMBO J.* **12**: 3061.

Grunwald D.J. and Streisinger G. 1992. Induction of recessive lethal specific locus mutations in the zebrafish with ethylnitrosourea. *Genet. Res.* **59**:103.

Grunwald D.J., Kimmel C.B., Westerfield M., Walker C., and Streisinger G. 1985. A neural degeneration mutation that spares primary neurons in the zebrafish. *Dev. Biol.* **126**:115.

Haffter P., Granato M., Brand M., Mullins M.C., Hammerschmidt M., Kane D.A., Odenthal J., van Eeden F.J.M., Jiang Y.-J., Heisenberg C.-P., Kelsh R.N., Furutani-Seiki M., Volgelsang E., Beuchle D., Schach U., Fabian C., and Nüsslein-Volhard C. 1996. The identification of genes with unique and essential function in the development of the zebrafish *Danio rerio*. *Development* **123**: 1.

Hamaguchi S. 1982. A light- and electron-microscopic study on the migration of primordial germ cells in the teleost *Oryzias latipes*. *Cell Tissue Res.* **227**:139.

Hay B., Jan L.Y., and Jan Y.N. 1988. A protein component of *Drosophila* polar granules is encoded by *vasa* and has extensive sequence similarity to ATP-dependent helicases. *Cell* **55**:577.

Hicks G.G., Shi E.G., Chen J., Roshon M., Williamson D., Scherer C., and Ruley H.E. 1995. Retrovirus gene traps. *Meth. Enzymol.* **254**: 263.

Hopkins N. 1993. High titers of retrovirus (vesicular stomatitis virus) pseudotypes, at last. *Proc. Natl. Acad. Sci.* **90**: 8759.

Hudson J.W., Alarcon V.B., and Elinson R.P. 1996. Identification of new localized RNAs in the *Xenopus* oocyte by differential display PCR. *Dev. Genet.* **19**: 190.

Izsvak Z., Ivics Z., and Hackett P.B. 1995. Characterization of a Tc1-like transposable element in zebrafish (*Danio rerio*). *Mol. Gen. Genet.* **247**: 312.

Jaenisch R. 1988. Transgenic animals. *Science* **240**: 1468.

Kane D. and Kimmel C.B. 1993. The zebrafish midblastula transition. *Development* **119**: 447.

Kimmel C.D., Kane D., Walker C., Warga R.M., and Rothman M.B. 1989. A mutation that changes cell movement and cell fate in the zebrafish embryo. *Nature* **337**: 358.

Lasko P.F. and Ashburner M. 1988. The product of the *Drosophila* gene *vasa* is very similar to eukaryotic initiation factor-4A. *Nature* **335**: 611.

Lin S., Long W., Chen. J. and Hopkins. N. 1992. Production of germ-line chimeras in zebrafish by cell transplantation from genetically pigmented to albino embryos. *Proc. Natl. Acad. Sci.* **89**: 4519.

Lin S., Gaiano N., Culp P., Burns J., Friedmann T., Yee J. -K., and Hopkins N. 1994. Integration and germ-line transmission of a pseudotyped retroviral vector in zebrafish. *Science* **265**: 666.

Mullins M., Hammerschmidt M., Haffter P., and Nüsslein-Volhard C. 1994. Large-scale mutagenesis in the zebrafish: In search of genes controlling development in a vertebrate. *Curr. Biol.* **4:** 189.

Rose J.K. and Bergmann J.E. 1982. Expression from cloned cDNA of cell surface and secreted forms of the glycoprotein of vesicular stomatitis virus in eucaryotic cells. *Cell* **30**: 753.

Schupbach T. and Wieschaus E. 1986. Maternal-effect mutations altering anterior-posterior pattern of the *Drosophila* embryo. *Roux's Arch. Dev. Biol.* **195**:302.

Streisinger G., Walker C., Dower N., Knauber D., and Singer F. 1981. Production of clones of homozygous diploid zebrafish (*Brachydanio rerio*). *Nature* **291**: 293.

Stuart G.W., Vielkind J.R., McMurray J.V., and Westerfield M. 1990. Stable lines of transgenic zebrafish exhibit reproducible patterns of transgene expression. *Development* **109**: 577.

Timmermans L.P.M. and Taverne N. 1989. Segregation of primordial germ cells: Their numbers and fate during development of *Barbus conchonius* (*Cyprinidae*, Teleostei) as indicated by 3H-thymidine incorporation. *J. Morphol.* **202**: 225.

Walker C. and Streisinger G. 1983. Induction of mutations by gamma-rays in pregonial germ cells of zebrafish embryos. *Genetics* **103**:125.

Weinberg E.S., Allende M.L., Kelly C.S., Abdelhamid A., Murakami T., Andermann P., Doerre O.G., Grunwald D.J., and Riggelman R. 1996. Developmental regulation of zebrafish *myoD* in wild type and *no tail* and *spadetail* embryos. *Development* **122**: 271.

Westerfield M. 1995. *The Zebrafish book: A guide for the laboratory use of zebrafish* (Danio rerio), 3rd edition. University of Oregon Press, Eugene.

Westerfield M., Liu D.W., Kimmel C.B., and Walker C. 1990. Pathfinding and synapse formation in a zebrafish mutant lacking functional acetylcholine receptors. *Neuron* **4**:867.

Wiley C., Ed. 1996. Zebrafish. *Development*, vol. 123.

Yoon C., Kawakami K., and Hopkins N. 1997. Zebrafish *vasa* homologue RNA is localized to the cleavage planes of 2- and 4-cell stage embryos and is expressed in the primordial germ cells. *Development* **124**: 3157.

Zavada J. 1972. Pseudotypes of vesicular stomatitis virus with the coat of murine leukemia and of avian myeloblastosis virus. *J. Gen. Virol.* **125**: 183.

Graded Sonic Hedgehog Signaling and the Specification of Cell Fate in the Ventral Neural Tube

J. Ericson,[1] J. Briscoe,[1] P. Rashbass,[2] V. van Heyningen,[2] and T.M. Jessell[1]
[1]*Howard Hughes Medical Institute, Center for Neurobiology and Behavior, Deptartment of Biochemistry and Molecular Biophysics, Columbia University, New York, New York 10032;* [2]*MRC Human Genetics Unit, Western General Hospital, Edinburgh, United Kingdom*

The patterning of cell types in vertebrate embryos depends on the function of organizing centers, specialized cell groups that direct the fate of adjacent cells through the secretion of inductive factors (Gurdon 1987). During early development of the vertebrate nervous system, the identity and pattern of cell types generated in the ventral half of the neural tube are controlled by signals provided initially by an axial mesodermal organizing center, the notochord (for review, see Tanabe and Jessell 1996). Assays of neural cell differentiation in vitro and analyses of mutant mouse and zebrafish embryos (for review, see Placzek 1995) have provided evidence that the notochord is the source of two operationally distinct inductive signals: a local signal that induces the differentiation of floorplate cells at the ventral midline of the neural tube and a longer-range signal that induces motor neurons (MNs) and ventral interneurons in more lateral positions. Although floorplate differentiation requires induction by the notochord, once induced, floorplate cells express the same short- and long-range signaling activities (Yamada et al. 1991; Placzek et al. 1993; Placzek 1995).

The best candidate as a mediator of both the short- and long-range inductive activities of the notochord is a secreted glycoprotein Sonic Hedgehog (Shh). Shh is expressed by the notochord and later by the floorplate over the period that these two midline cell groups exhibit their inductive activities (Echelard et al. 1993; Krauss et al. 1993; Riddle et al. 1993; Roelink et al. 1994). Gain- and loss-of-function studies of Shh signaling have provided strong evidence that Shh mediates the induction of floorplate differentiation by the notochord. In mice, floorplate differentiation does not occur when Shh signaling is eliminated by targeted mutation of the *Shh* gene (Chiang et al. 1996). Conversely, the biologically active amino-terminal fragment of Shh (Shh-N) (Lee et al. 1994; Porter et al. 1995) is able to induce the expression of floorplate markers, notably the transcription factor HNF3β (Hynes et al. 1995b; Roelink et al. 1995; Ericson et al. 1996). The induction of HNF3β in neural plate cells is an early and direct response to Shh signaling (Ruiz i Altaba et al. 1995b; Sasaki et al. 1997). Ectopic expression of HNFβ and related winged helix proteins can induce floorplate differentiation in neural tube cells (Ruiz i Altaba et al. 1993, 1995a; Sasaki and Hogan 1994; Hynes et al. 1995a). Moreover, HNF3β interacts directly with enhancer elements in the *Shh* gene (Chang et al. 1997) and may also activate genes such as *Netrin-1* (Serafini et al. 1994; Tanabe et al. 1995) and *F-Spondin* (Klar et al. 1992; Ruiz i Altaba et al. 1993) which are thought to contribute to the functions of the floorplate in axon guidance.

Many questions about the role of Shh signaling in the patterning of the ventral neural tube, however, remain unresolved. One issue is the precise contribution of Shh signaling to the specification of MN and ventral interneuron identity. Shh can induce MNs in vitro (Roelink et al. 1994, 1995; Marti et al. 1995; Tanabe et al. 1995), but MNs are generated well after neural tube closure and long after neural progenitor cells have first been exposed to notochord-derived signals (Ericson et al. 1992). It is therefore difficult to exclude the possibility that the generation of ventral neurons in response to Shh involves the induction of an intermediary factor. Thus, Shh could conceivably control cell pattern in the ventral neural tube entirely through local signaling with its long-range patterning activities dependent on a secondary diffusible factor. Precedent for such a relay mechanism has emerged from studies of *Drosophila* appendage formation where many of the patterning activities initiated by Hedgehog are mediated by the local induction of long-range signaling factors, notably, the transforming growth factor-β (TGF-β)-like protein Decapentaplegic (Dpp) (Zecca et al. 1995; Lecuit et al. 1996; Nellen et al. 1996).

A second unresolved issue is how Shh signaling is interpreted by ventral progenitor cells such that distinct cell types are generated at different dorsoventral positions within the ventral neural tube. Several key components involved in the cellular transduction of Hedgehog (Hh) signals have now been defined, revealing that the initial steps of Hh signal transduction in vertebrates and *Drosophila* are highly conserved. Signaling appears to be initiated by the binding of Hh ligands to transmembrane proteins of the Patched family (Chen and Struhl 1996; Marigo et al. 1996; Stone et al. 1996). In its ligand-bound form, Patched is no longer able to repress the activity of a second transmembrane protein, Smoothened (see Alcedo et al. 1996; Stone et al. 1996; van den Heuvel and Ingham 1996). As a consequence of this derepression, Smoothened is thought to transmit an intracellular signal that culminates in the activation of zinc finger transcription factors of the Gli/Ci family (Alexandre et al. 1996; Dominguez et al. 1996; Lee et al. 1997; Von Ohlen et al. 1997). The relevant targets of Gli signaling in vertebrate neural cells remain poorly characterized, although recent studies have provided evidence that *HNF*3β transcription in midline neural cells is activated directly by Gli proteins (Lee et al. 1997; Sasaki et al. 1997).

Several homeodomain transcription factors, notably, members of the Msx, Pax, and Nkx2 protein families, are expressed by neural plate and neural tube cells (Walther and Gruss 1991; Liem et al. 1995; MacDonald et al. 1995). At forebrain levels, Shh signaling represses the expression of *Pax6* and activates *Nkx.2* gene expression (Barth and Wilson 1995; Ekker et al. 1995; Ericson et al. 1995; MacDonald et al. 1995). The expression of *Pax6* in the spinal cord has also been shown to be regulated by notochord-derived signals (Goulding et al. 1993). In addition, mutations in *Pax6* result in pronounced defects in eye and forebrain development (Hill et al. 1991; Stoykova et al. 1996; Grindley et al. 1997; Warren and Price 1997). Taken together, these findings raise the possibility that members of the *Pax* and *Nkx2* gene families might contribute to the Shh-mediated control of neural cell identity and pattern.

In this paper, we summarize recent studies that have analyzed the mechanism by which Shh controls the identity and pattern of ventral cell types, focusing on caudal levels of the neuraxis that give rise to the spinal cord and hindbrain. A combination of in vitro assays of neural cell differentiation and studies of ventral patterning in mutant mice has resulted in two main findings: (1) The generation of different cell types at distinct positions in the ventral half of the neural tube is established by a gradient of Shh signaling activity and (2) Pax6 functions as a key intermediate in the regulation of progenitor cell identity and neuronal fate in response to Shh signaling.

Table 1. Ventral Cell Types Defined by Transcription Factor Expression

Cell type	Progenitor cell origin	Transcription factor code
V1 interneurons	$Pax6^{high}$	Enl, Pax2, Lim1/2
V2 interneurons	$Pax6^{intermediate}$	Chx10, Lim3, Gsh4
Dorsal (somatic) MNs	$Pax6^{low}$	Isl1, Isl2, Lim3(t), Gsh4(t)
Ventral (visceral) MNs	Nkx2.2	Isl1, Gsh4
Floorplate	—	HNF3β, HNF3α

This subdivision of motor neurons applies to the rostral spinal cord/caudal hindbrain level. (t) indicates transient expression. For details of homeobox genes used as markers, see Li et al. (1994) (Gsh4); Liu et al. (1994) (Chx10); Pfaff et al. (1996) (Isl1/2); Tsuchida et al. (1994) (Lim 3).

RESULTS

Shh Induces Different Ventral Neuronal Classes at Distinct Concentration Thresholds

To begin to examine how the identity and pattern of neuronal cell types in the ventral spinal cord are generated, we have focused on three major classes of neurons that can be defined by the expression of homeodomain proteins (Table 1). One class of interneurons (termed V1) is defined by coexpression of En1, Lim1/2, and Pax2 (Fig. 1A) and is generated in the dorsal-most region of the ventral spinal cord (Fig. 1A). A second class of interneurons (V2) is defined by coexpression of Chx10, Lim3, and Gsh4 (Fig. 1A) and is generated in the intermediate region of the ventral spinal cord, ventral to V1 interneurons (Fig. 1A). The third class (MNs) is defined by expression of Isl1 (Fig. 1A) and is generated ventral to V2 interneurons (Fig. 1A). Floorplate cells, defined by expression of HNF3β, differentiate ventral to MNs at the midline of the neural tube (Ruiz i Altaba et al. 1995b).

We first examined whether the differentiation of floorplate cells, MNs, V1, and V2 neurons is each induced by Shh and if so, whether the induction of each cell type is achieved at a distinct concentration threshold. To test this, stage-10 chick intermediate neural plate ([i]) explants (Yamada et al. 1993) were isolated and grown in vitro, alone or in the presence of recombinant Shh-N. When grown alone, [i] explants do not generate V1 or V2 neurons, MNs, or floorplate cells, whereas the generation of floorplate cells and all three neuronal classes is induced by Shh-N (Fig. 1B). The generation of V1 neurons requires the lowest concentration of Shh-N, V2 neurons an approximately twofold higher concentration, and MNs a further approximately twofold elevation (Fig. 1B) (Ericson et al. 1997). Floorplate cells require a two- to threefold increase in Shh-N concentration over that required for MN generation (Fig. 1B) (Roelink et al. 1995). These in vitro results suggest (Fig. 1C) that the identity and position of generation of different cell types in the ventral spinal cord are achieved by the exposure of progenitor cells to different Shh-N concentrations.

Requirement for Shh Signaling in Ventral Cell-type Generation

Antibodies that block the function of Shh-N (Marti et al. 1995; Ericson et al. 1996) have been used to address whether the generation of these ventral cell types requires Shh signaling. The requirement for Shh signaling by the notochord in the induction of floorplate differentiation has been tested in conjugates of notochord and [i] explants grown in the presence or absence of anti-Shh IgG. The induction of HNF3β expression by the notochord is blocked in the presence of anti-Shh IgG (Fig. 2a,b), showing that Shh signaling is required for the induction of floorplate differentiation. The induction of MNs by the notochord is also blocked in the presence of anti-Shh IgG (Fig. 2c,d) (Marti et al. 1995), indicating that the notochord does not secrete a factor that can induce MNs independently of Shh.

Because MNs and ventral interneurons differentiate well after the onset of Shh expression by the floorplate (Riddle et al. 1993; Ericson et al. 1996), we examined whether the generation of ventral neuronal cell types re-

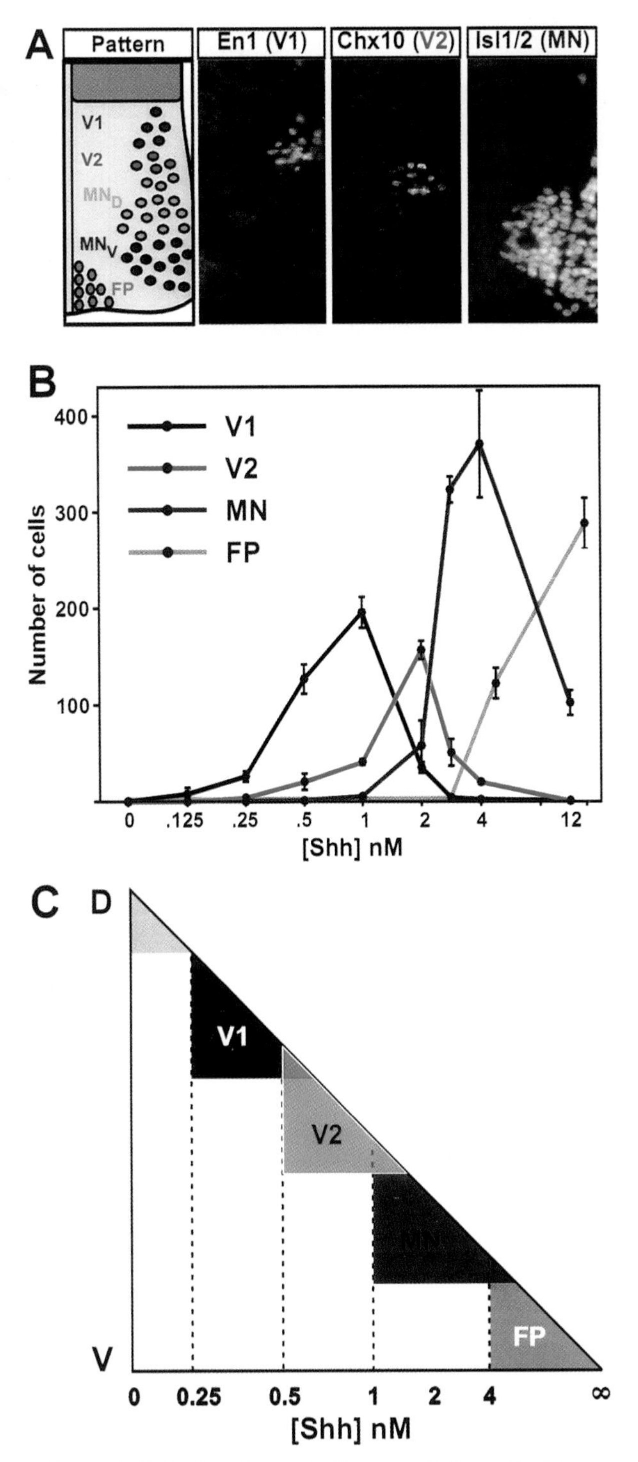

Figure 1. Induction of ventral cell types at different Shh-N concentration thresholds. (*A*) Neuronal subtype identity in the ventral spinal cord is defined by homeodomain protein expression. En1 expression defines V1 neurons, Chx10 expression defines V2 neurons, and Isl1/2 expression defines MNs. Presented here is a summary of the position of generation of floorplate cells and ventral neurons. (*B*) Quantitative analysis of the induction of En1 (V1), Chx10 (V2), and Isl1/2 (MN) neurons and HNF3β floorplate (FP) cells in stage-10 [i] explants at different Shh-N concentrations. Data for HNF3β induction are derived from Roelink et al. (1995). (*C*) Schematic diagram based on the French Flag model (Wolpert 1969) showing the relationship between Shh concentration, cell identity, and dorsoventral (D/V) position in the ventral neural tube.

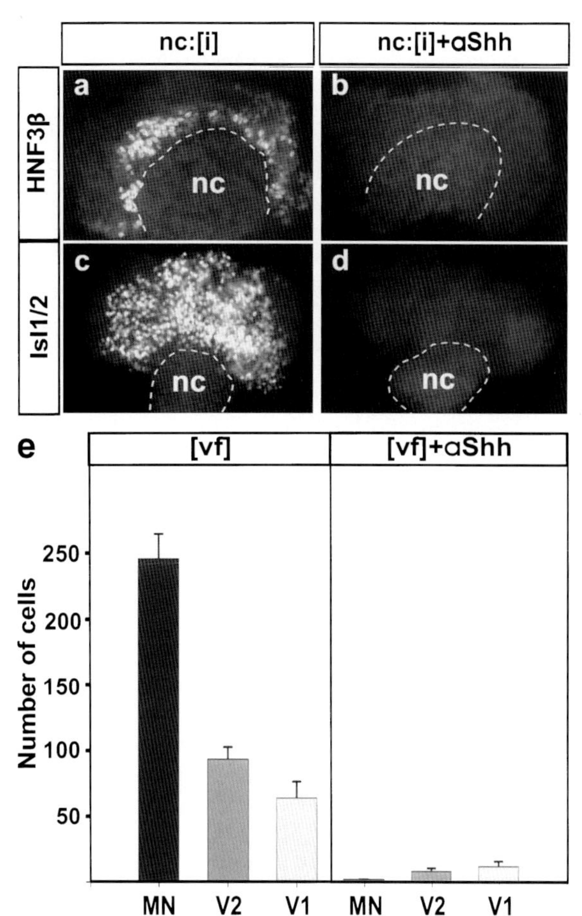

Figure 2. Requirement for Shh signaling in ventral neuronal differentiation. (*a,b*) Notochord (nc)-mediated induction of HNF3β cells (*a*) is blocked by anti-Shh IgG (*b*). (*c,d*) Notochord-mediated induction of Isl1/2 motor neurons (*c*) is blocked by anti-Shh (*d*). (*e*) Cells in stage-10 ventral neural tube/floorplate [vf] explants grown in the absence of anti-Shh IgG generate many MNs, V1, and V2 interneurons. In the presence of anti-Shh IgG (aShh), progenitor cells generate few if any MNs, V1, or V2 interneurons (for details, see Ericson et al. 1997).

quires Shh signaling from the floorplate. To test this, stage-10 neural tube explants comprising the floorplate and adjacent ventral tissue ([vf] explants) were grown alone or in the presence of anti-Shh IgG. When grown alone [vf] explants generated about 250 MNs, about 100 V2 neurons, and about 60 V1 neurons (Fig. 2e), whereas in the presence of anti-Shh IgG, less than 3 MNs, about 10 V1, and about 10 V2 neurons were generated (Fig. 2e). Thus, the differentiation of MNs, V1, and V2 neurons as well as floorplate cells requires Shh signaling from the floorplate.

Early Exposure of Neural Plate Cells to Shh Is Required for Floorplate and MN Differentiation

Soon after its formation, neural plate cells are exposed to Shh signals from the notochord (Liem et al. 1995), but the critical period for Shh signaling in the generation of ventral cell types has not been addressed. To determine

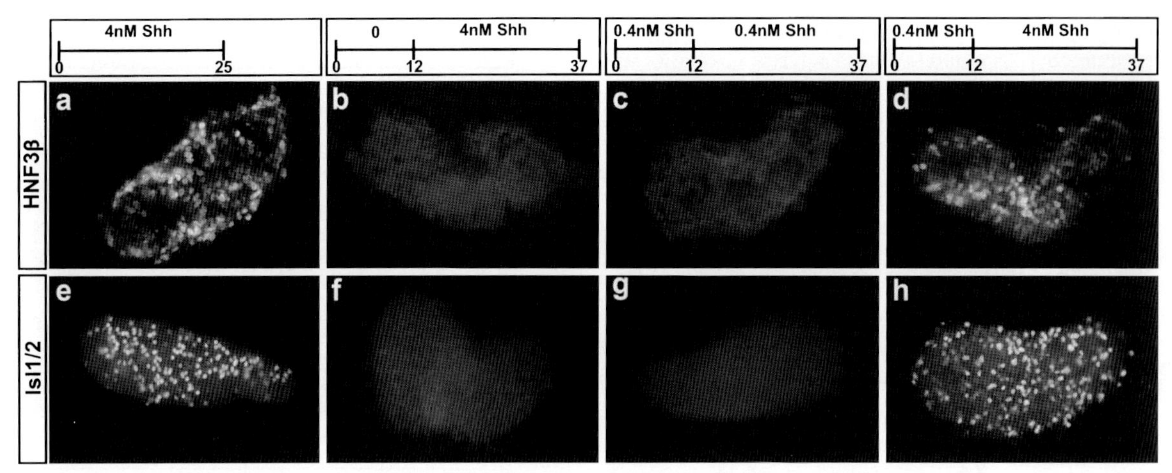

Figure 3. Early requirement for Shh signaling in floorplate and motor neuron generation. (*a,e*) HNF3β floorplate cells (*a*) and Isl1/2 motor neurons (*e*) are generated in stage-10 [i] explants grown in 4 nM Shh-N for 25 hr. (*b,f*) Neither HNF3β cells (*b*) nor Isl1/2 cells (*f*) are generated when [i] explants are grown for 12 hr in the absence of Shh-N followed by 25 hr in 4 nM Shh-N. (*c,g*) Neither HNF3β cells (*c*) nor Isl1/2 cells (*g*) are generated when [i] explants are grown for 37 hr in 0.4 nM Shh-N. (*d,h*) HNF3β cells (*d*) and Isl1/2 cells (*h*) are generated when [i] explants are grown for 12 hr in 0.4 nM Shh-N followed by 25 hr in 4 nM Shh-N.

whether early exposure of neural plate cells to Shh signals is required, stage-10 [i] explants were grown alone for 12 hours followed by the addition of 4 nM Shh-N for 25 hours. Cells deprived of early Shh signaling did not generate floorplate cells or MNs, despite their later exposure to a high concentration of Shh-N (Fig. 3b,f). This result provides evidence that early exposure to Shh is required for the generation of ventral cell types and shows also that neural plate cells rapidly lose their competence to generate floorplate cells and MNs when deprived of early Shh signals.

These findings raised the issue of whether the concentration of Shh-N necessary to maintain the competence of neural plate cells is different from that required later for the production of ventral cell types. To test this, stage-10 [i] explants were exposed continuously to a low concentration (0.4 nM) of Shh-N for 37 hours. At this Shh-N concentration, no floorplate cells or MNs were generated (Fig. 3c,g). In contrast, exposure of explants to 0.4 nM Shh-N for 12 hours followed by addition of 4 nM Shh-N for an additional 25 hours was sufficient to induce both floorplate cells and MNs (Fig. 3d,h). These findings show that the level of Shh signaling required during the early critical period is lower than that required later to generate specific ventral cell types.

Prolonged Requirement for Shh Signaling in MN Differentiation

During what period is Shh signaling required for the generation of ventral cell types? To address this question, stage-12 [v] explants, devoid of floorplate, were grown in vitro for 24 hours. Cells in such explants did not express HNF3β or *Shh* mRNA and did not contain detectable Shh immunoreactivity (Fig. 4c,d). Nevertheless, these explants generated approximately 50 MNs (Fig. 4a), raising the possibility that at this stage, some ventral progenitors are able to differentiate into MNs independent of further Shh signaling. However, this is not the case because MN generation in such explants is blocked by addition of anti-Shh IgG (Fig. 4b). The ability of anti-Shh antibodies to block MN generation in [v] explants provides evidence that the Shh protein required locally for the conversion of ventral progenitor cells into MNs is obtained by diffusion from the notochord or, more probably, from the floorplate. These results show that the differentiation of ventral progenitor cells into MNs requires a prolonged period of Shh signaling.

To define how long Shh signaling is required for the conversion of ventral progenitors into MNs, [vf] explants were isolated from stage-10–17 embryos, and MN differentiation was analyzed in vitro (Ericson et al. 1996). HNF3β and Shh expression is detected in these explants,

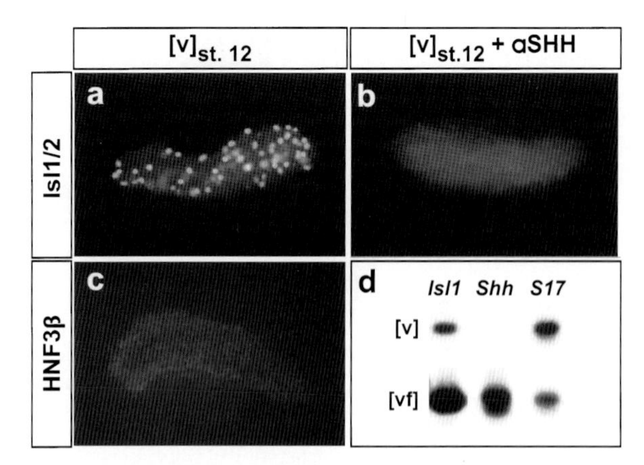

Figure 4. Ventral progenitor cells require prolonged Shh signaling for motor neuron generation. (*a*) Stage-12 [v] explants grown in the absence of floorplate cells generate Isl1/2 motor neurons after 36 hr. (*b*) Stage-12 [v] explants grown in the presence of anti-Shh IgG do not generate Isl1/2. (*c*) Stage-12 [v] explants do not contain HNF3β cells. (*d*) RT-PCR analysis of *Isl1*, *Shh*, and *S17* expression in stage-12 ventral [v] or [vf] explants grown for 24 hr in vitro.

and many MNs are generated (Fig. 5A,B). Nevertheless, the addition of anti-Shh IgG completely blocks MN generation in [vf] explants derived from stage-10–12 embryos (Fig. 5A,C). MN generation in explants derived from stage-13 embryos, although markedly reduced by addition of anti-Shh IgG, was not completely abolished (Fig. 5B,D). Thus, some ventral progenitor cells present at stage 13 are able to generate MNs in an Shh-independent manner. These Shh-independent progenitors appear to be the most advanced in their developmental program since the generation of MNs during the first 10-hour period in stage-13 [vf] explants is not inhibited by anti-Shh IgG, whereas the generation of MNs during the subsequent 26-hour period is completely blocked (Fig. 5D).

Shh-independent ventral progenitor cells are not detected until stage 13, shortly before the appearance of the first postmitotic MNs. The acquisition of Shh independence therefore probably occurs during the final cell division cycle of ventral progenitors (Langman et al. 1966). To determine the stage of the final cell cycle at which Shh dependence is lost, stage-13 [vf] explants were maintained in the presence of bromodeoxyuridine (BrdU), and the fraction of MNs that had incorporated BrdU was determined under conditions in which Shh signaling was either permitted or blocked. Of the MNs generated in the presence of anti-Shh IgG, the vast majority did not incorporate BrdU (Fig. 5F,G) and thus had completed their final round of DNA synthesis. In contrast, virtually all MNs generated in the absence of anti-Shh IgG did incorporate BrdU (Fig. 5E,G) and the few MNs that did not presumably derived from Shh-independent ventral progenitors. These results suggest that the differentiation of ventral progenitors into MNs depends on a prolonged period of Shh signaling that starts soon after the neural plate forms and persists late into the final S phase of the ventral progenitor cell.

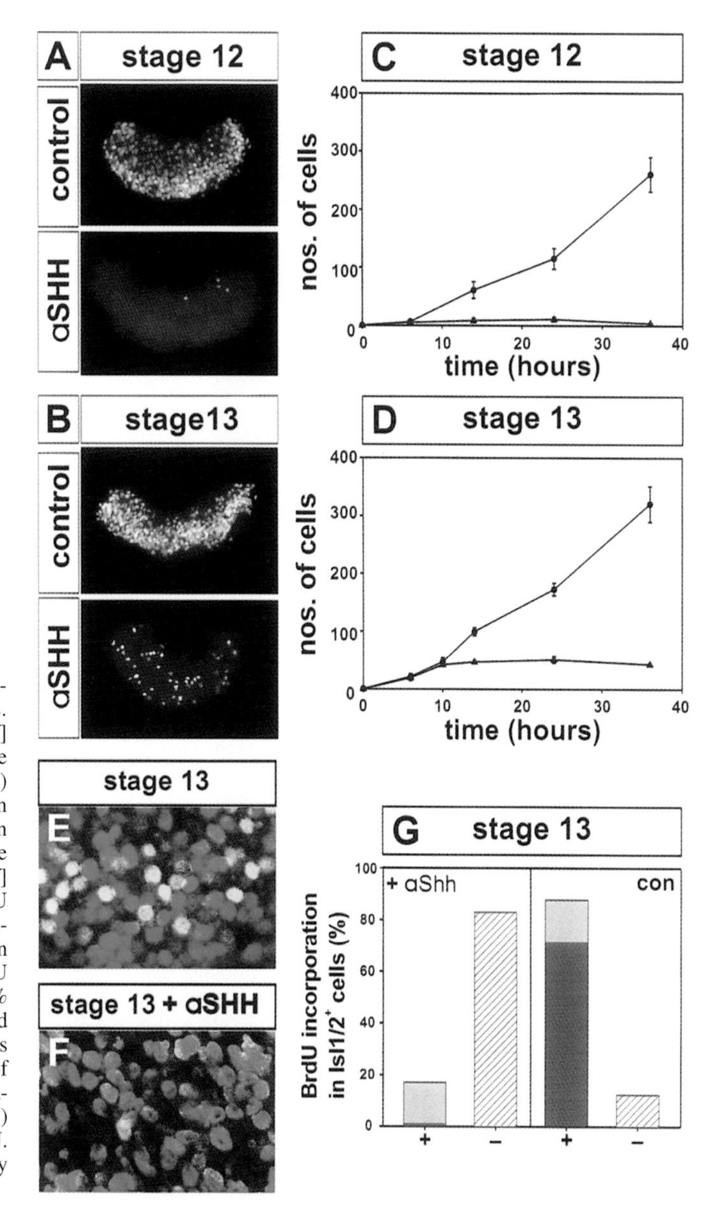

Figure 5. Prolonged requirement for Shh in the conversion of ventral progenitors into motor neurons. (*A,B*) Isl1/2 cells in stage-12 (*A*) and stage-13 (*B*) [f] explants grown in vitro for 34–36 hr, in the absence (control) or presence (aSHH) of anti-Shh IgG. (*C,D*) Temporal analysis of Isl1/2 motor neuron generation in stage-12 (*C*) and stage-13 (*D*) [vf] explants grown in vitro in the presence (*closed triangle*) or absence (*closed circle*) of anti-Shh IgG. (*E,F*) Stage-13 [vf] explants were grown in vitro in the presence of BrdU for 36 hr in the absence (*E*) or presence (*F*) of anti-Shh IgG. In the absence of anti-Shh IgG, more than 85% Isl1/2 motor neurons had incorporated BrdU (*yellow* cells). In the presence of anti-Shh IgG, 83% Isl1/2$^+$ motor neurons (*red*) had not incorporated BrdU (*green*), and the 17% of Isl1/2 motor neurons that were double labeled incorporated low levels of BrdU. (*G*) Quantitative analysis of BrdU incorporation into motor neurons in stage-13 [vf] explants. (+) BrdU labeled MNs; (−) MNs not labeled with BrdU. Intensity of shading in plus column indicates intensity of BrdU labeling.

Graded Shh Signaling and the Control of Progenitor Cell Identity in the Ventral Neural Tube

The studies described above leave unresolved how progenitor cells interpret graded Shh signals to generate different classes of ventral neurons. To begin to address this question, we have attempted to characterize the function of transcription factors whose expression by ventral progenitor cells is regulated by Shh signaling. We discuss here the potential roles of the Pax class of paired homeodomain proteins (Mansouri et al. 1996).

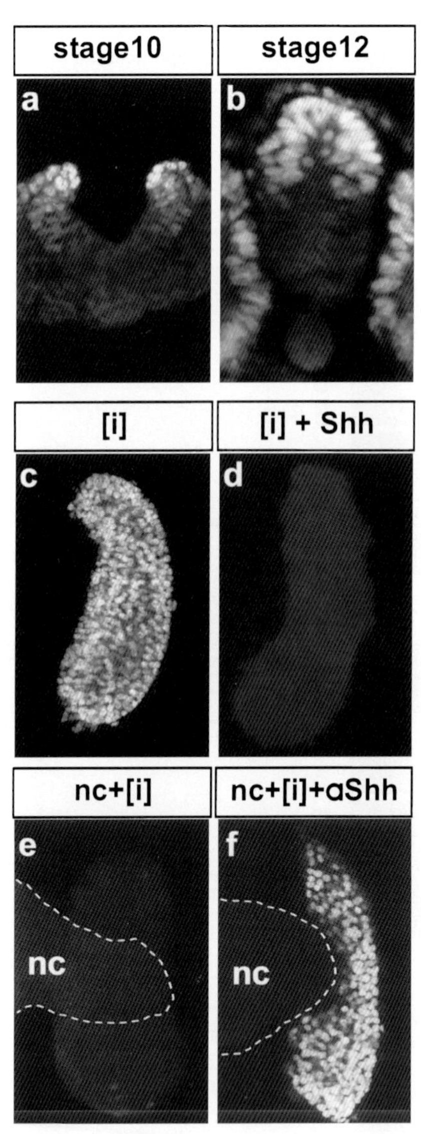

Figure 6. Notochord-derived Shh represses Pax7 expression. (*a*) Pax7 expression is extinguished from cells in the medial neural plate by stage 10. (*b*) Pax7 expression is restricted to cells in the dorsal neural tube at stage 12. Labeling of somite cells is also detected. (*c*) Pax7 is expressed by more than 95% of cells in [i] explants grown alone for 22 hr. (*d*) Shh-N (4 nM) represses Pax7 expression. (*e*) Notochord represses Pax7 expression. Border of notochord (nc) tissue is shown by the dotted line. (*f*) Anti-Shh IgG blocks the notochord (nc) repression of Pax7 expression. Pax7 is now detected in more than 90% of cells in neural plate explants.

The homeobox genes *Pax3* and *Pax7* are initially expressed by cells along the entire mediolateral extent of the neural plate, but their expression is rapidly extinguished from medial neural plate cells with the consequence that after neural tube closure, their expression is restricted to dorsal progenitor cells (Fig. 6a,b) (Liem et al. 1995; Ericson et al. 1997).

To test whether Shh signaling from the notochord is responsible for the repression of expression of these homeobox genes in medial neural plate cells, stage-10 [i] explants were grown alone or with the notochord, and the expression of Pax7 was analyzed. More than 95% of cells in [i] explants grown alone expressed Pax7, whereas expression was completely eliminated from neural plate cells that were grown in contact with the notochord or exposed to Shh-N (Fig. 6c–e). *Pax3* and Msx1/2 expression in such explants is also repressed by Shh-N (Liem et al. 1995). Importantly, addition of anti-Shh IgG blocks the repression of Pax7 expression by the notochord (Fig. 6f). These results, taken together with the detection of *Pax3* expression at the ventral midline of the neural tube in *Shh* null mice (Chiang et al. 1996), provide evidence that Shh mediates the notochord-induced repression of *Pax3* and *Pax7* in neural progenitor cells that populate the ventral half of the neural tube.

We also examined whether the $Pax7^{off}$ state of neural progenitors can be maintained independent of Shh signaling. To test this, stage-10 [v] explants were isolated after the expression of Pax7 had been extinguished and grown in vitro. Pax7 is reexpressed in more than 95% of progenitor cells in these explants (Ericson et al. 1996), indicating that neural cells require continued Shh signaling from the notochord to maintain their $Pax7^{off}$ state and thus the potential to generate ventral cell types. In contrast, cells in [v] explants derived from stage-12 embryos did not reexpress Pax7 when grown alone or with anti-Shh IgG and, consequently, can generate ventral cell types (Ericson et al. 1996). These findings indicate that between stages 10 and 12, naive neural plate cells are converted to ventral progenitors in response to Shh signaling.

Several distinct classes of neurons are generated from ventral progenitors at stages after Pax7 and Pax3 expression has been extinguished. This observation led us to define genes expressed by ventral progenitor cells that might be involved in the interpretation of graded Shh signaling. We focus here on two homeodomain proteins, Pax6 and Nkx2.2, both of which are expressed by dividing progenitor cells within the ventral neural tube (Walther and Gruss 1991; Price et al. 1992; Shimamura et al. 1995).

At neural plate and early neural tube stages, Pax6 is expressed by cells at all dorsoventral positions with the exception of floorplate cells at the ventral midline (Fig. 7a). Nkx2.2 is also detected at low levels at these stages, but its expression is restricted to ventral midline cells (Fig. 7b). From stages 12–16, however, the level of Pax6 in cells adjacent to the floorplate decreases below the limit of detection, and Nkx2.2 expression is initiated by cells that have extinguished Pax6 expression (Fig. 7c–f). From

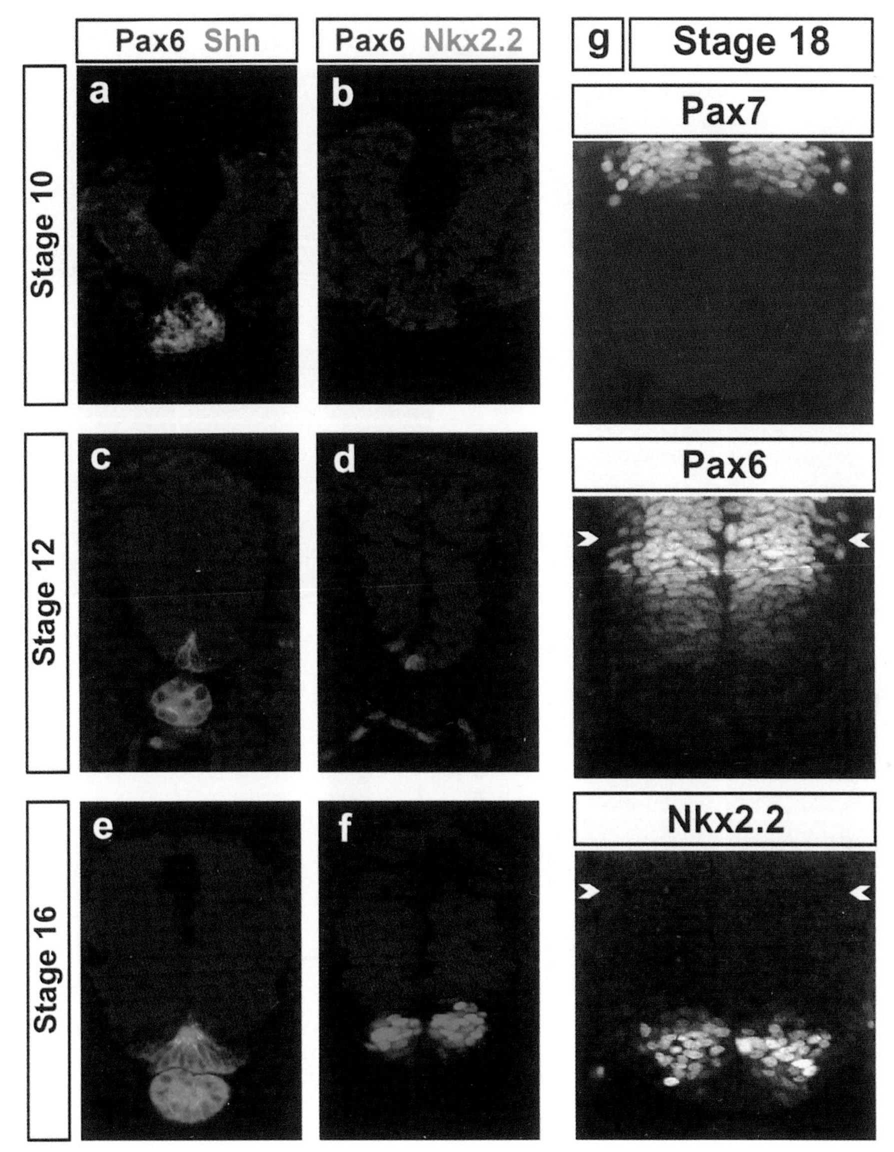

Figure 7. Pax7, Pax6, Nkx2.2, and Shh expression in the neural tube. (*a–f*) Expression of Shh, Pax6, and Nkx2.2 in chick cervical (12-somite level) neural plate (stage 10) (*a,b*), neural tube (stage 12) (*c,d*), and spinal cord (stage 16) (*e,f*). (*g*) Comparison of the domains of Pax7, Pax6, and Nkx2.2 in the ventral spinal cord at stage 18. Arrowheads indicate ventral boundary of Pax7.

stages 13–18, there is an extended ventral^{low}-dorsal^{high} gradient of Pax6 within the ventral neural tube, but the boundary of Nkx2.2 expression is more abrupt (Fig. 7g).

Both V1 and V2 neurons derive from Pax6 progenitor cells, whereas MNs appear to have a dual origin (Ericson et al. 1997). Dorsally positioned MNs derive from Pax6 progenitors, whereas ventrally positioned MNs derive from Nkx2.2 progenitor cells (see Table 1). These two populations of MNs can also be distinguished by their expression of LIM homeodomain proteins. Dorsal MNs coexpress Isl1, Isl2, and Lim3, whereas ventral MNs express Isl1 but not Isl2 or Lim3 (Table 1) (Ericson et al. 1997).

To examine whether the Pax6 gradient and the complementarity in Pax6 and Nkx2.2 expression are established by Shh signaling, stage-10 [i] explants were grown alone or with Shh-N. In the absence of Shh-N, more than 95% of cells coexpressed Pax7 and Pax6 and no cells expressed Nkx2.2 (Fig. 8a). Exposure of [i] explants to concentrations of Shh-N that generate predominantly V1 neurons results in the repression of Pax7 but in an elevation in the level of Pax6 (Fig. 8a). At a Shh-N concentration optimal for the generation of V2 neurons, Pax6 expression is still maintained by all cells but at a lower level and Nkx2.2 cells are not detected. At a twofold higher Shh-N concentration, which generates predominately MNs, Pax6 expression is extinguished and Nkx2.2 is now expressed by most cells (Fig. 8a). Thus, graded Shh signaling is sufficient to establish a gradient of Pax6 expression and to generate distinct Pax6 and Nkx2.2 progenitor cell populations.

To determine whether Shh signaling is required for the generation of these two ventral progenitor cell populations and whether the expression of Pax6 and Nkx2.2 is

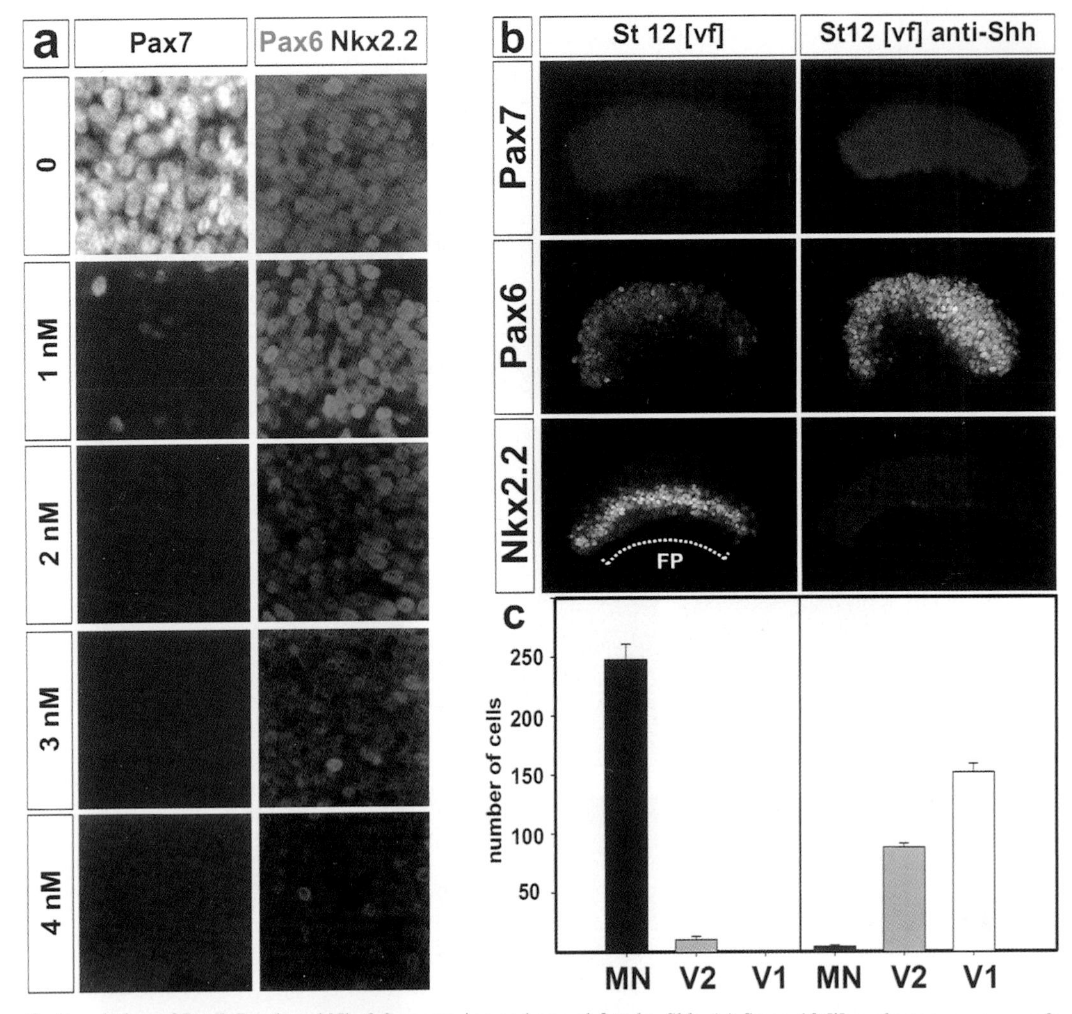

Figure 8. Regulation of Pax7, Pax6, and Nkx2.2 expression and neural fate by Shh. (*a*) Stage-10 [i] explants were grown for 22 hr in different concentrations of Shh-N, and the expression of Pax7, Pax6, and Nkx2.2 was determined. (*b,c*) Cells in stage-12 [vf] explants grown in the absence of anti-Shh IgG do not express Pax7, and at 22 hr express low levels of Pax6 in cells distant from the floorplate (FP) and express Nkx2.2 in cells close to the floorplate (*b*). At 38 hr, these explants generate many MNs, few V2 neurons, and no V1 neurons (*c*). In the presence of anti-Shh IgG, cells do not reinitiate Pax7 expression, express high levels of Pax6, and do not express Nkx2.2 (*b*). Cells in these explants generate few MNs but many V2 and V1 interneurons (*c*).

predictive of neuronal identity, we blocked Shh signaling in stage-12 [vf] explants. In [vf] explants grown in the absence of anti-Shh IgG, Pax6 is expressed at a low level by cells located at a distance from the floorplate, whereas Nkx2.2 is expressed at a high level by cells adjacent to the floorplate (Fig. 8b). These explants generate many MNs but few V2 or V1 neurons (Fig. 8c). In contrast, in [vf] explants grown in the presence of anti-Shh IgG, Pax6 is expressed at high levels by more than 95% of progenitor cells, and no cells express Nkx2.2 (Fig. 8b). These explants generate few MNs but many V2 and V1 neurons (Fig. 8c). Thus, the elimination of Shh signaling shortly before the onset of neurogenesis results in the maintenance of a high level of Pax6, in the absence of Nkx2.2 and in the generation of V1 and V2 neurons rather than MNs. Shh signaling is therefore required for the establishment of the distinct Pax6 and Nkx2.2 ventral progenitor cell populations. In addition, the state of Pax6 and Nkx2.2 expression by ventral progenitors is predictive of the identity of their neuronal progeny.

Control of Ventral Progenitor Cell Identity by Pax6

To determine whether *Pax6* functions in the establishment of ventral cell fates, we have analyzed the *Small Eye (Sey)* mouse that carries a presumed null mutation in the *Pax6* gene (Hill et al. 1991). We first address the contribution of Pax6 to the generation of distinct ventral progenitor cell populations, focusing on spinal cord and hindbrain levels.

In wild-type mouse embryos examined at e10 to e12, the expression of Pax6 in the ventral neural tube is graded in a manner similar to that observed in chick, and the complementarity in expression of Pax6 and Nkx2.2 is also evident (Fig. 9a). At all rostrocaudal levels of the spinal cord and hindbrain of *Sey/Sey* embryos, there is a

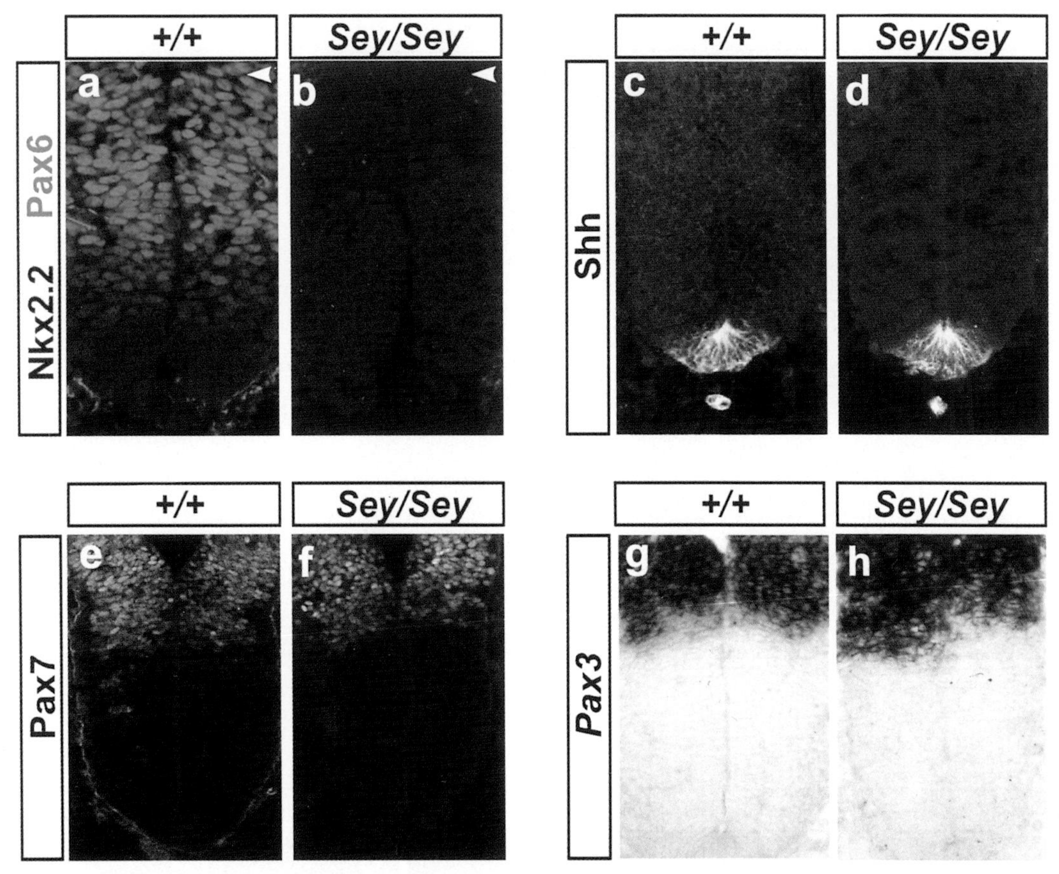

Figure 9. Alteration in progenitor cell domains in *Sey/Sey* embryos. (*a,b*) Pattern of Pax6 and Nkx2.2 at the C1-R7 level of e10 wild-type (*a*) and *Sey/Sey* (*b*) embryos. In wild-type embryos, Pax6 and Nkx2.2 are expressed in complementary domains in the ventral neural tube (*a*). In *Sey/Sey* embryos, Pax6 is not detectable and the Nkx2.2 domain is expanded dorsally (*b*). Arrowheads indicate ventral boundary of Pax7 expression. (*c–h*) The domains of Shh (*c,d*), Pax7 (*e,d*), and *Pax3* (*g,h*) expression are similar in the ventral neural tube of wild-type and *Sey/Sey* embryos.

marked approximately threefold increase in the number of Nkx2.2 cells and a dorsal expansion in the domain occupied by Nkx2.2 progenitor cells (Fig. 9b). The Nkx2.2 progenitor domain, however, never occupies the entire ventral neural tube (Fig. 9b). The location and level of Shh expression and the position of the ventral boundaries of *Pax3* and *Pax7* expression are not altered in *Sey/Sey* embryos, suggesting that the expanded Nkx2.2 domain is not the result of enhanced Shh signaling (Fig. 9c–h). These findings provide evidence that the domain occupied normally by Nkx2.2 progenitor cells is defined indirectly through the Shh-mediated repression of Pax6 and not by a direct requirement for high-level Shh signaling.

Impaired Generation of Ventral Interneurons in the Absence of Pax6 Function

We next examined whether the loss of Pax6 function influences the generation of specific ventral neuronal types, addressing first the fate of ventral interneurons. In wild-type mouse embryos, as in chick, V1 and V2 neurons are generated from Pax6 progenitors (Fig. 10a–c). In *Sey/Sey* embryos examined during the period e10 to e12, postmitotic neurons are generated at all dorsoventral positions as revealed by expression of the pan-neuronal marker Cyn1 (Fig. 10d,h), but V1 neurons are never detected (Fig. 10f). Moreover, V2 neurons are not detected in *Sey/Sey* embryos examined at e10 (Fig. 10g), although by e12, reduced (~50%) numbers of V2 neurons are generated (Ericson et al. 1997). Thus, Pax6 is required for the generation of V1 neurons and also has a profound influence on the generation of V2 neurons.

Control of MN Subtype Identity by Pax6

In *Sey/Sey* embryos, there are marked changes in the subtype identity of MNs, with the precise phenotype varying at different rostrocaudal levels. This variation appears to reflect distinctions in the progenitor cell population from which different MN classes derive. We illustrate the basic features of the role of Pax6 in the specification of MN fate by focusing on the C1-R7 level of the rostral spinal cord and caudal hindbrain.

In wild-type embryos, the C1-R7 level of the neural tube generates somatic MNs of the hypoglossal motor nucleus (hMN) and visceral MNs of the vagal motor nucleus (vMN) (Kuratani et al. 1988; Kuratani and Tanaka 1990). These two MN subclasses are generated in approximately

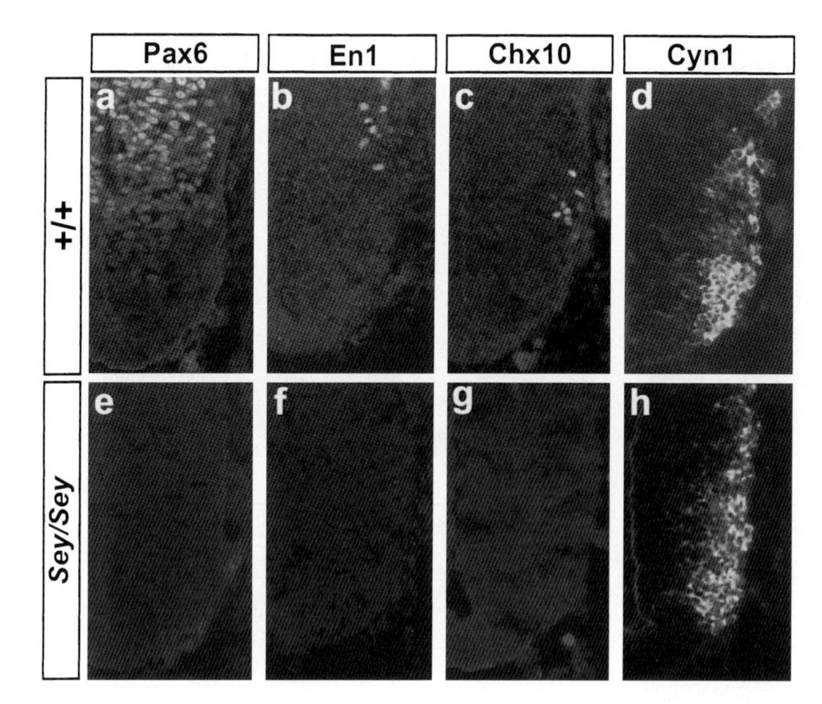

Figure 10. Alteration in ventral interneuron fate in *Sey/Sey* embryos. Shown are sections through the C1-R7 level of e10 (28 somite) wild-type (*a–d*) and *Sey/Sey* (*e–h*) embryos. In wild-type embryos, Pax6 is expressed in ventral progenitors (*a*) and En1 (V1) (*b*) and Chx10 (V2) (*c*) cells are detected. Cyn1 expression (*d*) defines postmitotic neurons. In *Sey/Sey* embryos, Pax6 is not detected (*e*) and no En1 (*f*) or Chx10 (*g*) cells are detected. Cyn1 neurons are generated within the domain normally occupied by En1 and Chx10 cells (*h*).

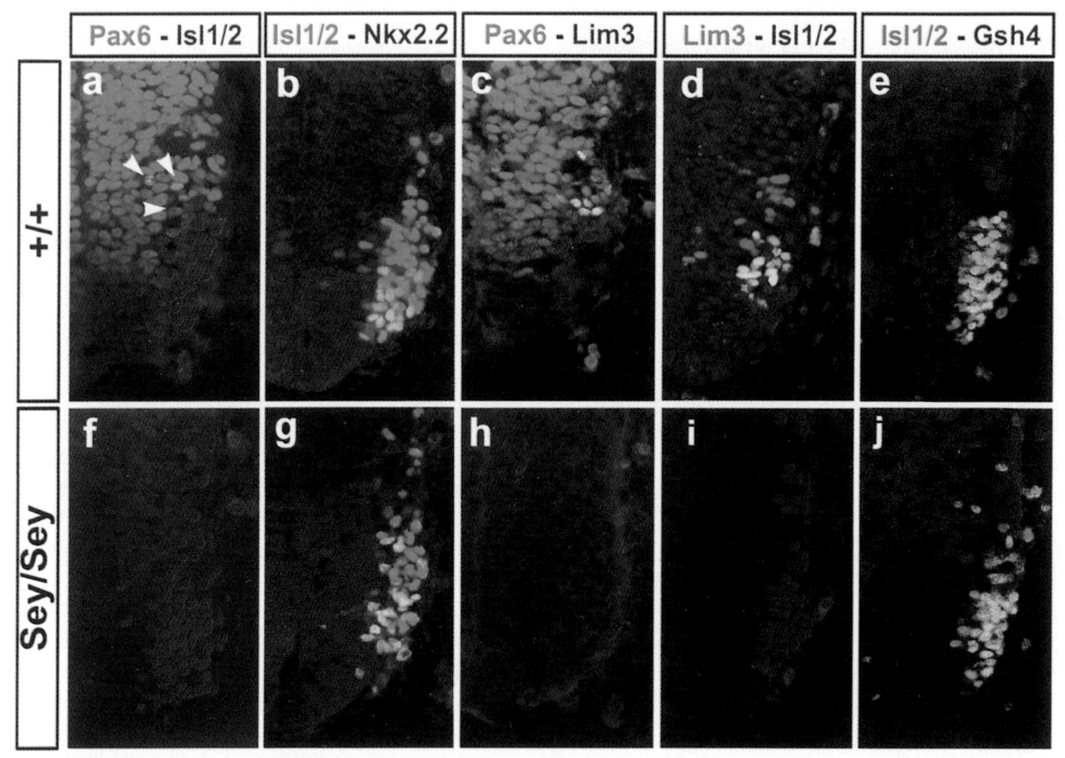

Figure 11. Alteration in LIM homeodomain protein expression by motor neuron *Sey/Sey* embryos. Sections show images of the C1-R7 level of the rostral spinal cord/caudal hindbrain in e10 wild-type (*a–e*) and *Sey/Sey* (*f–j*) embryos. (*a,b*) Dorsal Isl1/2 motor neurons are generated from Pax6 progenitor cells (*a*). Ventrally located Isl1/2 motor neurons derive from Nkx2.2 progenitors (*b*). (*c*) Lim3 neurons derive from Pax6 progenitor cells. (*d*) Lim3 is coexpressed only by dorsal Isl1/2 motor neurons. The dorsal group of Lim3 interneurons does not express Isl1/2 but coexpresses Chx10 (not shown). (*e*) Gsh4 is expressed by both dorsally and ventrally positioned MNs. Expression in ventrally located MNs occurs slightly after that in dorsal MNs. Medially located Gsh4 cells in a dorsal position have not begun to express Isl1/2. (*f,g*) In *Sey/Sey* embryos, Isl1/2 motor neurons are generated (*f*) and derive from Nkx2.2 progenitors (*g*). These cells do not express Isl2 (data not shown). Scattered dorsal Isl1 cells are evident (*f*). (*h*) In *Sey/Sey* embryos, no Lim3 cells are present. (*i*) In *Sey/Sey* embryos, Isl1 motor neurons do not coexpress Lim3. (*j*) In e10 *Sey/Sey* embryos, most MNs have begun to express Gsh4. Note the absence of dorsomedial Gsh4 cells that do not express Isl1/2.

equal numbers and can be distinguished in wild-type embryos by the ventral progenitor cell population from which they derive and later by their profile of LIM homeodomain protein expression, by the migration of their cell bodies, and by their axonal projections (Simon et al. 1994; Varela-Echavarría et al. 1996). hMNs derive from Pax6 cells and coexpress Isl1, Isl2, and transiently, Lim3 and Gsh4 (Fig. 11a–e; Table 1). At e12, hMNs retain expression of Isl1 and Isl2, but most cells no longer express Lim3 or Gsh4. The axons of hMNs project ventrally and exit the spinal cord via the ventral roots and coalesce to form the hypoglossal (XII) nerve (Fig. 12g,j–k). The vMN population is generated more ventrally from Nkx2.2 progenitor cells, and vMNs coexpress Isl1 and Gsh4 but not Isl2 or Lim3 (Fig. 11b–e; Table 1). Subsequently, vMNs migrate dorsally, settle in a dorsolateral position, and project axons via a dorsal exit point to form the vagus (X) nerve (Fig. 12h, j–l).

The differentiation of hMNs and vMNs is markedly altered in *Sey/Sey* embryos. Nkx2.2 cells occupied the entire ventral domain from which MNs are generated and all MNs derive from Nkx2.2 progenitor cells (Fig. 11f,g). In *Sey/Sey* embryos, the total number of MNs, defined by Isl1 expression, is not changed, but the patterns of LIM homeodomain protein expression, MN migration, and axonal projections are affected. From e10, all MNs coexpress Isl1 and Gsh4, but none express Isl2 or Lim3 (Fig. 11g–j). At e11, there is an approximately twofold increase in the number of MNs that migrate dorsally and these MNs settle in a dorsal position characteristic of vMNs (Fig. 12a–f). Moreover by e12, no MNs are detected in a ventral position characteristic of hMNs (Fig. 12e). A very low number of apoptotic cells are detected in the neural tube of both wild-type and *Sey/Sey* embryos between e10 and e12 (data not shown), indicating that the absence of hMNs does not result from their early death. The loss of hMNs has also been detected in rat embryos that carry a mutation in the *Pax6* gene (Osumi et al. 1997).

Analysis of the organization of cranial nerves at e10 and e11 reveals that the ventral roots that form the hypoglossal nerve are absent in *Sey/Sey* embryos, whereas the vagus nerve is present (Fig. 12j–l). Moreover, vMNs can still be identified in *Sey/Sey* embryos by retrograde labeling of the vagus nerve (Fig. 12i). Thus, at the C1-R7 level of the neural tube, the loss of Pax6 function results in a dorsal expansion in the Nkx2.2 progenitor cell population and in the transformation of hMNs into vMNs.

DISCUSSION

Several subclasses of neurons are generated at distinct dorsoventral positions in the ventral spinal cord and hindbrain. The studies described here provide evidence that the identity and position of differentiation of cell types in the ventral neural tube are initiated by the exposure of ventral progenitor cells to small differences in Shh

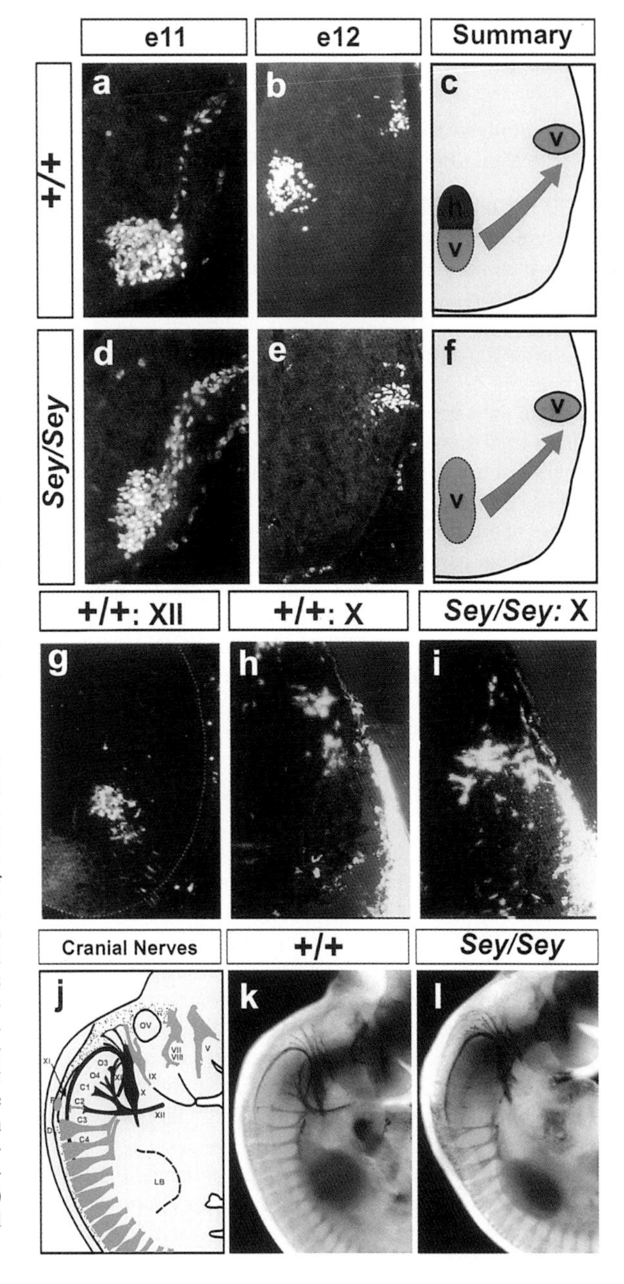

Figure 12. Alteration in motor neuron fate in the rostral spinal cord and caudal hindbrain of *Sey/Sey* embryos. (*a–c*) In wild-type embryos from e11 to e12, a population of Isl1 MNs (vMNs) migrate dorsally, whereas other Isl1 motor neurons (hMNs) remain in a ventromedial position. Summary of migration of vMNs is shown in (*c*). (*d–f*) In *Sey/Sey* embryos analyzed at e11, there is a twofold increase in dorsally migrating motor neurons (*d*). At e12, all MNs occupy a dorsolateral position and ventral MNs are missing (*e*). These dorsally positioned MNs coexpress Isl1 and Gsh4. A summary of MN migration in Sey/Sey embryos is shown in (*f*). (*g–i*) At the R8 level of e12 wild-type embryos, ventrally located motor neurons (hMNs) are retrogradely labeled from the hypoglossal (XII) nerve (*g*), whereas dorsally located motor neurons (vMNs) are retrogradely labeled from the vagus (X) nerve (*h*). vMNs can also be retrogradely labeled from the vagus nerve in *Sey/Sey* embryos (*i*). (*j–l*) Whole-mount labeling of e11 wild-type and *Sey/Sey* embryos using anti-neurofilament antibody reveals that the hypoglossal motor nerve (XII) is absent in *Sey/Sey* embryos, whereas the vagus (X) and spinal accessory (XI) nerves are present.

concentration. The Shh concentration threshold required to generate specific cell types in vitro is inversely related to the distance from the ventral midline in vivo at which each neuronal type is generated. One critical output of graded Shh signaling appears to be the regulation of *Pax* gene expression. In particular, the repression by Shh of *Pax6* expression results in the generation of a second population of ventral progenitor cells, defined by the expression of Nkx2.2. Elimination of Pax6 function results in a dorsal-to-ventral transformation in the identity of ventrally located progenitor cells and in a consequent change in MN fate. The lack of Pax6 function in dorsally located ventral progenitor cells results in the loss of certain ventral interneurons. Pax6 therefore appears to be a key intermediate in the Shh-dependent control of neuronal subtype identity in the ventral spinal cord and hindbrain. We discuss the implications of these findings for the signaling role of Shh, for the control of ventral progenitor cell identity, and for the specification of ventral neuronal subtypes.

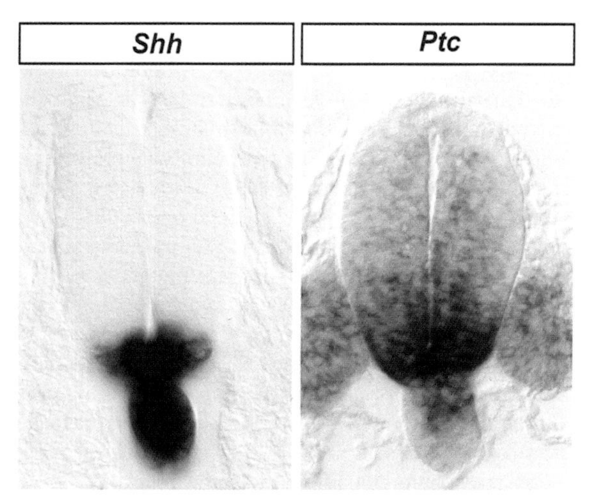

Figure 13. *Shh* and *Ptc* expression domains in the ventral neural tube. In stage-10 embryos, expression of *Shh* is restricted to floorplate cells, whereas *Ptc* is detected in a gradient in the ventral neural tube.

Prolonged Requirement for Long-range Shh Signaling in the Ventral Neural Tube

The present studies show that neural plate cells require exposure to Shh within a critical early period, soon after neural plate formation, in order to maintain their competence to generate ventral neurons. The requirement for Shh signaling in MN generation is, however, maintained long after naive neural plate cells have been converted into ventral progenitor cells. Indeed, ventral progenitors maintain their dependence on Shh until late into S phase of their final division cycle. This late dependence on Shh signaling sets a constraint on the time at which MN identity is determined, suggesting that it cannot be before late S phase of the final progenitor cell division. These observations have parallels in studies of neurogenesis in the mammalian cerebral cortex which have shown that the laminar identity of cortical neurons is determined late in the final progenitor cell division cycle (McConnell and Kaznowski 1991). The late requirement for Shh signaling in MN generation provides a potential reason for the inductive transfer of *Shh* expression from the notochord to the floorplate.

Shh secreted by the notochord and floorplate appears to act both locally and at a distance in the control of cell fate in the ventral neural tube. A local action of Shh from the notochord mediates the induction of floorplate differentiation (Placzek 1995), whereas long-range Shh signaling appears to be required for the generation of ventral neurons. It has been difficult to visualize the low levels of Shh protein that are presumed to diffuse from the notochord and floorplate. However, several lines of evidence indicate that Shh signals act at long range and in addition suggest that the range of Shh signaling initially extends throughout the ventral neural tube. First, the ability of anti-Shh antibodies to block MN differentiation in ventral neural tissue devoid of the floorplate indicates that the conversion of ventral progenitors into MNs depends on the diffusion of Shh from the floorplate. Second, during the early period of Shh signaling, the domain of the neural tube over which Pax7 expression is repressed extends approximately ten cell diameters from the notochord. Third, an elevated level of *patched* mRNA, diagnostic of the exposure of cells to Hh signaling in flies and vertebrates (Perrimon 1995; Goodrich et al. 1996), is detected in the medial neural plate and ventral neural tube over a domain similar to that in which Pax7 is repressed (Fig. 13) (Goodrich et al. 1996; Marigo and Tabin 1996). Finally, since the progenitors of MNs and ventral interneurons remain sensitive to the ambient Shh concentration until shortly before they leave the cell cycle, long-range Shh signaling is likely to involve direct action on ventral progenitor cells and not the activation of secondary inductive signals.

Control of Progenitor Cell Identity

Ventral progenitor cells respond to graded Shh signaling with the establishment of an inverse gradient of Pax6 protein expression. The gradient of Pax6 that is observed in vivo can be recapitulated in vitro by exposure of neural cells in vivo to an approximately tenfold range in Shh concentration. Thus, it is likely that a similar Shh concentration range is sufficient to specify the distinct cell types that are generated in the ventral neural tube in vivo. The conversion of Pax6 progenitors into Nkx2.2 progenitors in response to an approximately twofold change in Shh-N concentration indicates in addition that abrupt changes in gene expression and progenitor cell identity occur in response to graded Shh signals.

The differential sensitivity of Pax7, Pax6, and Nkx2.2 expression to Shh signaling defined in vitro may underlie the progression of progenitor cell specification within the ventral neural tube. The initial establishment of a general ventral progenitor cell population appears to involve the repression of Pax7 and *Pax3* expression (Liem et al.

1995; Ericson et al. 1996; Tremblay et al. 1996). The expression of Pax7 in vitro is repressed by Shh at a concentration approximately eightfold lower than that required to repress Pax6. Thus, the early exposure of neural cells to low concentrations of Shh provided by the notochord may be sufficient to extinguish expression of Pax7 from ventral cells without repressing Pax6. The stable inheritance of the early state of Pax7 expression (Ericson et al. 1996) may contribute to the establishment of a sharp boundary of Pax7 expression in the neural tube. Pax6 expression, in contrast to that of Pax7, remains sensitive to the ambient Shh concentration during the entire early period of ventral neurogenesis. This feature of Pax6 regulation may underlie the formation of the ventral^{low}–dorsal^{high} gradient of Pax6 in the ventral neural tube. The prolonged sensitivity of Pax6 to the local Shh concentration may also explain the late dependence of MN generation on Shh signaling.

The extinction of Pax6 expression at high Shh concentrations appears to result in the generation of a distinct progenitor cell domain, defined by expression of Nkx2.2 (Fig. 14). The marked dorsal expansion of Nkx2.2 expression in *Sey/Sey* mice indicates that the dorsal extent of the Nkx2.2 progenitor cell domain is defined indirectly by the Shh concentration required for extinction of Pax6. In *Sey/Sey* embryos, the dorsal limit of the expanded Nkx2.2 domain could reflect the position at which Shh activity falls below a threshold necessary to activate Nkx2.2 expression. Alternatively, Shh signaling could, in a Pax6-independent manner, define additional domains of gene expression and Nkx2.2 expression may be inhibited dorsally by such genes. The patterns of expression of the *Dbx1* and *Dbx2* genes (Lu et al. 1994; Shoji et al. 1996) make them candidates as inhibitors of the more dorsal expression of Nkx2.2.

Although the expression of Pax6 is graded within the ventral neural tube, it remains unclear whether this gradient is itself instructive in the control of neuronal fate or is an incidental consequence of the requirement for graded Shh signaling. The level of Pax6 expression is critical for appropriate eye development (Hill et al. 1991; Schedl et al. 1996) but in the spinal cord and hindbrain of *Sey*/+ embryos, the pattern of Nkx2.2 expression is not different from that of wild-type littermates, and no MN or interneuron patterning defects are detected (Ericson et al. 1997). The lack of a heterozygote phenotype may have its basis in the observation that in *Sey*/+ embryos, the level of Pax6 in cells in the ventral neural tube is close to the level detected in wild-type embryos (Ericson et al. 1997). Thus, in the spinal cord and hindbrain, it remains possible that Pax6 functions at a single threshold level, defining solely Pax6^{off} and Pax6^{on} progenitor cell states. If this is the case, Shh signaling may, independently of Pax6, establish additional distinctions in progenitor cell identity that contribute to the specification of ventral interneuron fate.

Pax6 and the Control of Neuronal Fate

Pax6 appears to function in a cell-autonomous manner in the ventral spinal cord and hindbrain, as in the control of eye development (Quinn et al. 1996). However, Pax6 appears to have somewhat different functions in the control of cell fate at different dorsoventral positions. In the dorsal region of the ventral neural tube, the loss of Pax6 function does not lead to a detectable ventral transforma-

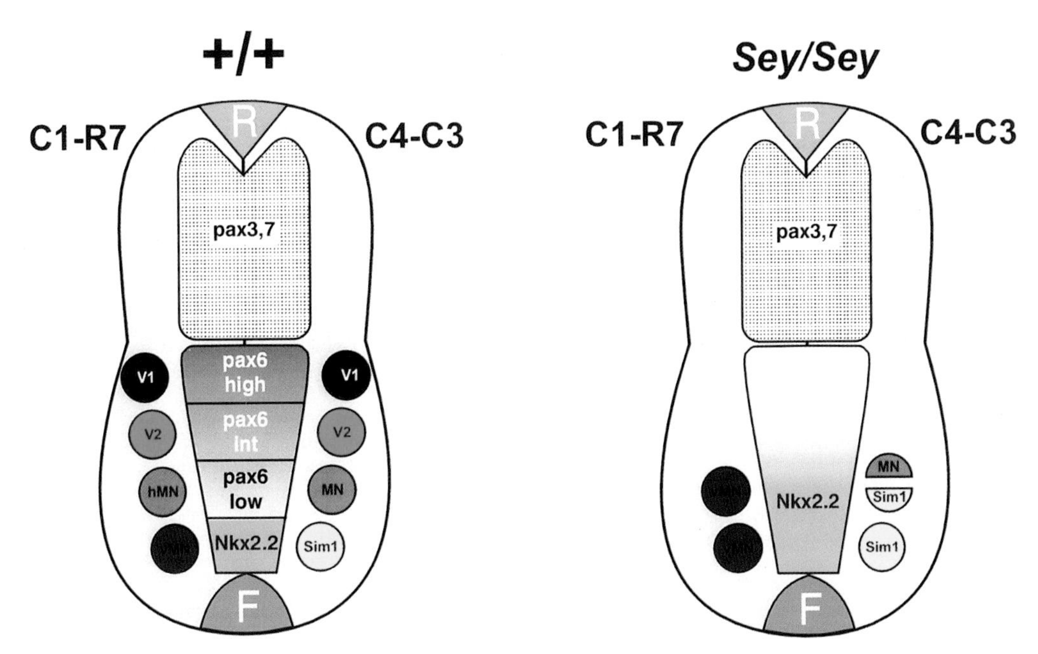

Figure 14. Change in ventral cell fate in the rostral spinal cord and caudal hindbrain of *Sey/Sey* embryos. Schematic representation of e10 caudal hindbrain (C1-R7) and cervical spinal cord (C4-C3) in wild-type (+/+) and Pax6 mutant (*Sey/Sey*) embryos. A consistent change in ventral progenitor cell identity is observed at both axial levels, but the consequences for MN differentiation vary at the two rostrocaudal levels. At the C1-R7 level, hMNs are transformed into vMNs, whereas at the C4–C3 level, there is a net loss of MNs and an expansion in the number of *Sim1* cells (for details, see text and Ericson et al. 1997).

tion in progenitor cell identity but nevertheless eliminates V1 interneurons. Since in *Sey/Sey* embryos neurons are generated within the dorsal domain that normally gives rise to V1 neurons, it remains possible that the loss of Pax6 causes V1 interneurons to assume a different neuronal fate. In contrast, the reduction in V2 neuron generation is likely to result from the dorsal encroachment of Nkx2.2 progenitor cells into the domain that normally generates V2 neurons.

The loss of Pax6 function more ventrally results in a clear dorsal-to-ventral transformation in progenitor cell identity. This transformation does not change the total number of MNs at rostral spinal cord/caudal hindbrain levels, indicating that Pax6 is not required directly for MN generation. Instead, MN identity is altered from somatic (hMN) to visceral (vMN) subtype (Fig. 14). At more caudal levels of the spinal cord, however, an equivalent progenitor transformation converts cells from a MN fate toward a more ventral cell fate, characterized by expression of *Sim1* (Fig. 14) (Ericson et al. 1997). The loss of Pax6 function therefore leads to a consistent transformation in the identity of ventral progenitor cells, but the consequences for MN fate differ according to position along the rostrocaudal axis of the neural tube (Fig. 14).

The transformation of somatic to visceral MNs detected in *Sey/Sey* embryos indicates that one important determinant of the subtype identity of MNs is the status of Pax6, and indirectly of Nkx2.2, expression by ventral progenitor cells. Since the state of Pax6 expression is controlled by Shh, it follows that graded Shh signaling controls not only the selection of MN and ventral interneuron identity, but also certain aspects of MN subtype identity. It is therefore likely that graded Shh signaling, through the establishment of distinct progenitor cell populations, controls MN subtype diversity at other levels of the neuraxis.

A Common Plan for Patterning the Ventral Neural Tube

The emergence of a link between Shh signaling, progenitor cell identity, and neuronal fate described at spinal cord and hindbrain levels may provide insight into the mechanisms of ventral patterning at more rostral levels of the neuraxis.

Pax6, Nkx2.2, and the related protein Nkx2.1 are expressed in ventral domains at forebrain levels (Walther and Gruss 1991; Price et al. 1992; Ericson et al. 1995; Shimamura et al. 1995). Shh has been shown to regulate the pattern of expression of these genes and neuronal fate in the developing forebrain (Barth and Wilson 1995; Ekker et al. 1995; Ericson et al. 1995; MacDonald et al. 1995). In addition, mice lacking Pax6 and Nkx2.2 function exhibit profound defects in eye development and other aspects of forebrain patterning (Hill et al. 1991; Kimura et al. 1996; Stoykova et al. 1996; Grindley et al. 1997; Warren and Price 1997). In particular, at diencephalic levels, markers of ventral progenitor cells are expanded dorsally in the *Sey*^{Neu}/*Sey*^{Neu} strain (Grindley et al. 1997). The role of Shh, Pax6, and Nkx2 proteins in the determination of progenitor cell identity and neuronal fate defined at spinal cord and hindbrain levels is therefore likely to be conserved at more rostral levels of the central nervous system.

ACKNOWLEDGMENTS

We thank C. Tabin for the chick *Ptc* cDNA. This work was supported by grants to T.M.J. from the National Institutes of Health, to J.E. from the Swedish MRC and the Swedish Institute, and to V.vH. from the MRC and the Howard Hughes Medical Institute International Scholars Program. J.B. was supported by an HFSP fellowship and P.R. by an MRC Clinical Training Fellowship. J.E. is a research associate and T.M.J. is an investigator of the Howard Hughes Medical Institute.

REFERENCES

Alcedo J., Ayzenozon M., Von Ohlen T., Noll M., and Hooper J.E. 1996. The *Drosophila smoothened* gene encodes a seven-pass membrane protein, a putative receptor for the hedgehog signal. *Cell* **86:** 221.

Alexandre C., Jacinto A., and Ingham P.W. 1996. Transcriptional activation of hedgehog target genes in *Drosophila* is mediated directly by the Cubitus interruptus protein, a member of the GLI family of zinc finger DNA-binding proteins. *Genes Dev.* **10:** 2003.

Barth K.A. and Wilson S.W. 1995. Expression of zebrafish nk2.2 is influenced by sonic hedgehog/vertebrate hedgehog-1 and demarcates a zone of neuronal differentiation in the embryonic forebrain. *Development* **121:** 1755.

Chang B.E., Balder P., Fischer N., Ingham P.W., and Strahle U. 1997. Axial HNF3β and retinoic acid receptors are regulators of the zebrafish sonic hedgehog promoter. *EMBO J.* **16:** 3955.

Chen Y. and Stuhl G. 1996. Dual roles for *patched* in sequestering and transducing Hedgehog. *Cell* **87:** 553.

Chiang C., Litingtung Y., Lee E., Young K.E., Corden J.L., Westphal H., and Beachy P.A. 1996. Cyclopia and defective axial patterning in mice lacking Sonic hedgehog gene function. *Nature* **383:** 407.

Dominguez M., Brunner M., Hafen E., and Basler K. 1996. Sending and receiving the hedgehog signal: Control by the *Drosophila* Gli protein Cubitus interruptus. *Science* **272:** 1621.

Echelard Y., Epstein D.J., St. Jacques B., Shen L., Mohler J., McMahon J.A., and McMahon A.P. 1993. Sonic hedgehog, a member of a family of putative signaling molecules, is implicated in the regulation of CNS polarity. *Cell* **75:** 1417.

Ekker S.C., Ungar A.R., Greenstein P., Von Kessler D.P., Porter J.A., Moon R.T., and Beachy P.A. 1995. Patterning activities of vetebrate hedgehog proteins in the developing eye and brain. *Curr. Biol.* **5:** 944.

Ericson J., Morton S., Kawakami A., Roelink H., and Jessell T.M. 1996. Two critical periods of Sonic hedgehog signaling required for the specification of motor neuron identity. *Cell* **87:** 661.

Ericson J., Thor S., Edlund T., Jessell T.M., and Yamada T. 1992. Early stages of motor neuron differentiation revealed by expression of homeobox gene *Islet-1*. *Science* **256:** 1550.

Ericson J., Muhr J., Placzek M., Lints T., Jessell T.M., and Edlund T. 1995. Sonic hedgehog induces the differentiation of ventral forebrain neurons: A common signal for ventral patterning along the rostrocaudal axis of the neural tube. *Cell* **81:** 747.

Ericson J., Rashbass P., Schedl A., Brenner-Morton S., Kawakami A., van Heyningen V., and Jessell T.M. 1997. *Pax6* controls progenitor cell identity and neuronal fate in response to graded Shh signaling. *Cell* **90:** 169.

Goodrich L.V., Johnson R.L., Milenkovic L., McMahon J.A., and Scott M.P. 1996. Conservation of the *hedgehog/patched* signaling pathway from flies to mice: Induction of a mouse *patched* gene by Hedgehog. *Genes Dev.* **10:** 301.

Goulding M.D., Lumsden A., and Gruss P. 1993. Signals from the notochord and floor plate regulate the region-specific expression of two *Pax* genes in the developing spinal cord. *Development* **117:** 1001.

Grindley J.C., Hargett L.K., Hill R.E., Ross A., and Hogan B.L.M. 1997. Disruption of PAX6 function in mice homozygous for the $Pax6^{Sey\text{-}1Neu}$ mutation produces abnormalities in the early development and regionalization of the diencephalon. *Mech. Dev.* **64:** 111.

Gurdon J.B. 1987. Embryonic induction—Molecular prospects. *Development* **99:** 285.

Hill R.E., Favor J., Hogan B.L.M., Ton C.C.T., Sauders G.F., Hanson I.M., Jordan J., Prosser J., Jordan T., Hastie N.D., and van Heyningen V. 1991. Mouse small eye results from mutations in a paired-like homeobox-containing gene. *Nature* **354:** 522.

Hynes M., Poulsen K., Tessier-Lavigne M., and Rosenthal A. 1995a. Control of Neuronal diversity by the floor plate: Contact-mediated induction of midbrain dopaminergic neurons. *Cell* **80:** 95.

Hynes M., Porter J.A., Chiang C., Chang D., Tessier-Lavigne M., Beachy P.A., and Rosenthal A. 1995b. Induction of midbrain dopaminergic neurons by Sonic hedgehog. *Neuron* **15:** 35.

Klar A., Baldassare M., and Jessell T.M. 1992. *F-Spondin:* A gene expressed at high levels in the floor plate encodes a secreted protein that promotes neural cell adhesion and neurite extension. *Cell* **69:** 95.

Kimura S., Hara Y., Pineau T., Fernadez-Salguero P., Fox C.H., Ward J.M., and Gonzalez F.J. 1996. The T/ebp null mouse: Thyroid-specific enhancer-binding protein is essential for the organogenesis of the thyroid, lung, ventral forebrain, and pituitary. *Genes Dev.* **10:** 60.

Krauss S., Concordet J. P., and Ingham P.W. 1993. A functionally conserved homolog of the *Drosophila* segment polarity gene *hh* is expressed in tissues with polarizing activity in zebrafish embryos. *Cell* **75:** 1431.

Kuratani S. and Tanaka S. 1990. Peripheral development of the avian vagus nerve with special reference to the morphological innervation of heart and lung. *Anat. Embryol.* **182:** 435.

Kuratani S., Tanaka S., Ishikawa Y., and Zukeran C. 1988. Early development of the hypoglossal nerve in the chick embryo as observed by the whole-mount nerve staining method. *Am. J. Anat.* **182:** 155.

Langman J., Guerrant R.L., and Freeman B.G. 1966. Behavior of neuroepithelial cells during closure of the neural tube. *J. Comp. Neurol.* **127:** 399.

Li H., Witte D.P., Branford W.W., Aronow B.J., Weinstein M., Kaur S., Wert S., Singh G., Schreiner C.M., Whitsett J.A., Scott W.J., and Potter S.S. 1994. *Gsh-4* encodes a LIM-type homeodomain, is expressed in the developing central nervous system and is required for early postnatal survival. *EMBO J.* **13:** 2876.

Lecuit T., Brook W.J., Ng M., Calleja M., Sun H., and Cohen S.M. 1996. Two distinct mechanisms for long-range patterning by Decapentaplegic in the *Drosophila* wing. *Nature* **381:** 387.

Lee J., Platt K.A., Censullo P., and Ruiz i Altaba A. 1997. Gli1 is a target of Sonic hedgehog that induces ventral neural tube development. *Development* **124:** 2537.

Lee J.J., Ekker S.C., von Kessler D., Porter J.A., Sun B.I., and Beachy P.A. 1994. Autoproteolysis in hedgehog protein biogenesis. *Science* **266:** 1528.

Liem K.F., Tremml G., Roelink H., and Jessell T.M. 1995. Dorsal differentiation of neural plate cells induced by BMP-mediated signals from epidermal ectoderm. *Cell* **82:** 969.

Liu I.S., Chen J.D., Ploder L., Vidgen D., van der Kooy D., Kalnins V.I., and McInnes R.R. 1994. Developmental expression of a novel murine homeobox gene (*Chx10*): evidence for roles in determination of the neuroretina and inner nuclear layer. *Neuron* **13:** 377.

Lu S., Wise T.L., and Ruddle F.H. 1994. Mouse homeobox gene *Dbx:* Sequence, gene structure and expression pattern during mid-gestation. *Mech. Dev.* **47:** 187.

MacDonald R., Barth K.A., Xu Q., Holder N., Mikkola I., and Wilson S.W. 1995. Midline signaling is required for *Pax* gene regulation and patterning of the eyes. *Development* **121:** 3267.

Mansouri A., Hallonet M., and Gruss P. 1996. *Pax* genes and their roles in cell differentiation and development. *Curr. Opin. Cell Biol.* **8:** 851.

Marigo V. and Tabin C.J. 1996. Regulation of *patched* by Sonic hedgehog in the developing neural tube. *Proc. Natl. Acad. Sci.* **93:** 9346.

Marigo V., Davey R.A., Zuo Y., Cunningham J.M., and Tabin C.J. 1996. Biochemical evidence that patched is the Hedgehog receptor. *Nature* **384:** 176.

Marti E., Bumcrot D.A., Takada R., and McMahon A.P. 1995. Requirement of 19K form of Sonic hedgehog for induction of distinct ventral cell types. *Nature* **375:** 322.

McConnell S.K. and Kaznowski C.E. 1991. Cell cycle dependence of laminar determination in developing neocortex. **254:** 282.

Nellen D., Burke R., Struhl G., and Basler K. 1996. Direct and long-range action of a DPP morphogen gradient. *Cell* **85:** 357.

Osumi N., Hirota A., Ohuchi H., Nakafuku M., Iimura T., Kuratani S., Fujiwara M., Noji S., and Eto K. 1997. *Pax-6* is involved in specification of the hindbrain MN subtype. *Development* **124:** 2961.

Perrimon N. 1995. Hedgehog and beyond. *Cell* **80:** 517.

Pfaff SL, Mendelsohn M., Stewart C.L., Edlund T., and Jessell T.M. 1996. Requirement for LIM homeobox gene *Isl1* in motor neuron generation reveals a motor neuron-dependent step in interneuron differentiation. *Cell* **84:** 309.

Placzek M. 1995. The role of the notochord and floor plate in inductive interactions. *Curr. Opin. Genet. Dev.* **5:** 499.

Placzek M., Jessell T.M., and Dodd J. 1993. Induction of floor plate differentiation by contact-dependent, homeogenetic signals. *Development* **117:** 205.

Porter J.A., von Kessler D.P., Ekker S.C., Young K.E., Lee J.J., Moses K., and Beachy P.A. 1995. The product of hedgehog autoproteolytic cleavage active in local and long-range signaling. *Nature* **374:** 363.

Price M., Lazzaro D., Pohl T., Mattei M.G., Ruther U., Olivo J.C., Duboule D., and DiLauro R. 1992. Regional expression of the homeobox gene *Nkx2.2* in the developing mammalian forebrain. *Neuron* **8:** 241.

Quinn J.C., West J.D., and Hill R.E. 1996. Multiple function for *Pax6* in mouse eye and nasal development. *Genes Dev.* **10:** 435.

Riddle R.D., Johnson R.L., Laufer E., and Tabin C. 1993. *Sonic hedgehog* mediates the polarizing activity of the ZPA. *Cell* **75:** 1401.

Roelink H., Porter J.A., Chiang C., Tanabe Y., Chang D.T., Beachy P.A., and Jessell T.M. 1995. Floor plate and motor neuron induction by different concentrations of the amino-terminal cleavage product of sonic hedgehog autoproteolysis. *Cell* **81:** 445.

Roelink H., Augsburger A., Heemskerk J., Korzh V., Norlin S., Ruiz i Altaba A., Tanabe Y., Placzek M., Edlund T., Jessell T.M., and Dodd J. 1994. Floor plate and motor neuron induction by vhh-1, a vertebrate homolog of hedgehog expressed by the notochord. *Cell* **76:** 761.

Ruiz i Altaba A., Roelink H., and Jessell T.M. 1995a. Restrictions to floor plate induction by *hedgehog* and *winged-helix* genes in the neural tube of frog embryos. *Mol. Cell. Neurosci.* **6:** 106.

Ruiz i Altaba A., Cox C., Jessell T.M., and Klar A. 1993. Ectopic neural expression of a floor plate marker in frog embryos injected with the midline transcription factor Pintallavis. *Proc. Natl. Acad. Sci.* **90:** 8268.

Ruiz i Altaba A., Placzek M., Baldassare M., Dodd J., and Jessell T.M. 1995b. Early stages of notochord and floor plate de-

velopment in the chick embryo defined by normal and induced expression of HNF3β. *Dev. Biol.* **170:** 299.
Sasaki H. and Hogan B.L. 1994. HNF-3β as a regulator of floor plate development. *Cell* **76:** 103.
Sasaki H., Hui C., Nakafuku M., and Kondoh H. 1997. A binding site for Gli proteins is essential for HNF-3β floor plate enhancer activity in transgenics and can respond to Shh in vitro. *Development.* **124:** 1313.
Schedl A., Ross A., Lee M., Engelkamp D., Rashbass P., van Heyningen V., and Hastie N.D. 1996. Influence of PAX6 gene dosage on development: Overexpression causes severe eye abnormalities. *Cell* **86:** 71.
Serafini T., Kennedy T.E., Galko M.J., Mirzayan C.M., Jessell T.M., and Tessier-Lavigne M. 1994. The netrins define a family of axon outgrowth-promoting proteins homologous to *C. elegans* UNC-6. *Cell* **78:** 409.
Shimamura K., Hartigan D.J., Martinez S., Puelles L., and Rubenstein J.L. 1995. Longitudinal organization of the anterior neural plate and neural tube. *Development* **121:** 3923.
Shoji H., Ito T., Wakamatsu Y., Hayasaka N., Ohsaki K., Oyanagi M., Kominami R., Kondoh H., and Takahashi N. 1996. Regionalized expression of the *Dbx* family homeobox genes in the embryonic CNS of the mouse. *Mech. Dev.* **56:** 25.
Simon H., Guthrie S., and Lumsden A. 1994. Regulation of SC1/DM-GRASP during the migration of motor neurons in the chick embryo brain stem. *J. Neurobiol.* **25:** 1129.
Stone D.M., Hynes M., Armanini M., Swanson T.A., Gu Q., Johnson R.L., Scott M.P., Pennica D., Goddard A., Phillips H., Noll M., Hooper J.E., de Sauvage F., and Rosenthal A. 1996. The tumor-suppressor gene patched encodes a candidate receptor for Sonic hedgehog. *Nature* **384:** 129.
Stoykova A., Fritsch R., Walther C., and Gruss P. 1996. Forebrain patterning defects in *Small eye* mutant mice. *Development* **122:** 3453.
Tanabe Y. and Jessell T.M. 1996. Diversity and pattern in the developing spinal cord. *Science* **274:** 1115.
Tanabe Y., Roelink H., and Jessell T.M. 1995. Induction of motor neurons by Sonic hedgehog is independent of floor plate differentiation. *Curr. Biol* **5:** 651.
Tremblay P., Pituello F., and Gruss P. 1996. Inhibition of floor plate differentiation by *Pax3:* Evidence from ectopic expression in transgenic mice. *Development* **122:** 2555.
Tsuchida T., Ensini M., Morton S.B., Baldassare M., Edlund T., Jessell T.M., and Pfaff S.L. 1994. Topographic organization of embryonic motor neurons defined by expression of LIM homeobox genes. *Cell* **79:** 957.
van den Heuvel M. and Ingham P.W. 1996. *Smoothened* encodes a receptor-like serpentine protein required for hedgehog signaling. *Nature* **382:** 547.
Varela-Echavarría A., Pfaff S.L., and Guthrie S. 1996. Differential expression of LIM homeobox genes among motor neuron subpopulations in the developing chick brain stem. *Mol. Cell. Neurosci.* **8:** 242.
Von Ohlen T., Lessing D., Nusse R., and Hooper J.E. 1997. Hedgehog signaling regulates transcription through Cubitus interruptus, a sequence-specific DNA binding protein. *Proc. Natl. Acad. Sci.* **94:** 2404.
Walther C. and Gruss P. 1991. *Pax-6,* a murine paired box gene, is expressed in the developing CNS. *Development* **113:** 1435.
Warren N. and Price D.J. 1997. Roles of *Pax*-6 in murine diencephalic development. *Development* **124:** 1573.
Wolpert L. 1969. Positional information and the spatial pattern of cellular differentiation. *J. Theor. Biol.* **25:** 1.
Yamada T., Pfaff S.L., Edlund T., and Jessell T.M. 1993. Control of cell pattern in the neural tube: Motor neuron induction by diffusible factors from notochord and floor plate. *Cell* **73:** 673.
Yamada T., Placzek M., Tanaka H., Dodd J., and Jessell T.M. 1991. Control of cell pattern in the developing nervous system: Polarizing activity of the floor plate and notochord. *Cell* **64:** 635.
Zecca M., Basler K., and Struhl G. 1995. Sequential organizing activities of *engrailed, hedgehog* and *decapentaplegic* in *Drosophila* wing. *Development* **121:** 2265.

Guidance of Developing Axons by Netrin-1 and Its Receptors

E.D. LEONARDO,[1] L. HINCK,[1] M. MASU,[1] K. KEINO-MASU,[1,2] A. FAZELI,[3] E. T. STOECKLI,[1]
S.L. ACKERMAN,[4] R.A. WEINBERG,[3] AND M. TESSIER-LAVIGNE[1]

[1]Howard Hughes Medical Institute, Departments of Anatomy and of Biochemistry and Biophysics, Programs in Cell and Developmental Biology and Neuroscience, University of California, San Francisco, California 94143; [2]Department of Physiology, National Defense Medical College, Saitama 359, Japan; [3]Whitehead Institute for Biomedical Research and Department of Biology, Massachusetts Institute of Technology, Cambridge, Massachusetts 02142; [4] The Jackson Laboratory, Bar Harbor, Maine 04609

The establishment of neuronal connections involves the accurate guidance of developing axons to their targets through the combined actions of attractive and repulsive cues in the extracellular environment. Accumulating evidence has indicated the importance of long-range mechanisms for axon guidance, involving diffusible chemoattractants secreted by target cells that attract axons to their targets, and diffusible chemorepellents secreted by nontarget cells that generate exclusion zones which axons avoid (Tessier-Lavigne and Goodman 1996). Two recently identified families of guidance molecules, the netrins and semaphorins, can function as diffusible attractants or repellents for developing axons, but the receptors and signal transduction mechanisms through which they produce their effects are poorly understood (Goodman 1996).

The netrins comprise a phylogenetically conserved family of long-range guidance cues related to the extracellular matrix molecule laminin, with members implicated in attraction and repulsion of axons in *Caenorhabditis elegans* (Ishii et al. 1992), in vertebrates (Kennedy et al. 1994; Serafini et al. 1994; Colamarino and Tessier-Lavigne 1995; Shirasaki et al. 1995; Varela-Echavarria et al. 1997), and in *Drosophila melanogaster* (Harris et al. 1996; Mitchell et al. 1996). In chicks, the netrin-1 and netrin-2 proteins have been implicated in guiding commissural axons in the spinal cord along a circumferential pathway from the dorsal spinal cord to floorplate cells at the ventral midline of the spinal cord. The two proteins were originally purified from embryonic chick brain on the basis of their ability to mimic an outgrowth-promoting effect of floorplate cells on commissural axons in collagen matrices in vitro (Serafini et al. 1994). In vivo, *netrin-1* is expressed in floorplate cells, and *netrin-2* is expressed in the ventral two thirds of the chick spinal cord, suggesting a decreasing ventral-to-dorsal gradient of netrin protein that functions to attract commissural axons to the ventral midline of the spinal cord (Kennedy et al. 1994). A netrin gradient may contribute to repelling some dorsally projecting hindbrain motor axons, including trochlear motor axons, away from the ventral midline, since netrin-1 can repel these axons in vitro (Colamarino and Tessier-Lavigne 1995; Varela-Echavarria et al. 1997). In *C. elegans*, UNC-6 is similarly thought to attract ventrally directed axons and to repel dorsally directed axons, since guidance of these axons is impaired in *unc-6* mutants (Hedgecock et al. 1990) and since UNC-6 appears to be concentrated in the ventral portion of the nematode (Wadsworth et al. 1996).

Insights into the mechanisms of action of netrins have come from *C. elegans*, where two genes, *unc-5* and *unc-40,* have been implicated in *unc-6*-dependent guidance of circumferential migrations of axons and mesodermal cells. Mutants in *unc-5* are primarily defective in dorsally directed migrations (presumed repulsions), whereas mutants in *unc-40* are primarily defective in ventrally directed migrations (presumed attractions) (Hedgecock et al. 1990). Both genes are known to encode transmembrane proteins, and both genes act cell-autonomously (Leung-Hagesteijn et al. 1992; Chan et al. 1996), raising the possibility that they encode netrin receptors.

The evidence suggesting that UNC-5 may be a netrin receptor extends beyond loss-of-function studies. Ectopic expression of UNC-5 in some neurons can redirect their migration away from a netrin source, showing that UNC-5 not only is required, but is also sufficient to redirect axons in an *unc-6*-dependent manner, at least in the neurons in which it has been tested (Hamelin et al. 1993). However, whether UNC-5 is a netrin receptor or simply an accessory to such a receptor has not been defined.

Mutations in the *unc-40* gene affect ventral migrations in the same way as do *unc-6* mutations, and *unc-6:unc-40* double mutants do not display any enhanced defects compared to the single mutants (Hedgecock et al. 1990). However, mutations in *unc-40* also affect dorsal migrations (although to a much lesser extent than *unc-6* or *unc-5* mutations), as well as several other patterning events in the nematode (Hedgecock et al. 1990), and *unc-40* appears to be expressed in some neurons whose axonal migrations are not affected in *unc-6* mutants (Chan et al. 1996). Thus, the precise function of UNC-40 in mediating responses to UNC-6 and, in particular, whether UNC-40 is an UNC-6 receptor, was not fully elucidated by these studies, although the evidence is consistent with a role for UNC-40 as an UNC-6 receptor involved in directing ventral migrations (Chan et al. 1996).

UNC-40 is a *C. elegans* homolog of vertebrate DCC (deleted in colorectal cancer) and neogenin (Fearon et al. 1990; Hedrick et al. 1994; Vielmetter et al. 1994; Chan et al. 1996), and *Drosophila* Frazzled (Kolodziej et al. 1996). Together, these proteins form a subgroup of the immunoglobulin (Ig) superfamily characterized by the presence of four Ig domains and six fibronectin type III repeats in their extracellular domains. The *DCC* gene was originally identified as a candidate tumor suppressor gene that is lost at high frequency in colorectal cancers (Fearon et al. 1990). *DCC* transcripts are present at low levels in almost all normal adult tissues, with highest levels in neural tissues (Reale et al. 1994; Cooper et al. 1995). *DCC* is also expressed in the developing nervous system in mouse, chick, and *Xenopus* (Chuong et al. 1994; Pierceall et al. 1994; Cooper et al. 1995). *neogenin* expression in chick correlates with the onset of neuronal differentiation and neurite extension, suggesting that neogenin is involved in terminal differentiation or axon guidance (Vielmetter et al. 1994). However, the actual functions of DCC and neogenin in the nervous system have not been identified.

In this paper, we describe evidence that DCC possesses netrin-binding activity and that antibodies to DCC can selectively block netrin-1-dependent outgrowth of commissural axons in vitro, suggesting that it is a netrin receptor or component of a netrin receptor. We show further that there is a family of vertebrate UNC-5-like proteins, that these proteins are expressed in the developing nervous system, and that they can bind netrin-1. Taken together, these studies suggest that the phylogenetic conservation of netrin function extends to netrin receptors and perhaps even to the signaling mechanisms through which responses to netrins are elicited.

EXPERIMENTAL PROCEDURES

Isolation of rat* DCC *and* neogenin *and of* unc-5 *homologs. For *DCC*, eight degenerate primers encoding conserved amino acid sequences between human DCC and chick neogenin were made. The sequences of the forward primers corresponded to the amino acid sequences KNG(D/E)VV, DEG(F/Y)YQC, KV(A/V)TQP, and DLWIHH. Those for the reverse primers corresponded to TGYKIR, MTVNGTG, NIVVRG, and EGLMK(Q/D). Polymerase chain reaction (PCR) was performed using cDNA from E12 rat spinal cord or brain. PCR products were subcloned and their sequences were determined. An E18 rat brain cDNA library was screened for *DCC* and *neogenin* with the PCR fragments as probes. For UNC-5 homologs, a search of the human expressed sequence tag (EST) databases revealed a small sequence (Genbank accession number R11880) with distant similarity to the carboxy-terminal portion of UNC-5. The corresponding cDNA fragment, amplified by PCR from an embryonic human brain cDNA library (Stratagene), was used to screen the library, resulting in the isolation of a 3.8-kb cDNA clone comprising all but the first 440 nucleotides of the coding region of the human homolog of *Unc5h1*. Probes from this cDNA were used to screen the E18 rat brain library, leading to the isolation of two distinct rat genes, *Unc5h1* and *Unc5h2*. Sequencing was done on a Licor (L4000) automated sequencer as well as by ^{33}P cycle sequencing. Searches of the databases were performed using the BLAST server, and sequence analysis was performed using Geneworks software (Intelligenetics). Genbank database accession numbers are *DCC* U68725, *neogenin* U68726, *Unc5h1* U87305, and *Unc5h2* U87306.

In situ hybridization. Cryostat sections (10 μm) were processed for in situ hybridization as described previously (Frohman et al. 1990). [^{35}S]UTP-labeled antisense riboprobes were synthesized using the PCR fragments for *DCC* and *neogenin* and using 3′UTR regions for *Unc5h1* and *Unc5h2*.

DiI labeling. DiI labeling was performed as described previously (Stoeckli and Landmesser 1995).

Purification of recombinant netrin-1. cDNAs encoding chick netrin-1 tagged with a myc-epitope at its carboxyl terminus (Serafini et al. 1994), or domains VI and V of chick netrin-1 fused to the constant (Fc) region of human IgG1, were subcloned into the expression vector pCEP4 (Invitrogen) and used to transfect 293-EBNA cells (Invitrogen). Cell lines permanently expressing either netrin-1 or netrin (VI·V)-Fc were established after drug selection (Shirasaki et al. 1996). Proteins were purified from conditioned media by heparin affinity chromatography to 85–90% homogeneity, as assessed by silver staining.

Binding experiments. Transfection of cDNAs encoding rat DCC, L1, TAG-1, UNC5H1, UNC5H2, or UNC5H3 (previously known as RCM) into 293-EBNA or 293T cells was performed using LipofectAMINE (GIBCO/BRL). Forty-eight hours after transfection, the cells were incubated with 2 μg/ml chick netrin-1 protein in phosphate-buffered saline (PBS) supplemented with 10% horse serum and 0.1% sodium azide in the presence or absence of 2 μg/ml heparin at room temperature for 90 minutes. After washing three times with PBS, the cells were fixed with methanol. Bound netrin was visualized with either an anti-netrin-1 antibody (affinity-purified rabbit polyclonal antibody [T.E. Kennedy and M.Tessier-Lavigne, unpubl.] used at 1:1000 dilution) or 9E10 (Evan et al. 1985), a monoclonal antibody to the carboxy-terminal myc-epitope tag.

Equilibrium binding experiments. For DCC, low-passage 293-EBNA cell lines permanently expressing rat DCC were used for binding studies, and untransfected and mock-transfected cells were used as negative controls. For UNC5H1, UNC5H2, and UNC5H3, 293T cells were transiently transfected with either a full-length *Unc5h2*, *Unc5h3*, or a truncated *Unc5h1* cDNA (lacking the sequence encoding the last 405 amino acids; the full-length *Unc5h1* appears to be toxic). The next day, cells were incubated in triplicate with different concentrations of netrin(VI·V)-Fc in PBS containing 2 μg/ml heparin and 1 mg/ml bovine serum albumin (BSA) on ice for 3 hours, rinsed with PBS (and for UNC5H proteins fixed briefly with methanol and 4% paraformaldehyde [PFA]), incubated with a ^{125}I-labeled conjugated anti-human IgG antibody (1 μCi/ml, NEN/Dupont) in PBS containing 10% horse serum for 30 minutes, rinsed again with PBS three times, and solubilized; the radioactivity bound on the cells was counted.

Explant cultures. Explants of E11 and E13 rat dorsal spinal cord were isolated and cultured in collagen gels as described previously (Tessier-Lavigne et al. 1988; Serafini et al. 1994). Outgrowth of commissural axons was elicited by addition of 300 ng/ml (for E13 explants) or 1.2 μg/ml (for E11 explants) of purified netrin-1. For blocking experiments, anti-DCC antibody (Oncogene Sciences, Inc.) and control mouse immunoglobulin solutions were dialyzed against F12 medium before being added to the culture.

RESULTS

A Family of Vertebrate Homologs of *C. elegans* UNC-5

Given the strong evidence implicating UNC-5 in mediating responses to UNC-6, we decided to search for vertebrate homologs to study their possible roles as netrin receptors. We isolated cDNAs encoding two rat homologs of UNC-5 (termed UNC5H1 and UNC5H2). The vertebrate UNC-5 family also comprises at least one additional member, the product of the mouse *Unc5h3* gene (formerly known as rostral cerebellar malformation [RCM]) (Fig. 1A) (Ackerman et al. 1997). UNC5H proteins show overall sequence similarity with UNC-5 (~30% identity) and possess two Ig-like domains and two thrombospondin type-1 repeats in their extracellular domains (Fig. 1A) (Leonardo et al. 1997). The cytoplasmic domains of the three UNC5H proteins do not contain any known catalytic domains, but they do possess two small regions of homology with known molecules. One small region shows homology with ZO-1, a protein that localizes to adherens junctions and is implicated in junction formation (Fig. 1B) (Itoh et al. 1993; Willott et al. 1993). Another region, at the carboxyl termini of both proteins, contains a divergent death domain, which is also found in UNC-5 (Fig. 1C) (Hofmann and Tschopp 1995). Death domains are protein interaction domains that have been found at the carboxyl termini of a number of receptors, including the low-affinity NGF receptor ($p75^{NGF\text{-}R}$) and the tumor necrosis factor receptor (TNF-R), and are thought to mediate the signaling response of these receptors (Rabizadeh et al. 1993; Hofmann and Tschopp 1995; Nagata and Golstein 1995). Together with UNC-5, the three UNC5H proteins define a new subfamily of the immunoglobulin superfamily.

DCC and UNC5H Family Members Are Netrin-binding Proteins

Although there was genetic evidence implicating UNC-40 and UNC-5 in mediating responses to UNC-6, there was no direct biochemical evidence that these molecules interact. To test the hypothesis that vertebrate homologs of UNC-40 and UNC-5 are netrin receptors, we examined whether netrin-1 would specifically bind cells expressing DCC, neogenin, UNC5H1, UNC5H2, or UNC5H3 (formerly RCM). Transfected human embryonic kidney 293 cells expressing these proteins showed binding of netrin-1 protein that was significantly above background as assessed using an antibody that specifically recognizes recombinant netrin-1 (Fig. 2A–D and data not shown). No such binding was seen with cells expressing TAG-1 or L1 (Fig. 2E and data not shown), two other members of the immunoglobulin superfamily. Thus, binding of netrin-1 to these proteins is specific and does not reflect a generalized interaction of netrin-1 with members of the Ig superfamily.

In these experiments, binding was performed in the presence of soluble heparin, which eliminates nonspecific binding of netrin-1 to the cells (Keino-Masu et al. 1996) but does not prevent binding to DCC or the UNC-5 homologs. Binding was similarly specific when cells were incubated with netrin-1 without added heparin and subsequently washed with only heparin-containing medium (data not shown), indicating that heparin is not required for the binding interaction.

The affinity of netrin-1 for several of these netrin-binding proteins was estimated in equilibrium binding experiments using netrin (VI·V)-Fc, a fusion of the amino-terminal two thirds of netrin-1 to the constant portion of human IgG (Keino-Masu et al. 1996). This chimeric netrin-1 molecule is bioactive, but unlike netrin-1, it does not aggregate at high concentrations. Specific binding of netrin (VI·V)-Fc to DCC, UNC5H1, UNC5H2, and UNC5H3 showed saturation, and these experiments yielded K_d values of 5.2 ± 0.2 nM, 19.8 ± 0.8 nM, 3.4 ± 1.0 nM, and 6.9 ± 1.8 nM for DCC, UNC5H1, UNC5H2, and UNC5H3, respectively (Keino-Masu et al. 1996; Leonardo et al. 1997). These values are consistent with the effective dose for the axon-outgrowth-promoting effects of netrin-1 (Serafini et al. 1994) and are of a similar order of magnitude to the dissociation constant for the interaction of the $\alpha_1\beta_2$ integrin with laminin-1 (Pfaff et al. 1994).

DCC and UNC5H Family Members Are Expressed in the Developing Spinal Cord

To act as netrin receptors in vivo, these netrin-binding proteins must have a spatio-temporal distribution that is consistent with known sites of netrin expression and function. To investigate potential sites of action for these candidate receptors, we performed in situ hybridization in rat embryos.

DCC, *neogenin*, *Unc5h1*, and *Unc5h2* are expressed in the developing spinal cord (Fig. 3). At E11, when the first

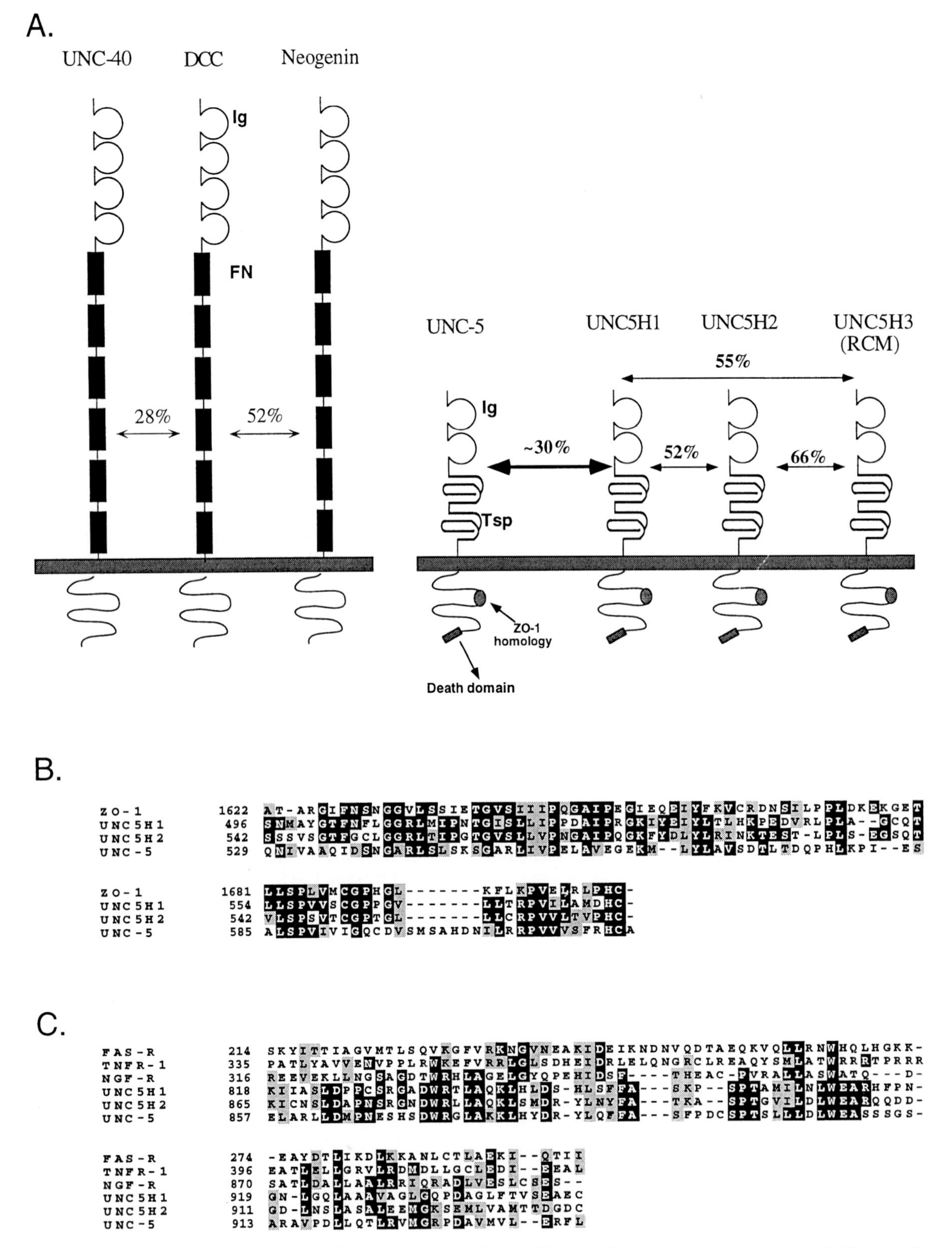

Figure 1. DCC and UNC-5 receptor subfamilies. (*A*) Members of the DCC subfamily possess four immunoglobulin-like domains (Ig) and six fibronectin type III repeats (FN) in their extracellular domains, and members of the UNC-5 family possess two Ig domains and two thrombospondin type-1 domains (Tsp). The cytoplasmic portions of members of the UNC-5 family show a domain of homology to ZO-1 (*shaded oval*) and a death domain (*shaded box*). (*B*) Alignment of sequences of UNC-5, UNC5H proteins, and ZO-1 in their regions of homology. (*C*) Alignment of the death domains of UNC-5 family members with those of FAS-R, TNFR-1, and p75^{NGF-R}.

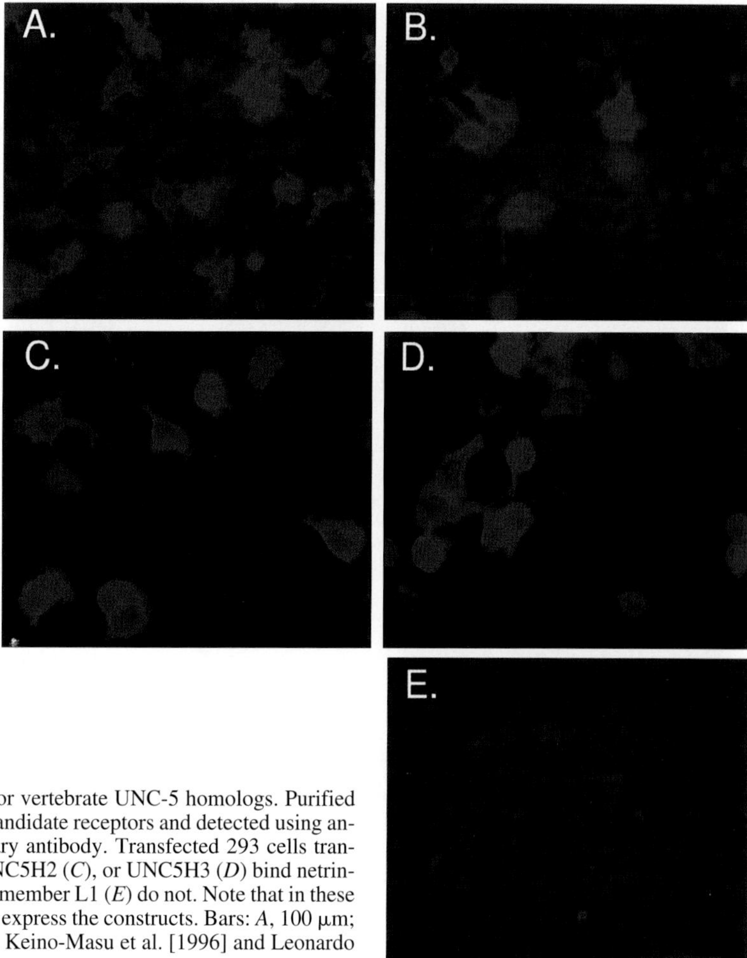

Figure 2. Interactions of netrin-1 with DCC or vertebrate UNC-5 homologs. Purified netrin-1 was incubated with cells expressing candidate receptors and detected using antibodies to netrin-1 and a fluorescent secondary antibody. Transfected 293 cells transiently expressing DCC (*A*), UNC5H1 (*B*), UNC5H2 (*C*), or UNC5H3 (*D*) bind netrin-1, whereas cells expressing the Ig superfamily member L1 (*E*) do not. Note that in these transient expression assays, ~10% of the cells express the constructs. Bars: *A*, 100 μm; *B–E*, 66 μm. (Adapted, with permission, from Keino-Masu et al. [1996] and Leonardo et al. [1997].)

neuronal populations are differentiating, *DCC* mRNA is expressed in the region of the developing motor column and is most highly concentrated in what appears to be a subpopulation of motor neurons that are most lateral. In addition, substantial hybridization is observed in the dorsal spinal cord over the cell bodies of commissural neurons (Fig. 3A). By E13, intense expression of *DCC* is observed in a pattern that corresponds to the cell bodies of commissural neurons, with weaker expression detected more ventrally in a subpopulation of cells in the developing motor columns (Fig. 3B). *neogenin* mRNA was observed in ventricular neuroepithelial cells at E11, with highest expression midway along the dorsal-ventral axis (Fig. 3C). The expression of *neogenin* becomes widespread in the E13 spinal cord and highest in the ventral third of the ventricular zone, but it is almost absent in commissural neurons, so that the expression patterns of *DCC* and *neogenin* at E13 are strikingly complementary (compare Figs. 3B and 3D). *Unc5h1* mRNA is first seen at E11 in the ventral half of the spinal cord excluding the floorplate (Fig. 3E). By E13, *Unc5h1* expression appears primarily confined to postmitotic cells in the ventral spinal cord, apparently including motoneurons, although weaker expression can be seen in dorsolateral regions (Fig. 3F and data not shown). Significant levels of *Unc5h2* transcripts in the spinal cord are not detected until E14, when they are found in the region of the roofplate (Fig. 3G). They are, however, detected in the developing sensory ganglia that flank the spinal cord, at low levels at E12 (data not shown) and at high levels by E14 (Fig. 3G).

Netrin Receptors Are Highly Expressed in the Developing Cerebellum

Expression of DCC subfamily members and of UNC-5 homologs is not limited to the spinal cord but can also be found at higher axial levels in the developing nervous system (Keino-Masu et al. 1996; Ackerman et al. 1997; Leonardo et al. 1997). One prominent site of expression for these receptors in the later-developing nervous system is the cerebellum. In situ hybridization analysis revealed a remarkable coexpression of these molecules during critical periods of cerebellar development. By P5, transcripts for *Unc5h1* and *Unc5h2* are abundant in the external germinal layer (EGL), where granule cell precursors differentiate before migrating to their final destination, the internal granule cell layer (IGL)—also a site of expression of both homologs (Fig. 4A,B). At a similar developmental stage in the mouse (P6), *Unc5h3* expression is seen in a more diffuse pattern encompassing cells in the EGL as

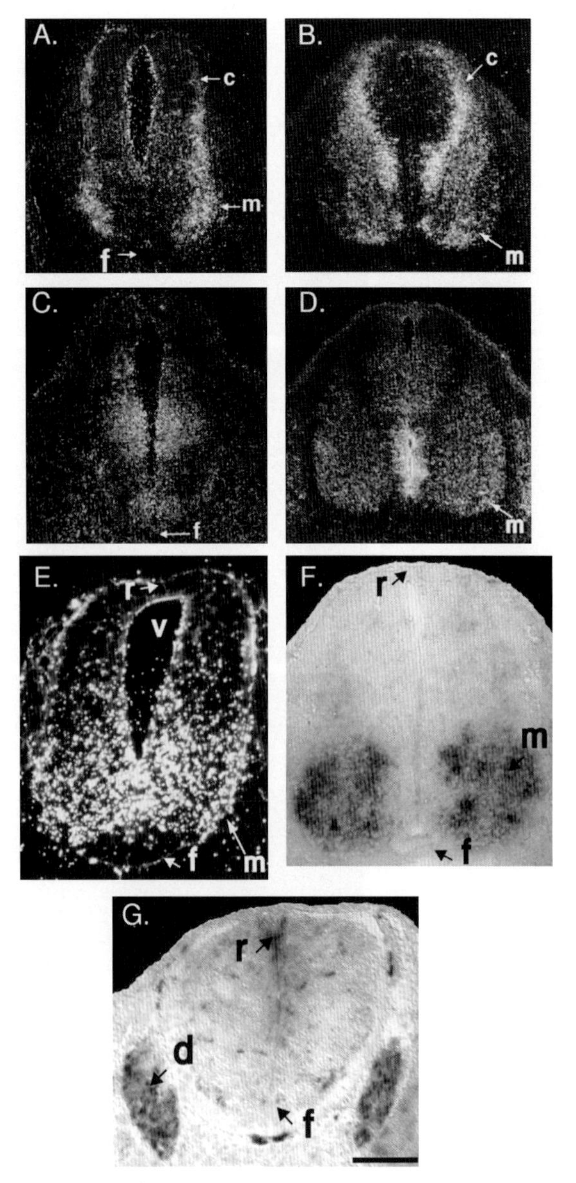

Figure 3. Sites of expression of *DCC, neogenin*, and *Unc5h* mRNAs visualized in transverse sections through the developing rat spinal cord at E11 (*A, C, E*), E13 (*B, D, F*), and E14 (*G*). At E11, *DCC* mRNA (*A*) is detected over the cell bodies of commissural neurons (c), and motoneurons (m); *neogenin* mRNA (*C*) is localized medially along the ventricular zone; and *Unc5h1* mRNA (*E*) is observed in motoneurons (m) in a pattern that is more ventrally restricted and more medial than *DCC* expression. At E13, intense expression of *DCC* mRNA (*B*) is detected over the cell bodies of commissural neurons (c), whereas *neogenin* expression is more restricted ventrally in a pattern that is roughly complementary to that of *DCC*. (*D*) *Unc5h1* at this age is detected primarily in the ventral horn in the region of motoneurons (m) (*F*). *Unc5h2* expression (*G*) is shown at E14, when it is found in the sensory ganglia and roofplate. Additional abbreviations: (d) dorsal root ganglia; (f) floorplate; (r) roofplate; (v) ventricular zone. Bars: *A,C*, 46 μm; *B,D*, 100 μm; *E*, 30 μm; *F*, 130 μm and *G*, 320 μm. (Adapted, with permission, from Keino-Masu et al. [1996] and Leonardo et al. [1997].)

well as in the molecular layer, IGL, and Purkinje cell layer (Fig. 4C) (Ackerman et al. 1997). *DCC* expression at this time can be detected in the EGL as well as the deep cerebellar nuclei (Fig. 4D) (Livesey and Hunt 1997), whereas *neogenin* expression is limited to the external germinal layer (Fig. 4E). Strikingly, *netrin-1* expression has also been localized to the outer aspect of the external germinal layer at this time (Livesey and Hunt 1997). As in the case of the spinal cord, coexpression of these candidate receptors in the vicinity of the netrin ligand at a critical developmental time raises the possibility that netrin-mediated signaling may play a role in the developing cerebellum.

DCC Function Is Required for Axon Outgrowth Evoked by Netrin-1

The finding that DCC was expressed on commissural neurons, a population known to respond to netrin-1 (Serafini et al. 1994), allowed us to test directly whether DCC is involved in mediating the effects of netrin-1. To do so, we examined the effect of perturbing DCC function on commissural axon outgrowth evoked by netrin-1 from explants of E11 (Fig. 5A,B) or E13 (Fig. 5E) rat dorsal spinal cord. Addition of a monoclonal antibody against the extracellular domain of DCC resulted in a reduction in the extent of commissural axon outgrowth in both types of cultures (Fig. 5D,F), whereas normal mouse immunoglobulin had no effect (Fig. 5C and data not shown). We also examined whether the anti-DCC antibody could interfere with the outgrowth of commissural axons from E13 dorsal explants that is evoked by floorplate cells (Fig. 5A,B) and that appears to be due to netrin-1 secreted by these cells (Kennedy et al. 1994; Serafini et al. 1996). The anti-DCC antibody completely abolished this outgrowth (Fig. 6D), whereas normal mouse immunoglobulin had no effect (Fig. 6C). Thus, the anti-DCC antibody can block the netrin-dependent outgrowth of commissural axons in vitro, whether netrin-1 is presented as pure protein or secreted from floorplate cells. Additional control experiments showed that the anti-DCC antibody blocks netrin-dependent but not netrin-independent outgrowth of commissural axons (Keino-Masu et al. 1996).

In addition to their ability to evoke outgrowth of commissural axons into collagen gels, both netrin-1 and the floorplate can reorient the growth of these axons within E11 dorsal spinal cord explants. The anti-DCC antibody was unable to block this reorientation of axons in response to either the floorplate or netrin-1, indicating that either DCC function is not required for the turning responses or that the antibody did not penetrate the tissue explant effectively (Keino-Masu et al. 1996).

Defects in Spinal Commissural Axon Projections and Brain Development in *Dcc*$^{-/-}$ Mice

To elucidate the functions of DCC in vivo, we have examined the effects of the inactivation of the mouse *Dcc* gene on the developing nervous system (Fazeli et al. 1997). Within the dorsal spinal cord of homozygous mice, we noted a reduction in the number of commissural axons, although those that did extend appeared to adopt a normal dorsal-to-ventral trajectory (compare Fig. 7A,C to 7B,D). Misrouting was also observed in the axons that extended into the motor column, with some projecting more

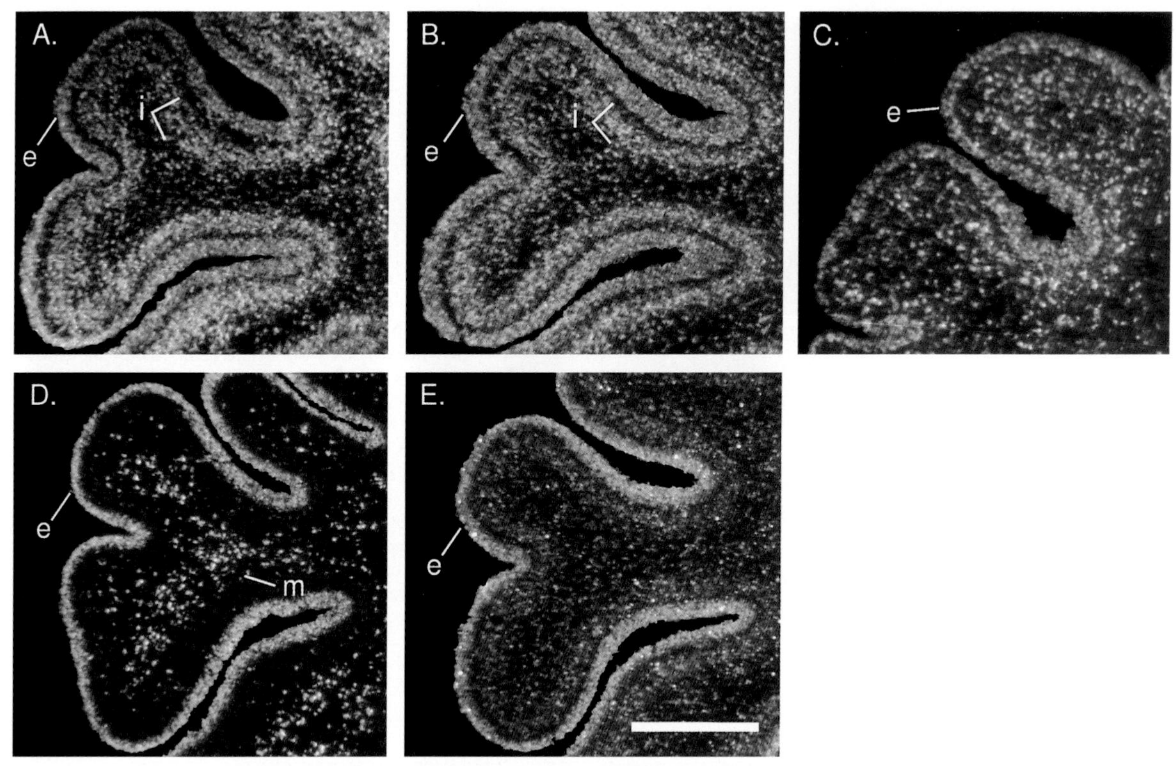

Figure 4. Expression of *DCC*, *neogenin*, and *Unc5h* mRNAs in parasaggital sections through individual folia of P5 rat (*A,B,D,E*) or P6 mouse (*C*) cerebella. (*A*) *Unc5h1* mRNA is seen expressed at high levels in the external germinal layer (e), as well as in the internal granule cell layer (i). No expression is seen in the Purkinje cell layer. (*B*) The pattern of *Unc5h2* expression is nearly identical to that of *Unc5h1*. (*C*) *Unc5h3* expression is more diffuse and can be seen in all layers. (*D*) *DCC* mRNA is expressed at high levels in the outermost region of the EGL as well as in scattered cells in the molecular layer. (*E*) *Neogenin* expression is restricted primarily to the EGL. Other abbreviations: (m) molecular layer. Bars: *A, B, C, D, E,* 415 μm; *C*, 340 μm.

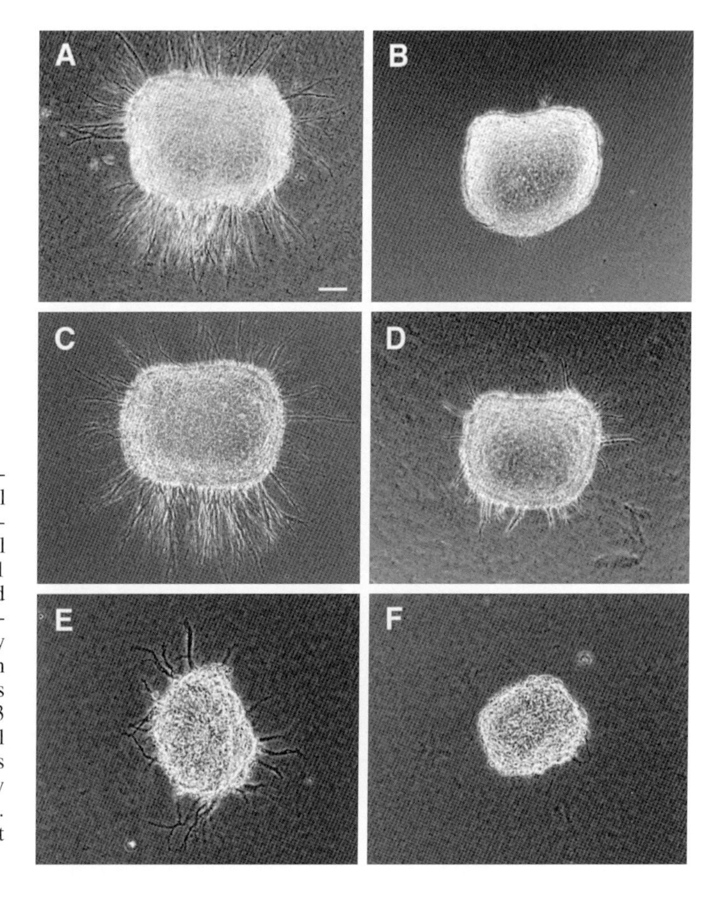

Figure 5. DCC function is required for axon outgrowth evoked by netrin-1. (*A–D*) E11 dorsal spinal cord explants cultured for 40 hr with 1.2 μg/ml netrin-1 (*A*), without netrin-1 (*B*), with netrin-1 and 10 μg/ml normal mouse immunoglobulin (*C*), or with netrin-1 and 1.0 μg/ml of the monoclonal antibody directed against the extracellular portion of DCC (*D*). Outgrowth evoked by netrin-1 (*A*) is markedly reduced by addition of anti-DCC antibody (*D*), but not by addition of control immunoglobulin (*C*). Little outgrowth is observed from explants cultured alone (*B*). (*E, F*) E13 dorsal explants cultured for 16 hr with 300 ng/ml netrin-1. Robust outgrowth elicited by netrin-1 (*E*) is blocked by addition of 10 μg/ml anti-DCC antibody (*F*). Little outgrowth is observed in controls (see Fig. 6*A*). (Reprinted, with permission, from Keino-Masu et al. 1996 [copyright Cell Press].)

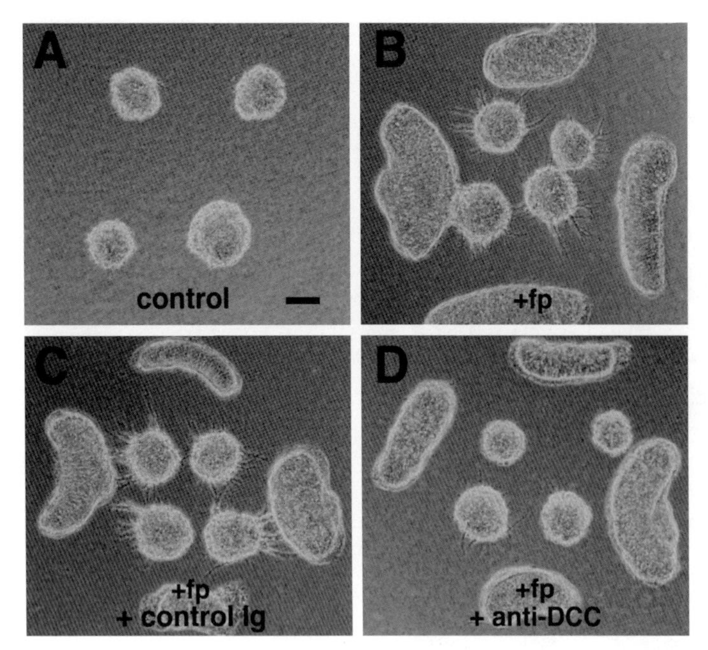

Figure 6. DCC function is required for axon outgrowth evoked by floorplate cells. Each panel shows four E13 dorsal spinal cord explants cultured for 14 hr in collagen matrices either alone (*A*) or surrounded by four explants of E13 rat floorplate (*B–D*). Explants were cultured either without antibody (*A*, *B*), with normal mouse immunoglobulin (10 μg/ml) (*C*), or with anti-DCC antibody (10 μg/ml) (*D*). Axon outgrowth evoked by floorplate cells (*B*) is reduced to control levels (*A*) by anti-DCC antibody (*D*) but not control immunoglobulin (*C*). Bar, 100 μm. (Reprinted, with permission, from Keino-Masu et al. 1996 [copyright Cell Press].)

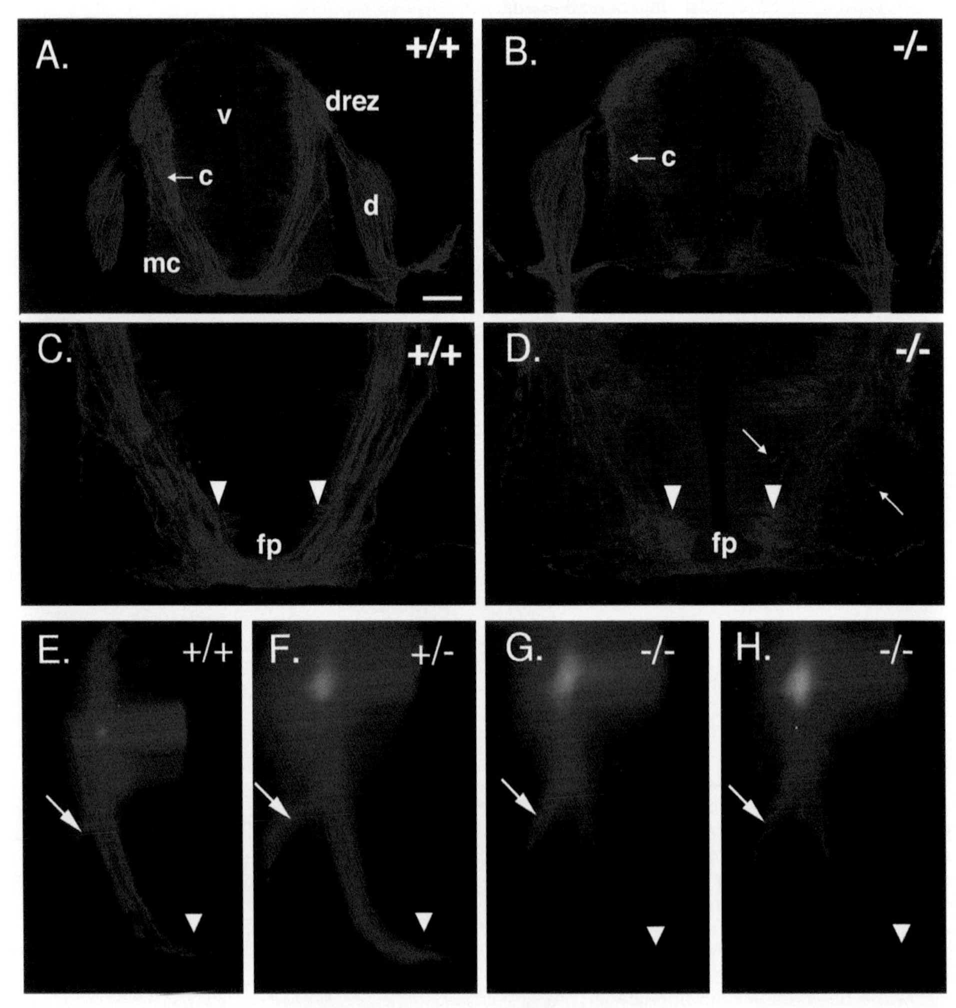

Figure 7. *(See facing page for legend)*

medially, and others more laterally (Fig. 7B,D). DiI labeling of commissural axons confirmed the misrouting of commissural axons in the ventral spinal cords of *Dcc*$^{-/-}$ embryos (Fig. 7E–H). The defects in axonal projections observed in and around the spinal cord appeared specific to commissural axons, inasmuch as the trajectories of sensory and motor axons appeared normal (Fig. 7B). Thus, *Dcc*$^{-/-}$ embryos show selective defects in commissural axon projections that are similar, but slightly more severe, than those seen in netrin-1-deficient mice (Serafini et al. 1996). In netrin-1-deficient mice, a similar misrouting of commissural axons is observed in the ventral spinal cord, but in the dorsal spinal cord the trajectory of these axons is normal (Serafini et al. 1996). Comparison of the two phenotypes suggests the possible existence of a second ligand for DCC in the dorsal spinal cord (Fazeli et al. 1997).

In addition to defects in the spinal cord, multiple defects were observed in the brains of *Dcc*$^{-/-}$ mice (Fazeli et al. 1997). The corpus callosum and hippocampal commissure appeared to be completely absent. The axons that normally form these commissures failed to cross the midline and remained ipsilateral, projecting to aberrant locations and forming tangles or "Probst bundles." The anterior commissure in *Dcc*$^{-/-}$ mice was severely reduced. In the brain stem of *Dcc*$^{-/-}$ mice, formation of an abnormal commissure was observed in the junctional region between hindbrain and midbrain. In addition, the pontine nuclei at the base of the rostral hindbrain appeared to be absent in the mutant animals. Thus, the defects observed in the brains of *Dcc*$^{-/-}$ animals were similar to those observed in netrin-1-deficient animals (Serafini et al. 1996). Unfortunately, both *netrin-1* and *Dcc* mutant animals die at birth, precluding examination of defects at later stages of development, in particular, during postnatal development of the cerebellum. It is interesting to note that mutations in murine *Unc5h3* result in specific cerebellar defects that are consistent with a role for UNC-5 homologs in axon guidance and cell migration events in the postnatal cerebellum (Ackerman et al. 1997).

DISCUSSION

DCC Is a Netrin-1 Receptor on Commissural Axons

We have shown that DCC is expressed on the axons and growth cones of spinal commissural neurons as they extend to and across the floorplate (Keino-Masu et al. 1996). Our studies also indicate that DCC is a netrin-1-binding protein that is required to mediate the outgrowth-promoting effects of netrin-1 on commissural axons in vitro. In vivo, netrin-1 has been directly implicated in guidance of commissural axons from the dorsal spinal cord to the ventral midline of the spinal cord (Serafini et al. 1996). Similarly, our results demonstrate that DCC is also required for normal guidance of commissural axons to the floorplate. The striking resemblance of the phenotypes of the two knockout mice, taken together with the in vitro data, strongly suggests that DCC is a receptor or a component of a receptor that is required for appropriate projections of commissural axons to their floorplate target.

Conservation of Netrin Signaling from Nematode to Rodents?

The finding of biochemical interactions between netrin-1 and members of the DCC and UNC5H families in vertebrates complements recent studies in *C. elegans*. In *C. elegans*, the DCC homolog UNC-40 is required for a subset of axon guidance events that are directed by UNC-6, the netrin-1 homolog. In particular, ventrally directed migrations, which are presumed attractive responses to UNC-6 (Wadsworth et al. 1996), are strongly affected by loss of UNC-40 function (Hedgecock et al. 1990). Moreover, *unc-40* appears to act cell-autonomously (Chan et al. 1996). Together, these studies have led to the suggestion that UNC-40 is a receptor or component of a receptor complex involved in mediating attractive effects of UNC-6 on ventrally directed axons.

Similarly, genetic studies have shown that UNC-5 is required for a subset of axon guidance events that are directed by UNC-6. In particular, UNC-5 is required cell-autonomously for dorsally directed migrations (Hedgecock et al. 1990; Leung-Hagesteijn et al. 1992), and ectopic expression of *unc-5* in neurons that normally project longitudinally or ventrally can steer their axons dorsally (Hamelin et al. 1993). Although consistent with the possibility that UNC-5 is an UNC-6 receptor, these results are also consistent with a role for UNC-5 in modifying the function of a different UNC-6 receptor. This modifier function is made more plausible by evidence that the DCC homolog UNC-40 is expressed by axons that project dorsally and is required for these projections (Hedgecock et al. 1990; McIntire et al. 1992; Chan et al. 1996), suggesting that UNC-5 might function by switching an attractive netrin receptor (UNC-40) into a repulsive netrin receptor. Although such a switching model is possible, our results suggest that UNC-5 might itself also

Figure 7. Defects in commissural axon projections in *Dcc*$^{-/-}$ embryos. (*A–D*) Trajectories of commissural axons are visualized using an antibody to TAG-1 in sections of wild type (*A,C*) and *Dcc*$^{-/-}$ (*B,D*) E11.5 embryos. In *Dcc*$^{-/-}$ embryos, TAG-1+ commissural neurons are present, but few axons extend into the ventral spinal cord (*B*), and those that do project along aberrant trajectories (arrows in *D*). Projections of sensory axons and motor axons in the ventral roots appear largely normal (*B*). Arrowheads in *C* and *D* indicate a population of TAG-1+ cells adjacent to the floorplate. Additional abbreviations: (d) dorsal root ganglia; (drez) dorsal root entry zone; (mc) motor column; (v) ventricle; (c) commissural axons; (fp) floorplate. Bars: *A,B*, 100 μm; *C,D*, 50 μm. (*E–H*) Trajectories of commissural axons revealed by DiI labeling in wild type (*E*), heterozygous (*F*), and *Dcc*$^{-/-}$ (*G,H*) E11.5 mouse spinal cord. In *F–H*, large arrowheads indicate floorplate and arrows indicate presumptive ipsilaterally projecting axons. (*E,F*) Normal trajectory of commissural axons in a wild type (*E*) and a heterozygous (*F*) embryo. (*G,H*) Trajectory of axons from two different *Dcc*$^{-/-}$ embryos. Many fewer axons project into the ventral spinal cord; some project ventromedially but are not particularly directed toward the floorplate, and many axons wander within the motor column. (Adapted, with permission, from Fazeli et al. 1997.)

function directly as a netrin receptor. A model in which UNC-5 and UNC-40 can form a receptor complex, but in which UNC-5 can also go it alone in transducing the UNC-6 netrin signal, would explain why loss of *unc-40* function results in a much less severe phenotype for dorsal migrations than does loss of either *unc-5* or *unc-6* function (Hedgecock et al. 1990; McIntire et al. 1992). In this context, it is notable that *Unc5h1* and *Unc5h2* are expressed in regions of the vertebrate nervous systems where the *Unc-40* homologs *DCC* and *neogenin* are expressed (e.g., in dorsal and ventral spinal cord, dorsal retina, dorsal root ganglia, and cerebellum) (Fig. 4) (Leonardo et al. 1997; Livesey and Hunt 1997; Keino-Masu et al. 1996). Thus, a model in which UNC-5 and UNC-40 form a receptor complex might also be applicable to vertebrates.

Ig Superfamily Members as Axon Guidance Receptors

Studies on Ig superfamily members expressed in the nervous system have focused for the most part on their homophilic or heterophilic interactions with other Ig superfamily members (for review, see Brummendorf and Rathjen 1994). There is, however, accumulating evidence that Ig superfamily members can also in some cases bind extracellular matrix (ECM) proteins: (1) Ng-CAM, a chick homolog of L1, can bind laminin-1 (Grumet et al. 1993); (2) Gicerin, a transmembrane Ig superfamily member with five Ig domains, binds the laminin-related molecule neurite outgrowth factor (NOF) (Taira et al. 1994); and (3) chick F11/contactin and its mouse homolog F3, which are glycosyl-phosphatidylinositol (GPI)-linked proteins with a structure similar to TAG-1, bind to members of the tenascin family and are implicated in mediating repulsive actions of these ligands (Norenberg et al. 1992; Zisch et al. 1992, Pesheva et al. 1993). Although netrins can function as diffusible chemoattractants and chemorepellents (Kennedy et al. 1994; Colamarino and Tessier-Lavigne 1995), in structure they are related to portions of the laminin molecules (Ishii et al. 1992; Serafini et al. 1994). Thus, the finding of interactions of netrin-1 with members of the DCC and UNC5H families parallels other observations on interactions between Ig superfamily proteins and ECM molecules.

The signal transduction mechanisms that are triggered by homophilic interactions between Ig superfamily members are beginning to be elucidated (for review, see Brummendorf and Rathjen 1994), but nothing is known yet about signaling triggered by binding of Ig superfamily members to ECM molecules. In this context, it is interesting to note that although the small cytoplasmic region of DCC family members is devoid of any known signaling motifs, the large cytoplasmic domain of the UNC-5 homologs contains two conserved regions that may provide clues to their signaling. The first is a region that shows homology with ZO-1, a protein that has been implicated in tight junction formation and may be involved in signaling at specialized cell-cell junctions (Willott et al. 1993). The second is the existence of a divergent death domain. Death domains are found in a number of transmembrane receptors that are involved in the apoptotic pathway (for review, see Hofmann and Tschopp 1995), and there is evidence to suggest that this domain is a protein-protein interaction domain involved in transmitting the apoptotic signal (Boldin et al. 1995; Chinnaiyan et al. 1995). More recently, several groups have identified a number of molecules that can specifically bind death-domain-containing receptors. Some of these molecules (e.g., Fadd) interact via a death domain of their own (for review, see Baker and Reddy 1996). It will be interesting to see what role (if any) these death domains play in transmitting netrin signals that are responsible for guiding axons. It also remains to be determined whether netrin-1 can regulate apoptosis in any context.

In conclusion, the identification of DCC and UNC-5 homologs as netrin-1-binding proteins provides a clear context for the elucidation of downstream components of DCC and UNC-5 signaling that are important for directing axon guidance and cell migration.

ACKNOWLEDGMENTS

This work was supported by grants to M.T.-L. from the International Spinal Research Trust, the American Paralysis Association, and the Howard Hughes Medical Institute, and to R.A.W., S.L.A., and M.T.-L. from the National Institutes of Health. K.K.-M. was supported by the National Defense Medical College, Japan; L.H. by a fellowship from the Jane Coffins Child Memorial Fund; E.D.L. by a UCSF chancellor's fellowship, and M.M. by a postdoctoral felowship from the Howard Hughes Medical Institute. R.A.W. is a research professor of the American Cancer Society, and M.T.-L. is an investigator of the Howard Hughes Medical Institute.

REFERENCES

Ackerman S.L., Kozak L.P., Przborski S.A., Rund L.A., Boyer B.B., and Knowles B.B. 1997. The mouse *rostral cerebellar malformation* gene encodes an UNC-5 like protein. *Nature* **386:** 838.

Baker S.F. and Reddy E.P. 1996. Transducers of life and death: TNF receptor superfamily and associated proteins. *Oncogene* **12:** 1.

Boldin M.P., Mett I.L., Varfolomeev E.E., Chumakov I., Shemer-Avni Y., Camonis J.H., and Wallach D. 1995. Self-association of the "death domains" of the p55 tumor necrosis factor (TNF) receptor and Fas/APO1 prompts signaling for TNF and Fas/APO1 effects. *J. Biol. Chem.* **270:** 387.

Brummendorf T. and Rathjen F. 1994. Cell adhesion molecules. 1. Immunoglobulin superfamily. In *Protein profile*, p. 1001. Academic Press, London.

Chan S.S.-Y., Zheng H., Su M.-W., Wilk R., Killeen M.T., Hedgecock E.M., and Culotti J.G. 1996. UNC-40, a *C. elegans* homolog of *DCC* (*Deleted in Colorectal Cancer*), is required in motile cells responding to UNC-6 netrin cues. *Cell* **87:** 187.

Chinnaiyan A.M., O'Rourke K., Tewari M., and Dixit V.M. 1995. FADD, a novel death domain-containing protein, interacts with the death domain of Fas and initiates apoptosis. *Cell* **81:** 505.

Chuong C.-M., Xiang T.-X., Yin E., and Widelitz R.B. 1994. cDCC (chicken homologue to a gene deleted in colorectal carcinoma) is an epithelial adhesion molecule expressed in basal cells and involved in an epithelial mesenchymal interaction. *Dev. Biol.* **164:** 383.

Cooper H.M., Armes P., Britto J., Gad J., and Wilks A.F. 1995. Cloning of the mouse homologue of the *Deleted in Colorectal Cancer* gene (mDCC) and its expression in the developing mouse embryo. *Oncogene* **11:** 2243.

Colamarino S.A. and Tessier-Lavigne M. 1995. The axonal chemoattractant *netrin-1* is also a chemorepellent for trochlear motor axons. *Cell* **81:** 621.

Evan G.I., Lewis G.K., Ramsey G., and Bishop J.M. 1985. Isolation of monoclonal antibodies specific for human c-*myc* proto-oncogene product. *Mol. Cell. Biol.* **5:** 3610.

Fazeli A., Dickinson S.L., Hermiston M.L., Tighe R.V., Steen R.G., Small C.G., Stoeckli E.T., Keino-Masu K., Masu M., Rayburn H., Simons J., Bronson R.T., Gordon J.I., Tessier-Lavigne M., and Weinberg R.A. 1997. Phenotype of mice lacking functional *Deleted in Colorectal Cancer (Dcc)* gene. *Nature* **386:** 796.

Fearon E.R., Cho K.R., Nigro J.M., Kern S.E., Simons J.W., Ruppert J.M., Hamilton S.R., Preisinger A.C., Thomas G., Kinzler K.W., and Vogelstein B. 1990. Identification of a chromosome 18q gene that is altered in colorectal cancers. *Science* **247:** 49.

Frohman M.A., Boyle M., and Martin G.R. 1990. Isolation of the mouse *Hox 2.9* gene: Analysis of embryonic expression suggests that positional information along the anterior-posterior axis is specified by mesoderm. *Development* **110:** 589.

Goodman C.S. 1996. Mechanisms and molecules that control growth cone guidance. *Annu. Rev. Neurosci.* **19:** 341.

Grumet M., Friedlander D.R., and Edelman G.M. 1993. Evidence for the binding of Ng-Cam to laminin. *Cell Adhes. Commun.* **1:** 177.

Hamelin M., Zhou Y., Su M.-W., Scott I.M., and Culotti J.G. 1993. Expression of the UNC-5 guidance receptor in the touch neurons of *C. elegans* steers their axons dorsally. *Nature* **364:** 327.

Harris R., Sabatelli L.M., and Seeger M.A. 1996. Guidance cues at the *Drosophila* CNS midline: Identification and characterization of two *Drosophila* netrin/UNC-6 homologs. *Neuron* **17:** 217.

Hedgecock E.M., Culotti J.G., and Hall D.H. 1990. The *unc-5*, *unc-6*, and *unc-40* genes guide circumferential migrations of pioneer axons and mesodermal cells on the epidermis in *C. elegans*. *Neuron* **2:** 61.

Hedrick L., Cho K.R., Fearon E.R., Wu T.-C,, Kinzler K.W., and Vogelstein B. 1994. The *DCC* gene product in cellular differentiation and colorectal tumorigenesis. *Genes Dev.* **8:** 1174.

Hofmann K. and Tschopp J. 1995. The death domain motif found in Fas (Apo-1) and TNF receptor is present in proteins involved in apoptosis and axonal guidance. *FEBS Lett.* **371:** 321.

Ishii N., Wadsworth W.G., Stern B.D., Culotti J.G., and Hedgecock E.M. 1992. UNC-6, a laminin related protein, guides cells and pioneer axon migrations in *C. elegans*. *Neuron* **9:** 873.

Itoh M., Nagafuchi A., Yonemura S., Kitani-Yasuda T., Tsukita S., and Tsukita S. 1993. The 220-kd protein colocalizing with cadherins in non-epithelial cells is identical to ZO-1, a tight junction-associated protein in epithelial cells: cDNA cloning and immunoelectron microscopy. *J. Cell Biol.* **121:** 491.

Keino-Masu K., Masu M., Hinck L., Leonardo E.D., Chan S.S.-Y., Culotti J.G., and Tessier-Lavigne M. 1996. *Deleted in Colorectal Cancer* encodes a netrin receptor. *Cell* **87:** 75.

Kolodziej P.A., Timpe L.C., Mitchell K.J., Fried S.R., Goodman C.S., Jan L.Y., and Jan Y.N. 1996. *frazzled* encodes a *Drosophila* member of the DCC immunoglobulin subfamily and is required for CNS and motor axon guidance. *Cell* **87:** 197.

Kennedy T.E., Serafini R., de la Torre J.R., and Tessier-Lavigne M. 1994. Netrins are diffusible chemotropic factors for commissural axons in the embryonic spinal cord. *Cell* **78:** 425.

Leonardo E.D., Hinck L., Masu M., Keino-Masu K., Ackerman S.L., and Tessier-Lavigne M. 1997. Vertebrate homologues of UNC-5 are candidate netrin receptors. *Nature* **386:** 833.

Leung-Hagesteijn C., Spence A.M., Stern B.D., Zhou Y., Su M.-W., Hedgecock E.M., and Culotti J.G. 1992. UNC-5, a transmembrane protein with immunoglobulin and thrombospondin type 1 domains, guides cell and pioneer axon migrations in *C. elegans*. *Cell* **71:** 289.

Livesey F.J. and Hunt S.P. 1997. Netrin and netrin receptor expression in the embryonic mammalian nervous system suggests roles in retinal, striatal, nigral and cerebellar development. *Mol. Cell. Neurosci.* **8:** 417.

McIntire S.L., Garriga G., White J., Jacobson D., and Horvitz H.R. 1992. Genes necessary for directed axonal elongation or fasciculation in *C. elegans*. *Neuron* **8:** 307.

Mitchell K.J., Doyle J.L., Serafini T., Kennedy T.E., and Tessier-Lavigne M. 1996. Genetic analysis of *Netrin* genes in *Drosophila:* Netrins guide CNS commissural axons and peripheral motor axons. *Neuron* **17:** 203.

Nagata S. and Golstein P. 1995. The Fas death factor. *Science* **267:** 1449.

Norenberg U., Wille H., Wolff J.M., Frank R., and Rathjen F.G. 1992. The chicken neural extracellular matrix molecule restrictin: Similarity with EGF-, fibronectin type III-, and fibrinogen-like motifs. *Neuron* **8:** 849.

Pesheva P., Gennarini G., Goridis C., and Schachner M. 1993. The F3/F11 cell adhesion molecule mediates repulsion of neurons by the extracellular matrix glycoprotein J1–160/180. *Neuron* **10:** 69.

Pfaff M., Gohring W., Brown J.C., and Timpl R. 1994. Binding of purified collagen receptors ($\alpha 1\beta 1$, $\alpha 1\beta 2$) and RGD-dependent integrins to laminins and laminin fragments. *Eur. J. Biochem.* **225:** 975.

Pierceall W.E., Reale M.A., Candia A.F., Wright C.V.E., Cho K.R., and Fearon E.R. 1994. Expression of a homologue of the *deleted in colorectal cancer (DCC)* gene in the nervous system of developing *Xenopus* embryos. *Dev. Biol.* **166:** 654.

Rabizadeh S., Oh J., Zhong L.T., Yang J., Bitler C.M., Butcher L.L., and Bredesen D.E. 1993. Induction of apoptosis by the low-affinity NGF receptor. *Science* **261:** 345.

Reale M.A., Hu G., Zafar A.I., Getzenberg R.H., Levine S.M., and Fearon E.R. 1994. Expression and alternative splicing of the *Deleted in Colorectal Cancer* (*DCC*) gene in normal and malignant tissues. *Cancer Res.* **54:** 4493.

Serafini T., Kennedy T.E., Galko M.J., Mirzayan C., Jessell T.M., and Tessier-Lavigne M. 1994. The netrins define a family of axon outgrowth-promoting proteins homologous to *C. elegans* UNC-6. *Cell* **78:** 409.

Serafini T., Colamarino S.A., Leonardo E.D., Wang H., Beddington R., Skarnes W.C., and Tessier-Lavigne M. 1996. Netrin-1 is required for commissural axon guidance in the developing vertebrate nervous system. *Cell* **87:** 1001.

Shirasaki R., Mirzayan C., Tessier-Lavigne M., and Murakami F. 1996. Guidance of circumferentially growing axons by netrin-dependent and -independent floorplate chemotropism in the vertebrate brain. *Neuron* **17:** 1079.

Shirasaki R., Tamada A., Katsumata R., and Murakami F. 1995. Guidance of cerebellofugal axons in the rat embryo: Directed growth toward the floorplate and subsequent elongation along the longitudinal axis. *Neuron* **14:** 961.

Stoeckli E.T. and Landmesser L.T. 1995. Axonin-1, Nr-CAM, and Ng-CAM play different roles in the in vivo guidance of chick commissural neurons. *Neuron* **14:** 1165.

Taira E., Takaha N., Taniura H., Kim C.-H., and Miki N. 1994. Molecular cloning and functional expression of Gicerin, a novel cell adhesion molecule that binds neurite outgrowth factor. *Neuron* **12:** 861.

Tessier-Lavigne M. and Goodman C.S. 1996. The molecular biology of axon guidance. *Science* **274:** 1123.

Tessier-Lavigne M., Placzek M., Lumsden A.G., Dodd J., and Jessell T.M. 1988. Chemotropic guidance of developing axons in the mammalian central nervous system. *Nature* **336:** 775.

Varela-Echavarria A., Tucker A., Puschel A.W., and Guthrie S. 1997. Motor axon subpopulations respond differentially to the chemorepellents netrin-1 and semaphorin D. *Neuron* **18:** 193.

Vielmetter J., Kayyem J.F., Roman J.M., and Dreyer W.J. 1994. Neogenin, an avian cell surface protein expressed during terminal neuronal differentiation, is closely related to the human tumor suppressor molecule *Deleted in Colorectal Cancer. J. Cell Biol.* **127:** 2009.

Wadsworth W.G., Bhatt H., and Hedgecock E.M. 1996. Neuroglia and pioneer neurons express UNC-6 to provide global and local netrin cues for guiding migrations in *C. elegans*. *Neuron* **16:** 35.

Willott E., Balda M.S., Fanning A.S., Jameson B., Van Itallie C., and Anderson J.M. 1993. The tight junction protein ZO-1 is homologous to the *Drosophila* discs-large tumor suppressor protein of septate junctions. *Proc. Natl Acad. Sci.* **90:** 7834.

Zisch A.H., D'Alessandri L., Ranscht B., Falchetto R., Winterhalter K.H., and Vaughan L. 1992. Neuronal cell adhesion molecule contactin/F11 binds to tenascin via its immunoglobulin-like domains. *J. Cell Biol.* **119:** 203.

The Many Faces of Fasciclin II: Genetic Analysis Reveals Multiple Roles for a Cell Adhesion Molecule during the Generation of Neuronal Specificity

C.S. GOODMAN, G.W. DAVIS, AND K. ZITO
Howard Hughes Medical Institute, Division of Neurobiology, Department of Molecular and Cell Biology, University of California, Berkeley, California 94720

The human brain contains trillions of neurons, each of which makes hundreds to thousands of synaptic connections with specific targets. The genome clearly does not contain a molecular blueprint or wiring diagram for the brain. Nature has solved this dilemma by generating brain circuitry over a series of stages allowing for a progressive refinement of the pattern of projections and connections (Goodman and Shatz 1993). The initial steps of growth cone guidance typically occur before neurons become functionally active and rely on molecular mechanisms of pathway and target recognition that are activity-independent (Tessier-Lavigne and Goodman 1996). Once the initial scaffold of connections has been established, however, the activity of emerging neural circuits drives the refinement and remodeling of the nervous system that continues throughout life (Katz and Shatz 1996).

An emerging principle of neuronal development is that certain molecules appear to be re-used at a number of different stages to serve novel functions. Neural activity appears to regulate synaptic remodeling by controlling the function of some of the very same molecular mechanisms that control selective growth and guidance during the earlier activity-independent stages of axon pathfinding and synapse formation. One of the molecular components common to both activity-independent and activity-dependent processes are cell adhesion molecules (CAMs). Here, we consider the multiple parts played by a single CAM during the development of synaptic specificity, as revealed by a detailed genetic analysis of its function.

CAMs come in a variety of shapes and sizes. Many neural CAMs belong to one of two large families—the immunoglobulin (Ig) and cadherin superfamilies—although other unrelated families of CAMs are also expressed in the developing nervous system (for review, see Goodman 1996; Tessier-Lavigne and Goodman 1996). Many CAMs can mediate homophilic adhesion, functioning as both a ligand on one cell and a receptor on another, and some members can also function as heterophilic ligands or receptors for distinct cell surface or ECM molecules. Here we consider the function of a member of the Ig superfamily: *Drosophila* Fasciclin II.

Ever since the molecular identification of the first neural cell adhesion molecule (NCAM) in the early 1980s (for review, see Edelman 1985), neurobiologists have suspected that CAMs must have important roles in the construction of brain circuitry. On the basis of in vitro functional analysis and their patterns of expression in vivo, CAMs have been implicated as potentially having roles in all phases of the generation of neural specificity, from neurite outgrowth to axon guidance, target recognition, synapse formation, and synaptic plasticity.

The initial wave of results from genetic analysis of CAM function was disappointing in that it did not fully confirm all of the earlier predictions. Loss-of-function mutations in the genes encoding *Drosophila* homologs of vertebrate L1 (Neuroglian; Bieber et al. 1989), NCAM (Fasciclin II; Grenningloh et al. 1991), and SC1/BEN (Irrec; Ramos et al. 1993), and mutations in the mouse NCAM gene (Tomasiewicz et al. 1993; Cremer et al. 1994), were shown to lead to much more subtle defects in guidance and connectivity (given their patterns of expression) than had been predicted. In each case, although specific defects were observed, the overall nervous system in the mutants looked remarkably normal. More recent genetic analysis of NCAM function in mouse has revealed defects in a variety of events of neuronal development ranging from cell migration to activity-dependent synaptic plasticity. However, many of these defects may be due in large part to the absence of a large carbohydrate on NCAM, polysialic acid, that is thought to modulate the function of other CAMs, rather than to the loss of NCAM itself (see, e.g., Becker et al. 1996; Hu et al. 1996; Muller et al. 1996). Thus, it remains an open question what part neural CAMs play during guidance, synapse formation, and synaptic remodeling.

These issues have begun to be resolved for one CAM using genetic analysis in *Drosophila*. A large body of literature now exists on the genetic analysis of Fasciclin II (Fas II) function in *Drosophila*, and this analysis has provided insights into the many functions played by this CAM during development. Fas II, an NCAM-like molecule in insects, was initially identified on the basis of its dynamic pattern of expression on a subset of fasciculating axons in the grasshopper embryo (Bastiani et al. 1987; Harrelson and Goodman 1988). Fas II is related to vertebrate NCAM (Cunningham et al. 1987) and *Aplysia* apCAM (Mayford et al. 1992) in structure (its ectodomain contains five C2-type Ig domains followed by two fibronectin type III domains), multiple isoforms (see below), sequence (~23% amino acid identity), and its ability to mediate homophilic cell aggregation (Harrelson and Goodman 1988; Grenningloh et al. 1990, 1991). The *FasII* mRNA, like that for apCAM and NCAM, is alternatively spliced to give rise to multiple isoforms of the Fas II protein, including a phosphotidylinositol (PI)-linked form and two transmembrane forms, one of which contains a PEST degradation sequence in its cytoplasmic domain (PEST$^+$ form) and the other does not (PEST$^-$ form) (Fig. 1).

The role of Fas II has been studied during axon guidance in the central nervous system (CNS) (Lin et al. 1994) and peripheral nervous system (PNS) (Lin and Goodman 1994; Fambrough and Goodman 1996), leading to a detailed understanding of its function in controlling selective fasciculation and the ways in which this function is modulated during selective defasciculation. Fas II has also been shown to function in synapse stabilization (Schuster et al. 1996a) and in the patterning of synapse formation (Davis et al. 1997). The carboxy-terminal amino acid sequences that control its synaptic localization, and its interaction with the DLG protein, have also been elucidated (Zito et al. 1997). Finally, we review the role of Fas II in synaptic growth and sprouting and show that neural activity regulates its level of expression (Schuster et al. 1996b).

ROLE OF FASCICLIN II DURING AXON GUIDANCE

Fas II Controls Selective Fasciculation

In the early 1980s, experiments in the grasshopper embryo showed that growth cones can distinguish one group of axons from another, leading to specific patterns of selective fasciculation (see, e.g., Raper et al. 1984; Bastiani et al. 1984; Goodman et al. 1984). These results led to the formulation of the labeled pathways hypothesis and to the subsequent search for axon pathway labels using monoclonal antibodies. Fas II was identified in such a monclonal antibody screen based on its expression on a subset of fasciculating axons in the grasshopper embryo (Bastiani et al. 1987; Harrelson and Goodman 1988). Thus, Fas II appeared to be a prime candidate as an axon pathway recognition molecule that controls the pattern of selective fasciculation.

Studies demonstrate that Fas II expression drives selective axon fasciculation. However, its function during axon outgrowth is not that simple. In *Drosophila*, Fas II is dynamically expressed on a subset of embryonic CNS axons, many of which selectively fasciculate in the pCC and MP1 pathways (Fig. 2) (Grenningloh et al. 1991; Lin et al. 1994). Fas II is also expressed on all motor axons in the periphery (Van Vactor et al. 1993) and on other cell types and tissues as well. Among the axons in both the CNS and PNS that normally express Fas II, some of them

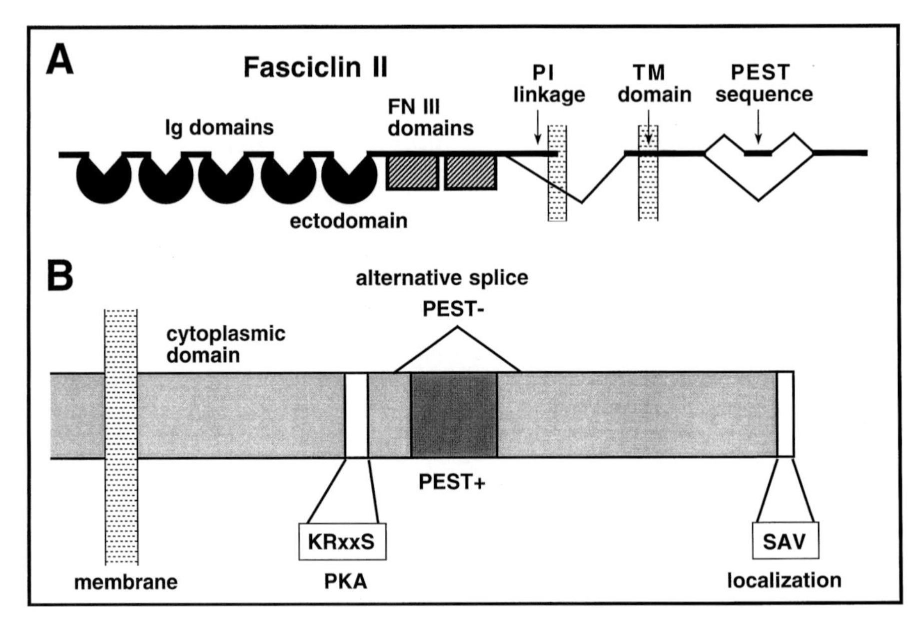

Figure 1. Structure, forms, and sequences of Fasciclin II. (*A*) Schematic diagram of Fasciclin II, showing its ectodomain consisting of five immunoglobulin (Ig)-type C2 domains followed by two fibronectin (FN) type III domains. The protein comes in three major forms as a result of alternative splicing. One form is linked to the membrane by a phosphotidylinositol (PI) anchor. The other two forms are transmembrane (TM) and differ by the splicing in or out of a cytoplasmic sequence that contains a PEST degradation sequence (Grenningloh et al. 1991; G. Helt et al., unpubl.). (*B*) The cytoplasmic domain of Fas II contains a number of functionally important sequences in addition to the PEST sequence. Shortly before the PEST sequence is a PKA consensus phosphorylation site. At the end of the carboxyl terminus is a three-amino-acid PDZ-interaction sequence that controls the synaptic clustering of Fas II protein. Fas II interacts with the MAGUK protein DLG via these three amino acids (Zito et al. 1997). See text for further details.

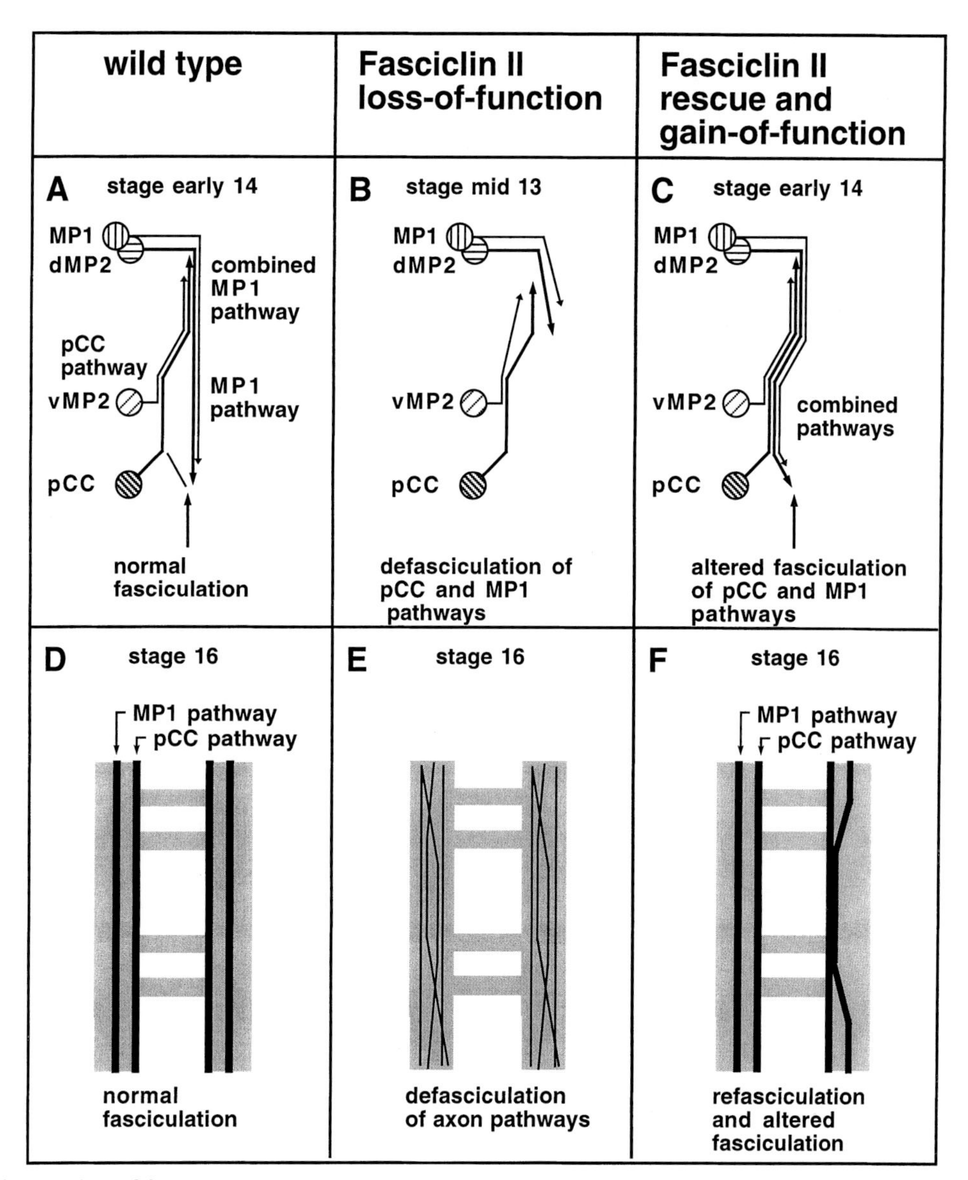

Figure 2. The *FasII* loss-of-function vs. gain-of-function conditions lead to complementary phenotypes in three axon pathways. Summary diagram showing the complementary *FasII* loss-of-function vs. gain-of-function fasciculation phenotypes for two different axon pathways (the pCC and MP1 pathways) at two different stages of development (stages 13/14 [*A–C*] and stage 16 [*D–F*]). (*A*) The wild-type pattern of axons in the pCC, MP1, and combined MP1 axon pathways at stage early 14. (*B*) The *FasII* loss-of-function defasciculation phenotype. (*C*) The *FasII* gain-of-function phenotype (ftz_{ng}*-GAL4* effector and *UAS-FasII* reporter transgenes) in which the pCC and MP1 pathways are fused together. (*D*) Wild-type pattern of axons in the pCC and MP1 axon pathways at stage 16. (*E*) *FasII* loss-of-function defasciculation phenotype. (*F*) *FasII* gain-of-function phenotype (ftz_{ng}*-GAL4* effector and *UAS-FasII* reporter transgenes in a *FasII* mutant background) in which the loss-of-function phenotype is rescued in that the axons in the pCC and MP1 pathways refasciculate. In some segments, the gain-of-function leads to altered patterns of fasciculation in which these two pathways are fused together, presumably due to the failure to properly defasciculate (see *C*). (Adapted from Lin et al. 1994.)

that initially fasciculate together later (at specific choice points) selectively defasciculate while still expressing Fas II. For example, in the CNS, the axons pioneering the pCC pathway (pCC and vMP2) transiently fasciculate with the axons pioneering the MP1 pathway (MP1 and dMP2). At a specific location, these axons selectively defasciculate from one another (i.e., pCC and vMP2 remain fasciculated, MP1 and dMP2 remain fasciculated, but each pair defasciculates from the other) to form two distinct axon pathways (Lin et al. 1994). Later, other changes take place in the pattern of fasciculation, leading to a switch in dMP2's association from MP1 to the pCC pathway (Hidalgo and Brand 1997). Similarly, as motor axons exit the CNS and extend into the periphery, the Fas-II-positive motor axons are initially fasciculated in one of two major motor nerves, but later subsets of these axons selectively defasciculate at specific choice points while still expressing Fas II (Van Vactor et al. 1993; see next section).

Genetic loss-of-function and gain-of-function studies provide strong support for the prediction that Fas II controls selective axon fasciculation. To test Fas II function, Grenningloh et al. (1991) generated a series of mutant alleles in the *FasII* gene and described the initial loss-of-function phenotypes. Lin et al. (1994) used these mutants to examine the axon guidance function of *FasII* in more detail. They found that when the levels of Fas II are decreased in *FasII* loss-of-function mutants, the axons in several CNS pathways that normally express Fas II defasciculate (Figs. 2 and 3). As a result, the longitudinal connectives and neuropil regions are disorganized. Nevertheless, these growth cones extend in the normal direction at a normal rate.

Two related phenotypes are observed under *FasII* gain-of-function conditions. First, transgenic constructs that specifically drive Fas II expression on the axons in these same pathways can rescue the defasciculation phenotype in a *FasII* loss-of-function background, thus creating a refasciculation of these major Fas-II-positive fascicles. Second, in both wild-type and *FasII* mutant backgrounds, these transgenic constructs can lead to a gain-of-function phenotype in which axons fasciculate incorrectly, in certain cases because they fail to defasciculate (Figs. 2 and 3). For example, pairs of pathways that normally begin together but later defasciculate (e.g., the pCC and MP1 pathways) now remain abnormally joined together (Lin et al. 1994). As analyzed in greater detail for motor axons (see next section), these results suggest that some mechanism must normally regulate Fas II function and thus allow Fas-II-positive axons to selectively defasciculate.

By increasing and decreasing the levels of Fas II in the developing organism using genetic analysis, Fas II has thus been shown to function in selective fasciculation and axon sorting. Moreover, the results define other aspects of growth cone initiation, outgrowth, and guidance in which Fas II function is not required. In this way, the function of this CAM during axon guidance can be viewed as one guidance force within the context of multiple forces impinging on the growth cone (for review, see Tessier-Lavigne and Goodman 1996).

Regulation of Fas II Function during Selective Defasciculation

The experiments described above show how the expression of Fas II can drive axon fasciculation. Yet, these same axons continue to express Fas II at times and places where they selectively defasciculate and leave a particular axon pathway. How is this accomplished? Insight has been provided by genetic analysis of the peripheral projections of motor axons in the embryo.

Fas II is normally expressed on all motoneuron growth cones and axons during the period of axon outgrowth and synapse formation. Many studies have focused on a single motor axon choice point: The motor axons of branch b of the intersegmental nerve (now called the ISNb) initially follow the intersegmental nerve (ISN) but then defasciculate from the ISN axons at a specific choice point and form a separate bundle which steers away (Van Vactor et al. 1993). Lin and Goodman (1994) studied the effects of increasing Fas II on these motor axons and found that increased Fas II can block the defasciculation of the ISNb motor axons at this choice point (Fig. 4). In some cases, increased Fas II caused the fasciculation of axons that would normally not bundle together (e.g., the SNa motor axons with the ISN). These effects of increasing Fas II on motor axons (Lin and Goodman 1994) are similar to the effects of increasing Fas II on CNS axons (Lin et al. 1994; described above); the common feature is that increased levels of Fas II lead to increased axon fasciculation and prevent defasciculation. These results, when combined with the observation that the Fas-II-positive ISNb axons normally defasciculate from the ISN without changing their levels of Fas II, lead to the suggestion that the selective defasciculation of these axons requires the modulation of Fas II function independent of changes in its expression.

Five genes have been identified that encode candidate negative regulators (or competitors) of Fas II function, as loss-of-function mutations in these genes give ISNb defasciculation phenotypes (also called bypass phenotypes) similar to those observed when Fas II levels are increased (Fig. 4). Three receptor protein tyrosine phosphatases (RPTPases: Dlar, DPTP69D, and DPTP99A) are expressed on motor axons, and mutations in the genes encoding them (either singly or in combination) give partially penetrant defasciculation phenotypes in which the ISNb fails to defasciculate from the ISN and thus does not enter its appropriate target region (Desai et al. 1996; Krueger et al. 1996). The strongest of the mutations in the genes encoding RPTPases is *Dlar* which displays an approximately 30% penetrant bypass phenotype.

Single mutations in two other genes—*beaten path* (*beat*; Fambrough and Goodman 1996; Bazan and Goodman 1997) and *sidestep* (*side*; Sink and Goodman 1994)—result in similar but more highly penetrant phenotypes: Virtually all ISNb axons fail to defasciculate and instead continue extending along the ISN. *beat* encodes a novel secreted protein of the Ig superfamily expressed by motoneurons. Beat contains two Ig domains and a cysteine-knot domain and is hypothesized to form a secreted dimer (Bazan and Goodman 1997). Genetic interactions between *beat* and *FasII* suggest that secretion of Beat by motor axons causes a decrease in adhesion of ISNb axons to ISN axons; this regulation of adhesion appears to be very specific, since the ISNb axons and the ISN axons remain tightly fasciculated within their own groups (Fambrough and Goodman 1996). Thus, these studies are beginning to identify candidate molecules that regulate the selective defasciculation of motor axons, possibly by modulating Fas II function.

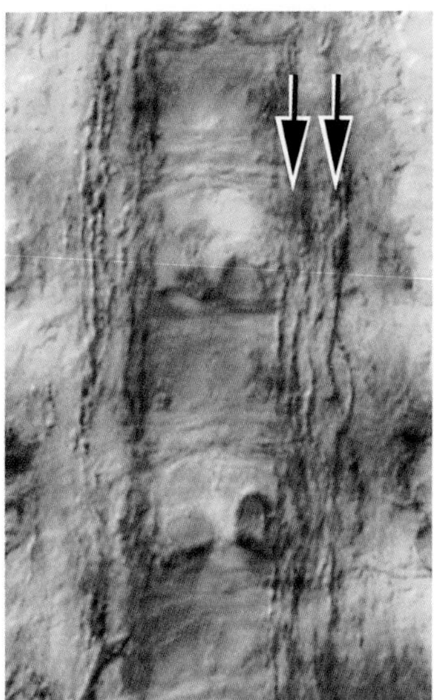

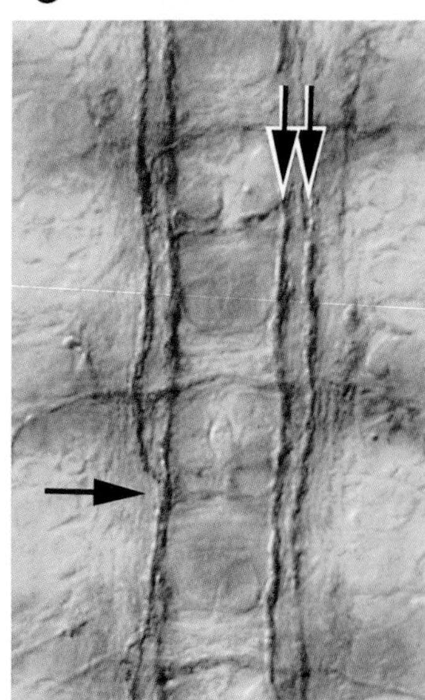

Figure 3. Defasciculation, refasciculation, and altered fasciculation of the pCC and MP1 pathways mediated by *FasII* loss and gain of function. (*Left*) Photomicrographs of the CNS in filleted, stage-16 wild-type embryo stained with anti-Fas II (MAb 1D4) antibody. Fas II is expressed by the fasciculating axons in three large dorsal axon pathways. The two pathways marked by the arrows are the pCC pathway axons (the more medial arrow) and the MP1 pathway (the more lateral arrow), as identified by Hidalgo and Brand 1997). (*Middle*) In *FasII* mutant (*FasII*^{eB112}) embryos, the pCC and MP1 pathway axons are defasciculated (Lin et al. 1994). Photomicrograph of embryo carrying the *ftz*$_{ng}$-*tau*-*β*-*gal* transgene marker and stained with serum anti-β-gal antibody and HRP immunohistochemistry. *ftz*$_{ng}$-*tau*-*β*-*gal* labels the pCC and MP1 pathway axons (*arrows*). (*Right*) The *FasII* gain-of-function condition as driven by the *ftz*$_{ng}$-*FasII* transgenes in a *FasII* ^{e76} (10% Fas II hypomorph) mutant background leads to the refasciculation of the pCC and MP1 pathways in stage-16 embryos. Moreover, the *FasII* gain of function leads to altered fasciculation in which the MP1 pathway abnormally fuses with the more medial pCC pathway. Axons are revealed with the 1D4 anti-Fas II monoclonal antibody and HRP immunohistochemistry. (*Two top arrows*) The refasciculated pCC and MP1 pathways. (*Bottom lateral closed arrow*) Locations where the two pathways are abnormally fused together. (Adapted from Lin et al. 1994.)

ROLE OF FASCICLIN II DURING SYNAPSE FORMATION AND STABILIZATION

Fas II Is Required for Synapse Stabilization

As described above, during the period of axon outgrowth in the *Drosophila* embryo, Fas II is expressed at high levels on motoneuron growth cones and axons where it functions to control selective fasciculation. It is also expressed at low levels by all muscles. However, during the period of synapse formation at the neuromuscular junction (NMJ), the pattern of Fas II expression dramatically changes as Fas II becomes localized to both the pre- and postsynaptic membranes and largely disappears from most of the muscle membrane and most motor axons.

Genetic analysis was used to show that Fas II expression at the synapse is required for synaptic stabilization and growth (Schuster et al. 1996a). In the absence of Fas II, the embryonic synapse forms and differentiates its initial complement of functional boutons. But, during the early stages of postembryonic development, the synapse fails to sprout and grow further. Rather, the boutons begin to retract, leading to synapse elimination at the NMJ and ultimately to death (Fig. 5). Both the synapse elimination and the resulting lethality are rescued by transgenes that drive Fas II expression both pre- and postsynaptically at the NMJ. Driving Fas II expression on either side alone is insufficient to either rescue the synapse or lethality.

It thus appears that a threshold amount of Fas II is required on both sides of the synapse to stabilize the synapse and allow it to further differentiate and grow. This conclusion is bolstered by synthetic mosaic analysis in which the expression of Fas II is separately controlled for two different motor axons synapsing on the same muscle. In these mosaic experiments, Fas II expression is rescued transgenically on certain neurons and muscles in an otherwise null mutant background. In these experiments, when Fas II is expressed on the postsynaptic target (muscle 3), but only on one of the two motor axons synapsing on that muscle, the synapse from the motoneuron expressing Fas II (MN 3a) is stabilized and persists, whereas the synapse from the motoneuron that does not express Fas II (MN 3b) is retracted.

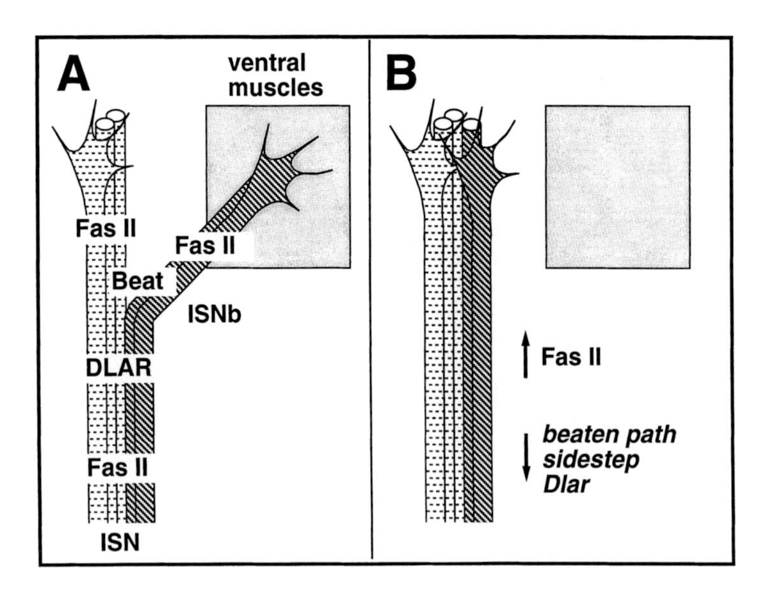

Figure 4. Regulation of Fas II and the selective defasciculation by motor axons.(*A*) In wild-type embryos, a subset of motor axons (called the intersegmental nerve b or ISNb) selectively defasciculate from the other axons in a major motor nerve (called the intersegmental nerve or ISN) at a specific choice point in the region of the ventral muscles (Van Vactor et al. 1993). All of the motor axons express high levels of Fas II. They also express several receptor protein tyrosine phosphatases, including Dlar. They secrete the Beat protein at the choice point required for the selective defasciculation of the ISNb axons (Fambrough and Goodman 1996). (*B*) Increased expression of Fas II (Lin and Goodman 1994) by motor axons results in the failure of the ISNb axons to defasciculate at the choice point. They continue extending distally along the ISN. Loss-of-function mutations in three other genes lead to the same lack-of-defasciculation phenotype: *Dlar* (Krueger et al. 1996), *beaten path* (Fambrough and Goodman 1996), and *sidestep* (Sink and Goodman 1994). See text for further discussion.

Differential Levels of Fas II Can Control the Pattern of Synaptic Connections

Fas II also functions during synapse formation to control the patterning of synaptic connections (Davis et al. 1997). When motoneuron growth cones explore their target muscle domain, they typically make filopodial contact with many different muscles. During normal synapse formation, many of these transient contacts are withdrawn, leaving most motor axons to form stable contacts with only one or a few muscles in a highly stereotyped pattern. Overexpression of Fas II on muscle leads to an increase in the number of filopodial contacts that are stabilized, resulting in ectopic synaptic connections that are fully functional and persist throughout larval development (Fig. 6). There is a critical period for this Fas-II-dependent remodeling of synaptic connectivity during the period of growth cone exploration in the embryo. Overexpression of Fas II on muscle after this period does not alter the pattern of synaptic connections.

The most dramatic respecification of connections is observed when Fas II is differentially overexpressed on subsets of muscle (Figs. 6 and 7). Differential overexpression of Fas II stabilizes ectopic synapses only at those muscles with increased Fas II levels, demonstrating that this rearrangement is target-specific. In addition, when Fas II is overexpressed on one of two muscles normally innervated by a single motoneuron, synapse formation by this neuron is biased onto the target with increased Fas II. This biased synapse formation is enhanced if the endogenous levels of Fas II are genetically reduced, suggesting that the growth cones are sensitive to the proportional difference of Fas II between neighboring muscles.

These results provide insight not only into the function of Fas II during synapse formation, but also into the molecular mechanisms that underlie the patterning of synapse formation. Growth cones behave as if they are not simply responding to the absolute level of a given molecule on a particular potential target, but rather they are comparing the relative levels of this molecule (and presumably other molecules as well) on neighboring target cells. These results suggest that target selection is not based on absolute attractants or repellents that either ensure or prevent synapse formation, but rather on the relative balance of these forces on any given cell in relationship to neighboring cells. This model of target selection (Davis et al. 1997) is similar to a model that views axon guidance in terms of a balance of forces (Tessier-Lavigne and Goodman 1996).

These results also provide experimental confirmation for an idea that was proposed by Sanes, Covault, and their colleagues a number of years ago (Covault and Sanes 1986; Covault et al. 1986, 1987; for review, see Hall and Sanes 1993). They examined the distribution of NCAM (and other CAMs and extracellular matrix molecules as well) on the surface of rat skeletal muscles prior to and after synapse formation. They described that NCAM was initially expressed at modest levels over the entire muscle surface but that after synapse formation, the level of NCAM increased at the synapse but decreased over the rest of the muscle. After denervation of adult muscle, they observed a dramatic increase in NCAM mRNA in the muscle and in the levels of NCAM protein across the muscle surface. On the basis of these observations, they proposed a model whereby NCAM made the surface of muscles attractive for growth cone exploration and synapse formation. Its decline after embryonic synapse formation left the muscle refractory as a potential target for further growth cone exploration. Upon denervation, up-regulation of NCAM makes the muscle surface attractive again so that regenerating motor axons can form synapses once again onto muscle.

Davis et al. (1997) provide experimental support for this hypothesis. The localization of Fas II during synapse formation appears to serve two functions. First, it increases the concentration of Fas II under certain growth cones, thereby stabilizing those contacts and facilitating their transformation into presynaptic terminals. At the

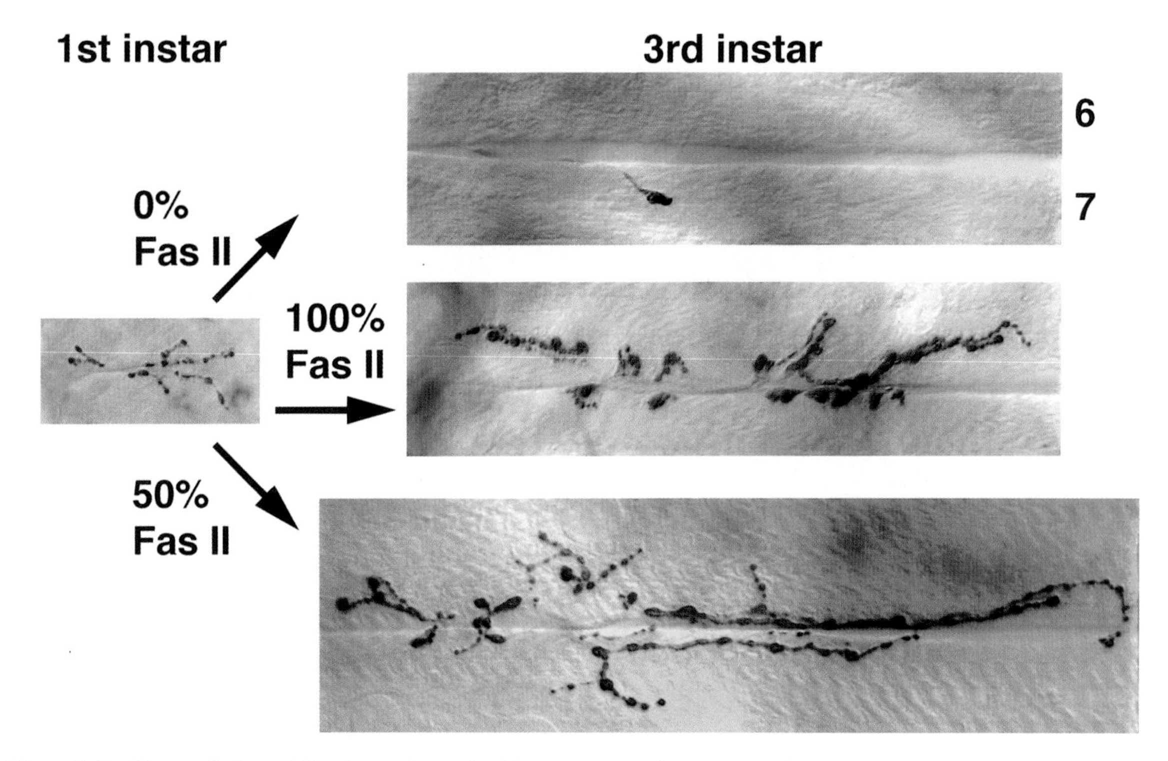

Figure 5. Fas II controls the stabilization and growth of the synapse. Anti-synaptotagmin immunostaining of a wild-type 1st instar larva (*left*) and 3rd instar larvae (*right*) of one of three genotypes: *FasII* synthetic mosaic null, A4 segment (*top*); wild type, A2 segment (*middle*); and *FasII* hypomorphic allele producing 50% Fas II, A2 segment (*bottom*). (*Left*) By late 1st instar, the morphology of the synapse on muscles 7 and 6 resembles that seen in the 3rd instar in terms of the number and pattern of terminal branches, except that it has many fewer boutons. (*Middle right*) Wild-type 3rd instar showing the synapse from MN RP3 and another motoneuron on muscles 7 and 6. (*Top right*) Several synthetic mosaic lines were able to rescue the 1st instar larval lethality of the mutant $FasII^{eB112}$. The synthetic mosaic $FasII^{eB112}$/Y; A51-GAL4; UAS-FasII expresses Fas II in a subset of motoneurons and muscles and is able to rescue the synapses at which it is expressed, thereby rescuing lethality. However, this synthetic mosaic does not express Fas II in MN RP3, MN 6/7b, or muscles 6 and 7 in segment A4. The normal innervation of both of these muscles has been completely retracted, and instead there is a pair of boutons from an ectopic branch of the transverse nerve innervating muscle 7. (*Bottom right*) The *FasII* hypomorph $FasII^{e86}$ produces about 50% the normal levels of Fas II protein. In this mutant, the neuromuscular junction grows more than in wild type, leading to a 50% or greater increase in the number of boutons on muscles 7 and 6. (Adapted from Schuster et al. 1996a.)

same time, a decrease in the concentration of Fas II over the rest of the muscle makes the muscle refractory to exploration and synapse formation by other growth cones. In this way, the first motor axon (or axons) to form a synapse with a muscle rapidly changes the way in which later growth cones can interact with that muscle. Previous results have shown that when the normal events of synapse formation are delayed or prevented, ectopic synapses form on these muscles (Keshishian et al. 1994; Halfon et al. 1995; Kopczynski et al. 1996). Davis et al. (1997) showed that when Fas II is overexpressed during embryogenesis, although the normal synapses form, ectopic synapses are also observed. These results suggest that there is normally a rate-limiting level of Fas II on muscles. The first synaptic contact rapidly concentrates Fas II to the synapse, which appears to lead to a concomitant decrease in Fas II across the rest of the muscle. As a result, additional growth cones are prevented from forming stable contacts.

To what extent does neuronal activity play a part in these events? A role for neuronal activity was shown by the demonstration that reducing activity, either pharmacologically or genetically, increases ectopic innervation of muscles (Jarecki and Keshishian 1995). Interestingly, ectopic innervation is twofold higher at uninnervated muscles compared with muscles that are normally innervated but by motoneurons with blocked activity. This suggests that some activity-independent signal associated with normal innervation may have a dominant role and activity-dependent synaptic transmission a more minor modulatory role in regulating ectopic synapse formation. Both activity-independent and activity-dependent mechanisms may control synapse formation by regulating the expression of the putative growth promoting signal emanating from the muscle.

We propose that Fas II on the muscle surface may represent this growth promoting signal that promotes ectopic innervation. Fas-II-dependent ectopic synapses are anatomically similar to those observed in experiments manipulating neuronal activity (Jarecki and Keshishian 1995) as well as in experiments either preventing or delaying normal synapse formation (Keshishian et al. 1994; Halfon et al. 1995; Kopczynski et al. 1996). Furthermore, reducing neuronal activity has the same critical period for ectopic synapse stabilization (Jarecki and Keshishian 1995) as does muscle overexpression of Fas II. The

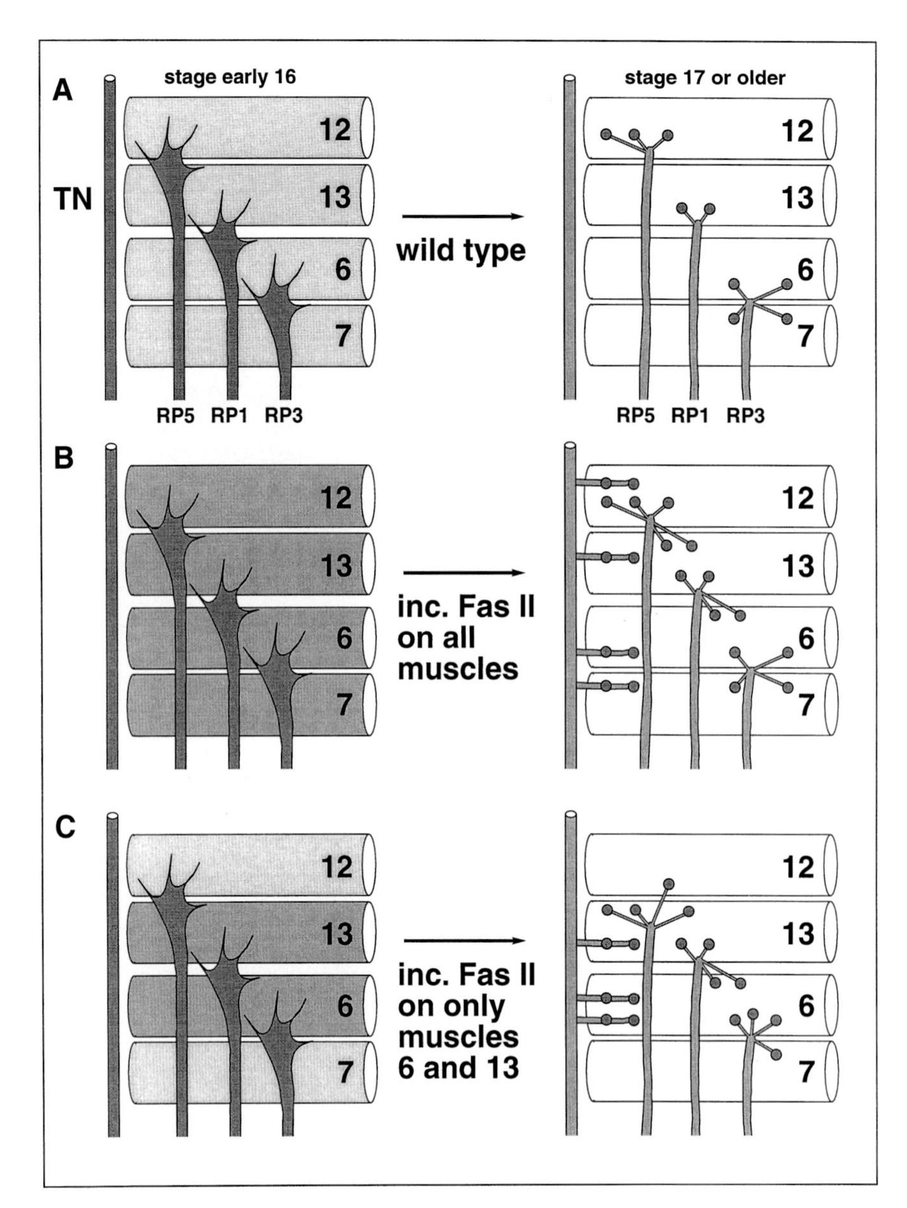

Figure 6. Fasciclin II-dependent regulation of synapse formation. The organization of muscles 7, 6, 13, and 12 and their innervation are diagrammed. Levels of Fas II expression are indicated by shades of gray. Embryonic stage 16 is diagrammed at left and the stereotyped pattern of synaptic connections from embryonic stage 17 through larval stages is diagrammed in the panels at right. In the wild-type stage early 16 embryo (*A, left*), Fas II is initially expressed at low levels on all muscles, and at higher levels on motoneurons including the segmental nerve, the intersegmental nerve, and the transverse nerve (TN). Growth cones contact multiple muscles, but withdraw many of these contacts and form stable synapses with specific muscles. In the wild-type larvae (*A, right*), the growth cones have been transformed into a highly stereotyped pattern of synaptic connections. Fas II expression both pre- and postsynaptically is localized to the synaptic boutons; Fas II disappears from the rest of the muscle surface. Overexpression of Fas II on all muscles in the embryo (*B, left*) stabilizes novel ectopic synaptic connections that are maintained at the mature larval neuromuscular junction (*B, right*). Novel synaptic connections are stabilized and normal innervation is also altered. Differential expression of Fas II on subsets of muscle (muscles 6 and 13) (*C, left*) induces muscle specific synapse stabilization (*C, right*). Novel ectopic synapses are stabilized and normal synaptic growth is biased onto muscles that express higher levels of Fas II. (Adapted from Davis et al. 1997.)

events of synapse formation drive the localization of Fas II to the synapse and the removal of Fas II from the rest of the muscle surface thus decreasing the probability that other contacts will stabilize and form synapses. Target selection in this system is thus a more competitive process than was previously suspected. The process is highly dynamic, and targeting decisions appear to be based both on the complement of receptors expressed on the growth cone surface and on the temporal relationship of whether the growth cone arrives first or after some other growth cone has already begun to form a synapse.

It follows from this model that anything that helps speed up or slow down these dynamic changes in Fas II levels will influence the likelihood that ectopic synapses will form. Neuronal activity may be able to modulate the rate of these events. For example, Schuster et al. (1996b)

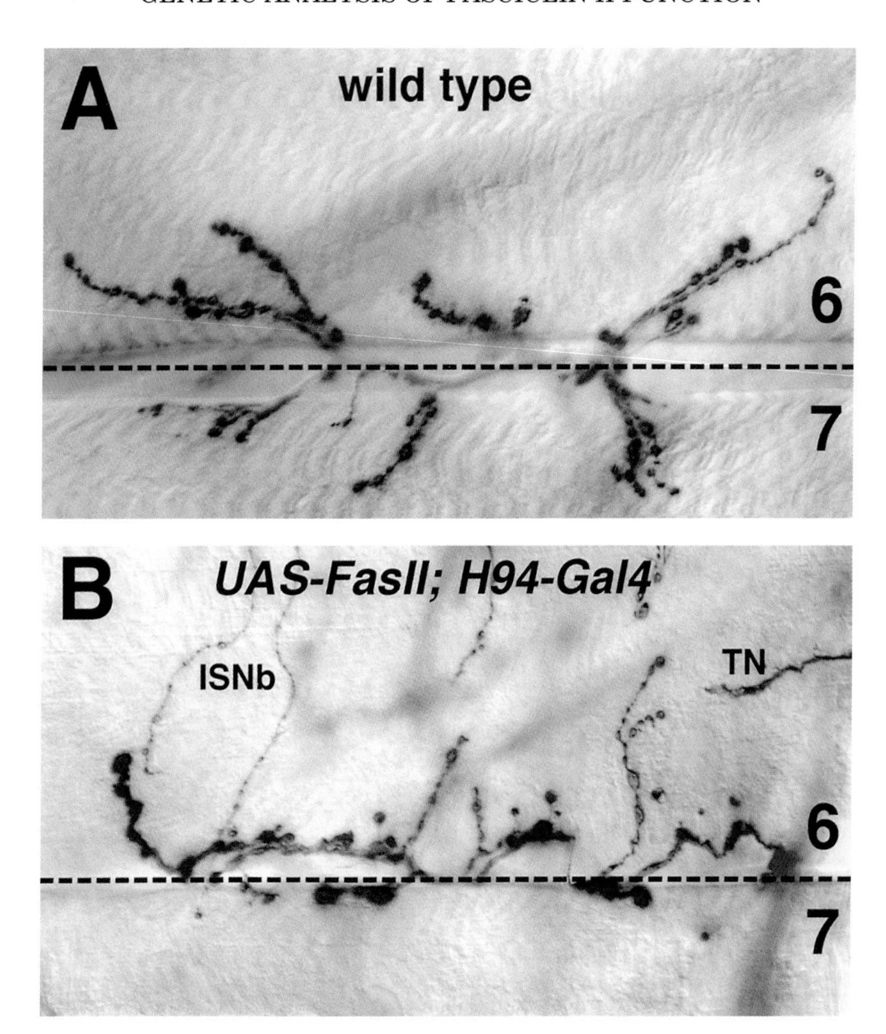

Figure 7. Differential Fas II expression directs both synapse formation and growth. (*A*) An example of the wild-type innervation of muscles 6 and 7. (*B*) Overexpression of *UAS-FasII* in muscle 6 but not muscle 7 by *H94-Gal4*. There is a significant increase in the percentage of boutons contacting muscle 6, and a concomitant reduction in the percentage of boutons contacting muscle 7 (Davis et al. 1997). ISNb indicates ectopic innervation from other motor axons in the ISNb nerve (presumably those that normally innervate muscles 13 and 12). TN indicates ectopic innervation from the transverse nerve which normally contacts muscles 7 and 6 but does not innervate it.

showed that Fas II expression at the synapse can be down-regulated by increased neuronal activity during larval development (as reviewed later in this paper). It is possible that neuronal activity during synapse formation helps drive the down-regulation of extrasynaptic Fas II, and if so, then the reduction in neuronal activity would slow down the removal of Fas II from the muscle surface and thus allow the stabilization of ectopic synapses.

Interactions of DLG with the Carboxyl Terminus of Fas II Controls Synaptic Clustering

The data described above suggest that the precise localization of Fas II at the pre- and postsynaptic membranes is required for the stabilization and specificity of synapse formation and, as described in the following section, for synaptic growth and plasticity.

A major insight into the mechanisms of protein localization at the synapse came with the identification of the MAGUK family of membrane-associated guanylate kinases and the discovery that these proteins appear to control the synaptic localization of many different ion channels and neurotransmitter receptors (for review, see Gomperts 1996; Sheng 1996). MAGUK proteins contain three PDZ domains at their amino terminus, an SH3 domain and a carboxy-terminal guanylate kinase domain. The PDZ domains are modular protein-protein interaction domains and appear to bind to membrane proteins by interacting with their extreme carboxy-terminal sequences (Doyle et al. 1996). Putative PDZ-interacting proteins were identified initially on the basis of a carboxy-terminal consensus motif, -S/T-X-V (Kornau et al. 1995). Fas II contains such a three-amino-acid sequence at its carboxyl terminus (Fig. 1) (Kornau et al. 1995), suggesting that it might interact with PDZ domains.

The MAGUK protein Discs-Large (DLG; Woods and Bryant 1991), a *Drosophila* homolog of PSD-95, has been implicated in the synaptic clustering of ion channels

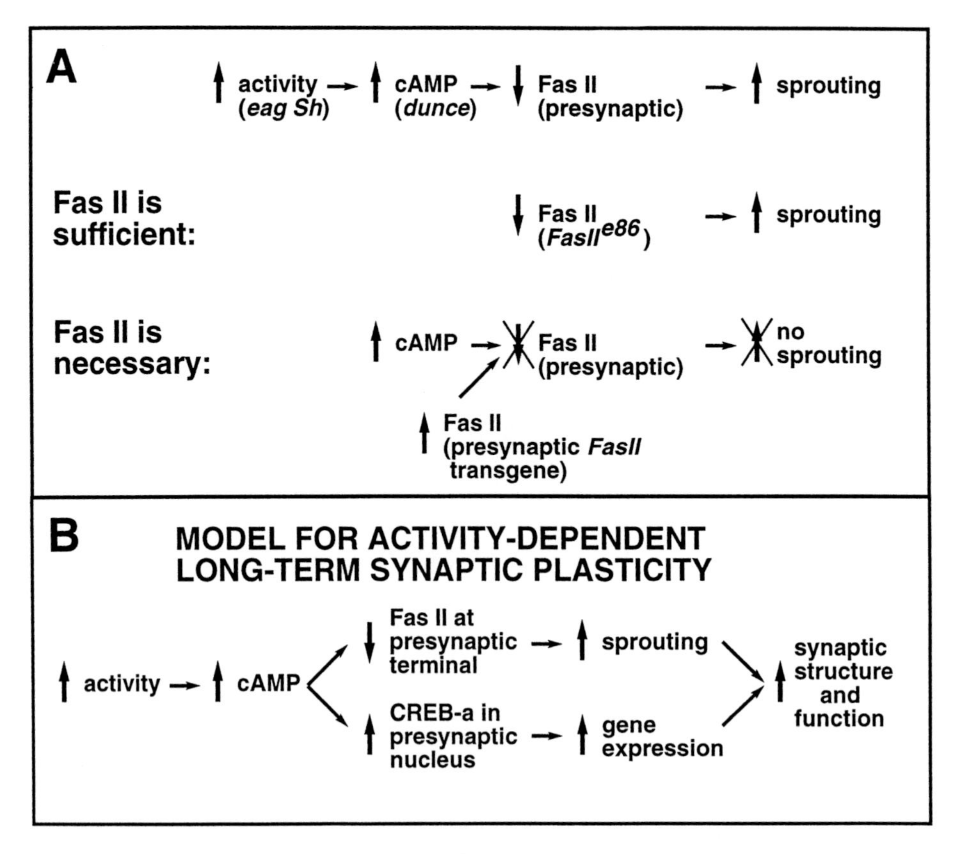

Figure 8. Model for Fas-II-dependent structural component of synaptic plasticity. (*A*) Schuster et al. (1996b) proposed that increases in neural activity (e.g., in *eag Sh* mutants) elevate cAMP levels (phenocopied in *dunce* mutants); they showed that these changes lead to a down-regulation of synaptic Fas II expression. Genetic analysis shows that the reduction in presynaptic Fas II is both necessary and sufficient to cause synaptic sprouting. Fas II is sufficient for structural plasticity (line 2) since genetic reduction of Fas II in the $FasII^{e86}$ hypomorphic mutant is sufficient to cause structural sprouting. Fas II is necessary (line 3) since maintenance of presynaptic Fas II expression with a *FasII* transgene blocks cAMP-dependent structural plasticity. (*B*) Increased cAMP (in *dunce* mutants) leads to an increase in the structure and function of the *Drosophila* neuromuscular junction. Synaptic Fas II controls this structural plasticity but does not alter synaptic function. Davis et al. (1996) showed that CREB, the cAMP-response-element-binding protein, acts in parallel with Fas II to cause an increase in synaptic strength. Expression of the CREB repressor in the *dnc* mutant blocks functional but not structural plasticity. Expression of the CREB activator increases synaptic strength only in *FasII* mutants that increase bouton number. This CREB-mediated increase in synaptic strength is due to increased presynaptic transmitter release. Expression of CREB in a *FasII* mutant background genetically reconstitutes this cAMP-dependent plasticity. Thus, cAMP initiates parallel changes in CREB and Fas II to achieve long-term synaptic enhancement.

at the NMJ (Tejedor et al. 1997) and is thus a good candidate for controlling the localization of Fas II. To determine whether these carboxy-terminal sequences in Fas II are responsible for synaptic targeting, and whether DLG controls this targeting, Zito et al. (1997) asked whether these sequences are capable of targeting a heterologous membrane protein (CD8) to the synapse. They showed that the carboxy-terminal sequence of Fas II is both necessary and sufficient for targeting to the subsynaptic muscle membrane at the larval NMJ of *Dro-sophila*. This localization depends on DLG and can be accounted for by an active clustering or selective retention of the protein at the synapse (Zito et al. 1997). Taken together with the results described above for the role of Fas II in synapse stabilization (Schuster et al. 1996a) and formation (Davis et al. 1997), these results suggest that the DLG-dependent localization of Fas II may have an important role in the stabilization of synaptic contacts and thus in the patterning of synapse formation.

ROLE OF FASCICLIN II DURING SYNAPTIC GROWTH

Down-regulation of Fas II Leads to Synaptic Sprouting

Schuster et al. (1996a) observed an enhanced growth of the neuromuscular synapse when Fas II levels are genetically decreased by approximately 50% (Fig. 5). Several different genetic conditions that lead to an approximately 50% reduction in Fas II levels lead to a 50% increase in the size of the synapse. Interestingly, similar kinds of enhanced synaptic growth have been seen at these same synapses in mutants with elevated neuronal activity (due to decreased potassium currents in *eag Sh*) or with increased cAMP levels (*dunce*) (Budnik et al. 1990; Zhong et al. 1992).

Might a down-regulation of Fas II at this synapse be involved in the activity-dependent control of synaptic growth? There is certainly a good precedent for thinking that Fas II might play a part in the enhancement of synaptic growth that is observed in *eag Sh* and *dunce* mutants. For example, when *Aplysia* sensory and motor neurons are grown in culture, applications of serotonin that induce presynaptic sprouting also lead to a down-regulation of presynaptic apCAM (Bailey et al. 1992; Mayford et al. 1992), the *Aplysia* homolog of Fas II. This work led to the hypothesis that perhaps activity-dependent events (involving neuronal activity and increased levels of cAMP) lead to a down-regulation of presynaptic apCAM, that this change in the level of apCAM leads to sprouting, and that this sprouting is the basis for the structural changes that accompany long-term memory. Genetic analysis in *Drosophila* allowed one step in this hypothesis to be confirmed, namely, that a down-regulation in the level of a synaptic Fas II leads to synaptic sprouting in the organism (Schuster et al. 1996a).

Neural Activity Leads to a Decrease in Synaptic Fas II

The results described above show that increased neural activity can lead to increased synaptic sprouting and that decreased levels of Fas II can do the same; but does activity regulate Fas II and is this a necessary step in its ability of neural activity to modulate sprouting? Schuster et al. (1996b) showed that *eag Sh* and *dunce* mutants do indeed lead to a down-regulation of synaptic Fas II and that this down-regulation of Fas II at the synapse is both necessary and sufficient for the cAMP-dependent enhancement of synaptic growth (Fig. 8). However, they showed this to be only part of the story, since decreases in synaptic Fas II lead to the structural but not the functional enhancement of the synapse, whereas an increase in cAMP leads to both. In another paper (Davis et al. 1996), they genetically dissected and then reconstituted the molecular mechanisms of cAMP-mediated synaptic enhancement. They showed that CREB, the cAMP-response-element-binding protein, acts in parallel with Fas II to control the cAMP-mediated enhancement in synaptic strength. Thus, an increase in cAMP appears to lead to two parallel changes: a decrease in synaptic Fas II and an elevation of CREB in the presynaptic nucleus (Fig. 8).

Decreased Fas II at the synapse drives its structural expansion, whereas increased activated CREB in the nucleus initiates the transcription of unknown rate-limiting components that are required for the functional expansion of the synapse. An interesting possibility, suggested by this work, is that the synapse-specific down-regulation of Fas II may prime specific synaptic terminals for functional modification. Increased CREB activation on its own does not influence synaptic function, whereas increased CREB activation in parallel with a down-regulation of synaptic Fas II leads to increased synaptic strength. Thus, an activity-dependent down-regulation of Fas II at selective synapses could prime those synapses for functional modification. In this manner, the regulation of Fas II could confer synaptic specificity onto the mechanisms of activity-dependent synaptic plasticity (Martin and Kandel 1996).

The mechanisms controlling the spatial and temporal regulation of cell adhesion molecules such as Fas II may be critical for synapse-specific plasticity. In *Aplysia*, MAP kinase, acting downstream from the serotonin receptor, phosphorylates apCAM, the *Aplysia* homolog of Fas II. This MAP kinase phosphorylation results in the internalization of apCAM from the cell surface (Bailey et al. 1997). Thus, MAP kinase phosphorylation appears to be an important component in the mechanism of 5-HT-dependent structural plasticity. A related type of mechanism appears to regulate Fas II function in *Drosophila*. Genetic evidence places Fas II downstream from cAMP in the molecular pathway leading to activity-dependent synaptic sprouting. How does cAMP regulate Fas II levels? Recent results suggest that increased cAMP acts through PKA to stimulate the internalization of Fas II from the cell surface (G.W. Davis and C.S. Goodman, unpubl.).

Differential Expression of Fas II Can Regulate Synaptic Growth

After synapse formation is complete, the differential expression of target-derived Fas II can have a dramatic input on the relative growth of the synapse (Davis et al. 1997). Muscles 7 and 6 are normally innervated by the same motoneuron, RP3. Overexpression of Fas II on muscle 6, but not muscle 7, biases the growth and development of this motoneuron onto muscle 6. In such animals, by the 3rd instar, the amount of synaptic growth on muscle 6 far exceeds what is normally seen in wild type (Fig. 7). It is differential levels of Fas II on different targets, and not simply the absolute level of Fas II on any one target, that appear to drive this bias in synaptic growth since (1) the distribution of the normal synapses between muscles 6 and 7 is not altered by a uniform elevation of Fas II expression, and yet (2) the biased growth on muscle 6 is enhanced beyond that observed by overexpression of Fas II on muscle 6 by genetically decreasing the level of Fas II on muscle 7, thus enhancing the proportional difference in Fas II levels between muscles 6 and 7. Thus, motoneurons respond to the differential expression of Fas II both by biasing their pattern of initial synapse formation and by biasing their growth onto the muscles with higher levels of Fas II.

CONCLUSIONS

In this paper, we have reviewed three faces of Fasciclin II and have shown that it functions to control selective axon fasciculation during axon guidance, the patterning and stabilization of synapses during growth cone exploration and synapse formation, and the growth of synapses during later periods of synaptic remodeling and plasticity. Some of these Fas II functions are controlled in an activity-independent fashion (e.g., axon fasciculation and growth cone guidance), whereas other functions are controlled by neural activity (e.g., synaptic growth and plasticity). Interestingly, the function of Fas II appears to be regulated in at least three different ways during these three stages.

During the period of axon guidance, growth cones fasciculate and defasciculate rather quickly, and thus there may be a requirement for rapid and transient changes in the function of CAMs. Although Fas II gets turned on to control axon fasciculation, it does not disappear from the surface to control defasciculation. Rather, axons can selectively defasciculate without apparently changing their levels of Fas II, suggesting that the function of the protein is regulated under these conditions. The secreted factor Beat is a good candidate as a potential regulator of Fas II function (Fambrough and Goodman 1996; Bazan and Goodman 1997). We imagine that Beat functions through some unknown receptor that somehow regulates Fas II function by interacting with its cytoplasmic domain, possibly by controlling one of the potential phosphorylation events on the Fas II cytoplasmic tail.

During the period of synapse formation and stabilization, the overall function of Fas II is controlled by the dynamic changes in its subcellular localization. Initially, there is a low level of Fas II across the entire muscle surface. During synapse formation, Fas II rapidly clusters under the forming synapse and disappears from the rest of the muscle membrane. Targeting sequences in Fas II have been identified (Zito et al. 1997) that control this synaptic clustering. Moreover, it has been hypothesized that this synaptic clustering results in the disappearance of Fas II from the rest of the muscle membrane and that this functions as a regulator of both synaptic stabilization and the patterning of synapse formation (Davis et al. 1997). The time course of synapse stabilization is slower that individual growth cone guidance decisions, and thus protein turnover at the plasma membrane, might work within a timeframe to modulate one but not the other.

Finally, during the period of synaptic growth and plasticity, increases in neural activity appear to lead to the presynaptic down-regulation of Fas II, which then leads to an increase in growth and sprouting of the synapse (Schuster et al. 1996a,b). In this case, the down-regulation appears to involve an active internalization that may be triggered by increased cAMP leading to increased activated PKA (G.W. Davis and C.S. Goodman, unpubl.).

Fas II thus appears to be regulated by functional modification, subcellular localization, and internalization. All three methods of regulation may be used in different development events or, in some cases, simultaneously, depending on the temporal and spatial constraints of the particular event. In this way, the organism has several ways to achieve a fine control of Fas II function. Fas II is called forth in a specific fashion during almost every stage in the generation of neuronal specificity, from the initial activity-independent events of guidance and synapse formation to the activity-dependent events of synaptic remodeling and growth. The variety of different mechanisms for the regulation of Fas II function may allow the organism a greater degree of freedom in how it uses this cell adhesion molecule. In addition, this work provides a clear example of how different events in the generation of neural specificity can reuse the same molecule by adjusting the ways it which the protein is expressed and its function regulated.

ACKNOWLEDGMENTS

We thank many present and previous members of the Goodman lab for their contributions to the experimental analysis of Fas II function reviewed here, including Michael Bastiani, Richard Fetter, Luis Garcia Alonso, Gabriele Grenningloh, Allan Harrelson, Greg Helt, Casey Kopczynski, David Lin, Jay Rehm, Christoph Schuster, Peter Snow, and Kai Zinn. We also thank our colleague Ehud Isacoff for his contributions. This work was supported by a National Institutes of Health grant (HD-21294) and by a grant from the Keck Foundation. K.Z. is a predoctoral fellow, G.W.D. is a postdoctoral fellow, and C.S.G. is an investigator with the Howard Hughes Medical Institute.

REFERENCES

Bailey C.H., Chen M., Keller F., and Kandel E.R. 1992. Serotonin-mediated endocytosis of apCAM: An early step of learning-related synaptic growth in Aplysia. *Science* **256:** 645.

Bailey C.H., Kaang B.K., Chen M., Martin K.C., Lim C.S., Casadio A., and Kandel E.R. 1997. Mutation in the phosphorylation sites of MAP kinase blocks learning-related internalization of apCAM in *Aplysia* sensory neurons. *Neuron* **18:** 913.

Bastiani M.J., Raper J.A., and Goodman C.S. 1984. Pathfinding by neuronal growth cones in grasshopper embryos. III. Selective affinity of the G growth cone for the P cells within the A/P fascicle. *J. Neurosci.* **4:** 2311.

Bastiani M.J., Harrelson A.L, Snow P.M., and Goodman C.S. 1987. Expression of fasciclin I and II glycoproteins on subsets of axon pathways during neuronal development in the grasshopper. *Cell* **48:**

Bazan J.F. and Goodman C.S. 1997. Modular structure of the *Drosophila* Beat protein. *Curr. Biol.* **7:** R338.

Becker C.G., Artola A., Gerardy-Schahn R., Becker T., Welzl H., and Schachner M. 1996. The polysialic acid modification of the neural cell adhesion molecule is involved in spatial learning and hippocampal long-term potentiation. *J. Neurosci. Res.* **45:** 143.

Bieber A.J., Snow P.M., Hortsch M., Patel N.H., Jacobs J.R., Traquina Z., Schilling J., and Goodman C.S. 1989. *Drosophila* neuroglian: A member of the immunoglobulin superfamily with extensive homology to the vertebrate neural adhesion molecule L1. *Cell* **59:**

Budnik V., Zhong Y., and Wu C.F. 1990. Morphological plasticity of motor axons in *Drosophila* mutants with altered excitability. *J. Neurosci.* **10:** 3754.

Covault J. and Sanes J. 1986. Distribution of N-CAM in synaptic and extrasynaptic portions of developing and adult skeletal muscle. *J. Cell Biol.* **102:** 716.

Covault J., Cunningham J.M., and Sanes J.R. 1987. Neurite outgrowth on cryostat sections of innervated and denervated skeletal muscle. *J. Cell Biol.* **105:** 2479.

Covault J., Merlie J.P., Goridis C., and Sanes J. 1986. Molecular forms of N-CAM and its RNA in developing and denervated skeletal muscle. *J. Cell Biol.* **102:** 731.

Cremer H., Lange R., Christoph A., Plomann M., Vopper G., Roes J., Brown R., Baldwin S., Kraemer P., Scheff S., Barthels D., Rajewsky K., and Wille W. 1994. Inactivation of the N-CAM gene in mice results in size reduction of the olfactory bulb and deficits in spatial learning. *Nature* **367:** 455.

Cunningham B.A., Hemperly J.J., Murray B.A., Prediger E.A.,

Brackenbury R., and Edelman G.M. 1987. Neural cell adhesion molecule: Structure, immunoglobulin-like domains, cell surface modulation and alternative RNA splicing. *Science* **236:** 799.

Davis G.W., Schuster C.M., and Goodman C.S. 1996. Genetic dissection of structural and functional components of synaptic plasticity. III. CREB is necessary for presynaptic functional plasticity. *Neuron* **17:** 669.

———. 1997. Genetic analysis of the mechanisms controlling target selection: Target-derived Fasciclin II regulates the pattern of synapse formation. *Neuron* **19:** 561.

Desai C.J., Gindhart Jr. G., Goldstein L.S., and Zinn K. 1996. Receptor tyrosine phosphatases are required for motor axon guidance in the *Drosophila* embryo. *Cell* **84:** 599.

Doyle D.A., Lee A., Lewis J., Kim E., Sheng M., and Mac Kinnon R. 1996. Crystal structures of a complexed and peptide-free membrane protein-binding domain: Molecular basis of peptide recognition by PDZ. *Cell* **85:** 1067.

Edelman G.M. 1985. Cell adhesion molecule expression and the regulation of morphogenesis. *Cold Spring Harbor Symp. Quant. Biol.* **50:** 877.

Fambrough D. and Goodman C.S. 1996. The *Drosophila beaten path* gene encodes a novel secreted protein that regulates defasciculation at motor axon choice points. *Cell* **87:** 1049.

Gomperts S.N. 1996. Clustering membrane proteins: It's all coming together with the PSD-95/SAP90 protein family. *Cell* **84:** 659.

Goodman C.S. 1996. Mechanisms and molecules that control growth cone guidance. *Annu. Rev. Neurosci.* **19:** 341.

Goodman C.S. and Shatz C.J. 1993. Developmental mechanisms that generate precise patterns of neuronal connectivity. *Cell* **72:** 77.

Goodman C.S., Bastiani M.J., Doe C.Q., du Lac S., Helfand S.L., Kuwada J.Y., and Thomas J.B. 1984. Cell recognition during neuronal development. *Science* **225:** 1271.

Grenningloh G., Rehm E.J., and Goodman C.S. 1991. Genetic analysis of growth cone guidance in *Drosophila*: Fasciclin II functions as a neuronal recognition molecule. *Cell* **67:** 45.

Grenningloh G., Bieber A., Rehm J., Snow P., Traquina Z., Hortsch M., Patel N., and Goodman C.S. 1990. Molecular genetics of neuronal recognition in *Drosophila*: Evolution and function of immunoglobulin superfamily cell adhesion molecules. *Cold Spring Harbor Symp. Quant.* **55:** 327.

Halfon M.S., Hashimoto C., and Keshishian H. 1995. The *Drosophila Toll* gene functions zygotically and is necessary for proper motoneuron and muscle development. *Dev. Biol.* **169:** 151.

Hall Z.W., and Sanes J.R. 1993. Synaptic structure and development: The neuromuscular junction. *Cell* **72:** 99.

Harrelson A.L. and Goodman C.S. 1988. Growth cone guidance in insects: Fasciclin II is a member of the immunoglobulin superfamily. *Science* **242:** 700.

Hidalgo A. and Brand A.H. 1997. Targeted neuronal ablation: The role of pioneer neurons in guidance and fasciculation in the CNS of *Drosophila*. *Development* **124:** 3253.

Hu H., Tomasiewicz H., Magnuson T., and Rutishauser U. 1996. The role of polysialic acid in migration of olfactory bulb interneuron precursors in the subventricular zone. *Neuron* **16:** 735.

Jarecki J. and Keshishian H. 1995. Role of neural activity during synaptogenesis in *Drosophila*. *J. Neurosci.* **15:** 8177.

Katz L.C. and Shatz C.J. 1996. Synaptic activity and the construction of cortical circuits. *Science* **274:** 1133.

Keshishian H., Chiba A., Chang T.N., Halfon M., Harkins E.W., Jarecki J., Wang L.S., Anderson M. S., Cash S., Halpern M.E., and Johansen J. 1994. Cellular mechanisms governing synaptic development in *Drosophila melanogaster*. *J. Neurobiol.* **24:** 757.

Kopczynski C.C., Davis G.W., and Goodman C.S. 1996. A neural tetraspanin, encoded by *late bloomer* that facilitates synapse formation. *Science* **271:** 1867.

Kornau H.C., Schenker L.T., Kennedy M.B., and Seeburg P.H. 1995. Domain interaction between NMDA receptor subunits and the postsynaptic density protein PSD. *Science* **269:** 1737.

Krueger N.X., Van Vactor D., Wan H.I., Gelbart W.M., Goodman C.S., and Saito H. 1996. The transmembrane tyrosine phosphotase DLAR controls motor axon guidance in *Drosophila*. *Cell* **84:** 611.

Lin D.M. and Goodman C.S. 1994) Ectopic and increased expression of Fasciclin II alters motoneuron growth cone guidance. *Neuron* **13:** 507.

Lin D.M., Fetter R.D., Kopczynski C. Grenningloh, G., and Goodman C.S. 1994. Genetic analysis of Fasciclin II in *Drosophila*: Defasciculation, refasciculation and altered fasciculation. *Neuron* **13:** 1055.

Martin K.C. and Kandel E.R. 1996. Cell adhesion molecules, CREB, and the formation of new synaptic connections. *Neuron* **17:** 567.

Mayford M., Barzilai A., Keller F., Schacher S., and Kandel E.R. 1992. Modulation of an NCAM-related adhesion molecule with long-term synaptic plasticity in *Aplysia*. *Science* **256:** 638.

Muller D., Wang C., Skibo G., Toni N., Cremer H., Calaora V., Rougon G., and Kiss J.Z. 1996. PSA-NCAM is required for activity-induced synaptic plasticity. *Neuron* **17:** 413.

Ramos R.G., Igloi G.L., Lichte B., Baumann U., Maier D., Schneider T., Brandstäter J.H., Frölich A., and Fischbach K.F. 1993. The irregular chiasm C-roughest locus of *Drosophila*, which affects axonal projections and programmed cell death, encodes a novel immunoglobulin-like protein. *Genes Dev.* **7:** 2533.

Raper J.A., Bastiani M.J., and Goodman C.S. 1984. Pathfinding by neuronal growth cones in grasshopper embryos. IV. The effects of ablating the A and P axons upon the behavior of the G growth cone. *J. Neurosci.* **4:** 2329.

Schuster C.M., Davis G.W., Fetter R.D., and Goodman C.S. 1996a. Genetic dissection of structural and functional components of synaptic plasticity: Fasciclin II controls synaptic stabilization and growth. *Neuron* **17:** 641.

———. 1996b. Genetic dissection of structural and functional components of synaptic plasticity: Fasciclin II controls structural plasticity. *Neuron* **17:** 655.

Sheng M. 1996. PDZs and receptor/channel clustering: Rounding up the latest suspects. *Neuron* **17:** 575.

Sink H. and Goodman C.S. 1994. Mutations in *side-step* lead to defects in pathfinding and synaptic specificity during the development of neuromuscular connectivity in *Drosophila*. *Soc. Neurosci. Abstr.* **20:** 1283.

Tejedor F.J., Bokhari A., Rogero O., Gorczyca M., Zhang J., Kim E., Sheng M., and Budnik V. 1997. Essential role for dlg in synaptic clustering of Shaker K^+ channels in vivo. *J. Neurosci.* **17:** 152.

Tessier-Lavigne M. and Goodman C.S. 1996. The molecular biology of axon guidance. *Science* **274:** 1123.

Tomasiewicz H., Ono K., Yee D., Thompson C., Goridis C., Rutishauser U., and Magnuson T. 1993. Genetic deletion of a neural cell adhesion molecule variant (N-CAM-180) produces distinct defects in the central nervous system. *Neuron* **11:** 1163.

Van Vactor D., Sink H., Fambrough D., Tsoo R., and Goodman C.S. 1993. Genes that control neuromuscular specificity in *Drosophila*. *Cell* **73:** 1137.

Woods D.F., and Bryant P.J. 1991. The discs-large tumor suppressor gene of *Drosophila* encodes a guanylate kinase homolog localized at septate junctions. *Cell* **66:** 451.

Zhong Y., Budnik V., and Wu C.F. 1992. Synaptic plasticity in *Drosophila* memory and hyperexcitable mutants: Role of cAMP cascade. *J. Neurosci.* **12:** 644.

Zito K., Fetter R.A., Goodman C.S., and Isacoff E.Y. 1997. Synaptic clustering of Fasciclin II and Shaker: Essential targeting sequences and role of DLG. *Neuron* (in press).

Cell Lineage Determination and the Control of Neuronal Identity in the Neural Crest

D.J. ANDERSON,[1] A. GROVES,[2] L. LO,[1] Q. MA,[1] M. RAO,[3] N.M. SHAH,[4] AND L. SOMMER[5]
[1]*Howard Hughes Medical Institute, Division of Biology 216-76, California Institute of Technology, Pasadena, California 91125*

The diverse cell types of complex tissues such as the blood and the brain are generated from self-renewing, multipotent progenitors called stem cells (for reviews, see Hall and Watt 1989; Potten and Loeffler 1990; Morrison et al. 1997). These stem cells must generate progeny of different phenotypes, in the correct proportions, sequence, and location. The manner in which this is accomplished is not well understood. It is clear that the local microenvironment of stem cells has an important influence on their development, as do transcription factors that act within the cells. However, the manner in which such signals and transcription factors interact to control lineage determination by multipotent stem cells is poorly understood. To address this issue, it is necessary to both alter the expression of transcription factors in stem cells and challenge the cells by altering their environment to determine their state of lineage commitment. There are relatively few experimental systems in which such combined genetic and cell biological manipulations of stem cells are feasible.

FATE AND POTENTIAL OF NEURAL CREST CELLS IN AVIAN EMBRYOS

We have studied the control of lineage commitment by stem cells in the neural crest. The neural crest is a migratory population of progenitor cells that detaches from the dorsolateral margins of the neural tube and migrates to distant locations throughout the embryo (Fig. 1). Fate-mapping experiments in amphibian and avian embryos have demonstrated that the crest generates a diverse array of neural and mesectodermal derivatives (for reviews, see Le Douarin 1982; Bronner-Fraser 1993a). These derivatives include the neurons and glia of the peripheral nervous system, melanocytes, smooth muscle cells of the cardiac outflow tracts, and the bones and cartilage of the face. Some crest derivatives are generated only at certain positions along the anteroposterior axis (A/P) (Le Douarin 1980). However, the crest also generates diverse derivatives at a single axial level. For example, neural crest cells in the thoracolumbar region of the trunk generate sensory neurons, sympathetic neurons, Schwann cells, adrenal chromaffin cells, and melanocytes (Le Douarin 1986).

Interspecific grafting experiments in avian embryos have revealed that the developmental potential of the crest is relatively homogeneous along the A/P axis, with the exception of craniofacial mesenchyme which apparently cannot be generated by transplanted trunk neural crest (Le Douarin 1982). This implies that the fate of neural crest cells is controlled by environmental signals. Such signals could act before, during, or after migration from the dorsal neural tube (Bronner-Fraser 1992). A major problem in neural crest cell biology, therefore, is to identify such environmental signals and their sources and to understand their mechanism of action on the crest (for reviews, see Stemple and Anderson 1993; Wehrlehaller and Weston 1997).

MULTIPOTENCY AND DEVELOPMENTAL RESTRICTION OF NEURAL CREST CELLS

The pleuripotency of the crest revealed by transplantation experiments could reflect a homogeneous population of pleuripotent cells, or a mixture of committed cells. In vivo lineage tracing experiments in chick have shown that many premigratory crest cells are multipotent (Bronner-Fraser and Fraser 1988). This observation is consistent with the results of in vitro clonal analysis of quail neural crest cells (Sieber-Blum and Cohen 1980; Baroffio, et al. 1988). The fact that the crest contains multipotent cells, and that the fate of these cells is influenced by the embryonic environment, leads to two extreme models for how lineage commitment is accomplished. On the one hand, specific environmental signals may instruct uncommitted cells to choose one fate at the expense of others. Such a mechanism would be termed "instructive." On the other hand, uncommitted cells may undergo lineage restriction by a cell-autonomous mechanism (either stochastic or deterministic), and environmental factors may permit the survival and proliferation of appropriate cells in the appropriate place. Such a mechanism would be termed "selective" (Morrison et al. 1997). One of our

Present addresses: [2]Division of Biology 139-74, Caltech, Pasadena, California 91125; [3]Department of Neurobiology and Anatomy, University of Utah, Medical School, 50 North Medical Drive, Salt Lake City, Utah 84132; [4]Columbia University, College of Physicians and Surgeons, 701 W. 168th St., New York, New York 10032; and [5]Institute of Cell Biology, Swiss Federal Institute of Technology, ETH-Hoenggerberg HPM E26, CH-8093 Zurich, Switzerland.

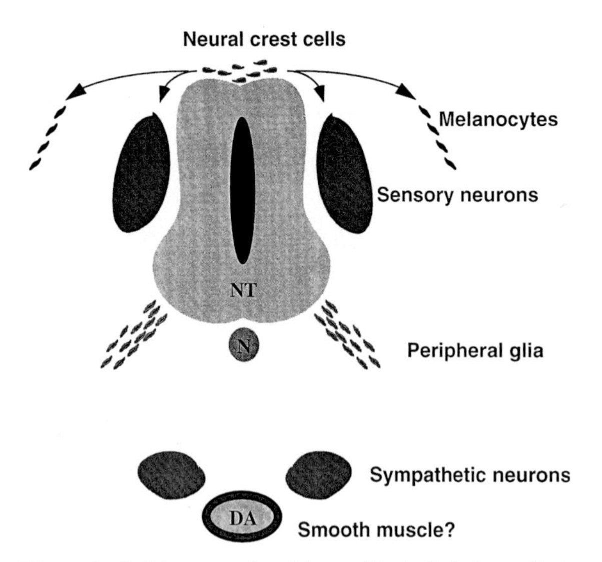

Figure 1. Cell types produced by multipotential stem cells in the trunk neural crest. The contribution to the vascular smooth muscle of the dorsal aorta is speculative. (Reprinted, with permission, from Anderson 1997.)

major objectives has been to determine which of these mechanisms is operative in the neural crest.

It is believed that the neural crest, like the hematopoietic system, undergoes progressive restrictions in the developmental potentials of individual progenitor cells (for discussions, see Anderson 1989; Sieber-Blum 1990; Le Douarin et al. 1991; Weston 1991). Subpopulations of crest-derived cells with apparently restricted developmental capacities have been detected in avian peripheral ganglia and other tissues colonized by crest cells (Sieber-Blum et al. 1993). However, rigorously documenting such restrictions requires the ability to challenge individual crest cells by exposing them to novel environments or to defined molecules capable of instructing alternative fates. With few exceptions (Sieber-Blum 1991; Lo and Anderson 1995) this has not been achieved: Either populations of crest cells have been challenged by transplantation in vivo (Weston and Butler 1966; Le Douarin 1986; Artinger and Bronner-Fraser 1992) or individual crest cells have been analyzed in vitro but not challenged with different instructive signals to assess their state of commitment (Duff et al. 1991; Sextier-Sainte-Claire Deville et al. 1992, 1994). The nature of the restricted sublineages that arise from the neural crest therefore remains poorly defined. Furthermore, the logic of how such restrictions emerge is not clear; various hierarchical, sequential, and stochastic models for lineage segregation have been put forth (for review, see Anderson 1993), and none have yet been experimentally validated.

We have approached the problem of neural crest cell lineage determination by isolating neural crest (Stemple and Anderson 1992) and crest-derived (Michelsohn and Anderson 1992; Lo and Anderson 1995) cells from embryos at different stages of development, and developing in vitro clonogenic assay systems (Stemple and Anderson 1992) where the developmental capacities of the cells, and their responses to environmental signals, can be assessed. This has allowed the identification of specific signals that can influence crest cell fate, and a determination of whether they act selectively or instructively (Shah et al. 1994, 1996). We have also begun to identify transcription factors important in the specification of different crest sublineages (Johnson et al. 1990; Ma et al. 1996; Sommer et al. 1996). By combining these cell biological and molecular approaches, we have begun to dissect the interplay between environmental signals and cell-intrinsic determinants in the control of neural crest lineage diversification. Here we review some of the concepts and particulars that have emerged from this experimental approach.

SELF-RENEWAL OF MULTIPOTENT MAMMALIAN NEURAL CREST CELLS

A defining characteristic of stem cells in other systems, such as the hematopoietic system, is their ability to self-renew; i.e., to divide to produce progeny with the same developmental capacities (Hall and Watt 1989; Potten and Loeffler 1990; Morrison et al. 1997). Although neural crest cells in avian embryos have been demonstrated to be multipotent (for review, see Bronner-Fraser 1993b), their ability to self-renew has not been addressed. We isolated a population of rat neural crest cells using antibodies to the low-affinity nerve growth factor (NGF) receptor ($p75^{LNTR}$) as a cell surface marker (Chandler Parsons et al. 1984) and have demonstrated that these cells, like their avian counterparts, are multipotent (Stemple and Anderson 1992). In clonal culture, they are able to generate at least three of the cell types that normally derive from the crest in vivo: autonomic neurons, glia, and smooth muscle. Moreover, by subcloning these cells, we have demonstrated that they are capable of self-renewal, at least for a limited number of generations (6–10) (Stemple and Anderson 1992). Thus, these neural crest cells exhibit multipotency and self-renewal, two properties of stem cells in the hematopoietic system (Davis and Reed 1996). More recently, similar studies have been performed for multipotent neural progenitors from the brain (Davis and Temple 1994; Gritti et al. 1996; Johe et al. 1996), supporting the idea that neural stem cells exist in the central nervous system (CNS) as well as in the peripheral nervous system (PNS).

The demonstration of self-renewal by multipotent neural progenitors in vitro raises the question of whether such self-renewal also occurs in vivo, and if so for how long. Multipotent neural progenitors have been recovered from the adult mammalian brain (Reynolds and Weiss 1992; Lois and Alvarez-Buylla 1993; Palmer et al. 1997; for review, see Aubert et al. 1995). Whether these adult stem cells reflect the self-renewal of embryonic stem cells, the persistence of such cells in a quiescent state, or rather de novo generation from a pre-stem cell remains to be determined. The existence of stem cells in adult neural-crest-derived structures has not yet been demonstrated. However, in avian embryos, multipotent cells can be recovered from some tissues colonized by the neural crest at late developmental stages (Sieber-Blum et al. 1993). These data are consistent with the idea that neural crest cells may undergo some self-renewal in vivo as well as in vitro.

IDENTIFICATION OF INSTRUCTIVE SIGNALS THAT INFLUENCE LINEAGE COMMITMENT BY NEURAL CREST STEM CELLS IN VITRO

We have identified several growth factors that can promote the differentiation of neural crest cells to specific lineages. Glial growth factor-2 (GGF2), a neuregulin (Marchionni et al. 1993; for review, see Burden and Yarden 1997), promotes the differentiation of peripheral glial (Schwann) cells (Shah et al. 1994). Members of the transforming growth factor-β (TGF-β) superfamily, TGFβ1-3 and BMP2/4, promote the differentiation of smooth muscle cells and autonomic neurons, respectively (although the BMPs also produce some smooth muscle cells in addition to neurons) (Fig. 2) (Shah et al. 1996). Clonal analysis and sequential observations of single clones have excluded the possibility that these factors act to promote the selective survival of a subset of founder cells pre-committed to a particular fate or that they selectively kill cells within clones that commit to the "wrong" fates (Shah et al. 1994, 1996). Thus, we have concluded that these factors act instructively rather than selectively.

The neural crest represents the first case in which growth factors have been shown to influence lineage determination by multipotent stem cells in an instructive rather than a selective manner. Similar data have recently been obtained for stem cells from the brain (Johe et al. 1996). In contrast, the available data in the hematopoietic system support selective rather than instructive actions for growth factors on lineage commitment (Fairbairn et al. 1993). Whether this reflects a fundamental difference in the way that the two systems utilize growth factors to control lineage decisions, or rather that instructive factors for hematopoietic stem cells have simply not yet been identified, remains to be determined.

We should point out that the developing nervous system also employs selective mechanisms in its development. For example, the neurotrophins (e.g., NGF, BDNF, and NT-3) act to promote the survival of subsets of peripheral sensory and autonomic neurons (for review, see

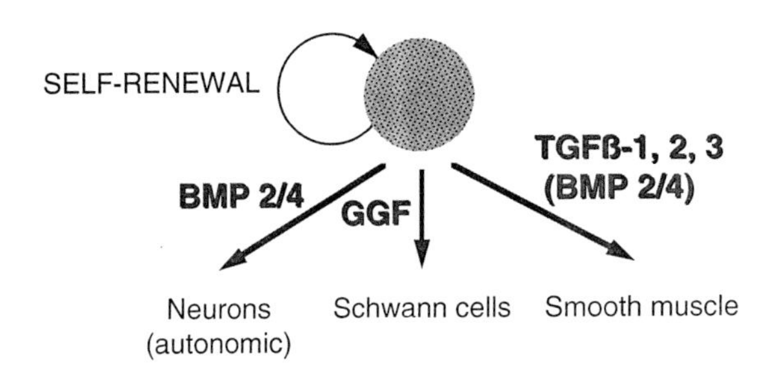

Figure 2. Instructive growth factors controlling lineage determination by neural crest stem cells. (Reprinted, with permission, from Shah et al. 1996 [copyright Cell Press].)

Thoenen 1991). Indeed, even GGF/neuregulin has been shown to promote the survival of lineage-committed Schwann cell progenitors (Dong et al. 1995), as well as the proliferation of mature Schwann cells (Lemke and Brockes 1984). Thus, the same factor may initially act as an instructive fate-determination signal and later as a selective survival factor or mitogen, at successive stages in the development of a particular lineage (Topilko et al. 1997).

Neural crest cells are likely to encounter multiple signals in their local environment in vivo. It is therefore important to understand how the cells integrate such opposing influences. Both GGF2 and TGF-β are able to antagonize the neurogenic influences of BMP2 on NC-SCs (Shah and Anderson 1997). However, the cells display very different sensitivities to these antagonistic interactions: TGF-β antagonizes BMP2 at saturating concentrations of the latter, whereas antagonism by GGF2 is only detectable at BMP2 concentrations 50–100-fold below saturation. These differences do not simply reflect different dosage sensitivities to GGF and TGF-β, they also reflect the fact that commitment to a glial fate in GGF2 occurs much more slowly (48–96 hours) than it does to a smooth muscle fate in TGF-β (≤24 hours) (N.M. Shah and D.J. Anderson, in prep.). The reasons for these differences in the kinetics of commitment are not yet clear.

The antagonistic interactions between TGF-β and BMP2 may reflect competition for limiting quantities of a shared signaling component, such as DPC4/Smad4 (A. Candia et al., pers. comm.). The antagonism between BMP2 and GGF2 is also interesting in light of recent data identifying antagonistic interactions between *Drosophila* homologs of these factors, DPP and D-EGF, in tracheal morphogenesis (see Shilo et al., this volume) as well as in follicle cell fate determination during oogenesis (L. Dobens and L. Raftery, pers. comm.). The molecular basis of this antagonism remains to be elucidated.

ROLE OF IDENTIFIED INSTRUCTIVE SIGNALS IN CONTROLLING NEURAL CREST DEVELOPMENT IN VIVO

The identification of GGF, TGF-βs 1-3, and BMP2/4 as instructive signals for neural crest lineage determination in vitro immediately raises several new questions. Do these growth factors play a role in determining neural crest cell fate in vivo, and if so what cells produce them and what regulates their production? How do these extracellular signals interact with transcriptional regulators to cause cells to commit to particular fates?

All of the identified factors are expressed in vivo at appropriate places and times to influence neural crest cell fate in a manner suggested by their actions in vitro. For example, several TGF-βs are found in the developing outflow tracts of the heart, where neural crest cells contribute to smooth muscle (for review, see Kingsley 1994). Knockouts in some of these growth factors lead to cardiac defects (for review, see Moses and Serra 1996), although whether these defects specifically affect crest-derived smooth muscle is not yet known. BMP2 and BMP4 are found in locations near or in which autonomic neurons develop: For example, the former is present in the dorsal aorta (Reissman et al. 1996; Shah et al. 1996) (Fig. 3A), the site of sympathetic gangliogenesis, whereas the latter is present in the gut (Bitgood and McMahon 1995; Lyons et al. 1995), the site of enteric neurogenesis. Unfortunately, knockouts in these genes die at a stage of embryogenesis too early to assess their requirement in the development of these neural-crest-derived autonomic ganglia (Winnier et al. 1995; Zhang and Bradley 1996; for review, see Hogan 1996). Therefore, tissue-specific knockouts of these factors or their receptors (Mishina et al. 1995) will be required to address their requirement for autonomic neurogenesis in vivo.

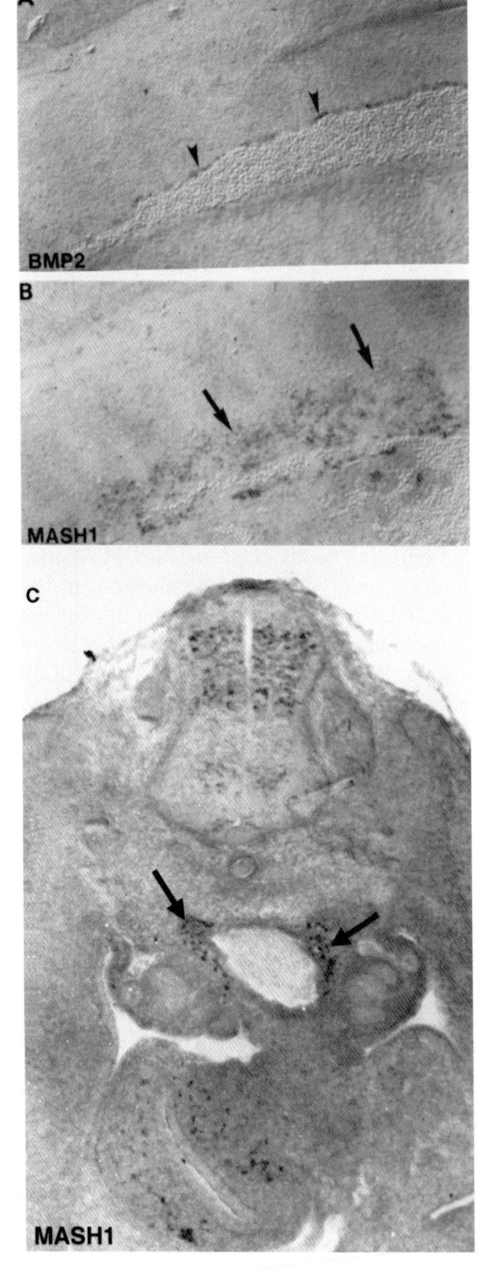

Figure 3. Coincident expression of BMP2 mRNA in the wall of the dorsal aorta (*A, arrowheads*) at the time that neural crest cells expressing MASH1 mRNA (*B, arrows*) or protein (*C, arrows*) can be detected adjacent to this structure. *A* and *B* are adjacent saggital sections, whereas *C* is a transverse section. $MASH1^+$ cells are also detected ventral to the sympathetic chain in the gut (*C*), which is known to contain BMP4 (Bitgood and McMahon 1995).

GGF/neuregulins are expressed by motor, sensory, and sympathetic neurons, near which peripheral glia develop (Marchionni et al. 1993; Meyer and Birchmeier 1994). We have suggested that this neuronal expression may constitute part of a negative feedback loop whereby neurons signal neighboring uncommitted stem cells to generate glia (Shah et al. 1994). The neuregulin knockout contains a reduced number of Schwann cells associated with peripheral nerve (Meyer and Birchmeier 1995), consistent with the in vitro actions of GGF2. However, whether the mutant phenotype reflects an in vivo action of the factor in lineage determination, progenitor cell survival (Dong et al. 1995), differentiation (Murphy et al. 1996), or proliferation (Lemke and Brockes 1984) is not yet known.

CONTROL OF AUTONOMIC NEURONAL LINEAGE DETERMINATION: INTERACTIONS BETWEEN BMP2/4 AND MASH1

How extracellular signals and transcription factors interact to control lineage commitment is beginning to emerge from studies of autonomic neurogenesis. Mammalian achaete-scute homolog-1 (MASH1) is a basic helix-loop-helix (bHLH) transcription factor that is specifically expressed in precursors of autonomic (but not sensory) neurons (Fig. 3C) (Lo et al. 1991). MASH1 is essential for the development of these neurons as shown by targeted mutagenesis in mice (Guillemot et al. 1993; Blaugrund et al. 1996). The fact that BMP2 and BMP4 promote expression of autonomic phenotypes in vitro (Varley et al. 1995; Reissman et al. 1996; Shah et al. 1996; Varley and Maxwell 1996) raised the question of whether these factors are also involved in the regulation of expression of MASH1. The available evidence suggests that they are. Purified recombinant BMP2 induces MASH1 expression in cultured neural crest cells within 6–12 hours (Fig. 4) (Shah et al. 1996) (a time by which many cells have not yet divided), suggesting that the growth factor acts directly on the cells to promote expression of the transcription factor. As mentioned earlier, BMP2 mRNA is expressed in the wall of the dorsal aorta (Fig. 3A), and this expression is detected at the time that $MASH1^+$ precursors of sympathetic autonomic neurons can first be observed adjacent to this structure (Fig. 3B,C) (Shah et al. 1996) Furthermore, explants of dorsal aorta tissue can induce MASH1 expression in cultured neural crest cells in a manner sensitive to inhibition by noggin (A. Groves and D.J. Anderson, unpubl.). These data suggest that neural crest cells begin to express MASH1 when they migrate to peripheral tissues that are local sources of BMP2 or BMP4.

What are the consequences of MASH1 expression for

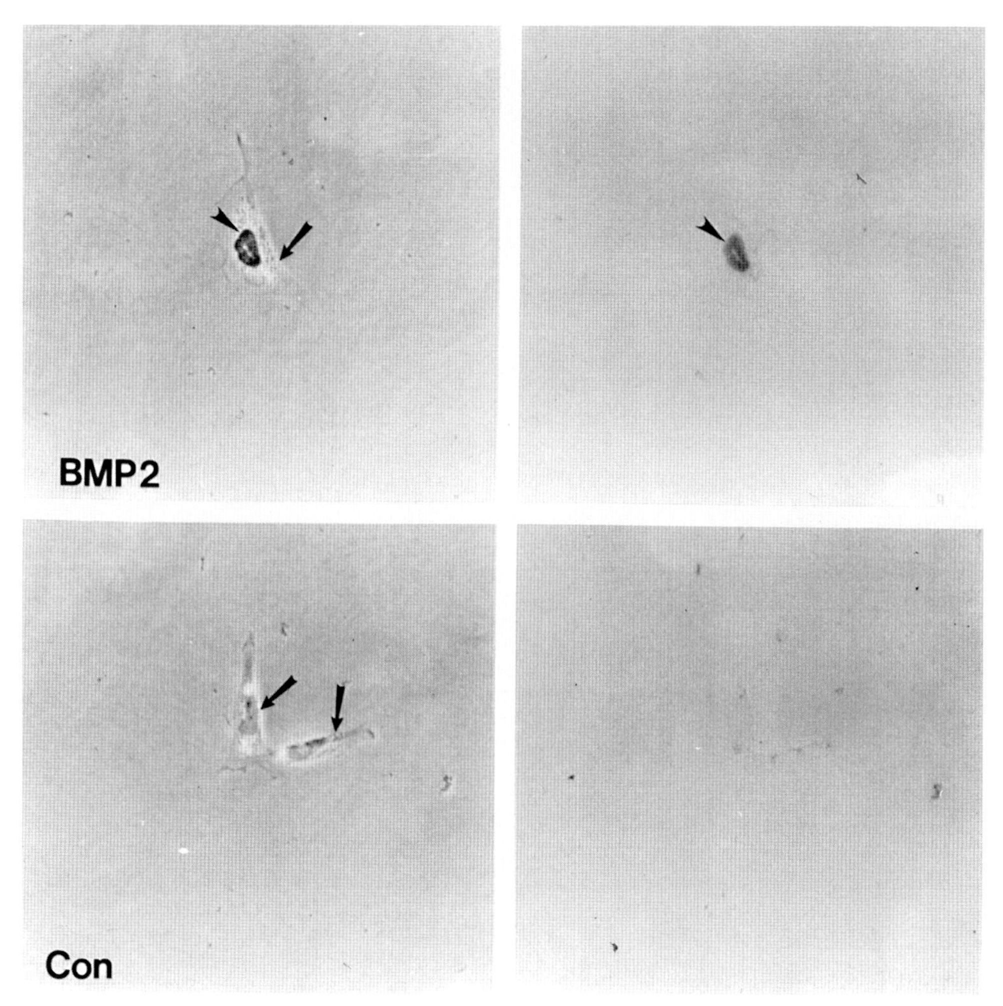

Figure 4. Induction of MASH1 protein in isolated neural crest stem cells by BMP2 in vitro. Arrowhead (*upper panels*) indicates the nucleus of a single MASH1-immunoreactive neural crest cell 12 hr after exposure to 50 ng/ml BMP2.

the fate of neural crest cells? Detailed analysis of *Mash1* mutant embryos has suggested that the gene is essential for the execution of terminal neuronal differentiation and is required only after the cells have already begun to express some neuronal genes (such as neurofilament 150-kD subunit) and have likely committed to a neuronal fate (Fig. 5) (Sommer et al. 1995). However, MASH1 expression is first detected prior to expression of such early neuronal markers (Fig. 4) (Lo et al. 1991; Sommer et al. 1995; Shah et al. 1996), raising the possibility that it has an earlier function in autonomic neurogenic lineage commitment not revealed by the null mutation.

To address this possibility, we have examined the relationship between expression of MASH1 and the commitment of neural-crest-derived cells to an autonomic neuronal fate. Postmigratory neural crest cells that are progenitors of autonomic enteric neurons can be isolated from fetal (E12.5–E14.5) rat gut using the receptor tyrosine kinase c-RET (Pachnis et al. 1993) as a cell surface marker (Lo and Anderson 1995). The majority (>85%) of such cells already express MASH1. Clonal analysis of these isolated c-RET$^+$ cells revealed that many of the cells (25–50%) are already committed to a neuronal fate (Lo and Anderson 1995). However, ≥50% of the cells are not committed to a neuronal fate, but rather generate non-neuronal derivatives, including glia and smooth muscle (Lo et al. 1997). These data suggest that, qualitatively at least, expression of MASH1 cannot be sufficient for commitment to a neuronal fate (although it may depend on the quantitative level of MASH1 expression, a parameter that has not yet been examined).

Does MASH1 have any function in the uncommitted postmigratory crest cells? A gain-of-function experiment suggests that MASH1 is required to maintain competence for neuronal differentiation in these cells. Virtually all c-RET$^+$ cells are initially competent for neurogenesis, since those cells which have not yet committed to a neuronal fate can be converted to neurons by BMP2 (Lo et al. 1997). This neurogenic competence is, however, lost with time in culture. In parallel, there is a loss of endogenous MASH1 expression, which can be prevented by maintaining the cells in BMP2. Constitutive expression of MASH1 from a retroviral vector, in turn, maintains competence for neurogenesis induced by BMP2 (Lo et al. 1997). These data suggest that expression of MASH1 in c-RET$^+$ postmigratory neural crest cells maintains com-

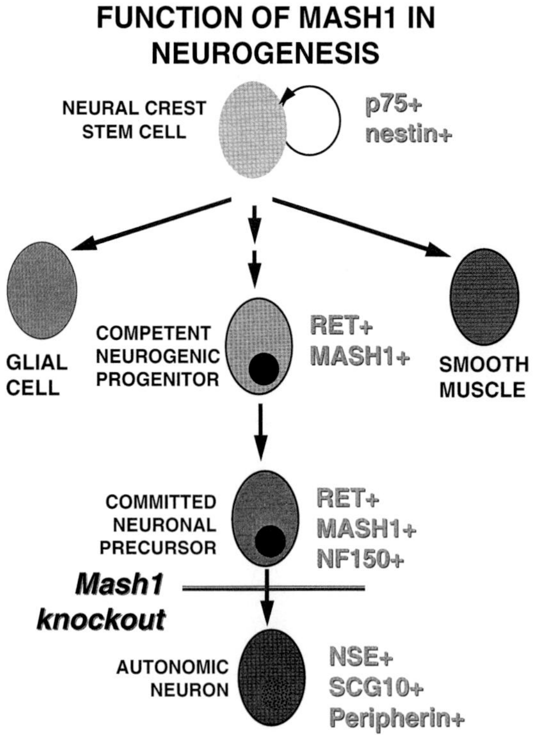

Figure 5. Essential function of MASH1 in autonomic neurogenesis as determined by targeted mutation in the mouse (Guillemot et al. 1993). The absence of *Mash1* function causes an arrest of autonomic neurogenesis at a stage when precursor cells already express a subset of neuron-specific genes (eg., NF150) and are likely committed to a neuronal fate (Sommer et al. 1995). However, MASH1 is expressed at earlier stages in the lineage, where it may have a nonessential function in maintaining competence for neurogenesis, as shown by overexpression experiments (Lo et al. 1997).

petence for neurogenesis in response to BMP2. Moreover, BMP2 is required not only to induce, but also to maintain MASH1 expression. In this way, MASH1 and BMP2 may participate in an indirect autoregulatory loop (Fig. 6). Whether this loop involves positive regulation by MASH1 of the BMP2 receptor, or some other component of the signaling pathway, remains to be explored.

These results suggest the following possible scenario for autonomic neuronal lineage commitment. Migrating neural crest cells arrive at peripheral target tissues (e.g.,

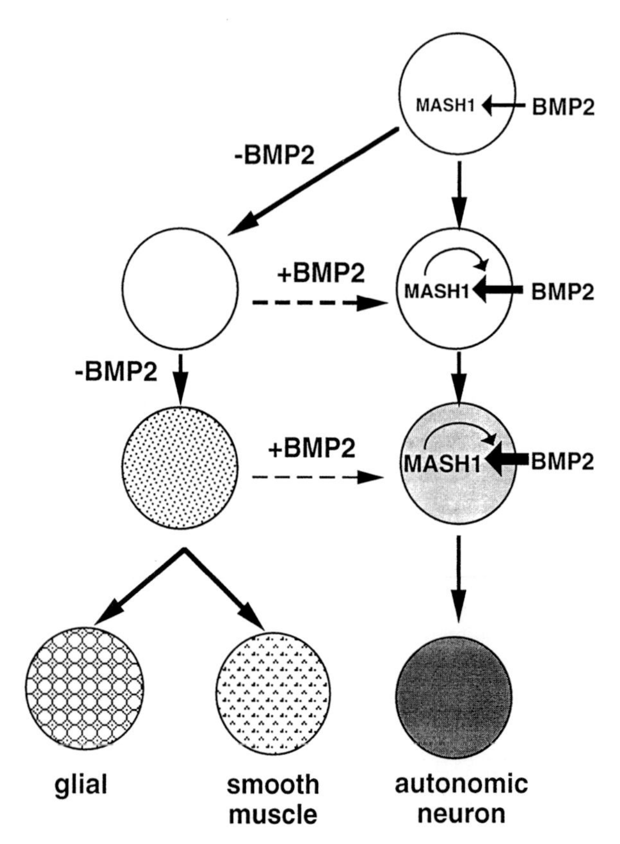

Figure 6. MASH1 participates in a positive autoregulatory loop with BMP2 that functions in maintaining competence for autonomic neurogenesis. The figure summarizes data obtained from analysis of MASH1 expression and function in a population of postmigratory neural crest cells isolated using monoclonal antibodies to c-RET (Lo and Anderson 1995). Maintenance of MASH1 expression in these cells in vitro requires continued exposure to BMP2; in turn, continued expression of MASH1 is required to maintain BMP2-responsiveness (Lo et al. 1997). If the cells are cultured in the absence of BMP2, expression of MASH1 is lost and the cells eventually irreversibly lose neurogenic capacity. (Reprinted, with permission, from Lo et al. 1997.)

dorsal aorta and gut), which produce a signal (BMP2 or BMP4) that induces expression of MASH1. Expression of MASH1 then confers competence to undergo further BMP2-mediated events that lead to commitment to a neuronal fate. This scenario raises several further questions that remain to be explored. What is the molecular basis of commitment to a neuronal fate? Does it involve a quantitative increase in levels of MASH1 expression, induction of cofactors that interact with MASH1, or a functionally distinct downstream factor? Does commitment to a neuronal fate require a higher level of BMP2 signaling than does the initial induction of MASH1, and if so why?

A further issue raised by this scenario is whether expression of MASH1 by migrating neural crest cells is solely dependent on their proximity to sources of BMP2 or BMP4. If that were the case, one might expect to observe MASH1-expressing crest cells just as they leave the dorsal neural tube and migrate near the ectoderm, which at early stages is a source of BMP4 (Liem et al. 1995). The fact that such cells are not observed in vivo suggests either that neural crest cells are initially not competent to express MASH1 in response to BMP2 or BMP4 when they first leave the dorsal neural tube or that the crest cells are protected from the effects of local BMP4 signaling by antagonists such as noggin (Smith and Harland 1992; Zimmerman et al. 1996), which is present in the roofplate at these stages (R.M. Harland, pers. comm.).

EXPRESSION OF A SYMPATHETIC PHENOTYPE REQUIRES THE INTEGRATION OF MULTIPLE EXTRACELLULAR SIGNALS

MASH1 is expressed in, and required for, the development of multiple autonomic neuronal subtypes. These subtypes can be distinguished by the kind of neurotransmitter they express. For example, sympathetic neurons synthesize norepinephrine, whereas parasympathetic neurons synthesize acetylcholine. Expression of MASH1 is not sufficient to specify these different autonomic subtypes, although it is necessary for this process (Guillemot et al. 1993). What other transcription factors are necessary to specify these different autonomic subtypes, and how is their expression controlled by extracellular signals?

Several transcription factors are specifically expressed in developing sympathetic ganglia besides MASH1, including Phox2a (Valarché et al. 1993), eHAND/Thg-1/Hxt (Cross et al. 1995; Cserjesi et al. 1995; Hollenberg et al. 1995), dHAND (Srivastava et al. 1995), and GATA-2 and GATA-3 (George et al. 1994; Tsai et al. 1994; Groves et al. 1995). Phox2a is a paired homeodomain protein; eHAND and dHAND are bHLH proteins; and GATA-2 and -3 are zinc finger proteins (which also function in the hematopoietic system [Briegel et al. 1993; Tsai et al. 1994]). Phox2a, eHAND, and dHAND are, like MASH1, expressed by all autonomic sublineages. GATA-2 and GATA-3, in contrast, are expressed by sympathetic neurons but not by enteric or parasympathetic neurons (Fig. 7) (Groves et al. 1995). These transcription factors appear to be expressed subsequent to MASH1 (Ernsberger et al. 1995; Groves et al. 1995), and the expression of eHAND (Ma et al. 1997) and Phox2a (M.-R. Hirsch et al.; L. Lo et al.; both in prep.) is dependent on MASH1 function. Although their functions in neurogenesis are not yet established, these transcription factors serve as useful markers to analyze the role of environmental signals in the specification of the sympathetic phenotype.

Evidence discussed earlier suggested that BMP2 (or BMP4) derived from the dorsal aorta induces expression of MASH1 in sympathetic precursors. However, in NCSCs grown at clonal density, neither BMP2 nor the dorsal aorta are sufficient to induce expression of GATA-2/-3 or tyrosine hydroxylase (TH), the rate-limiting enzyme in norepinephrine synthesis (Groves and Anderson 1996). In contrast, evidence from others indicates that BMP-2, -4, and -7 can induce TH expression in high-density mass cultures of avian neural crest cells, which, unlike NCSCs, are grown in the presence of fetal calf serum (Varley et al. 1995; Reissman et al. 1996; Varley and Maxwell 1996). The reason for this discrepancy is not clear, but it could reflect a requirement for density-dependent signals, additional signals provided by fetal calf serum, or a species difference.

In avian embryos, surgical ablation experiments have shown that the notochord and floorplate are required for the induction of catecholamine histofluorescence in neural crest cells aggregating near the dorsal aorta (Stern et al. 1991). We have shown by similar experiments that these structures are also required for expression of TH, Phox2, and GATA-2 by sympathetic precursors in vivo (Groves et al. 1995). Interestingly, they are not required for the induction of either CASH1 (the avian homolog of MASH1; Jasoni et al. 1994) or SCG10, a pan-neuronal marker (Stein et al. 1988). These observations are consistent with the finding that in cultured mammalian NCSCs, BMP2 leads to induction of MASH1, SCG10, and a neuronal morphology, but not to expression of TH or GATA-2 (Groves and Anderson 1996). (Although, as mentioned above, in the avian system, BMPs alone appear sufficient to induce expression of TH [Varley et al. 1995; Reissman et al. 1996; Varley and Maxwell 1996]; in most cases, these TH^+ cells do not coexpress neuronal markers [Christie et al. 1987].)

Taken together, these data suggest that the expression of pan-neuronal components of the sympathetic phenotype can be experimentally uncoupled from the expression of subtype-specific components, such as neurotransmitter synthesis. Expression of these different components may therefore be under the control of different genetic subprograms, which in turn may be under distinct environmental control. BMP2-like signals from the dorsal aorta and gut induce MASH1 and a subprogram leading to expression of pan-neuronal and some pan-autonomic properties, whereas additional signals from the notochord and floorplate may be required for expression of a subprogram leading to expression of the sympathetic neurotransmitter phenotype (Fig. 7). These subprograms are not completely independent, however, as MASH1 is required for expression of both the pan-neuronal and sub-

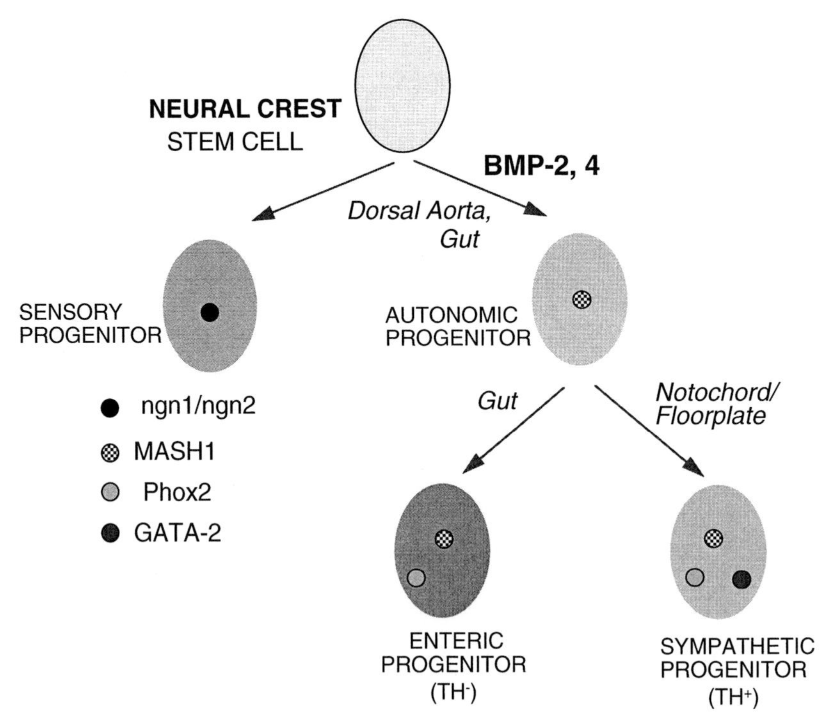

Figure 7. Different combinations of specifically expressed transcription factors are correlated with different autonomic neuronal identities. Progenitors in all three major autonomic sublineages (sympathetic, parasympathetic, and enteric) express MASH1, which is likely induced by BMP2 or BMP4 present in tissues such as the dorsal aorta and gut. These progenitors are likely exposed to additional signals in a location-specific manner, which cause the expression of sublineage-specific transcription factors such as GATA-2. Phox2 is also expressed by most or all autonomic sublineages, and it appears to be a downstream target of MASH1 (L.-C. Lo et al., in prep.).

type-specific components of the sympathetic phenotype (Guillemot et al. 1993; Sommer et al. 1995).

SPECIFICATION OF DIFFERENT SENSORY NEURON SUBTYPES MAY INVOLVE THE ACTION OF DISTINCT BUT RELATED bHLH PROTEINS

Although significant progress has been made in understanding the specification of autonomic lineages, much less is known about the specification of sensory lineages. The segregation of the sensory and autonomic lineages represents a major divergence point in the development of the neural crest (Le Douarin 1986), comparable to the segregation of lymphoid and myelo-erythroid lineages in hematopoiesis (Ikuta et al. 1992; Morrison et al. 1994). The logic and mechanisms underlying this lineage segregation event are not understood. Moreover, like the autonomic lineage, the sensory lineage comprises multiple neuronal subtypes, such as those mediating touch, pain, and posture (Snider 1994). In contrast to different subtypes of autonomic neurons, which develop in distinct embryonic locations, different subclasses of sensory neurons develop in a common location, the dorsal root glia, adjacent to the spinal cord. The development of sensory neurons thus poses a different set of problems than does that of autonomic neurons.

Recently, we identified a subfamily of related bHLH proteins, the neurogenins, that are likely to have a key role in the development of the sensory lineage (Ma et al. 1996; Sommer et al. 1996). Neurogenins (ngns)-1, -2, and -3 (also known as MATH4C, A and B, respectively; Gradwohl et al. 1996; Cau et al. 1997) are closely related to NeuroD and its relatives (for review, see Lee 1997), but form a distinct subfamily (Sommer et al. 1996). In vivo, *ngn1* and *ngn2* are expressed in developing sensory but not autonomic ganglia, in a manner complementary to MASH1 (Ma et al. 1996; Sommer et al. 1996). Interestingly, this complementarity extends to the central nervous system as well (Ma et al. 1997), suggesting that MASH1 and the *ngn*s have a broader role in the specification of different neuronal subtypes. (An exception is the olfactory placode, where MASH1, *ngn1,* and NeuroD have been shown to function in a cascade that likely acts within the same lineage [Cau et al. 1997].)

The function of neurogenins has been investigated by ectopic expression experiments in *Xenopus*. Injection of mRNA encoding *Xenopus* neurogenin-related-1 (Xngnr-

1) causes a massive induction of ectopic neurogenesis, both within the neural tube and in the flanking nonneurogenic ectoderm (Ma et al. 1996), similar to the phenotype caused by ectopic expression of XNeuroD (Lee et al. 1995). Ectopic expression of Xngnr-1 also induces ectopic expression of endogenous XNeuroD (Fig. 8) (Ma et al. 1996). In contrast, injection of XNeuroD mRNA fails to induce expression of endogenous Xngnr-1. These gain-of-function data suggest that Xngnr-1 can function as a neuronal determination gene and that it acts upstream of endogenous XNeuroD in a unidirectional cascade controlling neurogenesis (Ma et al. 1996). The early expression of endogenous Xngnr-1 and its interaction with the lateral inhibition machinery mediated by XNotch-1 and XDelta-1 (Fig. 8) (Chitnis et al. 1995; Chitnis and Kintner 1996; Ma et al. 1996) are consistent with the idea that Xngnr-1 normally does function in the process of neuronal determination, in vivo.

The function of the neurogenins in higher vertebrates is not yet established. However, the expression patterns of the genes are of interest in relation to the issue of sensory neuron subtype specification. In particular, *ngn1* and *ngn2* are expressed in mostly complementary subsets of cranial sensory ganglia: *ngn1* is most highly expressed in "proximal" ganglia whose neurons derive from the cranial neural crest and/or the otic and trigeminal placodes, whereas *ngn2* is prominently expressed in "distal" ganglia whose neurons derive from the epibranchial placodes (Gradwohl et al. 1996; Sommer et al. 1996). These data indicate a correlation between the expression of *ngns* and different sensory neuron subtypes. In trunk dorsal root sensory ganglia, both *ngn1* and *ngn2* are expressed; however, they are expressed sequentially (*ngn2* followed by *ngn1*) rather than simultaneously (Sommer et al. 1996). The expression of these two genes in distinct subsets of cranial sensory ganglia suggests, by analogy, that in the DRG, these same genes could specify distinct subtypes of trunk sensory neurons (Snider 1994). Targeted mutations in the *ngns* are in progress to examine this possibility.

SUMMARY

The molecular mechanisms underlying the determination of neuronal identity in the vertebrate peripheral nervous system are only just beginning to come into focus. Many of these mechanisms, such as the involvement of cascades of bHLH transcription factors and lateral inhibition via the Notch-Delta system, appear to have been conserved from *Drosophila* (Ghysen et al. 1993; Jan and Jan 1993). The way in which these genetic circuits are controlled by instructive growth factors, and the manner in which they lead to expression of a particular neuronal

NEUROGENIN BOTH ACTIVATES, AND IS INHIBITED BY, THE LATERAL INHIBITION MACHINERY

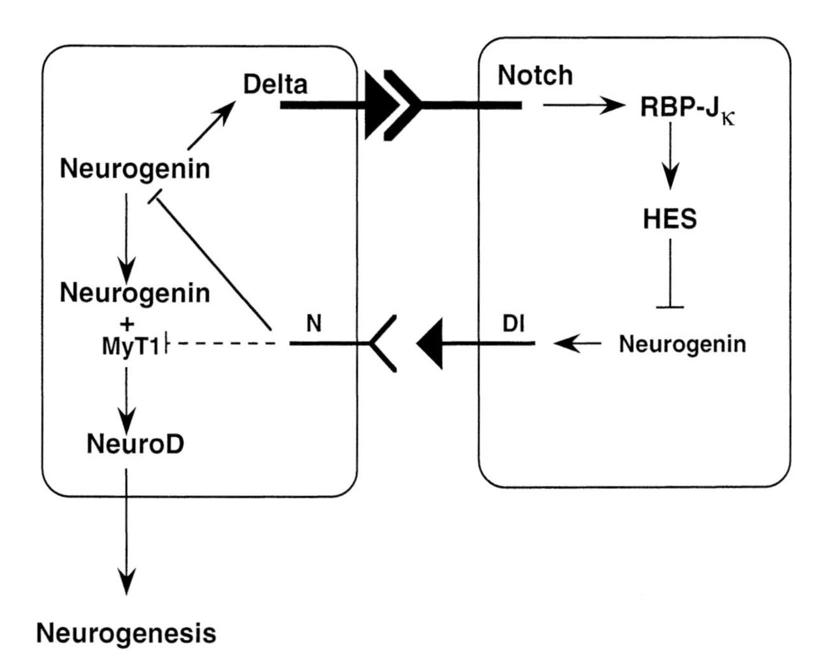

Figure 8. Role of neurogenin in neuronal determination as deduced from ectopic expression experiments in *Xenopus*. Neurogenin positively regulates expression of Delta and is in turn negatively regulated (at both transcriptional and posttranscriptional levels) by signaling through Notch. Neurogenin also activates expression of the zinc finger protein X-MyT1, which appears to collaborate with *ngn* to promote induction of XNeuroD and neuronal differentiation in a manner resistant to inhibition by Notch signaling (Bellefroid et al. 1996). Notch signaling likely involves the mammalian homolog of *Drosophila* Suppressor of Hairless, RBP-$J_κ$, (Artavanis-Tsakonas et al. 1995) as well as homologs of *Drosophila* hairy and Enhancer of Split (HES proteins) (Ishibashi et al. 1995). (Modified from Ma et al. 1996.)

identity, is far from clear. This process is being elucidated by studies of neurogenesis in the peripheral autonomic lineage, which is arguably the best-understood neurogenic lineage in vertebrates.

Emerging evidence is beginning to suggest that neuronal diversity within the autonomic and sensory lineages may be generated by related, but distinct, mechanisms. All autonomic progenitors express a common bHLH protein, MASH1, which appears to be induced by members of the BMP2 subfamily secreted by the tissues to which these progenitors migrate. Additional signals may then act on these progenitors in different locations to induce the expression of other transcription factors, which act in conjunction with MASH1 to specify the final phenotypes of the different autonomic neuron subtypes (sympathetic, parasympathetic, and enteric). Although different classes of autonomic neurons develop in very different locations within the body, different classes of sensory neurons are located together in dorsal root ganglia. The finding that distinct but related subtypes of bHLH proteins, the neurogenins, are expressed by different classes of sensory neuron precursors early in development suggests that sensory neuron diversity, in contrast to autonomic neuron diversity, may be pre-specified at or before the time neural crest cells begin their emigration from the neural tube.

ACKNOWLEDGMENTS

We thank the past members of the Anderson laboratory who have contributed to the work discussed in this paper, including A. Michelsohn, S. Birren, J. Johnson, T. Saito, D. Stemple, J. Verdi, and K. Zimmerman. We also thank our collaborators, including J.-F. Brunet, M.D. Gershon, C. Goridis, F. Guillemot, A. Joyner, C. Kintner, and M. Marchionni, for their important contributions. We are grateful to R. Axel, S. Fraser, A. Ghysen, T. Jessell, P. Patterson, M. Raff, H. Weintraub, B. Wold, and K. Zinn for important discussions, insights, and influences. Portions of the work described here have been supported by the National Institutes of Health Muscular Dystrophy Association, Thailand Foundation for Education and Welfare, Seaver Institute, Swiss National Science Foundation, and Human Frontiers Science Foundation. D.J.A. is an investigator of the Howard Hughes Medical Institute.

REFERENCES

Anderson D.J. 1989. The neural crest cell lineage problem: Neuropoiesis? *Neuron* **3:** 1.

———. 1993. Cell and molecular biology of neural crest cell lineage diversification. *Curr. Opin. Neurobiol.* **3:** 8.

———. 1997. Cellular and molecular biology of neural crest cell lineage determination. *Trends Genet.* **13:** 276.

Artavanis-Tsakonas S., Matsuno K., and Fortini M.E. 1995. Notch signalling. *Science* **268:** 225.

Artinger K.B. and Bronner-Fraser M. 1992. Partial restriction in the developmental potential of late emigrating avian neural crest cells. *Dev. Biol.* **149:** 149.

Aubert I., Ridet J.-L., and Gage F.H. 1995. Regeneration in the adult mammalian CNS: Guided by development. *Curr. Opin. Neurobiol.* **5:** 625.

Baroffio A., Dupin E., and Le Douarin N.M. 1988. Clone-forming ability and differentiation potential of migratory neural crest cells. *Proc. Natl. Acad. Sci.* **85:** 5325.

Bellefroid E.J., Bourguignon C., Hollemann T., Ma Q., Anderson D.J., Kintner C., and Pieler T. 1996. X-MyT1, a *Xenopus* C2HC-type zinc finger protein with a regulatory function in neuronal differentiation. *Cell* **87:** 1191.

Bitgood M.J. and McMahon A.P. 1995. *Hedgehog* and *Bmp* genes are coexpressed at many diverse sites of cell-cell interaction in the mouse embryo. *Dev. Biol.* **172:** 126.

Blaugrund E., Pham T.D., Tennyson V.M., Lo L., Sommer L., Anderson D.J., and Gershon M.D. 1996. Distinct subpopulations of enteric neuronal progenitors defined by time of development, sympathoadrenal lineage markers and *Mash-1*-dependence. *Development* **122:** 309.

Briegel K., Lim K.-C., Plank C., Beug H., Engel J.D., and Zenke M. 1993. Ectopic expression of a conditional GATA-2/estrogen receptor chimera arrests erythroid differentiation in a hormone-dependent manner. *Genes Dev.* **7:** 1097.

Bronner-Fraser M. 1992. Environmental influences on neural crest cell migration. *J. Neurobiol.* **24:** 233.

———. 1993a. Mechanisms of neural crest migration. *BioEssays* **15:** 221.

———. 1993b. Segregation of cell lineage in the neural crest. *Curr. Opin. Genet. Dev.* **3:** 641.

Bronner-Fraser M. and Fraser S. 1988. Cell lineage analysis shows multipotentiality of some avian neural crest cells. *Nature* **335:** 161.

Burden S. and Yarden Y. 1997. Neuregulins and their receptors: A versatile signaling module in organogenesis and oncogenesis. *Neuron* **18:** 847.

Cau E., Gradwohl G., Fode C., and Guillemot F. 1997. MASH1 activates a cascade of bHLH regulators in olfactory neuron progenitors. *Development* **124:** 1611.

Chandler C.E., Parsons L.M., Hosang M., and Shooter E.M. 1984. A monoclonal antibody modulates the interaction of nerve growth factor with PC12 cells. *J. Biol. Chem.* **259:** 6882.

Chitnis A. and Kintner C. 1996. Sensitivity of proneural genes to lateral inhibition affects the pattern of primary neurons in *Xenopus* embryos. *Development* **122:** 2295.

Chitnis A., Henrique D., Lewis J., Ish-Horowicz D., and Kintner C. 1995. Primary neurogenesis in *Xenopus* embryos regulated by a homologue of the *Drosophila* neurogenic gene *Delta*. *Nature* **375:** 761.

Christie D.S., Forbes M.E., and Maxwell G.D. 1987. Phenotypic properties of catecholamine-positive cells that differentiate in avian neural crest cultures. *J. Neurosci.* **7:** 3749.

Cross J.C., Flannery M.L., Blanar M.A., Steingrimsson E., Jenkins N.A., Copeland N.G., Rutter W.J., and Werb A. 1995. *Hxt* encodes a basic helix-loop-helix transcription factor that regulates trophoblast cell development. *Development* **121:** 2513.

Cserjesi P., Brown D., Lyons G.E., and Olson E.N. 1995. Expression of the novel basic helix-loop-helix gene *eHAND* in neural crest derivatives and extraembryonic membranes during mouse development. *Dev. Biol.* **170:** 664.

Davis A. and Temple S. 1994. A self-renewing multipotential stem cell in embryonic rat cerebral cortex. *Nature* **372:** 263.

Davis J.A. and Reed R.R. 1996. Role of Olf-1 and Pax-6 transcription factors in neurodevelopment. *J. Neurosci.* **16:** 5082.

Dong Z., Brennan A., Liu N., Yarden Y., Lefkowitz G., Mirsky R., and Jessen K.R. 1995. Neu differentiation factor is a neuron-glia signal and regulates survival, proliferation and maturation of rat Schwann cell precursors. *Neuron* **15:** 585.

Duff R.S., Langtimm C.J., Richardson M.K., and Sieber-Blum M. 1991. *In vitro* clonal analysis of progenitor cell patterns in dorsal root and sympathetic ganglia of the quail embryo. *Dev. Biol.* **147:** 451.

Ernsberger U., Patzke H., Tissier-Seta J.P., Reh T., Goridis C., and Rohrer H. 1995. The expression of tyrosine hydroxylase and the transcription factors cPhox-2 and Cash-1: Evidence for distinct inductive steps in the differentiation of chick sympathetic precursor cells. *Mech. Dev.* **52:** 125.

Fairbairn L.J., Cowling G.J., Reipert B.M., and Dexter T.M. 1993. Suppression of apoptosis allows differentiation and development of a multipotent hematopoietic cell line in the absence of added growth factors. *Cell* **74:** 823.

George K.M., Leonard M.W., Roth M.E., Lieuw K.H., Kioussis D., Grosveld F., and Engel J.D. 1994. Embryonic expression and cloning of the murine GATA-3 gene. *Development* **120:** 2673.

Ghysen A., Dambly-Chaudiere C., Jan L.Y., and Jan Y.-N. 1993. Cell interactions and gene interactions in peripheral neurogenesis. *Genes Dev.* **7:** 723.

Gradwohl G., Fode C., and Guillemot F. 1996. Restricted expression of a novel murine *atonal*-related bHLH protein in undifferentiated neural precursors. *Dev. Biol.* **180:** 227.

Gritti A., Parati E.A., Cova L., Frolichsthal P., Galli R., Wanke E., Faravelli L., Morassutti D.J., Roisen F., Nickel D.D., and Vescovi A.L. 1996. Multipotential stem cells from the adult mouse brain proliferate and self-renew in response to basic fibroblast growth factor. *J. Neurosci.* **16:** 1091.

Groves A.K. and Anderson D.J. 1996. Role of environmental signals and transcriptional regulators in neural crest development. *Dev. Genet.* **18:** 64.

Groves A.K., George K.M., Tissier-Seta J.-P., Engel J.D., Brunet J.-F., and Anderson D.J. 1995. Differential regulation of transcription factor gene expression and phenotypic markers in developing sympathetic neurons. *Development* **121:** 887.

Guillemot F., Lo L.-C., Johnson J.E., Auerbach A., Anderson D.J., and Joyner A.L. 1993. Mammalian achaete-scute homolog-1 is required for the early development of olfactory and autonomic neurons. *Cell* **75:** 463.

Hall P.A. and Watt F.M. 1989. Stem cells: The generation and maintenance of cellular diversity. *Development* **106:** 619.

Hogan B.L.M. 1996. Bone morphogenetic proteins—Multifunctional regulators of vertebrate development. *Genes Dev.* **10:** 1580.

Hollenberg S.M., Sternglanz R., Cheng P.F., and Weintraub H. 1995. Identification of a new family of tissue-specific basic helix-loop-helix proteins with a two-hybrid system. *Mol. Cell. Biol.* **15:** 3813.

Ikuta K., Uchida N., Friedman J., and Weissman I.L. 1992. Lymphocyte development from stem cells. *Annu. Rev. Immunol.* **10:** 759.

Ishibashi M., Ang S.-L., Shiota K., Nakanishi S., Kageyama R., and Guillemot F. 1995. Targeted disruption of mammalian *hairy* and *Enhancer of split* homolog-1 (*HES-1*) leads to upregulation of neural helix-loop-helix factors, premature neurogenesis, and severe neural tube defects. *Genes Dev.* **9:** 3136.

Jan Y.N. and Jan L.Y. 1993. HLH proteins, fly neurogenesis and vertebrate myogenesis. *Cell* **75:** 827.

Jasoni C.L., Walker M.B., Morris M.D., and Reh T.A. 1994. A chicken *achaete-scute* homolog (CASH-1) is expressed in a temporally and spatially discrete manner in the developing central nervous system. *Development* **120:** 769.

Johe K.K., Hazel T.G., Muller T., Dugich-Djordjevic M.M., and McKay R.D.G. 1996. Single factors direct the differentiation of stem cells from the fetal and adult central nervous system. *Genes Dev.* **10:** 3129.

Johnson J.E., Birren S.J., and Anderson D.J. 1990. Two rat homologues of *Drosophila achaete-scute* specifically expressed in neuronal precursors. *Nature* **346:** 858.

Kingsley D.M. 1994. The TGF-β superfamily: New members, new receptors, and new genetic tests of function in different organisms. *Genes Dev.* **8:** 133.

Le Douarin N.M. 1980. The ontogeny of the neural crest in avian embryo chimeras. *Nature* **286:** 663.

———. 1982. *The neural crest.* Cambridge University Press, Cambridge, United Kingdom.

———. 1986. Cell line segregation during peripheral nervous system ontogeny. *Science* **231:** 1515.

Le Douarin N., Dulac C., Dupin E., and Cameron-Curry P. 1991. Glial cell lineages in the neural crest. *Glia* **4:** 175.

Lee J.E., Hollenberg S.M., Snider L., Turner D.L., Lipnick N., and Weintraub H. 1995. Conversion of *Xenopus* extoderm into neurons by NeuroD, a basic helix-loop-helix protein. *Science* **268:** 836.

Lee J.F. 1997. Basic helix-loop-helix genes in neural development. *Curr. Opin. Neurobiol.* **7:** 13.

Lemke G.E. and Brockes J.P. 1984. Identification and purification of glial growth factor. *J. Neurosci.* **4:** 75.

Liem K.F., Tremml G., Roelink H., and Jessell T.M. 1995. Dorsal differentiation of neural plate cells induced by BMP-mediated signals from epidermal ectoderm. *Cell* **82:** 969.

Lo L.-C. and Anderson D.J. 1995. Postmigratory neural crest cells expressing c-*ret* display restricted developmental and proliferative capacities. *Neuron* **15:** 527.

Lo L., Sommer L. and Anderson D.J. 1997. MASH1 maintains competence for BMP2-induced neuronal differentiation in post-migratory neural crest cells. *Curr. Biol.* **7:** 440.

Lo L., Johnson J.E., Wuenschell C.W., Saito T., and Anderson D.J. 1991. Mammalian *achaete-scute* homolog 1 is transiently expressed by spatially-restricted subsets of early neuroepithelial and neural crest cells. *Genes Dev.* **5:** 1524.

Lois C. and Alvarez-Buylla A. 1993. Proliferating subventricular zone cells in the adult mammalian forebrain can differentiate into neurons and glia. *Proc. Natl. Acad. Sci.* **90:** 2074.

Lyons K.M., Hogan B.L.M., and Robertson E.J. 1995. Colocalization of BMP 7 and BMP 2 RNAs suggests that these factors cooperatively mediate tissue interactions during murine development. *Mech. Dev.* **50:** 71.

Ma Q., Kintner C., and Anderson D.J. 1996. Identification of *neurogenin*, a vertebrate neuronal determination gene. *Cell* **87:** 43.

Ma Q., Sommer L., Cserjesi P., and Anderson D.J. 1997. *Mash1* and *neurogenin1* expression patterns define complementary domains of neuroepithelium in the developing CNS and are correlated with regions expressing Notch ligands. *J. Neurosci.* **17:** 3644.

Marchionni M.A., Goodearl A.D.J., Chen M.S., Bermingham-McDonogh O., Kirk C., Hendricks M., Danehy F., Misumi D., Sudhalter J., Kobayashi K., Wroblewski D., Lynch C., Baldassare M., Hiles I., Davis J.B., Hsuan J.J., Totty N.F., Otsu M., McBurney R.N., Waterfield M.D., Stroobant P., and Gwynne D. 1993. Glial growth factors are alternatively spliced *erbB2* ligands expressed in the nervous system. *Nature* **362:** 312.

Meyer D. and Birchmeier C. 1994. Distinct isoforms of neuregulin are expressed in mesenchymal and neuronal cells during mouse development. *Proc. Natl. Acad. Sci.* **91:** 1064.

———. 1995. Multiple essential functions of neuregulin in development. *Nature* **378:** 386.

Michelsohn A. and Anderson D.J. 1992. Changes in competence determine the timing of two sequential glucocorticoid effects on sympathoadrenal progenitors. *Neuron* **8:** 589.

Mishina Y., Suzuki A., Ueno N., and Behringer R.R. 1995. *Bmpr* encodes a type I bone morphogenetic protein receptor that is essential for gastrulation during mouse embryogenesis. *Genes Dev.* **9:** 3027.

Morrison S.J., Shah N.M., and Anderson D.J. 1997. Regulatory mechanisms in stem cell biology. *Cell* **88:** 287.

Morrison S.J., Uchida N., and Weissman I.L. 1994. The biology of hematopoietic stem cells. *Annu. Rev. Cell Dev. Biol.* **11:** 35.

Moses H.L. and Serra R. 1996. Regulation of differentiation by TGF-β. *Curr. Opin. Genet. Dev.* **6:** 581.

Murphy P., Topilko P., Schneider-Maunoury S., Seitanidou T., Baron van Evercooren A., and Charnay P. 1996. The regulation of Krox-20 expression reveals important steps in the control of peripheral glial cell development. *Development* **122:** 2847.

Pachnis V., Mankoo B., and Costantini F. 1993. Expression of the *c-ret* proto-oncogene during mouse embryogenesis. *Development* **119:** 1005.

Palmer T.D., Takahashi J., and Gage F.H. 1997. The adult rat hippocampus contains primordial neural stem cells. *Mol. Cell. Neurosci.* **8:** 389.

Potten C.S. and Loeffler M. 1990. Stem cells—Attributes, cycles, spirals, pitfalls and uncertainties: Lessons for and from the Crypt. *Development* **110:** 1001.

Reissman E., Ernsberger U., Francis-West P.H., Rueger D., Brickell P.D., and Rohrer H. 1996. Involvement of bone morphogenetic protein-4 and bone morphogenetic protein-7 in the differentiation of the adrenergic phenotype in developing sympathetic neurons. *Development* **122:** 2079.

Reynolds B.A. and Weiss S. 1992. Generation of neurons and astrocytes from isolated cells of the adult mammalian central nervous system. *Science* **255:** 1707.

Sextier-Sainte-Claire Deville F., Ziller C., and Le Douarin N. 1992. Developmental potentialities of cells derived from the truncal neural crest in clonal cultures. *Dev. Brain Res.* **66:** 1.

———. 1994. Developmental potentials of enteric neural crest-derived cells in clonal and mass cultures. *Dev. Biol.* **163:** 141.

Shah N.M. and Anderson D.J. 1997. Integration of multiple instructive cues by neural crest stem cells reveals cell-intrinsic biases in relative growth factor responsiveness. *Proc. Natl. Acad. Sci.* **94:** (in press).

Shah N.M., Groves A., and Anderson D.J. 1996. Alternative neural crest cell fates are instructively promoted by TGFβ superfamily members. *Cell* **85:** 331.

Shah N.M., Marchionni M.A., Isaacs I., Stroobant P.W., and Anderson D.J. 1994. Glial growth factor restricts mammalian neural crest stem cells to a glial fate. *Cell* **77:** 349.

Sieber-Blum M. 1990. Mechanisms of neural crest diversification. In *Comments developmental neurobiology*, vol. 1 p. 225. Gordon and Breach, London.

——. 1991. Role of the neurotrophic factors BDNF and NGF in the commitment of pluripotent neural crest cells. *Neuron* **6:** 949.

Sieber-Blum M. and Cohen A. 1980. Clonal analysis of quail neural crest cells: They are pluripotent and differentiate in vitro in the absence of non-neural crest cells. *Dev. Biol.* **80:** 96.

Sieber-Blum M., Ito K., Richardson M.K., Langtimm C.J., and Duff R.S. 1993. Distribution of pluripotent neural crest cells in the embryo and the role of brain-derived neurotrophic factor in the commitment to the primary sensory neuron lineage. *J. Neurobiol.* **24:** 173.

Smith W.C. and Harland R.M. 1992. Expression cloning of *noggin,* a new dorsalizing factor localized to the Spemann organizer in *Xenopus* embryos. *Cell* **70:** 829.

Snider W.D. 1994. Functions of the neurotrophins during nervous system development—What the knockouts are teaching us. *Cell* **77:** 627.

Sommer L., Ma Q., and Anderson D.J. 1996. *neurogenins*, a novel family of *atonal*-related bHLH transcription factors, are putative mammalian neuronal determination genes that reveal progenitor cell heterogeneity in the developing CNS and PNS. *Mol. Cell. Neurosci.* **8:** 221.

Sommer L., Shah N., Rao M., and Anderson D.J. 1995. The cellular function of MASH1 in autonomic neurogenesis. *Neuron* **15:** 1245.

Srivastava D., Cserjesi P., and Olson E.N. 1995. A subclass of bHLH proteins required for cardiac morphogenesis. *Science* **270:** 1995.

Stein R., Mori N., Matthews K., Lo L.-C., and Anderson D.J. 1988. The NGF-inducible SCG10 mRNA encodes a novel membrane-bound protein present in growth cones and abundant in developing neurons. *Neuron* **1:** 463.

Stemple D.L. and Anderson D.J. 1992. Isolation of a stem cell for neurons and glia from the mammalian neural crest. *Cell* **71:** 973.

———. 1993. Lineage diversification of the neural crest: *In vitro* investigations. *Dev. Biol.* **159:** 12.

Stern C.D., Artinger K.B., and Bronner-Fraser M. 1991. Tissue interactions affecting the migration and differentiation of neural crest cells in the chick embryo. *Development* **113:** 207.

Thoenen H. 1991. The changing scene of neurotrophic factors. *Trends Neurosci.* **14:** 165.

Topilko P., Murphy P., and Charnay P. 1997. Embryonic development of Schwann cells—Multiple roles for neuregulins along the pathway. *Mol. Cell. Neurosci.* **8:** 71.

Tsai F.-Y., Keller G., Kuo F.C., Weiss M., Chen J., Rosenblatt M., Alt F.W., and Orkin S.H. 1994. An early haematopoietic defect in mice lacking the transcription factor GATA-2. *Nature* **371:** 221.

Valarché I., Tissier-Seta J.-P., Hirsch M.-R., Martinez S., Goridis C., and Brunet J.-F. 1993. The mouse homeodomain protein Phox2 regulates NCAM promoter activity in concert with Cux/CDP and is a putative determinant of neurotransmitter phenotype. *Development* **119:** 881.

Varley J.E. and Maxwell G.D. 1996. BMP-2 and BMP-4, but not BMP-6, increase the number of adrenergic cells which develop in quail trunk neural crest cultures. *Exp. Neurol.* **140:** 84.

Varley J.E., Wehby R.G., Rueger D.C., and Maxwell G.D. 1995. Number of adrenergic and islet-1 immunoreactive cells is increased in avian trunk neural crest cultures in the presence of human recombinant osteogenic protein-1. *Dev. Dyn.* **203:** 434.

Wehrlehaller B. and Weston J.A. 1997. Receptor tyrosine kinase-dependent neural crest migration in response to differentially localized growth factors. *BioEssays* **19:** 337.

Weston J.A. 1991. Sequential segregation and fate of developmentally restricted intermediate cell populations in the neural crest lineage. *Curr. Top. Dev. Biol.* **25:** 133.

Weston J.A. and Butler S.L. 1966. Temporal factors affecting localization of neural crest cells in the chicken embryo. *Dev. Biol.* **14:** 246.

Winnier G., Blessing M., Labosky P.A., and Hogan B.L.M. 1995. Bone morphogenetic protein-4 is required for mesoderm formation and patterning in the mouse. *Genes Dev.* **9:** 2105.

Zhang H.B. and Bradley A. 1996. Mice deficient for BMP2 are nonviable and have defects in amnion chorion and cardiac development. *Development* **122:** 2977.

Zimmerman L.B., Dejesus-Escobar J.M., and Harland R.M. 1996. The Spemann organizer signal *noggin* binds and inactivates bone morphogenetic protein-4. *Cell* **86:** 599.

Cadherins in Brain Patterning and Neural Network Formation

M. Takeichi, T. Uemura, Y. Iwai, N. Uchida, T. Inoue, T. Tanaka, and S.C. Suzuki
Department of Biophysics, Faculty of Science, Kyoto University, Kitashirakawa, Sakyo-ku, Kyoto 606-01, Japan

Development of the brain requires dynamic rearrangement of cells throughout the process. For example, in the earliest phase of neural development, the ectoderm is divided into two major domains: neural plate/tube and future epidermis. This process is achieved by physical separation of the two tissue domains. Then, during the growth of the neural tube, dividing neuroectodermal cells dynamically change their locations. To prevent them from random movement, however, regulatory mechanisms are present; e.g., cell migration is blocked by certain barriers localized between segmental units, such as neuromeres and rhombomeres, constituting the neural tube. Differentiating neuroblasts also undergo active migration, and, upon terminal differentiation, they cluster at appropriate positions to form specific "nuclei" or "layers." Finally, neurons extend axons to establish synaptic connections with their target cells, which are often located in remote positions. During this process, axons derived from multiple different neurons become fasciculated and migrate together, but upon reaching their targets, they become defasciculated and connect with distinct target cells. Neural networks are thus established.

A number of mechanisms should be involved in the regulation of such complex behaviors of developing neural cells. Cell adhesion molecules are thought to be essential components of these mechanisms. Among the many classes of cell adhesion molecules identified, we have been focusing on the cadherin family to determine their roles in neural morphogenesis. The cadherin family has divergent members, and its subfamily designated as "classic cadherins" has been best characterized as cell-cell adhesion molecules (Takeichi 1991, 1995). Classic cadherins, usually just called "cadherins," are important in the control of cell-cell adhesion from various aspects: About 20 members of this subfamily have been identified. Each member tends to bind preferentially to like molecules, and this property leads cells to associate preferentially with those expressing the same cadherins. Thus, cadherins are implicated in selective cell adhesion or cell sorting. At their cytoplasmic domain, cadherins are associated with "catenins." The catenins seem to be regulators of cadherin function, and they are even involved in signaling events in particular situations (Miller and Moon 1996; Fagotto and Gumbiner 1996). Thus, the cadherin-mediated adhesion is a dynamic system, which could directly control morphogenetic cell rearrangements. Indeed, for epithelial cells, cadherin-mediated cell-cell interactions are required not only for maintaining the static architecture of the epithelium, but also for dynamic cell rearrangement (Uemura et al. 1996). They are also essential for migration of particular cell types (Oda et al. 1997).

In this paper, we outline our recent findings on the possible roles of cadherins in the following aspects of neural morphogenesis: neural tube compartmentalization, axon bundling, synaptic connections, and neural network formation, all of which require regulation of cell-cell adhesion.

COMPARTMENTALIZATION OF EARLY BRAINS

The early neural tube is subdivided into compartments. Their longitudinal subdivisions are called "neuromeres," and these are also called "rhombomeres" specifically for the hindbrain subdivisions. Cell lineage or migration restriction is observed at the boundaries between these compartments. As a mechanism for such restrictions, we considered the possibility that cadherin-mediated cell sorting may have a role. For example, if a particular cadherin is expressed in restricted neuromeres, cells of these compartments may be sorted out from those not expressing this cadherin. Supporting this idea, we and other investigators found that a number of cadherins are expressed in restricted compartments or in their boundary regions (Espeseth et al. 1995; Redies and Takeichi 1996). For example, cadherin-6 (cad6) is transiently expressed in rhombomere 6 of E8.5 mouse brains (Inoue et al. 1997), and R-cadherin is expressed in the ventral thalamus and pretectum of E12.5 mouse brains (Fig. 1) (Matsunami and Takeichi 1995). All these expressions occur in undifferentiated neuroepithelial cells. To investigate the role of the R-cadherin expression in fetal brains, we isolated brain tissue fragments containing the R-cadherin-positive and adjacent negative regions, dissociated them into single cells, and allowed them to reaggregate under various conditions. The results showed that the R-cadherin-positive and -negative cells sorted out from each other in the aggregates that formed and that this segregation occurred in a cadherin activity-dependent manner. This finding supports the idea that the expression of a cadherin in restricted areas of the neuroepithelium has a role in maintaining the compartments, although the evidence is still indirect.

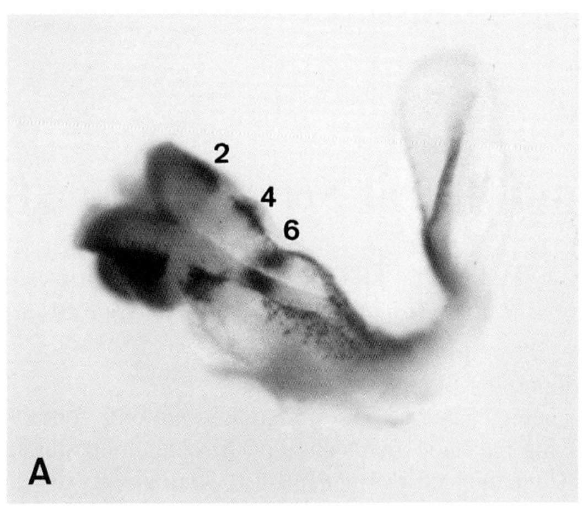

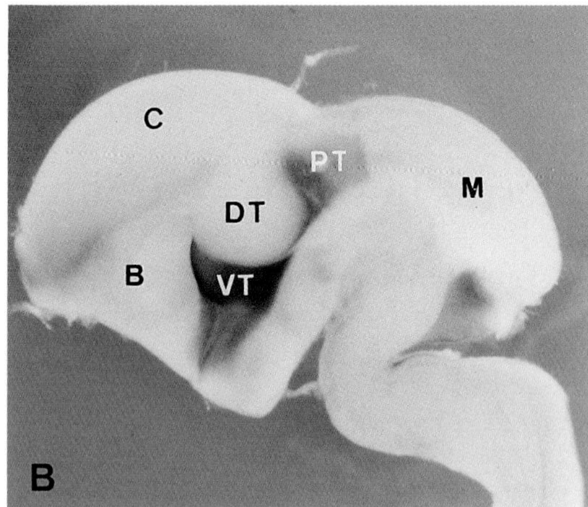

Figure 1. Compartment-dependent expression of cadherins in mouse embryonic CNS. (*A*) cad6 in a whole E8.5 embryo stained by RNA in situ hybridization. Note its expression in restricted rhombomeres (r): In r2 and r4, only the neural crest areas are positive, whereas in r6, the entire neural plate is positive. (*B*) R-cadherin expression in the ventricular area of an E12.5 brain visualized by staining with anti-R-cadherin antibodies. The brain was isolated, cut into halves sagittally to expose the ventricular surface, and stained with the antibodies. The R-cadherin expression occurs in the alar domain of VT, the contiguous hypothalamic region, and PT. Some other regions are also positive. (B) Basal ganglion; (C) cerebral cortex; (DT) dorsal thalamus; (M) midbrain; (PT) pretectum; (VT) ventral thalamus.

AXON PATTERNING

Growing axons, in general, express certain cadherins at their growth cones and lateral surfaces. Early studies showed that, in vitro, the N- or R-cadherin homophilic interactions mediate the migration of axons on the surface of other cells if both express the same cadherins (Matsunaga et al. 1988; Redies et al. 1992). In vivo, retinal ganglion cells transfected with a dominant-negative construct of N-cadherin cannot normally extend their axons (Riehl et al. 1996). These findings suggest that cadherin-mediated cell-cell interactions contribute to axonal outgrowth. Our following findings on a *Drosophila* neural cadherin have provided the first genetic evidence for the roles of cadherins in axon outgrowth and patterning.

We identified a novel *Drosophila* cadherin, *D*N-cadherin, that is intensely expressed by axons (Iwai et al. 1997). It has a large extracellular domain with 15 cadherin repeats, presenting a contrast to the four repeats in the vertebrate classic cadherins (Fig. 2), but it still can interact with catenins at its intracellular domain and induce cell aggregation as well. We therefore believe that the role of *D*N-cadherin in cell-cell adhesion is at least in part comparable with that of vertebrate classic cadherins. To investigate morphogenetic functions of *D*N-cadherin, we isolated null mutants for its gene and analyzed their phenotypes by use of various neuronal markers. As for the overall morphology of the central nervous system (CNS), only subtle phenotypic differences were detected between the mutant and wild types. However, significant phenotypes were observed when we focused on a subset of neurons expressing either Fasciclin II (Fas II) or Apterous (Lundgren et al. 1995). Anti-Fas II antibodies label a set of four neurons that pioneer the first two longitudinal axon pathways, vMP2 and MP1 (Goodman and Doe 1993; Lin et al. 1994). Loss of *D*N-cadherin did not appear to perturb the pathfinding by the pioneer growth cones. However, when follower neurons began Fas II expression and joined the pioneer tracks, various pattern alterations were recognized. For example, at the stage when Fas-II-positive axons were assembling into three longitudinal bundles, their medial pathways were often interrupted, and the bundles were locally bifurcated, which represents either defasciculation or abnormal fusion. These phenotypes can be interpreted as indicating reduced or altered interaction between axon fascicles, or misorientation of growth cones.

For examination of the axon patterning at a higher resolution, the transcription factor Apterous (Ap) was useful as a marker, as it is expressed only by three interneurons per abdominal hemisegment (Lundgren et al. 1995). The Ap-positive axons first extend medially; then their growth cones make right-angled turns and grow in an anterior direction. Subsequently, the transverse portion of each Ap axon starts a medial shift, and its turning point reaches the most medial surface of the longitudinal axon tracts. In the *D*N-cadherin mutants, the initial outgrowth and turning of the Ap axons looked normal. However, their turning points remained at the original positions in the axon scaffold, although most of the mutant growth

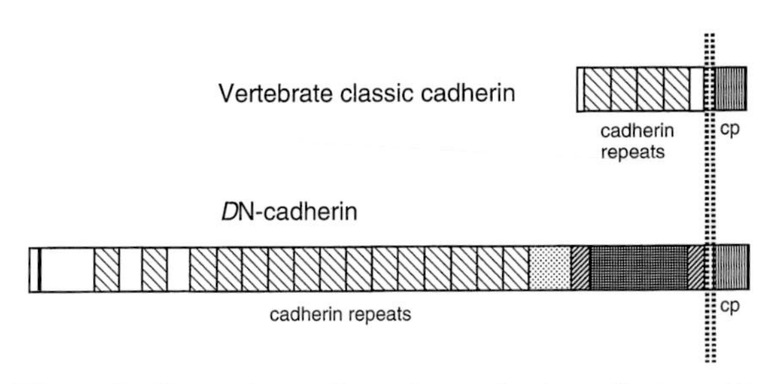

Figure 2. Comparison of vertebrate classic cadherins with *Drosophila D*N-cadherin. (cp) Cytoplasmic domain.

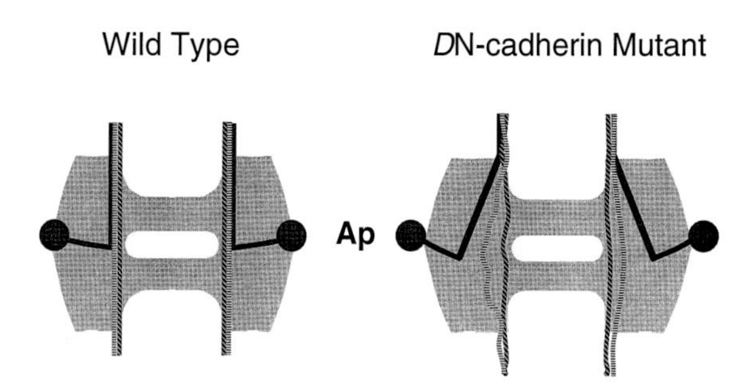

Figure 3. Aberrant axon fasciculation observed in the CNS of *D*N-cadherin mutants. Axons of apterous-positive neurons (Ap) cannot immediately fasciculate with other axons which themselves are also defasciculated. Only a subset of axons in one segment are illustrated.

cones of Ap neurons were able to migrate anteriorly and reach the adjacent anterior segments. Overall, Ap axons could not immediately fasciculate with the axons from the counterpart neurons of posterior segments (Fig. 3). Moreover, the mutants showed increased errors in the bundling of axons from the ventral and dorsal Ap cells; the dorsal and ventral axons sometimes failed to join, and this failure of fasciculation was occasionally accompanied by misorientation of growth cones. These findings suggest that axon patterning critically depends on *D*N-cadherin-mediated axon-axon interactions.

The above findings have been made for a unique type of cadherin identified in *Drosophila*. It remains to be determined whether vertebrate classic cadherins have a similar role.

SYNAPSE FORMATION

Upon reaching the target neurons, outgrowing axons form synapses with them. The synaptic junctions are the sites central for interneuronal communications. Despite such crucial functions, little is known about how the junctional structures are generated in synapses. We tested the possibility that the cadherin adhesion system may take part in this process. To this end, we examined the distribution of αN-catenin, a neuronal α-catenin, and β-catenin in adult mouse brains, assuming that they are associated with certain cadherins (Uchida et al. 1996). As both of these catenins gave a similar result, we refer to only that for αN-catenin below.

αN-catenin was found to be distributed widely in the adult brain as dotty signals. Close examination of these signals suggested that they may be localized in synaptic junctions, as they overlapped synaptic markers such as synaptophysin. Evidence to support this notion was obtained by electron microscopical immunolocalization for αN-catenin. As suggested above, αN-catenin was indeed present in synaptic junctions at their cytoplasmic portions, exhibiting a symmetrical distribution over the pre- and postsynaptic plasma membranes facing each other. Interestingly, these symmetrical αN-catenin signals were not present in the transmitter release zones, which can be identified by their associations with synaptic vesicles, but localized in areas bordering these zones. Furthermore, we examined the N-cadherin distribution in a certain region of the chick brain (see below) and found it to be similar to that of αN-catenin, except that the N-cadherin was located between the plasma membranes as expected. These observations demonstrated that the cadherin/catenin complex is a component of the synaptic junctions, which occupies a specific domain in the synaptic plasma membranes (Fig. 4).

In some regions of the adult brain, catenin signals were little detected, suggesting that this adhesion system is not always required for mature synaptic junctions. However, when such regions were examined at developmental stages, an abundance of catenin signals were detected at contact sites between axon terminals and their target neurons. Therefore, this adhesion system operates from the beginning of axon-target contacts, and it could be more generally important for initiating such contacts than for maintaining established synaptic junctions.

As mentioned above, N-cadherin was localized in certain synapses. Considering the general nature of cadherins to interact homophilically, N-cadherin should be expressed by both of the interconnected neurons. The expression of N-cadherin occurs only in restricted neurons in the newly hatched chick brain. In the midbrain, N-cadherin-positive neurons include those in the optic tectum

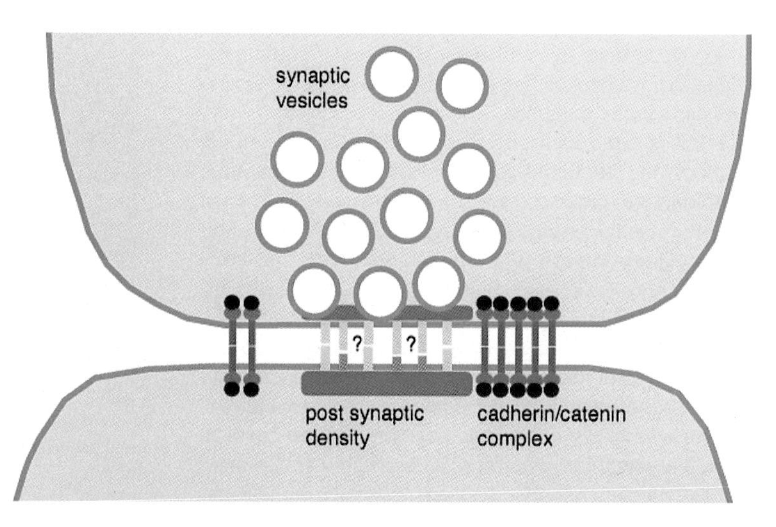

Figure 4. A schematic representation of synaptic junction focusing on junctional proteins. It is not known whether any adhesion proteins are present in the transmitter release zones.

and nucleus pretectalis, components of the visual circuits, and these are known to be connected with each other. The above synaptic localization of N-cadherin was found in this connection system. Notably, most of the other neuronal groups present in the midbrain were negative for N-cadherin expression, although they were positive for αN-catenin expression. This observation strongly suggests that other cadherins are expressed in the N-cadherin-negative neurons and prompted us to identify them.

NEURAL NETWORK FORMATION

It is known from biochemical studies that many cadherins are expressed in the brain. To address the above question, we chose three cadherins from them, cadherin-6 (cad6), cadherin-8 (cad8), and cadherin-11 (cad11), all of which belong to the type II classic cadherin subfamily, and determined their expression pattern in postnatal brains by mRNA in situ hybridization staining (Suzuki et al. 1997). We initially stained the cerebral cortex as whole-mount samples and found that each cadherin was distributed in a unique two-dimensional pattern on the surface of the cortex. Analysis of these patterns revealed that they were correlated with functional and anatomical subdivisions of the cortex. For example, cad6 was most strongly expressed in somatosensory and auditory areas (Fig. 5). On the other hand, cad8 and cad11 were intensely expressed in motor, cingulate/retrosplenial, and entorhinal areas. Thus, superficially, these cadherins delineated specific cortical areas. In transverse sections of the cortex, their distribution was more complicated; the cortical layers, divided into six laminae, were differentially delineated by these cadherin expressions. For example, in the auditory cortex, cad6 was expressed in layers II–IV and in part of layer V, and cad8 was expressed almost exclusively in layer V. In this cortex, the layers III and IV are the sites for afferent projections, and the layers V and VI are those for efferent projections. Thus, layers with afferent and efferent projections express distinct cadherins. Such layer-specific expressions of the three cadherins were observed in all cortical areas, although their distribution patterns differed from area to area. Overall, each cortical area displayed a unique combinatorial expression of the three cadherins.

As the cortical areas differentially expressed the three cadherins, we became interested in determining their expressions in the thalamus, whose subdivisions are selectively connected with different cortical areas. Examination of the thalamus revealed that each thalamic subdivision expressed a specific set of these cadherins; e.g., the medial geniculate body (auditory thalamic nucleus) expressed cad6 (Fig. 5); the ventroposterior thalamic nucleus expressed cad6 and cad11; and the anteroventral thalamic nucleus expressed cad6 and cad8. We extended such observations to the other regions of the brain and found that every region is subdivided into neuronal groups (nuclei or cortical layers), each of which expresses a unique set of cadherins. Although we have focused only on the above three cadherins, our preliminary studies indicate that other cadherin subtypes are also expressed in this way in postnatal brains. This illustrates that the postnatal brain is a collective of neurons expressing different cadherin subtypes in a complex mosaic pattern.

We then attempted to understand the role of the mosaic expression pattern of the three cadherins in the brain and noted a rule in the pattern. Each is correlated with a pattern of known neuronal circuitry. We can present numerous examples for such correlations, but only a few examples are shown below, focusing on cad6. As mentioned above, this cadherin expression occurs in layers II–IV of the auditory cortex and also in the medial geniculate body of the thalamus. The latter is known to be a thalamic relay nucleus in the auditory system and directly sends its axons to layer IV of the auditory cortex. In more caudal regions, cad6 is also expressed in the major, but not all, components of this system, including the inferior colliculus and dorsal cochlear nucleus. Thus, cad6 expression delineates a major subpathway of the auditory circuits. cad6 expression also occurs along subpathways of the somatosensory system; i.e., it is expressed in layer IV of the somatosensory cortex, ventroposterior thalamic nucleus, parts of the trigeminal nuclei, external cuneate nucleus, and so on. As noted for N-cadherin expression in the chicken visual system mentioned above, cad6 is thus detected in neurons interconnected with one another. cad8 and cad11 expressions delineate other neuronal pathways. Together with the finding that catenins are widely

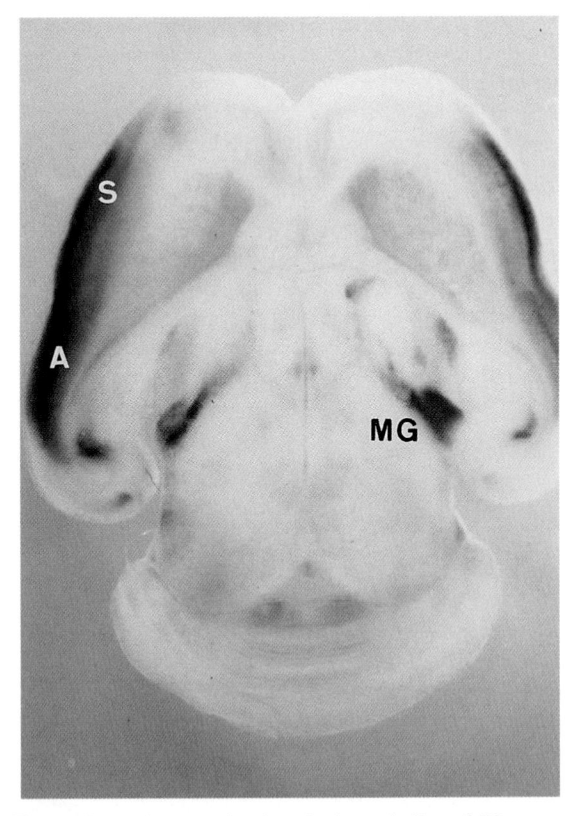

Figure 5. cad6 expression in a horizontal slice of P2 mouse brain. This cadherin is expressed along restricted neural circuits including the auditory pathway. Note intense signals in the auditory cortex (A) and medial geniculate body (MG), which are connected to each other. (S) Somatosensory cortex.

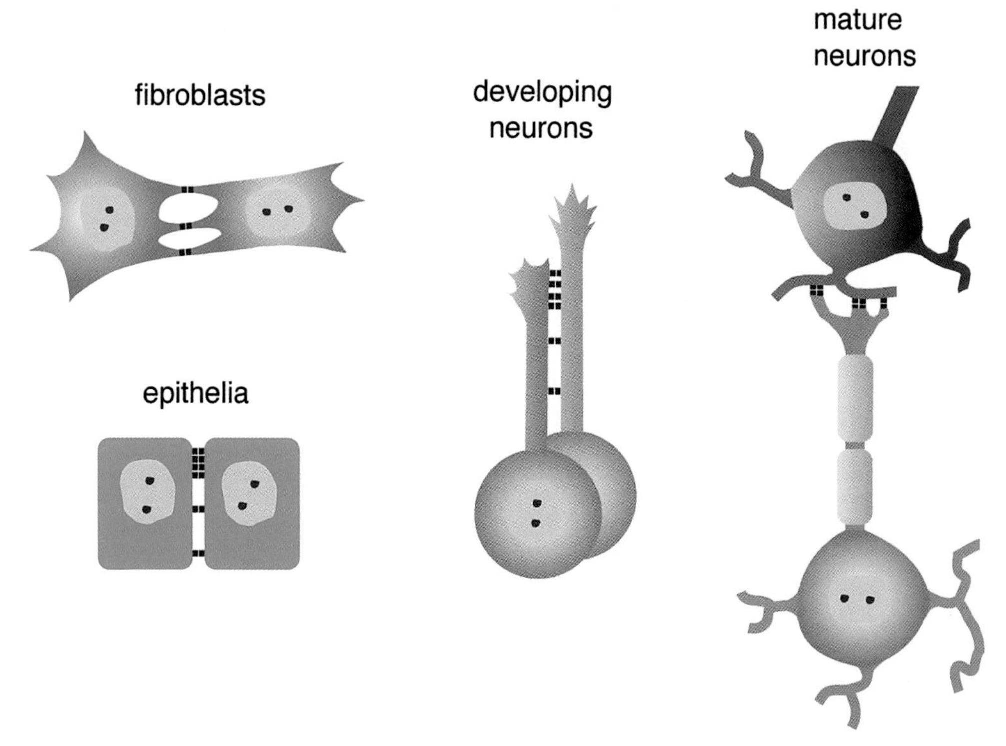

Figure 6. Various forms of cadherin-mediated cell-cell adhesion.

localized in synaptic junctions, the above observations suggest that not only N-cadherin, but also other multiple cadherin subtypes may be used for binding neurons to form synapses. The expression of each cadherin in distinct connection systems suggests that the selective binding nature of cadherins may be utilized for axons to be connected with specific partners.

CONCLUDING REMARKS

We have outlined our recent findings that suggest the involvement of cadherin cell-cell adhesion molecules in brain morphogenesis. Cadherin activity was first identified from a fibroblast cell line (Takeichi 1977). Then, cadherins were found to have a central role in epithelial organization; without them, epithelial layers are severely disorganized, losing their close lateral associations. In these fibroblastic and epithelial systems, a major role of cadherins is to connect like cells together through their homophilic interactions. For neuroepithelial cell associations, the same principle seems to exist, and therefore we proposed that the cadherin-mediated homotypic cell binding may have a role in maintaining the compartments that subdivide the neural tube.

Axons are connected together at their lateral surfaces for fasciculation (Fig. 6). This fashion of cell-cell interaction appears to be analogous to that of epithelial cell-cell association, as in both cases, like or similar cells are arranged in parallel, even though the axonal plasma membrane domain may not be equivalent to the epithelial lateral surfaces. Thus, the roles for cadherins may be essentially similar in both systems. By identifying the *Drosophila* axonal cadherin *D*N-cadherin, we could demonstrate that it is indeed essential for axon bundling; without this molecule, axons were at least in part defasciculated. Previously, we presented anatomical evidence that in sensory axon bundles, a subset of the axons are connected together with a particular cadherin (Shimamura et al. 1992; Uchiyama et al. 1994). On the basis of these observations, we proposed that cadherins expressed in such a pattern may play a part in segregating axons from one type to another within the fascicles. *D*N-cadherin expression occurs rather ubiquitously in the *Drosophila* CNS. Therefore, we could not test the proposed role of the differential cadherin expression in axonal segregation in the *Drosophila* system. It would be interesting to determine whether *Drosophila* has other cadherins expressed in a subset-specific manner in axons or if this animal does not require such cadherins for axonal organization.

In synaptic junctions, heterotypic cells are connected with each other by cadherins (Fig. 6), and this occurs in series to form neural networks. In contrast, in the fibroblastic and epithelial adhesion systems, the cadherin homophilic interactions mainly serve for collecting like cells. Thus, the synaptic neuronal connections can be regarded as a unique form of cadherin-mediated junctions. At the molecular level of cadherin interactions, however, nothing may be special for the synaptic junctions, as the formation of these functions probably simply depends on the rule that the same cadherins interact with each other, regardless of the cell types expressing them (Hirano et al. 1987).

In sum, we showed that cadherins have roles in various aspects of neuronal interactions by binding cells not only homotypically, but also heterotypically. We propose that the latter role could be especially important for wiring neurons to establish their complex networks. This is a novel view for the morphogenetic roles of cadherins.

ACKNOWLEDGMENTS

This work was supported by a grant-in-aid for Creative Fundamental Research from the Ministry of Education, Science, and Culture of Japan and by a grant from the Human Frontier Science Program. N.U. and S.C.S. are recipients of a fellowship of the Japan Society for the Promotion of Sciences for Young Scientists.

REFERENCES

Espeseth A., Johnson E., and Kintner C. 1995. *Xenopus* F-cadherin, a novel member of the cadherin family of cell adhesion molecules, is expressed at boundaries in the neural tube. *Mol. Cell. Neurosci.* **6:** 199.

Fagotto F. and Gumbiner B.M. 1996. Cell contact-dependent signaling. *Dev. Biol.* **180:** 445.

Goodman C.S. and Doe C.Q. 1993. Embryonic development of the *Drosophila* central nervous system. In *The development of* Drosophila melanogaster (ed. M. Bate and A. Martinez Arias), p. 1131. Cold Spring Harbor Laboratory Press, Cold Spring Harbor, New York.

Hirano S., Nose A., Hatta K., Kawakami A., and Takeichi M. 1987. Calcium-dependent cell-cell adhesion molecules (cadherins): Subclass-specificities and possible involvement of actin bundles. *J. Cell Biol.* **105:** 2501.

Inoue T., Chisaka O., Matsunami H., and Takeichi M. 1997. Cadherin-6 expression transiently delineates specific rhombomeres, other neural tube subdivisions and neural crest subpopulations in mouse embryos. *Dev. Biol.* **183:** 183.

Iwai Y., Usui T., Hirano S., Steward R., Takeichi M., and Uemura T. 1997. Axon patterning requires *D*N-cadherin, a novel neuronal adhesion receptor, in the *Drosophilia* embryonic CNS. *Neuron* **19:** 77.

Lin D.M., Fetter R.D., Kopczynski C., Grenningloh G., and Goodman C.S. 1994. Genetic analysis of fasciclin II in *Drosophila*: Defasciculation, refasciculation, and altered fasciculation. *Neuron* **13:** 1055.

Lundgren S.E., Callahan C.A., Thor S., and Thomas J.B. 1995. Control of neuronal pathway selection by the *Drosophila* LIM homeodomain gene *apterous*. *Development* **121:** 1769.

Matsunaga M., Hatta K., and Takeichi M. 1988. Guidance of retinal nerve fibers by N-cadherin adhesion molecules. *Nature* **334:** 62.

Matsunami H. and Takeichi M. 1995. Fetal brain subdivisions defined by R- and E-cadherin expressions: Evidence for the role of cadherin activity in region-specific, cell-cell adhesion. *Dev. Biol.* **172:** 466.

Miller J.R. and Moon R.T. 1996. Signal transduction through β-catenin and specification of cell fate during embryogenesis. *Genes Dev.* **10:** 2527.

Oda H., Uemura T., and Takeichi M. 1997. Phenotypic analysis of null mutants for *D*E-cadherin and *Armadillo* in *Drosophila* ovaries reveals distinct aspects of their functions in cell adhesion and cytoskeletal organization. *Genes Cells* **2:** 29.

Redies C. and Takeichi M. 1996. Cadherins in the developing central nervous system: An adhesive code for segmental and functional subdivisions. *Dev. Biol.* **180:** 413.

Redies C., Inuzuka H., and Takeichi M. 1992. Restricted expression of N- and R-cadherin on neurites of the developing chicken CNS. *J. Neurosci.* **12:** 3525.

Riehl R., Johnson K., Bradley R., Grunwald G.B., Cornel E., Lilienbaum A., and Holt, C.E. 1996. Cadherin function is required for axon outgrowth in retinal ganglion cells in vivo. *Neuron* **17:** 837.

Shimamura K., Takahashi T., and Takeichi M. 1992. E-cadherin expression in a particular subset of sensory neurons. *Dev. Biol.* **152:** 242.

Suzuki S.C., Inoue T., Kimura Y., Tanaka T., and Takeichi M. 1997. Neuronal circuits are subdivided by differential expression of type-II classic cadherins in postnatal mouse brains. *Mol. Cell. Neurosci.* **9:** 433.

Takeichi M. 1977. Functional correlation between cell adhesive properties and some cell surface proteins. *J. Cell Biol.* **75:** 464.

———. 1991. Cadherin cell adhesion receptors as a morphogenetic regulator. *Science* **251:** 1451.

———. 1995. Morphogenetic roles of classic cadherins. *Curr. Opin. Cell Biol.* **7:** 619.

Uchida N., Honjo Y., Johnson K.R., Wheelock M.J., and Takeichi M. 1996. The catenin/cadherin adhesion system is localized in synaptic junctions, bordering the active zone. *J. Cell Biol.* **135:** 767.

Uchiyama N., Hasegawa M., Yamashina T., Yamashita J., Shimamura K., and Takeichi M. 1994. Immunoelectron microscopic localization of E-cadherin in dorsal root ganglia, dorsal root and dorsal horn of postnatal mice. *J. Neurocytol.* **23:** 460.

Uemura T., Oda H., Kraut R., Hayashi S., Kataoka Y., and Takeichi M. 1996. Zygotic *D*E-cadherin expression is required for processes of dynamic epithelial cell rearrangement in the *Drosophila* embryo. *Genes Dev.* **10:** 659.

Retinoids and Posterior Neural Induction: A Reevaluation of Nieuwkoop's Two-step Hypothesis

P. J. KOLM AND H. L. SIVE

Whitehead Institute for Biomedical Research, Cambridge, Massachusetts 02142

A prevailing model of vertebrate neural regionalization, formulated by Pieter Nieuwkoop (1952), proposes that this process occurs in two steps. All presumptive neurectoderm initially goes through an "activation" step that specifies anterior neural fates. Subsequently, a graded "transforming" signal from the posterior mesoderm converts some of this tissue to more posterior fates, including hindbrain and spinal cord. Retinoids are able to posteriorize neurectoderm very efficiently and are therefore prime candidates for endogenous posteriorizing factors. We discuss here the role of retinoids in *Xenopus* hindbrain induction, with emphasis on the regulation and activity of the retinoid-regulated posterior patterning gene, *HoxD1*. Our data suggest several modifications to Nieuwkoop's two-step model. In particular, we propose that neural induction and posterior induction are independent events that can occur in any temporal order. The ventral parts of the hindbrain and spinal cord are induced to a neural state first and then later modified to form posterior neural tissue, as proposed by Nieuwkoop. More medial posterior neurectoderm is likely to be exposed to the posterior- and neural-inducing signals simultaneously. Finally, in the most dorsal regions of the hindbrain and spinal cord, the posterior-inducing signal would act before the neural inducing signal.

NEURECTODERMAL INDUCTION AND PATTERNING REQUIRES MULTIPLE SIGNALS

The molecular basis of regionalization of the vertebrate central nervous system (CNS) has been best characterized in amphibians. During *Xenopus* gastrulation, a series of inductive events that occur between the mesoderm and ectoderm and later within the dorsal ectoderm lead to nervous system formation. By molecular criteria, a complex anteroposterior (A/P) pattern has been specified in the nervous system by mid gastrula (H. Sive et al., unpubl.), with complexity increasing as neurula stages are reached (for review, see Slack and Tannahill 1992; Doniach 1993).

On the basis of specification and induction assays, Nieuwkoop formulated what has become the prevailing model for A/P neural induction (Nieuwkoop 1952) . He proposed that the induction and patterning of the neural plate involve two signals. An initial induction of anterior neural structures in the dorsal ectoderm was termed "activation." The presumptive posterior neurectoderm must necessarily pass through this anterior state so that a second signal can respecify the tissue to more posterior fates. This respecification was termed "transformation." A primary characteristic of this posterior-inducing signal is that it forms a gradient to induce different posterior fates along the A/P axis.

Several lines of evidence are consistent with the hypothesis that the anterior (forebrain) and posterior (hindbrain and spinal cord) parts of the nervous system are induced by different molecules. First, the timing of anterior and posterior neural induction is different, with anterior neural tissue specified before that of more posterior regions (Eyal-Giladi 1954; Sive et al. 1989). Second, where examined, it appears that more posterior neural regions are initially induced to an anterior-type fate that is later respecified to give rise to more posterior tissues (Eyal-Giladi 1954; Sive et al. 1989). This reprogramming can be recapitulated in conjugate assays between presumptive anterior neurectoderm and posterior mesoderm (Sive et al. 1989, 1990; Kolm 1997; Kolm et al. 1997).

Finally, neural-inducing molecules defined in *Xenopus* are able to activate formation of anterior neural tissues (forebrain) but not more posterior brain or spinal cord regions (Lamb et al. 1993; Hemmati-Brivanlou et al. 1994; Knecht et al. 1995). These molecules, including noggin, chordin, and follistatin, induce neural tissue by binding and inactivating bone morphogenetic proteins (BMPs) (Piccolo et al. 1996; Zimmerman et al. 1996; Fainsod et al. 1997). Several molecules have also been defined that can posteriorize anterior neural tissue. Preeminent among these is retinoic acid (RA), a factor that we and other investigators have shown can suppress head formation in *Xenopus* embryos and activate ectopic expression of posterior-specific genes (Durston et al. 1989; Sive et al. 1990; Ruiz i Altaba and Jessell 1991a,b; Sive and Cheng 1991; Papalopulu and Kintner 1996; Taira et al. 1997).

In this paper, we examine the function of retinoids in inducing posterior neurectoderm, particularly the hindbrain. We have approached this question by examining the regulation of *HoxD*1, a posterior gene activated by retinoids. Using insights gained from these studies, we reexamine the predictions of Nieuwkoop's two-signal model and propose a modified model for posterior neural induction.

METHODS

Embryo manipulation and microinjection were carried out as described in Sive et al. (1990) and Kolm and Sive (1995). Whole-mount in situ hybridization was performed as described previously (Harland 1991), with modifications as described by Bradley et al. (1996).

RESULTS AND DISCUSSION

Retinoid Treatment Posteriorizes the Nervous System

Retinoic acid treatment ablates the forebrain and expands the posterior hindbrain. In a search for modifiers of neural patterning, we and other investigators determined that retinoic acid was able to suppress anterior neurectodermal determination in *Xenopus* embryos. RA treatment causes the loss of forebrain, midbrain, and anterior hindbrain structures (Durston et al. 1989; Sive et al. 1990; Papalopulu et al. 1991; Ruiz i Altaba and Jessell 1991a,b; Drysdale and Crawford 1994), with concomitant expansion of the posterior hindbrain (compare A and B in Fig. 2) (Papalopulu et al. 1991; Ruiz i Altaba and Jessell 1991b; Manns and Fritzsch 1992). The hindbrain is the region of the nervous system most sensitive to RA treatment, since even low concentrations of RA cause loss of anterior hindbrain structures (rhombomeres 1–3), and concomitant expansion of posterior rhombomeres 4 and 5 (Papalopulu et al. 1991; Ruiz i Altaba and Jessell 1991b; Manns and Fritzsch 1992; Bradley 1993; Bradley et al. 1993). Retinoid treatment does not affect the total amount of neural tissue (Papalopulu et al. 1991; Ruiz i Altaba and Jessell 1991b); cells fated to become anterior ectodermal structures are not lost but form more posterior and ventral ectodermal structures (Agarwal and Sato 1993).

We and others also demonstrated that embryonic sensitivity to retinoids is maximal during gastrulation, at the time the A/P axis is specified (Durston et al. 1989; Sive et al. 1990; Ruiz i Altaba and Jessell 1991a; Sharpe 1991). This suggested that endogenous retinoids may play a role in neurectodermal patterning during this period of development. Supporting these findings, gastrulating *Xenopus* embryos have been shown to contain the proteins necessary for retinoid signal transduction, including receptors (RARs and RXRs) (Ellinger-Ziegelbauer and Dreyer 1991; Blumberg et al. 1992; Sharpe 1992; Marklew et al. 1994; Sharpe and Goldstone 1997; R. Sparks-Thissen et al., unpubl.), binding proteins (Dekker et al. 1994; Ho et al. 1994), and a variety of retinoids (Durston et al. 1989; Pijnappel et al. 1993; Creech-Kraft et al. 1994a,b; Blumberg et al. 1996).

Retinoic acid acts on both ectoderm and mesoderm. The loss of anterior ectodermal structures in RA-treated embryos could have been due to alterations in the inducing capacity of the dorsal mesoderm or a more direct effect on the ectoderm. RA can act upon both tissues, although the predominant effects of RA appear to be on the ectoderm. We showed that retinoids act directly on induced neurectoderm isolated from mid gastrula embryos to repress anterior marker genes and activate posterior markers (Sive et al. 1990; Sive and Cheng 1991; Kolm 1997; Kolm et al. 1997). Consistent results have been obtained using noggin-neuralized animal caps as a substrate for RA treatment (Papalopulu and Kintner 1996; Taira et al. 1997). We and other investigators also showed that RA-treated ectoderm is incapable of responding to untreated inducing mesoderm to activate anterior dorsal ectodermal genes (Durston et al. 1989; Sive et al. 1990; Sive and Cheng 1991). Additionally, RA can modify anterior mesoderm to activate expression of posterior genes in this tissue (Sive and Cheng 1991). Furthermore, mesoderm isolated from RA-treated embryos is somewhat defective at inducing anterior structures in einsteck (implant) experiments (Cho et al. 1991; Ruiz i Altaba and Jessell 1991a; Sive and Cheng 1991).

labial Hox Genes Are Posteriorly Expressed Targets of Retinoid Signaling

Xenopus labial Hox ***genes are activated as an immediate early response to retinoids.*** To better understand the role of endogenous retinoids, we decided to isolate potential retinoid target genes and analyze their normal regulation in *Xenopus* embryos. The *Hox* genes encode a family of homeodomain-containing DNA-binding proteins which provides positional information along the A/P axis, which are activated by retinoids in tissue culture (for review, see Boncinelli et al. 1991; Krumlauf 1994). *Hox* genes related to *Drosophila labial* (*HoxA1, HoxB1*, and *HoxD1*) had been shown to be rapidly and directly induced by retinoids in cell culture (LaRosa and Gudas 1988; Simeone et al. 1990; Stornaiuolo et al. 1990; Boncinelli et al. 1991). We isolated two *Xenopus labial*-like *Hox* genes, *HoxD1* and *HoxA1* (originally named *Xhox.lab1* and *Xhox.lab2*; Sive and Cheng 1991; Kolm and Sive 1995). We predicted that if these genes were endogenous direct retinoid targets, they would not only be retinoid-inducible, but this induction would not be dependent on protein synthesis in embryonic tissue. Consistent with this hypothesis, we showed that both *HoxA1* and *HoxD1* are strongly induced throughout the ectoderm and mesoderm of RA-treated gastrula embryos (Sive and Cheng 1991; Kolm and Sive 1994, 1995). This induction occurred in both ectoderm and mesoderm in the presence of the protein synthesis inhibitor cycloheximide, suggesting that these genes were part of the primary response to endogenous retinoids during early development (Kolm and Sive 1995).

Xenopus HoxD1 ***is expressed in a unique posterolateral domain.*** The expression pattern of retinoid-responsive genes should reflect domains of retinoid action in the embryo. If retinoids are involved in patterning the posterior neurectoderm, expression of retinoid response genes should be localized to the posterior ectoderm during gastrulation. We have shown that *HoxD1* is the first *Xenopus labial Hox* gene expressed, at early gastrula, followed by *HoxA1* and *HoxB1* at the end of gastrulation (Sive and Cheng 1991; Dekker et al. 1992; Godsave et al. 1994; Kolm and Sive 1995; Poznanski and Keller 1997). As

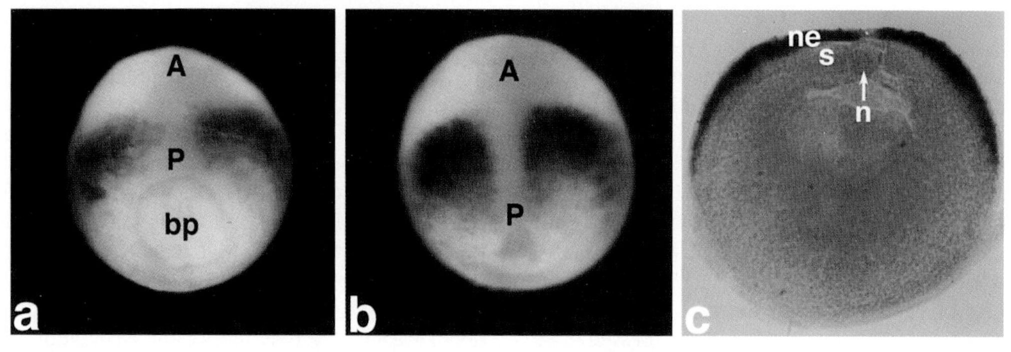

Figure 1. *Xenopus* HoxD1 is expressed in predominantly neurectoderm at gastrula and neurula stages. *HoxD1* expression was analyzed by whole-mount in situ hybridization of albino embryos (Methods). (bp) Blastopore; (A) anterior; (P) posterior; (ne) neurectoderm; (s) somitic mesoderm; (n) notochord. (*a*) Mid gastrula (stage 11.5), dorsovegetal view; (*b*) late gastrula (stage 12.5), dorsal view; (*c*) transverse section through an early neurula (stage 14) HoxD1-stained embryo.

predicted, by mid gastrula in *Xenopus* embryos, *HoxD1* RNA is restricted to the lateral and ventral posterior ectoderm and mesoderm (Fig. 1a) (Kolm and Sive 1995; Kolm 1997). This region of the ectoderm will primarily give rise to spinal cord and hindbrain (Fig. 4A) (Keller et al. 1992a). Expression is absent from the dorsal midline (Kolm and Sive 1994, 1995), which is fated to become floorplate and more ventral regions of the neural tube (Keller 1991).

By the end of gastrulation, *HoxD1* expression is localized to posterodorsal regions of the embryo (Fig. 1b). This domain is expanded anteriorly in RA-treated embryos (compare A and B in Fig. 2) (Kolm and Sive 1995). Examination of *HoxD1* expression at early neurula either in isolated ectoderm and mesoderm (Kolm and Sive 1995) or in transverse sections (Fig. 1c) shows that expression at this stage is predominantly neurectodermal. The anterior boundary of *HoxD1* expression at this stage lies in presumptive rhombomere 5 (Kolm 1997). Thus, *HoxD1* defines a domain of the posterior neurectoderm in the early neurula extending from the posterior hindbrain through the spinal cord.

Retinoid signaling is necessary for* HoxD1 *expression and hindbrain patterning. We reasoned that if endogenous retinoids are required for *HoxD1* activation, then disruption of retinoid signaling should ablate *HoxD1* expression. To test this, we used a truncated form of RARα2 (RARΔ) that efficiently blocks retinoid-induced gene expression in both isolated ectoderm (Blumberg et al. 1997; Kolm 1997; Sharpe and Goldstone 1997; Kolm et al. 1997) and cell culture (Damm et al. 1993) and also

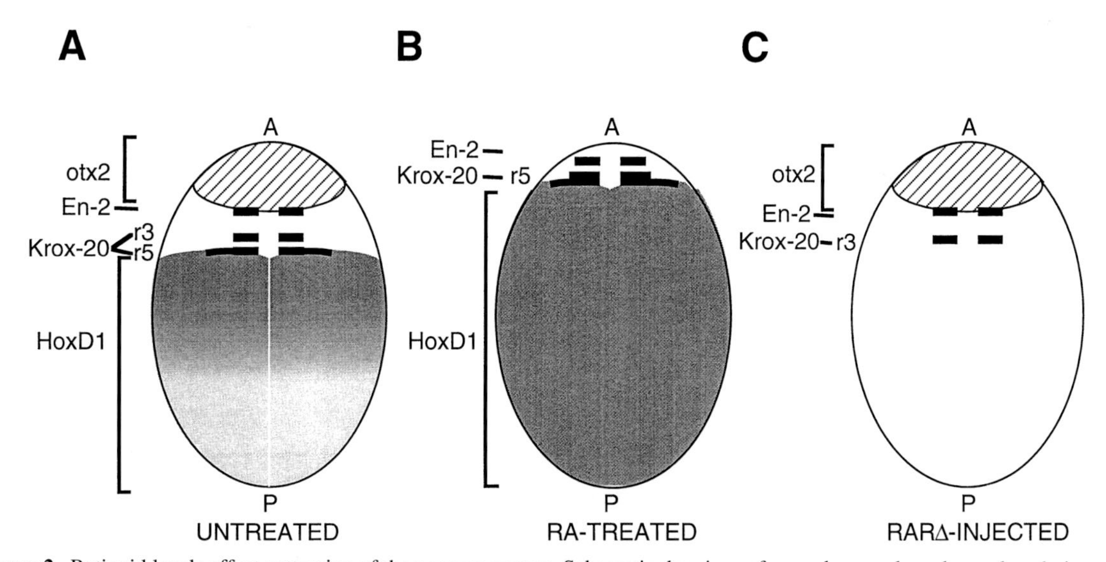

Figure 2. Retinoid levels affect patterning of the nervous system. Schematic drawings of an early neurula embryo, dorsal view; anterior at the top. (A) Anterior; (P) posterior. (*A*) Normal neurectodermal expression patterns. *Otx-2* (*hatched area*; expressed in cement gland [not shown], forebrain and midbrain; Blitz and Cho 1995; Pannese et al. 1995), *En-2* (midbrain-hindbrain boundary; Hemmati-Brivanlou et al. 1991), *Krox-20* (presumptive rhombomeres 3 (r3) and 5 (r5); Bradley et al. 1993), and *HoxD1* (*gray area;* Kolm and Sive 1995) are shown. (*B*) Gene expression in retinoic-acid-treated embryos. *HoxD1* expression is expanded (Kolm and Sive 1995), *En-2* expression shifts anteriorly (Sive et al. 1990), and *otx2* expression is repressed (Pannese et al. 1995; Gammill and Sive 1997). In the hindbrain, the r3 stripe of *Krox-20* is lost (Papalopulu et al. 1991), and the r5 stripe is expanded (Bradley 1993). (*C*) Gene expression in RARΔ-injected embryos. *HoxD1* expression is lost (Kolm 1997; Kolm et al. 1997), as is the r3 stripe of *Krox-20* expression (Blumberg et al. 1997; Kolm 1997; Kolm et al. 1997). *otx2* and *En-2* are unaffected (Kolm 1997; Kolm et al. 1997).

prevents the teratogenic effects of retinoids on *Xenopus* embryos (Blumberg et al. 1997; Kolm 1997; Kolm et al. 1997). High levels of the thyroid hormone receptor, c-erbA, can similarly inhibit retinoid signaling (Old et al. 1992; Banker and Eisenman 1993; Barettino et al. 1993; Kolm and Sive 1995). We asked what effect these dominant inhibitory receptors would have when expressed during early *Xenopus* development. We found that both RARΔ and c-erbA ablate expression of *HoxD1* (Kolm and Sive 1995; Kolm 1997; Kolm et al. 1997), consistent with a role for retinoids in the normal regulation of *HoxD1* expression (Fig. 2C). RARα2.2 is expressed in an expression domain very similar to that of *HoxD1* (Sharpe 1992; Kolm 1997; Kolm et al. 1997) and is likely to be one of the RARs normally involved in *HoxD1* regulation.

We further asked how disruption of retinoid signaling, and subsequent ablation of *HoxD1* expression, affects neural patterning. RARΔ blocks expression of *Krox-20* in presumptive rhombomere 5 of the posterior hindbrain while having little effect on either forebrain, midbrain, or spinal cord markers (Fig. 2C) (Kolm 1997; Kolm et al. 1997). Similar effects of dominant inhibitory RARs on hindbrain patterning have been shown by Blumberg et al. (1997).

labial Hox ***genes define a similar retinoid-dependent posterior domain in all vertebrates.*** The expression and regulation of *HoxD1* we described appear to be universal features of *labial Hox* genes. In all vertebrates, at least one *labial*-like *Hox* gene is expressed during gastrula and early neurula stages in a broad posterior neurectodermal domain. The precise *labial* paralog predominantly expressed during gastrulation is variable; *HoxD1* is expressed in this posterior domain in *Xenopus*, with *HoxA1* and *HoxB1* first expressed only weakly at the end of gastrulation (Dekker et al. 1992; Godsave et al. 1994; Kolm and Sive 1995; Poznanski and Keller 1997). A similar posterior ectodermal domain is defined by *HoxA1* in zebrafish gastrula embryos (Alexandre et al. 1996; C. Sagerstrom et al., unpubl.), *HoxB1* in chicken (Sundin et al. 1990; Sundin and Eichele 1992), and *HoxA1* and *HoxB1* together in the mouse (Duboule and Dollé 1989; Frohman et al. 1990; Sundin et al. 1990; Hunt et al. 1991; Murphy and Hill 1991). The anterior limit of *Xenopus HoxA1* and *HoxD1* at early neurula lies at the presumptive rhombomere 4/5 boundary (Kolm 1997), whereas in mouse, chick, and zebrafish, the *labial Hox* genes have an anterior limit at the rhombomere 3/4 boundary (Frohman et al. 1990; Sundin et al. 1990; Hunt et al. 1991; Murphy and Hill 1991; Alexandre et al. 1996; C. Sagerstrom et al., unpubl.), perhaps reflecting differences in hindbrain structure between amphibians and other vertebrates (Gilland and Baker 1993; Holland and Garcia-Fernàndez 1996). Several other non-*Hox* genes are expressed in a similar neurectodermal domain in both *Xenopus* (Demartis et al. 1994; von Bubnoff et al. 1996) and zebrafish (C. Sagerstrom et al., unpubl.), suggesting that this defines a general early division along the A/P axis.

Expression of *labial Hox* genes in early embryos of all vertebrates is strongly induced by retinoid treatment (Morriss-Kay et al. 1991; Sive and Cheng 1991; Conlon and Rossant 1992; Kessel 1992; Marshall et al. 1992; Sundin and Eichele 1992; Kolm and Sive 1995; Alexandre et al. 1996; Shimeld 1996). A more detailed analysis of *labial Hox* gene regulatory regions has shown that retinoic-acid-responsive elements (RAREs) lie 3′ to the mouse and human *HoxA1* (Langston and Gudas 1992; Langston et al. 1997) and mouse, human, chick, and pufferfish *HoxB1* genes (Marshall et al. 1994; Ogura and Evans 1995a,b; Langston et al. 1997). Consistent with our dominant-negative receptor expression data, functional RAREs are required for the early neurectodermal expression of these genes in mouse embryos (Marshall et al. 1994; Frasch et al. 1995; Dupé et al. 1997).

Finally, the posterior hindbrain is also sensitive to the loss of retinoid signaling in other vertebrates. This has been shown most strikingly in vitamin-A-deficient quail embryos that develop with deficiencies in rhombomeres 4 and 5 (Maden et al. 1996). Similar results have been shown in the mouse in which either expression of a dominant-negative RAR or double mutations in RARα and RARγ produces alterations in craniofacial patterning consistent with disruption of hindbrain patterning (Damm et al. 1993; Lohnes et al. 1994; for review, see Kastner et al. 1995).

Completing the Pathway: Downstream from Retinoid Targets

HoxD1 ***patterns the hindbrain in*** **Xenopus.** Since the *Hox* genes are thought to provide positional information along the A/P axis from the hindbrain to the posterior of the embryo (for review, see Krumlauf 1994), it was reasonable to predict that *HoxD1* has a direct role in A/P neural patterning. *HoxD1* expression precedes that of *Krox-20* in the future hindbrain, and the anterior boundary of *HoxD1* expression overlaps the posterior *Krox-20* stripe in presumptive rhombomere 5 (Bradley et al. 1993; Kolm 1997), suggesting that *HoxD1* is a good candidate for patterning the posterior hindbrain. Indeed, we have found that overexpression of *HoxD1* induces ectopic expression of the posterior hindbrain markers *Krox-20* and *Sek-1* (Bradley et al. 1993; Winning and Sargent 1994; Kolm 1997). Since elimination of retinoid signaling ablates both *HoxD1* expression and the posterior hindbrain, *HoxD1* activity may be a crucial step in hindbrain patterning in the *Xenopus* embryo.

We also asked whether the teratogenic effects of retinoids may in part be mediated by the ectopic *HoxD1* expression induced by retinoids. Our preliminary results show that high levels of *HoxD1* RNA repress expression of the anterior ectodermal markers, *XCG* (cement gland; Sive et al. 1989) and *otx2* (forebrain, midbrain and cement gland; Blitz and Cho 1995; Pannese et al. 1995) similar to RA treatment (Sive et al. 1990; Gammill and Sive 1997; Kolm 1997). However, the loss of forebrain is not complete in *HoxD1*-injected embryos. Perhaps an additional contribution of *HoxA1* and *HoxB1*, as well as other retinoid inducible genes, is required to complete repression of anterior structures.

labial Hox ***genes pattern the hindbrain in all vertebrates.*** Like *Xenopus HoxD1*, other vertebrate *labial* orthologs are also involved in hindbrain patterning (Fig. 3). Overexpression of *HoxA1* in the mouse causes transformation of anterior hindbrain (r2/3) into posterior hindbrain (r4) (Zhang et al. 1994), whereas in zebrafish, overexpression of a *HoxA1*-like *labial* paralog expanded the width of rhombomere 3, but did not otherwise alter hindbrain patterning (Alexandre et al. 1996). Similarly, mutations in mouse *HoxA1* and *HoxB1* cause deficiencies in the formation of rhombomeres 4 and 5 (Lufkin et al. 1991; Chisaka et al. 1992; Carpenter et al. 1993; Dollé et al. 1993; Mark et al. 1993; Goddard et al. 1996; Studer et al. 1996), consistent with their early expression domains.

Where Are Endogenous Retinoids?

A retinoid-dependent signal from the lateral mesoderm induces **HoxD1.** Although our data showed that retinoids have a critical role in *HoxD1* regulation, expression of the dominant-negative RAR throughout the embryo did not allow us to determine the source of retinoid signals. We therefore asked what tissues could induce *HoxD1* and whether this induction was dependent on retinoid signaling. We previously determined that mesoderm was required for *HoxD1* expression since the disruption of mesoderm formation by expression of dominant-negative fibroblast growth factor (FGF) (Amaya et al. 1991) or activin (Hemmati-Brivanlou and Melton 1992) receptors ablates *HoxD1* expression (Kolm and Sive 1995).

On the basis of the expression pattern of *HoxD1* during gastrulation, we predicted that the dorsolateral (D/L) mesoderm underlying ectodermal *HoxD1* expression at mid gastrula would be a source of *HoxD1*-inducing signals. Indeed, we found that D/L mesoderm induces *HoxD1* expression in ectoderm, whereas mesoderm from the dorsal midline cannot (Kolm 1997; Kolm et al. 1997). Expression of the dominant-negative RAR in ectoderm blocks this induction. This suggests either that retinoids are a component of the D/L mesodermal signal or that the D/L mesoderm induces a secondary retinoid signal within the ectoderm itself (Kolm 1997; Kolm et al. 1997).

Is there a retinoid gradient? In tissue culture, low levels of retinoids rapidly activate more anterior *Hox* genes (such as *HoxD1*), whereas higher retinoid levels activate more posterior *Hox* genes over a much longer time. This has led to the proposal that retinoids form a gradient in the embryo (Boncinelli et al. 1991; Chen et al. 1994; Sasai

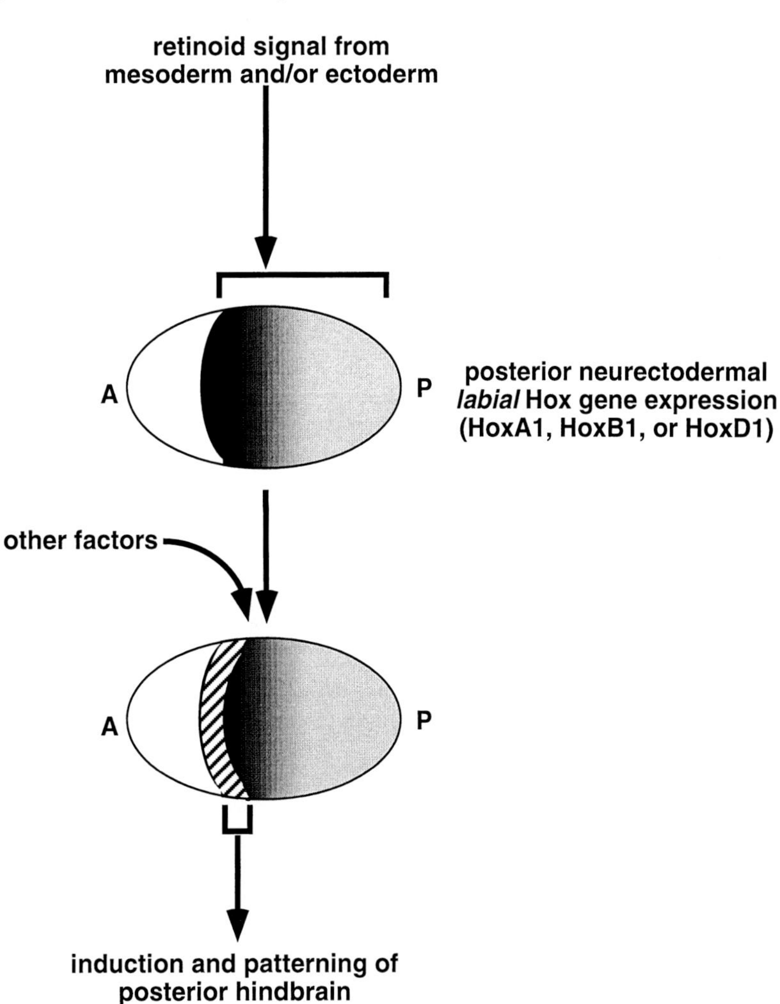

Figure 3. Retinoids pattern the hindbrain. Diagram of a dorsal view of a prototypical vertebrate embryo at early neurula. Anterior is to the left. Endogenous retinoids in the mesoderm and/or ectoderm activate expression of one or more *labial*-like *Hox* genes in a broad neurectodermal domain stretching from the posterior hindbrain through the spinal cord (*shaded area*). Additional factors (*hatched area*) cooperate with labial Hox proteins to restrict activity to the anterior-most region of their expression domains, such that this region is specified to form the posterior hindbrain.

and De Robertis 1997); however, this hypothesis is not supported by available data. By dissecting *Xenopus* embryos into three pieces, Chen and colleagues (1994) found high levels of retinoid activity at the posterior of the embryo, moderate levels in a central piece, and low levels at the anterior of the embryo. Although these data could indicate an endogenous retinoid gradient, they could equally reflect a high level of retinoids in the posterior and a low level in the anterior, with a sharp discontinuity in retinoid levels in the central region of the embryo. Assays of endogenous retinoid levels in rodents, using retinoid-sensitive reporter assays, showed uniform retinoid activity stretching from the node (future posterior hindbrain) to the posterior of the embryo (Wagner et al. 1990, 1992; Rossant et al. 1991; Hogan et al. 1992; Colbert et al. 1993; Horton and Maden 1995; Ang et al. 1996).

Are Retinoids the Only Posteriorizing Signals?

If retinoids are not present in a gradient, how are different posterior neurectodermal regions induced? One possibility is that other factors may play a part in posterior patterning. We propose that retinoid signaling is required for normal hindbrain patterning and *labial* Hox gene induction and that more posterior regions of the embryo may be induced by FGFs or wnts.

FGF has been suggested to be a posterior neural inducer, and in support of this, *eFGF* is expressed in the extreme posterior of the embryo during gastrula stages and later also in the notochord underlying the future spinal cord (Isaacs et al. 1992). Like RA, FGF can posteriorize artificially neuralized ectoderm (Cox and Hemmati-Brivanlou 1995; Kengaku and Okamoto 1995; Lamb and Harland 1995), and activates expression of genes that potentially pattern the posterior of the embryo in animal cap ectoderm (Green et al. 1992; Isaacs et al. 1992; King and Moore 1994; Northrop and Kimelman 1994; Kolm and Sive 1995; Pownall et al. 1996). However, we find that FGF is unable to repress anterior gene expression in isolated dorsal ectoderm and is only able to weakly induce a subset of posterior genes in this tissue (Kolm 1997; Kolm et al. 1997). Consistently, FGF signaling is necessary for some, but not all, gene expression in the spinal cord (Northrup and Kimelman 1994; Kroll and Amaya 1996; Pownall et al. 1996); whereas forebrain, midbrain, and hindbrain patterning is FGF-signaling-independent (Kroll and Amaya 1996; Pownall et al. 1996). These data suggest that FGF may normally pattern the spinal cord, whereas retinoids predominantly function to pattern the hindbrain.

Wnt-3A can also posteriorize neuralized animal cap ectoderm (McGrew et al. 1995). However, this gene is not expressed at high levels until early neurula, when it is localized to the anterior neural folds (Wolda et al. 1993), making it an unlikely endogenous posteriorizing factor. Wnt-8 is a potential candidate for a posteriorizing wnt molecule, since it has activity similar to that of Wnt-3A, is expressed in the lateral and ventral marginal zones, adjacent to posteriorly fated ectoderm, and can repress anterior neurectodermal patterning without reducing the total amount of neural induction (Christian and Moon 1993; Moon et al. 1993; Fredieu et al. 1997). Interestingly, murine Wnt-8 is RA-inducible in both cell culture and early embryos (Bouillet et al. 1995, 1996), suggesting that it may mediate retinoid effects.

Reevaluation of Nieuwkoop's Two-signal Model

These studies have given us new insight into the mechanisms of posterior neural patterning and allow us to reevaluate the prevailing two-signal model of neural induction. Nieuwkoop proposed that the initial "activation" signal (anterior neural inducer) spreads throughout the presumptive neural plate, from late blastula through late gastrula stages (Nieuwkoop and Albers 1990). Subsequent to this, the "transforming" signal is produced by the posterior mesoderm (Nieuwkoop and Albers 1990). This second signal has been proposed to function only after activation has occurred, beginning at late gastrula and continuing through mid neurula stages (Nieuwkoop 1952; Nieuwkoop and Albers 1990).

As Nieuwkoop had proposed, we agree that posterior neural determination requires both neural induction and posterior induction. However, we suggest that these are independent events that can and do occur in any order. Furthermore, we suggest that the posterior-inducing signal is not dorsally restricted but extends around the dorsoventral (D/V) circumference of the embryo. In contrast, the neural inducing signal is dorsally restricted. The considerations that have led us to these conclusions are discussed below.

The* Xenopus *fate map indicates that hindbrain and spinal cord derive from ectoderm extending into ventrolateral regions. The spinal cord and hindbrain are fated to form from lateral and ventral regions of the early gastrula embryo, spanning an arc greater than 240° (Fig. 4A) (Keller 1991; Keller et al. 1992a,b). This fate map remains relatively fixed until mid gastrula, when there is large-scale movement of cells toward the dorsal midline, which brings these cells to the dorsal side of the embryo by neurula stages (Keller 1991). Until late gastrula, much of the presumptive posterior neural plate lies well outside the neural-inducing region (organizer), suggesting that some parts of the neural plate are not induced as neural until after they have been exposed to a posterior-inducing signal (see below).

A posterior-inducing domain is present by early gastrula and extends around the circumference of the embryo. We can use expression of *HoxD1* and other genes expressed in a similar domain (such as *eFGF*) as indicators of a posterior domain in the embryo. The posterior domain extends initially almost 360° around the embryo, although for some genes, including *HoxD1*, exclusion of gene expression from the dorsal-most region (organizer) is observed. This indicates that a posterior-inducing signal is present around the entire embryonic circumference, both dorsally and ventrally (Figs. 1a and 4A). This do-

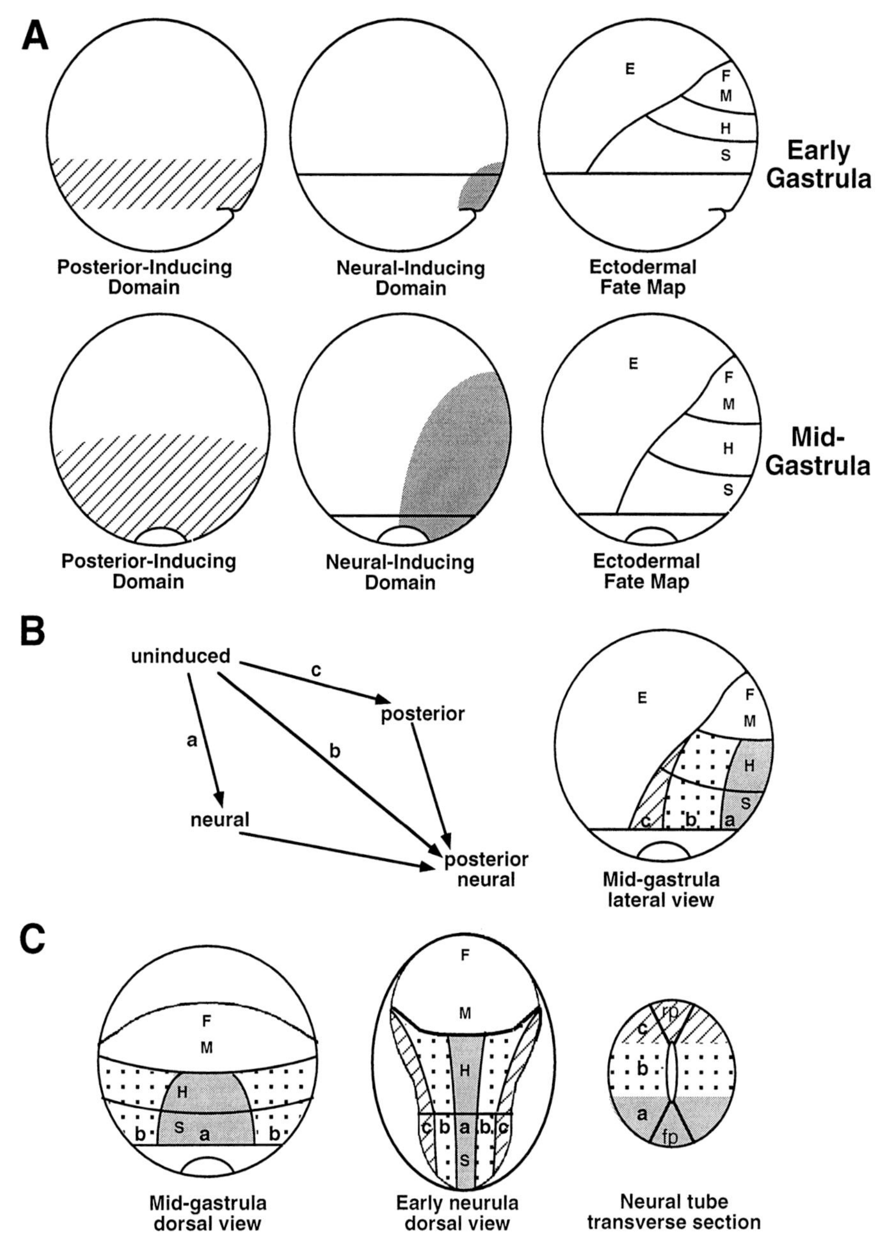

Figure 4. Model of posterior patterning. (*A*) Lateral views of gastrulating embryos. Dorsal is to the right. (E) Presumptive epidermis; (F) forebrain; (M) midbrain; (H) hindbrain; (S) spinal cord. (*Top row*) Early gastrula. A signal induces posterior patterning genes circumferentially (*hatched area*). The dorsal mesoderm (organizer)-derived neural inducer (*gray area*) at this stage is only in an arc of 90° centered on the dorsal midline. The presumptive neural plate is located in a wide arc spanning approximately 240°. (*Bottow row*) Mid gastrula. There is a broad posterior domain (*hatched area*), whereas the neural inducer (*gray area*) extends to an arc of nearly 180°. The neural fate map is largely similar to that at early gastrula, since most of the convergence to the dorsal side of the embryo occurs at late gastrula stages (Keller et al. 1992a). (*B*) Three pathways can lead to posterior neural determination. Uninduced ectoderm adjacent to the dorsal lip at early gastrula can first be induced to a neural fate, followed by respecification to a posterior fate (*a, shaded region*). Cells taking this route fit Nieuwkoop's model of initial anterior "activation" followed by a posteriorizing "transformation." Uninduced ectoderm can also receive posterior and neural inducing signals at about the same time (*b, stippled region*). Finally, the uninduced ectoderm can first be induced to a posterior fate, followed by neural induction (*c, hatched region*). Abbreviations are the same as those in *A*. (*C*) Cells exposed to posteriorizing and neuralizing factors at different stages of development make up different regions of the neural tube. Abbreviations and shading are the same as that in *B*. (*Left*) Dorsal view of a mid gastrula embryo; anterior at the top. (*Center*) Dorsal view of early neurula (open neural plate stage) embryo; anterior at the top. (*Right*) Transverse section through a late neurula neural tube; dorsal at the top. (rp) Roofplate; (fp) floorplate.

main includes both mesoderm, from which the posterior-inducing activity is likely to originate, and, by mid gastrula, the ectoderm spanning both future neural and non-neural tissues.

The neural-inducing region does not cover the entire presumptive hindbrain and spinal cord. At early gastrula, the neural-inducing signal is localized to the dorsal and D/L mesoderm (Jones and Woodland 1989; Mitani and Okamoto 1991). This neural-inducing capacity lies in an arc of only about 90°, with most of the activity in the central 30° (Stewart and Gerhart 1990; Gerhart 1996; Hamburger 1996). By examining the expression pattern of the BMP antagonist *chordin*, the region of mesoderm with neural inducing capacity can be followed during gastrulation (Fig. 4A) (Sasai et al. 1994). Consistent with induction assays, at early gastrula, *chordin* is expressed only near the dorsal lip of the blastopore. Expression follows the course of mesoderm involution, such that by mid gastrula, the neural inducing region is approximately 180° wide (Sasai et al. 1994). Since the width of the presumptive neural plate is considerably greater than the extent of *chordin* RNA expression at this stage, the data suggest that some of the presumptive posterior neural plate must be exposed to the neuralizing signal only when these cells have moved dorsally at late gastrula, after exposure to the posteriorizing signal (Fig. 4B).

Posterior induction can occur independently of neural induction. Retinoids are unable to induce either mesoderm or neural tissue (Durston et al. 1989; Sive et al. 1990; Ruiz i Altaba and Jessell 1991a; Tadano et al. 1993). Since RA induces *HoxD1* in naive animal cap or ventral ectoderm, *HoxD1* induction must be independent of neural induction. In particular, it shows that initial anterior neural specification is not necessary for expression of this posterior marker. RA can induce several other posterior-specific genes in animal caps, including *HoxA1*, more posterior *Hox* genes, and *Gbx-2* (Cho and De Robertis 1990; Sive and Cheng 1991; Kolm and Sive 1995; von Bubnoff et al. 1996), indicating that induction of posterior genes by RA in nonneural ectoderm is a general phenomenon. Not only are these posterior genes expressed in neurectodermal tissue, but they are also expressed in presumptive epidermis and, in some cases, the mesoderm. We suggest that such genes are involved in defining "posteriorness" of embryonic tissue, including the neurectoderm, and should be called "posterior patterning" genes. In contrast, genes such as N-tubulin, which are expressed exclusively in the posterior neurectoderm during neurula stages, are only expressed in cells that have received both neural and posteriorizing signals (see, e.g., Papalopulu and Kintner 1996; Taira et al. 1997).

Three routes for posterior neural induction. We suggest that there are three routes by which posterior neurectoderm can be induced (Fig. 4B). Using *HoxD1* expression as a marker of posterior identity, and chordin expression as representative of mesoderm with neural inducing capacity, the posterior neural plate can be divided into at least three populations of cells that correspond to these three routes.

At early gastrula, ectodermal cells adjacent to the dorsal lip are first exposed to the neural-inducing signal, followed by the posteriorizing signal (Fig. 4B, route a). This region will ultimately form the floorplate and ventral regions of the posterior neural tube (Fig. 4C). These cells initially express *otx2*, a gene that is later repressed in the posterior neurectoderm and is eventually specifically expressed in the anterior neural plate (Blitz and Cho 1995; Pannese et al. 1995; Gammill and Sive 1997), suggesting that these cells follow the classical Nieuwkoop model of anterior before posterior induction.

A second population of ectodermal cells, lying adjacent to region a, are exposed to both the posteriorizing and neuralizing signals during the short period between early and mid gastrula, when both signals spread into this presumptive neurectoderm (Fig. 4B, route b). It is likely that these cells receive the neuralizing and posteriorizing signals simultaneously; however, the order of exposure to these signals is difficult to determine. This region of the presumptive neurectoderm will form medial regions of the posterior neural plate (Fig. 4C).

Finally, a third population of cells, located in the early gastrula ventrolateral ectoderm, is first induced as "posterior," but does not come into contact with the neuralizing mesoderm until convergence movements at late gastrula move these located cells dorsally (Fig. 4B, route c). These cells form the more dorsal regions of the hindbrain and spinal cord (Fig. 4C).

This model emphasizes the independence of neuralizing and posteriorizing signals and suggests that posterior neurectoderm can be and is induced by exposure to these signals in any order. It is interesting to consider, however, that different D/V regions of the neural tube are not equivalent but express different genes and go on to make different kinds of neurons (for review, see Sasai and De Robertis 1997). It is possible then that part of the mechanism by which D/V patterning in the neural tube is achieved is a consequence of the order in which neural- and posterior-inducing signals are received. Using this model as a framework, we aim to continue defining the components and timing of posterior neurectodermal induction.

ACKNOWLEDGMENTS

We thank R. Harland, A. Hemmati-Brivanlou, C. Sharpe, J. Smith, and D. Wilkinson for gifts of plasmids. Many thanks also go to V. Apekin for helping to characterize the effects of RARΔ. P.J.K. would like to thank members of the Sive lab, especially M.E. Lane and B. Sun, for suggestions and insights regarding the model presented here. Additional thanks go to C. Sagerstrom and L. Gammill for suggestions on the manuscript, and L. Bradley for discussions about hindbrain patterning. This work was supported by the National Science Foundation and by the Council for Tobacco Research.

REFERENCES

Agarwal V. and Sato S. 1993. Retinoic acid affects central nervous system development of *Xenopus* by changing cell fate. *Mech. Dev.* **44:** 167.

Alexandre D., Clarke J., Oxtoby E., Yan Y., Jowett T., and Holder N. 1996. Ectopic expression of *Hoxa-1* in the zebrafish alters the fate of the mandibular arch neural crest and phenocopies a retinoic acid-induced phenotype. *Development* **122:** 735.

Amaya E., Musci T., and Kirschner M. 1991. Expression of a dominant negative mutant of the FGF receptor disrupts mesoderm formation in *Xenopus* embryos. *Cell* **66:** 257.

Ang H., Deltour L., Hayamizu T., Zgombic-Knight M., and Duester G. 1996. Retinoic acid synthesis in mouse embryos during gastrulation and craniofacial development linked to class IV alcohol dehydrogenase gene expression. *J. Biol. Chem.* **271:** 9526.

Banker D. and Eisenman R. 1993. Thyroid hormone receptor can modulate retinoic acid-mediated axis formation in frog embryogenesis. *Mol. Cell. Biol.* **13:** 7540.

Barettino D., Bugge T., Bartunek P., Ruiz M., Sonntag-Buck V., Beug H., Zenke M., and Stunnenberg H. 1993. Unliganded T3R, but not its oncogenic variant, v-*erbA*, suppresses RAR-dependent transactivation by titrating out RXR. *EMBO J.* **12:** 1343.

Blitz I. and Cho K. 1995. Anterior neurectoderm is progressively induced during gastrulation: The role of the *Xenopus* homeobox gene *orthodenticle*. *Development* **121:** 993.

Blumberg B., Bolado J., Moreno T., Kintner C., Evans R., and Papalopulu N. 1997. An essential role for retinoid signaling in anteroposterior neural patterning. *Development* **124:** 373.

Blumberg B., Mangelsdorf D., Dyck J., Bittner D., Evans R., and De Robertis E. 1992. Multiple retinoid-responsive receptors in a single cell: Families of retinoid "X" receptors and retinoic acid receptors in the *Xenopus* egg. *Proc. Natl. Acad. Sci.* **89:** 2321.

Blumberg B., Bolado J., Derguin F., Craig A., Moreno T., Chakravarti D., Heyman R., Buch J., and Evans R. 1996. Novel retinoic acid receptor ligands in *Xenopus* embryos. *Proc. Natl. Acad. Sci.* **93:** 4873.

Boncinelli E., Simeone A., Acampora D., and Mavilio F. 1991. HOX gene activation by retinoic acid. *Trends Genet.* **7**: 329.

Bouillet P., Oulad-Abdelghani M., Ward S., Bronner S., Chambon P., and Dollé P. 1996. A new mouse member of the *Wnt* gene family, *mWnt-8*, is expressed during early embryogenesis and is ectopically induced by retinoic acid. *Mech. Dev.* **58:** 141.

Bouillet P., Oulad-Abdelghani M., Vicaire S., Garnier J., Schuhbaur B., Dollé P., and Chambon P. 1995. Efficient cloning of cDNAs of retinoic acid-responsive genes in P19 embryonal carcinoma cells and characterization of a novel mouse gene, *Stra1* (mouse LERK-2/Eplg2). *Dev. Biol.* **170:** 420.

Bradley L. 1993. "Characterisation and functional analysis of *Xenopus* Krox-20." Ph.D. thesis. Open University, London.

Bradley L., Wainstock D., and Sive H. 1996. Positive and negative signals modulate formation of the *Xenopus* cement gland. *Development* **122:** 2739.

Bradley L., Snape A., Bhatt S., and Wilkinson D. 1993. The structure and expression of the *Xenopus* Krox-20 gene: Conserved and divergent patterns of expression in rhombomeres and neural crest. *Mech. Dev.* **40:** 73.

Carpenter E., Goddard J., Chisaka O., Manley M., and Capecchi M. 1993. Loss of Hox-A1 (Hox-1.6) function results in the reorganization of the murine hindbrain. *Development* **118:** 1063.

Chen Y., Huang L., and Solursh M. 1994. A concentration gradient of retinoids in the early *Xenopus laevis* embryo. *Dev. Biol.* **161:** 70.

Chisaka O., Musci T., and Capecchi M. 1992. Developmental defects of the ear, cranial nerves and hindbrain resulting from targeted disruption of the mouse homeobox gene Hox-1.6. *Nature* **355:** 516.

Cho K. and De Robertis E. 1990. Differential activation of *Xenopus* homeo box genes by mesoderm-inducing growth factors and retinoic acid. *Genes Dev.* **4:** 1910.

Cho K., Morita E., Wright C., and De Robertis E. 1991. Overexpression of a homeodomain protein confers axis-forming activity to uncommitted *Xenopus* embryonic cells. *Cell* **65:** 55.

Christian J. and Moon R. 1993. Interactions between the *Xwnt-8* and Spemann organizer signalling pathways generate dorsoventral pattern in the embryonic mesoderm of *Xenopus*. *Genes Dev.* **7:** 13.

Colbert M.C., Linney E., and LaMantia A.-S. 1993. Local sources of retinoic acid coincide with retinoid-mediated transgene activity during embryonic development. *Proc. Natl. Acad. Sci.* **90:** 6572.

Conlon R. and Rossant J. 1992. Exogenous retinoic acid rapidly induces anterior ectopic expression of murine Hox-2 genes in vivo. *Development* **116:** 357.

Cox W. and Hemmati-Brivanlou A. 1995. Caudalization of neural fate by tissue recombination and bFGF. *Development* **121:** 4349.

Creech-Kraft J., Schuh T., Juchau M., and Kimelman D. 1994a. Temporal distribution, localization, and metabolism of all-trans retinol, didehydroretinol, and all-trans-retinal during *Xenopus* development. *Biochem. J.* **301:** 111.

———. 1994b. The retinoid X receptor ligand, 9-*cis*-retinoic acid, is a potential regulator of early *Xenopus* development. *Proc. Natl. Acad. Sci.* **91:** 3067.

Damm K., Heyman R., Umesono K., and Evans R. 1993. Functional inhibition of retinoic acid response by dominant negative retinoic acid receptor mutants. *Proc. Natl. Acad. Sci.* **90:** 2989.

Dekker E., Vaessen M., van den Berg C., Timmermans A., Godsave S., Holling T., Nieuwkoop P., van Kessel A., and Durston A. 1994. Overexpression of a cellular retinoic acid binding protein (xCRABP) causes anteroposterior defects in developing *Xenopus* embryos. *Development* **120:** 973.

Dekker E., Pannese M., Houtzager E., Boncinelli E., and Durston A. 1992. Colinearity in the *Xenopus laevis* Hox-2 complex. *Mech. Dev.* **40:** 3.

Demartis A., Maffei M., Vignali R., Barsacchi G., and DeSimone V. 1994. Cloning and developmental expression of LFB3/HNF1β transcription factor in *Xenopus laevis*. *Mech. Dev.* **47:** 19.

Dollé P., Lufkin T., Krumlauf R., Mark M., Duboule D., and Chambon P. 1993. Local alterations of Krox-20 and Hox gene expression in the hindbrain suggest lack of rhombomeres 4 and 5 in homozygote null Hoxa-1 (Hox-1.6) mutant embryos. *Proc. Natl. Acad. Sci.* **90:** 7666.

Doniach T. 1993. Planar and vertical induction of anteroposterior pattern during the development of the amphibian central nervous system. *J. Neurobiol.* **24:** 1256.

Drysdale T. and Crawford M. 1994. Effects of localized application of retinoic acid on *Xenopus laevis* development. *Dev. Biol.* **162:** 394.

Duboule D. and Dollé P. 1989. The structural and functional organization of the murine HOX gene family resembles that of *Drosophila* homeotic genes. *EMBO J.* **8:** 1497.

Dupé V., Davenne M., Brocard J., Dollé P., Mark M., Dierich A., Chambon P., and Rijli F. 1997. In vivo functional analysis of the *Hoxa-1* 3′ retinoic acid responsive element (3′RARE). *Development* **124:** 399.

Durston A., Timmermans J., Jage W., Hendriks H., deVries N., Heidiveld M., and Nieuwkoop P. 1989. Retinoic acid causes an anteroposterior transformation in the developing central nervous system. *Nature* **340:** 140.

Ellinger-Ziegelbauer H. and Dreyer C. 1991. A retinoic acid receptor expressed in the early development of *Xenopus laevis*. *Genes Dev.* **5:** 94.

Eyal-Giladi H. 1954. Dynamic aspects of neural induction in amphibia. *Arch. Biol.* **65:** 179.

Fainsod A., Deissler K., Yelin R., Marom K., Epstein M., Pillemer G., Steinbeisser H., and Blum M. 1997. The dorsalizing and neural inducing gene follistatin is an antagonist of BMP-4. *Mech. Dev.* **63:** 49.

Frasch M., Chen X., and Lufkin T. 1995. Evolutionary-conserved enhancers direct region-specific expression of the murine *Hoxa-1* and *Hoxa-2* loci in both mice and *Drosophila*. *Development* **121:** 957.

Fredieu J., Cui Y., Maier D., Danilchik M., and Christian J. 1997. *Xwnt-8* and lithium can act upon either dorsal mesodermal or neurectodermal cells to cause a loss of forebrain in *Xenopus* embryos. *Dev. Biol.* **186:** 100.

Frohman M., Boyle M., and Martin G. 1990. Isolation of the mouse *Hox-2.9* gene; analysis of embryonic expression suggests that positional information along the anterior-posterior axis is specified by mesoderm. *Development* **110:** 589.

Gammill L. and Sive H. 1997. Identification of *otx2* target genes and restrictions in ectodermal competence during *Xenopus* cement gland formation. *Development* **124:** 471.

Gerhart J. 1996. Johannes Holtfreter's contributions to ongoing studies of the organizer. *Dev. Dyn.* **205:** 245.
Gilland E. and Baker R. 1993. Conservation of neuroepithelial and mesodermal segments in the embryonic vertebrate head. *Acta Anat.* **148:** 110.
Goddard J., Rossel M., Manley N., and Capecchi M. 1996. Mice with targeted disruption of *Hoxb-1* fail to form the motor nucleus of the VIIth nerve. *Development* **122:** 3217.
Godsave S., Dekker E., Holling T., Pannese M., Boncinelli E., and Durston A. 1994. Expression patterns of Hoxb genes in the *Xenopus* embryo suggest roles in anteroposterior specification of the hindbrain and in dorsoventral patterning of the mesoderm. *Dev. Biol.* **166:** 465.
Green J., New J., and Smith J. 1992. Responses of embryonic *Xenopus* cells to activin and FGF are separated by multiple dose thresholds and correspond to distinct axes of the mesoderm. *Cell* **71:** 731.
Hamburger V. (translator). 1996. Differentiation potencies of isolated parts of the anuran gastrula, by J. Holtfreter. *Dev. Dyn.* **205:** 217.
Harland R. 1991. In situ hybridization: An improved wholemount method for *Xenopus* embryos. *Methods Cell Biol.* **36:** 685.
Hemmati-Brivanlou A. and Melton D. 1992. A truncated activin receptor inhibits mesoderm induction and formation of axial structures in *Xenopus* embryos. *Nature* **359:** 609.
Hemmati-Brivanlou A., Kelly O., and Melton D. 1994. Follistatin, an antagonist of activin, is expressed in the Spemann organizer and displays direct neuralizing activity. *Cell* **77:** 283.
Hemmati-Brivanlou A., de la Torre J., Holt C., and Harland R. 1991. Cephalic expression and molecular characterization of *Xenopus* En-2. *Development* **111:** 715.
Ho L., Mercola M., and Gudas L. 1994. *Xenopus laevis* cellular retinoic acid-binding protein: Temporal and spatial expression pattern during early embryogenesis. *Mech. Dev.* **47:** 53.
Hogan B., Thaller C., and Eichele G. 1992. Evidence that Hensen's node is a site of retinoic acid synthesis. *Nature* **359:** 237.
Holland P. and Garcia-Fernàndez J. 1996. *Hox* genes and chordate evolution. *Dev. Biol.* **173:** 382.
Horton C. and Maden M. 1995. Endogenous distribution of retinoids during normal development and teratogenesis in the mouse embryo. *Dev. Dyn.* **202:** 312.
Hunt P., Gulisano M., Cook M., Sham M., Faiella A., Wilkinson D., Boncinelli E., and Krumlauf R. 1991. A distinct Hox code for the branchial region of the vertebrate head. *Nature* **353:** 861.
Isaacs H., Tannahill D., and Slack J. 1992. Expression of a novel FGF in the *Xenopus* embryo: A new candidate inducing factor for mesoderm formation and antero-posterior patterning. *Development* **114:** 711.
Jones E. and Woodland H. 1989. Spatial aspects of neural induction in *Xenopus laevis*. *Development* **107:** 785.
Kastner P., Mark M., and Chambon P. 1995. Nonsteroid nuclear receptors: What are genetic studies telling us about their role in real life? *Cell* **83:** 859.
Keller R. 1991. Early embryonic development of *Xenopus laevis*. *Methods Cell Biol.* **36:** 62.
Keller R., Shih J., and Sater A. 1992a. The cellular basis of the convergence and extension of the *Xenopus* neural plate. *Dev. Dyn.* **193:** 199.
Keller R., Shih J., Sater A., and Moreno C. 1992b. Planar induction of convergence and extension of the neural plate by the organizer of *Xenopus*. *Dev. Dyn.* **193:** 218.
Kengaku M. and Okamoto H. 1995. bFGF as a possible morphogen for the anteroposterior axis of the central nervous system in *Xenopus*. *Development* **121:** 3121.
Kessel M. 1992. Respecification of vertebral identities by retinoic acid. *Development* **115:** 487.
King M. and Moore M. 1994. Novel HOX, POU and FKH genes expressed during bFGF mesodermal differentiation in *Xenopus*. *Nuc. Acids Res.* **22:** 3990.
Knecht A., Good P., Dawid I., and Harland R. 1995. Dorsal-ventral patterning and differentiation of noggin-induced neural tissue in the absence of mesoderm. *Development* **121:** 1927.
Kolm P. 1997. "Patterning of the posterior neurectoderm by *labial*-like Hox genes and retinoids." Ph.D. thesis. Massachusetts Institute of Technology, Cambridge.
Kolm P. and H. Sive. 1994. Complex regulation of *Xenopus* HoxA1 and HoxD1. *Biochem. Soc. Trans.* **22:** 579.
———. 1995. Regulation of the *Xenopus* labial homeodomain genes, HoxA1 and HoxD1: activation by retinoids and peptide growth factors. *Dev. Biol.* **167:** 34.
Kolm P., Apekin V., and Sive H. 1997. *Xenopus* hindbrain patterning requires retinoid signalling. *Dev. Biol.* (in press).
Kroll K. and Amaya E. 1996. Transgenic *Xenopus* embryos from sperm nuclear transplantations reveal FGF signaling requirements during gastrulation. *Development* **122:** 3173.
Krumlauf R. 1994. Hox genes in vertebrate development. *Cell* **78:** 191.
Lamb T. and Harland R. 1995. Fibroblast growth factor is a direct neural inducer, which combined with noggin generates anterior-posterior neural pattern. *Development* **121:** 3627.
Lamb T., Knecht A., Smith W., Stachel S., Economides A., Stahl N., Yancopolous G., and Harland R. 1993. Neural induction by the secreted polypeptide noggin. *Science* **262:** 713.
Langston A. and Gudas L. 1992. Identification of retinoic acid responsive enhancer 3′ of the murine homeobox gene Hox-1.6. *Mech. Dev.* **38:** 217.
Langston A., Thompson J., and Gudas L. 1997. Retinoic acid-responsive enhancers located 3′ of the Hox A and Hox B homeobox gene clusters—Functional analysis. *J. Biol. Chem.* **272:** 2167.
LaRosa G. and Gudas L. 1988. An early effect of retinoic acid: Cloning of an mRNA (Era-1) exhibiting rapid and protein synthesis-independent induction during teratocarcinoma stem cell differentiation. *Proc. Natl. Acad. Sci.* **85:** 329.
Lohnes D., Mark M., Mendelsohn C., Dollé P., Dierich A., Gorry P., Gansmuller A., and Chambon P. 1994. Function of the retinoic acid receptors (RARs) during development. I. Cranofacial and skeletal abnormalities in RAR double mutants. *Development* **120:** 2723.
Lufkin T., Dierich A., LeMeur M., Mark M., and Chambon P. 1991. Disruption of the Hox-1.6 homeobox gene results in defects in a region corresponding to its rostral domain of expression. *Cell* **66:** 1105.
Maden M., Gale E., Kostetskii I., and Zile M. 1996. Vitamin A-deficient quail embryos have half a hindbrain and other neural defects. *Curr. Biol.* **6:** 417.
Manns M. and Fritzsch B. 1992. Retinoic acid affects the organization of reticulospinal neurons in developing *Xenopus*. *Neurosci. Lett.* **139:** 253.
Mark M., Lufkin T., Vonesch J., Ruberte E., Olivo J., Dollé P., Gorry P., Lumsden A., and Chambon P. 1993. Two rhombomeres are altered in Hoxa-1 mutant mice. *Development* **119:** 319.
Marklew S., Smith D., Mason C., and Old R. 1994. Isolation of a novel RXR from *Xenopus* that most closely resembles mammalian RXRβ and is expressed throughout early development. *Biochim. Biophys. Acta* **1278:** 267.
Marshall H., Nonchev S., Sham M., Muchamore I., Lumsden A., and Krumlauf R. 1992. Retinoic acid alters hindbrain Hox code and induces transformation of rhombomeres 2/3 into a 4/5 identity. *Nature* **360:** 737.
Marshall H., Studer M., Pöpperl H., Aparicio S., Kuiowa A., Brenner S., and Krumlauf R. 1994. A conserved retinoic acid response element required for early expression of the homeobox gene Hoxb-1. *Nature* **370:** 567.
McGrew L., Lai C., and Moon R. 1995. Specification of the anteroposterior neural axis through synergistic interaction of the *Wnt* signaling cascade with *noggin* and *follistatin*. *Dev. Biol.* **172:** 337.
Mitani S. and Okamoto H. 1991. Inductive differentiation of two neural lineages reconstituted in a microculture system from *Xenopus* early gastrula cells. *Development* **112:** 21.
Moon R., Christian J., Campbell R., McGrew L., DeMarais A., Torres M., Lai C., Olson D., and Kelly G. 1993. Dissecting *Wnt* signalling pathways and *Wnt*-sensitive developmental processes through transient misexpression analyses in embryos of *Xenopus laevis*. *Dev. Suppl.*, p. 85.

Morriss-Kay G., Murphy P., Hill R., and Davidson D. 1991. Effects of retinoic acid excess on expression of Hox-2.9 and Krox-20 and on morphological segmentation in the hindbrain of mouse embryos. *EMBO J.* **10:** 2985.

Murphy P. and Hill R. 1991. Expression of the mouse labial-like homeobox-containing genes, Hox 2.9 and Hox 1.6, during segmentation of the hindbrain. *Development* **111:** 61.

Nieuwkoop P. 1952. Activation and organization of the central nervous system in amphibians. III. Synthesis of a new working hypothesis. *J. Exp. Zool.* **120:** 83.

Nieuwkoop P. and Albers B. 1990. The role of competence in the craniocaudal segregation of the central nervous system. *Dev. Growth Differ.* **32:** 23.

Northrop J. and Kimelman D. 1994. Dorsal-ventral differences in Xcad-3 response to FGF mediated induction in *Xenopus*. *Dev. Biol.* **161:** 490.

Ogura T. and Evans R. 1995a. A retinoic acid-triggered cascade of HOXB1 gene activation. *Proc. Natl. Acad. Sci.* **92:** 387.

———. 1995b. Evidence for two distinct retinoic acid response pathways for HOXB1 gene regulation. *Proc. Natl. Acad. Sci.* **92:** 392.

Old R., Jones E., Sweeney G., and Smith D. 1992. Precocious synthesis of a thyroid homone receptor in *Xenopus* embryos causes hormone-dependent developmental abnormalities. *Roux's Arch. Dev. Biol.* **201:** 312.

Pannese M., Polo C., Andreazzoli M., Vignali R., Kablar B., Barsacchi G., and Boncinelli E. 1995. The *Xenopus* homologue of *Otx2* is a maternal homeobox gene that demarcates and specifies anterior body regions. *Development* **121:** 707.

Papalopulu N. and Kintner C. 1996. A posteriorising factor, retinoic acid, reveals that anteroposterior patterning controls the timing of neuronal differentiation in *Xenopus* neuroectoderm. *Development* **122:** 3409.

Papalopulu N., Clarke J., Bradley L., Wilkinson D., Krumlauf R., and Holder N. 1991. Retinoic acid causes abnormal development and segmental patterning of the anterior hindbrain in *Xenopus laevis*. *Development* **113:** 1145.

Piccolo S., Sasai Y., Lu B., and De Robertis E.M. 1996. Dorsoventral patterning in *Xenopus*: Inhibition of ventral signals by direct binding of chordin to BMP-4. *Cell* **86:** 589.

Pijnappel W., Hendriks H., Folkers G., van den Brink C., Dekker E., Edelenbosch C., van der Saag P., and Durston A. 1993. The retinoid ligand 4-oxo-retinoic acid is a highly active modulator of positional specification. *Nature* **366:** 340.

Pownall M., Tucker A., Slack J., and Isaacs H. 1996. *eFGF*, *Xcad3*, and Hox genes form a molecular pathway that establishes the anteroposterior axis in *Xenopus*. *Development* **122:** 3881.

Poznanski A. and Keller R. 1997. The role of planar and early vertical signaling in patterning the expression of Hoxb-1 in *Xenopus*. *Dev. Biol.* **184:** 351.

Rossant J., Zirngibl R., Cado D., Shago M., and Giguere V. 1991. Expression of a retinoic acid response element-hsplacZ transgene defines specific domains of transcriptional activity during mouse embryogenesis. *Genes Dev.* **5:** 1333.

Ruiz i Altaba A. and Jessell T. 1991a. Retinoic acid modifies mesodermal patterning in early *Xenopus* embryos. *Genes Dev.* **5:** 175.

———. 1991b. Retinoic acid modifies the pattern of cell differentiation in the central nervous system of neurula stage *Xenopus* embryos. *Development* **112:** 945.

Sasai Y. and De Robertis E. 1997. Ectodermal patterning in vertebrate embryos. *Dev. Biol.* **182:** 5.

Sasai Y., Lu B., Steinbeisser H., Geissert D., Gont L., and De Robertis E. 1994. *Xenopus chordin*: A novel dorsalizing factor activated by organizer-specific homeobox genes. *Cell* **79:** 779.

Sharpe C. 1991. Retinoic acid can mimic endogenous signals involved in transformation of the *Xenopus* nervous system. *Neuron* **7:** 239.

———. 1992. Two isoforms of retinoic acid receptor α expressed during *Xenopus* development respond to retinoic acid. *Mech. Dev.* **39:** 81.

Sharpe C. and Goldstone K. 1997. Retinoid receptors promote primary neurogenesis in *Xenopus*. *Development* **124:** 515.

Shimeld S. 1996. Retinoic acid, HOX genes and the anterior-posterior axis in chordates. *BioEssays* **18:** 613.

Simeone A., Acampora D., Arcioni L., Andrews P., Boncinelli E., and Mavillo F. 1990. Sequential activation of HOX2 homeobox genes by retinoic acid in human embryonal carcinoma cells. *Nature* **346:** 763.

Sive H. and Cheng P. 1991. Retinoic acid perturbs the expression of Xhox.lab genes and alters mesodermal determination in *Xenopus laevis*. *Genes Dev.* **5:** 1321.

Sive H., Hattori K., and Weintraub H. 1989. Progressive determination during formation of the anteroposterior axis in *Xenopus laevis*. *Cell* **58:** 171.

Sive H., Draper B., Harland R., and Weintraub H. 1990. Identification of a retinoic acid-sensitive period during primary axis formation in *Xenopus laevis*. *Genes Dev.* **4:** 932.

Slack J. and Tannahill D. 1992. Mechanism of anteroposterior axis specification in vertebrates: Lessons from the amphibians. *Development* **114:** 285.

Stewart R. and Gerhart J. 1990. The anterior extent of dorsal development of the *Xenopus* embryonic axis depends on the quantity of organizer in the late blastula. *Development* **109:** 363.

Stornaiuolo A., Acompora D., Pannese M., D'Esposito M., Morelli F., Migliaccio E., Rambaldi M., Faiella A., Nigro V., Simeone A., and Boncinelli E. 1990. Human HOX genes are differentially activated by retinoic acid in embryonal carcinoma cells according to their position within the four loci. *Cell Differ. Dev.* **31:** 119.

Studer M., Lumsden A., Ariza-McNaughton L., Bradley A., and Krumlauf R. 1996. Altered segmental identity and abnormal migration of motor neurons in mice lacking *Hoxb-1*. *Nature* **384:** 630.

Sundin O. and Eichele G. 1992. An early marker of axial pattern in the chick embryo and its respecification by retinoic acid. *Development* **114:** 841.

Sundin O., Busse H., Rogers M., Gudas L., and Eichele G. 1990. Region-specific expression in early chick and mouse embryos of Ghox-lab and Hox-1.6, vertebrate homeobox-containing genes related to *Drosophila* labial. *Development* **108:** 47.

Tadano T., Otani H., Taira M., and Dawid I. 1993. Differential induction of regulatory genes during mesoderm formation in *Xenopus laevis* embryos. *Dev. Genet.* **14:** 204.

Taira M., Saint-Jeannet J., and Dawid I. 1997. Role for the *Xlim-1* and *Xbra* genes in anteroposterior patterning of neural tissue by the head and trunk organizer. *Proc. Natl. Acad. Sci.* **94:** 895.

von Bubnoff A., Schmidt J., and Kimelman D. 1996. The *Xenopus laevis* homeobox gene *Xgbx-2* is an early marker of anteroposterior patterning in the ectoderm. *Mech. Dev.* **54:** 149.

Wagner M., Han B., and Jessell T. 1992. Regional differences in retinoid release from embryonic neural tissue detected by an in vitro reporter assay. *Development* **116:** 55.

Wagner M., Thaller C., Jessell T., and Eichele G. 1990. Polarizing activity and retinoid synthesis in the floor plate of the neural tube. *Nature* **345:** 819.

Winning R. and Sargent T. 1994. *Pagliaccio*, a member of the *Eph* family of receptor tyrosine kinase genes, has localized expression in a subset of neural crest and neural tissues in *Xenopus laevis* embryos. *Mech. Dev.* **46:** 219.

Wolda S., Moody C., and Moon R. 1993. Overlapping expression of *Xwnt-3A* and *Xwnt-1* in neural tissue of *Xenopus laevis* embryos. *Dev. Biol.* **155:** 46.

Zhang M., Kim H.-J., Marshall H., Gendron-Maguire M., Lucas D., Baron A., Gudas L., Gridley T., Krumlauf R., and Grippo J. 1994. Ectopic *Hoxa-1* induces rhombomere transformation in mouse hindbrain. *Development* **120:** 2431.

Zimmerman L.B., De Jesus-Escobar J.M., and Harland R.M. 1996. The Spemann organizer signal *noggin* binds and inactivates bone morphogenetic protein 4. *Cell* **86:** 599.

Genetic Analysis of Pattern Formation in the Zebrafish Neural Plate

W. DRIEVER, L. SOLNICA-KREZEL,[1] S. ABDELILAH,[2] D. MEYER, AND D. STEMPLE[3]
University of Freiburg, Department of Developmental Biology, Hauptstrasse 1, D-79110 Freiburg, Germany, and CVRC/MGH, Charlestown, Massachusettes 02129

The first attempts toward an understanding of the organization of the vertebrate brain were purely based on morphological analysis and led to the definition of the major domains of the central nervous system (CNS) along the rostrocaudal or anteroposterior (A/P) axis, as they are visible due to undulations and swellings (van Baer 1828; von Kupfer 1906). Common to vertebrates is the tripartite structure of the brain: forebrain (prosencephalon, later divided into telencephalon and diencephalon), the midbrain (mesencephalon, with the tegmentum and the tectum), and the hindbrain (rhombencephalon, with the metencephalon being the rostral most narrow domain and the myelencephalon being divided into rhombomeres). Caudal to the hindbrain extends the spinal cord. Along the dorsoventral (D/V) axis, a unifying concept of domains was not easy to obtain and was restricted to the naming of the anatomical domains within each rostrocaudal section of the brain.

During the last decade, analysis of expression patterns of transcription factors and other regulatory genes has indicated that some of the morphological parts might also represent developmental units. This has been most obvious in the hindbrain, where expression patterns of *hox* genes as well as various members of signaling pathways appear to be restricted to individual or defined subsets of rhombomeres (for review, see Lumsden and Krumlauf 1996). For the forebrain, such studies have suggested an organization in neuromeric domains, called prosomeres (Figdor and Stern 1993; Puelles and Rubenstein 1993; Rubenstein et al. 1993). However, clonal behavior and cell migration do question the functional significance of these domains (for review, see Lumsden and Krumlauf 1996). Within the spinal cord, *hox* genes have sharp anterior borders of expression and might define functional domains. However, within the midbrain, no clear neuromeric domains can be distinguished.

The combination of experimental embryology and genetic studies has provided the first insights into the control of pattern formation during development of the CNS. At present, it appears that in a first step, global A/P pattern is established in the neurectoderm, and as a second step, local signaling centers refine patterning in subregions of the brain. In the classic concept of neural induction and patterning, which is still subscribed to today, the organizer initially induces tissue of anterior character (activation) (Spemann 1938) that subsequently becomes posteriorized by signals from later involuting mesoderm (transformation; Nieuwkoop et al. 1952). However, candidate molecules for the activation signal, noggin and chordin, all bind and inactivate bone morphogenetic proteins (BMPs) and point toward a role in inhibition of inhibitors of neural development, BMP2 and BMP4, and thus neural induction might be permissive rather than instructive (Piccolo et al. 1996; Zimmerman et al. 1996). However, it has not been possible so far to separate induction of dorsal from induction of neural structures. Candidate molecules for the transforming signal are retinoic acid and basic fibroblast growth factor (bFGF) (Cox and Hemmati-Brivanlou 1995; Papalopulu and Kintner 1996).

Whereas other regions of the neural plate, like the zona *limitans* in the forebrain, are suspected to function as local signaling centers, such an activity has so far been demonstrated only for the isthmus, the region of the midbrain-hindbrain boundary. Transplantation of the isthmic region to forebrain or hindbrain induces mesencephalic or cerebellar structures, respectively (Bally-Cuif and Wassef 1994; Marin and Puelles 1994). A good candidate for this signaling activity is FGF8, which is expressed in the isthmic region and ectopically induces organized and polar midbrain structures (Crossley et al. 1996). Downstream from FGF8, the genes *wnt1*, *engrailed 2*, *pax-2*, *pax-5*, and *pax-8* appear to be involved in elaborating the pattern and in maintaining it (for review, see Joyner 1996), as judged from induced mutations in mice.

The initial mediolateral pattern of the neural plate and later D/V pattern of the neural tube appear to be influenced by at least two signaling sources. On the medial part (and later ventral side), axial mesoderm is the source of signals that pattern the ventral neural tube, inducing

Present addresses: [1]Department of Molecular Biology, Vanderbilt University, Nashville, Tennessee 37235; [2]Howard Hughes Medical Institute, UCSF, San Francisco, California 94143-0724; [3]Division of Developmental Biology, National Institute for Medical Research, The Ridgeway, Mill Hill, London NW7 1AA, Great Britain.

structures such as the floorplate. One of the signaling molecules in this pathway is sonic hedgehog (*shh*), which induces floorplate and motoneurons, among others (Echelard et al. 1993; Roelink et al. 1995). Mice with an inactivated *shh* gene are deficient in ventral CNS structures (Chiang et al. 1996). It must be noted that the axial mesoderm underlying the rostral brain, the prechordal plate, appears to have different anterior signaling capabilities (Adelmann 1936; for review, see Doniach 1995), when compared to the notochord in the caudal brain and spinal cord. At the lateral margin of the neural plate, and later in the dorsal neural tube, signals from the adjacent ectoderm, most likely BMP4 and BMP7 (Liem et al. 1995), are involved in specification of dorsal cell types and the roofplate. However, although it becomes apparent that during development, patterning of the neuroectoderm involves establishing a Cartesian coordinate system based on A/P and D/V signaling inputs, many questions remain open. For example, mechanisms that pattern the forebrain, as well as the mechanisms by which the local signaling centers like the midbrain hindbrain boundary are positioned in the neural plate, are unknown. To identify novel genes that are involved in as yet unknown steps of vertebrate neural patterning, we have performed a genetic screen in zebrafish and report here on the analysis of the role in neural patterning of some of the mutations isolated during this screen.

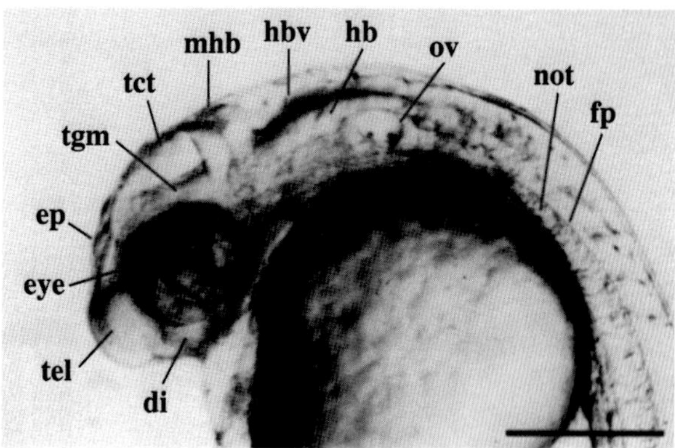

Figure 1. The zebrafish CNS at 28 hr of development. In a lateral view, many structures can be easily identified in living zebrafish embryos under the dissecting stereomicroscope: (tel) Telencephalon; (di) diencephalon with the ventrally located hypothalamus and the epiphysis (ep); (tgm) tegmentum; (tct) tectum and tectal ventricle; (mhb) midbrain-hindbrain boundary. (hb) Hindbrain; (hbv) hindbrain ventricle; (eye) eye with lens; (ov) otic vesicle with two otoliths. The otic vesicle is located lateral to rhombomere 5. (not) Notochord. The notochord extends anteriorly to the level of the otic vesicle. (fp) Floorplate. The floorplate extends anteriorly into the caudal diencephalon. Anterior is to the left and dorsal is up. Bar, 250 μm.

NEURAL DEVELOPMENT IN ZEBRAFISH

The shield region, which develops on the dorsal side of the early zebrafish gastrula, has been demonstrated to be the equivalent of the amphibian organizer (Oppenheimer 1936b; Shih and Fraser 1996). The exact location of the earlier Nieuwkoop center is unknown, but experiments indicate that its activity may reside within the dorsal yolk syncytial layer, underlying the shield region (Mizuno et al. 1996). During gastrulation, the neural plate is spread out on the dorsal side of the embryo, whereas neurulation starts only at later stages of gastrulation (yolk plug closure) and proceeds by infolding of the neural keel (Papan and Campos-Ortega 1994; Kimmel et al. 1995).

Neural patterning and morphogenesis proceed very fast in zebrafish. By the 18-somite stage (18 hours postfertilization), ten distinct swellings become obvious along the neural rod: rostrally the primordia of tel-, di-, and mesencephalon and caudally the seven rhombomeres. After 1 day of development (Fig. 1), the main domains of the brain can be easily recognized. The zebrafish embryo remains completely transparent during these stages, providing for easy identification of morphological features as well as for monitoring cellular behavior during normal development or in mutations affecting neural development.

GENETIC ANALYSIS OF NEURAL DEVELOPMENT

One of the great advantages of systematic genetic screens performed in invertebrates is that not only individual new loci, but also, in many cases, mutations in several or all genes constituting a regulatory pathway were identified, such that an in-depth understanding of molecular regulation is now possible for many aspects of invertebrate development (Nüsslein-Volhard and Wieschaus 1980). Such approaches have previously not been feasible in vertebrates, due to the inaccessibility of intrauterine embryos in mice, then the only well-developed vertebrate genetic system. However, the zebrafish system, with short generation time, ease of breeding, a large number of progeny from single crosses, and transparent embryos at all relevant developmental stages, made systematic large-scale screens for mutations possible (Driever et al. 1996; Haffter et al. 1996). During these screens, a total of more than 6000 embryonic or larval lethal mutations have been identified, about 2000 affecting specific aspects of development, rather than resulting in general growth retardation or degeneration. Most of the 2000 mutations were further characterized at a genetic level, and it appears that at least 600 developmentally important genes have been defined. There was a strong focus on the analysis of neural structures in the zebrafish embryo, and mutations affecting various aspects of neural development have been isolated: early patterning of the neural plate and neuraxis (Brand et al. 1996a,b; Hammerschmidt et al. 1996b; Heisenberg et al. 1996; Mullins et al. 1996; Schier et al. 1996), neurogenesis (Jiang et al. 1996), neuronal cell death (Abdelilah et al. 1996; Furutani-Seiki et al. 1996), sense organ development (Malicki et al. 1996a,b; Whitfield et al. 1996), neuronal pathfinding (Baier et al. 1996; Karlstrom et al. 1996; Trowe et al. 1996), behavior (Granato et al. 1996), and neurophysiology (Brockerhoff et al. 1995). Table 1 lists mutations that affect neural patterning at early stages.

Those first large screens were based on visual inspec-

Table 1. Mutations Affecting Early Neural Patterning

Class / Locus	Abr.	Main phenotype	References
Dorsoventral			
dino	*din*	ventralized, narrow neurectoderm	Hammerschmidt et al. (1996)
swirl	*swl*	dorsalized, expanded neurectoderm	Mullins et al. (1996)
somitabun	*sbn*	dorsalized, expanded neurectoderm	Mullins et al. (1996)
snailhouse	*snh*	dorsalized, expanded neurectoderm	Mullins et al. (1996)
Dorsoventral -axial mesoderm		all mutations in this group: ventral CNS affected various degrees of cyclopia	
bozozok	*boz*	prechordal plate, notochord and floorplate missing, neurectoderm anterior truncation	Solnica-Krezel et al. (1996)
one eyed pinhead	*oep*	prechordal plate and floorplate missing	Schier et al. et al. (1997)
cyclops	*cyc*	prechordal plate reduced and floorplate reduced	Hatta et al. (1991b); Thisse et al. (1994); Brand et al. (1996b); *unf* Schier et al. (1997)
schmalspur	*sur*	posterior part of prechordal plate reduced	
silberblick	*slb*	posterior part of prechordal plate reduced	Heisenberg and Nüsslein-Volhard (1997)
squint	*sqt*	prechordal plate reduced	Heisenberg and Nüsslein-Volhard (1997)
sleepy	*sly*	notochord differentiation, CNS defects	Schier et al. (1996)
bashful	*bal*	notochord differentiation, CNS defects	Schier et al. (1996)
grumpy	*gup*	notochord differentiation, CNS defects	Schier et al. (1996)
Dorsoventral -ventral CNS		all mutations in this group: floorplate is reduced, abnormal morphology	
chameleon	*con*	posterior cyclopia	Brand et al. (1996b)
you-too	*yoy*	posterior cyclopia	Brand et al. (1996b)
iguana	*igu*	posterior cyclopia	Brand et al. (1996b)
detour	*dtr*	posterior cyclopia	Brand et al. (1996b)
sonic-you	*syn*		Brand et al. (1996b)
schmalhans	*smh*		Brand et al. (1996b)
monorail	*mo*		Brand et al. (1996b)
Forebrain			
masterblind	*mbl*	telencephalon, eyes and nose absent; posterior forebrain expanded (epiphysis)	Heisenberg et al. (1996)
knollnase	*kas*	telencephalon: roofplate does not form	Heisenberg et al. (1996)
bozozok	*boz*	prechordal plate and notochord missing, neurectoderm anterior truncation	Solnica-Krezel et al. (1996)
Midbrain		mutations affecting the midbrain-hindbrain-boundary region	
acerebellar	*ace*	absence of isthmus and cerebellum	Brand et al. (1996a)
no isthmus	*noi*	absence of isthmus, tectum, cerebellum	Brand et al. (1996a)
spiel ohne grenzen	*spg*	absence of isthmus and cerebellum	Schier et al. (1996)
big head	*bid*	enlarged midbrain	Jiang et al. (1996)
Hindbrain			
valentino	*val*	required for rhombomere 5 and 6 formation	Moens et al. (1996)
flachland	*fll*	hindbrain neural tube thinner	Schier et al. (1996)
prachute	*pac*	hindbrain disorganized	Jiang et al. (1996)
atlantis	*atl*	extra neural bridges/folds in hindbrain	Jiang et al. (1996)
Neurogenesis			
mindbomb	*mib*	supernumerary primary neurons	Schier et al. (1996); *wit* Jiang et al. (1996)

This table presents a subset of mutations isolated during large-scale mutagenesis screens in Boston and Tbingen (Driever et al. 1996; Haffter et al. 1996) as well as from the Oregon zebrafish labs. Only mutations that might act during neural plate stages have been listed. Many other mutations have CNS phenotypes at later stages and are not shown here.

tion of living embryos and larvae and thus were limited in their ability to detect subtle phenotypes. Recently, several more specialized screens have been performed, in which mutations are identified by changes in gene expression patterns, as judged from whole-mount in situ hybridizations with probes indicative for neural patterning (Henion et al. 1996; Moens et al. 1996; B. Appel and J.S. Eisen; A.B. Chitnis; K. Artinger, and W. Driever; both unpubl.). A plethora of novel mutations is isolated in such screens, some with very subtle phenotypes, affecting only single types of neurons, and others with severe patterning defects, whose late visual phenotype would not obviously point to neural patterning abnormalities. The availability of these mutations and extensive means to study the function of the affected genes make zebrafish an organism of choice to study neural development in vertebrates.

AXIAL MESODERM AND D/V PATTERNING OF THE CNS

Head axial mesoderm—the prechordal mesoderm—and trunk and tail axial mesoderm—the notochord—have diverse and specific signaling activities in patterning the ventral CNS (Adelmann 1932, 1936; van Straaten et al. 1989; Yamada et al. 1991). In the anterior head, ventral diencephalon depends on prechordal plate signals, and in its absence, cyclopic embryos develop. In the trunk, notochord-derived signals induce floorplate and ventral neuronal cell types, motoneurons. The product of the *sonic hedgehog* gene appears to mediate some of these activities in the trunk and head (Echelard et al. 1993; Riddle et al. 1993; Roelink et al. 1994). Many zebrafish mutations have been identified that affect this patterning cascade at various steps.

Formation of both prechordal plate and notochord is affected in *bozozok (boz)* mutant embryos (Solnica-Krezel et al. 1996). *boz* mutant embryos display quite variable phenotypes, but most frequently, ventral CNS structures are deleted in head and trunk, with cyclopic eyes, or infrequently complete absence of eyes and forebrain structures (Fig. 2). *boz* is the only mutation isolated during the genetic screens to delete both floorplate and motoneurons in the trunk. Thus, a severe and early depletion of both head and trunk axial mesoderm appears to be one of the prerequisites to ablate motoneuron formation, which are not depleted in any of the mutations affecting prechordal plate or notochord only (described below).

Prechordal plate is absent in *one-eyed pinhead (oep)* and severely reduced in *cyclops (cyc)* mutants (Fig. 2) (Thisse et al. 1994; Schier et al. 1997). Interestingly, *oep* also affects formation of head endoderm, indicating a genetic link between the specification of head endoderm and axial mesoderm in patterning the anterior embryo. Predominantly, the posterior prechordal plate is affected in *schmalspur (sur*; formerly named *uncle freddy)* (Fig. 2) and *silberblick (slb)* embryos (Schier et al. 1996; Heisenberg and Nüsslein-Volhard 1997). A common feature of all zebrafish mutations affecting prechordal plate development is absence of or very fragmentary formation of the floorplate. Whether this reflects a requirement for the prechordal plate in induction of floorplate precursor cells, a common developmental pathway, or pleiotropic gene function, is still unknown. It should be noted that the function of *oep* does not appear to be restricted to anterior development: *no tail-oep* double-mutant embryos reveal a role for *oep* gene activity in the formation of somitic precursors as well as an interaction with *no tail* (*ntl*) in specifying chordamesoderm (Schier et al. 1997).

Notochord and chordamesoderm are absent in *floating head (flh)*, the zebrafish homolog of Xnot (Talbot et al. 1995), but, surprisingly, floorplate (albeit interrupted) and ventral neuronal cell types develop. This has been interpreted as being the result of very early induction of these cell types in the embryonic shield by axial mesoderm; indeed, *shh* is transiently expressed at the gastrula margin. Other genes affecting notochord differentiation appear to act downstream, with *ntl* affecting, among others, differentiation of early chordamesoderm precursor cells (Halpern et al. 1993; Schulte-Merker et al. 1994). Analysis of double mutants has suggested that Ntl may be important for early cell fate decisions for cells to form either chordamesoderm or floorplate precursor cells, Ntl antagonizing floorplate development and promoting notochord development (Halpern et al. 1997). Other mutations, *sleepy, bashful,* and *grumpy,* affect both notochord differentiation and brain patterning (Schier et al. 1996; Stemple et al. 1996).

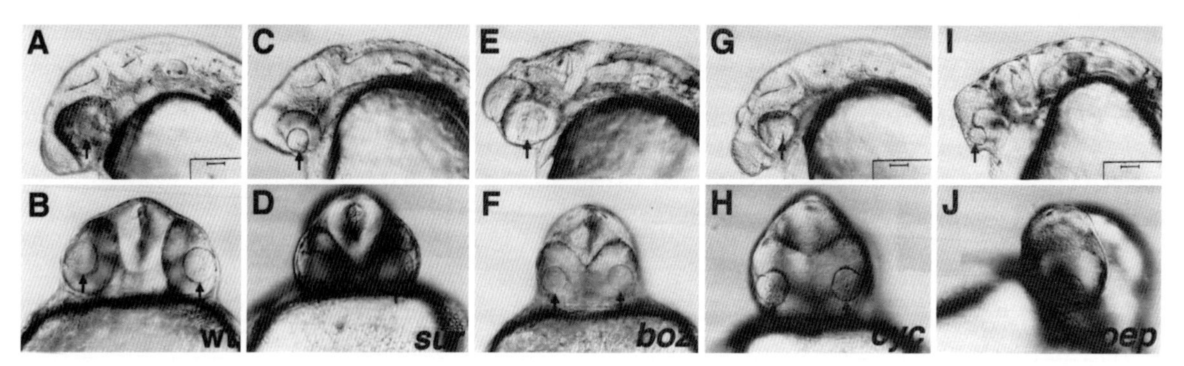

Figure 2. Mutations affecting D/V patterning. Phenotypes of mutations affecting the formation of ventral neuroectoderm on day 1 of development. Lateral (*A,C,E,G,I*) and A/V (*B,D,F,H,J*) views of wild-type (*A,B*), *schmalspur* sur^{m768} (*C,D*), *bozozok* boz^{m168} (*E,F*), *cyclops* cyc^{m122} (*G,H*), and *one eyed pinhead* oep^{m134} (*I,J*) mutant embryos at 28 hpf. Arrows indicate the position of the lens. (Modified from Schier et al. 1996.)

A further group of mutations has been isolated due to its developmental defects in floorplate development and formation of embryonic structures that depend on midline signaling. These include *chameleon, you-too, iguana, detour, sonic-you, schmalhans,* and *monorail* (Brand et al. 1996b). It has been suggested that these genes may act to maintain proper structure and inductive function of zebrafish midline tissues; as such, several of them may be components of the signaling pathway involving vertebrate *hedgehog* homologs.

THE ORGANIZER AND A/P PATTERNING OF THE NEURAL PLATE

Extensive experimental evidence establishes the role of the organizer in induction of neural structures in amphibians (see above). Several reports shed light on shield function in A/P patterning in teleosts, and possibly vertebrates in general (Oppenheimer 1936a; Ho 1992; Shih and Fraser 1996; D.L. Stemple and W. Driever, unpubl.). These experiments establish homology in organizer activity between the amphibian dorsal blastopore lip and the zebrafish shield. Transplantation of shield tissue into ventral margin results in formation of an ectopic axis; both donor and host tissue can contribute to the forming neuraxis. The extent to which a second neuraxis can form depends on the experimental procedure. Using surgical transplantation techniques similar to those used in amphibians, Shih and Fraser (1996) obtained incomplete secondary axes, which extended rostrally no further than the region of the hindbrain, with otic vesicles being the most anterior structures reported. In this experimental setup, experimental axis was also incomplete in that formation of hearts was observed in only 3 of 240 experimental embryos. In contrast, a new transplantation technique, developed by D. Stemple (D.L. Stemple and W. Driever, unpubl.), involves the use of specially prepared glass micropipettes to "stamp out" and transplant both superficial and deep parts of the shield. This technique yields a high percentage of experimental embryos with complete secondary axis, including forebrain and eyes (Fig. 3) (D.L. Stemple and W. Driever, unpubl.). The main difference between the two experimental approaches appears to be the amount of deep cells of the shield to be transplanted, cells that are fated to be mesendodermal precursors. It appears that the deep cells would be necessary for induction of anterior neurectoderm; these cells are more closely related to deep endodermal and mesendodermal cells in amphibians, based on their position during gastrulation, than to dorsal blastopore lip cells. Whether these deep mesendodermal cells are involved in induction of anterior neural structures requires further testing. A role for anterior primitive endoderm in patterning the anterior neural plate has recently been demonstrated in the mouse embryo (Thomas and Beddington 1996) and has been suggested in rabbits (Viehbahn et al. 1995). This area of the embryo is characterized by the expression of *Hesx1* in mice (Thomas and Beddington 1996), a homolog of XANF1 (Mathers et al. 1995) and XANF2 (Zaraisky et al. 1995) in *Xenopus*

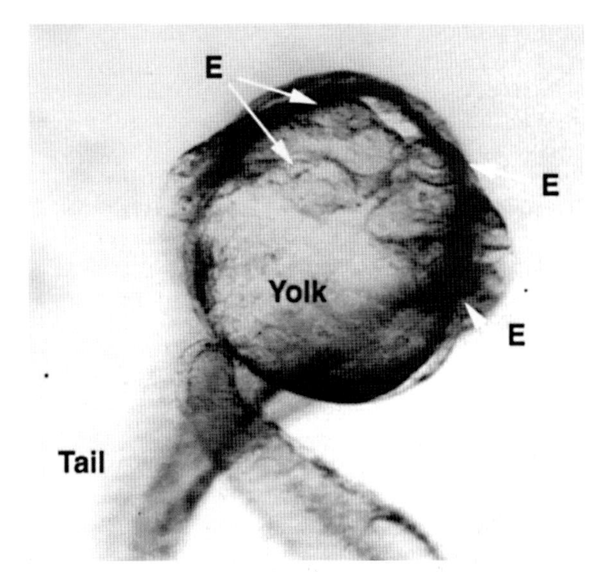

Figure 3. Axis duplications caused by shield transplantations. Isochronic transplantation of shield results in production of two nearly complete embryonic axes. Formation of the head in the secondary axis is normal (E = eye). The embryos are fused at the tail (Tail). The structure at the middle, bottom of image is a tail from a different embryo.

that are first expressed in deep cells of the organizer, deep yolky endomesodermal cells, which include progenitors of the anterior gut and do not have to ingress during gastrulation.

Informative also is the effect on development of experimental extirpation of the zebrafish shield. Both Shih and Fraser (1996) and W. Driever and D.L. Stemple (unpubl.) obtain similar results using either microsurgical removal or "stamp-out" of the shield regions: Experimental embryos develop all A/P elements of the neuraxis, but they lack derivatives of axial mesoderm in head and trunk. As such, they very much resemble intermediate phenotypes at the *boz* locus (Solnica-Krezel et al. 1996). Somewhat unexpectedly, embryos thus lacking both prechordal plate and notochord can form normal A/P pattern. It is possible that deep mesendodermal cells of the organizer region, which may contribute to anterior neural patterning as judged from the transplantation experiments, may not have been effectively removed by the extirpation techniques. Alternate explanations for this observation are that A/P pattern might already be initiated before shield formation in the very early gastrula or that the organizer activity is not as strictly localized as the morphological shield. Further refinement of pattern might depend on interactions within the neurectoderm, and/or with lateral tissues, rather than the presence of axial mesoderm. Support for this idea comes from the elaborate fate map of the CNS that can already be found at early gastrula stages (Woo and Fraser 1995). Evidence for a role of nonaxial signals in regionalization of the zebrafish CNS is also suggested by transplantation studies (Woo and Fraser 1997).

Why is it that so few mutations have been isolated during the zebrafish screens to affect establishment of early anterior neurectoderm? Is it that many of the interactions are performed by maternal gene products or are redundant genetic pathways involved in patterning the zebrafish CNS? On the basis of experimental analysis in amphibians, and gene expression patterns in various organisms, it is to be expected that many of the steps involved in A/P neural patterning will involve zygotic gene expression. The set of zygotic zebrafish mutations that set up initial D/V pattern in the zebrafish gastrula, analogous to the antagonism between BMPs and chordin/noggin/follistatin, affect the dorsolateral extent of the neural plate but have no distinct A/P patterning phenotype (Hammerschmidt et al. 1996a,b; Mullins et al. 1996; Fisher et al. 1996; Schulte-Merker et al. 1997; Kishimoto et al. 1997) However, although most of the neural patterning mutations isolated during the genetic screens were initially classified as affecting axial mesoderm formation and D/V patterning of the CNS, detailed genetic analysis reveals involvement of some of the mutated genes in initiating proper A/P neural patterning as well.

Mutations at the *boz* locus (Solnica-Krezel et al. 1996) display a phenotype of quite variable expressivity, ranging from cyclopia and notochord defects in the trunk (see Fig. 2) to complete absence of axial mesoderm and truncation of anterior neurectodermal structures (see Fig. 4) (L. Solnica-Krezel, unpubl.). As expected for embryos devoid of axial mesoderm, severe ventral CNS deficiencies develop in *boz* mutant embryos. The *boz* mutation is the only known zebrafish mutation that does not develop primary motoneurons. Very strong *boz* mutant embryos do not develop eyes or forebrain structures, as judged from the analysis of genes expressed in wild-type forebrain. Furthermore, midbrain tissue is reduced. It appears as if the A/P extent of the rhombencephalon is severely extended at the cost of rostral CNS. Analysis of genes expressed in a rhombomere-specific fashion (*sek1* in Fig. 4) indicates a vast anterior shift of rhombomere 1 and a strong broadening along the A/P axis and anterior shift of rhombomeres 2 through 5. Thus, the mutation behaves to some degree similar to that of mutations in the maternal morphogen *bicoid* in *Drosophila* (Driever and Nüsslein-Volhard 1988a,b). In mutant embryos, anterior fates are depleted, and the fate map at more posterior positions is expanded and shifted anteriorward.

The strong mutant phenotype occurs only in a small percentage of all embryos, but certain genetic backgrounds can stabilize and enhance the strong phenotype. In crosses of fish double heterozygous for *boz* and either cyc^{b16} or sur^{m768}, the strong phenotypes segregate at ratios close to those expected for the double mutants. Whether the strong phenotypes actually represent double homozygous mutant embryos, or a contribution (haploinsufficiency in the background of *boz/boz*) of heterozygosity at either *cyc* or *sur*, still needs to be determined. The earliest defects that can be detected in *boz* and *cyc* affect the expression of organizer genes (Fig. 5): *goosecoid* expression appears to be normal on one side of the embryo after mid blastula transition, but early during gastrulation, it is reduced in *cyc* (Thisse et al. 1994) and disappears completely in *boz* (Solnica-Krezel et al. 1996). In *sur*, *gsc* expression appears not to be affected, but *hlx-1* (Fjose et al. 1994) expressed in the wild-type posterior prechordal plate is detected at reduced levels only (D. Meyer and W. Driever, unpubl.), as is *fkd2* (Brand et al. 1996b). *gsc* expression is a direct response to maternal signals (Cho et al. 1991), and thus it appears that *boz* and *cyc* are involved in maintaining proper organizer function. Whether these genes act primarily in proper forma-

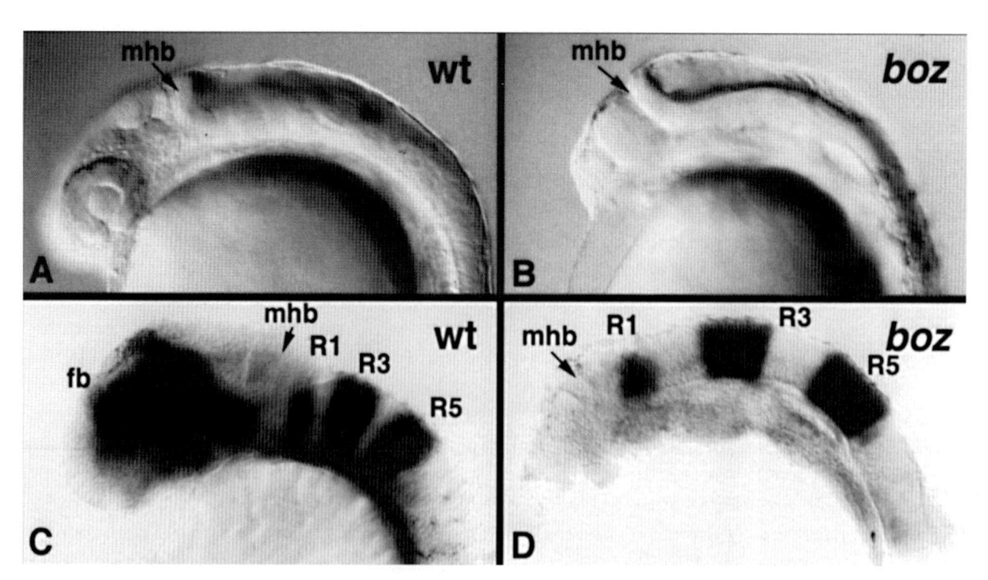

Figure 4. A/P neural patterning defects in mutations affecting the organizer. Defects in A/P patterning of the brain in *boz* embryos that express a strong mutant phenotype. (*A,B*) Nomarski images of live embryos at 30 hpf. Eyes do not develop, and most of the distance between the midbrain-hindbrain boundary and anterior end of an embryo is decreased in *boz* (*B*) relative to the wild-type embryo (*A*). (*C,D*) Expression pattern of a *sek1* gene in 30 hpf embryos detected by whole-mount in situ hybridization. The forebrain expression domains of the *sek1* gene (*C*) are missing in *boz* embryos (*D*). Furthermore, the hindbrain expression domains of this gene in rhombomeres 1, 3, and 5 appear enlarged.

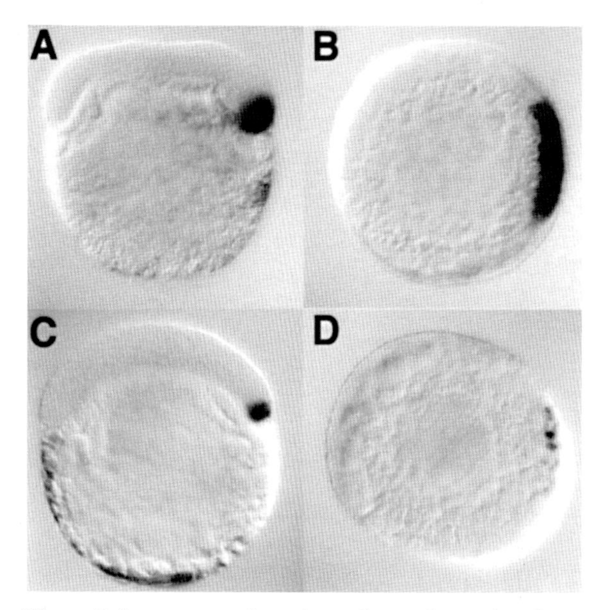

Figure 5. *boz* mutant embryos have abnormal organizer function. Expression of *goosecoid* is already affected at the onset of gastrulation in *boz* mutant zebrafish embryos. Lateral (*A,C*) and animal pole views (*B,D*) of wild-type (*A,B*) and *boz* mutant (*C,D*) embryos at 50% epiboly. *gsc* was visualized by whole-mount in situ hybridization with digoxygenin-labeled probes. Note that the number of *gsc*-expressing cells, and the expression level, are severely reduced even before formation of the embryonic shield.

tion of the shield organizer or are involved in the establishment of other signaling centers important for anterior neurectoderm remains to be determined.

EARLY ORGANIZING CENTERS IN THE NEURAL PLATE: THE MIDBRAIN HINDBRAIN BOUNDARY

One signaling center of the neural plate is the midbrain hindbrain boundary (MHB, see introduction). However, although it has been demonstrated that an isthmic organizing center can be induced by FGF8 (Crossley et al. 1996), it is currently unknown how FGF8 signaling activity is localized in the first place. During further development, *pax-2, -5, and -8, wnt-1,* and *en* appear to be mutually required to maintain the MHB and to specify tissue identity.

In zebrafish, three loci, *spiel ohne grenzen (spg), no isthmus (noi),* and *acerebellar (ace),* that affect the formation of the MHB have been described (Brand et al. 1996a; Schier et al. 1996). Mutations in *noi* have been linked to the *pax[zf-b]* locus. The neuroectodermal phenotype of the strongest *noi* alleles is a loss of the MHB constriction and the cerebellum. An analysis of the brain organization using various marker genes of the MHB indicates that both the dorsal and ventral aspects of the MHB are absent or respecified by the 20-somite stage (Brand et al. 1996a). Furthermore, the distance between cell populations within the prosencephalon and the first rhombomere is decreased, providing additional evidence for a reduced amount of mesencephalic tissues. At the 20-somite stage, the di-/mesencephalic boundary, as assessed by *pax[zf-a]* expression, appears to be normal (Brand et al. 1996a). In *noi* mutants, the tectum (dorsal mesencephalon) degenerates by an apoptotic mechanism during late somitogenesis. Outside of the brain, *noi* mutants lack a pronephric duct and develop a defective circulation.

A second locus, *acerebellar* (*ace*), results in a less severe mutant phenotype (Brand et al. 1996a). Marker analysis indicates that *ace* might be required later during development than *noi* and might therefore be a downstream target. Most importantly, the MHB constriction and the cerebellum are missing. However, a tectum is present, indicating that *ace* mutants possess activities that are sufficient to allow formation and survival of this tissue. In *ace* mutants, loss of the entire MHB is completed only later during somitogenesis (Brand et al. 1996a).

In *spg* mutant embryos, the MHB constriction and the cerebellum are lost except for some remaining dorsal rudiments (Fig. 6). Similar to *ace*, as assessed by morphological criteria in *spg* mutants, a tectum is present, suggesting that the dorsal MHB retains activities sufficient to allow the formation and survival of tectal tissues. Marker gene expression within the neighboring pros-, mes-, and rhombencephalic regions indicates that only *spg*, but not *noi,* mutants have distinct modulating effects on the positioning of the di-/mesencephalic boundary and the spatial organization within the hindbrain (Fig. 7). In contrast to *noi* mutant embryos, which develop an apparently correct spacing of the mesencephalic and anterior rhombencephalic region at this early stage (Brand et al. 1996a) in *spg* mutants, a characteristic ventroposterior shift of *pax[zf-a]* expression that marks the pros-/mesencephalic boundary (Macdonald et al. 1994) indicates a loss of ventral mesencephalic tissues. This phenotype might constitute an expansion of prosencephalic at the expense of ventral tectal and/or tegmentum tissues.

Within the hindbrain, the *spg* mutation has two effects that distinguish *spg* from *noi* mutant embryos. The first rhombomere is shifted anteriorward, and this effect is more pronounced on the ventral side, such that it obtains a tilted orientation in the dorsoposterior direction (S. Abdelilah et al., unpubl.). The effect can be so strong that ventral rhombomere tissue and ventral prosencephalic tissue appear to touch each other, with the ventral isthmus mostly or completely deleted. There are alterations in the size of subsequent rhombomeres. As indicated by expression patterns of *krx20*, rhombomeres r3 and r5 are shortened along the A/P axis, whereas the gap between the two *krx20* domains appears to be expanded. However, a final resolution to the exact spacing and size of the even-numbered rhombomeres requires an investigation using rhombomere-specific markers such as *Hoxb-1,* since the spacing of the gap between rhombomeres r3 and r5 does not necessarily reflect the size of the fourth rhombomere. Additionally, there is a loss of distinct *fgf8* expressing domains within the rhombencephalon (S. Abdelilah et al., unpubl.).

noi, *ace*, and *spg* all three affect the MHB. However, in contrast to *noi*, the *spg* mutation additionally affects the

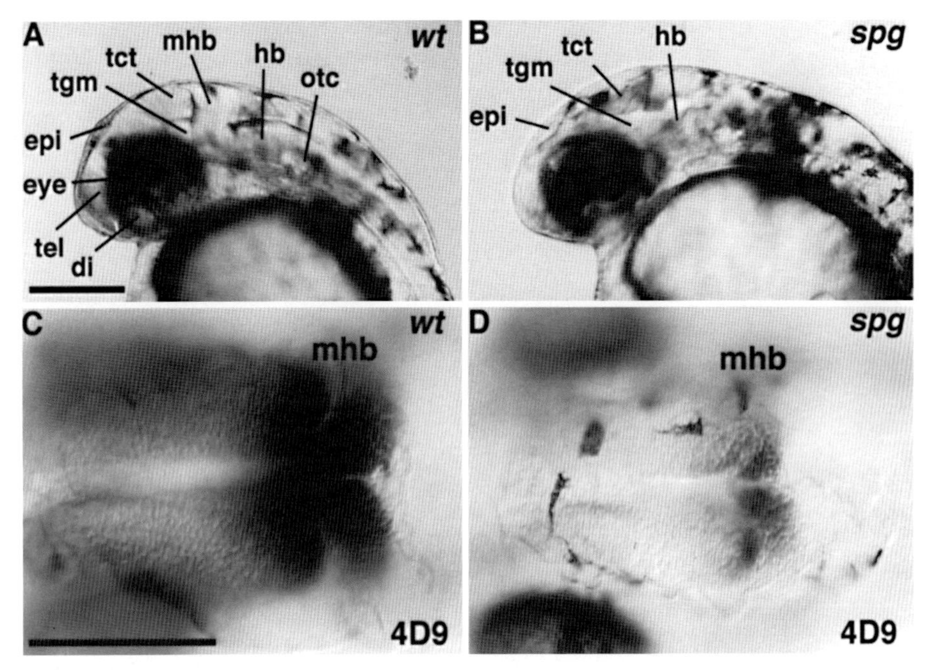

Figure 6. *spiel ohne grenzen*, a mutation affecting the midbrain-hindbrain boundary signaling center. The *spg* mutant phenotype. (*A,C*) Wild type at 30 hpf. (*B,D*) *spg* mutant phenotype at 30 hpf. The domain of the MHB is severely reduced in mutants. The tectum is present. (*C,D*) The constriction of the MHB visualized by the 4D9 antibody that recognizes all three engrailed proteins (Hatta et al. 1991a). (di) Diencephalon; (epi) epiphysis; (hb) hindbrain (met- and meyelencephalon); (mhb) midbrain-hindbrain boundary; (otc) otic vesicle; (tct) tectum; (tel) telencephalon; (tgm) tegmentum. Bars, 200 μm.

positioning of the boundaries between the prosencephalon and mesencephalon and between first and second rhombomere, as well as the correct spatial organization within the rhombencephalon. This phenotype is distinct from phenotypes of murine mutants of several genes that affect MHB formation, including *pax5* (Urbanek et al. 1994), *wnt-1 (*Thomas and Capecchi 1990; McMahon et al. 1992), and *engrailed-1* (Wurst et al. 1994) which display variable losses or reductions of the MHB and mesencephalic tissues. Similar to *noi* mutants, the size of the prosencephalon and the rhombomeric organization appears not to be altered in these murine mutants. The pleiotropic phenotype that distinguishes *spg* from other known mutants that affect the MHB formation raises several important issues: Does *spg* act at a higher level of genetic hierarchy? Where is spg function required? What is the function of *spg* during the regionalization of the anterior neural plate? In *Drosophila*, a genetic pathway that includes the genes *paired* and its downstream targets *engrailed* and *wingless* is required for the correct formation of the posterior compartment boundary within trunk segments (for review, see Lawrence 1992). As yet, it is not known whether the vertebrate homologs of these genes are arranged in a similar pathway during the formation of the MHB. In zebrafish, the expression of the *pax[zf-b]* gene precedes expression of *engrailed-2* and *wnt-1* (a vertebrate homolog of *wingless*) (Hatta et al. 1991a; Krauss et al. 1991a; Kelly and Moon 1995). Experimental evidence from injections of an antibody against *pax[zf-b]* (Krauss et al. 1992) and analysis of several mutant alleles of the zebrafish *noi* (*pax[zf-b]*) locus (Brand et al. 1996a) indicate that expression of the *engrailed-2* and *wnt-1* genes is reduced or absent without a functional pax[zf-b] protein. This evidence suggests that *engrailed-2* and *wnt-1* might be direct downstream targets of *pax[zf-b]*. Furthermore, recent work in mouse demonstrates that, similar to *Drosophila*, a major role of *wnt-1* signaling is the maintenance of *engrailed-1* (Danielian and McMahon 1996).

Several lines of evidence suggest that *spg* might act upstream of or parallel to *pax[zf-b]* with respect to the MHB formation. During normal development, expression of *pax[zf-b]* initiates at the late gastrula stages within the primordium of the MHB (Krauss et al. 1991a). We analyzed the expression of *pax[zf-b]* in *spg* mutant embryos during this initial phase. At the late gastrula stages, this expression is in a reduced and aberrant pattern in *spg* mutants. In comparison, the early *pax[zf-b]* expression is normal in *noi* mutants (Brand et al. 1996a), indicating that *spg* might be involved in the early regionalization within the neural plate and might precede *pax[zf-b]* activity. Loss of *spg* has similar effects on several markers that are expressed within the anlage of the MHB at the late gastrula stages, including *her5*, a zebrafish *Enhancer of split* homolog (von Weizsacker 1994; Müller et al. 1996). Similarly, in *spg* mutants, the presumptive rhombomeric region appears to be affected by the end of gastrulation as indicated by an altered *krx20* expression. The early phenotypes observed in *spg* mutants precede the expression of potential downstream factors of *pax[zf-b]* such as *engrailed-2* (Hatta et al. 1991a). Another line of evidence that argues for an upstream function of *spg* are the extended defects within the anterior neural plate, including the shifted pros-/mesencephalic boundary, that are not

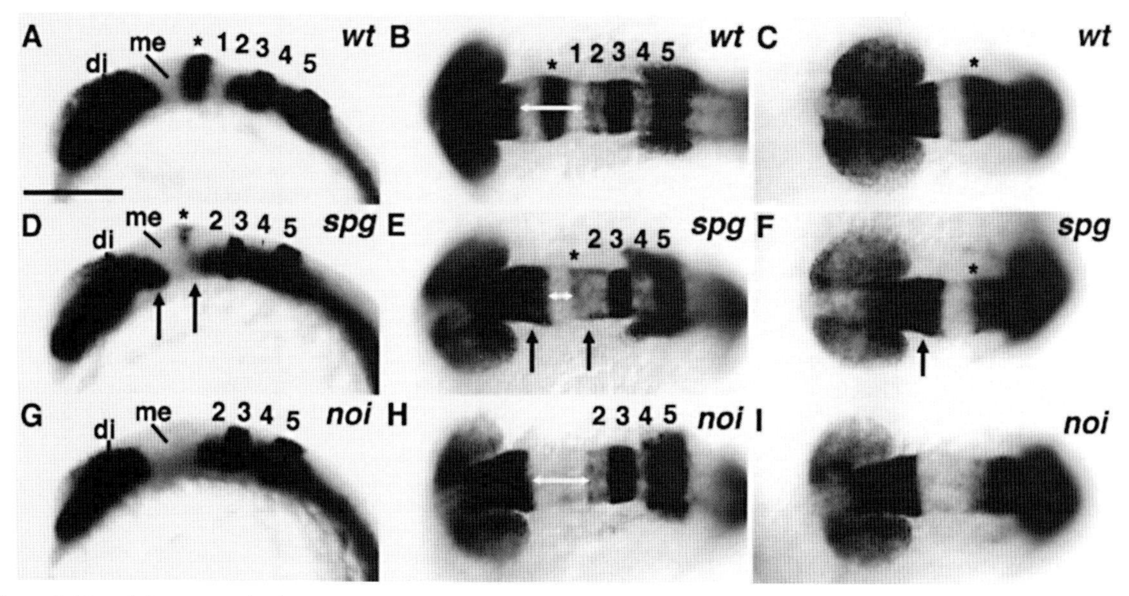

Figure 7. Neural plate patterning in *spg* mutant embryos. Comparative analysis of A/P pattern in wild-type, *noi-*, and *spg* mutant embryos at the 12–14-somite stage. Expression of *pax[zf-a]* and *[zf-b]* and *krx20* allows distinguishing the MHB (*asterisk*), the rhombomeric organization (*1–5*), and the diencephalic-mesencephalic boundary (between di and me). (*D–F*, *Closed arrows)* The expanded domains in *spg* mutant embryos. Note the severely reduced distance between the *pax[zf-a]* expression within diencephalon and rhombomere 2 in *spg* mutants (*open arrows*). (*A–C*) Wild type. (*D–F*) *spg* mutant phenotype; (*G–I*) *noi* mutant phenotype. (*A,D,G*) Lateral view; (*B,E,H*) dorsal view on mes-rhombencephalic regions; (*C,F,I*) dorsal view on mes-diencephalic regions. (di) Diencephalon; (me) mesencephalon. Bar, 200 μm

observed in *noi* mutants. Finally, the *spg* locus appears not to be a downstream element within the suggested *pax[zf-b]* pathway since the activation of several of the candidate downstream genes, including *engrailed-2*, is not disrupted per se in *spg* mutant embryos but rather restricted to domains that correspond to aberrant *pax[zf-b]* expression.

In summary, our data suggest that the *spg* locus is a partial regulator of *pax[zf-b]* and might be required for normal formation or specification of ventral isthmic tissues within the MHB primordium and mes-/rhombencephalic neural plate/tube. The inverse shifts of gene expression domains within rhombencephalic and ventral di-/mesencephalic tissues might indicate an abnormal pattern process, rather than a mis-specification as the underlying molecular defect. The phenotypes described for *spg* mutants are reminiscent of defects in gap genes in *Drosophila*. Loss of trunk gap gene function in *Drosophila* results in a deletion phenotype within the normal expression domain of this gene. In addition, there is an effect on the positioning of neighboring gap gene expression. For example, loss of activity of the anterior gap gene *hunchback* (*hb*) or the posterior gap gene *knirps* (*kni*) results in an anterior or posterior expansion of *Krüppel* (*Krü*) expression (Gaul and Jäckle 1987, 1991). These phenotypes reflect a mechanism of boundary formation that is enforced by mutual repression of gap gene expression (Pankratz and Jáckle 1993).

Whether or not the shifted expression domains of the zebrafish pair-rule homolog *pax[zf-a]* can be interpreted similar to the shifts of gap gene expression that have been described in *Drosophila* remains to be investigated. Future analysis of the *spg* mutation will provide important insights into the molecular mechanisms that are involved in the positioning of major subdivisions of the vertebrate neural plate/tube. We will have to await the cloning of the *spg* gene to investigate how *spg* regulates the regionalization of different regions within the anterior neural tube.

PROSPECTS FOR MOLECULAR ANALYSIS

Although the zebrafish mutations have already made important contributions to our understanding of vertebrate neural development by elucidating function of embryonic parts and mechanisms of interactions among parts, it will be important to characterize the mutated genes at a molecular level in order to tie their function into the network of genetic interactions that controls patterning of the nervous system. Extensive efforts are currently under way to map and clone zebrafish mutations.

Since a map of the zebrafish genome is available (Postlethwait et al. 1994; Johnson et al. 1996; Knapik et al. 1996; Postlethwait and Talbot 1997), it has been possible to determine relative locations of mutant genes and genes cloned by means of homology with other vertebrates or flies. The candidate gene approach has made it possible to assign genes to a number of mutant loci very quickly (see, e.g., Talbot et al. 1995). What is the prospect of those genes for which no candidates can easily be identified, since they might represent evolutionary young vertebrate-specific additions to the repertoire of developmental control genes? First, the evaluation of syntenic relationships between fish and the well-characterized mammalian genomes will provide a plethora of candidate genes whose functions are still unknown (Nadeau 1989; Postlethwait and Talbot 1997). Second, the tools are now in place to clone genes by positional means: Whereas the combined average distance between bins of genetic markers (RAPDs, SSLPs, STSs, cloned genes) is still 2–3 cM, methods are established to create systematically local high-density maps by identification of AFLP markers (Vos et al. 1995; D. Meyer and W. Driever, unpubl.). Furthermore, excellent genomic libraries are available: cosmids (H. Lehrach, German Resource Center), BAC, and PAC as well as YAC libraries (Zon, Fishman, Amemiya, distributed through Genome Systems, Inc., Research Genetics, Inc., and the German Genome Resource Center). Thus, it is not surprising that essentially all the mutant loci mentioned here have been mapped to defined regions of the zebrafish genome, and positional cloning efforts are under way for all mutations described in depth. For example, for *boz*, *spg*, and *unf*, regional high-density maps based on AFLP markers have been established. The ease with which large panels of mutant progeny can be generated (2500 for some of the crosses) makes very-high-resolution meiotic mapping possible, down to 0.1 cM and less, equivalent to only about 50–80 kb (D. Meyer et al., unpubl.). Following mapping to such resolution, genomic clones spanning the proximal markers are identified and sequenced as a whole over the closest interval containing the mutant locus, to identify open reading frames, or used in hybridization reactions to identify cDNA. The first positional cloning ever of a zebrafish gene, *one-eyed pinhead,* was reported at this CSH Symposium by the laboratory of Alexander Schier (unpubl.). It is to be expected that over the next years, the function of many interesting genes will be defined by zebrafish genetics and molecular biology.

ACKNOWLEDGMENTS

We thank Heinz-Georg Belting and Gieselbert Hauptmann for critical reading of the manuscript. This work was supported by grants from the National Institutes of Health to W.D. D.M. was an EMBO fellow, and D.S. was supported by the Helen Hay Whitney Foundation.

REFERENCES

Abdelilah S., Mountcastle-Shah E., Harvey M., Solnica-Krezel L., Schier A.F., Stemple D.L., Malicki J., Neuhauss S.C.F., Zwartkruis F., Stainier D.Y.R., Rangini Z., and Driever W. 1996. Mutations affecting neural survival in the zebrafish, *Danio rerio*. *Development* **123:** 217.

Adelmann H.B. 1932. The development of the prechordal plate and mesoderm of amblystoma punctatum. *J. Morphol.* **54:** 1.

_______. 1936. The problem of cyclopia, part II. *Q. Rev. Biol.* **11:** 284.

Baier H., Klostermann S., Trowe T., Karlstrom R.O., Nüsslein-Volhard C., and Bonhoeffer F. 1996. Genetic dissection of the retinotectal projection. *Development* **123:** 415.

Bally-Cuif L. and Wassef M. 1994. Ectopic induction and reorganization of *Wnt-1* expression in quail/chick chimeras. *Development* **120:** 3379.

Brand M., Heisenberg C.P., Jiang Y.-J., Beuchle D., van-Eeden F.J.M., Furutani-Seiki M., Granato M., Haffter P., Hammerschmidt M., Kane D.A., Kelsh R.N., Mullins M.C., Odenthal J., and Nüsslein-Volhard C. 1996a. Mutations in two zebrafish genes affect the formation of the boundary between midbrain and hindbrain. *Development* **123:** 179.

Brand M., Heisenberg C.-P., Warga R.M., Pelegri F., Karlstrom R.O., Beuchle D., Picker A., Jiang Y.J., Furutani-Seiki M., van-Eeden F.J.M., Granato M., Haffter P., Hammerschmidt M., Kane D.A., Kelsh R.N., Mullins M.C., Odenthal J., and Nüsslein-Volhard C. 1996b. Mutations affecting development of the midline and general body shape during zebrafish embryogenesis. *Development* **123:** 129.

Brockerhoff S.E., Hurley J.B., Janssenbienhold U., Neuhauss S.C.F., Driever W., and Dowling J.E. 1995. A behavioral screen for isolating zebrafish mutants with visual system defects. *Proc. Natl. Acad. Sci.* **92:** 10545.

Chiang C., Litingtung Y., Lee E., Young K.E., Cordon J.L., Westphal H., and Beachy P.A. 1996. Cyclopia and defective axial patterning in mice lacking *Sonic hedgehog* gene function. *Nature* **383:** 407.

Cho K.W.Y., Blumberg B., Steinbeisser H., and De Robertis E.M. 1991. Molecular nature of Spemann's organizer: The role of the *Xenopus* homeobox gene *goosecoid*. *Cell* **67:** 1111.

Cox W.G. and Hemmati-Brivanlou A. 1995. Caudalization of neural fate by tissue recombination and bFGF. *Development* **121:** 4349.

Crossley P.H., Martinez S., and Martin G.R. 1996. Midbrain development induced by FGF8 in the chick embryo. *Nature* **380:** 66.

Danielian P.S. and McMahon A.P. 1996. *Engrailed-1* as a target of the Wnt-1 signaling pathway in vertebrtae midbrain development. *Nature* **383:** 332.

Doniach T. 1995. Basic FGF as an inducer of anteroposterior neural pattern. *Cell* **83:** 1067.

Driever W. and Nüsslein-Volhard C. 1988a. The bicoid protein determines position in the *Drosophila* embryo in a concentration-dependent manner. *Cell* **54:** 95.

_______. 1988b. A gradient of bicoid protein in *Drosophila* embryos. *Cell* **54:** 83.

Driever W., Solnica-Krezel L., Schier A.F., Neuhauss S.C.F., Malicki J., Stemple D.L., Stainier D.Y.R., Zwartkruis F., Abdelilah S., Rangini Z., Belak J., and Boggs C. 1996. A genetic screen for mutations affecting embryogenesis in zebrafish. *Development* **123:** 37.

Echelard Y., Epstein D.J., St-Jacques B., Shen L., Mohler J., McMahon J.A., and McMahon A.P. 1993. *Sonic hedgehog*, a member of a family of putative signaling molecules, is implicated in the regulation of CNS polarity. *Cell* **75:** 1417.

Figdor M.C. and Stern C.D. 1993. Segmental organization of embryonic diencephalon. *Nature* **363:** 630.

Fisher S., Amacher S., and Halpern M.E. 1996. Loss of *cerebum* function ventralizes the zebrafish embryo. *Development* **124:** 1301.

Fjose A., Izpisúa-Belmonte J.C., Fromental-Ramain C., and Duboule D. 1994. Expression of the zebrafish gene *hlx-1* in the prechordal plate and during CNS development. *Development* **120:** 71.

Furutani-Seiki M., Jiang Y.-J., Brand M., Heisenberg C.P., Houart C., Beuchle D., van-Eeden F.J.M., Granato M., Haffter P., Hammerschmidt M., Kane D.A., Kelsh R.N., Mullins M.C., Odenthal J., and Nüsslein-Volhard C. 1996. Neural degeneration mutants in the zebrafish *Danio rerio*. *Development* **123:** 229.

Gaul U. and Jäckle H. 1987. Pole region-dependent repression of the *Drosophila* gap gene *Krüppel* by maternal gene products. *Cell* **51:** 549.

_______. 1991. Role of gap genes in early *Drosophila* development. *Adv. Genet.* **27:** 239.

Granato M., van-Eeden F.J.M., Schach U., Trowe T., Brand M., Furutani-Seiki M., Haffter P., Hammerschmidt M., Heisenberg C.-P., Jiang Y.-J., Kane D.A., Kelsh R.N., Mullins M.C., Odenthal J.J., and Nüsslein-Volhard C. 1996. Genes controlling and mediating locomotion behavior of the zebrafish embryo. *Development* **123:** 399.

Haffter P., Granato M., Brand M., Mullins M.C., Hammerschmidt M., Kane D.A., Odenthal J., van-Eeden F.J.M., Jiang Y.-J., Heisenberg C.-P., Kelsh R.N., Furutani-Seiki M., Warg R.M., Vogelsang E., Beuchle D., Schach U., Fabian C., and Nüsslein-Volhard C. 1996. The identification of genes with unique and essential functions in the development of the zebrafish, *Danio rerio*. *Development* **123:** 1.

Halpern M.E., Ho R.K., Walker C., and Kimmel C.B. 1993. Induction of muscle pioneers and floor plate is distinguished by the zebrafish *no tail* mutation. *Cell* **75:** 99.

Halpern M.E., Hatta K., Amacher S.L., Talbot W.S., Yan Y.-L., Thisse B., Thisse C., Postlethwait J.H., and Kimmel C.B. 1997. Genetic interactions in the zebrafish midline. *Dev. Biol.* **187:** 154.

Hammerschmidt M., Serbedzija G.N., and McMahon A.P. 1996a. Genetic analysis of dorsoventral pattern formation in the zebrafish: Requirement of a BMP-like ventralizing activity and its dorsal repressor. *Genes Dev.* **10:** 2452.

Hammerschmidt M., Pelegri F., Kane D.A., Mullins M.C., Serbedzija G., van-Eeden F.J.M., Granato M., Brand M., Furutani-Seiki M., Haffter P., Heisenberg C.P., Jiang Y.-J., Kelsh R.N., Odenthal J., Warga R.M., and Nüsslein-Volhard C. 1996b. *dino* and *mercedes*, two genes regulating dorsal development in the zebrafish embryo. *Development* **123:** 95.

Hatta K., Bremiller R., Westerfield M., and Kimmel C.B. 1991a. Diversity of expression of *engrailed*-like antigens in zebrafish. *Development* **112:** 821.

Hatta K., Kimmel C.B., Ho R.K., and Walker C. 1991b. The *cyclops* mutation blocks specification of the floor plate of the zebrafish central nervous system. *Nature* **350:** 339.

Heisenberg C.P. and Nüsslein-Volhard C. 1997. The function of *silberblick* in the positioning of the eye anlage in the zebrafish embryo. *Dev. Biol.* **184:** 85.

Heisenberg C.P., Brand M., Jiang Y.-J., Warga R.M., Beuchle, F.J.M. van-Eeden, M. Furutani-Seiki, M. Granato, P. Haffter D., Hammerschmidt M., Kane D.A., Kelsh D.A., Mullins M.C., Odenthal J., and Nüsslein-Volhard C. 1996. Genes involved in forebrain development in the zebrafish, *Danio rerio*. *Development* **123:** 191.

Henion P.D., Raible D.W., Beattie C.E., Stoesser K.L., Weston J.A., and Eisen J.S. 1996. Screen for mutations affecting development of zebrafish neural crest. *Dev. Genet.* **18:** 11.

Ho, R. 1992. Axis formation in the embryo of the zebrafish *Brachydanio rerio*. *Semin. Dev. Biol.* **3:** 53.

Jiang Y.-J., Brand M., Heisenberg C.-P., Beuchle D., van-Eeden F.J.M., Furutani-Seiki M., Granato M., Haffter P., Hammerschmidt M., Kane D.A., Kelsh R.N., Mullins M.C., Odenthal J., and Nüsslein-Volhard C. 1996. Mutations affecting neurogenesis and brain morphology in the zebrafish, *Danio rerio*. *Development* **123:** 205.

Johnson S.L., Gates M.A., Johnson M., Talbot W.S., Horne S., Baik K., Rude, S. Wong J.R., and Postlethwait J.H. 1996. Half-tetrad analysis in zebrafish II: Centromere-linkage analysis and consolidation of the zebrafish genetic map. *Genetics* **142:** 1277.

Joyner A.L. 1996. *Engrailed, Wnt* and *Pax* genes regulate midbrain-hindbrain development. *Trends Genet.* **12:** 15.

Karlstrom R.O., Trowe T., Brand M., Baier H., Crawford A., Grunewald B., Haffter P., Hammerschmidt M., Hoffman H., Klostermann S., Meyer S., Richter F.J.M., van-Eeden E., Vogelsang S., Nüsslein-Volhard C., and Bonhoeffer F. 1996. Zebrafish mutations affecting retinotectinal axon pathfinding. *Development* **123:** 427.

Kelly G.M. and Moon R.T. 1995. Involvement of *Wnt1* and *Pax2* in the formation of the midbrain-hindbrain boundary in the zebrafish gastrula. *Dev. Genet.* **17:** 129.

Kimmel C.B., Ballard W.W., Kimmel S.R., Ullmann B., and Schilling T.F. 1995. Stages of embryonic development of the zebrafish. *Dev. Dyn.* **203:** 253.

Kishimoto Y., Lee K., Zon L., Hammerschmidt M., and Schulte-Merker S. 1997. The molecular nature of zebrafish *swirl*: BMP2 function is essential during early dorsoventral patterning. *Developmant* **124:** 4457.

Knapik E.W., Goodmann A., Atkinson S., Roberts C.T., Shiozawa M., Sim C.U., Weksler-Zangen S., Trolliet M., Futrell C., Innes B.A., Koike G., McLaughlin M.G., Pierre L., Simson J.S., Vilallonga E., Roy M., Chiang P., Fishman M.C., Driever W., and Jacob H.J. 1996. A reference cross for zebrafish (*Danio rerio*). *Development* **123:** 451.

Krauss S., Johansen T., Korzh V., and Fjose A. 1991. Expression of the zebrafish paired box gene *pax[zf-b]* during early neurogenesis. *Development* **113:** 1193.

Krauss S., Maden M., Holder N., and Wilson S.W. 1992. Zebrafish *pax[b]* is involved in the formation of the midbrain-hindbrain boundary. *Nature* **360:** 87.

Lawrence P.A. 1992. *The making of a fly: The genetics of animal design.* Blackwell, Oxford, United Kingdom.

Liem K.F., Tremmi G., Roelink H., and Jessell T.M. 1995. Dorsal differentiation of neural plate cells induced by BMP-mediated signals from epidermal ectoderm. *Cell* **82:** 969.

Lumsden A. and Krumlauf R. 1996. Patterning the vertebrate neuraxis. *Science* **274:** 1109.

Macdonald R., Xu Q., Barth K.A., Mikkola I., Holder N., Fjose A., Krauss S., and Wilson S.W. 1994. Regulatory gene expression boundaries demarcate sites of neuronal differentiation in the embryonic zebrafish forebrain. *Neuron* **13:** 1039.

Malicki J., Neuhauss S.C.F., Schier A.F., Solnica-Krezel L., Stainier D.Y.R., Stemple D.L., Abdelilah S., and Driever W. 1996a. Mutations affecting development of the zebrafish retina. *Development* **123:** 263.

Malicki, J., Schier A., Solnica-Krezel L., Stemple D.L., Neuhauss S.C.F., Stainier D.Y.R., Abdelilah S., Rangini Z., Zwartkruis F., and Driever W. 1996b. Patterning mutations of the zebrafish ear. *Development* **123:** 275.

Marin F. and Puelles L. 1994. Patterning the embryonic avian midbrain after experimental inversions: A polarizing activity from the isthmus. *Dev. Biol.* **163:** 19.

Mathers P.H., Miller A., Doniach T., Dirksen M.-L., and Jamrich M. 1995. Initiation of anterior head-specific gene expression in uncommitted ectoderm of *Xenopus laevis* by ammonium chloride. *Dev. Biol,* **171:** 641.

McMahon A.P., Joyner A.L., Bradley A., and McMahon J.A. 1992. The midbrain-hindbrain phenotype of *Wnt-1*$^{-}$/ *Wnt-1*$^{-}$ mice results from stepwise deletion of engrailed-expression cells by 9.5 days post coitum. *Cell* **69:** 581.

Mizuno T., Yamaha E., Wakahara M., Kuroiwa A., and Takeda H. 1996. Mesoderm induction in zebrafish. *Nature* **383:** 131.

Moens C.B., Yan Y.-L., Appel B., Force A.G., and Kimmel C.B. 1996. *valentino*: A zebrafish gene required for normal hindbrain segmentation. *Development* **122:** 3981.

Müller M., von Weizsäcker E., and Campos-Ortega J.A. 1996. Transcription of a zebrafish gene of the *hairy*-Enhancer of *split family* delineates the midbrain anlage in the neural plate. *Dev. Genes Evol.* **206:** 153.

Mullins M.C., Hammerschmidt M., Kane D.A., Brand M., van Eeden F.J.M., Furutani-Seiki M., Granato M., Haffter P., Heisenberg C.-P., Jiang Y.-J., Kelsh R.N., Odenthal J., Warga R.M., and Nüsslein-Volhard C. 1996. Genes establishing dorsal-ventral pattern formation in the zebrafish embryo: The ventral specifying genes. *Development* **123:** 81.

Nadeau J.H. 1989. Maps of linkage and synteny homologies between mouse and man. *Trends Genet.* **5:** 82.

Nieuwkoop P.D., Boterenbrood E.C., Kremer A., Bloesma F.F.S.N., Hiessels E.L.M.J., Meyer G., and Verheyen F.J. 1952. Activation and organization of the central nervous system in amphibians. *J. Exp. Zool.* **120:** 1.

Nüsslein-Volhard C. and Wieschaus E. 1980. Mutations affecting segment number and polarity in *Drosophila*. *Nature* **287:** 795.

Oppenheimer J. 1936a. Transplantation experiments on developing teleosts (*Fundulus* and *Perca*). *J. Exp. Zool.* **72:** 409.

———. 1936b. Structures developed in amphibians by implantation of living fish organizers. *Proc. Soc. Exp. Biol. Med.* **34:** 461.

Pankratz M.J. and Jäckle H. 1993. Blastoderm segmentation. In *The development of* Drosophila melanogaster (ed. M. Bate and A. Martinez Arias), p. 467. Cold Spring Harbor Labora-

tory Press, Cold Spring Harbor, New York.
Papalopulu N. and Kintner C. 1996. A posteriorising factor, retinoic acid, reveals that anteroposterior patterning controls the timing of neuronal differentiation in *Xenopus* neuroectoderm. *Development* **122:** 3409.
Papan C. and Campos-Ortega J.A. 1994. On the formation of the neural keel and neural tube in the zebrafish *Danio (Brachydanio) Rerio. Roux's Arch. Dev. Biol.* **203:** 178.
Piccolo S., Sasai Y., Lu B., and De Robertis E.M. 1996. Dorsoventral patterning in *Xenopus*: Inhibition of ventral signals by direct binding of chordin to BMP-4. *Cell* **86:** 589.
Postlethwait J.H. and Talbot W.S. 1997. Zebrafish genomics: From mutants to genes. *Trends Genet.* **13:** 183.
Postlethwait J.H., Johnson S.L., Midson C.N., Talbot W.S., Gates M., Ballinger E.W., Africa D., Andrews R., Carl T., Eisen J.S., Horne S., Kimmel C.B., Hutchinson M., Johnson M., and Rodriguez A. 1994. A genetic linkage map for the zebrafish. *Science* **264:** 699.
Puelles L. and Rubenstein J. 1993. Expression patterns of homeobox and other putative regulatory genes in the embryonic mouse forebrain suggest a neuromeric organization. *Trends Neurosci.* **16:** 472.
Riddle R.D., Johnson R.L., Laufer E., and Tabin C. 1993. *Sonic hedgehog* mediates the polarizing activity of the ZPA. *Cell* **75:** 1401.
Roelink H., Porter J.A., Chiang C., Tanabe Y., Chang D.T. Beachy P.A., and Jessell T.M. 1995. Floor plate and motor neuron induction by different concentrations of the amino-terminal cleavage product of Sonic hedgehog autoproteolysis. *Cell* **81:** 445.
Roelink H., Augsburger A., Heemskerk J., Korzh V., Norlin S., Ruiz i Altaba A., Tanabe Y., Placzek M., Edlund T., Jessell T.M., and Dodd J. 1994. Floor plate and motor neuron induction by *vhh-1*, a vertebrate homolog of *hedgehog* expressed by the notochord. *Cell* **76:** 761.
Rubenstein J., Martinez S., Shimamura K., and Puelles L. 1993. The embryonic vertebrate forebrain: The prosomeric model. *Science* **266:** 578.
Schier, A.F., Neuhauss S.C.F., Helde K.A., Talbot W.S., and Driever W. 1997. The *one-eyed pinhead* gene functions in mesoderm and endoderm formation in zebrafish and interacts with *no tail. Development* **124:** 327.
Schier A.F., Neuhauss S.C.F., Harvey M., Malicki J., Solnica-Krezel L., Stainier D.Y.R., Zwartkruis F., Abdelilah S., Stemple D.L., Rangini Z., Yang H., and Driever W. 1996. Mutations affecting the development of the embryonic zebrafish brain. *Development* **123:** 165.
Schulte-Merker S., Lee K.J., McMahon A.P., and Hammerschmidt M. 1997. The zebrafish organizer requires *chordino. Nature* **387:** 862.
Schulte-Merker, S., Van Eeden F.J.M., Halpern M.E., Kimmel C.B., and Nüsslein-Volhard C. 1994. *no tail* (*ntl*) is the zebrafish homologue of the mouse *T* (*Brachyury*) gene. *Development* **120:** 1009.
Shih J. and Fraser S.E. 1996. Characterizing the zebrafish organizer: Microsurgical analysis at the early-shield stage. *Development* **122:** 1313.
Solnica-Krezel L., Stemple D.L., Mountcastle-Shah E., Rangini Z., Neuhauss S.C.F., Malicki J., Schier A., Stainier D.Y.R., Zwartkruis F., Abdelilah S., and Driever W. 1996. Mutations affecting cell fates and cellular rearrangements during gastrulation in zebrafish. *Development* **123:** 67.
Spemann H. 1938. *Embryonic development and induction.* Yale University Press, New Haven, Connecticut.
Stemple D.L., Solnica-Krezel L., Zwartkruis F., Neuhauss S.C.F., Schier A.F., Malicki J., Stainier D.Y.R., Abdelilah S., Rangini Z., Mountcastle-Shah E., and Driever W. 1996. Mutations affecting development of the notochord in zebrafish. *Development* **123:** 117.
Talbot W.S., Trevarrow B., Halpern M.E., Melby A.E., Farr G., Postlethwait J.H., Jowett T., Kimmel C.B., and Kimelman D. 1995. A homeobox gene essential for zebrafish notochord development. *Nature* **378:** 150.
Thisse C., Thisse B., Halpern M.E., and Postlethwait J.H. 1994. *goosecoid* expression in neurectoderm and mesendoderm is disrupted in zebrafish *cyclops* gastrulas. *Dev. Biol.* **164:** 420.
Thomas K.R. and Capecchi M.R. 1990. Targeted disruption of the murine *int-1* proto-oncogene resulting in severe abnormalities in midbrain and cerebellar development. *Nature* **346:** 847.
Thomas P. and Beddington R. 1996. Anterior primitive endoderm may be responsible for patterning the anterior neural plate in the mouse embryo. *Curr. Biol.* **5:** 1487.
Trowe T., Klostermann S., Baier H., Grunewald B., Hoffmann H., Granato M., Haffter P., Hammerschmidt M., van-Eeden F.J.M., Vogelsang, Meyer S., Crawford A., Stroh T., Richter S., Nüsslein-Volhard C., and Bonhoeffer F. 1996. Mutations disrupting the ordering and topographic mapping of axons in the retinotectal projection of the zebrafish, *Danio rerio. Development* **123:** 439.
Urbanek P., Wang Z.Q., Fetka I., Wagner E.F., and Busslinger M. 1994. Complete block of early B-cell differentiation and altered patterning of the posterior midbrain in mice lacking Pax5/BSAP. *Cell* **79:** 901.
van Baer K. 1828. *Über die Entwicklungsgeschichte der Thiere.* Bornträger, Königsberg.
van Straaten H.W.M., Hekking J.W.M., Beursgens J.P.W.M., Terwindt-Rouwenhoorst E., and J. Drukker. 1989. Effect of the notochord on proliferation and differentiation in the neural tube of the chick embryo. *Development* **107:** 793.
Viehbahn C., Mayer B., and Hrabe de Angelis M. 1995. Signs of the principlal body axis prior to primitive streak formation in the rabbit embryo. *Anat. Embryol.* **192:** 159.
von Kupfer K. 1906. Die Morphogenie des Central Nerven Systems. In *Handbuch der vergleichenden und experimentellen Entwicklungslehre der Wirbeltiere* (ed. O. Hertwigs), vol.2 (part 3), p. 1. Fischer Verlag, Jena.
von Weizsacker E. 1994. Molekulargenetische Untersuchungen an sechs Zebrafisch-Genen mit Homologie zur Enhancer of split Gen-Familie von *Drosophila*. Ph.D thesis, University of Cologne, Germany.
Vos P., Hogers R., Bleeker M., Reijans M., van de Lee T., Hornes M., Frijters A., Pot J., Peleman J., Kuiper M., and Zabeau M. 1995. AFLP: A new technique for DNA fingerprinting. *Nucleic Acids Res.* **23:** 4407.
Whitfield T., Granato M., van-Eeden F.J.M., Schach U., Brand M. Furutani-Seiki M., Haffter P., Hammerschmidt M., Heisenberg C.-P., Jiang Y.-J., Kane D.A., Kelsh R.N., Mullins M.C., Odenthal J., and Nüsslein-Volhard C. 1996. Mutations affecting development of the zebrafish inner ear and lateral line. *Development* **123:** 241.
Woo K. and Fraser S. 1995. Order and coherence in the fatemap of the zebrafish nervous system. *Development* **121:** 2595.
______. 1997. Specification of the zebrafish nervous system by nonaxial signals. *Science* **277:** 254.
Wurst W., Auerbach A.B., and Joyner A.L. 1994. Multiple developmental defects in *Engrailed-1* mutant mice: An early mid-hindbrain deletion and patterning defects in forelimbs and sternum. *Development* **120:** 2065.
Yamada T., Placzek M., Tanaka H., Dodd J., and Jessell T.M. 1991. Control of cell pattern in the developing nervous system: Polarizing activity of floor plate and notochord. *Cell* **64:** 635.
Zaraisky A.G., Ecochard V., Kazanskaya O.V., Lukyanov S.A., Fresnko I.V., and Duprat A.-M. 1995. The homeobox-containing gene XANF1 may control development of the Spemann organizer. *Development* **121:** 3839.
Zimmerman L.B., De Jesús-Escobar J.M., and Harland R. 1996. The Spemann organizer signal noggin binds and inactivates bone morphogenetic protein 4. *Cell* **85:** 599.

Cell Interactions in Patterning the Mammalian Midbrain

D.H. Rowitch,[1,2] P.S. Danielian,[2] S.M.K. Lee,[2] Y. Echelard,[3] and A.P. McMahon[2]
[1]*Joint Program in Neonatology, Harvard Medical School, Boston, Massachusetts 02115;* [2]*Department of Molecular and Cellular Biology, Harvard University, Cambridge, Massachusetts 02138;* [3]*Genzyme Integrated Genetics, Framingham, Massachusetts 01701*

Several lines of evidence indicate that the vertebrate brain develops on a segmental template. The forebrain is thought to be composed of six segments, or neuromeres, the midbrain of one segment, and the hindbrain of eight segments (Fig. 1) (Puelles and Rubenstein 1993; Rubenstein et al. 1994). To investigate the genetic pathways that control the establishment and elaboration of neuromeric development, we have focused our attention on the vertebrate midbrain. *Wnt-1*, encoding the murine ortholog of the *Drosophila* patterning signal, wingless (Rijsewijk et al. 1987), is expressed in the early neural plate, exclusively within the presumptive midbrain. In the absence of Wnt-1 signaling, the entire midbrain is lost (McMahon and Bradley 1990; Thomas and Capecchi 1990). Secondarily, there is a deletion of the neighboring first neuromere (rhombomere 1[r1]) of the hindbrain (McMahon et al. 1992; Mastick et al. 1996; Serbedzija et al. 1996). Thus, by understanding the molecular and cellular processes that regulate *Wnt-1* expression, we hope to elucidate how early patterning of the mammalian neural plate is regulated. Furthermore, by identifying targets of Wnt-1 signaling, we hope to shed light on the mechanisms responsible for subsequent development of the midbrain and anterior hindbrain. From available data, it is possible to model early mouse midbrain development from the mid-streak stage of gastrulation (7.0 dpc) to just after neural tube closure (9.5 dpc) as occurring in three phases, each with unique cell-cell interactions. These phases are discussed with reference to potential factors that may have roles in pattern formation.

METHODS

The production of *WEXP-En-1*$^{+}$, *Wnt-1*$^{-/-}$ transgenic mice and subsequent whole-mount in situ hybridization was carried out on 10.5 embryos as described previously (Danielian and McMahon 1996). Methods used in the generation of previously published data are indicated in legends of Figures 2–5 and Figure 7.

RESULTS

Pattern Formation in the Embryonic Midbrain

The neural plate is the first morphologically recognizable structure of the central nervous system (CNS), comprising a one-cell-layer-thick undifferentiated epithelium. It forms at approximately 7.25–7.5 dpc in the mouse, although fate mapping indicates that cells destined to contribute to the CNS can be identified at least 24 hours beforehand (Quinlan et al. 1995). The expression patterns and functions of certain genes in the gastrula further indicate that early patterning events take place before formation of the neural plate. *Otx-2* expression is ubiquitous at egg cylinder stages, and by 7.25 dpc, its expression is restricted to cells of the presumptive forebrain and midbrain. *Otx-2* null mutant embryos lack all CNS structures rostral to r3, demonstrating that it has an essential role in the early development of the anterior CNS (Acampora et al. 1995; Matsuo et al. 1995; Ang et al. 1996). Evidence from grafting and explant experiments indicates that signals from mesoderm to the dorsal epiblast (presumptive neuroepithelium) may be involved in induction and early pattern formation of the neural plate (for review, see Ruiz i Altaba 1994; Kelly and Melton 1995). Although the nature of inductive signals for the midbrain is unclear, the finding that an Fgf-8-soaked bead can induce an ectopic midbrain structure in prosomere 2 (Crossley et al. 1996) raises the possibility that the Fgf signal transduction pathway is involved (for further details, see Discussion).

The earliest known gene regionally expressed in the presumptive midbrain-hindbrain of the mouse is *Pax-2* (Nornes et al. 1990; Puschel et al. 1992; Rowitch and McMahon 1995), a paired domain-containing transcription factor (Nornes et al. 1990). *Pax-2* expression commences at 7.0–7.5 dpc, prior to neural plate formation (Fig. 2) (Rowitch and McMahon 1995). Genetic evidence from *Drosophila* has indicated that the *Pax* transcription factor, *paired*, is required for activation of *wg* and *en* (Ingham 1991). In zebrafish, *pax-b* is the only member of the *Pax* gene family expressed at the midbrain-hindbrain border (Krauss et al. 1991). The function of *pax-b* was studied by injecting antibody against pax-b into developing zebrafish embryos. In such injected embryos, morphological malformations and a reduction in the levels of *wnt-1, eng-2*, and *pax-b* at the midbrain-hindbrain junction were observed (Krauss et al. 1992). Furthermore, in the zebrafish mutant, *no isthmus (noi)*, the mid-hindbrain region is absent, a phenotype similar to that of null mutations of *Wnt-1* in the mouse (McMahon and Bradley 1990; Thomas and Capecchi 1990). *noi* mutations have

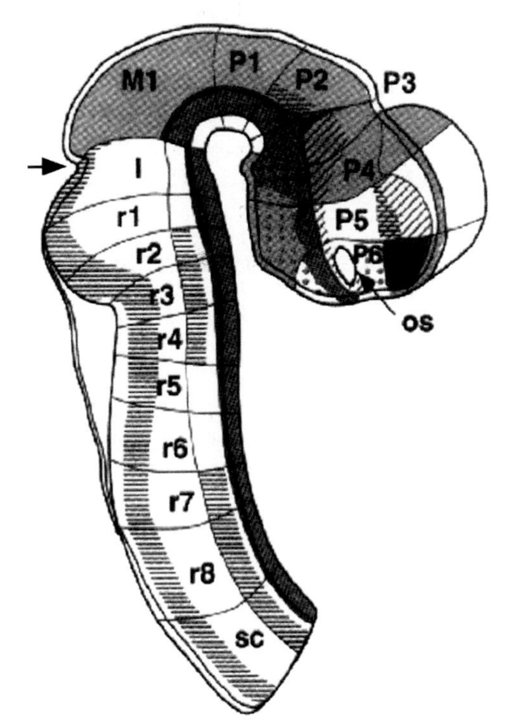

Figure 1. Neuromeric organization of the CNS. Scheme of 10.5 dpc mouse embryonic CNS showing division into six prosomeres (P1–P6), midbrain (M1), isthmus (I), eight rhombomeres (r1–r8), and the spinal cord (sc). The model is based on morphologic, grafting, and gene expression data from vertebrates at various stages of CNS development (Puelles and Rubenstein 1993; Rubenstein et al. 1994). The patterns of certain genes expressed in segments along the A/P axis (*Otx-2*, gray; *Gbx-2*, horizontal lines; *Dlx-2*, diagonal lines) are shown. The midbrain-hindbrain constriction (MHC) is indicated (*black arrow*). (Modified from Rubenstein et al. 1994.)

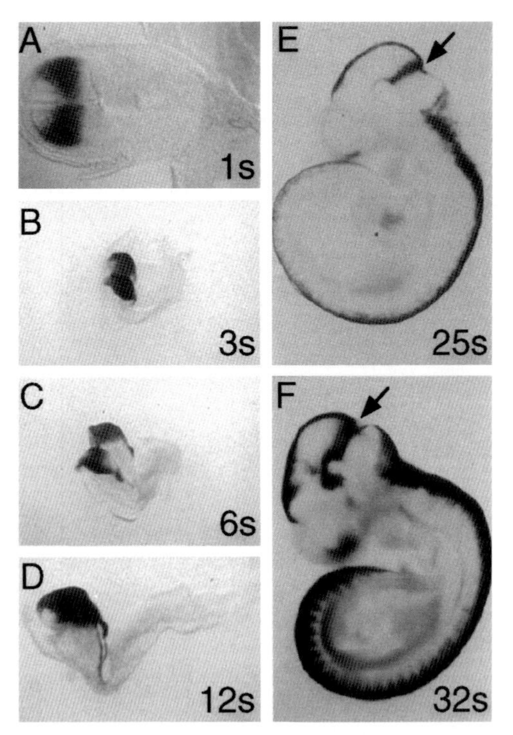

Figure 3. Expression of β-galactosidase in *Wnt-1* enhancer transgenic reporter embryos during development from 8.0–10.5 dpc (1–32-somites). In vivo transgenic mouse reporter assay. β-galactosidase histochemical staining of whole-mount transgenic embryos from the WZT9B line was carried out for 30 min to 2 hr. (*A–D*) Dorsal and (*E,F*) lateral views of embryos at the following somite stages: (*A*) 1s; (*B*) 3s; (*C*) 6s; (*D*) 12s; (*E*) 25s; (*F*) 32s. Note that expression directed by the Wnt-1 enhancer is widespread in the midbrain from the 1–12-somite stage (*A–D*). After neural tube closure (*E,F*), expression is restricted to the dorsal and ventral midbrain, the midbrain-hindbrain junction (*closed arrows*), and the roofplate of the spinal cord. (Modified from Echelard et al. 1994.)

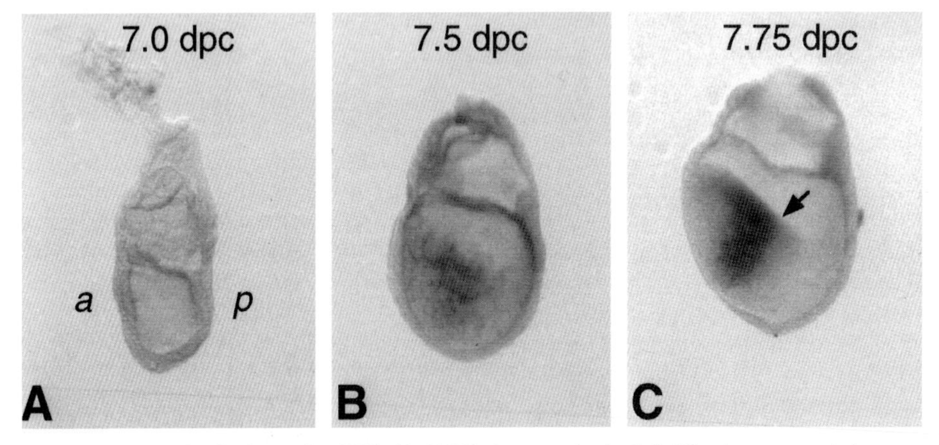

Figure 2. Onset of *Pax-2* expression in the murine CNS. (*A–C*) Whole-mount in situ hybridization was carried out on mid-late streak stage (7.0–7.5 dpc) mouse embryos using an anti-sense probe for *Pax-2*. (*A–B*) *Pax-2* expression commences at about 7.5 dpc in the anterior region of the mouse embryo. (*C*) After formation of a morphologically recognizable neural plate, *Pax-2* expression adopts a restricted segmental pattern with its posterior boundary corresponding to the future r1/r2 boundary. (Modified from Rowitch and McMahon 1995.)

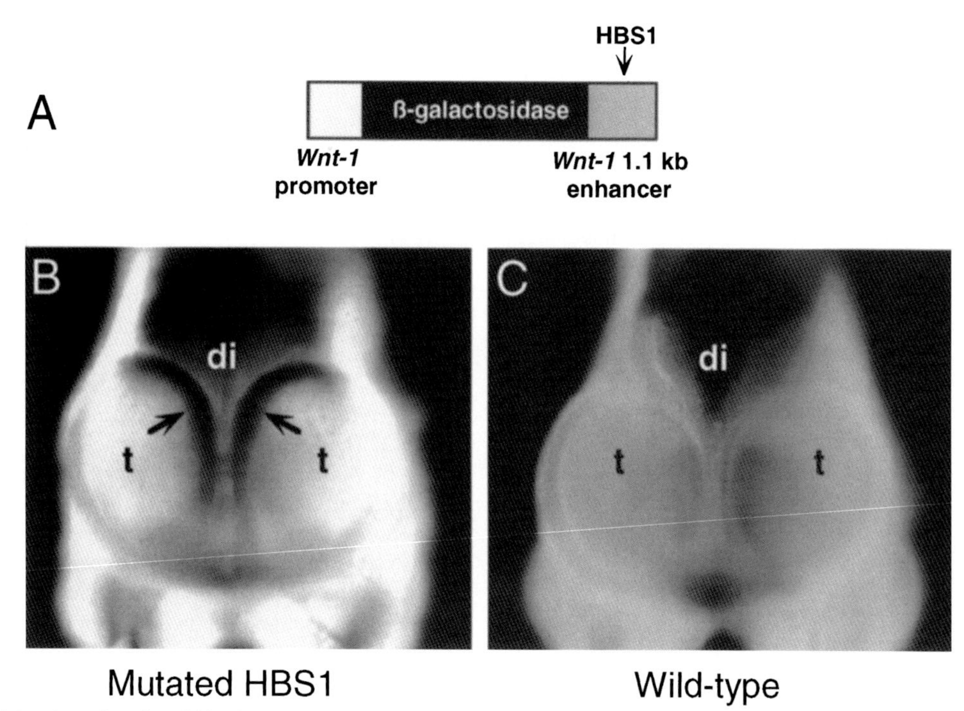

Figure 4. Mutation of HBS1 within the *Wnt-1* enhancer results in ectopic forebrain expression of a reporter transgene. (*A*) Schematic diagram of the reporter construct, pWTX1.1, comprising the *Wnt-1* promoter, β-galactosidase gene, and the minimal 1.1-kb *Wnt-1* enhancer. The position of HBS1 is indicated (*closed arrow*). (*B,C*) In vivo transgenic mouse reporter assay. β-galactosidase histochemical staining of whole-mount 10.5-dpc transgenic founder embryos was carried out for 2 hr. (*B*) Frontal view of representative embryo injected with pWTZ1.1-mHBS1. The region of ectopic dorsal-medial telencephalon staining is indicated (*closed arrows*). (*C*) Frontal view of representative embryo injected with pWTZ1.1. (di) Diencephalon; (t) telencephalon. (Modified from Iler et al. 1995.)

been mapped within *pax-b*, and analysis of *noi* embryos revealed an absence of *wnt-1* and *eng-2* expression in neural plate (Brand et al. 1996). Thus, it is of interest to consider a possible role for *Pax* genes in mouse neural plate patterning. Both *Pax-2* (Fig. 2) (Nornes et al. 1990; Puschel et al. 1992; Rowitch and McMahon 1995) and *Pax-5* (Asano and Gruss 1992; Urbanek et al. 1994) are expressed in overlapping regions of the mouse midbrain at neural plate stages. Null mutations of *Pax-5* result in abnormal development of the caudal tectum (Urbanek et al. 1994). Several *Pax-2* loss-of-function alleles have been described including *krd*, a deletion that includes the *Pax-2* locus together with a large surrounding region of chromosome 19 (Keller et al. 1994), *Pax-2*^{1neu}, a frameshift mutation within the paired domain (Favor et al. 1996), and finally, a null allele engineered by homologous recombination in embryonic stem (ES) cells (Torres et al. 1996). Interpretation of the *Pax-2* loss-of-func-

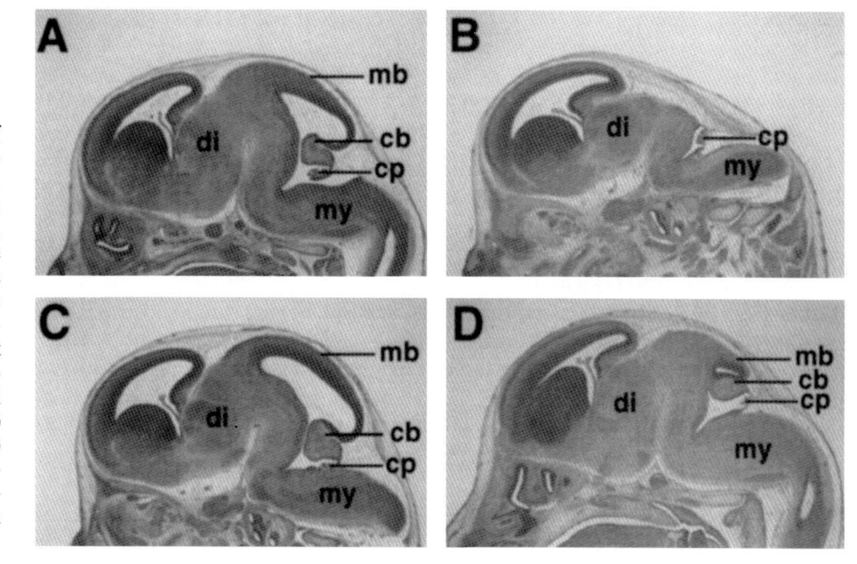

Figure 5. Histological analysis of CNS development in *Wnt-1*$^{-/-}$ embryos expressing *En-1* at 14.5 dpc. Sagittal views of *Wnt-1*$^{+/+}$ (*A*); *Wnt-1*$^{-/-}$ (*B*); and *Wnt-1*$^{-/-}$, WEXP-*En-1*$^{+}$ embryos (*C,D*). The posterior choroid plexus (cp) marks the metencephalic-myelencephalic junction. Embryos that were Wnt-$1^{+/+}$ or *Wnt-1*$^{+/-}$ with or without the transgene were indistinguishable at this stage. (Ch) Cerebral hemisphere; (di) diencephalon; (mb) midbrain; (cb) cerebellum; (cp) choroid plexus; (my) myelencephalon (caudal hindbrain). (Modified from Danielian and McMahon 1996.)

tion CNS phenotype is difficult to reconcile, however, since Torres et al. (1996) described a resultant neural tube defect (exencephally) with formation of a midbrain-hindbrain junction, whereas the *Pax-2*^{1neu} phenotype comprised failure of mid-hindbrain development (Favor et al. 1996). The inconsistent results obtained by these groups could be due to particularities of the alleles tested and/or the genetic background used to maintain the mutant lines. In any case, given the recent evidence for functional overlap between *Pax-2* and *Pax-5* (Urbanek et al. 1997), final interpretation of the *Pax* loss-of-function phenotype in the mid-hindbrain region awaits analysis of the *Pax-2*$^{-/-}$, *Pax-5*$^{-/-}$ compound mutant embryos. It will be of special interest to determine if the mid-hindbrain-specific genes *Wnt-1* and *En-1* are activated normally at the one-somite stage.

Regulation of *Wnt-1* can be divided into an activation phase (Fig. 3A–D), in which the mesencephalon broadly expresses the gene, and a maintenance phase (~15–20-somites and later; Fig. 3E–F), in which expression is restricted to the roofplate of the caudal diencephalon, mesencephalon, myelencephalon, and spinal cord, as well as a ring just anterior to the midbrain-hindbrain junction, and the ventral midline of the caudal diencephalon and mesencephalon (Wilkinson et al. 1987). Given its pivotal role in midbrain development, we have investigated the regulation of *Wnt-1*. Its enhancer lies downstream from the gene and was originally identified as a 5.5-kb region that was sufficient for directing expression of a reporter transgene in both the activation and maintenance phases of the *Wnt-1* pattern (Fig. 3) (Echelard et al. 1994).

To show that this region was necessary for endogenous *Wnt-1* expression, it was removed by gene targeting in ES cells and replaced by a *neo* cassette flanked by *loxP* sites. Subsequently, *neo* was removed by treating zygotes with *cre* recombinase. Embryos bred to the homozygous state demonstrated a phenotype indistinguishable from that of null mutants of *Wnt-1* itself; moreover, no expression of *Wnt-1* was detected in such mutant embryos (P.S. Danielian and A.P. McMahon, in prep.). We conclude from this that the 5.5-kb enhancer region contains the essential *cis*-acting regulatory sequences for activation of the gene.

To more precisely map the *cis*-regulatory region, a deletion analysis of the 5.5-kb enhancer was performed in transgenic mice. This analysis revealed a 110-bp regulatory element that was sufficient to drive *lacZ* expression in the midbrain portion of the *Wnt-1* pattern at the 1- and 25-somite stages (D.H. Rowitch et al., in prep.). To further investigate DNA sequences in the 110-bp enhancer, we performed a comparative analysis with a homologous region from the *wnt-1* locus of the pufferfish, *Fugu rubripes*. A 102-bp sequence was identified that has 69% homology with the mouse enhancer. The 102-bp pufferfish sequence accurately generated the activation phase pattern of *Wnt-1* at early somite stages in the transgenic reporter assay. However, at the 25-somite stage, staining was divergent from the pattern generated with the mouse regulatory element and was confined solely to the roofplate of the midbrain (D.H. Rowitch et al., in prep.). These results indicate that there are different regulatory sequences involved in the activation and maintenance of *Wnt-1*. Conserved sequences between mouse and pufferfish are involved in the activation phase of expression, whereas those DNA sequences present in the mouse but not present in the pufferfish enhancer are needed to recapitulate the full pattern of *Wnt-1* at the 25-somite stage (e.g., the staining at the midbrain-hindbrain junction; arrows in Fig. 3).

It is noteworthy that Pax-2-binding sites, identified on the basis of similarity to known consensus sequences (Czerny et al. 1993; Epstein et al. 1994), are not conserved in the pufferfish sequences. Indeed, our efforts at footprinting these sites with Pax-2 and Pax-5 have so far been inconclusive (P.S. Danielian and A.P. McMahon, unpubl.). Thus, we do not have evidence for direct interaction of Pax-2 with the *Wnt-1 cis*-acting regulatory sequences. On the other hand, a homeodomain core-binding consensus site, called HBS1, was identified that is conserved in the pufferfish sequences. *Wnt-1* reporter transgenes that carried mutations or deletion of the HBS1 site resulted in ectopic expression of *lacZ* in the dorsal telencephalon (Fig. 4) (Iler et al. 1995; Y. Echelard et al., unpubl.). The expression pattern of *Emx-2* is consistent with a role in mediating repression of the *Wnt-1* transgene (Boncinelli et al. 1993); moreover, Emx-2 has been found to bind HBS1 in vitro (Iler et al. 1995). Recent analysis of the *Emx-2* null mouse reveals that there is ectopic expression of *Wnt-1* in the dorsal telencephalon (Yoshida et al. 1997). Taken together, these experiments are consistent with a model in which Emx-2 normally functions to repress *Wnt-1* expression in the developing telencephalon.

The Role of *Wnt-1* in Early Development of the Midbrain

In *Wnt-1*$^{-/-}$ embryos, there is failure of midbrain development and subsequent loss of the anterior hindbrain, including the cerebellar anlagen (McMahon and Bradley 1990; Thomas and Capecchi 1990; Serbedjiza et al. 1996). Interestingly, axonal pathways that normally traverse the midbrain and midbrain-hindbrain boundary successfully navigate the new juxtaposition of diencephalon and r2 (Mastick et al. 1996). The phenotype of the *Wnt-1* null mutant embryo is clearly distinguishable from wild type by 8.75 dpc, demonstrating that *Wnt-1* has an essential role in brain development at early neural plate stages.

Expression of *Engrailed-1* is contemporaneous with *Wnt-1* at the 1-somite stage in a domain that is somewhat caudal to that of *Wnt-1*, including the adjacent first rhombomere of the hindbrain (Davis and Joyner 1988; McMahon et al. 1992; Rowitch and McMahon 1995). *En-1*$^{-/-}$mutant embryos lack the caudal tectum (inferior colliculus), cranial nerves III and IV, and the cerebellum, a less severe phenotype than that seen in *Wnt-1* null mutants (Wurst et al. 1994). Activation of *Engrailed-2* expression occurs approximately 6 hours after *En-1* at the 4-somite stage, coincident with the onset of *Pax-5* expression. Indeed, it has been suggested that *Pax* transcription factors may directly regulate *En-2* (Song et al.

1996). However, *En-2*$^{-/-}$ mice have subtle defects in cerebellar foliation (Davis et al. 1988; Joyner et al. 1991; Millen et al. 1994), whereas in *Pax-5*$^{-/-}$ embryos, there are abnormalities of the inferior colliculus, indicating that genes other than *En-2* are likely to be regulated by *Pax-5*. *En-1* and *En-2* appear to be functionally equivalent since (1) the *En-1*$^{-/-}$, *En-2*$^{-/-}$ double mutant has a more severe phenotype (similar to *Wnt-1*$^{-/-}$; W. Wurst and A. Joyner, pers. comm.), and (2) replacement of *En-1* by *En-2* via homologous recombination in ES cells results in normal homozygous animals (Hanks et al. 1995).

In *Wnt-1*$^{-/-}$ embryos, *En-1* and *En-2* are activated normally; however, there is a subsequent stepwise loss of *Engrailed* expression or the cells expressing *Engrailed* (McMahon et al. 1992). This observation is of particular interest because in *Drosophila* the ortholog of the Wnt-1 signal, Wg, is required for the maintenance of *en* expression in the embryonic cuticle, as well the corresponding region of each segment which gives rise to the naked cuticle (Ingham 1991). Remarkably, expression of *En-1* in the developing mesencephalon of *Wnt-1* null embryos is sufficient to rescue early mes- and metencephalic development, suggesting that a key role of *Wnt-1* signaling in the vertebrate CNS is the maintainance of *Engrailed* expression (Danielian and McMahon 1996). If so, what are the cellular roles that *Wnt-1* and *En-1* have? Ectopic expression of *Wnt-1* in the spinal cord of transgenic mice results in an increased number of proliferating cells (Dickinson et al. 1994), suggesting that Wnt-1 could have a mitogenic effect in the developing CNS. One intriguing model would be that *Wnt-1* may have a similar role in the midbrain through an *En-1*-dependent process.

Roles for *Wnt-1* in Later Midbrain Development

The *swaying* mutation is a spontaneous frameshift mutation in *Wnt-1* causing truncation of the carboxy-terminal half of the protein (Thomas et al. 1991). Surprisingly, mice homozygous for this mutation are viable and certain midbrain and anterior hindbrain structures, which are always absent in *Wnt-1* null embryos generated by gene targeting, are often present (McMahon and Bradley 1990; Thomas and Capecchi 1990; Thomas et al. 1991; Bally-Cuif et al. 1995). This would suggest that *swaying* may encode a hypomorphic allele, but this is by no means certain as mice with a *swaying* phenotype were occasionally found in a line generated by an identical gene targeting strategy to our own (Thomas and Capecchi 1990). One possibility is that a modifier in the genetic background may lead to variability in the phenotype.

Expression of *Otx-2* is normally confined to the fore- and midbrain, and its posterior boundary coincides with the midbrain-hindbrain junction at 10.5 dpc. However, in *swaying* mutants, the boundary is indistinct and *Otx-2* expressing cells are observed in r1, suggesting that the Wnt-1 signal may have a role in restricting *Otx-2* expression to the midbrain (Bally-Cuif et al. 1995). To further assess if this is the case, we examined *Otx-2* expression in *Wnt-1* null embryos which were substantially rescued by expression of *En-1* (Danielian and McMahon 1996). In these embryos, *Otx-2* expression appeared to be predominantly restricted to the developing midbrain and dorsal hindbrain (compare A and C in Fig. 6). Significantly, lateral (Fig. 6C) and dorsal (Fig. 6F) views showed that cells outside the normal area of *Otx-2* expression near the presumed midbrain-hindbrain junction were expressing *Otx-2* (compare A and C and D and F in Fig. 6; arrows in Fig. 6C,F). In *Wnt-1* null embryos, the midbrain and anterior hindbrain have failed to develop (Fig. 6B,E) (P. Danielian and A. McMahon, unpubl.). These results raise the possibility that Wnt-1 signal is required to restrict expression of *Otx-2* to the midbrain and are in keeping with the findings of Bally-Cuif et al. (1995). Interestingly, both *Wnt-1* null embryos substantially rescued by expression of *En-1* and mildly affected *swaying* mutant embryos lack cranial nerves III and IV (Bally-Cuif et al. 1995; Danielian and McMahon 1996). These results raise the possibility that the Wnt-1 signal may have several functions in addition to maintaining *En-1* expression.

Later Patterning of the Midbrain by *Fgf-8*

The mature midbrain has distinctive rostral-caudal polarity which can be defined in terms of gross morphology

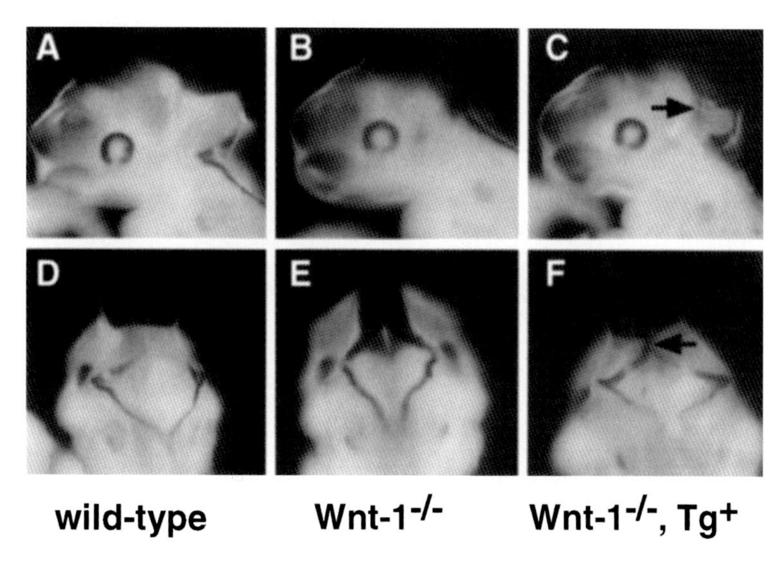

Figure 6. The Wnt-1 signal may be required to restrict the expression of *Otx-2* to the midbrain. (*A–F*) In situ hybridization demonstrating *Otx-2* expression at 10.5 dpc in (*A,D*) *Wnt-1*$^{+/+}$, (*B,E*) *Wnt-1*$^{-/-}$, and (*C,F*) *Wnt-1*$^{-/-}$ embryo which was substantially rescued by expression of *En-1* (Danielian and McMahon 1996). (*A,D*) In wild-type embryos, *Otx-2* expression is restricted to the developing midbrain and dorsal hindbrain caudal to r2. (*B,E*) In *Wnt-1* null embryos, note that the midbrain and anterior hindbrain have failed to develop. Lateral (*C*) and dorsal (*F*) views showed that cells outside the normal area of *Otx-2* expression near the presumed midbrain-hindbrain junction were expressing *Otx-2* (*arrows in C,F*).

and histological analysis. The dorsal midbrain (tectum) is divided into the superior colliculus (anterior) and inferior colliculus (posterior). In the chicken, a major role of the tectum is the processing of sensory input from the retina. Typically, neurons from the nasal retina synapse at posterior regions of the tectum, whereas temporal retinal neurons innervate the anterior tectum (for review, see Friedman and O'Leary, 1996a; Drescher et al. 1997). What are the molecular clues that guide these retinotectal connections? Recent work from the Flanagan and Bonhoeffer laboratories provides evidence that Eph receptor-ligand pairs, which are expressed in reciprocal gradients in the retina and tectum, are important determinants of axon guidance (Cheng et al. 1995; Drescher et al. 1995; Nakamoto et al. 1996).

An anteroposterior (A/P) organization of proliferation and differentiation is reflected in the developing embryonic midbrain. Neurons cease dividing and mature first in anterior regions, whereas the highest proliferative rates occur in the caudal areas of the midbrain which differentiates later on (see Lee et al. 1997). The expression of certain genes in the developing midbrain defines another aspect of A/P polarity. Genes encoding En-2 (Itasaki et al. 1991) and ELF-1, a ligand thought to be involved in retinotectal guidance (Cheng et al. 1995), are expressed in a decreasing caudal-to-rostral gradient in the tectum. Finally, the expression of *Wnt-1* rostral to the midbrain-hindbrain junction and *Fgf-8* caudal to this border serve as A/P markers of the isthmus region. Since Wnt-1 and Fgf-8 are secreted molecules with roles in cell signaling, they are candidate regulators of mid-hindbrain polarity. One may argue that Wnt-1 is unlikely to be involved since in the *Wnt-1* null rescue by *En-1*, aspects of mid-hindbrain polarity are maintained, e.g., the presence of tectal structures and cerebellar anlagen (Fig. 5) and expression of *Wnt-1, Pax-5*, and *Fgf-8* (Danielian and McMahon 1996). However, since the "rescued" midbrain lacks the occulomotor and trochlear nerves, it is possible that Wnt-1 is needed to specify certain cell types within the region.

To directly explore the role of Fgf-8 signaling in establishing mesencephalic rostrocaudal polarity, *Fgf-8* was ectopically expressed in the rostral mesencephalon of mouse embryos using the *Wnt-1* enhancer (WEXP-*Fgf8*; Lee et al. 1997). In 54% of WEXP-*Fgf-8* transgenic embryos, there was dramatic overgrowth of the mesencephalon and caudal diencephalon. Moreover, histological analysis revealed a severe reduction in differentiated cell types in the dorsal mesencephalon and diencephalon. The observed effects were likely due to direct effects of Fgf-8, since an ectopic midbrain-hindbrain (isthmus) organizing center was not observed in WEXP-*Fgf-8* transgenic brains. To investigate the possible role of Fgf-8 as a general regulator of posterior mesencephalon development, the distributions of several markers normally restricted to this region were characterized. Both *En-2* transcripts and the ELF-1 ligand were ectopically expressed in the rostral mesencephalon (Fig. 7). Taken together, these data argue that following mesencephalic induction, Fgf-8 is involved in a signaling loop which regulates several aspects of rostrocaudal polarity including later *En-2* transcription.

DISCUSSION

The concept of an organizer as a source of signals that pattern the vertebrate axis began with the discovery of Spemann and Mangold that transplantation of the blastopore lip to ventral regions of the frog embryo could induce a complete duplicate axis (for review, see Ruiz i Altaba 1994; Kelly and Melton 1995). Interestingly, this effect is mimicked by injection of *Xenopus* oocytes with

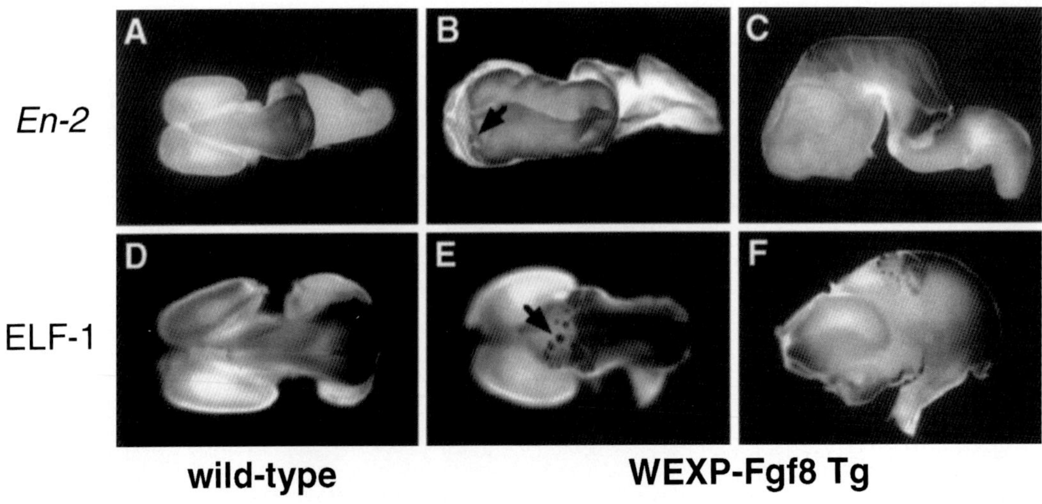

Figure 7. *En-2* and *ELF-1* are ectopically expressed in the rostral mesencephalon of WEXP-*Fgf8* transgenic mice. Isolated 12.5 dpc brains were stained as whole mounts using either in situ hybridization to detect *En-2* or the RAP in situ technique to detect ELF-1. In wild-type brains, mesencephalic expression of *En-2* (*A*) or *ELF-1* (*D*) is limited to caudal regions. In contrast, *En-2* is expressed throughout the alar region of the entire mesencephalon in transgenic brains (*B,C*). Similarly, ELF-1 is ectopically expressed in a very broad dorsal domain along the entire rostrocaudal extent of the mesencephalon and in a transverse stripe (limited to alar regions) in the rostral mesencephalon of transgenic brains (*E,F*). Note that clumps of *En-2* and ELF-1 (*closed arrows in B,D*) expressing cells are also present in dorsal p1. (Modified from Lee et al. 1997.)

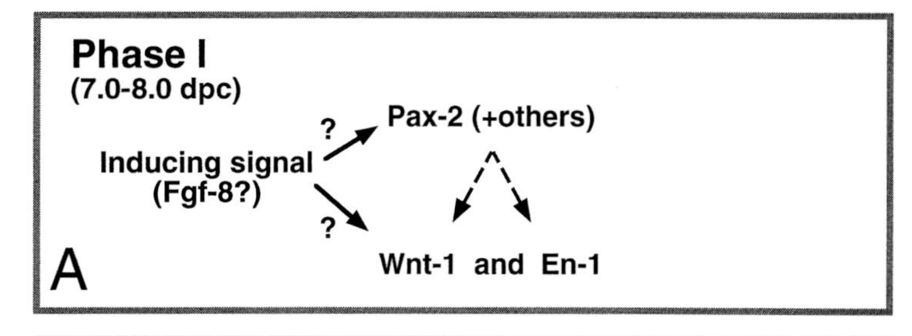

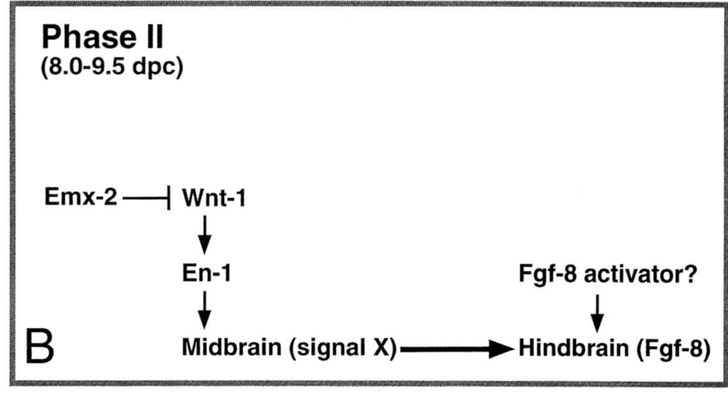

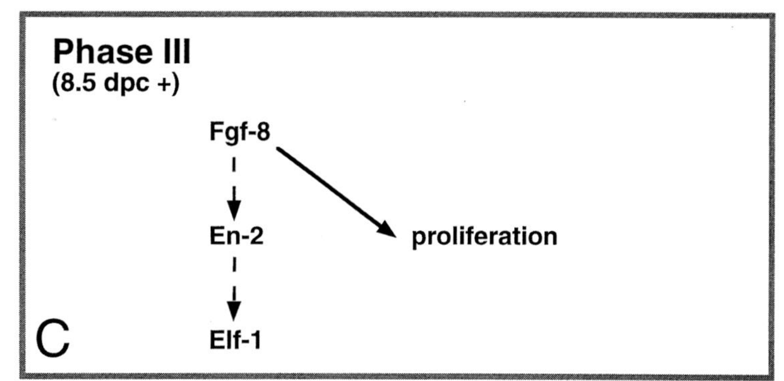

Figure 8. Proposed model summarizing cell interactions involved in pattern formation during midbrain and hindbrain development. For details see the text.

mRNA transcripts encoding the Wnt-1 signal (McMahon and Moon 1989), other *Wnt* gene transcripts, or putative components of the Wnt signal transduction pathway (for review, see Moon et al. 1997). A process closely linked to axis induction is regionalization of the neural ectoderm into the forebrain, midbrain, hindbrain, and spinal cord. We have focused our studies on pattern formation in midbrain development from the time of neural plate induction (~7.5 dpc) until 14.5 dpc, when midbrain polarity is reflected in A/P differences in gene expression patterns, the degree of cytoarchitectonic differentiation, and rates of cellular proliferation. On the basis of recent work, the nature of the regulatory signals involved in midbrain patterning is becoming somewhat clearer. In Figure 8, we propose a model that highlights and summarizes certain roles of the Wnt-1 and Fgf-8 signals in regional pattern formation.

Phase I—Neural Induction (7.0–8.0 dpc)

The earliest known marker of the midbrain-hindbrain region is *Pax-2*, which is expressed in the mouse embryo during gastrulation at 7.0–7.5 dpc (Fig. 2). As such, it could be a transcriptional target of midbrain inductive signaling and have a role in pattern formation of the neural plate. On the basis of genetic evidence from *Drosophila* and zebrafish, the possible conservation of regulatory interactions, and the temporal-spatial expression patterns of *Pax-2* and *Pax-5*, several investigators have proposed that *Pax* transcription factors may regulate *Wnt-1* and/or *Engrailed* genes (Rowitch and McMahon 1995; Joyner 1996). Indeed, Song et al. (1996) report the presence of Pax DNA-binding sequences in the *En-2* enhancer. We have identified a 110-bp *cis*-acting regulatory element that drives appropriate expression of a *Wnt-1* reporter transgene in the neural plate of transgenic mice (pattern as shown in Fig. 3). In contrast to the findings of Song et al. (1996), our data on Pax protein binding to the *Wnt-1* minimal enhancer are inconclusive. In any case, since the expression of *Pax-2* encompasses the distinct subdomains of *Wnt-1* and *En-1* (Rowitch and McMahon 1995), if it does have a regulatory role, it must act in concert with a coactivator or repressor. In this regard, it is interesting to note that Pax-5 cooperates with an Ets transcription factor in binding to the *mb-1* regulatory element (Fitzsimmons et al. 1996).

The midbrain-hindbrain constriction (MHC) has properties associated with an organizing center (for review, see Joyner 1996). Ectopic grafts of MHC to diencephalon can induce midbrain- and hindbrain-specific genes, such as *En-2*, in surrounding tissues. A clue to the nature of such cell-cell interactions is that an Fgf-8-soaked bead is capable of inducing an ectopic midbrain and expression of *Wnt-1* and *En-2* (Crossley et al. 1996). In this session, Huma Sheikh, Ivor Mason, and colleagues demonstrated expression of *En-1*, *Fgf-8*, and *Pax-2*, using a similar experimental paradigm (H. Sheikh et al., pers. comm.). *Fgf-8* is first expressed in the hindbrain at the 3-somite stage (Mahmood et al. 1995), following the onset of *Pax-2*, *Wnt-1*, and *En-1* expression. Thus, based on this sequence of midbrain-specific gene expression, Fgf-8 is unlikely to have a midbrain-inductive role in the setting of normal development (see also Lee et al. 1997). One possibility, therefore, is that an as yet undescribed *Fgf* family member expressed in the mesoderm could be involved in this aspect of midbrain induction and pattern formation. *Fgf-8* loss-of-function studies will be helpful in further clarifying this issue.

Phase II—Development of the Embryonic Midbrain (8.0–9.5 dpc)

Expression of *Wnt-1* in its normal pattern during embryogenesis is essential for CNS development, and one component of this is the repression of *Wnt-1* in the dorsal telencephalon, a function that is likely carried out in vivo by Emx-2 (Fig. 8B). It is not obvious why there should be a mechanism for repression of *Wnt-1* in the forebrain, given that five other *Wnt* genes (S. Lee and A. McMahon, unpubl.) are expressed in that region. Perhaps it is important to regulate the total dosage of the signal; alternatively, Wnt-1 could have specific effects on cell fate. One approach to this question is to ectopically express *Wnt-1* in the dorsal telencephalon using a *Wnt-1* enhancer construct that carries mutation of the HBS1 site (Fig. 4) and such studies are under way in our laboratory.

In the absence of the Wnt-1 ligand, there is failure of midbrain development and subsequent loss of r1. The finding that *En-1* can rescue significant aspects of this phenotype suggests that a major role of Wnt-1 is the maintenance of *Engrailed* expression (Fig. 5). Other roles for Wnt-1 are suggested by the finding that *Otx-2* is ectopically expressed in r1 of *Wnt-1* mutant embryos (Fig. 6) (Bally-Cuif et al. 1995). The midbrain supplies a second signal (signal ‘X’) to the anterior hindbrain, which prevents apoptotic cell death in r1. In turn, the anterior hindbrain is the source of the secreted factor, Fgf-8, which can then act on either side of the midbrain-hindbrain junction, for example, to induce gene expression and/or cellular proliferation in these tissues. These points are summarized in Figure 8.

Phase III—Later Patterning of the Midbrain (>9 dpc)

Some fibroblast growth factors are potent mitogens for CNS progenitors in vitro (Temple and Qian 1995 and references therein), and Fgf-8 is able to dramatically stimulate levels of proliferation in the midbrain of transgenic mice (Lee et al. 1997). *Fgf-8* is expressed in r1 immediately caudal to the midbrain-hindbrain junction, which places it in a position to stimulate proliferation in both the rhombic lip and tissues of the caudal midbrain. Indeed, rostral-to-caudal differences in proliferative rates in the midbrain (Lee et al. 1997) could be explained by exposure to diminishing concentrations of Fgf-8 ligand derived from r1. A “gradient” of *Engrailed* expression in the chicken tectum has been observed (Itasaki et al. 1991) and ectopic expression of *Engrailed* results in induction of the Eph-receptor ligand, ELF-1 (Friedman and O’Leary 1996b; Itasaki and Nakamura 1996; Logan et al. 1996). The finding that both of these molecules are ectopically induced in *Fgf-8* transgenic mice suggests a regulatory hierarchy in which Fgf-8 levels determine expression of *En-2* and *Elf-1* (Fig. 7) (Lee et al. 1997).

At the present time, we do not know whether Fgf-8 has a distinct role in regulating cell proliferation or whether this process is controlled through an *En/ELF-1* pathway. Mice ectopically expressing *En-1* in a pattern similar to that of the WEXP-*Fgf-8* trangenics do not show hyperplasia (Danielian and McMahon 1996; Lee et al. 1997), and misexpression of *En-1* widely in the chick tectum does not result in dramatic expansion of the midbrain (Friedman and O’Leary 1996b; Itasaki and Nakamura 1996; Logan et al. 1996). Thus, the simplest conclusion at this time is that the proliferative effect of Fgf-8 is independent of *En* regulation.

ACKNOWLEDGMENTS

D.H.R. is a Howard Hughes Medical Institute Physician postdoctoral fellow. P.S.D. is a recipient of a postdoctoral fellowship for the Human Frontiers Science Program. These studies were supported by grants NS-32691 and HD-30249 from the National Institutes of Health to A.P.M.

REFERENCES

Acampora D., Mazan S., Lallemand Y., Avantaggiato V., Maury M., Simeone A., and Brulet P. 1995. Forebrain and midbrain regions are deleted in $Otx2^{-/-}$ mutants due to a defective anterior neuroectoderm specification during gastrulation. *Development* **121:** 3279.

Ang S.L., Jin O., Rhinn M., Stevenson L., and Rossant J. 1996. A targeted mouse *Otx2* mutation leads to severe defects in gastrulation and formation of axial mesoderm and to deletion of the rostral brain. *Development* **122:** 243.

Asano M. and Gruss P. 1992. *Pax-5* is expressed at the midbrain-hindbrain boundary during mouse development. *Mech. Dev.* **39:** 29.

Bally-Cuif L., Cholley B., and Wassef M. 1995. Involvement of Wnt-1 in the formation of the mes/metencephalic boundary. *Mech. Dev.* **53:** 23.

Boncinelli E., Gulisana M., and Broccoli V. 1993. *Emx* and *Otx* homeobox genes in the developing mouse brain. *J. Neurobiol.* **24:** 1356.

Brand M., Heisenberg C.-P., Jiang Y.-J., Beuchle D., Lun K., Furutani-Seiki M., Granato M., Haffter P., Hammerschmidt M., Kane D., Kelsh R.N., Mullins M.C., Odenthal J., van Eeden F.J.M., and Nüsslein-Volhard C. 1996. Mutations in zebrafish genes affecting the formation of the boundary between midbrain and hindbrain. *Development* **123:** 179.

Cheng H., Nakamoto M., Bergemann A.D., and Flanagan J.G. 1995. Complementary gradients in expression and binding of ELF-1 and Mek-4 in development of the topographic retinotectal projection map. *Cell* **82:** 371.

Crossley P.H., Martinez S., and Martin G.R. 1996. Midbrain development induced by FGF8 in the chick embryo. *Nature* **380:** 66.

Czerny T., Schaffner G., and Busslingler M. 1993. DNA sequence recognition by Pax proteins: Bipartite structure of the paired domain and its binding site. *Genes Dev.* **7:** 2048.

Danielian P.S. and McMahon A.P. 1996. *Engrailed-1* as a target of the *Wnt-1* signalling pathway in vertebrate midbrain development. *Nature* **383:** 332.

Davis C.A. and Joyner A.L. 1988. Expression patterns of the homeo box-containing genes *En-1* and *En-2* and the proto-oncogene *int-1* diverge during mouse development. *Genes Dev.* **2:** 1736.

Davis C.A., Noble-Topham S.E., Rossant J., and Joyner A.L. 1988. Expression of the homeo box-containing gene *En-2* delineates a specific region of the developing mouse brain. *Genes Dev.* **2:** 361.

Dickinson M.E., Krumlauf R., and McMahon A.P. 1994. Evidence for a mitogenic effect of *Wnt-1* in the developing mammalian central nervous system. *Development* **120:** 1453.

Drescher U.,. Bonhoeffer F, and Muller B.K. 1997. The Eph family in retinal axon guidance. *Curr. Opin. Neurobiol.* **7:** 75.

Drescher U., Kremoser C., Handwerker C., Loschinger J., Noda M., and Bonhoeffer F. 1995. In vitro guidance of retinal ganglion cell axons by RAGS, a 25 kDa tectal protein related to ligands for Eph receptor tyrosine kinases. *Cell* **82:** 359.

Echelard Y., Vassileva G., and McMahon A.P. 1994. *cis*-acting regulatory sequences governing *Wnt-1* expression in the developing mouse CNS. *Development* **120:** 2213.

Epstein J., Cai J., Glaser T., Jepael L., and Maas R. 1994. Identification of a Pax paired domain recognition sequence and evidence for DNA-dependent conformational changes. *J. Biol. Chem.* **269:** 8355.

Favor J., Sandulache R., Neuhauser-Klaus A., Pretch W., Chatterjee B., Senft E., Wurst W., Blanquet V., Grimes P., Sporle R., and Schughart K. 1996. The mouse Pax2 (1Neu) mutation is identical to a human PAX2 mutation in a family with renal-coloboma syndrome and results in developmental defects of the brain, ear, eye, and kidney. *Proc. Natl. Acad. Sci.***93:** 13870.

Fitzsimmons D., Hodsdon W., Wheat W., Maria S.-M., Wasylyk B., and Hagman J. 1996. *Pax-5* (BSAP) recruits *Ets* proto-oncogene family proteins to form functional ternary complexes on a B-cell-specific promoter. *Genes Dev.* **10:** 2198.

Friedman G.C. and O'Leary D.D. 1996a. Eph receptor tyrosine kinases and their ligands in neural development. *Curr. Opin. Neurobiol.* **6:** 127.

———. 1996b. Retroviral misexpression of engrailed genes in the chick optic tectum perturbs the topographic targeting of retinal axons. *J. Neurosci.* **16:** 5498.

Hanks M., Wurst W., Anson-Cartwright L., Auerbach A.B., and Joyner A.L. 1995. Rescue of the *En-1* mutant phenotype by replacement of *En-1* with *En-2*. *Science* **269:** 679.

Iler N., Rowitch D.H., Echelard Y., McMahon A.P., and Abate-Shen C. 1995. A single homeodomain binding site restricts spatial expression of *Wnt-1* in the developing brain. *Mech. Dev.* **53:** 87.

Ingham P.W. 1991. Segment polarity genes and cell patterning within the *Drosophila* body segment. *Curr. Opin. Genet. Dev.* **1:** 261.

Itasaki N. and Nakamura H. 1996. A role for gradient *en* expression in positional specification on the optic tectum. *Neuron* **16:** 55.

Itasaki N., Ichijo H., Hama C. Matsuno T., and Nakamura H. 1991. Establishment of rostrocaudal polarity in tectal primordium: *Engrailed* expression and subsequent tectal polarity. *Development* **113:** 1133..

Joyner A. 1996. *Engrailed, Wnt* and *Pax* genes regulate midbrain-hindbrain development. *Trends Genet.* **12:** 15.

Joyner A.L., Herrup K., Auerbach B.A., Davis C.A., and Rossant J. 1991. Subtle cerebellar phenotype in mice homozygous for a targeted deletion of the *En-2* homeobox. *Science* **251:** 1239.

Keller S.A., Jones J.M., Boyle A., Barrow L.L., Killen P.D., Green D.G., Kapousya N.V., Hitchcock P.F., Swank R.T., and Meisler M.H. 1994. Kidney and retinal defects (Krd), a transgene-induced mutation with a deletion of mouse chromosome 19 that includes *Pax-2*. *Genomics* **2:** 309.

Kelly O.G. and Melton D. 1995. Induction and patterning of the vertebrate nervous system. *Trends Genet.* **11:** 273.

Krauss S., Johansen T., Korzh V., and Fjose A. 1991. Expression of the zebrafish paired box gene pax[zf-b] during early neurogenesis. *Development* **113:** 1193.

Krauss S., Maden M., Holder N., and Wilson S.W. 1992. Zebrafish pax[b] is involved in the formation of the midbrain-hindbrain. *Nature* **360:** 87.

Lee S.M.K., Danielian P.S.F.B., Fritzsch B., and McMahon A.P. 1997. Evidence that FGF8 signalling from the midbrain-hindbrain junction regulates growth and polarity in the developing midbrain. *Development* **124:** 959.

Logan C., Wizenmann A., Drescher U., Monschau B., Bonhoeffer F., and Lumsden A. 1996. Rostral optic tectum acquires caudal characteristics following ectopic *Engrailed* expression. *Curr. Biol.* **6:** 1006.

Mahmood R., Bresnick J., Hornbruch A., Mahony C., Morton N., Colquhoun K., Martin P., Lumsden A., Dickson C., and Mason I. 1995. A role for FGF-8 in the initiation and maintenance of vertebrate limb bud outgrowth. *Curr. Biol.* **5:** 797.

Mastick G.S., Fan C.M., Tessier-Lavigne M., Serbedzija G.N., McMahon A.P., and Easter S.E. 1996. Early deletion of neuromeres in the *Wnt-1*$^{-/-}$ mutant mice: Evaluation by morphological and molecular markers. *J. Comp. Neurol.* **374:** 246.

Matsuo I., Kuritani S., Kimura C., and Aizawa S. 1995. Mouse *Otx2* functions in the formation and patterning of the rostral head. *Genes Dev.* **9:** 2646.

McMahon A.P. and Bradley A. 1990. The *Wnt-1 (int-1)* proto-oncogene is required for development of a large region of the mouse brain. *Cell* **62:** 1073.

McMahon A.P. and Moon R.T. 1989. Ectopic expression of the proto-oncogene *int-1* in *Xenopus* embryos leads to duplication of the embryonic axis. *Cell* **58:** 1075.

McMahon A.P., Joyner A.L., Bradley A., and McMahon J.A. 1992. The midbrain-hindbrain phenotype of *Wnt-1*$^{-}$*Wnt-1*$^{-}$ mice results from stepwise deletion of *Engrailed*-expressing cells by 9.5 days post coitum. *Cell* **69:** 581.

Millen K., Wurst W., Herrup K., and Joyner A.L. 1994. Abnormal embryonic cerebellar development and patterning of postnatal foliation in two mouse *Engrailed-2* mutants. *Development* **120:** 695.

Moon R.T., Brown J.D., and Torres M. 1997. Wnts modulate cell fate and behavior during vertebrate development. *Trends Genet.* **13:** 157.

Nakamoto M., Cheng H.J., Friedman G.C., McLaughlin T., Hansen M.J., Yoon C.H., O'Leary D.D., and Flanagan J.G. 1996. Topographically specific effects of ELF-1 on retinal axon guidance in vitro and retinal axon mapping in vivo. *Cell* **86:** 755.

Nornes H.O., Dressler G.R., Knapik E.W., Deutsch U., and Gruss P. 1990. Spatially and temporally restricted expression of *Pax-2* during neurogenesis. *Development* **109:** 797.

Puelles L. and Rubenstein L.R. 1993. Expression patterns of homeobox and other putative regulatory genes in the embryonic mouse forebrain suggest a neuromeric organization. *Trends Neurosci.* **16:** 472.

Puschel A.W., Westerfield M., and Dressler G. 1992. Comparative analysis of *Pax-2* protein distributions during neuralation in mice and zebrafish. *Mech. Dev.* **38:** 197.

Quinlan G.A., Williams E.A., Tan S.-S., and Tam P.L. 1995. Neuroectodermal fate of epiblast cells in the distal region of the mouse egg cylinder: implication for body plan organization during early embryogenesis. *Development* **121:** 87.

Rijsewijk F., Schuermann M., Wagenaar E., Parren P., Weigel

D., and Nusse R. 1987. The *Drosophila* homologue of the mouse mammary oncogene *int-1* is identical to the segment polarity gene *wingless. Cell* **50:** 649.

Rowitch D.H. and McMahon A.P. 1995. *Pax-2* expression in the murine neural plate preceeds and encompasses the expression domains of *Wnt-1* and *En-1. Mech. Dev.* **52:** 3.

Rubenstein J.L.R., Martinez S., Shimamura K., and Puelles L. 1994. The embryonic vertebrate forebrain: The prosomeric model. *Science* **266:** 578.

Ruiz i Altaba A. 1994. Pattern formation in the vertebrate neural plate. *Trends Neurosci.* **17:** 233.

Serbedzija G.N., Dickinson M., and McMahon A.P. 1996. Cell death in the CNS of the *Wnt-1* mutant mouse. *J. Neurobiol.* **31:** 275.

Song D., Chalepakis G., Gruss P., and Joyner A.L. 1996. Two *Pax*-binding sites are required for early embryonic brain expression of an *Engrailed-2* transgene. *Development* **122:** 627.

Temple S. and Qian X. 1995. bFGF, neurotropins, and the control of cortical neurogenesis. *Neuron* **15:** 249.

Thomas K.R. and Capecchi M.R. 1990. Targeted disruption of the murine *int-1* proto-oncogene resulting in severe abnormalities in midbrain and cerebellar development. *Nature* **346:** 847.

Thomas K.R., Musci T.S., Neumann P.E., and Capecchi M.R. 1991. Swaying is a mutant allele of the proto-oncogene *Wnt-1. Cell* **67:** 969.

Torres M., Gomez-Pardo E., and Gruss P. 1996. *Pax2* contributes to inner ear patterning and optic nerve trajectory. *Development* **122:** 3381.

Urbanek P., Fetka I., Meisler M.H., and Busslinger M. 1997. Cooperation of *Pax2* and *Pax5* in midbrain and cerebellum development. *Proc. Natl. Acad. Sci.* **94:** 5703.

Urbanek P., Wang Z.-Q., Fetka I., Wagner E.F., and Busslinger M. 1994. Complete block of early B cell differentiation and altered patterning of the posterior midbrain in mice lacking Pax5/BSAP. *Cell* **79:** 901.

Wilkinson D.G., Bailes J.A., and McMahon A.P. 1987. Expression of the proto-oncogene *int-1* is restricted to specific neural cells in the developing mouse embryo. *Cell* **50:** 79.

Wurst W., Auerbach A.B., and Joyner A.L. 1994. Multiple developmental defects in engrailed-1 mutant mice: An early mid-hindbrain deletion and patterning defects in forelimbs and sternum. *Development* **120:** 2065.

Yoshida M., Yoko S., Matsuo I., Miyamoto N., Takeda N., Kuritani S., and Aizawa S. 1997. Emx1 and Emx2 functions in the development of dorsal telencephalon. *Development* **124:** 101.

Otx and *Emx* Functions in Patterning of the Vertebrate Rostral Head

I. MATSUO, Y. SUDA, M. YOSHIDA, T. UEKI, C. KIMURA, S. KURATANI, AND S. AIZAWA
Department of Morphogenesis, Institute of Molecular Embryology and Genetics, Kumamoto University School of Medicine, Kumamoto-860, Japan

HEAD IN ANIMAL BODY PLAN

All the animals that move for food have their heads in the dorsal front of the mouth. Phylogeny generally considers two major streams of animal evolution: protostomes and deuterostomes. The position of the blastopore confers the adult mouth in the former and the anus in the latter. Paradoxically, the anteroposterior (A/P) orientation is preserved between the two lineages: *HOM-C* code in *Drosophila* and *Hox* code in vertebrate have the same A/P orientation (Carroll 1995). Spiral cleavage in protostomes generates two sources of mesoderm: ectodermal and endodermal. As typically seen in development of annelid, the protocoel of trocophore larva is composed of ectodermal mesoderm, and the larva becomes the adult head in postlarval development. Deuterocoel of the adult trunk is made of endodermal mesoderm or teloblast posteriorly to blastopore. In contrast, radial cleavage in deuterostomes generates hollow blastula. Gastrulation occurs by invagination of archenteron anteriorly; the anterior part of the invaginant participates in head formation and the posterior part in trunk formation; the head organizer proceeds to trunk organizer. Thus, the bodies of protosomes are made posteriorly and of deuterostomes anteriorly from the blastopore with the same A/P orientation. Our mouth is not the anus of the insect. In both lineages, the head formation appears to advance to the trunk formation.

To increase body size, animals store yolk for development, and the cleavage and gastrulation are modified in a variety of ways by the amount of yolk they retain. *Drosophila* is unique in that it centralizes yolk for its development, and mesoderm invaginates as a furrow in the ventral. Nevertheless, in the insect body plan, as seen in short germ-band insects such as grasshopper, the early germ band consists only of the head lobes and a growth zone, and thoracic and abdominal segments are formed secondarily and posteriorly from the growth zone (Sander 1976; Tautz and Sommer 1995). In deuterostomes, the anterior tip of the archenteron generates mesoderm in echinoderm, and its dorsal side in *Amphioxus*. In vertebrates, the mesoderm production centers around the blastopore in amphibian and around node and primitive streak in avian and mammals. The basic plan for body formation noted above, however, is unchanged in both arthropoda and vertebrates. The neural system is formed as a ladder-like system in arthropoda and as a neural tube in vertebrates. The endodermal alimentary canal or archenteron that digests yolk is able to cross the ladder-like nervous system in the former, but not the neural tube in the latter, each mandatorily giving rise to the body plan illustrated in Figure 1. If arthropoda had made the yolk side ventral, this animal would have had the head in the ventral part of the mouth. Such an animal has never existed. To construct the head in the dorsal front of the mouth, protostomes had to make the yolk side dorsal and deuterostomes the yolk side ventral, as pointed out by G. Zaddach in 1854 (Nübler-Jung and Arendt 1994).

VERTEBRATE HEAD

The most extreme complexity achievable by the ladder-like neural system in protosomes that is comparable to the primitive neural system of vertebrates is seen in the octopus. The brain is most conspicuous in vertebrates, however, and may have become possible by the neural system as the neural tube. This is seen not only in vertebrates, but also in *Amphioxus* and tunicates. Its origin was once hypothesized to be the dorsal convergence of ciliary band with neoteny of *Auricularia* larva (Garstang 1894, 1929; Lacalli 1996) but remains totally uncertain. Tunicates have a brain vesicle and *Amphioxus* have a cerebral vesicle in the rostral tip of the neural tubes. Genes that are uniquely expressed in the vertebrate rostral head are also expressed in the anterior neural tube upon development of these vesicles (Holland and Garcia-Fernandez 1996; Wada et al. 1996; Williams and Holland 1997). Such genes are, however, also expressed in *Drosophila* head, as expected from the origin of all extant animals in the Cambrian Sea. The expressions of homologous genes or the existence of homologous gene cascades do not necessarily implicate the homology of the structure concerned. Vertebrates share a set of synapomorphs in the head region: possession of eyes, inner ear, skull, well-differentiated brain composed of forebrain, midbrain and hindbrain, neural crest derivatives, and so forth. No homologous structures are present in *Amphioxus* or tunicate, whereas the organization of the brain may have changed little from the earliest vertebrate lineage to today. *Pteraspid* in the *Silurian* period appears to have these structures (Janvier 1993).

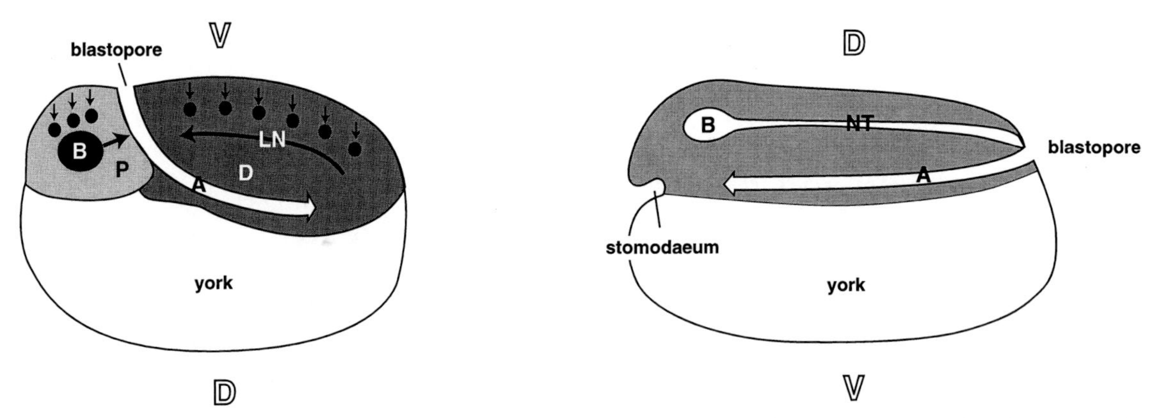

Figure 1. Schematic representation of lobster and avian body plan. The lobster is the animal with which Geoffroy St. Hilarie (1822) speculated on dorsoventral reversion between insects and vertebrates. (P) Protocoel made of ectodermal mesoderm; (D) deuterocoel made of endodermal mesoderm or teloblast; (A) endodermal alimentary canal or archenteron; (B) brain; (LN) ladder-like nervous system; (NT) neural tube.

HEAD ORGANIZER AND ANTERIOR VISCERAL ENDODERM

Development of the head has been considered to originate in the head organizer at gastrulation (Spemann 1931; Mangold 1933). Several genes are now identified as components of the head organizer: *Otx2, Lim 1*, and *cerberus*. (Acampora et al. 1995; Matsuo et al. 1995; Shawlot and Behringer 1995; Ang et al. 1996; Bouwmeester et al. 1996). In *Xenopus*, microinjection of these mRNAs causes a secondary axis with anterior structures and reduced trunk structures (Taira et al. 1994; Bouwmeester et al. 1996; Andreazzoli et al. 1997). *Otx2* as well as *Lim1* KO mice fail to develop structures anterior to rhombomere 3 (Acampora et al. 1995; Matsuo et al. 1995; Shawlot and Behringer 1995; Ang et al. 1996). The early dorsal blastopore lip or node that has head organizer activity generates the prechordal mesendoderm, whose role in development of the rostral brain has been suggested by many studies (Spemann and Mangold 1924; Slack and Tannahill 1992; Storey et al. 1992; Doniach 1993; Ang et al. 1994; Foley et al. 1997). *Otx2* and *Lim1* are both expressed in the anterior node derivatives. Of these, only *Otx2* becomes expressed in the anterior neural plate, but it apparently occurs when the prechordal mesoderm has not yet migrated into the head region. Several recent mutant studies suggest that the role of prechordal mesendoderm is not essential in A/P patterning of the brain (Ang and Rossant 1994; Chian et al. 1996; Schier et al. 1997); in the absence of prechordal mesoderm, the anterior neural tube is formed with A/P patterning. In addition, removal of the embryonic shield from zebrafish embryos does not prevent rostral brain development (Shih and Fraser 1996), and some A/P patterning occurs in *Xenopus* Keller explants (Mathers et al. 1995).

Determination of A/P axis in mammals appears to precede the primitive streak formation, and the role of anterior visceral endoderm in anterior specification has recently been suggested (Thomas and Beddington 1996; Varlet et al. 1997). The head organizer components stated above as well as *Rpx1/Hesx1* and *nodal* are also expressed in the anterior primitive endoderm (Bouwmeester et al. 1996; Hermesz et al. 1996; Thomas and Beddington 1996; Varlet et al. 1997). It is an intriguing question whether and how gene functions are coordinated between visceral endoderm and early node. Visceral endoderm may be homologous to avian hypoblast, but how did it evolve and how was it endowed with the anterior specification through vertebrate lineages? We are surveying an enhancer element that directs *Otx2* expression in anterior visceral endoderm. The enhancer will provide a precious starting point to examine the function of *Otx2* gene in visceral endoderm and thereby the function of the endoderm in anterior specification.

HEAD SEGMENT: *Otx* AND *Emx* GENES

The vertebrate body is believed to follow a metamerical plan, and the *Hox* cluster genes control segmental patterning in hindbrain and trunk regions (McGinnis and Krumlauf 1992). The *Hox* genes, however, are not expressed in the rostral head. Several neuromeric or prosomeric models have been proposed for the developmental patterning of rostral brain (Von Baer 1828; Figdor and Stern 1993; Puelles and Rubenstein 1993; Guthrie 1995), but it is yet uncertain whether this region is also patterned on a segmental plan. Far from that, until recently it had been a matter of dispute where the anterior end of the brain or of the sulcus limitans that divides the alar and basal plates is to be found. Analyses with molecular markers now indicate it is probably located around the optic chiasm (Shimamura et al. 1995).

Although cephalization may have occurred independently in the evolutional lineages leading to insects and vertebrates and no direct equivalency can be ascertained in any of the brain structures between them, molecules that function in the insect head should give us clues in analyses of molecules in the vertebrate head as the ancestors of both were born in the Cambrian Sea. In *Drosophila*, under the morphogenetic gradients of the maternal genes, *bicoid* and *hunchback*, the head is patterned by head gap genes such as *orthodenticle* (*otd*), *empty spiracles* (*ems*), and *buttonhead* (*btd*), whereas the trunk is patterned by a genetic cascade from trunk gap genes to homeotic selector (*Hom-C*) genes (Cohen and Jürgens 1990; Finkelstein and Perrimon 1990; Walldorf and Gehring 1992; Simpson-Brose et al. 1994). The mutational inactivation of *otd* and *ems* results in a loss of specific epidermal head segments and embryonic brain segments (Cohen and Jürgens 1990; Finkelstein and Perrimon 1990; Hirth et al. 1995).

Otx1 and *Otx2* and *Emx1* and *Emx2* genes were identified by Simeone and Boncineli's group as the mouse cognates of *otd* and *ems*, respectively (Simeone et al. 1992a,b, 1993). In mouse at the pharyngula stage, *Otx2* expression is first established in future rostral brain (Simeone et al. 1993). Thereafter, expressions of *Otx1*, *Emx2*, and *Emx1* genes occur in this order in more limited regions and in a nested pattern about 8–9 dpc (days postcoitus) (Simeone et al. 1992a,b, 1993), when regionalization of rostral brain takes place. Expression patterns of *Otx* genes are suggestive of their roles in the entire forebrain and midbrain, and those of *Emx* genes in delineation of the cerebral hemispheres out of the forebrain.

Probably coinciding with its nested expression, *Otx2* mutation displayed craniofacial defects by haploinsufficiency (Matsuo et al. 1995). In the skull, defects never extended caudal to the mandibular arch. Among the first arch elements, the defect was specifically found in the mandible, but not in other elements. In addition, *Otx2* heterozygous skull defects were also found in the trabecula cartilage. Together with the *Hoxa-2* mutant phenotype (Rijli et al. 1993), a transplantation experiment (Noden 1983), and the migration of midbrain neural crest cells into the first arch (Osumi-Yamashita et al. 1994), we propose that the distal part of the first arch and the more anterior region as well as midbrain and forebrain (see below) are under *Otx* control, whereas the second and more posterior arches are under *Hox* regulation (Matsuo et al. 1995). Such a scheme leaves a default state in rhombomere 1 (r1)/r2 and the proximal part of the first arch, which is destined to develop middle ear components (Fig. 2).

The vertebrate skull is generally classified into the neurocranium and viscerocranium. Viscerocranial components are of neural crest origin and metamerically organized with cranial nerves. The most anterior component of the viscerocranium is the first branchial arch, the mandibular arch, that belongs to the same head segment as the maxillomandibular branch of the trigeminal nerve. The neurocranium comprises mesodermal components and trabecular components of neural crest origin. However, *Otx2* heterozygous defects favor the view first presented by Huxley in 1874 that the trabecular is the most anterior component of the viscerocranium, the premandibular component, which belongs to the same head segment as the ophthalmic branch of the trigeminal nerve (de Beer 1931; Matsuo et al. 1995; Kuratani et al. 1997). The ophthalmic lobe originates from the midbrain neural crest, whereas the maxillomandibular portion originates from the hindbrain neural crest (Lumsden et al. 1991).

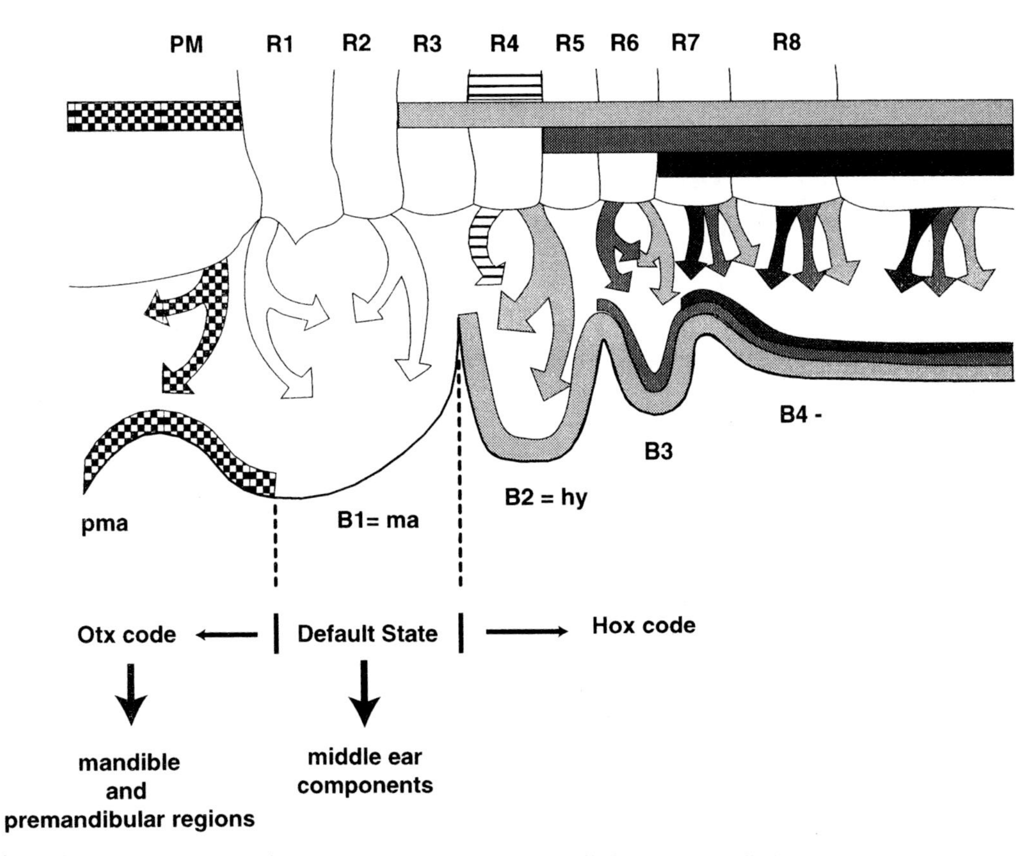

Figure 2. *Otx/Hox* code. (PM) Posterior mesencephalon; (pma) premandibular; (ma) mandibular; (hy) hyoid; (R1–8) rhombomere 1–8; (B1–4) branchial arch 1–4.

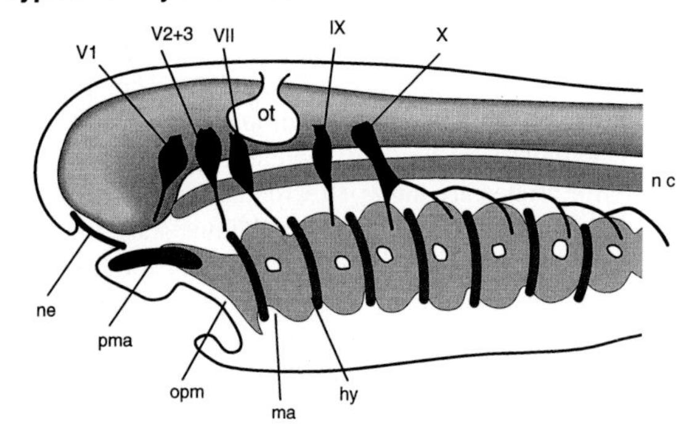

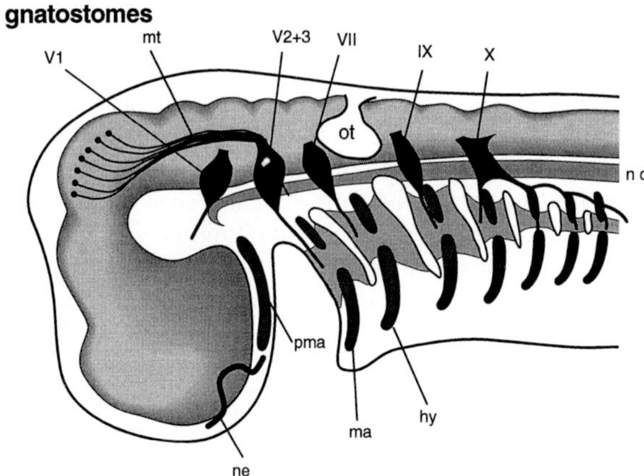

Figure 3. Transition from agnatha to gnathostomata based on the diagram by de Beer (1931). (hy) Hyoid arch skelton; (ma) mandibular arch skelton; (mt) mesencephalic trigeminal neurons; (nc) notochord; (ne) nasal epithelium; (opm) oropharyngeal membrane; (ot) otic vesicle; (pma) premandibular trabecular cartilage; (V1) profundus or opthalmic branch of the trigeminal nerve; (V2+3) maxillomandibular portion of the trigeminal nerve; (VII) facial nerve; (IX) glossopharyngeal nerve; (X) vagus nerve.

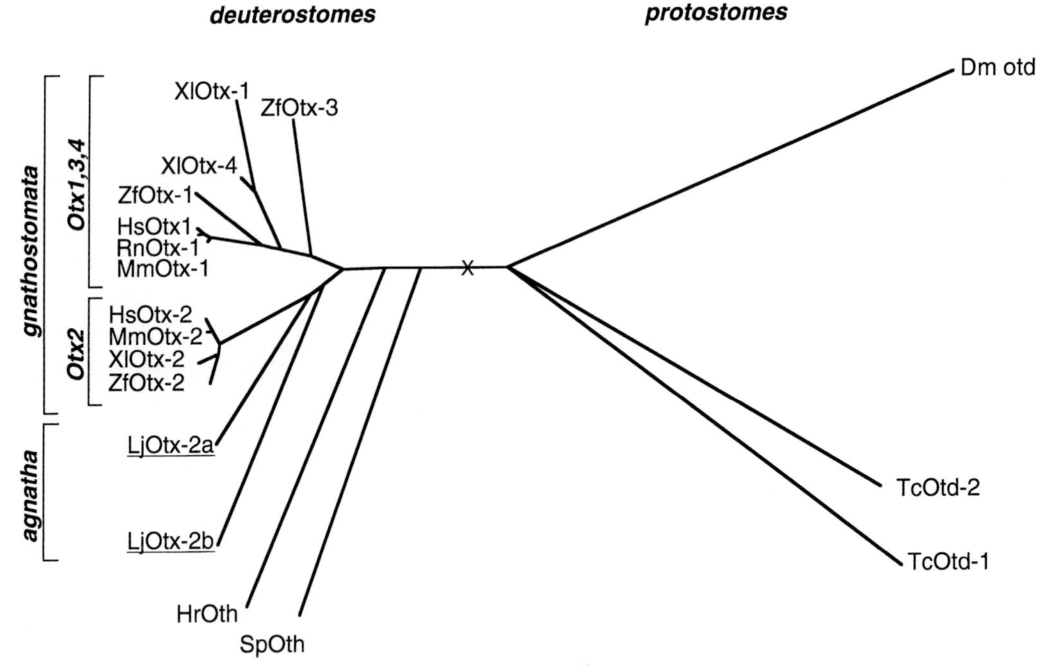

Figure 4. Phylogenetic tree of *Otx*/*Otd* cognates. Sequence data are from SWISSPROT and PIR protein database as follows: (Dm otd) *Drosophila orthodenticle* gene (Finkelstein and Perrimon 1990); (HsOtx1 and HsOtx2) human OTX-1 and OTX-2 genes (Simeone et al. 1993a); (MmOtx-1 and MmOtx-2) mouse *Otx-1* and *Otx-2* genes (Simeone et al. 1992); (RNOTX1) Rat *Otx-1* gene (Frantz et al. 1994); (XlOtx-1, XlOtx-2, XlOtx-4) *Xenopus Otx-1*, *Otx-2*, and *Otx-4* genes (Blitz and Cho 1995; Kablar et al. 1996); (ZfOtx-1, ZfOtx-2, and ZfOtx-3) zebrafish *Otx-1*, *Otx-2*, and *Otx-3* genes (Li et al. 1994; Mori et al. 1994); (SpOth) *Strongylocentrotus purpuratus Otx* gene (Gan et al. 1995); (HrOth) *Halocynthia roretzi orthodenticle* gene (Wada et al. 1996); (TcOtd-1 and TcOtd-2) *Tribolium castaneum Otd-1* and *Otd-2* genes (Li et al. 1996). Alignment and phylogenetic tree were made by neighbor joining method (Saitou and Nei 1987) using ClustalW software (Higgins et al. 1992).

The lobe is thought to be originally an independent nerve called the profundus nerve (Goodrich 1930; de Beer 1931). Indeed, *Otx2* heterozygotes exhibit defects of the ophthalmic branch, but never of the maxillary or mandibular branch.

FROM AGNATHA TO GNATHOSTOMATA

In the *Otx2* heterozygous nervous system, defects were also found in mesencephalic trigeminal (MT) neurons (Matsuo et al. 1995). The structures affected in *Otx2* heterozygotes, mandible, MT neurons, and trabecular components are those that were invented or revolutionized upon transition from agnatha to gnathostomata. They are also intimately related with midbrain neural crest (Fig. 3). Following this transition, the first arch made a masticating system; a pair of gill bars fringed the expanding mouth cavity and became armed with teeth. To control the force of bite, MT neurons may have been incorporated into this system. Agnatha that has no mandible also has no MT neurons (Sarnat and Netsky 1981). The neural tube beyond the anterior tip of the notochord expanded to develop the brain. No mesoderm was available in this region; instead, premandibular trabecular cartilage was used to develop the neurocranium to support the expanding brain.

We therefore searched for *Otx* homologs in an extant agnatha, the lamprey (T. Ueki et al., in prep.). In gnathostomes, *Otx2* cognates are highly conserved, whereas *Otx1* cognates are moderately diverged (Fig. 4). Although several invertebrates also have two *otd*/*Otx* homologs, the divergence into *Otx1* and *Otx2* lineages apparently coincides with the evolution of vertebrates. We identified two *Otx* homologs in lamprey. One, *LjOtx2a*, obviously belongs to the *Otx2* lineage, although it is diverged from ganthostome *Otx2* cognates. The other, *LjOtx2b*, belongs to *Otx2* lineage only with 42% probability, implying that the gene was diverged where *Otx1* and *Otx2* lineages branched. *LjOtx2b* was not expressed in brain neuroepithelium, but *LjOtx2a* is expressed in the fore- and midbrain. We believe that the comparison of *Otx2* functions in midbrain neural crest between agnatha and gnathostomes as well as the knowledge of whether or not the crest cells migrate into the first arch in the agnatha can provide us with valuable information on the transition from agnatha to gnathostomes.

In one of such studies, we searched for the *cis*-elements responsible for *Otx2* expression in midbrain neural crest cells of mouse and pufferfish, hypothesizing that they should be preserved through all gnathostomes. The tetraodontoid fish, *Fugu rubripes* (*Fugu*), has a compact genome of approximately 400 Mb, 7.5 times smaller than the human genome, whereas the two genomes have a similar number of genes (Brenner et al. 1993). This is due to smaller intergenic and intronic sequences and less-repetitive DNA sequences in the fish genome. With this advantage, the transcriptional enhancer region of the *Hoxb-4* gene was identified as having highly conserved intronic sequences between the mouse and *Fugu* genomes (Aparicio et al. 1995). In mouse, 5′-flanking 49-bp sequences were found to be essential and sufficient for *Otx2* expression in midbrain neural crest cells. By comparing the sequences with the *cis*-element of *Fugu Otx2* along with mutational analyses, two motifs, TAAATCTG and CTAATTA, conserved between the two species were found to be crucial for *Otx2* expression in midbrain neural crest cells (Kimura et al. 1997). The former has no known consensus binding sequences for transcriptional factors, but the latter contains the core motif for binding of the homeodomain proteins *MHox, Cart-1, Msx*, and *Dix*, of which mutations are reported to cause defects in cranial neural crest cell-derived structures (Satokata and Mass 1994; Martin et al. 1995; Zhao et al. 1996; Qiu et al. 1997). The obvious next step in the study is the production of mice that have a mutation in this *cis*-element to learn the effects of specific *Otx2* deficiency in midbrain neural crest cells.

REGIONALIZATION IN MIDBRAIN AND ISTHMIC ORGANIZER

At the pharyngula stage, the vertebrate rostral brain primordium consists of prosencephalic and mesencephalic vesicles. The mesencephalic vesicle comprises mesencephalon (mes) and anterior metencephalon (met) that later generates cerebellum (Martinez and Alvarado-Mallart 1989; Hallonet et al. 1990; Bally-Cuif and Wassef 1995; Bally-Cuif et al. 1995; Millet et al. 1996). Isthmus is known to function as an organizer for the development of midbrain (Marin and Puelles 1994). A series of gene candidates that would have functional roles in this regionalization have been suggested (Bally-Cuif and Wassef 1995). In particular, *Wnt1, En, Pax*, and *Fgf8* are thought to be key elements of a genetic cascade leading to mes/metencephalic development (Bally-Cuif and Wassef 1995; Crossley et al. 1996; Danielian and McMahon 1996; Joyner 1996; Lee et al. 1997; Urbanek et al. 1997). However, details of the molecular mechanisms to define the mes/metencephalic territory and to distinguish and specify each domain within it, allowing the correct formation of isthmus organizer, have not yet been elucidated.

In brain, *Otx2* heterozygous mutants exhibit defects in caudal midbrain structures such as the inferior colliculus and in the anterior forebrain structures of eyes, olfactory bulb, and hypophysis (Matsuo et al. 1995). Affected structures correspond to the most anterior and most posterior parts of *Otx2* expression where *Otx1* is not expressed. Defects are not found in the regions in which *Otx1* is expressed or in early brain patterning by *Otx1* mutation, suggesting that the *Otx1* and *Otx2* functions overlap in the regions in which both are expressed (Suda et al. 1996). The *Otx1* and *Otx2* double-mutant phenotype was then examined. Examinations were possible only in $Otx1^{+/-}Otx2^{+/-}$ state, but not in the $Otx1^{-/-}Otx2^{+/-}$, $Otx1^{+/-}Otx2^{-/-}$, or $Otx1^{-/-}Otx2^{-/-}$ state. Nevertheless, the perinatal double heterozygotes exhibited marked defects throughout fore- and midbrains where abnormalities were never found by a single mutation of either gene alone (Suda et al. 1996, 1997). The perinatal brain defects were primarily due to abnormalities in early regional patterning. Analyses with molecular markers at 9.5 dpc suggested the failure in development of prosomere 1 (p1)/p2 diencephalon and mesencephalon with the expansion of anterior metencephalon, r1. The isthmic genes *Pax2, Gbx2, Fgf8* as well as *Wnt1* exhibited a characteristic lateral stripe, but dorsally, *Fgf8* expression was expanded along with *Pax2*$^-$, *Gbx2*$^-$, *Wnt1*$^+$, *En1*$^+$, *En2*$^+$, *Otx1*$^+$, *Otx2*$^+$, *Wnt7b*$^-$, and *Emx2*$^-$ characteristics. In terms of these molecular markers, mutant defects were apparent at the six-somite stage, but not at the three-somite stage. Particularly broad *Fgf8* expression at the three-somite stage took place normally, but it did not concentrate into a spot corresponding to a future isthmus at the six-somite stage. We propose that regionalization of the mes/metencephalic domain takes place around the three- to six-somite stage with the onset of *Otx1* function. During this stage in mesencephalic vesicle, the caudal border of *Otx* expressions and isthmus may be established; the *Otx*-positive portion gives rise to mesencephalon and the *Otx2*-negative portion gives rise to the rostral r1 that corresponds to the mesencephalon-derived cerebellar primordia (Millet et al. 1996). Fgf8 does not participate in the establishment of the midbrain territory itself (Crossley et al. 1996; Lee et al. 1997). Instead, the Fgf8 signaling appears to give A/P polarity in the susceptible region, mesencephalon and diencephalon, by regulating growth and differentiation through its targets such as graded expression of *En-2* and *ELF-1* genes (Lee et al. 1997). Thus, the *Otx1* and *Otx2* genes may cooperate upstream of Fgf8 signaling for the establishment of mesencephalon and caudal diencephalon, allowing for the correct development of isthmus.

Double heterozygotes develop truncated structures in the anterior region of the lateral stripe characteristic of the isthmic genes, in the region corresponding to midbrain and/or p1/p2 diencephalon of wild-type brain, with the above molecular marker phenotype, and defects were milder in the ventral thalamus and telencephalon. However, it may not be extraordinary to imagine that the entire rostral brain can be lost when *Otx2* and *Otx1* genes are double-homozygously mutated with the rescue of the earlier *Otx2* function as head organizer. Of note is that *Otx2* heterozygous phenotype and *Otx1/Otx2* double-heterozygous phenotype are influenced by the genetic background of mice, C57BL/6 or CBA, suggesting that a modifier gene(s) has an important role in patterning of the rostral head. *Otx2* maps to mouse chromosome 14 [D14Mit3-8.4CM-Otx2-4.5CM-D14Mit5] (Oyanagi et al. 1997); near this locus, three developmental mouse mutations are known: pugnose (*pn*) and disorganization (*Ds*) are about 3 and 4 cM apart, respectively, whereas waved coat (*Wc*) is very close to *Otx2*. The defects in these mutants all appear to relate to body patterning and bear several common phenotypic features with *Otx2* heterozygous mutants (Hummel 1959; Kidwell et al. 1961; Diwan and Stevens 1974). Interestingly, *Otx2*, *Wc*, and *Ds* mutations are all semidominant, and the heterozygous phenotype displays variable phenotypes by the genetic background of mice.

REGIONALIZATION IN FOREBRAIN

The evolution of the cerebral hemisphere is one of the most spectacular stories in comparative anatomy (Romer and Parsons 1949). The brain tube initially develops as a single, unpaired structure all the way forward to the anterior end. Then, alar plates of the anterior prosencephalic vesicle protrude into distinct paired structures, cerebral hemispheres with lateral ventricles (Puelles et al. 1987). In Cyclostomata and Chondrichthyes, a single ventricular cavity, the third ventricle, bifurcates only anteriorly. In Cyclostomata, the hemisphere is mainly an olfactory lobe, paleopallium (Romer and Parsons 1949). In amphibians as well as fish such as shark and lungfish, basal ganglia develop ventrally, and the archipallium, dorsally and medially in the hemisphere. Paleopallium exists as a band of tissue along the lateral surface. In primitive reptiles, the basal ganglia are large and have moved to the inner part of the hemisphere. The first faint traces of mammalian cortical development, the neopallium, are to be seen between the paleopallium and archipallium in certain reptiles. The evolutional history of the mammalian brain is a story of the expansion and elaboration of this neopallium, with which the archipallium is folded into a restricted area on the median part of the hemispheres as the hippocampal region, and the paleopallium comes to constitute only a small ventral hemisphere region, the pyriform lobe. Basal ganglia, the corpus striatum, occupy much of the inside of the hemispheres. Almost nothing is known, however, about the molecular basis of the evolution or the regionalization of cerebral hemispheres.

As noted earlier, expressions of *Emx* genes are suggestive of their roles in delineation of the cerebral hemispheres out of the forebrain. In the *Emx2* mutant forebrain, the roof between cerebral hemispheres expanded, and the medial part of the pallium was greatly reduced, with a lateral shift of its boundary (Yoshida et al. 1997). This most medial pallium that was transformed into the choroidal roof by the *Emx2* mutation corresponds to the archipallium. In newborn *Emx2* mutants, the dentate gyrus was missing, and hippocampus and medial limbic cortex were greatly reduced in size (Pellegrini et al. 1996; Yoshida et al. 1997). When *Emx1* expression takes place around 9.0 dpc, *Emx1* and *Emx2* expressions largely overlap in dorsal telencephalon, but *Emx1* is not expressed in the most medial part of the pallium (Yoshida et al. 1997). There is an *Emx2*-positive but *Emx1*-negative domain in the most medial pallium; neither is expressed in the roof. Therefore, we speculated that *Emx2* determines the boundary between the pallium and the roof, and its disruption has shifted the boundary laterally to the site of the *Emx1* expression. *Emx2* expression, however, takes place in the prospective telencephalon around the two-somite stage at 8.0 dpc and is likely to correspond to the prospective archipallium when one refers to the model by Alvarez-Bolado et al. (1995). It is thus possible that the *Emx2* mutant phenotype resides in this earlier *Emx2* expression. On the other hand, defects were not found in the region where *Emx1* expression overlaps, nor in the early embryonic forebrain by the *Emx1* mutation. Analyses of the *Emx2* and *Emx1* double-mutant phenotype are preliminary, but the neopallium is meager in the double mutants, suggesting that *Emx1* and *Emx2* functions overlap for the development of the telencephalic region where both are expressed.

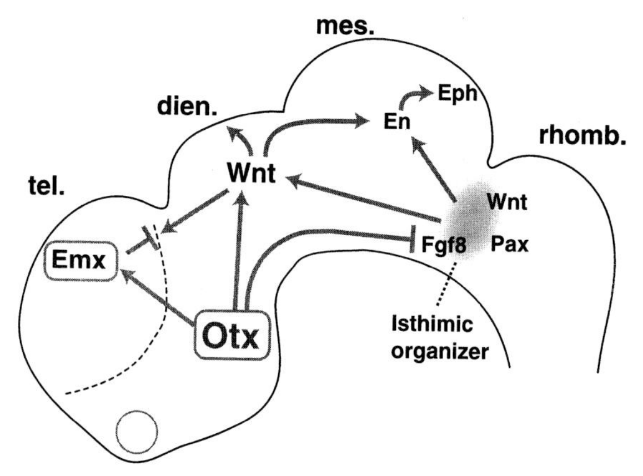

Figure 5. *Otx* and *Emx* functions in patterning of the vertebrate rostral brain. (dien) Diencephalon; (mes) mesencephalon; (rhomb) rhombencephalon; (tel) telencephalon.

Emx1 and *Emx2* mutants thus appear to present a clue to the molecular basis of the regionalization of cerebral hemispheres. Although the molecules that interact with *Emx* genes belong to future studies, it is noteworthy that in the enhancer of *Wnt1* gene, there exist sequences to which *Emx2* can bind (Iler et al. 1995). Transgenic analyses have indicated that the lack of these sequences leads to ectopic expression of *Wnt1* in the region that was affected in the *Emx2* mutant, suggesting that *Emx2* may function as a negative regulator of *Wnt1* expression. Supporting this, *Wnt1* expression extended anteriorly in the *Emx2* mutant at the two-somite stage (Yoshida et al. 1997; our unpublished results). Other *Wnt* cognates are also expressed in diencephalon (Parr et al. 1993). In *Drosophila* head, *wg* is hypothesized to be the target of the head gap genes *otd* and *ems* (Royet and Finkelstein 1996). It is possible at the time of brain regionalization that *Otx1* and *Otx2* genes cooperate to activate and maintain Wnt signaling for development of the forebrain and midbrain (Fig. 5). Caudally, *Otx* genes counteract *Fgf8* expression directly or indirectly, allowing the correct development of isthmus organizer. Rostrally, *Otx* genes may function upstream of *Emx* genes that counteract *Wnt* expression for development of cerebral hemispheres. An increasing number of genes have accumulated that function in the forebrain such as *Gli-3* (Franz 1994), *BF-1* (Xuan et al. 1995), *Nkx-2.1* (Kimura et al. 1996), and *Dlx* (Anderson et al. 1997), as well as those in cephalic neural crest. The molecular hierarchy in the patterning of rostral brain is thus emerging.

ACKNOWLEDGMENTS

We thank Dr. Kinya Yasui for helpful discussions and Mr. Naoki Takeda for production of chimeric mice. We are also grateful to the Laboratory Animal Research Center of Kumamoto University School of Medicine for housing of the mice. This work was supported in part by grants-in-aid from the Ministry of Education, Science and Culture of Japan (Specially Promoted Research), the Science and Technology Agency, Japan, and the Ministry of Public Welfare, Japan.

REFERENCES

Acamporar D., Mazan S., Lallemand Y., Avantaggiato V., Maury M., Simeone A., and Brûlet P. 1995. Forebrain and midbrain regions are deleted in $Otx2^{-/-}$ mutants due to a defective anterior neuroectoderm specification during gastrulation. *Development* **121:** 3279.

Alvarez-Bolado G., Rosenfeld M.G., and Swanson L.W. 1995. Model of forebrain regionalization based on spatiotemporal pattern of POU-III homeobox gene expression, birthdates, and morphological features. *J. Comp. Neurol.* **355:** 237.

Anderson S.A., Qiu M., Bulfone A., Eisenstat D.D., Meneses J., Pedersen R., and Rubenstein J.L.R. 1997. Mutations of the homeobox genes *Dlx-1* and *Dlx-2* disrupt the striatal subventricular zone and differentiation of late-born striatal neurons. *Neuron* **19:** 27.

Andreazzoli M., Pannese M., and Boncinelli E. 1997. Activating and repressing signals in head development: The role of Xotx1 and Xotx2. *Development* **124:** 1733.

Ang S.-L. and Rossant J. 1994. *HNF-3*β is essential for node and notochord formation in mouse development. *Cell* **78:** 561.

Ang S.-L., Conlon R.A., Jin O., and Rossant J. 1994. Positive and negative signals from mesoderm regulate the expression of mouse *Otx2* in ectoderm explants. *Development* **120:** 2979.

Ang S.-L., Jin O., Rhinn M., Daigle N., Stevenson L., and Rossant J. 1996. A targeted mouse *Otx2* mutation leads to severe defects in gastrulation and formation of axial mesoderm and to deletion of rostral brain. *Development* **122:** 243.

Aparicio S., Morrison A., Gould A., Gilthrope J., Chaudhuri C., Rigby P., Krumlauf R., and Brenner S. 1995. Detecting conserved regulatory elements with the model genome of the Japanese puffer fish, *Fugu rubripes. Proc. Natl. Acad. Sci.* **92:** 1684.

Bally-Cuif L. and Wassef W. 1995. Determination events in the nervous system of the vertebrate embryo. *Curr. Opin. Genet. Dev.* **5:** 450.

Bally-Cuif L., Cholley B., and Wassef M. 1995. Involvement of *Wnt-1* in the formation of the mes/metencephalic boundary. *Mech. Dev.* **53:** 23.

Blitz I.L. and Cho K.W. 1995. Anterior neurectoderm is progressively induced during gastrulation: The role of the *Xenopus* homeobox gene *orthodenticle. Development* **121:** 993.

Bouwmeester T., Kim S.-H., Sasai Y., Lu B., and De Roberitis E.M. 1996. Cerberus is a head-inducing secreted factor expressed in the anterior endoderm of Spemann's organizer. *Nature* **382:** 595.

Brenner S., Elgar G., Sandford R., Macrae A., Venkatesh B., and Aparicio S. 1993. Characterization of the pufferfish (*Fugu*) genome as a compact model vertebrate genome. *Nature* **366:** 265.

Carroll S.B. 1995. Homeotic genes and the evolution of arthropods and chordates. *Nature* **376:** 479.

Chiang C., Litingtung Y., Lee E., Young K.E., Corden J.L., Westphal H., and Beachy P.A. 1996. Cyclopia and defective axial patterning in mice lacking *Sonic hedgehog* gene function. *Nature* **383:** 407.

Cohen S.M. and Jürgens G. 1990. Mediation of *Drosophila* head development by gap-like segmentation genes. *Nature* **346:** 482.

Crossley P.H., Martinez Z., and Martin G.R. 1996. Midbrain development induced by FGF8 in the chick embryo. *Nature* **380:** 66.

Danielian P.S. and McMahon A.P. 1996. *Engrailed-1* as a target of the *Wnt-1* signalling pathway in vertebrate midbrain development. *Nature* **383:** 332.

de Beer G.R. 1931. On the nature of the *Trabecula cranii. Q. J. Microsc. Sci.* **74:** 701.

Diwan S. and Stevens L.C. 1974. *Mouse Newslett.* **51:** 24.

Donlach T. 1993. Planar and vertical induction of anteroposterior pattern during the development of the amphibian central nervous system. *J. Neurobiol.* **24:** 1256.

Figdor M. and Stern C.D. 1993. Segmental organization of embryonic diencephalon. *Nature* **363:** 630.

Finkelstein R. and Perrimon N. 1990. The *orthodenticle* gene is

regulated by *bicoid* and *torso* and specifies *Drosophila* head development. *Nature* **346:** 485.
Foley A.C., Storey K.G., and Stern C.D. 1997. The prechordal region lacks neural inducing ability, but can confer anterior character to more posterior neuroepithelium. *Development* **124:** 2983.
Frantz G.D., Weimann J.M., Levin M.E., and McConnell S.K. 1994. *Otx1* and *Otx2* define layers and regions in developing cerebral cortex and cerebellum. *J. Neurosci.* **14:** 5725.
Frantz T. 1994. Extra-toes (Xt) homozygous mutant mice demonstrate a role for the *Gli-3* gene in the development of the forebrain. *Acta Anat.* **150:** 38.
Gan L., Mao C.A., Wikramanayake A., Angerer L.M., Angerer R.C., and Klein W.H. 1995. An *orthodenticle*-related protein from *Strongylocentrotus purpuratus. Dev. Biol.* **167:** 517.
Garstang W. 1894. Preliminary note on a new theory of the phyologeny of the Chordata. *Zool. Anz.* **17:** 122.
———. 1929. The morphology of the Tunicata, and its bearings on the phylogeny of the Chordata. *Q. J. Microsc. Sci.* **72:** 51.
Goodrich E.S. 1930. *Structure and development of vertebrates.* Macmillan, London.
Guthrie S. 1995. The status of the neural segment. *Trends Neurosci.* **18:** 74.
Hallonet M.E.R., Teillet M.-A., and Le Douarin N.M. 1990. A new approach to the development of the cerebellum provided by the quail-chick marker system. *Development* **108:** 19.
Hermesz E., Mackem S., and Mahon A.K. 1996. *Rpx:* A novel anterior-restricted homeobox gene progressively activated in the prechordal plate, anterior neural plate and Rathke's pouch of the mouse embryo. *Development* **122:** 41.
Higgins D.G., Bleasby A.J., and Fuchs R. 1992. Clustal V: Improved software for multiple sequence alignment. *Comput. Appl. Biosci.* **8:** 189.
Hirth F., Therianos S., Loop T., Gehring W.J., Reichert H., and Furukubo-Tokunaga K. 1995. Developmental defects in brain segmentation caused by mutations of the homeobox genes *orthodenticle* and *empty spiracles* in *Drosophila. Neuron* **15:** 769.
Holland P.W.H. and Garcia-Fernandez J. 1996. *Hox* genes and chordate evolution. *Dev. Biol.* **173:** 382.
Hummel K.P. 1959. Developmental anomalies in mice resulting from the action of the gene, a semi-dominant lethal. *Pediatrics* **23:** 212.
Huxley T.H. 1874. On the structure of the skull and of the heart of *Menobranchus lateralis. Proc. Zool. Soc. Lond.* 1874 (cited by de Beer 1931).
Iler N., Rowitch D.H., Echelard Y., McMahon A.P., and Abate-Shen C. 1995. A single homeodomain binding site restricts spatial expression of *Wnt-1* in the developing brain. *Mech. Dev.* **53:** 87.
Janvier P. 1993. Patterns of diversity in the skull of jawless fishes. In *The skull* (ed. J. Hanken and B.K. Hall), vol. 2, p. 131. University of Chicago Press, Illinois.
Joyner A. 1996. *Engrailed, Wnt* and *Pax* genes regulate midbrain-hindbrain development. *Trends Genet.* **12:** 15.
Kabler B., Vignali R., Menotti L., Pannese M., Andreazzoli M., Polo C., Giribaldi M.G., Boncinelli E., and Barsacchi G. 1996. *Xotx* genes in the developing brain of *Xenopus laevis. Mech. Dev.* **55:** 145.
Kidwell J.F., Gowen J.W., and Stadler J. 1961. *pugnose*—A recessive mutation in linkage group 3 of mice. *J. Hered.* **52:** 145.
Kimura C., Takeda N., Suzuki M., Oshimura M., Aizawa S., and Matsuo I. 1997. *Cis*-acting elements conserved between mouse and pufferfish *Otx2* genes govern the expression in mesencephalic neural crest cells. *Development* **124:** 3929.
Kimura S., Hara Y., Pineau T., Fernandez-Salguero P., Fox C.H., Ward J.M., and Gonzalez F.J. 1996. The *T/ebp* null mouse: Thyroid-specific enhancer-binding protein is essential for the organogenesis of the thyroid, lung, ventral forebrain, and pituitary. *Genes Dev.* **10:** 60-69.
Kuratani S., Matsuo I., and Aizawa S. 1997. Developmental patterning and evolution of the mammalian viscerocranium: Genetic insights into comparative morphology. *Dev. Dyn.* **209:** 139.
Lacalli T.C. 1996. Landmarks and subdomains in the larval brain of *Branchiostomata:* Vertebrate homologies and invertebrate antecedents. *Is. J. Zool.* **42:** 131.
Lee S.M., Danielian P.S., Fritzsch B., and McMahon A.P. 1997. Evidence that FGF8 signalling from the midbrain-hindbrain junction regulates growth and polarity in the developing midbrain. *Development* **124:** 959.
Li Y., Allende M.L., Finkelstein R., and Weinberg E.S. 1994. Expression of two zebrafish *orthodenticle*-related genes in the embryonic brain. *Mech. Dev.* **48:** 229.
Li Y., Brown S.J., Hausdorf B., Tautz D., Denell R.E., and Finkelstein R. 1996. Two *orthodenticle*-related genes in the short-term beetle *Tribolium castaneum. Dev. Genes Evol.* **206:** 35.
Lumsden A., Sprawson N., and Graham A. 1991. Segmental origin and migration of neural crestcells in the hindbrain region of the chick embryo. *Development* **113:** 1281.
Mangold O. 1933. Über die induktionsfahigkeit der verschiedenen Bezirke der Neural von Urodelen. *Naturwiss.* **21:** 761.
Marin F. and Puelles L. 1994. Patterning of the embryonic avian midbrain after experimental inversion: A polarizing activity from the isthmus. *Dev. Biol.* **163:** 19.
Martin J.F., Bradley A., and Olson E.N. 1995. The *paired*-like homeo box gene *MHox* is required for early events of skeletogenesis in multiple lineages. *Genes Dev.* **9:** 1237.
Martinez A.A. and Alvarado-Mallart R.-M. 1989. Rostral cerebellum originates from caudal portion of the so-called "mesencephalic" vesicle: A study using chick/quail chimeras. *Eur. J. Neurosci.* **1:** 549.
Mathers P.H., Miller A., Doniach T., Dirksen M.-L., and Jamrich M. 1995. Initiation of anterior head-specific gene expression in uncommitted ectoderm of *Xenopus laevis* by ammonium chloride. *Dev. Biol.* **171:** 641.
Matsuo I., Kuratani S., Kimura C., Takeda N., and Aizawa S. 1995. Mouse *Otx2* functions in the formation and patterning of rostral head. *Genes Dev.* **9:** 2646.
McGinnis W. and Krumlauf R. 1992. Homeobox genes and axial patterning. *Cell* **68:** 283.
Millet S., Bloch-Gallego E., Simeone A., and Alvarado-Mallart R.-M. 1996. The caudal limit of *Otx2* gene expression as a marker of the midbrain/hindbrain boundary: A study using in situ hybridisation and chick/quail homotopic grafts. *Development* **122:** 3785.
Mori H., Miyazaki Y., Morita T., Nitta H., and Mishina M. 1994. Different spatio-temporal expressions of three *otx* homeoprotein transcripts during zebrafish embryogenesis. *Mol. Brain Res.* **27:** 221.
Noden D.M. 1983. The role of the neural crest in patterning of avian cranial skeletal, connective, and muscle tissues. *Dev. Biol.* **96:** 144.
Nübler-Jung K. and Arendt D. 1994. Is ventral in insects dorsal in vertebrates. *Roux's Arch. Dev. Biol.* **203:** 357.
Osumi-Yamashita N., Ninomiya Y., Doi H., and Eto K. 1994. The contribution of both forebrain and midbrain crest cells to the mesenchyme in the frontonasal mass of mouse embryos. *Dev. Biol.* **164:** 409.
Oyanagi M., Matsuo I., Wakabayashi Y., Aizawa S., and Kominami R. 1997. Mouse homeobox-containing gene, *Otx2*, maps to mouse chromosome 14. *Mamm. Genome* **8:** 292.
Parr B.A., Shea M.J., Vassileva G., and McMahon A.P. 1993. Mouse *Wnt* genes exhibit discrete domains of expression in the early embryonic CNS and limb buds. *Development* **119:** 247.
Pellegrini M., Mansouri A., Simeone A., Boncinelli E., and Gruss P. 1996. Dentate gyrus formation requires *Emx2. Development* **122:** 3893.
Puelles L. and Rubenstein J.L.R. 1993. Expression pattern of homeobox and other putative regulatory genes in the embryonic mouse forebrain suggest a neuromeric organization. *Trends Neurosci.* **16:** 472.
Puelles L., Amat J., and Martinez-De-La Torre M. 1987. Segment-related, mosaic neurogenetic pattern in the forebrain and mesencephalon of early chick embryos. I. Topography of

AChE-positive neuroblasts up to stage HH18. *J. Comp. Neurol.* **266:** 247.

Qiu M., Bulfone A., Martinez S., Ghattas I., Meneses J.J., Christensen L., Sharpe P.T., Presley R., Pedersen R.A., and Rubenstein J.L.R. 1997. Role of the *Dix* homeobox genes in proximodistal patterning of the branchial arches: Mutations of Dix-1, Dix-2, and Dix-1 and -2 alter morphogenesis of proximal skeletal and soft tissue structures derived from the first and second arches. *Dev. Biol.* **185:** 165.

Rijli F.M., Mark M., Lakkaraju S., Dierich A., Dolle P., and Chambon P. 1993. Homeotic transformation is generated in the rostral branchial region of the head by disruption of *Hoxa-2*, which acts as a selector gene. *Cell* **75:** 1333.

Romer A.S. and Parsons T.S. 1949. The nervous system. In *The vertebrate body*, 6th edition, p. 587. Harcourt Brace Jovanovich, Cleveland, Ohio.

Royet J. and Finkelstein R. 1996. *hedgehog*, *wingless* and *orthodenticle* specify adult head development in *Drosophila. Development* **122:** 1849.

Saitou N. and Nei M. 1987. The neighbor-joining method: A new method for reconstructing phylogenetic trees. *Mol. Biol. Evol.* **4:** 406.

Sander K. 1976. Specification of the basic body pattern in insect embryogenesis. *Adv. Insect Physiol.* **12:** 125.

Sarnat H.B. and Netsky M.G. 1981. *Evolution of the nervous system*, 2nd edition. Oxford University Press, United Kingdom.

Satokata I. and Maas R. 1994. *Msx1* deficient mice exhibit cleft palate and abnormalities of craniofacial and tooth development. *Nat. Genet.* **6:** 348.

Schier A.F., Neuhauss S.C., Helde K.A., Talbot W.S., and Driever W. 1997. The one-eye pinhead gene functions in mesoderm and endoderm formation in zebrafish and interacts with no tail. *Development* **124:** 327.

Shawlot W. and Behringer R.R. 1995. Requirement for *Lim1* in head-organizer function. *Nature* **374:** 425.

Shih J. and Fraser S.E. 1996. Characterizing the zebrafish organizer: Microsurgical analysis at the early shield stage. *Development* **122:** 1313.

Shimamura K., Hartigan D.J., Martinez S., Puelles L., and Rubenstein L.R. 1995. Longitudinal organization of the anterior neural plate and neural tube. *Development* **121:** 3923.

Simeone A., Acampora D., Gulisano M., Stornaiuolo A., and Boncinelli E. 1992a. Nested expression domains of four homeobox genes in developing rostral brain. *Nature* **358:** 687.

Simeone A., Gulisano M., Acampora D., Stornaiuolo A., Rambaldi M., and Boncinelli E. 1992b. Two vertebrate homeobox genes related to the *Drosophila empty* spiracles gene are expressed in the embryonic cerebral cortex. *EMBO J.* **11:** 2541.

Simeone A., Acampora D., Mallamaci A., Stornaiuolo A., D'Apice M.R., Nigro V., and Boncinelli E. 1993. A vertebrate gene related to *orthodenticle* contains a homeodomain of the *bicoid* class and demarcates anterior neuroectoderm of the gastrulating mouse embryo. *EMBO J.* **12:** 2735.

Simpson-Brose M., Treisman J., and Desplan C. 1994. Synergy between the *hunchback* and *bicoid* morphogens is required for anterior patterning in *Drosophila. Cell* **78:** 855.

Slack J.M.W. and Tannahill D. 1992. Mechanisms of anteroposterior axis specification in vertebrates. Lessons from the amphibians. *Development* **114:** 285.

Spemann H. 1931. Über den anteil von implantat und wirtskeim an der orientierungund beschaffenheit der induzierten embryonanlage. *Roux's Arch. Entwicklungsmech. Organ.* **123:** 389.

Spemann H. and Mangold H. 1924. Über Induction von Embryonanlagendurch Implantation Artfremder Organis Atoren. *Roux's Arch. Dev. Biol.* **100:** 599.

St. Hilarie G.E. 1822. Considérations générales sur la vertébre. *Mem. Mus. Natl. Hist. Nat.* **9:** 89.

Storey K.G., Vrossley J.M., De Robertis E.M., Norris W.E., and Stern C.D. 1992. Neural induction and regionalization in the chick embryo. *Development* **114:** 729.

Suda Y., Matsuo I., and Aizawa S. 1997. Cooperation between *Otx1* and *Otx2* genes in developmental patterning of rostral brain. *Mech. Dev.* (in press).

Suda Y., Matsuo I., Kuratani S., and Aizawa S. 1996. *Otx1* function overlaps with *Otx2* in development of mouse forebrain and midbrain. *Genes Cells* **1:** 1031.

Taira M., Otani H., Saint-Jeannet J.-P., and Dawid I.B. 1994. Role of the LIM class homeodomain protein Xlim-1 in neural and muscle induction by the Spemann organizer in *Xenopus. Nature* **372:** 677.

Tautz D. and Sommer R.J. 1995. Evolution of segmentation genes in insects. *Trends Genet.* **11:** 23.

Thomas P. and Beddington R. 1996. Anterior primitive endoderm may be responsible for patterning the anterior neural plate in the mouse embryo. *Curr. Biol.* **6:** 1487.

Urbanek P., Fetka I., Heisler M.H., and Busslinger M. 1997. Cooperation of *Pax2* and *Pax5* in midbrain and cerebellum development. *Proc. Natl. Acad. Sci.* **94:** 5703.

Varlet I., Collignon J., and Robertson E.J. 1997. *nodal* expression in the primitive endoderm is required for specification of the anterior axis during mouse gastrulation. *Development* **124:** 1033.

Von Baer K.E. 1828. *Über Entwicklungsgeschichite der Tiere.* 1. *Beobachtungen und Reflexionen.* Theil, Königsberg.

Wada S., Katsuyama Y., Sato Y., Itoh C., and Saiga H. 1996. *HROTH*, an *ortodenticle*-related homeobox gene of the ascidian, *Halocynthia roretzi:* Its expression and putative roles in the axis formation during embryogenesis. *Mech. Dev.* **60:** 59.

Walldorf U. and W.J. Gehring. 1992. *Empty spiracles*, a gap gene containing a homeobox involved in *Drosophila* head development. *EMBO J.* **11:** 2247.

Williams N.A. and Holland P.W.H. 1997. Old head on young shoulders. *Nature* **383:** 490.

Xuan S., Baptista C.A., Balas G., Tao W., Soares V.C., and Lai E. 1995. Winged helix transcription factor *BF-1* is essential for the development of the cerebral hemispheres. *Neuron* **14:** 1141.

Yoshida M., Suda Y., Matsuo I., Miyamoto N., Takeda N., Kuratani S., and Aizawa S. 1997. *Emx1* and *Emx2* functions in development of dorsal telencephalon. *Development* **124:** 101.

Zaddach G. 1854. Untersuchungen über die Entwickelung und den Bau der Gliederthiere I. Heft. *Die Entwicklung des* Phryganiden-*Eies.* Verlag von Georg Reimer, Berlin.

Zhao Q., Behringer R.R. and de Crombrugghe B. 1996. Prenatal folic acid treatment suppresses acranial and meroanencephaly in mice mutant for the *Cart1* homeobox gene. *Nat. Genet.* **13:** 275.

Summary: A Common Language

M.P. Scott
Department of Developmental Biology, Beckman Center, Stanford University School of Medicine, Stanford, California 94305

I Am not ignorant of what many Learned and Inquisitive Men, both at home and abroad, especially in this last Century, have performed in the Anatomy of Animals. After all whom, if it be demanded, what is left for me to do? I Answer in the words of *Seneca*: (Epist. 64), *Multum adhuc restat operis, multumq; restabit; nec ulli Nato, post mille Saecula, praecludetur occasio, aliquid adnuc adjiciendi.* [Much work still remains, and much will remain, nor will anyone born a thousand centuries from now be denied the opportunity to add something still.]

—Nehemiah Grew, M.D., F.R.S.
"The Comparative Anatomy of Stomachs and Guts Begun"
Lectures of the Royal Society (1676)

My central theme in this summary is that developmental biology has come so far so fast because the mechanisms of development are turning out to be unexpectedly simple. They are not simple on an absolute scale, but certain anticipated difficulties have not come to pass. Our field now has a common language. People working on vastly different organisms share principles and facts in a way that would have been unimaginable just 10 years ago. Proteins are in families, and many families are in pathways. Underlying the vast diversity of biology we increasingly see simple commonalities. A sense of optimism and excitement pervades the field, although we have a long way to go.

PAST

Monod and Jacob, at the 26th Cold Spring Harbor Laboratory Symposium in 1961, commented, "Unfortunately, in the face of formidable technical difficulties, the study of differentiation either from the genetic or biochemical point of view has not attained a state which would allow any detailed comparison of theory with experiment . . . Adequate techniques of nuclear transfer, combined with systematic studies of possible inducing or repressing agents, and with the isolation of regulatory mutants, may conceivably open the way to the experimental analysis of differentiation at the genetic-biochemical level."

At the 50th Symposium, "Molecular Biology of Development," in 1985, Gerald Rubin commented, "I believe we can be optimistic about understanding the basic features of development at the molecular level, certainly before the 100th Symposium, and perhaps by the 60th." At that last Cold Spring Harbor Symposium devoted primarily to developmental biology, where did we stand? We were four years past the first reports of the Nüsslein-Volhard and Wieschaus screen—the first functional genomics project for development. Their catalog of genes affecting the epidermis was to become a catalog of genes affecting all of us. The first segmentation genes had just been cloned. Localized maternal RNA molecules were being discovered in frog eggs. Mutations had been isolated that affect early nematode development. Most of what mattered about homeobox genes was known due to Ed Lewis's half century of work on the bithorax complex: colinearity of body and gene complex, combinatorial action, upstream regulators. The methods to get at the genes had come along at the same time as the discovery of most of the gene types. Chromosome walking, invented by David Hogness in the late 1970s, had opened the chromosomes to directed gene isolation and set in motion a flood of new projects. There was tremendous excitement, as well as substantial data, about similarities between transposons in plants, bacteria, and flies. Bulk genome analysis was to be replaced by an intimate knowledge of genes for abundant proteins, transposons, and all those mysterious genes that control development. Molecular biologists were showing that the homeotic and segmentation genes make related proteins and that fly homeotic genes are absolutely huge. The homeobox had been found in flies and had allowed the isolation of the mammalian genes. There was no evidence that these genes might have similar functions in mammals and insects and, indeed, little feeling that they do. (Rubin noted, "There is at present no evidence that any of the homeo-box-containing genes in vertebrates have a controlling function in embryonic development.") Many of the first developmental regulatory genes isolated encoded things that looked like transcription factors, but we were not sure. In situ hybridization had been invented by my thesis advisor Mary Lou Pardue, along with her thesis advisor Joe Gall, for looking at chromosomes. Thanks to Michael Akam and Walter Gehring, the technique was newly being applied to tissues, especially embryos. A first simplification had come to light: Transcription factors are often produced only in the tissues they affect. The elegant and original adapta-

All authors cited here without dates refer to papers in this volume.

tion of P elements for moving DNA into and around the genome of flies, by Gerald Rubin and Allan Spradling, was well established and being used to great benefit. Transgenic mouse methods had been developed by Ralph Brinster, Frank Costantini, Elizabeth Lacy, Erwin Wagner, Beatrice Mintz, and Frank Ruddle and were well represented at the meeting. The technique was still new enough to be mentioned as part of David Baltimore's title. Insertional mutations of mice were being isolated. Gene transfer in plants, sea urchins, nematodes, and *Dictyostelium* raised exciting prospects for progress, all since fulfilled. A small field, the cell cycle, was represented by yeast studies, mitotic factors and chromosome replication in frogs, and regulation of histone gene expression. The startling conservation of ras from mammals to yeast was discussed, and early studies of oncogene expression and function during development were presented. Gene regulation in yeast, flies, frogs, and mice was a major topic. Molecular developmental neurobiology was just getting going, with reports on neural crest gene expression, Notch (in flies only, of course; preposterous to imagine it might be elsewhere), circadian fly clocks, and insect neurogenesis. Major progress was reported in sex determination, particularly in the assembly of genetic pathways in flies and worms. Some sex determination genes had been cloned, and the existence of sex-specific RNA forms had been noted for flies, but the genetics was way ahead of the molecular biology.

Many people turning to the molecular biology of development for new opportunities had been intellectually raised on beautiful bacterial genetics. The special elegance of yeast mating type, arguably among the best understood genetic hierarchies in any organism, had extended sophisticated gene regulation to a eukaryote. The recognition of a relationship between bacterial repressors, mating-type repressors, and the homeodomain provided a satisfying unification of development with gene regulation in single-celled organisms. Now techniques used in bacteria could be applied, and *lacZ* and other gene fusions were beginning to be applied to developmental problems.

What was missing in 1985? Rubin commented, "The interactions that occur between cells to execute morphogenesis await our attention." The dearth of signaling studies at that meeting is really striking. Cell-cell interactions were discussed in the context of *Dictyostelium* cAMP signaling and *Myxococcus*; about the only higher eukaryotic signaling topic was hormonal signaling. In the 1997 meeting, about half of the talks dealt with signaling.

There was another difference between the earlier meeting and this one that cannot easily be seen by reading papers. The 1985 meeting encompassed several groups of scientists whose work seemed far apart. They were striving to understand each other. Sometimes they did not understand—sometimes they did not strive. There were strong sentiments that animal development could not be unified and that the "principles" being learned from bacteria, yeast, and small animals with easy genetics would eventually be eclipsed by biology more directly relevant to humans. I think that no one anticipated the actual outcome: that all the work being done on different organisms would be linked in ways that make everyone's efforts important. Within a few years of that 1985 meeting, we all had to be developmental biologists, rather than frog developmental biologists, mouse developmental biologists, or invertebrate developmental biologists. This transition is incomplete but has tremendous momentum. In a sense, there are no model organisms because the information learned is so directly applicable. We have a common language.

Mike Levine's tongue-in-cheek principle of a "statute of limitations" (see Zhou et al.), established during this 1997 meeting, stated that rediscoveries of results first reported more than 10 years ago could be described as new findings. In fact, the reality of this year's meeting has been an increasingly deep appreciation of seminal biological observations made over a period of centuries, with references to Leonardo da Vinci's illustrations of lungs, Ramon y Cajal's neuroanatomy, Cabry's 1887 experiments on ascidian mosaicism, and the rich biological findings of the present century that we still strive to explain at a molecular and cellular level. The old experiments can be approached once again, with new markers, pure genes, and pure proteins. Quantitative biology, in fact.

PRESENT

There are not very many kinds of proteins, and that has made all the difference in developmental biology. This is probably not a false result, because it is based on many years of extensive and varied genetic screens as well as other methods of gene discovery, like transducing viruses and disease genes. About one-third of the talks at this 1997 meeting focused on signals of just four types: Wnt, Hedgehog, TGF-β, and FGF, each of them at work in many contexts. Add in a dozen classes of transcription factors, a similar number of types of cell surface constituents, and machinery for transcription, translation, transport, and the cytoskeleton, and you would be well on your way to building a complex animal. It did not have to be this way. With perhaps 12,000 genes in the fly genome and perhaps 100,000 in a mammalian genome, there was room for far more complexity. Most of the proteins fall into families, and a quick look at a new protein sequence often tells you where to look in the cell—sometimes even where to look in the embryo. We always hope to discover new and exciting types of regulators, but a great many of them turn out to be in familiar groups. If we had to cope with hundreds of truly distinct signals our work would be very much harder. Instead, learning from fly genetics that β-catenin is involved in the Wnt pathway leads to a feast of good information coming from vertebrate systems, and eventually a direct connection with human colon cancer. Eddy De Robertis (De Robertis et al.) quoted Balzac (1842): "There is only one animal."

We have new tools, too. Since the previous meeting, we have knockout mice thanks to Mario Cappecchi and others, enhancer traps from the Gehring lab, the use of Flp recombinase for making clones lacking gene function from Kent Golic and Susan Lindquist, the use of Flp to make clones of cells expressing a gene from Gary Struhl, and the use of GAL4 for spatially restricted ectopic gene expression from Norbert Perrimon. An assortment of the rich array of zebrafish mutants was described for us by Driever et al. There are many and varied better ways of screening for differentially expressed genes. In the analysis of mice lacking FGF8 function (Lewandoski et al.), we saw advanced uses of Cre/Lox and Flp combined. Enhancer and gene traps have been used massively in *Arabidopsis* genetics to create a large array of useful mutants (Moore et al.; R. Martienssen, unpubl.), and viral mutagenesis in fish is allowing much more rapid gene isolation (Amsterdam et al.). Positional cloning is practical in the zebrafish (A.F. Schier, unpubl.). Many of these methods depend on transposons, a fitting tribute to Barbara McClintock and Cold Spring Harbor Laboratory.

Transcription in Development

Nathans described the complex features of expression and function for the Brn3 family of transcription factors. Each member of the family is differently expressed in neurons of the auditory, somatosensory, or optic systems. The presence of the same transcription factor in, for example, ganglion cells and auditory neurons may hint at functional parallels between cells previously viewed as entirely different. What looks different *to us* may not be so different.

Kenyon et al. showed how the beautifully precise pulsations of *mab5* (Hox) expression are regulated by Wnt signaling. They also demonstrated that when *mab5* is off, it's for a reason, and when it's on, it's for a reason. This work brings Hox gene functions to a new level of precision and opens the way to understanding the kinetics of their functions.

Levine (Zhou et al.) reported studies of transcriptional regulation in response to the dorsoventral regulator *Toll*, which was produced in activated form at the anterior of the embryo. This established compartmentalized transcription of dorsoventral regulators along the anteroposterior axis. The success of this experiment shows how firm is the understanding of the triggering influences. The roles of insulators and competition between enhancers for a promoter were described in reference to adjacent promoters in Hox complexes and engineered competitions. *Toll* was prominent again in a description of parallels between regulation of dorsoventral patterning and immunity (Wu and Anderson). Downstream events in the pathway are involved in immunity both in flies and mammals.

Homeobox genes play a prominent role in development in plants, as they do in fungi and animals. We learned of Knat1, a homeobox gene that affects meristem cell division and fate (R. Martienssen, pers. comm.). A plethora of homeobox genes in plants also provides a link to the animalia and protista.

At late stages of neurogenesis, large domains of the brain are laid out by a series of regulators, many of which are now known. They include Hox and Fgf in the hindbrain; Engrailed, Wnt, and Pax in the midbrain; and Lim1 (Wakamiya et al.), Otx, and Emx in the forebrain. Differential effects of one transcription factor are often attained by input from signaling pathways (Kolm and Sive). Rowitch (Rowitch et al.) and Aizawa (Matsuo et al.) described experiments manipulating or destroying these components, and a clear hierarchy is beginning to take shape. What functions are regulated by these signals and transcription factors further downstream is unknown.

The hindbrain is a prime location for studying regulation of Hox genes (Nonchev et al.). Kreisler, a MAF class transcription factor expressed early in the hindbrain, is an activator of Hox genes in rhombomere 5. The integration of the Kreisler-binding site with an Ets-binding site provides the complete pattern, with the Ets-binding site restraining where Kreisler is able to activate its Hox target. *Hoxb2* is regulated by *Krox20* in rhombomeres 3 and 5, but by *Hoxb1* in rhombomere 4, and so the complete pattern is built up. Pufferfish Hox complexes show distinctive changes compared to mammalian complexes; the conservation of sequences will lead us to a rich set of *cis* regulatory elements.

In muscle development the main emphasis was also on transcription (Baylies et al.). *twist* in flies is the "master of the universe," with the amount of Twist controlling whether cells will form somatic musculature, and high-level Twist converting ectoderm to muscle. Differentiation of the muscle requires the fly version of *Mef2*. Remarkably, each muscle derives from a founder myoblast that is determined by a process involving inhibition of other cells from following the same path. The founder then recruits fusion partners until an adequate population is built up.

The development of the heart is of direct clinical relevance; Harvey pointed out that 1/100 live births brings with it some type of heart anomaly. The heart arises from paired mesodermal precursors that migrate to the ventral midline and fuse. Schultheiss (Schultheiss and Lassar) described two signal sources involved in this process, one unknown source from the anterior endoderm, and BMP2 from the neural plate and notochord. The growth of the heart provides another remarkable case of evolutionary conservation, with both the Tinman/Nkx2.5 and the MEF2 transcription factors revealing common origins of the pump. The eHand and dHand bHLH proteins were a major subject this year, with elimination of dHand function in knockout mice leading to deletion of much of one of the ventricles (Lin et al.). Left–right asymmetry is prominent in multiple tissues, but most of all in the heart, and a large number of markers have been found to be asymmetrically expressed in heart precursors in a transcription and signaling cascade. The Nodal-related *Xenopus* gene Xnr1 is a signaling component that is able to effect situs inversus after injection into frog embryos (C. Wright, unpubl.); misexpression of Nodal in chicks has the same effect (Kengaku et al.). Evidence for the involvement of activins upstream of Nodal was provided by experiments in which endogenous activin-related molecules were inhibited by follistatin (Kengaku et al.), so we are moving from what could be to what is. eHand and a secreted cytokine called ANF may be important for the asymmetry (Biben et al.). The MEF2 family of regulators, four genes in mice, have all been knocked out and show the redundancy of some of the genes (Q. Lin et al.).

The effects of MEF2C are especially dramatic, with the right ventricle deleted and a failure of looping morphogenesis.

Nodal has also been implicated in primitive endoderm functions that appear to be crucial for determining anterior-ness in early vertebrate embryos (Varlet et al.; Thomas et al.).

Signals and Morphogens

Struhl (pers. comm.) described elegant experiments using the *Drosophila* abdomen to learn how Hedgehog and other signals control bristle polarity in the epidermis. The most important finding was that both graded signaling and signal relays seem to be involved, each one affecting a different readout of the initiating signal.

Jessell (Ericson et al.) described new work on dosage responses to Hedgehog signals in the neural tube. At least five different responses to Sonic hedgehog were observed depending on signal concentration, and only twofold changes were needed.

Fgf signaling analyzed with knockout mice employed some of the most advanced manipulations of endogenous mammalian genes (Lewandoski et al.; Rossant et al.). Alleles were created with partial function or specific splice forms eliminated, a tour de force of hard work. FGF8 has been implicated in cell migration events that are necessary for proper gastrulation. The cells are unable to move from the primitive streak (Lewandoski et al.). This may be a parallel with *C. elegans* FGF functions, which can also affect migration. Later roles for FGF8 were ascertained by doing tissue-specific gene elimination with Cre/*loxP*, revealing severe limb defects. A hypomorphic allele revealed the loss of the mid/hindbrain region (Lewandoski et al.). FGF receptor 1 has a role in controlling cell behavior in the primitive streak and is necessary for proper anteroposterior vertebral patterning, suggesting a role in activation or maintenance of Hox expression (Rossant et al.).

BMP signals play roles in many processes, among which are formation of the cartilage primordium, cartilage differentiation, and limb patterning by cell death. Their receptors can be dedicated to different purposes, raising the question of how downstream events take different directions (Zou et al.). But often BMPs have to be stopped in their tracks. The discovery of antagonistic activities in signaling has been highlighted here, in part for their role in Spemann's organizer. We had seen from Tam's experiments (Tam et al.) some special properties of the mouse organizer, as it can reappear after being deleted. A zebrafish mutant, now called *chordino*, has a substantially ventralized embryo (De Robertis et al.), in keeping with the view of Chordin as a ventralizer. The Noggin protein, a BMP antagonist, has been eliminated from the mouse genome, and the consequences include defects in skeletal development, chondrogenesis, and joint formation (R. Harland, unpubl.).

Wnts battling Frzbs, and chordins fighting TGF-βs, give us a new image of the checks and balances that produce animals. Why might these antagonistic activities have evolved originally? How about War!! One amusing idea is that antagonistic secreted proteins originally evolved not to coordinate growth and pattern, but as weapons. A protein produced by one organism would nullify a signal essential to the life or success of a competitor, like jamming enemy radio communications. How would the genes ever end up in the same animal? The most severe competition is usually between organisms that are related and occupy similar niches. A first step would be to evolve a variant signal, the second to produce an antagonist specific to the competitor, and the third to produce an antagonist to the home signal and use it in a well-regulated manner. Lobbing chordin at the enemy to dorsalize him—a colorful world of microbes.

Guidance Systems

Signaling systems were prominent in the investigations of branching morphogenesis. The initial determination of cell fates in the fly trachea primordia is key because no more cells are created as a vast network of tubes forms. Trachealess, a bHLH protein, initially determines the cells. The different types of cells depend on a battle between Dpp (a BMP2/4 relative) signaling and influences of fly EGF signaling. Outgrowth and branching of the trachea in flies is guided by FGF signals at one stage and by hypoxia in target tissues at a later stage (an exciting parallel to mammalian lung development, and some of the same genes are involved) (Shilo et al.). The tip cells that branch must prevent adjacent cells from doing so, and they do this by blocking the response to FGF while mysteriously being resistant to the blocking signal themselves (Krasnow). Sprouty, a cell surface protein which does this, may be a general regulator limiting FGF function, as a mouse homolog is expressed near FGF8 at multiple locations. The Sprouty-FGF relationship is reminiscent of the BMP-Chordin relationship, although in the former case the molecular mechanism is still to be learned.

Genetic and biochemical approaches have both contributed substantially to discovering the basis for axon guidance. The netrins, acting over long distances both to attract and to repulse, emerged from both nematode genetics and remarkable biochemical successes. Tessier-Lavigne (Leonardo et al.) described different receptors and how their different activities may lead a growth cone rushing toward or away from the nervous system midline. A battery of surface proteins with immunoglobulin-like domains and fibronectin-like domains has emerged from fly genetic studies reported by Goodman et al. A key finding in this arena is that the proteins are localized to specific parts of the cell surface, a phenomenon reminiscent of synaptogenesis itself. S.L. Zipursky (unpubl.) described the role of the Cdc42/Dck signal transduction system in guiding neuronal targeting in the visual system. This connection with cytoskeletal regulators was to me a particularly interesting step beyond the cell surface, beginning to lead us to the changes in cell shape and behavior that must surely underlie targeting. The cadherins also play a role in axon patterning, as revealed by *Drosophila* mutants (Takeichi et al.).

The argonauts of cells, the neural crest cells, are guided in their migration by Eph-class ligands made in the somites. Repulsion from Eph channels the moving cells. Crest cells are guided in their differentiation by transcription factors and signals, and Anderson et al. reported how transcription factors can make crest cells competent to see BMP signals and take on the fate of autonomic nerves. As the cells move to their targets, they become exposed to BMP, and the resulting induction of MASH1 transcription factor appears to make the cells still more responsive to BMP signals in a positive feedback loop.

Signal Transduction

This past year has equipped us with long-sought receptors for two major signaling families, Wnts and Hhs. The *Drosophila* Frizzled protein, studied for many years for its role in controlling bristle patterning, has a new family member, Frizzled2, and both can bind Wnt signals (Nusse et al.), as can related proteins in frogs (R.T. Moon, unpubl.). Perrimon et al. described the involvement of heparan sulfate proteoglycan-related functions in the Wg pathway, and he suggested the existence of a co-receptor that acts together with Fz2. Unity with the worm came from the finding of a mannose transporter implicated in Wg processing. This same transporter type was described by Horvitz (Herman and Horvitz) as being important for epithelial invagination. A debate over which of two transmembrane proteins, Patched and Smoothened, was most likely to be involved in Hh reception was decided in favor of using both. One (Ptc) appears to bind the protein and the other (Smo) to transduce the signal when released from the negative influence of Ptc. However, the molecular mechanics of this process are far from understood.

Rubin et al. showed the genetic pathway involving ras, derived from screens of 850,000 flies. Cascades and feedback, antagonistic functions, are evident in this pathway as well.

Hedgehog proteins are tethered through a remarkable modification, by cleaving the signal into two parts and attaching a cholesterol molecule to the amino-terminal signaling part. Mutations eliminating Sonic hedgehog function cause holoprosencephaly, as do inhibitors of cholesterol metabolism (Beachy et al.). Scott (Johnson and Scott) described the involvement of a kinesin-related protein in the Hedgehog signal transduction pathway. Hedgehog signaling has many roles, among them regulation of different muscle types in the zebrafish (Quirk et al.). A Sonic hedgehog mutation in the fish (Rauch et al.) causes altered somites and shortened pectoral fins; additional defects seen in the mouse knockout, such as cyclopia, may be eliminated in the fish due to the action of other Hedgehog family members. An array of other fish mutants holds great promise for being related to other Hedgehog signaling components.

Cascades of transcription factors and signal transduction systems in plants use components distantly related to animal regulators. No direct homologies with animal developmental mechanisms are apparent as yet, but 12 years from now . . . . A kinase has been implicated in meristem patterning, where it coordinates the process of cell division in different cell layers (Meyerowitz). We also learned about how few genes can be sufficient for dramatic morphological changes in plant breeding and therefore presumably evolution, such as from teosinte to modern corn (Doebley and Wang).

Cell Biology and Development

The increased focus on cell biology is part of a general trend away from pure developmental biology and to specific molecular biology of the cell questions. For example, fly sex determination became a study of RNA splicing, fly chromosome dosage compensation a study of chromatin, eye development a study of the ras pathway, segment polarity genes a study of Wnt and Hedgehog signal transduction, pair-rule genes a study of combinatorial actions of transcription factors, Hedgehog processing a study of cholesterol modification of proteins, and so on.

Certain experimental systems originally employed for global patterning studies are now being used for studies at the cellular level. This is a trend that seems particularly promising. Notable examples are the study of epithelial invagination in the worm vulva (Herman and Horvitz) and *Drosophila* oogenesis (Spradling et al.; Rongo et al.; Markussen et al.). Epithelial invagination requires a set of proteins involved in glycosylation. A screen for mutations affecting germ cell migration led to the identification of the Zfh1 transcription factor as important (Rongo et al.). Together with Tinman, it is required for the interaction between pole cells and the mesoderm into which they move. The fusome, a structure connecting the developing oocytes that is derived from mitotic spindles, appears to set up a transport system and may be involved in storing, modifying, or delivering molecules such as cyclins that control cell fates by controlling the cell cycle (Spradling et al.). The fusome is part of the system for producing "the equivalent of an 8-lb baby every three hours." (Daunting as this may have seemed to the XX members of the audience, the length of *Drosophila* sperm exceeds the length of the adult, giving pause to the rest of the audience.) Localized substances within the oocyte govern fates in the anteroposterior axis and regulate determination of germ cell identity. The regulation of *oskar*, one of these regulators, has been particularly interesting. The 3′ UTR of *oskar* is required to bring the RNA to the posterior pole. The bruno protein is required to block translation of *oskar* mRNA before it reaches the posterior pole, and Ephrussi (Markussen et al.) reports the identification of a new protein, p50, that may be required for translation after the mRNA reaches the pole.

The 3′ UTR has also been implicated in worm sex determination, in particular in the regulation of the decision to make sperm versus oocytes in the germ line. Two genes that act on the relevant UTR, *mog1* and *mog5*, both encode RNA-binding proteins with DEAH box motifs (Puoti et al.). The proteins are also related to the Pumilio protein previously implicated in related functions in flies. RNA localization, a major theme of the meeting, emerged again in the study of a fish *vasa* homolog that appears to be involved in germ cell determination (Amsterdam et al.). The remarkably limited precise localization of the fish *vasa* RNA early in development, and its preservation, is an indicator of much interesting biology to come.

Asymmetry

The talks on bacteria and yeast reminded us why these are the best experimental organisms. We would have been even more humbled by a phage talk. The "lowest life forms" were described by Hofmeister and Losick, who presented three mechanisms of asymmetric protein activity in *B. subtilis.* In an asymmetric division that creates a small spore, one sigma factor becomes active exclusively in the spore due to the mysterious activity of a membrane-bound protein localized to the septum between the two parts of the cell. A second sigma factor is activated only in the larger mother cell in two ways: (1) elimination of the protein from the forespore and (2) processing of the protein by a membrane-bound protease that mediates a signal across the septum from the forespore.

Asymmetry in yeast budding is inherited from the previous budding event, although the information about the previous division is interpreted according to the genotype of the cell (Herskowitz). Two GTPase systems control local assembly of the cytoskeleton for budding. Another form of asymmetry is the activation of HO in the daughter, but not mother, cell after division. A localized RNA, encoding an HO repressor, is found in the distal daughter cell. The budding site inheritance system controls localization of the RNA.

Proceeding to plants, we were reminded that *Fucus* responds to light with oriented growth. Remarkably, it can remember the direction of light for hours (Quatrano). The establishment of a cortical growth site allows asymmetric development, but how the light signal is transduced is not yet clear. The cortical site is established by microfilament-mediated movement of vesicles to the site.

Asymmetric cell division is a frequent event in biology and is especially prevalent in the fly nervous system. Several localized proteins (Prospero, Numb, Inscuteable, and A3) and at least one localized RNA (*inscuteable*) are involved. The localization of the RNA is separable from the localization of the protein (Knoblich et al.). A3 is a component critical for localizing Prospero protein, a transcription factor that spends part of its time hanging on to the membrane to end up in the right nucleus. Numb can also bind to A3. Numb, part of the output of the system and localized independently of Prospero, may act by binding to Notch and biasing cell-cell communication events to control cell fates. Chia et al. reported the exciting finding that a protein previously known to be involved in RNA localization during embryogenesis, Staufen, binds to *inscuteable* RNA, and *staufen* mutants affect *insc* RNA localization. Unity in biochemistry.

Evolution

The single most amazing advance during the past 12 years is, of course, the recognition of great similarities among animals that at first glance seem entirely different. The list of similarities is increasing all the time. Currently, there is evidence for the following: anteroposterior body plan controlled by Hox genes (Cappecchi; Kenyon et al.; J. Kim et al.), dorsoventral body plan controlled by TGF-β signals and their antagonists (De Robertis et al.; R. Harland; C. Wright; both unpubl.), heart development controlled by Tinman/Nkx2.5 (Biben et al.) and MEF2 (Lin et al.), eye development controlled by Pax6/eyeless/aniridia, CNS midline regulation of axon extension (Leonardo et al.; Goodman et al.), branching morphogenesis of lung and trachea (Krasnow; Shilo et al.), limb dorsoventral and distal-proximal patterning (Schwabe et al.; Kengaku et al.), limb anteroposterior patterning (Kengaku et al.), and Toll signaling in insect and mammalian immunity. In fairness, I must note apparent differences as well. No role for FGF in the outgrowth of insect limbs has been noted, in contrast to the vertebrate scenario.

We are increasingly aware that biological structure is stable, because it replicates, whereas geological structure is transient. Our meeting took place on a pile of gravel called Long Island, swept by glaciers off the surface of Maine within times measured in tens of thousands of years. The mosquitoes outside have scarcely changed in tens of millions of years, unfortunately.

The picture is therefore of commitment of certain gene systems to certain developmental purposes. When preformationists were retreating from their fully irrational stance to one slightly less so, they proposed that there was no complete homunculus, but instead each body part of the progeny inherited a representative bit from the corresponding parental organ. We now face a corrected version of that idea, of gene sets appropriate to certain tissues being inherited as working packages.

We learned that certain tooth structures are thought to have evolved more than 20 times despite the remarkable complexity of signaling proteins in the signaling center called the enamel knot (Thesleff and Jernvall). Eyes were thought to evolve more than 40 times. The idea is that selective pressure caused the same outcome on separate occasions. Another view seems more likely to me. The genetic program for a structure may have evolved once, and then was stashed away intact (perhaps used for other purposes?), only to become reactivated later. With this view, intermediate forms need not have retained all the structures common to their ancestors and descendants.

Hox complexes in anteroposterior patterning are not too hard to accept because the ancestors presumably had heads and tails and fronts and backs, but believing that our appendages evolved from a creature whose other progeny became insects or squids has been hard to believe. The whole appendage gene program may have existed for a primordial purpose that need not have been anything very fancy, for example, a simple outgrowth. Then as vertebrates or insects evolved, the easiest way to form an outgrowth was to recruit the set of gene systems intact and initiate them at a localized spot on the body (Kengaku et al.).

Following through this reasoning, I wonder whether we can dispense with the outgrowth altogether and think back to a still more primitive situation when the set of gene systems was integrated for another purpose altogether. We could even imagine events in the single cell, such as a shape change, requiring a certain genetic program. In speculating about this, we would be helped if we knew much more about the use of, for example, Wnt, Hh, FGF, and TGF-β signals in very simple organisms. In the same way that yeast mating-type studies have been directly informative in thinking about Hox genes, understanding how higher animal types of signals are used in single-celled organisms, or organisms with relatively few cells, might help us to understand how a sponge uses the genes that we use to form arms. Are organs related to organelles?

FUTURE

Developmental biology has at least two practical uses: (1) for learning how animals and plants grow in order to make it possible to change how they grow or regenerate and (2) for "functionating," i.e., taking novel proteins such as one important for a disease and finding out how they relate to other proteins and pathways.

Developmental biology has yet to make any serious impact on medicine, but this is rapidly changing. In the short term, the identification of inherited human disease genes has brought to light the involvement of familiar regulators of development. Putting them into genetic pathways and developmental contexts will permit the design of new therapies and diagnostics. Oncogenes and tumor suppressors, such as Ras (Rubin et al.) and Patched (Johnson and Scott), are now being viewed as development regulators gone awry. Their normal roles will be helpful in understanding their evil roles.

Regeneration has not been a major focus in recent years, but as normal development is better understood, the right questions will be asked about how complex structures can be reformed or healed. The current tests of bone morphogenetic proteins for helping fractures heal may seem, in years to come, the crudest beginning. The seemingly miraculous reformation of amphibian limbs may be extended to other tissues: to rebuilding a damaged spinal cord, making insulin from a regenerated pancreas, or repairing a heart or liver. These prospects are exciting.

The human genome project is bringing a new importance to work on animal development. Flies, frogs, and worms may well become test organisms for placing a mystery gene into some sort of conceptual framework. Rubin's description (Rubin et al.) of *kuzbanian* provides an example. A newly found gene was quickly recognized as belonging to the Notch pathway. The right biochemical tests (cleavage of Notch) could be immediately designed from this knowledge and the protein's sequence, which revealed a metalloprotease. "Model" organisms will become increasingly important for this sort of purpose; it is easy to imagine standardized ways of placing "new" proteins into established pathways, a sort of aptitude test for genes.

With All This Progress, We Still Have a Superficial Understanding

The underlying principles still evade us. We have an understanding of the instruments but not the orchestration. Why is FGF used in one instance and TGF-β in another? Was it history that gave us these choices, or is there logic to the application of certain protein types to certain purposes?

There is room for substantial technical advances. We do not have easy ways of identifying genes regulated by transcription factors. We do not know nearly enough about the inheritance of determined cell fates, or what makes cells responsive to a signal at one moment and refractory the next. We do not know how to reactivate embryonic programs. We are at the very beginning of applying developmental biology to human disease genes to create new treatments, or to other human needs such as dealing with pathogen resistance, pest control, and parasitology. We need to invent more ways to make dominant negative mutations, one of the best chances to overcome redundancy in vertebrate genomes. We had some good examples in this meeting, such as dominant negative BMP receptors (Zou et al.; C. Wright, unpubl.) and fusion of an Engrailed repression domain to the Lmx1 homeodomain (Schwabe et al.). We lack good methods for observing evolution in progress, although the rapidity of evolution that can be observed in field biology studies gives me optimism on this count (see, e.g., *The Beak of the Finch: A Story of Evolution in Our Time* [1994, Knopf, New York] by Jonathan Weiner, reviewing the work of Peter and Rosemary Grant).

How do signals move through tissues or between cells? Are there facilitating, as well as obstructive, factors? At this meeting, receptors were proposed as both facilitators and limiters to ligand movement. Roel Nusse (Nusse et al.) described experiments suggesting facilitation of Wingless movement by its receptor, Frizzled2, and G. Struhl (pers. comm.) described how the Patched receptor can restrict movement of its ligand, Hedgehog. In addition to these factors, extracellular matrix and transcytosis were mentioned as relevant to the movement of ligands through tissue. One practical value of better understanding how molecules move would be to allow the design of altered ligands with altered signaling range.

We need methods for looking at protein *activities* in space and time, providing information formally equivalent to knowing when and where genes are transcribed or where proteins are located. We had one nice example of this at the meeting from Benny Shilo (Shilo et al.): the detection of active MAP kinase with an antibody specific to the activated form.

We need to look at more animals. The wonders of nature provide tools as well as aesthetic pleasure in science. The butterfly patterning discussed by Carroll (J. Kim et al.) provides one example. Early development is really diverse, even within the vertebrates. What can be done to understand the biological meaning of the differences and the mechanisms used to generate those differences? Very often, interesting molecular biology is revealed by interesting biology, for example in antibody diversity in the immune system. R. Beddington (Thomas et al.) pointed out the potential value of looking at armadillo development, where four primitive streaks form in one embryo. The major hurdle to be overcome is that no one wants to do all the basic gene characterization for a whole new system, so shortcuts must be found.

With all of our optimism derived from the past dozen years, we must remember the remarkable complexity of the Organ of Corti shown to us by Nathans (Xiang et al.). He mentioned that whenever he feels we are really making progress, a look at the Organ of Corti is humbling. We have a very long way to go in understanding how global regulators of the sorts discussed here control cell morphology.

A Common Language

In groping for analogies that may help us to think about the common heritage of developmental mechanisms, I turned to linguistics. Here, I had heard, a revolution had come from the proposition that languages with seemingly nothing in common in fact had related underlying structures. This seems exactly like our awakening in developmental biology. The linguistic thesis, as I understand it, is that languages all have certain syntactical rules in common because language is constrained by the structure and organization of the brain. I referred to Noam Chomsky's more accessible writings (*Reflections on Language* [1975, Random House, New York]) to learn more and was startled to find that he had used developmental biology as an analogy for his ideas about language. He wrote:

> . . . "human cognitive systems, when seriously investigated, prove to be no less marvelous and intricate than the physical structures which develop in the life of the organism."
>
> "The idea of regarding the growth of language as analogous to the development of a bodily organ is therefore quite natural and plausible."
>
> "It is a curious fact about the intellectual history of the past few centuries that physical and mental development have been approached in quite different ways. No one would take seriously the proposal that the human organism learns through experience to have arms rather than wings."
>
> "More intriguing to me at least is the possibility that by studying language we may discover abstract principles that are universal by biological necessity and not mere historical accident, that derive from mental characteristics of the species."

If Chomsky is right, and language takes its structure from the working structure of the brain, then we are talking about one and the same thing: Genes that organize brain development organize language. Commonalities in language reflect common paths of development. Language becomes a phenotype of the genes we have discussed here. Let's find out. In the meantime, we can share and enjoy the new common language of developmental biologists.

ACKNOWLEDGMENTS

I am particularly grateful to Bruce Stillman for organizing a memorable meeting, and to the speakers whose work I have attempted to summarize for their inspiration and patience. I thank David Bilder for discovering, and translating, Nehemiah Grew's Royal Society comments. I have done my best to be accurate and, although some errors undoubtedly remain, I hope that the spirit and excitement of the meeting—its enduring value—are evident. Research in my own laboratory is supported by the Howard Hughes Medical Institute and by the National Institutes of Health.

Author Index

Subject Index

X

Z

WITHDRAWN